ENCYCLOPEDIA OF
EARTH
SYSTEM
SCIENCE

VOLUME 2

Cr-L

ADVISORY BOARD

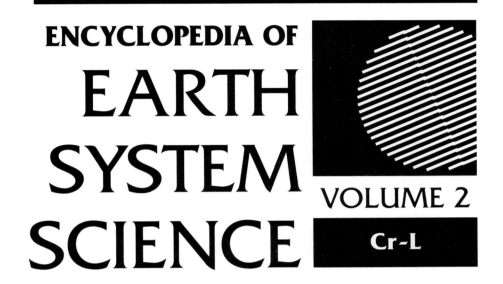

ENCYCLOPEDIA OF
EARTH
SYSTEM
SCIENCE

VOLUME 2

Cr-L

William A. Nierenberg, *Editor-in-Chief*

Scripps Institution of Oceanography
University of California, San Diego

Academic Press, Inc.
Harcourt Brace Jovanovich, Publishers
San Diego   New York   Boston   London   Sydney   Tokyo   Toronto

Academic Press, Inc.
San Diego, California 92101

United Kingdom Edition published by
Academic Press Limited
24–28 Oval Road, London NW1 7DX

Library of Congress Cataloging-in-Publication Data

Encyclopedia of earth system science / edited by William A.
  Nierenberg.
      p. cm.
    Includes bibliographical references and indexes.
    ISBN 0-12-226722-2 (v. 1). -- ISBN 0-12-226723-0 (v. 2). -- ISBN
  0-12-226724-9 (v. 3). -- ISBN 0-12-226725-7 (v. 4)
    1. Earth sciences--Encyclopedias.   I. Nierenberg, William Aaron,
  date.
  QE5.5E514   1991
  550'.3--dc20   90-29045
                                                          CIP

PRINTED IN THE UNITED STATES OF AMERICA
91 92 93 94   9 8 7 6 5 4 3 2 1

CONTENTS

v

■

■

GUIDE TO USING THE ENCYCLOPEDIA

The *Encyclopedia of Earth System Science* has been arranged in a format that enhances its usefulness to the reader. We would like to familiarize you with some of its features.

Each volume begins with a table of contents for that particular volume. The articles are ordered alphabetically using the word-by-word method (disregarding prepositions, conjunctions, and articles). For example, "Sediment Transport Processes" precedes "Sedimentary Geofluxes, Global," and "Continental Flood Basalts" precedes "Continental and Island Arcs." Article titles are arranged so that the main subject is listed first (e.g., "Clastic Sediments, Recycling and Evolution" rather than "Recycling and Evolution of Clastic Sediments").

All articles follow the same general organizational scheme. Each article begins with an outline that lists the main subject headings. This is followed by a paragraph or series of paragraphs that defines the primary subject of the article. The main text of the article is then presented. A list of bibliographic references follows the text. Finally, a glossary lists definitions of significant terms *in the context of their use in the article*. Therefore, a term may appear in more than one article and may be defined in a slightly different manner in each.

Cross-references to related articles are given in pertinent positions within the main text of each article. When more than one article is listed in a cross-reference, the titles are separated by semicolons.

Each volume contains a number of color plates. These plates are numbered consecutively within the volume. They are referred to in various articles within a volume. Below the caption of each plate, the article in which it is cited is listed.

Following the final article in Volume 4, a complete list of the contributing authors, their affiliations, and the titles of their articles is given (Volume 4, page 525).

The list of contributors is followed by a relational index that gives, for each article, the titles of articles about closely related subjects. Many of the article titles that are cross-referenced within the text of an article are listed here, along with any other closely related articles.

The final feature is a subject index that contains detailed entries for topics that are discussed within the *Encyclopedia*. Index entries appear with the volume number in boldface followed by a colon and the page number(s) in that volume on which the material is discussed. All terms that are listed in the glossaries of individual articles will appear as main entries in the subject index with the word "defined" either following them or as a subentry. As noted earlier, some terms occur in more than one glossary and therefore have multiple page citations in the subject index. In this way the reader is able to select the definition that fits a particular area of interest.

Crustal Fluids

Bruce E. Nesbitt and Barbara J. Tilley
University of Alberta

Crustal fluid regimes are composed of predominantly aqueous fluids that occupy pore spaces within oceanic and continental crust. Fluids in crustal regimes incorporate waters of marine, meteoric, magmatic, metamorphic, and mantle origins. The fluids vary widely in salinity, temperature, and composition of dissolved constituents, due to variations in composition of the original fluids and the nature and extent of water–rock interaction. Crustal fluids play major roles in the buffering of the chemistry of the oceans as well as the formation of mineral and petroleum resources.

I. Introduction

The earth's crust is defined as the outermost portion of the earth's lithosphere. It is composed of oceanic crust, which is predominantly gabbroic in composition, and continental crust, which is more felsic in composition. Occupying depressions in both oceanic and continental crust are sedimentary basins, which have hydrogeological characteristics distinct from those of both oceanic and continental crust.

The crust is nearly universally saturated with fluids of varying compositions. Sources for aqueous fluids involved in crustal regimes can be divided into two broad categories: surface (marine and meteoric) and subsurface (metamorphic, diagenetic, magmatic, mantle). Surface fluids comprise large and readily available fluid reservoirs, which dominate fluid regimes in the upper portions of the crust. In deeper portions of the crust, fluids of subsurface origins are more volumetrically important and probably the dominant source of crustal

fluids. In addition to aqueous fluids, fluids composed of hydrocarbons or igneous melts are significant in certain settings; however, due to the much greater abundance of aqueous fluids and the significance of water–rock interactions, the emphasis of this chapter will be on aqueous fluids. [See GROUNDWATER.]

The upper, cooler portions of the continental crust are relatively brittle with the deeper crustal units being more ductile. The transition between the regions of brittle and ductile rheological behavior is largely a function of temperature and in general occurs between 350 and 450°C and depths varying from 6 to 25 km, depending on geothermal gradients. The brittle–ductile transition has an important bearing on the fluid regimes of the crust. The typically fractured, upper crust is relatively permeable and, consequently, fluid pressures are generally near hydrostatic. Deeper, more ductile rocks have significantly lower permeabilities, and fluid pressures often approach or slightly exceed lithostatic pressures. This transition in characteristic fluid pressures leads to a transition in the hydrogeology of the systems as the fluid regimes in the near-surface, brittle rocks tend to be dominated by the circulation of surface waters (meteoric or marine) and fluid regimes of the deeper systems are dominated by fluids released from crystallizing magmas or metamorphism.

The most important driving force for fluid flow in the crust is fluid buoyancy resulting from density differences between deep, high-temperature, low-density fluids and overlying cooler, more dense fluids. In contrast in near-surface systems, differences in hydraulic head across an aquifer are of predominant importance in driving fluid flow. In regions where units are being compressed due to tectonic or sedimentary loading, fluids can be physically driven out of collapsing pore spaces. In sedimentary basins additional important driving forces for fluid flow include salinity contrasts and the presence of underpressured zones created by the erosion of overburden.

In addition to the driving force for fluid flow, rock porosity and permeability constitute important controls on the magnitude and direction of fluid flow in the crust. Porosity may be primary or secondary in origin. Primary porosity refers to the void space that a rock possesses at the end of its depositional phase. Secondary porosity is produced by post lithification fracturing and/or dissolution of the rock. Porosity can be reduced in brittle rocks by compression and the formation of new phases within the void space and by plastic flow in ductile rocks. Values for porosity are given in terms of volume percent occupied by void space and typically range from 5 to 50% for sandstones, karst limestone, shale, and fractured basalts down to 0 to 10% for crystalline rocks. Rock permeability depends on the connectivity of the pore spaces and depends largely on the size of the pore throats. Permeability values are given in terms of square meters ($m^2 = 1.01 \times 10^{12}$ darcies) and typically range for unfractured rocks from 10^{-13} to 10^{-16} m^2 for sandstones and limestones to 10^{-17} to 10^{-21} m^2 for shales and crystalline rocks. Fracturing can increase permeabilities by 2 to 6 orders of magnitude. Porosity and permeability along with the magnitude of the driving force causing fluid flow are the principle controlling variables that determine the values for fluid flux. The magnitude of fluid flux in a hydrogeological system is the key variable of a system in determining its capability to transport heat and dissolved constituents.

The chemical and physical interactions of fluids and rocks have important effects on the characteristics of both the fluids and the rocks. Circulating fluids often undergo extensive exchange of elements with rocks along the flow paths resulting in major chemical changes in both the fluids and the rocks. Physical effects of water–rock interaction include aspects such as the convective removal of heat from the crust and major changes in the electromagnetic properties of rocks. These and additional aspects of water–rock interaction are considered in detail below.

Fluid regimes within the crust can be divided into three categories — oceanic, sedimentary basin, and continental — based on distinct differences in the origins, chemistries, and circulation paths of fluids in the three settings. These subdivisions of crustal fluid regimes are used in the following sections describing the characteristics of fluids in the crust. For each category the sources of fluids, the driving mechanisms causing movement of those fluids, the nature and effects of water–rock interactions, and the impact on the earth's hydrosphere and lithosphere are examined.

II. Fluids in the Oceanic Crust

Oceanic crust comprises approximately 2/3 of the earth's crust. It is created at mid-oceanic spreading ridges at a rate of approximately 2 to 3 km^2/year and consumed in subduction zones (Figs. 1 and 2). An idealized section of oceanic crust is composed of four mafic to ultramafic igneous layers: an upper unit of pillowed basalts, underlain by a unit composed of sheeted mafic dikes, then a thick, massive gabbroic unit, and finally at the base, a large body of cumulate ultramafics. The overall thickness of the igneous units is estimated to be in the range of 5 to 10 km. Away from the spreading ridges, the pillow lavas are blanketed by a layer of deep-sea sediments composed of cherts, shales, and pelagic oozes. [See OCEANIC CRUST.]

Due to high thermal gradients and high permeabilities, vigorous convection of seawater through the basaltic units is common in rocks closely adjoining the spreading centers (Fig. 1). At the high temperature (up to 400°C) vent areas associated with these convection cells, large accumulations of sulfides, silicates, and oxides as well as deep-sea biological communities are commonly formed. Relative to seawater, vent fluids are enriched in Si, K, Rb, Li, Al, Fe, Mn, Cu, and Zn and depleted in Mg, as well as being significantly more reduced and at a lower pH (Table I). Similar less intense, lower temperature, seawater convection cells are believed to exist for up to hundreds of kilometers off the spreading axes. [See WATER GEOCHEMISTRY.]

At subduction zones, the downgoing slab dewaters, first due to compaction of sediments and later from devolatilization reactions. The resulting fluids rise toward the surface and have a significant impact on the chemical and physical characteristics of the overlying sedimentary wedge and the oceanic or continental crust.

A. Fluid Processes at Mid-Oceanic Spreading Centers

The convection of seawater at spreading centers was first suspected to be a major process in the mid to late 1960s, when it was noted that heat flow from the oceanic crust in the vicinity of the spreading centers was less than was expected, assuming conductive dissipation of heat from intruded magmas. Over the next several years, numerous geochemical anomalies in seawater and samples dredged from the ocean floor indicated that high-temperature fluids were venting onto the ocean floor in the vicinity of active spreading centers. In 1977, the first direct observation of an active hydrothermal vent and the associated biological community was made at the Galapagos Spreading Center on the East Pacific Rise. Subsequently, numerous discoveries of active sea-floor vent systems have been made and extensive investigations of the metalliferous deposits and biological communities associated with these vents have been conducted. [See HYDROTHERMAL SYSTEMS.]

Fluid flow in submarine hydrothermal systems is in direct response to thermal gradients existing in the oceanic crust. Near the spreading ridges, high-temperature molten and recently solidified oceanic crust raises the temperature and, consequently, lowers the density of fluids in the adjoining rocks. The low-density fluids are buoyant and rise rapidly through fractures in the oceanic crust, often venting onto the ocean floor. Fluid inflow occurs on the flanks of the system, and cold ocean water migrates down and in toward the heat source. Numerous computer models of the system have been developed. These models indicate that the two most significant factors controlling the size and intensity of the convection system are the size and temperature of the plutonic heat source and the nature and magnitude of rock permeability. Estimates of permeability in these systems vary considerably

Hydrogeological Regimes of the Crust: Extensional Margins

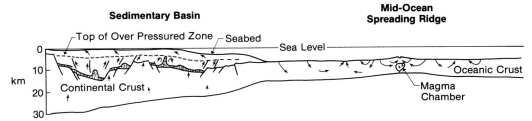

Fig. 1 Schematic diagram showing directions of fluid flow (arrows) in the crust at extensional margins. Subvertical lines indicate faults; the patterned areas in the sedimentary basins represent evaporite deposits.

but generally indicate widespread bulk permeabilities on the order of 10^{-15} m^2 with higher permeabilities (up to 10^{-9} m^2) in the upflow areas below sea-floor vents. Permeabilities are believed to decrease sharply at depths greater than 4 to 8 km due to the transition from brittle to ductile rock rheology.

Given the approximately 55,000-km linear extent of oceanic spreading ridges and the intensity of the convection process associated with these ridges, it is not surprising that the fluid flux associated with this process has a significant impact on the geophysical and geochemical characteristics of the oceans and oceanic crust. It is estimated that the convective heat loss produced by the circulation of sea water is between 11 and 15 $\times$ 10^8 cal/cm^2. Extrapolated over the extent of the ridge system, this results in 4 to 6 $\times$ 10^{19} cal/ year. This value is believed to represent approximately 80% of the heat lost by cooling of newly formed lithosphere and 20% of the annual heat loss of the earth.

Based on geophysical and geochemical data, the estimates of the worldwide flux of seawater through the convection systems are 1 to 9 $\times$ 10^{17} g/year, which is approximately 0.5 to 2.5% of the global river annual discharge. At this rate, the entire mass of the oceans cycles through the subsea-floor systems every 5 to 11 million years. The mass of sea water annually convected through the system is approximately 2 to 20 times the mass of newly created oceanic crust annually, indicating relatively high water/rock ratios in these systems. [See WATER CYCLE, GLOBAL.]

Due to the relatively high temperatures attained in these systems (up to 400°C) and the high flux rates, extensive water–rock interaction occurs. The processes involved and the resulting effects on water and rock chemistry have been studied by direct observation of fluids and rocks and by experiments and theoretical modeling.

Alteration of the oceanic crust can be divided into two categories: low-temperature alteration encompassing sea-floor weathering and water–rock interactions up to 200°C, and high-temperature alteration at temperatures greater than 200°C. Sea-floor weathering is the alteration process that takes place in the first few meters of the oceanic crust at temperatures of around 4°C. During this process, rocks are oxidized, basaltic glass devitrifies, and clays, hydroxides, and some zeolites form. At higher temperatures (50–150°C), the

Hydrogeological Regimes of the Crust: Compressional Margins

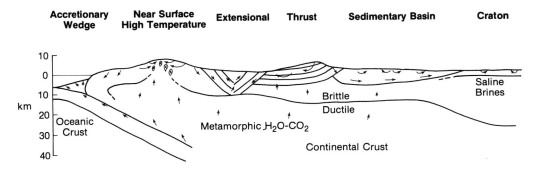

Fig. 2 Schematic diagram showing directions of fluid flow (arrows) in the crust at compressional margins. Enclosed shapes with X's represent igneous intrusions.

Table I Representative Water Analyses[a]

	Seawater	Sea floor vents[b]	Basinal brines[c]	Basinal brines[d]	Thermal springs[e]	Mountain springs[f]	Cratonic brines[g]
T (°C)	4	350	74	117	260		20
pH	7.8	3.5	6.5	6.2	8.6	6.8	6.4
Cl	19,350	18,282	121,711	142,000	1668	1.1	81,636
Na	10,760	10,488	48,222	77,000	980	6.0	14,000
Mg	1,290	0	2,606	1,129	0.02	1.7	631
Ca	411	647	20,892	10,148	2.4	10.4	31,273
K	399	939	5,431	1,054	200	1.6	166
Sr	8	7.1	786	911			715
Rb	0.12	2.5	9	3.4	2.2		
SO$_4$	2,710	0	340	6.4	6.5	2.4	185
H$_2$S	0	253		0.4	0		
CO$_3$ Total	142		396	446		54.6	31
Br	67		659	415	5.3		755
F	1.3			0.8	4.5	0.1	
Li	0.18	7.2	56	16	12.6		
SiO$_2$	10	1050		48	750	24.6	8
Fe	0.002	79.9	5	83	0.23	0.031	
Mn	0.0002	48.7	1.5		0.009		
Cu	0.0005	1.4	0.1		0.6		
Zn	0.002	5.5	1.8		0.6		
Pb	0.00003	54.1			0.8		

[a] Values are given in parts per million.

[b] Sea Floor Vent, 21°N, East Pacific Rise: From K. L. Von Damm et al. (1985). Geochim. Cosmochim. Acta 49, 2197–2220.

[c] Average of 15 samples from Upper Devonian carbonates in the Western Canadian sedimentary basin: From B. Hitchon et al. (1971). Geochim. Cosmochim. 35, 567–598.

[d] From the S-sand in south Louisiana: D. K. Gautier et al. (1985). Lecture Notes for Short Course No. 17, Society of Economic Paleontologists and Mineralogists, p. 123. SEPM, Tulsa, Oklahoma.

[e] Broadlands, New Zealand Well 13 (major elements), Well 2 (minor elements): From Henley and Ellis (1983).

[f] Perennial mountain springs, Sierra Nevada Mountains: From R. M. Garrels, and F. T. Mackenzie, (1967). In "Equilibrium Concepts in Natural Water Systems," pp. 222–242. American Chemical Society Advances in Chemistry Series, No. 67, Washington, D.C.

[g] Average of 22 brines from the Canadian craton: From S. K. Frape et al. (1984). Geochim. Cosmochim. Acta 48, 1617–1628.

alteration process is more reducing, and the alteration mineralogy is characterized by the formation of chlorite, laumontite, and wairakite. Low-temperature alteration processes persist for up to 400 to 700 km off the ridge axis and up to 50 million years after the formation of the basals.

High-temperature alteration (> 200°C) is restricted in its areal extent and duration to areas near the spreading centers and lasts for a million years or less. However, because of the high temperature of the system and the mass of fluid involved, such processes have a major impact on the chemistry of ocean water and oceanic crust. As cool ocean water penetrates into the crust on the flanks of the convection cell, it gradually increases in temperature and is reduced. Because anhydrite solubility decreases with increasing temperature, the fluid precipitates anhydrite when it reaches temperatures of 150 to 200°C, with the seawater supplying the sulfate and Ca coming from the basalts. At approximately the same time, substantial amounts of chlorite are formed from plagioclase in the basalts, as Mg from seawater is exchanged for Ca and

K. If water/rock values are less than 10 (a rock-dominated system), the Mg content of the circulating seawater will be exhausted and the resulting rock mineralogy will consist of chlorite, albite, epidote, and actinolite. If the water/rock values are greater than 10 (a water-dominated system), the rock mineralogy will be dominated by chlorite and quartz and the pH of the fluid will be substantially decreased due to the water–rock reactions. If the fluids undergo further reactions with basaltic or gabbroic rocks at high temperatures (300 to 400°C), epidote-rich rocks will be formed. The reactions that form epidote deplete the fluid in Ca, lower the pH further, and enrich the fluid in Na and K. In the upflow zone as the fluids cool and depressurize, it is common for large amounts of albite to be formed, depleting the fluid in Na.

As the fluids vent onto the ocean floor and mix with cool, pristine seawater, a series of major changes occur to the vented fluids including cooling, oxidation, increased pH, and the addition of sulfate. These chemical changes trigger the precipitation of a number of phases including sulfates, carbonates, Fe–Cu–Zn sulfides and Fe–Mn rich silicates, oxides, and hydroxides. The resulting accumulations of metals are generally economically insignificant, due to their small size and inaccessible location. However, occasionally, most notably in the Red Sea, the deposits are of sufficient size and tonnage as well as relative accessibility to be seriously evaluated as an economic resource. In addition to the formation of mounds of inorganic precipitates, thriving biological communities often develop. Life in these communities is based on chemosynthesis as opposed to photosynthesis. Microorganisms around the vents derive energy from the oxidation of reduced sulfur compounds in the vent fluids. These microorganisms constitute the first step in the food chain around the vents. [See MARINE MINERAL RESOURCES; ORE FORMATION.]

One of the most important aspects of convection of seawater through the oceanic crust is its effect on the mass balance characteristics of seawater and the crust. In the mid 1970s, it was recognized that the Mg content of sea water was substantially less than could be accounted for by consideration of the known sources of Mg input and extraction from the oceans. It is now recognized that this imbalance is the result of the nearly complete loss of Mg from seawater as it reacts with the oceanic crust during convection (Table I). Water–rock interaction during convection is similarly believed to account for the loss from seawater and consequent enrichment in the crust of significant amounts of K, B, and Rb and lesser amounts of H_2O and U. In contrast, seawater is enriched and the oceanic crust is depleted in Si, Ca, Ba, Li, and base metals during convection. In addition, seawater undergoes significant exchange of oxygen isotopes with the basaltic crust during circulation, which results in a long-term buffering effect on the $^{18}O/^{16}O$ ratio of the oceans.

B. Fluids in Zones of Subduction

The process of subduction of the oceanic plates beneath oceanic and/or continental material also has a major impact on the crustal fluid regime (Fig. 2). Since subduction zones are generally much less accessible than oceanic spreading ridges, the magnitude and details of this process are not as well understood as the processes occurring at ocean ridges. However, from the work done to date and the theoretical models of the process, it is clear that the fluid regime associated with subduction has a significant impact on the chemistry of the oceans and oceanic crust as well as on the continental crust.

As the oceanic crust is subducted, much of the sedimentary cover is scraped off the downgoing slab, forming an accretionary wedge. This results is the expulsion of a large amount of pore water from the unconsolidated material as it is compacted and deformed. Studies conducted in deep drill holes in modern accretionary wedges have shown that the expelled fluids are similar to seawater in their chemistries, with anomalously high values for CH_4 and low values for Cl^-. The remainder of the sedimentary material and the basalts are subducted. As the subducted plate increases in temperature, a variety of devolatilization reactions set in, producing a fluid that is largely H_2O, but with significant amounts of CO_2, CH_4, and SiO_2. The bulk of this fluid is believed to migrate back up the subduction zone through fracture permeability concentrated at the contact between the downgoing plate and the overlying crustal or mantle material. As the fluid migrates upward through this zone, it has the potential to transport heat—causing at least local thermal anomalies—and also dissolved constituents, most notably SiO_2. At deeper levels, it appears that some of the fluid migrates vertically into the overlying crust or mantle, causing melting.

Given our relatively limited knowledge of the quantitative aspects of this process, it is difficult to accurately estimate the fluxes involved. However, at present rates of subduction it appears that the mass of ocean water will cycle through the subduction zones every billion years. Oceanic carbonate cycles much faster, at approximately every million years.

Geophysical evidence for the involvement of fluids in subduction zones has been drawn from a variety of techniques. For example, seismic imaging and electromagnetic data obtained on and off the coast of British Columbia discerned several horizons above the subduction zones, which appear to be zones of 1 to 4% porosity, saturated with saline fluids.

III. Fluids in Sedimentary Basins

Sedimentary basins can form wherever a depression in oceanic or continental crust is coupled with a sediment supply. Varying rates of subsidence, supply of sediment, and environmental conditions throughout a basin's history can produce a unique stratigraphic profile with varied thicknesses and interlayered combinations of carbonate rocks, evaporites, coals, and clastic rocks including conglomerates, sandstones, and shales. [See SEDIMENTARY ROCKS.]

Sedimentary basins evolve with time through a stage of active basin filling followed by a cessation of subsidence and/or a stage of tectonic uplift. The flow dynamics and fluid chemistries of porewaters, which comprise about 20% by volume of most sedimentary basins, evolve as the basin evolves. Fluids in sedimentary basins move continuously, and only in rare cases of long-term stability does erosional leveling of basin topographic relief result in effectively stagnant-flow conditions. The possible sources of waters present in sedimentary basins include (1) meteoric water; (2) fresh or marine water buried with the sediments; (3) waters released during alteration of minerals at depth, and (4) waters released during devolatization of basement rocks during metamorphism.

With the rapid development of the oil and gas industry in the early part of the twentieth century, substantial new information on the nature of fluids in the subsurface became available. Geophysical borehole logs (such as spontaneous potential, resistivity, gamma ray, density, and sonic logs) have provided information on the nature of the rocks and the fluids they contain. Drill stem tests have provided pressure data, which have provided information on fluid flow; and well head samples have provided data on water chemistries. Seismic data have been used to evaluate the continuity of reservoir units and to detect the presence of fluid-saturated rocks versus dry rocks. The escalating costs of locating new resources and the need to isolate radioactive and chemical wastes from the biosphere for long periods of time have encouraged the development of models for subsurface flow in sedimentary basins.

A. Flow Regimes

Three types of flow regimes can operate in sedimentary basins: (1) gravity-driven, (2) pressure-driven, and (3) density-driven. Which of these flow regimes dominates largely depends on the stage of basin evolution — that is, whether the basin is depositionally active or inactive.

Gravity-driven flow is most important in depositionally inactive basins (the Paris basin and parts of the western Canadian sedimentary basin, Fig. 2), but can also occur in shallow, updip portions of depositionally active basins (parts of the Texas Gulf Coast basin that are above sea level). In gravity-driven flow, topographic relief along basin surfaces drives groundwater through deep strata. Meteoric water recharges the flow system at high elevations. Groundwater everywhere moves toward regions of lower fluid potential, eventually discharging at low elevation. Flow rates are on the order of meters per year. In a system with irregular relief, flow systems develop over a range of scales from local to regional. When the irregularity of the relief is large compared to overall relief and to the thickness of the flow system, the local flow systems can overwhelm regional flow. However, given a permeable aquifer at depth, regional-scale flows can occur even in the presence of highly irregular relief. A flow system generally develops so that most of the flow occurs in the direction of highest permeability and within the highest permeability units.

Pressure-driven flow can be important in depositionally active basins where the shale/sandstone ratio is high and the sedimentation rate is rapid (Fig. 1). Fluid is displaced up and out of the deepest parts of the basin as strata compact under the weight of the accumulating sediments and to a lesser extent from dehydration of clay minerals. Fluid flow that is driven by compaction is limited by the volume of pore fluid that is carried to depth within the sediments of the basin. Flow velocities in slowly subsiding basins are very slow (millimeters per year) and are independent of permeability.

In shaly basins where burial proceeds at rates of at least 0.1–1 mm/yr, fluid pressures greatly in excess of hydrostatic are likely to develop. In this case, burial proceeds so rapidly that enough fluids cannot escape to allow the sediments to compact fully. The pore fluid, then, becomes overpressured as it bears part of the overburden weight that would normally be supported by the sediment. Flow rates resulting from compaction in this type of basin are relatively slow, generally centimeters or tens of centimeters per year, but the large potential gradients in overpressured basins block meteoric water from circulating into deep sediments. Overpressured zones occur in the Viking graben of the North Sea basin and the Gulf of Mexico basin. In contrast, basins that subside slowly or have deep aquifers develop only small excess potentials by compaction, and meteoric recharge may dominate.

Another variation of pressure-driven flow can occur in basins where uplift and erosion are occurring. In strata with sufficiently small permeabilities, underpressures can develop in deep strata due to the diminishing weight of overlying sediments and, to a lesser extent, the thermal contraction of pore fluids. The origin and persistence of underpressured environments is of interest in waste disposal efforts because potential gradients can be expected to drive flow into deep strata and away from the biosphere. Such underpressured environments may also affect petroleum migration.

Fluid flow in sedimentary basins can also be produced by differences in fluid density due to variations in temperature or salinity. The process is similar to that of seawater convecting through the crust described earlier in this chapter. Possibilities for fluid convection are enhanced by high permeability, high thermal or salinity gradients, and great thickness of continuously permeable sediment. Rates of flow by convection in sedimentary basins are on the order of meters per year. The process of flow by convection has been invoked for some sedimentary basins because it is the only flow process that can explain the large volume of fluids necessary to account for the observed diagenetic alterations such as those in the Gulf Coast basin and the existence of some thermal anomalies. Thermohaline convection is due to variations in both solute concentration and temperature. The solutes enhance convection when it is the cooler fluids that are most saline and retard convection when it is the hotter fluids that are most saline. Complex convective patterns have developed surrounding salt domes in the Gulf of Mexico basin in Louisiana. There, convection is driven by salt dissolution at shallow depths, which causes warm flows that have ascended along the dome to descend (Fig. 2).

Fluid flow in sedimentary basins can result from a combination of gravity, pressure, and density-driven flow depending on the particular region and depth within the basin. The age, maturity, and tectonic history of the basin influences which kind of flow will dominate. A young, rapidly subsiding basin will likely be dominated by pressure-driven flow as shales compact (Fig. 1). Gravity-driven flow may occur in those parts of the basin that are above sea level. In such basins, compaction may drive saline fluids landward while gravity drives meteoric water basinward. The two flows converge and discharge vertically. In slowly subsiding basins where precipitation is high and the shale-to-sand ratio is low, gravity-driven flow may overcome compaction-driven flow. In such basins, the rate of gravity-driven flow is significantly greater than that of compaction-driven flow and the volume of available meteoric water is limited only by the amount of precipitation, whereas compaction water is a finite quantity.

Once sedimentation stops and structural uplift produces a topography, the character of flow gradually changes. Gravity-driven flow systems become more important as flow due to compaction ceases and overpressures dissipate (Fig. 2). Some examples of basins that have evolved to the point where gravity-driven flow dominates are the Aquitaine region of France, the Great Lowland Artesian Basin of Hungary, the Artesian Basins of the Central Plateau in Iran, and portions of the Western Canada Sedimentary Basin.

B. Chemistry of Basinal Fluids

Most subsurface fluids have chemical compositions unlike any surface waters. Waters from different sedimentary sequences have their own distinctive composition, although there are systematic variations. Subsurface waters of seawater salinity or less often show a depletion in Ca and Mg relative to seawater. With increasing total dissolved solids content, the proportions of Ca and Mg to Na increase. Chemical analyses of two basinal brines (Table I) illustrate the variability in Cl, Na, Mg, Ca, K, SO_4, and Fe concentrations in brines of similar salinity. [*See* Brines in Sedimentary Environments.]

The composition of a basinal brine depends on at least three interrelated factors: (1) its preburial and early diagenetic history, (2) the extent of water-rock interaction, and (3) membrane effects. At the preburial stage, subaerial evaporation can first increase the concentration of major dissolved species, and then, as saturation with respect to various mineral phases is reached, a decrease in concentration of certain solutes occurs. During the early stages of burial, the following modifications in proportion of major dissolved species can occur depending on the rate of burial and the amount of organic matter present: reduction of sulphate; production of bicarbonate; and removal of K, Ca, and Mg.

The drive toward thermodynamic equilibrium is one of the more important controls on the composition of sedimentary brines. Departure from equilibrium resulting in precipitation or dissolution of mineral phases and accompanying changes in porewater chemistry can be forced in a number of ways. Changes in temperature and pressure, an introduction of acids or mixing of waters of diverse composition are just a few. Migration of waters into a new host mineral assemblage, change in redox conditions, or a loss of volatiles can also force a departure from equilibrium. A fluid can be continually kept out of equilibrium by active convection through a thermal gradient or by continued pumping of NaCl into the sediment–porewater system as a result of dissolution of salt.

Composition of pore waters is also controlled to a lesser degree by membrane filtration effects. Clay minerals can act as membranes that create a partitioning of dissolved species between sands and adjacent shales. Mg, Ca, and HCO_3 can pass readily through clay membrane, whereas Na, K, SO_4, Cl, and Br pass through with increasing degrees of difficulty.

The sources of dissolved constituents in a basinal brine include the materials present in the basinal pore fluids at the time of their burial, materials exchanged with surrounding rock and organic matrix as a result of diagenetic exchange, and material imported into a basin by active mass transport. Redistribution of soluble constituents in brines of sedimentary basins can occur as a result of brine migration over short or long distances.

Most basinal brines are isotopically dissimilar from seawater. Trends in δD and $\delta^{18}O$ values are commonly interpreted in one of two ways: (1) All marine water molecules have been flushed out of the basin and have been replaced by meteoric water molecules. ^{18}O values of the water have been enriched by reaction with carbonates and D values have been preferentially enriched by membrane filtration. (2) The $\delta^{18}O$ and δD values of basinal brines represent the mixing of infiltrated meteoric water with diagenetically altered seawater. No conclusive evidence for either interpretation is yet available.

C. Economic and Environmental Impact

The migration and accumulation of hydrocarbons and the formation of ore deposits in sedimentary basins are largely the result of basinal fluid flow. Petroleum moves in response to buoyancy, the hydrodynamic drive of groundwater flow, and gradients in capillary pressure resulting from heterogeneous distribution of porosity or grain sizes. Meteoric water, which comes into contact with hydrocarbons in shallow parts of the basin, may cause degradation of the oil (e.g., tar sands and heavy oils of Alberta). Mississippi Valley type, Pb–Zn ore deposits, and many types of uranium deposits are the result of fluid flow in sedimentary basins. Gravity-driven flow of hot saline basinal brines is thought to be responsible for the leaching of metals from basinal sediments and their concentration at the basin margins.

Basin thermal budgets are dominated by heat conducted into the sedimentary pile from the underlying crystalline crust. In most sedimentary basins, the geothermal gradient is 20–45°C/km and heat flow values are 40–80 mW/m^2. Basinal heat flow moves to the surface by conduction (most common) or through advection of groundwater. When advective transfer dominates, discharge areas are warmed and recharge areas are cooled relative to a conductive gradient. For groundwater flow to create a positive thermal anomaly in the discharge area, the brines must move rapidly enough to avoid complete cooling by conduction to the surface.

Throughout much of the surface of a sedimentary basin

there is little interaction between surface or near-surface fluids and basinal brines. Shallow, local groundwater flow systems overlie the deeper, regional flow system involving basinal brines. Connection between the local and regional systems occurs at the recharge and discharge zones of the regional flow system. Recharge zones represent the beginning of the regional flow system, where fresh fluids enter the system. Geothermal gradients in these areas are lowered by the influx of cooler water. More significantly, discharge zones are areas where basinal brines rise to the surface of the basin and mix with any existing surface fluids. The effect on the surface fluids may be a rise in salinity and temperature. Surface accumulations of salt sometimes occur at discharge zones.

IV. Fluids in the Continental Crust

The continental crust comprises approximately one-third of the earth's crust and extends to 30 to 40 km depth. In comparison to the oceanic crust, it tends to be less dense and richer in Si, Al, and alkali elements. [*See* CONTINENTAL CRUST.]

Fluid regimes of the continental crust, excluding sedimentary basins, can be divided into four categories:

1. Near-surface, high-temperature ($T > 100°C$) hydrothermal systems associated with recent igneous intrusions or extrusions, such as Yellowstone or the geothermal fields of the North Island, New Zealand.
2. Fluid regimes in the brittle crust of tectonically active portions of the continents, such as the Himalayas or western coast of South and North America.
3. Fluids hosted by ancient, continental crust of the cratons.
4. Fluid regimes of the ductile portions of the continental crust.

A. Fluid Regimes in Near-Surface, High-Temperature Hydrothermal Systems

The emplacement of high-temperature ($T > 700°C$) plutons into relatively low-temperature ($T < 100°C$), shallow crustal settings creates substantial thermal gradients in the adjoining rocks. The resulting differences in density between the warm, low-density, near-pluton fluids and the more distant, cooler water compel the fluids near the pluton to rise. The resulting convection process creates some of the most spectacular examples of the dynamic nature of geologic processes, such as the geothermal fields of Yellowstone and the North Island, New Zealand. Such geothermal processes have a major influence on the fluid regimes in those areas with extensive water–rock interaction, leading to substantial changes in

rock and water chemistry and in some cases the formation of zones of economic mineralization.

Due to their association with areas of recent volcanism and/or plutonism, the distribution of high-temperature, near-surface systems is relatively restricted. Most of the presently active geothermal systems are distributed around the Pacific Rim above zones of active subduction. In addition, in locations where oceanic spreading ridges rise above sea level, such as Iceland, terrestrial geothermal systems are formed that are analogous to the submarine systems described earlier.

Due to the substantial interest in the exploration for and development of geothermal energy from terrestrial, high-temperature systems, there exists an extensive body of data on the geological, geochemical, and geophysical characteristics of such systems. Many of the areas have been thoroughly explored by drilling to depths of hundreds of meters with extensive investigations of the chemistries of rock and water samples. In addition, geophysical studies incorporating seismic, heat-flow, and electromagnetic techniques have been employed to document the characteristics of many of these systems.

Most of the active, near-surface systems are developed in intermediate to felsic volcanic or plutonic rocks in calderas or other zones of intense, extensive fracturing. The fluid convection systems vary in size but are typically 5 to 10 km in diameter and 4 to 6 km deep; the limit on depth of convection is provided by the brittle-to-ductile transition in rock rheology near the pluton. The heat to drive the systems is largely derived from the cooling of near-surface plutons. Most hydrothermal systems are active for periods up to 10^6 years and are driven by more than one pluton emplaced over that time span. Measured permeabilities in geothermal systems are on the order of 10^{-14} to 10^{-15} m², reflecting a dominance of fracture-controlled permeability. Estimated fluid flux for a typical system is on the order of 100 to 900 km³/10,000 years. Flow paths are similar to those described for convection associated with the oceanic ridges, with upflow directly over the plutons and inflow on the flanks. As the rising fluids approach the surface, they cool and interact with near-surface groundwater, resulting in a characteristic mushroom shape for the overall flow patterns in these systems.

The fluids in these systems are dominantly composed of meteoric water with little or no detectable contribution of fluid from the igneous rocks. The fluids generally have temperatures in the range of 100 to 300°C, are low in total dissolved salts, and have pH's close to neutral. At some distance below the surface, the fluids often boil, releasing a gas phase with CO_2, N_2, H_2S, etc., in addition to steam. The gas phase either escapes through fumaroles at the surface or condenses near the surface forming low pH, sulfate, bicarbonate fluids. Water–rock interaction during fluid convection results in the formation of a suite of minerals including quartz, albite, clays, calcite, chlorite, sericite, and epidote. In the zones altered by the low pH fluids, kaolinite, alunite, pyrophyllite, and opal are common.

While these systems constitute an important source of energy in a few locations around the world, the overall impact of circulation in near-surface, high-temperature geothermal systems on the mass balance of the atmosphere and surface waters is probably limited due to the restricted distribution of such systems. However, near-surface high-temperature systems are important as a concentrating mechanism for economically significant trace elements such as Au, Ag, and Hg.

B. Fluid Regimes in Tectonically Active Regions

The continual movement of crustal plates creates a number of zones within the continental crust of compression, such as the Himalayas, and extension, such as the East African rift. These zones of major crustal tectonic activity have a substantial impact on the fluid regimes of the continental crust and, due to the areal extent of these regions, the potential impact of the fluid regimes on the chemistry of the crust and surface fluids is significant. [See CONTINENTAL DEFORMATION.]

Data on the chemistry of the hydrological regimes associated with tectonically active zones are limited, due to the depths to which these systems are active (15 km or more) and the lack of an economic incentive to drill deep research holes. Available data are largely composed of chemical analyses of fluids from springs and from the few deep holes drilled within these areas. In addition, information on the chemistry of fluids present during deformation and metamorphism is obtainable from the study of fluid inclusions occurring in rock samples and the chemical changes produced in the rocks as a result of water–rock interaction. During the last few years, a considerable contribution to the understanding of fluid regimes in this setting has also been provided by electromagnetic, seismic, and heat-flow geophysical studies.

In extensional tectonic regimes, such as in eastern Africa, the deep penetration of vertical structures to several kilometers into the crust creates relatively continuous, highly permeable zones. The presence of these zones of high permeability keeps fluid pressures approximately equal to the hydrostatic pressure gradient and permits the deep circulation of surface (marine or meteoric) fluids. Due to the high thermal gradients, which are characteristic of extensional zones, convection cells are established that move the surface fluids through the brittle units with a lower boundary at brittle-to-ductile transition. The geochemical effects of this process have been clearly documented by fluid inclusion and stable isotope studies in a variety of these complexes. Electromagnetic evidence for the existence of fluids in this setting have been obtained from both the western United States and eastern Africa.

The movement of a large mass of fluid through extensional areas results in significant chemical effects on both the rocks and the fluids. Fluids in these systems tend to increase in temperature. CO_2 content, and $\delta^{18}O$ values. In areas of high fluid flux, the rocks are visibly altered to chlorite and other low-temperature phases. In terrestrial settings, the formation of both Cu and Au deposits have been linked to this process; and in marine settings, the formation of syngenetic Pb–Zn sulfide deposits often results from the convection of seawater in extensional zones in the continental crust.

In compressional regimes, such as the Himalayas or southern Europe, thrust faulting is the predominant style of deformation. Permeability in such regions tends to be more variable and discontinuous vertically, as the principal orientation of faults and stratigraphy is roughly horizontal. Consequently, fluid pressures are probably more variable than the generally hydrostatic values believed to be common in extensional regimes. In addition, during thrust faulting, fluid pressures significantly in excess of hydrostatic are required to facilitate movement on the faults. As a result of these features, the fluid regimes in compressional zones are probably much more heterogeneous than those present in extensional terranes. [See FAULTS.]

An important source of fluids in compressional terranes is fluid expelled from partially lithified units during compression. These fluids are believed to be essentially squeezed out of sediments in front of the thrust sheet, as the sediments are being overridden by thrust sheets. In addition, there is some evidence for the infiltration of surface waters into the upper layers of a thrust complex. These fluids often recharge the hydrological systems supplying water to warm springs (Table I), which are common in these tectonically complex compressional regimes. Waters issuing from springs in such regions generally have temperatures in the 25 to 90°C range and pH's near neutral. The spring waters often contain substantial amounts of CO_2, H_2S, N_2, or CH_4 and variable chloride and carbonate contents. Fluid inclusion studies of veins from deeper portions of such systems indicate similar chemistries for the higher-temperature, deeper fluids.

The extensional and compressional environments characterized above represent the end members of what are in reality highly complex tectonic environments. Even within a single tectonic zone, it is not unusual to have regions with active extensional faulting adjoining areas of active thrust faulting. Furthermore, in many areas early compressional phases of deformation are often followed by later rebound and extensional faulting. Consequently, the hydrological regimes of tectonically active areas are most likely also complex, transitory, and dynamic. It should be expected that during a prolonged deformational-metamorphic event, the fluid pressure at any one site might fluctuate between hydrostatic and lithostatic and that at different times in an orogenic event a given rock unit may be exposed to the influx of surface-derived fluids as well as the presence of a fluid regime dominated by metamorphic devolatilization reactions.

C. Fluid Regimes of Continental Cratons

Cratons are composed of stable, generally, Precambrian-age continental crust, which has not undergone significant deformation for extended periods of time. Due to the long-term stability of cratons as well as their generally low relief and low geothermal gradients, it is predictable that the fluid regimes within cratons should be relatively uniform both within a particular craton and among cratons worldwide. Studies of fluids and rocks from deep drill holes and mines, as well as geophysical results, indicate that there is indeed a very consistent and uniform pattern to the hydrological regimes of cratons.

The upper 200 m of most cratonic fluid regimes is dominated by relatively pristine meteoric water of local origin. These fluids generally have low total dissolved salt contents, and the chemistries of the fluids are strongly dependent on the composition of the host rocks. Below a transitional zone from 200 to 1000 m, the fluids chemistry is radically different. The deep fluids are highly saline with total dissolved salt contents up to 300,000 mg liter^{-1} and very high Ca, Na, and Cl contents. Isotopically these fluids are also unusual, with anomalously low $\delta^{18}O$ values. The details of the origins of these fluids are not well understood but they appear to be a product of prolonged, low-temperature interaction of surface fluids with the cratonic units.

D. Fluid Regimes of the Deep Continental Crust

At temperatures in excess of approximately 400°C, most rocks become increasingly ductile. As open fractures are eliminated by ductile sealing of the rock, the permeability of the rocks decline sharply to values probably less than 10^{-18} m². Such low permeabilities essentially seal off the rocks from the influx of surface fluids; as a result, the fluid regimes of the deeper portions of the continental crust are most likely dominated by fluids of a metamorphic or magmatic origin.

During prograde metamorphism, rock units undergo a virtually continuous series of devolatilization reactions, which yield large volumes of H_2O and CO_2 and minor amounts of N_2, CH_4, and H_2S. The amount of fluid ($H_2O + CO_2$) released during metamorphism is between 3 to 5% by weight, depending on the initial volatile content of the rock. Fluid inclusion and phase equilibria studies indicate that generally fluids in greenschist and amphibolite metamorphism are low salinity and water-rich. However, in the granulite facies the activity of H_2O in the fluid phase is low, and CO_2 often comprises the bulk of the fluid phase. Estimates of fluid flux rates in metamorphic regimes are on the order of 10^{-11} g/cm² s. Geophysical evidence of the presence of fluids in deep, ductile rocks has been obtained from both seismic and electromagnetic studies in the western United States. [See META-MORPHIC PETROLOGY.]

The contribution to the deep-crustal fluid regime by devolatilization of cooling plutons is not well understood. As a pluton crystallizes, fluids that were not incorporated into hydrous phases are expelled into the surrounding rock. A granitic melt can contain 5% (weight) or more H_2O depending on composition of the magma, temperature, and pressure; consequently, in regions with large batholithic intrusions, the contribution of magmatic fluid to the regional fluid budget can be substantial. Remnant samples of magmatic fluids have been recognized in the study of fluid inclusions from igneous plutons and closely associated ore deposits. The results of these studies generally indicate that the magmatic fluids are highly saline with up to 60% equivalent weight NaCl. [See IGNEOUS PETROGENESIS.]

V. Summary

It is clear from the discussions above that the movement of fluids in the crust is a product of the complex interaction between the various driving forces for fluid movement, crustal permeability, and the availability of fluids. Likewise, the chemistry of fluids in the crust is the result of the interaction of the initial chemistry of the fluids and the chemical evolution of the fluids as they react with rocks along their flow paths. Our understanding of the movement and chemistry of crustal fluids has evolved tremendously in the last 10 years; however, a large number of fundamental questions still remain to be answered. As documented in this presentation, it is essential to improve our understanding of crustal fluids, since processes active in the crust clearly have a significant impact on the chemistry of the oceans and atmosphere and on the understanding of many aspects of global environmental change.

Bibliography

Barrett, T. J., and Janbor, J. L., eds. (1988). Seafloor hydrothermal mineralization. *Can. Mineral.* 26, 429–888.

Bethke, C. M. (1989). Modeling subsurface flow in sedimentary basins. *Geol. Rundsch.* 78, 129–154.

Hanor, J. S. (1987). "Origin and Migration of Subsurface Sedimentary Brines." SEPM Short Course No. 21.

Henley, R. W., and Ellis, A. J. (1983). Geothermal systems ancient and modern: A geochemical review. *Earth Sci. Rev.* 19, 1–50.

Nesbitt, B. E., ed. (1990). "Fluids in Tectonically Active Portions of the Continental Crust." Mineralogical Association of Canada, Short Course Notes Vol. 17.

Rona, P. A. (1984). Hydrothermal mineralization at seafloor spreading centers. *Earth Sci. Rev.* 20, 1–104.

Rona, P. A., Bostrom, K., Laubier, L., and Smith, K. L., Jr., eds. (1983). "Hydrothermal Processes at Seafloor Spreading Centers." Plenum Press, New York.

Glossary

Fluid flux Mass of fluid passing through a given cross-sectional area per unit time.

Hydraulic head Difference in elevation between parts of the upper surface of the zone of saturation.

Hydrostatic Pressure due to the weight of an overlying column of water.

Lithostatic Pressure due to the weight of an overlying column of rock.

Water/rock ratio Amount of fluid passing through a given rock volume or mass, integrated over the lifetime of the fluid flow system.

Cyclic Sedimentation, Climate and Orbital Insolation Changes in the Cretaceous

Robert J. Oglesby and Jeffrey Park

Yale University

Cyclic sedimentation refers to the alternating occurrence of two or more distinct types of layers, or beds, in sediments deposited on the ocean floor and subsequently preserved. The alternating sedimentary beds can be caused by either a change in the type of material being deposited or a change in the sediments subsequent to deposition. Frequently, the alternating beddings are visible to the naked eye, typically as alternating light and dark bands in an outcrop or deep-sea core. In other cases, the cycles are revealed only after chemical analysis. Many alternating sedimentary beddings are thought to be due to changes in climate, which can affect the nature of sedimentary materials or the depositional environment. These changes in climate may be due to cyclic changes in insolation received by the earth; the insolation changes are due to periodic fluctuations in the earth–sun orbit called Milankovitch cycles. Many striking examples of cyclic sedimentary beddings, with periodicities similar to those of Milankovitch orbital cycles, are found during the Cretaceous, a geologic period that lasted from about 65–130 million years ago.

I. Introduction

The earth's geologic record contains many examples of cyclic sedimentary bedding, caused either by periodic or quasi-per-iodic shifts in the depositional environment (primary cycles) or by diagenetic alteration of sediments subsequent to deposition (secondary cycles). Perhaps the best known examples of these sequences are the cyclical signatures of the Pleistocene Ice Ages recorded in deep-sea sediments; this cyclicity can be subtle and revealed only after isotopic analysis. An important feature of these cyclical sequences is that they are related presumably to large changes in continental ice; the implication is that the changes in ice volume in some way help cause the sedimentary sequences. The quantity, young age, and reasonably accurate dating of these records has made possible the recognition that correspondences exist between the principal periods of the Pleistocene cycles and the quasi-periodic perturbations to the earth's orbit known as Milankovitch cycles. [*See* Ice Age Dynamics.]

Cyclic stratigraphic sequences are also common in earlier geologic epochs, including the Precambrian (the period prior to 600 Ma). From the Cretaceous period (65–135 Ma) there are many good examples of cyclical carbonate sequences involving limestone–marl and limestone–black shale alternations. Paleoclimatic and paleoenvironmental reconstructions indicate that Pleistocene and Cretaceous climates were considerably different. However, there is strong evidence from some Cretaceous sequences of periods associated with Milankovitch cycles. This argues that, even under widely divergent "mean states" of global climate, the earth system is sensitive to the small insolation perturbations associated with natural fluctuations of the earth's orbit. The Pleistocene was a cold period dominated by the cyclic growth and decay of continental ice sheets. The changes in albedo and topography caused by ice sheets have been proposed as amplifying the effect of orbital insolation changes. There is no evidence for widespread continental ice during the Cretaceous, which is thought to have been about 6°C warmer than at present. Evidence from sediment sequences associates Milankovitch rhythms instead with such factors as widespread changes in ocean bottom-water redox state and regional changes in the precipitation minus evaporation balance. A particular challenge is posed to account for Cretaceous cyclic sedimentary sequences without resorting to the feedback effects of large ice sheets. [*See* Global Cyclostratigraphy.]

In this article we focus on the Cretaceous period, although our comments are in many cases applicable to earlier Mesozoic periods and later warm Cenozoic periods (e.g., the Eocene). We describe some examples of Cretaceous cyclical sequences, concentrating on where they are found, of what materials they are composed, and possible theories as to how they may have been formed. The results of spectrum analysis of selected records are described, emphasizing what information this approach can and cannot yield. We also describe attempts to model Cretaceous climate cycles with general circulation models. [See EOCENE PATTERNS AND PROCESSES.]

II. Examples of Cretaceous Bedding Cycles

The occurrence of climate cyclicity in the Pleistocene is supported primarily by geochemical analysis of sediment cores taken from pelagic, shelf, and lacustrine environments. Often such sequences are not visibly cyclic. Evidence for climate cyclicity in the Cretaceous, however, is found most often from carbonate sequences with alternating lithologies. Pelagic sequences often consist of limestone–black shale and limestone–marl beds. Carbonate cycles can be found as well from shallow-water continental shelf environments. Well-studied examples of these shallow-water cycles include chalk–marl sequences from the present-day western United States (e.g., the Greenhorn Formation and Bridge Creek Limestone Member) and many locations in Europe (including the Upper Cretaceous Middle and Upper Chalk of southern England and Cenomanian deposits of southern France). Black shale sequences in shelf environments have been drilled in the South Atlantic. Deep-water cycles are somewhat more difficult to uncover, largely because most of the oceanic crust extant during the Cretaceous has since been subducted. There are some good examples of intermediate water-depth cycles that have since been uplifted above sea level (e.g., the Scisti a Fucoidi Formation of central Italy and the Annona Formation of Arkansas). Deep-sea cores obtained by the Ocean Drilling Program (ODP) reveal deep-water sequences from unsubducted crust in the Pacific and South Atlantic oceans. Though not exhaustive, Table I lists many of the well-known Cretaceous bedding cycles. The origins are those proposed by the key workers on each bedding.

Primary limestone–marl cycles are thought to be generated by cyclic variation of four factors: (1) biogenic carbonate productivity, (2) detrital input from terrigenous runoff, (3) carbonate dissolution at or near the sediment–water interface and (4) preservation of organic matter. Each of these factors can cause the relative concentration of calcium carbonate ($CaCO_3$) in the deposited sediment to fluctuate about a critical value that divides the lithologic types. The primary

influence of the detrital material is direct if its proportion in the sediment layer reflects only fluctuations in its input. The influence of the detrital material can under certain circumstances be indirect if it modulates the preservation of carbonate input in the sediment by affecting the chemistry of the sediment–water interface. Changes in oceanic circulation (e.g., the location and intensity of upwelling regions) can affect the oceanic nutrient supply, which can affect carbonate productivity. Changes in productivity can affect bedding sequences by changing the amount of carbonate tests formed and by affecting the carbonation of deep and bottom waters and hence retention of carbonate. [See OCEAN CIRCULATION.]

Mesozoic sedimentary cycles from shallow, subtidal environments are thought to result from cyclical changes in sea level, though how this can occur on a global scale at Milankovitch time scales without continental ice sheets is not clear. It is possible that sea-level fluctuations were confined to restricted basins. When bedding planes are lifted above sea level no more sediments are deposited, and lithification of lime mud can occur. These cycles stand out as sharp boundaries in the stratigraphic record. [See SEA LEVEL FLUCTUATIONS; SEDIMENTARY GEOFLUXES, GLOBAL.]

A key factor in the formation of deep-ocean sediment cycles is the degree of dissolution of carbonate tests prior and/or subsequent to deposition on the ocean bottom. Dissolution is controlled by the degree of carbonate saturation and alkalinity of the ocean. Above a certain depth (3.7–4.0 km in the present-day ocean) negligible dissolution of calcite occurs. Below this depth, called the lysocline, is a transition zone in which partial dissolution occurs. Below the calcite compensation depth (CCD), 4.2–5.0 km in the present-day oceans, virtually all carbonate tests in the form of calcite are dissolved. A similar, though shallower, set of dissolution levels can be found for aragonite, another carbonate mineral. Cyclical changes in the depths of the lysocline and CCD, related to changes in the oceanic carbon budget and thermohaline circulation, can account for limestone–marl sequences by affecting the amount of carbonate actually deposited and preserved.

Black shale deposition is associated with oxygen-depleted conditions at the sediment–water interface and/or high biogenic productivity. The organic component need not be entirely marine. For example, wind-blown (aeolian) terrigenous organic material is thought to be a major contributor to the black shales of the Scisti a Fucoidi Formation of Italy; these periods of enhanced organic deposition alternate with periods of enhanced carbonate deposition and preservation. The deposition of marine black shales in shallow water is often associated with phosphorous deposits, suggesting an oceanic upwelling region. In such a scenario, upwelled nutrient-rich waters from oxygen-depleted intermediate or deep waters fuel productivity in the photic zone, which leads to

Table I   Examples of Well-Studied Cretaceous Cyclical Sedimentary Beddings

Sedimentary sequence	Age	Type of alternation	Proposed origin
Greenhorn Limestone, central United States	Throughout Cretaceous	Limestone with marlstone	Dilution (oxic/anoxic)
Scisti a Fucoidi, central Italy	Mid Cretaceous	Limestone; black shale	Oxic anoxic
Walvis Ridge, south Atlantic	Mid to late Cretaceous	Carbonate; black shale	Changes in dissolution
Northern Spain	Late Cretaceous	Limestone; marlstone	Dilution
Lower Chalk, southeast England	Early Cretaceous	Chalk; marlstone	Dilution
Upper Chalk, southeast England	Late Cretaceous	Chalk; turbidities	Bottom currents
Arcola Limestone, south central United States	Mid Cretaceous	Limestone; marlstone	High–low productivity
South France	Mid Cretaceous	Chalk; marlstone	Dilution

greater oxygen depletion and a greater supply of organic carbon. In deep water, the deposition of black shales requires bottom water that is anoxic or nearly so. In a silled basin like the Eastern Mediterranean, cyclic anoxia can be produced by an alternation of circulation at the sill, induced by fluctuations in the precipitation–evaporation balance over the basin's drainage area. The organic-rich Cenozoic sapropels in the Eastern Mediterranean are thought to have been deposited in periods where excess precipitation created a low-density cap on the basin, causing an outflow of surface waters across the Strait of Sicily and an inflow of nutrient-rich intermediate waters from the Western Mediterranean. This "nutrient trap" model leads to the rapid uptake of the remaining oxygen by organisms and hence to bottom-water anoxia. Similar silled basins in the Cretaceous Tethys and South Atlantic may have been governed by this scenario. According to geochemical modeling studies, open-ocean bottom-water anoxia appears to require special behavior in the region of deep-water formation. High productivity in this region can lead to oxygen-depleted bottom waters. This scenario fits well with the hypothesis that warm, saline Cretaceous bottom waters were generated at low latitudes, where increased sunlight enhances biogenic productivity. Given this model, cyclic anoxia in the global ocean could be generated by a cyclic shift of bottom-water formation regions from low- to high-productivity environments.

The above discussion concerns sedimentary bedding cycles that are of *primary* origin; that is, the cycles are due to changes in deposition environment. However, diagenetic processes may also play a key role in producing bedding cycles, although few specific mechanisms have yet been identified. The production of carbonate bedding cycles by diagenesis can occur through "rhythmic unmixing" whereby carbonate dissolves preferentially in one bed and redeposits in another. Until compaction has driven off pore water, calcite tends to accrete in beds with high $CaCO_3$ concentrations at the expense of beds with lower concentrations. Thus diagenesis can amplify a primary bedding rhythm involving

carbonate concentration. The dissolution–accretion mechanism is unstable and could create carbonate cycles out of random fluctuations in the sediment column. If this instability has a characteristic length or bedding thickness, bedded sequences that appear equivalent to orbitally induced primary bedding cycles may be created.

There are five main methods by which to distinguish primary from secondary carbonate rhythms: (1) Degree of dissolution of foraminifera and nannofossil tests. Undissolved tests imply little diagenetic unmixing since the diagenesis involves migration of dissolved carbonate. This method is important because several key Cretaceous bedding cycles show little dissolution of tests. Apparent dissolution does not imply necessarily that bedding is diagenetic rather than primary in origin, but it signals that a simple interpretation in terms of primary sediment inputs may be biased. (2) Differential compaction within different lithologies, as evidenced by relative distortion of macrofossils. Diagenetic unmixing causes carbonate-poor beds to undergo significantly greater compaction. (3) Preservation of fine structures (e.g., microbeddings in the transition of one lithology to another) argues that little or no diagenesis has occurred (4) Presence of burrow holes (dug by benthic organisms) that cross different lithologies indicates that significant diagenesis could not have occurred. (5) Evidence of secondary concretion (e.g., nonbiogenic carbonate in pore spaces of limestone) is direct evidence of diagenesis.

Another secondary effect is bioturbation, the reworking of sediments by bottom-dwelling, burrowing organisms. Substantial bioturbation can destroy primary lithologic cyclicity partially or completely. Avoiding this effect means generally that the sedimentary environments must meet one of two requirements: Either the bottom waters must be anoxic, in which case the bottom dwellers are not present, or the sedimentation rate must be high enough that structural beddings survive even with the occurrence of bioturbation. In summary, no single factor distinguishes whether Cretaceous bedding cycles are of primary or secondary origin. The relative contribution of primary and secondary factors will vary with the chosen sequence, making the construction of a global record of Cretaceous climate cyclicity more difficult.

III. Earth Orbital Insolation Changes

If the earth–sun orbit were circular and the earth's rotational axis not tilted with respect to the axis of the orbital plane, the insolation received by any point on the earth's surface would be a function only of latitude and the time of day. There would be no dependence on the time of year, and hence no seasons. This of course is not the case. Elliptical orbits are solutions to the gravitational two-body problem for the earth–sun system. The ellipticity of the orbit is quantified by its eccentricity. The gravitational forces of additional bodies (the moon and planets) in the physical system renders the earth's orbital motion chaotic, and induces quasi-periodic fluctuations in orbital eccentricity. Eccentricity variations have dominant quasi-periods near 95, 125, and 413 kyr, as well as periods greater than 1 Myr.

The obliquity angle δ is defined as the angle made by the earth's rotation axis to the axis of the earth–sun orbital plane. If $\delta = 0$, the varying earth–sun distance associated with orbital eccentricity would provide only a slight seasonality. The axis of rotation currently is tilted approximately 23.5° with respect to the perpendicular, yielding the familiar strong seasonality at mid and high latitudes. The obliquity, or "tilt" angle δ also varies in a quasi-periodic manner, within a range $\approx 2.5°$ and with dominant periods near 41 kyr; δ determines the location of the Arctic and Antarctic Circles and the Tropics of Cancer and Capricorn. Obliquity has no effect on the insolation received when integrated over the total earth surface, but it does affect how insolation is distributed. A warmer summer in one hemisphere will coincide with a cooler winter in the other hemisphere. Even with no net globally averaged insolation anomaly, climatic feedbacks triggered by obliquity may have a strong net global effect. Also, an increase in obliquity will increase annual-average insolation at mid and high latitudes, with the surplus compensated by a decreased annual-average insolation at low latitudes.

Earth rotation induces an elliptical bulge at its equator, and the combined gravitational forces of the sun, moon, and planets attempt to pull the associated mass anomaly into alignment with the orbital plane. In response the axis of earth rotation precesses about the perpendicular to the orbital plane, similar to the precession of a spinning top. This precession, with dominant quasi-periods near 16, 19, and 23 kyr, is referred to commonly as the precession of the equinoxes. This terminology applies because precession describes the relationship between the equinoxes and solstices, defined by relation of the tilt with the directed earth–sun distance (e.g.,

summer solstice occurs in the northern hemisphere when the north pole is tilted directly toward the sun) and the perhelion and aphelion of the earth's orbit (i.e., minimum and maximum earth–sun distances). For example, if perhelion occurs in summer (in the hemisphere tilted toward the sun), the insolation and hence seasonal warming will be enhanced. If perhelion occurs at an equinox, precession has a relatively small winter/summer effect.

Figure 1 shows the latitude-dependent seasonal insolation changes associated with precession and obliquity. Orbital eccentricity has a relatively small annual-average global effect; however, it plays a more important role in modulating the precessional cycle. The combined effect of precession and eccentricity is expressed via the precessional index $p = e \sin \omega$ where e is the eccentricity and ω is the celestial longitude of perhelion, measured with respect to the vernal (spring) equinox. The longitude of perhelion determines the relation between the maximum and minimum earth–sun distance and the summer–winter seasonality of the tilt cycle. Clearly the index is zero (no effect of precession) if e is zero (circular earth–sun orbit) or if perhelion occurs at one of the equinoxes ($\omega = 0$ or π). A maximum effect requires both a large eccentricity and a perhelion at one of the solstices ($\omega = \pi/2$ or $3\pi/2$). Precession thus affects the total insolation received by the earth. A relatively colder winter follows a warmer summer, while a warmer winter follows a cooler summer. On an annual-average basis there is no net insolation effect from precession either globally or at any latitude, but there is a net global effect at any instant in time. [See SOLAR RADIATION.]

Although astronomers of the eighteenth century realized that the earth's orbit underwent fluctuations and had some idea of what the dominant periods of variability were, much work continues on the precise determination of the dominant periods and associated insolation anomalies. Culminating in the recent work of A. D. Vernekar and of A. Berger, we now have integrations of the celestial system considered reliable from the present to 2 Ma. The earth–sun orbital cycles are quasi-periodic, so that any finite expansion of periodicities fails to represent celestial motion after a sufficiently long time interval. Therefore, a direct extrapolation of present-day orbital motions to the Cretaceous is not advised. However, if the relative positions of planetary orbits remain constant, the dominant quasi-periods of the earth–sun orbit will not change, save for small perturbations to precession and obliquity periods associated with changes in the earth's rotation rate caused by tidal friction. This is consistent with spectrum analysis by P. E. Olsen and by J. Park and T. D. Herbert, which suggests that the ratios of the major orbital periods in the Mesozoic are approximately the same as today. In fact, in the Park and Herbert study, the obliquity period appeared shifted relative to eccentricity periods, with sign and magnitude of the shift consistent with the current rate of

Insolation Perturbation at Solstice

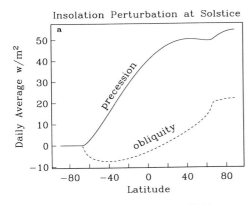

Avg Perturbation over Halfyear

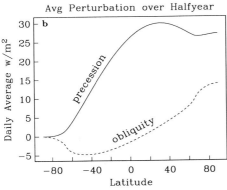

Fig. 1 Comparison of insolation anomalies associated with precession and obliquity. Graphed is anomalous energy, averaged over a day (in $W\ m^{-2}$) as a function of latitude for an orbit in which the earth's north pole tilts toward the sun at closest approach. For comparison, typical daily mean insolation values range from 200 to 400 $W\ m^{-2}$ in summer and at low latitudes, and from 0 to 100 $W\ m^{-2}$ in winter at mid and high latitudes. Negative latitude corresponds to the southern hemisphere. The solid lines show insolation anomalies appropriate for a 10% increase in the solar constant at constant present-day obliquity, appropriate for eccentricity $e \approx 0.05$ relative to a circular orbit. The dashed lines show insolation anomalies for a circular orbit given a 1° increase in the obliquity angle δ from its present-day value of 23.45°. (Both perturbations are close to the maximum total deviations from the mean values of the orbital parameters.) (a) Insolation anomaly at the closest earth–sun distance (northern hemisphere summer). (b) Daily-average insolation anomaly for the halfyear centered on the closest approach of the earth to the sun (i.e., from fall to spring equinox in the northern hemisphere).

tidal friction. This suggests that analysis and interpretation of cyclical bedding cycles may provide information on the past history of earth rotation.

In the nineteenth century both J. Croll and J. A. Adehmar speculated on the effect orbital insolation changes have on the earth's climate, with particular attention to implications for Pleistocene glaciation. In the early part of the twentieth century, M. Milankovitch quantified the insolation changes described above and theorized about the importance of mid-summer mid-latitude insolation changes for glaciation. If Cretaceous bedding cycles are indeed coincident with orbital periods, then the task remains of explaining how the insolation changes could pace or drive the sediment cycles, especially in the absence of large-scale continental ice. Furthermore, the apparent dominance of 100-kyr variability in many Cretaceous cyclical sediments requires explanation since insolation changes are much smaller in general at 100 kyr than at 19 and 23 kyr and at 41 kyr. This key question will be explored further in subsequent sections.

IV. Spectrum Analysis of Cretaceous Bedding Cycles

The inference of orbital periods in Cretaceous bedding cycles depends on several assumptions about the record. In the best of circumstances these assumptions can be checked against other geologic evidence. However, in many cases one is left with conclusions no more specific than that the sedimentary record is *consistent* with Milankovitch cyclicity, but that other models are possible. Many hypotheses about the record can be tested with the tools of spectrum analysis. A time series (e.g., carbonate percentage sampled down a deep-sea drill core) can be decomposed into Fourier components that yield information on any periodic variations about the *mean* of the time series. Preferred periods of variation stand out as peaks above the background spectrum, which is assumed to be generated by random, nonperiodic behavior and measurement noise. A narrow peak can indicate a very regular periodicity (such as the annual cycle of solar insolation). Time windows of orbital parameter calculations exceeding 1-Myr duration typically return narrow spectral peaks, despite the quasi-periodic behavior of the celestial system. A broad spectral peak can indicate either a strongly quasi-periodic process, such as an instability in the climatic system that may be amplified by orbital insolation changes, or a lack of sufficient time control in the time series. The latter problem is common in analyses of sedimentary sequences where variations in the sediment accumulation rate are poorly known. Figure 2 shows the spectral decomposition for a well-preserved Cretaceous core. The challenge to the analyst is to discriminate

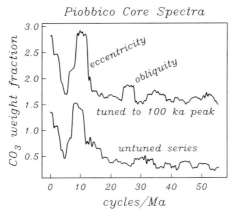

Fig. 2 Smoothed-spectrum estimate for the mid-Cretaceous core drilled near Piobbico, Italy. Plotted is the carbonate fraction by weight versus cycles per million years. The lower curve is the spectrum of a 522-point time series of roughly 1.4-Myr duration, and the estimate in the upper curve uses the same series after tuning to the 100-kyr eccentricity peak. This latter peak is a dominant feature of both spectra. The obliquity peak shows up clearly in the tuned spectra but is not readily apparent in the untuned spectra. Precession peaks are not readily apparent in either spectra, although this is probably because the resolution of the core is inadequate to resolve them.

whether a broad observed spectral peak is a distorted version of a Milankovitch cycle or represents another periodic or quasi-periodic process (e.g., diagenetic unmixing) or is due to an instability in the climatic system.

Paleo-environmental proxy variables in sedimentary sequences can be chemical (e.g., oxygen isotopes or $CaCO_3$ concentration) or lithologic (limestone, marl, sandstone, black shale). Oxygen isotope measurements, useful in Pleistocene analyses, are often biased by lithification in older carbonate sequences. Measurements of proxy variables are not made against time but against depth within the sequence. Accurate dating of the time series is critical to an accurate determination of the cycle periods inferred by spectrum analysis. Biostratigraphic time constraints often establish that observed sedimentary cycles have principal periods within the range of orbital time scales. Dating Cretaceous sedimentary beddings is uncertain, however, as accurate dates are restricted usually to a few stratigraphic markers with intermediate dates interpolated. In fact, if it is established that cyclic Cretaceous sequences are indeed causally linked to orbital periods, the cycles themselves would be a powerful dating tool (as with the Pleistocene record). In the absence of adequate absolute time constraints, the analyst can test whether spectral peaks revealed by time series analysis have periods in the same proportion as the orbital cycles (e.g., ratios of 19, 23, 41, 95, 125, and 413 kyr). Sedimentologists note that

many Cretaceous sequences possess a "bundling" of bedding cycles, in which four or five lithologic alternations are grouped into larger-cycle sets. This is associated commonly with the ratio of precession and eccentricity periods, and is plausible because eccentricity modulates the precession insolation anomaly. Spectrum analysis often, but not always, confirms these observations by revealing spectral peaks that correspond roughly to orbital period ratios.

A more detailed comparison requires that corrections be made for variations in the sediment accumulation rate before spectrum analysis. This is called tuning the time–depth function. The analyst chooses a frequency associated with a prominent spectral peak and varies the time–depth function to preserve that peak's period and, perhaps, phase-coherence. This tuning procedure, however, involves the assumption that at least one of the spectral peaks revealed by analysis of the untuned data series is truly periodic. If we use spectrum analysis of the tuned series to argue for a link with orbital insolation cycles, this may lead to a circular argument. It is helpful if analysis of the tuned series reveals more behavior consistent with orbital cycles than is assumed in the tuning procedure. In analysis of carbonate content sampled in a drill core of mid-Cretaceous carbonate limestone–black shale cycles from central Italy, Park and Herbert tuned the time–depth function using a prominent spectral peak, which was constrained by biostratigraphy to be near 100 kyr period. After tuning, this peak resolved into two peaks with frequency ratio indistinguishable from that of the two major eccentricity periods at roughly 95 and 125 kyr. In addition, a smaller, but highly periodic, spectral feature was revealed whose observed period, when scaled relative to the presumed eccentricity cycles, was consistent with the obliquity cycle after correction for 100 myr of tidal friction effects. Results such as these argue against circularity in the procedure, but may not be achievable in all cases. For instance, sedimentation rates in that sequence appear not to have been high enough to avoid substantial distortion of shorter-period precession cycles by bioturbation.

The existence of eccentricity cycles in many Cretaceous sequences is particularly intriguing, both because the late Pleistocene glaciations also show a dominant 100-kyr periodicity and because the direct insolation effect of eccentricity at 100 and 400 kyr is quite small. Because eccentricity modulates the precession insolation signal, nonlinear earth system response can give rise to eccentricity cycles in the geologic record. A forced, nonlinear response of the atmospheric–ocean system to insolation variations, analogous to that hypothesized for the 100-kyr ice sheet cycles in the Pleistocene, might lead to primary eccentricity bedding cycles. Carbonate deposition depends in part on biologic productivity in the oceans, whose dependence on insolation and nutrient supply need not be linear. Secondary processes, such as bioturbation and diagenetic unmixing, may produce eccentricity cycles

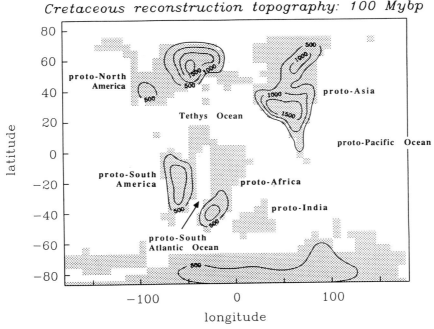

Fig. 3 Mid-Cretaceous (100 Ma) paleogeographic and topographic reconstruction. The precursors to the modern continents are shaded and labeled; topographic height contours are every 500 m.

from primary precessional bedding cycles. Then again, it is possible that the 100-kyr periodicity might itself be primarily due to internal instabilities in the climate system, which can be "paced" by orbital insolation changes. In this case, a roughly 100-kyr cycle could occur even in the absence of Milankovitch orbital insolation changes.

V. Establishing Links between Insolation Changes and Sedimentation

A. Cretaceous Climate as Inferred from the Geologic Record

Figure 3 shows the reconstructed geography and topography of the mid-Cretaceous. The dominant feature of the reconstructed Cretaceous climate is its warmth compared to the present day. One key indicator of this is the apparent absence of widespread continental glaciation throughout the entire period. Another key indicator is the presence at relatively high latitudes of fossil plant remains that appear to be from frost-intolerant species. Dinosaur remains have been found at Arctic sites that appear to have been at 70–80°N paleolatitude during the Cretaceous. It has been estimated that globally and annually averaged surface temperatures during the

Cretaceous were at least 6°C warmer than at present. Though much of this warming appears to have occurred at mid and high latitudes, it is likely that the tropics were somewhat warmer as well. Warm temperate forest remains have been found at 50–60° reconstructed latitude, while cool temperate forests appear to have occupied near-polar locations. In general, the Cretaceous also appears to have been somewhat drier than at present, as suggested by the lack of widespread tropical and subtropical rainforests and by the widespread presence at low and mid latitudes of evaporite deposits. Many workers, notably D. I. Axelrod, have suggested that the Cretaceous climate was more equable—that is, demonstrated less seasonality than at present. Paleofloral evidence suggests widespread frost-free winters. The existence of extensive continental seaways could also act to moderate climate. By extension, the influence of orbital insolation changes would likewise be moderated, since obliquity and precession both are seasonal effects. However this notion of equability has been challenged by other workers on both observational and theoretical grounds.

The meridional atmospheric and oceanic surface temperature gradients were likely to have been smaller than at present, suggesting less stormy conditions at mid-latitudes. There is some indication of this from records of aeolian deposition and precipitation patterns, although in general it is difficult to make inferences about anything other than surface temperature and precipitation from records as ancient as those from the Cretaceous. There is also some indication that the Cretaceous climate may have had an important monsoonal (land–sea) circulation. This may be because of a reduction in the number and/or intensity of cyclonic storms at mid-latitudes, especially since the continents in the Cretaceous were generally smaller than at present. This relative importance of monsoonal circulations will play an important role in many of the scenarios described below that relate climatic changes to sedimentation cycles. [*See* ATMOSPHERIC CIRCULATION SYSTEMS.]

Inferences about the Cretaceous oceanic circulation are more problematical than those regarding the atmospheric circulation. One observation that seems certain is that the ocean as a whole was up to 10°C warmer than at present, both at the surface and at depth. The explanation of this warmth revolves around one of three scenarios, none of which have yet been clearly supported by interpretations of geologic data. The first holds that Cretaceous oceanic thermohaline circulation was driven in a manner similar to the present, which implies that warm, high-latitude atmospheric conditions must have driven Cretaceous high-latitude ocean temperatures, that, while relatively cold, were warm compared to the present. The atmosphere would warm the entire ocean by warming the deep-water formation zones. The second possibility is that the oceanic circulation was the reverse of the present circulation, with warm deep waters formed at low latitudes, probably through the influence of excess salinity on the density of ocean waters (warm saline deep water). The warm deep waters are transported poleward and upwell to warm the high latitudes. In this case the oceanic state drives the high-latitude atmospheric warmth. The third possibility is that there was no localized source of deep-water formation; instead, the ocean warmed through purely diffusive means to be close to the globally and annually averaged mean surface temperature of the earth, currently 15°C. Though we cannot yet adequately discriminate between these scenarios, they are crucial to possible scenarios relating insolation changes to changes in oceanic sedimentation. [*See* AIR–SEA INTERACTION.]

To summarize, the Cretaceous climate appears to have been warmer and drier than the present-day climate. The warmth appears to have been most prevalent at mid and high latitudes, while drier conditions were prevalent at low and mid latitudes. Seasonal extremes are thought to have been less than at present with much evidence for frost-free winters. The atmospheric circulation is thought to have been more sluggish, with a reduced importance of mid-latitude cyclonic disturbances and enhanced importance of monsoonal circulations. The oceanic circulation is more problematical and could reflect the influence of a warm high-latitude atmosphere on the ocean, or the influence of a warm high-latitude

ocean on the atmosphere. A major challenge remains to gather enough geologic data and develop enough physical understanding to resolve these uncertainties.

B. Possible Scenarios

1. Changes in Biologic Productivity

Biologic productivity can change because of changes in the riverine input of nutrients, or because of changes in the oceanic circulation that affect the distribution of nutrients. Cyclic climate change can influence both factors. Systematic changes in precipitation and evaporation patterns would lead to systematic changes in the amount and/or location of riverine runoff. Changes in the oceanic thermohaline circulation can result from regional changes in evaporation and precipitation patterns affecting the formation of deep-ocean waters, or through atmospheric circulation changes that affect the location and intensity of upwelling regions such as those found off the west coasts of continents. W. S. Broecker and T.-H. Peng have noted that the residence time of oceanic phosphorus, a critical nutrient, is close to 100 kyr. This long nutrient residence time may introduce inertia into the ocean biosphere response. If so, rapid precession insolation changes could be damped nonlinearly so that only the low-frequency 100 kyr eccentricity envelope is apparent.

The same types of climatic changes can induce shallow-water sedimentary cycles, though in a different fashion. In this situation, the changing distribution of runoff due to regional climate changes would change the amount of terrigenous material delivered to near-shore waters, possibly accounting for dilution–productivity cycles.

The above scenarios require changes in climate that are largely manifested by changes in the hydrologic cycle. One possibility is that insolation changes affected the Cretaceous climate by affecting primarily its *monsoonality* (i.e., land–ocean circulations). Shifts in the location and intensity of regions of precipitation and aridity that result from the changing monsoons could provide the required runoff or net evaporation changes required.

2. Oxic versus Anoxic Bottom-Water Conditions

Under oxygen-depleted (anoxic) conditions, black shale or marl is preferentially deposited, representing preservation of organic materials; under oxygenated (oxic) conditions carbonate deposits can dominate. Warmer water can hold less dissolved oxygen, so the warmer pelagic realm of the Cretaceous was probably less oxygenated than the present-day ocean. In addition, the degree of oxygenation of ocean waters is related to the vertical stratification. Under highly stratified conditions, little vertical mixing occurs and deep waters can become oxygen-starved. Such a situation can occur in near-shore environments or restricted basins during times of heavy precipitation and runoff, when a light, fresh-water cap can inhibit vertical mixing. During times of light precipitation and excess evaporation, vertical mixing can be facilitated by the formation of warm saline surface water. Deep waters of the open ocean can be affected on a larger scale if changes in evaporation and precipitation occur over critical regions of deep-water formation. In general, greater deep-water production increases the oxygenation of the deep ocean.

3. Changes in Aeolian Deposition

As described in Section II, changes in aeolian input to the ocean can play a role in forming sedimentary bedding cycles. Climate changes can affect aeolian input to the ocean (a) by affecting the size and the location of source regions (e.g., arid regions) and (b) by affecting the distance that airborne particles are transported through the intensity of the surface and low-level winds.

Information on dust-grain size has provided information on wind intensity during the Pleistocene. W. L. Prell and colleagues argue for a relationship between the aridity of northern Africa and the Middle East, and aeolian deposits in the Indian Ocean, presumably related to changes in the Asiatic monsoon. The waxing and waning of the Pleistocene ice sheets likely played a major role in these changes. Implications for the Cretaceous are more problematical, since there were no major continental ice sheets; but it is quite possible that cyclical changes in monsoon circulations affected both aridity and wind intensity patterns, hence playing a role in explaining the evidence for cyclic changes in aeolian input that has been reported from some Cretaceous bedding cycles. [See AEOLIAN DUST AND DUST DEPOSITS.]

4. Changes in Sea Level

Changes in sea level, both local and global, can produce sedimentary cycles in near-shore environments (see Section II). Two methods of explaining cyclical changes in Cretaceous sea levels resulting from orbital insolation changes have been proposed. One method involves local sea-level changes resulting from cyclical changes in sedimentary loading resulting from basinwide climatic changes (the weight of sediments depresses the ocean floor). The second method questions the inference that the Cretaceous was free of significant continental ice cover. In this approach, it is assumed that an ice cover did exist over Antarctica and possibly limited regions in the northern hemisphere, with temperate, ice-free conditions elsewhere. Cyclical waxing and waning of the ice cover in part influenced by orbital insolation changes would have resulted in *global* fluctuations in sea level, with consequent changes in the deposition and preservation of near-shore sediments.

5. Changes in Carbonate Supply to the Ocean

A final scenario involves changes in the supply of carbonate to the ocean. These fluctuations can be important, as they change the chemistry of the ocean and hence the levels at which carbonate dissolution occurs. Lack of a thorough understanding of the global carbon cycle hampers the exploration of this scenario, although one important factor could be changes in the net carbonate supply to the ocean via climatically induced changes in runoff. Another factor could be changes in the deep-ocean circulation, which affect the age and hence alkalinity of the waters. An additional factor is the occurrence of turbidities, which are the debris left from underwater landslides that frequently contain a large fraction of sand and silt. Layering associated with turbidities can add distinct, in some cases very thick, layers to sediment sequences. Turbidite flows are thought to occur fairly randomly in time, so that possible links between their frequency and insolation cycles are more difficult to establish in the geologic record. Much work will be required to clarify these mechanisms. [See CHEMICAL MASS BALANCE BETWEEN RIVERS AND OCEANS.]

VI. Mathematical Modeling of Cretaceous Climate

The theme that most of the above scenarios share is a cyclic climatic response involving elements of the hydrologic cycle. Mathematical models of climate can be used to help evaluate the potential role and significance of each of these scenarios. A hierarchy of models, ranging from simple energy-balance models to more complex statistical–dynamical models, to the very complex general circulation models have been applied to theoretical investigations of Cretaceous climate. A key goal of most studies to date has been to account for the inferred Cretaceous warmth relative to the present, a task that has been only partially successful. A few modeling studies (T. J. Glancy, E. J. Barron, and M. A. Arthur; R. J. Oglesby and J. Park) have directly addressed the effects of orbital insolation changes on Cretaceous climate. [See ATMOSPHERIC MODELING.]

Energy-balance models suggest that albedo changes resulting from changed continental locations are insufficient to account for the Cretaceous warmth and that seasonality, especially summertime warmth, may play an important role in accounting for ice-free Cretaceous conditions. This mechanism, however, is at odds with the idea of an "equable" Cretaceous climate. Studies by R. J. Oglesby and B. Saltzman with a more sophisticated statistical–dynamical model that includes a computation of the zonally symmetric atmospheric circulation and hydrologic cycle suggest that a warm deep ocean can play a crucial role in accounting for Cretaceous warmth, aridity, and ice-free conditions, but leaves open the question of how the oceanic warmth was maintained. These modeling studies did not make direct use of orbital insolation changes. [*See* LAND–ATMOSPHERE INTERACTIONS.]

The most detailed modeling studies to date have employed general circulation models of the earth climatic system. These models consist of computer codes designed to solve for present-day weather patterns (forecast models), but they are executed with Cretaceous boundary conditions. This modeling was pioneered by Barron and colleagues at the National Center for Atmospheric Research (NCAR). Using an increasingly advanced series of NCAR global circulation models (GCMs), these workers investigated the effects of continental location and surface ocean heat transport on Cretaceous warmth and ice-free conditions. The studies yielded significant global warming; but they failed to predict ice-free, frost-free conditions, as cold mid-winter continental temperatures occurred at mid and high latitudes. Speculation regarding other warming factors has centered around elevated levels of atmospheric CO_2, a greenhouse gas. A recent GCM simulation with present-day geography and 1000 ppm CO_2 (three times the control value) has a globally averaged surface temperature about 6.5° warmer than the present-day control, consistent with the inferred Cretaceous record, with considerably warmer sea-surface temperature (SST) at all latitudes. However, this simulation still has below-freezing mid and high latitude temperatures with extensive snowcover in winter, as well as seasonal (but little perennial) sea ice in the Arctic Ocean. Figure 4 shows July and January surface temperatures for a mid-Cretaceous GCM simulation. The cold winter temperatures in both hemispheres appear to contradict previous interpretations of Cretaceous paleofloral data, which suggest frost-free conditions on the land masses.

Some workers are skeptical, however, that continental interiors are sufficiently well-represented in the paleofloral record to sustain these interpretations. Figure 5 shows January and July sea-level pressures. A key feature here is the seasonal alternation of pressure patterns. In the summer hemisphere, low pressures are found over the continents, with high pressure over adjacent oceans. In winter the continents are regions of high pressure, with low pressure over the oceans. These land–ocean pressure differences and their seasonal shift drive the monsoons that dominate the simulated Cretaceous climate.

The studies of Glancy, Barron, and Arthur and of Oglesby and Park have investigated by response of Cretaceous climate to changes in insolation. Both studies used versions (but different ones) of the NCAR GCM run in perpetual season mode at midwinter and midsummer. This execution mode allows climate statistics to be estimated in short (150-day) model simulation, but requires that the sea surface temperature be prescribed *a priori* rather than calculated. The goal of the Glancy *et al.* study was to model the formation of the

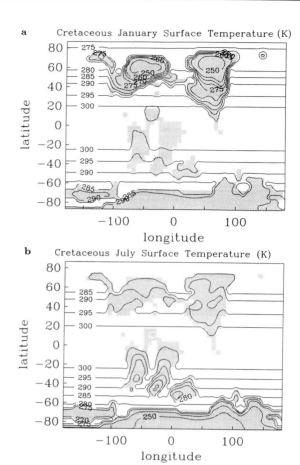

Fig. 4 Mid-Cretaceous simulated surface temperature, in degrees Kelvin, for (a) January and (b) July. Contour interval is 5 K.

Bridge Creek Member of the Colorado Greenhorn formation, a typical shallow-water limestone–marl sequence. They examined the scenario in which fluctuations in rainfall over adjacent land regions influenced sediment deposition in the North American Interior Seaway. In particular, they examined the hypothesis that an increase in the maximum seasonal range (i.e., warm summers with cool winters versus cool summers with warm winters) could account for appropriate climatic changes. At maximum seasonality, warmer summers could enhance monsoonal precipitation by increasing the flow of moist air from the ocean to the adjacent land, while the cooler winters could enhance storminess because of the increased land–ocean temperature contrast, thereby increasing precipitation. A combination of orbital parameters yielding a minimum seasonal range was compared with a combination yielding a maximum seasonal range. Precipitation over North America, the key model variable in their study, showed a fairly clear response in January, increasing with decreasing insolation (cooler winter), consistent with the hypothesis

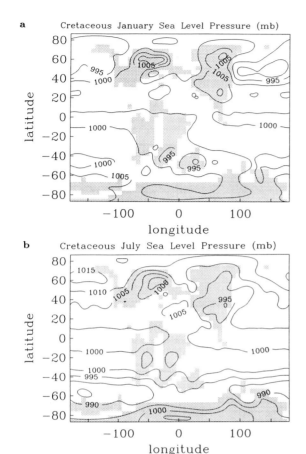

Fig. 5 Mid-Cretaceous simulated sea-level pressure, in millibars, for (a) January and (b) July. Contour interval is 5 mbar.

they were examining. However, July precipitation showed no trend with orbital parameters, in large part because there was little July precipitation over North America in either simulation. [*See* CLIMATIC CHANGE: RELATIONSHIP BETWEEN TEMPERATURE AND PRECIPITATION.]

Oglesby and Park used a Cretaceous geographic reconstruction and prescribed SST regime quite similar to that used by Glancy *et al.*, but they used a later version of the NCAR GCM. A large model response to precession was inferred from their study, especially with regard to monsoonal circulations. A weaker response to obliquity was apparent. Model control simulations for the Cretaceous demonstrated the dominance of monsoonal circulations, both because of large continent–ocean heating asymmetries and because of a reduced importance of mid-latitude cyclonic activity. Because it contains the larger land masses, the northern hemisphere monsoon dominates the relatively large overall model response to precession. As typical examples of the model response, Fig. 6 shows the response of surface temperature,

sea-level pressure, atmospheric specific humidity, and precipitation area-averaged over south proto-Asia for January (winter); and Fig. 7 shows the same quantities area-averaged over northeast Africa for July (summer). For south proto-Asia in winter, surface temperature and sea-level pressure show a systematic linear sensitivity, although significant model noise is apparent; specific humidity and precipitation show less systematic responses. In the summertime northeast Africa, however, all four variables show a clear linear response. Maximum–minimum precessional pairs (warm summer–cool winter *or* cool summer–warm winter) yield stronger or weaker monsoons, respectively. The response of the summer monsoon to precession is somewhat more systematic than that of winter, probably reflecting the relative importance of surface heating in summer and storminess in winter.

These modeling results suggest that the response to changes in obliquity is everywhere smaller than that of precession, and that obliquity response is generally most significant at low latitudes. The significant sensitivity to obliquity at low latitudes is somewhat unexpected, since its greatest relative insolation anomalies occur at high latitudes. The explanation may be that the inherent model variability for surface pressure is generally large at high latitudes and small at low latitudes. In model runs, insolation anomalies at high latitudes do not lead to a systematic pressure response, while smaller, low-latitude anomalies are more frequently significant because model variability is less there. Systematic surface-pressure responses to obliquity appear to modulate the longitudinal (east–west) range of the precession-enhanced continental monsoons.

Of particular interest is the response of the model hydrologic cycle to precession and obliquity changes. The model results demonstrate sensitivity of the evaporation minus precipitation (E–P) balance to orbital cycles in a number of regions. Most interesting is the response of the E–P balance over the newly rifted South Atlantic Ocean. Figure 8 graphs linear sensitivity of average moisture transport divergence $\nabla \cdot \overline{VQ}$ to changes in the precession and obliquity parameters. Regions where $\nabla \cdot \overline{VQ} < 0$ experience excess precipitation relative to evaporation. Regions where $\nabla \cdot \overline{VQ} > 0$ experience excess evaporation relative to precipitation. The moisture divergence in the perpetual-January and perpetual-July control simulations (which correspond to a circular orbit and tilt angle $\delta = 23.45$) are plotted for reference. Other plots represent the linear response of the climate system to the total perturbation range associated with precession and obliquity. Note that the overall response in July is larger than in January. Formal error analysis of the linear sensitivity coefficients suggest that the January response of $\nabla \cdot \overline{VQ}$ over the South Atlantic is poorly determined relative to the July response, which shows large (and opposite) responses to precession at low and mid latitudes in the narrow ocean basin. This restricted ocean basin was the site of cyclic black shale/

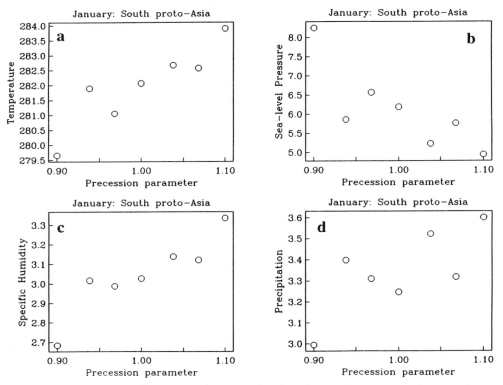

Fig. 6 Simulated response of Cretaceous climate as a function of precession area averaged over south proto-Asia for January for (a) surface temperature in degrees Kelvin; (b) sea level pressure in millibars; (c) atmospheric specific humidity in grams per kilogram; and (d) precipitation in 10^{-8} ms^{-1} (1×10^{-8} ms^{-1} is about 2.5 cm per month).

limestone sediment deposition during the Cretaceous, and it has also been cited as a possible location for the formation of warm saline deep water for the wider Cretaceous ocean. The modeling results suggest that the cyclical evaporation/precipitation changes could occur in response to precessional changes, further suggesting cyclical changes in the formation of warm saline deep water and hence both local and global oxic/anoxic conditions. [See WATER CYCLE, GLOBAL.]

The results of Oglesby and Park also bear on the question of whether the aeolian transport and subsequent deposition of terrigenous materials shows any systematic response to precession. Of particular interest would be results pertaining to changes in continental aridity, surface wind direction, and wind mean-intensity and variances, the latter reflecting the mean and episodic potential of the wind to sweep terrigenous material from arid regions. There is some indication of a model response over proto-Africa and the South Atlantic, where wind variances increase as the wet/dry annual cycle over land intensifies (max precession case). This tendency, however, occurs during the wet season, providing apparently contradictory results. In general, the model results do not

yield a clear determination of sensitivity to changes in aeolian entrainment, transport, and deposition. In large part, this is likely due to inadequacies in existing climate models, especially their representation of the atmospheric boundary layer.

Over all regions the model response could not be considered largely different from linear plus model noise (due to the variability inherent in the model). A linear response to precession and obliquity forcing will yield responses only near 19–23 and 41 kyr. Studies with current general circulation models do not exhibit the nonlinear response to insolation required for the explanation of 100-kyr eccentricity periods seen in some Cretaceous bedding cycles.

VII. Summary: Where Do We Go from Here?

Analysis of Cretaceous bedding cycles suggests a relationship between the primary periods of cyclical sedimentary deposition and those of orbital insolation changes. However, our knowledge of the occurrence of sedimentary cycles is re-

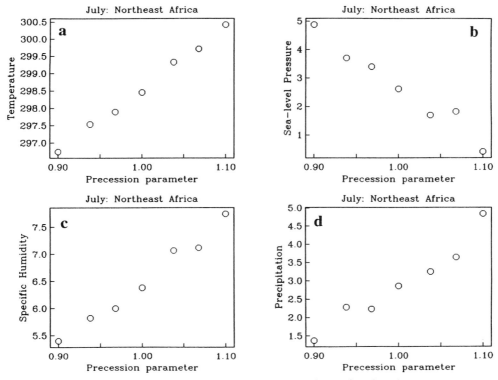

Fig. 7 Same as Fig. 6 but area-averaged over northeast Africa for July.

stricted to those locations where they are accessible and/or apparent. Large gaps presumably exist in our knowledge of the spatial and temporal extent of these cycles, which hampers attempts at a global reconstruction of cyclic climate changes. Many of the Cretaceous cyclical sequences have been sampled at outcrops. Better depth/age constraints and less bias due to chemical weathering can be obtained by drilling formations, but this has been done at relatively few formations to date. A more serious deficiency is that we usually have only crude and potentially inaccurate means of dating, both relatively and absolutely, the sedimentary sequences. Careful biostratigraphy and/or radiometric dating is necessary for absolute time constraint. The analysis of drill cores, rather than outcrops, reduces considerably an important source of relative time scale uncertainty. [See STRATIGRAPHY—DATING.]

The modeling results suggest linkages between insolation changes, Cretaceous climate, and bedding cycles. However, these results are subject to large uncertainty. The studies suggest that changes in surface heating and consequent changes in the atmospheric circulation, especially involving the hydrologic cycle, result from changes in orbital insolation. However, inferences based on these results are limited for three reasons: (1) These are coarse-resolution models known to contain many deficiencies in replicating the present-day observed climate. To date, only atmospheric GCMs (sometimes coupled to simple oceanic thermodynamic models for computing SST) have been used to address directly these linkages. (2) Surface boundary conditions and the atmospheric composition (e.g., CO_2 concentration) must be prescribed in these models and are not well known for the Cretaceous, especially the spatial coverage of vegetation. These uncertainties may explain in part why GCM studies have had difficulty in accounting fully for the inferred Cretaceous warmth. Because the inferred mean Cretaceous climate has not been simulated successfully, model results regarding the effect of insolation perturbations relative to the mean state must be treated with caution. (3) Sediment deposition and preservation involves all components of the land–atmosphere–ocean system. Full explanation of bedding cycles will thus require coupling of atmospheric GCMs, ocean GCMs, biogeochemical models (e.g., models of ocean productivity), models of sedimentary deposition, and models of diagenetic alteration of sediments. To date, no attempt has been made to synthesize all of these factors. Thus, the best modeling studies can offer at present is to suggest plausible links between insolation changes and bedding cycles. It will remain for future work to uncover more precisely the physical mechanisms involved.

Finally, there remains the problem of explaining the appar-

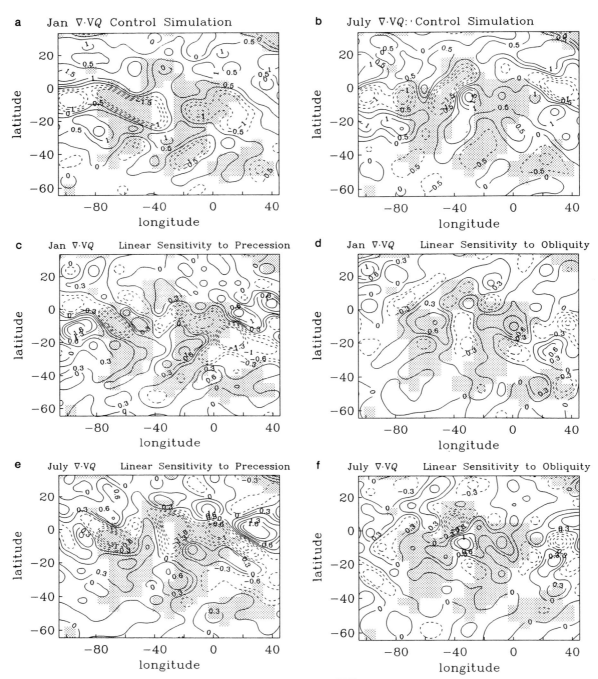

Fig. 8 Linear sensitivity of average moisture transport divergence $\nabla \cdot \overline{VQ}$ to changes in the precession and obliquity parameters. The southern continents are shown, with the proto-South Atlantic separating South America and Africa. Regions where $\nabla \cdot \overline{VQ} < 0$ experience excess precipitation relative to evaporation. Regions where $\nabla \cdot \overline{VQ} > 0$ experience excess evaporation relative to precipitation. Contour units are 10^{-8} m/s. The moisture divergence in the perpetual-January and perpetual-July control simulations (which correspond to a circular orbit and tilt angle $\delta = 23.45$) are plotted for reference. Other plots represent the linear response of the climate system to the total perturbation range associated with precession and obliquity.

ent preponderance of 100-kyr cycles in the sedimentary record, despite the small amount of direct orbital forcing at that eccentricity period, compared to the shorter periods of obliquity and precession. The modeling results of Oglesby and Park suggest a largely linear response of the atmosphere to precessional and obliquity insolation changes. If the modeling results are not misleading (possible, given the crudeness of existing models), the linear atmospheric response may be followed by nonlinear response within either the biologic, oceanic, or sedimentary systems. An alternative option posits that some internal instability in the climatic system generates the 100-kyr cycles, so that they would be evident to some extent even in the absence of Milankovitch orbital forcing.

Bibliography

Arthur, M. A., and Garrison, R. E., eds. (1986). *Paleoceanography* 1 (4), 369–586. Special issue devoted to Milankovitch cycles throughout earth history. (Includes GCM modeling study by Glancy, Barron, and Arthur.)

Barron, E. J., and Washington, W. M. (1984). The role of geographic variables in explaining paleoclimates: Results from Cretaceous climate model simulations. *J. Geophys. Res.* **89,** 1267–1279.

Barron, E. J., Arthur, M. A., and Kauffman, E. G. (1985). Cretaceous rhythmic bedding sequences: a plausible link between orbital variations and climate. *Earth Planet. Sci. Lett.* **72,** 327–340.

Berger, A., Imbrie, J., Hays, J., Kukla, G., and Saltzman, B., eds. (1984). "Milankovitch and Climate, Part 1." D. Riedel, Dordrecht, The Netherlands.

Dean, W. E., Arthur, M. A., and Stow, D. A. V. (1984). Origin and geochemistry of Cretaceous deep-sea black shales and multicolored claystones, with emphasis on Deep Sea Drilling Project Site 530, southern Angola Basin. *Int. Rep. Deep Sea Drill. Proj.* **75,** 819–844.

Einsele, G., and Seilacher, A., (1982). "Cyclic and Event Stratification." Springer-Verlag, New York.

Oglesby, R. J., and Park, J. (1989). The effect of precessional insolation changes on Cretaceous climate and cyclic sedimentation. *Geophys. Res.* **94,** 14793–14816.

Olsen, P. E. (1986). A 40-million-year lake record of early Mesozoic orbital climate forcing. *Science* **234,** 842–844.

Park, J., and Herbert, T. D. (1987). Hunting for paleoclimatic periodicities in a geologic time series with an uncertain time scale. *J. Geophys. Res.* **92,** 14027–14040.

Ricken, W. (1986). "Diagenetic Bedding: A Model for Marl–Limestone Alternations." Springer-Verlag, Berlin.

ROCC (1986). Rhythmic bedding in Upper Cretaceous pelagic carbonate sequences: Varying sedimentary response to climate forcing. *Geology* **14,** 153–156.

Sarmiento, J. L., Herbert T. D., and Toggweiler, J. R. (1988). Mediterranean nutrient balance and episodes of anoxia. *Global Biogeochem. Cycles* **2,** 427–444.

Glossary

Aeolian deposits Sedimentary deposits consisting of wind-blown dust.

Anoxia Situation where water is depleted in oxygen; most living organisms cannot survive under such conditions.

Bioturbation Disturbance of oceanic sediments near the water interface by bottom-dwelling (benthic) organisms, for example, burrowing worms.

Diagenesis Progressive alteration of sediments over time; especially important for the progression from carbonate ooze through chalk to limestone through time and continued burial.

Eccentricity Measures how elliptical the earth–sun orbit is, with consequent effects on solar radiation received by the earth. If e is the eccentricity, then $(1 + e)/(1 - e)$ is the ratio of the farthest and closest earth–sun distances.

Monsoonal circulations Atmospheric circulations induced by differential land–ocean heating. In summer, the flow is from the ocean to land; in winter it is from land to ocean, but with modification by the effects of the earth's rotation.

Obliquity Measures the tilt of the earth's axis of rotation with respect to the plane of the earth–sun orbit; provides the mid and high latitudes with seasonality.

Pelagic Zone of free swimming or floating organisms in the open ocean (as opposed to bottom-dwelling or tidal-zone organisms).

Perihelion Time of the closest earth–sun approach (presently occurs in early January). Aphelion is the time of the farthest earth–sun approach (presently occurs in early July).

Precession Motion of the earth's axis of rotation about a perpendicular to the earth–sun orbital plane, in response to the earth rotation and the combined gravitational forces of the sun, moon, and planets. Affects the relationship of the winter and summer seasons to the earth–sun distance during an annual orbit.

Sapropels Lithified organic-rich marine sediments deposited under anoxic conditions. If the organic content is very high, the sediments lithify to become black shale; if dilution by clays or other materials occurs, the sediments lithify to become marlstone.

Thermohaline circulation Density-driven circulation of the deep ocean. The density variations are due to temperature and/or salinity changes.

Deep Ocean Seismic Noise

A. E. Schreiner

Scripps Institution of Oceanography

Seismic noise is the ambient ground motion that exists independently of major earthquakes or artificial sources. Seismometers are used to measure earth motion in the frequency range 0.0003–100 Hz (wave period 3000–0.01 s), encompassing phenomena ranging from the earth's free oscillations caused by very large earthquakes to the sound from a nearby dynamite explosion. The noise that affects ocean floor seismic measurements is mostly caused by motions in the water at various scales, and these in turn are caused by the wind. At some frequencies, noise generated on the ocean floor can be measured at land stations thousands of miles from the nearest ocean. Deep ocean noise can be a nuisance to those whose interest is sensing distant events, but it also contains a wealth of information that can be used to study the ocean environment.

I. Description of the Noise Spectrum

The seismic noise spectrum can be divided into several frequency bands of characteristic energy. Figure 1 shows an example spectrum for an ocean bottom site and also, for comparison, a continental site. The physical mechanisms that generate the noise in each band are quite distinct.

At very low frequencies (<0.02 Hz) the noise level on the deep ocean floor is high and rises toward low frequency. This band of noise is primarily due to the passage of ocean surface waves with very long periods. Between 0.02 and 0.1 Hz, there is a notch in the deep ocean floor noise. Filtering by the ocean of shorter-wavelength fluctuation cuts off the effect of the long waves. Above 0.1 Hz, there is a distinct peak, widely known as the microseism peak. This peak is widely present, even observed on land seismometers. The microseism peak is due to ocean swell and wind waves, but through a nonlinear interaction between sets of waves which propagate in different directions. Structure in the high-frequency side of the peak is due to the growth and decay of local wind wave systems. Noise above 10 Hz is due to the action of breaking waves. [*See* SEISMOLOGY, THEORETICAL.]

II. Historical Measurements of Noise

Nearly as soon as sensitive measurements of earth motion were made, it was discovered that there is a permanent background signal. In the middle of the twentieth century it was realized that some of this seismic noise was related to atmospheric storms. The background level at land stations often increases when there are major storms at nearby coastlines. The correlation is good enough that it was thought for a time that seismic noise measurements could be used to give early warning of hurricanes. That approach was superseded by the advent of weather satellites, but the association of microseisms with oceans and waves remained clear.

In the 1960s, the desire to monitor nuclear testing led to renewed efforts to understand and characterize seismic noise. The background level sets a lower limit on the size of an earthquake or explosion that can be detected by seismic methods. Two of the most useful initiatives were the construction of large arrays of seismometers on land and the development of seismometers which could be placed on the deep ocean floor. [*See* NUCLEAR TESTING AND SEISMOLOGY.]

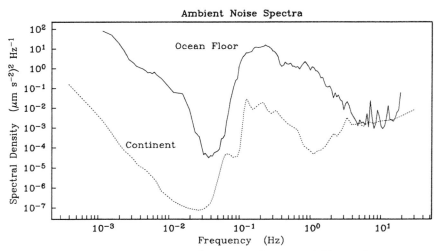

Fig. 1 Example noise spectrum measured at an ocean floor site (solid line) and at a continental site far from a coastline (dashed line). The peak between 0.1 and 0.5 Hz is known as the microseism peak.

Measurements made with sensor arrays can be analyzed in terms of spatial frequencies. Seismic signals are found at a variety of wavelengths, which are controlled by the type of propagation and the seismic medium through which the signals propagate. In addition, one can make a direct measurement of the signal delay across the array and thus determine the direction and the speed of propagation of the recorded signal. Measurement of microseisms with these arrays confirmed that increased noise levels could be correlated with specific ocean storms. Velocity measurements of the noise in these frequencies identified the microseisms as Rayleigh waves, a type of seismic wave which travels along the earth's surface. [*See* SEISMIC DATA, RECORDING AND ANALYSIS.]

Because two-thirds of the earth's surface is ocean, the absence of measurement in these regions creates an extreme undersampling. Placement of seismic sensors on the ocean floor thus provides some potential advantages. It was thought originally that the ocean floor would be a very quiet place to make seismic measurements, being removed from cultural activity and noises associated with air turbulence. However, the earliest ocean bottom seismic measurements proved this assumption wrong. Noise levels in the microseism frequency band are higher on the ocean floor than on land. It was confirmed, though, that microseism noise propagates along the ocean floor as Rayleigh waves.

Until the 1980s, only a single deep ocean bottom seismometer had made low-frequency measurements of noise. Data from this instrument was telemetered to a shore station via cable, a technique that is not practical for sensors far from land. Low-frequency seismic measurements require higher electrical power, which has not generally been available with battery-powered autonomous instruments. However, low-frequency pressure sensors and electric field sensors near the ocean bottom have been employed to make indirect inferences of ocean floor motions at periods between 10 and 200 s.

Ocean bottom noise studies have largely been point measurements. In the 1980s some small arrays of seismometers were deployed in the deep ocean to study the spatial frequencies of deep ocean noise. Direct measurement of the seismic noise wavelength showed that Rayleigh waves are important but that at higher frequencies the seismic waves have very short wavelengths and propagate in the sediments of the ocean floor.

III. Sources of Deep Ocean Seismic Noise

At low frequencies, seismic noise on the ocean floor is generated by pressure fluctuations in the water associated with the passing of waves on the ocean surface. At high frequencies, a more important source is the generation of bubbles by breaking waves and whitecaps. Other, more localized sources are geological phenomena such as microearthquakes and hydrothermal circulation associated with the generation of ocean crust at midocean spreading centers.

There is a continuous spectrum of wavelength and period of water waves, varying from tidal periods and planetary scales to tiny ripples. The wavelength and period of sea surface ocean waves are interrelated through the water wave dispersion relation. In the linear wave theory, for an ocean of

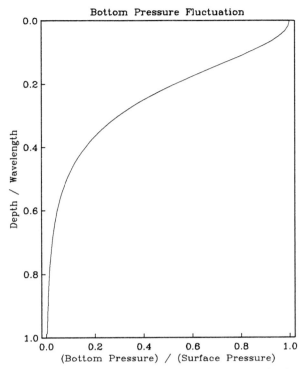

Bottom Pressure Fluctuation

(vertical axis) Depth / Wavelength

(horizontal axis) (Bottom Pressure) / (Surface Pressure)

Fig. 2 Ratio of the linear-theory pressure fluctuation amplitude at the ocean floor to that just below the water surface plotted against the ratio of the wavelength to the water depth.

depth D, the dynamic water pressure beneath a passing sinusoidal wave is given by

$$p(f, z) = \rho g A \frac{\cosh 2\pi(z + D)/\lambda}{\cosh 2\pi D/\lambda} \cos 2\pi\left(\frac{x}{\lambda} - ft\right)$$

where p is the pressure, f the frequency, ρ the density of water, g the gravitational acceleration, A the wave height, $-z$ the distance from the surface with downward negative, D the depth of the bottom, and λ the wavelength of the passing wave. The frequency of the pressure oscillation is the same as the frequency of the wave at the surface. The pressure fluctuation in the water depends on both the distance below the surface of the point of observation and the total depth of the water layer. The amplitude of the pressure fluctuation on the seafloor ($z = D$) diminishes nearly exponentially as the depth of the seafloor increases beyond one wavelength (Fig. 2).

The degree to which ocean waves affect the bottom of the ocean depends on the wavelength and the depth, and the amount of seismic noise that they generate further depends on the material properties of the ocean crust. Each of the major sources of noise will be described.

A. Very Long Period Ocean Surface Waves

Most ocean wave processes have their origin in the transfer of momentum from the wind to the water, but not all frequencies are generated directly. Very long period waves result from a nonlinear process involving breaking waves at coastlines, which is sometimes referred to as "surf beat." Most of this energy is trapped near the shore, but some radiates into the deep ocean. Once in deep water, the waves propagate freely across the ocean and are attenuated very slowly.

The amplitude of very long period waves is very small but the wavelength is very large, about 20 km for a 100-s wave in a 4-km deep ocean basin. Their passage has a direct effect on the ocean bottom. With increasing frequency, the wavelength gets shorter until at 50 s (0.02 Hz) it is down to about 4 km, comparable to the average ocean depth, and the exponential decay begins to have an effect, resulting in strongly reduced noise levels. The deep ocean effectively filters short-wavelength fluctuations. In the low-noise-level notch, the energy that is present may be due to turbulence associated with currents on the bottom.

B. Wind Waves

The peak of the ocean swell and wind wave spectrum occurs in the 8–16-s period range. This is the swell that is plainly visible on the open seas, with amplitudes between a fraction of a meter in very light seas to 10 or more meters in very rough conditions. At these frequencies, the wavelengths are on the order of 150–600 m, too short to affect the deep ocean bottom directly. However, when there are two sets of swell of the same wavelength traveling in opposite directions, a standing wave pattern develops. The wind-generated swell is large enough that when two sets interact it is necessary to include the nonlinear terms of the hydrodynamic equations to model the effect fully. An expansion of the nonlinear equations into the linear and quadratic terms has proved useful. The pressure fluctuation from the linear term has already been described. The quadratic term predicts three additional effects: (1) that the resulting pressure oscillation beneath the standing wave has double the frequency of the originating waves, (2) that the amplitude of the pressure fluctuation depends the square of the surface wave amplitude, and (3) that there is no explicit dependence on depth. Opposing swell of the same frequency can occur near a coastline, where the wave train reflects back on itself, or in the open ocean when a fast-traveling storm overtakes the waves and generates swell in its wake.

The wave–wave interaction can be generalized beyond the case of purely opposing waves. Observations of the directional spectrum of wind waves show components in nearly all directions. The wave number describes the direction and length scale of a wave. The magnitude of the wave number is proportional to the reciprocal of the wavelength. A rigorous analysis of the quadratic term with the full distribution of wave numbers leads to an expression for the subsurface pressure spectrum with sums of wave numbers and differences of wave numbers. The sum terms lead to large wave numbers and therefore do not contribute to the pressure spectrum at large depths. The difference terms become important when the wave numbers of the interacting waves are nearly equal in magnitude but nearly opposite in direction. The resulting wave number then becomes very small, and the wavelength thus very large. The pressure spectrum of acoustic energy incident from above can be approximated by

$$p(f) = A f^3 S^2 \left(\frac{f}{2}\right) \int_{-\pi}^{\pi} G(\theta) G(\theta - \pi) \, d\theta$$

where f is the frequency, S the wave amplitude spectrum, G the wave directional spectrum, θ the azimuth, and A a constant of proportionality. The integral computes the effect of directional spectrum at any azimuth multiplied by the directional spectrum at the opposite azimuth. It does not suffice for the waves to be large; they must also be distributed in azimuth to have opposing components and thus to create a substantial ocean bottom pressure oscillation.

Subsidiary peaks on the limb of the microseism peak can be associated with the growth of the local wind wave field. Wind waves tend to arise at high frequencies first and then extend to lower frequencies as long as the prevailing wind direction lasts. The directional spectrum of ocean waves at higher frequencies is difficult to measure, requiring radar or side-looking sonar. The wave–wave interaction can occur to arbitrarily high frequencies, but for a significant effect on the deep ocean floor, the source at the surface must be correlated over a wide area.

C. Breaking Waves

Noise levels at subsonic frequencies (2–20 Hz) are known to rise steadily with increasing wind speed at the ocean surface. At a wind speed of about 25 m/s, the noise level below 10 Hz no longer increases, but above 10 Hz the noise jumps to higher levels. This is due to the onset of whitecaps. When waves break, there is impact of water against water and bubble formation and decay. Both these processes are high-frequency sound sources. These disturbances can be detected with seismometers on the ocean floor.

D. Geological Phenomena

Most of the ocean floor is geologically stable, but near active sites, such as oceanic spreading ridges, ocean bottom noise can come from within the crust. Submersible divers at ridge crests have observed very active hot-water circulation. Hydrothermal flow may produce a broad frequency range of acoustic energy. Movement of lava within rock generates a narrowband signal called harmonic tremor. Such signals at 1–5 Hz have been measured by ocean bottom seismometers, but the noise does not propagate great distances in the ocean crust.

E. Coupling between the Ocean and the Crust

A variety of mechanism, then, can generate pressure fluctuations or acoustic waves at different frequencies. Some, but not all, of this energy is coupled into the solid ocean floor and then propagates away as elastic waves, generally to be recorded as noise.

The ocean floor, and indeed the whole earth, is an elastic system with certain resonant frequencies. The elastic properties and density of the earth vary with depth. These properties control the velocity at which seismic waves travel through the earth. An example of velocity structure for the upper 8 km of oceanic crust is shown in Fig. 3. In general, seismic velocities increase with depth and, since propagating waves refract toward the slower medium, the seismic energy is kept close to the surface of the earth and a waveguide is created. Seismic surface waves such as Rayleigh waves propagate efficiently in the waveguide in horizontal directions. Waveguides support mode propagation; only certain combinations of frequency and velocity propagate in the waveguide without loss, and these discrete combinations are known as modes. The relationship between velocity and frequency is known as the dispersion curve. The result of this is that the pressure waves whose frequency and velocity match one of the seismic modes in the ocean floor result in seismic waves.

The very long period ocean waves are detected by seismometers on the ocean floor because they flex the crust as they pass overhead. However, their speed does not match a seismic mode, so the disturbance does not propagate away in the crust. At the microseism frequency, the resulting disturbance matches seismic modes that propagate efficiently and therefore are detected all over the globe.

At higher frequencies, above 0.5 Hz, the waveguide that controls propagation is the one in the upper 50 m of the sediments on the ocean floor. These sediments are water-

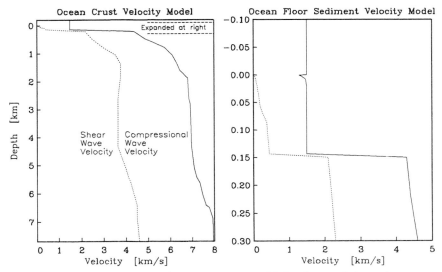

Fig. 3 Seismic velocity structure for an ocean crust model. Depths are relative to the ocean floor. Both compressional wave velocity (solid line) and shear wave velocity (dotted line) increase with depth. The right-hand frame shows the ocean floor sedimentary layer at an expanded scale, emphasizing the very low shear wave velocity in the sediments.

logged and very soft, and seismic waves propagate very slowly through them. The pressure fluctuations at these higher frequencies do not propagate at a velocity which is matched by the slowly propagating modes in the sediments. Nevertheless, some of the energy does couple into the sedimentary layer by scattering at the rough ocean bottom. The resulting seismic waves in the sediments propagate over short distances because they are attenuated by the soft material.

Bibliography

Kerman, B. R., ed. (1987). "Sea Surface Sound." Kluwer Academic Publishers, Dordrecht, The Netherlands.

Kibblewhite, A. C., and Ewans, K. C. (1985). Wave–wave interactions, microseisms, and infrasonic ambient noise. *J. Acoust. Soc. Am.* 78, 981–994.

Urick, R. J. (1985). "Principles of Underwater Sound." McGraw-Hill, New York.

Glossary

Directional spectrum Distribution of wave energies over direction.

Microseism Peak in the seismic noise spectrum at 6–7 s period.

Rayleigh wave Type of seismic wave which is vertically polarized and travels along the earth's surface rather than through the interior.

Seismic waveguide Low-velocity channel which confines seismic energy.

Wave number Spatial frequency, proportional to the reciprocal of the wavelength.

Deglaciation, Impact on Ocean Circulation

Eystein Jansen
University of Bergen

The most striking features of the earth's climate during the last 600–700 thousand years of the planet's history are the dramatic deglaciations that occurred roughly every 100,000 years. During these phases, which lasted about 10,000 years, the northern hemisphere was transformed from maximum ice-age cooling to maximum interglacial warmth, and the large ice sheets over North American and Europe disappeared. It is now considered proven that the large-scale glacial cycles are fundamentally driven by variations in the geographical and seasonal distribution of solar insolation. Variations in incoming solar energy are again caused by cyclic changes in (1) the angle of the earth's axis with respect to the plane of the earth's orbit around the sun (orbital tilt); (2) the time of the year when the earth is located closest to the sun (perihelion) (precession of the equinoxes); and (3) the eccentricity of the earth's orbit, which varies at distinct periods of 41, 23, and 100 thousand years, respectively. This confirmation of the Milankovitch theory for ice ages was done in the late 1970s by J. Imbrie, N. Shackleton, and J. Hays based on new astronomical calculations by A. Berger. Although the fundamental cause for glacial cyclicity has been found, we do not understand precisely how and why the climate system responds so strongly to these relatively moderate variations in energy input. Most remarkably we do not know why the climate response is strongest at the 100,000-year period (defined by major deglaciations) when the insolation variations are least in this particular frequency band. Another puzzling factor is that while the rising solar insolation associated with deglaciations is a gradual process, the climatic response can be abrupt and step-wise, and large changes take place within as short time as decades. This behavior documents that the climate system responds in a nonlinear fashion to the climatic forcing imposed by the increased insolation during deglaciations. The ocean is an important factor for determining this response. To understand the way climate changed during these transitions, one needs to understand the role of the oceans as a mediator of both gradual and abrupt climate change.

I. Role of Ocean Circulation for Global and High-latitude Climates

In the present world the surface currents of the North Atlantic modify the strong zonal equator-to-pole temperature gradient by advecting heat to the Arctic and sub-Arctic regions. This produces a temperature anomaly in Northern Europe with much warmer climates that the zonal mean. The excess heat given off by the ocean adds about 30% to the heat directly received from the sun in the North Atlantic. The cause of this heat advection lies partly in the wind field of the North Atlantic, which is driven by the meanders of the tropospheric jet and the prevailing westerlies that pump atmospheric heat and warm water northeastward. It partly depends on the global thermohaline ocean circulation. When warm and salty North Atlantic waters are exposed to the cold fall and winter atmosphere in the Nordic seas (Norwegian–Greenland and Iceland seas) they are cooled by complex interactive processes and release heat to the atmosphere, which warms up the surrounding region. This cold water becomes dense, and due to its saltiness it will ultimately convect and sink toward the bottom. A minor portion of deep water is additionally formed in the Arctic from sea-ice formation on the continental shelves. Sea-ice formation extracts fresh water from the surface waters and leaves the remaining water enriched in salt. This brine can become so dense that it starts sinking to the bottom. [See Air–Sea Interaction; Sea Ice.]

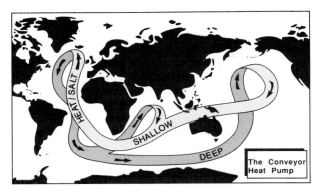

Fig. 1 The global oceanic conveyor circulation by which deep water formation in the Nordic seas leads to export of deep water and the return of surface water, which brings salt and heat into the region.

The newly formed deep water in the Nordic seas is well oxygenated and ventilated because it has just been in contact with the atmosphere. It flows southward across the sills of the Greenland–Scotland ridge and forms North Atlantic Deep Water (NADW) after mixing with intermediate waters. Within about 100 years NADW has traveled the entire length of the Atlantic as a deep-water mass, easily characterized by its high oxygen content. It mixes with circum-Antarctic deep and intermediate waters, which form from brines from freezing processes around Antarctica, and finds its way into the deep Indian and Pacific oceans. There is no similar vertical overturn in the North Pacific because the surface waters there are 0.6‰ less saline than their North Atlantic counterparts due to a net export of moisture was water vapor out of the Atlantic. The Pacific surface waters thus do not become dense enough after cooling to overturn. The net export of deep waters out of the North Atlantic is being compensated by a net inflow of surface water, which transports heat across the equator from the southern to the northern hemisphere. This global circulation, named the "Ocean Conveyor Belt" by W. Broecker, is shown schematically in Fig. 1. Since the rate and mode of deep-water formation also determines many of the chemical characteristics of the ocean waters, it will influence the ocean's ability to absorb and store carbon dioxide and thus influence the magnitude of the greenhouse effect and the planet's climate. [*See* OCEAN CIRCULATION.]

The ocean circulation of the high latitudes consequently influences climate as a heat source that modifies regional climates and determines on a global basis to a large degree the natural magnitude of the greenhouse effect. This system is likely to be easily affected by changes in atmospheric circulation and the salinity distribution of the North Atlantic and the Nordic seas. It can thus shift as a result of climatic

change; but because it is such a key element of the coupled global climate system, such changes will by themselves have strong climatic impact and serve as an initiator of climatic change or as feedback mechanisms that amplify or moderate climate change. Knowledge about the stability of ocean circulation and how sensitive it is to perturbations is therefore pivotal for attempts to understand and estimate climate change. Such knowledge can be gathered from studies of past changes — especially from periods where large climatic fluctuations took place, such as during deglaciations. [*See* CLIMATE; POLAR REGIONS, INFLUENCE ON CLIMATE VARIABILITY AND CHANGE.]

II. Methods for Reconstructing Past Ocean Circulation

A. Surface Ocean Characteristics

During the last 20 years our improved ability to reconstruct the physical and chemical state of the ocean has revolutionized the study of climate change and has led to the general acceptance of the orbital, or Milankovitch, theory for explaining the ice ages. Widespread use of chemical and microfossil techniques have made these advances possible. Reconstructions of past sea-surface temperature are performed by analysis of fossil microfauna or microflora studied in deep-sea sediment cores. Based on statistical methods that link abundance of floral or faunal planktonic assemblages to present sea-surface temperatures, one is able to calculate temperatures back through time. The reconstruction of the global ocean temperature field for the last glacial maximum 18,000 ^{14}C years ago by the CLIMAP project has been particularly useful. This reconstruction has been extensively used as boundary conditions for numerical climate model experiments on the climate of the glacial world and for testing and developing global climate models. [*See* PALEOCLIMATIC EFFECTS OF CONTINENTAL COLLISION AND GLACIATION.]

The stable isotopic composition of carbonate shells of foraminifers have been extensively utilized in paleoceanography. The $\delta^{18}O$ of carbonate is determined by the temperature of formation and the $\delta^{18}O$ of the water in which the shells form. The preferential inclusion of ^{16}O relative to ^{18}O in fresh water and hence in glacial ice, due to the distillation of "light" from heavy water, leads to an enrichment of ice-age ocean water in ^{18}O as ^{16}O is distilled out and preferentially stored in the ice sheets. When the glacial ice is stored on land and subsequently returned to the ocean as meltwater on deglaciation, it produces a global ice-volume $\delta^{18}O$ signal that reflects the magnitude of ice stored on land, but also produces local isotopic excursions (negative $\delta^{18}O$ spikes) in the areas that directly receive the meltwater. Both these processes can be recorded by the $\delta^{18}O$ of planktonic foraminifers. Increasing temperatures will also produce lower $\delta^{18}O$ values.

B. Deep Water Characteristics

Primary production in the euphotic zone of the oceans produces organic matter that is depleted in $\delta^{13}C$ by about 20‰, compared to the $\delta^{13}C$ of dissolved CO_2. This leaves the remaining CO_2 enriched in $\delta^{13}C$. When this organic matter is oxydized, recycled, and remineralized in deeper water masses, carbon dioxide with low $\delta^{13}C$ is returned. A positive correlation thus exists between $\delta^{13}C$ and oxygen content, and hence there is an inverse correlation between $\delta^{13}C$ and ocean-water nutrient and carbon-dioxide content. This forms carbon isotopic gradients between surface and deep waters and basin-to-basin isotope fractionation where "young" deep water such as NADW is characterized by higher $\delta^{13}C$ than "older" water. The oldest waters in the North Pacific also have the lowest $\delta^{13}C$ values. There is therefore a close tie between the world's ocean circulation and the $\delta^{13}C$ distribution. The $\delta^{13}C$ of the carbonate in foraminiferal shells reflect these gradients, and the $\delta^{13}C$ of benthic and planktonic foraminifers extracted from sediment cores are utilized to reconstruct changes in circulation patterns back in time. [See CARBON DIOXIDE TRANSPORT IN OCEANS; OCEAN PRODUCTION EXPORT: HYDRODYNAMICS AND ECOSYSTEMS.]

For unknown reasons there is a close similarity between ocean water nutrient content (and $\delta^{13}C$) and the dissolved cadmium (Cd) content. Since the calcium (Ca) content does not vary much, studies of the Cd/Ca ratio in foraminiferal shells will reflect past Cd and nutrient distributions. This technique developed by E. Boyle is another way of reconstructing past ocean chemistry and circulation changes.

A third method was established in recent years by the evolution of tandem accelerator mass spectrometers (AMS), which made it possible to measure the radiocarbon (^{14}C) content of small samples of foraminiferal shells from deep-sea cores. This has led to a much more refined dating of paleoceanographic events during the last 30,000 years, which is the time limit of the radiocarbon method, than was previously possible. Radiocarbon is also a powerful tracer utilized by oceanographers to study the present ocean circulation and to estimate its magnitude and speed by measuring ocean-water radiocarbon ages. Although there are problems connected with the way the different foraminiferal species incorporate carbon isotopes into their shells (this is also true for $\delta^{13}C$), the AMS method can be used to directly study past radiocarbon gradients between surface and deep-water (benthic-to-planktonic gradients), ocean-to-atmosphere gradients, and basin-to-basin gradients between the different ocean basins. Compared with the present ^{14}C distribution, one can by this method obtain information about the scale of past circulation changes. [See RADIOCARBON DATING.]

The combination of these techniques has dramatically improved ability to accurately describe past oceanic changes in time and space, and enhanced the ability of paleoceanographers to provide insight into the role and response of oceans during deglaciations and climate change in general.

III. Circulation of the Glacial Ocean

There is firm evidence to prove that the circulation of the world ocean during the last glacial maximum was very different from the present mode of circulation. Cd/Ca and $\delta^{13}C$ data document that the production of NADW was drastically diminished, and that in the deep North Atlantic less ventilated Antarctic Bottom Water (AABW) was more widespread than today. This loss of oxygenated northern component water diminished the chemical gradients between the North Atlantic and the Pacific, and the reduced flux of NADW made the $\delta^{13}C$ of the Southern Ocean deep water shift closer to the low Pacific values. At the last glacial maximum, the radiocarbon age difference between the Pacific and the North Atlantic was less than today; but the North Atlantic remained better ventilated, and the radiocarbon age of Pacific deep waters was about 2000 years, which is 300 to 500 years higher than today, reflecting the reduced oxygenation of the global deep sea.

The glacial-to-interglacial amplitude of oxygen isotope variations in benthic foraminifers is in most deep-ocean basins about 0.4 to 0.5‰ larger than the amplitude expected from reconstructions of the global ice volume change from estimates of the corresponding sea-level fluctuations. This suggests that the deep waters were 1 to 3 degrees colder in glacials compared to the present. The strongest cooling is observed in the North Atlantic, which today has the warmest deep waters of the major ocean basins.

The reduction in NADW production led to increased nutrient content (witnessed by higher Cd/Ca and lower $\delta^{13}C$) in the glacial deep ocean, but at the same time the intermediate waters, which today display a strong nutrient maximum, became depleted in nutrients. That is, nutrients were moved to the waters at the ocean floor during glacials. This is observed in the North Atlantic, the Carribean, and the Indian Oceans. It is unclear where these nutrient-depleted intermediate waters came from. A number of mechanisms have been proposed: Increased circulation rates of intermediate waters that strip them of nutrients more effectively than today, increased flux of Mediterranean outflow, higher production of ventilated intermediate waters in the North Atlantic and the Circum-Antarctic.

Why was the glacial circulation so different from

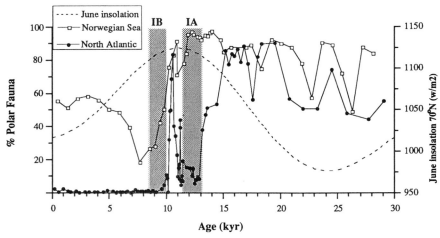

Fig. 2   The abundance of polar planktonic foraminifers, an indicator of cold surface water, plotted against radiocarbon age. The Norwegian Sea record comes from core HM52-43 collected from 2600 m water depth in the central southern Norwegian Sea. The North Atlantic record comes from core V23-81 collected from 2400 m water depth off Ireland. The arrows indicate the direction of increased ventilation and deglaciation. The shaded intervals correspond to the two phases of most rapid sea-level rise and deglacial meltwater transfer (see text).

the present? The North Atlantic heat pump did not operate as today. Slightly less saline surface waters probably impeded vertical overturn. Reduced overturning reduced the cross-equatorial conveyor-belt transport of heat. Studies by W. Ruddiman and A. McIntyre have shown that the entire North Atlantic north of 40°N was covered by cold Arctic or Polar surface waters, and that the present Arctic front which today stretches SW–NE off Greenland was located W–E across the North Atlantic between northeast America and Portugal. This is also shown in Fig. 2 by the high content of polar species of planktonic foraminifers at 18–20,000 YBP both in the North Atlantic and the Norwegian Sea. Experiments with global atmospheric circulation models indicate that when the Laurentide ice sheet over North America grew high enough, it split the atmospheric Jet Stream into two branches, with a northerly branch advecting cold arctic air masses to the North Atlantic through the Greenland–Labrador gap. Such a wind field effectively cools the surface of the ocean and reduces the meridional flux of surface water in the area, thereby reducing the advection of warm salty water and heat to the Nordic seas, which today is a prerequisite for deep-water formation. [*See* ATMOSPHERIC MODELING.]

This situation shows that many key elements of the global climate system operated very differently from today during glacial times. This scenario with enhanced nutrient content of deep waters and a dramatic shift in the nutrient gradient between intermediate and deep waters was suggested by Boyle and Broecker to have made the glacial ocean able to

absorb more CO_2 than today. This mechanism is probably responsible for bringing the glacial atmospheric CO_2 level down by 30% below the preindustrial level—or 45% lower than the present level. Such a reduction is documented from measurements of the carbon dioxide content of air trapped in bubbles in ice from Antarctic and Greenland ice cores. The ocean was apparently responsible for forcing a reduction in the greenhouse effect. This coupling demonstrates the intimate interplay between ocean circulation and the atmosphere.

When the ocean subsequently changed from the glacial type of circulation into the type we know as the present ocean through the deglaciation, it affected and interacted strongly with the other elements of the climate system. [*See* CLIMATE–ICE INTERACTIONS; ICE AGE DYNAMICS.]

IV. Evidence for Meltwater and Salinity Anomalies during Deglaciation

Deep convection and deep-water formation are intimately linked with surface water salinities and temperatures. Since deglaciation means the transfer of huge amounts of fresh water, equal to 120 m of global sea level, mainly to the North Atlantic sector of the world ocean, this transfer forms the most probable way deglaciation would affect ocean circulation. Evidence bearing on the timing of meltwater transfer comes from two sources: estimates of sea-level change and

$\delta^{18}O$ records from areas that directly receive the meltwater, because the low oxygen isotope values of the meltwater can be traceable as light isotope excursions in the $\delta^{18}O$ of the foraminiferal shells. An extremely detailed sea-level record spanning the last 17,000 YBP was recently published by R. Fairbanks. This was derived by dating surface-dwelling corals drilled from submerged reefs off Barbados. He showed that the glacial sea level was about 120 m lower than present and the deglacial rise in sea level started slowly at 16,000 YBP radiocarbon; but the deglaciation was concentrated in two major steps, one between 13,000 and 11,500 YBP and another at 10,000–8,500 YBP, during which the rate of meltwater transfer was about five times higher than during the period in between. The disintegration of the large northern hemisphere ice sheets thus took place in two major steps, and the transfer of meltwater was to a large extent concentrated in those intervals of most rapid sea-level rise. [See SEA LEVEL FLUCTUATIONS.]

How is this reflected by the $\delta^{18}O$ evidence for salinity anomalies? The available data are summarized in Table I. In the Gulf of Mexico, which received a major amount of meltwater from the Laurentide ice sheet through the Mississippi, anomalously low $\delta^{18}O$ values are recorded 13,000–11,000 YBP, closely corresponding with the first deglacial step. The same anomaly is observed in different places in the North Atlantic and in the Norwegian and Iceland seas. In the Fram Strait, the gateway between the Nordic seas and the Arctic and in the northeast Norwegian Sea, there is a local early anomaly at 13,500–14,000 YBP, marking the early deglaciation of marine-based ice sheets in the sub-Arctic. The second deglacial step 10,000–8,500 YBP is also marked by salinity anomalies, but less so than the first (see Table I). The major phases of deglaciation and meltwater transfer can thus be recorded as surface-water salinity anomalies.

Benthic $\delta^{18}O$ records from the Norwegian Sea and the NE Atlantic off Ireland also document this two-step transfer of the deglacial meltwater signal (see Fig. 2). In particular the Norwegian Sea $\delta^{18}O$ record shows a remarkable coherence in timing with the two meltwater steps from the Barbados sea-level record (named IA and IB in Fig. 2). The reduced deglaciation rate between 11,000 and 10,000 YBP (Younger Dryas period) when sea level only rose about 10 m, is clearly seen as a return to high $\delta^{18}O$ in both the North Atlantic and the Norwegian Sea, indicative of a reduced transfer of isotopically light water to the deep ocean. The figure also shows that while the whole deglaciation occurred during a period of maximum summer isolation in the northern hemisphere, the existence of the steps and the intervening period of reduced deglaciation rate is clearly unrelated to orbital forcing. This is further discussed in Section VII.

The fact that the deglacial signal is seen almost instantly at 2500 m water depth (within the uncertainty of the dating method) shows that the isotope spike received from the meltwater is rapidly transported into the deep-water despite the reduction in vertical overturn one would expect to result from the transfer of low salinity meltwater to the high northern latitudes. The mechanism responsible for this transport is further discussed in the next section.

Recent discussion on the origins of rapid climatic change during the deglaciation has focused on the probable effects of *meltwater drainage patterns* on ocean circulation and climate. Broecker has suggested that the diversion of meltwater drainage from the Mississippi to the St. Lawrence at 11,000 YBP and catastrophic discharge of water from ice-dammed glacial lakes draining out this new route, wiped out vertical overturn, canceled the conveyor-belt heating, and caused a swift climatic deterioration (the Younger Dryas cooling). It has not yet been possible to see any clear meltwater anomaly in the open ocean related to this event, which took place when the bulk meltwater transfer was about 20% of that of the preceding period. This could have canceled out the effect of the drainage shift (see Section VII).

V. Deglacial Fluctuations of Deep Water Formation

Because the production of oxygenated water masses through vertical overturn is very sensitive to salinity change, one would expect that the episodes of large-scale meltwater dispersal would lead to a stabilization of the surface water through the formation of a low-salinity layer, thereby preventing winter overturning. Such responses have been documented from the deep-sea Cd and $\delta^{13}C$ records. A detailed record was produced from a core collected from 4000 m on the Bermuda Rise in the western North Atlantic, close to the present boundary between NADW and AABW. This record documents four episodes of reduced oxygenation (i.e., when AABW replaced NADW at this depth). The first at 14.5 took place during the initial deglaciation. The next two happened during the first major deglacial step. The midpoint of the last one corresponds to the second deglacial step, but it appears to stretch over the time span 10,007 to 9000 YBP. The dating control on the initiation of this event is not precise, however.

In the Northeast Atlantic off Ireland the benthic $\delta^{13}C$ record shows that the most marked episodes of reduced ventilation were at the last glacial maximum at about 18,000 YBP and at 13,000–14,000 YBP (see Fig. 3). The latter might correspond to the first deglacial step, since the accuracy of the dating is approximately 1000 years. In the Norwegian Sea, the deep waters were well ventilated until the first deglacial step (see Fig. 3), indicating that vertical overturn took place both at the last glacial maximum and in the first, slow deglacial phase, and that only the major deglacial meltwater releases at 13,000–12,000 YBP was capable of turning off the

Table I Ocean Events during Deglaciation

Area	Time[a]	Evidence
Meltwater events		
Global	13–11.5 kyr	Barbados sea level
Global	10.2–8.5 kyr	Barbados sea level
Gulf of Mexico	13–11 kyr	Planktonic $\delta^{18}O$
Gulf of Mexico	10.1 kyr	Planktonic $\delta^{18}O$
Bermuda Rise	13.5 kyr	Planktonic $\delta^{18}O$
Bermuda Rise	12 kyr	Planktonic $\delta^{18}O$
NE Atlantic	14–12 kyr	Planktonic $\delta^{18}O$
Norwegian Sea	14–12 kyr	Planktonic $\delta^{18}O$
Iceland Sea	13–12 kyr	Planktonic $\delta^{18}O$
Norwegian Sea	10–9 kyr	Planktonic $\delta^{18}O$
Fram Strait	14–13 kyr	Planktonic $\delta^{18}O$
Reduced oxygenation events (evidence of reduced NADW)		
Bermuda Rise	14.5 kyr	Cd/Ca, benthic $\delta^{13}C$
Bermuda Rise	13.5 kyr	Cd/Ca, benthic $\delta^{13}C$
Bermuda Rise	12 kyr	Cd/Ca, benthic $\delta^{13}C$
Bermuda Rise	10.7–9 kyr	Cd/Ca, benthic $\delta^{13}C$
NE Atlantic	13.5 kyr	Benthic $\delta^{13}C$
Norwegian Sea	13–11.5 kyr	Benthic $\delta^{13}C$
Norwegian Sea	10–8.5 kyr	Benthic $\delta^{13}C$
Warming of surface water		
NE Atlantic	14–11 kyr	Planktonic foraminifers
NE Atlantic	10.2–9 kyr	Planktonic foraminifers
Norwegian Sea	12–11 kyr	Planktonic foraminifers
Norwegian Sea	10.2–7 kyr	Planktonic foraminifers/diatoms
Southern Ocean	~13 kyr	Planktonic foraminifers/diatoms
Southern Ocean	10.5–8 kyr	Planktonic foraminifers/diatoms
Cooling of surface water		
NE Atlantic	11–10.2 kyr	Planktonic foraminifers
Norwegian Sea	11–10.2 kyr	Planktonic foraminifers
Norwegian Sea	7–4 kyr	Plantonic foraminifers/diatoms
Reduction in sea-ice cover		
NE Atlantic	14–13 kyr	Planktonic foraminifers/carbonate flux
Norwegian Sea	13 kyr	Planktonic foraminifers/carbonate flux
Iceland Sea	9 kyr	Planktonic foraminifers/carbonate flux

[a] Ages given in 1000 ^{14}C years before present.

formation of oxygenated deep water. We also note from Fig. 3 that the deep waters of both the Norwegian Sea and the Northeast Atlantic became better oxygenated in the period 11,000–10,000 YBP (Younger Dryas), when the total deglacial meltwater release was markedly reduced.

Figure 3 also shows the evolution of $\delta^{13}C$ gradients between the Norwegian Sea and the North Atlantic. During the Holocene we see low gradients and well-oxygenated waters resulting from the intense formation of NADW. The large gradients before 13,000 YBP are clearly an indication of formation of oxygenated waters north of the Scotland–Greenland ridge and presence of less ventilated AABW south of the ridge. The magnitude of this formation apparently was much lower than today. Perhaps the density of these waters also were slightly lower than today, due to slightly reduced salinities of the source waters. Thus the outflow of these waters may have formed intermediate waters rather than deep waters after flowing over the Greenland–Scotland ridge. They thereby probably contributed to the formation of the low-nutrient high $\delta^{13}C$ intermediate waters in the glacial ocean instead of sinking to the deep ocean that was occupied by low $\delta^{13}C$ AABW.

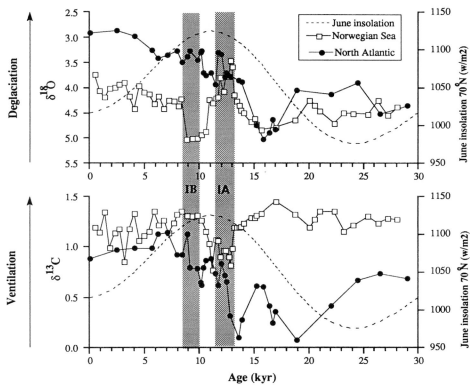

Fig. 3 Benthic oxygen and carbon isotope records plotted against radiocarbon age for the same cores as presented in Fig. 2. Symbols and time scale are the same as in Fig. 2.

When evaluating both the $\delta^{13}C$ and $\delta^{18}O$ gradients of Fig. 3, we can see that only during the first deglacial step were the characteristics of the water masses on each side of the Scotland–Greenland ridge similar. This shows that during this phase it is possible to construct scenarios whereby the meltwater release had as a consequence that the flow of deep waters was reversed compared to the modern mode. Before we can conclude whether this really happened, more data is needed, especially Cd/Ca and benthic radiocarbon data.

If deep-water formation dropped during the first deglacial step, how can it be that the deglacial $\delta^{18}O$ anomaly so rapidly was transferred to the deep-water realm? Oxygenated deep waters are primarily formed by open-ocean convection but also from sea-ice freezing and brine formation (see Section I). This process might still continue while open-ocean convection is closed off by the formation of a low-salinity lid. During sea-ice formation the remaining salt water retains its isotopic composition, while its salt content increases. The deep water that is eventually formed by this process will carry the deglacial $\delta^{18}O$ spike to the deep ocean. The Norwegian Sea data also shows 0.3‰ lower glacial-to-glacial interglacial $\delta^{18}O$ amplitude than expected from the global ice-volume/

sea-level record. This can be accounted for by the same mechanism — namely, that during glacials, the ratio of deep waters produced by brine formation to those formed by open-ocean convection was higher than in the present situation.

Available evidence clearly shows that the deep-ocean circulation is highly sensitive to changes in high-latitude surface salinities. During the 1970s a low-salinity anomaly in the North Atlantic, originating from increased fresh-water flow out of the Arctic, stopped or reduced vertical overturn in some years. On a larger scale, during the last deglaciation, all major low-salinity anomalies detected by planktonic $\delta^{18}O$ records apparently led to a decrease in NADW formation and a northward and upward migration of AABW in the Atlantic. Complete shutdown of vertical overturn, however, appears to take place only during the major deglacial steps, when meltwater release to the ocean reached peak levels.

Some radiocarbon and $\delta^{13}C$ data from the Pacific indicate that there was a local source for deep or intermediate water ventilation in the Pacific during the deglaciation. These records need to be supported by a larger data set before firm conclusions can be drawn, and the available data more strongly suggest that convection might only have formed intermediate water in this region during glacial times, and perhaps later also.

Newly formed oxygenated northern source waters may, together with intermediate waters formed in the Antarctic, be a major source for the oxygenated, low nutrient bearing intermediate waters of the glacial ocean.

VI. Surface Ocean Warming and Ocean Fronts

Figure 2 shows that during the deglaciation, polar planktonic foraminiferal faunas were replaced by subpolar faunas in both the Norwegian Sea and the Northeast Atlantic. Polar faunas indicative of Arctic and Polar water masses spread as far south as 40°N at 20,000–18,000 BP, defining a strong temperature front between the U.S. East Coast and Portugal. The figure also shows that the transition into a warm ocean did not happen regularly. Although it happened when northern hemisphere solar insolation was at a maximum, it took place in sudden steps, interrupted by a sudden shift to cold conditions 11,000–10,200 YBP (the Younger Dryas cooling). There was also significant differences in the timing of the warming events between the Northeast Atlantic and the Norwegian Sea to the north.

In the present Nordic seas low carbonate flux and about 90% dominance of polar planktonic foraminifers characterize *Polar Water* with almost year-round sea ice, whereas the combination of high carbonate flux and a uniform polar foraminiferal fauna characterize *Arctic Water*, which is only ice-covered during parts of the winter season or only in severe ice years. These indices correspond with the *Polar* and *Arctic Fronts*, respectively; the *Polar Front* defines the mean to summer sea-ice limit and surface waters with reduced salinities, while the *Arctic Front* is a temperature front between warm Atlantic and cold Arctic waters, close to the extreme maximum sea-ice boundary. It is thus useful to distinguish between these two fronts when interpreting core data.

Figure 2 shows that the first warming episode started after 15,000 YBP and lasted until 11,000 YBP, and estimated temperatures (estimated from the foraminiferal assemblages) indicate a summer warming of 7 degrees, from 6 to 13°C off Ireland. This warming coincided with the rise in sea level and the first phase of deglaciation, but the surface of the North Atlantic had started to warm prior to the first major deglacial step. In the Norwegian and Iceland seas there was a strong increase in the carbonate flux at 14,000–13,000 YBP signaling a transition from perennial sea-ice cover into a situation with Arctic water and limited seasonal sea ice. This happened when the North Atlantic warmed and when significant

warming was recorded in Western Europe. Temperature estimates based on fossil beetles indicate a summer temperature increase of 9°C in England at 13,000 YBP. The Norwegian Sea remained relatively cold until about 11,800 YBP, well into the first major deglacial step (Fig. 2).

Synchronous return to cold conditions took place at 11,000 YBP, lasting about 800 years, during the Younger Dryas cold phase. Summer sea-surface temperatures in the North Atlantic dropped by about 7°C. This was, however, not a return to glacial conditions. Faunal data and very high carbonate flux indicate that both the Norwegian Sea and the North Atlantic south to about 45°N were covered by *Arctic* rather than *Polar* waters. This indicates a situation that over large areas, was very similar to the situation that today prevails in the areas with most active vertical overturn in the Iceland and Greenland seas. It is therefore not surprising that the carbon isotope data displayed in Fig. 3 show a strong increase in oxygenation, and that oxygen isotopes document extremely cold deep waters formed by open-ocean convection during this cold phase. After the Y. Dryas, sea-surface temperatures rose quickly again during the next deglacial step.

After the ice sheets had vanished and their influence on regional climates had gone away, the continued high summer insolation produced surface waters about 2°C warmer than at present until about 6,000 YBP in the central Norwegian Sea. Note the low content of polar fauna between 9,000 and 6,000 YBP in Fig. 2. Evidence from cores off East Greenland also show that the sea-ice extent was less than today in the early part of postglacial time. This is in accordance with results from climate model experiments, which indicate that high summer insolation in the northern hemisphere leads to a reduction in the thickness and extent of the Arctic sea-ice cover.

In summary, the warming of the North Atlantic and the movements of the Polar and Arctic fronts followed the main deglacial steps. The phases of strongest warming took place roughly in course with the most rapid ice-sheet diminution, but with a delay of about 1200 years in the Norwegian Sea during the first deglacial step. Thus the warm surface of the ocean provided a positive feedback that enhanced the deglaciation. The Younger Dryas cooling marked the return of Arctic waters and coincided with a period of reduced meltwater flux. During this period the cold ocean provided a negative feedback to deglaciation, which helped reduce the deglaciation rate.

For some time it was thought that the warming of the circum-Antarctic areas in the Southern Ocean took place earlier than in the North Atlantic. In a general sense, the sea-surface temperature variations of the Antarctic lead the North Atlantic variations during the Quaternary. During the last deglaciation AMS-dated records show the first marked warming of the Southern Ocean at 13,000 YBP, the same time as in the North Atlantic (Table I). It is not yet quite clear to what extent the deglacial warming and polar front movements in the Southern Ocean also was step-like. Some evidence suggest it, but more data are needed before we are able to reach a conclusion.

VII. Abrupt Climatic Change and Possible Ocean-Climate Feedbacks

We have seen that both the surface and deep-water circulation and the physical and chemical characteristics of ocean water changed dramatically during deglaciation. Rapid changes in both surface and deep-ocean properties as well as the reversal of the warming trend during the Y. Dryas cooling, show that the ocean circulation changes are not linearly related to orbital forcing. It appears from the work of J. Imbrie that they are fundamentally caused by orbital variations and that the period of most rapid change coincides with maximum summer insolation in the northern hemisphere. Uranium series dating of corals performed by high precision mass spectrometry has now documented that the radiocarbon time scale probably is compressed during the deglacial interval due to changes in the production rate of ^{14}C. Radiocarbon ages may consequently be 1500 years younger than true calendar years at 10,000 YBP and as much as 3000 years younger at 20,000 YBP. Since solar insolation is calculated for calendar years, this discrepancy accordingly shifts the relation between the data plotted on the radiocarbon time scale in Figs. 2 and 3 and insolation. This will slightly alter the phasing between ocean response and orbital forcing, but will not change the main relationship between forcing and response.

The complexity and rapidity of the oceanic changes during the deglaciation makes it difficult to provide a coherent explanation and a clear picture of the relationships between causes and effects of changing ocean circulation. Due to the rapidity of the events, some of the changes occur over time spans shorter than the accuracy of the best available dating methods and the stratigraphic resolution of proxy climate records. It is thus difficult to establish the right chronological chain of events.

Some elements seem to emerge from recent research. The main differences between deglacial, peak glacial, and present circulation patterns of the North Atlantic region are outlined in Fig. 4. The picture that emerges from the presently available data is that the ocean circulation was significantly different from the present and from that of the last glacial maximum during most of the deglaciation. This includes the Y. Dryas cold phase, which cannot be envisaged as a return to glacial conditions. Due to its ventilated deep waters and corresponding cold surface waters it was also dissimilar to both the peak deglacial situation and present ocean circulation.

The most dramatic alteration of the circulation regime took place during the two steps of rapid deglaciation, but meltwater anomalies observed by isotopic means appear in general to reduce the strength of NADW formation by producing a stable low-salinity surface layer. Since the reduction in oxygenation is most marked during the two deglacial steps, it follows that the strength of the North Atlantic heat conveyor was at its lowest during those times. It is thus difficult to understand what source of heat provided both the heat to explain the ocean warmth during those phases and the heat needed to disintegrate the ice sheets. A large portion of the disintegration is, however, probably related to calving rather than ablation. The following is a sketch of a tentative scenario that could account for the missing heat:

The initial reduction in the height of the Laurentide ice sheet must have led to a reduction in the advection of cold polar winds to the North Atlantic. At the same time summer insolation was about 7–8% higher than today. The huge amounts of meltwater and a more stratified upper ocean would make this heat more effective than present summer heating. It would be easier for summer warming to heat up a thin, more stable surface layer than in the present well-mixed ocean. The large amounts of fresh water that were produced at peak deglaciations also led to a net fresh-water export from the high and middle latitudes. To compensate for this, one needs a compensation current flowing underneath in the opposite direction. This would bring relatively warm thermocline waters northward which could, by upward mixing, provide additional heat in high-latitude areas. Such a situation is indicated in Fig. 4 and could provide a positive feedback to enhance deglaciation.

This does not explain why there was a sudden shift to glacial temperatures in the Y. Dryas. Since the cooling is most clearly pronounced in and around the North Atlantic, it is very tempting to assign it to on-off switches in the North Atlantic heat conveyor. Such an elegant model has been advocated strongly by, among others, W. Broecker. He argued that the location of the meltwater discharge may be more important than the absolute rate of meltwater release, and specifically that the diversion of meltwater flow to the St. Lawrence river system at 11,000 YBP effectively turned off deep convection because it directly hit the main flow of North Atlantic drift water. As discussed above, recent evidence documenting the changes in deep-water formation are exactly opposite to what this model prescribes. Oxygenation was low before and after but rose during the Y. Dryas (Fig. 2). Also, recent evidence for a more global nature of the cooling may point to causes outside of the Atlantic heat conveyor. It

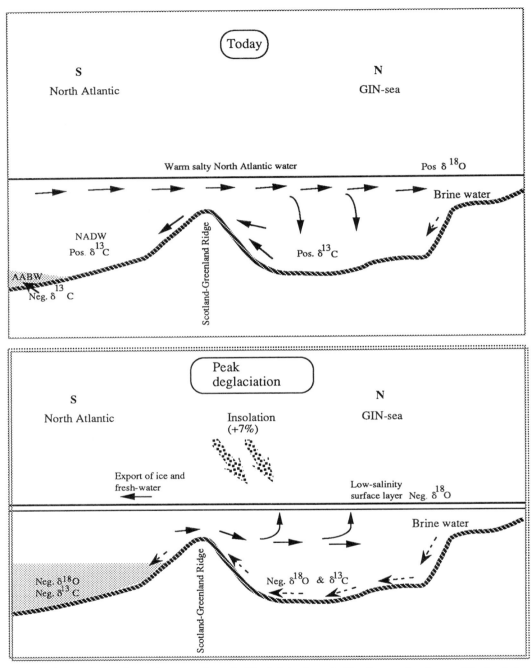

Fig. 4   A tentative north–south transect from the North Atlantic to the Nordic seas, showing possible circulation patterns during the glacial maximum 18,000 years ago, peak deglacial times, and the present ocean circulation. AABW is Antarctic Bottom Water, NADW is North Atlantic Deep Water, and NAIW is North Atlantic Intermediate Water.

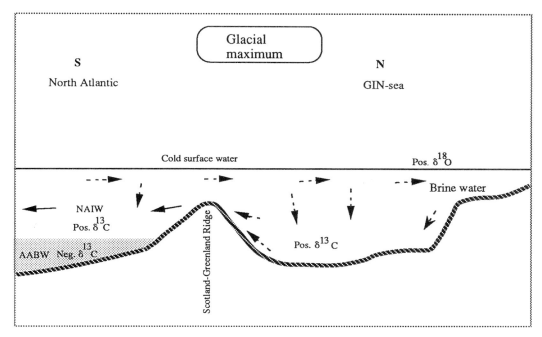

Fig. 4   *(Continued)*

thus appears that the explanation for this sudden cooling has to lie elsewhere in the complex ocean–atmosphere–ice system. [*See* AIR–SEA–ICE EXCHANGE PROCESSES.]

It cannot be explained by a rejuvenated influence of ice height on the jet-stream pattern, since the ice sheets continued to diminish, at a lower rate, during the cold phase. A sudden shift in atmospheric CO_2 content is a possible consequence of the rapid shifts in ocean circulation and ocean chemistry. At the moment the ice-core records are not accurate enough to document shifts in CO_2 content on the short time scales of the deglacial steps. It is also not yet clear if such shifts in deep-water circulation necessarily would lead to rapid changes in the carbon cycle, because the deep ocean circulates slowly and responds globally on time scales of 1000 rather than 100 years. It has also been proposed that negative albedo feedbacks brought about by increased flow of icebergs to the critical regions of the North Atlantic could account for the cooling. Icebergs would tend to cool the area due to their reflectivity and insulate the ocean from solar insolation. The problem with these explanations is that the evidence to support that such events actually took place at the right time, the initiation of the cooling, is not yet very strong. Discussions of the cause of the two-step character and the events of rapid shifts between warm and cold climates undoubtedly will continue to go on in the years to come. At present no theory satisfactorily explains the evidence. [*See* ICEBERGS.]

Bibliography

Aagaard, K., and Carmack, E. C. (1989). The role of sea ice and other fresh water in the Arctic circulation. *J. Geophys. Res.* **94**(C10), 14485–14498.

Bard, E., Labeyrie, L. D., Pichon, J.-L., Labracherie, M., Arnold, M., Duprat, J., Moyes, J., and Duplessy, J.-C. (1990). The last deglaciation in the Southern and Northern Hemispheres: A comparison based on oxygen isotope, sea surface temperature estimates and accelerator ^{14}C dating from deep-sea sediments. *In* "The Geologic History of the Polar Oceans: Arctic versus Antarctic" (U. Bleil and J. Thiede, eds.), pp. 405–416. Kluwer Academic Publ., Dordrecht, The Netherlands.

Berger, W. H., and Keir, R. S. (1984). Glacial-Holocene changes in atmospheric CO_2 and the deep sea record. *Geophys. Monogr., Am. Geophys. Union* **29**, 337–351

Berger, W. H., and Vincent, E. (1986). Sporadic shutdown of North Atlantic deep-water production during the Glacial–Holocene transition? *Nature (London)* **324**, 53–55.

Boyle, E. A. (1988). Vertical oceanic nutrient fractionation and glacial/interglacial CO_2 cycles. *Nature (London)* **331**, 55–56.

Boyle, E. A., and Keigwin, L. D. (1987). North Atlantic thermohaline circulation during the past 20,000 years linked to high-latitude surface temperature. *Nature (London)* **330**, 35–40.

Broecker, W. S., Peteet, D. M., and Rind, D. (1985). Does the ocean atmosphere system have more than one stable mode of operation? *Nature (London)* **315**, 21–26.

Broecker, W. S., Andree, M., Wolfli, W., Oeschger, H., Bonani, G., Kennett, J., and Peteet, D. (1988). The chronology of the last deglaciation: implications to the cause of the Younger Dryas event. *Paleoceanography* **3**, 1–19.

Broecker, W. S., and Denton, G. H. (1989). The role of ocean–atmosphere reorganization in glacial cycles. *Geochim. Cosmochim. Acta* **53**, 2465–2501.

COHMAP Project Members (1988). Climatic changes of the last 18,000 years: observations and model simulations. *Science* **241**, 1043–1052.

Fairbanks, R. G. (1989). A 17,000-year glacio-eustatic sea level record: influence of glacial melting rates on the Younger Dryas event and deep ocean circulation. *Nature (London)* **342**, 637–642.

Jansen, E., and Veum, T. (1990). Evidence for two-step deglaciation and its impact on North Atlantic deep-water circulation. *Nature (London)* **343**, 612–616.

Oppo, D. W., and Fairbanks, R. G. (1987). Variability in the deep and intermediate water circulation of the Atlantic Ocean during the past 25,000 years: Northern Hemisphere modulation of the Southern Ocean. *Earth Planet. Sci. Lett.* **86**, 1–15.

Ruddiman, W. F. (1987). Northern oceans. *In* "North America and Adjacent Oceans during the Last Deglaciation." (W. F. Ruddiman and H. E. Wright, eds.), pp. 137–154. Geological Society of America, Denver, Colorado.

Ruddiman, W. F., and McIntyre, A. (1981). The North Atlantic Ocean during the last deglaciation. *Palaeogeogr., Palaeoclimatol., Palaeoecol.* **35**, 145–214.

Shackleton, N. J., Duplessy, J. C., Arnold, M., Maurice, P., Hall, M. A., and Cartlidge, J. (1988). Radiocarbon age of last glacial Pacific deep water. *Nature (London)* **335**, 708–711.

Glossary

Benthic foraminifers Benthic protozoans (i.e., live on the sea floor), which may form carbonate tests. The tests are preserved in sediments and form an important basis for paleoceanographic reconstructions.

^{14}C Radioactive isotope of carbon with a half-life of 5.568 years, used to date geological matter and as a tracer for ocean circulation.

δ^{13}C Ratio of heavy (^{13}C) to light (^{12}C) carbon isotopes expressed as the per mil difference from a standard.

δ^{18}O Ratio of heavy (^{18}O) to light (^{16}O) oxygen isotopes expressed as the per mil difference from a standard.

Paleoceanography Discipline of the geosciences aimed at reconstructing past ocean circulation by analyzing sediments from the sea floor.

Planktonic foraminifers Planktonic protozoans that may form carbonate tests. The tests are preserved in sediments and form an important basis for paleoceanographic reconstructions.

Depositional Remanent Magnetization

Kenneth P. Kodama

Lehigh University

Depositional remanent magnetization is the magnetic record carried by clastic sediments and sedimentary rocks of the geomagnetic field at the time of deposition. It is acquired by a series of processes that start with the settling of sediments in the water column and end with their burial by subsequent sedimentation. The magnetic record provided by the depositional remanent magnetization of sediments is important to understanding the generation of the earth's magnetic field, dating and correlating sedimentary rocks globally and locally, and tracking the motion of tectonic plates through the earth's history.

I. Introduction

A. Remanence of Sediments and Sedimentary Rocks

Sediments can acquire a remanence, or permanent magnetization, by three different processes; physical, chemical, and biological. The physical process involves mechanical alignment of the magnetic moment of micron-sized ferromagnetic particles with the earth's magnetic field either during or soon after their deposition. The ferromagnetic particles are in low concentrations (typically less than 1% by weight) and are dispersed through the nonmagnetic minerals that make up the bulk of the sediment. This type of magnetism is the depositional remanent magnetization that is the main focus of this article. It is also known as detrital remanent magnetization, assuming that the magnetic particles are fragments of eroded rocks; but the terms are used interchangeably. They are often simply called DRM. This is the process by which sediments acquire a primary magnetization. [See GEOMAGNETISM.]

Once the sediments have been deposited and buried by other sediments, they are changed by diagenesis and lithified into sedimentary rocks. Therefore, depositional remanent magnetization (DRM) is the primary magnetization of clastic sedimentary rocks. Clastic rocks are made up of fragments of other rocks that have been eroded, transported, and deposited in a basin. The other major category of sedimentary rocks is chemical sedimentary rocks, such as carbonates formed by either the biologic or inorganic precipitation of calcite or dolomite or evaporites formed by the precipitation of salts. DRM plays a lesser role in this type of sedimentary rock since the erosion of other rocks typically provides the ferromagnetic mineral particles necessary for a strong DRM. [See SEDIMENTARY ROCKS.]

Depositional remanence is usually the strongest and most accurate in the finest-grained clastic rocks. Siltstones, shales, and mudstones form in low-enough energy environments so that the alignment of the magnetic particles can take place. Sedimentary rocks formed from lake, river, continental shelf, and deep sea sediments typically have a DRM.

Sediments and sedimentary rocks may also acquire a remanence by the growth of magnetic minerals after deposition and sometimes after lithification. When ferromagnetic minerals grow through a certain size, usually in the submicron range, they will become spontaneously magnetized parallel to the earth's magnetic field. This is termed a chemical remanent magnetization, or CRM, and is always a secondary magnetization. Although a CRM may occur very early in the post-depositional history of a sediment, it can occur long enough after deposition so that the time of magnetization and the time of deposition are significantly different. [See CHEMICAL REMANENT MAGNETIZATION.]

Magnetotactic bacteria, which produce chains or clusters of submicron-sized ferromagnetic particles, are found in both marine and nonmarine sediments. These bacteria use their magnetism to swim down along the earth's magnetic field lines so that they may avoid the oxygen-rich surface waters. When they die their magnetic particles are released to the sediment, which then may be aligned with the earth's magnetic field by the physical process mentioned above. In this case the source of the magnetic particles is not from eroded rocks, but from biological production. However, since these organisms live near the sediment – water interface the magnetization may still be considered to be primary.

The two most common naturally occurring ferromagnetic minerals are magnetite, Fe_3O_4, and hematite, Fe_2O_3; however, magnetite is usually the primary magnetite mineral in sediments and sedimentary rocks and therefore most likely to be associated with DRM. Hematite typically is a secondary mineral in sediments and so is usually associated with a CRM; however, hematite may carry a primary DRM in some red sediments.

B. Sedimentary Remanence and Global Change

The primary remanence of sedimentary rocks allows the measurement of past global changes in several ways. It can provide a continuous time record of the normal directional and relative intensity changes (the secular variation) of the earth's magnetic field and the more dramatic changes that occur during the interval when the field switches from one polarity to the other (polarity transitions). This type of information is important for constraining models of the field's generation in the earth's core and the processes that may lead to a field polarity reversal. Sediments and sedimentary rocks may be dated and correlated regionally using the remanence record of the field's secular variation and globally using the record of polarity reversals. This information, coupled with the environment of deposition deduced from the sediment's characteristics, allows correlation and dating of paleoenvironments on regional and global scales that can give a better measure of past environmental changes. [*See* Magnetic Stratigraphy; Paleomagnetism.]

Sedimentary remanence may also be used for plate tectonic reconstructions. By using the average magnetic direction from sedimentary units of different ages and assuming that the earth's magnetic field is best approximated by a magnetic dipole at the earth's center, the movement of tectonic plates with respect to the earth's rotation axis may be measured. If this information is available for all of the earth's major tectonic plates, their position with respect to each other at different times in the past can be reconstructed. Since the position of the continents affects oceanic circulation and oceanic circulation affects climate, computer simulations of past climates could ultimately be obtained from plate tectonic reconstructions.

One of the most intriguing observations linking sedimentary remanence to measurement of recent global changes has been the unexpected detailed correlation between the intensity and sometimes the direction of sedimentary remanence and paleoclimatic indicators such as oxygen isotope ratios, the abundance of certain foraminifera, or calcium carbonate concentration. Although this was originally thought to indi-

cate a fundamental link between changes in the geomagnetic field and climate, it is more likely that changes in climate have caused changes in magnetic mineralogy and the concentration of magnetic minerals. The investigation of these relationships is the focus of an important new field, environmental magnetism. [*See* Climate.]

II. The DRM Acquisition Process

A. Predepositional Processes

Before a ferromagnetic mineral grain can contribute to sedimentary remanence it must become magnetized. For sedimentary rocks this process occurs in the provenance or source area of the sediments that eventually become the sedimentary rock. If the provenance area is an igneous terrane then the ferromagnetic minerals, usually an oxidized form of magnetite or titanomagnetite, are primary and have acquired a thermal remanent magnetization or TRM. A TRM is acquired when a ferromagnetic mineral cools down through its Curie point, the temperature at which exchange interactions (see ferromagnetism in the Glossary) can overcome thermal oscillations and magnetic order at the atomic scale sets in. If the provenance area is a metamorphic terrane, ferromagnetic minerals may acquire either a TRM or a chemical remanent magnetization as new mineral phases grow in elevated temperature and/or pressure conditions. If the provenance area is a sedimentary terrane, the ferromagnetic minerals in the rock have already been magnetized in their source area.

For the magnetization acquired by these ferromagnetic grains in the provenance area to be important to DRM it must be stable enough to last from the formation of the source terrane at least until deposition. If the DRM is to be useful as a record of the earth's field it must last until the present, when the DRM is measured. Magnetite grains ranging in size from about 0.05 microns to several microns are single-domain or pseudo-single-domain grains, meaning that they either have or behave as though they have their whole volume magnetized in one direction. Single-domain grains cannot lose or have their intrinsic magnetism easily reset at the temperature and pressure conditions near the earth's surface; hence they are stably magnetized.

When the provenance area is eroded, magnetic and nonmagnetic grains are transported to the depositional basin, where the sediments then acquire a DRM by the processes detailed below.

B. Alignment in the Water Column

As the ferromagnetic grain settles through the water column of the depositional basin, a torque attempts to rotate its

Fig. 1 Angle θ between the magnetic moment m of a ferromagnetic grain and the earth's magnetic field H. As the grain settles through the water column the field exerts an aligning torque on m [see Eq. (1) in text].

magnetic moment into alignment with the earth's magnetic field. This torque has the magnitude:

$$mH \sin \theta \qquad (1)$$

where m is the magnetic moment of the grain, H is the intensity of the earth's magnetic field, and θ is the angle between the grain's magnetic moment and the earth's magnetic field lines (Fig. 1). This aligning torque is opposed by the inertia of the grain given by

$$(\pi \, d^5 \rho / 60) \, d^2\theta / d^2t \qquad (2)$$

where the quantity in the parentheses is the moment of inertia I of a spherical grain, d is the grain diameter, and ρ is the grain density. The aligning torque is also opposed by the viscous drag between the particle and the surrounding water:

$$(\pi \, d^3 \eta) \, d\theta / dt \qquad (3)$$

where the expression in the parentheses is a damping parameter λ, and η is the viscosity of the water. These terms may be combined to write an equation of motion for the grain:

$$I \, d^2\theta / d^2t + \lambda \, d\theta / dt + mH \sin \theta = 0 \qquad (4)$$

For the very small ferromagnetic grains important to DRM, the inertial term is not important and the system quickly approaches small values of θ; so the small-angle approximation can be used and $\sin \theta \approx \theta$. This leads to a simplified equation:

$$d\theta / dt = -(mH\theta) / (\pi \, d^3 \eta) \qquad (5)$$

with the solution

$$\theta(t) = \theta_0 \exp(-t/t_0) \qquad (6)$$

where θ_0 is the initial angle and t_0 is given by

$$t_0 = (\pi \, d^3 \eta) / (mH) \qquad (7)$$

and is the characteristic time for θ to decrease from θ_0 to θ_0 / e. When appropriate values for the moment of a ferromagnetic grain, the viscosity of water, and the intensity of the earth's magnetic field are substituted into the expression for the characteristic alignment time, t_0 is found to have a value of about one second. Based on this simple theory all the ferromagnetic grains in a depositing sediment should be perfectly aligned with the earth's magnetic field. However, observations in the laboratory and in nature show that the intensity of DRM is directly proportional to the earth's field intensity (Fig. 2a). This indicates that the ferromagnetic grains in sediments cannot be perfectly and completely aligned with the field.

One possible misaligning influence would be the random Brownian motion of water molecules bumping into the ferromagnetic grains. This effect would be more important for the very small submicron-sized grains; however, they carry the most stable magnetic signal and are most important to DRM. Langevin theory from statistical thermodynamics can be applied in this case. It essentially states that the proportion of ferromagnetic grains aligned with the field will depend exponentially on the ratio of aligning energy to thermal energy, which provides the misaligning influence. If the ferromagnetic grains' magnetic moments are assumed to be uniformly distributed from 0 to some maximum value, m_{max}, Langevin theory yields:

$$\text{DRM}/\text{DRM}_0 = (1/x) \ln(\sinh(x)/x) \qquad (8)$$

where DRM_0 is the completely saturated remanence that would result if the ferromagnetic grains were completely and perfectly aligned. The parameter x is given by

$$x = m_{max}H/kT \qquad (9)$$

where k is Boltzmann's constant and T is the absolute temperature. Note that $m_{max}H$ is the aligning energy and kT is the misaligning thermal energy. Equation (8) is plotted in Fig. 2b. Its form is very similar to the variation of DRM intensity with field for redeposited glacial sediments shown in Fig. 2a. Furthermore, laboratory experiments in which the magnetization of slurries made from deep sea sediments is measured while a field is applied supports the Langevin description of the misaligning effects of Brownian motion.

Other misaligning influences during settling could come from water currents near the sediment–water interface rotating the ferromagnetic grains just before touch-down, or from

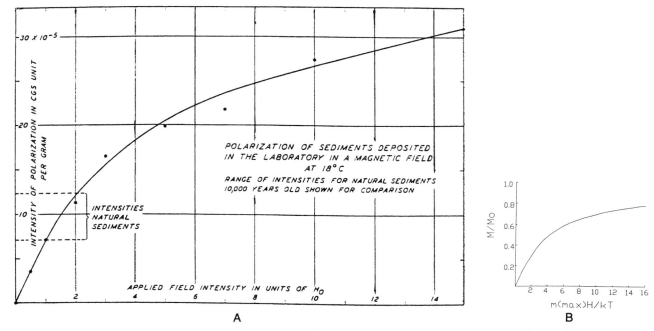

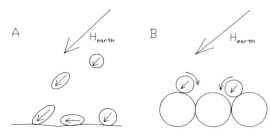

Fig. 2 (A) Intensity (polarization) of DRM as a function of field strength during laboratory redeposition of a varved clay sample. [From E. A. Johnson, T. Murphy, and O. W. Torreson (1948). Pre-history of the earth's magnetic field. *Terr. Magn. Atm. Electr.* **53**, 349.] (B) Increase in DRM intensity (M/M_0) as a function of field strength (H) assuming that the misaligning influence of Brownian motion can be modeled using the Langevin formula. [From F. D. Stacey and S. K. Banerjee (1974). "The Physical Principles of Rock Magnetism." Elsevier, Amsterdam.]

the irregular shape of a grain causing it to tumble as it falls through the water column.

C. Misalignment during Touch-Down, the Classical Inclination Error

In the 1950s and early 1960s a great amount of laboratory experimental work was done to try to better characterize DRM. In these experiments natural and artificial sediment slurries with magnetite as the ferromagnetic mineral were quickly deposited and allowed to dry in the earth's magnetic field. Invariably when these sediments were sampled and measured their magnetic inclination, the angle that the magnetic vector in the sediments makes with the horizontal, was 10° to 20° shallower than the inclination of the earth's field during deposition. This is what has become known as the classical inclination error and has led to some concern that sediments and sedimentary rocks could not accurately record the ancient earth's field.

This early experimental work also led to several theoretical models to explain the inclination error. One model postulated that a certain proportion of the ferromagnetic grains in

a sediment were plate-like and would lie flat on touch-down due to gravitational forces (Fig. 3a). Since the magnetic moment of these grains would be parallel to their flat surface to minimize their magnetostatic energy, the magnetic inclination recorded by these grains would be flattened. In the model the rest of the magnetic grains were spherical and would maintain their magnetic alignment with the earth's field during touch-down. Unfortunately, using this model to explain the labora-

Fig. 3 (A) Spheres and plates model of the inclination error. Arrows are individual grains' magnetic moments. Large arrow indicates the earth's magnetic field (H_{earth}). (B) Rolling spheres model of the inclination error.

tory results led to an unrealistically high proportion of plate-like grains. A second model postulated that all the ferromagnetic grains were spherical, a reasonable assumption for the cubic crystal symmetry of magnetite, and were aligned with the field as they settled toward the bottom. As they touched down these grains would roll into the nearest hollow on the sediment's surface created by the previously deposited grains (Fig. 3b). This is modeled mathematically as rotation through some angle around randomly oriented horizontal axes. To explain the laboratory results, however, implausibly large rotations ($\approx 75°$) were needed. Recent very slow redeposition experiments do not indicate any inclination error and suggest that the inclination error may have been an artifact of the earlier experimental work, either due to the very fast rate of deposition or the very short drying time. However, the disparity between the early laboratory work showing large inclination errors and observations that recently deposited natural sediments apparently did not have large inclination errors led workers to different models of DRM acquisition.

D. Post-Depositional Remanent Magnetization and Its Lock-In

While some workers were trying to characterize DRM by conducting redeposition experiments to see the accuracy of alignment during settling and touch-down on the sediment–water interface, other workers were investigating the possibility that the ferromagnetic grains in a sediment could realign with the earth's field after they were deposited. One study investigated a slumped bed in a sandstone. The bed had apparently slumped soon after deposition, while its water content was still quite high; and it had subsequently gone through diagenesis and become a sedimentary rock. The remanence throughout this slumped bed was constant in direction, suggesting that it had become magnetized after the bed had slumped. Variations in the character of the remanence with different grain-size sandstone beds suggested that the remanence was related to deposition and not a late-stage chemical remanent magnetization. This was the first example of post-depositional realignment of ferromagnetic grains in a sediment. Subsequently several laboratory experiments confirmed that the post-depositional realignment of ferromagnetic grains can take place.

In one experiment a mixture of quartz and magnetite grains was flooded with water in the presence of the earth's field. This sediment was able to accurately record the magnetic field direction. In another experiment small amounts of a slurry made from deep-sea sediments were deposited daily. The laboratory field was reversed in direction during the experiment, and it was found that the sediments didn't record the field reversal until 10 days after it occurred. Clearly the magnetic grains were able to realign and follow

changes in the ambient field for 10 days after they were deposited. Recent observations of different kinds of natural sediments show that post-depositional realignment of ferromagnetic grains leading to what is termed a post-depositional remanent magnetization (PDRM) is very important in nature. This realization reduced the importance of misalignment at touch-down or the classical inclination error for natural sediments and suggested that in most cases sedimentary remanence would be an accurate recorder of the earth's magnetic field.

If ferromagnetic grains can realign with the ambient magnetic field in a sediment after deposition, yet sediments are also known to carry a stable remanence, there must be some point after deposition when the ferromagnetic grains become "locked-in" and can no longer follow changes in the magnetic field. This occurs when burial compaction has forced enough water out of the sediment so that the pore spaces between the nonmagnetic matrix grains of the sediment are too small to let the ferromagnetic grains move. A great deal of work has been done to try to characterize the "lock-in" of a PDRM.

Several different types of experiments have been done to determine the "critical water content" when the ferromagnetic grains become locked-in. Soil consolidometers, centrifuges, drying, and burial compaction resulting from long, slow redeposition have all been used to slowly remove water from sediments in the presence of either constant or changing laboratory magnetic fields. The sediments' magnetism is measured, and the critical water content is that at which either the magnetic intensity of the sediment reaches a maximum value or the magnetic grains can no longer follow magnetic field changes. When these results are applied to natural sediments they suggest that lock-in should occur in the top 1 to 2 m of the sediment column; however, variations within this range are caused by the relative size of the nonmagnetic matrix grains and the ferromagnetic grains.

Remanence measurements of cores taken in the deep sea or near-shore marine environments show that the magnetization intensity of these sediments reaches a maximum value at a depth of about 10 cm below the sediment–water interface (Fig. 4). This would suggest that all the ferromagnetic grains are locked in by this depth and contribute to the sedimentary remanence. At shallower depths some of the ferromagnetic grains are still free to move and cannot carry a stable DRM.

Theoretical treatments of DRM lock-in have essentially approached it from two different perspectives: either a continuously increasing viscosity impeding the motion of the ferromagnetic grains or more discontinuous models. One continuous model introduces the important concept of a lock-in zone. The model predicts that the ferromagnetic grains in a sediment will have increasingly longer characteristic times for movement [similar to the characteristic time in

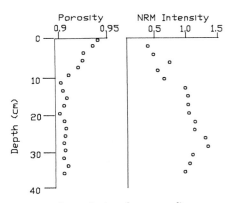

Fig. 4 Decrease of porosity in a deep-sea sediment to a constant level and simultaneously an increase in NRM (natural remanent magnetization) intensity to a constant level suggests lock-in of a PDRM at a depth of about 10 cm. [From T. Yamazaki (1987). Fixation of remanent magnetization with compaction in deep-sea environment. *J. Geomagn. Geoelectr.* **39**, 307.]

Eq. (7)] as they are buried deeper and deeper, since the viscosity resisting their motion will increase as the sediment grains move closer together during burial. At any given time after deposition grains deeper in the sediment column will be more locked-in than others closer to the sediment–water interface. A lock-in zone, a range in depth where grains can respond to magnetic field changes to varying degrees, will result. Those at the top of the zone can respond completely to field changes while those at the bottom of the zone cannot reorient at all. This process will tend to produce a remanence record that has smeared out the actual changes in the earth's magnetic field. Although this model only considers ferromagnetic grains all of the same size, a range in magnetic grain sizes would also lead to a lock-in zone, since larger grains would tend to be locked-in at shallower depths than smaller grains. By contrast a discontinuous model suggests that the torque acting on a magnetic grain by the earth's field [see Eq. (1)] is opposed not by a continuously varying viscosity but by friction with the surrounding nonmagnetic matrix grains. As the sediment compacts, the matrix grains will reorient and may shift to a different structure that will allow the magnetic grain to quickly move into closer alignment with the magnetic field. Thus post-depositional mobility occurs in short, discrete steps as the sediment compacts. When the magnetic grain is close to being aligned with the field the magnetic torque is small; it is more difficult for the grain to overcome the increasingly larger frictional forces imposed by the compacting matrix grains, and it becomes locked in. Whether the forces resisting ferromagnetic grain mobility increase continuously or in discontinuous steps, as compaction proceeds the forces grow and the ferromagnetic grains become less mobile until all the grains in the sediment cannot respond at all to

magnetic field changes. The resulting lock-in zone will have different depths and thicknesses depending on nonmagnetic and magnetic grain size and shape, but the PDRM of the sediment will be a smeared record of geomagnetic field changes. This effect will be considered further in Section III.

E. Post-Depositional Disturbances

The two main post-depositional disturbances of PDRM are bioturbation, which occurs very soon after deposition, and the compaction of the sediments that occurs at depths on the order of hundreds of meters.

Bioturbation is the disturbance of approximately the top 10 cm of the sediment column by the burrowing of animals. Bioturbation could potentially be a very destructive force for PDRM, causing randomization of the ferromagnetic particles; however, observations of natural sediments and laboratory experiments suggest just the opposite effect. In several experiments where the remanence of bioturbated, present-day sediments is monitored the PDRM is not destroyed but is actually enhanced (Fig. 5). These at first puzzling results can be explained by laboratory experiments in which an artificial sediment is stirred in a magnetic field. As the sediment is stirred the intensity of PDRM increases and accurately

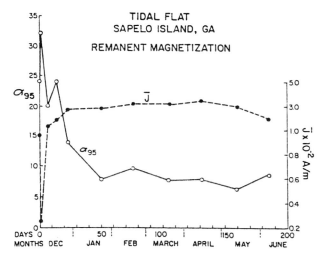

Fig. 5 The results of an experiment in which the magnetization of sediments placed in a tidal flat environment is monitored as they are increasingly bioturbated. α_{95} is a measure of the scatter of directions from several samples; α_{95} is seen to decrease with time indicating that the PDRM is actually enhanced by increased bioturbation. J, the magnetization intensity, is seen to increase early in the experiment, suggesting that bioturbation allows the magnetic grains to become better aligned with the field. [From B. B. Ellwood (1984). Bioturbation: minimal effects on the magnetic fabric of some natural and experimental sediments. *Earth Planet. Sci. Lett.* **67**, 367.]

records the direction of the magnetic field. It is thought that stirring and, by analogy, bioturbation actually increases the porosity of the sediment, decreasing the viscosity opposing grain mobility, and allows the ferromagnetic grains to align more easily with the geomagnetic field. But bioturbation, along with a lock-in zone, tends to smear or smooth the geomagnetic signal recorded by the sediments since the ferromagnetic grains are remobilized over a finite depth range.

As marine sediments are buried to depths of several hundreds of meters they typically lose about 50% of their volume through the expulsion of water from their pore spaces. There is the possibility that this volume loss will cause a shallowing of the magnetic inclination of the PDRM or DRM carried by the sediments since the volume loss occurs by vertical shortening. Until fairly recently cores taken of deep-sea sediments only penetrated several meters down into the sediment column, and no compaction-caused inclination shallowing had been observed. Recent technological advances have allowed sediment cores about a hundred meters long and down-core inclination shallowing, probably caused by compaction, has been observed. Laboratory experiments also show that volume losses caused by vertical shortening can cause inclination shallowing and that 5–10° of shallowing could be expected for a 50% volume loss.

The most obvious model of compaction-caused inclination shallowing is that as pore spaces close during burial compaction, magnetic grains are forced into more horizontal orientations. Since the magnetic vector in a grain tends to lie along its long axis, magnetic inclination is shallowed. This model would predict that the largest ferromagnetic grains would be most affected by compaction; however, some experiments with clay-rich sediments have indicated that the finest magnetic grains are affected the same or more than the larger magnetic grains. To explain these results it is suggested that submicron to micron-sized magnetite grains are sticking to the clay particles by electrostatic or Van der Waals forces and rotating with the compacting clay matrix. This would suggest that the magnitude of compaction-caused inclination shallowing would be dependent on sediment type, perhaps more of a factor in clay-rich sediments. However, the decreasing-pore-space model of compaction inclination shallowing would also predict that clay-poor sediments can be affected by compaction inclination shallowing. It should be noted that compaction-caused inclination shallowing has only been recognized in several long deep-sea sediment cores and has not been observed in many shorter cores taken from lake sediments, so the extent to which compaction shallowing of inclination affects sedimentary remanence is not presently known.

The magnetite–clay sticking model of compaction shallowing may also have implications for PDRM lock-in. If the finest grained magnetite grains stick to clay particles, they may become locked-in in clay-rich sediments at much shallower depths than would be predicted from simple decreasing pore space models of lock-in. Since the finest magnetite grains tend to carry the most stable remanence, this would affect models of how lock-in zones tend to smear or smooth the record of geomagnetic field changes (see Section III).

F. DRM Processes in Hematite-Bearing Sediments

Since magnetite is typically the primary ferromagnetic mineral in clastic sediments, the syndepositional and post-depositional processes discussed above usually are assumed to apply to magnetite-bearing sediments. However, hematite-bearing, red-colored, clastic sedimentary rocks (red beds) frequently occur in the geologic record and are very important paleomagnetically, since they are strongly magnetic and can carry a very stable remanence. A still unresolved question is whether these rocks carry a primary DRM/PDRM signal or a secondary CRM. The very fine-grained pigmentary hematite that gives these sediments their red color is usually secondary, and its CRM contribution to the total remanence of the rock can be removed by heating and cooling samples in very low magnetic fields (thermal demagnetization); however, coarser-grained detrital hematite grains could carry a depositional remanence. The best evidence for depositional remanence in red beds comes from samples collected in different parts of a sedimentary structure called cross-bedding. The bedding in cross beds was deposited with different orientations with respect to the horizontal at various locations in the structure. Since the magnetic inclination can be correlated with bedding tilt, apparently the orientation of the hematite grains was affected by the dip of the sediment surface at touch-down and the magnetization is a primary DRM.

Evidence that the remanence in hematite-bearing rocks has been affected by bioturbation may also indicate it is a DRM. Redeposition experiments with hematite-bearing sediments give similar results to experiments with magnetite-bearing sediments. A syndepositional classical inclination error is observed, but post-depositional reorientation of the hematite grains when the sediment slurry is shaken gives an accurate record of the earth's magnetic field direction. The limited data suggest that red beds may carry a depositional remanence, but general statements about red bed remanence are difficult to make since hematite is usually a secondary magnetic mineral and assumed to indicate a CRM.

III. Resolving the Earth's Past Magnetic Field Variations from the DRM Signal

A finite-width lock-in zone acts as a low-pass filter of geomagnetic field variations. The effect of a zone of DRM

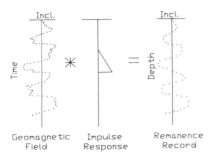

Fig. 6 Hypothetical example of how the smearing effects of a lock-in zone or bioturbation, quantified by an impulse response function, can affect the true variations of a magnetic field parameter, in this case inclination (incl.). Note that the high-frequency content of the signal has been lost in the remanence record. If the impulse response function is known, one can work backwards from the remanence record to the geomagnetic field variations. * indicates the convolution operation.

acquisition is to average the variations of the earth's field over the time it takes to deposit that thickness of sediments. Bioturbation can have a similar effect by remobilizing all the ferromagnetic grains over a depth range in the sediment column. By using linear systems theory we can mathematically model this effect with convolution:

$$r(t) = g(t) * l(t) \qquad (10)$$

where $g(t)$ is the input, in this case the actual variations of the geomagnetic field with time; $r(t)$ is the output, the remanence record of the geomagnetic field variations; and $l(t)$ is the impulse response function representing the filtering characteristics of a lock-in zone (Fig. 6). Any filter may be mathematically described from its impulse response, essentially how it would respond to a spike, an input with infinite amplitude that is infinitely short in duration. The * represents the convolution operation, which for Eq. (10) would be the integral:

$$r(t) = \int_{-\infty}^{+\infty} g(\tau)l(t - \tau)\, d\tau \qquad (11)$$

If each of the DRM processes that occurs during deposition and after deposition could be described in terms of its impulse response, the convolution operation could be used to fully characterize the sedimentary remanence in terms of the geomagnetic field variations. For example, if $a(t)$, $t(t)$, $m(t)$, $l(t)$, $b(t)$, and $c(t)$ represent the impulse responses describing the effects of alignment in the water column, misalignment at touch-down, post-depositional mobility and alignment, lock-in, bioturbation, and compaction, respectively, then the remanence could be given in terms of the geomagnetic field variations by

$$r(t) = g(t) * a(t) * t(t) * m(t) * l(t) * b(t) * c(t) \qquad (12)$$

Experiments with natural and synthetic sediments can be used to determine these impulse responses so that the remanence record can be deconvolved to provide the true geomagnetic field variations.

$$\begin{aligned} g(t) = r(t) &* 1/a(t) * 1/t(t) * 1/m(t) * 1/l(t) \\ &* 1/b(t) * 1/c(t) \end{aligned} \qquad (13)$$

Some promising work has already been done using this approach, particularly in determining the lock-in zone thickness of deep-sea sediments given the known sequence of geomagnetic field reversals. The linear systems theory approach has the greatest promise for providing the most accurate record of the geomagnetic field from DRM, the ultimate goal of all DRM studies.

Bibliography

Lund, S. P., and Karlin, R. (1990). Introduction to the special section on physical and biogeochemical processes responsible for the magnetization of sediments. *J. Geophys. Res.* 95 (B4), 4353.

Tarling, D. H. (1983). "Palaeomagnetism, Principles and Applications in Geology, Geophysics, and Archaeology." Chapman and Hall, London.

Tucker, P. (1983). Magnetization of unconsolidated sediments and theories of DRM. *In* "Geomagnetism of Baked Clays and Recent Sediments" (K. M. Creer, P. Tuchulka, and C. E. Barton, eds.), pp. 9–19. Elsevier, Amsterdam.

Verosub, K. L. (1977). Depositional and postdepositional processes in the magnetization of sediments. *Rev. Geophys. Space Phys.* 15(2), 129.

Glossary

Diagenesis Chemical and physical processes occurring after deposition that change a sediment into a sedimentary rock.

Ferromagnetic Spontaneous magnetism that arises in some minerals from the coupling, through quantum mechanical exchange interactions, of adjacent atomic magnetic moments. If the adjacent moments are equal and opposite in direction, it is termed antiferromagnetism; if the opposing moments are not equal, it is ferrimagnetism. In this paper ferromagnetism is used to indicate spontaneous magnetism in general and refers to both ferrimagnetic and antiferromagnetic minerals.

Remanence Permanent magnetism of rocks or sediments that remains when they are no longer in the presence of a magnetic field.

Desertification

Wesley M. Jarrell

Oregon Graduate Institute of Science and Technology

Deserts are usually considered to be dry, abandoned, useless land areas hostile to humans. Desertification is the conversion of land from significant productivity, in terms of human values, to low productivity, again in human terms. For example, the conversion of land from grasses that are palatable to cattle to land dominated by shrubs or trees that are not palatable, may be considered desertification by those who wish to raise cattle, even though the amounts of biomass produced in both systems may be similar.

I. Introduction

Most areas are considered to be deserts because of insufficient rainfall to produce agricultural crops or forage for livestock. Approximately one-sixth of the earth's surface falls into this category. Other deserts may have soils low in nutrients or high in toxic elements or salts and are considered desert-like because they support relatively little plant life, or only slow-growing species. [See ARID AND SEMIARID ECOSYSTEMS.]

As the effect of human activities intensifies across the globe, the risk of desertification increases because soil and vegetation properties can change rapidly as a result of human impacts, compared with rates of changes due to natural processes alone. Thus, as human populations and demands on the earth resources increase, the ability of the ecosystems to meet these demands decreases. Without high levels of input in the form of irrigation waters, nutrients, soil amendments, or other management factors, land loses productivity and can no longer support human populations.

II. Historical Significance

The following discussion focuses on dry lands, but there are similarities to degraded humid systems as well.

A. Historical Climate Changes

We know that many areas of the world now considered deserts were formerly well vegetated. In the 10,000 years since the last Pleistocene glaciation, changes in the global climate, particularly precipitation, led to the gradual replacement of humid-zone forests by savannahs; savannahs were replaced by grasslands; and finally grasslands were replaced by the widely spaced shrubs and succulents that characterize many deserts. Some deserts also include extensive areas completely devoid of vegetation. Fossils of rainforest trees from the Sahara clearly show that it was once a humid environment.

B. Agriculture and Grazing

Agriculture and grazing are often viewed as key components of soil degradation leading to desertification. Natural, adapted vegetation is removed from the land and replaced with food and fiber crops. Disturbances to the soil surface, like tillage and animal traffic, and prolonged exposure when soils are completely bare (even years in some fallow-crop rotations), lead to enhanced erosion by wind and water.

Great civilizations grew up in part because of the organizational requirements of complex water distribution systems. But because many arid-land waters contain more salts than the plants can take up, these salts may accumulate in the soil. Poor management of even slightly salty irrigation water can lead to soil salinization, loss of soil structure, and waterlogging. In many cases this is accompanied by the accumulation of toxic elements such as boron, selenium, molybdenum, or arsenic. When these salts or trace elements exceed critical levels of bioavailability, plant or animal health is affected. Even after the land is abandoned, it often will not return to its original state, because the basic resource—the soil—has been too drastically changed. [*See* SOIL SALINITY.]

C. Deforestation

Even in dry lands, trees are often present in areas where water accumulates along streams or arroyos, or may be widely scattered across the landscape. Removal of these trees for fuel wood, construction, or other purposes exposes more soil to erosional processes and added heat. Dry land trees also exploit deep water that many other plant types cannot reach and may provide foods, fibers, and soil improvement functions that non-tree species cannot replace. Under natural conditions, trees often become established only infrequently, during years of high rainfall. With abandonment of the destructive practices, trees may not return naturally for many years or even decades.

III. Soil, Water, and Vegetational Changes

A. Vegetation

Many changes start with removal of natural vegetation. This may occur through over-grazing, clearing for agriculture, burning (or at times unnatural control of burning), and deforestation. Once the vegetation is removed, the soil surface is exposed, eventually disturbed, and natural nutrient and water cycles are interrupted. There can be total loss of vegetation or conversion of vegetation to unusable types.

1. Total Loss of Vegetation

Occasionally the vegetation is completely lost from the landscape. In this case, the land surface may be converted to bare rock or shifting sand dunes.

2. Vegetational Changes, Conversion to Unusable Types

In many cases, one vegetation type is converted to another. The new plants are often not valuable to humans for fuel, forage, or other uses.

B. Soil

Soil is a fundamental, but often underappreciated, resource for global plant production. Soil physical, chemical, and biological properties usually change during the process of desertification. There can be loss of nutrients, loss of physical structure, and loss of organic matter.

1. Loss of Nutrients

When vegetation growth is disrupted, plants take up less nutrients. This often leaves more nutrients available to be washed from the soil, lost as gases, or blown away in dust.

2. Loss of Physical Structure

A productive, stable soil usually allows water to enter it and pass through it rapidly. In addition, the small particles in the soil often stick to each other to form aggregates, which makes them heavier and less susceptible to wind and water erosion. When plants are removed, the surface is exposed directly to the forces of wind, raindrops, and flowing water. This power may sweep away particles and even aggregates of soil. It may cause the surface to become more compact and dense, with small pores that allow water to move through them very slowly. Because water can only move into the soil slowly, much of it runs off the surface. This action removes nutrient-rich top soil and also results in less water storage in the soil profile for plants. [*See* SOIL PHYSICAL CONDITION AND PLANT GROWTH.]

3. Loss of Organic Matter

Organic matter is an important resource in surface soils. It is a long-term source of plant nutrients; it helps bind together soil particles into aggregates; and it provides an energy source for beneficial soil microorganisms and other soil biota. In natural systems, this organic matter comes directly from plants or from animal manures in most cases. When the plants are removed, and erosion occurs, soil organic matter content usually decreases. Once it is lost, it is usually hard to restore soil organic matter levels, unless large amounts of organic materials are supplied from outside sources. [*See* SOIL BIOCHEMISTRY.]

C. Water Resources

1. Reduced Storage

Water that flows along the surface of the soil (runoff) will not infiltrate or percolate. Runoff therefore represents a loss of water that might otherwise be stored in the soil for plant use. The water that runs off of one area of the soil may accumulate in others. In these cases, dry river beds quickly become raging torrents. Some of the redistributed water infiltrates into the soil, making it available to vegetation along the stream bank. In other cases the water may percolate to greater depths and recharge groundwater levels and aquifers.

Portions of the landscape that do not receive water from other areas experience a shorter period of plant-available soil water. Some plants cannot complete their life cycles in this shortened period and are likely to disappear from the ecosystem. Animals will suffer during extended dry periods as well.

2. Soil Salinization

One classic direct effect of agriculture in many arid lands is the excess accumulation of soluble salts in the soil. These not

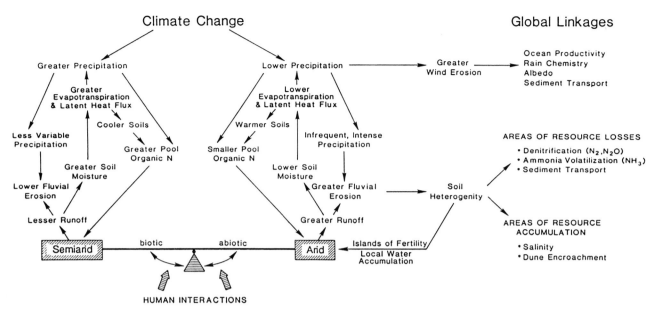

Fig. 1 Linkages between global change and human activities during desertification in dry lands. [From Schlesinger *et al.* (1990).]

only include table salt (sodium chloride), but also calcium, magnesium, and sulfate salts such as gypsum.

Some desert soils are naturally salty. Others become saline through poor irrigation practices. Many arid zone waters contain more salts than plants can take up. As a result, the plant removes the water and leaves behind the salt. Salt accumulates in the soil until the osmotic potential becomes so low that plants are stunted or they die.

Several solutions to this problem have been tested. One is to grow salt-tolerant plants. However, if salts are still accumulating, even the most salt-tolerant plants will eventually die.

The long-term, sustainable solution to salinization is to remove salts in the drainage water. Usually this is done by installing subsurface drain lines that carry the water to ditches or other surface drains, eventually to rivers and the sea. However, if the water is retained anywhere short of the sea, some will evaporate, leaving behind even saltier water. Kesterson Reservoir in Central California is a classic example of this phenomenon. Salts accumulated here and trace elements such as selenium eventually reached toxic concentrations.

IV. Global Linkages

The model of desertification proposed by W. H. Schlesinger and co-workers illustrates linkages between local processes and global impacts (Fig. 1). Increased erosion or over-grazing leads to increased atmospheric dust, trace gas emissions, ground reflectance, and sediment transport. Each of these

potentially affects the quality of other ecosystems, including the oceans, as well as the desertified region.

A. Atmospheric Changes

Changes in the atmosphere may occur due to increased dust and/or increased reflectance/albedo.

1. Increased Dust

When more of the soil surface is exposed to the sun, wind, and rain, dust storms are likely to increase. In the 1930s widespread drought in the Great Plains, combined with agricultural practices that left the soil disturbed and bare, resulted in huge clouds of dust that blew out into the Atlantic Ocean. The direct effects of air pollution caused by these clouds of particulates is obvious. In addition, the dust particles may disturb other ecosystems, for example, by "fertilizing" the ocean with elements like iron. In some large areas of ocean, plants may be iron deficient. When dust settles in these areas, it releases iron, which then stimulates the growth of these planktonic plants. [*See* ATMOSPHERIC CYCLE OF DESERT DUST.]

2. Increased Reflectance/Albedo

Fewer plants means that more sunlight hits the soil directly. The warm soil increases the air temperatures and thermal circulation of the atmosphere. As a consequence, lower humidity and precipitation result. Higher soil temperatures can also result in accelerated losses of soil organic matter through decomposition.

Desertification can also change the timing of water release back into the atmosphere. If water enters the soil uniformly over the landscape instead of being lost as runoff, plants will slowly extract it from the soil depths and release it. Without plants, rainfall is more likely to run off and only wet the soil to a shallow depth. As a result, water is released from the soil for short periods of time after rains, resulting in reduced thunderstorm activity or hotter environments overall.

V. Current Trends in Global Desertification

A. Extent of Desertification Worldwide

Estimates of desertification vary a great deal, because it is difficult to know what to measure and because many of the areas most affected are poorly described or characterized. However, wherever dry lands are being affected by human activity, desertification-like processes are potentially able to change land productivity.

About one-third of the earth's surface is currently dry lands, including deserts and semiarid grasslands and woodlands. It has been estimated that global change could result in a 17% expansion in the area of the region classified as dry lands. When combined with increased pressure from humans, this could lead to acceleration in the rates of desertification that are now occurring.

B. Means for Reversing Desertification

The United Nations and many other institutions have attempted for many years to reverse effects of desertification processes. Lowering the intensity of land use, or increasing the inputs of energy and materials into a system, are typically required. To be more effective in reversing desertification, we must understand the basic changes in the resource base that in turn produce vegetation and soil changes. Once these are well understood, we can assess the best means for restoring the productivity of degraded areas.

Bibliography

Batisse, M. (1985). Progress and perspective: a look back at the UNESCO arid zone activities. *In* "Arid Lands: Today and Tomorrow" (E. E. Whitehead, C. G. Hutchinson, B. N. Timmerman, and R. G. Varady, eds.), pp. 21–30. WestView Press, Boulder, Colorado.

Presser, T. S., and Barnes, I. (1984). Selenium concentration in waters tributary to and in the vicinity of Kesterson National Wildlife Refuge, Fresno and Merced Counties, California. *Water-Resour. Invest. (U.S. Geol. Surv.)* **84–4122.**

Schlesinger, W. H., Reynolds, J. F., Cunningham, G. L., Huennecke, L. F., Jarrell, W. M., Virginia, R. A., and Whitford, W. G. (1990). Biological feedbacks in global desertification. *Science* **247,** 1043–1048.

Verstraete, M. M. (1986). Defining desertification. *Clim. Change* **9,** 5–18.

Glossary

Soil productivity Ability of the soil to produce vegetation.
Soil salinization Accumulation of soluble salts (predominantly sodium, calcium, and magnesium chlorides and sulfates) in plant root zone; eventually plant growth diminishes.
Water use efficiency Effectiveness of irrigation or rainfall in producing plant biomass.

Doppler Radar

Peter S. Ray

Florida State University

Radar is an acronym for RAdio Detection And Ranging. An antenna is used to focus a beam of electromagnetic waves generated by the transmitter. The receipt of backscattered energy is used to detect discrete targets such as aircraft, or distributed targets such as rain. A Doppler radar compares the change in phases of the transmitted and received signals to determine the Doppler frequency shift. This is the scatterer's component of motion toward or away from the radar (i.e., the radical velocity). For weather, most radars transmit in the 3-cm (X-band) to 10-cm (S-band) wavelength band. The letter codes were developed in response to wartime security needs. Figure 1 shows the portion of the spectrum considered. Most meteorological radars operate between 3-cm and 6-m wavelength. The measured signal strength (converted to reflectivity) and radial velocity patterns enable us to identify storm types and to predict their course and severity. Radars operating at longer wavelengths receive more intense echoes from refractive index fluctuations in the optically clear air. This is used to monitor three components of the winds throughout the depth of the atmosphere in time.

I. Introduction

Radar was developed during World War II to identify and track warships and aircraft. It was greatly advanced by the British through the development of the cavity magnetron. The use of radar was a crucial advantage for the Allies during World War II. For example, it played an integral role in the destruction of one-third of the German U-boats during four months, March–June, in 1943. After Germany discovered the equipment on an aircraft shot down over Holland, U-boats were equipped with microwave listening devices and the success was reduced. But the few months of tactical advantage for the allies unarguably made a tremendous difference for the battle of the Atlantic. Similar contributions of the role of radar were repeated in other applications in other war theaters.

The role of radar in meteorological studies began shortly after World War II. Initial observations were descriptive. Only recently has it become possible to directly infer the dynamical and microphysical properties of convection from radar; this opens a new era of investigation in convection dynamics.

II. Operating Principles

Electromagnetic waves can be described by their amplitude, phase, and polarization. Weather radar is based on the interaction of electromagnetic waves with hydrometeors or other atmospheric inhomogeneities. A small fraction of energy that is incident on a refractive index inhomogeneity or particle is scattered, some backward in the direction of the antenna that normally is used for both transmitting and receiving.

There are many radar designs; only the main elements are described here. Refer to Fig. 2 for a schematic diagram of a radar. A transmitter produces power at a known frequency. Most radars use either a magnetron or a klystron to amplify the signal to be transmitted. The advantage of the magnetron is that it is much lighter and less costly. However, until recently, klystrons were necessary to preserve phase coherence (or the same phase) over many pulses. The principal difference between conventional (incoherent) and Doppler radar is that the conventional radar does not provide phase information—it only measures the amplitude of the received power.

In either case, the high-power pulse goes through the T/R (transmit, receive) switch (or duplexer). The purpose of this switch is to protect the receiver from damage by the high-power transmitted pulse. It connects the antenna to the

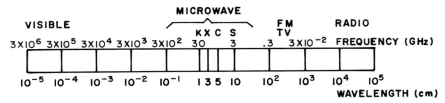

Fig. 1 Frequency spectrum from visible to radio frequencies and corresponding scale in wavelength. The microwave region, important to meteorological radar, is highlighted. Most meteorological radars operate between 3 and 10 cm wavelength.

transmitter when it is "on" and to the receiver when it is "off."

An antenna focuses the transmitted waves to a beam, usually about 1° wide, thereby both increasing the power density by a factor of 10^4 or so and indicating direction. Additionally, the antenna receives the fraction of the power backscattered from the target.

The receiver detects, amplifies, and converts the reflected microwave single into a low frequency, where it is interfaced to some type of indicator on which the signal can be displayed.

Most meteorological radars are pulsed radar, transmitting a burst of energy for about $\tau = 10^{-6}$ s (1 μs), every 10^{-3} s. The interval between pulses is the interpulse period, and the reciprocal is the pulse repetition frequency, or simply the PRF. Both of these parameters are widely varied.

To determine range, the return signals are sampled periodically in bins called "gates," typically sampling a depth every 150 m or so. This continues until the time for the next pulse.

The returned Doppler-shifted frequency is denoted by $f_t + f_d$, where f_t is the transmitted frequency and f_d is the Doppler-shifted component. When $F_d > 1/\tau$, the Doppler signal may be easily described from the information in a single pulse. But an S-band radar with a pulse width of 1 μs implies velocities greater than 10^4 m s^{-1}, or $\sim 30,000$ miles h^{-1}, to obtain the requisite frequency shift. To obtain speeds using microwave frequencies, many successive samples at each range gate must be used. Typically, discrete Fourier transform methods are used to process the complex time series samples.

A. Scattering

When a particle intercepts a radio wave, some of the energy is absorbed and some is scattered. For a single antenna, the portion of greatest interest is that portion scattered backward, towards the transmitting and receiving antenna. The amount scattered depends on the dielectric properties of the scatterer (refractive index), and the ratio of the wavelength to the size of the scatterer. Surface structure and other effects also play a role. Although scattering mechanisms are not completely understood, the complete solution for scattering

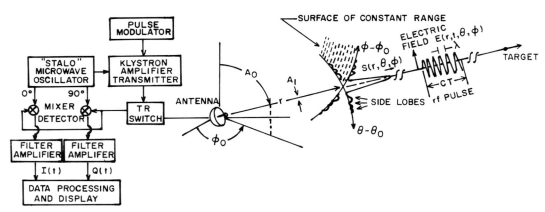

Fig. 2 Schematic of a radar system, showing main electronic components and fundamental characteristics of the transmitted pulse. [From Doviak and Zrnic (1984).]

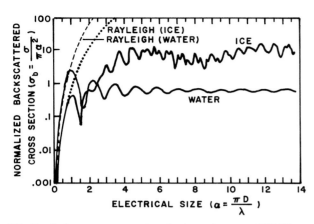

Fig. 3 Backscatter cross sections for ice and water at 0°C. The Rayleigh approximation is indicated for both cross sections. The electrical size has been determined with a constant wavelength of 10 cm.

by spheres was given by G. Mie in 1908. For the backward direction in the far field, it takes the form:

$$\sigma = \frac{\pi D^2}{4\alpha^2} \left| \sum_{n=1}^{\infty} (-1)^n (2n+1)(a_n - b_n) \right|^2$$

where D is the drop diameter, $\alpha = \pi D / \lambda$ and is called the electrical size, λ is the incident radiation wavelength, and a_n and b_n are coefficients of the scattered field involving spherical Bessel and Hankel functions. These are functions of the scattering angle, electrical size, and complex index of refraction, $m = n - ik$, where n is the ordinary refractive index and k the absorption coefficient. The variation of backscattered energy with electrical size is shown in Fig. 3.

If the diameter is small compared to the wavelength, the complete Mie solution can be simplified to

$$\sigma = \frac{\lambda^2}{\pi} \alpha^6 \left| \frac{m^2 - 1}{m^2 + 2} \right|^2$$

or

$$\sigma = \frac{\pi^5}{\lambda^4} \left| \frac{m^2 - 1}{m^2 + 2} \right|^2 D^6$$

Thus, the backscattered energy is proportional to the sixth power of the particle's diameter, as expected for a radiating dipole.

B. The Radar Range Equation

The radar range equation relates the received power to a knowledge of the transmitted power P_t, the radar system, the range r, and the characteristics of the target. J. R. Probert-Jones and others considered the antenna gain G, the beam dimensions ϕ, θ, and h, and the sum of the scatterers in the volume to obtain the relation

$$\overline{P}_r = \frac{P_t G^2 \lambda^2 \theta \phi h l}{512 (2 \ln 2) \pi^2 r^2} \frac{1}{\Delta V} \sum_{vol} \sigma_i$$

Typical values in storms range from 10^2 to 10^6 (mm)^6 ms^{-3}. It is conventional to express reflectivity in decibels, so that a reflectivity of 10^5 (mm)^6 m^{-3} is expressed as 50 dBZ.

As the energy propagates through the air or through precipitation, it will experience some loss, here denoted by l. This loss is wavelength dependent and is most pronounced for shorter wavelengths. The attenuation for 3.2-cm wavelength is about six times greater than at 5-cm and 40 times greater than at 10-cm wavelength.

C. Signal Processing

Through phase detection of the backscattered signal, when the transmitted signal is used as a reference, the bipolar video, or the I and Q (or in-phase and quadrature phase) components are isolated. The radial wind speed (but not direction) is contained in either the I or Q component. If the Doppler shift is positive (indicating motion toward the radar), Q will lead I by $\pi/2$; if the motion is away from the radar, Q will lag I by $\pi/2$. A succession of I and Q pairs form a complex time series. Using standard Fourier analysis techniques, the power density spectrum can be found. In general, the longer the sampling time (the dwell time) the more resolution in the frequency (velocity) domain. The more frequent the samples (higher PRF), the larger the resolvable velocity.

The spectral moments can be expressed as

$$f_n(V) = \frac{\displaystyle\int_{-\infty}^{\infty} (V - \overline{V})^n S(V) \, dV}{\displaystyle\int_{-\infty}^{\infty} S(V) \, dV}$$

where n is the moment of the spectrum. The integral of the spectrum is the reflectivity. The first moment is the mean velocity, and the second moment is the spectrum variance. The spectrum width is related to sampling parameter, the strength of the signal, turbulence in the sampling volume, the spread of the scatterers' terminal fall speeds, the shear of the wind along or across the beam, and the antenna rotation

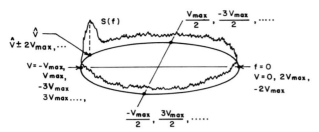

Fig. 4 Circular power density spectrum from a Doppler radar.

rate. Spectrum width has found limited utility in meteorological analysis.

One difficulty arises when the Doppler frequency shift is greater than the Nyquist frequency, $f_{max} = \pm 1/(2PRF)$. These frequencies (velocities) are aliased and interpreted as arising from a lower frequency. This is illustrated in Fig. 4. The peak in the power density spectrum could be from velocities $\hat{V}, \hat{V} + 2V_{max}, \hat{V} - 2V_{max}, \hat{V} + 4V_{max}, \hat{V} - 4V_{max},$ etc.

In the early 1970s, Doppler radar signal processing was revolutionized by the implementation of the autocovariance or pulse-pair processing algorithms. This greatly reduced the computational requirements to calculate the mean velocity (first moment of the Doppler power spectral density) over the Fourier transform methods. For the first time, the full scanning volume could be processed and displayed, illuminating the interior flow structure of storms. Recently, advances in computer technology have made the Fourier transform methods competitive.

III. Uses of Reflectivity Information

Reflectivity information is used for tracking storms, identifying them by characteristic structures or features, and estimating rainfall based on the strength of the returned signal.

Tracking storms is straightforward and generally is just a linear extrapolation from past to future time. An example of a storm structure is the hook echo that is usually found with tornadic and other severe thunderstorms. This reflectivity feature is formed by the advection of precipitation particles by the rotating flow in which the tornado is embedded. Other storms show a distinctly linear or another structure that indicates a possible threat or probable evolution.

A. Rainfall Estimation and Phase Discrimination

It has already been shown that power backscattered to the radar is approximately related to the sixth power of the

drop diameter, biasing the signal to the contribution of the largest drops, which also contribute the most to the mass of rainwater. More specifically, the reflectivity factor is given by

$$Z = \int N(D)D^6 \, dD$$

Many studies have established that the distribution of drop sizes is well represented by the exponential form

$$N(D) = N_0 e^{-\Gamma D} \quad (\text{cm}^{-4})$$
$$\Gamma = 41R^{-0.21} \quad (\text{cm}^{-1})$$
$$N_0 = 0.08 \quad (\text{cm}^{-4})$$

Here, R refers to the rainfall rate and is usually measured as the depth of water per unit time. Rainfall rates are usually of the form

$$R = \frac{\pi}{6} \int_0^\infty D^3 N(D) V_t(D) \, dD$$

where $V_t(D)$ is the terminal velocity for each diameter. Since N_0 and Γ cannot both be known from the measurement of Z alone, the $Z-R$ relationship is not unique. Generally, it is expressed in the form

$$Z = AR^b$$

where A and b are empirically determined. There are many solutions for this relation, but a commonly used one is $Z = 200 \, R^{1.6}$. In many places, rainfall estimation is improved by calibrating the reflectivity by the use of rain gauges. It has been shown that significant improvement can result from as few as one gauge per 1600 km². [*See* WEATHER PREDICTION, NUMERICAL.]

B. Multiparameter Techniques

As previously mentioned, reflectivity alone cannot uniquely determine two parameters of a rainfall distribution. However, additional information is available through multiparameter techniques. For example, if the reflectivity from two orthogonal polarizations are compared, additional information is given. Since large drops are distorted, they scatter more energy backward when illuminated with a horizontally polarized wave than they do when illuminated with a vertically polarized wave. The difference indicates their size. A common measure of this effect is called differential reflectivity, or

$$Z_{DR} = 10 \log\left(\frac{Z_H}{Z_V}\right) \quad (\text{dB})$$

Z_{DR} is proportional to raindrop size, allowing a unique solution to an estimation of the drop-size distribution and, therefore, the rainfall rate.

Perhaps the most useful application of Z_{DR} is in phase discrimination. High values of reflectivity can come from either heavy rain or hail. Typically, reflectivities in excess of 55 dBZ are associated with hail. But, whereas rain associated with high reflectivities will have a large (i.e., 3.5 dB) positive value of Z_{DR}, hail will have values near zero. [See HAIL CLIMATOLOGY.]

IV. Uses of Doppler Velocity Information

In a pulsed Doppler radar system, the pulses are separated at a distance R that is related to the pulse repetition frequency (PRF), by

$$R = \frac{c}{PRF}$$

where c is the radio wave propagation speed (3×10^8 m s^{-1}). At the scattering volume, a portion of the pulse is returned to the radar. The time between pulses is $1/PRF$, so the maximum range, R_{max}, that light can travel from the radar to the scattering volume and return is given by

$$R_{max} = \frac{c}{2\ PRF}$$

For a PRF of 1000 s^{-1}, R_{max} corresponds to an ambiguous range of 150 km. However, for any range R, echoes at ranges $R + nR_{max}$, where n is the number representing previous pulses, will be received at the same time. Without employing polarization or phase discriminating techniques, there is no way to unambiguously discriminate between the returns from different ranges. In a radar display, range-folded echoes may be distinguished by a typically elongated appearance, since their angular width is unchanged when they are displayed at their apparent range. This is illustrated in Fig. 5.

A fundamental sampling theorem states that to measure a frequency f_d, it is necessary to sample at a frequency of at least $2f_d$. Since the sampling rate is the PRF, the maximum frequency that can be resolved is PRF/2.

The Doppler theorem relates the observed velocity to the frequency shift by

$$V = -\frac{f\lambda}{2}$$

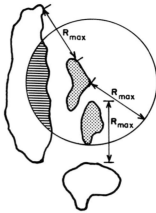

Fig. 5 Schematic illustration of range-ambiguous echoes. The heavy line defines the extent of the storm. The PRF-determined maximum range is labeled R_{max}. The echo area denoted by hatching is in its correct position. However, the radar processes the return so that echoes are displayed in the hatched and stippled areas. The correct position of the stippled area is radially outward a distance R_{max}.

It follows that the maximum ambiguous Doppler velocity that can be defined is

$$V_{max} = \pm \frac{(PRF)\lambda}{4}$$

Thus, an S-band (10 cm) Doppler radar operating with a PRF of 1000 s^{-1} has a maximum unambiguous velocity of Nyquist interval of ± 25 m s^{-1}. An X-band Doppler radar operating at the same PRF would have an unambiguous velocity interval of only ± 8 m s^{-1}. All velocities exceeding these bounds are aliased and are not distinguished from the fundamental if conventional Doppler processing is used.

Increasing the PRF to extend V_{max} decreases R_{max}. This is shown in the relationship

$$V_{max}R_{max} = \pm \frac{\lambda c}{8}$$

and in Fig. 6.

V. Multiple Doppler Radar Analysis

A powerful use of radars is to use them in combination, or in a network. Since each radar provides an independent estimate of the vector wind field, it is possible, with two or three radars, to describe the wind field within storms. This enables new insight into the processes that govern storm develop-

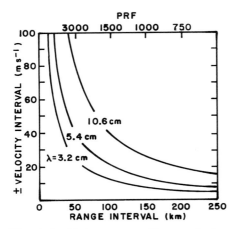

Fig. 6 The relationship between unambiguous velocity, range, wavelength, and PRF for three meteorologically important wavelengths.

ment. There are two principal approaches to the wind field synthesis. They share some common features.

The location of the ith radar is denoted by the coordinate triple (x_i, y_i, z_i), and a grid point is denoted by (x, y, z). If u, v, and W $(=w + V_t)$ represent the particle motion in the x, y, and z directions, where w is the vertical wind component, then the measured radial velocity (V_i) is related to the Cartesian wind components by

$$\frac{u(x - x_i)}{R_i} + \frac{v(y - y_i)}{R_i} + \frac{W(z - z_i)}{R_i} = V_i \qquad (1)$$

Here V_t represents particle terminal velocity, and the range is given by

$$R_i = [(x - x_i)^2 + (y - y_i)^2 + (z - z_i)^2]^{1/2}$$

Generally, radar observations are interpolated to a grid point in either a Cartesian or a conical section. Different methods have been used, but a distance-weighted average usually is employed.

A. Direct Solution

Three independent observations were sufficient to determine the (u, v, W) components directly. At the ground, vertical velocity estimates become unbounded. To overcome this restriction, one can use Eq. (1) to obtain the horizontal wind components and obtain w from the solution of the mass continuity equation.

$$\frac{\partial u}{\partial x} + \frac{\partial v}{\partial y} + \frac{\partial w}{\partial z} - \kappa w = 0$$

B. Dual-Doppler Solution

In Cartesian coordinates with observations from at least two radars and the mass continuity equation, there are sufficient observations to deduce all wind components. This is solved by using an estimate of W, reducing the problem to two equations (observations) and two unknowns, and then using the estimates to provide a refined estimation of W through the continuity relation. The updated W is then used in a new calculation of u and v. The iteration continues until convergence is reached. A covariance is introduced between the horizontal and vertical velocity errors, but it is small in most applications.

An example of the horizontal wind field at w km above the ground in a tornadic storm is given in Fig. 7. Here the 40-dBZ contour is highlighted and downdraft regions are stippled. Regions of updraft are also contorted in the region near the "hook echo." [See CLOUD DYNAMICS.]

VI. Airborne Doppler Radar

Airborne Doppler radar systems allow interrogation of storms where they occur and for longer periods of time as they advect to new locations. Size and weight considerations commend X-band systems for aircraft. One such system is onboard the National Oceanic and Atmospheric Administration (NOAA) Orion P-3, as illustrated in Fig. 8. The antenna scans in a direction normal to the flight path, prescribing a helical sampling pattern as the aircraft moves through the air. Pseudo dual-Doppler data can be obtained by flying L-shaped patterns. Another scanning method on the P-3, called "fore-aft" scanning, provides pseudo dual-Doppler data by alternately scanning in a direction somewhat less than or greater than 90° relative to the flight path.

Another implementation is the National Center for Atmospheric Research (NCAR) Eldora radar. This concept is illustrated in Fig. 9. This radar has two antennas and two receivers, one pointed 70° forward and the other 70° backward from the flight direction. As the airplane flys past the area of interest, it will view from angles that are separated by 40°. This is sufficient to use the dual-Doppler methodology as outlined above.

VII. National Doppler Radar Networks Profilers

A. NEXRAD

In the early 1990s, the nation will be largely covered with a network of NEXRAD (NEXt generation weather RADars) radars. When operational these radars will be designated as

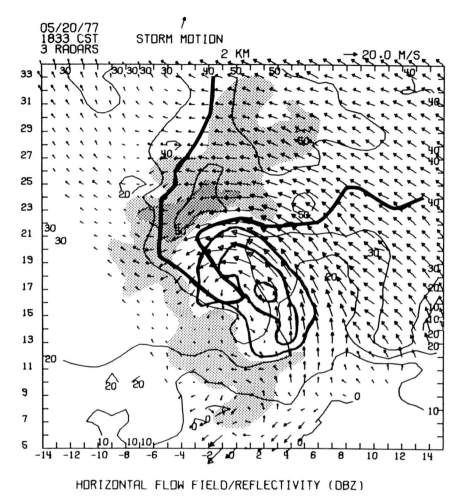

Fig. 7 Flow field and reflectivity at 2 km above the ground in a tornadic storm that occurred near Oklahoma City, Oklahoma on May 20, 1977. The 40-dBZ contour is heavy and the regions of updraft are also heavier in intervals of 10 m/s. Areas of descending motion are stippled. Tornado formed at coordinates (2, 16) about 7 min after these data were obtained.

WSR-88D. These S-band radars will have a beamwidth of less than a degree and other characteristics that would qualify them as first-rate research radars, as well as important tools in the operational meteorologist's inventory. Some of these will be equipped with the capability of recording and archiving the "base level" data, not just the products. Base level would include mean reflectivity, velocity, and spectrum-width data. This opens up new opportunities for researchers to combine other facilities with the NEXRAD and profiler systems to more easily examine the geographic diversity of storms, as well as reduce the costs of special research field observations.

The combination of NEXRAD and airborne Doppler radar is particularly attractive, since it would combine the spatial flexibility of aircraft with an extensive ground-based network. Long-lived systems such as squall lines could be followed through their life cycle.

B. Profiler

Radars operating in the very high frequency (VHF) band (40–50 MHz) to the upper ultra-high frequency (UHF) band (3 GHz) can detect echoes from Bragg scatter from refractive

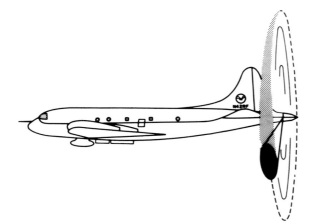

Fig. 8   Schematic of NOAA P-3 airborne Doppler radar.

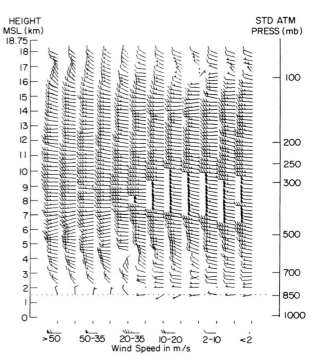

Fig. 10   An example of hourly averaged winds measured by the Platteville prototype profiler. A weak frontal passage is evident in the lower troposphere and ridge axis passage at middle levels. Winds plotted on the site elevation line are from standard surface measurements taken at the profiler site.

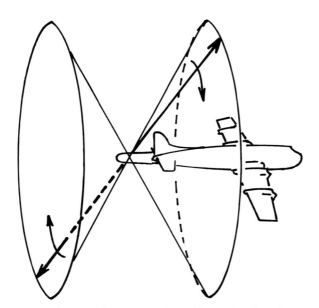

Fig. 9   Schematic of NCAR ELDORA airborne Doppler radar scanning pattern.

index structure due to variations in humidity and temperature in the visually clear atmosphere. With a vertically pointing antenna and two others pointing in orthogonal directions and about 15° from the vertical, it is possible to deduce the three wind components. For example, assume one of the antennas is pointing eastward and 15° from the vertical. Then the eastward component of wind is $u = 3.86V - 3.73w$, where w is obtained directly from the vertically pointing beam and defined as negative downward, and V is the measured radial wind from the eastward pointing beam. An example of Profiler data is shown in Fig. 10.

A network of wind profilers is being deployed across the central United States, adding to an already existing dispersed network of research and operational profilers. This augmentation of over 30 additional profilers will take this technology from the research arena and from the single-unit site to implementation of fields of data into our most fundamental forecast tools. These two networks (NEXRAD and the Profiler) are illustrated in Fig. 11.

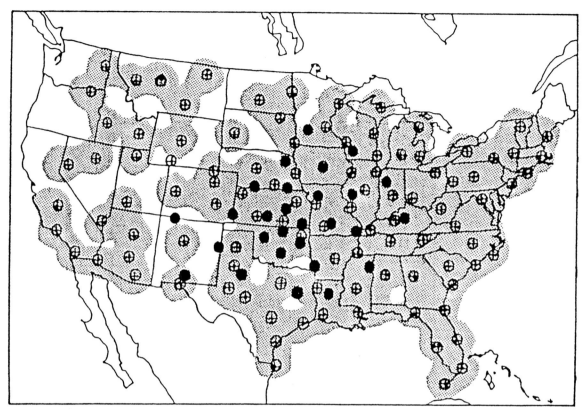

Fig. 11 Location of NEXRAD and Profiler networks. Coverage area of the NEXRAD system is indicated by the circular stippled regions.

Bibliography

Armijo, L. (1969). A theory for the determination of wind and precipitation velocities with dual-Doppler radars. *J. Atmos. Sci.* **26,** 570–573.

Atlas, D, ed. (1990). "Radar in Meteorology." American Meteorological Society, Boston.

Battan, L. (1973). "Radar Observations of the Atmosphere." Univ. of Chicago Press, Chicago.

Doviak, R. J., and Zrnic, D. S. (1984). "Doppler Radar and Weather Observations." Academic Press, Orlando, Florida.

Mie, G. (1908). Beitrage zur optik trüber medien, speziell Kolloidaler Metallösungen. *Ann. Phys. (Leipzig)* **30,** 377–442.

Probert-Jones, J. R. (1962). The radar equation in meteorology. *Q. J. R. Meteorol. Soc.* **88,** 485–495.

Ray, P. S., ed. (1986). "Mesoscale Meteorology and Forecasting." American Meteorological Society, Boston.

Ray, P. S., Ziegler, C. L., Bumgarner, W., and Serafin, R. J. (1980). Single- and multiple-Doppler radar observations of tornadic storms. *Mon. Weather Rev.* **108,** 1607–1625.

Skolnik, M., ed. (1990). "Radar Handbook." McGraw-Hill, New York.

Glossary

Direct solution Combined use of three or more Doppler radars to determine the three dimensional wind. No other information is required.

Doppler radar Radar configured such that the phase of the transmitted signal can be compared to the received signal so that a frequency shift can be measured. This is used to compute the component of scatter motion in the direction toward or away from the radar.

Dual-Doppler solution Method of deducing the three-dimensional winds within storms that requires at least two radars and an auxiliary relation. Usually, the requirement of continuity of air entering and leaving a volume is used.

NEXRAD Acronym NEXt generation Doppler RADar. Approximately 144 ten-centimeter wavelength radars will form the component of the National Weather Service's radar storm identification network.

Profiler Multibeam fixed-antenna Doppler radar that points nearly vertically and measures the wind up to at least 15 km height. Operates from 900 to 50 MHz.

Radar Acronym RAdio Detection And Ranging. Developed largely during World War II, it is the use of focused transmitted electromagnetic waves to identify scatterers in the atmosphere. From the direction of the receiving antenna, and the time between transmission and receiving, direction and range are found.

Reflectivity Range-independent measure of the energy returned to the radar from the scatterer. A single millimeter drop in a cubic meter has a reflectivity of $1 \text{ mm}^6 \text{ m}^{-3}$. Frequently reflectivity is expressed in decibels.

Downhole Measurements beneath the Oceans

Paul F. Worthington

BP Research

Downhole measurements of the physicochemical properties of geological formations are part of a family of sensing techniques that range from micro to global scales and are directed at quantitative investigations of the drillable subsurface as an integral part of the earth's physicochemical system. Tools or probes are lowered into a drillhole and continuous measurements are made for the duration of their deployment. Although the output data may be chemical or physical in nature, the sensing techniques themselves are almost exclusively physical. The primary downhole-measurement technique is wireline logging.

I. Introduction

Downhole measurements can be divided into two groups, those deployed for spatial surveys and those directed at time-variant studies. There are several different types of downhole measurements within the category of spatial surveys. These include wireline logging, formation testing, measurement-while-drilling, borehole geophysics, and stress probing. Of these technologies, wireline logging is by far the most widely used, and it is given primary consideration here. Time-variant studies essentially use downhole observatories that comprise sensors emplaced in a drillhole from which they provide continuous data over a period of time. However, this category of measurement also includes the so-called

time-lapse surveys whereby a borehole is reentered for separate deployments of the same downhole tool at different times. The essential difference between the data outputs from a downhole observatory and a time-lapse survey is that the former are temporally continuous and spatially discrete, whereas the latter are temporally discrete and spatially continuous.

II. Borehole Logging

Borehole logs are spatially continuous records of key physicochemical properties of the formations penetrated by a drillhole. Logging tools are conventionally deployed by means of a conducting cable or "wireline" either shortly after drilling or later through wireline reentry (see Fig. 1). In addition to providing a spatially continuous characterization of the subsurface in the form of physicochemical properties, logs sense at a measurement-scale that is intermediate relative to core analysis and borehole geophysics, and that is central within the range of earth-science measurements. Figure 2 illustrates that log measurements are made with a higher resolution than is achievable through geophysics and yet sample a volume that can be 100 times greater than the volume of recovered core. Log measurements are made at *in-situ* conditions and are usually designed to be minimally affected by any alteration due to the drilling of the borehole itself.

The majority of logging tools have been developed commercially for application in hydrocarbon reservoirs. Therefore logging-tool technology, which has evolved steadily since the first documented well log was run by the Schlumberger Brothers in France in 1927, is matched to conventional oilfield environments. When an exploration well is abandoned, logs constitute the only continuous record of the subsurface characteristics at that site. As such, they are a commercial legacy with a trading value.

In addition to the commercial logging tools, there are also available a limited number of special tools that have been developed to address specific scientific objectives. The deployment of logging tools for scientific as opposed to commercial purposes provides a scientific legacy of high-quality

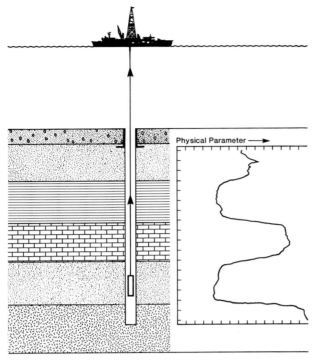

Fig. 1 Scheme of borehole logging at sea.

data that will benefit not only present but also future generations of geoscientists. In the particular case of ocean science, targets include a better understanding of the distribution, structure, composition, permeation properties, stress regime and history of ocean crust. Attainment of this improved understanding would provide important new insights into how the earth really works as a physicochemical system. A knowledge of global mechanisms is, of course, a prerequisite for predicting global change.

III. Principles of Log Measurements

The principal arena for deploying both commercial and special tools in subocean boreholes for scientific purposes is the international Ocean Drilling Program (ODP). A summary of these tools is presented in Table I where it can be seen that virtually every branch of physics is represented.

The logging tools of Table I are usually run in combination to save time. Conventional tool combinations have evolved from the oilfield requirement to have a failsafe log run initially, in case the hole is lost, followed by more detailed measurements to evaluate porosity (the fractional volume of pores) and hydrocarbon saturation (the fraction of the pore space that is filled with hydrocarbons). In scientific drilling, where hydrocarbons are avoided on safety grounds, the prac-

tice has been to substitute the hydrocarbon-sensing tool string with one directed at formation chemistry. The following description relates to the practices of the Ocean Drilling Program, which has the most technically advanced downhole measurements capability that is being deployed routinely in the world today. The logging tools are described in groups according to their purpose. Principles of operation and target parameters are contained in Table I. [*See* BASIN ANALYSIS METHODS.]

A. Seismic-Stratigraphic

This tool combination provides borehole calibration for surface seismics in terms of elastic wave velocity, and indirectly measures two variables that influence velocity: porosity and clay-mineral content. Constituent tools are

- Full waveform sonic tool
- Dual induction tool
- Spherically focused resistivity tool
- Natural gamma tool
- Caliper

A natural gamma tool is included within each tool string to facilitate depth correlation.

B. Litho-Porosity

The litho-porosity combination of nuclear tools is primarily used for the determination of porosity. An important intermediate parameter is density, which can be combined as a product with elastic wave velocity to yield acoustic impedance, an important parameter for the cross-scale integration of acoustic data. The determination of porosity and density is lithology-dependent, and therefore some lithological data are needed. This need is partly met by the provision of estimates of the concentrations of the primary radioactive elements (potassium, thorium, uranium). Constituent tools are

- Neutron porosity tool
- Spectral density tool
- Natural gamma spectral tool

C. Geochemical

The value of the geochemical tool string lies in its ability to measure relative concentrations of twelve elements: aluminum, calcium, chlorine, gadolinium, hydrogen, iron, silicon, sulfur, titanium, potassium, thorium, and uranium. Of these, two elements (chlorine and hydrogen) are fluid related. The remaining ten provide a geochemical signature of the formations penetrated by a drillhole. Constituent tools are

- Aluminum activation clay tool
- Induced gamma spectral tool
- Natural gamma spectral tool

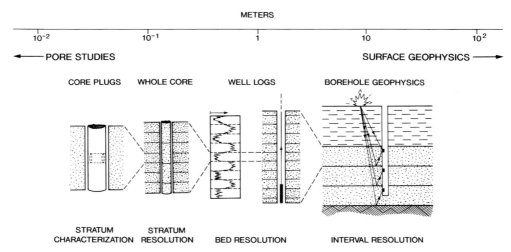

Fig. 2 Scale and resolution in petrophysics.

In practice, tool string B can be integrated into A and C with some data sacrifice.

D. Formation Microscanner

The formation microscanner is a pad-type tool that performs multiple measurements of formation resistivity at a small scale (<1 cm). Subsequent processing allows these data to be presented as an electrical image of the borehole wall. Benefits include the recognition of fractures and possibly stress-driven breakouts, diagnosis of structure, and recognition of sedimentary facies. The tool is run in combination with a natural gamma tool and with a general purpose inclinometer that records the vector components of the downhole magnetic field and thereby allows the microscanner pads to be oriented.

The tools within strings A–D have been developed primarily for oilfield application and they are available off the shelf. However, the oil industry does not run tool strings C and D on a routine basis. In contrast, tool strings A–D, sometimes run in different combinations, have come to be regarded as "standard" for scientific logging of subocean boreholes within the Ocean Drilling Program. Examples of "standard" well logs are shown in Fig. 3.

E. Special Tools

In addition to the standard logs, various special tools are available. These are deployed according to either environmental constraints or scientific needs. The tools described below are available commercially. Others, which have been developed specifically for scientific application, are indicated in Table I.

1. Dual Laterolog

Induction logs do not always perform well in highly resistive formations such as oceanic basalts. The dual laterolog, which is a contact-resistivity device, is sometimes used to provide resistivity measurements with a high depth of current penetration in these formations. This, in turn, can lead to an improved estimate of porosity.

2. Well Seismic Tool

Sonic logs do not measure at seismic frequencies. The well seismic tool does provide a determination of seismic velocity by measuring the travel time from a surface seismic shot to clamped single-component wellbore geophones. By moving the geophones to different positions, a vertical seismic profile can be built up. This experiment allows estimates of the depths to predominant reflectors, which can be below the base of the borehole (cf. Fig. 2). The primary benefit is therefore the control of seismic structure and stratigraphy. [*See* SEISMIC DATA, RECORDING AND ANALYSIS.]

3. Seismic Acquisition Tool

This tool comprises a three-component array of geophones. These allow a much wider range of applications than the one-axis recordings achievable through the well seismic tool. The benefits include a directional diagnostic capability for feature identification, the separation of compressional and shear waves, and the extraction of Poisson's ratio and its relation to lithology.

4. Borehole Televiewer

The borehole televiewer (BHTV) transmits an ultrasonic pulse from a centralized position in a wellbore and receives a pulse reflected from the borehole wall. The received signal is

Table I Shipboard Tools for Downhole Measurements[a]

Branch of physics	Tool[b]	Principle of operation	Parametric objective	Principal area of application
Acoustics	Full waveform sonic	Travel time of sound (8 receivers)	Sound velocity and attenuation (compressional and shear)	Crustal composition and structure
	Well seismic tool	Travel time of sound between surface shot-point and wellbore geophones Single component waveform	Sound velocity Depths to reflecting horizons VSP (single-component)	Crustal composition and structure
	Seismic acquisition tool	As above but with three-component waveform	VSP (three-component)	Crustal composition and structure
	Borehole televiewer (BHTV)	Ultrasonic reflection	Hole diameter Reflectivity (image)	Global stress regime
	*Multichannel sonic tool	Travel time of sound (12 receivers)	Sound velocity and attenuation (compressional and shear)	Crustal composition and structure
Electricity	Dual induction	Induced current (deep/medium sensing)	Conductivity (resistivity)	Hydrogeology
	Dual laterolog	Alternating current (deep/medium sensing)	Resistivity	Hydrogeology
	Spherically focused	Alternating current (shallow sensing)	Resistivity	Hydrogeology
	*Large-scale resistivity	Direct current (very deep sensing)	Resistivity	Hydrogeology
	Formation microscanner	Alternating current (micro sensing)	Microresistivity (image)	Crustal composition and structure
Magnetics	General purpose inclinometer	Orientated magnetic field and inclination	Vector components of magnetic field Hole azimuth and direction	Distribution and history
	*Geomagnetic	Orientated magnetic field	Vector components of magnetic field, susceptibility	Distribution and history

[a] Based on the technology currently deployed in the Ocean Drilling Program.
[b] The asterisk denotes scientific specialty tools.

(continues)

analyzed for amplitude, which indicates the acoustic reflectivity of the borehole wall. The BHTV data are processed to yield a 360° picture of the borehole wall in terms of acoustic reflectivity. Benefits are similar to those of the formation microscanner but with a more complete image and a better capability for recognizing stress breakouts. The BHTV can also operate as a 360° acoustic caliper.

IV. Applications of Log Measurements

The scope of downhole measurements is summarized in Fig. 4, which illustrates the primary contribution(s) of the respective measurements to the identified research targets associated with oceanic lithosphere. Examples of the application of downhole measurements to these targets is given below.

A. Distribution and History

This all-encompassing subject includes such topical issues as global environmental change. The history of environmental change can be preserved in sedimentary records, such as arid-zone or humid-zone clay minerals and grain-size distribution of wind-blown sands reflecting wind strength. A key environmental factor is climate. In addition to major changes in

Table I *(Continued)*

Branch of physics	Tool[b]	Principle of operation	Parametric objective	Principal area of application
Mechanics	Caliper	Electrical monitoring of mechanical gauge	Hole diameter	Environmental conditions
	*Drillstring packer (rotatable/straddle)	Single/double element assembly for isolating borehole intervals	Pore pressure, permeability	Hydrogeology
	*Water sampler	Pressure intake	Pore water samples Pore water chemistry	Crustal composition and structure Hydrogeology
	*Wireline sampler	Double element assembly for isolating borehole intervals	Pore water samples Pore water chemistry	Crustal composition and structure Hydrogeology
Nuclear	Neutron	Slowing down/ absorption of neutrons from moderate energy source	Porosity	Hydrogeology
	Spectral density	Scattering/attenuation of gamma rays	Bulk density (porosity) Photoelectric absorption factor	Crustal composition and structure
	Natural gamma spectral	Natural gamma-ray emissions	Total gamma count rate plus constituent count rates for uranium/ potassium/thorium	Crustal composition and structure
	Induced gamma spectral	Gamma-ray emissions from slowing-down/ absorption of neutrons from downhole accelerator	Gamma-ray energy spectrum (elemental analysis)	Crustal composition and structure
	Aluminum activation	Gamma-ray emissions from slowing-down/ absorption of neutrons from low energy californium source	Gamma count rate windowed for aluminum (elemental analysis)	Crustal composition and structure
Thermodynamics	Temperature	Thermistor	Formation temperature	Distribution and history

lithology due to tectonic movement through different climatic belts, there are also shorter term cyclic changes associated with astronomically driven climatic variations derived from cyclic changes in the eccentricity of the earth's orbit (95 and 410 ka), the obliquity of the elliptic (41 ka), and the precession of the equinoxes (19–23 ka) as postulated by the Yugoslav scientist, M. Milankovitch. Borehole logs can recognize this cyclicity provided that the post-compaction sedimentation rate is sufficient for the events to be spatially resolvable by the logging tools. Unlike core data, which are rarely continuous, logs provide a continuous record, which is essential for cycle recognition. The ultimate prospect is one of time-logging with a temporal resolution (40–100 ka and possibly even finer) that exceeds the resolution achievable through biostratigraphy alone. For a typical logging-tool spatial resolution of 0.6 m, a temporal resolution of 100 ka

would require a sedimentation rate of at least 6 m/Ma. [See GLOBAL CYCLOSTRATIGRAPHY.]

Figure 5 shows an example of a log product that indicates sediment cyclicity. This example is from ODP Hole 693A on the Weddell Sea margin of East Antarctica (70°49.89' S, 14°34.41' W). Fourier power spectra have been determined for a resistivity log of overlapping intervals of 39-m thickness. Several amplitude peaks are evident, and these can be correlated from interval to interval. The predominant peaks are believed to be a manifestation of Milankovitch cyclicity. The assignment of periodicities to the peaks requires a knowledge of sedimentation rate(s). However, in the absence of this information, the relative positions of the peaks can be used to identify provisionally those that correspond to the Milankovitch postulates of 19–23, 41, 95, and 410 ka. Thus, for example, peaks A and B within the 108–147 m interval might

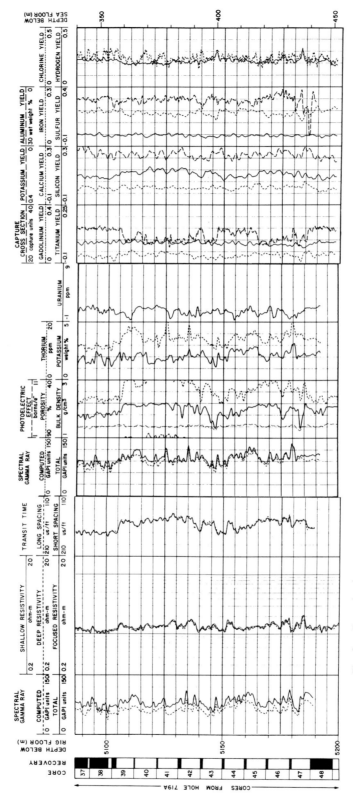

Fig. 3a Example of a composite standard ODP well log from ODP Hole 719B, Distal Bengal Fan (0°57.646′ S, 81°23.967′ E). This hole was drilled prior to the availability of the slimhole formation microscanner [From J. R. Cochran, D. A. V. Stow, et al. (1989). *Proc. ODP, Int. Rep.* 116.]

74

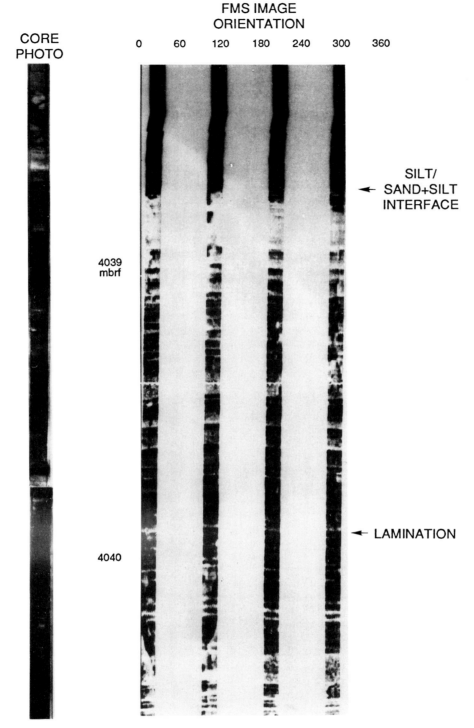

Fig. 3b Formation microscanner (FMS) images from ODP Hole 793B, Bonin Arc-Trench System (31°6.33′N, 140°53.27′E). The more resistive a formation, the lighter its color. (Data are courtesy of Borehole Research Group, Lamont-Doherty Geological Observatory, Palisades, New York.)

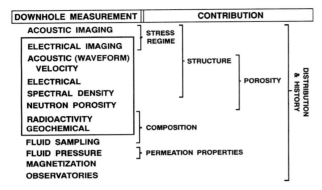

Fig. 4 Contribution of downhole measurements to identified research targets associated with oceanic lithosphere.

correspond to the 41 and 95 ka periods, which would imply that they are respectively caused by obliquity and eccentricity cycles associated with orbital variations of the earth. If this assignment is correct, a sedimentation rate of 50 m/Ma can be identified with subsequent scope for iterative refinement. Successful completion of this process would allow cycle counting to define key chronological markers in the stratigraphic succession.

The relative manifestations of the cyclic events associated with the three causative factors (eccentricity, obliquity, and precession) are known to vary from place to place, both laterally across the globe and vertically within a succession. A proper understanding of the response of the climate to orbital drives is a prerequisite for separately identifying perturbations due to human activity. [*See* CYCLIC SEDIMENTATION, CLIMATE AND ORBITAL INSOLATION CHANGES IN THE CRETACEOUS.]

B. Crustal Composition and Structure

A knowledge of crustal composition and structure is a prerequisite for understanding the origin and evolution of oceanic crust. Although the latter is seismically layered, the causes of these acoustic contrasts were not appreciated until logs were run in Layers 2A, 2B, and 2C of oceanic crust. An important input to the understanding of crustal structure is the variation in chemical composition indicating, for example, different magma sources. Yet again, the standard logs can be deployed with the borehole televiewer with its capability for detecting fractures. Thus, the conjunctive use of different logs allows a comprehensive interrogation of crustal composition and structure. [*See* OCEANIC CRUST.]

An example of the conjunctive use of logs to infer crustal characteristics is afforded by ODP Hole 735B drilled in a fracture zone on the Southwest Indian Ridge (32°43.395′ S, 57°15.959′ E). Figure 6a reveals a zone of anomalously high-

density and high-capture cross section [956–1004 mbrf (meters below rig floor)], which is associated with gabbros rich in iron–titanium oxide and forms part of crustal Layer 3. The geochemical logs (Fig. 6b) detail the layered structure of these iron–titanium oxide-rich gabbros and clearly differentiate them from the olivene gabbros above.

Immediately below the iron–titanium oxide-rich zone is a brecciated zone of alteration within an olivene-gabbro unit. The logs show an increase in gadolinium, thorium, and neutron porosity together with reduced density and iron concentration (Figs. 6a and 6b). This zone possibly represents a pathway for high-temperature fluids associated with the intrusion into the (now frozen) magma chamber of the iron–titanium oxide-rich gabbros.

Although vertical seismic profile (VSP) data indicate that seismic velocities around 735B are uniform on the (VSP) scale of 50–200 m, there is some evidence that localized decreases in compressional wave velocity observed in the (higher-resolution) sonic log are associated with fractures detected by the borehole televiewer in the brecciated zone (Fig. 6c). Some of the fractures seen by the BHTV are actually pyrite-rich veins, as evidenced by the sulfur peaks in the geochemical log.

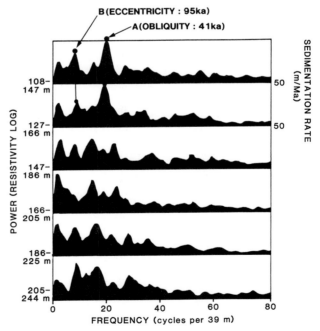

Fig. 5 Provisional log-based recognition of astronomically driven sediment cyclicity at ODP Hole 693A, Weddell Sea. Depths are meters below sea floor. (Data are courtesy of Borehole Research Group, Lamont-Doherty Geological Observatory, Palisades, New York.)

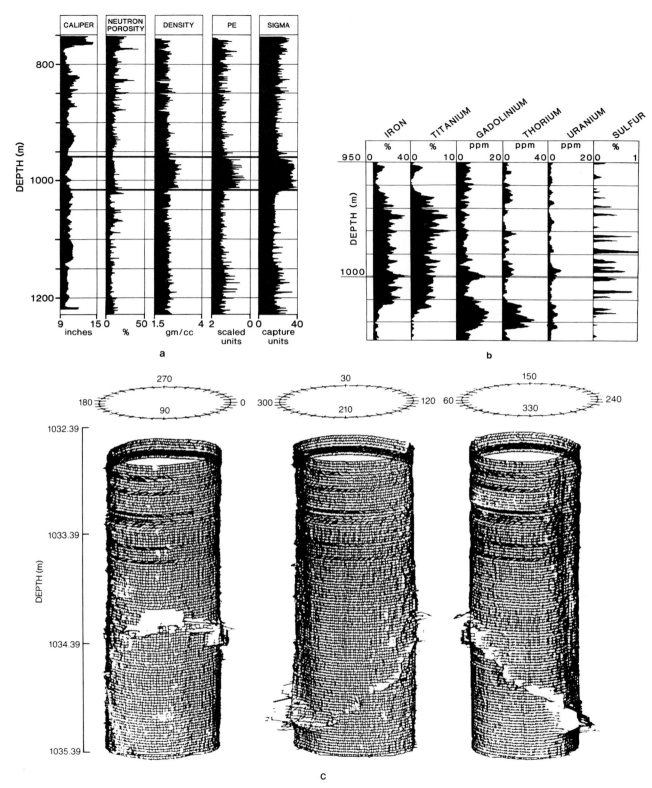

Fig. 6 Borehole logs of ODP Hole 735B, Southwest Indian Ridge: (a) Identification of dense gabbros; (b) rich in iron–titanium oxide and underlain by a brecciated zone of alteration; (c) processed borehole televiewer data indicating the fractured nature of the brecciated zone. Depths are meters below rig floor. [From P. T. Robinson, R. P. von Herzen, *et al.* (1989). *Proc. ODP, Init. Rep.* **118.**]

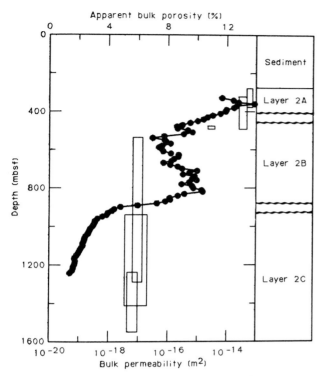

Fig. 7 Depth plot of apparent porosity, calculated from a resistivity log, and permeability from packer tests in ODP Hole 504B, Costa Rica Rift. The rectangles indicate the locations and vertical extents of the test intervals and the estimated uncertainties in the derived permeabilities. [From K. Becker, H. Sakai, *et al.* (1988). *Proc. ODP, Init. Rep.* 111.]

C. Hydrogeology

Crustal fluid flow is an essential component in the transfer of matter from the mantle to the oceans and atmosphere. A knowledge of fluid circulation is therefore a prerequisite for a proper understanding of the global geochemical budget. Two key parameters govern fluid storage and flow, porosity and permeability. Porosity can be evaluated from density, sonic, neutron, and resistivity logs. Permeability has to be measured by downhole pressure tests using packers, devices that allow a target zone of downhole investigation to be isolated hydraulically from the rest of the borehole column. [*See* CRUSTAL FLUIDS.]

Figure 7 shows depth plots of apparent porosity, calculated from the (continuous) resistivity log, and permeability, determined from (discrete) packer tests, in ODP Hole 504B located on the southern flank of the Costa Rica Rift (1°13.611′ N, 83°43.818′ W). Figure 7 also indicates the seismic stratigraphy at this site. The data suggest that the

lower pillow lavas of Layer 2B, which geochemical logs have indicated to be largely sealed by alteration products, and the sheeted dykes of Layer 2C are essentially impermeable. These account for over 1 km of hole. The only significant permeability is associated with the upper pillow lavas of Layer 2A. This suggests that hydrothermal circulation around Hole 504B is now confined to shallow basement. These conditions have presumably prevailed since the fracture-sealing by the alteration products, which indicate paleo-porosity and permeability.

D. Global Stress Regime

Although the theory of plate tectonics has been accepted for more than 20 years, we do not understand the forces that drive the plates and determine their motions. Our understanding can be advanced through information about the stress field within plates, since the driving forces give rise to measurable stress orientations and magnitudes. These orientations, and possibly the magnitudes, can be identified using the borehole televiewer. When regional tectonic stresses exceed the strength of the rock immediately around a drillhole, and the horizontal principal stresses are anisotropic, rock failure will occur. The cavities so formed, known as "breakouts," can be detected by the borehole televiewer. The orientation of the breakouts is perpendicular to the directions of the larger horizontal principal compressional stress. [*See* PLATE TECTONICS, CRUSTAL DEFORMATION AND EARTHQUAKES.]

Figure 8 shows breakout orientations imaged with a borehole televiewer in Deep Sea Drilling Project (DSDP) Hole 597C, located in crust of age 35 Ma within the Pacific Plate about 1500 km west of the East Pacific Rise (18°48.43′ S, 129°46.22′ W). Breakouts are oriented north–northeast to

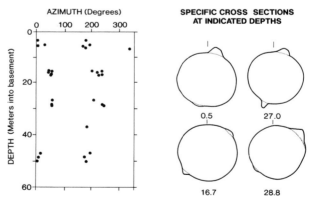

Fig. 8 Breakout orientations identified by the borehole televiewer in DSDP Hole 597C, Central Pacific Plate. (Data are courtesy of Borehole Research Group, Lamont-Doherty Geological Observatory, Palisades, New York.)

south–southwest. Interestingly, this orientation is perpendicular to horizontal convective "rolls" in the mantle imaged from satellite measurements of gravity, supporting the view that the larger horizontal principal compressional stress is oriented west–northwest to east–southeast. An ultimate goal is to build up a network of observed stress orientations to produce a global horizontal stress azimuth map.

V. Future Scenario

In the past, developments in logging technology have been driven by oilfield requirements. The oil industry benefited from the revolution in data acquisition technology in the 1970s through advances in digital telemetry, which allowed spectral and multichannel logging tools to be deployed, and in computer power, which allowed the introduction of well site computers. The oil industry is currently asking the question "What do all these data mean?" The industry seeks less empiricism, fewer overdeterminations, and greater accuracy and precision. It is the charge of science to provide the necessary understanding.

There are two key aspects to seeking this understanding. One is the reconciliation of (core, log, borehole geophysical, and surface geophysical) data measured at different scales. Downhole measurements are a fulcrum for this process, which involves the progressive calibration of low-resolution larger-scale data by high-resolution smaller-scale data and, through this procedure, the cross-scale compatibility of interpretative models. The second key aspect is the integration of disciplines. Downhole (physical) measurements are providing an increasing input to conventional subdisciplines such as geochemistry, stratigraphy, sedimentology, and structural geology. This trend will continue with earth scientists making a far greater use of the downhole-measurement capability that is available to them. For example, the approaching development of a sediment magnetometer, with a capability to recognize field reversals in weakly magnetized sediments, could revolutionize event stratigraphy. [See MAGNETIC STRATIGRAPHY.]

Scientific borehole logging is entering a very exciting era, one in which commercial logging instrumentation will not meet perceived needs. These needs are governed by ambitions to drill at crustal spreading centers where temperatures of 350–400°C are anticipated. Most commercial logging tools are rated to 260°C or lower. Therefore, new tools will have to be developed if comprehensive logging in these hazardous environments is to be realized. Thus, future developments in (high-temperature) logging technology will be driven by scientific programs and not by commercial considerations as in the

past; in this light, scientific borehole logging appears to be at a watershed. The scientific community is adapting positively to this new situation.

Acknowledgments

The author acknowledges those ODP scientists and engineers who contributed to the acquisition and reporting of the data cited herein.

Bibliography

Bateman, R. M. (1984). "Log Quality Control." D. Reidel, Dordrecht, The Netherlands.

Desbrandes, R. (1985). "Encyclopedia of Well Logging." Graham & Trotman, London.

Dewan, J. T. (1983). "Essentials of Modern Open-Hole Log Interpretation." PennWell, Tulsa, Oklahoma.

Worthington, P. F., Anderson, R. N., Bell, J. S., Becker, K., Jarrard, R. D., Salisbury, M. H., and Stephen, R. A. (1987). Downhole measurements for the Ocean Drilling Programme. In "Report of the Second Conference on Scientific Ocean Drilling (COSOD II)" (G. B. Munsch, ed.), pp. 131–136. European Science Foundation, Strasbourg, France.

Worthington, P. F., Anderson, R. N., Jarrard, R. D., Becker, K., Bell, J. S., Salisbury, M. H., and Stephen, R. A. (1989). Scientific applications of downhole measurements in the ocean basins. *Basin Res.* 1, 223–236.

Glossary

Borehole geophysics Science of (seismic or electrical) physical measurement between boreholes or between a borehole and the surface.

Formation testing Application of pressure tests to downhole intervals of formation whose response is used to infer permeability.

Measurement-while-drilling Acquisition of petrophysical borehole logs using sensors positioned within the drillstring.

Stress probing Measurement of *in-situ* stresses in a dedicated small-diameter hole drilled ahead of the primary drill bit.

Wireline logging Measurement of depth variations (logs) of physical properties using a tool or sonde deployed by means of a conducting cable (wireline).

Drought

D. A. Wilhite
University of Nebraska

Scores of definitions of drought exist, reflecting different applications and regions of concern. Common to all types of drought is the fact that they originate from a deficiency of precipitation that results in water shortage for some activity (e.g., plant growth, transportation) or some group (e.g., farmer, water suppliers). Drought can be defined as a deficiency of precipitation from expected or "normal" that, when extended over a season or longer period of time, is insufficient to meet the demands of human activities. Drought must be considered a relative, rather than absolute, condition. The ultimate results of these precipitation deficiencies are, at times, enormous economic and environmental impacts as well as personal hardship. Scientists speculate that the frequency and severity of droughts may increase if projected changes in climate occur because of increasing concentrations of CO_2 and other atmospheric trace gases.

I. Introduction

Throughout human history, drought has been a threat to our existence, often altering the course of history itself. Drought should not be viewed as merely a physical phenomenon. It is the result of an interplay between a natural event (precipitation deficiencies due to natural climatic variability on varying time scales) and the demand placed on water supply by human-use systems. Literature is replete with references showing how extended periods of drought have resulted in food supply disruptions, famine, massive migrations of people, and wars. In the United States, for example, the droughts of the 1890s and 1930s significantly altered the settlement of the western frontier. [*See* Water Cycle, Global.]

The impact of drought is often exacerbated by human beings. The earth's rapidly expanding population is placing an ever-increasing demand on local and regional water resources and, in many areas, accelerating environmental degradation. In the past several decades we have been continuously besieged by reports of drought and its impacts on natural and anthropogenic ecosystems. Recent droughts in developing and developed countries and the concomitant impacts and personal hardships that resulted have underscored the vulnerability of all societies to this natural hazard. It is difficult to determine whether it is the frequency of drought that is increasing, or simply societal vulnerability to it.

II. Drought: An Overview

Drought differs from other natural hazards (e.g., floods, hurricanes, earthquakes) in several ways. First, it is a "creeping phenomenon," making its onset and end difficult to determine. The effects of drought accumulate slowly over a considerable period of time and may linger for years after the termination of the event. Second, the absence of a precise and universally accepted definition of drought adds to the confusion about whether or not a drought exists and, if it does, its severity. Third, drought impacts are less obvious and are spread over a larger geographical area than are damages that result from other natural hazards. Drought seldom results in structural damage. For these reasons the quantification of impacts and the provision of disaster relief is a far more difficult task for drought than it is for other natural hazards.

Drought is a normal part of climate for virtually all climatic regimes. It is a temporary aberration that occurs in high as well as low rainfall areas. Drought therefore differs from aridity, since the latter is restricted to low rainfall regions and is a permanent feature of climate. The character of drought is distinctly regional, reflecting unique meteorological, hydrological, and socioeconomic characteristics. Many people associate the occurrence of drought with the Great Plains of North America, Africa's Sahelian region, India, or Australia; they may have difficulty visualizing drought in Southeast Asia, Brazil, Western Europe, or the eastern United States, regions normally considered to have a surplus of water.

Drought should be considered relative to some long-term average condition of balance between precipitation and evapotranspiration in a particular area, a condition often perceived as "normal." It is the consequence of a natural reduction in the amount of precipitation received over an extended period of time, usually a season or more in length, although other climatic factors (such as high temperatures, high winds, and low relative humidity) are often associated with it in many regions of the world and can significantly aggravate the severity of the event. Drought is also related to the timing (such as the principal season of occurrence, delays in the start of the rainy season, occurrence of rains in relation to principal crop growth stages) and the effectiveness of the rains (i.e., rainfall intensity, number of rainfall events). [See CLIMATIC CHANGE: RELATIONSHIP BETWEEN TEMPERATURE AND PRECIPITATION; EVAPORATION AND EVAPOTRANSPIRATION.]

A. Definitions and Types of Drought

Because drought affects so many economic and social sectors, scores of definitions have been developed by a variety of disciplines. In addition, because drought occurs with varying frequency in nearly all regions of the globe, in all types of economic systems, and in developed and developing countries alike, the approaches taken to define it also reflect regional differences as well as differences in ideological perspectives. Impacts also differ spatially and temporally, depending on the societal context of drought. A universal definition of drought is an unrealistic expectation.

Definitions of drought can be categorized broadly as either conceptual or operational. Conceptual definitions are of the "dictionary" type, generally defining the boundaries of the concept of drought, and thus are very generic in their description of the phenomenon. Operational definitions attempt to identify the onset, severity, and termination of drought episodes. Definitions of this type are often used in an "operational" mode to detect the onset, continuation, severity, and termination of drought. These definitions can also be used to analyze drought frequency, severity, and duration for a given historical period. An operational definition of agricultural drought might be one that compares daily precipitation to evapotranspiration (ET) rates to determine the rate of soil-water depletion and then expresses these relationships in terms of drought effects on plant behavior at various stages of development. The effects of these meteorological conditions on plant growth would be reevaluated continuously by agricultural specialists as the growing season progresses. [See EVAPORATION, TRANSPIRATION, AND WATER UPTAKE OF PLANTS.]

Many disciplinary perspectives of drought exist. Each discipline incorporates different physical, biological, and/or socioeconomic factors in its definition of drought. Because of these numerous and diverse disciplinary views, considerable confusion often exists over exactly what constitutes a drought. Research has shown that the lack of a precise and objective definition in specific situations has been an obstacle to understanding drought, which has led to indecision and/or inaction on the part of managers, policy makers, and others. It must be accepted that the importance of drought lies in its impacts. Thus definitions should be impact and region specific to be used in an operational mode by decision makers.

Drought can be grouped by type as follows: meteorological, hydrological, agricultural, and socioeconomic. Meteorological drought is expressed solely on the basis of the degree of dryness (often in comparison to some "normal" or average amount) and the duration of the dry period. Definitions of meteorological drought must be considered as region specific, since the atmospheric conditions that result in deficiencies of precipitation are highly variable from region to region. For example, some definitions of meteorological drought differentiate periods on the basis of the number of days with precipitation less than some specified threshold. Extended periods without rainfall are common for many regions; such a definition is unrealistic in this case. Other definitions may relate actual precipitation departures to average amounts on monthly, seasonal, or annual time scales. Definitions derived for application to one region but applied to another often create problems, since meteorological characteristics differ. Human perceptions of these conditions are equally variable. Both of these points must be taken into account to identify the characteristics of drought and make comparisons between regions.

Hydrological droughts are concerned more with the effects of periods of precipitation shortfalls on surface or subsurface water supply (such as stream flow, reservoir and lake levels, ground water) rather than with precipitation shortfalls. Hydrological droughts are usually out-of-phase or lag the occurrence of meteorological and agricultural droughts. Meteorological droughts result from precipitation deficiencies while agricultural droughts are largely the result of soil moisture deficiencies. More time elapses before precipitation deficiencies show up in components of the hydrological system. As a result, impacts are out of phase with those in other economic sectors. Also, water in hydrological storage systems (e.g., reservoirs, rivers) is often used for multiple and competing purposes, further complicating the sequence and quantification of impacts. Competition for water in these storage systems escalates during drought, and conflicts between water users increase significantly.

The frequency and severity of hydrological drought is often defined on the basis of its influence on river basins. Since low-flow frequencies have been determined for most streams, hydrological drought periods can be of any specified length. If the actual flow for a selected time period falls below a certain threshold, then hydrological drought is considered to be in progress. However, the number of days and the level of probability that must be exceeded to define a hydrological drought period is somewhat arbitrary. These criteria will vary between streams and river basins.

Agricultural drought links various characteristics of meteorological drought to agricultural impacts, focusing on precipitation shortages, differences between actual and potential evapotranspiration, soil-water deficits, and so forth. A plant's demand for water is dependent on prevailing weather conditions, biological characteristics of the specific plant, its stage of growth, and the physical and biological properties of the soil. An operational definition of agricultural drought should account for the variable susceptibility of crops at different stages of crop development. For example, deficient subsoil moisture in an early growth stage will have little impact on final crop yield if topsoil moisture is sufficient to meet early growth requirements. However, if the deficiency of subsoil moisture continues, a substantial yield loss may result.

Finally, socioeconomic drought associates the supply and demand of some economic good with elements of meteorological, hydrological, and agricultural drought. Some scientists suggest that the time and space processes of supply and demand are the two basic processes that should be included in an objective definition of drought. For example, the supply of some economic good (such as water, hay, electric power) is weather dependent. In most instances, the demand for that good is increasing as a result of increasing population or per capita consumption. Therefore, drought could be defined as occurring when the demand exceeds supply as a result of a weather-related supply shortfall. This concept of drought supports the strong symbiosis that exists between drought and human activities. For example, poor land use practices such as overgrazing can decrease animal carrying capacity and increase soil erosion, which exacerbates the impacts of and vulnerability to future droughts.

B. Drought Characteristics and Severity

Droughts differ in three essential characteristics — intensity, duration, and spatial coverage. Intensity refers to the degree of the precipitation shortfall and/or the severity of impacts associated with the shortfall. It is generally measured by the departure of some climatic index from normal and is closely linked to duration in the determination of impact. The simplest index in widespread use is the percent of normal precipitation. With this index, actual precipitation is compared to "normal" or average precipitation (defined as the most recent 30-year mean) for time periods ranging from one to 12 or more months. One of the principal difficulties with this approach is the choice of the threshold of precipitation deficiency (e.g., 75% of normal) to define the onset of drought. Thresholds are usually chosen arbitrarily, but they should be linked to impact. Actual precipitation departures are normally compared to expected or average amounts on a

monthly, seasonal, annual, or water year (October–September) time period.

The most widely used method for determining drought severity in the United States is the Palmer Drought Severity Index (PDSI). Developed in the mid-1960s, the PDSI is a meteorological index that evaluates prolonged periods of abnormally wet or abnormally dry weather. The index can be thought of as a hydrological accounting system. The input to the system is precipitation. Outputs include evapotranspiration, runoff, soil infiltration, and deep percolation through the soil layer to the ground water. The PDSI relates accumulated differences of actual precipitation to evapotranspiration, runoff, and soil infiltration to average precipitation for individual climatic regions. PDSI values generally range from +4 (extreme wetness) to −4 (extreme drought), although values above or below these thresholds are not unusual. For example, during the severe drought of 1976–1977 in the United States, PDSI values exceeded −4 for extended periods of time for portions of the Pacific Northwest and the upper Midwest. The PDSI has been used to classify and compare historical drought periods from 1895 to the present.

Another distinguishing feature of drought is its duration. Droughts usually require a minimum of two to three months to become established but then can continue for several consecutive years. The magnitude of drought impacts is closely related to the timing of the onset of the precipitation shortage, its intensity, and the duration of the event. An analysis of the sequence of monthly PDSI values for southeast Nebraska from 1931 to 1978 indicates seven drought periods exceeding 10 months in length. These occurred in the 1930s, 1950s, and 1970s. The duration of the longest drought in that period of record began in May 1936 and extended through August 1941, sixty-four consecutive months of PDSI values below −1.0. During that drought period, 61 months were calculated to have had PDSI values less than −2.0 (moderate drought). Of these 61 months, 21 and 24 months, respectively, were in the extreme (−4.0) and severe (−3.0) category.

Droughts of equal and longer duration are common in the Great Plains. Unfortunately, weather records for this region seldom exceed 100 years. To ascertain a clearer picture of the occurrence of drought over the last several hundred years, scientists must rely on other sources of data to extend the weather record. The most notable source of data is tree rings. Trees respond to wet or dry periods by producing wider or narrower growth rings. These growth rings are calibrated with weather records and then extended back in time. Early studies of tree rings taken from samples in western Nebraska back to nearly A.D. 1200 reveal numerous drought episodes ranging in length from five to 38 years. More recent studies conducted by scientists at the University of Arizona's Laboratory of Tree-Ring Research confirm drought as a normal part of the western United States' climate back to A.D. 1700.

Each drought has unique spatial characteristics. The percentage of the total area of the contiguous United States

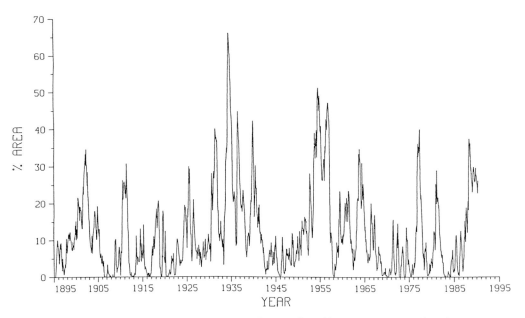

Fig. 1   Percentage of the area of the contiguous United States affected by severe to extreme drought, 1895–1990.

affected by severe to extreme drought has been highly variable over the past century (Fig. 1). The largest area affected by drought occurred during the 1930s—particularly 1934, when more than 65% of the country was experiencing severe or extreme drought. Using the percent of total area to define major drought episodes, significant areas were also affected in the 1890s, 1910, 1925–1926, 1953–1957, 1964–1965, 1976–1977, 1983, and 1988–1990.

The spatial dimensions of several major drought episodes of this century in the United States illustrate the uniqueness of each drought. The drought years of the 1930s, commonly referred to as the Dust Bowl or dirty thirties, affected nearly all parts of the United States to some degree during the decade. The most widespread drought conditions occurred in 1934 and 1936. The 1934 drought was concentrated primarily in the northern part of the country, extending from New York to the West Coast. A significant area of prolonged drought also occurred in the southern Great Plains states. During 1936 the principal drought area was concentrated mainly in the northern and central Great Plains. By contrast, the pattern of extreme drought in July 1956 was centered mainly in the Southwest and in the southern and central Great Plains states.

In 1976–1977 the far western, northern Great Plains, and upper midwestern states were most seriously affected. Although this drought equaled previous droughts in intensity for some parts of the United States, neither duration nor spatial extent was comparable. The very recent severe drought of 1988 (Fig. 2) was somewhat similar in spatial extent to the drought of 1976–1977, concentrating in the far

western and upper midwestern states. In addition, it also affected significant portions of the northern plains and Corn Belt states. This drought continued into 1989, with the principal area affected extending from the western Corn Belt through the Central Rocky Mountain states to California. The western half of the United States continued to be affected through the summer of 1990, with California being the hardest hit. The southeastern portion of the country was also affected in 1990.

III. Causes and Predictability

Empirical studies conducted over the past century have shown that drought is never the result of a single cause. Rather, it is the result of many causes, and these are often synergistic in nature. Some of the causes may be the result of influences that originate far from the drought-affected area. A great deal of research has been conducted in recent years on the role of interacting systems or *teleconnections* in explaining regional and even global patterns of climatic variability. These patterns tend to recur periodically with enough frequency and with similar characteristics over a sufficient length of time that they offer opportunities to improve our ability for long-range climate prediction, particularly in the tropics. One such teleconnection is the El Niño/Southern Oscillation (ENSO). [*See* EL NIÑO AND LA NIÑA.]

The immediate cause of drought is the predominant sinking motion of air (subsidence) that results in compressional warming or high pressure, thus inhibiting cloud formation

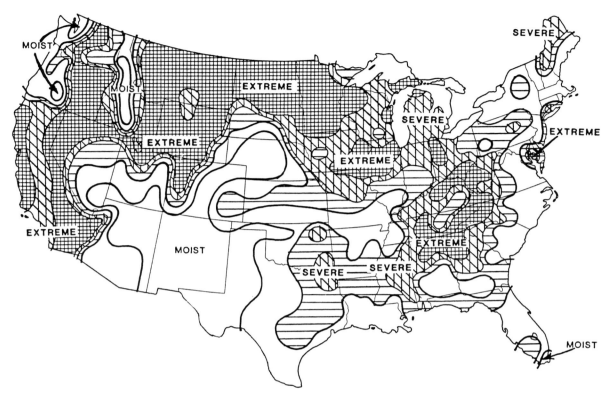

Fig. 2 The pattern of drought severity in the United States (July 23, 1988) according to the Palmer Drought Severity Index.

and resulting in a lowered relative humidity and less precipitation. For regions under the influence of semi-permanent high pressure during all or a major portion of the year, desert (arid) conditions result (e.g., Sahara and Kalahari of Africa, Gobi desert of Asia). Most climatic regions, however, experience varying degrees of dominance by high pressure, often depending on the season. Prolonged droughts occur when large-scale anomalies in atmospheric circulation patterns become established and persist for periods of months, seasons, or longer. The extreme drought that affected the United States and Canada during 1988 is a good example of a large-scale atmospheric circulation anomaly. [*See* ATMOSPHERIC CIRCULATION SYSTEMS.]

The drought of 1988 was one of the most extensive droughts to occur in North America in many years. During the peak of the drought in July, nearly 40% of the contiguous states and a substantial portion of the Prairie Provinces of Canada were experiencing severe to extreme drought. In addition, another 30% of the United States was experiencing moderate drought conditions. Figure 2 illustrates the extent of the drought in the United States during late July 1988. Because of the complexity of the temporal and spatial pattern represented in this illustration, the origins of drought cannot be traced to a single cause. A common explanation for the

drought that set up quickly in the spring and continued through most of the summer months was the displacement of the jet stream to the north of its normal position so that storm tracks were similarly displaced. However, to fully understand the origins of the drought, one must investigate the reasons for the displacement of the jet stream.

Several years of drought for portions of the United States preceded the extremely dry conditions of 1988. The southeastern United States, for example, had experienced a severe to extreme drought during 1986, the beginning of which can be traced to the fall of 1985. In the spring of 1987, drought conditions continued in the southeast and had also developed along the West Coast and in the Pacific Northwest. These conditions persisted into the spring of 1988, spreading across the Prairie Provinces and the northern and midwestern portions of the United States during the spring and summer months, connecting the substantial drought areas of the west and southeast that had existed before the spring.

The West Coast drought of 1987 was associated with the occurrence of El Niño conditions in the tropical Pacific Ocean. Associated with an El Niño event are major alterations in atmospheric circulation, which in turn result in conditions favorable to the development of an unusually strong high-pressure ridge near the West Coast of the United

States and lower pressure over the north Pacific Ocean. In 1987, this resulted in a split jet stream. The southerly branch was not very active and did not result in much precipitation in southern California; the northerly branch was displaced far to the north. The end product of this pattern was that the high-pressure ridge blocked the passage of precipitation-producing low-pressure systems and cold fronts into the western states and the northern Great Plains states. The establishment and persistent recurrence of an atmospheric system such as a ridge of high pressure can dominate a region for a month, season, year, or period of years and thus set the stage for the persistent subsidence of air and drought.

Very little skill currently exists to predict drought for a month or more in advance. What are the prospects that these predictions can be improved significantly in the near future? The potential predictability differs by region, season, and climatic regime. Recent technological advancements make prospects somewhat better today than a decade ago for some regions. In the tropics, for example, meteorologists have made significant advances in their understanding of the climate system. Specifically, it is now known that a major portion of the atmospheric variability that occurs on time scales of months to several years is associated with variations in tropical sea-surface temperatures. Major global meteorological experiments are underway to investigate these questions further. An improvement in the predictability of ENSO episodes, for example, would have a profound influence on seasonal predictions in the tropics. To date, empirical relationships have been developed for some tropical and near-tropical regions such as the Indian Peninsula and Australia. Significant advancements beyond what has been achieved will require major breakthroughs in the use of dynamical models that couple the ocean–atmosphere systems. [See HYDROLOGIC FORECASTING.]

In the extratropical regions, current long-range forecasts are of very limited skill. The National Weather Service of the United States periodically issues 30-day and 90-day forecasts of temperature and precipitation for regions north of 30°N latitude. The skill that does exist is primarily the result of empirical and statistical relationships. In the tropics, empirical relationships have been demonstrated to exist between precipitation and ENSO events, but few such relationships have been confirmed above 30°N. Therefore, meteorologists do not believe that highly skilled forecasts are attainable for all regions a season or more in advance.

IV. Impacts of Drought

The impacts of drought are diverse and often ripple through the economy. Thus, impacts are often referred to as direct or indirect, or they are assigned an order of propagation (i.e., first-, second-, or third-order). Conceptually speaking, the more removed the impact from the cause, the more complex the link to the cause. In other words, a loss of yield resulting from drought is a direct or first-order impact of drought. However, the consequences of that impact (e.g., loss of income, farm foreclosures, outmigration, government relief programs) are secondary or tertiary impacts. First-order impacts are usually of a biophysical nature, whereas higher-order impacts are usually associated with socioeconomic valuation, adjustment responses, and long-term "change."

Because of the number of affected groups and sectors associated with drought, the geographic size of the area affected, and the difficulties connected with quantifying environmental damages and personal hardships, the precise determination of the financial costs of drought is an arduous task. Scientists have estimated the direct losses of drought in the United States to be on the order of $1.2 billion annually. Although drought occurs somewhere in the country each year, such figures are misleading since significant or major episodes often occur in clusters. Therefore, direct and indirect losses may be extremely large for one or two consecutive years and then negligible for several years. Government estimates of recent losses associated with the droughts of 1976–1977 and 1988 were $36 and $40 billion, respectively. These estimates include direct losses broadly grouped into foodstuffs, transportation, energy, production, and sales.

The impacts of drought can be classified into three principal areas: economic, environmental, and social. Table I presents a simplified illustration of the impacts associated with each of these areas. Economic impacts range from direct losses in the broad agricultural and agriculturally related sectors, including forestry and fishing, to losses in recreation, transportation, banking, and energy sectors. Other economic impacts would include added unemployment and loss of revenue to local, state, and federal government. Environmental losses are the result of damages to plant and animal species, wildlife habitat, and air and water quality; forest and range fires; degradation of landscape quality; and soil erosion. Although these losses are difficult to quantify, growing public awareness and concern for environmental quality has forced public officials to focus greater attention on these effects. Social impacts mainly involve public safety, health, conflicts between water users, and inequities in the distribution of impacts and disaster relief programs.

As with all natural hazards, the economic impacts of drought are highly variable within and between economic sectors and geographic regions, producing a complex assortment of winners and losers with the occurrence of each disaster. For example, decreases in agricultural production result in enormous negative financial impacts on farmers in drought-affected areas, at times leading to foreclosure. This decreased production also leads to higher grain, vegetable, and fruit prices. These price increases have a negative impact on all consumers as food prices increase. However, farmers

Table I Classification of Drought-Related Impacts

Problem sectors	Impacts
Economic	• Loss from dairy and livestock production
	Reduced productivity of range land
	Forced reduction of foundation stock
	Closure/limitation of public lands to grazing
	High cost/unavailability of water for livestock
	High cost/unavailability of feed for livestock
	Increased predation
	Range fires
	• Loss from crop production
	Damage to perennial crops; crop loss
	Reduced productivity of cropland (wind erosion, etc.)
	Insect infestation
	Plant disease
	Wildlife damage to crops
	• Loss from timber production
	Forest fires
	Tree disease
	Insect infestation
	Impaired productivity of forest land
	• Loss from fishery production
	Damage to fish habitat
	Loss of young fish due to decreased flows
	• Loss from recreational businesses
	• Loss to manufacturers and sellers of recreational equipment
	• Loss to energy industries affected by drought-related power curtailments
	• Loss to industries directly dependent on agricultural production (fertilizer manufacturers, food processors, etc.)
	• Unemployment from declines in drought-related production
	• Strain on financial institutions (foreclosures, greater credit risks, capital shortfalls, etc.)
	• Revenue losses to state and local governments (from reduced tax base)
	• Revenues to water supply firms
	Revenue shortfalls
	Windfall profits
	• Loss from impaired navigability of streams, rivers, and canals
	• Cost of water transport or transfer
	• Cost of new or supplemental water source development
Environmental	• Damage to animal species
	Wildlife habitat
	Lack of feed and drinking water
	Disease
	Vulnerability to predation (e.g., from species concentration near water)
	• Damage to fish species
	• Damage to plant species
	• Water quality effects (e.g., salt concentration)
	• Air quality effects (dust, pollutants)
	• Visual and landscape quality (dust, vegetative cover, etc.)
Social	• Public safety from forest and range fires
	• Health-related low-flow problems (diminished sewage flows, increased pollutant concentrantions, etc.)
	• Inequity in the distribution of drought impacts/relief

outside the drought-affected area with normal or above-normal production or those with significant grain in storage reap the benefits of these higher prices. Similar examples of winners and losers could be given for other economic sectors as well.

V. Drought Response and Preparedness

With the occurrence of any natural disaster come appeals for disaster assistance from the affected area. During the twentieth century, governments have typically responded to drought by providing emergency, short-term, and long-term assistance to distressed areas. Emergency and short-term assistance programs are often reactive, a kind of "Band-Aid" approach to more serious land and water management problems. Actions of this type have long been criticized by scientists and government officials, as well as by recipients of relief, as inefficient and ineffective. Long-term assistance programs are far fewer in number, but they are proactive. They attempt to lessen a region's vulnerability to drought through improved management and planning.

Governmental response to drought includes a wide range of potential actions to deal with the impacts of water shortages on people and various economic sectors. In the United States, agencies of the federal government and Congress typically respond by making massive amounts of relief available to the affected areas. Most of this relief is in the form of short-term emergency measures to agricultural producers, such as feed assistance for livestock, drilling of new wells, and low-interest farm operating loans. Few, if any, of these assistance measures in recent years have been aimed at reducing future vulnerability. The drought program passed by the U.S. Congress in April 1977 is a good example (Table II). The intent of this program was to reduce the immediate, short-term effects of drought. The President's proposed program totaled $943.8 million; however, Congress appropriated only $843.8 million of that requested. In the 1974–1977 drought period, the federal government provided in excess of $7 billion in relief, principally to the agricultural sector. An additional $6 billion was provided in 1988. Until recently, states have traditionally played a passive role in drought assessment and response efforts, relying largely on the federal government to come to their rescue.

Because of the unique character of drought, governments have been less inclined to invest resources to develop well-conceived mitigation programs and contingency plans. This reactive approach to natural disasters is commonly referred to as crisis management. In crisis management the time to act is perceived by decision makers to be short. Research has demonstrated that reaction to crisis often results in the implementation of hastily prepared assessment and response procedures that lead to ineffective, poorly coordinated, and untimely response. An alternative approach is to initiate planning between periods of drought, thus developing a more coordinated response that might more effectively address longer-term issues and specific problem areas. Also, the limited resources available to government to mitigate the effects of drought could be allocated in a more beneficial manner.

A. Drought Planning

Drought planning is defined as actions taken by individual citizens, industry, government, and others in advance of drought for the purpose of mitigating some of the impacts and conflicts associated with its occurrence. From an institutional or governmental perspective, drought planning should include, but is not limited to, the following activities:

1. A monitoring/early-warning system to provide decision makers at all levels with information about the onset, continuation, and termination of drought conditions and their severity.
2. Operational assessment programs to reliably determine the likely impact of the drought event in a timely manner.
3. An institutional structure for coordinating governmental actions, including information flow within and between levels of government, and drought declaration and revocation criteria and procedures.
4. Appropriate drought assistance programs (both technical and relief) with predetermined eligibility and implementation criteria.
5. Financial resources to maintain operational programs and to initiate research required to support drought assessment and response activities.
6. Educational and public awareness programs designed to promote an understanding and adoption of appropriate drought mitigation and water conservation strategies among the various economic sectors most affected by drought.

To be successful, drought planning must be integrated between levels of government.

B. Drought Policy and Planning Objectives

Before the development of a contingency plan for more effectively assessing and responding to drought, government officials should first define what they hope to achieve by that plan. Thus, a drought policy statement should be prepared in advance of a plan. The objectives of a drought *policy* differ from those of a drought *plan*. A clear distinction of these differences must be made at the outset of the planning process. A drought policy will be broadly stated and should express the purpose of government involvement in drought

Table II President Carter's Proposed Drought Program, 23 March 1977.

Title	Purpose/description	Amount ($)
Emergency Loans Program (FmHA)	5% loans to cover prospective losses to farmers and ranchers	100,000,000
Community Program Loans (FmHA)	$150 million in 5% loans and $75 million in grants to communities of less than 10,000 for emergency water supplies	225,000,000
Emergency Conservation Measures Program (ASCS)	Soil conservation cost-sharing grants	100,000,000
FCIC Insurance	Increase in FCIC Capital Stock	100,000,000
Drought Emergency Program (BuRec)	Creation of water bank, protection of fish and wildlife, grants to states, 5% for water supply and conservation measures	100,000,000
Emergency Fund (BuRec)	Emergency irrigation loans	30,000,000
Emergency Power (SWPA)	Purchase of emergency power supply	13,800,000
Community Emergency Drought Relief Program (EDA)	$150 million in 5% loans and $75 million in grants to communities of more than 10,000 for emergency water supply	225,000,000[a]
Physical Loss and Economic Injury Loans (SBA)	Low-interest loans for small businesses (including farmers)	50,000,000[b]

Total requested 943,800,000
Total appropriated 843,800,000

[a] Only $175 million of this amount was finally appropriated.

[b] Action on this proposal resulted in the lowering of interest rates for physical loss and economic injury loans (both ongoing, funded programs) but none of the additional appropriation originally requested was granted.

assessment, mitigation, and assistance programs. Drought plan objectives are more specific and action-oriented. Typically, the objectives of drought policy have *not* been stated explicitly by government. What generally exists is a *de facto policy*, one defined by the most pressing needs of the moment. Ironically, under these circumstances, it is the specific instruments of that policy (i.e., assistance measures), particularly at the federal level, that define the objectives of the policy.

The objectives of drought policy will differ between levels of government. Generally speaking, these objectives should encourage or provide incentives for agricultural producers, municipalities, and other water-dependent sectors or groups to adopt appropriate and efficient management practices that help to alleviate the effects of drought. Past relief measures have, at times, discouraged the adoption of appropriate management techniques. Assistance should also be provided in an equitable, consistent, and predictable manner to all without regard to economic circumstances, industry, or geographic region. Assistance can be provided in the form of technical aid or relief measures. Whatever the form, those at risk would know what to expect from government during drought and thus would be better prepared to manage risks. An objective should also seek to protect the natural and agricultural resource base. Degradation of these resources can result in spiraling economic, environmental, and social costs.

To develop drought policy objectives government officials should consider many questions. A basic question that must be addressed is the purpose and role of government involvement in drought mitigation efforts. Other questions should address the scope of the plan and identify geographic areas, economic sectors, and population groups that are most at risk. The principal environmental concerns must also be identified. Government officials must also determine what human and financial resources are available to invest in the planning process. Answers to these and other questions should help to determine the objectives of drought policy and therefore focus the drought planning process.

C. Impediments to Drought Planning

Identifying the principal obstacles or impediments to drought planning may be the first step in any attempt to initiate the development of a drought plan. Impediments include an inadequate understanding of drought, uncertainty about the economics of preparedness, lack of skill in drought prediction, variability in societal vulnerability to drought, information gaps and insufficient human resources, inadequate scientific base for water management, and difficulties in identifying drought impact sensitivities and adaptations.

In the United States, the most significant impediments to drought planning are an inadequate understanding of drought and uncertainty about the economics of preparedness. Drought is often viewed by government officials as an extreme event that is, implicitly, rare and of random occurrence. Officials must understand that droughts, like floods, are a normal feature of climate. Their recurrence is inevitable. Drought manifests itself in ways that span the jurisdiction of numerous bureaucratic organizations (agricultural, water resources, health, and so forth) and levels of government (e.g., federal, state, and local). Competing interests for scarce government resources and institutional rivalry impede the development of concise drought assessment and response initiatives. To solve these problems, policy makers and bureaucrats, as well as the general public, must be educated about the consequences of drought and the advantages of preparedness. Drought planning requires input by several disciplines, and decision makers must play an integral role in this process.

Planning, if undertaken properly and implemented during nondrought periods, can improve governmental ability to respond in a timely and effective manner during periods of water shortage. Thus, planning can mitigate and, in some cases, prevent some impacts while reducing physical and emotional hardship. This, in turn, could improve the constituents' perception of government. Planning should also be a dynamic process that reflects socioeconomic, agricultural, and political trends.

It is sometimes difficult to determine the benefits of drought planning versus the costs of drought. There is little doubt that drought preparedness requires financial and human resources that are, at times, scarce. This cost has been and will continue to be an impediment to the development of drought plans. Preparedness costs are fixed and occur now, while drought costs are uncertain and will occur later. Further complicating this issue is the fact that the costs of drought are not solely economic. They must also be stated in terms of human suffering and the degradation of the physical environment, items whose values are inherently difficult to estimate.

Post-drought evaluations have shown assessment and response efforts of state and federal governments with a low level of preparedness to be largely ineffective, poorly coordinated, untimely, and economically inefficient. Unanticipated expenditures for drought relief programs can also be devastating to state and national budgets. For example, during the droughts of the mid-1970s in the United States, specifically 1974, 1976, and 1977, the federal government spent more than $7 billion on drought relief programs. The federal government has expended similar amounts during subsequent drought periods. Between 1970 and 1984, state and federal government in Australia expended nearly $1 billion on drought relief. The Republic of South Africa has spent approximately $1.5 billion for drought relief in the past decade. When compared to these expenditures, a small investment in mitigation programs in advance of drought would seem to be a sound economic decision.

Drought plans should be incorporated into general natural disaster and/or water management plans wherever possible. This would reduce the cost of drought preparedness substantially. Politicians and many other decision makers simply must be better informed about drought, its impacts, and alternative management approaches and how existing information and technology can be used more effectively to reduce the impact of drought at a relatively modest cost.

D. Status of Drought Planning

Governments worldwide have shown increased interest in drought planning since the early 1980s. Several factors have contributed to this interest. First, the widespread occurrence of severe drought over the past several decades and, specifically, the years during and following the extreme ENSO event of 1982–1983 focused attention on the vulnerability of all nations to drought. Second, the costs associated with drought are now better understood by government. These costs include not only the direct impacts of drought but also the indirect costs (i.e., personal hardship, the costs of response programs, and accelerated environmental degradation). Nations can no longer afford to allocate scarce financial resources to short-sighted response programs that do nothing to mitigate the effects of future droughts. Finally, the intensity and frequency of extreme meteorological events such as drought are likely to increase, given projected changes in climate associated with increasing concentrations of CO_2 and other atmospheric trace gases. Droughts are a climatic certainty and recent events worldwide have highlighted the importance of preparing now for future episodes. From an institutional point of view, learning today to deal more effectively with climatic events such as drought may serve us well in preparing proper response strategies to long-term climate-related issues.

Governmental interest in and progress toward the development of drought plans worldwide has increased significantly in the past decade. The greatest progress has been made at the state level in the United States. In 1982, three states had developed drought plans: South Dakota, Colorado, and New York. At present, 23 states have drought plans. These plans differ considerably in their structure and comprehensiveness, but at least these states have taken a first step to address the unique and complicated assessment and response problems associated with drought. Considerable progress is also being made in Canada, Brazil, Australia, and many drought-prone African countries.

The challenge of changing the perception of policy makers and scientists worldwide about drought is a formidable one. The typical mode of operation for government in dealing with natural hazards is crisis management. It is indeed a difficult task for government to engage in long-range planning. However, the progress made toward planning in recent years demonstrates a new awareness and improved understanding of drought and its impacts on individual citizens, economic development, and the environment.

VI. Conclusions

Drought is a pervasive natural hazard that is a normal part of the climate of virtually all regions. It should not be viewed as merely a physical phenomenon. Rather, drought is the result of an interplay between a natural event and the demand placed on water supply by human-use systems. Drought should be considered relative to some long-term average condition of balance between precipitation and evapotranspiration.

Many definitions of drought exist; it is unrealistic to expect a universal definition to be derived. Drought can be grouped by type or disciplinary perspective as follows: meteorological, hydrological, agricultural, and socioeconomic. Each discipline incorporates different physical, biological, and/or socioeconomic factors in its definition. It must be accepted that the importance of drought lies in its impacts. Thus definitions should be impact and region specific to be used in an operational mode by decision makers.

The three characteristics that differentiate one drought from another are intensity, duration, and spatial extent. Intensity refers to the degree of precipitation shortfall and/or the severity of impacts associated with the departure. Intensity is closely linked with the duration of the event. Droughts normally take two to three months to become established but may persist for months or years, although the intensity and spatial character of the event will change from month to month or season to season.

Drought has many causes, which may be synergistic in nature. Some of the causes may be the result of influences that originate far from the drought-affected area. Prolonged droughts occur when large-scale anomalies in atmospheric circulation patterns become established and persist for periods of months, seasons, or longer. Recent droughts in the United States (1988–1991) are a good example.

The skill to predict meteorological drought for a month or season in advance is very limited. The potential for improved forecasts differs by region, season, and climatic regime. Significant advances have been made in understanding the climate system in the tropics. Much of this improvement is the result of a better understanding of the fact that a major portion of atmospheric variability that occurs on time scales of months to several years is associated with variations in tropical sea-surface temperatures. In the extratropical regions, current long-range meteorological forecasts are of very limited skill and are not likely to improve significantly in the next decade.

The impacts of drought are diverse; they ripple through the economy and may linger for years after the termination of the period of deficient precipitation. Impacts are often referred to as direct or indirect. Because of the number of groups and economic sectors affected by drought, its geographic extent, and the difficulties in quantifying environmental damages and personal hardships, the precise calculation of the financial costs of drought is difficult. Drought years frequently occur in clusters, and thus the costs of drought are not evenly distributed between years. Drought impacts are classified as economic, environmental, and social.

Government response to drought includes a wide range of potential actions to deal with the impacts of water shortages on people and various economic sectors. The types of actions taken will vary considerably between developed and developing countries and from one region to another. Few, if any, actions of government attempt to reduce long-term vulnerability to the hazard. Rather, assistance or relief programs are reactive and address only short-term, emergency needs; they are intended to reduce the impacts and hardship of the present drought.

Developing a drought policy and contingency plan is one way that governments can improve the effectiveness of future response efforts. A drought policy will be broadly stated and should express the purpose of government involvement in drought assessment, mitigation, and response programs. Drought plan objectives are more specific and action oriented and will differ between levels of government. The development of a drought contingency plan results in a higher level of preparedness that can mitigate and, in some cases, prevent some impacts while reducing physical and emotional hardship. An increasing number of governments in the United States and elsewhere are now developing policies and plans to reduce the impacts of future periods of water shortage associated with drought.

Bibliography

Glantz, M. H., ed. (1987). "Drought and Hunger in Africa: Denying Famine a Future." Cambridge Univ. Press, New York.

Karl, T. R., Quinlan, F. T., and Ezell, D. S. (1987). Drought termination and amelioration: Its climatological probability. *J. Clim. Appl. Meteorol.* **26,** 1198–1209.

Klemeš, V. (1987). Drought prediction—A hydrological perspective. *In* "Planning for Drought: Toward a Reduction of Societal Vulnerability" (D. A. Wilhite and W. E. Easterling, eds.), pp. 81–94. Westview Press, Boulder, Colorado.

Riebsame, W. E., Changnon, S. A., and Karl, T. R. (1990). "Drought and Natural Resources Management in the

United States: Impacts and Implications of the 1987–90 Drought." Westview Press, Boulder, Colorado.

Wilhite, D. A., and Easterling, W. E., eds. (1987). "Planning for Drought: Toward a Reduction of Societal Vulnerability." Westview Press, Boulder, Colorado.

Wilhite, D. A., and Glantz, M. H. (1985). Understanding the drought phenomenon: The role of definitions. *Water International* **10**, 111–120.

Wilhite, D. A., Rosenberg, N. J., and Glantz, M. H. (1986). Improving federal response to drought. *J. Clim. Appl. Meteorol.* **25**(3), 332–342.

Glossary

El Niño Invasion of warm surface water from the western equatorial part of the Pacific Basin to the eastern equatorial region and along the west coast of South America. El Niño events occur about twice every ten years, although the interval between two events is irregular.

ENSO Combination of El Niño and Southern Oscillation events.

Evapotranspiration (ET) Total process of water transfer into the atmosphere by transpiration from vegetation and evaporation from the soil surface.

Jet stream Strong zonal current of air, usually near the 500-mbar constant pressure surface in each hemisphere, that encircles the earth. Referred to as the jet stream because of its high concentration and great speed, often up to 500 km h^{-1}.

Southern Oscillation Out-of-phase relationship between atmospheric pressure over the southeast Pacific and the Indian Ocean. When pressure is high in the Indian Ocean, it is lower than usual in the south Pacific Ocean. Rainfall varies in the opposite direction. The Southern Oscillation is closely linked to El Niño events and was first observed during the latter part of the nineteenth century.

Teleconnections Regional or global patterns of atmospheric variability that reappear with considerable frequency in roughly the same form and often persist or recur throughout a month or season. ENSO is a good example of a teleconnection.

Dynamic Hydrology of Variable Source Areas

Rao S. Govindaraju and M. Levent Kavvas
University of California, Davis

During rainfall events, expanding and shrinking overland flows are observed over hillslope surface regions near streams. These regions are called variable source areas (VSAs). Expansion and shrinkage of VSAs are due to saturation of the soil from below due to rainfall. As such, VSAs occur even when the rain intensity is small. The expansion/shrinkage characteristics of VSAs are governed by the interaction of atmospheric processes with land surface hydrologic processes at the hillslope scale. The contribution from these surface areas to the stream hydrograph is practically instantaneous compared to the subsurface response.

I. Introduction

A. Scope and Occurrence of Variable Source Areas

A hillslope, in hydrologic terms, is a three-dimensional complex with soil, vegetation, bedrock, and a stream that transports the water out of the system. Even if all these hillslope parameters could be adequately sampled (which they cannot), the hydrologic response of a hillslope to a precipitation event is complicated due to the nonlinear flow processes involved. A hillslope is the smallest unit at a catchment or watershed scale that incorporates the three flow components: overland flows, stream flows, and subsurface saturated–unsaturated groundwater flows. These flow components are in a state of dynamic equilibrium and interact continuously through their common boundaries.

Figure 1 shows the three flow components in a physical hillslope setting. When rainfall starts, water infiltrates into the soil, migrates downward in response to gravity, and feeds the stream as baseflow or subsurface stormflow. In regions adjacent to streams, wetter conditions are maintained due to shallow water tables. These regions are more responsive to precipitation events and are likely to develop overland flow even when the rainfall intensity is less than the saturated hydraulic conductivity of the soil. Under prolonged rainfall, the saturated regions adjacent to streams, expand and so does the surface overland flow region. When the rain stops, water drains out from these near-stream saturated regions and the overland flow region diminishes. These hillslope areas, which develop overland flows and contribute to nearby streams, and which expand and shrink in response to rainfall and the moisture status within the soil, are called variable source areas (VSAs). An accurate modeling of this phenomena is through the study of overland flow on moving boundaries. The degree to which saturated regions and consequent overland flow contributing VSA expansion/shrinkage occurs depends on factors like intensity, duration, and spatial location of rainfall, antecedent soil moisture conditions, and other soil properties. [*See* HYDROLOGIC FORECASTING.]

It is claimed that in most natural catchments and hillslopes, the major portion of storm runoff is produced by overland flows on saturated areas near streams. The lateral inflow to these wetland areas is supplied by water escaping from the ground surface and reaching the stream as overland flow. The most important transition between the flow systems thus appears to be at the soil surface, when water is released from the high damping effect of subsurface flow and travels 100 to 500 times faster as overland flow. In investigating the stream hydrograph, the response time of any water in the system is therefore controlled by how far it has to travel to get to the stream (slope length) and the mechanics of its transfer (pathway). Methods of predicting VSA expansion and shrinkage

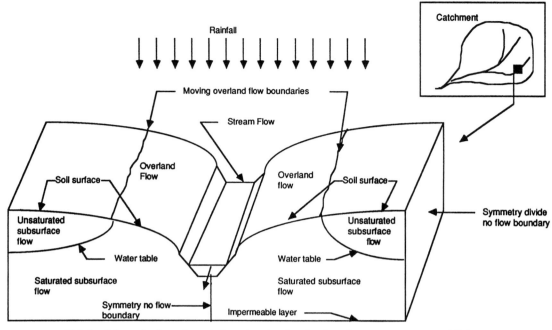

Fig. 1 Schematic view of a typical section showing the three flow components on a hillslope.

during rainfall events are therefore useful, since overland flow routes from the near-channel contributing areas have practically zero time lag (as compared to subsurface flow contributions) in reaching the receiving waters downstream.

B. Interconnection of Atmospheric, Surface, and Subsurface Processes

Study of the hydrology of variable source areas is important for understanding the characteristics of the interactions of land moisture dynamics along hillslopes and stream channels with atmospheric processes and their role in causing floods. The nature of the dynamic interaction between the hillslope overland flows and hillslope subsurface flows is very much dependent on the atmospheric processes that act as the boundary conditions on the hydrology of VSAs. The evaporation process, which is an outcome of the interaction of atmospheric processes (like temperature and wind fields) with the hydrologic soil moisture dynamics, dictates the antecedent soil moisture conditions prior to the onset of rainfall events. As shall be seen in the following, the antecedent soil moisture conditions have a significant influence on whether the outcome of a rainfall event will produce flooding or not. Therefore, the hydrology of VSAs is a product of the interaction of atmospheric processes with the land moisture move-

ment dynamics. [*See* FLOODS; LAND-ATMOSPHERE INTERACTIONS.]

C. Brief Review of Related Literature

R. A. Freeze was the first to address the effect of hillslope parameters on the stream hydrograph at the outlet of a hillslope section. He found that on concave slopes with low soil permeabilities and on most convex slopes, the dominating contribution to the stream hydrograph comes from overland flows on transient variable source areas neighboring the streams. K. Beven extended on the work of Freeze by using a finite element model for a two-dimensional subsurface flow coupled to a one-dimensional stream flow. His results clearly demonstrate the nonlinear response of the hillslope hydrograph to rainfall and parameter variations. He has further demonstrated that the initial conditions, particularly in the unsaturated zone, are of primary importance in determining the time and magnitude of the hillslope hydrograph peaks.

Later, Freeze studied the problem of parameter uncertainties in the rainfall-runoff process occurring on a hillslope through an event-based Monte Carlo technique. In this technique, many computer simulations of the hillslope rainfall-runoff process are performed under varying parameter values. Freeze has shown that variation in soil hydraulic conductivities along a hillslope has a profound impact on the runoff response of a hillslope to rainfall. R. E. Smith and R. H. B. Hebbert discuss the interaction present in the surface and subsurface flow processes. They use a kinematic wave approximation for surface flows in conjunction with a simplified but analytical infiltration model that is applicable to cases when rainfall exceeds infiltration capacity of the soil. Their description of the interaction between overland flow and subsurface flow processes is very interesting, and a similar procedure is adopted here. More recently, A. Binley and others have studied the problem of stochastic runoff generation by using a physically based heterogeneous hillslope model. Their study concentrates on the effect of randomness in hydraulic conductivities. Their surface-flow routing is based on a linearized approach instead of the solution of the partial differential equations governing unsteady flows.

Traditionally, two distinct mechanisms for producing overland flow have been identified. The Horton type flow (infiltration excess) occurs when the rainfall intensity is greater than the soil infiltration capacity and the rain excess flows off as surface runoff. If there is an impeding layer within the soil horizons, then water fills up from this impeding layer upward, finally causing saturation at the soil surface. Ponding of water occurs here, eventually resulting in overland flow. Overland flow from such a mechanism is referred to as saturation excess flow here. In the literature, it is also called the Dunne mechanism of overland flow generation. These two mechanisms of flow generation do not occur independently in nature. It has been concluded that, depending on the rainfall characteristics and hydrogeologic properties, a hillslope may generate overland flow by both mechanisms simultaneously.

II. Mathematical Formulation of the Hydrologic System

The mathematical representation of the flow processes over a hillslope leads to a set of nonlinear partial differential equations. Analytical solutions of such systems are not available and numerical techniques are used to obtain the behavior of the quantities of interest. The equations used in this study for describing the flow processes are presented in this section.

A. Stream Flow Equations

The equations of continuity and momentum in a channel are

$$V \frac{\partial A}{\partial x} + A \frac{\partial V}{\partial x} + \frac{\partial A}{\partial t} = Q \tag{1}$$

$$S_f = S_0 - \frac{\partial y}{\partial x} - \frac{V}{g} \frac{\partial V}{\partial x} - \frac{1}{g} \frac{\partial V}{\partial t} - \frac{QV}{gA} \tag{2}$$

These are the Saint–Venant equations describing the propagation of flood waves in a channel. In the above equations, x is the distance along the horizontal direction, V is the mean velocity of water in the channel cross section, A is the area of the flow cross section normal to the horizontal direction, t is the time, Q is the lateral inflow entering the channel per unit length, S_0 is the channel bed slope, y is the flow depth normal to the flow direction, g is acceleration due to gravity, and S_f is the friction slope or the slope of the total energy line along the channel.

There are no known analytical solutions to the Saint–Venant equations. Even their numerical solutions are fraught with problems of convergence and stability due to their highly nonlinear nature. Hence investigators have used simplified versions of these equations whenever the physical situation justified them. Common simplification techniques involve the modification of the momentum equation (2). Some of these simplified models are (a) the momentum equation is not considered at all, leading to the storage routing Muskingum method; (b) only the first term of the right-hand side of Eq. (2) is considered, leading to kinematic wave routing; and (c) the first and second terms on the right-hand side of Eq. (2) are considered, resulting in the general diffusion approximation. W. S. Gonwa and M. L. Kavvas have demonstrated that the diffusion wave approximation to the full Saint–Venant equations is adequate for cases of practical interest. The diffusion wave approximation is used here in analyzing the stream flow component in this study. Substituting the diffusion approximation of the momentum equation into the continuity equation (1) leads to the following equation describing the flood propagation in trapezoidal channels:

$$\frac{\partial y}{\partial t} + C_w \frac{\partial y}{\partial x} + \delta = K \left(\frac{\partial^2 y}{\partial x^2} - \frac{\partial S_0}{\partial x} \right) \tag{3}$$

For trapezoidal channels, the cross-sectional area A is

$$A = by + zy^2 \tag{4}$$

where b is the bottom width and z is the inverse of the side slope. The generalized friction law is represented as

$$V = DR^m \left(S_0 - \frac{\partial y}{\partial x} \right)^j \tag{5}$$

where R is the hydraulic radius of the flow section. With appropriate choice of D, m, and j, the familiar Manning's law, Chezy's law, or the Darcy–Weisbach friction relationship can be obtained. In Eq. (3) the wave celerity C_w is

$$C_w = V(1 + m\Omega) \tag{6}$$

where Ω is given by the expression

$$\Omega = 1 - \frac{2R}{T}(1 + z^2)^{1/2} \tag{7}$$

where T is the top width of the flow section. δ in Eq. (3) is

$$\delta = \frac{Vy}{T}\frac{\partial b}{\partial x} + \frac{Vy^2}{T}\frac{\partial z}{\partial x} + \frac{VAm}{TR}\xi - \frac{Q}{T} \tag{8}$$

The unknown term here is ξ, which is defined as

$$\xi = \frac{1}{P}\left(y\frac{\partial b}{\partial x} + y^2\frac{\partial z}{\partial x}\right)$$
$$- \frac{A}{P^2}\left\{\frac{\partial b}{\partial x} + 2zy(1 + z^2)^{-1/2}\frac{\partial z}{\partial x}\right\} \tag{9}$$

Here P is the wetted perimeter, which for trapezoidal sections is

$$P = b + 2y(1 + z^2)^{1/2} \tag{10}$$

The diffusion coefficient K in Eq. (3) is defined as

$$K = \frac{VAj}{T(S_0 - \partial y/\partial x)} \tag{11}$$

Equations (3) to (11) constitute the nonlinear diffusion wave approximation for the description of flood propagation in trapezoidal channels.

B. Overland Flow Equations

Overland flows may be treated as flows per unit width on very wide rectangular channels. Since the width is very much greater than the depth, the hydraulic radius is equal to the depth of flow. Using the procedure discussed above results in the following equation for the diffusion wave approximation for overland flows:

$$\frac{\partial h}{\partial t} + C_{wo}\frac{\partial h}{\partial x} + \psi_o = K_o\left(\frac{\partial^2 h}{\partial x^2} - \frac{\partial S_{0o}}{\partial x}\right) \tag{12}$$

and the generalized friction law for overland flows becomes

$$V_o = D_o h_{mo}\left(S_{0o} - \frac{\partial h}{\partial x}\right)^{jo} \tag{13}$$

where h is the depth of flow, V_o is the depth-averaged overland flow velocity, S_{0o} is the slope of the overland flow bed, x is the coordinate along the horizontal direction, and t

is time. In Eq. (13), D_o, mo, and jo are constants. The value assigned to these constants depends on whether the flow is laminar or turbulent, and an appropriate friction relationship can be specified. For overland flow applications, the definitions of terms appearing in Eq. (12) are

$$C_{wo} = V_o(1 + mo) \tag{14}$$

$$\psi_o = -Q_o \tag{15}$$

$$K_o = \frac{V_o hjo}{S_{0o} - \partial h/\partial x} \tag{16}$$

where Q_o is the net lateral inflow to the overland flow section per unit length of flow.

C. Saturated – Unsaturated Groundwater Flow Equations

The subsurface flow of water takes place in the pores or voids of the soil medium. This component is complicated by the presence of two distinct regions—the saturated and the unsaturated zones. There exists a region from the ground surface to some varying depth within the soil (the water table elevation) where the pore spaces are only partially filled with water. This zone comprises the unsaturated region of the soil. The water table is a theoretical surface at which the pressure in the soil equals the atmospheric pressure as the soil is assumed to be perfectly aerated. The zone below the water table comprises the saturated region of the soil.

In the unsaturated zone (also called as the zone of aeration) flow is driven by two primary forces. These are the gravitational potential and the negative hydrostatic potential due to capillary suction. Other less significant forces include those resulting from osmotic pressure due to charged soil particles, adsorption forces between water and soil, and thermal potential due to temperature differences. The voids and pore spaces are completely filled with water below the water table in the saturated zone except for small negligible volumes of entrapped gases. In this region, the flow is governed by the piezometric head or the total energy head, which is composed of the elevation head and the positive pressure potentials. Some of the basic assumptions made in modeling subsurface flows are (a) only liquid flow is considered even though the flow may occur both in liquid and vapor form; (b) the porous medium is incompressible; and (c) Darcy's law is valid. Within these limitations, subsurface flow may be described by a continuity equation expressed as

$$\frac{\partial}{\partial t}[\rho\theta] + \text{div}(\rho q) = s \tag{17}$$

where q is the specific discharge of water, ρ is the density of water, θ is the moisture content by volume of the soil, s is a source/sink term (incorporating the effects of evapotranspiration, rain infiltration, etc.). The continuity equation (17) is combined with Darcy's law, which may be stated as

$$q = -K(\theta) \, \text{grad} \, \phi \qquad (18)$$

where $K(\theta)$ is the hydraulic conductivity of the soil, which varies with the moisture content in some nonlinear fashion, and ϕ is the piezometric head. The piezometric head may be expressed as

$$\phi = \psi + z \qquad (19)$$

where ψ is the capillary potential or suction head of the soil and z is the elevation above some fixed datum. Combining Eqs. (17), (18), and (19) leads to a single equation for a two-dimensional slice in the vertical (x, z) plane in terms of ψ as

$$\left[\frac{\theta}{\eta} S_s + C(\psi) \right] \frac{\partial \psi}{\partial t} = \frac{\partial}{\partial x} \left\{ K_x(\psi) \frac{\partial \psi}{\partial x} \right\}$$
$$+ \frac{\partial}{\partial z} \left\{ K_z(\psi) \left(\frac{\partial \psi}{\partial z} + 1 \right) \right\} + s \qquad (20)$$

where $C(\psi)$ is called the specific moisture capacity of the soil medium and is expressed as

$$C(\psi) = \frac{d\theta}{d\psi} \qquad (21)$$

In Eq. (20), η denotes the soil effective porosity and S_s is the specific storage of the soil defined as

$$S_s = \rho g \eta \, (C_w + C_f) \qquad (22)$$

where g is the acceleration due to gravity, C_w is the compressibility of water, and C_f is the compressibility of the soil medium. The soil porosity is taken as 0.3 and the specific storage for the soil matrix is taken as 0.0001 m in all examples presented in this chapter. Thus, under isotropic conditions, only two functional relationships are required to solve for Eq. (20): $\theta(\psi)$ and $K(\psi)$. These functions are called characteristic curves and are shown for a typical soil in Fig. 2. The hydraulic conductivity of the soil is usually expressed as

$$K(\psi) = K(\text{sat}) \cdot K_r(\psi) \qquad (23)$$

where $K(\text{sat})$ is the saturated hydraulic conductivity of the soil and is a constant for a given soil type and $K_r(\psi)$ is the relative hydraulic conductivity of the soil. Figure 2 shows single-valued relationships for the characteristic curves of a

soil, but in general these functions are hysteretic and hence non-unique specification for $K(\theta)$ and $\psi(\theta)$ may result for a given θ depending on the previous wetting and drying history of the soil. [*See* GROUNDWATER; SOIL PHYSICS; SOIL WATER: LIQUID, VAPOR, AND ICE.]

D. Initial and Boundary Conditions

The equations for stream flows, overland flows, and subsurface flows in the previous sections need to be combined with appropriate initial and boundary conditions for a complete specification of the problem. Figure 3 shows the side hillslope section investigated here. The subsurface section is 30 m long on each hillslope and is 1 m deep at the no-flux boundary. The stream section is 100 m long and 2 m wide. The overland and subsurface flow phenomena occur on the two side hillslopes draining into the stream. Such a section would occur along the length of a natural catchment (see inset in Fig. 1) or where there are two neighboring agricultural plots separated by a stream. Freeze has discussed the end conditions applicable to such sections. Referring to Fig. 3, along the basal boundary GF, there is no flux.

$$\frac{\partial}{\partial z} (\psi + z) = 0 \qquad (24)$$

Along the imaginary, vertical boundaries AG and EF impermeable conditions are imposed invoking symmetry arguments as

$$\frac{\partial \psi}{\partial x} = 0 \qquad (25)$$

On the stream bottom ABC, the boundary condition is dictated by the time-dependent stream elevation at that particular section (for any point κ along ABC in Fig. 3) Therefore

$$\psi = y(\kappa, t) \qquad (26)$$

The boundary conditions along CDE are particularly interesting. The location of the point D determines the extent of ponded domain, and the region CD is the variable source area. Along the region CD the subsurface nodes are specified by a time–space varying head boundary condition dictated by overland flow conditions occurring on the surface as

$$\psi = h(x, t) \qquad (27)$$

The net flux $I(x, t)$ (which is positive for rainfall and negative for evapotranspiration) is equal to the amount of water infiltrating/exfiltrating into/from the soil. Hence the boundary condition along DE from Eq. (18) is given as

$$I(x, t) = K(x, t) \left[\frac{\partial \psi(x, t)}{\partial z} + 1 \right] \qquad (28)$$

K(sat) = 0.264 cm/min., Por. = 0.3

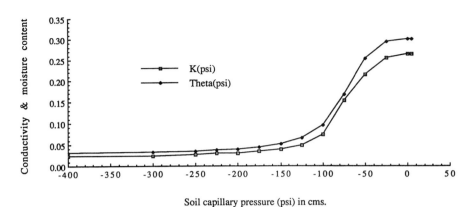

Fig. 2 Typical soil characteristic curves for a loose sandy soil [similar to Freeze (1972a).]

The value of $K(x,t)$ in Eq. (28) will depend on the value of the matric potential $\psi(x,t)$ and the nature of the characteristic curve $K_r(\psi)$ for the soil along the boundary DE.

The upstream boundary condition of the channel is taken as the zero inflow condition. It is usually assumed that the channel achieves uniform flow at the downstream. The downstream boundary condition of zero-depth-gradient has been adopted for overland flows in the examples given here. However, one could also utilize a critical flow condition at the overland flow downstream section. The upstream moving boundary condition for overland flow is taken as the zero

depth condition at the transient point D (see Fig. 3). The lateral inflow into the stream at any section is the sum of the rainfall, the contribution from the overland flow section at the outlet point C (note that a symmetric section exists on both sides of the stream) minus/plus the quantity infiltrating/exfiltrating from/onto the base of the stream surface. Thus the lateral inflow into the channel at any channel node is

$$Q = R(t) + Q(\text{over}) \pm Q_s \qquad (29)$$

ABC = Region in contact with stream (head condition)

CD = Overland flow region (head condition)

DE = Region with flux boundary condition

EF = Impermeable boundary from symmetry conditions

FG = Impermeable base

AG = Impermeable boundary from symmetry conditions

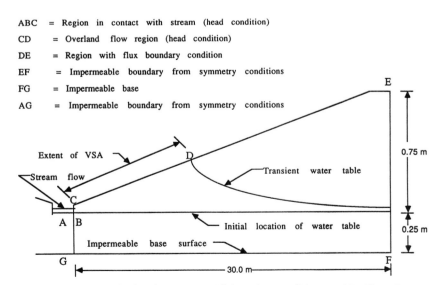

Fig. 3 A typical subsurface section with boundary conditions used in this study.

where $R(t)$ is the time-varying rain falling over the width of the stream section (positive for rainfall and negative for evaporation), $Q(\text{over})$ is the overland flow contribution and Q_s is the quantity infiltrating into (or exfiltrating from) the soil over the width of the channel section.

Similarly the lateral inflow into the overland flow nodes are determined by the algebraic sum of precipitation and infiltration into the soil. Mathematically this is represented as

$$Q_o(x, t) = I(x, t) - K(\psi)\left[\frac{\partial\psi(x, t)}{\partial z} + 1\right], \quad Q_o > 0$$
$$= 0, \qquad\qquad\qquad Q_o \leq 0 \quad (30)$$

where $I(x,t)$ represents the rainfall intensity contribution to the lateral inflow and $\psi(x, t)$ represents the capillary pressure on the subsurface boundary nodes along CE in Fig. 1. If $Q_o(x, t)$ as determined from Eq. (30) is negative, then there is no net lateral inflow contribution to that particular overland flow node for that time instant. The situation of partially ponded areas CD in Fig. 3 occurs when the rain intensity does not exceed the saturated soil hydraulic conductivity.

Note that Eq. (30) essentially dictates what portion of the rainfall enters the soil through infiltration and what portion of the rainfall contributes to overland flow. Overland flows are also governed by a continuity equation similar to Eq. (1) as

$$\frac{\partial h}{\partial t} + \frac{\partial}{\partial x}(hV_o) = Q_o(x, t) \quad (31)$$

where the variables are defined as in Eq. (12) except that the net lateral inflow to overland flow $Q_o(x, t)$ is obtained from Eq. (30). The second term on the right-hand side of Eq. (30) [call it $I_2(x, t)$ for convenience] represents the rate of water infiltrating into the soil medium. Equation (30) holds even for ponded conditions when the quantity of water infiltrating into the soil depends on the rainfall, overland flow depth on the soil surface, and the status of soil moisture at the surface of the soil. At initial times, the soil surface is dry and $I_2(x, t)$ is greater than $I(x, t)$; thus $Q_o(x, t)$ is zero for this point in time and space. As the water table rises and the initial moisture deficit is satisfied, $I_2(x, t)$ becomes gradually smaller, but still there will be no contribution to overland flow if the rain intensity does not exceed the saturated hydraulic conductivity of the soil. In such instances, overland flow will have positive contributions only when the groundwater table practically reaches the soil surface. This is the case of saturation excess overland flow, which occurs due to subsurface saturation from below and forms the VSAs. [See SOIL KINETICS.]

Due to the sparsity of data in most hydrological situations, two sets of physically meaningful initial conditions for the subsurface component are in common use. These are (1)

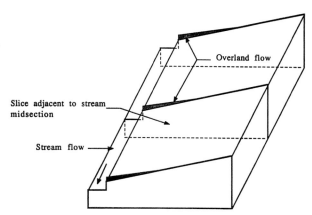

Fig. 4 An example showing orientation of three slices for representing the 3-D subsurface and the 2-D overland flow.

static or no flow condition and (2) the steady-state condition. Under static initial condition, the total hydraulic head $\phi(x, z, 0)$ is constant for all spatial locations (x, z) with a horizontal water table at level with the stream depth. The moisture contents and pressure heads in the subsurface region are at equilibrium. For steady-state initial conditions the time derivatives of the flow equations disappear, and the equations may then be solved subject to appropriate boundary conditions that include a special constant input rate. Static initial conditions have been chosen for the subsurface domain in the simulation efforts of this study. The initial condition for overland and stream flow components, dictated by static initial conditions for the subsurface, is that of a dry section.

E. Solution Methodology for Conjunctive Modeling

In our demonstration studies, implicit centered finite-difference techniques have been used for all flow components. Details of this scheme can be obtained elsewhere and are not repeated here. Small time steps (in the order of seconds) were used in the beginning of each simulation because of the highly nonlinear nature of the equations.

The equations used in the subsurface are two-dimensional with a one-dimensional overland flow description on the surface. Such a unit of overland and subsurface flow is called as a "slice," as indicated in Fig. 4. The channel flow is perpendicular to the orientation of these slices (see Fig. 1). The time–space varying channel-flow depth serves as a boundary condition for the subsurface nodes attached to the channel bottom (region ABC in Fig. 3). To simulate this situation, many slices are chosen along the channel reach, and each slice interacts with the channel depending on its loca-

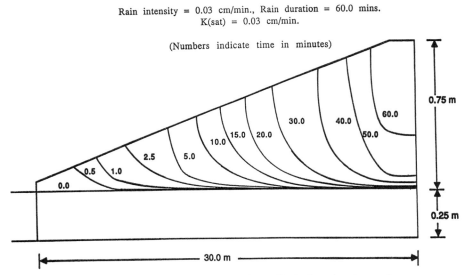

Fig. 5   Time varying location of the rising water table profiles in the subsurface at the stream midsection.

tion along the channel reach. This approximates the three-dimensional nature of flow without actually using a three-dimensional subsurface flow setup. This leads to considerable savings in computer effort, and the problem is simpler to formulate.

The problem now reduces to the simultaneous solution of a one-dimensional stream section and various slices. These components are internally coupled, since solving one component modifies the boundary conditions or the lateral inflow/outflow conditions for the other components. The stream depths and the overland flow depths at each surface node and the capillary potentials at each subsurface node are used as the convergence criterion quantities during the iteration procedure. Convergence is achieved when the maximum differ-

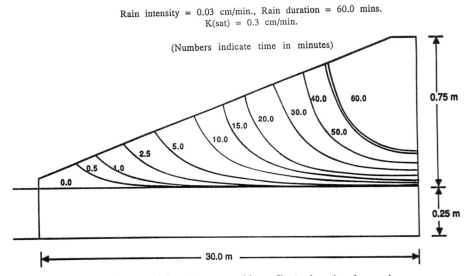

Fig. 6   Time varying location of the rising water table profiles in the subsurface at the stream midsection.

Rain intensity = 0.03 cm/min., Duration = 60.0 mins.

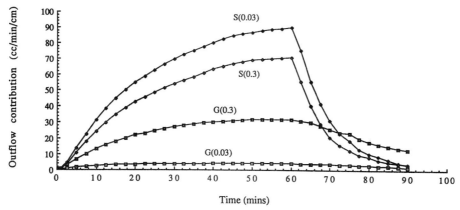

Fig. 7 Comparison of the relative contributions to the stream from the surface and subsurface flow for the case in Figs. 5 and 6.

ence for any nodal value of these quantities between two successive iterations is less than some preset tolerances. Each sweep starts with the numerical solution of the stream flow section using the net lateral inflow from rainfall, and base-flow and overland flow discharge from each slice (estimated from solution at the previous time step). The new stream solution thus obtained prescribes a new head condition at each slice along region ABC for the subsurface (see Fig. 3). Using this new boundary condition, obtained from the time–space varying stream depth, the subsurface flow components are solved for all slices. The new subsurface solutions pre-scribe new values of net lateral inflows to overland flows along CD or change the capillary potentials along DE for each slice. The overland flow component is now solved for each slice. The new overland flow solutions for each slice deter-mine new boundary conditions along CDE for the subsurface flow and provide new lateral outflow for each slice to a particular stream location. This starts off a new sweep cycle, and the stream flow is solved for new lateral inflows from the subsurface and new overland flow inputs. These cycles con-tinue until all the nodes for each component in the system converge within preset tolerances.

III. Dynamic Response of Variable Source Areas

For the examples studied here, it is found that the channel merely drains the water from the side hillslopes and does not play an active part in the overland flow dynamics. Therefore our efforts shall be concentrated on modeling the location

and movement of the dynamic variable source areas. Simula-tion results will be shown for the slice adjacent to the stream midsection (see Fig. 4).

A. Effect of Soil Hydraulic Conductivity

Hydraulic conductivity has a strong influence on the subsur-face moisture status. Figures 5 and 6 show the location of the water tables for different times when the $K(\text{sat})$ values are varying by an order of magnitude. These figures also indicate the extent of the variable source areas, since the water table intersects the soil surface just where overland flow is starting (the dynamic point D in Fig. 3). In both Figs. 5 and 6, the rainfall intensity is 0.03 cm/min and the rain duration is 60.0 min. The $K(\text{sat})$ values are 0.03 and 0.3 cm/min in Figs. 5 and 6, respectively. Till the first 30 min, Figs. 5 and 6 show similar results as far as the VSA extent is concerned. But the shape of the water table profile at identical times is very much steeper for $K(\text{sat}) = 0.03$ cm/min and much flatter for $K(\text{sat}) = 0.3$ cm/min in Figs. 5 and 6. After 30 min, the contributing saturated regions are quite different for these figures. In Fig. 5, the water table profiles are still steep and are rising at 60 min when the rain stops. There is little difference in the water table profiles for 50 and 60 min in Fig. 6 indicating that equilibrium conditions are approaching. Yet, even after 60 min of rain, not all the slope is contributing to overland flow for the case of $K(\text{sat}) = 0.3$ cm/min in Fig. 6. To transport the same amount of water to the stream, the soil with greater $K(\text{sat})$ develops a smaller overland flow region. A greater share of the water will be transported through the subsurface region. Thus different values of soil hydraulic

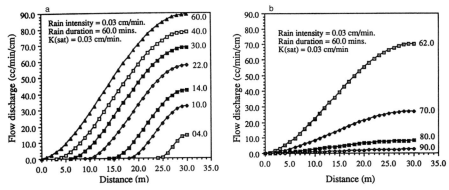

Fig. 8 Time-space varying overland flow discharges (depth × velocity). Numbers indicate time in minutes. (a) Rising overland flow profiles. (b) Receding overland flow profiles.

conductivity influence both the slope length and the mode through which water is transported to the stream.

These conclusions are illustrated more explicitly in Fig. 7, where the relative outflow contributions to the stream mid-section are shown for different cases of hydraulic conductivity. In this figure, the numbers in brackets are the saturated hydraulic conductivities (cm/min) and S and G stand for surface and groundwater contributions, respectively. This figure shows that when the hydraulic conductivity of the soil is smaller, the surface contribution is more dominant. But the subsurface contribution is more pronounced when the hydraulic conductivity is greater. Increasing the hydraulic conductivity of the soil increases the relative importance of the subsurface component as a contributor to the stream hydrograph. Figure 7 shows that the surface response time is much smaller than the subsurface response time. The surface contributions rise more rapidly when rain starts and recede faster when rain stops.

Figures 8a and 8b show the spatially varying flow discharge (depth × velocity) profiles at different times for the rising and recession stages of overland flow. In these figures, the 30.0-m location is adjacent to the stream and indicates the downstream end of overland flow region. The gradual increase in the contributing area is brought out clearly in Fig. 8a for the rising stage. Figure 8b, for the recession stage, shows that the overland flow water drains off rapidly after the rain stops.

The conclusions of Figs. 5 to 7 are very dependent on the geologic conditions. For a wider and more deeply incised channel, subsurface contributions would increase as the area of the subsurface in contact with the stream increases. But there would be no such increase in the overland flow contribution. This agrees with the conclusion of Freeze that the

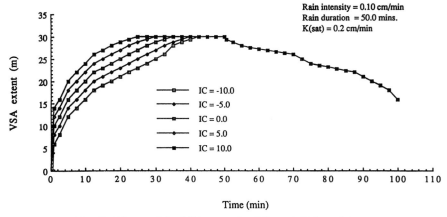

Fig. 9 Time evolving VSA extent for different initial conditions.

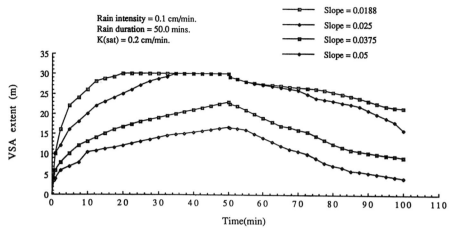

Fig. 10 Time evolving VSA extent for different slopes.

role of subsurface flow increases for more deeply incised channels and for soils exhibiting larger hydraulic conductivities.

B. Effect of Initial Conditions

The effect of initial conditions (or antecedent soil moisture conditions) is manifested in the change of response time of a catchment. The VSA extent is used as an indicator of the response time for different initial conditions (IC) in this study. Figure 9 shows the time variation of the location of the moving overland flow boundary for five different initial positions of the water table. Figure 3 shows the location of the water table for IC = 0.0. An IC value of 5.0 indicates that the water table has been raised by 5.0 cm over the initial location in Fig. 3. The initial conditions over the rest of the flow domain are determined through the static initial conditions, ensuring a no-flow situation. Figure 9 shows that the closer the water table is to the soil surface, the faster the saturation regions expand, since there is less region where the moisture deficit has to be satisfied. All cases in Fig. 9 were obtained under a rainfall intensity of 0.1 cm/min and rain duration of 50 min. The K(sat) value for the soil is 0.2 cm/min. During the rising stages, for any time, the VSA extent is maximum when IC = 10.0 and least when IC = −10.0. By the time the rain stops, all the solutions have reached the same equilibrium state, and thus the recession curves for all cases of IC are similar in Fig. 9. This figure also illustrates that for different starting conditions (all other conditions remaining the same) the equilibrium solution is the same. The time taken to reach this equilibrium depends on how close the initial condition is to the steady-state solution.

C. Effect of Slope Variation

Changes in slope of the hillslope were achieved by keeping the horizontal extent of the slope constant and varying the height of the section of the initially unsaturated subsurface zone. Thus a greater slope not only affects the topography of the hillslope but also increases the unsaturated region, which in turn increases the moisture deficit.

Figure 10 compares the extent of the contributing variable source areas as they evolve in time for four different slope conditions. For each case, the K(sat) value is 0.2 cm/min, the rainfall intensity is 0.1 cm/min, and rain duration is 50.0 min. The curves for the rising portion in Fig. 10 indicate that the milder the slope, the faster the zone of saturation increases with time. Thus for a slope of 0.0188, the whole 30 m contributes to overland flow in about 40 min, while for a slope of 0.05 barely 15 m of the slope gets a chance to develop overland flow in 50.0 min. The recession curves, after the cessation of rain, also exhibit a strong dependence on slope. The greater the slope the faster the recession as the overland flow drains off at a faster rate. The recession of the overland flow region is fastest for a slope of 0.05 in Fig. 10. Thus for steeper slopes, smaller overland flow domains occur for both rising and recession parts. The overland flow contribution to stream flows would therefore recede very rapidly after cessation of rain where steep slopes exist.

IV. Concluding Remarks

The dynamic response of variable source areas over hillslopes has been discussed here through the concept of moving boundaries for the overland flow domain. The upstream overland flow boundary moves back and forth in response to infiltration–exfiltration processes (see Figs. 8a and 8b). A physics-based distributed model was used to study this phenomena by internally coupling the nonlinear partial differential equations for the flow components. Using this model, the sensitivity of the variable source areas to some hillslope parameters was described above.

The hydrologic response in the subsurface to variation in hydraulic conductivity is shown in Figs. 5 and 6. An increase in the hydraulic conductivity of the soil increases the relative importance of the subsurface as a contributor to stream flows (see Fig. 7). This figure also indicates that the subsurface response is considerably damped and sluggish, while the surface response is more rapid to external rain events. This is why flashy stream response is attributed to VSA contributions and the low dry-weather flows are maintained by subsurface contributions.

The closer the water table is to the ground surface, the faster is the response and increase in VSA extent (Fig. 9). This is because there is a smaller unsaturated region where the moisture deficit needs to be satisfied in case of more shallow initial water tables. It appears that, for all other conditions remaining the same, different initial conditions lead to the same steady-state solutions. The results for various slopes, shown in Fig. 10, suggest that the steeper slopes exhibit slower expansion in VSA extent during rising phase and faster shrinking of the wetland areas when the rainfall stops. The overland flow drains away faster on steeper slopes, and thus the surface-flow response time is smaller for steeper slopes.

There are many more combinations and ranges of hillslope parameters that could provide interesting results and throw more light on hillslope response. For other descriptions on the topic, the reader is referred to the bibliography. For all the cases considered here, the stream plays a dormant role and merely drains off the water from the side hillslopes. However, in many flooding situations, the stream depth increases beyond the downstream overland flow elevation at a hillslope section, thus forcing the overland flow to retreat against the slope. Thus the dynamic upstream flow boundary moves back due to downstream effects. Further studies of such situations could improve our understanding of the fundamental flooding mechanisms on river systems.

Bibliography

Beven, K. (1977). Hillslope hydrographs by the finite element method. *Earth Surf. Processes* **2**, 13–28.

Binley, A., Elgy, J., and Beven, K. (1989). A physically based model of heterogeneous hillslopes, 1, runoff production. *Water Resour. Res.* **25**(6), 1219–1226.

Binley, A., Beven, K., and Elgy, J. (1989). A physically based model of heterogeneous hillslopes, 2, effective hydraulic conductivities. *Water Resour. Res.* **25**(6), 1227–1233.

Dunne, T., and Black, R. D. (1970). An experimental investigation of runoff production in permeable soils. *Water Resour. Res.* **6**(2), 478–490.

Dunne, T., and Black, R. D. (1970). Partial area contributions to storm runoff in a small New England watershed. *Water Resour. Res.* **6**(5), 1296–1311.

Freeze, R. A. (1972a). Role of subsurface flow in generating surface runoff 1, base flow contribution to channel flow. *Water Resour. Res.* **8**(3), 609–623.

Freeze, R. A. (1972b). Role of subsurface flow in generating surface runoff 2, upstream source areas. *Water Resour. Res.* **8**(5), 1272–1283.

Freeze, R. A. (1978). Mathematical models of hillslope hydrology. *In* "Hillslope Hydrology" (M. J. Kirkby, ed.). pp. 177–225. Wiley (Interscience), New York.

Freeze, R. A. (1980). A stochastic-conceptual analysis of rainfall-runoff processes on a hillslope. *Water Resour. Res.* **16**(2), 391–408.

Gonwa, W. S., and Kavvas, M. L. (1986). A modified diffusion wave equation for flood propagation in trapezoidal channels. *J. Hydrol. (Amsterdam)* **83**, 119–136.

Govindaraju, R. S., and Kavvas, M. L. (1989). "On the Physics-Based Hydrology of Floods in First Order Basins." Report to Natural and Manmade Hazards Mitigation Program, National Science Foundation, Washington, D.C.

Smith, R. E., and Hebbert, R. H. B. (1983). Mathematical simulation of interdependent surface and subsurface hydrologic processes. *Water Resour. Res.* **19**(4), 987–1001.

Woolhiser, D. A. (1974). Unsteady free-surface flow problems. *Proceedings of Institute on Unsteady Flow in Open Channels*, pp. 195–213. Colorado State University, Fort Collins.

Glossary

Hydraulic conductivity Ability of the soil to transport water.

Hydrograph Depth (or discharge) versus time curve used as the hydrologic response of some flow section.

Infiltration Process enabling applied surface water to enter the soil.

Matric potential Affinity with which the soil holds the water.

Piezometric head Total hydraulic head driving flow within soil.

Dynamic Topography

Michael Gurnis

The University of Michigan

Dynamic topography is the pushing up and the pushing down of the earth's surface in response to the forces of thermal convection inside the earth's mantle. Although the observational evidence is indirect, global dynamic topography exerts a fundamental control on sea-level change and continental flooding over hundreds of millions of years. By influencing sea level, dynamic topography is an important force influencing continental erosion, global sedimentation, the chemistry of seawater, and climate.

I. Introduction

A. What Is Dynamic Topography?

Topography is the elevation of the solid surface of the earth; on a global scale topography is dominated by the dichotomy between oceans and continents. This rather sharp difference is caused by variations in crustal thickness—the continents have a thickness of about 30 km on average and the oceans have a uniform thickness of 6 km. This sharp distinction between continents and oceans does not show up in the earth's long-wavelength geoid, the equipotential surface of the gravitational field; this means that at large lateral dimensions, the crust of the earth is isostatically compensated—the mass anomalies causing the topography balance the mass of the topography.

Dynamic topography is unlike this familiar isostatically compensated topography—the mass anomalies driving the topography are buried deep inside the planet. To understand this process, one must appreciate that the mantle, the outer 3000-km solid layer of the earth, behaves as a very viscous fluid over geologic time scales (in this case, for times greater than about 10^3 years). In a viscous medium, fluid flow can be generated by density differences within. Fluid rises where the density is low and sinks where it is high. This pattern of fluid flow transmits stress to the top of the fluid, where the vertical hydrostatic stress on the top surface of the fluid is balanced by the deflection of the surface. This deflection is the dynamic topography. Positive mass anomalies within the viscous mantle cause the interface between the earth's solid part and its fluid parts (the ocean and atmosphere) to be deflected downward. Within the mantle, mass anomalies and flow are driven by thermal convection associated with plate tectonics and the cooling of the earth. The observational evidence for this dynamic topography is indirect, as we shall see, but on a global scale recent estimates for its magnitude range from less than 100 m to about 1 km. [*See* MANTLE.]

B. Why Is Dynamic Topography Important for the Earth System?

In the context of the earth system, dynamic topography is important for two reasons: (a) Observational estimates of dynamic topography provide important constraints on mantle dynamics, and (b) dynamic topography can result in profound changes in the face of the earth as the pattern of mantle convection evolves with time.

Estimates of dynamic topography for the present day as well as for past geological ages provide boundary conditions for models of mantle dynamics. The record of platform flooding over periods of tens to hundreds of millions of years provides observational constraints on how dynamic topography has changed over time and therefore provides important information about the dynamics of the mantle—where we have few additional constraints. Most geophysical observations (like heat flow, present-day topography, and the seismic structure of the deep mantle) constrain the current physical state of the mantle but provide few clues as to how the pattern of convection has evolved over time. Only in the last year have earth scientists started to quantitatively exploit platform flooding data to constrain mantle convection. [*See* HEAT FLOW THROUGH THE EARTH.]

Dynamic topography is also important because variations in topography strongly influence interactions between the solid part of the earth and the oceans and atmosphere. As the pattern of mantle convection changes, the resulting dynamic topography can push continents up above the sea surface and can result in an increased erosion rate and higher continental denudation; dynamic topography can bow continental surfaces below the sea surface to result in increased rates of platform and margin sedimentation. Variations in the amount of continental surface exposed influences the amount of atmospheric carbon dioxide and consequently the efficiency of the atmospheric greenhouse. Furthermore, dynamic topography can be greatly modified when supercontinents form and perturb mantle convection, resulting in profound changes in climate. Some scientists have speculated that the longest cycle of mantle convection, the 300-million-year cycle of supercontinent aggregation and dispersal, is the primary driver of the longest cycle of climate change. [*See* SUPERCONTINENTS.]

II. Estimates of Global Dynamic Topography

A. Direct Estimates of Present-Day Dynamic Topography

Terrestrial topography is dominated by crustal thickness variations within the upper 30 to 50 km of the solid part of the planet, and this makes it difficult to distinguish dynamic from the isostatically compensated topography. One approach at extracting dynamic from total topography would be to strip away the topography resulting from crustal thickness and crustal density variations; unfortunately, this is difficult given the poor understanding of the structure of continental crust. [*See* CONTINENTAL CRUST.]

The structure of the oceanic crust and lithosphere is considerably less complex than the continents. The oceanic topography is simple because new oceanic lithosphere is continuously created at mid-ocean ridges and the lithosphere symmetrically moves away from the ridges. The surface of the earth is divided into tectonic plates, which move with constant angular velocity with respect to the center of the earth's mass; the oceanic lithosphere is one type of plate. As new and hot oceanic lithosphere moves away from the mid-ocean ridges it cools by diffusion. The oceanic lithosphere acts as a simple thermal boundary layer for a large-scale mode of mantle convection. Because this process is well understood, the resulting topography can be easily predicted and removed from the total topography of the ocean floor. [*See* LITHOSPHERIC PLATES; OCEANIC CRUST.]

Independently, two French research groups have tried to use this approach of stripping away the topography caused by the subsidence of the oceanic lithosphere. From the observed total topography, Anne Cazenave and her colleagues extracted a global residual topography of approximately 300 m amplitude. The residual topography is dominated by a degree two pattern, strongly correlated with the geoid. (By degree two we mean that there are two topography highs and two lows as we move around the equator.) Interestingly, Cazenave *et al.* found that the residual topography signal associated with the continents contributed very little power to the long-wavelength components. Recently, another French group led by Luce Fleitout has been unable to detect such a significant degree two anomaly in residual oceanic topography. But because isostatic topography has a much larger amplitude compared to dynamic topography, such a discrepancy is not surprising and many believe the discrepancy will be resolved soon.

B. Indirect Estimates of Present-Day Dynamic Topography

An indirect method at estimating global dynamic topography, first developed by Bradford H. Hager, is independent of the observed topography. This method uses the observed seismic structure of the mantle and the observed long-wavelength geoid; the former observation is poorly known, while the latter is well known at long wavelengths. Hager developed a viscous earth model wherein both the interior density contrasts and the dynamic topography control the geoid; the flow is driven by density contrasts constrained by seismic observations of the lower mantle. This approach was developed after the observational detection and mapping of deep-mantle density anomalies.

These dynamic geoid models embody the principals mentioned above: Interior density contrasts drive the flow (often referred to as buoyancy-driven flow) and cause areas above less-dense fluid to be pushed upward. The new models have been developed for a spherical viscous shell with radial variations in viscosity. The geoid that results from such a density system is complex, because the gravitational field is not only influenced by the driving density anomaly but also by the dynamic topography since it creates a mass anomaly at the surface. The process is rather complicated because a low-density driving anomaly is a mass deficit and results in a gravity low or a depression in the geoid, but the bowed-up surface above the light fluid is a mass excess and results in a geoid high. The resulting geoid is the difference between these two contributions. Moreover, the sign of the geoid above upwelling fluid depends on the viscosity structure of the medium. In a constant viscosity medium, there are geoid highs above upwellings because the positive geoid anomaly associated with the dynamic topography is larger in magnitude than the negative geoid associated with the low-density fluid driving the flow. Interestingly, the magnitude of the dynamic topography can be reduced by a low-viscosity layer above the mass anomaly so that the resulting total geoid can become negative! This is a primary reason why geodynamicists are interested in dynamic topography and the geoid:

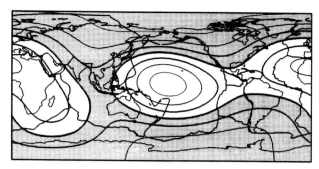

Fig. 1 Dynamic topography calculated from a viscous spherical earth model. The buoyancy forces that drive the flow pushing up the earth's surface are constrained by the lateral seismic structure obtained from seismic tomography. [Reproduced by permission from Hager *et al.* (1985). *Nature (London)* 313, 541. Copyright © 1985 Macmillan Magazines Limited.]

Constraints can be placed on the viscosity structure of the interior.

In 1984, the first maps of the lower mantle were obtained using a seismic tomography method. Large earthquakes send seismic (essentially acoustic) waves through the earth's interior, and these seismic waves speed up and slow down as they travel through the mantle with differing material properties. An important factor influencing the propagation of seismic waves is the temperature of mantle rocks: Seismic velocity is slower in hot solids and is faster in cool solids. By mapping out lateral variations in seismic velocity, seismic waves essentially map out the structure of mantle convection. The pattern of warm and cool regions through the lower mantle has broad hot regions below the western Pacific and also under Africa. The pattern that emerged from the seismic tomography was consistent with the distribution of hotspots. Hotspots are areas of volcanism relatively fixed with respect to the fast-moving lithospheric plates; hotspots give rise to linear island chains such as Hawaii. Most hotspots lie above the long-wavelength distribution of hot, seismically slow, mantle. [*See* SEISMOLOGY, THEORETICAL.]

One of the first models of global dynamic topography is shown in Fig. 1. The flow is driven by observed mantle density anomalies, and the predicted geoid matches the observed geoid. This global-scale dynamic topography has an amplitude of about 600 m. Interestingly the pattern of dynamic topography is similar to the pattern of residual topography that was obtained by removal of oceanic lithosphere subsidence. The notable difference between the direct and the indirect methods of dynamic topography extraction is a factor of two in amplitude; the direct methods have the smaller amplitude.

In 1982, Don L. Anderson showed that the supercontinent Pangea would have been located over the degree two geoid high now centered over Africa. As we can see in Fig. 1, Africa

is also the site of the dominant high-elevation area in dynamic topography. Interestingly, if the present-day continents that fragmented from Pangea had moved with respect to this pattern of dynamic topography, they would have been flooded by seas that were 300 to 600 m deep! Our greatest uncertainty in making such a speculation is that the two methods just discussed do not constrain how dynamic topography and the geoid vary over geologic time.

C. Fluctuations in Dynamic Topography

Mantle convection evolves with time and has complicated time-dependent interactions with other components of the earth system. The two methods to constrain dynamic topography that we just discussed cannot be used to determine how the pattern and amplitude of global dynamic topography has evolved with time. Although the technique is not direct, we can use the sequence of sedimentary rocks laid down onto continent platforms during periods of locally high sea level to place some constraints on time-evolving dynamic topography. The primary problem with using the spatial coverage and temporal variations of marine deposits is that the sedimentary record is blurred by erosion, by subsequent burial, and by metamorphism. This means that we may underestimate both the spatial extent and the amplitude of dynamic topography. [*See* SEDIMENTARY ROCKS.]

One fundamental piece of information that we can immediately extract from the continental record of marine deposits is the maximum amount by which the continents were flooded during the Phanerozoic. The maximum value is not well known but is probably somewhere between 30 and 40%; and these high sea levels occurred during the Ordovician, Devonian–Mississippian, and Cretaceous periods. If we assume that this flooding was caused by the motion of continents over a global pattern of dynamic topography, then we should be able to put an upper bound on the amplitude of dynamic topography. Implicit in the hypothesis of Anderson is that continental fragments moved off of a dynamic topography high and into lows. This phenomenon has been studied with one-dimensional models where continents move from dynamic topography (and geoid) highs to lows and are partially exposed and then flooded. One such model is shown in Fig. 2, where we have a long-wavelength component of dynamic topography with an amplitude of 200 m and a geoid with an amplitude of 100 m. The surface of the continent is shaded, and the sea surface is shown by the horizontal line. Interestingly, the continent is exposed at the geoid high (Fig. 2, top) because the dynamic topography pushes the continent above the sea surface and is more flooded at the geoid low (Fig. 2, bottom). The models tell us that if dynamic topography is greater than about 150 m, the model continent is flooded by more than 30%; if flooding exceeded 45%, dynamic topography must be greater than 250 m. This simple kinematic model suggests that a large-scale pattern of dy-

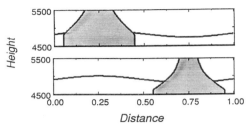

Fig. 2 Total topography and sea surface. The topography of the continent (shaded) is the result of isostatic topography (caused by crustal thickness variations), dynamic topography, and seawater loading. The continent becomes more flooded when positioned over a dynamic topography and geoid low (*bottom*). The original hypsometry, without dynamic topography and water loading, would follow the observed hypsometry of continents. [Reproduced by permission from M. Gurnis (1990). Bounds on global dynamic topography from Phanerozoic flooding of continental platforms. *Nature (London)* **344**, 754–756. Copyright © 1990 Macmillan Magazines Limited.]

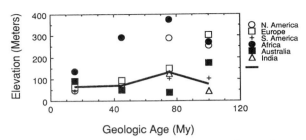

Fig. 3 Relative sea-level elevations that correspond to the intersection of the fraction of flooding with the continent's hypsometric curve. The solid curve is an inferred eustatic sea level. [Modified from Bond (1978).]

namic topography must have an amplitude less than about 250 m; this estimate is consistent with that estimated with the direct methods of extracting dynamic topography, but it is inconsistent with the indirect method using seismic tomography.

A more direct method can be used to extract time variations in dynamic topography. For each continent, we can find the elevation difference between the sea surface and the continent by determining the intersection of the fraction of continent covered by marine deposits with the continent's distribution of elevation versus area. The distribution of elevation versus area is called the hypsometric curve. If over geologic time there were only globally uniform fluctuations in sea level (eustatic fluctuations), then at each geologic time each continent would have the same elevation difference. But in the mid-1970s Gerard Bond showed that the apparent sea level as seen by individual continents differed enormously. Bond applied the flooding–hypsometry technique individually to North America, Africa, South America, Europe, Australia, and India and found that the intersection points with their hypsometric curves (their relative sea level) varied by as much as an inferred eustatic signal (Fig. 3). The large scatter between individual relative sea levels that we see in Fig. 3 requires that the hypsometry of continents have changed. The problem to be resolved is whether these changes in hypsometry are due to changes in crustal thickness or to what geologists have called epeirogenic uplift and subsidence — essentially what earth dynamists call dynamic topography. In general, the sediments were laid down on flat and stable platforms that have not been subject to tectonism — strongly suggesting that the changes in hypsometry were due to changes in dynamic topography. When relative sea level falls

for a given continent, the continent is essentially being uplifted; since the Cretaceous, for example, Africa has risen by about 300 m and its sea level has fallen by about 300 m (Fig. 3). [*See* SEA LEVEL FLUCTUATIONS.]

III. Summary

Only in the last few years has dynamic topography been recognized as an important component of the earth system. Unfortunately, direct detection of dynamic topography, and especially its corroboration by independent methods, has remained elusive. However, various methods suggest that global dynamic topography may be somewhere between 100 m and 1 km. Study of the record of marine deposits on continental platforms suggests that dynamic topography is time dependent. Studies of dynamic topography brings together the disparate fields of marine geology, stratigraphy, seismology, and geodynamic modeling.

Bibliography

Bond, G. (1978). Speculations on real sea-level changes and vertical motions of continents at selected times in the Cretaceous and Tertiary Periods. *Geology* **6**, 247.

Cazenave, A., Souriau, A., and Dominh, K. (1989). "Global coupling of the earth surface topography with hotspots, geoid and mantle heterogeneity. *Nature (London)* **340**, 54.

Fischer, A. G. (1984). The two Phanerozoic supercycles. *In* "Catastrophes and Earth History" (W.A. Berggren and J. A. Van Couving, eds.), pp. 129–150. Princeton Univ. Press, Princeton, New Jersey.

Gurnis, M. (1990). Continental flooding and mantle-lithosphere dynamics. *In* "Glacial Isostasy, Sea-Level, and Mantle Rheology" (R. Sabadini and K. Lambeck, eds.), pp. 445–491. Kluwer Academic Publishers, Dordrecht, The Netherlands.

Hager, B. H., Clayton, R. W., Richards, M. A., Comer, R. P., and Dziewonski, A. M. (1985). Lower mantle heterogeneity, dynamic topography and the geoid. *Nature (London)* **313**, 541.

Glossary

Epeirogeny Large-scale, vertical movement of the crust; this is a gradual process that does not lead to folding or faulting. Usually, one continent is thought of as vertically moving relative to all other continents.

Eustatic change Sea-level change that is global in extent and uniform in magnitude.

Hotspots Sources of volcanism that are relatively fixed with respect to the faster-moving tectonic plates. Indirect evidence suggest that the sources of hotspots may be on the core–mantle boundary.

Lithosphere Upper 50 to 100 km region of the mantle, which has been cooled by conduction and is a layer capable of supporting small crustal loads on geologic time scales. The oceanic lithosphere defines the upper thermal boundary layer of a large-scale mode of mantle convection.

Earthquake Prediction Techniques

Chi-Yu King
U.S. Geological Survey

Various geologic, geophysical, and geochemical techniques are currently being tested for possible use in predicting the time, location, and magnitude of future earthquakes.

I. Introduction

Earthquakes pose a major natural hazard that affects many parts of the world. It is estimated that more than 20,000 lives have been lost to earthquakes on average each year since the beginning of the twentieth century. One possible way to reduce this hazard is earthquake prediction. The possibility that some damaging earthquakes may be predicted has been recognized for centuries, particularly in China and Japan, where many earthquakes were preceded and accompanied by various unusual phenomena in and near the epicentral areas. Such phenomena include changes in the color, taste, and level of groundwater; appearances of mysterious rainbows, fog, and light; and various aberrant behavior of animals. They were used by people, including astrologers, soothsayers, and individual scientists, in attempts to predict earthquakes.

Systematic scientific research on earthquake prediction began only a few decades ago. Large-scale national efforts were initiated in such countries as China, Japan, the United States, and the Soviet Union in response to some recent earthquake disasters. The Soviet program evolved from stud-

ies of seismic activity in Central Asia in the 1950s, using a wide range of geophysical and geochemical methods. Japan initiated its official program in 1965, shortly after the 1964 Niigata earthquake. China mounted a large-scale effort shortly after the 1966 Xingtai earthquake, involving not only professional scientists in various disciplines but also many amateurs. The Chinese have successfully predicted several large earthquakes, including the magnitude 7.3 event in Haicheng in 1975, but failed to predict others, including the magnitude 7.8 event in Tangshan in 1976. In the United States, the research effort was intensified after the 1964 Alaska earthquake, and an official program was authorized in 1977. The U.S. program emphasizes seismologic and crustal-deformation methods, especially along several segments of the San Andreas fault (such as the Parkfield segment in central California, where a moderate earthquake is expected to occur within a few years), and includes a significant effort in laboratory and theoretical model studies.

Recent research findings indicate that a successful prediction strategy requires multidisciplinary fundamental research aimed at a better understanding of the mechanisms of earthquakes and the associated premonitory changes. Also, it is found to be desirable to divide the practice into several stages, such as long-, intermediate-, and short-term predictions.

It is commonly accepted that most earthquakes are generated by elastic rebound, as proposed by H. F. Reid in 1910: When a crustal region is gradually deformed to such a degree that the accumulated stress exceeds the fracture or friction strength of the rock, rupture occurs, and the opposite sides of the fault rebound to a new equilibrium position. During this process, the released strain energy is partly consumed in heating and crushing of rock materials in the slip zone, and partly converted to seismic waves. The distribution of earthquakes in time and space in the earth and the slip directions had not been well understood until the theory of plate tectonics was developed in the 1960s, mainly on the basis of worldwide seismologic and paleomagnetic data. According to this theory, the earth's surface layer (about 100 km thick) is composed of a comparatively few gigantic rigid plates moving relative to one another at a rate of several centimeters per year, possibly driven by convection currents generated by nonuniform distribution of temperature in the deeper interior. These lithospheric plates ride on a weaker layer called the asthenosphere.

Most tectonic deformation and slippage occur in the regions along the plate boundaries, of which there are three kinds. Along divergent or spreading boundaries, the plates move away from each other; along transform or strike-slip boundaries, the plates pass horizontally by each other; and along convergent or subduction boundaries, one plate collides with and may be thrust beneath the other. Although the theory of plate tectonics provides a framework useful for long-term prediction of most large earthquakes along the plate boundaries, a significant number of earthquakes also occur within the plates, such as the 1811–1812 New Madrid earthquakes along the Mississippi River in Missouri. The occurrence of intraplate earthquakes indicates that the plates are not ideally rigid or free of internal faults, and this makes the task of prediction more difficult. [*See* LITHOSPHERIC PLATES; PLATE TECTONICS, CRUSTAL DEFORMATION AND EARTHQUAKES; SEISMOLOGY, THEORETICAL.]

Long-term prediction efforts seek to identify areas along active faults where damaging earthquakes may occur during the next few decades, including estimates of magnitude and probability of occurrence. These estimates, which are based on various geologic, seismic, and geodetic methods, may be used as a guide for other mitigation and preparedness measures. Focused efforts in intermediate- and short-term predictions seek to predict (in addition to magnitude and location) the time of earthquakes to a precision of months to years and days to weeks, respectively. Such predictions are based mainly on instrumental recordings of the various possible premonitory phenomena (such as uplift of land and increase in the radon content of groundwater) of different durations. Although some of these phenomena were observed in some places before certain earthquakes, none has been found to consistently precede every earthquake. The recorded data also commonly contain variations due to many concurrent and commonly unavoidable environmental changes unrelated to earthquakes, such as rainfall, barometric-pressure changes, artificial electromagnetic emissions, and groundwater pumping. Thus, finding precursors that can be universally and reliably used for intermediate- and short-term predictions has been extremely difficult.

This article gives an overview of the geologic, geophysical, and geochemical techniques used in the various stages of earthquake prediction. Other topics related to earthquake-hazard reduction, such as earthquake-caused ground and engineering failures and the socioeconomic impact of predictions, are not discussed.

II. Techniques for Earthquake Prediction

A. Fault Mapping

Because most large earthquakes occur repeatedly along certain segments of active faults, it is important for long-term prediction to determine their locations and to estimate their movements in the past. In some seismic regions, such as California and Japan, active faults have been mapped in much more detail than previously known from the analysis of global seismic data, on which plate tectonics is based. The techniques include remote sensing from aircraft and satellites, measurement of such geomorphologic and anthropogenic features as offset river channels and fences and uplifted coastal terraces, and accurate determination of the hypocenter locations and first motions of small earthquakes using densely deployed regional or local networks of seismic stations. Since the mid-1970s, exploratory trenches have been dug across and along some faults to look for evidence of earthquake-related movements in the past. By measuring the amount of fault offsets at different depths and by dating the offset materials with such techniques as radiocarbon, the recurrence intervals of large earthquakes in the past hundreds or thousands of years can be estimated. For example, the average recurrence interval on the segment of the San Andreas fault in southern California where a magnitude 8 earthquake occurred in 1857 is estimated to be about 150 years since about A.D. 260. [*See* FAULTS.]

B. Pattern of Earthquake Activity in Space and Time

The spatial and temporal distributions of past earthquakes of different sizes recorded historically and by seismic networks (in existence for only about a century), as well as many aspects of earthquake source and near-source characteristics, have been studied in attempts to discover some patterns that might be useful for long- to short-term predictions of future earthquakes. [*See* SEISMIC DATA, RECORDING AND ANALYSIS.]

1. Seismic Gap

One concept that has been found useful for long-term prediction is called the seismic gap: A segment of an active fault where large earthquakes have occurred more than several decades ago, while the adjacent segments are seismically active, is a likely site for such an event in the future (decades). The magnitude of the event may also be estimated from the magnitudes of the past events and from the gap length, on the basis of empirical relations that show increasing rupture length and offset with increasing magnitude. The occurrence time cannot be predicted precisely by this concept because the recurrence intervals of past earthquakes generally vary considerably. However, from known recurrence intervals and the time elapsed since the previous event (together with crustal-deformation data to be discussed later), the probability for occurrence of a future event in the gap during a specified period can be estimated. As an early example, in 1928, A. Imamura expected a disastrous earthquake to occur soon in the Nankaido–Tokaido area of southwestern Japan, because large earthquakes had occurred there every 100 to 150 years during the previous 600 years, and the last one occurred about 75 years earlier. His long-term prediction was fulfilled by the occurrence of two magnitude 8 earthquakes in 1944 and 1946 in this area. Many seismic gaps were identified in the 1960s and 1970s throughout the circum-Pacific region, a number of which have since been ruptured by large earthquakes.

In 1980, K. Shimazaki and T. Nakata noted that, for a seismic gap in Japan, the recurrence interval between two successive large earthquakes was correlated with the coseismic fault displacement (or magnitude) of the preceding earthquake, and so the occurrence time of the following earthquake was better predictable than previously thought possible, but its displacement was not. Since then, this "time-predictable" model has been found applicable to some other seismic gaps. Nonetheless, few gaps have been found to fit a complementary "slip-predictable" model, for which the coseismic displacements are correlated with the preceding recurrence intervals.

2. Earthquake Migration

Another concept potentially useful for long- and intermediate-term prediction is migration of earthquake epicenters along seismic zones, presumably due to stress redistribution caused by interaction of adjacent fault segments or propagation of aseismic movement at subseismogenic depths. Earthquake migration was first recognized for the North Anatolia fault in Turkey: Between 1939 and 1967, earthquakes of magnitude 7 or larger showed a westward migration trend, with most successive rupture zones abutting one another and leaving some seismic gaps filled in by later earthquakes. Since then, the possibility of earthquake migration has been noticed along many seismic zones, including many regions of the circum-Pacific belt and several parts of China. The estimated migration speeds range from 10 to 270 km/yr.

3. Anomalous Changes in Small Earthquake Activity

Some large earthquakes are preceded by changes in the pattern of small earthquakes in or near the epicenter areas. Such changes include a decrease or increase in seismicity, changes in magnitude distribution and source characteristics, and

horizontal or vertical migration of hypocenters. They have been studied for possible application to intermediate- and short-term predictions. K. Mogi recognized in 1969 that before several large earthquakes in Japan, the regions surrounding the future rupture zones were seismically active, while the rupture zones themselves remained quiescent. The appearance of such doughnut-shaped patterns of seismicity has since been used for long- to intermediate-term predictions of some earthquakes. In a typical case, short-term foreshocks tend to occur in the doughnut-hole region, where the main shock also takes place. Aftershocks then fill in the doughnut-hole region, and eventually the entire region returns gradually to a state of broadly distributed seismicity.

The occurrence of foreshocks or the increase of preearthquake seismicity is potentially useful for short-term prediction of the occurrence time of some earthquakes. But foreshocks occur only before some large earthquakes (about 44%, notably the 1975 Haicheng earthquake in China), and they are generally difficult to recognize soon enough to be useful. Several methods have been proposed to distinguish foreshocks from background seismicity, with mixed results. One of these methods involves the magnitude distribution of small earthquakes. Foreshocks have been reported to have distributions favoring larger magnitudes, or smaller b values in the following empirical relation by B. Gutenberg and C. F. Richter:

$$\log N = a - bM,$$

where N is the number of earthquakes of magnitude M or greater, and a and b are empirical constants.

Another group of methods attempt to find differences in earthquake-source characteristics detectable on seismograms. Parameters studied include amplitude ratio between different types of seismic waves, wave spectrum, stress drop, and stress orientation as indicated by first motions. Still other methods attempt to discover some subtle horizontal or vertical migration patterns of foreshocks, such as increasing and then decreasing focal depths. Such studies require a dense seismic network.

4. Anomalous Changes in Source Medium

As discussed below, under increasing tectonic stresses, the rock medium in and near the earthquake-source volume may start to undergo some inelastic changes, such as a volumetric increase called dilatancy due to the creation of an increasing number of small cracks, long before the earthquake occurrence. Several methods have been proposed to detect such changes by analyzing seismic waves generated within or passing through this preparation volume. One of these methods looks for anomalous changes in the velocity ratio between longitudinal and shear waves. This method was tested extensively in several countries during the 1960s and 1970s, but with mixed results. Another method looks for anomalous polarization of shear waves, because most of the expected dilatancy cracks may have a preferred direction and tend to split shear waves into components with different velocities. Recently, seismic coda waves (tail parts of the wave trains of local earthquakes, probably produced by backscattering) have been analyzed to uncover dilatancy-caused changes in amplitude and duration.

5. Application of Pattern-Recognition Techniques

Since the 1960s, pattern-recognition techniques have been applied, especially in the Soviet Union, to identify premonitory features in seismic data. Computer algorithms are formally defined and used to objectively search the data for specific patterns. Some studies also included consideration of geomorphologic characteristics, such as structural lineaments and their intersections, topographic relief, fault distribution, and sedimentary cover. Other types of data can be included in more extensive studies.

C. Crustal-Deformation Measurement

Because earthquakes are sudden mechanical failures resulting from gradual strain accumulation in the earth, it is important to monitor crustal deformation in seismic regions to detect and understand all earthquake-related processes. Many techniques have been developed to measure horizontal and vertical deformation repeatedly (by detecting relative position changes between well-installed and well-located benchmarks) at different intervals in time and space. Some of these techniques are primarily useful for long-term, and others for shorter-term, predictions. To be recognized, premonitory strain changes must be greater than about 10^{-7}, which is the amplitude of background variation due to other ongoing geophysical processes, mainly solid-earth tides. [See CONTINENTAL DEFORMATION.]

1. Geodetic Measurements

Repeated geodetic measurements can detect average horizontal-strain changes to tidal precision over distances of 10 km or more. Before about 1960, geodetic surveys were mostly done with triangulation methods, measuring changes in angle between survey lines to an accuracy of about 5 parts in 10^6. Since then, geodimeters, especially those using lasers as light source, have been increasingly used to detect changes in line length of as much as 10 km or so between the instrument and a reflector to 10^{-6} by measuring travel times of the light beams. The precision has been further improved in some seismic regions by contemporaneously measuring temperature, pressure, and humidity along survey lines by aircraft to correct for the refractivity effect of the air. Re-

peated geodetic measurements may be useful only for long-term to intermediate-term predictions because, being time-consuming and expensive, they are carried out infrequently (about once a year in some seismic regions).

Since the mid-1970s, a special kind of geodimeter with a two-color laser as light source has been used in California. Because of better correction for refractive index, this instrument has a precision of 10^{-7}, or about an order of magnitude higher than that of a single-color instrument. Also, measurement can be carried out more frequently (usually once a day), potentially useful for short-term prediction.

Two space-based geodetic techniques became available in the 1970s, enabling measurement of relative horizontal and vertical position changes of earth stations separated by thousands of kilometers with an accuracy of a few centimeters. The results of these large-scale measurements have largely confirmed our knowledge of plate motions during the past several million years derived from earth-based measurements, and may provide rapid and frequent monitoring of the strain field over a large region. One of these techniques, namely satellite laser ranging (SLR), uses short pulses of light from lasers at a network of ground-based stations toward retroreflectors on the surface of a satellite in orbit around the earth. Relative station locations can be determined by precise timing of the round-trip travel of the pulses and from a precise knowledge of the orbit. The other technique, very long baseline interferometry (VLBI), uses stations equipped with radio antennas to observe a fixed celestial radio source (quasar). As the earth rotates, the varying signals received at the stations can be electronically corrected to determine the delays in arrival times and thus the relative station positions.

A more recently developed technique, global positioning system (GPS), is similar to VLBI in principle but uses orbiting transponders as signal sources and smaller, portable receiver systems on the ground. The GPS technique can measure distances over baselines of 150 km or less with an accuracy of about 1 cm. Because of its portability and lower cost, this technique may be used to monitor movements in a complex region involving many blocks and faults, by frequent measurements over a dense network of stations.

2. Leveling Surveys

Vertical ground movement can be measured by repeated leveling and gravity surveys and by monitoring sea and lake levels. A leveling survey is carried out at many bench marks along a route by sequentially reading vertical scales (made of some low-thermal expansion materials) placed between the marks with a precisely leveled telescope from the marks in both forward and backward directions. The relative heights of any two adjacent marks are obtained by subtracting the readings of the same scale from the marks. Leveling surveys are time-consuming and expensive. The random errors due to observational uncertainties are relatively small, but systematic errors due to unusual refraction associated with elevation differences and to rod miscalibration can be significant and difficult to detect. For surveys along a long route, it is desirable that the route form a closed loop and the data be checked against some other data, such as by tide gauges if in a coastal area. The measurement error is about 1 mm in 1 km.

3. Strainmeters and Tiltmeters

In addition to the above-mentioned techniques for large-scale but infrequent measurements of crustal deformation, various types of strainmeters and tiltmeters are used for frequent or continuous measurements of local deformation at individual stations. An ordinary strainmeter is an extensometer, measuring extension and contraction of the ground against a length standard that may be a tube, rod, or wire of a low-thermal-expansion material, such as quartz or invar. One end of the standard is fixed to a pier in the ground; the other is free to move relative to a reference point on another pier. The relative movement is sensed with some device, such as an optical lever, or a variable inductance or capacitance transducer. Strainmeters of this kind are commonly used in a group of three arranged in different directions at a station to determine the strains completely in a horizontal plane. To reduce noise, they are ordinarily setup in underground vaults, where temperature variation is small. The stability of such meters is about 10^{-6} per year, being affected by thermoelastic strains in the surface materials and by post-installation instrument settling. Recently, a new type of strainmeters using laser interferometry has been developed; it has an intrinsic stability of 10^{-12} in an ideal environment. Another type of strainmeter sensitive to volume changes was developed by I. S. Sacks and D. W. Evertson. It consists of a stainless-steel cylinder, 4 m long and 11 cm in diameter, filled with degassed silicon oil and ordinarily buried in a deep borehole in such a way that it is bonded to the surrounding rock by a cement that expands after solidification. The thickness of the cylinder wall is such that the apparent cylinder rigidity approximately matches that of the rock. When the cylinder volume is compressed by the rock, the oil is pushed through a fine pipe at the cylinder top to an expansion bellows. The expansion of this bellows is then measured with a sensor, such as a differential transformer.

Various types of tiltmeters have been used to continuously record small local tilt of the ground surface to a precision of a few microradians. For example, the borehole tiltmeter is basically a weight hung with a fine string in a borehole. A small ground tilt causes a change in the position of the weight relative to the wall, which can be measured with a variable-inductance transducer. The horizontal-pendulum tiltmeter basically consists of a pendulum with a weight oscillating slowly about an almost-vertical axis in an almost-horizontal plane. A small tilt change of the ground and the axis causes a relatively large change in the angle of oscillation, which change is used as a sensitive measure of the tilt of the ground. The water (or some other fluid such as mercury) tube tiltmeter consists of two water-filled pots fixed on piers and interconnected with a pipe. The ground tilt is sensed by height changes of some reference marks in the pots relative to the water surface. The bubble tiltmeter has a bubble under a concave quartz lens, floating on the surface of a conducting electrolyte in a container. The position of the bubble is sensed by two ac bridge circuits and is used to measure two tilt components.

Strainmeters and tiltmeters need to be stably attached to the ground to eliminate spurious signals. Those installed near the ground surface are affected by many meteorologic and geologic factors. Those placed in mines and tunnels are more expensive and subject to "cavity" effects. Data recorded by short-base meters may be only local and not representative of the regional deformation related to earthquakes.

4. Stress Measurement

Since 1980, information about tectonic-stress distributions has been obtained for many regions of the world from various sources, such as seismic fault-plane solutions, young geologic data on fault slip and volcanic alignment, and *in situ* measurements by different techniques. For example, the hydraulic-fracture technique involves drilling a borehole and then hydraulically inducing a tensile fracture in a section of the hole; the least compressive stress and its direction can be obtained from the maximum fluid pressure and the location of the fracture, respectively. The overcoring technique involves installing strain gauges at the bottom of a borehole in a competent rock unit and then overcoring it; stress information can be inferred from the gauge readings. Information can also be obtained from the wellbore breakouts that occur naturally in many boreholes when the compressive hoop stress exceeds the strength of the rock.

5. Creepmeters and Alignment Arrays

Along certain fault segments, displacement occurs not only suddenly during earthquakes but also slowly without generating earthquakes. This type of movement, called fault creep, has been monitored continuously with creepmeters and intermittently by surveying marks along lines across the fault (alignment arrays) since the mid-1960s in several regions. Some moderate earthquakes in California were found to be preceded by changes in creep rate for periods of months in the respective epicenter areas, suggesting the possible usefulness of creep measurements for intermediate-term prediction. Note that similar movements may occur at subseismogenic depths, where the material is more ductile. Most creepmeters in California are extensometers (like some of the strainmeters described above), spanning a fault trace at approximately a 45° angle to measure strike-slip motion.

D. Hydrologic and Geochemical Monitoring

The earth's crust contains numerous pores and fractures filled with water, gas, and other fluids that have different chemical compositions and can move under changing tectonic stress, resulting in hydrologic and chemical changes at affected locations. Such changes, which were first observed a long time ago, have been instrumentally monitored in search of earthquake precursors since the 1960s in many seismic regions. Parameters studied include water level at wells, flow rate at springs, concentration of various ions and dissolved gases, and components of soil gas in shallow holes. Intermediate- to short-term premonitory changes have reportedly been observed before some earthquakes. They tended to be widely distributed in epicentral distance (hundreds of kilometers for large events) and to cluster along active faults, possibly because of strain concentration and higher fluid content in the fault zones. However, like all other measurements made near the ground surface, noise caused by other environmental variables needs to be distinguished from precursors.

1. Water Level

Water levels in wells (preferably deep or tapping confined aquifers) can be measured simply with a tape or monitored continuously with meters that use floats or pressure transducers. The water level at each well responds in its own way to barometric pressure and tidal-stress changes, which, if known, can be used to calibrate the well's response to crustal strain changes. A calibrated well may thus be considered a type of strainmeter, and the data recorded there can be digitally filtered to search for anomalous changes of a few centimeters in amplitude.

2. Geochemical Monitoring

Geochemical measurements may contribute to earthquake prediction in two general ways: locating active faults and showing premonitory changes. Many active fault zones, even some that are not obvious at the surface, are characterized by higher concentrations of various terrestrial gases in ground water and soil air, including radon, helium, hydrogen, mercury vapor, and carbon dioxide. This observation suggests that such faults, like volcanoes, may be paths of less resistance for gases at depth to escape to the atmosphere.

The possibility that radon content may show premonitory changes was noted as early as 1927 in Japan. However, substantial research did not begin until such changes were detected before the 1966 Tashkent earthquake in the Soviet Union. Currently, some 40 geochemical parameters are being monitored *in situ* or by taking water and gas samples for laboratory analysis in many seismic regions, especially the Soviet Union, China, Japan, and the United States. These parameters include temperature, electric conductivity, isotopic ratios of hydrogen and oxygen, and concentrations of various ions and gases. Many anomalous changes of different durations (hours to years) have been recorded before some large and moderate earthquakes, mostly at sites along faults at distances of as much as hundreds of kilometers, suggesting their usefulness for intermediate- and short-term predictions. Care must be taken, however, to distinguish true anomalies from noise caused by other environmental variables. Also, the anomaly and earthquake occurrences do not appear to show one-to-one correspondence. Most of these anomalies appear to result from some local processes in response to a broad-scale stress variation associated with the corresponding earthquakes. Several mechanisms have been proposed to explain the various observed anomalies; they include increased upward flow of deep-seated fluids to the monitored aquifers or shallow holes, squeezing of gas-rich pore fluids out of rock matrix into the aquifers, mixing of water from other aquifers through tectonically created cracks in the intervenient barriers, increased water/rock interactions, and increased gas emanation from newly created crack surfaces in the rock to the pore fluids. Some of these possible explanations have been tested by crushing rock samples in the laboratory and by analyzing similar changes resulting from such field operations as underground explosions and groundwater pumping.

E. Geoelectric and Geomagnetic Measurements

Anomalous electromagnetic changes observed before earthquakes include those in resistivity, telluric current, geomagnetic field, and electromagnetic emission. Care must be taken to exclude changes caused by other geophysical and anthropogenic processes, such as lightning, meteors, and arcing power lines. Possible resistivity anomalies commonly are recorded at large distances (as far as hundreds of kilometers from earthquakes of magnitude 7–8) with unexpectedly large amplitudes (e.g., 10^{-4} where the strain is calculated to be 10^{-8}). Telluric currents are generally observed by measuring the electric potential between two electrodes buried in the ground separated by hundreds of meters. Because these currents are affected by geomagnetic variations of extraterrestrial origin, it is difficult to attribute any changes to earthquakes.

The geomagnetic field may change locally by as much as 10^{-4} in total intensity or in the ratio of different components as a result of crustal-stress changes because of the piezomagnetic effect of magnetized rocks. Such changes may be detected by repeated surveys or continuous recordings, and by comparing the observed values near earthquake epicenters with those at a remote (hundreds of kilometers away) standard observatory to approximately eliminate background variations. In repeated surveys, it is important to reoccupy observation points in exactly the same way each time to minimize errors. Since about 1960, data have been reliably recorded with proton-precession magnetometers, which measure absolute magnetic-field intensity by counting the frequency of free precession of protons in water. Because they rely on an atomic constant (magnetic moment of a proton), such instruments are drift free and unaffected by environmental changes.

Unusual electromagnetic emissions in various frequency bands up to many kilohertz have been observed hours to weeks before some moderate and large earthquakes at epicentral distances as far as several hundred kilometers. Such signals were detected by some specially designed instruments, as well as by regularly used radio-communication receivers and household radios. Possible mechanisms include charge separation and recombination produced by fractures of rock materials, piezoelectric effects of quartz-containing rocks, and streaming potentials produced by groundwater movement.

III. Mechanisms of Earthquake Precursors

Many laboratory and theoretical studies have been conducted for the purpose of understanding the mechanisms by which earthquakes and their various precursors are generated. Most of these studies involve the prefailure phenomena of dilatancy and stable sliding, as observed in many soil and rock specimens. A laboratory specimen under increasing compressive uniaxial or triaxial stress commonly begins to show nonelastic increase in volume (dilatancy) at a stress level of only a fraction of the corresponding failure stress, because of the increasing number and size of microcracks created in the specimen. The same phenomena presumably occur at earthquake sources and may be responsible for the generation of the various geophysical and geochemical precursors.

The return to normal values may be explained in two different ways: flow of ground water into the cracks, weakening the rocks and leading to earthquakes; and concentration of crack growth along the eventual faulting zone, reducing stress and crack activity on either side of the fault where measurements are made. To account for the occurrence of precursors at large epicentral distances and at only certain locations, the inhomogeneity of the crust must be considered and a broad-scale but low-amplitude seismotectonic-deformation field must be invoked. Under such circumstances, strain may concentrate along some weak zones, such as active

faults, and local dilatancies may occur only at locations where the augmented strained are sufficiently high.

IV. Conclusion

Scientific research during the past few decades has made considerable progress in developing techniques to identify the probable locations of large future earthquakes and to estimate their magnitudes. However, to estimate the time of occurrence precisely enough to be useful for intermediate and short-term predictions remains a challenge because few, if any, geophysical or geochemical parameters are found to have shown repeatable premonitory changes. The difficulties include the noise inevitably caused by many environmental variables, the probably intrinsic absence of one-to-one correspondence between earthquake and precursor occurrences, the lack of long-term stability or time resolution for the monitoring equipment, and insufficient speed in data analysis and synthesis. To overcome such difficulties, a prediction effort should adopt a multistage, multisite, and multidisciplinary approach; use telemetered instruments of good long-term stability and fine time resolution; and deploy them at sensitive locations long enough to recognize normal background noises with the help of appropriate statistical methods for data analysis.

Bibliography

Asada, T. (1982). "Earthquake Prediction Techniques." Univ. of Tokyo Press, Tokyo.

Guha, S. K., and Patwardhan, A. M., eds. (1986). "Earthquake Prediction, Present Status." Univ. of Poona, Pune, India.

Isikara, A., and Vogel, A., eds. (1982). "Multidisciplinary Approach to Earthquake Prediction." Friedr. Vieweg and Sohn, Braunschweig/Wiesbaden, Germany.

Mogi, K. (1985). "Earthquake Prediction." Academic Press, Tokyo.

Rikitake, T., ed. (1983). "Current Research in Earthquake Prediction." Center for Academic Publications, Tokyo, and D. Reidel Publishing Company, Dordrecht, The Netherlands.

Shimazaki, K., and Stuart, W. D., eds. (1985). "Earthquake Prediction." Birkhauser Verlag, Basel, Switzerland.

Simpson, D. W., and Richards, P. G., eds. (1981). "Earthquake Prediction—An International Review." American Geophysical Union, Washington, D.C.

Stuart, W. D., and Aki, K., eds. (1988). "Intermediate-Term Earthquake Prediction." Birkhauser Verlag, Basel, Switzerland.

UNESCO (1984). "Earthquake Prediction—Proceedings of the 1979 International Symposium, Paris." Terra Scientific Publishing Company, Tokyo.

Vogel, A., and Brandes, K., eds. (1988). "Earthquake Prognostics." Friedr. Vieweg and Sohn, Braunschweig/Wiesbaden, Germany.

Glossary

Dilatancy Inelastic volume increase caused by the occurrence of small cracks in rock or soil under stress.

Epicenter Point on the earth's surface directly above the focus of an earthquake.

Fault Large fracture in the ground with two sides displaced relative to each other in directions parallel to the fracture.

Geodetic Pertaining to measurement of the shape and dimensions of the earth.

Geomorphology Branch of geology that studies landforms.

Seismic Pertaining to earthquake activity.

Tectonic Pertaining to the deformation of the earth's crust.

Telluric currents Natural electric currents that flow in the ground.

Plate 1   (*Top*) During El Niño of 1983 in the Pacific Ocean, the trade winds over the Atlantic were unusually strong, and La Niña conditions prevailed in that ocean. The presence of a tongue of cold surface waters in the Atlantic and its absence in the Pacific are evident in the satellite-measured sea-surface temperatures shown here. (*Bottom*) In 1984 the conditions were reversed from those in the top photo, with the Atlantic experiencing El Niño and the Pacific experiencing La Niña. (Courtesy of O. Brown, University of Miami, and R. Legeckis, NOAA.)
(Citation appears in the article EL NIÑO AND LA NIÑA.)

El Niño and La Niña

S. George Philander
Princeton University

The Southern Oscillation is an irregular interannual climatic fluctuation between El Niño (a period of weak trade winds and warm, wet conditions in the eastern tropical Pacific) and La Niña (a complementary period of intense trade winds and dry, cold conditions).

I. Introduction

Although the coastal zone of Peru is a barren desert, the adjacent cold waters of the Pacific are one of the most productive regions of the world ocean, teeming with fish and other forms of marine life that support a huge bird population. A warm southward current moderates the low surface temperatures during the early months of the calendar year. The current was named *El Niño* (the child Jesus) because it appears around Christmas. Every few years it is more intense than normal, penetrates unusually far south, is exceptionally warm, and is accompanied by very heavy rains. Such years were known as *años de abundancia,* or years of abundance, when, according to an early observer, "the sea is full of wonders, the land even more so. First of all the desert becomes a garden. . . . The soil is soaked by the heavy downpour, and within a few weeks the whole country is covered by abundant pasture. The natural increase of flocks is practically doubled and cotton can be grown in places where in other years vegetation seems impossible."[1] The wonders in the sea can sometimes include long yellow-and-black water snakes,

[1] R. C. Murphy (1926). Oceanic and climatic phenomena along the west coast of South America during 1925. *Geogr. Rev.* **16,** 26–54.

bananas, and coconuts carried southward by the warm current from the coastal rain forests farther north. At the same time, however, the birds and marine life that are usually abundant disappear temporarily.

Not until the 1960s did oceanographers realize that the warm surface water along the coast of Peru during *años de abundancia* extends thousands of kilometers offshore and is but one aspect of unusual conditions throughout the tropical Pacific Ocean. These conditions arise from changes in the circulation of the entire ocean basin, in response to changes in the surface winds that drive the ocean. To understand El Niño it is necessary to explain how the ocean adjusts to the changes in the surface winds. [*See* OCEAN CIRCULATION.]

Efforts to describe the changes in surface winds and more generally to document interannual variations in the circulation of the tropical and global atmosphere started toward the end of the nineteenth century. The motivation was not El Niño but the occasional failure of the monsoons. Sir Gilbert Walker, who became director general of observatories in India in 1904, was probably unaware that the catastrophic Indian droughts in 1877 and 1899, which caused severe famine, coincided with *años de abundancia* along the coast of Peru, but he realized that the monsoons are part of a global phenomenon. He set out to document its full scope, and by the 1930s had demonstrated the existence of an irregular interannual fluctuation, which he named the Southern Oscillation, involving major changes in the rainfall patterns and wind fields over the tropical Pacific and Indian oceans. [*See* ATMOSPHERIC CIRCULATION SYSTEMS.]

Figure 1 illustrates the remarkably coherent low-frequency variations associated with the Southern Oscillation. Walker was unable to provide theoretical support for the statistical relations he discovered among the various parameters, probably because data about sea-surface temperatures like those in Fig. 1 were unavailable. Only in the 1960s were such data sufficiently plentiful to permit Jacob Bjerknes to conclude that large-scale changes in sea-surface temperatures in the tropical Pacific cause the Southern Oscillation. He pointed out that *años de abundancia* coincide with one phase of the Southern Oscillation, when sea-surface temperatures are high, the trade winds are weak, and the differences in surface pressure across the tropical Pacific are small. The term El Niño is now reserved for these interannual occurrences.

Not only has our use of the term changed, but our view of

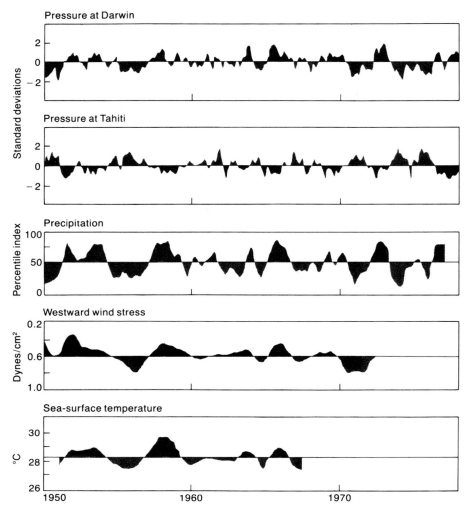

Fig. 1 Beginning with the observation of an annual warming of normally cold Pacific Ocean waters off the coast of Peru, meteorologists and oceanographers have gradually delineated a vast system of ocean–atmosphere exchanges covering the tropics. The complementary fluctuations in atmospheric pressure at Darwin, Australia and Tahiti illustrate the so-called Southern Oscillation, which is closely related to variations in a number of other atmospheric and oceanic parameters. El Niño is a period of high pressure near Darwin, low pressure near Tahiti, high rainfall near the date line at Ocean and Nauru islands (5°S, 167°E), weak trade winds as sampled near the equator west of the date line, and high sea-surface temperatures at Canton Island (3°S, 171°W). During La Niña, tendencies are just the opposite. The data have been filtered to eliminate fluctuations with time scales of a few weeks or less. Units for pressure are standard deviations; precipitation is given in a percentile index.

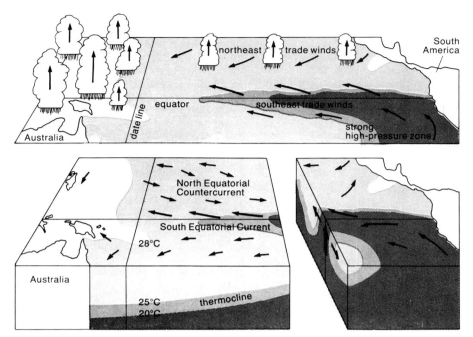

Fig. 2   A schematic view of the tropical Pacific shows atmospheric conditions *(top)* and oceanic conditions *(bottom)* toward the end of a year in which La Niña occurs. Exceptionally intense trade winds come together in the Intertropical Convergence Zone (which is relatively far north) and the huge convective zone to the west of the date line, where sea-surface temperatures exceed 28°C. The eastward surface current between 3°N and 10°N, which is known as the North Equatorial Countercurrent because it flows counter to the prevailing winds, is relatively weak. The westward South Equatorial Current is extremely strong, especially near the equator, where divergent motion causes intense upwelling and hence low sea-surface temperatures. The thermocline, the layer of large vertical temperature gradients that separates the warm surface waters from cold water at depth, slopes steeply to the west, where its depth is approximately 150 m.

El Niño has become negative. El Niño is now associated primarily with ecological and economic disasters, which include devastating droughts over the western tropical Pacific, torrential floods over the eastern tropical Pacific, and unusual weather patterns over various parts of the world. This phase of the irregular Southern Oscillation precedes and follows generally benign periods, when oceanic and atmospheric conditions are complementary to those during El Niño. The term *La Niña* (the girl) is apposite for this other phase, when sea-surface temperatures in the central and eastern Pacific are unusually low and the trade winds are very strong. La Niña has recently attracted attention because the occurrence of exceptionally intense La Niña conditions in the tropical Pacific Ocean during 1988 may have contributed to the drought over North America that summer. Figures 2 and 3 contrast the conditions that prevail under La Niña and El Niño. [*See* DROUGHT; FLOODS.]

Interannual variations in sea-surface temperatures in the tropical Pacific cause the Southern Oscillation. But from an oceanographic point of view the variations are themselves caused by the surface wind fluctuations associated with the Southern Oscillation. From these circular arguments Bjerknes inferred that interactions between the ocean and atmosphere are at the heart of the Southern Oscillation. He described how an initial change in the ocean could affect the atmosphere in such a manner that the altered meteorological conditions would in turn induce further oceanic changes to reinforce the initial one. For example, a slight relaxation of the trade winds, which drive the warm surface waters westward and expose cold water to the surface in the east, can cause a modest warming of the central and eastern tropical Pacific. This in turn can cause a further relaxation of the winds and further warming, so that El Niño gradually develops. These arguments can also be reversed to explain the evolution of La Niña. Bjerknes envisioned "a never-ending succession of alternating trends by air–sea interactions in the equatorial belt." [*See* AIR–SEA INTERACTION.]

Attempts to quantify these air–sea interactions received a

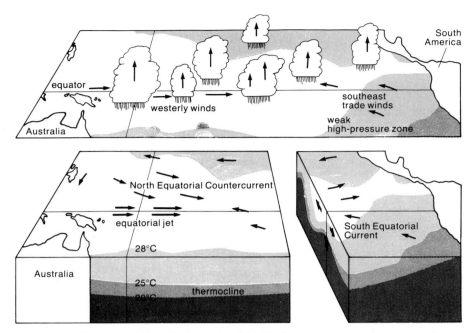

Fig. 3 Conditions toward the end of a year in which El Niño occurs are just the opposite of those prevailing during La Niña (Fig. 2). The conditions depicted here correspond to a very intense El Niño such as that of 1982. The trade winds have collapsed, to be replaced in the west by westerly winds. The eastward movement of the convective zone is associated with an eastward expansion of warm surface waters, a thermocline that is elevated in the west and depressed in the east, an intensified eastward North Equatorial Countercurrent, a weakened westward South Equatorial Current that is replaced by an eastward equatorial jet in the west, and very weak equatorial upwelling.

strong impetus from the extremely severe El Niño of 1982–1983. That event caught oceanographers and meteorologists by surprise; even in late 1982 nobody was aware that the most severe episode of the past century was occurring at that moment. By the time of the next El Niño in 1987, however, the National Meteorological Center in Washington was issuing a monthly *Climate Diagnostics Bulletin* describing in detail Southern Oscillation conditions in the ocean and atmosphere during the previous month. Not only was it possible to follow the erratic development of El Niño of 1987 as it occurred, but models devised since 1982 succeeded in predicting the event.

The models are still relatively crude. They isolate some of the physical processes that cause the Southern Oscillation and thus provide a better understanding of this phenomenon. They can also be used to anticipate whether or not a warm event is imminent, but they are less successful at predicting how it will evolve or what amplitude it will attain. Sophisticated numerical models capable of realistic solutions are currently under development and should be available soon for routine predictions. [*See* ATMOSPHERIC MODELING.]

II. The Southern Oscillation

Atmospheric motion in the tropics corresponds to direct thermal circulations in which moisture-laden air converges on the warmest regions of the earth's surface, where rising motion causes condensation in towering cumulonimbus clouds that extend through the entire troposphere to a height of approximately 15 km. These convective zones, which are characterized by heavy precipitation and low surface pressures, are shown in Fig. 4. In the upper troposphere the air diverges from the convective zones and, drained of moisture, subsides over regions with minimal rainfall, high surface pressures, and low surface temperatures. While subsiding, the air radiates heat to space. The tropical atmosphere can therefore be viewed as a heat engine that operates between the relatively high temperature of the ocean surface, where latent heat is gained, and the low temperature of the upper troposphere and lower stratosphere, where heat is lost. [*See* CLOUD DYNAMICS.]

The convective zones move seasonally in response to changes in the surface temperatures, as shown in Fig. 4. These

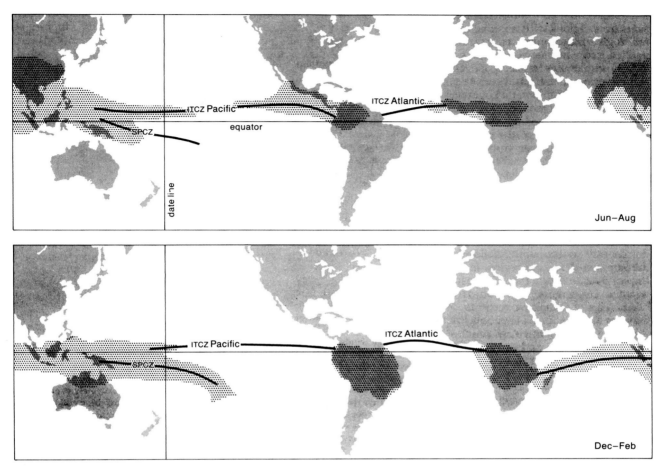

Fig. 4   Atmospheric motion in the tropics corresponds to an immense thermal circulation. Moisture-laden air converges on the warmest regions of the earth's surface, rises, and condenses in towering cumulonimbus clouds. This map shows selected satellite readings of outgoing long-wave radiation, which indicates the extent of cloud cover. A reading of less than 240 W/m² implies widespread coverage by high, cold clouds, upward motion, and heavy convective rainfall. The tropical regions with such readings *(shaded areas)* are predominantly in the hemisphere where summer is occurring, the Northern Hemisphere from June through August *(top)* and the Southern Hemisphere from December through February *(bottom)*. The most extensive area is found over the Australian monsoon region, with an eastward extension along the Intertropical Convergence Zone (ITCZ) north of the equator and a southeastward extension along the South Pacific Convergence Zone (SPCZ) toward the Southwest Pacific. During El Niño, an even more pronounced eastward shift of convective zones occurs. [After Rasmusson (1985).]

movements are predominantly north–south, following the sun, so that each region in the tropics has a distinct rainy season when a convergence zone is overhead. The seasonal movements of the convergence zones, especially the one over the "maritime continent" of the western Pacific and southeastern Asia, can have a large east–west component too. This is in response to changes in the area covered by water warmer than approximately 27.5°C, which is the favored area for convection over the oceans. Figures 2 and 3 illustrate how

sea-surface temperature patterns can change between El Niño and La Niña. The eastward expansion of the area of warm surface water during El Niño results in an eastward drift of the convective zone of the western tropical Pacific. This causes the eastern tropical Pacific to experience an increase in rainfall, a decrease in surface pressure, and a relaxation of the trade winds, while to the west of the date line rainfall decreases and surface pressure increases.

Figure 5 depicts the development of these anomalous con-

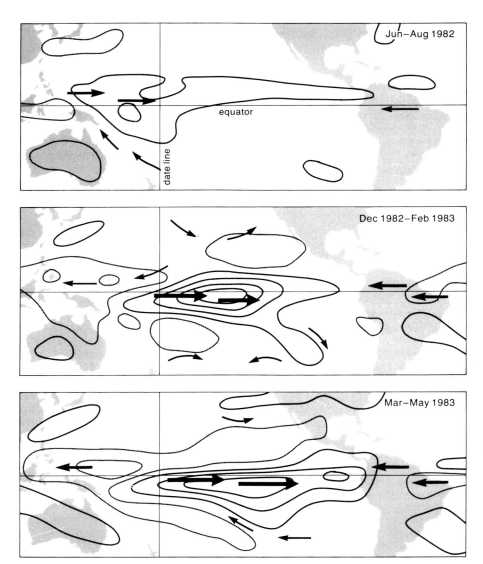

Fig. 5   In 1982 El Niño conditions first appeared to the west of the date line and gradually expanded eastward. The arrows, which indicate winds near the ocean surface, show how westerly winds penetrated farther and farther eastward as the trade winds over the Pacific collapsed. The trades intensified over South America and the Atlantic Ocean. The contours represent anomalous atmospheric convection (outgoing long-wave radiation at intervals of 20 W/m²). Convection (and precipitation) is enhanced in regions with darker contours and is suppressed in regions with lighter contours. [After Rasmusson and Wallace (1983).]

ditions during El Niño of 1982–1983. That event had an exceptional amplitude, but, as is evident from the data in Fig. 1, similar conditions characterize all El Niño episodes. During cold La Niña events, the area of warm surface waters contracts toward the western tropical Pacific, so that heavy rainfall and low surface pressures are confined to that region. The trade winds are intense during such periods. The hypothesis that changes in sea-surface temperatures influence the location of atmospheric convective zones clearly provides a persuasive explanation for the high correlations evident in Fig. 1 and hence for the Southern Oscillation.

Why does large-scale atmospheric convection occur over the warmest surface waters? Several reasons have been proposed. One possibility is that high evaporation rates over very warm regions give rise to local condensation of water vapor and a local heating of the atmosphere. This causes rising motion and hence convergent low-level winds. The moisture carried by these winds is fuel that intensifies the heating, thus causing even stronger low-level convergence. Changes in the thermal structure of the atmosphere limit this positive feedback, for which several models exist. A second possibility is that convection occurs preferably over the warmest waters because the vertical stability of the atmosphere, which depends on its moisture content, is minimal in such regions. A third possibility is that if the heating of the atmosphere in a convective region takes place at too high an altitude, it is unlikely to influence the winds immediately above the ocean surface. It has been proposed that, because of mixing processes in the lower atmosphere, sea-surface temperature gradients give rise to atmospheric pressure gradients that drive motion above the ocean surface. The resultant winds converge on regions with maximum sea-surface temperature.

Different simple dynamical models based on the three different mechanisms that relate atmospheric convergence to sea-surface temperatures give approximately the same surface winds in response to a given temperature anomaly. At this time it is not possible to decide which of the three mechanisms is dominant in the atmosphere; it is conceivable that each of the three plays a role.

The most realistic and sophisticated models of the atmosphere, known as general circulation models, solve the equations that govern atmospheric motion. It is not possible to resolve all the scales of motion, however, so that small-scale phenomena, such as clouds and mixing processes, are parametrically related to resolved fields. These parameterizations are reasonably successful because the models are capable of realistic simulations of the earth's weather and climate. Of special interest is a set of calculations that focuses on atmospheric variability over a 15-year period. When climatological sea-surface temperatures are specified as a lower boundary condition, so that the temperature for each January is the same and similarly for the other months, the model reproduces a realistic seasonal cycle but has no Southern Oscilla-

tion. When the sea-surface temperatures observed in the tropical Pacific between 1961 and 1976 are specified, the model reproduces the Southern Oscillation as observed during that period. Only the seasonal and interannual variations and not the superimposed high-frequency fluctuations are simulated. These results imply that if future sea-surface temperatures were known, it would be possible to make perfect predictions of El Niño and La Niña.

III. How the Ocean Adjusts

To explain and simulate variations in sea-surface temperatures such as those shown in Fig. 1, it is necessary to take into account that the patterns reflect subsurface thermal gradients. Along the equator, for example, decreasing sea-surface temperatures from west to east are associated with isotherms that slope upward as shown in Fig. 2. The slope of the thermocline, the layer of large vertical temperature gradients that separates the warm surface water from the cold water at depth, is maintained by the trade winds, which drive the warm water westward while exposing cold water to the surface in the east. If the winds relax, as they do during El Niño, some of the warm water flows back eastward, so that the thermocline rises in the west and falls in the east, as is evident in Fig. 3. Sea level, which is a measure of the vertically integrated temperature of the water column, therefore falls in the west while it rises in the east. During El Niño episodes, the warm eastward surface current between 3° and 10°N intensifies, while the westward currents weaken and, near the equator, sometimes reverse direction. Equatorial upwelling is reduced during such periods, but it intensifies during La Niña, when the westward current strengthens and the eastward surface current weakens.

The seasonal reversal of the Somali Current in response to the reversal of the monsoons over the Indian Ocean was the basis of a seminal theoretical study of an ocean's adjustment to changes in the wind. Studies over the past two decades of the similarities and differences among the three tropical oceans have contributed to considerable progress in our understanding of the dynamics of the oceans in low latitudes. The Atlantic and Pacific oceans are forced by winds that are similar, but their geometries are very different. The Indian and Atlantic oceans have similar dimensions, but the prevailing winds are very different.

The abrupt changes in the winds over the Indian Ocean come closest to the much-studied idealized case in which spatially uniform winds suddenly start to blow over an ocean initially at rest. At first the motion is independent of longitude, and the winds drive accelerating currents. This acceleration has stopped by the time a state of equilibrium is reached, and the winds then maintain zonal pressure gradients associated with sloping isotherms. The gradients are established by wave fronts (or bores) that propagate across the ocean basin after being excited initially when the winds started to blow. The waves are therefore of central importance in the ocean's adjustment to a change in the winds.

If the ocean were a nonrotating basin of fluid of constant density, the relevant waves would be surface gravity waves. The ocean, however, is a rotating spherical shell of stratified fluid, and the waves have complex properties. For example, their speeds increase rapidly with decreasing latitude, and some are trapped near the equator, which is a waveguide. The dominant waves within a few hundred kilometers of the equator travel from west to east, taking approximately three months to cross the Pacific Ocean. The slower off-equatorial waves propagate from east to west. Since all these waves affect the adjustment to a state of equilibrium, the time it takes to reach such a state is far shorter in low than in high latitudes. Generating the Gulf Stream in an ocean initially at rest would take several years, but generating the Somali Current takes a matter of weeks. Large-scale changes in the oceanic circulation such as those shown in Figs. 2 and 3 can occur on far shorter time scales in low than in high latitudes.

Oceanographers have developed a hierarchy of models with which to study oceanic adjustment. The simpler models clarify the roles various physical processes play in a certain phenomenon and are used to analyze results from more complex and realistic models. Sophisticated general circulation models are capable of realistic simulations of low-frequency variability associated with the Southern Oscillation. One such model, which is forced with the observed winds, provides a detailed description of the surface and subsurface thermal fields and currents of the tropical Pacific Ocean. The description, which is published each month, is impossible on the basis of measurements only — they are too sparse — and so data generated by the model supplement them. If future wind fluctuations were known, this model could be used to predict El Niño and La Niña very accurately.

IV. Interactions between Ocean and Atmosphere

The development of El Niño of 1982–1983, shown in Fig. 5, started with the appearance of anomalies in westerly winds over the far western tropical Pacific Ocean. This caused an eastward expansion of the pool of warm surface waters, which in turn caused an eastward migration of the convective zone and of the wind anomalies. In response, the warm surface waters progressed even farther eastward. This positive feedback between the ocean and atmosphere, which is confined to the tropics, where, as we have seen, the ocean adjusts rapidly to changes in the winds, not only characterizes El

Niño of 1982–1983 but more generally causes the Southern Oscillation.

The interactions between the ocean and atmosphere support gradually amplifying oscillations, with a period near three years, between warm El Niño and cold La Niña states. Analyses with simple couple ocean–atmosphere models reveal that a variety of modes of oscillations exists: Some have eastward phase propagation similar to that observed in 1982, some have westward phase propagation, and some are stationary. In the case of a mode with eastward phase propagation, modest El Niño conditions appear in the west and gradually expand eastward so that westerly winds begin to prevail. At a certain stage in this evolution, modest easterly wind anomalies appear in the west, as shown in the bottom panel of Fig. 5. These are the seeds that develop into the eastward-expanding La Niña that succeeds El Niño. The speed of propagation of a mode, which is determined by air–sea interactions, is usually slower than that of the oceanic waves mentioned earlier. The oceanic waves, although they are of central importance, are not explicitly evident in the air–sea modes. The exception is a certain type of stationary mode in which oceanic waves that reflect off the western boundary of the ocean basin can play a critical role in the turnabout from El Niño to La Niña.

The properties of a mode — its direction of propagation, for example — depend on the processes that determine variations in sea-surface temperatures. If changes in the rate of upwelling of cold water dominate surface temperature variations, there is eastward phase propagation. If horizontal advection is dominant, the direction of propagation depends on the sign of surface temperature gradients and can be westward. Sea-surface temperature is the most complex of oceanic parameters, and the processes that determine it vary with space and time, so that each El Niño and La Niña episode tends to develop in a different manner. In 1982, El Niño conditions first appeared in the western tropical Pacific and then expanded eastward. In 1972, unusually warm surface water appeared off the coast of Peru, whereafter anomalous conditions migrated westward.

A model that captures an ocean–atmosphere mode succeeds in reproducing a Southern Oscillation with several realistic features — it has the correct time scale, and the oscillation is between warm El Niño and cold La Niña states — provided the model takes into account dissipative processes or nonlinearities that limit the amplitude. (For example, the rapid rate at which evaporation from the ocean surface increases as sea-surface temperature increases prevents the surface waters from becoming much warmer than 30°C.)

An example of a simulated Southern Oscillation appears in Fig. 6. It was excited in the coupled ocean–atmosphere model by imposing westerly winds over the western tropical

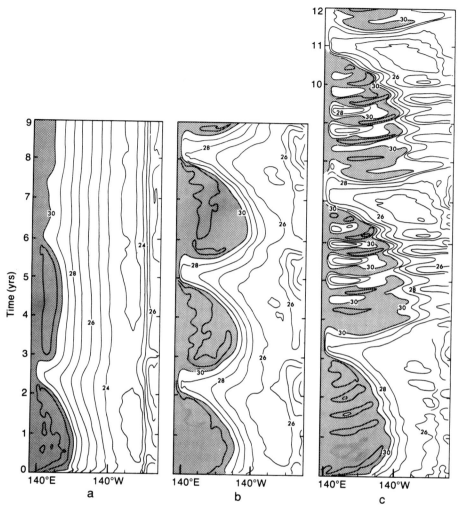

Fig. 6 In the coupled ocean–atmosphere model developed by J. D. Neelin, interactions between the two media are initiated by westerly winds that are imposed over the ocean to the west of the date line for a period of one month. The diagrams show the movement of isotherms (°C) along the equator over a period of years. Temperatures exceed 30°C in shaded regions. Initially the eastward winds drive eastward oceanic currents that advect warm water in that direction, evoking El Niño. In (a) the coupling between the ocean and atmosphere is very weak, and a damped oscillation is excited. An increase in the strength of the coupling in (b) results in a sustained, regular oscillation between El Niño and La Niña states, with a period of approximately three years. A further increase in the strength of the coupling in (c) introduces a secondary oscillation with a period of approximately six months, superimposed on the low-frequency oscillation. Further increases in the strength of the coupling lead to chaotic fluctuations.

Pacific for a month. Subsequent developments in the model depend on the strength of the coupling between the ocean and atmosphere. If it is weak, the interannual oscillations are damped, as in Fig. 6a. Sufficiently strong coupling permits self-sustaining oscillations between El Niño and La Niña states, as in Fig. 6b. In Fig. 6c the coupling is even stronger, so that secondary instabilities with a period near six months appear. Further increases in the strength of the coupling results in completely chaotic fluctuations with a broad spectrum of variability. This last parameter range is not believed to be realistic, because the Southern Oscillation has well-defined temporal and spatial scales, as is evident in Fig. 1. The appropriate strength of the coupling between the ocean and atmosphere, or the degree to which interactions between them are unstable, is believed to cause damped or self-sustaining oscillations similar to those in Figs. 6a and 6b.

The strength of the coupling between the ocean and atmosphere depends on a number of parameters and varies with time. Consider, for example, the meteorological conditions that are necessary for unusually warm surface waters in a certain region to initiate unstable ocean–atmosphere interactions that lead to El Niño. The warm waters must cause a local heating of the atmosphere so that winds converge onto that region. The local heating requires that there be rising air over the warm surface waters. Hence, if large-scale atmospheric motion is such that there is subsidence over the unusually warm surface waters, unstable interactions will not ensue. Rising motion in the atmosphere occurs in the convergence zones shown in Fig. 4, so that the seasonal movements of those zones cause the strength of the coupling between the ocean and atmosphere to vary seasonally.

One unrealistic feature of the model that produces the results shown in Fig. 6 is the absence of a seasonal cycle. Another is the absence of "weather." The atmospheric component of the coupled model is a very simple one that describes the anomalous steady winds in response to a given pattern of sea-surface temperatures. The introduction into such a model of weather—that is, random high-frequency fluctuations that are associated with instabilities of the mean atmospheric motion and that are independent of changes in sea-surface temperature—results in a realistically irregular interannual oscillation. The extent to which weather affects a periodic ocean–atmosphere mode depends on the degree to which the coupled system is unstable. Random disturbances influence the self-sustaining oscillation in Fig. 6b less than the damped oscillation in Fig. 6a. In fact, for the parameter range of Fig. 6a, a continual, irregular oscillation is entirely dependent on a constant source of disturbances that develop into El Niño on some occasions and into La Niña on others. If a seasonal cycle were present in the model, then, as explained above, interactions between the ocean and atmosphere would be weak during certain seasons (conditions corresponding to those in Fig. 6a) and strong during others (condi-

tions corresponding to those in Fig. 6b). This means that in the presence of a seasonal cycle, random disturbances would disrupt the air–sea mode of oscillation more readily in some seasons than in others. The Southern Oscillation can therefore be attributed to unstable ocean–atmosphere interactions that are modulated by the seasonal movements of the atmospheric convergence zones. It would be perfectly regular and predictable were it not for the disruptive effect of random disturbances.

V. Predicting El Niño

The Southern Oscillation has a long time scale, on the order of a few years. If the superimposed high-frequency fluctuations can be eliminated by filtering the data, it should be possible to make predictions by extrapolating the low-frequency trend. This can be done either statistically or by means of a dynamical model. A research group at the Scripps Institution of Oceanography uses statistical methods to exploit the correlations between low-frequency surface wind and fluctuations in sea-surface temperatures in the tropical Pacific. Using past records, the group's members identify wind patterns that correlate with future increases in sea-surface temperatures in the central and eastern tropical Pacific. The wind variations usually lead by a few months. On the basis of such variations, they are able to predict whether El Niño will occur during the next few months. [See WEATHER PREDICTION, NUMERICAL.]

A dynamical model can be used to integrate the equations that govern oceanic and atmospheric motion forward in time, thus providing a forecast. If initial conditions are similar to those in the model of Fig. 6b, energetic, regular oscillations are likely to ensue, so that future developments are highly predictable. If, on the other hand, initial conditions are similar to those in the model of Fig. 6a, predictability is very limited because random perturbation will strongly influence the behavior of the coupled system.

To determine whether conditions at time T correspond to those in Fig. 6a or 6b, it is necessary to make repeated predictions, starting from slightly different initial conditions on each occasion. The rate at which different predictions—of sea-surface temperature in a certain region, for example—diverge provides a measure of the predictability of the coupled system at time T. Forecasts that do not diverge indicate that the coupled system is predisposed to certain developments. Oceanic data with which to initiate the calculations are unavailable. Therefore the available wind data is used to force the oceanic component of the coupled model, thus generating the necessary initial oceanic data. These data are reasonably accurate because the predictions are at least as skillful as those by means of statistical methods.

Evidence that the Southern Oscillation involves natural

modes of oscillation of the coupled ocean–atmosphere system is persuasive. It is unclear to what degree the details of the picture provided by the models, a picture of a perfectly regular oscillation made irregular by the disruptive effects of weather, are accurate. The models developed thus far are highly idealized, neglecting a number of processes that could contribute to a far more complicated picture. They cope in an inadequate manner with the seasonal cycle, which is simply specified as part of the basic state of the model so that it does no more than merely modulate the natural modes that correspond to the Southern Oscillation. This amounts to assuming that the response of the ocean–atmosphere system to the seasonal variations in solar radiation is strictly an annual signal that is independent of interannual oscillations. In reality, matters are more complicated: The existence of a biennial oscillation in the tropics suggests that the response to seasonal forcing is nonlinear and could involve interannual fluctuations too. The extent to which the Southern Oscillation is part of that response is yet to be determined. [See SOLAR RADIATION.]

The Atlantic Ocean experiences a phenomenon very similar to El Niño. One occurred in 1984, as is evident from Plate 1. On such occasions rainfall is very heavy along the normally arid coast of southwestern Africa. Whereas El Niño in the Pacific is associated primarily with an eastward movement of the atmospheric convergence zone in the west, the Atlantic phenomenon is associated with a southward displacement of the Intertropical Convergence Zone shown in Fig. 4. (Apparently the convective zone over the Amazon Basin is not easily dislodged.) Interactions between the ocean and atmosphere are again responsible, but they are poorly understood. All the unstable ocean–atmosphere modes that have been studied have such large east–west scales that they can be realized in the Pacific only, not in the Atlantic. This is but one of the many puzzles that remain to be solved in the study of interactions between ocean and atmosphere around the world.

Bibliography

Battisti, D. S., and Hirst, A. C. (1989). Interannual variability in the tropical atmosphere-ocean system: The influence of the basic state and ocean geometry. *J. Atmos. Sci.* **46**, 1687–1712.
Cane, M. A., Dolan, S. D., and Zebiak, S. E. (1986). Experimental forecasts of the 1982/83 El Niño. *Nature (London)* **321**, 827–832.
Neelin, J. D. (1989). Interannual oscillations in an ocean GCM-simple atmospheric model. *Philos. Trans. R. Soc. London A* **329**, 189–205.
Philander, S. G. H. (1990). "El Niño, La Niña, and the Southern Oscillation." Academic Press, San Diego.
Rasmusson, E. M. (1985). El Niño and variations in climate. *Am. Sci.* **73**, 168–177.
Rasmusson, E. M., and Wallace, J. M. (1983). Meteorological aspects of the El Niño/Southern Oscillation. *Science* **222**, 1195–1202.
Schopf, P. S., and Suarez, M. J. (1988). Vacillations in a coupled ocean-atmosphere model. *J. Atmos. Sci.* **45**, 549.
Wyrtki, K. (1975). El Niño—the dynamic response of the equatorial Pacific Ocean to atmospheric forcing. *J. Phys. Oceanogr.* **5**, 572–584.
Zebiak, S. E., and Cane, M. A. (1987). A model ENSO. *Mon. Weather Rev.* **115**, 2262–2278.

Glossary

El Niño Periods of warm surface waters and heavy rainfall off the coast of Peru that coincide with relaxed trade winds over the Pacific.
La Niña Conditions that are complementary to El Niño—cold surface waters off Peru and intense trades over the Pacific.
Southern Oscillation Irregular interannual fluctuation between El Niño and La Niña conditions that in the atmosphere has a scale larger than the tropical Pacific and extends into the Indian Ocean.
Ocean–atmosphere interactions Interactions that are invoked to explain why certain phenomena, such as the Southern Oscillation, are attributable to changes in the ocean from a meteorological perspective and are attributable to changes in atmospheric conditions from an oceanographic perspective.

Electrification in Winter Storms

Earle R. Williams
Massachusetts Institute of Technology

The formation of summer thunderstorms and their electrical structure are described as background for the examination of the electrification in winter storms. Important differences in the electrical behavior of summer and winter storms are attributed to large-scale meteorological differences. The prevalence of cloud-to-ground lightning in the winter is discussed in this context.

I. Introduction

The electrification of clouds in the earth's atmosphere and their most spectacular manifestation, lightning, are common phenomena in summer months at mid-latitude. Perhaps less well recognized is the occurrence of electrification and lightning in winter snowstorms. This article is mainly concerned with the latter phenomenon. However, because the conditions surrounding the electrification of summer storms have been investigated more extensively and because there are important connections between winter and summer storm electrification, we shall begin with a discussion of the summer case.

II. Summer Thunderstorms

It is well recognized that electrification and lightning in summer thunderstorms is closely associated with vigorous convection and the formation of precipitation. Convection involving water substance is initiated by the lifting of parcels of air containing water vapor. Under ordinary conditions, the highest concentrations of water vapor are found adjacent to the earth's surface, which is the source of the water vapor

through processes of evaporation (from water surfaces) and transpiration (from the surfaces of vegetation). The clear air containing water vapor is largely transparent to the visible radiation from the sun, and the earth's surface is strongly heated by absorption in the morning hours. So-called sensible heat is transferred from the earth's surface to adjacent air parcels, thereby reducing their density and endowing them with upward buoyancy. The buoyant parcels, or "bubbles," are the thermals that sailplane pilots rely on to remain aloft. As the parcels rise, they cool by adiabatic expansion. According to the Clausius–Clapeyron relation, air can hold less water vapor at lower temperature. As a rough rule of thumb, the amount of water vapor an air parcel can hold doubles with every 10°C of parcel temperature. Eventually the parcel's water vapor begins to condense on cloud condensation nuclei, the generally invisible particulate naturally present in the earth's atmosphere. This cloud condensation level defines the cloud base, and because the water vapor and temperature distribution in the atmosphere are often laterally uniform, the cloud base is often a reasonably well-defined horizontal surface. [*See* EVAPORATION AND EVAPOTRANSPIRATION.]

Continued condensation above cloud base and the attendant release of latent heat endows air parcels with additional positive buoyancy relative to the cloud-free environment and can lead to cloud vertical development to great heights above cloud base. Figure 1 illustrates a small cumulus cloud containing no precipitation and exhibiting no lightning and a larger cumulonimbus cloud that has developed to thunderstorm status. Figure 1 also includes information about the atmosphere's temperature structure. When the rising air parcels traverse the 0°C level in the atmosphere, a nucleation process may occur to produce ice crystals directly from the water vapor phase. The rarity of ice nuclei in the atmosphere tends to inhibit the transition to the ice phase however, and supercooled water droplets are usually more prevalent than ice crystals at subfreezing temperatures in early stages of cloud development. [*See* CLOUD DYNAMICS.]

The formation of precipitation, a feat achieved by only a small fraction of clouds in the atmosphere, is initiated when particles of liquid water or ice achieve sufficient size and attain sufficient fall speed to collide with other particles and thereby form still larger particles. In the warm part of the cloud this process is called coalescence, and in clouds in the tropics whose temperature is everywhere greater than 0°C,

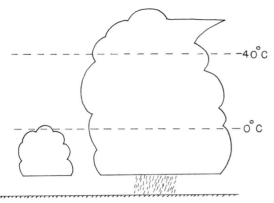

Fig. 1   Shallow cumulus cloud (left) and deeper thundercloud (right), showing temperature structure and the mixed-phase region.

submillimeter-sized raindrops may appear within 15–30 min of the initial appearance of cloud. In the cold part of the cloud, the Bergeron process plays an important role and is somewhat more involved than coalescence. Because the equilibrium vapor pressure of ice is systematically less than that of liquid water at the same temperature (again following the Clausius–Clapeyron relation), ice crystals tend to grow by vapor diffusion at the expense of supercooled water droplets, which undergo evaporation. The growing ice crystals attain larger fall speeds and collide with slower-moving supercooled droplets. These droplets freeze on contact with the ice crystal, a process known as riming. The continuation of riming leads to the formation of irregular assemblages of frozen supercooled droplets — which themselves are called soft hail, or graupel, with sizes of millimeters to centimeters and fall speeds of several meters per second. Graupel particles are the most prevalent precipitation type in the cold parts of thunderclouds. In cases of still more vigorous convection with sustained updrafts of several tens of meters per second, graupel may transition to hailstones. These largest of precipitation particles may fall to earth without completely melting. [*See* Hail Climatology.]

The Bergeron process and its evolution to the graupel stage is fundamental to mixed-phase microphysics, and it requires the simultaneous presence of water in all three phases (solid, liquid, and vapor). Considerable evidence suggests that mixed-phase microphysics is also fundamental to the cloud electrification process. This is not to say that the physical details at the molecular scale are well understood; indeed they are not. Furthermore, theories exist for cloud electrification that make no reliance on mixed-phase microphysics, and there are a few observations (all in the tropics) of lightning in clouds judged to be 'warm' and therefore not containing ice particles of any kind. Yet on balance, both thunderstorm observations and laboratory simulations of the mixed-phase

microphysical processes believed to occur in thunderstorms strongly support the role of mixed-phase microphysics in the electrification of both summer and winter storms.

As shown in Fig. 2, the gross electrical structure of summer thunderstorms in a wide variety of geographical locations has been shown to be a positive dipole, with dominant positive charge uppermost. The main negative charge region is laterally extensive (5–10 km) and has a vertical thickness of only 500–1000 m. Observations of the location of the negative charge region have invariably shown it in the mixed-phase region where the cloud temperature lies between 0°C and −40°C. In addition to the main positive and negative regions, a subsidiary region of lower positive charge has been noted, frequently in association with localized shafts of precipitation.

According to the currently most popular mechanism for cloud electrification, the electrostatic structure shown in Fig. 2 is formed by collisions between graupel particles and ice crystals (or small ice fragments) in the presence of supercooled water. Without supercooled water, the charge transfer caused by these ice particle collisions is substantially less, according to the results of laboratory experiments. Negative charge is selectively transferred to the faster-falling graupel particles (for reasons not currently well understood) and opposite positive charge to the ice crystals. Subsequent differential particle motions under gravity result in a positive dipole structure. It is important to note that such motions can result in quite shallow zones of net charge.

Despite the apparent simplicity of this process, the foregoing mechanism cannot account for (1) the distinct lower boundary on the negative charge region, (2) the subsidiary lower positive charge, nor (3) the observation of positively charged precipitation arriving at the earth's surface. Laboratory simulations of collisions between graupel particles and

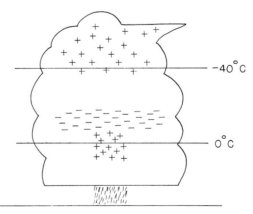

Fig. 2   Electrostatic structure of an isolated thunderstorm, showing upper positive, main negative, and subsidiary lower positive charge regions.

ice crystals also suggest that if the temperature in the mixed-phase region is sufficiently high (i.e., sufficiently close to 0°C), and if the cloud liquid-water content is sufficiently large, then selective positive charge is transferred to the graupel particles and negative charge left on the ice crystals. Such a charge-reversal phenomenon, though poorly understood at the molecular scale, helps explain the aforementioned observations at the thundercloud scale.

The electrostatic structure shown in Fig. 2 and maintained by active charge separation in the mature stage of the storm, gives rise to large differences of electric potential ($\sim10^8$ V) and two principal types of lightning, intracloud (IC) and cloud-to-ground (CG). IC lightning is an electrical discharge between the upper positive charge and the main negative charge. CG lightning is a discharge between the main negative charge and the (locally) positively charged earth. In summer thunderstorms, well over 90% of CG lightning transfers negative charge to earth. The amount of charge transferred in lightning discharges in isolated summer storms ranges from a few to several tens of coulombs.

Observations show that the lightning-flash rate of summer storms increases dramatically with their size and convective vigor. Ordinary mid-latitude thunderstorms with cloud depths of 10–12 km may produce only a few flashes per minute, whereas the energetic storms that attain heights of 18–20 km, both at mid-latitude and in the tropics, may achieve flash rates as large as a few per second. Recent observations suggest that cloud height alone is often not a reliable indicator of instantaneous flash rate; the development of intense precipitation in the mixed-phase region, often well correlated with cloud height, is equally important in affecting the lightning activity. [See MID-LATITUDE CONVECTIVE SYSTEMS.]

III. Winter Storm Electrification

Although winter storm electrification is less well understood than summer thunderstorms, observations suggest that mixed-phase microphysics is operating in both situations. However, large-scale meteorological differences between summer and winter give rise to important differences in cloud electrical behavior. We now address these important meteorological differences.

By definition, winter storms occur at latitudes farther from the sun's position than do summer storms. As one result, winter storms form in an environment with lower surface temperature. In summer storms, the surface temperature of moist air that supplies the storm with water vapor in the form of thermals is 20–30°C, whereas for winter storms the surface temperature is in the range of −10 to +10°C. Recalling the Clausius–Clapeyron result that a 10°C temperature change is a doubling of saturation water-vapor content, and

that water vapor is the energy source for moist convection, it is immediately apparent that the energy available (per unit horizontal area) to form a deep cloud and drive vigorous convection may be less in winter by as much as an order of magnitude.

A second important meteorological effect for storms at high latitude is the tendency for larger latitudinal temperature gradients and for the prevalence of so-called frontal zones between cold dry air masses (from polar regions) and warm moist air masses (from equatorial regions). The size of these air masses may be half the size of the continental United States. In frontal situations, the lifting process essential for condensation and the ultimate formation of precipitation and promotion of mixed-phase microphysics has a different physical basis than in isolated summer thunderstorms. In the case of a warm front in a winter storm, for example, warm moist air is forced up an extensive wedge of cold dry air by differences in atmospheric density at a still larger scale. As a consequence, the area over which precipitation may form (tens of thousands of square kilometers) is far greater than the area of an isolated thunderstorm covering perhaps one hundred square kilometers. [See ATMOSPHERIC FRONTS.]

The meteorological effects of temperature and temperature gradient give rise to summer storms that are tall and narrow and to winter storms that are shallow and broad. In effect, the conditions on temperature in the winter move the ice-containing region of the storm to the surface observer. For example, snow is produced by ordinary summer thunderstorms but it does not survive the fall through the subfreezing portion of the atmosphere. Because of the smaller convective energy and the shallow slopes (100 to 1) of frontal boundaries, the vertical air motions in winter storms are only a few meters per second or less. These motions are sufficient to provide for supercooled water in the mixed-phase region (necessarily closer to earth because of the winter atmosphere's temperature structure) and to promote a riming process in the presence of ice crystals, but are usually insufficient to promote the formation of large graupel and hail. Nonetheless, the microphysical quantities believed to be responsible for the electrification of summer storms (i.e., supercooled water, ice particles, differential particle motions) are also present in winter storms, but with generally smaller magnitudes. This point is consistent with the observations that lightning rates in winter storms are significantly smaller than in summer storms. A rate of one flash per minute is considered large for a winter storm; a few flashes per hour is more typical. The charge transferred in these events tends to be large — in some cases several hundred coulombs and therefore more than an order of magnitude larger than in isolated summer storms. Evidently a large reservoir of charge is made possible by the large area associated with winter storms.

Some snowstorms produce no lightning at all. Such observations provide additional indirect evidence for the role of

DEGREES WEST LONGITUDE

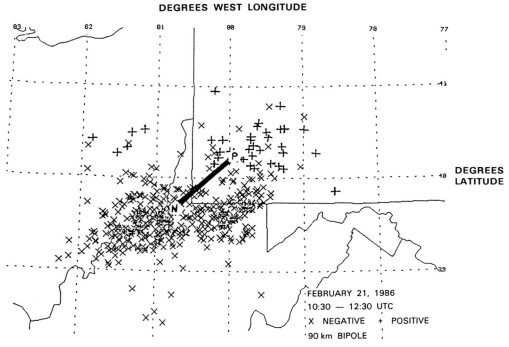

Fig. 3 Example of a bipolar lightning pattern in a winter storm over Ohio, Pennsylvania, and West Virginia. X indicates location for cloud-to-ground lightning of negative polarity, + indicates positive polarity.)

mixed-phase microphysics. So-called "dry" snowstorms, in which the surface temperature is below the freezing point (0°C) and in which the riming process is largely inactive, do not generally exhibit lightning. Wet snowstorms, whose convective motions are usually more vigorous, are far more likely to reach the lightning stage.

Consistent with the large areas of precipitation associated with winter storms, the electric field at the ground beneath is far more widespread than beneath the more compact summer storms. The field usually exhibits considerable variability in polarity and is less easily interpreted than the summer case. Occasionally, a pronounced oscillatory character is observed at single stations, with periods varying from a few minutes to large fractions of an hour. It has not been established whether

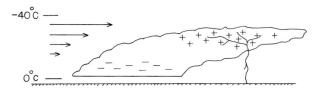

Fig. 4 Illustration of the possible effect of wind shear on the production of positive discharges to ground in a winter storm.

these oscillations are the result of convection-induced changes in mixed-phase microphysics or the result of advection of alternating regions of positive and negative charge, or some other effect.

A distinct electrical characteristic of winter storms, identified within the last two decades in several geographic locations, is the prevalence of cloud-to-ground lightning of positive polarity. In contrast with summer storms, which often exhibit exclusively negative flashes, winter storms most often show a mixture of polarities and occasionally all positive flashes. The reasons for this behavior are still poorly understood, but some hints are provided by the spatial organization of the location of flashes to ground of positive and negative polarity. Cloud-to-ground flash locations with specified polarity in a winter storm are shown in Fig. 3. It can be seen that the positive flashes tend to be displaced from the negative flashes in a direction aligned with the cloud top winds in the atmosphere. The centroid of positive locations and the centroid of negative locations together constitute a so-called lightning bipole.

One explanation for bipoles in winter storms and their alignment with the upper-level wind is based on the tilting of originally vertical cloud dipoles (recall Fig. 2) by the change in wind speed with height in the atmosphere, as shown in Fig. 4.

This latter quantity, the wind shear, results in the lateral displacement of upper positive charge away from lower negative charge, to a position where direct positive charge transfer to ground can occur. According to a well-known relation in dynamical meteorology, the thermal wind relationship, the vertical wind shear at any location is directly proportional to the local horizontal temperature gradient. For reasons discussed earlier, winter storms are generally characterized by larger horizontal temperature gradients than summer storms and hence are more prone to produce positive cloud-to-ground flashes according to the picture in Fig. 4. Furthermore, because of the shallow nature of winter storms, the mixed-phase region and the upper positive charge are closer to the earth's surface, a factor that may promote the occurrence of positive cloud-to-ground lightning. This explanation cannot account, however, for the relative infrequency of negative cloud-to-ground lightning in a winter storm with electrostatic structure shown in Fig. 4. [*See* ZONAL AND MERIDIONAL WINDS IN THE EARTH'S ATMOSPHERE.]

An alternative explanation for positive flashes rests on the possible reversal of polarity in ice particle collisions, or the rearrangement of charge by vertical air motions and the large-scale creation of vertical dipoles of inverted polarity — that is, negative charge uppermost and positive charge at lower levels. These simple structures, the inverted dipole and the tilted dipole, are not easily distinguished in real winter storms and require further observations. The observers' challenge here is the simultaneous acquisition of information on both the storm scale (ten to hundreds of kilometers) and the particle scale (microns to centimeters).

While certain systematic features of winter storm electrification have been identified, no comprehensive understanding of the electrification of either winter or summer storms exists. Part of the reason for the lack of understanding is the incomplete picture of the air and particle motions within clouds, but perhaps the most notable gap involves the behavior of ice and the physics of the mixed-phase. Water substance deserves further study toward elucidating cloud electrification. [*See* WATER, PHYSICAL PROPERTIES AND STRUCTURE.]

Bibliography

Brook, M., Nakano, M., and Takeuti, T. (1982). The electrical structure of the Hokuriku winter thunderstorms. *J. Geophys. Res.* 87, 1207–1215.

Chalmers, J. A. (1982). "Atmospheric Electricity." Pergamon Press, Oxford.

Engholm, C. D., Williams, E. R., and Dole, R. M. (1990). Meteorological and electrical conditions associated with positive cloud-to-ground lightning. *Mon. Weather Rev.* 118, 470–487.

Krider, E. P., and Roble, R., eds. (1986). "The Earth's Electrical Environment." National Academy Press, Washington, D.C.

Vonnegut, B. (1963). Some facts and speculations concerning the origin and role of thunderstorm electricity. *Meteorol. Monogr., Am. Geophys. Union* 5, 224–241.

Williams, E. R. (1989). The tripole structure of thunderstorms. *J. Geophys. Res.* 94, 13151–13167.

Glossary

Adiabatic expansion Increase in volume of parcels of atmospheric air without any exchange of heat with their surroundings.

Bergeron process Process of precipitation formation based on the growth of ice crystals at the expense of supercooled water droplets, and the subsequent capture and freezing of the droplets by the falling crystals.

Bipole Organized pattern of locations for cloud-to-ground lightning of positive and negative polarity.

Clausius–Clapeyron relation Physical law relating the temperature and the equilibrium vapor pressure (or vapor density) of water.

Graupel Precipitation particles composed of ice and formed by the accretion and freezing of supercooled water droplets.

Lightning Electrical discharge in thunderstorms caused by the separation of electric charge of positive and negative polarity.

Mixed-phase region Zone in the atmosphere that contains water substance in all three of its phases (vapor, liquid, and solid) and whose temperature is generally between $-40°C$ and $0°C$.

Riming Process of freezing of supercooled water droplets when they come into contact with ice particles.

Thermal wind relation Theoretical relationship in dynamical meteorology stating that the horizontal gradient in atmospheric temperature is proportional to the vertical gradient in horizontal wind.

Electromagnetic Methods in Geophysics

Ajit K. Sinha

Geological Survey of Canada

Application of electromagnetic (EM) methods in geophysics involves the excitation of the ground by an alternating-current primary field and recording the resultant secondary electrical and magnetic fields by suitably placed receivers. The secondary fields are generated by the presence of conductive materials such as ore bodies or conductive geological materials such as clays or sandstones containing water. Since the electrical resistivity of various earth materials span many orders of magnitude, EM methods can be successfully used for mapping and detecting resistivity variations inside the earth from shallow to large depths. The EM methods include a wide range of techniques with various types of transmitters and receivers using frequencies from extremely low (0.001–10 Hz), as in magnetotelluric methods, to very high (several megaherz), as in radar systems. The primary field is generated artificially in most EM methods, but some methods utilize natural EM fields (e.g., magnetotellurics) or fields that have been generated artificially at large distances from the receiver (as in the very low frequency, or VLF, method). The depth penetration of EM systems is controlled by ground resistivity, strength of the primary field, noise, and frequency—the lower frequencies penetrate deeper into the ground. EM

This is Geological Survey of Canada Contribution No. 53890.

methods are used in mineral exploration, geological mapping, geothermal exploration, environmental studies, permafrost mapping, and in deep crustal studies. EM methods can be used on the ground, in air, and in boreholes.

I. Sources of Electromagnetic Fields

Any discussion of the electromagnetic (EM) methods in geophysics must contain a short review of the nature and sources of EM fields that are used for unraveling the nature of the material below the ground. Broadly speaking, the sources can be divided into two categories: natural and artificial. The existence of natural EM fields has been known to scientists for well over a century. The fields are generated by natural electromagnetic disturbances in the atmosphere caused by fluctuating ionospheric currents with frequencies ranging from 10^{-5} to 10^4 Hz. The fields travel with very small attenuation between the lower ionosphere and the ground surface by bouncing back and forth between these surfaces. At a large distance from the source, the field appears like a plane wave. Atmospheric electric disturbances or sferics also contribute to natural EM fields. Worldwide thunderstorm activity propagates EM energy in the earth–ionosphere interspace, in the frequency range of 1 to 10^4 Hz. Four EM methods, magneto-variational method, telluric, magnetotelluric, and audio-frequency magnetics (AFMAG) use natural EM sources as their primary fields. The first method will be discussed elsewhere in this encyclopedia; the last three methods will be described in this article. [*See* IONOSPHERE; SOLAR–TERRESTRIAL INTERACTION.]

Other EM methods use artificial excitation sources. Although this causes an extra complication, since a transmitter is needed, it provides greater control over the nature of the primary field for optimization of response from targets. The transmitters are often portable, so that they may be carried by one or two people even in remote rough terrain. In most cases, the energization is by induction; that is, the ground is energized by passing an alternating current, most often in the frequency range between 100 and 10,000 Hz, through a loop of wire without making direct contact with the ground. The ground can also be energized by conduction—that is, by electrodes making direct contact with the ground. An advan-

Encyclopedia of Earth System Science, Volume 2

Source: Energy, Mines and Resources Canada. Reproduced with the permission of the Minister of Supply and Services Canada, 1990.

137

tage of the inductive mode of energizing is that EM surveys can be performed from an aircraft.

Most portable EM systems use a small loop as the excitation source. However, when deeper penetration is desired, stronger primary fields are provided by large square or rectangular loops of wire lying on the ground, or by linear cables several kilometers long, carrying large currents (10–100 A). In the so-called very low frequency or VLF method, the source is one of several powerful Navy transmitters with radiating power in the range of 100 to 1000 kW, which may be located more than a thousand kilometers from the survey area. Because of the large distance between the transmitter and the receiver, the primary field behaves like a plane-wave field, similar to natural-source EM methods.

II. Electrical Properties of Geological Materials

The behavior of EM fields in any medium is governed by Maxwell's equations, which may be written in vector form as

$$\nabla \times \mathbf{H} = \mathbf{J} + \frac{\partial \mathbf{D}}{\partial t} \tag{1}$$

$$\nabla \times \mathbf{E} = -\frac{\partial \mathbf{B}}{\partial t} \tag{2}$$

$$\nabla \cdot \mathbf{D} = \rho' \quad \text{and} \quad \nabla \cdot \mathbf{J} = 0 \tag{3}$$

where $\mathbf{H}$ and $\mathbf{E}$ are magnetic and electric fields, $\mathbf{J}$ is the current density, $\mathbf{D}$ is the electric displacement, $\mathbf{B}$ is the magnetic induction vector, and ρ' is the charge density. In most isotropic media, one can write

$$\mathbf{B} = \mu \mathbf{H}, \quad \mathbf{D} = \epsilon \mathbf{E}, \quad \text{and} \quad \mathbf{J} = \sigma \mathbf{E} \tag{4}$$

where μ is the magnetic permeability, ϵ is the dielectric permittivity, and σ is the electrical conductivity. These are the three basic electrical constants of a medium; they determine the way the electric and magnetic fields behave inside a medium.

A. Electrical Conduction in Geological Materials

Electrical conduction in rocks and minerals takes place in three ways: electronic, electrolytic or ionic, and dielectric. In the first case, current flows through the movement of free electrons in materials such as metals. Electrolytic conduction takes place through the movements of ions such as those present in liquids containing salt. Dielectric conduction is

important for poor conductors or insulators that have few free electrons or ions. The atomic electrons in dielectric materials are displaced slightly under the influence of an external EM field. The separation of positive and negative charges is known as dielectric polarization and normally becomes important at frequencies greater than 50 kHz.

1. Electrical Conductivity

Electrical conductivity σ of a material defines the ease with which electrical currents can flow through that material. Mathematically, it may be written as

$$\sigma = \frac{L}{R \cdot A} \quad \text{(in Siemens/m)} \tag{5}$$

where L, R, and A represent the length in meters, resistance in ohms, and cross-sectional area in square meters of a solid cylinder of the material, respectively. The resistance R can be measured by applying a voltage V across the ends of the cylinder and measuring the current I passing through it, since by Ohm's law

$$R = V/I \tag{6}$$

The reciprocal of conductivity is resistivity, indicated by ρ. Conductivity is a basic electrical property of all materials and controls the conduction of electrical currents at low to moderate (10–10,000 Hz) frequencies, the frequency range normally used in EM exploration. Both electronic and ionic conduction contribute to the conductivity of a material, the latter increasing in importance as the porosity and presence of electrolytes in pore fluids increase. Rocks and minerals show a wide range of bulk electrical resistivity (Table I). Generally, igneous rocks have the highest electrical resistivities, followed by metamorphic and sedimentary rocks.

2. Dielectric Permittivity

The dielectric permittivity ϵ, defined in Eq. (4), determines the amount of dielectric current in a medium caused by displacement of positive and negative charges under the influence of a time-varying EM field. In contrast to conductivity, the dielectric permittivity of most materials varies within a narrow range. The permittivity for free-space has a value of $\epsilon_0 = 8.854 \times 10^{-12}$ F/m. The ratio of the dielectric permittivity of a medium to that of free space is called the dielectric constant k of the material. Table II lists the dielectric constants of some common rocks and minerals. The dielectric constant of water is high, and consequently the percentage of pore fluid in rocks determines their dielectric constants. However, the displacement currents caused by dielectric conduction are much smaller than conduction currents for most earth materials in the frequency range commonly used in EM methods.

Table I Resistivities of Common Rocks and Minerals

Material	Resistivity range (ohm-m)
Rocks	
Granite	$3 \times 10^2 - 10^6$
Gabbro	$10^3 - 10^6$
Basalt	$10 - 1.3 \times 10^7$ (dry)
Diorite	$10^4 - 10^5$
Gneiss	$6.8 \times 10^4 - 3 \times 10^6$
Schists	$20 - 10^4$
Limestone	$50 - 10^7$
Sandstone	$1 - 6.4 \times 10^8$ (dry)
Clays	$1 - 100$
Oil sands	$4 - 800$
Minerals	
Argentite, Ag_2S	$2 \times 10^{-3} - 10^4$
Chalcopyrite, $CuFeS_2$	$1.2 \times 10^{-5} - 3 \times 10^{-1}$
Pyrite, FeS_2	$2.9 \times 10^5 - 1.5$
Pyrrhotite, Fe_nS_m	$6.5 \times 10^{-6} - 5 \times 10^{-2}$
Galena, PbS	$3 \times 10^{-5} - 3 \times 10^2$
Sphalerite, ZnS	$1.5 - 10^7$
Chromite, $FeCr_2O_4$	$1 - 10^6$
Hematite, Fe_2O_3	$3.5 \times 10^{-3} - 10^7$
Magnetite, Fe_3O_4	$5 \times 10^{-5} - 5.7 \times 10^3$
Cassiterite, SnO_2	$4 \times 10^{-4} - 10^4$
Quartz, SiO_2	$4 \times 10^{10} - 2 \times 10^{14}$

Source: W. M. Telford, L. P. Geldart, R. E. Sheriff, and D. A. Keys (1976). "Applied Geophysics." Cambridge Univ. Press, London.

3. Magnetic Permeability

Equations (2) and (4) show that for a time-varying EM source, the fields are determined not only by conductivity σ and dielectric permittivity ϵ, but also by magnetic permeability μ. Magnetic permeability is defined as the ratio of magnetic induction $\mathbf{B}$ to the inducing field $\mathbf{H}$ [Eq. (4)]. The magnetic permeability of most materials is close to 1. Three exceptions are the magnetic minerals magnetite, pyrrhotite, and ilmenite, which have permeability values of 5.0, 2.55, and 1.55, respectively.

It should be noted that most earth materials are anisotropic; in other words, the electrical properties of materials are different along orthogonal directions. Thus the three electrical constants are, strictly speaking, tensors, not scalars.

III. Basic Theory of Electromagnetic Method of Exploration

There are three basic elements in EM method of exploration: first, a primary EM field to excite the ground; second, a target

body with an anomalous conductivity, generally with a higher conductivity than the material surrounding it (host rock); and third, a receiver located some distance away from the transmitter to record the secondary and primary fields. The primary field excites the ground by electric and magnetic current lines. When the current lines hit the anomalous conductor, they induce eddy currents to flow in it. This gives rise to a secondary field, which is normally much smaller than the primary field and is shifted in phase in relation to the primary field.

To measure the secondary field, some method must be devised to cancel or neutralize the much larger primary field at the receiver; or the secondary field must be recorded when the primary field is zero, as is done in transient EM methods. In most cases of EM exploration, the ground is energized by induction without making direct contact with the ground. In some cases, particularly when deep penetration is required, the energy is introduced into the ground by conduction, that is, by direct contact. The receiver records the secondary field by induction in almost all cases.

A. Vector Wave Equation

Assuming a sinusoidal time dependence in the form of $\mathbf{H} = He^{i\omega t}$ and $\mathbf{E} = Ee^{i\omega t}$, where $\omega = 2\pi \times$ frequency, Maxwell's equations may be rewritten in the form of vector wave equations.

$$\nabla^2 \mathbf{E} = i\omega\mu(\sigma + i\omega\epsilon)\mathbf{E} \tag{7}$$

$$\nabla^2 \mathbf{H} = i\omega\mu(\sigma + i\omega\epsilon)\mathbf{H} \tag{8}$$

where ∇^2 is a vector Laplacian operator. Equations (7) and (8) are called Helmholtz equations. Electric and magnetic fields can be determined by solving these equations along three coordinate axes (i.e., solving a total of six equations). A

Table II Dielectric Constants of Some Rocks and Minerals

Rocks	Dielectric constant	Minerals	Dielectric constant
Granite	4.8 – 18.9	Galena	18
Gabbro	8.5 – 40	Sphalerite	7.9 – 69.7
Basalt	12	Hematite	25
Gneiss	8.5	Calcite	7.8 – 8.5
Sandstone	4.7 – 12	Quartz	4.2 – 5
Soil	3.9 – 29.4	Water (20°C)	80.4
Clays	7 – 43	Ice	3 – 4.3

Source: W. M. Telford, L. P. Geldart, R. E. Sheriff, and D. A. Keys (1976). "Applied Geophysics." Cambridge Univ. Press, London.

simplification may be introduced in the form of a vector potential **F** defined by

$$\mathbf{E} = -\nabla \times \mathbf{F} \tag{9}$$

$$i\omega\mu\mathbf{H} = \nabla(\nabla \cdot \mathbf{F}) - \gamma^2\mathbf{F} \tag{10}$$

where $\gamma = [i\omega\mu(\sigma + i\omega\epsilon)]^{1/2}$ is the propagation constant of the medium.

Since the frequency used during most EM surveys is below 10 kHz, the effect of displacement currents is much smaller than that of conduction currents for most geological materials ($\sigma \gg \omega\epsilon$). Hence, Eqs. (7) and (8) can be simplified as

$$\nabla^2 \begin{matrix} \mathbf{E} \\ \mathbf{H} \end{matrix} = 0 \qquad \text{for poor conductors} \quad (\sigma \rightarrow 0) \tag{11}$$

$$\nabla^2 \begin{matrix} \mathbf{E} \\ \mathbf{H} \end{matrix} = i\omega\mu\sigma \begin{matrix} \mathbf{E} \\ \mathbf{H} \end{matrix} \qquad \text{for good conductors} \tag{12}$$

Neither Eq. (11) nor Eq. (12) represents the propagation of steady waves. Equation (11) is obeyed by static electric and magnetic fields, whereas (12) represents a vector diffusion equation. In other words, when the frequency is low and the conductivity is finite, the wave slowly diffuses into the medium instead of propagating.

B. Boundary Conditions

When an EM wave meets an interface between two electrically distinct media, it undergoes changes in amplitude and phase. The field components must satisfy some boundary conditions at each interface, in addition to satisfying the field equations in the different regions. The boundary conditions can be stated as follows:

1. The tangential electric field is continuous at the interface.
2. The tangential magnetic field is continuous at the interface.
3. The current density normal to the interface is continuous.
4. The magnetic flux normal to the interface is continuous.

C. Phase Relationships

A primary EM field impinging on a conductive inhomogeneity will produce a secondary field with the same frequency as the primary field but different phase. A receiver (Rx) placed near the transmitter (Tx) will record the combined effects of the two fields. Figure 1a shows a typical situation where a transmitter loop produces a primary field $\mathbf{H}_p$ at a receiver located in the survey grid. The eddy currents produced inside the conductor by its interaction with the primary field pro-

duces a secondary field $\mathbf{H}_s$, which is recorded along with $\mathbf{H}_p$ at the receiver. Figure 1b illustrates the relationship between the primary field shown by the vector $\mathbf{H}_p$, the secondary field by the vector $\mathbf{H}_s$, and the resultant field $\mathbf{H}_r$.

If R_c and L_c indicate the resistance and inductance of the conductor, the secondary induced field will lag behind the primary field by $\pi/2 + \phi$, where $\tan\phi = \omega L_c/R_c$. The phase lag of $\pi/2$ is caused by the coupling between the Tx and the conductor circuitry, while the lag of ϕ is due to the properties of the conductor as an electrical circuit. For a good conductor, $R_c \rightarrow 0$ and $\phi \rightarrow \pi/2$ whereas for poor conductors $R_c \rightarrow \infty$ and $\phi \rightarrow 0$. In most real cases, ϕ will lie between 0 and $\pi/2$, so that the phase difference between $\mathbf{H}_p$ and $\mathbf{H}_s$ will range from $\pi/2$ to π.

As shown in Fig. 1b, the component of $\mathbf{H}_s$, 180° out of phase with the primary field ($\mathbf{H}_s \sin\phi$), is called the *real* or *in-phase* component and the component 90° out of phase ($\mathbf{H}_s \cos\phi$) is called the *quadrature* or the *out-of-phase* component. In general, the superposition of two fields that differ in phase by anything other than 0 or π will result in elliptic polarization of the field, in which the resultant vector is finite at all times and rotates in space with continuous amplitude change. The extremity of the vector $\mathbf{H}_r$ describes an ellipse.

The secondary voltage induced in the receiver coil (Fig. 1a) can be written as

$$\mathbf{E}_R = K\left(\frac{Q^2 + iQ}{1 + Q^2}\right) = K(A + iB) \tag{13}$$

where K is a constant and Q is defined by $Q = \omega L_c/R_c$. The secondary field is thus a complex quantity, and A and B represent the real and quadrature parts of the response function. The parameter Q is called the response parameter of the conductor, since it is composed of the equivalent inductance and resistance of the conductor. Figure 2 shows a plot of A and B versus the response parameter Q.

At low frequencies ($\omega \rightarrow 0$), or when the resistance is very high ($R_c \rightarrow \infty$), $Q \rightarrow 0$. In such a case, both A and B are infinitely small. This is known as the resistive limit. As frequency increases and/or R_c becomes finite, Q increases in value. In the range $0 < Q < 1$, the quadrature component B is greater than the real component A and they become equal at $Q = 1$. With increasing Q, the parameter A increases, but B begins to decrease. When Q is very large, A stabilizes at a value of 1 and the so-called *inductive limit* is reached. Parameter B tends toward zero in the inductive limit. Figure 2 also shows the importance of frequency in EM exploration. At low frequencies, both real and quadrature responses are small; whereas at very high frequencies, the real part saturates, making distinctions between good and poor conductors difficult, and the quadrature component becomes zero. This shows the importance of using several frequencies for

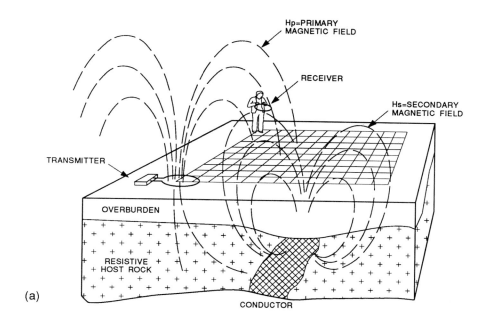

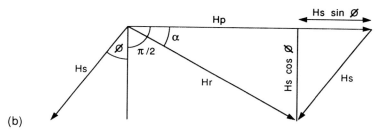

Fig. 1 (a) A typical loop-loop EM system over a ground containing a conductor under an overburden. (b) Vector diagram illustrating the relative amplitudes and phase shifts between the primary field H_p and secondary field H_s.

identification and categorization of conductors in EM surveys.

D. Depth Penetration

An important consideration for all EM systems is their depth penetration capability. The primary field in passing from the surface to the conductor gets attenuated as it travels through the earth. A similar attenuation is suffered by the secondary field as it travels from the conductor to the receiver. In addition, the primary field also decreases with distance from the source (geometric decrease), and the rate of decrease depends on the source itself. In case of plane-wave EM fields, for which there is no geometric decrease, the attenuation of

EM energy in the earth can be described in terms of the *skin effect,* in which alternating currents tend to concentrate close to the surface of a conductor. In the case of a plane EM wave incident over a homogeneous earth, the amplitude of the wave is attenuated exponentially due to the skin effect. The skin depth δ is defined as the depth at which the amplitude of the incident field has dropped to $1/e$, or about 37% of the surface value. The skin depth for a nonmagnetic earth for low to moderate frequencies can be written as

$$\delta = 159.15\sqrt{10\rho/f} \qquad (14)$$

where δ is in meters, ρ is the resistivity in ohm–meters, and f is frequency in hertz.

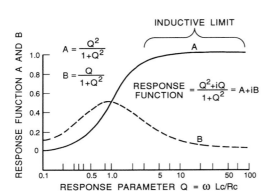

Fig. 2 Variation of the response function of a conductor with change of the response parameter. [Reproduced with permission from F. S. Grant and G. F. West (1965). "Interpretation Theory in Applied Geophysics." McGraw-Hill, Toronto.]

Figure 3 shows the variation of the skin depth with variations of resistivity and frequency as well as the values of δ for some common geological materials. At high frequencies, when displacement currents are important, the values will be somewhat different. Apart from skin depth, another term, "practical depth of penetration," is used in EM terminology, which defines the maximum depth at which a conductor may be detected. This depth depends not only on the skin depth but also on the strength of stray anomalies from near-surface conductivity variations (geological noise) and instrumental sensitivity. It appears from experience that this depth varies from 1 to 5 times the separation between the transmitter and the receiver if geological noise is small.

IV. Classification of Electromagnetic Prospecting Systems

The large number of available EM systems and their great diversity makes the task of classifying them difficult. The methods may be classified, for example, based on the origin of the primary fields or on the nature of the primary field waveform. It is also possible to classify them based on whether the primary field is a plane wave or an inhomogeneous wave from a finite source. Another way of classifying them may be on the basis of where it is used — whether it is used on the ground, in the borehole, or from an aircraft. Figure 4 illustrates a classification scheme based on all these factors.

EM methods are used for either profiling or sounding types of survey. In horizontal profiling, one is interested in detecting lateral inhomogeneities in the ground. In depth sounding systems, one is more interested in determining the variation of structure of the ground with depth by changing either

geometry or frequency. Various systems belonging to the different groups mentioned in Figure 4 will be described in Sections V–VII, with the exception of the first natural EM method, which will be dealt with elsewhere in this encyclopedia in the article "Geomagnetism." Radar methods, which are used to obtain information about the surface using reflections of very high frequency EM waves, will also not be discussed in this article.

V. Ground-Based Systems

A. Natural-Source EM Systems

Three natural-source EM methods, namely, the telluric, the magnetotelluric, and the AFMAG methods will be discussed in the following sections.

1. Telluric Current Method

Telluric currents are natural, low-frequency alternating currents that are induced in the earth mainly by the interaction of the electrically charged solar winds with the magnetic field of the earth. These low-frequency currents are present everywhere and at all times, although their intensity and orientation change with time. [*See* Geomagnetism; Magnetospheric Currents.]

The existence of telluric currents have been known to scientists for well over a century. Telluric currents were first used in exploration in the 1920s in Europe, North Africa, and the Soviet Union mainly as a tool for oil exploration. The method involves the measurement of the potential gradient or electric field at a site using two electrodes placed on the ground (and occasionally in drill holes). The distance between the electrodes is usually a few hundred meters. Nonpolarizing electrodes are generally used for potential measurement to minimize the effects of contact potentials. Any static potential between the electrodes, which may be several hundred millivolts, must be balanced out using a voltage compensator or any similar device. Since the source of the telluric fields is located at a large distance from the observation point, the incident field is assumed to be uniform, and the current lines are parallel to the earth's surface for homogeneous or layered ground.

Since magnetic fields are not recorded in this method, variations in the electric field recorded at any site can be ascribed to either a change of the inducing magnetic field or a change in the electrical properties of the subsurface. To separate the two effects, observations are made at two locations simultaneously, one at a base station, and the other at a field station, which may be several kilometers apart. Since the telluric field signals vary both in amplitude and azimuth with time, the base and field stations normally have two pairs of

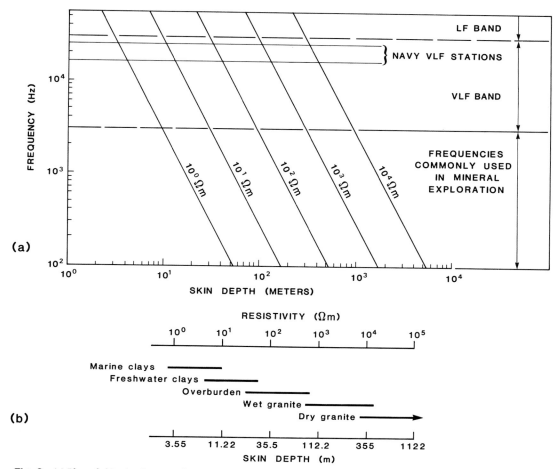

Fig. 3 (a) Plot of skin depth versus frequency for several resistivity values. (b) Resistivity ranges of common surficial materials with their corresponding skin-depth values at 20 kHz.

electrodes, one N–S and the other E–W. In this way, continuous records of two horizontal electric field components are obtained at each station. It is important that precise time correlation (to a fraction of a second) be established between the field and base stations. This is done by using radio time signals, direct radio communication, or synchronized crystal clocks.

With a setup like this, in theory one can compare the horizontal components of electric field variations between the base and field stations with regard to frequency, amplitude, and phase. When the electric field intensities are normalized by the intensity values at the base, the resultant field at the field station generally describes an ellipse. The interpretation of the telluric data is based on the characteristic dimensions of the elliptical patterns traced out by the normalized electric fields at the field stations. However, since no absolute parameters are measured, the method is used pri-

marily as a reconnaissance mapping tool and provides no direct information on the depths of anomalous bodies.

2. Magnetotelluric Method

The magnetotelluric (MT) method, developed in France and the Soviet Union in the 1950s uses the same natural EM fields as the telluric method. The method uses simultaneous measurements of orthogonal horizontal electric and magnetic field components of the natural EM field as a function of frequency to determine the resistivity distribution inside the earth. French geophysicist Louis Cagniard in a classic paper in 1953 showed that for a plane EM wave propagating vertically into the earth, the ratio of the orthogonal horizontal electric and magnetic fields at any frequency is directly proportional to the square root of apparent resistivity of the ground at that frequency. Since the frequency spectrum of MT fields varies from 10^{-5} to 10^4 Hz, depth soundings from very large to very

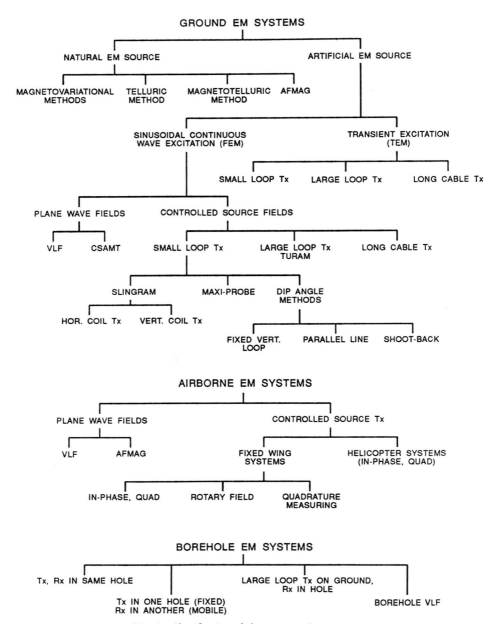

Fig. 4 Classification of electromagnetic systems.

shallow depths can be achieved with this method, although frequencies greater than 100 Hz are not normally used in standard MT surveys.

The early models used in MT sounding interpretations were one-dimensional (i.e., the earth was modeled as a layered medium); and hence, the resistivity variations were considered to be functions of depth only. In such a case, the impedance and the apparent resistivities are independent of direction of measurement. However, such simple models are rarely encountered in reality, and the earth in most areas is laterally inhomogeneous, which makes the apparent resistivity direction-dependent. In such cases, two horizontal orthogonal telluric fields and three components of the orthogonal magnetic field are measured.

The measurement system for MT soundings consists of electric and magnetic field sensors, amplifiers, filters, and recorders. While the recording of electric field parameters is straightforward, the technology for accurate measurement of small magnetic fields is more complex. Two types of magnetometers are used for measurement of magnetic fields in the entire frequency range of MT measurements, namely, induction coils and cryogenic magnetometers called Squids (*Super*conducting *qu*antum *i*nterference *d*evices).

The field data are used to determine the impedance of the ground in orthogonal directions in a wide frequency band. From that, apparent resistivity amplitude and phase parameters may be plotted against time. The data may be interpreted graphically for layered ground or by an inversion in the general case. Two recent variations of the standard MT are called AMT (audio magnetotellurics) and CSAMT (controlled source audio magnetotellurics). The first refers to a method in which higher frequencies in the audio frequency range ($10 - 10^4$ Hz) are used for determining the earth's structure at medium to shallow depths. The CSAMT method is described in Section V, B, 1, a.

3. AFMAG Method

The audio-frequency magnetics or AFMAG method is an inductive EM technique where the source is audio-frequency natural EM fields arising from atmospheric discharges (sferics) mainly from worldwide thunderstorm activity. Anthrosogenic noises from many sources, for example those due to urban electrification or jet aircraft vapor trails, also contribute to the AFMAG energy. The sferic fields arise in three principal thunderstorm centers of Central Africa, South-Central America, and the East Indies and are propagated in a spherical waveguide bounded by the ground surface and the atmosphere. The sferic energy is strongest in equatorial regions during summer and weakest in polar regions, especially in winter.

Since the source of AFMAG energy is at great distances from the survey area, the field is essentially plane with a horizontal magnetic field. Near a zone of lateral resistivity change, especially in the vicinity of a highly conductive zone, the plane of polarization of the magnetic field is changed from horizontal due to the presence of vertical secondary magnetic fields. The azimuth of the field will also be less random and will be directed normal to the orientation of the conductive zone. The field measurements consist of the tilt angle and azimuth of the natural magnetic field at two widely separated frequencies, such as 150 and 510 Hz.

AFMAG surveys are conducted using traverses normal to the assumed geological strike. In common with all natural EM methods, this method has the advantage that no transmitters are needed and the depth penetration can be large, since low frequencies are used and the geometrical decrease factor is zero (plane waves). Its disadvantages include the effects of large random changes in the amplitude and direction of the inducing field and weak strength of the field at high latitude, especially in winter months. In common with all plane-wave inducing fields, the method preferentially excites long linear conductors of large strike extent, such as faults and shear zones. Hence the method is more useful for geological mapping than for exploration.

B. Artificial-Source EM Systems

Most EM systems use controlled sources in geophysical exploration. While this adds to the complexity of the systems, it also enables the geophysicist to choose an appropriate source for investigating a particular target and to select the parameters of the transmitter (Tx) so that the coupling between the primary Tx field and the target is optimum.

1. Sinusoidal Continuous-Wave (CW) Techniques

Ground-based EM systems using sinusoidal CW technique were the first to be developed for exploration purposes. The CW techniques are divided into two groups: those that use plane waves and those that use an inhomogeneous primary field from a controlled source close to the receiver.

a. Techniques Using Plane Waves

Two methods that use plane or quasi-plane sinusoidal EM fields from artificial sources are the very low frequency (VLF) and the controlled source audio magnetotelluric (CSAMT) techniques.

The *VLF method* uses plane-wave EM fields from distant Navy transmitters located strategically around the earth in the frequency range of 15 – 25 kHz (VLF band). The method is based on measurement of the variations of the induced magnetic fields produced by lateral inhomogeneities relative to the horizontal primary magnetic fields. Thus only a portable receiver is required to measure the secondary field in this method, making it one of the popular reconnaissance techniques. The name of the method is somewhat misleading, since the frequency of operation (15 – 25 kHz) is much higher than the ones used in other induction systems, and the depth penetration of VLF systems is somewhat limited.

Early VLF receivers measured the tilt of the major axis of the ellipsoid of polarization and the vertical quadrature component. More recent versions can also record the total VLF field in addition to those two parameters, using three mutually orthogonal coils. Modern VLF receivers can tune in from one to three frequencies simultaneously. The VLF method can also be used to measure the apparent resistivity of the ground by measuring ground impedance; this is done by simultaneously recording orthogonal horizontal electric and magnetic field components. The VLF tilt angle measurement is similar to the tilt measurement in the AFMAG system with the advantage that the primary field direction is fixed and the signal strength is fairly uniform. As with other plane-wave EM systems, the system detects a conductor when its strike is normal to the primary magnetic field. In places where such a VLF field is not available, a large loop, called a local-loop VLF transmitter, may be used to simulate a VLF field several kilometers away from the loop. In conjunction with the field from a Navy VLF station, a properly positioned local-loop VLF transmitter can help in producing a total conductor map of an area.

The *CSAMT method* simulates the magnetotelluric method using an artificial source in the audio frequency range. The transmitter is generally a long line source (2–4 km) carrying alternating current in a large frequency range (4–10,000 Hz). The fields are measured normal to the line several kilometers from the line such that the measurement points are in the far zone — that is, when the distance is greater than several skin depths in the medium. The field is quasi-plane in the far zone and hence simulates the natural AMT method. When the criteria of far zone field is not met, the primary field is not a plane wave. Complicated corrections are required to convert the near-zone information to far-zone data in such cases.

b. Techniques Using Controlled Sources

Techniques using controlled sources and sinusoidal CW excitation include those that use a small loop, a large loop, or a long cable as transmitters. The methods using small loops as transmitters are numerous; techniques using a large loop or a long cable as transmitters are used when deep penetration is required.

The *slingram method* is one of the oldest and most popular small-loop transmitter systems for EM exploration. It uses two small loops, one as the transmitter (Tx) and the other as the receiver (Rx) with a connecting cable between them. The Tx and Rx are moved in tandem, keeping a constant spacing between them. Frequently, the coils are horizontal and co-planar, which is ideal for detecting horizontal layerings and semi-vertical conductors at depth. The coils are maximum-coupled in this configuration. A variation in which the Rx is vertical to produce minimum coupling is also used sometimes.

The Rx coil receives both primary and secondary field components, out of which the primary field is removed, or "bucked out," by comparing the total field with a reference primary field carried by the connecting cable. The secondary field is then split into in-phase and quadrature components, using a phase-comparator circuitry, and recorded. Most modern slingram systems can operate at several frequencies, transmitted sequentially, and also with several Tx–Rx separations. The accuracy of measurements depends greatly on the correct coil separation and orientation. For example, the error in in-phase can be ±1% for an error of ±30 cm for 100-m coil separation. A misorientation error of 4% in level-

ing can produce an error of ±1% in the in-phase response. A recently developed slingram system called Genie has eliminated the need for the connecting cable by measuring the ratio of the field parameters at two frequencies simultaneously.

Maxi-probe is a multifrequency dipole–dipole system in which the Tx consists of two or three horizontal loops, approximately 10 m in diameter, connected in parallel so that fields from all loops reinforce each other. The Rx consists of a pair of orthogonal vertical and horizontal coils located at a distance of 200 m to 1 km from the Tx. The system, powered by a gasoline generator, can transmit up to 128 frequencies from 1 Hz to 60 kHz sequentially. The Rx measures the ratio of the vertical and horizontal magnetic field components and their phase differences. The results are interpreted in terms of the parameters of a layered ground. The data may also be converted to tilt angle and ellipticity of the ellipsoid of polarization, which may be interpreted by numerical inversion for layered earth or for detection of lateral inhomogeneities. Similar systems have been developed in France with fewer frequencies. The system was used successfully for deep EM exploration over the permafrost terrain in the arctic regions of Canada and over unfrozen terrain down to depths of 300–600 m.

The *dip angle method* measures the dip or the inclination of the total magnetic field about a horizontal axis at a number of stations. There are generally three variations of this method. The fixed vertical loop method, developed in the 1920s, is still used occasionally because of the simplicity of operation. The Tx coil is vertical with several turns of wire and free to rotate about a vertical axis. The Rx coil has several hundred turns wound on a ferrite core. The Tx is placed in the center of the survey area and the Rx is moved along traverse lines normal to the geological strike. For each Rx location, the Tx is rotated so that the planes of both lie along the line joining them. When so arranged, the Rx is tilted about the axis formed by joining the Tx and Rx for minimum signal. Operations can be speeded up by having another Rx traverse a parallel line so that their planes lie along the same line. The dip angle is zero in the absence of a conductor. When a steeply dipping conductor is present, a typical crossover type of response is obtained; the inflection point determines its location.

In the parallel line or *broadside method,* the Tx is portable and moved along with the Rx, the two moving along parallel lines. The Tx always points toward the Rx, which is horizontal and is rotated about the Tx–Rx axis to obtain a null. The line joining the Tx and Rx is maintained parallel to the general strike for maximum coupling; their separation is approximately 150–200 m. More than one receiver can be used simultaneously to speed up the survey, but they must all be along lines parallel to the strike.

The *shoot-back method* is used over hilly or rough terrain

where the slingram method has difficulty maintaining correct alignment. However, the Tx and Rx coils are interchangeable in this method. Thus, when coil 1 is transmitting, its axis is directed toward the Rx (coil 2), but at a finite angle, about 15° below the horizon. Coil 2 is rotated about a horizontal axis normal to the traverse to get a minimum. Next, coil 2 becomes the Tx at the same location; its axis points the same amount (15°) above horizon and toward coil 1. Coil 1 is now rotated to obtain a minimum. The difference between the two dip values is plotted at the middle point between the Tx and Rx. This value will be zero in the absence of a conductive inhomogeneity even if there is an elevation difference between them. A maximum response is obtained over a semi-vertical conductor.

In *large-loop transmitter systems* a large-loop Tx carrying a large alternating current is used, as in the so-called *Turam* method. Usually a large rectangular loop is laid on the ground and traverses are taken normal to the longer side, which is parallel to the geological strike. In one variation, the Rx is a vertical coil with its axis along the traverse to measure the horizontal component of the magnetic field. The measured values are then normalized with precomputed values of field over a half-space at various positions. The plotted normalized values provide indications about the presence of lateral inhomogeneities. In another variation, two small horizontal coils are used as receivers. The spacing between the coils is maintained constant, and the amplitude ratios and phase differences between the fields recorded at the two receivers are plotted. These parameters can also be normalized with precomputed values, so that the ratio stays at 1 and the phase difference at zero when no conductors are present. A finite anomaly is obtained when anomalous conductors are traversed.

In the *long cable method*, a long cable grounded at both ends is laid parallel to the geological strike. A large alternating current is sent through the cable and traverses are taken normal to the wire in a fashion similar to that described with the large-loop transmitter method. The measurement parameters are also similar to that method. Often two frequencies are used (220 and 660 Hz) to help discriminate between deeper conductive features and shallow overburden effects. The last two techniques have the advantage of large depth penetration and a wider coverage with one setup of the transmitter. A disadvantage of the two methods is that long, moderately conductive features like shear zones are accentuated in the response at the expense of smaller, potentially more economic conductors. In this method, conduction currents are involved in the excitation, which complicates the total response and the interpretation.

2. Transient EM Methods

Transient EM (TEM) methods use repetitive current pulses instead of sinusoidal currents for excitation. TEM methods have gained popularity in the last two decades because of their ability to measure the secondary field in the absence of the primary field. Furthermore, the TEM decay recorded over a finite length of time has the same information as the frequency domain EM (FEM) response over a range of frequency that must be recorded sequentially. In most TEM systems, a steady current is passed through a loop or long wire and then abruptly terminated in a controlled fashion. The rapid decrease of the primary field flux generates an electromotive force (emf) in nearby conductors proportional to the rate of change of flux. This emf causes eddy currents to flow in conductors, which decay in a characteristic manner depending on the conductivity, size, and shape of the conductor. The decay generates a secondary field, the time rate-of-change of which may be recorded by an Rx coil when the primary field is zero.

Theoretically, TEM and FEM measurements are equivalent. However, practical considerations make TEM measurements easier and more reliable since it is possible to measure the decay a few tens of milliseconds after the current drops to zero. This is equivalent to taking measurements at extremely low frequencies, which is difficult to do in FEM. The TEM systems have the biggest advantage in areas with high surficial conductivity (clay or salt). It can be shown that to detect a finite conductive target under conductive overburden or host rock, observations must be taken over a finite time range or *window* where the target response is greater than the background response. Another advantage of the TEM system is that the spatial position of the Rx does not have to be absolutely correct since the late time response (when the current distribution has stabilized) is insensitive to the distance between the Tx and Rx. There are three basic variations of the ground TEM systems; they are discussed below.

a. Moving Source Small-Loop Tx

In this arrangement, the Tx and Rx are both moved in tandem, somewhat like that in the slingram system. The Tx is normally a horizontal loop on the ground as in the Crone pulse EM (PEM), or a square loop (50–100 m on a side) as in the Sirotem system. The Rx may be separated as in the Crone PEM or coincident with the Tx as in the Sirotem. Geonics Ltd., Toronto has recently introduced a system somewhat similar to the Crone system. Unlike the FEM slingram systems, small errors in coil separations or elevation differences between the Tx and Rx do not produce significant errors in the decay amplitudes. The depth penetration is somewhat limited with this type of system.

b. Fixed Source Large-Loop Tx

In this method, a large rectangular loop with dimensions of several hundred meters on each side is used as the Tx, similar to that used in Turam. The survey lines are taken normal to the longer side, which is generally taken parallel to the geological strike. The Rx consists of a small multiturn loop, which may be oriented in any horizontal or vertical direction. Before commencing the survey, the timing circuitry of the Tx and Rx are synchronized using temperature-controlled crystal clocks or reference cables. In all cases of TEM survey, the receiver amplifies and stacks the signals over many cycles to improve the signal/noise ratio, and then the signals are recorded. Most commercial TEM systems use rectangular pulses of excitation with base frequencies in the range of 0.75 to 300 Hz. A system called UTEM uses a precision sawtooth pulse in the Tx; the secondary field decay is measured in the presence of the primary field, which is removed from the total field at each channel during processing. Large-loop TEM systems with low base frequencies have large depth penetration and can cover a substantial area for each Tx loop position. These methods have been used successfully for exploration of minerals (uranium, sulfide conductors) at large depths and for geological mapping over frozen and unfrozen ground down to depths of 700 m or more.

c. Long Grounded Cable Tx

In this method, a grounded cable 1 to 3 km long is injected with a square-wave electric current, which induces both galvanic and eddy currents into the ground. The currents can range from 10 A to 100 A or more when very deep penetration is desired. The Tx is kept fixed while the receiver, which can detect both electric and magnetic fields, can be placed anywhere around the Tx up to a distance of 20 km away from the Tx. The decays are recorded up to 10 s after the current is reduced to zero, thus providing large depth penetration of the order of the Tx–Rx separation. This type of survey is therefore suitable for hydrocarbon detection or deep crustal studies. It can be shown that resistive inhomogeneities at depth can best be detected using electric field transients from a long grounded source, whereas good conductors are best detected using an inductive loop source. This method is expensive to use and can be justified only when great depth penetration is needed.

VI. Airborne Systems

The variations in airborne EM systems are almost as great as in ground-based EM systems. The principle of airborne EM methods are similar to their ground counterparts with the additional complication that the systems must be more portable, capable of orienting themselves in the air, and able to record measurements fast — of the order of 6 to 20 measurements per second. The development of airborne EM was pioneered in Canada for locating economic mineral deposits, but significant developments were also made in Scandanavian countries (Sweden, Finland). [*See* REMOTE SENSING OF EARTH RESOURCES FROM AIR AND SPACE.]

A. Airborne Systems Using Plane Waves

Two airborne EM systems use plane-wave primary EM fields. The airborne VLF method uses VLF fields transmitted by Navy VLF transmitters. However, it is impossible to measure the tilt of the ellipsoid from a moving platform. Hence, airborne VLF systems measure the total magnetic field using three orthogonal coils and the vertical quadrature components. The airborne VLF systems are normally used in conjunction with other geophysical sensors. The latest airborne VLF system can record the data at two VLF frequencies simultaneously. Because higher frequencies are used, the depth penetration achieved with airborne VLF system is not large. The system also preferentially detects long linear conductors whose strikes are normal to the direction of the primary VLF field. Airborne VLF fields were used in the early 1970s to measure the electric wavetilt (ratio of horizontal to vertical electric field), which provided a measure of the apparent resistivity of the ground in a system called E-Phase. The system is no longer in use.

Airborne AFMAG systems that use the natural EM fields in the audio frequency range have been used in exploration for some time. The sensor consists of a pair of orthogonal coils mounted in a "bird" such that their axes are along the direction of flight, each 45° off the horizontal. The direction of flight is generally normal to the general strike. Signals from the two coils are compared at two frequencies, such as 140 and 510 Hz, in such a way that the variation is proportional to the tilt angle. The depth penetration by this system is good because of the low operating frequency; the system is particularly useful in geological mapping, especially for long linear conductors. However, the airborne AFMAG system suffers from the same problem of low field strengths at high latitude, especially in the winter months, and the random nature of the primary field that affect ground AFMAG surveys. The system is not currently in use.

B. Airborne Systems Using Controlled Sources

Most of the currently available airborne EM systems use small coils (magnetic dipoles) as transmitters for convenience and greater control over the nature of the primary field. Early systems in this group used fixed-wing aircraft, which have low operating costs and permit the use of larger transmitter loops. However, the recent trend is toward using helicopter-borne systems at closely spaced lines, which permits flexibility of use and excellent signal/noise values.

1. Fixed-Wing Airborne EM Systems

The earliest fixed-wing airborne EM systems used coils wound around the aircraft as Tx and small coils to measure the in-phase and quadrature components of the secondary magnetic fields using sinusoidal excitation. The Tx coils were mounted on either the wing-tips or the nose and tail of the aircraft. A transient EM system, called INPUT (INduced PUlse Transients) appeared in the 1960s in which the Tx was wound around the aircraft (horizontal loop) and the receiver coil was placed in a bird trailing below and behind the aircraft with its axis along the survey direction (vertical coil). This has been one of the most successful airborne EM systems in detecting hidden orebodies in the past 20 years because of its large depth penetration and good signal/noise ratio. Modified versions of this system, as well as a similar system with a digital receiver called GEOTEM, are still in extensive use in exploration and mapping.

Most fixed-wing airborne EM systems used one frequency to measure the in-phase and quadrature components of the secondary field in parts per million (ppm) of the primary field. A system called Tridem was developed in the 1970s that used three coil pairs at three frequencies simultaneously. However, poor signal/noise values led to the abandonment of this system in North America in the 1980s.

In the Rotary system, developed in Sweden, two aircraft are used for the survey, one carrying the Tx and the other the Rx. The planes fly with constant separation. The Tx uses two identical orthogonal coils with a 90° phase difference to produce a rotary field. The total field recorded by a similar orthogonal coil pair is passed through phase-shifting circuitry to reduce the differential output to zero over barren ground. A finite signal indicates the presence of a buried conductor. This system is also not currently in use because of the high cost of operation.

In the quadrature measuring system, the Tx is strung between the wing tip and tail (horizontal coil) and the Rx is vertical and carried in a bird such that its axis is along the survey direction. Generally, two frequencies (400 and 2300 Hz) are transmitted simultaneously and the ratio of quadrature responses (Q_{400}/Q_{2300}) provides a rough criterion for the conductivity of an anomalous body, ratios greater than 1 indicating a good conductor. The system is no longer in use.

2. Helicopter EM Systems

Helicopter EM (HEM) systems have become the systems of choice for airborne EM surveys in the last 10 years. In almost all HEM systems, both Tx and Rx are placed inside a nonmetallic cylindrical shell (bird) and towed below and behind the helicopter. The systems measure the in-phase and quadrature components with excellent resolution and accuracy. Because of the excellent aerodynamic features of the bird and the maneuverability of the helicopter, HEM systems can fly close to the ground at slower speed, thus increasing the resolution and usefulness of the data. Presently, all HEM systems are flown in the multifrequency mode, with three or four frequencies and with several coil combinations for detection of

horizontal layerings and steeply dipping sulfide conductors. The art of building airborne EM systems has probably reached its zenith in the design of modern HEM systems, which have nominal noise levels of less than 1 ppm in several systems. The systems can fly closely spaced lines (100 m apart) at elevations of 40 m or less, thus approaching the accuracy and resolution of ground EM surveys, at substantial savings in time and expense.

VII. Borehole Systems

Borehole EM systems, although not new, have not reached the diversity and sophistication achieved by ground-based and airborne EM systems. Induction logging systems, with both Tx and Rx placed inside the same hole a fixed distance apart, became available many years ago and have been used in mineral and hydrocarbon exploration. A variation of this technique, called hole-to-hole survey, involves placing a Tx in one hole and recording the response at several nearby holes. Presence of a conductive inhomogeneity between the holes will cause an anomalous response at the receiver.

Recently, large-loop borehole EM systems using frequency and transient excitation have become available. In such systems, a large-loop transmitter is laid on the ground and excited either by frequency or, more commonly, a transient EM source, and the receiver is moved in one or more holes to record the response. A borehole VLF system has recently been developed for engineering purposes, such as for detection of weak fracture and shear zones in restrictive plutonic rocks, which may be used as repositories for spent nuclear wastes. The technique is still experimental, and the interpretation is largely empirical. [*See* Nuclear Waste Disposal.]

VIII. Application of Electromagnetic Methods

Electromagnetic methods are used for solving a wide variety of problems requiring depth penetrations ranging from shallow (1–10 m) to depths up to 1 km or more for deep mineral or crustal exploration. But all the methods are applied in either one of the two basic modes, namely, profiling (detection of lateral contrasts in electrical conductivity) or sounding (investigation of the conductivity distribution at one position with depth). Sometimes both modes of operation are combined.

A. Profiling

Electromagnetic profiling can be performed using both natural and controlled-source techniques. It can be done using either a fixed-source (Turam) or a moving (slingram) transmit-

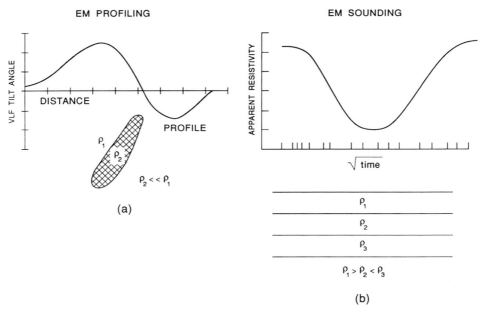

Fig. 5 (a) A typical VLF tilt-angle profile over a sheet-like conductor. (b) A typical transient EM sounding plot of apparent resistivity versus square root of time.

ter; the receiver is mobile in both cases. Figure 5a shows the essential features of a profile survey, in this case a VLF tilt-angle profile, over a sheet-like conductor. The Tx is fixed and the Rx is moved along a line normal to the strike of the conductor. The response shows a typical crossover type anomaly; the inflection point indicates the location of the conductor.

B. Sounding

Multifrequency sounding can be performed either by changing the frequency while keeping the separation between Tx and Rx constant or by changing the coil spacing keeping the frequency constant, or a combination of the two. In frequency-domain sounding, it is preferable to conduct the sounding by changing the frequency rather than changing the separation. Decreasing the frequency while keeping Tx–Rx constant permits greater depth penetration without bringing in the effects of lateral inhomogeneities, as is the case when greater depth penetration is achieved by increasing the separation between Tx and Rx. In transient EM sounding, greater depth penetration is achieved by taking decay measurements over a longer period of time; longer decay time is equivalent to lower frequencies in FEM. Figure 5b shows the sounding data in TEM measurements. The computed late-time apparent resistivities plotted against square root of time shows the influences of deeper layers as the decay time increases. There is however a limit to how low a frequency can be or how long

a decay time can be recorded, since the signal/noise ratio decreases rapidly at low frequency or long decay-time intervals. In the past few years, several modern TEM systems have become available that permit EM sounding down to depths of several hundred meters.

IX. Electromagnetic Interpretation Methods

Electromagnetic data are normally interpreted by comparing the field data with precomputed response over anomalous bodies of various shapes and sizes that are likely to be present in that area. Such response can be obtained analytically in the case of bodies of regular shapes, such as spheres, cylinders, or plates; by numerical modeling on a digital computer for bodies of complex shape; or by physical scale-model experiments in the laboratory for bodies of even more complex shape. Analytical solutions are also available for horizontally layered ground excited by a plane-wave, dipolar source or by a long cable carrying alternating current.

The interpretation can be accomplished by graphical or analytical techniques. In the first case, master curves or interpretation diagrams are prepared for the interpretation of anomalous bodies of regular shape from the analytical or scale-model responses of such bodies for variables such as depth, conductivity, thickness, etc. Many such interpretation diagrams are available in geophysical literature. However, the method may be tedious and the interpreter does not get an idea of the cumulative error made in the interpretation in the form of any least-square errors. In the analytical approach, after the nature of the anomalous body has been guessed, some initial parameters for it are entered into a computer program. The program computes the response of the initial model, compares that to the field data, and decides which parameters of the body should be changed to bring the computed response closer to the field response. The process continues until a model is found for which the least-square error between the computed and field data set is minimum. At this stage, the interpreter can view the result and change one or more parameters of the model to bring the response closer to the field response. The advantage of the analytical method is that it avoids the personal bias of the interpreter. But it still needs a good initial model in order to converge to a geologically sound model.

Bibliography

Fitterman, D. V., ed. (1990). Developments and applications of modern electromagnetic surveys. *U.S. Geol. Surv. Bull.* **1925.**

Nabighian, M. N., ed. (1988). "Electromagnetic Methods in Applied Geophysics — Theory," Vol 1. Society of Exploration Geophysicists, Tulsa, Oklahoma.

Palacky, G. J., ed. (1986). "Airborne Resistivity Mapping." Geological Survey of Canada, Paper 86-22.

Sinha, A. K. (1988). EM soundings for mapping complex geology in the permafrost terrain of northern Canada. *In* "Permafrost: Proceedings of the Fifth International Conference on Permafrost" (K. Senneset, ed.), pp. 994–999. Tapir Publishers, Trondheim, Norway.

Sinha, A. K. (1990). Interpretation of ground VLF-EM data in terms of inclined sheet-like conductor models. *Pure Appl. Geophys.* **132,** 733–756.

Sinha, A. K. (1990). Stratigraphic mapping of sedimentary formations in southern Ontario by ground electromagnetic methods. *Geophysics* **55,** 1148–1157.

Sinha, A. K., and Hayles, J. G. (1988). Experiences with a local-loop VLF transmitter for geological studies in the Canadian Nuclear Fuel Waste Management Program. *Geoexploration* **25,** 37–60.

Van Zijl, J. S. V., and Köstlin, E. O., ed. (1985). "The Electromagnetic Method — The Field Manual for Technicians No. 3." South African Geophysical Association, Johannesburg, South Africa.

Vozoff, K., ed. (1986). "Magnetotelluric Methods," Geophysics Reprint Series No. 5. Society of Exploration Geophysicists, Tulsa, Oklahoma.

Glossary

AFMAG Abbreviation for audio-frequency magnetic technique, in which natural EM fields in the audio-frequency range are used to study lateral changes in earth resistivity.

AMT Abbreviation for audio magnetotelluric method, in which magnetotelluric measurements are made in the audio frequency range ($10–10^4$ Hz) for medium to shallow depth investigation.

Bird Sensor suspended from an aircraft by a cable to make geophysical measurements.

CSAMT Abbreviation for controlled source audio-magnetotelluric method in which an artificial source, generally a long cable, is used to simulate the audio-frequency magnetotelluric fields.

FEM Abbreviation for frequency electromagnetic method in which the primary field is sinusoidal in character.

Host rock Rock surrounding an economic ore deposit, which usually has much greater resistivity than the deposit.

Impedance Apparent resistance to flow of alternating current; it is analogous to electrical resistance in direct current method and is a complex quantity. For plane wave EM sources, the impedance of the ground is equal to the ratio of the orthogonal horizontal electric and magnetic fields.

Induction Process by which a body becomes magnetized or electrified by merely placing it in a magnetic field or electric field. Also refers to the process under which electric currents are initiated in a conductor by merely placing it in an EM field.

Sferics Natural atmospheric fluctuations of the EM fields, caused mainly by thunderstorm activity, in the range of $1–10^5$ Hz.

Skin depth Measure of the effective depth of penetration of EM fields. It is defined as the depth at which the amplitude of an EM field drops to 37% of its surface value.

TEM Abbreviation for transient electromagnetic method, in which the primary energizing field is a repetitive pulse.

Turam Electromagnetic survey method in which the energizing source is a long, grounded, insulated cable, or a large, rectangular, horizontal loop excited by one or more frequencies in the range of 100–1000 Hz.

Environmental Injury to Plants

F. R. Katterman
University of Arizona

Environmental injury to plants is a study of the various facets of the environment such as drought, heat, chill, freezing, high salinity, and oversaturation of the soil with water that initiate a high-enough level of stress within the plant to cause an observable or hidden injury. Such damage can either be irreversible or countered by an alteration of the physiological and/or metabolic reactions that normally take place within the plant cell. Such studies are of importance in that they determine the environmental limit of valuable agronomic or horticultural plant population boundaries. In addition such studies can also help to engineer various strategies that enhance some of the coping mechanisms of the plant to stave off injury.

I. Introduction

In the past decade there has been a gradual change in the earth's climatic patterns. These changes are thought to be attributable in part to the large-scale interference and influence of mankind on the global environment. Massive production of carbon dioxide from industrial factories and greater use of the internal combustion engine are causing a gradual increase of the mean global temperature. This temperature increase is thought to occur when radiant solar energy from the sun cannot be easily reflected back into the outer atmosphere due to a gradual increase of a carbon dioxide blanket in the upper atmosphere. This trapped high energy is then converted into a lower energy form represented by heat. Such a process is known as the "greenhouse effect." The latter is

also enhanced by the excessive release of fluorinated hydrocarbons into the atmosphere. These compounds react with atmospheric ozone to destroy portions of the ozone layer that protects the earth from excessive high-energy radiation (such as ultraviolet light). Thus more heat results from the extra radiation that is allowed to pass through the partially fragmented ozone cover from the sun. [See ATMOSPHERIC OZONE; OZONE, ABSORPTION AND EMISSION OF RADIATION; SOLAR RADIATION.]

Normally, much of the accumulated carbon dioxide can either be absorbed into the ocean or photosynthetically converted into oxygen via the vast areas of rain forest foliage scattered throughout the world. However, with the increased population of mankind and the obvious need to clear land for settlement and grazing, much of the primeval rain forest areas are being systematically destroyed. Therefore, the carbon dioxide blanket can exist as a stable entity or imperceptibly increase over a finite amount of time. [See PHOTOSYNTHETIC CARBON METABOLISM AND ATMOSPHERIC CO_2.]

With such a significant and stable rise in the mean global temperature, the stage is set for gradual but dramatic shifts and changes in climatic patterns that can enhance several kinds of environmental stress responsible for a large-scale injury and death of important horticultural and agronomic plants that supply food and fiber for humanity. In this article, the following types of stress that reflect these global changes of climatic patterns will be emphasized: chill and freezing injury due to the abnormal onset of winter in regions not normally affected by stable climatic conditions; drought and heat stress above that normally encountered in regions now becoming more arid in nature; osmotic stress that occurs after an extensive accumulation of salts on irrigated arable land located in the arid regions; and finally, anaerobic stress produced by soaking and flooding of fertile areas of soil subjected to greater-than-normal rains and ice thaws during the spring season. [See ATMOSPHERIC CO_2 AND TREES, FROM CELLULAR TO REGIONAL RESPONSES; VEGETATION, EFFECT OF RISING CO_2.]

II. Chill and Freezing Injury

Plants that originate from warm climes are susceptible to low chilling temperatures (from 0 to 15°C) in that they show signs of injury, such as wilt and bleaching, as well as changes in physiology that affect seed germination, growth, and photosynthesis. In some cases even reproductive development is affected. Examples of valuable crop plants that are cultivated in warm weather are cotton, soybeans, tomato, rice, corn, sorghum, and cucumbers. Besides the obvious visual appearance of chilling damage, a more subtle level of damage is evidenced by a multiple array of biochemical and cellular changes. Included are changes in rates of respiration, protein synthesis, protoplasmic streaming, altered membrane permeability, and inefficient photosynthesis. In addition, chilling exposure in the light is more deleterious than under darkened conditions.

In contrast to chilling, freezing damage is much more severe. In plant tissues, crystallization of cellular water as the temperature drops below 0°C can occur anywhere inside the cell and cause lethal damage of a physical nature to the cell. If many cells in the plant are affected in this manner, death of the plant can occur on thawing. In addition to ice formation within the cell (intracellular), the formation of ice can also take place in the intercellular spaces at subzero temperatures. When such an event takes place, liquid water will move from inside of the cell (due to a vapor pressure difference between the water and the ice crystals) and contribute to the growing ice crystals on the outside of the cell. As this process continues, the cell or cells will become dehydrated and undergo a condition similar to that of osmotic shock induced by a high salt concentration outside the cell.

A component of the cell that is initially acted upon by both chilling and freezing events is viewed by most investigators to be the plasmalemma membrane (the membrane on the exterior of the protoplast, in turn surrounded by the plant cell wall); this is known as the "membrane hypothesis." This is where the common ground of the two events (chilling and freezing) ends, however. The actual mechanism of lesion formation or membrane alteration differs in both cases. For chilling, the original concept of the membrane hypothesis proposed thermotropic lipid phase transitions of the membrane that is induced by low temperatures (0 to 15°C) in chill-sensitive plants for all classes of lipids. Such a temperature-induced liquid-gel phase was thought to initiate conformational changes in membrane-associated enzymes and membrane permeability that resulted in a cascade effect on cellular-biochemical processes. Notable among these processes were a cessation of protoplasmic streaming, respiration rate changes, alterations in photosynthesis, and modifications in protein synthesis. The initiating process was shown to be reversible as opposed to the time-dependent and nonreversible secondary or cascading events.

Recently, the membrane hypothesis became refined to a certain extent when the proposed role of the lipids as a whole in chilling sensitivity became more focused on one molecular species of the polar lipid fraction — namely, phosphatidylglyc-

erol. The influence of this small subpopulation of the membrane lipids on the phase separation event (and thus the relative sensitivity to chilling temperatures) depends largely on the phosphatidylglycerol molecular species composition (i.e., on the specific ordering of saturated and unsaturated fatty acid carbon chains on the phosphatidylglycerol molecule).

It has recently been proposed that the first important and perhaps controlling factor of the "cascading effect" is initiated by the altered permeability of the chill biological membranes. Under normal conditions, both chill-tolerant and chill-sensitive plant cells maintain a homeostatic level (about 10^{-7} M) of calcium (Ca^{2+}) in the cytosol. Recent studies show that chilling temperatures may increase Ca^{2+} by reducing the rate at which the chill-sensitive plant cells exclude Ca^{2+} from their cytosol; or alternatively, chill temperatures may activate voltage-dependent cation channels that transport Ca^{2+} into the chill-sensitive cell. A higher-than-normal level of Ca^{2+} in the cytosol is generally toxic and can lead to the death of the chill-sensitive cell due to an imbalance of the calcium/calmodulin activating system for inactive proteins or enzymes.

As with chilling injury, destabilization of the plasma membrane is also a primary site of freezing injury; it too plays a prominent role in cellular-biochemical behavior during a freeze–thaw cycle. Different opinions, however, are presented as to the cause and nature of plasma membrane destabilization, since many of the investigations measure the expression of injury after the freeze–thaw event. Recent investigations of a direct and less inferential nature, in which the plasma membrane of the plant protoplasts was observed during a freeze–thaw cycle, were able to characterize the conditions that lead to the destabilization of the protoplast plasma membrane. Supercooling at a specific temperature range in a given plant-cell cytosol is a function of the rate of heat transfer, which is then relative to the mass transfer of water. Water flux is determined by the water permeability of the plasma membrane. Generally, the greater the cooling rate, the greater is the incidence of intracellular ice formation, which is usually disruptive and lethal to the cell. Thus a plasma membrane with a relatively lower water permeability would tend to negate or slow down the formation of ice crystals in the cytosol.

Under conditions of nonintracellular ice formation, injury or plasma membrane destabilization during a freeze–thaw cycle seemed to be caused by two main factors: (a) osmotic shifts from the inside to the outside of a cell and vice versa, and (b) disorganization of the plasma membrane components induced by severe dehydration during freezing. The extreme changes in osmotic pressure during a typical freeze–thaw cycle usually cause lysis or cell bursting, which is one of the predominant forms of cell injury. Osmotic contraction results in the formation of large unilamellar vesicles or exocytotic extrusions from the plasma membranes of both nonacclimated and acclimated protoplasts. However, during osmotic expansion, the extruded material that is drawn back into the plasma membrane behaves differently between both types of protoplasts. The cold-acclimated protoplasts produced extrusions that were continuous with the parent bilayer during the contraction phase so that when expansion took place, the extruded material was smoothly drawn back into the bilayer. The extrusion withdrawal of nonacclimated protoplasts, on the other hand, does not proceed as smoothly as the acclimated protoplasts, and lysis generally results. The differential behavior of the plasma membrane observed between both types of protoplasts is attributed to changes in lipid composition. Recently, it was demonstrated that alteration of the proportion of a single lipid species in the plasma membrane was responsible for the difference in stability between both types of protoplasts.

If the freezing conditions are such that no ice crystals are formed in the cytosol of protoplasts frozen to temperatures below $-5°C$, more than 90% of the osmotically active water is removed from these cells. Under such conditions, lamellar to hexagonal rearrangements are induced in the ultrastructure of the plasma membrane, which results in a loss of semipermeability. The protoplasts are then in an osmotically unresponsive state, which leads to biochemical and physiological imbalances preparatory to injury or death. The exact mechanism that causes dehydration to induce the particular type of membrane rearrangements mentioned above has not yet been resolved. Several attractive theories have been advanced; however, most of them have to do with forces of hydration repulsion between the two bilayer components of the membrane and the pressures required to overcome the hydration forces so as to induce several different types of change in bilayer structure. Although these pressures are large, they do occur from the osmotic potential increase during freezing of the aqueous solutions that surround and are included within the protoplasts at temperatures ranging from $-10°C$ to $-20°C$.

Cold acclimation for many species of plants in the temperate and alpine regions involves an initial exposure over a given period of time to low nonfreezing temperatures ($0-10°C$) so that the plants will be able to alter the freezing tolerance of their cells. Although intensive study has been given to the cold acclimation process, not much is really known about the mechanisms that lead to increases in freezing tolerance, especially with regard to distinguishing between the specific biochemical changes that assume the main role in freezing-tolerance mechanisms from those that take place in adjustments to temperature in the nonfreezing or chilling ranges. In spite of such a lack of definitive knowledge in this area, there is increasing evidence that implicates the involvement of a genetic component connected with the plant's ability to acclimate to chilling temperatures followed by the induction of freezing tolerance. Plants grown and evolved in tropical regions do not have this ability, as seen by experiments in which a plant hormone, abscisic acid, is unable to turn on genes that cause cell culture of freeze-sensitive species to become more freeze tolerant. On the other hand, cell cultures of cold-hardy species can respond to this plant hormone and develop tolerance to chilling and freezing temperatures.

Although widely accepted that cold acclimation follows a well-defined inheritance pattern, there appears to be a disagreement as to whether cold acclimation is a multigenic process or whether a few genes control cold acclimation as seen with the heat shock and anaerobiosis models. Since cold acclimation is an inducible trait that depends on changes in cellular physiology, cold acclimation genes can be studied by means of standard molecular and DNA recombinant methods. Most of these studies have dealt with nucleic acid and protein changes in cold-hardy plants. At the onset of acclimation DNA remains constant while an accumulation of RNA occurs. Most of the RNA increase was attributed to ribosomal RNA, which made up the increased level of ribosomes and polysomes. The latter remain intact and are three times higher in content than that of nonacclimated tissues. This increase can be attributed to the need to synthesize more protein, since enzymatic processes are slowed down significantly at the lowered temperature. It would be of interest to know which mRNAs are being translated by the polysomes with regard to freezing tolerance. Recent attempts to clone low-temperature-induced genes revealed that three low-temperature specific cDNAs were isolated from the poly(A) RNAs (active messenger RNAs) induced by low temperatures.

The soluble proteins accumulate at the start of acclimation and closely follow the changes in tissue freezing tolerance. The appearance of a glycoprotein component in the soluble protein function was thought to be a factor for resistance to ice formation inside the cell. Recent protein labeling studies in plants exposed to low temperatures provide additional evidence for induction of changes in protein content by low temperatures. The pattern of protein synthesis differed significantly from that of plants exposed to warm temperatures. Part of this change demonstrated the appearance of new protein polypeptide bands not present in non-freeze-tolerant plant species. It is not clear whether these protein bands are similar to those that appear as heat and osmotic shock proteins and whether they appear in many species of cold-acclimated plants with enough consistency to be termed "cold-shock proteins."

In certain cold-acclimated plants there exists a group of soluble proteins of high molecular weight ($10,000-20,000$) capable of protecting membranes against freezing damage. Enzymes from cold-acclimated plants (such as esterases, phosphatases, amylase, and lactate dehydrogenase) also differ in isozyme patterns from those shown by nonacclimated plants.

Thus it appears that the mass of information gained in the study of cold tolerance is still in the descriptive stage and more work needs to be done to demonstrate whether the cold acclimation process is initiated by a few genes as seen in heat shock and anaerobioses.

III. Drought Stress, Injury and Adaptation

Lack of water is a stress factor that has been the source of much investigation, since this factor often limits agricultural productivity in arid and semiarid regions of the earth. Most of the important agricultural crop plants are either classified as hydrophytes (growth where water is superabundant such as rice) or mesophytes (where water is intermediate, such as cotton). Thus with changing global air movement and increasing heat, the plant population boundary with regard to drought is continually changing in certain parts of the earth. Drought injury manifests itself in plants that do not have built-in mechanisms to ward off the effects of severe lack of water (as opposed to Xerophytes, where water is scarce) such as bound water, small transpiring surfaces, leaf abscission, deep and wide-spreading root systems, etc. Irreversible drought-induced injury in a typical plant cell appears to be caused by mechanical forces when the vacuole shrinks during drying with a resultant inward pull and a simultaneous outward pull on the cytoplasm as exerted by the cell wall to which the cytoplasm is attached. The cell is contracted with a pulling in of the wall and a visible torn protoplasm (cytoplasm plus plasmalemma membrane). [See Arid and Semiarid Ecosystems; Drought; Evaporation, Transpiration, and Water Uptake of Plants.]

With a less severe drought where relatively smaller amounts of water are withheld, however, there is a slowing down or lag in cellular enlargement and subsequently in root and shoot growth in the plant. This lag period is followed by an adaptation period that consists of a sequence of events that cause the plant to resume normal operations on a less efficient but somewhat effective plane. During the lag period there is an alteration of biochemical and physiological processes ranging from protein synthesis to photosynthesis and solute accumulation. Throughout the stress period as a whole, the water and osmotic potential are reduced by equal amounts so that turgor pressure remains the same in the growing regions of the plant, even though growth first stops and finally continues after the lag period is over.

$$\Psi_p = \Psi - \Psi_\pi$$

where Ψ_p is turgor pressure, Ψ is water potential, and Ψ_π is osmotic potential. In addition, decreases in water potential seem to coincide with the occurrence of growth retardation caused by drought stress. The factors involved in osmotic potential adjustment primarily involve molecules of a molecular weight of about 200, such as the hexoses glucose and sucrose. Both of these moieties take part in rapid osmotic adjustment while proline, an amino acid, is not involved until a longer period of continuous stress has taken place. Stress-caused growth reductions seem to be highly correlated with a reduction of polyribosomes to single ribosomes in the growing areas of the plant. Thus growth seems to be affected at the translational level of protein synthesis.

The anatomy of a plant, particularly in the expanded and growing portions of the leaf, are critically related to the stress response. Mature and open vessel elements that transport water are present throughout the growing and expanded leaf blade regions. Transpirational water will move throughout the organizational pattern of the mature and open xylem vessels. As noted above, a sudden osmotic stress causes an almost immediate growth cessation. Such suggests there must be a lateral transmission of the stress signal from the vessel elements to all of the intervening mesophyll cells. The signal consists of a reduction in the hydrostatic pressure of the xylem cells. The pressure change is probably transmitted through the pits in the vessel walls onto the continuous mass of water in the mesophyll cell walls of the growing region. The only difference between the growing versus the nongrowing expanded regions (which are insensitive to the hydrostatic pressure change) is that there are more mesophyll cells between the intervening transpiration vessels in the expanded leaf blade than in the area of leaf growth. As a result much of the transpirational water in the expanded leaf tends to bypass most of the mesophyll cells on its way to the stomatal regions. Such is the path of least resistance, since the apoplastic water (water between and in the cell walls of the contiguous mesophyll cells) does not pass easily across the plasma membrane and into the cells. Eventually, the mesophyll cells in the expanded portions of the leaf blade will lose turgor as opposed to the growing areas. The blades turgor reduction occurs because the water loss through the stomata exceeds its ability to replace lost water and because much of the limited water is removed by the cells in the growing regions. It is proposed that the apoplastic water potential controls the direction of water flux. Since the growing region maintains positive turgor, normally and under mild stress, it is possible that the net water influx might control the release of factors necessary for cell wall extension. If the stress is great, on the other hand, adequate solute accumulation to reduce the cellular loss of water must occur.

One of the most critical factors that the plant uses to respond to low soil-water availability is the role of abscisic acid (ABA) influence on stomatal closure to control the transpirational loss of water. Rather than increased amounts of ABA in the vicinity of the guard cells to induce closure of the latter as formerly hypothesized, it is now believed that increased sensitivity of the guard cells to ABA is responsible for this phenomena. The initial event is a release of ABA from storage areas, such as the mesophyll cells and roots, to the extracellular space (apoplast). Once in this apoplastic region, the ABA can be carried by the transpiration stream toward the guard cells. It is known that guard cells do not produce their own ABA and in addition do not have any plasmodesmata connections to other cells. Some reports have appeared that give evidence for binding sites specific for ABA on the guard cell membranes. Thus, the sensitivity of the guard cells toward ABA is related to the number and binding ability of these sites to the modest concentration of ABA in the surrounding apoplastic water.

The biochemistry of ABA synthesis in higher plants is much more difficult than that observed in fungi and involves a more indirect pathway in addition to the usual derivation from mevalonic acid through the isoprenoid pathway. Water stress, however, stimulates both biosynthesis and degradation of ABA, but the rate of biosynthesis usually exceeds that of catabolism during the drought period. This excess of ABA is generally high enough to maintain levels necessary to keep the stomata closed. Other uses for the high concentration of ABA that are concerned with drought (but which have no relation to stomatal closure) are the senescence and defoliation of older leaves on the plant. This action indirectly aids the plant in conserving much-needed water.

Mathematical models of the relationship between transpirational water loss and carbon dioxide (CO_2) uptake for photosynthetic reduction to starch precursors show that although both processes take place through the same stomatal openings in the epidermis of the leaf, their pathways vary within the leaf and are subject to different constraints. As a result, stomatal closure during drought stress will not affect these two processes in the same manner. It can be shown that stomatal closure diminishes water evaporation more than CO_2 diffusion into the leaf. However, under more severe drought conditions, this relationship does not hold because of the direct damaging effect on chloroplast metabolism. Since factors controlling the ABA release from the mesophyll are determined partially by photosynthetic reactions (water stress inhibits electron transport and disrupts the proton transfer process in the chloroplasts, causing a change in pH that in turn facilitates the release of ABA from the mesophyll cells), it is suggested that ABA could function as a "messenger" from the mesophyll to the stomates to also provide a fine-tuning effect on gas exchange in the leaves under normal drought conditions.

Recent investigations, using molecular biological techniques, show that applications of ABA caused the appearance of a new set of messenger RNAs that almost matched the set induced by wilting. Such results indicate that nearly all drought-initiated events are responses due to ABA.

IV. Heat Shock Responses in Plants

Drought is often accompanied or caused by high-temperature conditions, especially in the arid regions throughout our planet. Plants must alter their cellular biochemistry and physiology to survive such an environmental change that often rapidly occurs. Death of a plant exposed to high temperature is caused mainly from two interactive factors. Coagulation of the protein within the plant cell and the resultant disruption of the protoplasmic structure is most likely to be induced by a rapid rise in temperature. On the other hand a gradual temperature increase causes a protein degradation that releases ammonia, which is toxic to the cell. Plants resistant to high temperatures (Xerophytes) possess high amounts of bound water and high protoplasmic viscosity. Unlike drought or freezing temperatures, plant acclimation (aside from the Xerophytes) to high temperatures is not as extensive or efficient.

In the past decade probably the most investigated of all stress responses is that induced by heat shock factors. The most easily noted response to heat stress, in plants as well as other organisms, is the initiation of heat shock proteins (hsp). It is thought that the stimulus perception for the induction of the heat shock response may involve several changes in the physiology or biochemistry of the heat-shocked cells. A change in the oxidative–reductive environment of the cell toward a predominantly oxidative state as induced by the rise in temperature somehow activates the heat-shock transcription factor so that it binds to the promoter region of the genes. Another possible alteration would be a resultant denaturation of proteins in the cell, which are then combined with a specific protease (known as ubiquitin) for degradation. Such a preferential binding of ubiquitin to the denatured proteins lowers the level of ubiquitin so that less is available to bind and inactivate the heat-shock transcription factor. A third possibility is the perturbation of the membrane ion pumps by heat shock so that the exchange of protons and potassium ions or the sequestration of calcium ions are in an imbalanced state, especially with regard to calcium, which is now available for the calcium/calmodulin activation of kinases (enzymes that add phosphate groups to proteins). Some of these kinases could probably be instrumental in the activation of the heat-shock transcription factor (through the addition of phosphate) which can now bind efficiently to the heat-shock protein (hsp) gene promoter. Even though the heat-shock proteins (in large, intermediate, and low molecular weight classes) are coordinately expressed under heat stress, the induction of these proteins is transient and lasts only for several hours in spite of a continuous omnipresence of the elevated temperatures. During this brief transient period of hsp induction, transcription of normal cellular proteins is momentarily suspended. In addition, the messenger RNAs (mRNAs) for the hsp are preferentially translated. The normal cellular mRNAs, however, remain stable and become reactivated during the recovery process.

In addition to a change in direction of physiological processes, heat shock also changes some aspects of cell ultrastructure that probably plays a role or is linked to some of the overall physiological changes. In certain plant systems, heat shock causes a definite disassociation of the endoplasmic reticulum (phospholipid membranes associated with ribosomes, polysome formation, and protein synthesis). As to the cause of this dissociation, it is proposed by some investigators that heat shock lowers the activity of enzymes responsible for the synthesis of endoplasmic reticulum (ER) or possibly enhances the activity of phospholipases a and d which break down certain phospholipids within the ER membrane. Membranes, as discussed in the sections on chilling, are very sensitive to temperature changes; and membrane fluidity may be a liability at higher temperatures, especially with regard to protein synthesis. It is probable that disassociation of the ER is an attempt to decrease the ER fluidity and thus maintain a reasonable balance between membrane integrity and protein synthesis at higher temperatures. Heat shock also produces a change in the ultrastructural appearance of the nucleoli. The latter exhibit less of a granular appearance and a lower heterochromatin content than under normal temperature conditions. This observation is also consistent with the heat-shock inhibitory action on RNA processing and resultant ribosome assembly.

The function of the heat-shock proteins is not yet clear, but several hsp have had certain roles ascribed to them in cell protection. One of these roles appear to be involved with the stabilization of important cellular constituents. Studies by means of specific poly and monoclonal antibodies in conjunction with cell fractionation procedures have been helpful in localizing the position of specific hsp in the cell. There is a strong association of these proteins with nuclear pellets. Other tests with this pellet fraction suggest that these proteins function in a structural capacity either inside or outside the nucleus as nucleoskeletal elements. Other hsp are localized in the cytoplasm, where they are believed to function in the protection (by sequestering) of normal cellular protein mRNA from turnover; they also stabilize important heat-labile proteins during heat shock. No hsp synthesis has been observed in plastids or mitochondria, but there are observations that hsp produced in the nucleus are transported to these particulates as well as to the chloroplasts. These transported proteins might be involved in the important physiological processes of the chloroplast, such as gene activation or photosynthesis. In addition to a stabilization role as discussed above, some of these proteins are involved in cell cleanup or recovery where protein aggregates are dismantled along with a continuation of proper RNA splicing.

Thus the heat-shock response to high temperatures aids the cells to survive and recover from an environmental stress that could do irreversible damage. Additional information on the role and significance of as yet unidentified hsp is an active on-going phase of research.

V. Osmotic (Salt) Stress and Trace Metal Tolerance

Along with drought, salt (or osmotic stress) is another widespread factor in the restriction of growth on large tracts of essential agricultural land on our planet. As salt accumulates in the soil from irrigation water, the plant is confronted with two problems—namely, obtaining water from a soil of moderate to high negative osmotic potential and coping with higher than normal concentrations of potentially toxic chloride and carbonate ions. Although salt tolerance or hardiness can be increased in some measure by exposure to saline conditions, it is the accumulation of these ions that prevents salt hardiness to be on a par with drought or cold acclimation. [See SOIL SALINITY.]

Essentially, there are two basic classifications of plants with regard to the study of salt hardiness. Halophytes (such as *Atriples* and *salicornia*) grow only where there are high levels of salt in the soil, such as brackish waters or on near sea coasts. There are some halophytes that actually accumulate salt within their tissues. Glycophytes on the other hand perform better at lower levels of salt and include many of the important crop plants, such as alfalfa, cotton, barley, and tomatoes. These plants are usually termed salt-tolerant species.

A similar factor in common between plants that are exposed to either a dry or saline environment is that of growth inhibition. For many decades scientists attributed this growth reduction to a loss in turgor pressure from an imbalance in plant–water relations. Normally, a plant cell is surrounded by water (with a minimal amount of mineral components) as a water potential (Ψ) close to zero. The cell's water potential on the other hand is determined by the sum of its osmotic potential (π) plus its pressure potential (P), which is related to the cell turgor pressure. Given that the membrane permeability is constant, the rate of water movement into the cell can be enhanced by the lowering of either the osmotic potential (osmotic adjustment) or the pressure potential. The π becomes more negative by salt uptake or by production of organic solutes in the cytoplasm by either synthesis of certain organic components (such as proline) or by the breakdown of macromolecules (such as hydrolysis of starch to glucose). A decrease in P by cell-wall loosening through auxin-stimulated growth occurs via cell expansion. In this instance, the cell wall is stretched in an irreversible manner that release pressure and enhances water uptake. It must be noted that decreases in either π or P during cell growth are more or less negated by the resulting water uptake, so that well-watered plants essentially retain their original π and P values. The relationship between turgor itself and growth has been approximated by plant physiologists with the following expression:

$$G = \phi(P - Y)$$

where G is growth, ϕ is the extensibility coefficient, P is turgor pressure, and Y is the threshold turgor that must be exceeded preparatory to any growth or cell expansion. This expression shows that constant growth can still proceed even at lowered turgor pressure provided that ϕ is increased and Y reduced.

During the past decade, however, studies on long-term effects of osmotic stress on growth have definitely shown that even though osmotic adjustment had led to a complete restoration of turgor pressure in the presence of continuous osmotic stress, inhibition of growth rates still continued. It was then thought that this was the plant's method of conserving water uptake from its nearly moistureless environment; increased growth meant more water would be needed to keep the turgor pressure constant. With the advent of osmotic stress studies wherein cell suspension cultures (small clumps of undifferentiated plant cells shaken in a liquid media) were utilized, the cultured cells that had adapted to the added NaCl by means of osmotic adjustment still showed a reduced growth response even when the turgor pressure was above that in prestress conditions. This finding indicated that stressed cells being surrounded by adequate amounts of water was not a factor in restoring normal rates of growth. A reassessment involves a concept of restricted cell expansion as a necessary adjunct to the high salt environment; this concept postulates that a reduced rate of cell expansion might be the mechanism by which a manageable ion flux rate through the cytoplasm to the vacuole or an adequate organic solute synthesis rate in the cytoplasm is achieved. A rapid rate of cell expansion of adapted cells on the other hand would necessitate a rapid influx through a small cytoplasmic pool (to prevent toxicity) into a large vacuolar pool to provide the necessary osmotic adjustment to the high salt environment outside the cell. Since it is best to obtain as large a size difference between both pools as possible, a greater amount of osmotic adjustment is required. In turn, a greater influx must be available to adequately drive the vacuolar osmotic adjustment. Such an ion flux rate through the cytoplasm would most likely be unattainable, hence the need for a more

efficient and adaptable system such as a reduced rate of cell expansion.

Studies on a molecular basis of adaptation to salinity stress with regard to mechanisms of altered or retarded cell expansion have recently focused on the involvement of cell wall biochemistry in cell growth. The cell wall itself is a well-organized structure of very large protein and carbohydrate polymers. The synthesis of these polymers occurs at different stages of the cell's development. Certain polymers originate only during cell division, while others appear only during cell expansion. After cell expansion stops, other polymers are made that are responsible for a permanent locked-in configuration of the cell wall pending further differentiation of the cell itself. On comparison of unadapted and salt-adapted cells, the latter fail to gain fresh weight as rapidly as the unadapted cells. Although decreased cell division could account for a lower rate of fresh weight growth in adapted cells, it can be demonstrated that reduced cell division does not account for all the observed reduction in fresh weight gain. In addition, the slowly expanding salt-adapted cells contain as much or more turgor as the unadapted cells. From these observations, it is apparent that the salt-adapted cells have much less cell wall extensibility (ϕ) than the nonadapted cells.

The nature of reduced cell wall extensibility or cell wall stiffness is not as yet clearly understood. There are, however, some observations on physical and biochemical alterations in the cell walls that somewhat pave the way to a more definitive understanding of this phenomena. Initial investigations showed that osmotic stress-adapted cells partitioned more carbon toward osmotic adjustment (via organic solute synthesis) than toward cell wall polymer synthesis. Despite this preferential diversion of the available carbon, decreased extensibility still took place. Recently this anomaly was resolved in part by studies on the tensile strength of plant cell walls. The loss of tensile strength of salt-adapted cell walls as compared to the nonadapted cells seemed to correlate with a diminished mass of the cell wall's cellulose-extension framework as a result of preferential use of carbon for osmotic adjustment. As a result the adapted cells remained in a locked configuration wherein no microfibril separation resulting in cell wall expansion could take place. Certain proteins (extensions) that are ionically bound to the cell wall control the extension of the wall. Much of this protein is released into the high-salt media of the salt-adapted cells as compared to the unadapted cells in the same media. Loss of this protein may then result in an inhibited extension process of the salt-adapted cell well. It is of interest that plants regenerated from the salt-adapted cell cultures have leaf cells that exhibit lower rates of expansion. The leaves of these plants are smaller in size than those of normal plants.

As with heat, drought, and water stress, the plant hormone abscisic acid (ABA) plays a prominent role in plant tissue subjected to water stress by a high-salt osmoticum. Endogenous ABA levels increase in plant tissue subjected to this type of stress and are involved in the induction of proteins termed as osmotins in the presence of salt. These proteins, specifically synthesized and accumulated in cells that undergo gradual osmotic adjustment to salt stress, are localized mostly in the vacuoles. Although the exact role of these proteins is not fully defined, most of them are believed to play a role for strategies of defense in the plant against the type of environmental injuries mentioned above. The main activity of these proteins seems to be linked to the reduced growth discussed previously and to stress adaptation. Osmotin synthesis along with all of the other metabolic adjustments mentioned previously consumes additional energy that is reflected by increases in respiration.

Recent investigations on the molecular genetics of salt stress have dealt mainly with the following areas: (a) a limited number of metabolic pathways that are involved in the stress response of the affected plants and that are involved with the plasmalemma. These include the control of salt uptake in the roots, salt transport in the xylem, regulatory control of pathways for organic osmolyte synthesis, maintenance of photosynthesis and growth (cell cycle control and biosynthesis of necessary cell wall components). (b) Mechanisms used by plants to transfer signals from the perception of osmotic changes in their environments to initiate changes in gene expression. Changes in turgor pressure or the osmotic state outside of the cell initiate physical and chemical changes in the plasmalemma, which in turn brings about the rapid activation of specific genes. (c) Studies of molecular mechanisms that control gene expression, which in turn finally result in the necessary metabolic adjustments to saline stress. Thus it is postulated that salt stress acts through a chain of events: perception, signal transfer, and gene expression.

Along with salt stress is often the presence of toxic trace-metal ions present in the soil from a number of sources. These ions are generally introduced into the soil from emissions of fuel-burning power plants and automobiles, from addition of industrial and municipal sewage sludge to the soil as a fertilizer, and from metallurgical mining and processing. These toxic trace-metal ions can be released from the soil in free ionic form through changes in soil acidity. Such a pH change has in some instances led to the contamination of valuable groundwater. Airborne distribution from emissions has also insured a widespread deposit of these contaminants within the soil. Recently information has been gathered as to the biochemical and molecular aspects involved in plant tolerance of these toxic metal ions.

Metal ions have been categorized as to their affinity for different binding sites found within organic molecules synthesized by higher plants. Class A metals (such as aluminum) form stable complexes with organic molecules containing oxygen. Class B metals (such as cadmium and copper) form

complexes with compounds that contain a nitrogen or sulfur center.

Aluminum, although not a trace metal, is a common element in certain parts of the earth's crust. Changes in soil pH cause small amounts of the free ion to be taken up by plants. Toxicity occurs as a result of several interactions that have an adverse effect on some aspects of the plant's physiology. Exposure to aluminum results in the reduction of calcium, magnesium, and phosphorus uptake to levels below that needed for tissue survival. Aluminum tends to be concentrated in the nucleus, where it interferes with DNA synthesis and thus inhibits the cell cycle. Several strategies are used by the aluminum-tolerant plant to overcome the toxic effect of this metal. Changes in the ammonium versus the nitrate ion tend to raise the pH in the cell and thereby render the aluminum ion insoluble and ineffective as a toxic agent. Aluminum-tolerant plants also produce different chelating agents, which form a metal complex that again is no longer toxic to the plant. Most of these chelating agents are an overproduction of organic acids, such as citric or malic acid which are normally in the plant at lower concentrations.

Cadmium normally enters the soil from emissions of fossil fuel–burning power plants and mining activity. Cadmium concentrations and soil pH are major factors that influence the uptake of this metal into plants. Cadmium is known to replace copper and zinc in critical enzymes that use these metals as cofactors. This ion then interferes with several metabolic pathways that are critical for cellular as well as whole plant physiology. Such pathways include formation of chlorophyll, critical steps in the Calvin cycle, and respiratory reactions. Cadmium tolerant plants are able to synthesize polypeptides that have metal binding properties and are able to form cadmium–polypeptide complexes. Although some cadmium-sensitive plants are able to also produce metal-binding polypeptides, the latter are unable to form complexes with cadmium. Once taken into the cell as a complex, cadmium is then accumulated and stored as an insoluble form (as the insoluble cadmium phosphate) and is not released into the environment.

Under normal conditions copper, manganese, and zinc are necessary micronutrients for plants. In quantities above micronutrient levels, however, these metals are toxic. Plants tolerant to high levels of these metals are able to synthesize sulfur-containing polypeptides and in some cases also accumulate high levels of organic acids. Thus, either complexing or chelation reactions of these metals take place depending on which metal is being considered.

As with aluminum, selenium is present as a natural component of certain soils. Selenifrous soils exist in a broad section of the central part of the United States and contain plants that accumulate high levels of this element initially in its anionic form. Most of these plants belong to the genus *Austragalas* (milk vetch). The main reason that selenium is toxic to plants is its similarity in structure to sulfur. Even though selenium replaces sulfur in many biological compounds, it cannot carry out the same functions provided by sulfur. For example, selenium accumulators (which are tolerant to selenium) are capable of excluding the seleno amino acids and thus do not incorporate them into the critical proteins (enzymes, etc.) during synthesis. Otherwise, defective key enzymes would adversely affect several important metabolic pathways that control the plant's growth and development.

More knowledge of the mechanisms of metal tolerance in higher plants is important, because without such knowledge it would be difficult to predict the transport of these toxins through the biological chain. Thus, the storage of toxic ions in plant tissues would have a different impact on human health than if the plants could altogether exclude metal ions from their system by a different mechanism.

VI. Anaerobic Stress and Adaptation

During an excessive rainy season or a rapid spring thawing of winter snow build-up, much flooding of ordinary arable land where plants are under cultivation takes place. Many of these economically important plants are nonaquatic in nature and in some cases suffer from oxygen deprivation to their root cells. Such an extreme condition is termed anoxia and results in either death or cessation of root cell growth where complete fermentative metabolism is taking place. Some of these plants however are able to partially counter the extreme effects of anoxia by providing limited amounts of oxygen (hypoxia) from intercellular air spaces (aerenchyma) present in the nonsubmerged tissues.

There are basically two types of reactions from plants under anaerobic stress. The first is a production of ethylene by the roots, which causes structural changes within this organ to allow more oxygen diffusion through the roots and also causes the root tips to grow upward through the soil toward a more aerobic environment (as opposed to their natural inclination to grow downward toward a gravitational pull). Secondly, within half a day after flooding, a set of nearly 20 soluble proteins are produced that eventually replace all the proteins produced by the root cells under normal aerobic conditions. The replacement or more accurately the repression of the preexisting aerobic protein synthesis is accompanied by a disassociation of nearly all the polysomes that were involved in aerobic protein translation. This event is somewhat reversible, since 70–90% of the mRNAs originally involved with aerobic protein synthesis are reassembled back into translatable polysomes without synthesis of new mRNAs after restoration of aerobic conditions. Anaerobic protein synthesis takes place in two stages: (1) There is an immediate production of transition proteins. (2) After a lag period, anaerobic proteins appear. Several of the anaerobic proteins are linked to fermentation reactions (fermentative glycolysis). One of these enzymes, alcohol dehydrogenase (ADH) is necessary for the survival of plants under anaerobic conditions. There are two possible end products of glycolytic fermentation—ethanol or lactic acid. To prevent acidosis, which would be toxic for root cells during anaerobiosis, there is a mechanism of pH control. During a short increase of lactate production, the cellular pH drops below 7.0 and activates enzymes that preferentially lead toward formation of ethanol and simultaneously inactivate enzymes that enhance lactic acid formation. Thus if the pH remains stable and slightly below pH 7.0, the formation of ethanol prevails. The importance of ADH was demonstrated when mutant plants lacking the ADH gene did not survive flooding more than 3–5 days.

Under hypoxic conditions, ADH activity increases in root tissue; but as the oxygen supply diminishes the glycolytic flux or rate of reaction of glucose metabolism must increase to keep the ATP (adenosine triphosphate—the "energy currency" of the cell) at least at a 50–70% level of that present under normal aerobic conditions when the Kreb's cycle of glucose oxidation is operative. ATP concentrations substantially below this level will not allow the adaptive biosynthetic responses to take place in the root cells. Finally, both of these coping mechanisms appear to be linked to one another in that when the ATP levels are maintained at minimum efficiency, the root ethylene levels decrease and vice versa.

It is of interest that the induction of heat-shock proteins is immediate and similar to that of the transition proteins involved in anaerobic stress. Both heat-shock and transition proteins then show a decelerated rate of synthesis. This is where the similarity ends, however, for there are only one or two protein polypeptides that overlap with regard to similar molecular weight. In addition, it should be mentioned in closing that most of the stress topics discussed in this article (osmotic, chill, drought, heat, and anaerobiosis) can be shown to have a few of the induced stress proteins in common with one another. Future research in the area of environmental stress will probably clarify a set of physiological or biomolecular reactions that are directly related to the shock proteins in common with each of these stress categories.

Acknowledgments

The author wishes to acknowledge the following persons whose concepts on the various aspects of environmental stress on plants were utilized to provide the essentials needed for this summary article: Drs. R. A. Bressan, Purdue University, Indiana; M. R. Brodl,

Knox College, Illinois; K. Cornish, USDA, Albany, California; J. C. Cushman, University of Arizona; C. Guy, University of Florida; T. O. Ho, Washington University, Missouri; P. J. Jackson, Los Almos National Laboratory, New Mexico; D. V. Lynch, Williams College, Massachusetts; K. Matsuda, University of Arizona; M. M. Sachs, Washington University, Missouri, and P. L. Steponkus, Cornell University, New York.

Bibliography

Atkinson, B. G., and Walden, D. B., eds. (1985). "Changes in Eukaryotic Gene Expression in Response to Environmental Stress." Academic Press, New York.

Cherry, J. H., ed. (1989). "Biochemical and Physiological Environmental Stress Tolerance." Springer-Verlag, Berlin.

Cherry, J. H., ed. (1989). "Environmental Stress in Plants." Springer-Verlag. Berlin.

Ho, T. D., and Sachs, M. M. (1989). Stress-induced proteins: characterization and the regulation of their synthesis. *In* "The Biochemistry of Plants," Vol. 15 (A. Marcus, ed.), pp. 347–378. Academic Press, San Diego.

Katterman, F. R., ed. (1990). "Environmental Injury to Plants." Academic Press, San Diego.

Key, J. L., and Kosuge, T., eds. (1985). "Cellular and Molecular Biology of Plant Stress." Liss, New York.

Li, P. H., and Sakai, A., eds. (1987). "Plant Cold Hardiness." Liss, New York.

Monti, L., and Proceddu, E., eds. (1987). "Drought Resistance in Plants: Physiological and Genetic Aspects." Commission of the European Communities, Luxembourg.

Glossary

Anaerobiosis Stress to plant root cells, either under low oxygen level or no oxygen, caused by excess of water in the soil. Stress changes direction of glucose metabolism.

Cytosol Soluble cytoplasm of the cell.

Exocytosis Vesicular mechanisms for discharge of materials to exterior of cell.

Osmotic stress Passing of water through selectively permeable membrane from region of low solute to region of high concentration of solute. Stress occurs on passage of water from plant cell to environment of high salt concentration.

Plasmodesmata Fine cytoplasmic threads that pass through openings in cell walls and connect the protoplast from one cell to that of an adjacent cell.

Protein polypeptides Subunits of a protein or enzyme that are separated from one another by a buffered detergent and resolved by gel electrophoresis.

Thermotropic Physical phase change induced by a specific and narrow range of temperature.

Transpirational water Water that passes through the xylem cells through the stomata and raised pores of the leaf.

Turgor pressure Force that builds up in a cell caused by absorption of water.

Vacuole Spaces within the cytoplasm that contain air and a solution of various solutes.

Eocene Patterns and Processes

Brian McGowran

The University of Adelaide

The Eocene epoch is a part of the Tertiary period or, as we now prefer, a part of the Paleogene period of the Cenozoic era. The record of what happened during the Eocene epoch is to be found in the rocks of the Eocene series. Fossil molluscan assemblages in Europe, studied by the French conchologist Gerard Deshayes, were used by Charles Lyell in the 1830s to subdivide the Tertiary. Eocene assemblages had only about 3% living species, a proportion that increased through the Miocene and Pliocene toward the present. It is not difficult to imagine the problems encountered in recognizing the Eocene series far from the Paris basin, where fossils of the characteristic shallow-marine molluscan species do not occur—in strata deposited in the deep ocean, for example, or containing the bones of terrestrial animals, as in North America.

I. Introduction

Eocene strata were being traced around the world in the nineteenth century. This procedure is one of correlation with the stratotypes, meaning those strata on which the Eocene series is defined, using fossils of many kinds and representing many environments. Building that pattern of the global Eocene record is still an active research program on which much other research must rely. If we get our stratigraphical correlations wrong, then not much else can be correct. [*See* STRATIGRAPHY—DATING.]

The next step is calibration. Numerical estimates of Eocene time based on isotopic decay have not reached consensus but suggest that the Eocene extended roughly from 57–54 to 36–34 Ma (million years before present). That is about 20 million years (m.y.). The most rapid recent development has occurred in the piercing together of the record left in the rocks of the repeated reversals through geological time of the earth's magnetic field. We have, as a result, a geomagnetic stratigraphy which, cross-checked with the fossils and the numerical ages, greatly strengthens our geological time scale.

In dealing with patterns in Eocene earth history, accurate correlation and cross-checking are usually more important than time in millions of years—even though we need those numerical estimates for scaling our thinking and for sustaining a healthy skepticism about our accuracy. We need resolution as well as accuracy. That means we can "see" into time at the right level—it is no use asking questions about global changes at, say, scales of decades or centuries when the geological record does not break down to less than hundreds of thousands or millions of years. There is a 3-m.y. uncertainty about the age of the close of the Eocene at about 35 Ma; but in fact we can see into the record in the rocks at finer resolution than that, using fossils and other techniques, as discussed later.

A widely used scale for the Paleocene and Eocene is shown in Fig. 1. It integrates the various components referred to above. The ages are based on the type sections in Europe. The zones are based on the distribution of microfossils through successive sedimentary strata. The flip-flop behavior of the geomagnetic field gives the numbered events which in turn give the chronological succession of so-called chrons. All of this is calibrated; that is, it is put against a numerical time scale in millions of years.

II. What Is Special about the Eocene?

The Eocene epoch mostly was a warm time. The nineteenth-century geologists believed, on uniformitarian grounds, that

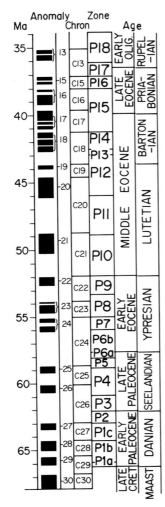

Fig. 1 Integrated time scale for the Paleocene and the Eocene. Note the cross-correlation between age (based on original designations in NW Europe); geomagnetic reversal scale expressed in chrons (black: times of earth's "normal" — like today's — field; blank: reversed"); fossil zones (in this case, planktonic foraminifera, where zones are the so-called P-zones); and estimated numerical ages (Ma: millions of years). [Based on W. A. Berggren *et al.* (1985).]

plained away as an effect of continental drift, for example. Figure 2 shows three mutually confirming trends of global, post-Eocene cooling. Such cross-checking is perhaps the most powerful scientific method available to geo- and biohistorians.

Also well known are the higher sea levels of those times as manifested by more extensive distribution of marine strata across continental margins than subsequently. Warmth and high sea levels suggest high humidity, helping to explain another phenomenon — the preservation of vast amounts of organic material as coal and as oil shale (again, in both hemispheres). A global curve of economic phosphate abundance peak twice in the Cenozoic: once in the early Eocene and again the Miocene.

Eocene geography was different from today's in the most fundamental sense — the distribution of continents and oceans. During the Mesozoic era the supercontinent, Pangea, broke up and the continental fragments drifted apart. One important consequence was that the Tethys, the equatorial embayment extending westward from the superocean, Panthalassa, closed progressively as the fragments now known as Africa, Arabia, India, and Australia drifted northward toward Europe and Asia. Such phenomena provide modern explanations for hypotheses advanced by earlier workers. Thus, the Eocene has been long noteworthy for being a time of "orogenic climax" — compression, metamorphism, igneous activity, and thrust-faulting — distributed along the narrow tracts of crustal instability and mostly high relief known as the Alpine mountain belt. The belt mostly marks the vanished Tethys (the Alps themselves are only a small segment) (see Fig. 3).

Finally, we can generalize usefully about Eocene life. The archaic appearance of the mollusks described by Deshayes is found repeatedly through the fossil record — in the plants, the mammals of the northern continents, the other invertebrates of the shallow seas, the skeletonized microplankton of the open oceans. As we look back from the present, we see organisms not too dissimilar from today's back to the Miocene, whereas the Oligocene was a time of transition and reorganization after the disappearance of the older-looking Eocene life forms from the face of the earth.

III. Some Modern Generalizations about the Eocene

The plates of the earth's crust are rearranged from time to time in response to boundary collisions. The Eocene was a time of extensive plate tectonic rearrangement and episodic sea-floor spreading. In particular, Chron 24r at the Paleocene/Eocene boundary marks an intensive episode of change in the pattern of spreading. The change was accompanied by

the world was warmer in Eocene times. Thus, fossils of crocodiles, palms, marine mollusks, and corals were known from higher northern latitudes than their closest living counterparts ever attain. Subsequent research has confirmed that marine and terrestrial biotas of tropical or subtropical aspect have been retreating toward the equator since the Eocene. The retreat has not been smooth or uniform but a retreat just the same; it is recognized in the southern hemisphere as well. It has left a record of climatic change that cannot be ex-

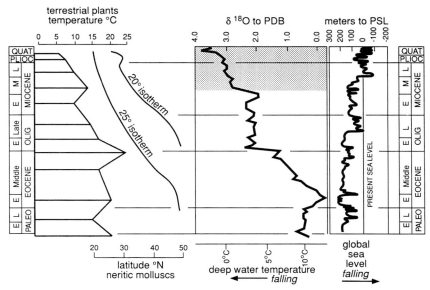

Fig. 2   Three independent strategies for chronicling the climatic deterioration since the Eocene. From the left: A curve derived from the analysis of terrestrial floras in North America by Erling Dorf in the 1950s. The floras indicate cooling since the Eocene, as estimated on a scale of temperatures. Next, isotherms constructed by J. Wyatt Durham in the 1950s retreat equatorwards through geological time—showing the same cooling. The isotherms are based on assemblages of shallow-marine mollusks. Then, an oxygen isotopic curve for the deep ocean, based on a recent curve by N. J. Shackleton (stippling indicates time of major ice growth on the poles, distorting the temperature scale which applies until the middle Miocene). The three interpretations drawn from three realms of the biosphere agree quite well. The global sea level at the right is based on Haq *et al.* (1987). Note that sea level is highest when climate is warmest in the early Eocene and that temperature and sea level both fall after that.

an outpouring of volcanic lavas and ash in the North Atlantic region—the Thulean volcanic episode. It recalled the Deccan episode in India of 10 m.y. previously, at the end of the Cretaceous, which was suggested by some to be implicated in the demise of the dinosaurs and the ammonites. [*See* PLATE TECTONICS, CRUSTAL DEFORMATION AND EARTHQUAKES.]

But the most spectacular event at Chron C24r was the collision of northward-drifting India with Asia—an initial collision, so-called, involving a volcanic island arc and the emplacement of granites in the crust. In reaction to those great crustal events, crustal rearrangements were appearing at Chron 19 about 13 m.y. later. We see the record of that global plate reorganization in the basaltic crust of all ocean basins. Suffice it here to mention only the change in motion of the Pacific ocean floor from more northerly to more westerly (as seen in the great bend in the Hawaiian–Emperor chain of islands and seamounts), the rapid separation of Australia from Antarctica, and the growth of the Southern Ocean.

It will become apparent that Chrons C24r and C19 are magic numbers in the global patterns and problems of the

Eocene. Some workers see the history of the earth's crust as a succession of states analogous to convenient divisions or time slices of, say, European history. Thus one configuration of plates and of orogenies at the colliding plate boundaries is succeeded by another configuration. The Laramide global tectonic regime was succeeded between Chrons C24 and C19 by the Himalayan regime.

Those changes in the global land and seascapes were dramatic and far-reaching. The Tethys seaway was largely destroyed (destruction was completed in the Miocene). The circum-Antarctic oceanic passageway was opened. Whereas warm oceanic water could flow around the globe in the equatorial region during much of the Mesozoic era and during the early Cenozoic era, the largest comparable ocean system today is the cold circum-Antarctic current or westwind drift. That switch in the distribution of oceans and continents was the underlying cause of the climatic changes that have brought on the Neogene (including the present) ice age. The Eocene epoch is the time of crucial transformation —it intervenes between the world of the dinosaurs and our own world.

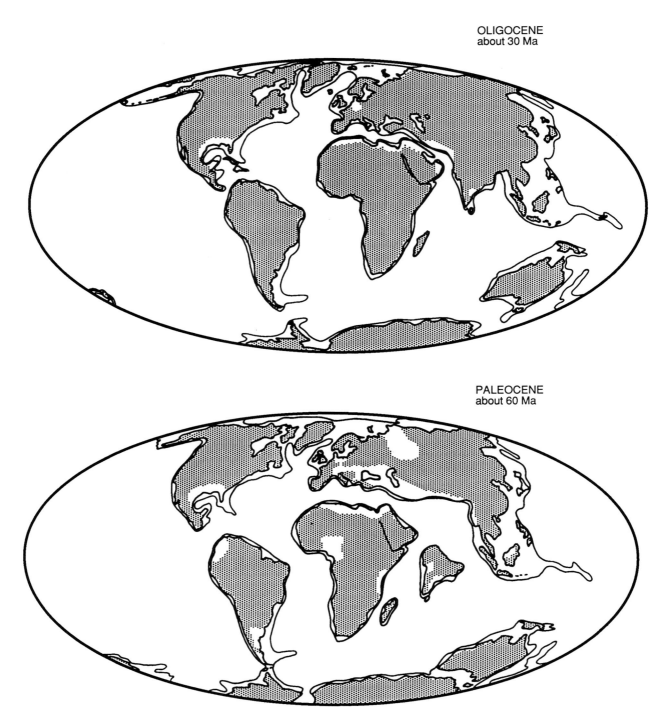

OLIGOCENE
about 30 Ma

PALEOCENE
about 60 Ma

Fig. 3 Paleocene and Oligocene global geography, to demonstrate Eocene changes. The changes in the distribution of ocean basins and continents, essential to the development of today's icehouse world, were the destruction of the equatorial oceanway Tethys and the opening of the circum-Antarctic ocean. Those developments were completed by the Miocene but these maps show that they occurred mostly in the interim—that is, in the Eocene. [Adapted from reconstructions published by B. U. Haq and F. W. B. van Eysinga (1987). "Geological Time Table," 4th ed. Elsevier, New York.]

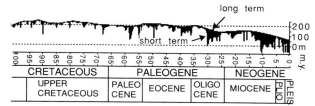

Fig. 4 The Exxon sea level curve, showing both long-term and short-term configuration. There are both a long-term fall and an increase in the intensity of short-term fluctuations toward the present. Scales are in meters and millions of years. [Based on B. U. Haq *et al.* (1987).]

IV. Time Series in the Eocene

As in all stratigraphy, the rock record of the Eocene is built up by correlating numerous bits and pieces—the local outcrops and drilled sections—into regional and global frameworks. This gives us the overview, the big stratigraphic picture, that we need. The signals of what was going on in the Eocene world are found in the more sensitive parts of the system. They include the fossil record, of course, and those sediments that respond far and wide to changes in the environment. Most effort has gone into investigating carbonates and the results of cycling carbon through the biosphere—photosynthesis and the consumption of its products.

Time series are constructed to reveal changes in the exogenic systems, which include the hydrosphere, the cryosphere (ice), the atmosphere, the biosphere, and the reactive lithosphere, or weathering zone. If they have one characteristic in common, it is that a clear signal usually has a less-than-clear cause. Different causes can give the same result. Curves are based on stable-isotope ratios, especially of oxygen and of carbon, or on changes in the balance of calcium carbonates accumulating in the ocean basins, among other geochemical signals. Curves are often based on some part of the fossil record that might reveal an aspect of evolutionary change or a response to environmental perturbation.

We begin with a modern quest for a Holy Grail, the Exxon global sea-level curve (Fig. 4). That curve is a synthesis of data from the world's continental margins that bear on the advance and retreat of the sea across those margins. The synthesis was made by researchers at Exxon Production Research Company. Advances, or marine transgressions, signal either local subsidence of the continental crust or global displacement of water from the ocean basins that spills onto the continents, or both. There are subtleties and complications to this interpretation, but they need not detain us here. The curve for 100 m.y. shows two strong characteristics. One is an overall fall toward the present day, its highest level during the

Cenozoic era being in the early Eocene. The other is the sharp fluctuations superimposed on the overall trend.

There are two main problems. First, is the Exxon curve a real signal? Or is it a chimera, built of local up-and-down movements of the crust (known as isostasy) that are deceptively like a global pattern (eustasy)? That is a problem of pattern. The majority of workers probably would agree that the curve can be improved rather than rejected outright. Second, what controls the pattern? That is a process problem. Note that the spilled-paint effect intensifies toward the present but the width of the dribbles does not change. The long-term trend can be attributed to crustal processes acting to increase or to decrease the capacity of the ocean basins. But the short-term pattern is generally considered to be too fast for that, leaving essentially the waxing and waning of polar icecaps as the control. That belief forces us to confront the problem of a change (as is widely believed to occur) from a greenhouse earth in the Cretaceous and Paleogene to the icehouse earth of the Neogene. Now how can that be, if tectonic processes are too slow and ice effects are often trivial? Either we look again at this notion of an ice-free earth during some times past (pattern), or we look at whether global tectonism is faster than we thought (process).

Let us consider now the stable isotopes as seen in changing ratios, oxygen-16 to ^{18}O and carbon-12 to ^{13}C (Fig. 5). We are interested in isotope fixing during the biomineralization as calcite in skeletons of planktonic and benthic microorganisms. Since the numbers are clumsy as simple ratios, they are given as a delta notation in positive or negative deviations in parts per thousand from the accepted standard.

An increase in $\delta^{18}O$ through time indicates cooling of the seawater. The increase through the Cenozoic, then, suggests cooling as we approach the present day. A complication arises in that the organism producing the shell responds not only to the temperature of the reservoir (the ocean) but also to its isotopic composition, which gets heavier as ^{16}O is locked preferentially into icecaps. There has been agreement but not unanimity that we can read Eocene ratios as a paleothermometer, whereas the ice effect becomes prominent during the Neogene. Note the warming after level C and the coolings at level D and especially at level F (Fig. 5).

Carbon has a different response. ^{12}C is taken up preferentially during photosynthesis, leaving the reservoir enriched in ^{13}C. Thus a marked increase in $\delta^{13}C$ in a time series suggests increased biological productivity—removal of ^{12}C. A positive trend in $\delta^{13}C$ in the calcite of oceanic organisms sustained on a scale of millions of years indicates removal of light carbon from the reservoir. That can happen by locking organic matter into sediments as source material for oil and gas, or as coal. Over shorter times the advance and retreat of the global rainforest will affect the oceanic signal in biominerals. Note the sudden decrease in $\delta^{13}C$ at level A, suggesting to some a catastrophic collapse in photosynthesis and production of

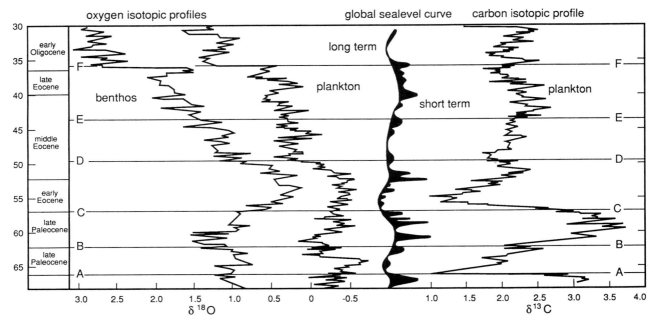

oxygen isotopic profiles   global sealevel curve   carbon isotopic profile

Fig. 5   The isotopic curves for oxygen (bottom waters at left, surface waters at right) and for carbon (surface waters only) are based on a compilation by N. J. Shackleton. The global sea level curve with both long- and short-term changes is based on the Exxon curve of B. U. Haq *et al.* (1987). Levels A to F are for ease of reference to time series in subsequent figures.

plankton, therefore the extinction of ammonites and other higher organisms — the "strangelove ocean" of the end of the Mesozoic era. Note too the great positive spike in $\delta^{13}C$ between levels B and C.

The global significance of these stable signals is established by comparing time series from different regions of the world and finding close similarities in broad patterns and in many details.

V. Paleobiological Time Series in the Eocene

Fossils can be thought of as documents of organic evolution, and of course that is true. But the rich potential of the fossil record lies in revealing ancient systems through organisms' responses to paleoclimatic, paleogeographic, and paleocean-ographic changes. "Evolution" refers not only to changes that can be measured in fossils but also by using a species in its true role as a basic unit in nature. By making counts, we see changes in the fundamental evolutionary phenomena of spe-ciation (origin of species) and extinction. This is the taxic approach to evolution; the other is the morphological ap-proach.

Figure 6 presents 10 time series drawn from the fossil record. They are plotted against the horizons identified in the stable isotopic series. The curves include several kinds of synthesized data:

1. Simple taxic diversity refers to the number of kinds of organisms counted from the published descriptions of faunas or floras. Here, the mollusks of the U.S. Gulf coastal plain include the species of bivalves (clams and others) and gastropods (snails) described in monographs over many years. Note the great drop in diversity at the beginning of the Paleocene, the partial recovery among the gastropods at level C, and more strongly, together with the bivalves, at level E. The extinction curve for marine animals in general clearly is relevant to these big fluctuations in diversity.

2. Paleoclimatic curves are based on changes in the fossil assemblages of sensitive organisms. The examples in Fig. 6 are analyses taken of terrestrial floras in the United States and southern England, and North Atlantic plank-tonic foraminifera. Another proxy for paleoclimatic change is found in the large foraminifera that host pho-tosynthesizing microorganisms in their cytoplasm. Pre-ferring the warm, well-lit shallow seas of the tropics, they

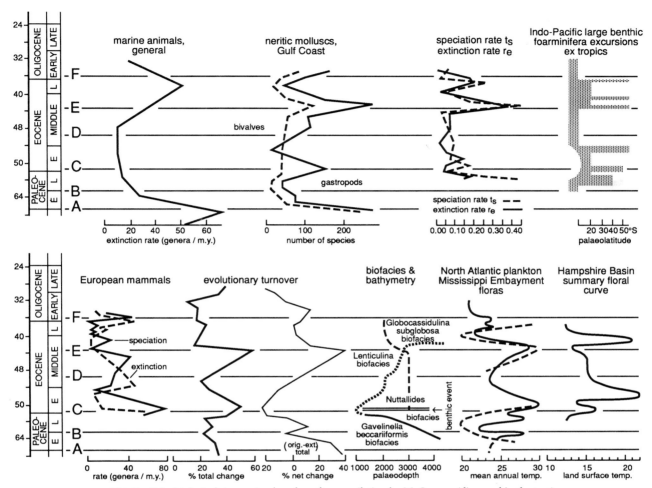

Fig. 6 Paleobiological time series, based on the compilation by McGowran (discussed in the text).

expanded briefly to higher latitudes when the world warmed.

Terrestrial plants are very sensitive to environmental changes. The characteristics that allow one to generalize about modern tropical, temperate, and cold floras can be used to place fossil assemblages environmentally and to draw regional paleoclimatic curves. The two curves in Fig. 6 show pronounced warming in the early Eocene and again above level E (the slight mismatch in time is almost certainly a shortcoming in correlation).

3. In the deep sea environment, the envelope describes the upper depth limit of a deep-oceanic assemblage of foraminifera named after the genus *Nuttallides*. The greatest rates of change involve 3000 m toward the shallow peak at level C and 1500 m above level E.

4. Metrics of taxic evolution include rates of speciation and rates of extinction. They are displayed (using different

measures) for planktonic foraminifera from the global ocean and for terrestrial mammals from western Europe. Rates of change in deep-oceanic foraminifera peak at the turning points on the *Nuttallides* envelope. Finally, the general extinction curve for marine animals is actually based on those with skeletons and therefore presents a reasonable fossil record.

This assemblage of time series is very wide in its coverage of organisms, environments, and kinds of data and synthesis. The organisms are marine and terrestrial; animals, plants, and single-celled organisms; planktonic and benthic; and neritic and deep-oceanic. The data are evolutionary and environmental. Any perceived parallels might be trying to tell us something quite fundamental and pervasive about the biosphere.

Figure 7 presents a modern compilation of data on the

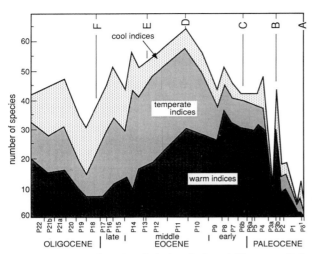

Fig. 7 Diversity of species of planktonic foraminifera, compiled by I. Premoli Silva and A. Boersma. The three groups are "climatic indices" according to their temperature preferences in the surface waters of the global ocean.

diversity of planktonic foraminifera in the Paleogene. Note especially the pronounced change across the Paleocene/Eocene boundary, the peaking by warm-water species, and then the rise in numbers of those preferring cooler-water habitats. Note too the rise and fall of late Paleocene numbers following closely the trajectory of the $\delta^{13}C$ spike (Fig. 5) from levels B to C.

Figure 8 presents a compilation of taxic evolutionary data on the warm-water planktonic components (Fig. 7) together with two benthic groups, the alveolinids and the nummulitids. They are large (although single-celled, their shells can measure in millimeters and even in centimeters) and structurally more complex than are the great majority of foraminiferans. Both the large benthic groups and the warm-water planktonics carried populations of photosynthesizing "algae" in their cytoplasm, exchanging CO_2 and other wastes for carbohydrates and oxygen. That was (and is) a splendid adaptation for surviving in warm, shallow waters low in nutrient—the oligotrophic state (Fig. 9). That is one way to characterize the ecological specialists, known as k-mode strategists, and to contrast them with the generalists or opportunists (r-mode strategists), which flourish in high-nutrient situations. Note that the warm-water specialists, interpreted as occupying oligotrophic environments in the pelagic and neritic realms, respectively, are parallel in their taxic evolution—in origination, extinction, and their summation, overturn. The importance of those parallels is discussed here.

VI. Are There Eocene Patterns That Make Sense in Terms of Process?

Inspection of the fossil-based curves suggest that the paleobiological history of the Eocene world divides naturally into three successive chunks. They are the early Eocene, the earlier part of the middle Eocene, and the later middle plus late Eocene. There is, as it were, heightened evolutionary activity in the first and the last of those time intervals on the measures of speciation, extinction, diversity, and evolutionary turnover. That activity seems to correlate with paleobiological evidence for global warming. Intervening, there seems to have been a pause, an interregnum, a time of subdued activity in the biosphere.

Are such grand generalizations useful and valid? Searching for correlations with other time series is one way of showing that these patterns over millions of years are real and interesting. It points the way as well toward some explanation in terms of process. We will scrutinize these Eocene time slices in turn.

VII. The Warm World of the Early Eocene

There is now an abundance of mutually supporting evidence that the early Eocene was the warmest time of the Cenozoic era and that extraordinarily warm conditions prevailed to high latitudes. The evidence is found in terrestrial floras and faunas to 80°N latitude, in marine fossils, and in $\delta^{18}O$ values. It was also very wet at that time, as evidenced in part by deep chemical weathering of the landscape at high latitudes. The analysis of eolian particles in oceanic sections indicates a pronounced drop in wind strengths across the Paleocene/Eocene boundary. The deep ocean warmed, with the very important corollary that temperature gradients, both vertical and latitudinal, were considerably flatter than today's (Fig. 10). Earth's fluid systems, then, were less vigorous than they are now. The key problem becomes: If modern oceanic thermohaline circulation together with a vigorous atmosphere are unsuitable as a vehicle for transporting heat poleward in the early Eocene—as they surely are—then what was the vehicle?

It has been postulated that the oceanic thermohaline circulation of an icecap-prone world—such as today's—is replaced during global greenhouse times by a salinity-driven circulation. That comes about through the intensification of subtropical evaporation, generating dense saline water that sinks and moves poleward at depth. We have both evidence and a plausible mechanism for a reversal of circulation happening at the Paleocene/Eocene boundary. The major event in the fossil record of the deep global ocean is not at the

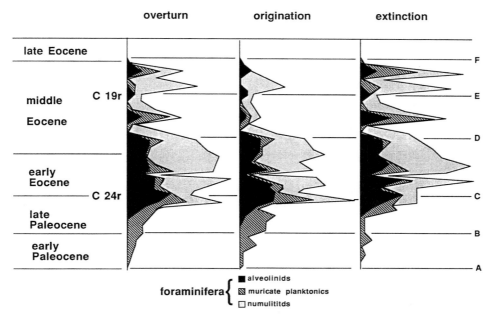

Fig. 8 Taxic evolutionary rates in three groups of foraminifera interpreted to be K-mode specialists (see Fig. 9). The alveolinids and the nummulitids are large benthic forms. The muricate planktonics (including the morozovellids of Fig. 9) occupied very shallow niches in the planktonic community. All of these forms are inferred to have relied heavily on their private microfloras. (Adapted from the compilation of the first and last appearances and their summation, overturn, is by P. Hallock, I. Premoli Silva, and A. Boersma.)

"mass extinction" horizon at the end of the Mesozoic, nor at the next such, claimed for the end of the Eocene. It is at the end of the Paleocene, when there was a major overturn in the benthic foraminiferal community. Note that this is a striking event in both of the deep-ocean foraminiferal indices in Fig. 6, is the level C event in Fig. 11, and is at the onset of intense bottom-water warming on $\delta^{18}O$ evidence.

Can this remarkable confluence of diverse observations and inferences be tied together? Recall the "magic number," Chron C24r, at the Paleocene/Eocene boundary. That is precisely the time of an intensification of global crustal tectonism. A recent hypothesis suggests that the accompanying intensified hydrothermal and volcanic activity, clearly seen in ocean drilling, included an injection of CO_2 into the exogenic system, inducing global warming and increasing evaporation in "evaporation pans" at lower latitudes. That change reversed circulation in the ocean and reduced the vigor of atmospheric circulation. All of this impacted rapidly on the biosphere, not only in the deep ocean, but also in the evolution of terrestrial mammals. The modern orders of mammals mostly appear in the fossil record in a narrow interval of time that correlates with Chron C24r. The evidence for that conclusion is seen most clearly in the strata representing the Clarkforkian and early Wasatchian ages in the western interior of North America, but it correlates very nicely with the peak of evolutionary activity in western Europe. The lush forests at high latitudes survived the long polar nights by being cushioned in a warm and very moist environment—this would also explain the intensive and deep chemical weathering and the anomalous accumulation of coals at high latitudes.

VIII. The Early Middle Eocene: First Climatic Deterioration

Although the precise shape of a climatic curve based on a $\delta^{18}O$ series is arguable, it is well established that the first substantial cooling in the Cenozoic era occurred across the early/middle Eocene boundary, bottoming at level D (Fig. 5). Consistent with that is some evidence for an early glaciation (the "Krakow glaciation") on Antarctica dated as early middle Eocene. Note the short-term drops near level D on the Exxon curve. They seem to lead, not lag, the temperature decline, and they correlate with the cutting of canyons into exposed continental margins. That suggests that the sea level

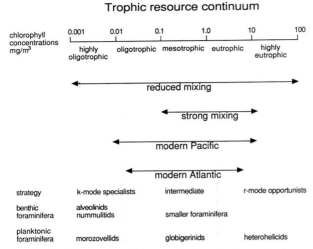

Trophic resource continuum

Fig. 9 Broad but very useful generalizations can be made about oceans in terms of mixing and the effects that derive from that, as discussed in the text. Oceans of the Cretaceous and the Paleogene mostly spanned the continuum; Neogene oceans mostly were more restricted. (Compiled by P. Hallock, I. Premoli Silva, and A. Boersma.)

was lowered before cooling, which was caused by forcing a continental climate. Or are these correlations wrong, and cooling lead cutting, which was forced by glacio-eustatic lowering of the sea level? Stating the problem of pattern and process so baldly makes us acknowledge that we really do not know the answer.

What might have caused the end of the early Eocene warm period? There are at least two possibilities. In one, we have a

reaction to the Chron C24r crustal rearrangements, during which the global oceanic crust subsided during ridge detumescence (some evidence available), withdrawing oceanic water from the continental margins and stimulating continentality in climate. Again, there is the Monterey effect (Fig. 10). That is an hypothesis based on isotopic time series patterns in the Miocene (during the most recent time of postulated oceanic circulation reversal, as it happens). Strongly positive benthic and planktonic gradients in $\delta^{13}C$ imply withdrawal of organic carbon from the system. That brings on cooling by a reversed-greenhouse effect during CO_2 drawdown. We see the cooling in the $\delta^{13}O$ curve, beginning just as the positive $\delta^{13}C$ trend peaks. In both the time scales and the isotopic shifts involved, the early Eocene pattern is remarkably similar to the Miocene Monterey effect, giving us an hypothesis by pattern analogy.

At any rate, there was a global cooling, perhaps with a brief accumulation of polar ice. We do not see here the diverse clues of warmth and humidity so evident in the early Eocene record. Instead, there are indications of aridity, although the age constraints are a bit loose. This is the interregnum in the Laramide/Himalayan changeover. It is also the interregnum in evolutionary activity and in pantropical adventuring by warm-adapted organisms.

IX. The Khirthar Restoration

We come to the "magic number" of Chron C19, the time of the new global tectonic regimen enforced by the events of Chron C24r. In northwest Europe this is the Lutetian/Bartonian boundary. In the Indo–Pacific region, it is a major transgression which I have named the Khirthar transgression,

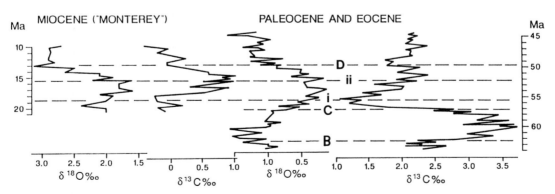

Fig. 10 The Monterey hypothesis applied to the early Eocene by pattern analogy. In the Miocene (19–13 Ma) the arguments are that $\delta^{13}C$ became strongly positive (levels i–ii), implying withdrawal of ^{12}C from the system. By reversed greenhouse, that triggered global cooling at level ii, and the cooling is monitored by the rise in $\delta^{18}O$ (levels ii–D). Comparing the aligned early Eocene pattern (55–49 Ma), we see closely similar $\delta^{13}C$ (i–ii) and $\delta^{18}C$ (ii–D) shifts over equivalent spans of time. (The pre-i and post-D patterns are ignored here.) [Modified from B. McGowran (1989).]

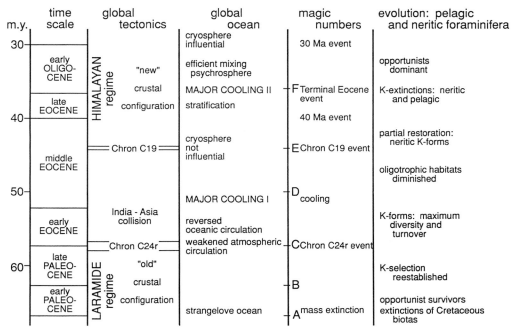

Fig. 11 Summary of major patterns and inferences for the early Cenozoic, including the Eocene. The table builds from the control—the crust of the earth—to the global ocean and climate, and finally to events in the biosphere signaled by the abundantly preserved and environmentally sensitive foraminifera.

after the middle Eocene stage in India and Pakistan in which it occurs.

The paleobiological time series demonstrate a rather widespread return to the situation in the early Eocene (Fig. 6). Note particularly the evidence for warming in some of the curves, and indirect evidence in others (e.g., the sharp increase in molluskan diversity could be a proxy for warming). That sense of the restoration of a former regime is heightened by evidence of increased humidity and renewed exogenic activity: deep weathering; large coal deposits and oil shales at mid-high latitudes, as in Australia. But now, inspect the oceanic $\delta^{18}O$ profiles again (Fig. 5). It is widely accepted now that the deep ocean temperature never again returned to the old levels of warmth after the early-middle Eocene deterioration. The surface-water values held a bit steadier, but that is all. Again, the global plot of planktonic foraminiferal diversities in Fig. 7 shows a progressive replacement of warmer water kinds with temperate and cool water kinds. In short, there is a decoupling of the oceanic evidence from the continental and shallow marine evidence for global warming—a decoupling that we did not see in the early Eocene. That difference between the early and later Eocene is a significant problem.

The later Eocene intervenes between the presumed termination of the episode of reversed circulation and the develop-

ment of the modern-type ocean, as we shall see. There is some evidence of oceanic stratification. It is seen in the periodic blooms among opportunistic planktonic foraminifera, which take advantage of an increased nutrient supply accompanying an expanded oxygen minimum zone. That, in turn, requires decreased mixing, accomplished (in one hypothesis) by the progressive cooling of the deep global ocean through the middle and late Eocene while surface cooling lagged. There is also a "silica window" in the late Eocene when siliceous sediments were more common. (The Eocene ocean was a silica-prone ocean. The chert-rich oceanic sediments known as Horizon Ac are associated with the earlier cooling that stimulated delivery of increased nutrient to the photic zone. Silica will become a major signal when we understand its behavior better.)

X. The Terminal Eocene Event

The modern ocean can be regarded as a large volume of cold water, the psychrosphere, covered by a thin skin mostly of warmer water. It can also be regarded as a large volume of nutrient-rich water covered by a thin skin in which that nutrient is needed, because the skin is in the photic zone. The modern ocean is a vigorously circulating ocean because of its and the atmosphere's strong gradients. It was not ever thus. It was proposed in the 1970s that the psychrosphere originated in the late Eocene on the evidence of widespread, deep-oceanic ostracodes. That seemed to be confirmed by the discovery at the same time (mid-1970s) of the very sharp increase in δ^{18}O in both surface- and deep-water profiles—the most spectacular cooling event in the Cenozoic.

On land, it was realized by 1909 that the vertebrate faunas, specifically the artiodactyls and the perissodactyls (e.g., modern cattle and horses, respectively) showed a radical change, a replacement by immigration after earlier disappearances. Thus there was a pronounced discontinuity in the fossil record, a *Grande Coupure*, at the Eocene/Oligocene boundary. A recent compendium of time ranges among marine invertebrates revealed an episode of heightened extinction at the same time (Fig. 6). These and still more phenomena have become known collectively as the Terminal Eocene Event (TEE). Discovery of microtectites in different oceans was invoked in support of an extraterrestrial impact as the cause of the perceived mass extinction. That extinction, in turn, became a critical item in the claimed 26-m.y. (or thereabouts) cycle of mass extinctions.

The literature on the TEE, while not rivaling the sheer quantity of writing on the end-Cretaceous event, is copious. Perhaps the following general statements would be acceptable to most workers with no axe to grind.

The *Grand Coupure* is indeed a time of profound change among terrestrial vertebrates, the change being stimulated by sweeping climatic changes. The oceanic δ^{18}O signal is very sharp and occurs at the base of Chron C13r. There is plenty of evidence for major changes in the biotas of the oceans and the continental shelves at that time. However, there was no mass extinction. Rather, the TEE consists of a succession of steps beginning with cooling, habitat shifts, extinctions at the middle-late Eocene boundary and proceeding for about 3 m.y. until the big drop. The microtectite have been confirmed; there may be up to six layers within the late Eocene. But their presumably extraterrestrial cause had no catastrophic impact on the biosphere.

XI. Overview of the Eocene

History, including global geohistory and biohistory, can be characterized as law plus contingency. We should of course violate no physical laws. However, we should not apply them in a machine-like way, because contingencies are never quite the same twice. The search for the big event that extinguished the dinosaurs and the ammonites is a case in point; it has stimulated the whole field of biotic extinction. But we soon

found that similar evidence for an ET cause at the TEE did not match a similar effect. Again, the Thulean volcanism at C24r has been identified as part of a profound impact on the exogenic system. The Deccan volcanism of 10 m.y. before the Thulean was even more voluminous but there is not the evidence available to justify a scenario comparable to our story for the early Eocene. (The critical difference is that the Deccan is a hot-spot effect whereas the Thulean is part of a crustal reorganization.)

We find the same thing in the other turning points of the Eocene. But first, what are those turning points? The world does not change gradually during its history. Gradualism arose out of an entirely justifiable insistence on the operation of physical laws, in opposition to the notion of wholesale creation and extinction by a Creator from time to time during earth history. So far, so good, except that geologists tended to abjure all notion of sudden or catastrophic change, and to react strongly to such ideas as short, sharp global episodes of mountain building, or the evolution of organisms in big jumps.

At present our worldview is somewhere in between. It seems to be clear that global crustal change is indeed episodic. It seems clear too that the global exogenic system moves from one mode to another in steps, which means that the shift is much more brief than the occupancy. In using these fertile notions we must beware two traps in particular. One is time scale. "Greenhouse" is a good example. We use it at several scales from our immediate social and cultural scales, through the "ice ages" of history and prehistory into the Pleistocene, all the way to the major icehouse/greenhouse alternations of the past billion years. That is acceptable so long as discussants are acutely aware of their time scales. If not, then confusion reigns.

The other trap is the either/or situation. Again, greenhouse and icehouse can exemplify. A strong present view has it that the present icehouse world began with the psychrosphere at the TEE. That does not imply that the change in mode was due to a concentration of causes entirely at the TEE; nor does it imply that the world has not reverted on occasion to—or toward—the greenhouse mode. Polarizations are valuable and perhaps indispensable conceptual tools, but they must be disciplined.

Figure 11 offers some major generalizations about the Eocene. It suggests that there is a rhythm in the patterns of Eocene history that extends from crustal events to organic evolution. Chrons C24r and C19 clearly are critical times—turning points—as are the climatic events labeled as major coolings I and II. The 40-Ma event is signaled by the strongest kick in the Exxon curve for more than 10 m.y. and by the biggest change in the deep-ocean microbiota since the drama of C24r. The 30-Ma event is beyond the scope of this discussion; however, there are biospheric changes at that time associated with the greatest interpreted fall in sea level for the entire Paleogene.

Notwithstanding the caveats expressed above, we might tentatively generalize that the TEE marks the development of the psychrosphere and the emergence of the cryosphere as a powerful influence on the exogenic systems of this planet. There probably was Eocene ice—the Krakow glaciation—but glacioeustatic effects were muted. That changed in the Oligocene. An interesting new hypothesis recognizes a difference in the seismically profiled shapes of canyons cut into the Atlantic continental margin of North America. Eocene canyons are U-shaped whereas in the Oligocene and Neogene they are V-shaped. Pre-Oligocene cuts were caused when global spreading rate changes—tectonoeustasy, detumescence of the global spreading system—lowered the sea level. The younger cuts were caused by faster, glacioeustatic lowerings.

The next generalization is that the prepsychrospheric ocean was not so well mixed, at least for most of the time, as the psychrospheric ocean of the Oligocene and the Neogene, which rarely reverted to the earlier state (it did during the Monterey event of the Miocene). Reversed circulation probably collapsed at the cooling I. In this mood of global generalization, we may substitute "greenhouse state" and "icehouse state" for prepsychrospheric and psychrospheric, respectively.

We now look again to the biosphere, this time for hints of process in the strange Eocene ocean. Consider the broad statements in Fig. 9. The "trophic resource continuum" describes a spectrum of resources available for consumption in terms of the simple but powerful variable, the chlorophyll concentration in water in the photic zone. Note especially that the continuum contracts at both ends as mixing improves with better circulation. That contraction impacts more strongly on the specialists at the oligotrophic end—among the foraminifera, the symbiont-bearing forms discussed earlier. The story being told by the graphs of diversity and taxic evolution is essentially twofold. First, the specialists in the nutrient-poor Eocene seas at high sea level yield progressively to the generalists as mixing improves and the sea level falls. Second, those changes are not smooth, but pulsating.

Perhaps the boldest of all generalizations about the Eocene arises from drilling in the Weddell Sea. J. P. Kennett and L. D. Stott have suggested that the Eocene ocean was two-layered, with warm saline waters formed at low latitudes and overlain by cooler waters formed at high latitudes. They have named that halothermal ocean "Proteus." The modern thermohaline ocean, formed in the Miocene, is "Oceanus." The transition took place in the Oligocene—"Proto-oceanus." Those names crystallize the differences in the global ocean during its Cenozoic history.

The global systems operating in the Eocene were rather different from today's systems. While scrutinizing the present for clues to the past, we must avoid over-reliance on the present. The story of the early Eocene would never have been revealed in that way. The pattern correlations simply are not represented in today's world. It is possible that organic evolution is different in a greenhouse world in that more species are susceptible to environmental perturbation than is the case in the icehouse mode. To be able to test such suggestions is an exciting possibility for the near future.

Bibliography

Berger, W. H., and Mayer, L. A., eds. (1987). Cenozoic paleoceanography. *Paleoceanogr.* **2**, 613–761.

Berggren, W. A., Kent, D. V., and Flynn, J. J. (1985). Paleogene geochronology and chronostratigraphy. *In* "The Chronology of the Geological Record" (N. J. Snelling, ed.), Geological Society of London Special Paper 10, pp. 141–195.

Boersma, A., Premoli Silva, I., and Shackleton, N.J. (1987). Atlantic Eocene planktonic foraminiferal paleohydrographic indicators and stable isotope paleoceanography. *Paleoceanogr.* **2**, 287–331.

Hallock, P., Premoli Silva, I., and Boersma, A. (1991). Similarities between planktonic and larger foraminiferal evolutionary trends through Paleogene paleoceanographic changes. *Palaeogeogr. Palaeoclimatol. Palaeoecol.* **83**, 49–64.

Haq, B. U. (1983). Paleogene paleoceanography: early Cenozoic oceans revisited. *Oceanologica Acta* **4** (suppl.), 71–82.

Haq, B. U., Hardenbol, J., and Vail, P. R. (1987). Chronology of fluctuating sea levels since the Triassic. *Science* **235**, 1156–1167.

Hsü, K. J., ed. (1986). "Mesozoic and Cenozoic Oceans." American Geophysical Union Geodynamics Series, Vol. 15. Washington, D.C.

Kennett, J. P. (1982). "Marine Geology." Prentice-Hall, Englewood Cliffs, New Jersey

Kennett, J. P., and Stott, L. D. (1990). Proteus and Protooceanus: ancestral Paleogene oceans as revealed from Antarctic stable isotope results; ODP Leg 113. *In* "Proceedings of the Ocean Drilling Program, Scientific Results," Vol. 113 (P. F. Barker, J. P. Kennett, *et al.*, eds.). College Station, Texas.

McGowran, B. (1989). Silica burp in the Eocene ocean. *Geology* **17**, 857–860.

McGowran, B., (1990). Fifty million years ago. *Am. Sci.* **78**, 30–39.

Premoli Silva, I., and Boersma, A., guest editors (1991). The oceans of the Paleogene. *Palaeogeogr. Palaeoclimatol. Palaeoecol.* 83, 1–263.

Rea, D. K., Zachos, J. C., Owen, R. M., and Gingerich, P. D. (1990). Global change at the Paleocene–Eocene boundary: climatic and evolutionary consequences of tectonic events. *Palaeogeogr., Palaeoclimatol., Palaeoecol.* 79, 117–128.

Shackleton, N. J., guest editor (1986). Boundaries and events in the Paleogene. *Palaeogeogr., Palaeoclimatol., Palaeoecol.* 57(1).

Glossary

Circulation, oceanic Today, the ocean transports heat to high latitudes in its surface waters; Antarctic freezing leaves a residue of both cold and saline, therefore especially dense, water that underplates much of the global oceanic fill as Antarctic bottom water and, together with cold North Atlantic deep water, returns to low latitudes. Neglecting some important matters, that modern pattern is *thermohaline* circulation. During heightened evaporation and presumably less significant or small polar icecaps, slugs of warm, saline, relatively dense water may escape from their shallow, low-mid latitude generating pans into the ocean, where they will sink. Thus we have reversed, salinity-dominated, *halothermal* circulation.

Evolution Often related to *transformational* changes—the number of horses' toes, the size of our ancestors' brain—evolution is actually a more fertile notion in its *taxic* sense. That means changes in the rates of speciation (origin of species), of extinction, of turnover (the two combined), and of diversity (species richness). We are now in an exciting period wherein global environmental changes can be (or cannot be!) related to taxic evolutionary changes as preserved in the fossil record, suggesting cause and effect.

Exogenic system Earth's interacting envelopes: the *hydrosphere* (mostly the ocean), *atmosphere, biosphere, reactive lithosphere* (that part of the rocky crust within reach of processes of weathering). The processes of deep seawater circulation at spreading ridges overlap with *endogenic* (within the earth) processes. Also relevant are *extraterrestrial* processes. Refractory problems such as mass extinction usually become arenas for competing theories drawn from the three sources.

Greenhouse and **Icehouse** Useful polarized concepts of the state of the planet. "Greenhouse" collects together global warmth, flattened thermal gradients, increased humidity, low relief, high sea level, and reduced vigor of oceanic circulation with a tendency toward enhanced anoxia in some parts of the ocean. "Icehouse" is essentially the reverse state in that assemblage of generalizations. It is important to add not-so-icecap-prone and icecap-prone, respectively, without implying necessarily that the world ever was ice-free.

Integrated time scale The Eocene rock record initially was and still is recognized by its characteristic *fossils;* fossils are cross-checked against the *geomagnetic reversal scale.* But rates of change and other quantifications demand *numerical calibration*—time in millions of years—which is founded ultimately in isotopic decay in the potassium–argon and other decay series. The geomagnetic scale consists of chrons, of which Chron C24 and Chron C19 are important here. Some fossil successions have their own names; thus the Clarkforkian and Wasatchian are North American "land mammal ages" whose mutual boundary is within C24r and close in age to the Seelandian/Ypresian boundary in the "standard," established in NW Europe (Fig. 1).

Isotopes in paleoceanography Stable isotopes of some light elements, occurring in nature in given ratios, fractionated by the biosphere, and thereby providing informative signals about the latter. Most useful ratios are $^{16}O/$ ^{18}O; $^{12}C/^{13}C$. Marine organisms biomineralize using oxygen in their calcite, etc., and in doing so record the temperature of the reservoir (i.e., the ocean) in the $^{16}O/$ ^{18}O ratio, which is expressed as $\delta^{18}O$, a departure from an agreed standard. But evaporation removes a bit more ^{16}O—a lighter fraction—from the reservoir. That does not matter at geological time scales until that very light fraction is locked up as an icecap, whereupon the reservoir is distorted in the heavier direction. We believe—although not all of us—that that ice effect is not a great distorter in the Eocene; it becomes assertive in the Oligocene. Carbon behaves differently. Photosynthesis removes ^{12}C preferentially, so that calcite skeletons are enriched in ^{13}C. Thus, high productivity is signaled as an increased difference between the planktonic record (from the photic zone) and the benthic record (where the light fraction is returned to the system). But when both signals go positive through time, that shift signals a removal of organically fixed carbon from the reservoir altogether—into oceanic sediments, becoming a potential source of oil, or into increased terrestrial plant biomass and thence into coals.

Paleogeography (1) Distribution of continents and oceans, changing as basaltic *oceanic* crust is emplaced at spreading ridges and lighter granitic *continental* crust is moved around. Main Eocene effects: closure of low latitude oceanic embayment Tethys, isolation of Antarctica, opening between NW Europe and NE America. (2) Distribution of land and sea, changing as the sea level changes (higher in the Eocene); also topography.

Erosion and Weathering

William W. Hay
University of Colorado

Erosion and weathering are the processes by which geological materials are modified and transported at the earth's surface. Weathering involves physical disintegration and chemical dissolution of rock materials *in situ* in response to the environmental conditions prevailing at the earth's surface. Erosion is the transport of the material from the site of weathering to a site of deposition.

I. Physical and Chemical Weathering

The term *weathering* originally referred to the effects of physical and chemical processes in causing the disintegration of rock exposed to the weather on land. Physical weathering takes place as a result of temperature changes. It is caused by differential expansion and contraction of mineral grains of different composition or by the expansion of ice in cracks and crevices. Chemical weathering involves attack by active agents at the exposed rock surface or at the soil–rock inter-face. Physical weathering breaks down rock by causing disaggregation of the constituent mineral grains; chemical weathering dissolves susceptible grains and the cement that holds them together. Biological agents contribute to both physical and chemical weathering. The expansion of roots as plants grow may widen cracks, and the humic acids released into the soil by plants are an important agent in dissolution of rock materials. Weathering releases material into solution and leaves a residue of detrital particles that can be transported by water, ice, and wind to a new site of deposition.

The major processes involved in chemical weathering on land are hydrolysis, oxidation, and carbonation and, to a lesser extent, reaction with sulfuric, nitric, and humic acids. These processes affect all rocks and sediments, but the rates at which the weathering reactions proceed are strongly dependent on composition of the material being weathered.

Recently, the term *weathering* has been expanded to include analogous processes that occur subaquatically, referring particularly to the interaction of seawater and ocean crustal rocks in hydrothermal systems. [*See* HYDROTHERMAL SYSTEMS.]

A. Weathering of Silicate Minerals

Igneous and many metamorphic rocks consist chiefly of silicate minerals. The products of weathering of igneous and metamorphic rocks are sediments that, on compaction and induration, become sedimentary rocks. It is the chemical weathering of silicate minerals that alters minerals that are stable at the high temperatures and pressures prevailing in the interior of the earth to minerals stable under the conditions existing on the surface of the earth. Weathering of silicate minerals involves breakdown of the parent mineral to release cations and silica. The cations and silica may migrate into new minerals or be removed in solution. [*See* CHEMICAL WEATHERING CONTROLS ON SAND COMPOSITION; SILICATES.]

The major weathering reactions are hydrolysis, oxidation, and carbonation, shown in the following examples:

Hydrolysis: $2KAlSi_3O_8 + 11H_2O \rightarrow$
Orthoclase

$$Al_2Si_3O_8 + 4H_4SiO_4 + 2K^+ + 2OH^-$$
Kaolinite

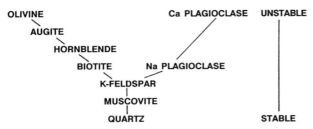

Fig. 1 Stability of minerals found in igneous rocks under conditions prevailing at the earth's surface. The order of increasing stability is the inverse of their order of crystallization (Bowen's reaction series) and corresponds to decreasing temperature of formation.

Oxidation: $2Fe_2SiO_4 + 4H_2O + O_2 \rightarrow$
Fayalite

$2Fe_2O_3 + 2H_4SiO_4$
Hematite

Carbonation: $Mg_2SiO_4 + 4H_2O + 4CO_2 \rightarrow$
Forsterite

$2Mg^{2+} + 4HCO_3^- + H_4SiO_4$

Orthoclase is the common feldspar of granitic rocks. Kaolinite is a clay mineral. Fayalite and forsterite are olivines, found in ultrabasic igneous rocks and very susceptible to weathering at the earth's surface. Hematite is the rust-colored mineral that often stains the surface of ultrabasic rocks. H_4SiO_4 is silicic acid, the soluble form in which silica is transported by surface- and groundwater. All of these reactions can affect all silicate minerals; rates depend on temperature, presence of water, pH, and Eh.

The increasing resistance of silicate minerals to weathering is shown in Fig. 1. The order is the same as that of Bowen's reaction series describing the order of crystallization of minerals from a silicate melt and corresponds to the increasing number of Si–O–Si bonds, from zero in olivine to four in quartz.

Mechanical weathering, the result of differential expansion and contraction in response to temperature changes, freezing and thawing of intergranular water, growth of plant roots, etc. causes the mineral grains to separate. This both greatly increases the surface area for chemical reaction and enhances the likelihood that the grains will be moved by transporting media.

B. Alteration and Dissolution of Sediments

Sedimentary minerals are derived from silicate minerals by alteration to become more stable under conditions at the earth's surface and from other minerals produced during the weathering process. Igneous silicate minerals become clay minerals that are progressively leached of all soluble cations. The clay mineral products of silicate weathering depend on the climate. Thus albite, the Na-plagioclase feldspar, weathers to kaolinite under wet tropical conditions but to montmorillonite under less humid, cooler subtropical conditions.

$2NaAlSi_3O_8 + 2CO_2 + 11H_2O \rightarrow$
Albite

$Al_2Si_2O_5(OH)_4 + 2Na^+ + 2HCO_3^- + 4H_4SiO_4$
Kaolinite

$2NaAlSi_3O_8 + 2CO_2 + 6H_2O \rightarrow$
Albite

$Al_2Si_4O_{10}(OH)_2 + 2Na^+ + 2HCO_3^- + 2H_4SiO_4$
Montmorillonite

Some sedimentary minerals are the ultimate residual products of weathering [e.g., gibbsite, $Al_2(OH_6)$] and are virtually insoluble and immune to further decomposition. The leached materials are carried in solution to the sea or to playas where, under evaporation conditions, they are precipitated as salts [e.g., halite (NaCl)]. The sedimentary minerals thus formed may be highly soluble. Sedimentary minerals are affected by hydrolysis, oxidation, and carbonation reactions, but exhibit an even greater range of sensitivity than do igneous silicates.

C. Development of Residuum

Weathering of rock results in solution of some components and their removal by percolating waters. Disaggregated rock that is little altered chemically is termed "gruss." The smaller particles formed as the rock disintegrates and disaggregates increase the surface area available for chemical reaction. As weathering proceeds, soluble material goes into solution and is removed. The insoluble material that remains behind is termed "residuum" and forms a mantle over the parent material. The nature and thickness of this weathering mantle varies with climate; in high latitudes and the arid zone the absence of liquid water inhibits chemical weathering, so that it is largely rubble resulting from physical weathering. In the humid belts chemical weathering dominates, and the residuum consists of those minerals that are most stable and resistant to solution. In the tropics the warmth makes the chemical reactions proceed rapidly, so that the weathering

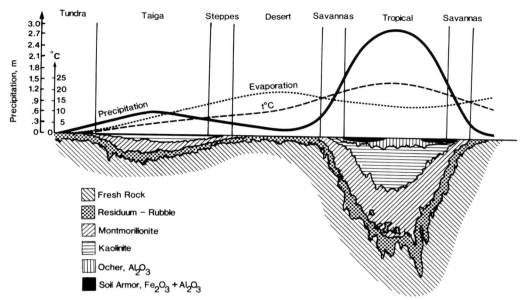

Fig. 2   Schematic diagram of the latitudinal distribution temperature, precipitation, evaporation, and different types of material produced by weathering on the surface of the earth today. [Modified from N. M. Strakhov (1963). "Types of Lithogenesis and Their Evolution in the History of the Earth" (in Russian). Gasgeoltekhizdat, Moscow.]

mantle is often very thick and leaching is extreme. Figure 2 shows the changes in thickness of the weathering mantle with latitude.

D. Soil Formation

Soil is a complex material containing not only the residues of rock weathering but also moisture, air, decomposing organic matter and the associated organic compounds, along with living bacteria, plants, and animals. In the upper part of the soil, the zone of eluviation, chemical reactions result in leaching and mobilization of material that is carried downward by percolating waters. In the lower part of the soil, or zone of illuviation, some of the material dissolved from the upper part of the soil is deposited. Other soluble material is flushed from the soil, infiltrating the underlying rock or sediment, and is carried away in solution in groundwater.

Soil formation is strongly dependent on the nature of the parent materials, slope of the land, climate, and the vegetation cover. Weathering processes take place within the soil, as well as within the weathering mantle beneath the soil and at the rock interface. Figure 2 shows the variability of weathering mantle and soil types with latitude and climate.

II. Environmental Factors Influencing Weathering Rates

A. Climate

Climatic factors are significant in influencing weathering rates. Mechanical erosion depends on rapid temperature changes. Temperature variations are greatest at the rock–air interface but decrease rapidly downward. The daily cycle of temperature change penetrates only a few centimeters, and the annual cycle is usually damped out below a meter depth. Thus the insulating effect of a thickness of soil or weathering mantle of more than a few centimeters greatly reduces physical weathering. Chemical reactions proceed more rapidly at higher temperatures and virtually stop at the freezing point of water. Precipitation–evaporation balance is important in determining the moisture content of the weathering mantle. If precipitation exceeds evaporation, soluble material will be flushed away with water infiltration to the groundwater system; if evaporation exceeds precipitation, solutes are drawn with the evaporating water to the soil surface and form a crust. Seasonal rainfall may result in an alternation of hydrolysis and carbonation with the oxidation reaction.

B. Atmospheric and Soil-Atmospheric Compositions

Although soil contains atmospheric gases, the soil-atmosphere composition differs from that of the atmosphere itself. Soil atmosphere is enriched in CO_2 supplied from decomposing organic matter. Rain water always contains some carbonic acid (H_2CO_3), from solution of atmospheric CO_2 in rain drops, and is hence at least weakly acidic. However, the water percolating through the soil into the deeper weathering mantle becomes enriched in carbonic acid due to the higher CO_2 pressure in the soil and is more strongly acidic. Oxygen is also present in soil atmosphere, but its partial pressure is less than that of the atmosphere because of oxygen consumption in the decomposition of organic matter. Oxygen is also consumed in the weathering of sulfide minerals at the rock interface, producing sulfuric acid as one of the weathering products.

$$4FeS + 4H_2O + 9O_2 \rightarrow 2Fe_2O_3 + 4H_2SO_4$$
Pyrite   Hematite

The atmosphere contains some SO_2 from natural volcanic as well as anthropogenic sources; hence rainwater already contains small amounts of H_2SO_4, sulfuric acid. However, the sulfuric acid content of waters percolating through weathering mantle or rock containing sulfides increases greatly. Sulfides occur in most mineral deposits, hence the waters draining from mine tailings are often highly acid. Sulfides also occur in some sediments, notably black shales, making the waters percolating through these sediments acid. [*See* ACIDS AND BASES IN WATERSHEDS; LAND–ATMOSPHERE INTERACTIONS.]

C. Rainwater–Soil–Groundwater Compositions

Average rainwater contains salts in proportions that resemble very dilute (0.02%) seawater. Groundwater differs markedly from rainwater because it contains the solutes derived from the weathering process. River water is a mixture of about two-thirds direct runoff, water that has run over or through the soil, and one-third groundwater. The average groundwater has a content of dissolved solids about 20 times that of rainwater and about 0.4% that of seawater; average river water is slightly more dilute. The carbonation reaction dominates the composition of groundwater, so most groundwaters and river waters are bicarbonate solutions, with Ca^{2+} the major cation. Limestone is readily dissolved by carbonic acid and transported to be redeposited in the sea mostly through the intermediate agency of organisms that precipitate $CaCO_3$ as either calcite or aragonite skeletal material.

$$CaCO_3 + H_2CO_3 \rightarrow Ca^{2+} + 2HCO_3^{2-} \rightarrow$$
Limestone

$$CaCO_3 + H_2O + CO_3$$
Shells

The CO_2 for the carbonic acid came from the atmosphere either directly or indirectly through the intermediate agency of fixation of carbon by plants and subsequent decomposition of the organic matter. It returns to seawater and the atmosphere when calcium carbonate is deposited in the sea. The weathering of limestone (and dolomite) uses atmospheric CO_2 as a catalyst to transport $CaCO_3$ from one site to another; it does not consume atmospheric CO_2. Atmospheric CO_2 is consumed only by the weathering of silicates, which proceeds much more slowly and uses bicarbonate for transport in solution.

The composition of groundwater is usually dominated by Ca^{2+} and HCO_3^-, but a variety of other cations and anions are present, reflecting particularly the abundance of soluble sedimentary rocks in the drainage basin. Na^+ and Cl^- are present because of their introduction as salt recycled from the sea through rainwater, but their concentrations become significant where circulating groundwaters bathe halite deposits. Significantly, human activities, such as bringing brine from great depths to the surface in petroleum extraction and in direct mining of salt, has increased the Na^+Cl^- content of many rivers. Similarly, the SO_4^{2-} content of rivers is high if the river drains areas underlain by gypsum or anhydrite ($CaSO_4$). Weathering of dolomite, leaching of clay minerals, and weathering of igneous rocks contributes Mg^{2+}; and K^+ is derived from weathering of igneous rocks and leaching of clay minerals. Because of the differences in solubility of these ions in silicates and sedimentary rocks, the composition of rivers is largely controlled by the availability of soluble sedimentary rocks in the drainage basins.

D. Elevation and Slope

The regional elevation and local slope of the ground both significantly affect the rates of weathering. Slope is a major factor contributing to the removal of mechanically separated detrital particles. Steep slopes may not develop a residuum adequate to retain moisture or a soil that can supply acids for chemical weathering. Also, once in transport, chemical decomposition is curtailed, so that steep slopes promote physical disintegration not chemical decay. Higher elevations have lower temperatures, reducing the rates of chemical weathering, but are drained by rivers steep enough to remove mechanically derived particles.

E. Vegetation

Vegetation plays a major role in determining the nature of the soil, and hence the composition of waters percolating down to the horizons where active chemical weathering proceeds. Coniferous forests develop soils rich in organic acids. Vegetation also modifies the way in which rainfall reaches the soil surface. Forests that develop canopies intercept much of the rain and both slow the rate of rainfall onto the forest floor and prolong the time of throughfall as the canopy continues to drip after the rain has stopped. In many trees, the rainfall intercepted by the canopy is directed to the forest floor by flowing along the branches and trunk, where the water may become enriched in organic acids. Finally, vegetation plays a major role in binding the soil and weathering mantle with its roots, retarding mechanical erosion.

III. Weathering as the Rate-Limiting Process in Denudation of the Continents

Weathering is the rate-limiting process in denudation of the continents. Chemical weathering dissolves the cement that holds the particles of rocks together. Chemical and mechanical weathering prepare the rocks for transport by water, wind, and ice. The surface of the land cannot be lowered by erosion and transport of the material until weathering has reduced the detrital particles to a size small enough to be moved by ice, water, or wind.

IV. Global Change and Weathering Rates

Natural changes, such as mountain building and climatic change, will affect rates of weathering. Weathering processes go to completion only after they have operated for very long periods of time, probably millions of years. During the past few million years most of the earth has been subjected to repeated rapid climate change, and as a consequence there are very few areas where the weathering process has developed stable soil profiles. Australia is one of the few continents where the climate has remained nearly constant, and soils are quite ancient. In the regions of North America and Europe covered with debris left by the glaciers, soil formation has just begun, and the soils are especially fertile.

As an agent for exposing rock and sediment to weathering, human activity exceeds natural processes. Farming and ranching have altered the natural vegetation cover and exposed bare soil to erosion. Plowing churns the soil and exposes it to the oxygen-rich atmosphere much more rapidly than the natural activity of burrowing animals. Construction activity pulverizes and exposes rock that would otherwise remain buried. Mining activity has brought to the surface, pulverized, and spread out as tailings minerals containing rare metals that would otherwise reach the weathering zone at a much slower rate.

Global warming will change weathering processes in a variety of ways. Mechanical weathering would be reduced by lessened seasonal temperature variation in high latitudes, but it may be increased if climate change involves a greater frequency of extreme events. Chemical weathering will increase with increased temperature. Increased atmospheric CO_2 may have some effect on increasing the rates of chemical weathering, but this is moderated by the already high CO_2 content of soil atmosphere. Increased SO_2 and NO_2 in the atmosphere, causing "acid rain" may become more important as chemical weathering factors. [*See* ACID RAIN.]

V. Weathering of Rocks in the Ocean

Submarine weathering is the result of the chemical reaction of seawater with igneous rocks forming the ocean floor. Near the mid-ocean ridges where new sea floor forms by volcanic activity, seawater percolates into the hot rock through cracks and clefts, and the reactions lead to hydrothermal alteration of the rock. The basalt of the sea floor consists mostly of plagioclase feldspars, amphiboles, and pyroxenes—minerals that are rich in Ca and Fe. In reaction with seawater, Ca is replaced by Mg, Na by K, less abundant metals are mobilized, and silica goes into solution. The result is conversion of the minerals typical of basalt into clay minerals. Some of the cations carried in solution in the hot waters precipitate on contact with cold seawater; but some other cations, such as Ca^{2+}, are introduced into the ocean in quantities that are of the same order of magnitude as the inputs from land. This large submarine input acts to stabilize the ocean chemical system, so that perturtabtions from the delivery of material from land may be less important in changing ocean composition than was previously thought.

Reverse weathering is a term used to describe the silicate mineral-forming processes that occur in the sea, removing solutes from seawater to form clay and other minerals that would be unstable if exposed to the atmosphere and rain water.

VI. Erosion

Erosion is removal of weathered material or sediment and its transport to a new site of residence or deposition. The major processes of erosion on land involve transport by ice, water, or air. Erosional processes are usually measured in terms of transport of material to the sea. At present the total amount of material eroded and delivered to the sea is about 250×10^{14} g/yr, equivalent to lowering the land surface by about 0.07 mm/yr. Of this, the particulate load of rivers accounts for about 73% of the total, that of glaciers 8%, the dissolved load of rivers 16%, the dissolved load of groundwater 2%, erosion on coasts 1%, and atmospheric dust only about 0.2%. The comparison of processes in terms of transfer from land to sea is misleading because some transport processes are more effective in moving material from erosional to depositional sites on land. Also, erosion, transport, and redeposition of sediment may occur entirely subaqueously; erosion and redeposition are important in both shallow seas and the deep ocean basin. [*See* SEDIMENT TRANSPORT PROCESSES; SEDIMENTARY GEOFLUXES, GLOBAL.]

A. Glacial Erosion

Glacial erosion is the result of moving ice. Because the temperatures in glaciated regions are usually below the freezing point of water, chemical weathering is reduced, and physical weathering dominates. The expansion of water as it freezes to ice is also an important factor in producing debris that can be moved by glaciers. Glacial transport can accommodate large boulders, but the effectiveness of glaciers as erosive agents is dependent on the slopes down which they flow and whether the flow is confined. Mountain glaciers carry large amounts of debris plucked from the valley walls and floors. Continental ice sheets may be much less effective because the flow of the ice takes place above its floor. Although glacial transport today accounts for about 8% of the total of material delivered to the ocean, glaciers play a larger role in moving material from higher to lower elevations in mountainous regions. Glaciers are effective in pulverizing rock, reducing it to grain sizes more readily transported by rivers. Glaciers often oversupply sediment to their meltwater; so that the streams emerging from them are usually braided, unable to fix a channel because of the deposition of excess sediment. Glaciers can move detritus uphill as well as downhill. The Pleistocene ice cap scooped the basins of the Great Lakes from the soft sediments surrounding the crystalline rocks of the Canadian Shield, depositing the excavated sediment as the ground moraine of the midwestern states of the United States and the prairie provinces of Canada.

B. Fluvial Erosion

Fluvial erosion dominates transport of material from land to sea. The chemical load of rivers is dependent on the area drained and on the rocks, sediments, and soils exposed to the percolating and moving waters. The concentration of solutes is inversely proportional to the runoff; rivers draining wet areas are dilute, those draining arid regions have higher concentrations of dissolved solids, reflecting the rate-limiting nature of weathering processes. The dissolved load of rivers is easily measured and well known, but analysis of the solutes does not reveal all of the cations being transported. Significant cation transport may take place by their absorption on the surface of clay minerals with a high exchange capacity.

These adsorbed cations may be exchanged and released where the river enters the sea, but there are very few data on these processes. Most of the load of rivers is transported as fine detrital particles, mostly clay minerals and quartz, in suspension. Kaolinite, which has a low exchange capacity, is the dominate clay mineral in tropical rivers; at intermediate and higher latitude, where weathering proceeds under cooler conditions, it is replaced by montmorillonite, which has a higher exchange capacity. [*See* CHEMICAL MASS BALANCE BETWEEN RIVERS AND OCEANS.]

The bed load of rivers consists of coarser sand and gravel moved by traction, averaging 10% of the suspended load; but the proportion of bed load varies greatly among rivers. The particulate loads of rivers are highly dependent on water volume and flow rate. If these are adequate to move all of the sediment supplied, the river is said to be competent. Evaporation from the surface of the river may reduce the volume flow so that the river is no longer able to transport the sediment introduced in its upper reaches; the river becomes incompetent, and deposition occurs. Global warming may result in major changes of the flow regimes of rivers and change the competence of rivers to carry loads. Human modifications of rivers have greatly altered their sediment transport characteristics. The construction of dams, from those used for farm ponds to major barriers, has completely changed the way in which detrital sediment is transported. Although agricultural activity has increased erosion rates, dams trap sediment, so that less sediment reaches the sea than under natural conditions. Dams change the competence along the length of the river. Sediment-free water released below a dam has a high erosive capacity, and the still waters of the lakes have low transport competence. [*See* RIVERS.]

C. Aeolian Erosion

Under natural conditions, aeolian erosion is limited to the subtropical desert regions and to periglacial areas. In the desert regions, lack of moisture results in sparse or nonexistent vegetative cover to bind the weathering mantle and soil. In periglacial regions, glacial outwash plains are devoid of vegetation, and when dry the fine sediment is readily transported by the wind. Human agricultural practices seasonally destroy vegetation cover and expose soil to erosion by the winds in regions where aeolian erosion would not normally occur. In times of drought, when the vegetative cover is not regenerated wind erosion may become severe. [*See* AEOLIAN DUST AND DUST DEPOSITS.]

D. Coastal Erosion

Coastal erosion is the result of wave action that mobilizes particles and currents that transport them to a new site of deposition. On rocky coasts, wave energy is concentrated on promontories and material removed from them is deposited in the adjacent embayments. On coasts bordered by a coastal plain the effects of coastal erosion may be more pronounced. Erosion occurs where wave energy is concentrated, and the mobilized sediment is then transported, usually longshore, to a deposition site. In many regions the balance between sediment supply by rivers and longshore transport determines the position of the coast. Human activities often upset this balance by changing the river supply or longshore transport. If the longshore transport is interrupted by an artificial feature, such as a dredged harbor entrance, the result is downstream erosion of the coast.

VII. Environmental Factors Affecting Erosion Rates

A. Climate

Cold temperatures affect erosion rates because frozen ground cannot be readily eroded and frozen rivers do not transport sediment. The sediment loads of Arctic rivers are only a few percent of the loads of rivers in warmer parts of the world, because significant parts of their drainage basins are underlain by permafrost and their flow is limited to a few months of the year. As global warming melts permafrost and allows rivers to flow for longer periods of time, sediment transport budgets in high-latitude regions can be expected to increase markedly.

The precipitation–evaporation balance is very important in sediment transport. High erosion rates occur at the margin between savannas and desert areas, where rainfall occurs but is infrequent, and vegetation cover is sparse. Temporal distribution of rainfall is also important. Exceptional storms are responsible for mobilizing weathered material by slope wash and transporting it downslope into local streams, where it is usually redeposited but remains available for transport by lesser flood events.

B. Atmospheric Composition

The short-term effects of changes in atmospheric composition on erosion are largely indirect, through climate change. If increases in atmospheric greenhouse gases induce global warming, the precipitation–evaporation balance will favor evaporation, expanding regions of sparse vegetation and high fluvial and aeolian erosion rates.

C. Elevation and Slopes

On the long term, increases in atmospheric CO_2 must result in increased weathering of silicate minerals of igneous and metamorphic rocks. The formation of mountain ranges and their erosion modulate atmospheric CO_2 content. Volcanic activity during mountain building releases CO_2 into the atmosphere, and the later weathering and erosion of silicate exposed in the mountain ranges consumes CO_2 from the atmosphere.

D. Vegetation

Detrital erosion and denudation rates vary with the nature of the vegetation cover that determines how rainfall reaches the ground. Where vegetation is sparse and the plant cover incomplete, soil or weathering mantle is directly exposed to the rain. At the other extreme, rain forest and jungle canopies prevent most rainfall from reaching the ground directly, reducing its erosive effect. Chemical erosion and denudation rates also vary with the nature of the vegetation cover. The rates of production of humic acids and CO_2 in the soil are ultimately controlled by the vegetation.

VIII. Transport, Storage, Delivery to Endoric Regions, Delivery to the Sea

Most of the global information on loads of rivers, atmospheric dust transport, etc., refers to the flux from land to sea. The land–sea flux does not represent the global flux nor does it correspond to the denudation rate of specific land areas. Measurements are also made in other parts of the transport systems and show that eroded material is temporarily stored within drainage basins as a result of decreasing competence of rivers. Changes in river competence—their ability to transport sediment—is a function of the gradient and amount of water in the river.

Some rivers do not reach the sea but drain into closed basins in the interior of the continents, termed "endoric regions." These basins have no external drainage because evaporation rates within them exceed both precipitation and supply of water from rivers and the groundwater system. The increased evaporation expected with global warming may expand existing endoric regions and create new ones.

IX. Subaquatic Erosion, Transport, and Redeposition of Sediment

Erosion, transport and redeposition of sediment within lakes and the sea are not as well known as for land areas. Of the material brought to the sea in solution, much of the calcium and carbon are removed from the seawater and deposited by organisms as $CaCO_3$ shell material. The remaining carbon that is not recycled back to the atmosphere as CO_2 or other gases is buried as organic carbon. Most of the silicic acid brought to the sea is precipitated by organisms as opal skeletal material. Other salts are removed in playas, marine evaporite basins, and along the shore in very arid regions.

Detrital material brought to lakes and the sea is sorted by waves, with coarser material forming a moving band on and near the shore, kept in motion by longshore currents. Finer detrital material settles to the bottom further offshore. Deposition on the continental shelves may be only temporary, and during storms material is mobilized into dense flows of turbid water that flows down submarine canyons onto the continental rise and abyssal plains. Mass movements, slumps, and slides also transport large quantities of material to the deep-sea floor. Very little is known about the relative importance of transport processes in the sea.

X. Global Change and Erosion

Natural global change affects erosion rates by changing elevations through mountain building and the distribution and amounts of rainfall, and vegetative cover is changed through climatic change. All of these changing conditions alter rates of slope erosion and the competence of rivers. Permafrost is extensively developed around the Arctic, and it effectively shields weathered material from erosion. If the permafrost melts, it frees large amounts of particulate debris for erosion and transport.

Human agricultural activity has destabilized the natural balance; farming practices have increased soil erosion rates, but dams trap the eroded material in reservoirs. Although denudation rates have generally increased, sediment delivery to the sea has decreased; and because of the lower sediment supply, coastal erosion has increased. Construction activities expose natural materials to erosion, and paving and buildings increase runoff. As a result of construction requirement, man has become a major agent in moving sand and gravel from one place to another on the surface of the earth.

Global warming will result in higher evaporation off reservoirs and rivers, reducing the amount of water and the competence of rivers, increasing storage of sediment in the drainage basins.

Human activity has also been significant in altering the budget of material near the shore; obstructions slow longshore currents and cause sediment to be deposited, and downcurrent the lost sediment supply results in coastal erosion. Warming of the ocean might cause destabilization of the continental slopes. Clathrates, ices formed by water and gases such as methane at high pressures and temperatures above the freezing point of water, are present under extensive areas of the continental slopes. They act as cement within the sediment, holding it in place. Rising oceanic temperatures may cause the clathrates to decompose to water and gas, and the resulting loss of cement may cause large areas of the continen-

tal slopes to fail catastrophically as massive subsea slides. [*See* CLATHRATES.]

XI. Summary

Weathering involves mechanical and chemical processes that transform rocks and minerals to materials more stable under conditions prevailing at the surface of the earth. Weathering produces a mantle of residuum over the parent rock. Organisms modify the residuum, adding organic compounds and facilitating penetration of gases and moisture to produce soil. Erosion is the movement of weathered material to a new site of deposition. Detrital material is supplied to the sea from land. Solutes having a mass of about one-fourth that of the detrital material are also supplied from land, but there are additional supplies of solutes from submarine weathering in hydrothermal systems and from waters being squeezed out of thick sediment wedges. The magnitude of these subsea sources of solutes is not known, but for some ions may be as large as or larger than the flux from rivers and groundwater. Human activity modifies the weathering and erosion system in many ways, affecting rates of weathering, transport and deposition.

Bibliography

Garrels, R. M., and Mackenzie, F. T. (1971). "Evolution of Sedimentary Rocks." Norton, New York.

Gregor, C. B., Garrels, R. M., Mackenzie, F. T., and Maynard, J. B., eds. (1988). "Chemical Cycles in the Evolution of the Earth." Wiley (Interscience), New York.

Hay, W. W., and Southam, J. R. (1977). Modulation of marine sedimentation by the continental shelves. *In* "The Fate of Fossil Fuel CO_2 in the Oceans" Marine Science Series, Vol. 6, (N. R. Anderson and A. Malahoff, eds.), pp. 569–604. Plenum Press, New York.

Holland, H. D. (1978). "The Chemistry of the Atmosphere and Oceans." Wiley, New York.

Lerman, A., and Meybeck, M., eds. (1988). "Physical and Chemical Weathering in Geochemical Cycles." NATO ASI Series, Series C: Mathematical and Physical Sciences, Vol. 251.

Loughnan, F. C. (1969). "Chemical Weathering of the Silicate Minerals." American Elsevier, New York.

Meybeck, M. (1987). Global chemical weathering of surficial rocks estimated from river dissolved loads. *Am. J. Sci.* **287**, 401–428.

Milliman, J. D., and Meade, R. H. (1983). Worldwide delivery of river sediment to the oceans. *J. Geol.* **91**, 1–21.

Glossary

Denudation Lowering of the land surface through erosion.
Erosion Removal of residuum or soil by the action of running water, ice, or wind.
Residuum Material remaining *in situ* after weathering.

Reverse weathering Reactions with aquatic agents that reverse the ordinary weathering reactions.
Sedimentation Process of deposition of sediment.
Soil Residuum admixed with decaying organic matter and containing moisture, atmospheric gasses, and living animals and plants.
Storage Temporary deposition of material during transport from its source to its ultimate site of repose.
Weathering Destructive action of atmospheric or aquatic agents on rocks and sediments.

Evaporation and Evapotranspiration

John R. Mather
University of Delaware

Natural evaporation and evapotranspiration are much more than the reverse of rainfall, the return of moisture to the atmosphere. Tremendous amounts of energy are required to produce evaporation and evapotranspiration, and this energy is also transferred to the air with the water vapor as latent heat. The measurement of moisture loss from a water-filled pan does not provide a reliable value of evaporation from lakes or other water bodies because of problems of exposure and energy transfers. Evaluation of evaporation or evapotranspiration based on theoretical considerations often offers reasonable estimates, but some of the expressions are so complex that they cannot be widely applied. Values of evaporation and evapotranspiration are most often used in conjunction with precipitation in water budget computations to obtain quantitative information on such factors as streamflow, soil moisture storage, leaching, and drought.

I. Introduction

Evaporation and evapotranspiration play fundamental roles in earth system science because these processes supply not only the atmospheric moisture needed for precipitation (often at places far removed from where the initial evaporation occurs) but they are also a significant contributor to the energy balance of the earth–atmosphere system. In fact, nearly as much energy is transferred as latent heat to the atmosphere by evaporation and evapotranspiration as is absorbed into the atmosphere directly from solar radiation. Lack of a full understanding of the role of evaporation and evapotranspiration contributes to our present debate over possible global warming due to increasing CO_2 and other trace gases in the atmosphere. Current global circulation models (GCMs) do not adequately represent evaporation and evapotranspiration transfers, and thus they fail to recognize possible feedbacks that might negate or accentuate the role of increased CO_2 in future climates.

II. Definitions

A. Evaporation

Evaporation represents the total movement of water molecules away from the surface in the form of vapor. There can also be some downward movement of moisture to the surface, a recondensation of water vapor. It is, thus, the difference between the total evaporation and the recondensation, the net evaporation, that is loosely referred to as the evaporation. It is usually given as a depth of moisture so that it can be compared directly with the depth of precipitation.

B. Transpiration

Transpiration refers to the loss of water from living plants (as perspiration refers to water loss from humans and other animals). The amount of water lost by transpiration varies in partial response to rooting depth and leaf area index. The principal role of transpiration is to cool the plant and to keep its temperature within the limits for life of the plant.

C. Actual Evapotranspiration

The combined evaporation from both moist land surfaces and water bodies and the transpiration from the vegetation cover over the ground is identified as "evapotranspiration." It represents the net water loss from the total plant, soil, and water surfaces in an area. When the ground is well covered with plants, the principal loss of water is by transpiration. Evaporation actually constitutes only a small fraction of the total water loss from a well-vegetated forest or pasture area, since little energy for evaporation can reach the moist soil surface.

Actual evapotranspiration depends on (a) various climatic factors (e.g., net radiation, wind velocity, humidity); (b) type of soil; (c) soil moisture content; (d) type of vegetation; and (e) land management practices. It can never exceed the value of precipitation at a place or in an area (unless there is an inflow of water by surface or near-surface flow from a nearby area). Any relation using precipitation and actual evapotranspiration to express the relative moistness or dryness of a climate fails, since evapotranspiration will not exceed the value of precipitation. The true degree of dryness of an area will not be revealed. [*See* CLIMATE.]

D. Potential Evapotranspiration

One further term that was originally defined by the American climatologist C. W. Thornthwaite in the early 1940s, and used extensively by him and others in evaluating the climatic water budget is *potential evapotranspiration*. It is the water loss from an extensive, closed (no bare soil areas visible), homogeneous (all vegetation growing to the same height), cover of vegetation that never suffers from a lack of water. A later addition to that definition specified that the vegetation should reflect between 20 and 25% of the solar radiation falling on it.

Potential evapotranspiration as defined by Thornthwaite has been found to depend almost entirely on the single climate factor of net radiation. As long as the evaporating area (and the buffer area around it maintained in the same condition as the evaporating area) is sufficiently large, wind and humidity factors have been found to be of limited importance. With adequate moisture always available, type of soil, land management practices, and soil moisture content are unimportant. Even the type of vegetation or its depth of rooting becomes less significant under conditions of always adequate soil moisture content. When the precipitation supply at a place or in an area is compared with the climatic demand for water (potential evapotranspiration), the relative dryness or moistness of a climate becomes clearly defined. For example, in a desert area, the annual precipitation supply may be 125 mm whereas the potential evapotranspiration may be 2000 mm. The actual evapotranspiration cannot exceed 125 mm. Comparison of the precipitation supply with the climatic demand for water, rather than with the actual loss of water by evapotranspiration, shows more realistically by how much the precipitation fails to supply the climatic water needs in the area.

III. Methods of Evaluation and Their Limitations

A. Direct Methods

Because of its climatic significance, considerable effort has been spent on the development of instruments to measure evaporation or evapotranspiration directly. Since size and exposure of the evaporating surface influences the rate of moisture loss, it is first necessary to distinguish between small

evaporating surfaces surrounded by markedly different surfaces and evaporating surfaces surrounded by essentially similar surfaces. The former case, similar to an *oasis* exposure, is of real interest to the climatologist concerned with the representativeness of desert observations of evaporation or the validity of pan evaporation. The latter case, described as a *mid-ocean* exposure, assumes a similar environment over such a large area that humidities will not vary horizontally and advection of energy will be negligible. In that situation, it can be shown theoretically that evaporation rate is proportional to the product of the wind speed and vapor pressure gradient.

Water-filled pans such as the standard National Weather Service Class A evaporation pan (1.2 m in diameter, 0.25 m deep, exposed 0.15 m above the ground surface) provide only rough estimates of the evaporating power of the air. Because of its unique exposure, with air passing all around it and radiation falling on the walls of the pan, the resulting values of water loss seldom approximate either potential or actual evapotranspiration.

Inflow–outflow measurements from lakes also provide only estimates of evaporation. The number of observations is quite limited and it is not possible to obtain short-period values of evaporation from such observations. Problems of seepage through the lake bottom and unmeasured inflows and diversions make errors likely. The thermal inertia of a large, deep lake as opposed to a small, shallow lake or an evaporating pan will result in great time discrepancies in the patterns of evaporation from each of these water bodies.

Soil-filled tanks (lysimeters), covered with a continuous stand of vegetation that is in every way similar to the vegetation that surrounds the tanks and having a water supply fully adequate to the needs of the vegetation, have been used for many years to measure potential evapotranspiration water loss. As long as the soil in and around the tanks is always kept moist, they can provide reasonable values of potential evapotranspiration. If the soil is allowed to follow normal changes in moisture content due to rainfall, actual evapotranspiration can be determined as long as some method of evaluating daily or monthly changes in soil moisture content in the lysimeter is available. Weighing lysimeters are able to provide the needed information on changes in soil moisture content in the tank.

The definition of potential evapotranspiration requires that the only source of energy for evapotranspiration must come from solar energy. To eliminate advection as an additional source of energy for evapotranspiration, the lysimeters must be exposed within a field planted to the same vegetation as is on the tanks. This vegetation must receive the same watering treatment as the tanks themselves. The size of the buffer area needed will depend on the climate; in a moist climate, a square 100 m on a side might be sufficient, but in a desert, a square 1000 m on a side might not be large enough. The need for reliable observations in arid areas is great, but these are the very areas in which it becomes most difficult to establish and maintain the exacting conditions required by the definition of potential evapotranspiration.

The question of the advection of sensible heat and its effect on evaporation rates from moist tracts such as irrigated areas has been studied in some detail without satisfactory solution. There is no question that if the moist tract is maintained in the midst of much drier fields (the oasis experiment), the total energy for evapotranspiration will include both solar receipts and the energy advected to the tract from the dry surroundings. The resulting evapotranspiration from the tract may be anywhere from 20 to 50% greater than the potential evapotranspiration. The actual amount of evapotranspiration from the tract will be in great part a function of the particular exposure and relative moisture conditions of the moist tract and its surroundings. Potential evapotranspiration, which requires the previously mentioned mid-ocean exposure, could not be measured as a result of such exposure.

Failure to maintain proper conditions both within the evapotranspirometer tank and in the field outside results in either too much or too little water loss. When the soil in the tanks is too dry, the evapotranspiration rate drops below the potential. When the soil outside is drier than the soil in the tank, the evapotranspiration rate will exceed the potential because of the advection of dry air. The condition of the vegetation is also important. If the vegetation in the tanks is taller than that outside, the water loss will be excessive. A height difference of just 2 or 3 cm may make a serious difference. If the vegetation in the tanks stands up prominently above the surroundings, the observations of water use are probably worthless and should be discarded. Tall-growing crops will inevitably give excessive values of water use if they are grown in association with lower plants. For that reason, it is almost impossible to compare the potential evapotranspiration of different types of vegetation covers on adjacent tanks at the same time. [See EVAPORATION, TRANSPIRATION, AND WATER UPTAKE OF PLANTS.]

Pan evaporation is strongly influenced by the moisture content of the air passing over the pan, and so it is not possible to determine potential evapotranspiration from pan evaporation. The same criticism, of course, applies to any soil-filled, vegetation-covered tank not operated under the proper conditions. In dry periods, pan evaporation is always higher than potential evapotranspiration. Any formula for determining potential evapotranspiration that contains a humidity term or that is based on pan evaporation will give excessive values in dry areas or during dry periods. This discrepancy has resulted in the habitual use of too much water in dry-land irrigation when evaporation pans are used to determine irrigation need, wasting precious water and damaging valuable land at the same time.

B. Indirect Methods

Measurements of evaporation or evapotranspiration are not available at many places, so it is necessary to consider alterna-

tive methods to evaluate these important factors on the basis of other data. A large number of indirect methods for the quantitative determination of evapotranspiration exist. These may be categorized as (1) mass transport techniques, (2) aerodynamic or profile techniques, (3) eddy correlation techniques, (4) energy-budget techniques, and (5) empirical (generally bookkeeping) techniques.

Mass transport expressions have generally developed from J. Dalton's early relation, $E = C(e_0 - e_a)$, where C is an empirical constant, usually containing a wind speed term, while e_0 and e_a are the saturation vapor pressure at the surface and in the air above. There is great difficulty in measuring the vapor pressure exactly at the surface. Various methods of expressing the effect of wind speed have been introduced into recent mass transport expressions. Finally, the height at which the wind or the vapor pressure in the air must be measured is not fully specified.

Aerodynamic or profile techniques require making certain assumptions concerning the turbulent diffusion of heat and water vapor in the atmosphere. Diffusion of moisture in the vertical is assumed to be proportional to the product of the gradient of moisture content with height and a turbulent-diffusion coefficient that is a function of the intensity of air turbulence in the surface layer. It is assumed to be dependent on the wind speed profile. One limitation is that atmospheric buoyancy influences the diffusion coefficient when the air temperature profile is not near neutral stability.

The eddy correlation techniques recognize that the net diffusion of water vapor upward can only occur if upward-moving turbulent eddies are more moist than downward-moving eddies. The magnitude of the flux is determined from simultaneous observations of vertical wind speed and moisture content of the air. The average product of these two terms, when multiplied by air density and specific heat, gives the moisture flux due to turbulent eddies. Sensitive and fast-response instruments are required for those measurements, so reliable observations at many different places over long periods of time are difficult to obtain.

Energy-budget techniques are based on the partitioning of available net radiation R_n into its different categories of use at the earth's surface. Neglecting the energy used in photosynthesis and other minor exchanges, the energy budget can be written as

$$R_n = S + H + LE$$

where S is the soil heat flux, H is the atmospheric heat flux, and LE is the energy going into the evaporation of water. Evaporation may be determined as a residual if the other terms are measured.

The percentage of the total net radiation used in evaporation has been shown to be proportional to the amount of moisture in the soil profile expressed as a percentage of field capacity. Thus net radiation might be used to compute evapo-

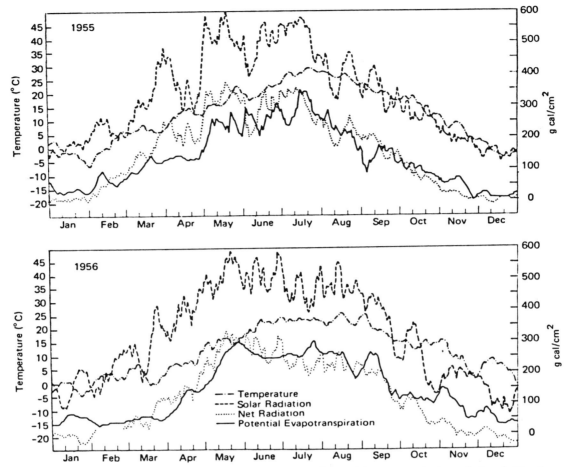

Fig. 1 Annual course of temperature, radiation (solar and net), and potential evapotranspiration at Seabrook, New Jersey for 1955 and 1956.

ration or evapotranspiration as long as the horizontal advection of energy and the role of stored heat are in some way eliminated. Figure 1 shows the close relation between measured net radiation and potential evapotranspiration for two years at Seabrook, New Jersey along with values of air temperature and solar radiation. The data are presented as 10-day running means to smooth daily fluctuations. Potential evapotranspiration is given in gram calories per square centimeter, the energy needed to evaporate the measured water loss. The curves of net radiation and potential evapotranspiration are, thus, directly comparable. The close correspondence between these two curves indicates that for Seabrook, at least, practically all the net radiation is utilized in evaporating water when the soil is moist. In March, April, and May not all the net radiation is used for evapotranspiration—for slightly more net radiation goes into heating the cold air and soil in

springtime than at other times of the year. In October, November, and December, when there is some transfer of stored heat from the soil to the air, evapotranspiration is greater than one would expect from net radiation alone. However, at Seabrook, advection and stored soil heat play only minor roles in the energy budget of this relatively moist area. [See LAND–ATMOSPHERE INTERACTIONS; SOLAR RADIATION.]

Energy-budget and aerodynamic approaches have been joined in what may be called combination techniques in an effort to eliminate certain unmeasured terms. One such expression (from the British climatologist H. L. Penman) relates evapotranspiration from a moist, short green cover (E_T) to the slope of the saturation vapor-pressure curve versus temperature, the net radiation, the latent heat of vaporization for water, a wind function, and the saturation vapor-pressure difference obtained from the air temperature and the dew-point temperature both taken at the height of the temperature observation shelter or screen. Since observations of duration of bright sunshine, mean air temperature, mean vapor pressure, and mean wind speed are required, this expression is difficult to evaluate at a large number of places. The inclusion of more factors active in the evaporation process, however, may add to the validity of the expression under a wide range of meteorological conditions. Many ways to simplify the Penman equation have been suggested, but certain accuracy is lost in these simplifications. The original expression has proved useful in different types of water budget studies.

Various empirical methods to determine potential evapotranspiration exist, the most familiar, possibly, being due to Thornthwaite, who originally defined the term. Using data from lakes and irrigated plots, he developed the following expression for monthly unadjusted potential evapotranspiration (in centimeters):

$$e = 1.6\left(\frac{10t}{I}\right)^a$$

where t is monthly temperature (°C), I is an annual heat index determined from the sum of the 12 monthly heat index values, $I = \Sigma\, i$, where $i = (t/5)^{1.514}$; and a is a nonlinear function of the heat index equal to

$$a = 6.75 \times 10^{-7}I^3 - 7.71 \times 10^{-5}I^2 + 1.79 \times 10^{-2}I + 0.49$$

These complicated and mathematically inelegant expressions can be evaluated readily with the use of available computer programs. Unadjusted potential evapotranspiration e is the water loss for a 30-day month with each day 12 hours long. This value is adjusted by a factor that expresses how the actual day and month length differ from these values.

Thornthwaite's expression for potential evapotranspiration is useful because it is simple to evaluate at many places in the world, requiring as it does only data of temperature and daylength. It has been found to provide reasonably reliable values of monthly evapotranspiration, especially if marked monthly changes in humidity (as in a monsoon climate) or in wind speed do not occur. The Thornthwaite expression is less effective on a daily basis because daily variations in wind speed and humidity are not included in the expression for potential evapotranspiration.

In selecting a method for estimating evapotranspiration, investigators must decide between ease and simplicity of evaluation, on the one hand, and increased accuracy through the inclusion of more factors that are related to evapotranspiration, on the other. The method of Penman, for example, appears to provide reliable values of daily or monthly evapotranspiration under a wide range of environmental conditions. But it requires considerable data and so is neither easy

to use nor of widespread applicability. Estimates are often used for some of the generally unmeasured terms but, when this is done, some of the advantages of using the more physically sound expression are lost. The simple Thornthwaite expression can be readily applied worldwide, and while it may be less accurate on a monthly basis (and surely is less accurate on a daily basis), its wide usefulness often outweighs its inability to reflect short-period changes in wind or humidity. Although many studies have sought to demonstrate the advantages of one method over another, there is still no universally agreed-upon method for computing evapotranspiration.

IV. Distribution

There are a number of existing maps of the areal distribution of evaporation from Class A pans or lakes or the evapotranspiration as computed by one of the various empirical formulas. Thornthwaite has provided a map of mean annual potential evapotranspiration in the United States, which, while having the same general pattern as National Weather Service maps of lake evaporation, differs appreciably in absolute values in arid areas. Annual evaporation based on estimates from lakes exceeds 86 in. (2180 mm)/year along the California–Arizona border, while Thornthwaite shows just 60 in. (1520 mm) of potential evapotranspiration. Throughout the northern Great Plains up into Montana, annual lake evaporation varies from 30 to 40 in. (760 to 1020 mm)/year, while Thornthwaite indicates no values over 30 in. (760 mm) and most below 24 in. (610 mm). The lake evaporation map should, of course, indicate somewhat higher values, since it reflects the influence of both solar and advective energy, whereas Thornthwaite's potential evapotranspiration assumes no influence of advection.

The differences can, in large measure, be explained by the relation that exists among rate of water loss, moisture content of the air, and size of the evaporating surface (Fig. 2). When a psychrometer is used to determine the humidity of the air, a thermometer bulb is moistened and becomes the evaporating surface. The surface area of the bulb is small and the amount of water vaporized from it is very small. Heat flows into the water film on the bulb from the warmer surrounding air, and the evaporation process will reach equilibrium at a rate and at a wet-bulb temperature such that the energy obtained from the air is just sufficient to maintain the evaporation. Solar radiation contributes little energy to this process. The water from the wet bulb moistens the air, but the amount is so small that the effect on the moisture content of the air is completely negligible.

The Weather Service Class A evaporation pan on a summer day in a dry situation may evaporate 8 liters of water. Solar radiation contributes an important share of the energy for

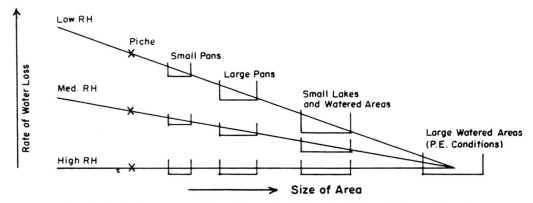

Fig. 2 Schematic relation between rate of water loss under different humidity conditions and size of evaporating area.

evaporation, the amount depending on the turbidity of the water and on the amount of reflection from the pan, which varies greatly with type, age, and condition of the material used. Additional energy for evaporation is available from the air. The amount of water evaporated from the pan does little to modify the moisture content of the air; but immediately over the water surface the humidity is raised, the moisture gradient reduced, and the evaporation impeded. The extent of this influence depends on the rate at which air is advected across the evaporating surface from outside.

If the area of the evaporating surface is large, the influence of the moisture condition of the air passing over the evaporating surface becomes small and solar radiation is the primary source of energy for evaporation. The influence of the evaporation on the atmospheric humidity is very important because this influence cannot fail to have a reciprocal effect on the evaporation. In moist air, the temperature of the evaporating surface will rise to a point above the dewpoint of the air such that the evaporation is in accord with the available energy. Similarly, in dry air, rapid evaporation will lower the temperature of the evaporating surface until the evaporation is in accord with the available energy. The size of the water body or moist land area necessary to ensure potential evapotranspiration is difficult to determine. The size of the area under high moisture conditions has to be large enough so that the evapotranspiration from the area is not affected by external factors such as the advection of moist or dry air masses and their modification by local conditions. Water pans that vary from one another only in size are influenced differently by such conditions and, thus, result in different values of evaporation. Most land evaporation pans are certainly too small to be uninfluenced by external conditions and, in some cases, even soil pans surrounded by swamps and marshes appear to be too small.

Figure 2 shows that water loss is fairly independent of the size of the measuring instrument when the air humidity is high, but as the air becomes drier, the size of the evaporating surface greatly influences the rate of evaporation or evapotranspiration from it. Because of this limitation, pans or other instruments like the Piché evaporimeter (13 cm^2 evaporating surface) do not give reproducible measures of evapotranspiration except under conditions of high humidity.

The well-known German climatologist Rudolph Geiger has prepared a world map of actual evaporation or evapotranspiration based on energy-budget considerations and estimates of moisture storage available for evaporation. Coupling these with world maps of precipitation, it has been possible to obtain values of precipitation and evaporation over the surface of the globe by 10° squares of latitude and longitude. The results are summed by latitudinal belts in Table I. Precipitation in each latitudinal belt on land always equals or exceeds evapotranspiration, while evaporation from the ocean areas can often greatly exceed precipitation in the same latitudinal belt because of the continuous availability of water for evaporation.

The latitudinal values of precipitation and evaporation given in Table I have been plotted graphically in Fig. 3. The upper two curves represent the latitudinal distribution of precipitation and evaporation over the earth as a whole, and the lower two curves represent precipitation and evapotranspiration on the land areas alone. The nonlinear latitudinal scale represents the actual surface area contained within adjacent latitudinal belts. Thus, the area between upper and lower evaporation curves represents the actual volumes of evaporation from the ocean areas of the world. [See WATER CYCLE, GLOBAL.]

In both hemispheres, evaporation is found to exceed precipitation in the belt from 10 to 40° latitude. These corre-

Table I Estimates of Annual Precipitation (P) and Evapotranspiration (E) from Land and Ocean Areas of the Globe by 10° Latitudinal Belts[a]

Latitude	P(Ocean)	E(Ocean)	P(Land)	E(Land)	Runoff (P − E)
80–90°N					
70–80	183	218	62	40	22
60–70	314	271	486	227	259
50–60	1,115	604	763	417	346
40–50	1,639	1,258	851	586	265
30–40	1,876	2,812	850	606	244
20–30	1,897	3,846	890	537	353
10–20	3,700	4,925	939	714	225
0–10	7,148	4,796	1,596	1,085	511
10–0°S	4,408	5,024	2,021	1,144	877
20–10	3,095	5,683	1,142	782	360
30–20	2,306	4,692	523	396	127
40–30	3,016	3,801	235	171	64
50–40	3,722	2,263	59	38	21
60–50	2,749	1,184	19	6	13
70–60	879	462	37	21	16
80–70	95	67	152	90	62
90–80					
Total	38,142	41,906	10,625	6,860	3,765

Source: Mather (1969), with permission of the American Water Resources Association.
[a] All values ×10 km³

spond generally to the worldwide subtropical high-pressure belts of reduced precipitation. Precipitation exceeds evaporation in the other latitude belts. In interpreting these results, it must be remembered that the values have been summed by 10° latitude belts so that the results in the 30 to 40° belt may be more influenced by the difference between precipitation and evaporation found in the 30 to 35° belt than in the 35 to 40° belt where precipitation may actually exceed evaporation.

Average values of water budget factors by continents as determined by several investigators are given in Table II. The methods used to achieve the continental values are somewhat different since J. R. Mather worked from Geiger's large maps of precipitation and evaporation, M. I. Budyko from energy balance considerations, and M. I. L'vovich from his own differentiated water-balance equations.

Mather found more precipitation and evaporation over Africa and less over Australia than did the others. Largest discrepancies in the three sets of results are found in Australia and South America, where L'vovich has much higher values of precipitation and evaporation and nearly three times more runoff from Australia than the others. South America would seem to be the best-endowed continent in terms of water resources and either Africa or Australia the least

well-endowed, based only on continent-wide summaries. [*See* GLOBAL RUNOFF.]

V. Uses of Evaporation/Evapotranspiration Information

Graphical or tabular comparisons of daily or monthly values of precipitation and evapotranspiration are fundamental for understanding the nature, distribution, and extent of the water resources of a place or area. Figure 4 graphically represents a long-term average climatic water budget, the month-by-month march of water supply (precipitation) and climatic demand for water (potential evapotranspiration), at Berkeley, California. Comparing precipitation with potential evapotranspiration on a monthly basis, one finds that they never coincide. There is too much precipitation in winter and too little in summer. In November, the precipitation over and above that needed for evaporation is stored in the soil and begins to recharge the upper layers of the soil. After the soil reaches field capacity (in February), any precipitation not needed for evapotranspiration is considered to be surplus, and ultimately it is lost as runoff or deep percolation to the water table.

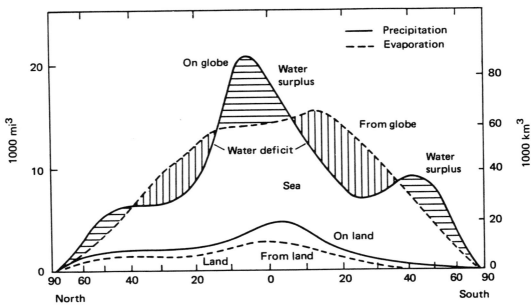

Fig. 3   Latitudinal variation of evaporation and precipitation on the globe as well as on the land and ocean areas. [From Mather (1969), with permission of the American Water Resources Association.]

Table II   Summary of Continental Values of Precipitation, Evaporation, and Runoff Obtained by Three Investigators

	Precipitation			Evaporation			Runoff		
	Mather (1969)	Budyko (1956)	L'vovich (1973)	Mather (1969)	Budyko (1956)	L'vovich (1973)	Mather (1969)	Budyko (1956)	L'vovich (1973)
Values in km³/yr									
Africa	21,450	20,190	20,780	17,590	15,370	16,555	3,860	4,820	4,225
Asia	30,390	26,920	32,690	18,440	17,210	19,500	11,950	9,710	13,190
Australia	3,560	3,620	6,405	2,830	3,160	4,440	730	460	1,965
Europe	6,380	5,980	7,165	3,730	3,590	4,055	2,650	2,390	3,110
North America	15,550	16,370	13,910	9,430	9,780	7,950	6,120	6,590	5,960
South America	27,030	23,990	29,355	15,470	15,280	18,975	11,560	8,710	10,380
Antarctica	1,890			1,110			780		
Total	106,250	97,070	110,305	68,600	64,390	71,475	37,650	32,680	38,830
Values in mm depth/yr									
Africa	704	670	686	577	510	547	127	160	139
Asia	691	613	726	421	392	433	270	221	293
Australia	457	480	736	363	419	510	94	61	226
Europe	638	603	734	373	362	415	265	241	319
North America	636	668	670	386	399	383	250	269	287
South America	1,522	1,350	1,648	871	860	1,065	651	490	583
Total	775	725	834	500	481	540	275	244	294

Source: J. R. Mather (1984). "Water Resources: Distribution, Use, and Management." Copyright © 1984 by John Wiley and Sons, Inc.

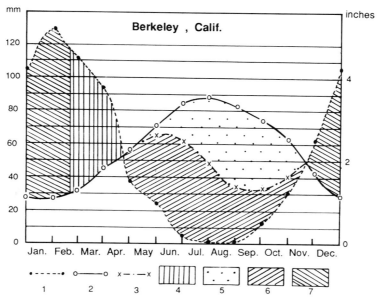

Fig. 4 Graphical representation of the average climatic water budget for Berkeley, California showing the average course of precipitation (1), potential evapotranspiration (2), actual evapotranspiration (3), as well as period of water surplus (4), water deficit (5), soil water utilization (6), and soil water recharge (7).

April is the first month in which the rapidly rising climatic water needs exceed the falling precipitation supply. The unmet needs for water are satisfied in part by water removed from the soil. The actual evapotranspiration equals the precipitation plus that water that is removed from the stored soil moisture by the vegetation cover. Water deficit is the difference between potential and actual evapotranspiration—the climatic demand for water that is not satisfied by the two available sources, the precipitation and the stored soil moisture.

The climatic water budget is, therefore, a straightforward way to determine quantitatively from an accounting of precipitation and evapotranspiration over appropriate time and space dimensions, information on streamflow as well as on the fluctuations in the height of the groundwater table. Information on the water deficit (the water demands of the vegetation unmet by precipitation and water stored in the soil) will be the irrigation need. The daily or monthly relation between precipitation and evapotranspiration also provides quantitative information on soil moisture content, a factor closely related to the ability of vehicles or humans to move over unpaved surfaces (soil trafficability). Finally, the relation of the actual evapotranspiration at a place to the potential evapotranspiration has been found to be directly related to the ratio of the actual yield of vegetation to its potential yield. This relation holds for both agricultural crops and forest

productivity. [See GROUNDWATER; SOIL WATER: LIQUID, VAPOR, AND ICE.]

The aim of most past evaporation and evapotranspiration studies has been to find reliable estimates of this factor that can be compared with values of precipitation in a water budget to permit evaluation of the various factors of water supply, water surplus, water deficit, and water storage in the soil. Because of the seeming simplicity of measuring precipitation (a measuring container and a ruler), many users feel we have good knowledge of the distribution of precipitation but that we have only inexact information on the return flow of water to the atmosphere—the evaporation or evapotranspiration. Thus, numerous conflicting studies have provided a range of expressions and instrumental techniques by which on can estimate potential or actual evapotranspiration or over-water evaporation. Often, strongly worded differences of opinion have appeared in the literature concerning which instrument or which empirical expression for evapotranspiration to use in water-budget analyses. There seems to be little recognition of the fact that the most significant error in water-budget input terms may be in precipitation rather than in evapotranspiration. While errors of 20% may occur in values of evapotranspiration obtained from different estimating techniques, such errors are commonly found in precipitation observations due to wind effects, exposure problems, interpolation from point observations, and other instrumental problems. In any evaluation of the water budget, one must look carefully at the problems and limitations inherent in precipitation data before uncritically accepting these data as "correct" and attributing all inconsistencies to problems in understanding evapotranspiration.

Bibliography

Budyko, M. I. (1956). "The Heat Balance of the Earth's Surface." Gidrometeoizdat, Leningrad. (Translated by N. A. Stepanova, Office of Climatology, U.S. Weather Bureau, 1958).

L'vovich, M. I. (1973). The water balance of the world's continents and a balance estimate of the world's freshwater resources. *Sov. Geogr.* **14**(3),135–152.

Mather, J. R. (1969). The average annual water balance of the world. *In* "Proceedings of the Symposium on the Water Balance of North America," American Water Resources Association, Series 7, pp. 29–40.

Mather, J. R. (1984). "Water Resources: Distribution, Use, and Management." Wiley, New York.

Mather, J. R., and Ambroziak, R. A. (1986). A search for understanding evapotranspiration. *Geogr. Rev.* **76**(4),355–370.

Miller, D. H. (1977). "Water at the Surface of the Earth." Academic Press, New York.

Oliver, J. E., and Fairbridge, R. W., eds. (1987). "The Encyclopedia of Climatology." Van Nostrand Reinhold, New York.

Rosenberg, N. J., Hart, H. E., and Brown, K. W. (1968). "Evapotranspiration: Review of Research." Publication 20, University of Nebraska, College of Agriculture and Home Economics and Nebraska Water Resources Research Institute, Lincoln, Nebraska.

Glossary

Dewpoint Temperature of saturation of the air or of 100% humidity.

Field capacity Amount of water remaining in the capillaries of a previously saturated soil after all possible water has been removed by gravity.

Latent heat Heat added to liquid or ice without change in temperature as it passes into a vapor state or is later released as the vapor is recondensed.

Net radiation Difference between all incoming long (atmospheric) and shortwave (solar) radiation and all outgoing long and shortwave radiation from the earth's surface.

Saturation deficit Difference between the saturation vapor pressure at a given temperature and the actual vapor pressure at that same temperature.

Saturation vapor pressure Pressure exerted by the water vapor in the air in a saturated space.

Evaporation, Transpiration, and Water Uptake of Plants

G. S. Campbell
Washington State University

Evaporation, in the context of this chapter, will refer to the loss of water from plant and soil surfaces as it is converted from liquid water to water vapor and transported from the surface. Transpiration will refer to the special case of evaporation from the substomatal cavities of plant leaves, which results in the transport of water through the vascular system of the plant and absorption of water from the soil by plant roots. Both processes play key roles in the hydrologic cycle and the overall energy balance of the earth's surface.

I. Overview

The processes of evaporation and transpiration have profound terrestrial importance. At the global scale, 90% of the annual heat transfer from oceans and about 50% of the heat transfer from land surfaces is as latent heat. Water vapor, as it enters the atmosphere, is by far the most important of the greenhouse gases. Since human activities can influence both the partitioning of heat at the surface and the amount of water vapor entering the atmosphere, anthropogenic alteration of evaporation may be a more important issue in global climate change than increasing CO_2 and other trace atmospheric gases. Perhaps more important in the long term, however, is the fact that evaporation supplies the water to drive the hydrologic cycle that renews and shapes the land,

creates rivers and lakes, and cleanses the atmosphere. [*See* ATMOSPHERIC TRACE GASES, ANTHROPOGENIC INFLUENCES AND GLOBAL CHANGE; LAND–ATMOSPHERE INTERACTIONS; WATER CYCLE, GLOBAL.]

On a much smaller scale, living organisms, including ourselves, are mainly water. Living cells must be maintained in a highly hydrated state. The equilibrium relative humidity of living cells is around 99%, while the average humidity of the terrestrial environment in which they live is around 50%. Necessary exchange of oxygen and carbon dioxide between organisms and their surroundings therefore results in unavoidable losses of water by evaporation. This evaporation is an important component of the water economy of organisms, but it is also important in energy exchange and in driving the circulation systems of plants. Nature's only method of cooling in a hot environment is by evaporation, so without this important process, life could not exist in many environments where it now thrives.

Partitioning of the sun's energy, on a global scale, between heating the atmosphere and evaporating water depends on physical processes at the micro-scale of individual plant leaves and soil clods. The influence our activities have on absorption of radiation and its partitioning are easily understood at a micro-scale; but this influence has generally been ignored at the global scale. While our focus in this chapter will be on the microclimate of soil and crop surfaces, it is important to realize that global energy exchange is the summation of the energy exchange occurring on individual square meters of the earth's surface. Alteration of any part of the surface alters the global exchange. The processes of desertification, deforestation, irrigation, denudation by grazing, etc., which are occurring in many parts of the world today could have energetic consequences on a global scale comparable to those being considered from alteration in greenhouse gas concentrations in the atmosphere. [*See* DESERTIFICATION; SOLAR RADIATION.]

The soil is the main reservoir from which evaporation and transpiration occur. Precipitation that falls on the soil surface infiltrates into the soil under the influence of capillary and gravitational forces. As evaporation occurs at the soil surface, water from beneath the surface is brought back to the surface by capillary conduction to evaporate. With continued evaporation, a drying front moves into the soil, and water moves to the atmosphere by diffusion through the soil pores. Eventually, the movement of water by diffusion and capillary flow becomes so slow that the evaporation rate becomes negligibly small. The depth of soil dried by evaporation varies depending on soil properties, but it is in the range of 10 to 20 cm. Plant roots, however, may reach to a depth of several meters and can withdraw water from the entire rooted volume for transpiration. Plants are therefore a major factor in determining how much of the soil water reservoir is depleted, and the rooting depth of the species growing in the soil will often determine how much total water is extracted. This, in turn, determines both the evaporative cooling available to a region and the yield of groundwater from a watershed. Deep-rooted species will evaporate more water and leave less water to percolate down to replenish groundwater than will shallow-rooted species. [*See* GROUNDWATER; SOIL WATER: LIQUID, VAPOR, AND ICE.]

An additional effect of plants on evaporation is through their effect on surface roughness. Vapor is transported away from an evaporating plant canopy or soil surface by atmospheric turbulence. Rougher surfaces have more turbulence, and therefore more effective transport. A tall canopy, such as a forest, may loose almost twice as much of the intercepted water that wets the leaves during a rain as do leaves of a short grass canopy. This difference is due to differences in turbulent transport between the two canopies. [*See* EVAPORATION AND EVAPOTRANSPIRATION.]

Our detailed consideration of evaporation and transpiration will be most useful if we first understand the atmospheric factors that determine potential evaporation rate. We can then consider how potential evapotranspiration is modified by limitations in soil and plant water supply. Finally, a few examples will show how the principles apply to practical problems.

II. Evaporative Demand of the Atmosphere

Fick's first law of diffusion has direct application to predicting evaporation rates. The law, in integrated form, says that the flux density (mass per unit area, per unit time) of water vapor from a surface is directly proportional to the difference in concentration of water vapor between the surface and the air around it, and inversely proportional to the diffusion resistance for water vapor between the surface and the surroundings. If the concentration of vapor at the evaporating surface and in the atmosphere are known, as well as the diffusion resistance, the rate of evaporation can be calculated. Unfortunately, none of these quantities can be mea-

sured very well, let alone be predicted or modeled for cases where measurements are unavailable. The surface concentration depends on the temperature at the surface and on the availability of water (which we will discuss in more detail later). The atmospheric concentration is strongly dependent on the evaporation rate of the surroundings, the energy supply, and the downward mixing of dry air from higher in the atmosphere; and the resistance depends on wind speed, surface roughness, and atmospheric stability. Direct application of Fick's law, while attractive in principle, is clearly not useful in practice.

The most successful approaches to predicting potential evapotranspiration (PET) have been those that recognize the important role played by energy supply and calculate evapotranspiration as the residual of an energy balance for a surface. The energy balance is a statement of the law of conservation of energy for a representative unit area of soil or vegetation. It can be written as

$$R_n - G - C - \lambda E = 0 \qquad (1)$$

where R_n is the net radiation (solar and thermal radiation absorbed by the surface, less the thermal radiation emitted), G is the flux density of heat into the soil and vegetation, C is the flux density of sensible heat into the atmosphere, E is the flux density of water vapor away from the surface, and λ is the latent heat of vaporization for water. Rearranging the equation gives the rate of evaporation.

$$E = (R_n - G - C)/\lambda \qquad (2)$$

Equation (2) is more useful than Fick's law for predicting PET because R_n is more easily measured and modeled and because λE tends to be a fairly large fraction of R_n. However, large uncertainties still exist because of the unpredictability of C. Further progress can be made by investigating the question of how R_n is partitioned between C, G, and λE. The flux density of energy into the soil G is usually small compared to R_n and is well correlated with R_n. We will therefore combine $R_n - G$ into a single term H, the enthalpy flux density, which represents the total energy available for evaporating water or raising the temperature of the air.

Using integrated forms of Fick's and Fourier's laws for latent and sensible heat transport in the atmosphere, the following can be established:

$$C = c_p g_h (T_s - T_a) \qquad (3)$$

and

$$\lambda E = \lambda g_v (x_s^* - x_a) \qquad (4)$$

where c_p is the molar specific heat of air, g_h and g_v are atmospheric and surface conductances for heat and vapor, T_s and T_a are surface and air temperatures, and x_s^* and x_a are water vapor concentrations (mole fractions) at the surface and in the air. We will assume that the evaporating surface is at 100% relative humidity, so the concentration of vapor at the surface, for a given atmospheric pressure, depends only on the surface temperature. The asterisk indicates saturation concentration.

Equation (4) can be expanded by adding and subtracting the saturation concentration of vapor at air temperature:

$$\lambda E = \lambda g_v(x_s^* - x_a^*) + \lambda g_v(x_a^* - x_a) \qquad (5)$$

Since x^* is a function only of temperature, the first term in parenthesis can be approximated as $s(T_s - T_a)$, where s is the slope of the saturation vapor concentration function. The second term in parenthesis is the saturation deficit of the atmosphere, which we will give the symbol D. Making these substitutions into Eq. (5), and combining Eqs. (5) and (3) to eliminate $(T_s - T_a)$ gives

$$C = (\gamma^*/s)(\lambda E - \lambda g_v D) \qquad (6)$$

Here, γ^* is the apparent psychrometer constant, given by $\gamma^* = c_p g_h/(\lambda g_v)$. Equation (1) can be rewritten as $C = H - \lambda E$, which can be combined with Eq. (6) to eliminate either C or λE. Eliminating C gives

$$\lambda E = \frac{s}{s + \gamma^*}[H + (\gamma^*/s)\,\lambda g_v D] \qquad (7)$$

which is the Penman–Monteith formula for calculating evapotranspiration. The parallel formula for sensible heat is obtained by eliminating λE between Eqs. (1) and (6) to give

$$C = \frac{\gamma^*}{s + \gamma^*}[H - \lambda g_v D] \qquad (8)$$

These two equations specify how a given quantity of heat H will be partitioned between sensible and latent heat at the surface. The conditions that determine this partitioning, besides H, are the vapor deficit of the air D, the surface conductances for heat and vapor (determined by wind speed, surface roughness, and atmospheric stability), and the slope of the saturation vapor concentration function s.

The implications of Eqs. (7) and (8) can be made clearer by considering some limiting cases. One case occurs for evaporation into a saturated atmosphere (sometimes referred to as a mid-ocean condition) where the atmospheric vapor deficit $D = 0$. In this case, $\lambda E = sH/(s + \gamma^*)$ and $C = \gamma^* H/(s + \gamma^*)$. If the surface is saturated, so that the ratio $g_h/g_v = 1$, then $\gamma^* = \gamma$, the thermodynamic psychrometer constant (6.6 $\times$

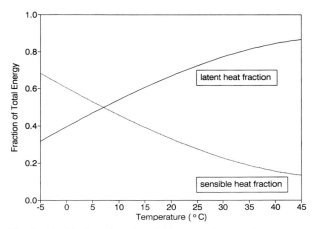

Fig. 1 Partitioning of total available energy between latent and sensible heat in a mid-ocean condition. The solid line is $s/(s + \gamma)$ and the dashed line is $\gamma/(s + \gamma)$.

$10^{-4}\,°C^{-1}$). Under these conditions, the partitioning of H between λE and C depends only on temperature (since s is temperature dependent). This temperature dependence is shown in Fig. 1. At low temperatures, the available energy is partitioned mainly to sensible heat, but as temperatures approach 40°C, almost 90% of the available energy goes to evaporation.

To predict potential evapotranspiration when saturation deficit is not zero (the normal daytime condition for soils and vegetation), Eq. (7) is used, but with $\gamma = \gamma^*$. This makes the PET dependent only on available energy and atmospheric conditions. Further simplification is possible if one recognizes that the first and second terms in parentheses in Eq. (7) are likely to be strongly correlated (large available energy will result in substantial heating of the atmosphere and, therefore, a large vapor deficit). This simplification results in the Priestly and Taylor formula for PET.

$$\lambda E = \frac{\alpha s H}{s + \gamma} \qquad (9)$$

The term α is 1 plus the ratio of the vapor deficit term to the available energy term; it has a value around 1.3 for humid climates and perhaps 1.5 for arid climates. For the mid-ocean condition mentioned earlier, $\alpha = 1$.

Equation (9) provides a simple but reliable basis for estimating atmospheric demand for moisture. The strong dependence on available energy is clear through the H term (net radiation minus soil heat), and the effect of temperature on partitioning of the available energy is also clear. Figure 2 shows the available energy for a typical crop or soil surface in a temperate climate, how that energy is partitioned between

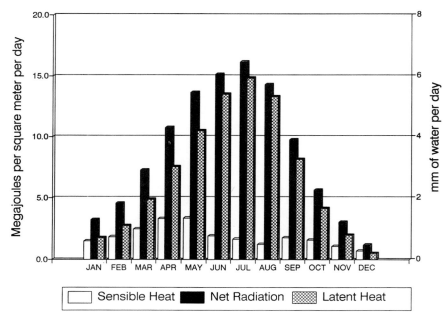

Fig. 2 Monthly averages of daily net radiation, sensible heat flux, and potential evapotranspiration (latent heat) for Pullman, Washington. The graph shows that most of the energy available at the surface is used for evaporation when water is available. Shifts in partitioning throughout the year are due to changes in temperature. Values are expressed in both energy units (megajoules per square meter) and depth of evaporated water. Approximately 2.5 MJ of energy is required to evaporate 1 mm of water.

sensible and latent heat loss and the resulting potential evapotranspiration. Clearly, evapotranspiration is a major component of the energy balance and is strongly correlated with available energy.

III. Evaporation from Soils

In the following discussion it will be useful to think of evaporation in terms of supply and demand. Equation (9) can be thought of as specifying the demand, or the highest rate at which evaporation can proceed. If the soil can supply water to the atmosphere at that rate, then Eq. (9) specifies the actual as well as the potential evaporation rate. If the soil is not capable of supplying water at that rate, then the surface of the soil dries, the conductance (part of g_v) increases, and evaporation rate decreases. Actual evaporation rate is then determined by the characteristics of the soil, through g_v. This limitation by the soil can be thought of as the maximum supply rate, and it varies depending on how deep and how porous the layer of dry soil is at the surface. You can always tell whether soil or atmosphere is determining the evaporation rate. If the soil surface is wet, then evaporation is proceeding at the potential, or atmosphere-limited rate. If the

soil surface is dry, then actual evaporation rate is below potential and the soil is controlling the rate of evaporation. Equation (7) is valid for either case because the surface condition is properly accounted for in either case through g_v. It is not very useful, however, for the dry surface case because the atmospheric demand is not controlling water loss in this case.

Figure 3 illustrates several of the principles involved in soil evaporation. Two distinct stages of drying are evident in Fig. 3. The first, or constant-rate stage of drying is atmosphere-limited, and the rate is predicted by Eq. (9). The second, or falling-rate stage is soil-limited, and its rate is determined by the ability of the soil to transmit water to the drying front. The length of first-stage drying and the evaporation rate during second-stage drying are dependent on the evaporative demand and the texture of the soil. Conditions for high and low evaporative demand with sand and loam soils are illustrated in the Figure. [See SOIL PHYSICS.]

Figure 4 shows the cumulative evaporation over time for the soils and evaporative demands in Fig. 3. The soil with the high evaporative demand sustains the highest total evaporative loss when both soils are in first-stage drying, but the low-demand soil stays in first-stage drying longest. If both soils are allowed to evaporate for a long period of time, the

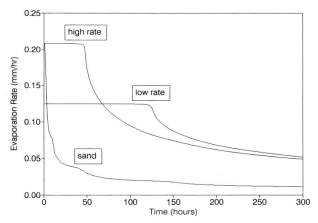

Fig. 3 Evaporation rate over time for a loam soil and a sand exposed to a constant potential evaporation environment. The surface of the sand dries quickly, causing the evaporation rate to decrease rapidly. The surface of the loam exposed to high potential evaporation dries after about 50 h; at the low rate, first-stage evaporation lasts for almost 150 h.

low-demand soil eventually will loose almost as much water as the high-demand soil.

Soil texture, however, is important in determining total evaporation. Fine-textured soils tend to maintain much higher water contents after wetting than sands, and they therefore hold more water close to the surface and available for evaporation. In Fig. 4, the total evaporation for the loam soils is much higher than for the sand. The lower the storage capacity of the surface material, the lower the evaporative loss. Minimum evaporative loss occurs with a gravel surface, which has a very low storage capacity. The surface dries quickly, forming a vapor barrier and minimizing water loss. First-stage evaporation is limited to the small amount of water used to wet the surfaces of the stones. A natural gravel surface, called a pebble pavement, is created in some deserts where interstitial material is blown away leaving only stones. This surface minimizes water loss to evaporation and maximizes storage efficiency.

The total amount of evaporation from soil is determined primarily by four factors: the frequency of wetting, the amount of precipitation received in a storm, the near-surface storage capacity of the soil for water, and the potential evaporation rate. Frequent, small storms keep the soil surface wet and in first-stage evaporation much of the time. Total evaporation is therefore maximized. Large storms provide enough water so that much of the water infiltrates below the surface layers and out of the reach of evaporation. This reduces the fraction of the storm that is lost to evaporation. The larger the surface storage capacity of the soil, the more water is held near the surface and available for evaporation;

and the higher the potential evaporation rate, the more water is lost during periods of infiltration and redistribution.

Potential interactions among these factors are obvious. For example, frequent small rains during the summer when evaporative demand is high (such as in the Great Plains of the United States) result in almost total loss of the applied water. Those same rains in winter, with low evaporative demand (as in the Pacific Northwest) result in substantial moisture storage. In the winter, the new moisture comes before the surface layers are depleted, and so part of the moisture moves on down into deeper storage.

The total amount of water evaporated from soil and vegetation (not including transpiration) ranges from 100% of precipitation in bare, arid areas with frequent summer rain to 10 to 20% of potential evapotranspiration in humid areas with full vegetative cover and abundant moisture. From the plant's point of view, evaporation is the worst form of regressive tax; it is by far the heaviest on those least able to pay. In arid, vegetated areas, evaporation can easily account for 50% or more of total water loss, and it is taken out before the plant has access to the water. Thus, on years when water is short, the absolute amount of evaporation stays about constant for a location, but its fraction of precipitation increases, leaving proportionately much less for plants. This is illustrated by measurements made on desert soils and vegetation in central Washington. Precipitation during the 1972–1973 water year was 14.6 cm, well below normal, and during the 1973–1974 year it was 35.7 cm, well above normal. Evaporation during the dry year was 6.5 cm, or 45% of precipitation, and in the wet year it was 8.2 cm, only 23% of precipitation.

The implications on vegetative production of controlling evaporation through management become clear when we

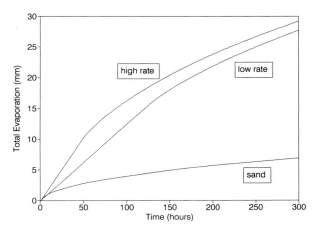

Fig. 4 Cumulative evaporation for the soils in Fig. 3. Total evaporation for the sand is low because it stores so little water near the surface, while the cumulative water loss for the loam soil is about the same whether the potential evaporation rate was high or low.

recognize the direct relationship between transpiration and evaporation and the relationship between transpiration and crop production. As previously mentioned, water that evaporates is not available for transpiration. Research has shown that, for a particular climate, there is a direct relationship between the amount of water transpired and the amount of photosynthesis occurring in a plant. Reducing evaporation increases the amount of water available for transpiration, and should therefore increase production.

Previous discussions should suggest at least two methods for altering evaporative loss through management. One is to alter the storage capacity of the surface layer and the other is to alter the potential evaporative demand. In fact, both amount to the same thing. Crops and crop residues cover the soil surface and prevent net radiation from penetrating to the surface. This reduces the potential evaporation rate at the soil surface, thus decreasing total evaporation. The crop or residue surface intercepts some water, but it has little storage capacity so it dries quickly. The reduction in evaporation that can occur through surface coverings such as crops and crop residues varies greatly depending on climatological factors, but it can have a significant effect on crop production.

IV. Plant Water Uptake and Transpiration

The bulk of the water vapor that enters the atmosphere in terrestrial environments evaporates through stomates in the leaves of plants. In arid areas, perhaps 50% of evapotranspiration is transpiration; for surfaces covered by vegetation, transpiration accounts for 90% of the water loss. The factors that control transpiration are therefore important in determining evapotranspiration.

As with soils, it is useful to consider transpiration in terms of supply and demand. The maximum demand is set by Eq. (9), as with evaporation. When the surface is not fully covered with vegetation, potential transpiration can be estimated as the fraction of net radiation intercepted by the canopy multiplied by the potential evapotranspiration rate.

The potential supply rate is determined by the maximum rate at which the plant can remove water from the soil and supply it to the evaporating surfaces in the leaves. When the demand exceeds the supply, stomates in the leaves close, increasing g_v and reducing transpiration. A number of factors, however, in addition to soil-water availability, influence stomatal opening. The important factors are those that influence photosynthesis including light, temperature, nutrition, and hormonal balance. Our discussion here will limit itself to the direct effects of water availability on the ability of the plant to supply water to the atmosphere.

The pathway for water transport through a plant is shown in Fig. 5. On the left is a schematic, magnified view showing some of the cells and tissues through which water moves. Liquid water flows through the soil to the root in capillary films along the surfaces of soil particles. The water enters the plant through the epidermis, the surface of which is often greatly enlarged by small protrusions called root hairs, which are 1–2 mm long. The water moves through the cells and cell walls of the cortex until it encounters the endodermis, where a special layer called the Casparian strip forces all of the water to flow through living cell membranes. This allows the plant to selectively take up and exclude solutes in the water, rejecting, for example, sodium ions and actively concentrating potassium, but greatly increasing the resistance to water flow. Most of the resistance to liquid water movement in the plant is at the endodermis. After the endodermis, water enters the vascular system of the plant, where it is transported through the xylem vessels to the leaves. After leaving the vascular system of the leaf, the water flows through cells and along cell walls until it evaporates in the substomatal cavities. In the gas phase, the water vapor diffuses through the air spaces within the leaf and out to the atmosphere through the stomates.

The driving force that causes water to flow in the soil–plant system is a gradient in the water potential. Water potential is defined as the potential energy per unit mass of water, with the potential energy of pure, free water taken as zero. It is equivalent to the partial specific Gibbs free energy in thermodynamics. Water in soils and plants is subject to adhesive and capillary forces and contains solutes, so the potential of this water is less than the potential of free water. Water potentials in soils and plants are therefore negative numbers. To give an idea of typical values, plant leaves, in the day, have water potential between −1000 and −2000 J/kg. At night, when transpiration stops, typical leaf water potentials are around −200 J/kg. Significant long-range liquid water movement in the soil stops at potentials below −20 to −50 J/kg. Thus, when a soil is initially wetted, water moves into the soil through infiltration and redistribution until the water potential of the wetted zone is around −30 J/kg. This is often termed "field capacity." Roots extract water until the soil reaches a potential of around −1500 J/kg, often called the "permanent wilt point." The water held between these two potentials within the root zone of the plant is called plant available water.

Differences in water potential between different parts of the soil–plant system cause water to flow within the system — just as voltage differences in a circuit cause current to flow, or pressure differences in a hydraulic system cause fluid to flow. The current or flux of water is directly proportional to the potential difference between points and inversely proportional to the resistance. The potential in the leaves must be lower than the potential in the soil for water to flow through the plant. On a typical day with high evaporative demand and adequate soil moisture, the soil water potential might be

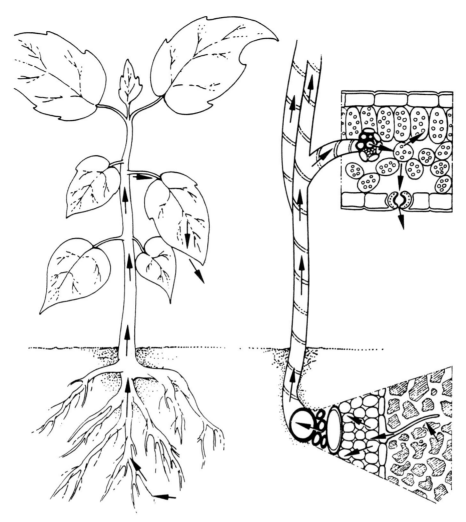

Fig. 5 Diagram of the path for water movement through the plant. The diagram on the right shows a schematic of the cells and tissues through which water moves. [From R. M. Devlin (1966). "Plant Physiology." Reinhold, New York.]

−100 J/kg while the leaf water potential is −2000 J/kg. This potential difference provides the driving force that moves water through the plant.

Ultimately, it is the atmosphere that provides the sink for water vapor, allowing evaporation to occur in substomatal cavities to sustain the potential gradient and thus the liquid flow within the plant. It is possible to calculate the water potential of the atmosphere using a relationship from thermodynamics.

$$\psi = (RT/M) \ln a_w \qquad (10)$$

where ψ is water potential, R is the gas constant, T is kelvin temperature, M is the molecular weight of water, and a_w is the water activity or relative humidity of the vapor phase. If the relative humidity of the atmosphere is 0.5, the water potential, from Eq. (10), is −100,000 J/kg, 50 to 100 times that inside the leaf. Using the inverse of Eq. (10), we can calculate the humidity or water activity inside the leaf. If leaf water potential is −1500 J/kg, the humidity inside the leaf is 0.99 or 99%. Both calculations clearly indicate that the main resistance to water flow is in the vapor phase and that no resistance in the liquid phase can significantly alter the rate of water loss. The only way the plant can control its water loss is by closing stomates or reducing leaf area. Any significant control through reduction in the humidity within the leaf would be lethal to the plant.

While the potentials within the plant have no direct effect on transpiration, they do have an indirect effect through their effect on stomatal resistance. The guard cells that open and close the stomates are hydraulically connected to the leaf xylem, so when leaf water potential decreases, guard cell turgor pressure decreases and stomates can close. Other, more subtle effects result when hormonal balance is altered by lowered leaf water potentials. Change in the hormone abscisic acid causes changes in potassium accumulation in guard cells, which can also result in stomatal closure. Our knowledge of these processes is incomplete, but their net result is an elegant control system that adjusts stomatal resistance and leaf area to balance transpirational supply and demand. This control is exercised, in Eq. (7), through adjustments in g_v. Typically, as soil moisture is depleted, leaf water potentials will decrease, resulting in senescence of older leaves that are most shaded and least capable of photosynthesis. As the moisture deficit develops, leaf area decreases to balance water uptake and loss. Stomatal closure protects the plant from desiccation during short-term episodes of high evaporative demand.

Over the long term, total transpiration depends on evaporative demand, timing of recharge, and size of the soil moisture reservoir. The soil texture and characteristics of the soil profile determine the field capacity and permanent wilt points for the reservoir, and the rooting depth determines the total volume of soil in the reservoir. In the absence of a water table, there is little upward movement of water into a drying root zone, so water that is not used as it travels through the root zone goes on down to recharge groundwater. When storms come in amounts and intervals that recharge the soil moisture reservoir during times of high evaporative demand, then almost all of the water delivered to the root zone will be used for transpiration. If recharge of the root zone occurs during winter when vegetation is dormant and evaporative demand is low, then water can move below the root zone and will go on to groundwater. Such contributions to groundwater are possible even in quite arid areas, where annual potential evapotranspiration greatly exceeds precipitation, if the precipitation occurs during times of low evapotranspiration.

Water loss by transpiration is more easily managed than is evaporation. Removal of vegetation results in complete elimination of transpiration. Reduced transpiration is achieved just by reducing plant cover. If vegetative cover is desirable, but transpiration needs to be minimized, for example, to maximize ground water recharge, shallow-rooted species can be chosen.

Considerable effort has gone into methods that control g_v directly through antitranspirants sprayed on the plants. These have been relatively unsuccessful, partly because of the expense of achieving even modest decreases in g_v through spraying things on plants, and partly because the antitranspirants increase leaf temperature, increasing the vapor concentration difference between the leaf and the atmosphere, which offsets part of the benefit gained by decreased vapor conductance.

V. Applications and Examples

We will close with three examples using applications of the principles just discussed. An important application of Eq. (9) and similar equations for predicting evaporative demand is in planning for irrigation of agricultural crops and scheduling irrigation on those crops. Water has been plentiful and cheap in irrigated areas, and producers generally have not had to bear the costs of environmental degradation from overirrigation. Irrigation is therefore often scheduled with a calendar or according to the convenience of the farmer or water authority with little regard to variations in water use by crops. Overirrigation usually occurs during times of low evaporative demand in spring and fall, and this extra water percolates through the soil carrying fertilizer, salts, and pesticides with it. The waste of water by these methods is serious, but not as serious as the damage to the environment.

With time, this picture is changing. Water resources are becoming scarcer, and competition between urban centers and irrigation for scarce water resources is increasing. Increased awareness of environmental degradation is also resulting in stricter regulation of non-point-source pollution. As water and the environmental costs of poor water management become more expensive, the attractiveness of sound irrigation scheduling methods is increasing. Ideally, irrigation water would be supplied frequently, and in amounts that would just meet the evapotranspiration demands of the crop. Careful water management would allow salts to accumulate and precipitate just below the root zone, and salt, fertilizer, and pesticide discharge to surface and groundwater would become negligibly small.

At present, we cannot measure either the water applied or the water used with enough precision to reach this ideal, but the knowledge exists to do much better than we are doing. Hopefully, in the near future, the knowledge we have will be fully applied and new techniques will become available so that ideal irrigation scheduling will become a reality.

A second example illustrates the hazard to the environment of failure to recognize principles discussed here and then shows an application of these principles to protect the environment. During and after World War II facilities were set up at various locations in the United States for processing radioactive materials to be used for weapons. One such location is Hanford, near Richland, Washington. Precipitation there is less than 20 cm per year, and potential evapotranspiration is over 100 cm per year; so it was assumed that no recharge to groundwater was possible. Wastes from processing were therefore often stored directly in the soil, without concern for potential groundwater contamination. Some waste sites were even treated to kill all vegetation, so that the only upward water loss from these sites was by evaporation from the soil surface.

Eventually, the fallacies of the no-recharge argument were brought to the attention of those who manage the waste. It was shown that perhaps 25 to 50% of the precipitation percolates through the waste to groundwater in the bare sites; and even with vegetation, some recharge may occur. Most of the precipitation occurs during winter months when plants are not able to use the water, so some goes below the root zone when the storage capacity of the soil for winter storms is insufficient. Once past the root zone, the water continues on, through the wastes and to groundwater.

Recently, scientists have begun to design covers for these waste sites that consider the principles we have discussed. Plants are used in the cover to maximize evaporation and transpiration, and the cover is designed to maximize water storage capacity. The cover is designed, taking climate into consideration, so that no water percolates below the root zone. Such covers should provide relatively safe and economical storage for radioactive wastes in the Hanford area.

As a final example we will consider the effect of urbanization on energy balance and evapotranspiration. In urban areas, paved streets and buildings provide a low-storage surface covering that reduces evaporation. Precipitation is intercepted and channeled into storm sewers, where evaporative loss is very small. The paved surfaces and built-up areas then quickly dry with minimal evaporation and no transpiration. Energy intercepted by these surfaces must all be dissipated as sensible and radiant heat, resulting, at times, in very high surface temperatures. As the size of the area increases, this "urban heat island" effect increases. The importance of green areas in cities is now widely recognized. They serve many purposes, but an important one is the partitioning of incoming solar energy into latent as well as sensible heat.

VI. Conclusion

The evapotranspiration that occurs at the earth's surface is tightly coupled to the availability of energy, on one hand, and the availability of water on the other. Energy limitations apply when plants and soils are well supplied with water. Water availability limits evapotranspiration when soil surfaces dry or when plants are not able to withdraw water from

their root zones rapidly enough to meet evaporative demand. Part of the available energy always goes to heating the air, but as water supply starts to limit evapotranspiration, the fraction of available energy going to heating the air increases.

Bibliography

Campbell, G. S. (1977). "An Introduction to Environmental Biophysics." Springer-Verlag, New York.

Campbell, G. S. (1985). "Soil Physics with BASIC: Transport Models for Soil–Plant Systems." Elsevier, Amsterdam.

Hillel, D. (1980). "Fundamentals of Soil Physics." Academic Press, New York.

McNaughton, K. G. (1989). Regional interactions between canopies and the atmosphere. *In* "Plant Canopies: Their Growth, Form and Function" (G. Russell, B. Marshall, and P. G. Jarvis, eds.), pp. 63–81. Cambridge Univ. Press, Cambridge.

Monteith, J. L. (1981). Evaporation and surface temperature. *Q. J. R. Meteorol. Soc.* **107,** 1–27.

Monteith, J. L., and Unsworth, M. H. (1990). "Principles of Environmental Physics." Edward Arnold, London.

Glossary

Evaporation Conversion of liquid water to water vapor at an air–water interface with the concomitant absorption of the latent heat of vaporization of water.

Latent heat Heat transported as a result of a phase change of water. The latent heat of vaporization of water (2430 J or 580 cal for each gram of water evaporated) is absorbed as water evaporates at a surface. This energy is transported from the surface with the vapor, resulting in a heat loss from the surface. Condensation of the water vapor releases the latent heat.

Sensible heat Heat transported by convection or diffusion as a result of a temperature difference.

Stomates Openings or pores between guard cells in the leaf epidermis through which vapor diffuses to the atmosphere. The plant controls the opening and closing of stomates through changes in guard-cell turgor and can therefore control the rate of vapor diffusion.

Transpiration Evaporation from the substomatal cavities of plant leaves (implies diffusion of the transpired water through stomates to the atmosphere and uptake of water from soil to replace transpired water).

Evaporites

Peter Sonnenfeld
University of Windsor

Evaporites are rocks composed of minerals precipitated from naturally occurring brines of marine or continental provenance. These brines are concentrated to saturation by evaporation or freeze-drying. To concentrate brines, annual evaporation rates must exceed rates of atmospheric precipitation; and freshwater inflow (atmospheric precipitation, river and groundwater discharge) declines with increasing evaporation rates.

Evaporation of a brine requires the partial water pressure of air to be lower than that of the fluid; this difference declines as the brine concentrates. Moreover, the surface tension increases in a concentrating brine and it becomes more difficult for water molecules to escape. The evaporation rate from NaCl-saturated and KCl-saturated solutions drops by about 30% and about 50%, proportionally, of that from a freshwater surface.

I. Introduction

Groundwater-derived crystals and crusts of evaporite minerals are known from the dawn of earth history, the Early Precambrian era. Bedded marine evaporites produced by evaporation of an exposed brine surface have formed only since the Proterozoic era. The earliest gypsum or anhydrite beds are about 1.7–1.8 Ga (billion years) old, the earliest halite and potash beds only about 0.7–0.8 Ga. A worldwide hiatus in evaporite deposition, of about 40 Ma (million years), occurred in the mid-Ordovician, another equally long one at the Cretaceous/Tertiary boundary.

Cooler periods in the earth's history coincided with a widening of the subtropical dry belt and high rates of evaporite accumulation. In the last ice age, maxima in aridity lagged slightly behind maxima in glaciation. More evaporation from the oceans in warm periods caused copious rains over wider areas.

II. Evaporite Minerals

Evaporite rocks are oligomineralic; that is, they are composed of only a few minerals. Each bed commonly contains no more than one or two minerals. Evaporite minerals crystallize in a temperature range between about $-55°C$ (the eutectic point) and $55°C$ (the eutonic point; the brine density has reached 1.3995 g/cm^3).

Evaporite minerals precipitate because of a deficiency of free water molecules to act as a solvent. Clusters of water molecules grow with decreasing brine temperature and as the amount of hydrogen bonding available to other cations increases. The lower the temperature of formation, the greater is the number of water molecules attached to a crystal lattice. Hydrated minerals are not very soluble and precipitate faster. Rising brine salinity increases the hygroscopy and fosters precipitation of anhydrous minerals—minerals containing no water in the crystal lattice.

Evaporite minerals form in four environments:

1. In almost all climatic regions as encrustations, rarely more than a few centimeters thick, in microenvironments such as mine outlets, tunnels, and lava caves.
2. As groundwater-derived crusts and efflorescences in semiarid soil horizons of any latitude.
3. As lacustrine deposits in semiarid and arid regions.
4. As precipitates in open waters of marine embayments in semiarid subtropical climates. These comprise the bulk of ancient evaporitic rocks. Associated with them are interstitial precipitates in surrounding coastal sands and silts.

Concentrating groundwater or lake waters first precipitate Na- and Ca-carbonates (Table I), seawater calcite or aragonite; Na- or Na-Ca- or Mg-sulfates follow in lakes, gypsum in soils or in marine environments. If a chlorine source is avail

Table I Lacustrine Carbonates[a]

Mineral	Formula	Specific gravity[b]
Anhydrous sodium carbonates		
Nahcolite	$NaHCO_3$	2.16
Wegscheiderite	$Na_5H_3(CO_3)_4$	2.14
Nyerereite	$Na_2Ca(CO_3)_2$	2.47
Shortite	$Na_2Ca_2(CO_3)_3$	2.50
Eitelite	$Na_2Mg(CO_3)_2$	2.73
Compound anhydrous sodium carbonates		
Northupite	$Na_3Mg(CO_3)_2Cl$	n.det.
Burkeite	$Na_6CO_3(SO_4)_2$	n.det.
Tychite	$Na_4Mg_2(CO_3)_2(SO_4)_3$	n.det.
Bradleyite	$Na_3MgCO_3PO_4$	n.det.
Hanksite	$Na_{22}K(CO_3)_2(SO_4)_9Cl$	n.det.
Burbankite	$Na_2(Ca,Sr,Ba,La,Ce)_4(CO_3)_5$	
Anhydrous potassium carbonates		
Kalicinite	$KHCO_3$	2.17
Fairchildite	$K_2Ca(CO_3)_2$	n.det.
Hydrated sodium carbonates		
Thermonatrite	$Na_2CO_3 \cdot H_2O$	2.20
Natron (soda)	$Na_2CO_3 \cdot 10H_2O$	1.44
Trona	$Na_3H(CO_3)_2 \cdot 2H_2O$	2.11
Pirssonite	$Na_2Ca(CO_3)_2 \cdot H_2O$	2.35
Gaylussite	$Na_2Ca(CO_3)_2 \cdot 5H_2O$	1.99
Compound hydrated sodium carbonates		
Dawsonite	$NaAl(OH)_2CO_3$	n.det
Hydrated potassium carbonates		
Buetschliite	$K_2Ca(CO_3)_2 \cdot 2H_2O$	n.det
Anhydrous iron-magnesium carbonates		
Magnesite	$MgCO_3$ (Fe : Mg > 20)	3.00
Breunerite	$(Mg,Fe)CO_3$ (Fe : Mg = 3–20)	3.10
Mesitite	$(Mg,Fe)CO_3$ (Fe : Mg = 1–3)	3.23
Pistomesite	$(Fe,Mg)CO_3$ (Fe : Mg = 0.3–1)	3.45
Sideroplesite	$(Fe,Mg)CO_3$ (Fe : Mg = 0.05–0.3)	3.63
Siderite	$FeCO_3$ (Fe : Mg < 0.05)	3.90
Compound anhydrous magnesium carbonates		
Dolomite	$CaMg(CO_3)_2$	2.87
Huntite	$CaMg_3(CO_3)_4$	2.70
Norsethite	$BaMg(CO_3)_2$	3.84
Hydrated magnesium carbonates		
Hydromagnesite	$Mg_4(OH)_2(CO_3)_3 \cdot 3H_2O$	2.16
Dypingite	$Mg_5(OH)_2(CO_3)_4 \cdot 3H_2O$	2.15
Artinite	$Mg_3(OH)_2(CO_3)_2 \cdot 3H_2O$	2.02
Barringtonite	$Mg(CO_3) \cdot 2H_2O$	2.11
Nesquehonite	$Mg(CO_3) \cdot 3H_2O$	1.95
Lansfordite	$Mg(CO_3) \cdot 5H_2O$	1.73
Anhydrous calcium carbonates		
Ankerite	$CaFe(CO_3)_2$	3.80
Aragonite	$CaCO_3$	2.93
Calcite	$CaCO_3$	2.71
Vaterite	$CaCO_3$	2.56
Strontianite	$SrCO_3$	3.70
Benstonite	$Ba_6Ca_7(CO_3)_{13}$	3.60

(continues)

Table I *(Continued)*

Mineral	Formula	Specific gravity[b]
Hydrated calcium carbonates		
Monohydrocalcite	$CaCO_3 \cdot H_2O$	n.det
Trihydrocalcite	$CaCO_3 \cdot 3H_2O$	n.det
Pentahydrocalcite	$CaCO_3 \cdot 5H_2O$	n.det
Ikaite	$CaCO_3 \cdot 6H_2O$	1.77

[a] Ordered according to declining carbon dioxide content.
[b] n. det., not determined.

able, halite is the next precipitate, followed by K- and Mg-chlorides.

Rubidium, cesium, thallium, and ammonium do not form independent evaporite minerals, but they substitute for potassium in the crystal lattice in some evaporite minerals. Bromide substitutes for chloride and the amount varies with brine concentration, Mg-availability, distance from shore, brine depth and temperature, degree of recrystallization of any precipitate, sorption by clays, scavenging by algae or bacteria. Fluorine, iodine, lithium, and many trace elements remain in the residual brine to be soaked up in interstitial spaces of the precipitates. Poorly soluble barite either occurs in nests in limestones or it replaces gypsum.

Borates form many hydrated minerals precipitating together with gypsum in both lacustrine and marine settings, or remain in the terminal brine. Boron comes from nearby volcanic sources or from muscovite flakes swept into aquatic environments. Authigenic tourmaline slivers form occasionally.

Phosphates are rare in evaporitic environments and are mutually exclusive with sulfates; iodides and chlorates are unknown. Nitrates and iodates occur only in surface exposures in extremely arid areas (<10 mm annual rainfall); they are unknown as ancient deposits.

III. Continental Evaporites

A. Groundwater-Derived Evaporites

Evaporating groundwater rising to the surface precipitates crusts and crystals in soil. More soluble components leapfrog over less soluble ones: A calcite cement is overlain by a gypsum crust that marks the depth of wetting in a leached soil horizon; halite efflorescing at the surface is largely blown away. Where seawater at high tide wets an arid soil and then ebbs away, seepage will form a reverse sequence of halite encrustations beneath gypsum crusts. [*See* GROUNDWATER.]

Sebkhas (or sabkhas) are shallow depressions—filled episodically by rains, rising groundwater, or storm tides—that precipitate evaporites. Significant are the dominance of detrital fragments, lattice imperfections in Sr-enriched anhydrites, replacive or displacive gypsum in diagenetic nodules or enterolithic shapes, and a calcite deficiency.

Continental sebkhas, also called playas, are either periodically flooded by surficial waters or fed by groundwater rising to the surface. Brines seeping into a lake-filled depression precipitate evaporite minerals in concentric belts, the most soluble minerals closest to the lake shore. The slower the groundwater movement and the more dilute the initial brine, the larger are the crystals. Paralic sebkhas are prograding peritidal plains subject to storm tides or seawater interaction with discharging groundwater.

B. Lacustrine Evaporites

Lacustrine evaporites accumulate in endorheic basins. Ions supplied by the drainage area and their interactions determine the nature of precipitates. Due to a low Ca/Na ratio in continental runoff, brines in alkaline lakes preferentially precipitate Na-carbonates, Na-sulfates, compound Na-Ca-carbonates, or compound Na-Ca-sulfates. All the hydrated species are unstable and either convert in time to anhydrous varieties or redissolve. The anhydrous $CaCO_3$ variety is normally calcite.

Soda lakes form mainly over bedrock composed of crystalline rocks or of polymictic, feldspathic, or argillaceous sandstones; Mg-rich brines form where carbonated groundwater leaches basic igneous rocks.

Gypsum follows, but not where $NaHCO_3$ is present; calcite and mirabilite or glauberite (Table II) precipitate instead. If Mg ions are present, bloedite will form rather than epsomite, which dominates in efflorescences of Mg-rich brines. Sulfide oxidation supplies the sulfate ions. Most potassium replaces sodium in clays, turning them into illites. K-Mg-solutions can alter gypsum to polyhalite. Mirabilite occurs wherever the brine approaches freezing temperatures in the winter; at 32.4°C it dehydrates to thenardite.

Many hypersaline lakes develop an alkalinity and a pH high enough to dissolve swept-in silicates. They precipitate as hydrated framework silicates in the zeolite group (e.g., hydrated Na-A1, Ca-A1, and Ca-Na-A1 silicates found in several lakes in the sahel of Africa). However, these minerals eventually dehydrate and convert to authigenic feldspar crystals, or turn into quartz or chert when their cations are leached out. Complex hydrated silicates are only known from

Table II   Evaporitic Sulfate Minerals

Mineral	Formula	Specific gravity
Lacustrine sulfates		
Celestite	$SrSO_4$	3.97
Polyhalite	$K_2SO_4 \cdot 2CaSO_4 \cdot MgSO_4 \cdot 2H_2O$	3.00
Anhydrite	$CaSO_4$	2.98
Glauberite	$Na_2SO_4 \cdot CaSO_4$	2.85
Bassanite	$CaSO_4 \cdot \frac{1}{2}H_2O$	2.76
Thenardite	Na_2SO_4	2.70
Aphthitalite	$Na_2SO_4 \cdot 3K_2SO_4$	2.66
Bruckerite	$2Na_2SO_4 \cdot CaSO_4 \cdot 2H_2O$	
Hanksite	$2Na_2CO_3 \cdot 9Na_2SO_4 \cdot KCl$	2.57
Burkeite	$Na_2CO_3 \cdot 2Na_2SO_4$	2.57
Uklonskovite	$NaOH \cdot MgSO_4 \cdot 2H_2O$	2.45
Gypsum	$CaSO_4 \cdot 2H_2O$	2.33
Bloedite	$Na_2SO_4 \cdot MgSO_4 \cdot 5H_2O$	2.28
Hydroglauberite	$5Na_2SO_4 \cdot 3CaSO_4 \cdot 6H_2O$	1.51
Mirabilite	$Na_2SO_4 \cdot 10H_2O$	1.49
Marine sulfates		
Ca-sulfates (decreasing calcium content)		
Anhydrite	$CaSO_4$	2.98
Bassanite	$CaSO_4 \cdot \frac{1}{2}H_2O$	2.76
Gypsum	$CaSO_4 \cdot 2H_2O$	2.33
Goergeyite	$K_2SO_4 \cdot 5CaSO_4 \cdot H_2O$	2.77
Glauberite	$Na_2SO_4 \cdot CaSO_4$	2.85
Polyhalite	$K_2SO_4 \cdot 2CaSO_4 \cdot MgSO_4 \cdot 2H_2O$	3.00
Syngenite	$K_2SO_4 \cdot CaSO_4 \cdot H_2O$	2.60
Sr-sulfates (decreasing strontium content)		
Celestite	$SrSO_4$	3.97
Kalistrontite	$K_2SO_4 \cdot SrSO_4$	3.32
K-Mg sulfates (decreasing potassium content)		
Kainite	$4KCl \cdot 4MgSO_4 \cdot 11H_2O$	2.19
Langbeinite	$K_2SO_4 \cdot 2MgSO_4$	2.83
Picromerite	$K_2SO_4 \cdot MgSO_4 \cdot 6H_2O$	2.15
Leonite	$K_2SO_4 \cdot MgSO_4 \cdot 4H_2O$	2.20
Ammonium sulfates		
Boussingaultite	$(NH_4)_2SO_4 \cdot MgSO_4 \cdot 6H_2O$	1.72
Mg-sulfates (decreasing magnesium content)		
Kieserite	$MgSO_4 \cdot H_2O$	2.57
Sanderite	$MgSO_4 \cdot 2H_2O$	
Leonhardtite	$MgSO_4 \cdot 4H_2O$	
Allenite	$MgSO_4 \cdot 5H_2O$	
Sakiite	$MgSO_4 \cdot 6H_2O$	1.76
Epsomite	$MgSO_4 \cdot 7H_2O$	1.68
Na-sulfates (decreasing sodium content)		
Thenardite	Na_2SO_4	2.70
D'ansite	$9Na_2SO_4 \cdot MgSO_4 \cdot 3NaCl$	2.65
Vant'hoffite	$3Na_2SO_4 \cdot MgSO_4$	2.69
Loeweite	$6Na_2SO_4 \cdot 7MgSO_4 \cdot 15H_2O$	2.42
Bloedite	$Na_2SO_4 \cdot MgSO_4 \cdot 5H_2O$	2.28
Aphthitalite	$Na_2SO_4 \cdot 3K_2SO_4$	2.66

dried-out Neogene lakes, but not from more ancient sediments.

Igneous and metamorphic rocks contain minimal amounts of chlorides. Wherever groundwater carries chloride ions, these are generally due to a discharge of $CaCl_2$-enriched formation waters or to dissolution of ancient marine chloride deposits; salt creeks or salt rivers are common even outside the arid belt. Halite can then precipitate, or hydrohalite below $-23°C$, and remain stable to about $+1.5°C$. Reprecipitation leads to secondary nonmarine halite deposits that display all aspects of marine evaporites. However, they are chemically purer, because they are stripped of any ions that were lattice substitutions in the original deposit; in other words, they display a negligible Br/Cl ratio. Commonly they do not contain a complete marine succession of evaporites, either being devoid of $CaSO_4$ or of potash deposits. Such halite accumulations are common on all continents; examples are the deposits in North Dakota and Arizona. Great Salt Lake, Utah, derives its salt content from leached Jurassic beds in the Wasatch Range. The salt content of the Dead Sea has been leaching out of Quaternary salt domes around and under it; it is not a primary residue of a severed arm of the Red Sea.

IV. Marine Evaporites

Bedded marine evaporites have only formed in subtropical lagoonal embayments. In the open ocean, currents thoroughly mix the waters and prevent the formation of increasingly saline brines. However, individual gypsum or halite crystals occur where concentrated brines are discharged into the sea, or in basaltic crevices on the ocean floor, or in beds underneath. [*See* Brines in Sedimentary Environments.]

A. The Entrance Barrier

Carl Ochsenius suggested in 1877 that evaporite precipitation in a lagoon requires an entrance sill to restrict water circulation and to foster brine concentration. Bays with unrestricted access maintain only marginally higher salinities than the open sea. He envisaged a continuous inflow of brine preventing the bay from drying out.

In 1917 Otto Krull augmented the Ochsenius model of a barred basin by assuming a concurrent bottom outflow that retards the solute removal and accounts for the shortfall in more soluble salts compared to those less soluble.

The configuration of the entrance strait determines the rate of bay water concentration and ultimately dictates whether the surface input of solute equilibrates with the export by bottom brines. The outflow depletes the shelf in front of the entrance sill of all nutrients by washing them down the continental slope. Bottom dwellers no longer migrate from the open ocean, and a rising salinity and osmotic pressure in the bay gradually kills resident burrowers and terminates bioturbation. A rising salinity is also deadly to many planktonic organisms and deposits organic films.

Reef builders continue to flourish in the oxygenated surface inflow, as their natural enemies die out. Their oxygen consumption is high; they produce more organic matter than they consume. Reef growth and reef debris in the entrance strait reduce the cross-sectional area, accelerate an inflow determined by evaporation losses, slow down outflow, and thus promote concentration. When sediments block the entrance strait, seawater seepage alone cannot match evaporation losses, unless the entrance barrier is extremely thin compared to the surface area of the bay. The permeability of algal limestones is too small to allow seepage to maintain a water level over wide areas.

The interface between the inflow and the outflow is a pycnocline. Storm energies are dissipated along the interface by working against buoyancy without overturning the density gradient. The depth of this interface is a function of the length of a strait to its total depth and the velocity of the inflow — the lower the interface, the less brine can escape. The Coriolis effect deflects the inflow in the northern hemisphere anticlockwise along the shore, tilts the interface both westward and poleward, and controls the facies distribution.

B. Ion Imbalance

The uppermost 10 cm of inflowing seawater absorbs 45% of solar energy, or about 1.35 kW/m^2; the next 90 cm absorbs nearly 35%. The interface acts as a one-way mirror to entrap almost all the remaining radiation in bottom brines of higher refractive index. The shelves heat up faster, as less brine needs to be heated per unit area and the interface is shallower. The warming in concentrating brines is helped by lowered specific heat and evaporation rates.

The Soret coefficient decreases with concentration but increases with temperature. Most marine evaporitic compounds are thermophile and increase the stability of density stratification; cryophile solutes decrease it. Some solutes change their character; for example, NaCl becomes cryophile below 12°C. Halite or sylvite precipitation releases heat; gypsum absorbs heat to 37.5°C and evolves heat above that temperature.

The hydrostatic pressure gradient of overlying brines causes downward solute diffusion — that is, in the direction of increasing salinity, dynamic viscosity, and density gradient. A steep vertical gradient impedes vertical fluid mixing and prevents convection. Large density differences are, however, only possible with a shallow interface between inflow and resident brine; the differences decrease with increasing depth of the inflow. Thermal diffusivity and dynamic and kinematic viscosities increase with concentration, but viscosities decrease with temperature.

A gradient of 600–800 mV between oxygenated surface waters of positive redox potential and anaerobic bottom waters of a negative one drives cations into the brine and extracts H_2S. To preserve a charge balance, the cations combine with water, generating hydroxyl ions and free hydrogen atoms for bacterial growth.

A brine saturated for $CaSO_4$ contains less than 4 ppm dissolved oxygen. Anaerobic sulfate-reducing bacteria (e.g., *Desulfovibrio*) then scavenge oxygen out of sulfate ions. If enough organic matter is present, the anaerobes are capable of destroying more than $\frac{2}{3}$ of incoming sulfate ions and converting them to hydrogen sulfide; halite-saturated brines no longer contain any sulfate ions. A cation surplus develops of at least 19.3 mol/m³ Mg^{2+}. Although seawater contains three times as much sulfate as calcium, a sulfur deficiency develops early in the concentration cycle, and not all calcium is taken up. Some is retained and seeps out as $CaCl_2$-rich formation water or precipitates as tachyhydrite.

CO_2 dissolved in seawater is taken up by bacteria and blue-green algae and is used in aragonite precipitation, or it escapes when increasing salinity or heating decrease its solubility. Sulfate-reducing anaerobes can utilize inorganic carbonate provided there is a source of hydrogen, such as derived from iron oxidation or clay chloritization. An increase in hydrogen ions and an excess of cations resulting from hydrogen sulfide removal raise the pH and drop the $CaCo_3$ solubility. However, anaerobes prefer metabolizing organic matter to using inorganic carbon. Macerated organic compounds thus accelerate sulfate reduction.

A partial destruction of gypsum laminae around the rim of marine lagoons produces ^{12}C-enriched low-Mg calcite laminae or, by a further bacterial conversion, magnesite. However, if the bacteria produce Na-carbonates, these react with seawater to form either huntite or hydromagnesite as possible magnesite precursors.

Where the interface is shallow enough for bottom waters to reach into the photic zone, one or several pink horizons (called "bacterial plates") develop within the mixing zone between oxygenated surface waters and bottom brines. They are mainly composed of thiobacteria that reoxidize hydrogen sulfide into sulfate ions. These sulfate ions attack precipitated $CaCO_3$ and convert it back to gypsum. Completely gypsified algal stromatolites are common. Bluegreen algae flourish in them, because gypsum is transparent to ultraviolet radiation.

C. Precipitation Sequence

The sequence of primary precipitation is identical in all Phanerozoic marine evaporite deposits. The laws of physics and chemistry of brines have thus remained constant, biogenic influences on precipitation did not change, and the ionic composition of seawater (whatever its salinity might have been) has never deviated far from present values.

In concentrating brines, $CaCO_3$ precipitates first. Hydrous carbonates do not occur in marine evaporites: Rapidly precipitating aragonite is the most common form in Mg-rich seawater, and slowly crystallizing calcite predominates in lacustrine environments. With rising brine temperature fewer Mg ions are needed to subdue calcite precipitation and to foster aragonite formation; Na ions counteract this effect. After deposition, aragonite eventually converts to calcite. Mesozoic aragonite is very rare, pre-Permian aragonite is unknown. Ca-carbonates as rocks are called limestones, but this includes also rocks containing some dolomite.

As the inflow concentrates, carbonate sedimentation retreats to the entrance or to supratidal parts of the shore as beach rock, and gypsum precipitation takes over. Primary gypsum may contain significant amounts of Sr, Zn, Pb, Mn, Fe, and other metals either built into the crystal lattice or finely disseminated. Gypsum dehydrates to anhydrite in brines saturated at least for NaCl, but only if organic compounds are present. Admixtures are released during this conversion: Strontium reprecipitates as celestite nests and other metals migrate with the crystal water. Holocene anhydrite occurs only where organic decomposition products are present; even water salinities of 200,000 ppm do not alter gypsum to anhydrite in sterile environments.

Gypsum deposition initially spreads throughout the marine bay. If synsedimentary subsidence drops deeper parts of the basin out of the photic zone, gypsum precipitation continues only on shallow shelves and shoals. In 1957 G. Richter-Bernburg aptly named these "saturation shelves." Stripped of sulfates the brine glides toward the deeper, most rapidly subsiding parts of the basin, be they graben structures or other depressions.

Further concentration leads to the deposition of chlorides (Table III) of progressively greater hygroscopicity. Gypsum is followed by halite, when at least 91% of the water has been removed. Halite is in turn followed by sylvite in brines covered by inflow, or carnallite in those exposed to the atmosphere. In the distal parts of a basin, Fe^{2+} may substitute for Mg^{2+} and NH_4^+ for K^+ in the carnallite lattice. The final precipitate is tachyhydrite at 300-fold concentration of seawater. Further precipitation must cover both carnallite and tachyhydrite before the bay dries out, otherwise they redissolve in their own crystal water. Tachyhydrite is stable only above 20°C and is thus found in localities that have never experienced lower temperatures — specifically in Lower Cretaceous evaporites of the Cuanza Basin of Angola, the Sergipe Basin of Brazil, the Sakonnakon Basin of Laos, and the Khorat Basin of Thailand.

D. Depositional Fabrics

Proteins, naphthenic acids, amino acids, resins, and sugars delay gypsum precipitation or prevent it. The higher the pH during gypsum crystallization, the stubbier are the crystals; alkanes, phenols, and fatty acids foster the formation of more

Table III Chloride Minerals

Mineral	Formula	Specific gravity
Continental environments		
Halite	NaCl	2.16
Hydrohalite	$NaCl \cdot 2H_2O$	1.54
Antarcticite	$CaCl_2 \cdot 6H_2O$	1.71
Sylvite	KCl	1.99
Carnallite	$KMgCl_3 \cdot 6H_2O$	1.60
not found:	$MgCl_2 \cdot 8 - 12H_2O$	
Marine environments		
Halite	NaCl	2.16
Sylvite	KCl	1.99
Rinneite	NaK_3FeCl_6	2.35
Carnallite	$KMgCl_3 \cdot 6H_2O$	1.60
Tachyhydrite	$CaMg_2Cl_6 \cdot 12H_2O$	1.66
Bischofite	$MgCl_2 \cdot 6H_2O$	1.60
Rare accessory minerals		
Sal Ammoniac	NH_4Cl	1.53
Baeumlerite	$KCaCL_3$	2.30
Molysite	$FeCl_3$	3.04
Chloraluminite	$AlCl_3 \cdot 6H_2O$	2.40
Kremersite	$(NH_4)_2FeCL_5 \cdot H_2O$	1.82
Erythrosiderite	$K_2FeCl_5 \cdot H_2O$	2.37
Douglasite	$KFeCL_3 \cdot 2H_2O$	2.15
Rokuehnite	$FeCl_2 \cdot 2H_2O$	2.36

tabular, equidimensional, or discoid-lenticular crystals. Elongate, prismatic gypsum crystals grow in low-pH environments or in the presence of dissolved SiO_2.

Sylvite nucleates much more slowly than carnallite. Alcohols or nitrogen compounds derived from decaying organic matter, such as urea and complex cyanides, greatly reduce the KCl-solubility and are protected by brine inflow against oxidation. Urea also increases the solubility of $MgCl_2$, and so a yellowish sylvite crystallizes. Where the brine is exposed to the atmosphere, carnallite precipitates and contains hematite platelets, derived from the oxygenation of Fe^{2+}-organic complexes. Even small amounts of $CaCl_2$, derived from an influx of formation waters, reduce the solubility of KCl and more so of $MgCl_2$, thus forcing carnallite precipitation or a conversion of precipitated sylvite to carnallite.

Carnallites deposited in open brines alternate with halite beds. Carnallites deposited into a halite slush form disseminated textures and are called "carnallites"; correspondingly, sylvite in a halite crystal matrix is called "sylvinite."

Anhydrite crystals forming on halite crystal faces tend to inhibit halite crystal growth; halite develops enlarged crystals only in the absence of anhydrite. Alcohols and other organic compounds decrease halite and sylvite solubility. Both crystallize normally as cubes, but they form octaeders if some organic compounds are present.

Hopper crystals can form at a brine surface, if the brine was not density stratified and did not contain soapy surfactants cutting the surface tension.

Current-deposited, cross-bedded gypsum sand (gypsarenite) or halite sand (haloarenite or halolite), turbidites, gravity flows, and submarine slides occur in a dense brine over very gentle slopes (for example, in Miocene anhydrite beds of the northern Appennines in Italy, Permian anhydrite beds in East Germany, Miocene halite beds from Caltanissetta in Sicily or Verotyshchensk in the Ukrainian Carpathians, and Holocene channels in the southern Dead Sea).

Nodular anhydrite is either a supratidal alteration of a gypsum crust precipitated in the capillary zone of an ancient groundwater table, or an early diagenetic alteration of a semifluid gypsum/brine mush beneath the lagoonal floor, or a compaction texture of soft gypsum, or a recrystallization of gypsum laminations in percolating halite-saturated hygroscopic brines.

Fluid inclusions in evaporites are voids, commonly negative crystals, filled with gaseous or liquid phases of fluids, mainly NaCl-brines, hydrocarbons, CO_2, and organogenic N_2. Fluid inclusions migrate, especially under uniaxial stress, because of solubility differences along warmer and colder cavity walls. Those with less than 10 vol % of gas move toward warmer temperatures, while those with a larger gaseous fraction move in the opposite direction. Along the way they may coalesce or be contaminated by intercrystalline moisture. Continued migration is hindered by grain boundaries that can permanently entrap the inclusion.

E. Siliciclastic Content

Lands surrounding an evaporite basin commonly possess a gentle slope with a shoreward fining of clastics; onshore sediments frequently are only red silty clays. A slight oscillation in sea level then results in a large change in the water surface.

Leaves, pine cones, fern pinnules, and tree trunks, found well preserved in halites, indicate an at least discontinuous vegetation cover nearby and a not very arid climate with annually variable rainfall and occasional flash floods.

Clastic intercalations are scarce because of scant runoff in semiarid areas. The amount of swept-in clastics is very subordinate, except close to the shoreline of an abutting hill chain (for example, the Miocene Transcarpathian Basin or the Permian Fore-Ural Basins). Without evaporite precipitation such basins would have been "starved basins" (subsiding depressions with meagre sedimentation as, for example, the Plio–Pleistocene and Holocene Mediterranean Sea).

Quartz sand and silt occur embedded in marginal gypsum shelves. Discrete quartz sand stringers encased in anhydrite are very rarely present in halites, never in the potash sequence. They mark a temporary freshening.

Detrital quartz or feldspar grains do not occur in halites, sylvites, carnallites, or tachyhydrites. A few detrital grains of staurolite, zircon, or tourmaline have been found in halites; they suggest that ancient dust storms did exist, but that both quartz and feldspar grains have been removed. The solution of blown-in quartz and the leaching of SiO_2 attest to a very alkaline pH. Even a few mg/m³ of ammonia in the brine (derived from protein breakdown) suffice to raise the pH substantially, while dissolved Mg^{2+} and Al^{3+} reduce silica solubility. Euhedral quartz and feldspar grains become neoformations.

Ocean surface waters supply seawater free of suspended clay particles. All clays in evaporites are brought in from nearby shores by flash floods. They spread out along the interface between inflow and resident brine and flush an unabraded fauna far into the bay.

The clays settle very slowly as floccules that retain water in their pore spaces; as the solute concentrates, the floccules grow and retain more brine. When halite precipitation resumes, the brine entrapped in the floccules keeps the clays permeable. Thin clay laminae can later act as conduits for brines or meteoric waters entering from the basin margin. Hygroscopic brines induce shrinkage cracks in clays, called "syneresis cracks." They are indistinguishable from subaerial desiccation cracks and increase the permeability.

All clays are efficient ion exchangers and in a hypersaline brine are ultimately transformed into mixed-layer varieties of the Mg-chlorite family by inserting $Mg(OH)_2$ pillars, liberating hydrogen ions, and leaching various cations and some SiO_2. Halloysites, kaolinite, Fe-chlorite, muscovite, and montmorillonite are restricted to nearshore sites, palygorskite is present offshore, replaced by sepiolite farther out. K-rich brines can later alter Mg-chlorites to illites.

F. Organic Matter and Metals

Evaporite basins generate organic matter up to three orders of magnitude faster than the open ocean. In a Na-enriched brine the organic matter goes either into solution or forms chloride-organic complexes with Mg, Fe, and other metals. Colloidal transport peaks at 70°C; above 85°C the hydrocarbons form an independent phase.

Calcium sites in gypsum lattices can break up organic chains; clays polymerize them into molecules too large to migrate through available pores. Dissolved organic compounds, organometallic complexes and even broken-up polymers migrate as finite globules or as gas in hypersaline brines seeping through a porous substrate of gypsum or halite slush into formation waters.

Reefs and other porous strata around ancient evaporite basins are prime exploration targets for hydrocarbons, while shales and micritic limestones are commonly bituminous because of trapped organic matter. Oil shales, however, are of lacustrine origin, associated with trona and thenardite. A statistical evaluation of paleolatitudes of hydrocarbon occurrences shows a prevalent correspondence to the latitudinal belt of marine evaporites.

Base metals concentrated in the brine by plankton are entrapped in gypsum and halite lattices; the metals are released in the conversion of gypsum to anhydrite or the recrystallization of halite. Lead-zinc deposits commonly surround ancient evaporite beds or occur in caprocks of salt diapirs. The organic fraction of metal-organic complexes turns into liquid petroleum that appears in fluid inclusions or in potash mine seeps.

G. Facies Distribution

In analogy to vertical and horizontal distribution of evaporites in lakes, evaporite sequences in marine embayments are also zoned. Marine anhydrites commonly grade abruptly into near-shore limestones and the limestones in turn into red beds with gypsum crusts in the soil. The thickness of marginal anhydrites varies inversely with the halite thickness offshore. Thickest halites mark subsidence centers.

All potash deposits were originally encased in halite, indicating a gradual rise and eventual drop in the brine concentration. They occur on slopes of ancient evaporite basins, because KCl-NaCl mixtures on cooling decrease both their specific gravity and solubility. Potash deposits generally do not reach the shores because the nearshore rainwash and redissolution (Fig. 1). They occur preferentially on western slopes (in paleo position) of the basin or on eastern slopes of sills and shoals within the basin; this could be related to the westward tilt in the interface, or a greater inertia against earth rotation.

Brine concentration is not continuous; it is interrupted by freshenings due to excessive rains, flash floods, or storm tides. Basin-wide dissolution and desiccation surfaces signify a temporarily halted salt accumulation. Sedimentation resumes with a sharp basal contact of an anhydrite bed or a carbonate bank, giving the halite a banded character. Each clay lamina, gypsum wafer, or dolomite stringer represents some corrosion parallel to bedding planes. The ratio between elapsed times of deposition and nondeposition probably exceeds 1:10.

A moderate freshening of the brine allows the marginal gypsum platform to expand onto precipitated halite as a wedge-shaped intercalation. More pronounced freshenings, such as caused by cyclonic storms, may even allow marginal algal and reefal limestones to cover the whole basin in nearly uniform thickness. They prove that the bay floor remained within the photic zone. Continued subsidence eventually allows for further thick halites.

Every deposit is capped by a reverse sequence of precipitates from progressively less concentrated brines. This reverse

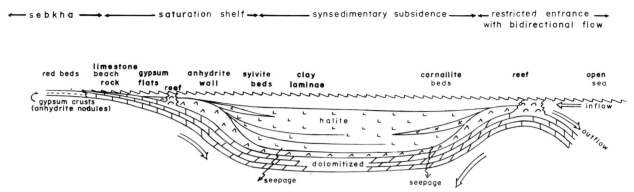

Fig. 1 Facies relationships in an evaporite basin.

sequence protects the salts from dissolution, unless parts of it later erode. Deposited from a progressively freshening brine, the sequence is thinner than the one precipitated from the concentrating brine. The ultimate dilution is thus faster than the original concentration.

During slow calcium carbonate and gypsum precipitation a subsiding basin deepens. Calcium carbonate precipitation rates do not reach 1 mm/yr, those of gypsum rarely exceed 1–2 mm/yr, and subsidence rates can be as high as 10–12 cm/yr. Halite precipitation rates vary between 3.5–14 cm/yr, but halite preservation or accumulation rates may be as low as 0.1–4.0 mm/yr.

A bull's eye distribution of facies, common in lakes, does not occur in marine evaporites. Even in perfectly round basins (such as the Silurian Michigan Basin) anhydrites surround the halites, but the sylvites are displaced to one flank (Fig. 2A). In elongate basins (for example, the Devonian Elk Point Basin in western Canada) the bottom currents redistribute the sediments into an alignment with the basin axis and still displace potash deposits to one margin. Where a shallow basin is broken up by marginal grabens, halites are restricted to them (as in the Miocene Tyrrhenian or Aegean Seas).

H. Synsedimentary Subsidence

Inflow and outflow tend to equilibrate, and the salinity in the embayment stabilizes. At prevailing deficits in the water budget of any subtropical sea, the brine cannot concentrate to halite saturation unless continued subsidence of the basin floor or a constriction of the entrance channel disturb the equilibrium between inflow and outflow. This occurred in the Gulf of Kara Bogaz Gol east of the Caspian Sea, first tectonically in the early part of this century and then artificially in the latter part of the century.

In ancient environments, the ratio between subsidence rates of gypsum precipitating shelves and halite precipitating deeps has commonly been 1:5–1:7. The deposition of several hundred meters of rock salt on top of shallow-water gypsum beds mandates such synsedimentary subsidence. The thicker these units, the longer was the time interval in which tectonic forces and climatic conditions remained relatively stable.

Synsedimentary subsidence alters thickness ratios between shelf and basin and allows the brine to transgress over its former shoreline. Distension faults and slumping mark ancient hinge lines. In Permian evaporites of Germany the axis of greatest subsidence shifted northwest during sedimentation; the greatest expanse of Triassic halite in North Africa is offset against the greatest thickness of the evaporite sequence.

A sequential arrangement of interconnected basins encourages brine preconcentration. Salina operators lead seawater through numerous concentrator pans before draining the brine into a crystallizer pan to harvest halite; the water surface exposed to evaporation must be 2–4 times larger than the surface of halite precipitation and even greater for sylvites or carnallites. Triassic evaporites of North Africa and Lower Cretaceous evaporites of Brazil and Angola display a pronounced deficiency of $CaCO_3$ and $CaSO_4$, possibly due to the destruction of their anterior basins by an opening and widening Atlantic Ocean.

When halite starts precipitating, the brine weight has increased nearly 20%, and the combined weight of brine and precipitate soon exceeds 170% of a seawater column. This must affect the isostatic equilibrium of the earth's crust.

Basin depth estimates based on studies of halites commonly differ by two orders of magnitude or more from those of intercalated potash deposits. Equating wave bases in KCl-saturated brines (<5 m) with those in seawater (>30 m) leads to overestimating brine depths. From Br/Cl studies and from

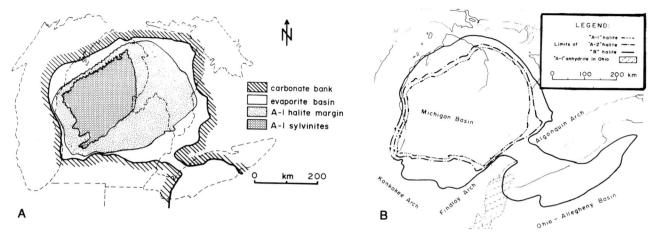

Fig. 2 The Silurian Michigan Basin. (A) Halite and sylvite distribution in the first evaporite cycle, the A-1 unit, the only potash bearing cycle. (B) Halite distribution in the first three evaporite cycles, the A-1, A-2, and B units. Note progressively increased areal extent. Later evaporite cycles expand farther into the Ohio Basin, even as far as West Virginia.

rates of freshening to produce anhydrite intercalations, estimates of brine depth in evaporite basins yielded a maximum of 100–150 m.

The larger the brine volume in a basin, the longer it takes after a major freshened influx to reach saturation; in other words, the rate of salination is inversely proportional to the brine depth. A longer time required to saturate a brine makes it less probable that evaporation or subsidence rates remain constant; saturation of deep basins is thus difficult to achieve.

I. Cyclicity of Deposition

Cyclic changes of brine concentration or episodic agitation cause varve-type laminations on a millimeter to centimeter scale. They extend over more than 100 km and mark basin-wide synchronous salinity fluctuations. Laminations form only in perennially submerged parts of the basin, where they are not destroyed in bioturbation or a vertical brine discharge. They cannot occur in deep brines; annual, seasonal, or daily fluctuations from supersaturation to undersaturation for the more soluble salt are only possible in shallow water.

Alternating calcite/gypsum, gypsum/halite, or halite/sylvite couplets generally have a sharp basal contact of the less soluble member of the pair and an irregular contact on top. Calcite/gypsum couplets can also be due to episodic gypsum destruction by anaerobic bacteria. Commonly, every eighth to tenth couplet is thicker, suggesting a record of major storm frequency related to sunspot cycles. Since each storm can remove up to 80% of the varve couplets, their count does not give a realistic estimate of elapsed time.

All evaporite basins contain alternating cycles of brine concentration and dilution caused by oscillations of environmental conditions. Primary marine evaporite deposits display from three or four to a score of cycles, alternating between several hundred meters of evaporites and deposits of normal salinity.

The transgressive nature of evaporite basins is best demonstrated by the growing volume and areal extent from one cycle to the next one (Fig. 2B). Frequently, the first cycle is restricted to a small basin; additional subsiding depressions develop over former shelves in subsequent cycles (for example, in the Silurian Michigan Basin, the Devonian Elk Point Basin in western Canada, the Permian Delaware Basin of west Texas, or the Permian Zechstein Basin of northwestern Europe).

J. Epigenetic Alterations

All chloride precipitates are initially slushy, with a porosity of 10–50%. Compaction by a growing overburden converts the slush into a solid rock, expels interstitial brines and those trapped in clay partings. The weight of overburden plasticizes rock salt and, to a lesser extent, anhydrite. Rock salt begins to expand into voids at an overburden of 300 m, potash minerals somewhat earlier. Percolating undersaturated waters increase the plasticity of the salt.

Before compaction destroys the permeability of the sediment, circulating waters seep in from above, diffuse into formation waters, and on the way recrystallize the underlying precipitates. Evidence for downward seepage is provided by preserved microfissures filled with salts of greater solubility than the wall rock and cylindrical vertical voids.

Horizontal brine movements are facilitated because horizontal stresses in salt average only 80% of vertical stresses. Waters entering laterally ascend as artesian brines rising through cracks and joints to find a hydrostatic equilibrium. Vertical movements cause crystal growth at right angles to bedding planes, such as herringbone (chevron) crystals of gypsum or halite druses stretched out vertically, separated by individual cylindrical walls. Brines percolating before the precipitate has been consolidated lead to recrystallization; brines percolating later produce pseudomorphs, crystal molds, and casts.

Undersaturated brines dissolve sylvite or carnallite under concurrent salting out of halite. They leave behind a recrystallized barren zone, called a "salt horse." Descending brines produce "roof horses" or "back horses"—barren areas along the hanging wall. Ascending brines produce "floor horses," barren areas along the footwall. Waters rising through an evaporite deposit saturate quickly and produce brine chimneys or pipes, salt horses of white halite rimmed by clays, organic films, and dolomites; all banding is destroyed.

Chloride-saturated hygroscopic brines lower the temperature range at which anhydrous minerals are stable. They extract the crystal water from gypsum at ambient temperatures and convert it to anhydrite. Rising geostatic temperatures at depth also convert gypsum to anhydrite. Gypsum is thus rare at depth, but common near the surface where anhydrite has been rehydrated by atmospheric humidity or groundwater.

Dehydration of gypsum increases the pore pressure; the combined volume of anhydrite and released water is more than 10% greater than that of the original gypsum. This pore pressure is the driving force to dissipate the waters and with it the trace elements leached out of crystal lattices. Overlying beds commonly show undisturbed laminations indicating that gypsum dehydration happened before these became consolidated.

Determinations of the transition temperature from gypsum to anhydrite cover a wide range with clusters around 38–42°C and 55–58°C, possibly because gypsum from different sources contains different lattice imperfections and surface-active organic matter.

Meteoric waters are required for a conversion of anhydrite to selenite, a variety of gypsum. Twinned selenite crystals, with fishtail fins pointing up, are taken to have grown into an open brine column as subaqueous exit points of slowly moving interstitial brines. A porphyroblastic texture denotes very early volume-for-volume replacement under near-equilibrium conditions, while an alabastine texture forms at a later stage. Rehydration generates a large amount of heat that must be rapidly disseminated, lest it brings the interstitial brine to the boiling point.

Primary K-, Mg-, or Na-sulfates do not occur in marine evaporite sequences. Worldwide, however, all Permian and Neogene potash deposits have largely been altered to secondary complex K-, Mg-, or even Na-sulfate minerals (Table II). Ca-Mg-sulfate minerals do not exist. Evaporite beds of other ages are similarly affected only where they have been tectonically disturbed and opened to meteoric water influx (for example, Cambrian evaporites in the Salt Range, Pakistan, but not undisturbed coeval ones in Siberia). A seawater incursion or a conversion of basal gypsum beds to anhydrite provide insufficient amounts of sulfate ions. Each meter of resulting K-Mg-mineral would have required many hundred meters of basal gypsum to generate the needed sulfatic waters; such thicknesses were not available.

An early exposure of gypsum shelves to meteoric waters could have supplied the required amounts of Ca^{2+} and SO_4^{2-}. Pre- and post-glacial oscillations in sea level exposed the gypsum shelves to runoff. Even a slight dissolution of marginal halites nearly quadrupled the gypsum solubility. Where waters, saturated for $CaSO_4$ and containing some NaCl, entered along clay partings and anhydrite laminae into incompletely consolidated evaporites, they soon reached potash seams that occurred on the shelf and slope.

A transformation of primary sylvite and carnallite into K-Mg-sulfates proceeds along a complex path of multiple temperature-dependent alterations. Either kainite or langbeinite develops, of which the latter can rehydrate to picromerite or leonite nests. Local nests of aphthitalite, loeweite, vanthoffite, d'ansite, and finally bloedite and thenardite indicate a growing surplus of sodium ions and further leaching of potassium. Multiple retrograde crystallizations are also known.

Anhydrite beds are converted into polyhalite or kieserite, if $MgCl_2$ solutions are present, or into syngenite or (rarely) goergeyite if not. Minimal amounts of kieserite occur also in Pennsylvanian, Triassic, and Eocene evaporite basins. Kieserite probably forms by the interaction of anhydrite laminae with percolating $MgCl_2$ brines, or by magnesite decomposing in hot $CaSO_4$-bearing brines. Kieserite corrodes anhydrite or forms pseudomorphs after it, and is more common at basin-margin sites—which are precisely the sites of gypsum precipitation. Kieserite and polyhalite thicken and become more frequent toward shores, just as their anhydrite precursor. Both contain little anhydrite and do not possess clay inclusions even where adjacent to clay-bearing anhydrite. Finally, both minerals fluoresce, a property not shared with primary salt minerals.

J. Usiglio published in 1849 a now famous and commonly repeated experiment of evaporating seawater in his well-oxygenated laboratory. He coprecipitated first iron oxide and gypsum, then halite and $MgSO_4$—probably as epsomite, because kieserite cannot be produced at temperatures below 110°C. In anaerobic brines, iron remains dissolved to late stages of concentration, and halite does not normally contain about 5% $MgSO_4$.

Reduced anhydrite/halite ratios in both Permian and Neogene evaporites, compared to other ages, confirms a loss of anhydrite. The volume reduction by sulfatization and progressive potassium removal leads to very contorted bedding planes and intraformational breccias. $CaCl_2$ brines derived from formation waters destabilize the secondary minerals and revert them to a new generation of chlorides and salted-out anhydrite.

Sulfatization of carnallite liberates some $MgCl_2$ and bischofite, poorly soluble in sulfatic brines, precipitates. It occurs in several Permian potash horizons, but has hitherto been detected in Neogene ones only in Sicily. Bischofite also forms during tachyhydrite disintegration.

A byproduct of the generation of a hypersaline brine is the eventual production of Mg-rich, SO_4-deficient bitterns that seep into the subsurface. Magnesium is utilized to alter the swept-in continental clays into Mg-chlorites with brucite pillars or to dolomitize surrounding limestones. With decaying organic matter, especially amines, present, magnesium ions can turn a $CaCO_3$-substrate into dolomite. All carbonate intercalations, most surrounding reefs and downdip limestone banks show evidence of extensive secondary dolomitization by a brine that was devoid of sulfate ions and probably carried Mg-organic complexes.

These dolomites possess a fine to coarse crystallinity and a sucrosic texture (in other words, all crystals are nearly euhedral and nearly of equal sizes). The crystals are smallest near the source of Mg-rich brines and become coarser along a paleohydraulic gradient. Dolomites are also found near the interface between groundwater discharge and seawater influx as cryptocrystalline soft paste that eventually hardens into a very fine-grained rock, quite different in appearance from sucrosic dolomites that are secondary after limestones. An elimination of $CaSO_4$ from the brine seems to be an essential step in dolomite formation in either case.

Primary hematite-bearing carnallite converts into a red sylvite by leaching of $MgCl_2$, at nearly 80% loss in volume. In turn, sylvite can be carnallitized at a 475% volume increase; carnallite commonly forms a rind on a sylvite core. Recrystallization of a red hematite-bearing sylvite displaces impurities to crystal boundaries; in carnallite it expels lattice substitutions.

Rinneite can occur in the distal part of a basin as a nest in white sylvite; it is then commonly overgrown with sylvite, and contains carnallite, anhydrite, halite, or fluid inclusions. It never accompanies pure carnallite, but may display pseudomorphs after it. This suggests a recrystallization in sterile anoxic brines of very low pH (<4.6). Erythrosiderite, douglasite, and rokuehnite seem to be rare later alteration products.

K. Salt Domes

Salt domes or diapirs are stocks of recrystallized halite, capped by gypsified anhydrites and dolomites. Examples include salt domes along the margin of the Gulf of Mexico, the Arctic Archipelago of Canada, the southeastern margin of the North Sea, and eastern Iran. A Quaternary downwarp of the Mediterranean Sea produced diapirs of Upper Miocene salt in the northwestern Balearic Sea and along the southern slope of the Mediterranean Rise in the Levantine Sea, but not in the deepest parts of either of these salt basins or the Ionian Sea. Deeply buried salts in the Silurian Michigan and Ohio Basins, the Devonian Elk Point Basin of western Canada, and many other intracratonic basins have not been mobilized to pierce overlying strata.

Halite becomes ductile under 12 km of overburden, sylvite under 10 km, or less if exposed to tectonic stresses or to a regional downwarp. Mobilized salt first produces salt pillows that later grow into swarms of salt domes. Overlying beds are commonly pierced, upended, and cut by normal faults dipping into the salt stocks. Meteoric waters entering along fault planes can produce a salt karst. Bacteria can also enter, scavenge some of the available organic matter, and transform the gypsum in the caprock into native sulfur.

Gypsum, dolomite, and shale intercalations wedge toward the basin center; their aggregate thickness increases toward the margin. The more competent dolomite, shale, and anhydrite fractions combine in rising salt domes into caprocks thinning away from ancient basin margins.

Bedded evaporites are unknown in metamorphic rock suites. Chlorides and sulfates are squeezed out well before pressures and temperatures rise to recrystallize silicates. Only fluid inclusions in the metamorphic rocks then bear witness to the former presence of evaporites. Many tectonic glide planes in both the Alps and the Proterozoic Grenville Province in Canada show evidence of lubrication by traces of anhydrite.

Bibliography

Borchert, H., and Muir, R. O. (1964). "Salt Deposits; The Origin, Metamorphism and Deformation of Evaporites." Van Nostrand-Reinhold, Princeton, New Jersey.

Braitsch, O. (1971). "Salt Deposits, Their Origin and Composition." Springer-Verlag, Berlin.

Mattox, R. B., Holser, W. T., Ode, H., McIntire, W. L., Short, N. M., Taylor, R. E., and van Siclen, D. C. (1968). Saline deposits. *Spec. Pap. Geol. Soc. Am.*, No. 88.

Sonnenfeld, P. (1984). "Brines and Evaporites." Academic Press, Orlando, Florida.

Sonnenfeld, P., and Perthuisot, J. P. (1989). Brines and evaporites. *28th Intl. Geol. Congr. Short Course Notes* 3.

Warren, J. K. (1989). "Evaporite Sedimentology. Importance in Hydrocarbon Accumulation." Prentice-Hall, Englewood Cliffs, New Jersey.

Warren, J. K., and Kendall, C. St. C. (1985). Comparison of sequences formed in marine sebkha (subaerial) and salina (subaqueous) settings; modern and ancient. *AAPG Bull.*, 69(6), 1013–1023.

Zharkov, M. A. (1981). "History of Paleozoic Salt Accumulation." Springer-Verlag, New York.

Zharkov, M. A. (1984). "Paleozoic Salt Bearing Formations of the World." Springer-Verlag, New York.

Glossary

Cryophile Solute diffuses into cooler surface waters.

Endorheic or closed basins Basins with insufficient drainage.

Eutectic point Both solvent and solute solidify.

Eutonic point Brines dry out and the solute crystallizes.

Pycnocline Layer of density changes.

Soret effect (thermal diffusivity) Solute flux due to a temperature gradient.

Thermophile Solute diffuses into warmer bottom brines.

Faults

Kenneth Watkins Hudnut
California Institute of Technology

Faults are fractures in the earth's surface across which a finite displacement has accumulated through time. They are relatively weak surfaces that serve to localize tectonic strain. Faults form along boundaries between plates of the earth's lithosphere, both in the ocean basins and in the continents. Most of the earth's active faulting, and its related earthquake activity, is concentrated near plate boundaries, although earthquakes also occur within the stable cratons. Geologists use information on the orientation and sense of displacement along a fault and on the amount and timing of displacement episodes to investigate prehistoric tectonic motions. This approach can yield reconstructions of either large motions occurring over millions of years or details in the record of earthquakes during the past few thousand years along an active fault. Such studies have been crucial to testing plate tectonic theory and quantifying earthquake hazards.

I. Introduction

Faulting of the earth's crust is of interest to the pragmatic scientist because the process often generates destructive earthquakes; hence there is a societal imperative to understand the process. Much of the earth's population is concen-trated along belts of earthquake activity, and in the future increasing numbers will surely continue to crowd these hazardous regions. It is important to estimate regional hazards and to describe uncertainties in these estimates. Planning and regulation of construction need to consider earthquake hazard in order to avoid repetitions of the recent disasters in Mexico City and in Soviet Armenia.

Study of faulting has also led to important advances in plate tectonic theory and in documentation and analysis of relative motion between the earth's great tectonic plates. Furthermore, the association of earthquakes with surficial tectonic faulting occasionally provides dramatic evidence of how tectonic geomorphology develops. One can combine observations of a single surface-rupturing earthquake with models of repeated occurrences over long time intervals to model the development of entire mountain ranges.

This article mostly examines these prominent aspects of active faulting, emphasizing the major contributions to plate tectonics and the analysis of earthquake hazards. The reader should note, however, that this topic forms only one specialty within an expansive range of fault studies. Much of what is known or inferred about the faulting process has been learned instead through investigation of ancient, exhumed faults. Inactive uplifted faults expose deep structural levels of the earth's surface, showing structures that form when faults move at higher pressures and temperatures. From this, one can estimate the mechanisms and stress conditions responsible for the observed active structures. Also, study of inactive faults can provide evidence toward understanding previous tectonic motions.

II. Mechanics and Dominant Styles of Faulting

The style and pattern of faulting reflect both the tectonic stress acting on a region and the material that is being faulted.

219

Fig. 1 Example of surface rupture along a strike–slip fault. The 1940 Imperial Valley earthquake produced about 6 m of lateral offset near the U.S.–Mexico border. Here it is shown traversing an orange grove just north of the border. Motion on this fault is part of the distributed deformation along California's continental transform fault system, dominated by the San Andreas fault. (Photograph by C. R. Allen.)

Oceanic crust and continental crust behave differently, for instance. Also, crust containing inactive faults will generally tend to respond to new stresses by reactivation of the older generation of faults.

Global motion of large pieces of the earth's crust relative to one another is known to occur, as is more localized motion of plates with respect to the underlying mantle. This motion is thought to be driven by large-scale flow in the mantle. In both the continents and oceans, the crustal material yields at higher deviatoric stress than does the underlying mantle. As a result, the earth's crust acts as a stress guide. Often, at structural breaks in this crustal stress guide, such as at the interfaces of continental crust and oceanic crust, a fault or broad zone of deformation may develop. Faults serve to localize tectonic strain, since a fault surface is weak compared to the surrounding rocks. The relative motion of large plates occurs almost entirely in a series of active belts surrounding the earth. In between these belts the crust behaves relatively rigidly, while supporting and transmitting stress across the large distances between plate boundaries. [See LITHOSPHERIC PLATES; OROGENIC BELTS.]

Stress may be represented by a tensor, and a fault can generally be thought of as planar. For any orientation of the stress tensor in the earth's crust, one can readily envision how

Fig. 2 Example of surface rupture along a reverse fault. The 1988 earthquake in Soviet Armenia involved both reverse and strike–slip motion. Here, northwest of the town of Spitak, the surface break involved a large reverse component. (Unattributed photograph.)

that stress would be resolved onto a given fault plane. If there is a resultant shear stress across a fault, then there is impetus to cause a displacement. Thus, the stress tensor orientation and the orientation of existing fault planes determine the mode of strain within the crust. In places where either exceedingly homogeneous crustal material exists or pervasive fractures of any conceivable orientation exist in the crust, crustal strain occurs as displacements on Coulomb fractures. The mechanics of such systems can be explained by the Anderson theory of faulting.

E. M. Anderson's theory states that, because axes of the stress ellipsoid are orthogonal and there may never be shear stress at the earth's surface, there should be one vertically oriented stress axis. Faulting will be favored on conjugate Coulomb fracture planes that intersect along the intermediate principal stress axis. As a generality, this view can be useful for broadly categorizing styles of deformation. Deforming

regions can often be simplified as dominantly extensional, compressional, or transcurrent. Each of these regimes relates simply to one of the cases Anderson portrayed.

In a transcurrent or transform zone, strike–slip faulting tends to dominate. An example of a strike–slip rupture from California is shown in Fig. 1. Regions of tectonic compression, such as subduction zones and intercontinental collision zones, are most often dominated by thrust and reverse faulting. An example of a reverse-oblique rupture from the Caucasus Mountains of Soviet Armenia is shown in Fig. 2. Along midocean ridges and where continental extension occurs, as in the Basin and Range Province of the United States, normal faulting dominates, as seen in Fig. 3. [See CONTINENTAL DEFORMATION; CONTINENTAL RIFTING.]

Faulting of the oceanic crust demonstrates both successes and limitations of this theory. Midocean ridges are commonly paralleled by faults with orientations and senses of

Fig. 3 Example of the fault surface exposed in a fresh break along a normal fault. The 1954 Fairview Peak–Dixie Valley earthquake sequence in Nevada produced mostly vertical displacements. As seen here from striations etched in the fault surface, however, some oblique slip also occurred. (Photograph by H. Benioff.)

displacement just as predicted by Anderson's theory for an extensional setting. Similarly, transcurrent slip on faults between midocean ridges is predicted for the situation of a vertically oriented intermediate stress axis. At subduction zones, shallow-dipping thrusts occur. Essentially, all of these common observations are as predicted. [*See* Oceanic Crust.]

Yet, in J. Tuzo Wilson's explanation of the interrelation of all these features, he noted that the Anderson theory fails to explain continuity between the earth's mobile belts. In order to do so, Wilson pointed out, one must consider crustal discontinuities and "large areas of crust must be swallowed up in front of an advancing continent and re-created in its wake." It was necessary to explain faults that "transformed" ridges to subduction zones, for example. Wilson did so by invoking the creation and destruction of crust and by allowing the sense of slip on tear faults between these zones of divergence and convergence to be determined by relative

motion of the plates involved. Thus, although Anderson's theory may explain any one fault, it does not do well in broad application to relative motion of the earth's plates.

Many mobile tectonic belts in the continents are not easily categorized with Anderson's model. Instead, these belts involve a combination of faulting styles. In Fig. 4, for example, an earthquake rupture in Mongolia is shown to involve both large strike–slip and reverse components. Similarly, strike–slip and reverse slip both occurred coseismically in the October 1989 Loma Prieta earthquake south of San Francisco, California. Interestingly, large-scale faults in regions of oblique deformation often display partitioning between faulting styles rather than faults with oblique slip. [*See* Plate Tectonics, Crustal Deformation and Earthquakes.]

Furthermore, active faults in extensional and compressional regions often develop as delaminations between layers of varied strengths. An example of this behavior occurs in some continental crust, in which there lies a relatively weak layer between about 10 and 15 km depth. Where such a situation occurs, a double-tiered stress guide may develop with the material between layers shearing easily and eventually forming a thick subhorizontal shear zone (e.g., Fig. 5). Such shear zones can develop in extensional as well as compressional deformation of the continental crust and may also occur below broad transform zones in the upper crust.

Fault zones that cut through the upper lithosphere occur in all types of tectonic settings. In extensional regions, block faulting in the upper crust can be underlain by a shallow-dipping shear zone that widens with depth as it passes through the lower crust. Such a style of faulting is seen in exposures called *metamorphic core complexes*, as shown in Fig. 5. Such exhumed mid- to lower-crustal shear zones are also exposed throughout the Basin and Range along the margins of the Colorado Plateau. Similarly, compressional and transcurrent boundaries often show a transition with depth from brittle faulting in the upper crust to increasingly plastic deformation with greater depth.

The transition in fault behavior with depth is related to changes in temperature, pressure, rock type, fluids, and strain rate. In the shallowest few kilometers, faults can undergo stable sliding commonly called *creep*. This is usually associated with a well-developed zone of weak fault *gouge*, which is very finely pulverized rock that develops a weak fabric through continuous shear. Between depths of a few kilometers and about 15 km, the fault typically is in an unstable regime characterized by stick–slip behavior. This is the depth range in which crustal earthquakes typically initiate and propagate, (although some earthquakes initiate and propagate as deep as 700 km in oceanic subduction zones). Fault zone rocks from these depths undergo *abrasive wear*, leading to development of *cataclastite*, which is coarsely pulverized rock often shot through with veins and typically lacking any well-

Fig. 4 The displacement vector commonly varies along a fault surface rupture. For example, at the location shown here the Gobi–Altai earthquake of 1957 was associated with more reverse slip (4 m) than strike slip (2 m). At another location, the surface break was purely strike–slip and involved nearly 9 m of offset. Such variations often can be related to discontinuities such as major stepovers or to significant changes in orientation of the fault surface. [Photograph by N. A. Florensov. Reproduced from N. A. Florensov and V. P. Solonenko, eds. (1963). "The Gobi-Altai Earthquake." Academy of Sciences of the USSR, Moscow.]

developed fabric. Deeper still, and perhaps extending down past 20 km in some cases, the fault rock undergoes *adhesive wear*. At these depths, and increasingly at greater depths, plastic flow of the rock occurs and rocks in the shear zone develop a strong fabric called *mylonite*. These mylonites, when exposed at the surface, are the vestiges of mid- to lower-crustal shear zones (as shown in Fig. 5).

Much of the current theory of fault mechanics is based on rock friction. A fault is conceptualized as two rough surfaces in contact, with stress concentrations at contact points. In the intermediate depth ranges where large crustal earthquakes occur, unstable friction results in stick–slip. That is, the surface contacts become stuck until enough stress accumulates to shear through them. This is the simple mechanical

basis for the earthquake cycle. Note that in this model only the load orthogonal to the fault and the real area of the fault in contact (which depends on the normal load) are important factors for determining friction and, hence, fault strength.

III. Earthquakes as Increments in the Faulting Process

The examples of surface faulting shown in Figs. 1–4 demonstrate a clear process by which faulting affects geomorphology. Through time, repeated earthquakes along these faults will progressively displace juxtaposed sides of the fault. For

Fig. 5   At deep levels in the continental crust, shear zones can develop between upper and lower crustal stress guides. Such shear zones can be several kilometers thick and can occur in extensional, compressional, or transcurrent continental plate boundaries. In rare cases, as shown here, deep levels are rapidly uplifted and exposed at the earth's surface. This photograph shows the several-kilometers-thick extensional shear zone on Naxos Island in the Greek Cyclades. The shear zone developed during post-Alpine orogenic collapse and has since been exhumed through ongoing extension. (Photograph by the author.)

strike–slip faults, linear features such as stream channels crossing the fault will be diverted along the fault trace, forming a kink in the channel. Normal and reverse faults, because of the vertical component of slip, will develop pronounced escarpments through time. Impressive mountain range fronts such as the eastern Sierra Nevada in California and the eastern face of the Teton Range in Wyoming are the result of repeated earthquakes over a long time interval. We can therefore consider each earthquake as producing an increment of displacement and the summation of these increments through time as producing the tectonically induced landforms so commonly seen along active fault zones.

Consideration of the *earthquake cycle* is also useful for quantitative estimation of earthquake hazard. Models of the earthquake cycle are used to infer the time of the next large earthquake along a given fault segment, based usually on very little data. Three models bearing on the earthquake cycle are widely applied. First among these is H. F. Reid's perfectly periodic model, in which the interval between events is always the same and each earthquake has the same displacement vector as the last and relieves the same amount of stress. Variations on this are special cases in which either the failure stress is always the same (*time-predictable* model) or earthquakes always drop the stress to the same value (*slip-predictable* model). The time-predictable model is used in earthquake forecasting when the slip in each event and the time since the last event can be estimated, but there is no record to allow estimation of an *average recurrence interval*. [See EARTHQUAKE PREDICTION TECHNIQUES.]

Strongly related to models of the earthquake cycle are

models of slip distribution along a fault. Some such models allow the long-term slip at a given site to accumulate through many events to the same value as anywhere else along the fault. Other models hold the slip at a given site to be the same in every earthquake, so that variable amounts of displacement accumulate through time along a fault. Another important issue in considering the earthquake cycle is the influence on a main fault of coupled or independent stress cycling on intersecting or nearby secondary faults.

Most of the recent research on and discussion of the earthquake cycle has emphasized reasons to be skeptical of the simplest models. This skepticism arises because there has been no credible occurrence of periodic fault behavior. It remains unclear how best to model the few existing well-documented observations of nonperiodic fault behavior.

In the face of such problems, perhaps it seems overzealous to attempt even long-term earthquake forecasting. In some cases in which data are relatively good, however, it is possible to make earthquake forecasts that are necessarily couched in a probabilistic estimate of the errors involved. To formulate earthquake forecasts, one calculates a conditional probability from a probability density function representing the estimated recurrence interval and associated error. Serious limitations of this approach can be pointed out at all levels. First, as previously discussed, models of the earthquake cycle vary from extremely simple, perfectly periodic repetitions of ruptures along the identical fault segment all the way to extremely complicated, nearly aperiodic or chaotic patterns of ruptures that have uneven recurrence times and frequently overlap along the fault or are strongly influenced by ruptures on neighboring faults. Second, it is not clear how best to estimate errors in determining parameters such as the displacement vector in each earthquake or the average interval of time between earthquakes. And finally, it is difficult to decide what type of statistical distribution should be used to represent these uncertainties. Despite shortcomings, such earthquake forecasts are being attempted for areas like California, Alaska, and Japan, where we have at least fragmentary data on sizes and dates of previous earthquakes along the major faults.

Bibliography

Cox, A. (1973). "Plate Tectonics and Paleomagnetic Reversals." Freeman, San Francisco.

Reid, H. F. (1910). The mechanism of the earthquake. *In* "Report of the State Earthquake Investigating Committee, the California Earthquake of April 18, 1906," pp. 1–192. Carnegie Institution of Washington, Washington, D.C.

Scholz, C. (1990). "The Mechanics of Earthquakes and Faulting." Cambridge Univ. Press, London.

Suppe, J. (1985). "Principles of Structural Geology." Prentice-Hall, Englewood Cliffs, New Jersey.

Vita-Finzi, C. (1986). "Recent Earth Movements." Academic Press, Orlando, Florida.

Wallace, R. E., *et al.* (1986). "Active Tectonics." National Academy Press, Washington, D.C.

Glossary

Displacement Vector defining offset on a fault that may represent slip in a single event or the accumulation of slip over a much longer time.

Normal fault General term for a fault dipping between vertical and near-horizontal, and with rock above the fault (the *hanging wall*, on which coal miners would hang their lamps) moving downward with respect to rock below the fault (the *foot wall*, similarly, derived from the mining term).

Paleoseismology Study of prehistoric earthquakes, often by excavation across an active fault to examine evidence of past surface-rupturing events.

Reverse fault General term for a fault dipping between 30° and vertical, and with the hanging wall moving up with respect to the foot wall.

Separation Single component of the fault displacement vector, usually either vertical or horizontal.

Strike–slip fault General term for a vertical fault with horizontal displacement vector, producing lateral relative motion of the rock on either side.

Thrust fault General term for a fault dipping between horizontal and 30°, and with the hanging wall moving up with respect to the foot wall.

Transcurrent fault One class of vertical faults with a horizontal displacement vector that *does not* link two or more plate tectonic scale boundaries.

Transform fault Another class of vertical faults with a horizontal displacement vector that *does* link two or more plate tectonic scale boundaries.

Floods

R. Craig Kochel
Bucknell University

Floods occur when the height or stage of a body of water rises above some datum designated as the normal water level. Floods result from a wide variety of causes such as snowmelt, landslide dams, ice blockages, precipitation, and tidal and storm surge along coastal regions. This article will focus on floods in riverine environments, defined as flows that overtop the river banks (overbank flow) and spill onto the valley bottom or floodplain. Floods along rivers are complex and interact with a variety of other systems. This discussion will be on floods produced by precipitation and will emphasize aspects of flood geomorphology, including (1) flood generating processes, (2) flood erosional and depositional processes, (3) geomorphic work done by floods, (4) paleofloods, and (5) interrelationships between flooding and global climate change.

I. Introduction

A. Extreme Flood Events Perspective

The largest flood episode known on earth occurred between 17,000 and 12,000 years before present in the northwestern United States — the catastrophic failure of Pleistocene Glacial Lake Missoula. Erosion from these Missoula floods carved an impressive array of channels and left behind extensive deposits of flood sediment in eastern Washington, an area known as the Channeled Scabland. One of the greatest geological controversies raged over the origin of the Scabland channels for decades following the seemingly outrageous hypothesis of J Harlen Bretz in the 1920s that the channels were formed by cataclysmic floods unlike any experienced in historic times. The Scabland channels are readily visible from spacecraft imagery of the earth. Viking orbiter photographs of Mars taken in the late 1970s revealed evidence of extensive floods earlier in that planet's history.

Interest in floods has seen an exponential increase during the last decade because catastrophic flooding is now viewed as part of the uniformitarianistic perspective of geomorphic processes and is significant in shaping fluvial landscapes. Three major texts on floods have appeared since 1987. The last major text on riverine flooding was written in 1955 by W. G. Hoyt and W. B. Langbein, entitled "Floods."

B. Frequency of Flooding

Research during the 1950s and 1960s demonstrated that on the average, bankfull flows (discharges just at the verge of spilling onto the floodplain) occur in rivers about once every 1.5 years. The concept of flood frequency illustrates the chances or probability that floods of different magnitude will occur over a specified period of time. Flood frequency is the basis for managing floodplain development and setting standards for design of engineering projects such as culverts, bridges, and dam spillways.

1. Flood Frequency Curve

Flood frequency is the probability of a selected discharge being equaled or exceeded in any one year. A flood frequency curve (Fig. 1) is a cumulative distribution that relates flood magnitude (discharge) to the frequency of its occurrence. The reciprocal of the flood probability is the recurrence interval (normally expressed in years). The recurrence interval (R.I.) of a flood is determined by $1/P = $ R.I., where P is the probability of a flow being equaled or exceeded in any year. The probability of a flood of x years R.I. being equaled or exceeded over an n-year period is computed by the relation $1 - [1 - 1/x]^n$. For example, a flow having a 50% chance of being equaled or exceeded any year has a recurrence interval of 2 years. The area inundated by this flood would be defined as the 2-year floodplain. The chance of experiencing the 100-year flood during any year is 1%. As the recurrence interval of a flood increases, its chances of occurring during any year are reduced. However, it is important to remember that its chances of occurrence are never zero.

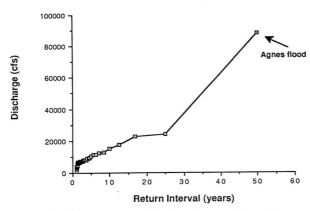

Fig. 1 Flood frequency curve for the Conestoga River at Lancaster, Pennsylvania, for 1929–1977. (Data from the U.S. Geological Survey.)

2. Frequency Analyses and Problems

Methodologies for constructing flood frequency curves fall into two basic categories: (1) theoretical distributions and curves derived (plotted) from mathematical relationships, and (2) graphical curve fitting using no predetermined distribution. Both approaches utilize the record of annual maximum peak flows from a river gaging station to develop the frequency curve. Numerous types of distributions can be selected for application of the annual flood series to theoretical-mathematical models. The most common distributions employed are the log Pearson III method (a three-parameter distribution adopted by United States governmental agencies) and the Gumbel extreme value theory. These methods assume that floods of variable magnitude occur as populations explained by one of these distributions, thereby permitting extrapolation to large-magnitude events beyond the observed record.

The graphical approach assumes no special frequency distribution and simply assigns a flood ranking according to magnitude when annual flows are plotted according to the formula:

$$\text{R.I.} = (N + 1)/M$$

where R.I. is recurrence interval (years), N is number of years of record, and M is magnitude rank of floods. Extrapolation from frequency curves developed using the graphical method is difficult beyond the length of the period of direct observations at the gaging station.

Depending on the method selected for flood frequency analysis, considerable differences can be obtained in the estimates of recurrence intervals for specific flood discharges and in the estimates of discharges associated with floods of specified frequency (Fig. 2). These problems generally arise from either the short period of direct measurement (generally less than 100 years) or from the fact that floods result from complex generating processes, each process having its own distinctive population and associated frequency characteristics.

C. Relevance of Flood Studies

1. Flood Geomorphology

Interest in the scientific study of floods has greatly increased since 1970. Prior to that time, most fluvial studies focused on the mechanics of bankfull or lower-magnitude flows. It is impossible to separate studies of flood processes from the landscape affected by the flood. The science of flood geomorphology attempts to integrate studies of flood causes and flood processes and to assess the role of floods in shaping the landscape through time.

Over the past few decades, geomorphologists have amassed observations that suggest prediction of the effects of large floods is exceedingly complex. The amount of flood erosion, flood sedimentation, or landscape change induced by a flood depends on many interconnected factors such as (1) the type of sediment carried by the river (its load), (2) the geometry of the channel, (3) the size of the drainage area contributing to the flood, (4) the stresses exerted on the stream bed by the flowing water, (5) the nature of the channel banks and associated vegetation, and (6) the temporal ordering of flood events. Recent studies have shown that riverine landscapes can preserve a record of past floods. This information becomes critical in risk estimation and application to environmental hazards. [*See* RIVERS.]

II. Framework of Flood Generation

Floods are generally triggered either by episodes of intense and/or long-duration rainfall or rapid melting of extensive snow and ice accumulations. Our discussion of flood-generating processes will focus on rainfall and associated runoff.

A. Global Flood Climates

The atmosphere serves as the reservoir for floodwater associated with rainfall-runoff floods. The available water in the atmospheric column combined with the intensity and duration of rainfall determine the flood potential for any given region. From a global perspective, there are three major types of flood reservoirs: (1) circum-tropical barotropic air masses, (2) mid-latitude baroclinic air masses, and (3) circum-polar air masses. [*See* ATMOSPHERIC CIRCULATION SYSTEMS; WATER CYCLE, GLOBAL.]

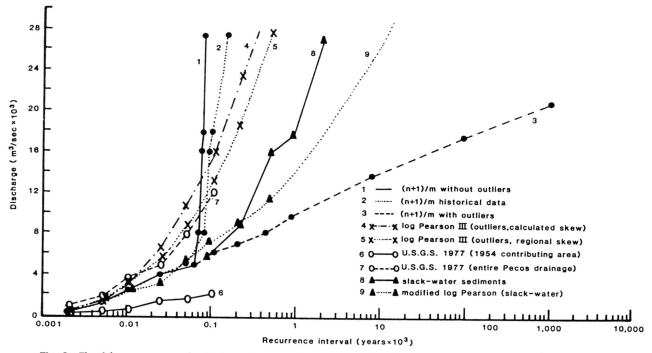

Fig. 2 Flood frequency curves for the Pecos River near Langtry, Texas. Note the wide range of interpretations of the data possible using conventional analysis techniques. Curves 8 and 9 have been developed using a 10,000 year record of paleofloods along the river. [From R. C. Kochel and V. R. Baker (1982). Paleoflood hydrology. *Science* **215**, 353–361. Copyright © 1982 by the AAAS.]

Barotropic conditions are typified by storm systems that obtain their energy from the latent heat of evaporation, usually over warm oceans. Convective thunderstorm activity predominates in these settings, evidenced by the omnipresent bright band of high clouds along the equatorial Intertropical Convergence Zone visible in satellite images. Convectional thunderstorms result in localized, intense rainfall, which is effective in generating large-magnitude floods in small drainage basins. Large cyclonic storms of tropical origin have produced some of the greatest floods from rainfall known on Earth. Recent examples of tropical storm-induced floods in the United States were the flood from Hurricane Camille in mountainous central Virginia in 1969 (over 150 lives lost) and the regional mid-Atlantic states flooding in June 1972 from Hurricane Agnes rainfall, which caused damages of over $3.5 billion and the loss of 118 lives.

Due to the dramatic differences in temperature and pressure between air masses in the mid-latitudes, extratropical cyclones tend to form along fronts that delineate the boundary between polar and tropical air masses. These cyclones typical of baroclinic mid-latitudes are predominantly fueled by the mechanical movement of air masses from regions of high pressure to areas of low pressure and act as large sole-noids. The intersection of temperature and pressure gradients from contrasting air masses creates powerful wind circulation systems. Significant rains can be produced along these frontal zones when tropical moisture is pumped over the airmass boundary into the extratropical systems. Typically, however, large extratropical cyclones produce lower-intensity precipitation widespread over large drainage areas and result in more regionalized flooding. [*See* ATMOSPHERIC FRONTS.]

Polar air masses contain less moisture. Rarely do these systems have access to moist air masses sufficient to generate precipitation capable of producing large floods. Polar region floods are usually associated with rapid and widespread snowmelt events and/or ice jams along rivers.

In cold regions, typically termed *nival* climates, the hydrologic cycle is predominantly influenced by the accumulation of snow and ice and its associated meltwater runoff. Rainfall-generated floods are less common and usually of low magnitude due to the infrequency of incursion of moisture-laden circulation systems into high-latitude and high-altitude regions. Nival floods are typically seasonal, associated with spring and summer melt, and often have protracted duration due to the mechanics involved with melting the snow and ice and concentrating runoff into downstream channels. Additionally, maximum flood peaks associated with runoff are typically lower per unit drainage area than the flashier floods produced from intense precipitation in mid-latitude temperate and low-latitude tropical environments. [*See* GLOBAL RUNOFF.]

The two primary processes responsible for large-magnitude and damaging floods in cold regions are ice blockage and the catastrophic failure of proglacial lakes. Ice jams are common along major rivers during spring runoff. Ice blockades can result in significant stage rises upstream followed by catastrophic floods downstream upon their disintegration. Resulting flood waters can be exceptionally erosive because the flows carry large ice blocks capable of scouring floodplains, destroying vegetation, and damaging structures.

Glacial floods are among the most catastrophic of flows ever recorded on earth. Most glacial floods result from catastrophic failure of temporary proglacial lakes impounded by sediment dams or ice blockades proximal to the glacier. The most catastrophic of these were the cataclysmic Spokane or Missoula Floods during the Pleistocene Epoch that carved a maze of channels in basalts of eastern Washington known as the Channeled Scablands; these channels are visible even in satellite views of the earth. [*See* GLACIAL LAKES.]

B. Flood Hydroclimatology

Flood hydroclimatology is a relatively new approach that attempts to place floods in context with their climatological context or mechanism of generation. This approach aims to provide physical connection between the flood and the processes responsible for generating rainfall. Floods occur over a wide range of spatial and temporal domains (Fig. 3). This can be explained in part because of the wide range of drainage basins and their integration, which is responsible for taking the precipitation and transferring the excess runoff through the hierarchy of a drainage system. In this manner, small basins experience floods of short duration, and large basins commonly experience longer floods proportional to their respective basin areas. However, within basins of similar size, there is likely to be a range of space–time domains associated with flooding that can be linked to meteorological and climatological processes responsible for generating the rainfall.

Most floods can be linked to immediate or local causes such as a specific convective storm cell. At the same time, the immediate cause can often be linked to a more regional or global framework of atmospheric conditions that facilitated its development. Hydroclimatology integrates the disciplines of meteorology, climatology, hydrology, geomorphology, and statistics to focus on identifying the special combinations of circumstances associated with flood events. These analyses have shown us that floods within any given region may originate from various mechanisms during different times of the year. Thus, a series of flood observations may be com-

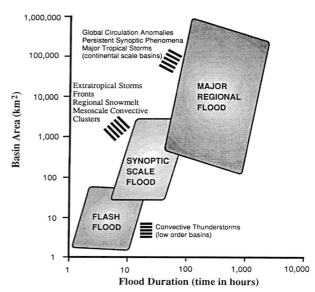

Fig. 3 Temporal and spatial domain for floods. The shaded areas represent three scales of flooding. Likely associations with atmospheric phenomena are shown with each flood category.

posed of multiple separate flood populations, each with its own magnitude and frequency relationship. Hydroclimatologic studies, in combination with analyses of the interaction between air masses and topography, promise to improve our ability to forecast flood conditions in the future. [*See* MID-LATITUDE CONVECTIVE SYSTEMS; MOUNTAINS, EFFECT ON AIRFLOW AND PRECIPITATION.]

C. Flood Catchment Characteristics and Runoff

Drainage basins are the fundamental compartments of the fluvial landscape for collecting water and sediment associated with rainfall-runoff events. Runoff occurs when the capacity of the soil and/or bedrock to allow precipitation to infiltrate is exceeded or when the soil reaches saturation, yielding runoff by saturation excess flow. Much of the infiltrated water entering is transferred as interflow to surface channels through pore spaces and macropores (large cavities from features such as fractures along adjacent soil peds, roots, or burrows) above the water table, while the remainder enters the groundwater system (Fig. 4). [*See* GROUNDWATER; SOIL PHYSICS.]

1. The Flood Hydrograph

A hydrograph shows the time history of the variation in discharge at a gaging station on a specific stream. The flood hydrograph in Fig. 5 shows the history of flood runoff from a

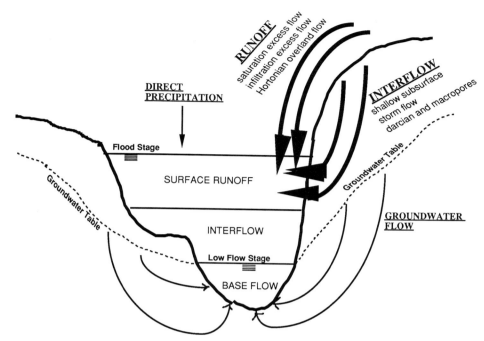

Fig. 4 Hydrologic components of a flood. Schematic shows the primary sources contributing to the flood discharge. The size and thickness of arrows relates to their respective importance. Shallow subsurface flow becomes more prominent in humid regions with deep regolith.

rainfall event in a drainage basin as a function of time. The area under the curve represents the volume of runoff once it is separated from the base flow and interflow contributed from subsurface sources during the flood. Lag time, measured between the time of the center of mass of the rainfall and the center of mass of the runoff, provides an index of the flashiness of the stream. Basins with high runoff-collecting efficiency are considered flashy because they exhibit sharply peaked hydrographs with short lag times.

The shape of a flood hydrograph is influenced by the temporal and spatial distribution of rainfall and by the physical character of the drainage basin. Sophisticated models of runoff have been developed in recent years that couple drainage basin parameters with equations designed to account for floodwater storage within the basin. These same models may be used to predict the movement of flood discharge and stage along channels through the drainage network, referred to as flood routing. [*See* HYDROLOGIC FORECASTING.]

Geomorphically, the flood hydrograph represents the summation of the effects of the physiography of the basin as these factors influence the collection, routing, and summation of runoff from a precipitation event. Unit hydrographs are generated by using a selected depth of precipitation (typically one inch) distributed uniformly over the drainage basin over a specified interval of time (typically one inch over one hour).

They illustrate the predicted runoff-time relationships for standardized precipitation events over the watershed. These effects can be observed when hydrographs of similar-sized basins within climatic environments are compared.

2. Physiographic Controls on Flooding

While the ultimate control on the magnitude and frequency of flooding rests in the meteorological factors, the physical characteristics of the drainage basin (e.g., basin shape, relief, drainage density) can exert substantial control over the concentration of runoff in stream channels. Numerous studies in recent decades have used multivariate regression techniques to demonstrate the connection between basin morphology and flood discharge. Although these studies have failed to develop a comprehensive predictive model for peak flood discharge, they have shown that flood discharge can be related to parameters such as (1) basin area; (2) ruggedness number—the dimensionless product of relief and drainage density; (3) basin magnitude—the number of first-order collector tributaries in the basin; and (4) less quantifiable aspects of soil thickness, vegetation density, and basin shape. The ultimate goal of these studies is to develop a model that can predict peak discharge from ungaged watersheds based on regression analyses of gaged basins within the same general

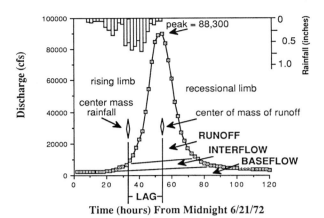

Fig. 5   Flood hydrograph for the Agnes rainfall on the Conestoga River at Lancaster, Pennsylvania in June 1972. (Data from the U.S. Geological Survey and National Climate Data Center.)

physioclimatic region. [*See* DYNAMIC HYDROLOGY OF VARIABLE SOURCE AREAS.]

III. Flood Processes

Regardless of whether the impetus for understanding floods is toward forwarding our understanding of the evolution of fluvial landscapes or aimed at the design of structures designed to pass flood flows, the understanding of the processes of flood erosion and deposition will be paramount in successfully achieving these goals.

A. Flood Flow Types

Research over the past few years has shown that flood processes are very complex and linked to aspects of the hillslope system in headwater regions of watersheds. Sediment transport in small, steep basins often mobilizes enormous quantities of sediment during floods, which results in flow phenomena different from the normal (Newtonian) water flows. The standard equations for predicting sediment transport operate only for water flows; hence, the rheology of the flood flow will regulate the kind of erosional and depositional behavior that accompanies the event.

The rheology of the flow type depends on the relative proportion of water and sediment. Flow types range from water flows to hyperconcentrated sediment flows to debris flows, in order of increasing sediment concentration. Water flows are "normal" (Newtonian) flows where energy is dissipated dominantly by turbulence. Sediment concentrations in water floods are less than 20% by volume, but often less than

1%. Water floods transport coarse bedload (sand and gravel) along channel floors by various tractive forces, including saltation, rolling, and sliding. Finer-grained materials (typically sand and mud) are distributed throughout the water column during flood as suspended load, maintained by turbulence within the flow. Debris flows occur when sediment concentrations exceed 45–50%. The fluid develops internal shear strength and behaves as a Bingham material. Debris flows may exhibit a range of flow behaviors temporally during an event and spatially along their path as they accept water and sediment from tributary flows. Because of their Bingham properties, debris flow are capable of transporting large particles on their surfaces and have extremely high impact forces. Hyperconcentrated flows fall in between and possess attributes of both ends of the flow spectrum. [*See* SEDIMENT TRANSPORT PROCESSES.]

Detailed observations of the morphology and sedimentology of flood deposits can be used to determine the nature of the flood (Table I). These distinctions are important where flood control structures and other engineering projects must be designed to accommodate floods that can have notably different characteristics.

B. Flood Erosion and Sedimentation Processes and Landforms

Research has demonstrated that it is easier to understand depositional landforms and processes compared to erosional features. This is because deposits are readily preserved after a flood and are available for study in the field. Much of the evidence of erosional processes is removed by the erosional processes themselves. Flood erosion and deposition can best be considered from a perspective of compartmentalizing river channels according to their overall framework of resistance. In this regard, erosion and sedimentation will be considered for resistant boundary channels—namely, bedrock rivers—and for alluvial channels characterized by temporally adjustable cross sections. [*See* EROSION AND WEATHERING.]

C. Flood Processes in Adjustable Alluvial Channels

Alluvial channels can respond to major floods by adjusting a large number of variables within their flow hydraulics and the geometry of their channels. This ability to adjust their cross-sectional boundaries during flood separates them from their bedrock channel counterparts whose only option is to adjust the hydraulics of their flow to accommodate changes in discharge and energy.

Numerous studies have documented that alluvial channels tend to scour their beds during rising stage and high flow,

Table I Comparison of Water Floods and Debris Flows[a]

Flow type	Characteristics	Deposits, landforms, and sedimentology
Debris flow	50% + sediment by volume; high strength; Bingham material; laminar	Terminal lobes; lateral levees (often bouldery); U-shaped channel; erosional or depositional in channel; sharp cutoff of erosion along channel margins; inverse grading or no grading; no imbrication or very weak; weak or no bedding; very poorly sorted; matrix-supported
Water flood	<20% sediment by volume; insignificant strength; Newtonian material; turbulent	Bars; minor levees; variably shaped channels; well-bedded; imbricated gravels; point bars; may be normally graded; better sorted; channel fills; clast-supported

[a] *Note:* Hyperconcentrated deposits fall in between these end-members but have more deposit affinities with water floods.

while filling generally occurs during flood recession. Many times, the net effect of transmitting a flood wave at a cross section will be negligible. Erosion of channel banks and floodplains depends on the complex interrelationship between the erosive potential of the flow and the resistance of the banks and floodplain. Floodwaters deficient in suspended sediment tend to have excessive erosive potential and may cut large chutes into floodplain surfaces, commonly across the insides of meander bends. Turbulence may be dampened when floodwaters carry large amounts of suspended sediment, which may result in deposition of sediments on the floodplain. Floodplain sediments are composed largely of clay, silt, and sand. However, cases have been observed in gravel-bed rivers where gravel lobes have been deposited on the floodplain from suspension or ramped up onto the floodplain by tractive transport processes.

Bank erosion can occur during rising flood stage by shearing forces in hydraulic action that remove sediments on a grain-by-grain basis. Alternatively, bank erosion may occur late in the flood or during flood recession by a host of mass movement processes resulting in the failure of large blocks of bank material. Bank failures may be triggered by undercutting during hydraulic activity earlier in the flood or may be related to elevated pore pressures and sapping as saturated floodplain sediments drain back into channels whose water has returned rapidly to levels below bankfull stage.

Response of alluvial channels to a given flood event can vary greatly between rivers. Low gradient streams in humid regions with luxuriant vegetation on their cohesive, fine-grained banks and floodplain surfaces may undergo insignificant adjustments during major floods. On the other hand, floods in less cohesive, sand- and gravel-bed channels with sparse vegetation in semiarid regions may experience extensive changes during a flood of similar magnitude and frequency. Observations of a major flood in Tucson, Arizona in 1983 documented extensive channel erosion and channel migration during that event. Channel migration resulted in considerable damage to homes and other structures located well above the level of flood inundation (and well above the floodplain) due to undercutting and slumping related to large-scale lateral adjustments in the planform of those desert channels.

D. Flood Processes in Resistant Boundary Channels

Channels with resistant flow boundaries are developed in bedrock or some type of extremely resistant indurated sediment. Only very recently have fluvial geomorphologists begun serious investigation of flood processes in bedrock channels. Flood flows in bedrock channels are typically narrow and deep, with dramatic increases in stage with discharge (Fig. 6). Because of their inability to adjust channel geometry, bedrock channels experience floods characterized by large adjustments in flow hydraulics. Extraordinarily high values of flow velocity, shear stress, and stream power per unit area of channel bed are achieved in these channels during large floods; values are commonly more than an order of magnitude above those experienced in alluvial channels.

In resistant channels, sediment availability is generally limited with respect to the large increases in stage. The resulting deep, swift flows contain excess energy beyond that which can be dissipated by roughness or adjustments in the immobile flow boundaries. This excessive energy typically permits the development of macroturbulent flow characterized by exceptionally intense vortex activity and resultant scour. Powerful subfluvial vortexes called kolks commonly form, which are capable of plucking enormous blocks from jointed bedrock floors. Plucking, combined with cavitation (powerful collapses due to pressure differentials developed within the flow), can exert enormous shear stresses along the boundaries of bedrock channels during major floods. Erosional features created during such flood flows are listed in Table II.

Competence values are generally high during floods in bedrock channels, while the volume of sediment transported is limited by the supply to the channel. Gravel moves through

Fig. 6 Bedrock channel of the lower Pecos River near Langtry, Texas. Floods experience dramatic increases in stage, promoting development of macroturbulence. The record 1954 flood stage exceeded 30 m along this reach and spilled out onto the plateau surface.

these channels as large-scale bars and gravel-wave bedforms (Fig. 7), sometimes being moved in suspension during exceedingly large floods. Sand flushes through the system as washload. Much of the sand is exported out of the system; however, some accumulates locally in zones where flood velocity experiences sudden deceleration. These areas of sand accumulation are referred to as slackwater deposits and are most common in (1) tributary mouth junctions, (2) at sites of major channel expansions, (3) in the lee of obstacles and meander bends, and (4) in shallow caves in channel walls.

E. Magnitude and Frequency of Flooding

Since a seminal article by M. G. Wolman and J. Miller in 1960, geomorphologists have debated the relative role of rare, large-magnitude floods versus frequent, smaller floods in shaping the landscape and in accomplishing geomorphic work. Geomorphic work was originally defined as a measure of the suspended sediment transported by a river. As such, studies revealed that many rivers (particularly those in humid regions of the United States carried as much or more sediment by cumulative high-frequency flows as they did during the uncommon catastrophic floods. Researchers working on gravel-bed rivers in semiarid settings later showed that frequent, low-magnitude floods were incapable of entraining the coarse sediment loads or modifying channel and floodplain landforms. The studies of gravel-bed rivers indicated that less-frequent large flows were required to initiate motion of the coarse-grained sediment load; thus, they were responsible for most of the geomorphic work.

This controversy becomes even more complex when issues of geomorphic effectiveness are considered. Effectiveness

Table II Flood Erosional Features

	Macroscale features	Location/occurrence
Bedrock channels	Cataracts	Mid-channel, slope breaks
	Inner-channel	Mid-channel
	Grooves	Mid-channel, lip of cataracts
Alluvial channels	Chutes	Inside meander bends
	Irregular scour	Floodplain surfaces
	Channel scour	Constrictions

	Mesoscale features	Location/occurrence
Bedrock channels	Potholes	Mid, side channel
	Caves	Outer meander bends
Alluvial channels	Bank scour	Irregular along banks

**Flood Sediment Bedforms
in Gravel-Bed and Bedrock Rivers**

Macroscale Bedforms	Location/Occurrence
point bars	meander bends
lateral bars	channel margins
pendant bars	lee of obstacles
expansion bars	channel expansions
overbank gravel lobes	floodplain surfaces
chute bars	inside meander bends

Mesoscale Bedforms	Location/Occurrence
gravel ripples	bars, floodplains
irregular gravel mounds	bars, floodplains

Sand Deposits	Location/Occurrence
slackwater deposits	tributary mouths, caves,
	lee of obstructions
sand waves	bars, floodplains
levees	channel banks

Fig. 7   Common bedforms formed by floods in gravel-bed rivers, found in bedrock and alluvial channels. (a) gravel bar with advancing foresets, (b) ripples in gravel, and (c) pendant bar formed in the lee of a flow obstacle. The table shows a range of other flood-produced bedforms common in these rivers.

refers to the ability of a flood to reshape or modify channel and floodplain landforms. Studies of floods in a variety of climates during the past few decades have illustrated that rules of thumb for predicting the effects of rare, large-magnitude floods have little usefulness. Catastrophic response to large floods is common in small, low-order watersheds where access to coarse-grained sediment is common along with high stream gradients. These attributes often combine to produce high shear stresses and high values of stream power per unit of channel bed, resulting in wholesale destruction of floodplain features (Fig. 8). Much more research aimed at quantifying the observations of river response to floods is needed before reasonably useful models for predicting response can be constructed. We do know, however, that individual rivers are likely to respond very differently to floods of similar magnitude. This suggests caution for engineering projects, because a structure designed for transmitting floods on one

river may not function well for another river of similar size, but having different geomorphic character.

Additional complexity has been added to the problem of predicting flood response in rivers from studies in the past 10–15 years. Rivers sometimes exhibit dramatically different responses to successive floods of similar magnitude. This can be explained with concepts of landscape recovery and permanence of features produced by flooding. Flood deposits or adjustments in the fluvial landscape made during floods typically are modified slowly by subsequent flows of lower magnitude until the river readjusts to a low-flow equilibrium state. If there is insufficient time between successive high-magnitude floods along a river, the later flows may have little apparent effect, because the system is still adjusted to the high-flow conditions from the preceeding floods. The time required for fluvial systems to recover from changes experienced during a major flood varies between sites, typically as a

Fig. 8 Catastrophic destruction of channel and floodplain surfaces along small stream in Nelson County, Virginia during the Hurricane Camille flood of August 1969. Channel enlargement exceeded 75% in many areas affected by upstream debris avalanching. (Photo courtesy of Virginia Division of Mineral Resources.)

function of climate. Recovery times in arid climates are generally much longer than recovery periods in humid settings. In this manner, the temporal ordering of flood events may be critical in governing the responses of a river or watershed to a rainfall-runoff event of a selected magnitude.

IV. Paleofloods

Paleohydrology is the investigation of flow characteristics of streams of the past, using combined methodologies from geomorphology, sedimentology, botany, hydraulics, and stratigraphy. Paleohydrologic research has generally followed one of two major approaches. One approach aims at reconstructing conditions of mean flow or bankfull flow for ancient rivers. The other approach focuses on reconstructing discharges of discrete flood events—called paleoflood hydrology.

Methods for reconstructing mean flow conditions of ancient channels apply one or more empirically derived relationships between stream discharge and attributes of channel geometry, meander geometry, sediment characteristics, or other features that can be observed in modern channels. This regime-based paleoflow approach appears to work best when applied to alluvial channels because of their ability to adjust their geometric configuration in accord with the forcing flow conditions. Numerous studies in the last 25 years have

presented equations useful in estimating paleoflow conditions from observations of paleochannel geometry, sedimentary architecture, and grain size.

A. Paleoflood Hydrology

Paleoflood hydrology is the science of the study of floods that have occurred prior to the time of direct measurement, using modern hydrologic procedures or documentation by historical records. Paleoflood discharges and frequency are reconstructed using the coupling of hydraulics–hydrology principles with sedimentological, geobotanical, and geomorphological methods of analyzing floodplain and channel deposits. Two basic approaches have been used in paleoflood research as the field experienced extensive development over the past 20 years: (1) Paleocompetence estimates and (2) paleostage-based slackwater deposit studies.

B. Paleoflood Stage, Discharge, and Frequency

1. Paleocompetence Estimators

Paleocompetence techniques are based on theoretical and empirical relationships between parameters relating to the ability of flood flows to entrain particles and the dimensions of the particles observed in a flood deposit. The most common relationships involve regressions between shear stress, flow velocity, and stream power and the transported particles. Values of paleoflood discharge made from these estimators of paleocompetence generally have significant uncertainty in the estimates; however, trends in recent research have narrowed these potential errors and are beginning to develop separate equations applicable to the variety of geomorphic types of rivers that exist.

The attractive feature of the paleocompetence approach is that the requirements for its use are generally preserved in the geologic record—namely, the flood sediments and sometimes an indication of the paleoflow slope (or a reasonable surrogate provided by the gradient of a river terrace). The primary limitation is that only rarely can this approach provide data on more than one flow event. These sediments rarely are preserved in stratigraphic sequences from which a history of flooding can be extracted. However, they provide useful data on the maximum paleoflood of record for a stream.

2. Paleostage Indicator—Slackwater Deposit Method

The most accurate and detailed reconstructions of paleofloods over time periods in the range of hundreds to thousands of years have been achieved in the past 10 years by investigations in bedrock channel reaches containing slackwater deposits. Sites of slackwater deposition receive new strata of flood sediment with each flood that achieves a stage capable of overtopping the accumulating stratigraphic sequence. Extensive reconstructions of paleofloods over 10,000 years have been made in semiarid bedrock channels using tributary-mouth slackwater sites and shallow cave sites in west Texas, Arizona, Utah, and parts of Australia (Fig. 9).

Paleoflood stage is obtained by tracing each slackwater flood sedimentation unit up tributary mouths until they pinch out. This maximum stage is then surveyed back to the main channel to provide a paleostage high-water mark associated with that cross section of the main channel. In this manner, multiple sites, each with records of multiple paleoflood events, can be used to establish a series of paleostage levels along the main river channel. These stages are then inserted into a flow-modeling program such as the Step-Backwater model for computation of paleoflood discharge for the floods preserved in the slackwater record.

3. Paleoflood Frequency

Before a flood frequency analysis can be performed based on the paleoflood record, the chronology of paleoflooding must be determined. Slackwater sediments commonly contain a host of organic materials that can be sampled for age determination using radiocarbon dating techniques. Table III shows the variety of materials generally available for dating and a ranking based on their potential to provide an accurate estimate of the age of the flood. In general, finely disseminated organics often deposited at the top of flood layers provide the best estimate for the timing of the flood. Standard radiocarbon techniques can date organic materials as old as 50,000 years. Recent advances in radiocarbon technology using accelerator mass spectrometry have permitted extensions of this technique beyond 75,000 years; this method can work with minute amounts of organic materials, such as individual seeds. Recent developments in radiocarbon interpretation have also shown that floods which have occurred after A.D. 1945 may be dated by comparison of the amount of elevated radiocarbon in relation to a calibrated curve of atmospheric radiocarbon elevation caused by nuclear bomb testing. [See RADIOCARBON DATING.]

C. Applications of Paleoflood Studies

1. Geomorphology and Paleoclimatology

Paleoflood analysis provides a viable means of calibrating true magnitude and frequency values for catastrophic floods. Historical data sets may be far too short to adequately assess the return intervals of infrequent floods. With paleoflood data, studies aimed at assessing the role of rare, large floods on fluvial landscape development can be made. Extended paleoflood records for selected rivers can also provide evidence of shifts in the frequency and character of flooding over long time periods spanning the entire Holocene and into the late Pleistocene. These kinds of data may serve as proxy data for evaluating changes in paleoclimate, because we have already seen that the atmospheric dynamics regulate flood generation. For example, new studies are emerging that are using paleoflood records to reconstruct histories of global cyclic atmospheric phenomena such as the El Niño. [See EL NIÑO AND LA NIÑA.]

Paleoflood records derived from sedimentary evidence can often be combined with flood reconstructions made from analysis of the interaction of flood flows and trees growing in riparian settings. Dendrogeomorphic studies in recent years have extended flood records for several hundreds of years along some rivers. Figure 10 shows an example of the burial of a desert tree in California and its subsequent exhumation, documenting fluvial response along this river in response to short-term (decadal) cyclic climatic changes. Urban development in such an area prone to cyclic climatic perturbations may be adversely affected by subsequent climate shifts. During excursions into wet periods, rainfall may increase to levels over 100% above dry periods, causing normally stable hillslopes to be mobilized and causing widespread flooding, erosion, and sedimentation in downstream alluvial fans and river valleys.

2. Hydrology and Engineering

One of the most problematic aspects plaguing flood frequency analyses is the short period of historical, systematic

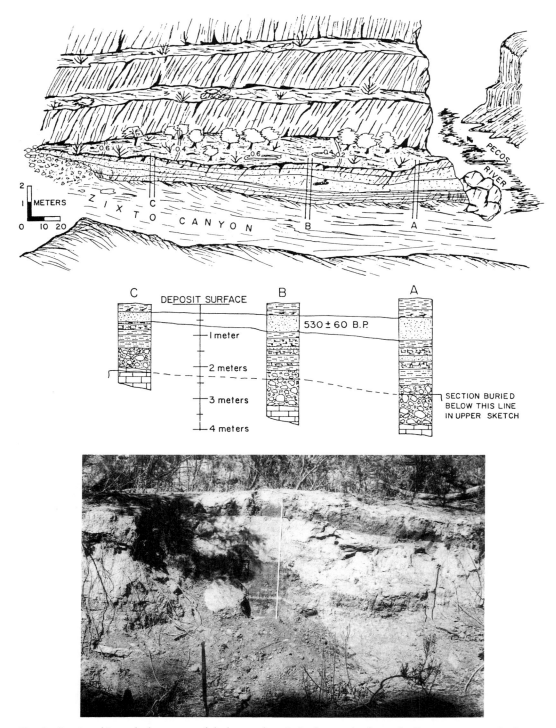

Fig. 9 Stratigraphic stacked sequence of slackwater deposits in the mouth of a small tributary canyon to the Pecos River in southwest Texas. Sketch shows the interpretation of numerous paleoflood units and an event radiocarbon dated at approximately 500 years B.P. [From R. C. Kochel and V. R. Baker (1982). Paleoflood hydrology. *Science* **215**, 353–361. Copyright © 1982 by the AAAS.]

Table III Dating Paleoflood Sediments

Type of material	Dating technique	Problems, range, and comments
Living and buried trees	Dendrochronology	Coring or slabs; accurate to the year; range to several hundred years
Prehistoric artifacts	Archeological studies	Mixing and time transgression between cultures is possible — error within a few hundred years is common
Trees, logs	Radiocarbon dating	Often older than flood — reworked, slow decomposition in arid areas; error possibly tens to hundreds of years
Charcoal	Radiocarbon dating	Same problems as logs, often worse
Paleosols	Radiocarbon dating	Buried soil organics yield minimum age by mean residence time; errors likely to be hundreds of years
Fine-grained organics	Radiocarbon dating	Best date for flood; errors likely to be within zero to tens of years

data on floods available for use in the various analytical techniques. Additionally, many rivers lack any gaging data or contain exceedingly short records. Paleoflood data can be readily applied to flood frequency analyses by blending its long-term record of major flow events with the shorter historical data set. In this manner, better estimates of flood frequency can be obtained for high-risk projects requiring knowledge of the expected return interval of major floods. Many engineers and agencies responsible for flood control are beginning to modify their procedures to include paleo-

flood information where it can be obtained. For example, the state of Colorado passed a bill in 1986 requiring the inclusion of paleoflood data in the design of reservoirs.

V. Global Climate Change and Flooding

Evidence is currently being assimilated to suggest that changes in patterns of flooding are due to a very complex interchange between naturally variable climatic phenomena

Fig. 10 Base of a small tree which has been buried and subsequently exhumed within the last few decades due to episodic aggradation and erosion in this stream channel due to alternating wet and dry climate cycles in southern California. This tree, which has been dated at less than 25 years old by tree ring studies, was buried up to the level of the sediment line on its trunk that correlates to the surface of the low terrace in the background. Aggradation accompanied the last wet episode between 1978–1983. Since 1983, erosion has exposed the tree as the terrace sediments are being removed during the dry period.

and change in the watershed induced by anthropogenic activity. Our present concern over the potential for global climate changes caused by human-related activity requires that environmental scientists must be able to distinguish between natural variation and that related to human-induced changes.

Variations in global hydrologic phenomena such as flooding represents one of the major emphases in the accelerating research in global systems studies. Flood geomorphologic studies will be able to gather data on natural hydrologic variations set within a temporal framework of millenia. This will enable scientists to develop models of natural variations so we can appropriately assess irregularities attributed to anthropogenic disturbance. The intimate linkage between river flooding and the dynamics of the atmosphere will afford opportunities to use the geologic record of flooding as proxy data for interpreting the temporal and spatial history of climate. This approach promises to work especially well for rivers whose flood history is primarily dependent on excursions of major climatic phenomena into their watersheds, such as the El Niño and tropical storms into otherwise arid regions.

Bibliography

Baker, V. R., and Nummedal, D., eds. (1978). "The Channeled Scablands." NASA, Washington, D.C.

Baker, V. R., Kochel, R. C., and Patton, P.C., eds. (1988). "Flood Geomorphology." Wiley, New York.

Beven, K., and Carling, P., eds. (1989). "Floods: Hydrological, Sedimentological and Geomorphological Implications." Wiley, New York.

Mayer, L., and Nash, D., eds. (1987). "Catastrophic Flooding." Allen and Unwin, Boston.

Schumm, S. A. (1977). "The Fluvial System." Wiley, New York.

Singh, V. P., ed. (1987). "Regional Flood Frequency Analysis," 1 of 4 vols. from Proceedings International Symposium on Flood Frequency and Risk Analysis 1986. Reidel, Dordrecht, The Netherlands.

Thorne, C. R., Bathhurst, J. C., and Hey, R. D., eds. (1987). "Sediment Transport in Gravel-Bed Rivers." Wiley, New York.

Wolman, M. G., Leopold, L. B., and Miller, J. P. (1964). "Fluvial Processes in Geomorphology." Freeman, San Francisco.

Glossary

Dendrogeomorphology Application of tree-ring studies (dendrochronology) to the interpretation of geomorphologic processes. Often supplemented by related botanical observations.

Flow rheology Deformational character of the flow when stress is applied. Rheology varies due to differences in fluid strength, viscosity, density, and sediment concentration.

Landscape recovery Amount of time necessary for stream landforms (channel, bank, and floodplain features) modified by a flood to return to their pre-flood configuration, which was adjusted to low-flow conditions.

Macroturbulence Flood flows of great depth and high velocity are characterized by special turbulent flow phenomena that are exceedingly powerful. Macroturbulent phenomena include powerful upward vortices called kolks, intense shocks resulting from sudden pressure change known as cavitation, and powerful roller vortices oriented parallel to flow. These features provide enormous shear stresses and stream power to flood flows.

Paleocompetence Competence is a measure of a stream's ability to entrain sediment particles, measured by the maximum grain size that can be transported. Empirical relationships between grain size and flow velocity or stream power are applied to paleoflood deposits to derive estimates of these same paleoflow parameters.

Forest Ecosystems

K. Lajtha
Boston University

Perhaps the most diverse and widespread (at least before anthropogenic influences) vegetation classification on the surface of the earth is the *forest ecosystem*. Composed of a dominant canopy cover of trees, the largest of all terrestrial organisms, forest ecosystems differ in general physiognomy, in the relative diversity of both overstory and understory species, and in the degree of canopy cover. Forests may also differ in the relative composition of deciduous versus evergreen species. While climate, particularly temperature and water regime, will determine both the presence or absence of forests as well as the general physiognomy and composition of the tree species, local geology and soil nutrient availability will modify this climate control. As will be discussed below, forest ecosystems will also affect the climatic regime of their environment.

I. Introduction

A forest *community* consists of the particular assemblage of biological organisms that inhabit the forest, including the plants and the animals, as well as soil organisms such as bacteria, fungi, and soil-dwelling invertebrates. An *ecosystem,* on the other hand, consists of the populations of species that inhabit the community taken together with the abiotic environment, such as the soils, atmospheric inputs, and the transfers and circulation of energy and nutrients.

Distinct forest ecosystems integrade not only with each other over environmental gradients but with other vegetation community types as well. For example, temperate deciduous forests change in composition and integrade with spruce–fir forests with increasing elevation in the mountains of the northeastern United States; with increasing elevation in southwestern states, dry deserts and desert grasslands integrade with juniper grasslands, pinyon–juniper woodlands, ponderosa pine forests, and finally montane spruce–fir forests. Although these communities are generally continuous with one another and species composition can (but does not always) change gradually over gradients, these communities have been classified by a variety of different criteria into distinct groupings for the sake of convenience and comparison.

II. Forest Communities

The two most common and general community classification systems are the *physiognomic* approach and the *dominant species* approach. The physiognomic approach names community types by their overall structural appearance and the growth form of the dominant plant species. The rationale for this approach is based on the general assumption that whenever similar climatic environments occur across continents on the earth, the same growth forms exist, although often with very different species. One example of this would be the temperate woodland community type that occurs in western North America, the Mediterranean, and in parts of Australia, characterized by open-growth, usually short trees with a shrub or grass understory. Although these woodlands have similar physiognomic traits and perhaps similar functional traits, the dominant trees can be needle-leafed, deciduous, sclerophyll, or any combination of these in the different regions. The dominant species approach is perhaps the most direct and natural approach, particularly in communities dominated by one or two major species. However, this approach will not be as effective in communities that are extremely diverse with many overstory dominants, such as in tropical rain forests.

Another more general approach to classifying ecological communities is to use the *biome* or *biome-type* approach. A biome comprises a grouping of plants and their associated fauna on a given continent that are easily recognizable based on general physiognomy, but that might consist of species that differ greatly in actual taxonomic affinity. A biome-type is a grouping of convergent biomes across different continents. The needle-leafed conifer forest biome-type is found across northern Eurasia and North America, for example, yet encompasses the temperate rain forest of the Pacific Coast of North America that is composed of the redwood and sequoia forests of California and the Douglas fir–western hemlock forests of the northwest, the relatively less dense and colder boreal forests of spruce and balsam fir, and the warmer and drier Ponderosa pine forests farther south.

Different authors have recognized different groupings of terrestrial ecosystems into formations, biotic provinces, and ecoregions and have used either very broad categories of biomes or else have subdivided terrestrial ecosystems into many separate categories. The system used here follows the general vegetation classification scheme of R. H. Whittaker, a recent pioneer of plant community ecology. The geographic extent of the forest community types defined here are outlined in Fig. 1.

A. Woodlands and Tree Savannas

Although usually classified as being distinct from forest ecosystems, *tree savannas* and *woodlands* share much in common with forests in that the dominant overstory vegetation is composed of trees. Woodlands are loosely defined as communities of small trees that generally occur in climates too dry for true forests, and thus canopy coverage tends to be significantly lower than in true forests. Well-developed stands of grasses or shrubs occupy the canopy interspaces, and compete successfully with the overstory for soil water. Because of these grasses, woodlands support a varied fauna of naturally occurring grazers, but they are often used for domesticated grazers such as cattle that can degrade the soils and vegetation. [*See* SHRUBS: ADAPTATIONS AND ECOSYSTEM FUNCTIONS.]

Temperate woodlands in the United States are usually transitional between upper elevation, moist forest, and drier grasslands, or shrublands at lower, drier elevations. The oak–sagebrush communities of Utah, the evergreen–oak woodlands of California, and the oak–pine woodlands of Mexico are all such transitional communities. One of the most widespread woodland communities of the western United States is the pinyon–juniper ecosystem, in which different species of *Pinus* and *Juniperus* co-occur. Similar communities occur in Australia, where tall evergreen *Eucalyptus* dominates the wandoo woodland.

Tropical broadleaf woodlands occur in both dry areas and over very infertile soils. The cerrado communities of central Brazil occur between seasonal forest and thorn shrubland in acid, sandy soils with low nutrient availability. Low nutrient availability may also explain the occurrence of the miombo of interior southern Africa, and the semievergreen to evergreen woodlands found in Burma and elsewhere in southeast Asia.

Savannas generally represent a continuum community between true woodlands and grasslands. Savannas are most extensive in central and southern Africa, but also occur in Australia, southern Asia, and the llano and campos of tropical South America. Many of the African and Australian savannas occur in climates too dry for forests, while other savannas, particularly the South American savannas, occur in less arid climates and exist due to either soil conditions or fire history. Savannas generally receive most of their annual rainfall during only one brief period, and thus species must be tolerant of a long, dry season and are prone to periodic, frequent fires. While most savannas are natural, others are created by human-caused disturbance such as fire or grazing. The dominant grass understory supports a large and rich diversity of large grazing and browsing herbivores, particularly in Africa, such as the rhino, and including ungulates such as the zebra, gazelle, wildebeeste and antelope, as well as their mammalian carnivores such as lions, hyenas, and wild dogs. In terms of absolute consumption, however, the dominant herbivores in these ecosystems are invertebrate consumers including grasshoppers, termites, and ants. Tropical savannas, much like temperate woodlands, are currently under assault from overgrazing of cattle, which degrade the system due to daily trekking to watering holes, as they require more water more frequently than do native herbivores such an antelope.

B. Temperate Deciduous Forests

The temperate deciduous forest, composed of broadleaved tree associations, grow in moderately humid continental climates, characterized by moderate summer temperatures, no extended period without precipitation, and temperatures that are below freezing during winter months. This forest type is found throughout much of the eastern United States, in western Europe, in eastern Asia, and in limited areas of southern Chile. American deciduous forests include the species-rich forests of the southern Appalachians, including the oak–hickory, oak–pine, and oak–chestnut dominated ecosystems. These forests are excluded from climates that are either very cold or very dry, and from soils that are very nutrient-poor, where coniferous forests predominate. Unlike most woodland ecosystems, and even unlike most coniferous temperate or subarctic–subalpine forests, these forests are rarely dominated by only one or two canopy species, and shifts between community dominants over environmental gradients tend to be more gradual than in more water-limited systems. In mature deciduous forests leaf area index is greater than one, and the canopy is generally closed, although gaps of various sizes open small areas of the forest floor to light. There are generally several subdominant layers to the temperate forest, composed either of understory trees that are adapted to low light conditions, light-suppressed younger dominants, or understory shrubs. There is often a rich her-

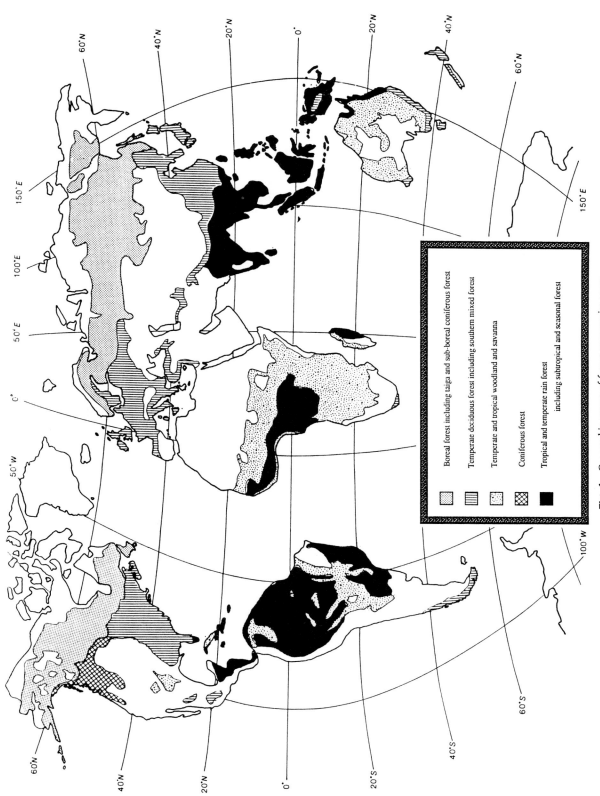

Fig. 1 Geographic extent of forest community types

Legend:
- Boreal forest including taiga and sub-boreal coniferous forest
- Temperate deciduous forest including southern mixed forest
- Temperate and tropical woodland and savanna
- Coniferous forest
- Tropical and temperate rain forest including subtropical and seasonal forest

baceous understory vegetation of annual plants or small perennials that can take advantage of the sunlight and warm temperatures that reach the forest floor in the early spring before the canopy trees leaf out.

In temperate deciduous forests of the United States, characteristic mammals include squirrels, raccoons, deer, and black bears. Both the mammalian and the avian fauna necessarily adapt to the strong seasonality of the forest, either by hibernating or remaining inactive during the coldest months (bears and squirrels) or by being migratory (many birds), or by being efficient at obtaining food resources even in deep snow (deer). Reptilian fauna is represented by snakes, frogs, and salamanders, although not in as great diversity as in tropical ecosystems, and lizards are generally not well represented in temperate forests when compared, for instance, to desert ecosystems.

Early plant geographers noted that the Asian and American deciduous forests are significantly more diverse than their European counterparts, and postulated that extinction effects were more severe here during Pleistocene glaciation than in North America or Asia. This could occur due to differences in recolonization success during late Pleistocene and early Holocene post-glacial warming, caused by differences in continental topography, rather than post-glacial climate. For instance, mountain chains in central and western Europe are generally west-to-east trending, and few mountain passes are below 2000 m in elevation, and thus northward expansion of tree populations from southern refugia might have been limited.

In contrast, the eastern U.S. Appalachian Mountains trend southwest-to-northeast, and have significantly lower topographic relief than the European mountains, thus allowing for easier northward dispersal and expansion during late-glacial climatic amelioration. However, recent reanalysis suggests that forest diversity, measured as species richness in all three continents, is related to productivity of the forests, which is itself related to climate, and that current climate conditions can adequately explain differences in species diversity across the continents. However, forests in Asia are more diverse than their American and European counterparts at the very highest productivities, and thus species impoverishment due to low recolonization after glaciation may be affecting the diversity of both American and European temperate deciduous forests. There are several indications that glaciation has had long-lasting effects on species diversity in temperate forests. Within the United States, Southern Appalachian cover or mixed mesophytic forests were presumably less affected by glacial extinctions and are more species-rich both in the tree and understory layers than their northeastern counterparts; they are thought to be most similar to the temperate forests that were of wide extent in the northern hemisphere during the Tertiary.

C. Tropical Rain Forests

Tropical rain forests and tropical monsoon forests occur in the humid tropics, forming a disjunct and uneven belt around the equator worldwide. The largest continuous rain forest is in the Amazon Basin of South America, and rain forest covers much of Central America, western and central equatorial Africa, and extends from Southeast Asia through Indonesia to northwest coastal Australia, as well as to various islands in the Indian and Pacific oceans. The tropical rain forest climate is significantly less seasonally variable than that of the temperate deciduous forest. Average annual rainfall is more than 240 cm, spread relatively evenly throughout the year; temperatures stay well above freezing even during winter months.

Tropical rain forests are the most species-rich of all forest ecosystems, with tree species often numbering in the thousands within a given forest. Usually no one single species is dominant, and it is commonly noted that conspecific individuals are often widely spaced from each other, perhaps as a defense mechanism against specialized insect feeders. Leaf area index is higher than in temperate forests, with very little light reaching the forest floor except in treefall or crown gaps; and thus species diversity and plant density in the understory is reduced compared to temperate forests. Thus there are not the corresponding complex subdominant canopy layers that are seen in more open, temperate forests. Trees of the rain forest are often tall and similar in appearance if not in taxonomic affinity, and support many climbing plants (lianas) and epiphytes, including orchids and bromeliads. Although most trees form a fairly even canopy over the forest floor, certain individuals of certain species are emergent, or overtop the rest of the canopy. Leaf shapes of the different species are often quite similar, with only minor lobing if present at all; most have terminal drip tips, presumably so that rain water is removed quickly and completely from the surface of the leaf, thus discouraging the growth of epiphytes and microorganisms.

Animal diversity, especially that of birds and insects, is also high in tropical rain forests. However, species tend to be quite distinct among the tropical forests of Asia, the Americas, and Africa, with little overlap even at the generic level in many cases.

D. Coniferous Forests

Within the very general class of coniferous forest ecosystems can be included forests that are often separated into temperate rain forests, temperate evergreen forests, and taiga forests of the northern hemisphere. However, although climate varies greatly among the continents and locales where these forests are found, conifer forests share in common the fact that the dominant tree species have narrow, sclerophyll leaves or needles that can persist for several years, rather than broad deciduous leaves. These modified leaves can be advantageous either in dry climates or in extreme cold; because of the unique morphology and often the unique secondary chemistry of these needles, nutrient cycling and ecosystem function can be quite different in conifer forests compared to broadleaf forests.

Along the northwest coast of the United States, temperate rain forests extend from the Olympic Peninsula to the coastal redwoods of California, where tree height can extend over 100 m. Although considered rain forests, many of the coastal forests do not receive much direct precipitation during the summer, but the foliage can collect abundant cloud and fog water. Species diversity in these evergreen temperate rain forests is significantly lower than in tropical rain forests and are usually dominated by anywhere from one to four tree species. This forest type encompasses the redwood and sequoia forests of California as well as the Douglas fir–Sitka spruce forests with western hemlock understory forests farther north.

Montane coniferous forests, well represented in the Rocky Mountain regions of the western United States, are made up of several associations that vary with elevation. At the highest elevations are subalpine forests, characterized by Engelmann spruce and alpine fir. Lower in elevation, where winter snowpack is less, winters are less severe, and summers tend to be drier, are extensive stands of Douglas fir and pines such as Ponderosa pine and lodgepole pine. Both the density of trees and total production in these ecosystems tend to be limited primarily by water availability. At yet lower elevations and drier, warmer climates, Ponderosa pine forests intergrade with pinyon–juniper woodlands, juniper grasslands, or shrubland ecosystems.

The taiga, or boreal evergreen forest, is separated from the temperate evergreen forest as an ecosystem of the cold edge of the climate range of forests. The boreal forest is circumpolar, extending from Canada and the high Appalachians of New England to Alaska and northern Eurasia. Species diversity in taiga is low, with either white or black spruce as a codominant of balsam fir. Due to the short growing seasons of 1–3 months and maximum summer temperatures in the 20°C to less than 30°C range, these conifer forests are significantly lower in stature than their rain forest or montane conifer forest counterparts. Moving south from the taiga, there is a general integradation with the subalpine zone with other spruce–fir codominants, such as red spruce and fraser fir in New England mountains.

Other needle-leaf evergreen forests, composed mostly of pine, occur in Asia, western Europe, and the southeastern United States. In some areas, such as Australia and the southeastern United States, pines have either been planted extensively for forestry purposes or else occur naturally but only as early successional stages—yet they are maintained for their value as lumber. Pines are generally considered to be a successional stage after a severe disturbance, such as an extreme fire or a harvest for lumber or to convert the land to agricultural use, in the eastern U.S. deciduous forest ecosystem. Curiously enough, it is a deciduous broadleaf species, quaking aspen, that is a common invasional successional species in many western conifer ecosystems after disturbance.

III. Forest Ecosystem Function

Biogeochemical cycles and transformations within forests can be separated into intersystem versus intrasystem transfers (Fig. 2) Intersystem transfers are those processes of chemical gains and losses to other ecosystems on the surface of the earth, including precipitation, dryfall, and gaseous absorption inputs, and streamflow, groundwater, and gaseous emission losses. The uptake of ions, their return to the soil, and the chemical transformations of these ions within the system constitute the intrasystem cycle. In general, the intrasystem circulation of essential nutrient ions through the biota is significantly greater than the annual inputs or outputs to the ecosystem, suggesting that vegetation receives most of its annual requirements from recirculation through the soil rather than through new inputs.

With the exception of nitrogen, most nutrient ion inputs to a forest ecosystem come ultimately from the weathering of parent rock material or from dust deposition. Rates of chemical weathering and nutrient release from bedrock depend not only on the hydrologic regime but also on the specific rock type, and thus inputs can vary widely. Inputs of nitrogen, however, come solely from atmospheric deposition and from *in situ* nitrogen fixation by microbes, as there is no fixed, usable form of nitrogen in rocks. Although there is much variation in rates of N fixation, atmospheric deposition, and bedrock weathering among sites, primary production in most temperate forest ecosystems has been found to be limited by the availability of nitrogen. In contrast, many tropical forest ecosystems are phosphorus limited, in part due to the very old, weathered soils that have lost much of the original P content and that contain minerals with a high P adsorption capacity. Boreal forest soils have been shown to be N, P, or N and P limited, perhaps due to the slow turnover of organic matter in boreal soils that contain much of the ecosystem's organically bound nutrient capital. [*See* NITROGEN IN SOIL; PHOSPHORUS, BIOGEOCHEMISTRY.]

Recently, a hypothesis of nitrogen saturation of forest ecosystems in highly industrialized northern temperate regions has been proposed. Atmospheric deposition of nitrogen compounds is greatly increased in forests near industrial

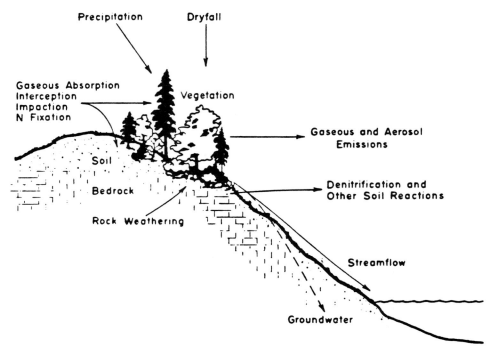

Fig. 2   A model of the pathways of nutrient movement through a forest ecosystem. [From Waring and Schlesinger (1985).]

centers, from as low as 2 kg N ha^{-1} yr^{-1} to more than 40 kg N ha^{-1} yr^{-1} in high-elevation forests of New England. Nitrogen saturation is defined as inputs of N that exceed N requirements of plant and microbial need for maximum production, until growth is limited by some other nutrient. Because most of these forests are currently limited by rates of N input and intrasystem turnover, added N deposition should initially have a fertilization effect, increasing forest productivity. However, preliminary measurements suggest that many forests in New England may soon reach their limits for biotic uptake and storage of nitrogen, and several negative consequences could occur. High levels of N often cause shoot growth to occur over the growth of roots, thus decreasing the uptake of other nutrients or water that are now limiting production. High levels of nitrate in leaves could reach toxic levels, and high levels of nitrate in forest soils may cause the leaching of critical cations (see Section IV, B on forest decline) and influence quality of streamwater draining high-elevation forest ecosystems. However, there is little evidence that current levels of N accumulation in forested ecosystems has switched N economy from a limiting situation to one of saturation. [See NITRIFICATION.]

Climatic regime and the chemistry and physiology of the dominant tree species control such ecosystem processes as production of organic matter, decomposition of organic litter, and nutrient remineralization. Consequently, world forests differ greatly in the quantity of organic matter and nutrients stored both aboveground and in the forest floor (Table I). Although local temperature and moisture regime should theoretically limit maximum net primary production (NPP) of any given forest, correlations between NPP and temperature and moisture variables are generally significant but weak, due to differences in species' rooting depths, length of foliage display, physiological efficiencies, and site nutrient and soil characteristics. Over a wide range of ecosystem types, NPP has been found to be more correlated with the maximum site leaf area than to direct climatic variables.

Temperature and moisture exert strong controls over litter decomposition in forests, with temperature of primary importance. Rates of decomposition of fresh litter can be accurately predicted using site actual evapotranspiration, a variable that combines the effects of temperature and moisture availability. However, the relationship is significantly improved if characteristics of litter chemistry are included, such as N content and concentrations of resistant, carbon-rich chemicals. Because soil organic-matter accumulation reflects the difference between primary production and litter decomposition, soil organic-matter content is also strongly con-

Table I Storage of Organic Matter and Nutrients in World Forest Ecosystems[a]

Ecosystem	Aboveground vegetation						Forest floor					
	Biomass	N	P	K	Ca	Mg	Mass	N	P	K	Ca	Mg
Boreal coniferous	51,300	116	16	44	258	26	113,700	617	115	109	360	140
Temperate												
Coniferous	307,300	479	68	340	480	65	74,881	681	60	70	206	53
Deciduous	151,900	442	35	224	557	57	21,625	377	25	53	205	28
Tropical	292,000	1,404	82	1,079	1,771	290	27,300	214	9	22	179	24

Source: Waring and Schlesinger (1985).
[a] Data are in kilograms per hectare.

trolled by regional differences in temperature and moisture. Thus cold boreal forests, with relatively low biomass in aboveground tree components, accumulate a great deal of carbon in the forest floor, while tropical forests accumulate relatively low amounts of carbon and nutrients in the forest floor compared to the aboveground standing biomass (Table I).

Because climate and anthropogenic inputs to forest ecosystems affect much of ecosystem functioning, clearly even small-scale alterations of climate can affect such factors as maximum annual production, the balance between production and decomposition, nutrient limitation, and nutrient exports to streams and rivers.

IV. Disturbance, Forest Decline, and Deforestation

A. Disturbance and Equilibrium

There is much debate among scientists as to whether forest ecosystems can be considered structurally and functionally to be equilibrium communities. The recent geologic record clearly shows that the earth's climate at any given location has not remained constant; and paleographic evidence, both from pollen records in preserved sediments and from leaves preserved in fossil structures such as packrat middens, has shown that distributions of forest species change dramatically as climate changes. Pollen records from the eastern U.S. forests during the last 10,000 years document the movement northwards of individual tree species following the retreat of glaciers, and show that movement of each species is individualistic, rather than as a conglomerate flora moving together. For instance, 18,000 years ago spruce grew in association with a variety of sedges in the American midwest; and by 10,000 years ago they formed a closed monospecific canopy in southern Canada. The spruce–birch association of the modern boreal forest is no more than 6000 years old and is

probably as sensitive to variations in climate as its community predecessors.

Even at smaller spatial scales and over shorter temporal scales there is much heterogeneity and lack of equilibrium in natural communities. Although the term *disturbance* is generally thought of as an event or series of events that are abnormal or unusual to a system, forest ecologists now view disturbance, at least at some scale, as being both commonplace and essential for maintaining the structure of forests and the coexistence of species. Thus two kinds of ecological disturbances can be defined, although these two types intergrade and are merely extremes of a continuum. Disturbance can be disruptions of ecosystem, population, or community structure that are commonplace or recurring, discrete events defined by environmental fluctuation; but disturbance can also be unusual destructive events that occur rarely or perhaps only once, from which ecosystem recovery is only gradual or perhaps incomplete. The death of a canopy tree or several trees, forming a gap, can be considered to be a commonplace disturbance in most forests, allowing the release of suppressed saplings or new seedling germination to occur. Windthrow, hurricane damage, and drought occur infrequently but repeatedly in eastern deciduous forests, and affect larger areas at any given time. In general, eastern moist forests tend to be dominated by such small-scale disturbances characterized by the formation of many small gaps. As one moves across the country to the west, however, and to drier forests, disturbance regimes change. In the Allegheny Plateau of Pennsylvania, for example, sites dominated by dry, sandy soils with a pine overstory are more subject to larger-scale fires and blowdowns than are forest communities on upland moist sites. In the northern hardwoods section of Minnesota and Canada, as well as in dry southwestern coniferous forests, widespread fires were prehistorically more important to community structure than local gap formation events.

In all of these discussions, however, it is important to remember that anthropogenic disturbances now commonly override and overshadow these natural disturbance regimes. Certainly logging has changed the face of the western U.S.

landscape, as it has for the Amazonian and Southeast Asian tropics (see Section IV, C on deforestation). Clearing for agriculture has affected almost all of the world's forest types, and effects include not just forest destruction, but forest fragmentation as well. Even such practices as forest fire suppression in populated areas where repeated fire was a natural disturbance regime have affected community structure, by ensuring that when such systems do finally burn, fire is much more widespread, intense, and devastating—as the forest fires of 1988 in Yellowstone National Park demonstrated. Finally, the introduction of exotic species by man to a new environment has often profoundly altered species composition or ecosystem function. In Hawaii, for instance, invasion of young volcanic sites by an exotic nitrogen-fixing tree species has strongly increased the biological availability of nitrogen, and thus has altered the nature of ecosystem development after volcanic eruptions. Many exotic pest species have had a similar effect. The chestnut blight, *Endothia parasitica*, is a fungus that is parasitic on chestnut trees in Eurasia. When introduced to North America and to a chestnut species not accommodated to it, virtually all of the chestnuts *(Castanea dentata)*, one of the most important species in the eastern deciduous forest, were killed.

B. Forest Decline

In recent years ecologists have noted a widespread decline in vigor and death of forest trees in many parts of the world. Such decline and dieback has been observed in industrialized areas as well as in areas with relatively low air or land pollution. Decline is usually characterized by a progressive deterioration in the health of older and larger trees as well as stand-level mass mortality. While the ultimate cause of death is often fungal, microbial, or insect disease, initial causes of stress that lead to an increased susceptibility to disease are not always completely understood. Vegetation and forest community composition on any given site is rarely static over long periods of time, and natural occurrences of decline and dieback are known from pollen records of eras unaffected by anthropogenic pollution. For example, there was a widespread decrease in the eastern hemlock in eastern North American forests about 4800 years ago, but not in other co-occurring species with similar habitat requirements; there was a similar decline of elm species in northwest Europe about 5000 years ago. It is unlikely that a sudden climate change could account for these species-specific mortality events, yet disease agents for ancient decline occurrences are also unknown. Modern plant geographers have also noted sudden decreases in the abundances of specific forest tree species in the last century, with the primary causes of decline unknown.

In the early 1970s, a general decline in certain conifers of northern Europe, particularly older Norway spruce in West Germany, was noted. By the early 1980s, large-scale mortality was seen in stands of spruce, beech, and oak in much of industrialized Europe. Similar decline symptoms were being observed in high-elevation red spruce stands of New England. By 1989, more than half of the West German forest areas and 15% of the total European forest area showed at least some signs of decline.

Because much of the documented forest mortality was in areas potentially exposed to urban and industrial air pollution, anthropogenic stresses such as ozone, sulfur dioxide, toxic heavy metals, excess nitrogen deposition, and acid rain were implicated in this massive new decline. However, it is difficult to determine the cause of mortality in any natural system, as experiments usually cannot be done with appropriate pollutant-free controls. To date there is no proof that any one single factor is the primary cause of forest decline. Levels of ozone in West Germany, for instance, have been measured as high as 400 $\mu g/m^3$, 2–3 times as high as concentrations known to cause injury, and yet decline symptoms do not match known ozone injury symptoms. Similarly, although elevated levels of sulfur dioxide were implicated in European forest decline, the major areas exhibiting forest dieback did not necessarily experience elevated SO_2, decline symptoms did not necessarily match known symptoms of SO_2 damage, and often other species such as lichens that are known to be extremely sensitive to elevated SO_2 were not affected in these areas. Although such toxic gases may not be the primary cause of decline, it is quite possible that an interaction among factors, or cumulative stresses, might be allowing trees to succumb to disease. [*See* ENVIRONMENTAL INJURY TO PLANTS.]

Acid rain has also been implicated in causing decline, both through direct effects on foliage and through indirect effects such as elevated Al levels in soils, which can inhibit Ca uptake by roots. Acidity in high-elevation mountain fog and cloud water is known to be greatly elevated, and it is in such high-elevation forests in the eastern United States that decline symptoms are greatest. In addition, the two species that initially showed decline symptoms, Norway spruce and red spruce, have been shown to be highly sensitive to elevated levels of Al in the root medium in culture experiments. [*See* ACID RAIN.]

Most recently, nutrient imbalances have been implicated as a factor in forest decline. Many of these high-elevation forests are subject to high rates of atmospheric nitrogen inputs from wetfall, particulate dryfall, and gas and fog inputs. Sources of this increased nitrogen include emissions of NO_x from cars and power plants as well as N_2O losses from fertilized agricultural fields. Excess nitrogen fertilization could cause stand-level injury symptoms through several mechanisms: (1) increased N leads to increased plant growth, primarily of shoots rather than of roots; (2) increased growth leads to an increased demand for other nutrient elements as well, such as Ca or Mg that are not part of the atmospheric fertilization; (3) these elements become limiting to photosynthesis and new tissue growth, and are prematurely withdrawn from old needles to supply new needles, thus causing the common chlorosis and premature needle-drop symptoms; (4) this element deficiency is exacerbated by excess NO_3 deposition to the forest floor, because as excess NO_3 is leached from the soil, cations are removed at the same time, keeping the soil solution electrically neutral. In addition, excess nitrogen fertilization has been shown to lead to delayed frost hardiness in conifers, as well as a decrease in allocation to defensive chemistry against insects and pathogens. Support for this new decline hypothesis has come from fertilizer trial with K and Mg, where decline symptoms have been reversed.

To date, the relative roles of natural vegetation variations due to climatic variations versus anthropogenic pollution effects have not been determined. Although declines of red spruce in the Adirondacks occurred over a century ago near where the modern decline is occurring, this alone is not evidence that the same factors caused both events. Nor is there evidence that all modern-day forest species declines are occurring for the same reason, as species' tolerances to air pollution and toxic element stresses vary considerably. Several researchers have suggested that a mix of elevated ozone and cloud acidity are causing spruce and fir decline through direct impairment of photosynthetic tissues, nutrient leaching, and fine-root growth impairment, whereas ozone alone is causing decline symptoms in beech. In any case, it will be critical to understand the role, if any, of anthropogenic pollution in causing dieback symptoms in world forests.

C. Deforestation

The greatest loss of forest land from known human disturbance is through the clearing of tropical forests. Although the tropics contain over 40% of the forests of the world and are estimated to contain 64% of the earth's potentially cultivatable land, tropical forest soils are generally characterized as being of marginal quality for repeated cropping, due to low fertility or by being too steep or shallow. However, as population increases in the rural tropics lead to increased demand for fuelwood and new arable land, deforestation is accelerating. Estimates of the rate of deforestation in the tropics vary widely, but current estimates suggest that in 1989 almost 8 million hectares were logged or burned in tropical America, primarily in the Amazon, and 1.6 and 4.4 hectares of closed forest were cleared in tropical Africa and Asia, respectively. If open forest types are included in this estimate, nearly 22 million hectares were cleared worldwide, representing nearly a doubling in rate since 1980. Since 1950, forest loss in Africa is estimated at over 200 million hectares; most of the Atlantic coastal forest of Brazil has disappeared, with the exception of small forest reserves.

There are many possible global consequences of this large-scale deforestation that is taking place in the humid and subhumid tropics. The loss of species is perhaps the most obvious consequence of forest destruction. Of the estimated quarter million species of plants inhabiting the earth, roughly two-thirds are found in tropical regions, many of which have not yet been described. The diversity of insect species is equally as great in tropical regions, and yet they are even less well known and described. Of concern to human societies is not only the loss of the aesthetic benefits of a diverse flora and fauna, but also plants potentially useful in medicine, as the genetic basis for new crop species, or as raw materials for new fibers, chemicals, or fuels. Less well known is the effect of loss of genetic diversity for ecosystem function. [*See* BIODIVERSITY AND HUMAN IMPACTS.]

Other ecological effects include losses of soil through erosion, changes in microclimate and, potentially, regional climate, alterations of the hydrologic cycle, and effects on greenhouse gases. Because vast areas are often felled or burned at any given time, soil stability is greatly reduced. Once the forest canopy is removed and the soil surface is disturbed by felling equipment, raindrop impact on the soil surface is greater, infiltration rate is reduced, and runoff velocity is greater, leading to massive erosion of soils, degrading soil fertility and reducing water quality in local rivers and streams. Such erosion further increases the need to clear additional forest land; it is estimated that almost half of the acreage cleared annually in tropical forests can be attributed to the need to replace degraded agricultural soils.

Another significant impact of deforestation or forest conversion is on local, regional, and potentially global climate. Within the Amazon basin, which contains about half of the earth's tropical forests, it is estimated that 50% of the rainfall is derived from local evaporation, including transpiration from vegetation. When simulations of deforestation and forest conversion to pasture effects have been run using three-dimensional atmospheric global climate models in this area, results predicted a weakened hydrologic cycle, with lower precipitation and evaporation and an increase in surface temperature. Such a disturbance of the regional climate could lead to an irreversible loss of the tropical forest ecosystem. Perhaps the strongest effect on the global climate would be through the increase in albedo from forest conversion. Locally, increases in soil and near-ground air temperature and decreases in relative humidity have been observed in pastures compared to intact forests. On a regional scale, increases in streamflow and total water yield have been observed in areas with high rates of deforestation, again demonstrating strong changes in regional water balance. [*See* CLIMATIC CHANGE: RELATIONSHIP BETWEEN TEMPERATURE AND PRECIPITATION.]

The global climate is also potentially affected by deforestation through changes in emissions or uptake of chemically or radiatively active greenhouse gases such as carbon dioxide, water vapor, methane, and nitrous oxide. When forests are cleared or burned, CO_2 is released to the air from decaying or burned wood, leaf litter, and oxidizing soil organic matter.

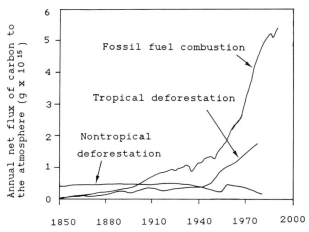

Fig. 3 Annual net flux of carbon to the atmosphere from nontropical deforestation, tropical deforestation, and fossil fuel combustion. [Modified from: R. A. Houghton (1990). The global effects of tropical deforestation. *Environ. Sci. Technol.* **24**, 414–422.]

Current estimates suggest that deforestation is now contributing 26–33% of the worldwide annual emissions of CO_2, with the rest being largely from the combustion of fossil fuels (Fig. 3). Similarly, CH_4 is released to the atmosphere both as a direct effect of deforestation and fuelwood burning and as an indirect effect of forest conversion, from such activities as cattle ranching and the annual burning of pastures and grasslands. Although total emissions of CH_4 are several orders of magnitude lower than emissions of CO_2, the radiative effects of CH_4 are 25 times greater than that of CO_2, and thus emissions could be a significant source of global warming. Although absolute values of increases in such trace gases as N_2O from forest conversion are less well known, N_2O has a radiative effect 250 times that of CO_2, and preliminary measurements show that new pastures and disturbed soils may contribute significantly to global increases in N_2O concentrations. When differences in radiative forcing of these three gases are taken into account and estimated releases are tallied, tropical deforestation could account for up to 25% of total greenhouse gas emissions globally. [*See* ATMOSPHERIC TRACE GASES, ANTHROPOGENIC INFLUENCES AND GLOBAL CHANGE; METHANE, BIOGEOCHEMICAL CYCLE.]

V. Forest Vegetation and Global Climate Change

With the build-up of greenhouse gases such as CO_2, the global climate is predicted to change by as much as 1–4°C by the middle of the next century. Although the earth's climate has alternately cooled and warmed between glacials and interglacials in the past, such a rapid change is

unprecedented. It is not known if forest vegetation can track this predicted climate change. Pollen analysis shows that past climate changes caused species to migrate at rates of approximately 10 to up to 200 km per century under unusual circumstances, and yet models of predicted climate change suggest that species will be required to move at a rate of up to 500 km per century to maintain life in their same approximate thermal environment. This entails very high dispersal rates, rates that are too high for many if not most plant species, if dispersal must be made latitudinally to compensate for the temperature increase. For forested communities in a mountainous region an altitudinal shift can compensate for climatic shifts, yet this leads to a concomitant reduction in total area of habitat, and many species could become extinct. Thus future forest communities may be composed of quite different species assemblages, and slow-dispersing species with narrow physiological tolerances or large area requirements may face extinction.

Increases in global temperatures will affect not only the species composition of forests, but ecosystem function as well. Increased mean temperatures can affect the balance between production and decomposition, as decomposition is generally more sensitive to and limited by low temperatures than is NPP. Thus boreal forests, currently net carbon sinks that contain about 13% of the earth's soil carbon pool, could have a net release of CO_2 to the atmosphere and have a positive feedback effect on global temperatures. Carbon stored in temperate forest soils could also face a slow release, although this will depend in part on how natural disturbance regimes are affected by climatic change. If the frequency of extreme disturbance events such as hurricane and drought increase as has been suggested, then forests will be constantly in a cycle of growth and tree loss, decreasing the capacity of the ecosystem to store carbon in reduced organic matter. Changes in disturbance frequently will also affect the rate at which vegetation can respond to climate change, with increased disturbance frequency overcoming lags in vegetation response. The nature of disturbance in specific regions should change with climate change as well; fire may become a more significant factor in many regions where it had not formerly been a frequent disturbance event.

The increase in atmospheric CO_2 can directly affect vegetation as well. Because plants use CO_2 as a building block for growth, there are predictions that a CO_2 increase will serve as fertilizer for the vegetation of the earth and forest productivity will increase. Not all vegetation if any is predicted to respond, however, as other factors such as water and nutrients may overwhelmingly limit plant production. The physiology of certain species may also change in the face of increased CO_2; water-use efficiency, or the ratio of carbon gain in photosynthesis to water loss in transpiration, is predicted to increase, thus perhaps overcoming water limitations in certain areas. Plant allocation to reproduction versus growth may change, thus affecting dispersal ability of individual species. Because many climate factors are expected to change with increased concentrations of greenhouse gases in the atmosphere, such as temperature, CO_2 concentrations, and rainfall regime, it is difficult to make predictive models of direct changes in tree species' physiology with climate change. [*See* ATMOSPHERIC CO_2 AND TREES, FROM CELLULAR TO REGIONAL RESPONSES; VEGETATION, EFFECT OF RISING CO_2.]

Perhaps the greatest difficulty in predicting vegetation response to climate change results from vegetation feedbacks to the global system. Individual species and vegetation type and density can affect such system properties as albedo, surface roughness, snow cover, and ground frost. The models that predict future forest ecosystem composition and function must incorporate not only direct effects of each climate factor on species physiology and dispersal, but also feedbacks of altered forest ecosystems to the global climate system.

Bibliography

Hutchison, B. A., and Hicks, B. B., eds. (1985) "The Forest–Atmosphere Interaction." D. Reidel, Dordrecht, The Netherlands.

Waring, R. A., and Schlesinger, W. H. (1985). "Forest Ecosystems: Concepts and Management." Academic Press, Orlando.

Whittaker, R. H. (1975). "Communities and Ecosystems." Macmillan, New York.

Glossary

Biogeochemistry Study of the cycling, transformations, and transport of chemicals in and between landscapes and ecosystems.

Deciduous Tree or shrub in which all leaves senesce before a dormant season, caused either by drought or low temperature.

Ecosystem function Ecological processes such as primary production, decomposition, and nutrient cycling that occur at the scale of a system and are affected by a variety of organisms and the environment's microclimate.

Evergreen Coniferous, needle-leaved, or sclerophyllous plant that does not exhibit a marked seasonal leaf fall.

Net primary production Net accumulation of organic matter in an ecosystem in plant or soil biomass; the difference between gross primary production and respiration.

Geodesy

R. Rummel
Delft University of Technology

Geodesy is the scientific discipline concerned with the measurement and determination of the figure of the earth and parts of the earth. The geometric figure is represented by coordinates of points and by maps. The shape of the earth, as expressed by its gravitational field, is represented in the form of level surfaces relative to a global, spherical, or ellipsoidal reference body. The directions to the stars and the orbits of satellites provide an external reference. The changing orientation of the earth with respect to the directions to stars is expressed by the earth's slightly variable spin rate, polar motion, and precession-nutation. Advanced terrestrial and space measurements show the variations of these factors with time.

I. Introduction

The earth is an almost perfect sphere. Deviations from its spherical shape are less than 0.4% of its radius (6371 km) and arise from its rotation, producing an ellipsoidal shape symmetric about the rotation axis. Relative to an ellipsoid the deviations are less than 10 km. The measurement and representation of these deviations are the object of geodetic activity.

The topographic features, mountains and valleys, are very irregular. They can only be followed in any detail by direct measurement. No simple generic mechanism exists that permits inference from a few selected points for the topography as a whole. In this respect the shape of the earth as defined by its gravity field, is more representative.

Gravity as observed on the earth's surface is the combined effect of the gravitational mass attraction and the centrifugal force due to the earth's rotation. The force of gravity provides a directional structure to the space above the earth's surface. It is tangential to the vertical plumb lines and perpendicular to all level surfaces. Any water surface at rest is part of a level surface. As if the earth were a homogeneous, spherical body, gravity is almost constant all over the earth's surface, the well-known 9.81 m s^{-2}. The slightly curved plumb lines are directed toward the earth's center of mass, and this implies that all level surfaces are nearly spherical too and that the density distribution inside the earth must consist of approximately spherical layers. The gravity decreases from the poles to the equator by about 0.05 m s^{-2}. This is caused by the flattening of the earth's figure and the negative effect of the centrifugal force, which is maximal at the equator. In contrast to the topographic surface, the level surfaces are smooth and always convex. Each surface is uniquely determined by its gravity potential value, as shown in Fig. 1. This makes them ideal reference surfaces, for example for heights. The deviations of the ocean surface from a level surface, due to wind, waves, or storms, are small, usually less than 2 m. Thus almost 70% of the earth's surface is practically a level surface. The level surface at mean sea level is called the geoid. As expressed by Newton's law of gravitation, the trajectory of any object in free fall is determined by the earth's mass distribution, whether it is the orbit of a satellite or the fall of the test mass of an absolute gravity apparatus. It is the objective of geodesy to determine the earth's surface and the gravity field at its surface and outside it, employing both terrestrial and space observations.

Encyclopedia of Earth System Science, Volume 2

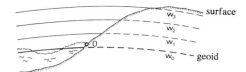

Fig. 1 A set of level (equipotential) surfaces with potential values W_0 to W_4; the geoid is at sea level.

II. Geodetic Concepts

A. Bruns' Polyhedron

The classical method of determination of the topographic surface is by triangulation. Characteristic terrain points are connected by triangles. The measurement of angles or, equivalently, of distance ratios gives the form of the triangles. Observing more than the minimally required number of elements permits, after adjustment, an evaluation of the precision and reliability of the network of triangles. The plumb line through a triangulation point serves as a reference for the measurement of the vertical angles with the sides to all adjacent network points. Thus the geometric form of a network is determined horizontally and vertically.

For larger point configurations it is convenient to introduce a coordinate system. In three dimensions this implies that the three coordinates of a fundamental point are chosen. The orientation is determined by fixing two coordinates of a second and one coordinate of a third point. One arbitrary side length defines the scale of the network. Independent networks (e.g., separated by oceans) can be positioned and oriented on the globe by means of a common external reference, the stars. Measurements are made at selected points of the vertical angles of stars and the horizontal angle between the direction to a star and to a terrestrial point.

However, these angles continuously change due to the earth's rotation, polar motion, and precession-nutation. Consequently these effects have to be carefully monitored and modeled. Only then can the star observation be converted into astronomical latitude and longitude and azimuth (angle of a point with respect to north). They are defined with respect to a conventional terrestrial coordinate system relative to the equator plane, the Greenwich meridian, and the North Pole, respectively. For mapping purposes a projection surface is chosen (e.g., a best-fitting ellipsoid) and the point coordinates are expressed, for example, by their geographical latitude, longitude, and height above the ellipsoid.

So far the gravity field has not played any real part, except for its role in defining horizontal and vertical. Angles and distances provide the geometric shape of point configurations, the astronomically determined latitude, longitude, and azimuth provide the proper location on the earth's surface. Figure 2 shows the basic situation. One drawback is that geometric heights (e.g., above a chosen ellipsoid) carry no

physical information relating to the flux of water. For this purpose the difference in gravity potential between the level surfaces passing through each of the terrain points has to be determined. The potential difference between two points P and Q is

$$W_P - W_Q = -\int_Q^P g\,dh \approx -\sum_i \Delta h_i g_i \qquad (1)$$

where W_P and W_Q are gravity potential in P and Q, respectively; g is scalar gravity; dh height increment; Δh_i leveled height increment; and g_i measured gravity at points i of the leveling line between P and Q. For general purposes the potential differences are converted into metric height differences (e.g., orthometric or normal heights). The heights of all points of a certain area are referred to one fundamental benchmark, the height datum. As the datum points are usually chosen close to sea level, one can speak of heights above sea level.

This concept of the determination of the figure of the earth was formulated by H. Bruns in 1878 and uses all measurement types available at that time. It is still valid today, although with significantly improved measurement techniques. The concept suffered in the past from two limitations. Vertical angles are usually distorted by atmospheric refraction and no direct connection between the geodetic networks of different continents was possible. Both drawbacks have been overcome by means of modern space techniques. A truly global polyhedron, covering the entire earth, can now be established. Crustal deformation studies in the horizontal and vertical directions require repeated application of the above principles.

B. Geodetic Boundary Value Problem

An alternative concept aims at the combined determination of the figure of the earth and its exterior gravity field. It is based on the assumption that all over the earth the four compare Eq. (1)], astronomical latitude Φ and longitude Λ, and scalar gravity g are available. The latter three quantities determine the direction and length of the gravity vector. Let us also assume that the gravitational effects of the sun and moon and of the earth's atmosphere are accounted for by

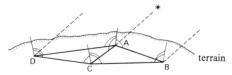

Fig. 2 Two triangles ABC and ACD of the Bruns' polyhedron with vertical angles referring to the plumb lines and with an external star reference.

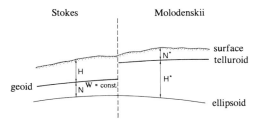

Fig. 3 The two fundamental approaches to the solution of the geodetic boundary-value problem: Stokes' method with the geoid and the method of Molodenskii with the telluroid.

means of corrections. The gravitational part of the gravity potential can then be regarded as a harmonic function. Hence the determination of the gravity potential in the exterior of the earth from its boundary function (i.e., from W on the earth's surface) could be considered a classical boundary-value problem of potential theory. However, as the boundary surface itself is also unknown, one is faced with a complicated nonlinear problem. For linearization purposes a globally best-fitting ellipsoidal reference gravity field is introduced.

The choice of the approximate surface characterizes the two alternative approaches to the solution of this so-called geodetic boundary-value problem (GBVP). One was conceived by G. G. Stokes in 1849, the other by M. S. Molodenskii in 1960. Stokes proposes reducing the given data from the earth's surface to the geoid. As the geoid is a level surface its potential value is constant. The difference between the reduced gravity on the geoid and the reference gravity on the ellipsoid is called gravity anomaly Δg. The difference T between the actual and the reference potential, the so-called disturbing potential, and the geoid height N (i.e., the height difference between geoid and ellipsoid) are unknown. The relation between Δg and T is, in spherical approximation,

$$\Delta g = -\left(\frac{\partial T}{\partial r} + \frac{2}{r} T\right) \qquad (2)$$

where $\partial T / \partial r$ is the derivative of T in a radial direction. T and N are connected by

$$N = \frac{T - \Delta W_0}{\gamma} \qquad (3)$$

where ΔW_0 is the unknown constant potential difference between geoid and ellipsoid and γ is the reference gravity.

In the case of Molodenskii the heights derived from the potential differences are directly planted on the ellipsoid to define an approximate surface that closely follows the earth's surface. It is called telluroid. In this case the geometric unknown is the height anomaly N^*, the deviation between the earth's surface and the telluroid. The geometric situation of both approaches is shown in Fig. 3. The disadvantage of the Stokes approach is that the reduction to the geoid re-

quires the introduction of assumptions concerning the unknown mass distribution between the earth's surface and the geoid. The solution of the GBVP is Stokes' formula

$$T = -\frac{\delta(GM)}{R} + \frac{R}{4\pi} \int_{\sigma} \mathrm{St}(\psi)\{\Delta g + G_1 + G_2 + \cdots\}\, d\sigma$$

$$(4)$$

It yields the disturbing potential at any point on the earth's surface and outside it. The first term on the right represents the uncertainty in the product gravitational constant times mass of the earth. The integration is carried out over a unit sphere; $\mathrm{St}(\psi)$ is the Stokes integral kernel where ψ is the spherical distance between computation and integration point. The terms G_1 and G_2 depend on the topographic heights of the terrain and appear only in the Molodenskii approach. Consequently, in moderately undulating terrain the Stokes and Molodenskii solutions practically coincide. The geoid height or height anomaly, respectively, are obtained by introducing (4) into (3). They complete the determination of the earth's figure, as seen from Fig. 3. The uncertainty $\delta(GM)$ is very slight; ΔW_0 is more of a problem, as each continent has its own fundamental benchmark, each very likely on a slightly different level surface.

This shows the conceptual vagueness of the geoid definition. With precise positioning by satellites, intercontinental height links can probably be established in the future. Recently solutions of the GBVP have been investigated that include more than the four boundary functions mentioned or that can accept different types of observable boundary functions on land and at sea.

The major assumption of the GBVP is that the boundary functions are given all over the earth. In fact observations are only available at discrete terrain points, and there are large areas, particularly at sea, where no data are given at all. The first problem is approached using methods such as numerical integration or least-squares collocation; the second limitation can be partly overcome by combining terrestrial gravity material with gravity field information deduced from satellite orbit analysis.

III. Geodesy and Earth System Processes

Geodesy provides a representation of the geometric shape of the earth, of its gravity field (expressed for example by the geoid), and of the variable orientation of the earth's body in space. Since the surface of the earth is directly accessible for measurement, more and better measurements, together with a suitably improved mathematical model, enable a better approximation of the earth's geodetic characteristics. In principle this improvement can be continued to almost any degree of perfection. At the moment, geodesy typically approaches a 10^{-8} relative precision, expressed, for example, relative to the earth's radius or gravity.

The parts of the earth system relevant to geodetic activity are the solid earth with its fluid outer core, the oceans, and the atmosphere. Since the earth's interior is not directly accessible to measurement, the physics have to be inferred indirectly from surface observations. Consequently, in contrast to the geodetic approach, the approach of solid earth physics is of a much more hypothetical character. In general it is agreed that our planet developed about 4.6 thousand million years ago and that during its first hundred million years it became segregated into core, mantle, and lithosphere with a preliminary hydrosphere and atmosphere. Heat accumulated in its interior. As the mantle material is of poor thermal conductivity, the most effective process of heat exchange is convection. Mantle convection, with a typical time scale of hundreds of millions of years, is the major process in the earth's interior. Plate tectonics, volcanic activity and earthquakes at the plate boundaries, and mountain-building are the visible results of the mechanism. Gravitation, on the other hand, is counteracting this process, trying to attain a smooth homogeneous zonation of core and mantle material according to its chemical and physical properties. At the earth's surface the smoothing mechanism is strongly supported by solar heat and weathering and includes every possible form of erosion. [See DYNAMIC TOPOGRAPHY; HEAT FLOW THROUGH THE EARTH; PLATE TECTONICS, CRUSTAL DEFORMATION AND EARTHQUAKES.]

All changes in the topographic surface related to this process can be monitored using geodetic methods. The gravity field reflects all mass inhomogeneities, the long spatial wavelengths being mainly due to mantle convection. Earth rotation and polar motion time series reveal the global response of the earth to the gravitational attraction of the sun and moon and to internal and external mass rearrangement.

The dynamics of the ocean and atmosphere represent much shorter time scales. Measurement of the dynamic topography of the oceans is therefore more complicated and requires a high precision and repetition rate. Only by means of precise altimetric measurements from satellites can these requirements be met. A secular change of sea level, which also affects ocean circulation, is caused by the melting of the ice caps. The post-glacial uplift due to the missing weight of the ice not only changes the topographic surface but also results in a secular variation of gravity.

The effect of the atmosphere on geodetic observables is more indirect. Half of the mass of the atmosphere is found below 6000 m altitude. Thus the topographic relief significantly interferes with the atmospheric circulation. The resulting exchange of angular momentum is clearly visible in earth rotation observations. The gravity field also contains a component from the gravitational attraction of the atmosphere. However, in contrast to the effect of atmospheric refraction

Table I   Geodesy and Earth System Components

Class[a]	Type	Component	Geophysical phenomenon
Ge	Leveling and point positioning	Lithosphere	Plate motion and deformation
Ge	Altimetry	Ocean	Variability, tides
Ge	Tide gauges	Ocean/ice/land	Sea level fluctuation
Ge/Gr	Altimetry and geoid	Ocean	Ocean circulation
Gr	Geopotential, gravimetry	Mantle and lithosphere	Convection, plate tectonics, isostasy
E.r.	VLBI, SLR, moon laser ranging	sun-moon, atmosphere, ocean, solid earth	Global response to angular momentum changes

[a] Geodesy divided into geometry (Ge), gravity field (Gr), and the earth's rotation (E.r.).

on geodetic measurements, these phenomena can be modeled very well. A summary of the connection between geodesy and the earth system components is given in Table I.

IV. The Contribution by Geodesy to Geodynamics

The possibilities for geodesy changed fundamentally with the advent of satellites. Only then did it become truly three-dimensional and global. A satellite can serve in geodesy as

- A measurement beacon at a high altitude with which points far apart can be connected
- A gyroscope, the orbit plane of which is stable enough to serve as an external direction reference
- A gravity probe in free fall, from the orbit of which the structure of the earth's gravitational field can be extracted
- A remote sensing platform, for measurement from space globally and with a high repetition rate

Since most or all of these features can be combined in one satellite mission, the division of geodetic concepts into figure, gravity field, and orientation determination becomes less appropriate. Nevertheless it makes sense to discuss the contribution to geodynamics along these lines.

A. Tectonic Deformations

The relative drift rates of lithospheric plates, based on paleomagnetic evidence, range typically from 3 to 5 cm year^{-1} to

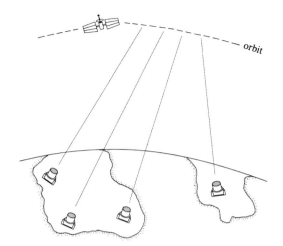

orbit

Fig. 4 Nonsimultaneous tracking of a satellite by four laser stations, three on one tectonic plate with their positions known and a fourth one on a second plate, the motion of which has to be determined.

maximum values of 18 cm year^{-1}. Despite the small rates and usually large distances between points on different plates, two geodetic measurement techniques are capable of providing *in situ* verification. These are laser ranging to artificial satellites and very long baseline interferometry (VLBI).

In the case of satellite laser ranging (SLR) the distance from ground station to satellite is derived from the travel time of emitted laser pulses. Reflectors mounted on the satellite direct the pulses back to the receiver at the ground station. The polyhedron principle can be followed again here. Assume that the configuration of three stations, located on one plate is known. Simultaneous distance measurements to a satellite yield its position. Inclusion of a fourth station located on a second plate permits the determination of the relative movement of the latter with respect to the other three.

At the present moment SLR attains centimeter precision. A complication arises from the fact that direct visibility is required. Clouds, for example, prohibit measurement. This means that simultaneous measurement of a satellite from four stations is hardly ever possible. Consequently, nonsimultaneous measurements have to be used with the computed orbit as intermediary, as shown in Fig. 4. As the dynamics of the motion of the satellite also play a part, the simple geometric concept is lost.

VLBI, operating at radio frequencies, is not weather-dependent. Radio telescopes located on different plates track signals from quasi-stellar objects (quasars). The signal emitted arrives at the earth as a sequence of parallel, planar wavefronts traveling with the velocity of light c. The signal is received at slightly different times at the individual stations.

Correlation analysis of the signal recorded at two stations gives the time delay τ. The product $s = c \cdot \tau$ represents, after correction for the earth's rotation, the length of projection of the baseline l between two stations in the direction of the radio source, as shown in Fig. 5. Again, the tracking of several quasars or the same source with different orientations enables the precise determination of the baseline and, after some years, its change with time.

Rates of plate motion, as derived so far from VLBI and SLR, show an overall agreement with geophysical tectonic rate models. For some more complex regions, such as the Mediterranean and Southern California, more stations and more sophisticated modeling are required. Transportable VLBI and SLR stations (see Fig. 6) provide more flexibility in this respect. It is expected that the station density and ease of measurement will improve significantly with the Global Positioning System (GPS). GPS will consist of 24 satellites in its operational phase and will enable the simultaneous tracking on radio frequencies of four or more satellites anywhere on earth. In this instance the geometric principle as described for the case of SLR is inverted, with the ground station taking the place of the satellite. The precision of absolute positioning is still limited, as it depends on the precision of the satellite orbit ephemerides. Relative positioning, by comparing the phase of the carrier waves received, yields precisions better than 10^{-7} of the measured baseline length. [*See* GLOBAL POSITIONING SYSTEM.]

At distances of 100 km and less, traditional geodetic triangulation and leveling are still very successful in monitoring horizontal and vertical crustal movements. Long measurement records have been established in this way for certain tectonically active zones. An example of such an area is Iceland, which forms part of the Reykjanes ridge.

B. The Earth's Rotation

Observations of stars and quasars and also the orbit plane of the moon and artificial satellites are used in geodesy to determine the orientation of networks and to help reduce distortions. For this purpose the variable orientation of the earth in space must be known. More precisely, a celestial coordinate system, represented by the positions of stars, has to be related to a terrestrial coordinate system linked to some fundamental points on earth. In view of today's high measurement standards even such small effects as plate tectonic motions have to be taken into account in the definition of these systems.

At the moment the connection between the two systems is established by VLBI, laser ranging to the moon, and SLR to very high and compact satellites. In particular VLBI provides earth-rotation parameters with unprecedented precision and data density. However, the older and less accurate

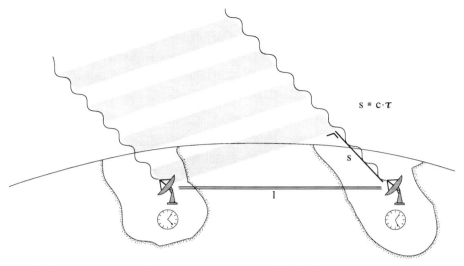

$$s = c \cdot \tau$$

Fig. 5 The principle of very long baseline interferometry (VLBI): The parallel wavefronts emitted from a quasar are received at one station with a time delay of τ relative to the second station. The time delay multiplied by the velocity of light c yields the distance s.

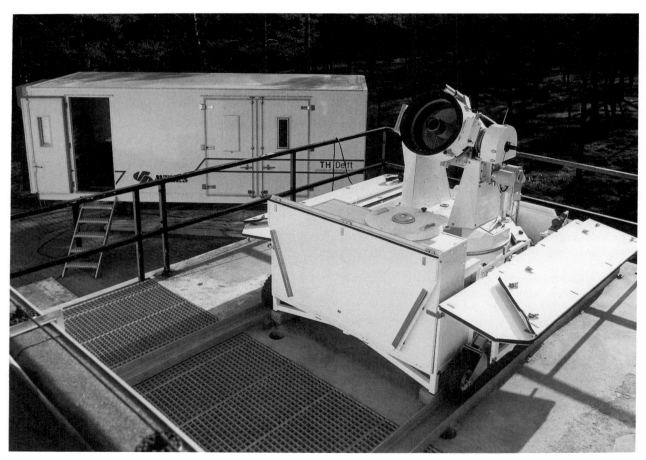

Fig. 6 The transportable laser ranging equipment of the satellite observatory of the Faculty of Geodetic Engineering, Delft University of Technology.

optical observations and historical records of astronomical phenomena are also important for the study of long periodic and secular effects.

The variable orientation between the two fundamental coordinate systems is split into three: precession-nutation, polar motion, and angular velocity of the rotation axis. Precession and nutation define the orientation of the earth's spin axis with respect to the celestial coordinate system. The equator plane is inclined by 23.5° with respect to the ecliptic and is not very different from that of the moon's orbit. Owing to its rotation the oblate earth reacts gyroscopically to the gravitational attraction of the sun and moon. The resulting precession of the equator plane has a period of 25,600 years. The complex interaction of the orbits of the sun and moon result in much smaller, shorter-period nutations. The earth's response to these gravitational torques depends on its internal structure. The precision of recent models, based on a rotating, elliptically stratified, elastic and oceanless earth with a fluid outer and solid inner core, already exceeds that of all older data. Small unmodeled annual nutations could be related to the free nutation mode, the pressure-coupling of the fluid core with the elliptical mantle boundary. [*See* CORE-MANTLE SYSTEM.]

Polar motion is the deviation of the instantaneous spin axis from some chosen mean pole position or simply the orientation of the spin axis with respect to the terrestrial coordinate system. The changes in the length of day (l.o.d.) are the variations in the earth's angular velocity. Polar motion, split into a north–south and an east–west component, and the changes in the l.o.d. are treated together as three elements of a vector. Euler's theory shows that a rigid body can wobble freely when its spin axis is displaced from either its axis of greatest or least rotational inertia. The free oscillation of 304 days, however, is very different from the so-called Chandler period, actually observed, of 428 days.

Modification of Euler's equations to apply to a nonrigid planet describes reality much more closely. These are the Liouville equations. They relate in linear approximation the changes in angular momentum to the three vector components. The major part of the Chandler wobble can be modeled employing an elastic earth model with fluid outer core and an ocean in equilibrium. The remaining differences may be attributable to inelasticity in the mantle. The important issue is how this oscillation is maintained or excited in the presence of dissipation. The three possible mechanisms currently under discussion are the integral effect of all earthquakes, atmospheric forcing, or electromagnetic torques between core and mantle. It seems doubtful whether this question can be resolved from polar motion data alone, despite their accuracy. Apart from the Chandler period, the polar motion spectrum shows an annual term. The latter can be clearly explained by the seasonal redistribution of air and water masses. Finally, there is a secular drift of the pole, the

polar wander. Although still uncertain, it could be due to a post-glacial readjustment of masses in the earth's mantle.

The spectrum of l.o.d. data displays a large variety of amplitudes. The secular and long-periodic part of 1–2 ms per century is very likely due to tidal friction and sea level changes. Long-period tidal components change the earth's polar moment of inertia and cause changes in l.o.d. with periods of 6 months to 14 days. Exchange in angular momentum between topography and the atmosphere accounts for more than 90% of the seasonal, nontidal amplitudes. A spectacular example of this type is the El Niño event of 1983, which can clearly be identified in the l.o.d. records. [*See* EL NIÑO AND LA NIÑA.]

In conclusion, precession-nutation, polar motion, and the changes in l.o.d. can now be measured and modeled very accurately. However, the relation between observable effect and physical cause is indirect. An observed effect is the integral response of the rotating earth to the exchange of energy between the main system components of the earth. Inversion, the determination of the individual causes, may therefore prove very complicated. The effects of the earth's rotation are summarized in Table II. [*See* POLAR MOTION AND EARTH ROTATION.]

C. Gravity Field

Assume the earth to be a rotating, perfectly homogeneous sphere. In such a case the orbit of a satellite would be an ellipse. The inclined orbit plane would remain fixed in space. In the case of an oblate earth the orbit plane would precess around the polar axis and the ellipse in the orbit plane around the earth's center of mass. Even in the late 1950s the analysis of the orbit precession of satellites led to very reliable estimates of the earth's flattening. The orbit of a satellite in the actual gravity field of our planet, in the presence of the atmosphere and of the sun and moon, is more complicated. However, the oblate earth situation still serves as a good approximation, in particular for linearization.

The link between tracking observations, such as ranges, range rates, or Doppler measurements and the earth's gravity field is established by Newton's second law and his universal law of gravitation. The gravity field is for this purpose usually expressed in a series of spherical harmonics, with the series coefficients as unknowns. A complete set, up to a certain degree and order, is called a geopotential model. From the analysis of a very large number of observations involving many different satellites with different orbit characteristics, detailed descriptions of the earth's gravity field have been derived. A recent example is the GEM-T1 model developed by NASA Goddard Space Flight Center. It is complete up to degree and order 36, comparable to a spatial resolution of about 500 km. The geoid, computed relative to a best-fitting ellipsoid (flattening $f = 1:298.247$) displays variations up to

Table II   The Earth's Variable Rotation

Phenomenon	Time scale	Physical cause	Unsolved
Precession/ nutation	18.6 yr – 4.7 days	Sun and moon on oblate and stratified earth	Free nutation mode
Polar motion	Secular (wander)	Viscous long-term stress (e.g., post-glacial)	
	428 days (Chandler)	Free wobble of a nonrigid planet	Effect of inelasticity, excitation, and maintenance
	Annual	Seasonal redistribution air/ water	Contribution of ground and surface water
Changes in l.o.d (variable spin)	Secular	Tidal dissipation	
	Decade	Core – mantle coupling	
	Seasonal (nontidal)	Zonal winds	
	6 months – 13 days	Change in earth's moment of inertia due to sun and moon	

100 m; see Fig. 7. Comparison of the geoid with a hydrostatic equilibrium figure ($f = 1 : 299.63$) reveals about a 200-m nonhydrostatic signal. At wavelengths greater than 4000 km 90% of the observed geoid can be explained by the effect of density variations in the mantle, based on the results of seismic tomography and modeling of subducted slabs. At shorter spatial wavelengths the gravity field contributes significantly to the study of (1) the structure of the oceanic lithosphere and its interaction with the asthenosphere and (2) the thermal structure of the continental lithosphere and the driving forces for continental tectonics. The resolution of geopotential models is not sufficient for this purpose. One has to

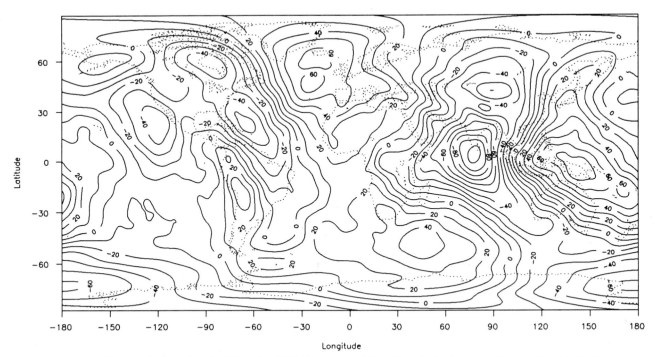

Fig. 7   Geoid heights (in meters) derived from the GEM-T1 geopotential model and referring to a reference ellipsoid with flattening 1 : 298.247.

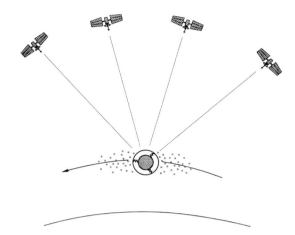

Fig. 8 Concept of simultaneous tracking of a satellite in a low orbit by four GPS satellites. The low satellite is drag compensated as indicated by the inner proof mass (the cloud symbolizes the perturbing influence of the atmosphere).

rely on terrestrial and shipborne gravimetry or gravity indirectly derived from satellite altimeter measurements above the oceans.

The attenuation of the gravitational attraction with altitude imposes a natural limit on any geopotential model improvement from satellites. Currently, further progress is expected from continuous tracking of satellites or from gradiometry. The Global Positioning System or similar future navigation systems will enable continuous tracking of a low-flying satellite serving as a gravity probe by several high-orbiting satellites simultaneously. In contrast, ground stations typically cover only 10% of a full revolution. Consequently, short periodic gravity effects in the orbit can be determined.

The use of drag-free compensation for the low-flying spacecraft also eliminates the effects of atmospheric drag. The principle is shown in Fig. 8. A satellite gradiometer, on the other hand, measures the derivatives of the components of ∇V. Geometrically speaking it measures the gravitational curvature in space. In principle the relative motion of neighboring, free-falling test masses is observed. Gradiometer measurements are highly sensitive to the fine structure of the earth's gravity field.

LAGEOS is a geodetic satellite specially designed for crustal-deformation and earth-rotation studies. It is compact, equipped with reflectors, and circles in a high orbit (6000 km). Time variations in the earth's gravity field have been derived from several years of laser ranging of LAGEOS and careful orbit analysis. There is a secular drift in the earth's flattening, which can very likely be attributed to post-glacial mass readjustment. The solid earth and the ocean tides also have an effect. This may help to improve our understanding

of the dissipation of tidal energy and, if the frequency dependence of the Love numbers can be shown, provide a measure of mantle inelasticity. Traditionally the tidal response of the deformable earth is studied on its surface with precise gravimeters, tilt meters, and strain meters. [*See* TIDES.]

D. Geodesy and the Oceans

Any deviation of the topographic surface from the geoid implies imbalance, and this imbalance results in dynamic processes. This applies both for land topography and for the topography of the oceans. Naturally, dynamic or sea-surface topography (SST) is much smaller and the time-scales of the variations are much shorter.

Before the space age the oceans were scarcely accessible to geodetic activity. Not until the 1970s, with the launch of the first altimeter satellites, did the situation change profoundly. Now satellite radar altimeters measure with centimeter precision the distance from the spacecraft to the ocean surface. Combined with the accurately determined orbit, the geometric height h of the instantaneous sea surface above a chosen reference ellipsoid is obtained, a shown in Fig. 9. The choice of the orbit elements determines the spatial and temporal coverage of the oceans by measurements and their northern and southern latitude bounds.

What type of information can be extracted from the measured sea surface heights? First, they approximate the geoid, since the SST is small and can be partly corrected for. Second, repeated measurement of the same ground track configuration provides the temporal variations of the ocean surface, which is useful for studies of changes in global circulation patterns, meandering of ocean currents and eddy motion and also for tidal modeling in open oceans and shelf seas. Finally, combination of sea surface heights with an independently determined precise geoid yields the SST—a measure of the

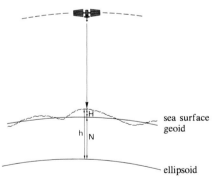

Fig. 9 Principle of satellite altimetry: The distance from the satellite to the ocean surface is measured. The geometric height h of the sea surface consists of the geoid height N and the sea surface topography H.

surface dynamics of the oceans that is essential, for example, for the analysis of heat transport. The full capability of the latter for ocean modeling will become apparent when geoid models are further improved. [*See* OCEAN CIRCULATION.]

Sea-level rise or fall has been studied in the past in conjunction with height system definition and research on regional recent crustal movement. The tide-gauge measurement records used for this purpose extend back to more than one hundred years for some stations. Currently there is a growing awareness that anthropogenic climate changes might lead to a more rapid melting of the remaining ice caps. The implications of this are not yet well understood. Any study of the consequences, however, must rely on highly precise and reliable measurement records of a worldwide sea-level monitoring system. Only then can the global pattern be studied, the crustal motion be separated from oceanic effects, and trend changes recognized early. The integration of tide gauges into one system will be achieved by precise satellite positioning and VLBI. [*See* DEGLACIATION, IMPACT ON OCEAN CIRCULATION; SEA LEVEL FLUCTUATIONS.]

Bibliography

Baker, T. F. (1984). Tidal deformations of the earth. *Sci. Progr.* **69**(274), 197–233.

Carter, W. E., and Robertson, D. S. (1986). Studying the earth by very-long-baseline interferometry. *Sci. Am.* **255**(5), 44–52.

Hager, B. H., and Richards, M. A. (1989). Long-wavelength variations in earth's geoid: physical models and dynamical implications. *In* "Seismic Tomography and Mantle Circulation" (R. K. O'Nions and B. Parsons, eds.), pp. 309–327. The Royal Society, London.

Kovalevsky, J., Mueller, I. I., and Kolaczek, B., eds. (1989). "Reference Frames in Astronomy and Geophysics." Kluwer Academic Publishers, Dordrecht, The Netherlands.

Lambeck, K. (1988). "Geophysical Geodesy." Clarendon Press, Oxford.

Moritz, H. (1980). "Advanced Physical Geodesy." Abacus Press, Tunbridge Wells, England.

Moritz, H., and Mueller, I. I. (1987). "Earth Rotation." Ungar, New York.

Mueller, I. I., and Zerbini, S., eds. (1989). "The Interdisciplinary Role of Space Geodesy." Springer-Verlag, Berlin.

Rummel, R. (1984). Geodesy's contribution to geophysics. *ISR, Interdiscipl. Sci. Rev.* **9**(2), 113–122.

Sansó, F., and Rummel, R., eds. (1989). "Theory of Satellite Geodesy and Gravity Field Determination." Springer-Verlag, Berlin.

Vaniček, P., and Krakiwsky, E. (1982). "Geodesy: The Concepts." North-Holland, Amsterdam.

Glossary

Geoid Level (or equipotential) surface at mean sea level.

Geopotential model Set of coefficients of a series representation of the earth's gravitational field in terms of spherical harmonics.

Gravity Combined effect of the gravitational mass attraction of the earth and the centrifugal force caused by the earth's rotation.

Polar motion Path of the earth's spin axis relative to an earth fixed-coordinate system

Sea-surface topography Deviation of the ocean surface from the geoid.

Geomagnetic Jerks and Temporal Variation

M. G. McLeod

Naval Oceanographic and Atmospheric Research Laboratory, Stennis Space Center

Geomagnetic temporal variations, or time variations of the earth's magnetic field, cover a wide spectrum that extends over more than 20 orders of magnitude in frequency. The spectrum extends from frequencies greater than 1000 Hz to periods of more than 100 million years. Sources of the geomagnetic field and its time variations are electric currents located both internal and external to the surface of the earth. The major portion of this article is concerned with periods from 1 year to several hundred years.

I. Introduction

The natural magnetic field observed at the earth's surface is approximately the field of a magnetic dipole. This field is believed to be generated by a self-excited dynamo that depends on the earth's rotation. Motion of the electrically conducting fluid outer core of the earth drags magnetic field lines along with the fluid and results in the formation of higher multipoles in the geomagnetic field. Magnetic diffusion causes the decay of these higher multipoles so that there is a balance between regeneration and decay. [*See* CORE; GEOMAGNETISM; GEOMAGNETISM, ORIGIN.]

A solar wind of ionized particles emanates continually from the sun. This wind interacts with the geomagnetic field and confines the field to a cavity. The boundary of the cavity is called the magnetopause, and the region within the cavity is called the magnetosphere. The magnetopause is located at about 12 earth radii from the earth's center on the side of the earth that faces the sun. The magnetosphere extends over 1000 earth radii from the center of the earth in a long tail (the magnetotail) in the antisolar direction. Electric currents flow on the magnetopause. Because the solar wind flows past the earth at a speed greater than the speed at which plasma waves propagate in the ionized solar wind, a shock front is formed a few earth radii from the magnetopause in the solar direction. Electric-current systems are present within the magnetosphere, including a ring current at about 3 earth radii from the center of the earth near the plane of the geomagnetic equator, ionospheric currents a few hundred kilometers above the surface of the earth, and field-aligned currents that originate in the magnetotail and flow through the ionosphere in the auroral regions. All of the external current systems vary with time because of varying solar activity. Time variations of the external current systems produce time-varying magnetic fields that induce electric currents within the earth by electromagnetic induction. [*See* IONOSPHERE; MAGNETOSPHERIC CURRENTS; SOLAR–TERRESTRIAL INTERACTION.]

The frequency spectrum of that part of the geomagnetic field produced by external-current systems extends from the sunspot frequency (11-year period) to more than 1000 Hz. There is also an average (or DC) component on time scales of a few hundred years and probably some power in the range from DC to the sunspot frequency as well, although this energy has not been demonstrated. The spectrum of that part of the geomagnetic field due to internal sources extends from periods of about 1 year to 100 million years. This article is primarily concerned with periods from 1 year to several hundred years.

The geomagnetic field and its time variations are of interest, not only for their own sake but also for the knowledge that can be acquired about some of the earth's physical properties, such as electrical conductivity of the mantle and motion of core fluid. Benefits and applications of this knowledge include improved magnetic charts and magnetic field models. These charts and models are valuable tools for accurate navigation and related uses, such as determining an earth satellite's orientation. [*See* CORE–MANTLE SYSTEM; MANTLE.]

II. Geomagnetic Measurements

Approximately 180 permanent magnetic observatories are currently in operation (1990). New observatories are estab-

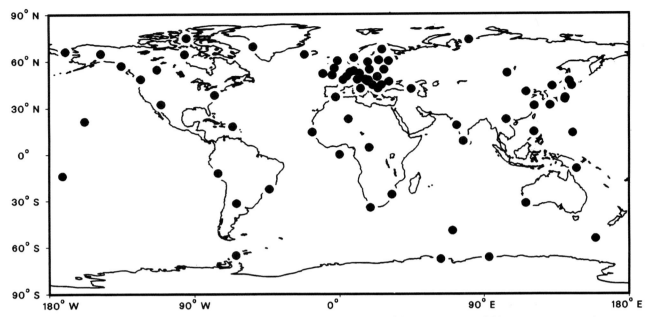

Fig. 1 Locations of 73 magnetic observatories that provided annual means of three geomagnetic-field vector components for each year from 1961 through 1983.

lished occasionally and old ones sometimes shut down. Between 50 and 100 observatories are used for most current global geomagnetic analyses. Locations of 73 observatories that had vector data for each year from 1961 through 1983 are shown in Fig. 1. Distribution of magnetic observatories is uneven—many are located in Europe and Japan, and relatively few are located in the southern hemisphere. A large region of the South Pacific has no observatories with continuous vector data for 1961–1983.

All-day annual means of vector magnetic-field components at magnetic observatories are commonly used as input data for models of the global geomagnetic field that cover the frequency spectrum from periods of about 1 year to 100 years. These all-day means are averages of field components for all hours of the day and all days of the year. Other types of annual means are also available for some observatories; for example, quiet-day annual means, which are averages of the five magnetically quietest days for each month of the year, averaged over the year. Magnetically quiet days are those days for which fluctuations of the field about its nominal value are minimal; the precise way in which these fluctuations are defined and measured is rather complex.

First time differences of the all-day annual means for two field components at the magnetic observatory in Eskdalemuir, Scotland are shown in Fig. 2. The two time series were filtered with a simple lowpass filter for which the weights are 0.25, 0.50, and 0.25. This filter has a zero at the Nyquist frequency of 0.5 cycle/year, so that frequencies for which

temporal aliasing errors are most severe have been highly attenuated. The east component of the first-difference field shows a pronounced change of slope at 1969.

Similar large slope changes of first differences of annual means were observed near 1969 at observatories located worldwide. This sudden slope change was designated a secular variation impulse, or geomagnetic jerk, and is discussed later in this article. The north component of the first-difference field shows an oscillation with about 11-year period. This oscillation is largely due to external current systems that are modulated by the sunspot cycle frequency. The 11-year oscillation is also evident for the east component of the first-difference field, but is much smaller for the east component than for the north component.

In addition to permanent magnetic observatories, measurements are made on a regular basis at relocatable stations, also called repeat stations. For example, measurements are made every 5 years at a French array of about 30 repeat stations. Measurements are also made from ships and aircraft.

Earth satellites have been used in recent years to measure the global magnetic field. The most recent and complete such satellite was MAGSAT (Magnetic Field Satellite, launched by the National Aeronautics and Space Administration), which obtained high-quality data for seven months during 1979–1980. MAGSAT's orbit was nearly polar, so that data were obtained for nearly the entire earth. In time, data from magnetic-field measuring satellites should increase our understanding of geomagnetic time variations, if more such

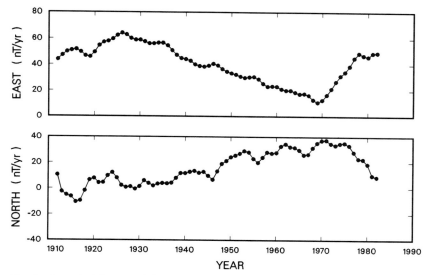

Fig. 2 Smoothed first time differences of annual means of X (north) and Y (east) magnetic field components at Eskdalemuir, Scotland.

satellites are put into orbit. Current satellite data are not available for sufficiently long periods to contribute significantly to our knowledge of geomagnetic time variations in the frequency range covered by this article.

There were not enough magnetic observatories before the twentieth century to permit construction of highly accurate global models of the geomagnetic field. In 1986, ship records were used to model the field for the eighteenth and nineteenth centuries in much greater detail than had been previously possible. These models were used to study time variation of the field.

The geomagnetic field also varies on a much longer time scale than a few hundred years. The field is nearly always approximately that of a magnetic dipole aligned roughly in the direction of the spin axis of the earth; however, the dipole axis reverses direction from time to time. The time between reversals varies tremendously; this time is typically a few hundred thousand years, but a period of 50 million years with no reversals has been observed. Our knowledge of geomagnetic behavior before instrumental measurements became available was derived by methods of archaeomagnetism and paleomagnetism. Human artifacts (e.g., bricks or hearths) and natural rocks (e.g., lava flows) that were once heated above Curie temperature recorded strength and direction of the geomagnetic field when these objects cooled below Curie temperature. The past geomagnetic field is also recorded by the process of sea-floor spreading, wherein molten material from the earth's mantle rises at the mid-Atlantic ridge to form the sea floor where the European and North American plates are separating from each other. Lake sediments also form a record of past geomagnetic behavior; these sediments are

slightly magnetized, and when they were deposited they aligned themselves in the direction of the geomagnetic field. [*See* DEPOSITIONAL REMANENT MAGNETIZATION; PALEOMAGNETISM.]

III. Geomagnetic Modeling

A. Spherical Harmonic Analysis

In a closed region that contains no magnetic sources (i.e., no electrical currents), the geomagnetic field can be derived from a potential function that satisfies Laplace's equation. For most purposes, it can be assumed that there are no significant currents in the region between the earth's surface and the bottom of the ionosphere. The magnetic potential for that region can be expanded by the separation-of-variables method into a series of orthogonal functions, represented as

$$V = a \sum_{n=1}^{\infty} \sum_{m=0}^{n} (a/r)^{n+1}(g_n^m \cos m\phi + h_n^m \sin m\phi) P_n^m(\cos \theta)$$
$$+ a \sum_{n=1}^{\infty} \sum_{m=0}^{n} (r/a)^n(q_n^m \cos m\phi + s_n^m \sin m\phi)P_n^m(\cos \theta) \quad (1)$$

where (r, θ, ϕ) are spherical coordinates, a is the mean radius of the earth (6371.2 km), $P_n^m(\cos \theta)$ are seminormalized Legendre polynomials, and g_n^m, h_n^m, q_n^m and s_n^m are the spherical harmonic coefficients. The magnetic field vector is the negative gradient of V. The first summation with terms containing the factor $(a/r)^{n+1}$ arises from sources within the earth; the second summation, with terms containing the factor $(r/a)^n$, arises from sources outside the earth. Equation (1) is valid in the region between sources where the magnetic field is a potential field. The terms $\cos m\phi\, P_n^m(\cos\theta)$ and $\sin m\phi\, P_n^m(\cos\theta)$ are called spherical harmonics. Spherical harmonics for which $m = 0$ are called zonal harmonics. Index n is called the degree of the harmonic, and index m is called the order of the harmonic.

Since the summations in Eq. (1) extend to infinity, the expression for V must be approximated by a finite set of terms to compute the spherical harmonic coefficients from vector field measurements at a finite set of measurement sites, such as magnetic observatories. Field components at each site can be computed from a finite approximation for Eq. (1) as a function of the unknown spherical harmonic coefficients. These unknown coefficients can then be estimated by fitting the computed field to the measured field in some fashion, such as ordinary least squares or weighted least squares.

Carl Fredrich Gauss developed the method of spherical harmonic analysis for the magnetic potential function in 1838 to separate magnetic fields due to sources internal to the earth from magnetic fields due to sources external to the earth. Gauss was able to show that substantially all of the geomagnetic field is due to sources internal to the earth, within the accuracy achievable at that time. Spherical harmonic coefficients for the geomagnetic potential are often called Gaussian coefficients in his honor.

The portion of the mean-squared geomagnetic field at the earth's surface associated with spherical harmonics of degree n can be computed from the Gaussian coefficients as

$$R_n = (n+1) \sum_{m=0}^{n} [(g_n^m)^2 + (h_n^m)^2]$$
$$+ n \sum_{m=0}^{n} [(q_n^m)^2 + (s_n^m)^2] \qquad (2)$$

where the first summation is the portion due to internal sources and the second summation is the portion due to external sources. The best determinations of the values for R_n are based on data obtained from MAGSAT in 1980. Only degree-one spherical harmonic coefficients could be determined for the portion of the field due to external sources. The degree-one external field was found to be about 20 nT on magnetically quiet days. Values of R_n were found to be given approximately by the equations

$$R_n = 1.349 \times 10^9 (0.270)^n \qquad (2 \le n \le 12) \qquad (3a)$$

and

$$R_n = 37.1(0.974)^n \qquad (16 \le n \le 23) \qquad (3b)$$

where the units for R_n are $(nT)^2$.

The ratio $(c/a)^2$ is 0.299, where c is the radius of the earth's core and a is the mean radius of the earth. Equation (3a) implies that for $2 \le n \le 12$, the mean-squared field at the core–mantle boundary is approximately the same for each degree; for $16 \le n \le 23$, the mean-squared field at the earth's surface is approximately the same for each degree. This observation has been interpreted to mean that values of R_n given by Eq. (3a) are due to core sources and that values of R_n given by Eq. (3b) are due to crustal sources. The mantle has been treated as an insulator in the discussion, so it has been assumed that Eq. (1) is valid within the earth down to the core–mantle boundary.

When making a spherical harmonic model from magnetic-observatory annual means by an ordinary least-squares or a weighted least-squares procedure, it is necessary to decide which spherical harmonics to include in the model. External-source terms are not included in most models and, because values of R_n decrease with increasing n according to Eq. (3a), the usual approach is to truncate the expression given by Eq. (1) at some maximum degree N. If N is chosen too small, the resolution as well as the accuracy of the model suffers. If N is chosen too large, the model tends to take on unrealistically large values in regions where the data density is low. Because the distribution of magnetic observatories is very uneven, the value of N appropriate for one region of the earth is not the value appropriate for another region. One analysis method that does not suffer from this limitation is stochastic inversion (see the following section). This method uses assumed *a priori* statistical properties of the magnetic field and measurement errors.

The same principles that apply to making spherical harmonic models of magnetic-observatory annual means apply to making spherical harmonic models of first differences of the magnetic-observatory annual means. Such models amount to filtered versions of the first time derivative of the geomagnetic field. These models are useful for studying geomagnetic temporal variation.

B. Stochastic Inversion

The term stochastic inversion refers to estimating the parameters of a physical model from a set of measurements using the statistical approach pioneered by Norbert Wiener and Andrey Nikolayevich Kolmogorov. Let $\mathbf{y}$, $\boldsymbol{\beta}$, and $\boldsymbol{\epsilon}$ be random vectors with zero mean and assume that

$$\mathbf{y} = W\boldsymbol{\beta} + \boldsymbol{\epsilon} \qquad (4)$$

where $\mathbf{y}$ is a known m-dimensional data vector, $\boldsymbol{\beta}$ is an unknown n-dimensional parameter vector, W is a constant $m \times n$ matrix, and $\boldsymbol{\epsilon}$ is an unknown m-dimensional error vector. For example, $\mathbf{y}$ could represent magnetic field component annual means at a set of magnetic observatories, $\boldsymbol{\epsilon}$ could represent errors in the annual means, $\boldsymbol{\beta}$ could be coefficients of a spherical harmonic expansion of the magnetic potential given by Eq. (1) truncated at sufficiently high degree that truncation errors are negligible, and W could be a matrix derived from Eq. (1) and magnetic observatory locations.

Let Q be the data-error covariance matrix and let R be the *a priori* parameter covariance matrix so that

$$Q = E[\boldsymbol{\epsilon}\boldsymbol{\epsilon}^T] \qquad (5a)$$

and

$$R = E[\boldsymbol{\beta}\boldsymbol{\beta}^T] \qquad (5b)$$

where $E[\ ]$ is the average or expected value and superscript T indicates the transpose of a matrix. Wiener and Kolmogorov have shown that the best linear estimate $\hat{\boldsymbol{\beta}}$ of $\boldsymbol{\beta}$ is given by

$$\hat{\boldsymbol{\beta}} = (W^TQ^{-1}W + R^{-1})^{-1} \, W^TQ^{-1}\mathbf{y} \qquad (6)$$

and that the parameter-error covariance matrix P is given by

$$P = (W^TQ^{-1}W + R^{-1})^{-1} \qquad (7)$$

Estimate $\hat{\boldsymbol{\beta}}$ is the best linear estimate of $\boldsymbol{\beta}$ in the sense that individual diagonal elements of the parameter-error covariance matrix P are minimized.

For the special case that R^{-1} is negligible in comparison to $W^TQ^{-1}W$, Eq. (6) becomes the usual weighted least-squares estimate of parameter vector $\boldsymbol{\beta}$. If, in addition, Q is a multiple of a unit matrix, then the solution given by Eq. (6) is the usual ordinary least-squares estimate.

Because of the presence of the parameter covariance matrix R in Eq. (6), a solution to Eq. (6) will generally exist even if the number of unknowns exceeds the number of measurements, unless some of the diagonal elements of R^{-1} are zero. Therefore, Eq. (1) can usually be truncated at a sufficiently high degree that truncation errors are negligible. This contrasts with the requirement that the number of measurements exceed the number of unknown parameters for an ordinary least-squares estimate.

The best estimate (in the sense of minimum expected error) for any linear function of $\boldsymbol{\beta}$ is that same function of $\hat{\boldsymbol{\beta}}$. Therefore, field components computed from Eq. (1) at any location where Eq. (1) is valid, using coefficients given by $\hat{\boldsymbol{\beta}}$, will be the best estimate of these field components.

The method of stochastic inversion can be extended to the nonlinear case. An example where the nonlinear case is useful is modeling of the geomagnetic field from measurements that include scalar field measurements, that is, measurements of the field vector magnitude. The field vector magnitude is the square root of the sum of the squares of the field components and is, therefore, a nonlinear function of the spherical harmonic coefficients for the magnetic potential. Stochastic inversion, both linear and nonlinear cases, has been used by

several scientists to model the geomagnetic field and geomagnetic secular variation. Examples of secular variation at various magnetic observatories, computed from models produced using stochastic inversion, are described in the following section of this article.

IV. Geomagnetic Secular Variation

A. Westward Drift

The astronomer Edmund Halley was first to notice westward drift of some features of the geomagnetic field, near the end of the seventeenth century. Drift rates on the order of 0.2° per year have been estimated in the twentieth century.

Secular variation of the geomagnetic field is considerably more complex than a simple rotation of the field about the earth's rotation axis, such as would be expected if secular variation were due only to a simple rotation of the earth's fluid core relative to the solid mantle and crust. For example, the magnetic dipole moment has been decreasing for the last two centuries at a rate of about 5% per century. If that rate of decrease were to continue, the dipole moment would vanish in about 2000 years. Some scientists, however, believe this phenomenon indicates an impending magnetic field reversal.

Numerical values determined for westward drift depend on how this drift is defined and measured. The rotation rate for individual spherical harmonics can be defined and determined independently; different harmonics are found to rotate at different rates. Some scientists have defined an overall rotation rate as a particular weighted average of rotation rates for individual harmonics. Other scientists define rotation rate as that value of angular rotation rate that produces a secular variation field that best fits the observed secular variation field. With this definition the numerical value of rotation rate depends on whether the best fit is found at the earth's surface or at the surface of the core. The numerical value of rotation rate also depends on the specific feature of the secular variation field that is fitted; for example, the scalar field, the vector field, or a particular field component can be fitted. Largest values of rotation rate are found when the east–west field component is fitted. The numerical value of rotation rate, as determined by fitting secular variation of the east–west field component, varied during 1950–1980 from a maximum value of nearly 0.2° per year to a minimum of about 0.1° per year. Minimum occurred near 1969 at the time of the geomagnetic jerk displayed in Fig. 2.

Westward drift of the geomagnetic field can be observed in global contour plots of field components for different epochs. For example, a comparison of contour plots of the vertical field component at the earth's surface for the years 1715, 1777, 1842, 1905, and 1980 shows that the zero vertical-field contour has moved progressively westward. This contour is called the magnetic dip equator. Besides westward drift, some changes in shape of the contour are also apparent. A general westward drift is evident in some but not all of the other features of the geomagnetic field at the earth's surface. Contour plots of field components at the core–mantle boundary, computed from field models, also reveal general westward drift of the geomagnetic field; however, different contours drift westward at different rates, and some contours drift eastward. Contour drift is generally believed to reflect the motion of core fluid; contour plots of field components at the core–mantle boundary reveal this fluid motion with a finer spatial resolution than similar plots for the earth's surface.

B. Geomagnetic Jerks

Three French scientists, Vincent Courtillot, Joel Ducruix, and Jean-Louis LeMouel, reported in 1978 that a sudden change in slope of the secular variation had occurred in 1969 for the east–west component of the geomagnetic field in Europe. They called this slope change a secular variation impulse. The slope change was termed the "geomagnetic jerk of 1969" by a number of other scientists. The geomagnetic jerk of 1969 is apparent on the graph of the east–west component of secular variation shown in Fig. 2. Observatories worldwide noted a similar sharp change of slope near 1969 for other vector magnetic-field components. However, the slope change was not detected for each field component at each observatory. These slope changes appear to take place in a year or two at some observatories.

Research on geomagnetic jerks is continuing. There is some controversy in the scientific community concerning time duration and worldwide extent of the geomagnetic jerk of 1969. The jerk is generally believed to be due to sources within the earth's core; however, some scientists have suggested that the apparent short time duration may be due to superposition of fields from sources external to the earth. Some research at the U.S. Naval Oceanographic and Atmospheric Research Laboratory has been directed toward this subject; some results of that research are described in following paragraphs.

In Fig. 3, the magnetic north component of the first time derivative of the geomagnetic field is shown plotted against time. Magnetic north is defined in terms of the magnetic dipole axis, in contrast to geographic north, which is defined in terms of the earth's rotation axis. The first column of Fig. 3 shows the magnetic north component of the first derivative of the field at three magnetic observatories: Honolulu, Hawaii; Huancayo, Peru; and Hermanus, South Africa. Annual means of the magnetic-north field component were computed from all-day annual means of the three vector field components. First time differences of magnetic-north annual means are plotted against time in the figure. The data were

not filtered beyond the filtering implicit in finding annual means and taking first differences of annual means.

Time-variation fields due to internal sources and external sources were separated from each other and from noise—to the degree that separation is possible. First time differences of annual means of magnetic-field components from a set of worldwide observatory (see Fig. 4) data were analyzed by the method of stochastic inversion. Inversion was for each year individually, within the 1911–1983 interval. Not all observatories had data for each year, but the distribution of observatories was approximately the same for each year. To carry out the stochastic inversion, a data-error covariance matrix and an *a priori* parameter covariance matrix were estimated. Both matrices were taken to be diagonal.

First time differences of the magnetic north component of the geomagnetic-field annual means, as computed from the stochastic inversion models, are plotted against time (Fig. 3) for three magnetic observatories. Portions of the first differences due to external sources (column 2) and internal sources (column 3) correspond to the two summations in Eq. (1). Column 4 is the difference between column 1 and the sum of columns 2 and 3.

The external source portion of the first derivatives has nearly the same waveform at the three observatories, because most of the model external-source field derivative is due to a single term in the spherical harmonic expansion of Eq. (1). This single term is the external first-degree zonal harmonic in geomagnetic coordinates. Model internal-source field derivatives are considerably smoother than "measured" derivatives or model external-source derivatives. This observation is good evidence that the stochastic inversion model effectively separates internal-source and external-source field derivatives. Sharp changes of slope for internal-source field derivatives are evident in "measured" derivatives.

Figure 5, which is similar to Fig. 3 in format, illustrates the first time derivatives of three vector field components at two magnetic observatories near Washington, D.C. Data before 1956 are from the observatory at Chaltenham, Maryland, and data after 1956 are from the observatory at Fredericksburg, Virginia. External-source and internal-source field derivatives, as well as residuals, are shown; these quantities are computed from the same set of stochastic inversion models described. As in Fig. 3, the magnetic-north internal-source derivative of Fig. 5 is considerably smoother than the corresponding external-source derivative or "measured" derivative. The vertical internal-source derivative is also smoother than the corresponding external-source or "measured" derivatives for years after 1956. Before 1956, larger residuals suggest that greater roughness of the internal-source derivative for this period is due to "noise" still present in the model. The model has no external-source magnetic-east component.

The magnetic-east component of the field derivative is shown in Fig. 6 in the same format as used for Fig. 3. Figure 6

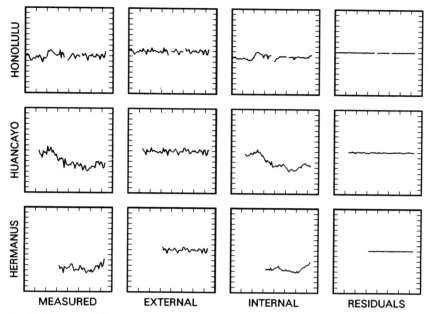

Fig. 3   First time differences of annual means of geomagnetic field magnetic-north component at three different magnetic observatories. Horizontal axes extend from 1910 to 1990. Vertical axes extend from −160 nT/yr to +160 nT/yr.

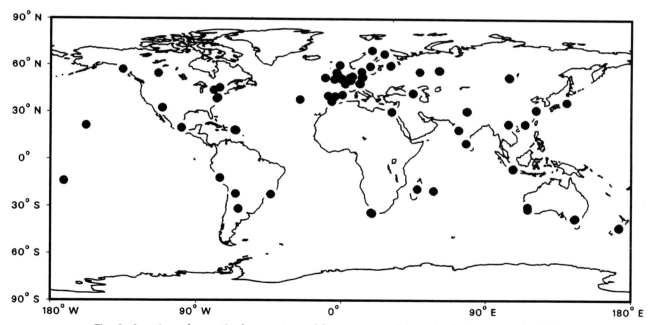

Fig. 4   Locations of magnetic observatories used for geomagnetic field models for interval 1911–1983.

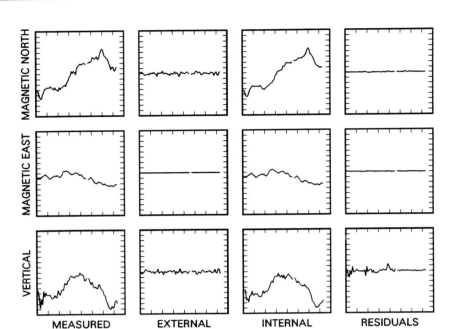

Fig. 5 First time differences of annual means of geomagnetic field near Washington, D.C. Horizontal axes extend from 1910 to 1990. Vertical axes extend from −160 nT/yr to +160 nT/yr.

is for three observatory locations different from the three locations used for Fig. 3. The Washington location refers to two nearby observatories used for Fig. 5; the Ottawa location refers to two observatories near Ottawa, Canada; Agincourt, Ontario (data before 1970) and Ottawa, Ontario (data after 1970). The third observatory location is Eskdalemuir, Scotland.

The structure of waveforms at Washington and Ottawa is very similar for both "measured" and internal-source model derivatives. This observation is a reason for confidence in the physical reality of the waveform structure; the structure is unlikely to be due to errors or "noise." The geomagnetic jerk of 1969 is apparent in the Eskdalemuir waveform; a similar change of slope is evident near 1978.

Vertical-field derivatives are shown (Fig. 7) in the same format and for the same observatory locations used for Fig. 6. Residuals for the vertical component of field derivative are considerably larger than Fig. 6 residuals for the magnetic-east component and residuals for the magnetic-north component seen in previous figures. Larger residuals for vertical-field derivatives (Fig. 7) are especially prominent for years before 1960. Relatively large residuals for vertical-field derivatives before 1960 are found at many other magnetic observatories.

The geomagnetic jerk of 1969 is evident in many but not all internal-source waveforms shown in previous figures. Similar prominent changes of slope, or jerks, can be seen in many of these figures at approximately 1925, 1940, and 1978.

C. Multilinear Secular Variation Models

Much, but not all, of the internal-source secular variation seen in the previous figures and also at other magnetic observatories (not used for the figures) can be modeled by straight lines with changes of slope at 1925, 1940, 1969, and 1978. Much of the time variation for time derivatives of spherical harmonic coefficients of degrees 1, 2, and 3 representing the internal-source field can also be modeled by straight lines with changes of slope at 1925, 1940, 1969, and 1978. These observations led some scientists to propose that such a multilinear secular variation model can be used as a reference to extract other less regular phenomena. The scientific community disagrees as to whether the slope change at 1925 should be included in such a model. They also disagree as to whether such a model is useful, since some phenomena due to internal sources are not described by the model. Geomagnetic secular variation is the subject of continuing research, and utility of multilinear secular variation models remains to be demonstrated.

V. Variations Due to External Sources

A. Magnetopause Current

A solar wind of ionized particles emanates continually from the sun. This wind confines the geomagnetic field to a cavity

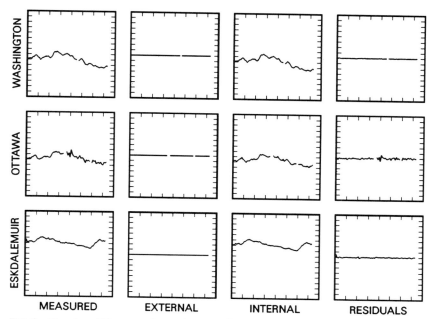

Fig. 6 First time differences of annual means of geomagnetic field magnetic-east component at three different magnetic observatories. Horizontal axes extend from 1910 to 1990. Vertical axes extend from −160 nT/yr to +160 nT/yr.

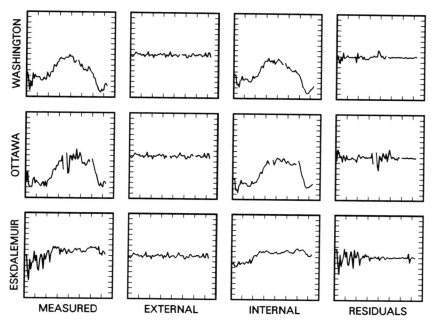

Fig. 7 First time differences of annual means of geomagnetic field vertical component at three different magnetic observatories. Horizontal axes extend from 1910 to 1990. Vertical axes extend from −160 nT/yr to +160 nT/yr.

that extends about 12 earth radii from the center of the earth in the solar direction. The cavity extends over 1000 earth radii from the earth's center in the antisolar direction, beyond the orbit of the moon, and is about 30 earth radii in diameter at the earth in the direction perpendicular to the solar direction. Dimensions of the cavity are determined by pressure balance between solar wind pressure and pressure attributable to the geomagnetic field. This latter pressure is proportional to the square of the geomagnetic field magnitude at the boundary. A smaller contribution to pressure balance is made by the solar magnetic field, also known as the interplanetary magnetic field. The cavity boundary is called the magnetopause. Electric currents flow on the magnetopause because the geomagnetic field deflects ionized solar wind particles; these currents produce a jump in magnetic field magnitude at the magnetopause. Field magnitude is nearly zero external to the cavity, while field magnitude internal to the cavity is increased from what it would be if magnetopause currents were not present. Magnitude of the jump is typically about 50 nT.

The jump in magnetic field magnitude at the magnetopause has been observed many times by spacecraft. Magnetopause location varies with variation in solar wind pressure, so that a spacecraft moving away from the earth to interplanetary space may cross the magnetopause several times on a single trajectory as the magnetopause moves back and forth past the spacecraft. Average solar-wind pressure varies with the sunspot activity; consequently, the portion of the average magnetic field at the earth's surface produced by magnetopause currents varies with the solar cycle, which has an approximate 11-year period.

B. Ring Current

A ring current is located at about 3 earth radii from the center of the earth. The ring current encircles the earth near the plane of the geomagnetic equator. The ring current is due to electrically charged particles trapped in the geomagnetic field.

A charged particle that moves in a uniform magnetic field in a direction perpendicular to field lines will travel in a circle. The circle's radius is inversely proportional to charge and field strength and is directly proportional to mass and speed. Any velocity component in the direction of the magnetic field is unaffected by the field.

A charged particle that moves in the geomagnetic field will tend to circle the field lines while simultaneously moving along the field lines. The geomagnetic field lines converge as they approach the earth so that a charged particle spiraling along a field line is slowed as it approaches the earth, finally stops its motion along the field line, and is reflected back along the field line away from the earth. Thus, a charged particle in the geomagnetic field spirals about field lines from one hemisphere to the other, unless the particle has sufficient momentum to reach the earth's atmosphere and collide with atmospheric particles. Because the geomagnetic field is weaker at higher altitudes, the radius of the spiral is tighter at low altitudes than at high altitudes; therefore, the particle drifts either eastward or westward during each rotation about a field line. Positively charged particles drift in the opposite direction from negatively charged particles. This drift of charged particles trapped in the geomagnetic field is the source of the ring current. Particles are continually being lost and new ones inserted into the ring current through mechanisms not completely understood. Average ring current varies with solar activity and with the sunspot cycle.

The magnetic field produced by the ring current at the earth's surface is approximately in the opposite direction to the field produced by magnetopause currents. Both current systems are sufficiently far from the earth's surface that the field they produce can be well represented by degree-one zonal spherical harmonics.

The spherical harmonic coefficient of the degree-one zonal external-source magnetic field is plotted against time in the upper-left graph of Fig. 8. The first time derivative of this coefficient plotted against time is shown in the upper-right graph. The spherical harmonic coefficient for the derivative was obtained from spherical harmonic models described in the previous section of this article. The spherical harmonic coefficient for the field was obtained from the spherical harmonic coefficient for the field derivative by digital integration. The lower two graphs of Fig. 8 show the sunspot number and its derivative, plotted against time. Seven sunspot cycles are shown. Each of the time series shown in Fig. 8 has been filtered with a simple lowpass filter by using weights (0.25, 0.50, 0.25).

Field and sunspot-number waveforms in Fig. 8 are highly correlated but are not identical. Derivatives of these waveforms are also highly correlated, although the derivative of the field usually has two peaks for each peak in the derivative of the sunspot number. Therefore, the field derivative has a large frequency component at the second harmonic of the sunspot cycle.

Time variation of the external source field seen in Fig. 8 is most likely mainly due to time variations in the magnetopause currents and ring current previously described, since most of this time variation is described by a first-degree zonal spherical harmonic, as would be expected for fields due to current systems whose distance is large in comparison to the earth's radius. The geometry of current systems nearer to the earth is such that larger magnitude, higher degree spherical harmonic coefficients should be expected if these current systems were a major contributor to the field shown in Fig. 8.

C. Ionospheric Currents and Field-Aligned Currents

Currents flow in the ionosphere due mainly to tidal effects and to ionospheric heating by the sun. Both mechanisms

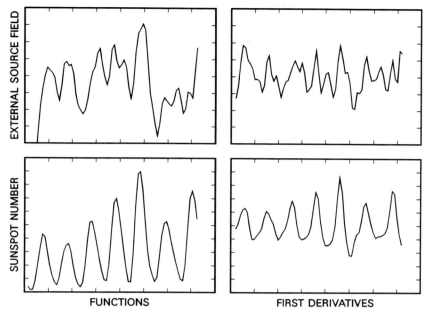

Fig. 8 Sunspot number and external source degree-one zonal spherical harmonic coefficient for geomagnetic field. Derivatives of these functions are also shown. Horizontal time axes extend from 1910 to 1990. Vertical axes for geomagnetic field and derivative extend from −20 nT to +20 nT and from −20 nT/yr to +20 nT/yr, respectively. Vertical axes for sunspot number and derivative extend from 0 to 200 sunspots and from −100 to +100 sunspots/yr, respectively.

cause ionized particles to move in the earth's magnetic field; the field causes the charged particles to be deflected from what would otherwise be their trajectories, and the deflection of the charged particles gives rise to electric currents. The ionospheric current systems are approximately symmetric about the geomagnetic equator and about the Sun–Earth line. These current systems consist largely of current loops centered near the noon meridian at about ±30° geomagnetic latitude. As the earth rotates beneath these current systems, the magnetic field components at a fixed location on the earth's surface exhibit a time variation with a 1-day period. This time variation is called daily variation. Amplitude and waveforms of daily variation are different for different geomagnetic latitudes and different magnetic field components; a typical amplitude would be ±30 nT.

Another external current system consists of currents that flow along magnetic field lines in the earth's magnetotail. The currents flow toward the earth on one side of the magnetotail and away from the earth on the other side of the magnetotail. These two current flows are connected by currents flowing in the ionosphere near auroral latitudes.

When fields are averaged over all times of day and all days of the year at magnetic observatories to obtain annual means, contributions of ionospheric and field-aligned current systems to the resulting annual means must be small in comparison to contributions to annual means from magnetopause and ring currents. Evidence for this conclusion is the observation that nearly all of the external-source magnetic-field annual means can be represented by degree-one zonal harmonics.

D. Electromagnetically Induced Currents

Electrical currents are induced in the earth's interior by time-varying magnetic fields produced by time-varying external current systems. These induced currents produce magnetic fields that can be observed at the earth's surface. Spherical harmonic coefficients of the part of the internal-source magnetic field due to induced currents are related to corresponding spherical harmonic coefficients of the external-source field, provided a linear relationship exists between induced and inducing fields. If spherical harmonic coefficients, expressed as functions of time, are Fourier analyzed into frequency components, then the amplitude and the phase of the transfer function that relates an induced coefficient to inducing coefficient are found to depend on frequency, as well as on degree and order of the coefficients.

Transfer functions for different degree and order coefficients are related to conductivity of the earth's core, crust, and mantle. To the extent that the earth can be approximated by a spherically symmetric model Earth, transfer functions between coefficients depend on the degree but not on the order of the coefficients. These transfer functions can be used to study the earth's conductivity as a function of radial distance from the earth's center, as discussed in the next section of this article.

VI. Interpretation and Applications

A. Physical Properties of the Earth

Since direct-measurement instruments cannot reach far below the earth's surface, investigation of the earth's interior requires sophisticated remote-sensing methods. Geophysical inverse theory has been used to obtain information about the earth's interior from seismological, electromagnetic, geothermal, and gravitational measurements.

Seismological measurements have been particularly valuable. Seismic waves generated by earthquakes and explosions have been used to determine laterally averaged density and seismic velocities to better than 1% accuracy.

Geomagnetic measurements have played a key role in the discoveries of plate tectonics, convection of core fluid, and the more sluggish convection of the silicate mantle. Temporal variations of the geomagnetic field have been used to study mantle electrical conductivity as a function of depth, as well as to study motion of the fluid core surface.

1. Mantle Conductivity

Time-varying magnetic fields that originate from current systems external to the earth induce electric currents within the earth. These induced currents in turn produce time-varying magnetic fields. The transfer function between induced and inducing fields for a given degree spherical harmonic can be used to study electrical conductivity of the earth as a function of depth. Lateral homogeneity is usually assumed as a reasonable approximation, so that the transfer function depends on the degree but not the order of the spherical harmonic.

Penetration depth of a time-varying magnetic field depends on frequency and conductivity profile as well as on degree of spherical harmonic. Periods that have been used to study mantle conductivity range from less than 1 day to the 11-year solar cycle period. For periods of a few seconds, the earth can be considered an ideal conductor; at the earth's surface, therefore, induced and inducing vertical-field components must be equal in magnitude for signals with periods of a few seconds, since a time-varying field cannot penetrate into an ideal conductor. For the solar-cycle period, some scientists have reported that the mantle behaves as a nearly ideal insulator while the core behaves as a nearly perfect conductor. Other scientists have reported larger apparent values for mantle conductivity. Determining a transfer function between degree-one induced and inducing spherical harmonics for this period is difficult because of "noise" in magnetic observatory measurements and because part of the signal from internal sources is not induced but originates in the earth's core. Determining a transfer function between induced and inducing fields for this frequency range is the subject of continuing research.

Constraints can be placed on mantle conductivity from studies of geomagnetic jerk propagation. If the jerk were an ideal impulse in the third time derivative of the magnetic field at the core surface, spherical harmonics of the third time derivative of the magnetic field at the earth's surface would be impulses of finite width delayed from the time of the impulse at the core surface. The amount of time delay and impulse broadening would depend on the degree of spherical harmonics and on the mantle conductivity profile. Attempts to determine mantle conductivity profile from the geomagnetic jerk of 1969 have not been fully successful, possibly because the jerk is more complex than the idealized model and possibly because of the presence of other signals from the core surface. The presence of noise and external-source signals contributes to the difficulty of determining mantle conductivity in this manner. Because the jerk seems to occur within a few years at some observatories, some scientists believe that the mantle must be a nearly perfect insulator at the sunspot-cycle period of 11 years.

2. Core Fluid Motion

Observations of the geomagnetic field and its secular variation can be used to derive information on motions of the upper portion of core fluid. At the core–mantle boundary, the radial portion of fluid velocity must be zero. Information on horizontal components of fluid velocity below a thin kinematic boundary layer can be obtained with the aid of the reasonable approximation that the core is an ideal conductor and the mantle is a perfect insulator. Approximation of the mantle as a perfect insulator allows extrapolation of the geomagnetic field and secular variation observed at the earth's surface to the core–mantle boundary. Approximation of the core fluid as an ideal conductor means that magnetic flux diffusion can be ignored, so that the radial component of magnetic field at the core–mantle boundary changes only as a result of fluid motion. However, this approximation, called the frozen-flux approximation, is not sufficient to permit determination of fluid velocity at the core–mantle boundary without introducing additional assumptions.

One approach that has been used to derive fluid velocity at the core–mantle boundary is to assume that the horizontal component of Coriolis force is balanced by the horizontal pressure gradient. This assumption is called the geostrophic approximation; the approximation fails at the geographic equator where Coriolis force is zero. The geostrophic approximation, in conjunction with the frozen-flux approximation, permits determination of fluid velocity over most of the core–mantle boundary.

Maps for 1980 of fluid motion at the core–mantle boundary, derived with aid of the frozen-flux and geostrophic approximations, show an upwelling and a downwelling near the geographic equator. The upwelling is centered beneath the Indian Ocean, and the downwelling is centered beneath a point off the western coast of Peru. The upwelling and downwelling are in the poloidal portion of fluid flow that results from exchanges of fluid with the deeper core. The toroidal flow shows a westward drift that is strongest near the equator in the Atlantic hemisphere in the region from 90° W to 90° E. This region also has a westward fluid motion due to poloidal flow; therefore, net westward drift is much greater in the Atlantic hemisphere than in the Pacific hemisphere.

3. Other Global Geophysical Phenomena

Some scientists have reported correlations between geomagnetic secular variation and other global geophysical phenomena, such as irregularities in the earth's rotation and climate.

It is reasonable to expect a westward acceleration of outer-core fluid to be associated with an eastward acceleration of the mantle because of conservation of momentum. The mechanism for momentum transfer is not known, but the possibility of "bumps" on the core–mantle boundary has been suggested. An eastward acceleration of the mantle corresponds to a decreasing length of day. A strong correlation between secular variation of the east magnetic-field component in Europe and length of day from 1900 to 1985 has been reported.

Chandler wobble is a global geophysical phenomenon that may be related to geomagnetic secular variation. The earth's rotation axis precesses about its mean position on the earth's surface. The precession period is about 435 days. The path is roughly circular and has a radius of a few meters. This precession is called Chandler wobble. The path radius decays because of energy dissipation; therefore, the wobble must be excited by some mechanism. Some scientists have suggested that the exciting mechanism is an electromagnetic torque between the core and the mantle. Correlations between the geomagnetic jerk of 1969 and the radius of the Chandler wobble path have been reported.

Some scientists have suggested a correlation between geomagnetic secular variation and long-term climatic changes. Climatic changes have been reported to lag secular variation by 20 years.

B. Navigation

Models, maps, and charts of the geomagnetic field and its secular variation are useful for navigation and related applications.

1. World Magnetic Charts

The United States Navy is responsible within the Department of Defense for producing global geomagnetic-field models. These models are used for navigation and related applica-

tions. The Naval Oceanographic Office accumulates the necessary magnetic-field measurement data and produces such models every five years. Because the geomagnetic field changes with time, a secular variation model is also produced for updating the field model in the period between production of field models.

In recent years the British Geological Survey has collaborated with the Naval Oceanographic Office in the production of field models. Other organizations throughout the world also produce global field models. These organizations include the United States Geological Survey, the National Aeronautics and Space Administration (NASA), and the Soviet Union. A model is selected from candidate models by the International Union of Geomagnetism and Aeronomy and is called the International Geomagnetic Reference Field (IGRF). Sometimes the IGRF model is obtained by averaging corresponding coefficients of candidate models.

2. Satellite Orientation

One use for global geomagnetic models is to aid in determining the orientation of artificial Earth-orbiting satellites. Provided the satellite carries a vector magnetometer, direction of the magnetic field vector can be determined in a satellite-fixed frame of reference. To determine the orientation of the satellite-fixed reference frame relative to an inertial reference frame, it is sufficient to determine the orientation in the satellite frame of two nonparallel vectors whose orientation is known in the inertial frame. Therefore, if satellite orientation is fixed in the inertial frame, then determination of the magnetic field vector direction in the satellite frame at two points along the orbit, for which the two corresponding magnetic field vectors are nonparallel, is sufficient to determine satellite orientation in the inertial reference frame, provided location of the two points is known and an accurate global geomagnetic field model is available.

If satellite orientation is not fixed in the inertial frame, orientation can still be determined with use of another vector in addition to the magnetic field vector. Direction to a star determined from a star tracker is often used for this purpose.

For ship or aircraft navigation, direction of the horizontal component of magnetic field is sufficient for determination of vehicle orientation in the horizontal plane. Orientation of the vehicle relative to the horizontal plane is determined from the gravity vector orientation relative to the vehicle. Thus, the gravity vector serves the same purpose in this application as the direction vector to a star served in the previous application.

Bibliography

Backus, G. (1988). Comparing hard and soft prior bounds in geophysical inverse problems. *Geophys. J.* **94**, 249–261.
Bloxham, J., Gubbins, D., and Jackson, A. (1989). Geomag-

netic secular variation. *Philos. Trans. R. Soc. London,* A329, 415–502.

Courtillot, V., and LeMouel, J. L. (1988). Time variations of the earth's magnetic field: from daily to secular. *Annu. Rev. Earth Planet. Sci.* **16,** 389–476.

Langel, R. A., Kerridge, D. J., Barraclough, D. R., and Malin, S. R. C. (1986). Geomagnetic temporal change: 1903–1982, a spline representation. *J. Geomagn. Geoelectr.* **38,** 573–597.

McLeod, M. G. (1986). Stochastic processes on a sphere. *Phys. Earth Planet. Inter.* **43,** 283–299.

Glossary

Earth's core Central part of the earth; radius is approximately 3485 km. The outer portion of the core is an electrically conducting fluid.

Earth's crust Outer part of the earth composed essentially of crystalline rocks. Crustal thickness is approximately 50 km and outer radius is approximately 6371 km.

Earth's mantle Region of the earth between the core and the crust.

Gaussian coefficients Coefficients of a spherical harmonic expansion of the potential function of the geomagnetic field.

Magnetic jerk Change in the slope of the geomagnetic secular variation. The change occurs within a few years and is global in extent.

Magnetopause Boundary between the geomagnetic field and interplanetary space. Electrical currents flow on this boundary surface.

Ring current Electrical current at about three earth radii from the center of the earth. The current results from charged particles trapped in the geomagnetic field.

Secular variation That portion of the temporal variation of the geomagnetic field due to sources internal to the earth. Induced current sources due to time-varying external current systems are excluded from this definition.

Stochastic inversion Method for computing Gaussian coefficients (or equivalent parameters) that is based on *a priori* estimates of parameter and noise covariances.

Geomagnetism

Paul H. Roberts

University of California, Los Angeles

The earth was known to be magnetic long before historical records were kept, and that knowledge was translated into a valuable navigational tool, the compass needle. Today much is known about the structure of the *magnetic field* of the earth as it was not only during the recent past but also, through the study of paleomagnetism, during geological time. This new knowledge has already proved to be a valuable geophysical research tool. The magnetic elements describing the geomagnetic field are defined in this article, and the harmonic analysis proving that it is mainly of internal origin is explained. A general description of the broad spectrum of geomagnetic field variations is provided.

I. The Magnetic Elements

The magnetism of the earth was discovered long before careful records were kept, and the directional property of the magnet has been known in China for perhaps more than four thousand years. Chinese mythology describes a battle in 2634 B.C. between the emperor Hoang-Ti and the rebellious prince Tchi-Yeou: Hoang-Ti was guided through a smoke screen, laid by the retreating prince, by the figure of a man riding on the emperor's chariot, whose arm constantly pointed to the south, enabling the emperor to overtake and put to death the unfortunate prince. Whether the figure on the chariot really made use of geomagnetism will never be known for certain, and to describe Hoang-Ti's supremacy as an early example of how superior technology wins battles would be unconvincing. Nevertheless Chinese geomancers of the first century A.D. certainly knew that when a spoon made of lodestone was spun on a smooth table, it always came to rest pointing in the same direction. The earliest discussion of a compass is found in the encyclopedia written in China by Shen Kua in the eleventh century.

The first account in Europe of the directional property of the compass needle was provided about a century later by Alexander Neckam, a monk from St. Albans in England. The compass needle was used by Arabian mariners early in the thirteenth century, and it was in regular use on ships of the British Navy in the fourteenth century. By the fifteenth century, a traveler might equip himself with a portable sundial whose gnomen could be correctly orientated with the help of a compass needle on its base (Fig. 1a). The makers of many of these sundials included an offset on the compass dial; apparently they knew, as Shen Kua had known before them, that even a perfect needle will not point due north. The angle of deviation is known as magnetic *declination D*; it varies over the earth's surface, a fact first noted by Columbus in the log of his famous voyage of discovery. The *Tabula Nautica* of Edmund Halley, the first of many magnetic charts of *D* constructed to help seamen in navigation, was published in 1701 (Fig. 2a). These charts collated the observations he had made on the sloop *Paramour* (made available to Halley by King William III of England from 1698 to 1700 — the world's first geophysical vessel).

A horizontally pivoted magnetic needle dips downward. This was first remarked in 1576 by Robert Norman, an English hydrographer, who saw this as evidence that the orientation of the compass is controlled from within the earth. This magnetic *inclination I*, or *dip* from the horizontal, was carefully studied by the chief court physician of Queen Elizabeth the First of England, William Gilbert, who published in 1600 the first modern scientific treatise, *De Magnete*, containing the fruits of long years of thought and experiment. Probably the most famous of his experiments today was essentially a repetition of one performed as early as 1269 by Petrus Peregrinus. Peregrinus had mapped out the orientation taken up by a magnetic needle in the vicinity of a roughly spherical lodestone, which he called, like Gilbert after him, a *terrella*, or little earth (Fig. 3). He observed that there were two *magnetic poles* on its surface where the needle pointed directly to the center of the sphere. Gilbert stressed the striking similarity between the variations in *I* over the surfaces of terra and terrella. "The earth," he wrote, "is a magnet."

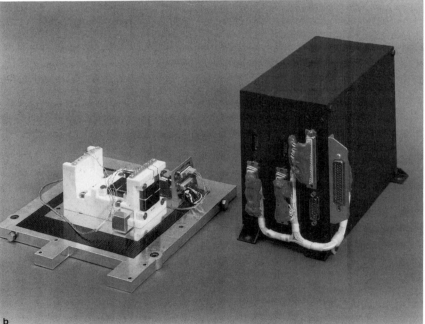

Fig. 1 (a) A portable sundial in the possession of the author. (b) A modern satellite magnetometer similar to the one that flew on MAGSAT. (Photograph (b) courtesy of Mario Acuna, Goddard Space Flight Center.)

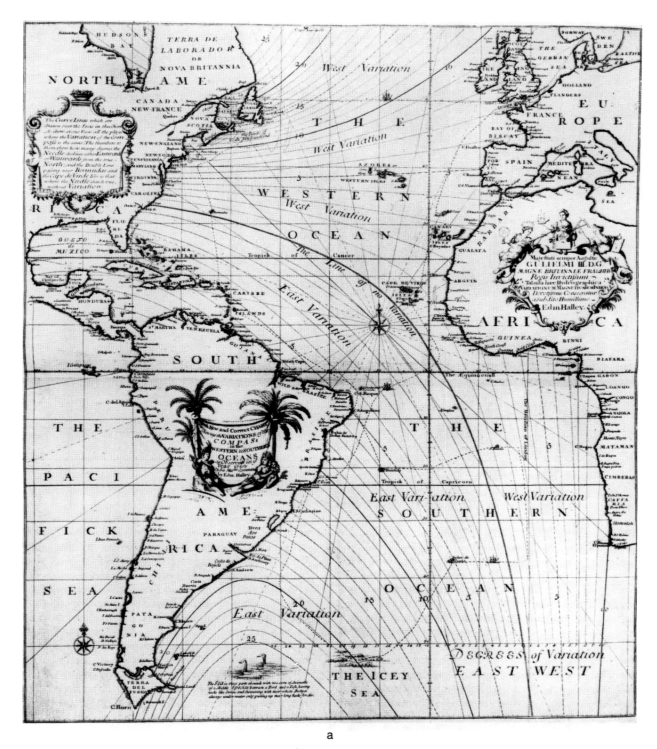

a

Fig. 2 (a) Reproduction of Halley's 1701 map of D. Note the position of the "Line of no variation," $D = 0$. (b) Isolines of D for epoch 1983. Note how far the contour $D = 0$ has moved west relative to (a). (Figures courtesy of R. A. Langel.) (*Figure continues.*)

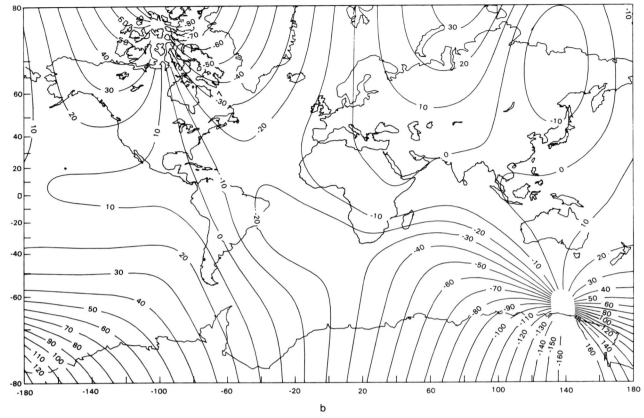

Fig. 2 (*Continued*)

Gilbert knew that the strength with which the earth attracts a compass needle varies with position. As a famous London clock maker, George Graham, realized as early as 1723, the period of oscillation of a freely suspended dip needle decreases with the strength of the earth's attraction. Alexander von Humboldt, during his famous American journey of 1799–1803, counted at a number of Peruvian sites the number of swings a dip needle made in 10 minutes. He found that that number decreased both to the north and to the south of the magnetic equator and greatly exceeded the number of swings in the same time interval at higher latitudes. He regarded this proof of the decrease in *magnetic intensity F* from poles to equator as his most important discovery in the Americas. In collaboration with Gay-Lussac, Humboldt also measured during a journey to Italy in 1807 the oscillation period of a vertically pivoted needle; he demonstrated that this increased with latitude, showing that the *horizontal intensity H* increases from poles to equator. The famous mathematician Carl Friedrich Gauss realized that magnetic intensity can be expressed in units of mass, length, and time; and with the help of the eminent physicist Wilhelm Weber he

devised an experiment with which in 1832 they measured F at Göttingen.

Today's world is that of the artificial satellite, and the earth's magnetism is described with a detail and accuracy never before achieved. Figure 1b shows a magnetometer of the type used to measure F on *MAGSAT*, the only satellite dedicated exclusively to geomagnetic measurements. Figure 2b shows a modern chart of D that makes extensive use of satellite observations. Similar maps can be drawn for all the magnetic elements, I, H, D, and F, depicting different facets of the earth's magnetic field, or the geomagnetic field. "Field" is the word used to describe a quantity that depends on position $\mathbf{r}$. This particular quantity is a vector; that is, it has at each $\mathbf{r}$ a magnitude F and a direction defined D and I. It is theoretically more convenient to represent this vector field $\mathbf{B}$ by its northward, eastward, and downward components X, Y, and Z, so that, in spherical coordinates (r, θ, ϕ),

$$\mathbf{B} = (-Z, -X, Y)$$
$$= F(-\sin I, -\cos I \cos D, \cos I \sin D)$$

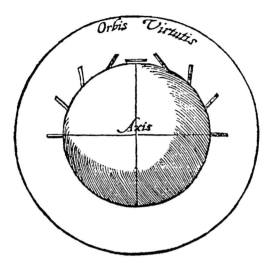

Fig. 3 Gilbert's terrella, showing the inclinations taken up by iron spikes placed on its surface at different latitudes. [After Chapman and Bartels (1940).]

and

$$H = F \cos I \qquad (1)$$

Here $r = |\mathbf{r}|$, θ is colatitude, ϕ is east longitude from Greenwich, and the geocenter O is the origin of $\mathbf{r}$.

Despite the similarity of the fields of terra and terrella, one must not too hastily conclude that the earth is a permanently magnetized body like the spherical lodestone. In fact, the geomagnetic field is created by electric currents flowing deep within the earth. [See GEOMAGNETISM, ORIGIN.]

II. Representation of the Magnetic Field

A. Multipoles

In 1835 Gauss was able to separate the geomagnetic field into parts of internal and external origin and to demonstrate that $\mathbf{B}$ is almost entirely produced within the earth. Gauss's success followed his realization that there are no important sources of $\mathbf{B}$ near the earth's surface, $r = a$, so that $\mathbf{B}$ is a *potential field* to high accuracy. In other words, it may be written near $r = a$ as the gradient of a scalar field V,

$$\mathbf{B} = -\nabla V \qquad (2)$$

where the *geomagnetic potential* V satisfies Laplace's equation,

$$\nabla^2 V = 0 \qquad (3)$$

We shall now develop the general solution of Eq. (3). The simplest solution is that for an isolated magnetic pole of strength q:

$$V_0 = q r^{-1} \qquad (4)$$

Although such poles do not exist (since $\nabla \cdot \mathbf{B} = 0$), it is often convenient to develop solutions of (3) from pairs of equal and opposite poles. For example, when the separation d of two such poles q and $-q$ tends to zero as $q \rightarrow \infty$ with $m = qd$ held finite, another solution of (3) is obtained:

$$V_1 = -\mathbf{m} \cdot \nabla r^{-1} \qquad (5)$$

This "dipole" solution is physically meaningful; the magnetic field created by a closed system of electric currents is, at great distances from that system, predominantly dipolar. The limiting process $d \rightarrow 0$, $q \rightarrow \infty$ that led to (5) is evidently equivalent to differentiation:

$$\begin{aligned} V_2 &= (\mathbf{m}_1 \cdot \nabla)(\mathbf{m}_2 \cdot \nabla) r^{-1} \\ V_3 &= -(\mathbf{m}_1 \cdot \nabla)(\mathbf{m}_2 \cdot \nabla)(\mathbf{m}_3 \cdot \nabla) r^{-1} \\ V_4 &= \cdots \end{aligned} \qquad (6)$$

corresponding to a quadrupole (four poles brought together at O), an octupole (eight poles brought together at O), and so on. An arbitrary potential field created by sources internal to radius r can be represented by a "multipole expansion"— that is, a linear combination of solutions (5) and (6); this also satisfies (3). Following Gauss, this sum is equivalently, but more economically, written as

$$V_{\text{int}} = a \sum_{n=1}^{\infty} \sum_{m=0}^{n} [g_n^m \cos m\phi + h_n^m \sin m\phi]$$
$$\cdot \left(\frac{a}{r}\right)^{n+1} P_n^m(\theta) \qquad (7)$$

where P_n^m denotes the Legendre function of degree n and order m. [In geomagnetism, these are usually "Schmidt normalized." This means that the square of any of the $2n + 1$ "spherical harmonics" Y_n^m of degree n (i.e., the set of functions $P_n^m \cos m\phi$ and $P_n^m \sin m\phi$) integrates to $4\pi/(2n + 1)$ over the unit sphere (i.e., over $0 \leq \theta \leq \pi$ and $0 \leq \phi < 2\pi$). Products of two different Y_n^m vanish in such an integration. These properties are needed in deriving (14) below.]

To relate (5) to (7), let $Oxyz$ be Cartesian coordinates, with Oz parallel to the earth's rotation axis ($\theta = 0$), while Ox ($\phi = 0$) and Oy lie in the equatorial plane ($\theta = \frac{1}{2}\pi$); let $\hat{\mathbf{x}}$, $\hat{\mathbf{y}}$, and $\hat{\mathbf{z}}$ be unit vectors along those axes. Then the *centered dipole* is

$$\mathbf{m} = \frac{4\pi a^3}{\mu_0} (g_1^1 \hat{\mathbf{x}} + h_1^1 \hat{\mathbf{y}} + g_1^0 \hat{\mathbf{z}}) \quad \text{A m}^2 \qquad (8)$$

and the magnitude of the dipole moment (currently 7.835×10^{22} A m^2) is

$$m = \frac{4\pi a^3}{\mu_0}\sqrt{[(g_1^1)^2 + (h_1^1)^2 + (g_1^0)^2]} \quad \text{A m}^2 \quad (9)$$

We have here introduced the factor $4\pi/\mu_0$, where μ_0 is the permeability of free space; in this way, $\mathbf{m}$ is converted from T m^3 to A m^2, the appropriate unit of magnetizing force $\mathbf{H} = \mathbf{B}/\mu_0$ substituting for $\mathbf{B}$. [To convert to the unrationalized cgs unit more commonly employed—Gauss cm^3—multiply (8) and (9) by 10^3.] In a similar way, V_n can be related to the order-n terms in (7). For example, for the quadrupole ($n = 2$), we have (again introducing the $4\pi/\mu_0$ factor)

$$g_2^0 = \frac{\mu_0}{4\pi a^4}(3m_{1z}m_{2z} - \mathbf{m}_1 \cdot \mathbf{m}_2)$$

$$g_2^1 = \frac{\mu_0\sqrt{3}}{4\pi a^4}(m_{1x}m_{2z} + m_{1z}m_{2x})$$

$$g_2^2 = \frac{\mu_0\sqrt{3}}{4\pi a^4}(m_{1x}m_{2x} - m_{1y}m_{2y})$$

$$h_2^1 = \frac{\mu_0\sqrt{3}}{4\pi a^4}(m_{1y}m_{2z} + m_{1z}m_{2y})$$

$$h_2^2 = \frac{\mu_0\sqrt{3}}{4\pi a^4}(m_{1x}m_{2y} + m_{1y}m_{2x})$$

B. Solutions of Internal and External Origin

The most general solution of (3) is not (7) but

$$V = V_{\text{int}} + V_{\text{ext}} \quad (10)$$

where V_{ext} is an arbitrary potential field created by sources exterior to radius r and consists of a linear combination of "exterior harmonics"

$$\hat{V}_n = r^{2n+1}V_n \quad (11)$$

where V_n is expressed as in (5) or (6). This is equivalently, but more economically, written as

$$V_{\text{ext}} = a\sum_{n=1}^{\infty}\sum_{m=0}^{n}[q_n^m \cos m\phi + s_n^m \sin m\phi]$$

$$\cdot \left(\frac{r}{a}\right)^n P_n^m(\theta) \quad (12)$$

When Gauss fitted (2) and (10) to the observations, he found that, to the accuracy of the data available to him, q_n^m and s_n^m and therefore V_{ext} were zero, so establishing mathemat-

ically what Norman could only conjecture: the sources of $\mathbf{B}$ are within the earth, and

$$V = V_{\text{int}} = a\sum_{n=1}^{\infty}\sum_{m=0}^{n}[g_n^m \cos m\phi + h_n^m \sin m\phi]$$

$$\cdot \left(\frac{a}{r}\right)^{n+1} P_n^m(\theta) \quad (13)$$

Although the terms "dipole," "quadrupole," "octupole," and so on abound in the geophysical literature, and although (13) is equivalent to a sum of these, it is important to realize that such multipoles have no physical reality. The core is metallic (and even the mantle is slightly conducting). Strong electric currents flow in the core (and weak currents flow in the mantle, not to mention the permanent magnetism present in the crust—see below). Therefore (2) and (3) are totally false in the core (and are not precisely correct even in the mantle). A magnetic field is necessarily nonsingular everywhere, even at $r = 0$. Infinitely many sources of $\mathbf{B}$ in $r < a$ can be constructed to duplicate the one $\mathbf{B}$ observed on and above $r = a$. Nevertheless, the main sources of $\mathbf{B}$ are probably deep within the earth, for reasons we will now discuss. The description of $\mathbf{B}$ as a sum of multipoles is therefore useful in geomagnetism. [See CORE; MANTLE.]

C. Energy Spectra

With increasing r, the dipole field decreases in strength as r^{-3}, but the quadrupole and higher terms in expansion (13), which define the earth's nondipole field, decrease as r^{-4} or more rapidly. The greater r, the weaker the geomagnetic field, but the better that field is represented by a dipole of strength $\mathbf{m}$. The fact that the dipole field dominates the nondipole field at the earth's surface *indicates* that the sources of field lie deep within the earth. It does not *prove* this; because of the nonuniqueness described above, it is always possible to recreate $\mathbf{B}$ by shallow sources, provided these are distributed in a sufficiently bizarre fashion. Disregarding such fantasies, it is more natural to suppose, in a first approximation, that no sources of $\mathbf{B}$ exist above $r = b$, the core–mantle boundary (CMB).

To judge the plausibility of this hypothesis, consider the average over a sphere of radius r of the magnetic energy density. This is $(2\mu_0)^{-1}\Sigma W_n$, where W_n is the contribution from the harmonics of degree n, which by (13) is

$$W_n(r) = (n+1)\left(\frac{a}{r}\right)^{2n+4}\sum_{m=0}^{n}[(g_n^m)^2 + (h_n^m)^2] \quad (14)$$

If we confine attention to $2 \le n \le 12$, W_n is well represented at the earth's surface by $\log_{10} W_n(a) = 9.13 - 0.569n$, where W_n is given in units of nT^2. If our hypothesis is correct, (13) will hold right down to the CMB. The energy spectrum (14) decreases much more slowly with n at $r = b$, but still apparently converges as $n \to \infty$; in fact, $\log_{10} W_n(b) = 10.18 - 0.024n$. Thus far, the hypothesis survives.

The hypothesis fails when the $n > 12$ harmonics are added in. The best empirical fit for $16 \le n \le 23$ is $\log_{10} W_n(a) = 1.57 - 0.0114n$, and the corresponding spectrum at the CMB increases rapidly with n. This is easily explained: The slow decrease of $W_n(a)$ with n for $n \ge 16$ betrays the existence of sources close to the earth's surface, principally crustal magnetic anomalies. At $r = b$, these anomalies are exterior sources of V. Strictly therefore, expansion (13) is untenable at $r = b$, and it diverges if adopted. By truncating (13)—that is, by using (13) with ∞ replaced by some "cut-off" N (≤ 12)—we obtain an apparently converged solution throughout $r \ge b$, simply because the contamination introduced by the crustal anomalies (and by currents flowing in the mantle) are, for $n \le 12$, swamped by fields produced below the CMB. In this sense, the $N = 12$ field might be called the *main geomagnetic field*, although it is deficient in two respects. First, the anomalies and currents above $r = b$ create $n \le N$ harmonics that contaminate the terms retained in (13); second, the sources beneath the CMB produce $n > N$ harmonics that have been discarded in (13). It is estimated that, because of these deficiencies, the $N = 12$ solution differs at the earth's surface from the true field created by sources beneath the CMB by about 5–10 nT. Because of anomalies, it differs from data at observatories by about 100 nT. (Data taken prior to 1950 is so sparse and inaccurate that $N = 6$ is the practical truncation limit.)

Recently, a technique was developed to remove the crustal field prior to fitting the data to (13). At spacecraft altitudes, crustal fields are some 20–30 nT smaller than on $r = a$. By factoring this into the analysis of data gathered from ground-based observations, it is possible to estimate the bias introduced by local crustal fields into measurements at each geomagnetic observatory. These biases, being associated with geological inhomogeneities of the crust, are for our purposes time-independent and can be subtracted from the observatory data prior to determining g_n^m and h_n^m. The resulting **B** replicates the de-biased observatory data with an accuracy of 5–10 nT.

It should be pointed out that, even if a technique could be devised that eliminated the effects of all sources in crust and mantle, there would still be a practical limit to N. If terms beyond that N were retained, they would be magnified to such an extent in extrapolation to $r = b$ that series (13) would be meaningless. This well illustrates the inherent instability of the extrapolation process; mathematically, extrapolation is not a well-posed problem.

The possibility of extrapolating **B** to the CMB has been increasingly exploited during the last decade in an attempt to learn more about processes at work in the core, and tech-

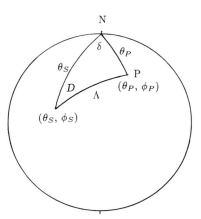

Fig. 4 Spherical triangle for finding the colatitude θ_P and longitude ϕ_P of a VGP from the D and I measured at a site at latitude θ_S and longitude ϕ_S.

niques of minimizing errors have been devised. Nevertheless, not surprisingly, the degree to which **B** can be resolved on the core surface is currently a matter of some dispute.

D. Different Kinds of Magnetic Poles

Geomagnetic poles are usually defined as the points where the dipole axis (8) meets the earth's surface. Owing to the nondipole terms in (13), these do not coincide with the (generally nonantipodal) *dip poles*, where the field is vertical ($H = 0$); there could be more than two such poles. The axisymmetric ($m = 0$) component, $\mathbf{m}_z = 4\pi a^3 g_1^0 \hat{z}/\mu_0$, of (8) is called the *axial dipole*; the remaining perpendicular component, $4\pi a^3 (g_1^1 \hat{x} + h_1^1 \hat{y})/\mu_0$, is the *equatorial dipole*. Together they form the *centered dipole* and, by moving this away from the geocenter to create the *eccentric dipole*, the quadrupole terms $n = 2$ in (13) can be partially allowed for.

The geomagnetic and dip poles are site independent; not so the *virtual geomagnetic poles*, or VGPs, which are defined from D and I at a particular site (θ_s, ϕ_s) on the earth's surface as the points where the geomagnetic poles would be situated if the field were that of a centered dipole. According to (5), the magnetic colatitude Λ of the north VGP (θ_P, ϕ_P), which is denoted by P in Figure 4, is given by

$$\cot \Lambda = \tfrac{1}{2} \tan I$$

The cosine rule of spherical trigonometry,

$$\cos \theta_P = \cos \theta_S \cos \Lambda + \sin \theta_S \sin \Lambda \cos D$$

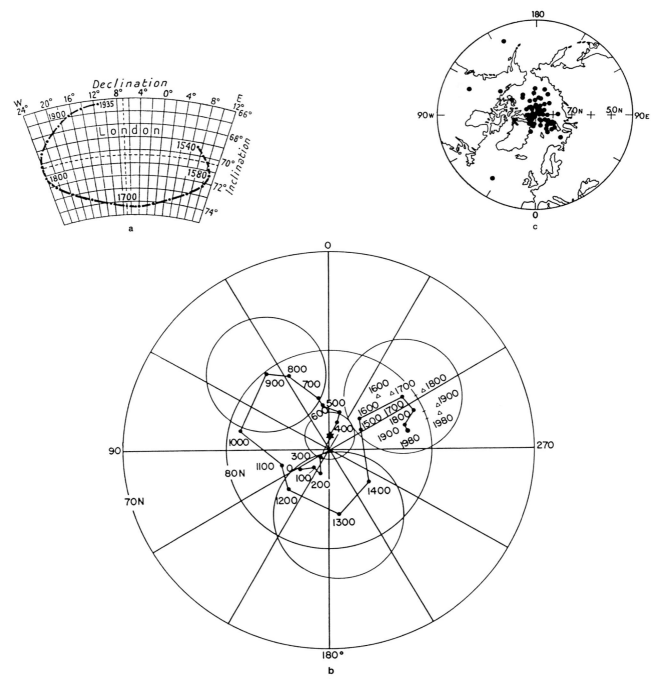

Fig. 5 (a) Bauer plot, i.e., the curve tracing the evolution of (D, I) with time for one site, here London. [After Chapman and Bartels (1940).] (b) The path of the north geomagnetic pole during the last 2000 years. [After Merrill and McElhinny (1983).] (c) Paleomagnetic poles for the past 5 Myr; X indicates the present position of the north geomagnetic pole. [After McElhinny (1973).]

determines θ_P, and the sine rule,

$$\frac{\sin \delta}{\sin \Lambda} = \frac{\sin D}{\sin \theta_P}$$

then yields δ, where $0 \le |\delta| \le \frac{1}{2}\pi$; from this ϕ_P may be determined with the help of another cosine rule:

$$\cos \Lambda = \cos \theta_P \cos \theta_S + \sin \theta_P \sin \theta_S \cos(\phi_P - \phi_S)$$

This shows that

$$\phi_P = \phi_S + \begin{cases} \delta, & \text{if} \quad \cos \Lambda > \cos \theta_P \cos \theta_S \\ \pi - \delta, & \text{if} \quad \cos \Lambda < \cos \theta_P \cos \theta_S \end{cases}$$

(Paleomagnetic poles will be defined in Section III,B.)

III. Variability in Time

A. Secular Variation and Westward Drift

Perhaps Gilbert had too much faith in his analogy between terra and terrella, for he opined in *De Magnete* that the earth's field, like that of the terrella, does not change with time t. Following work by William Borough in 1580 and his predecessor Edmund Gunter in 1624, Henry Gellibrand, a professor of astronomy at Gresham College in London, announced in 1635 that D had decreased from 11.3°E in 1580 to 4.1°E in 1634; **B** is a function not only of **r** but also of t. In 1722, Graham, while making observations with his new and greatly improved instrument for measuring D, noted that its compass needle was in continuous motion. This marked the birth of a now enormous and significant subject: *aeronomy*. [*See* AERONOMY.]

We shall in this article ignore externally created fast changes of **B** and, with them, the whole field of aeronomy. We shall be interested only in the slow *secular variations* (SV) that originate within the earth. The key phrase here is *originate within the earth*; the distinction cannot be made on the basis of the adjectives *fast* and *slow*. On the one hand, the 11-year solar cycle creates variations in the geomagnetic field of the same period; but we shall ignore these because they are externally produced. On the other hand, according to LeMoüel and his associates, the SV is sometimes subject to sudden worldwide *impulses* that are completed in only about one year; these appear to originate in the core and are therefore relevant to our topic. Gauss's method, which applies also to time-dependent fields, allows us to determine $V_{int}(t)$ as a function of time. Three contaminants to $\mathbf{B}_{int}$ will, however, be disregarded here, namely, fields induced by ocean currents; those induced in the mantle and crust by changing current systems above the earth's surface; and those pro-

duced by crustal anomalies—we shall suppose that these have been removed from **B**. [*See* GEOMAGNETIC JERKS AND TEMPORAL VARIATION.]

The strength of the geomagnetic field has varied over the recent past. At the time Gauss made his analysis, m was nearly 9% larger than at present. It achieved its last maximum, of approximately 1.2×10^{23} A m² (i.e., about 50% greater than m today), about 600 B.C. If it decreased at its present rate, the centered dipole would disappear in about 1200 years; but on past form it is more likely to reach a minimum before then and to start increasing. The average of m over geological time has been estimated as 9×10^{22} A m²—close to its present strength.

Both Gellibrand and Halley are variously credited with having discovered the westward drift of **B**, a tendency for the geomagnetic field patterns to move slowly in longitude at an angular speed of about 0.1–0.2° yr⁻¹ but in the opposite direction to the prevailing motion of the world's weather patterns. This drift results in the clockwise sense of the path shown in Fig. 5a, which is a so-called Bauer plot of the field direction, that is, a record of D versus I at one site, here London. The westward drift is also evident when the isolines of D in 1983 (Fig. 2b) are compared with those of 1701 (Fig. 2a). More detailed study reveals that westward drift is far from uniform over the globe and may even be negative (i.e., eastward) in some places at some times.

By downward extrapolating the field in the way described above, J. Bloxham and D. Gubbins have concluded that, during the recent historical past, the radial field emerges mainly from two patches on the southern hemisphere of the CMB and returns to the core mainly in two patches at similar (north) latitudes and longitudes; these four patches have moved little during historical times. Other high-flux areas on the southern hemisphere of the CMB, accompanied by reverse field patches, have wandered slowly west and south, so providing an explanation for the westward drift of the earth's surface field. We now turn to variations on a longer time scale. [*See* CORE–MANTLE SYSTEM.]

B. Paleomagnetism

Paleomagnetism is the study of the geomagnetic field from Precambrian times to the present, as revealed by the field fossilized into rocks and sediments at their birth. The subbranch *archeomagnetism* provides the same information for historic and prehistoric times from the field imprinted on artifacts such as shards of pottery and bricks from the kilns that made that pottery. We shall not describe here the processes by which ancient fields are trapped, nor the techniques and pitfalls of extracting reliable geomagnetic information from samples. We shall focus on a few findings that are particularly significant for Section IV, following, and are also needed to explain the origin of geomagnetism. It should be

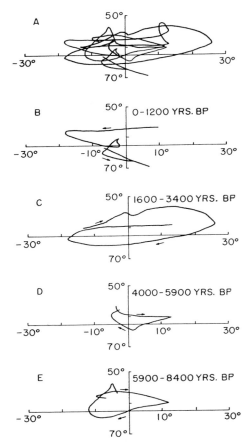

Fig. 6 Westward and eastward drift during the past 8400 yr at Lake St. Croix, Minnesota as determined from sediments; a clockwise sense in these Bauer plots indicates westward drift; anticlockwise indicates eastward. The entire record is shown in A, and this is broken down into different periods in B–E. [After Olson and Hagee (1987).]

noticed that paleomagnetism is mainly useful in inferring the direction of the geomagnetic field in the past and that, because of the difficulty in assessing the degree to which a rock loses its magnetization over geologically long periods of time, it is extremely hard to determine the strength of an ancient field. [*See* DEPOSITIONAL REMANENT MAGNETIZATION; PALEOMAGNETISM.]

Figure 5b indicates by filled circles (•) the location of the north geomagnetic pole over the past 2000 years, as deduced from archeomagnetic data gathered from many sites; the 95% confidence circles are also indicated for the epochs A.D. 900, 1300, and 1700. (Statistically, there is only a 5% probability that the mean of a number of observations would lie outside such a circle by chance.) Also shown, by open triangles (△), are the pole positions deduced from (8) using spherical har-

monic analyses of **B** data taken over the last 400 yr; the oldest of these are directional measurements made mainly by mariners. Unlikely though it appears from Fig. 5a, which spans too short a period, Fig. 5b makes it plausible that the geomagnetic poles coincide with the geographic poles when averaged over the past 2000 yr. Such an averaging is hard to avoid when delving farther into the past. Rock samples taken from the same geological epoch may well be separated in age by as much as 10^6 yr. The VGP, averaged over an interval of 10^5 yr or more, from rocks from one geographical location and epoch, is called the *paleomagnetic pole* for that location and epoch. Such poles are shown in Fig. 5c for rocks younger than 5 Myr. Their scatter is convincingly centered on the north geographic pole.

Lake-sediment data confirm that, although westward drift has been the norm in the recent past, short periods of eastward drift recur spasmodically. This is illustrated in Fig. 6 (in which BP stands for "before present" and "present" = A.D. 1950. Clockwise motions in these Bauer plots indicate westward drift, counterclockwise eastward.

C. Reversals

As early as 1906 the French physicist Bernard Brunhes found that some recently formed volcanic rocks, and some clays that had been contemporaneously heated and baked when those rocks were formed, were magnetized in almost exactly the opposite direction to that of the present geomagnetic field. He concluded that, at the time of their formation, the polarity of the field was reversed with respect to its present "normal" polarity. Subsequent study revealed that the geomagnetic field has changed its polarity many times in the past.

Polarity states of the geomagnetic field have been broadly divided into two categories: polarity epochs and polarity events. In the latter, a sense of polarity lasts for a geologically short period, say 10^5 yr or less; in the former, the change in polarity endures for a much longer period, although that period may be interrupted by a number of polarity events. The last four polarity epochs are named after Brunhes, Matuyama, Gauss, and Gilbert, and commenced approximately 0.72, 2.4, 3.4 and 4.9 Myr BP, respectively. The polarity in the Brunhes and Gauss epochs is normal; in the Matuyama and Gilbert epochs it is reversed. Within all but the Brunhes epoch there are clear indications of polarity events. For example the Jaramillo and Olduvai normal polarity events occurred in the Matuyama epoch, the former lasting about 70,000 yr and the latter somewhat longer. The situation is summarized in Fig. 7.

Since the late Cretaceous, polarity epochs have on average lasted only about a million years. There have however been "superchrons" in which the reversal frequency was considerably less, notably the normal polarity epoch that existed for more than 30 Myr in the mid-Cretaceous (see Fig. 8) and a

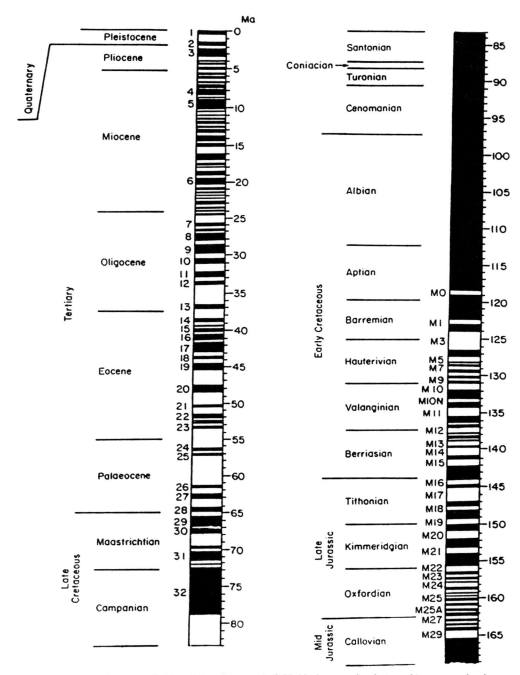

Fig. 7 Summary of recent polarities of the geomagnetic field: black, normal polarity; white, reversed polarity. [After Merrill and McElhinny (1983).]

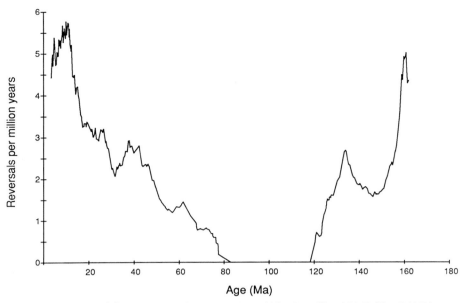

Fig. 8 Reversal frequency over the past 165 Myr. [After Merrill and McFadden (1990).]

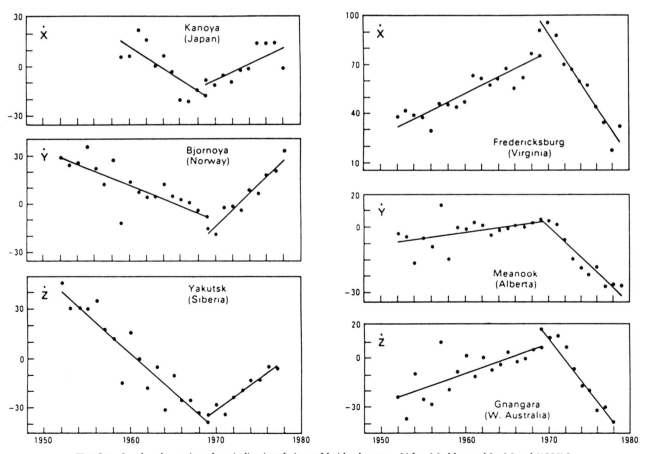

Fig. 9a Secular-change impulses, indicating their worldwide character. [After Madden and Le Mouël (1982).]

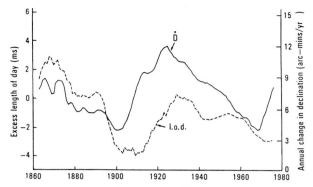

Fig. 9b Comparison of 5-yr means of D at Paris with the variation in the length of the day (l.o.d.). [After Le Mouël and Courtillot (1981).]

reversed polarity epoch lasting 70 Myr in the Permian. There is evidence that the scatter of field directions is less during superchrons of infrequent reversals than during superchrons of frequent reversals. This is another indicator of the greater geomagnetic stability of field directions.

The actual time taken by the field to make the transition from one polarity state to the other appears to be rather short, perhaps of the order of 5000 yr. One may ask, "Which of the following is closer to reality: (a) The direction of the geocentric dipole **m** rotates through 180° without changing its strength m; or (b) m decreases to zero with the direction of **m** fixed, but when m grows again **m** has exactly reversed its direction?" Undoubtedly the truth lies between these ex-

tremes, but it appears that (b) is more plausible than (a). There is clear evidence that F decreases during a reversal to 5–10 mT (i.e., to $\frac{1}{10}$–$\frac{1}{3}$ of its normal level).

In addition to the polarity epochs and events, "polarity excursions" have been reported, in which during a short period the VGP at a site moves by perhaps as much as 90° and then returns. Sometimes the field may actually reverse at one site but not at another, suggesting that the nondipole field dominates the dipole field during this time.

D. Geomagnetic Time Scales, a Summary

The geomagnetic field exhibits variations on many different time scales, ranging from perhaps 1 to 10^9 years. The shortest is that of the secular-change impulses shown in Fig. 9a; the longest are illustrated in Fig. 9f. Between these extremes, SV on the decade variations are shown in Fig. 9b; these are linked to the decade variations in the length of the day. The existence of a 60-year periodicity has been claimed (see Fig. 9c), and time scales in the range 200–2000 years certainly exist (Fig. 9d). The presence of a fundamental period of about 7000 years seems probable (Fig. 9e). Reversals occur much less frequently (see Fig. 8).

IV. The Geomagnetic Field as a Tool

A. Earthquake Prediction

Tectonomagnetism is the study of the variations in the geomagnetic field associated with tectonic processes within the

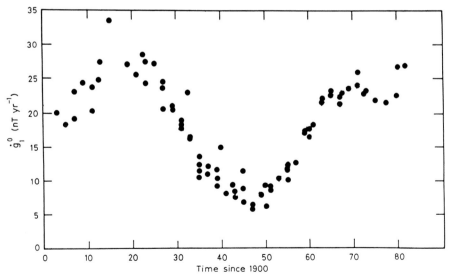

Fig. 9c Indication of a periodicity near 60 yr. [After Langel (1987).]

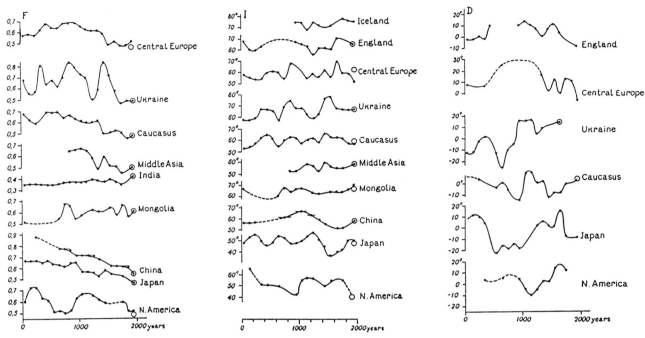

Fig. 9d Variation of *F*, *I*, and *D* over the past 2000 years at a number of sites. [After Bucha (1983).]

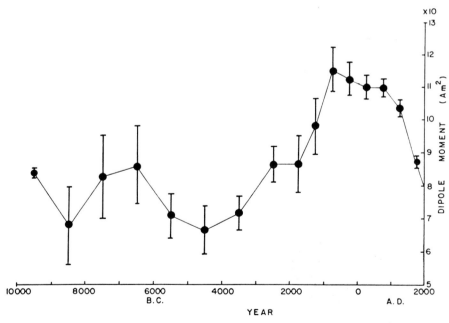

Fig. 9e Variation of the dipole moment *m* over the last 12,000 years obtained from archeomagnetic data; error bars at 95%. [After McElhinny and Senanayake (1982).]

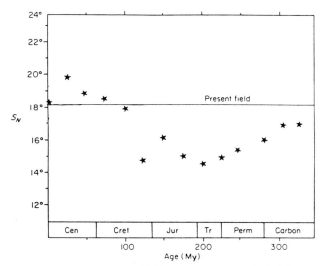

Fig. 9f Angular variance of field directions over geological time; geological epochs are indicated by the abbreviations at the bottom. [After Irving and Pullaiah (1976).]

earth. These variations are mainly due to alterations in the magnetic properties of crustal rocks created by changes in the temperatures and especially in the mechanical stresses of their environments. The latter distort the crystal lattice of magnetized minerals and therefore alter the fields they produce, a phenomenon known as the "piezomagnetic effect." It is thus the converse of the better-known phenomenon of magnetostriction, which is the change in dimensions of such a crystal or magnetization.

Tectomagnetic changes are produced in localized volumes of the crust and can be regarded as variations in the "stationary" crustal field, contained in the large n tail of (7). Tectomagnetic change is distinguished from the secular variation ($n \leq 12$) by its short length scale and, when the difference in fields measured at sites only 10 km apart alters, stress changes in the crust may be the cause. Unfortunately, however, such a difference is typically only of order 1 nT, a signal difficult to separate from random noise. Claims have been made that differences occur of as much as 20 nT before and after an earthquake. These have led to the hope that such seismomagnetic effects may be used to monitor stress in the crust and to assist in the prediction of earthquakes. [See EARTHQUAKE PREDICTION TECHNIQUES.]

B. Sea-Floor Spreading

The mid-ocean ridges that run along the floor of nearly every ocean are also regions of anomalously high heat flux from the earth. Below the sedimentary surface layer of the ocean floors, whose thickness increases with distance from a ridge,

lies a layer of basalt in the form of pillow lavas and dykes. It has been realized since the early 1960s that the ocean floors are growing through the continuous extrusion of this layer along the ocean ridges, newly extruded material pushing to each side of the ridge the material that had been extruded earlier. (See Fig. 10.)

When the proton precession magnetometer became available to map the magnetic anomalies over the oceans, a remarkable pattern emerged over nearly all the oceans. Parallel strips of anomalies run, sometimes almost without interruption, roughly parallel to the ridges for sometimes thousands of miles. This is consistent with the Vine–Matthews hypothesis (VMH) that, as the molten material solidifies at the ridges and cools below its Curie point, it becomes permanently magnetized along the direction of the field at that time, the alternation in pattern being a manifestation of the succession of normal and reversed polarities that the geomagnetic field has undergone during the extrusion.

The duration of polarity epochs is quite variable (see Section III), and correspondingly the widths of the successive anomaly strips should be variable also (assuming that the rate of sea-floor spreading does not change). Moreover, if the VMH is correct, the sequence of broad and narrow strips on one side of a ridge should be repeated (and reflected) on the other side (see Fig. 10). Further, that pattern should be the same for all ridges. Also, the central anomaly, the one over the ridge itself, should be normal because it corresponds to the geomagnetic field since the beginning of the Brunhes epoch. The VMH survives all these tests convincingly. And the field magnitude of the anomalies is consistent with the idea that a basaltic layer beneath the sea floor has been magnetized in a field of roughly the same strength as today's geomagnetic field.

Accepting the VMH as true, one can use it to learn more about the geological past. From the known durations of the Brunhes, Matuyama, Gauss, and Gilbert epochs, and the horizontal scales of the four anomaly strips born during those epochs, the rates of sea-floor spreading can be assessed. These vary from about 1 cm/yr near Iceland to about 8–9 cm/yr at the East Pacific Rise. Assuming that the spreading rate (though position dependent) is constant in time and applies during even earlier epochs, those epochs can be dated from their widths. The answers agree well with estimates made by drilling the ocean crust and by radiometric dating. Evidence for the Cretaceous superchron of normal polarity is found.

The ridges and anomaly patterns show breaks where they are crossed by the long *transform faults*, where the great lithospheric plates slip past one another. The magnetic anomalies provide a key to decipher the past and present motion of plates. In this way geomagnetism has provided an important tool for unlocking the geological past. [See LITHOSPHERIC PLATES.]

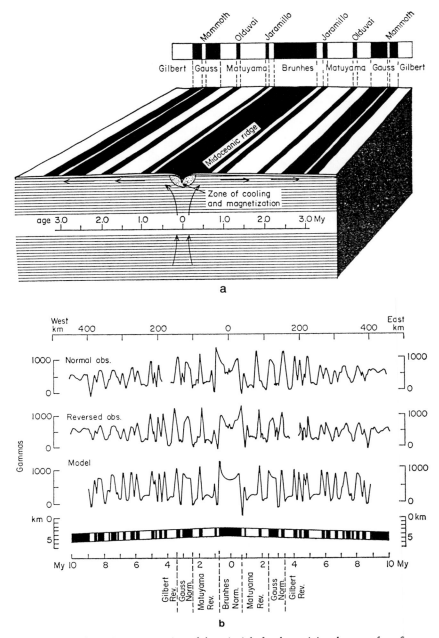

Fig. 10   (a) Schematic representation of the principle for determining the rate of sea-floor spreading and the history of the oceans. (b) Actual magnetic profiles along 1000-km transverse to the Pacific–Antarctic ridge. (1 gamma = 1 nT.) [After Allan (1969).]

Fig. 11 (a) Comparison of apparent polar wandering paths for America (•) and Europe (■). The letters indicate geological epochs. (b) The paths shown in (a) after subjecting them to the same relative rotation as that given to Africa and South America in (c), the Bullard *et al.* (1965) fit of the North Atlantic. [Parts (a) and (b) after McElhinny (1973).]

C. Continental Drift

It was seen in Section III that the average VGP positions inferred from rocks younger than 5 Myr coincide more closely with the geographic poles than do even the present geomagnetic poles: paleomagnetic poles coincide with the geographic poles during the recent geological past. It is natural to make the hypothesis that this coincidence holds throughout geological time. It was found however that, for earlier geological epochs, the paleomagnetic poles do not coincide with the present geographic poles; there has been relative movement between the site and the geographic poles. The path of the paleomagnetic pole in time defines for that site an *apparent polar wander path*. And it was found that the polar wandering paths for sites in different continents do not coincide, indicating that these sites have also moved relative to one another during the history of the earth. (See Fig. 11). For example, in the Carboniferous the south pole was in south central Africa, but relative to South America it was only 30° off the cost of Argentina. It appears that at that time the South Atlantic must have been small or absent, and that the two continents must then have "fitted together." (Fig. 11c).

The construction and comparison of the polar wander paths of all regions of the globe have led to a detailed picture of the motion of the major land masses during the history of the earth, a phenomenon known as continental drift. It has brought about a reconciliation and unification of diverse geological facts that prior to the study of paleomagnetism had appeared to be irreconcilable.

Acknowledgments

I am grateful to S. Banerjee, S. Braginsky, M. Kono, R. A. Langel, and R. T. Merrill for their help and advice.

Bibliography

Allan, T. D. (1969). Review of marine geomagnetism. *Earth Sci. Rev.* **5**, 217.

Bucha, V., ed. (1983). "Magnetic Field and the Processes in the Earth's Interior." Academia, Czechoslovak Academy of Sciences, Prague.

Bullard, E., Everett, J. E., and Smith, A. G. (1965). The fit of the continents around the Atlantic. *Philos. Trans. R. Soc. London,* A **258**, 41.

Chapman, S., and Bartels, J. (1940). "Geomagnetism. I. Geomagnetic and Related Phenomena." Clarendon Press, Oxford, England.

Irving, E., and Pullaiah, G. (1976). Reversals of the geomagnetic field, magnetostratigraphy, and relative magnitude of paleosecular variation in the Phanerozoic. *Earth Sci. Rev.* **12**, 35.

Langel, R. A. (1987). The main field. *In* "Geomagnetism," Vol. 1 (J. A. Jacobs, ed.), p. 249. Academic Press, London.

Le Mouël, J.-L., and Courtillot, V. (1981). Core motions, electromagnetic core–mantle coupling and variations in the earth's rate of rotation: new constraints from secular variation impulses. *Phys. Earth Planet. Inter.* **24**, 236.

Madden, T., and Le Mouël, J.-L. (1982). The recent secular variation and motions at the core surface. *Philos. Trans. R. Soc. London,* A **306**, 271.

McElhinny, M. W. (1973). "Paleomagnetism and Plate Tectonics." Cambridge Univ. Press, Cambridge, England.

McElhinny, M. W., and Senenayake, W. E. (1982). Variations in the geomagnetic dipole 1, the past 50,000 years. *J. Geomagn. Geoelectr.* **34**, 39.

Merrill, R. T., and McElhinny, M. W. (1983). "The Earth's Magnetic Field. Its History, Origin and Planetary Perspective." Academic Press, London.

Merrill, R. T., and McFadden, P. L. (1990). Paleomagnetism and the nature of the geodynamo. *Science* **248**, 345.

Olson, P., and Hagee, V. L. (1987). Dynamo waves and palaeomagnetic secular variation. *Geophys. J. R. Astron. Soc.* **88**, 139.

Parkinson, W. D. (1983). "Introduction to Geomagnetism." Scottish Academic Press, Edinburgh and London.

Glossary

Declination Angle between true north and the direction (magnetic north) taken by a magnetic compass needle free to turn in the horizontal plane.

Inclination Angle between the horizontal and the direction taken by a magnetic compass needle free to turn in the vertical plane containing magnetic north.

Main geomagnetic field Part of the geomagnetic field that is produced deep within the earth.

Secular variation Slow change in the magnitude and direction of the main magnetic field of the earth.

Geomagnetism, Origin

Paul H. Roberts

University of California, Los Angeles

Everyone knows that the magnetic compass needle points approximately to the north, but the reason for this has not been understood until comparatively recently. It is now recognized that the *geomagnetic field* is generated by fluid motions in the electrically conducting core of the earth. These motions are strongly affected by the rotation of the earth, which imparts to the field a rough north–south symmetry. This *geomagnetic dynamo* is a subtle mechanism. For example, the deviation of the compass needle from geographic north is no accident; it is essential if the geodynamo is to work successfully. Despite such subtleties, the secrets of the geodynamo are beginning to yield to theory, in a way summarized below.

I. Background to Modern Theory

A. Principal Questions

A description of the magnetism of the earth, and its variation in space and time, is given elsewhere in this encyclopedia. [*See* GEOMAGNETISM.]

Implicit in that account of the so-called *geomagnetic field* are a number of serious challenges for the theoretician, of which we may particularly mention the following:

- Q1. Why is the earth magnetic?
- Q2. Why has its magnetic field persisted throughout geological time?
- Q3. Why is it predominantly dipolar?
- Q4. Why has its strength changed so little during geological time?
- Q5. What determines that strength?
- Q6. Why does the compass needle point north? In other words, why do the geomagnetic and geographic axes so nearly coincide?
- Q7. Why, when averaged over geological time, are they apparently actually coincident?
- Q8. What causes the irregular polarity reversals of the geomagnetic field?
- Q9. Can one explain the behavior of the geomagnetic field during a reversal?
- Q10. Why does the frequency of those reversals vary so greatly over geological time?

There are in addition numerous secondary questions, many connected with the wide spectrum of frequencies exhibited by the field. We may mention:

- Q11. How can one explain the occurence of polarity events and excursions?
- Q12. What causes the slow secular changes in the field, that is, those on a global scale and with time scales of (say) 200–2000 yr?
- Q13. What significance does the slow westward drift of geomagnetic field patterns have?
- Q14. What causes the short-period secular variations (SPSVs) of field, having time scales of (say) 20–200 yr?
- Q15. What is the explanation of the secular variation "impulses" that last only of the order of a year?
- Q16. Can the earth be fitted into a general scheme into which the magnetism of other planets also fits?

Frequent reference is made in this article to the subject of magnetohydrodynamics (or MHD for short) which is covered elsewhere in this encyclopedia. Readers unfamiliar with the subject may find it useful to consult that article while reading this article. [*See* MAGNETOHYDRODYNAMICS.]

B. Early Theories

The existence of the geomagnetic field was initially accounted for in primitive ways, for example by a supposed affinity of the compass needle for the pole star. The first serious explanation followed William Gilbert's famous pronouncement of 1600 that the earth is a great magnet: Perhaps (it was argued) it *is* a great magnet; perhaps it is permanently magnetized. This proposal has the merit not only of answering Q1 and Q3 but also of obviating the need to answer Q2 and Q4. Answers to the questions concerning the variability of the field with time, and especially Q8, are embarrassingly lacking, and the idea completely fails to answer Q5. Every magnetic material loses its permanent magnetization when heated, at its so-called *Curie point,* which is at most several hundred degrees Celsius. The temperature increases with depth into the earth, and no material can be permanently magnetized at depths greater than about 30 km.

The crustal magnetization required to explain the dipole moment, $m \approx 7.8 \times 10^{22}$ A m^2, of the earth is about 6×10^3 A m^{-1}, which exceeds the saturation magnetization of all known minerals, including magnetite. It is also hard to imagine how the magnetization of a thin crust could by accident produce a field having a scale comparable with the radius a of the earth. Moreover, to answer Q6 by saying that the dipole axis produced by crustal magnetism coincides with the polar axis by accident is hardly satisfying! In short, permanent magnetism cannot explain geomagnetism, and can be ignored as insignificant. (We may therefore also assume that μ, the magnetic permeability of the earth, is that of free space, $\mu_0 = 4\pi \times 10^{-7}$ H m^{-1}.)

In 1947 P. A. M. Blackett made a revolutionary proposal that went far toward answering Q1–Q7. Every self-gravitating body has, he claimed, a magnetic field in virtue of its rotation. He argued that this field is predominantly dipolar, with moment:

$$\mathbf{m} = k \frac{4\pi G(\mathcal{M}\mathcal{R})^{3/2}}{c^2 \mu_0^{1/2}} \mathbf{\Omega} \tag{1}$$

where $\mathcal{R}$, $\mathcal{M}$, and $\mathbf{\Omega}$ are the radius, mass, and angular velocity of the body, c is the speed of light, G is the universal constant of gravitation, and k is a dimensionless constant that we shall set to unity. In the case of the earth, (1) gives $m \doteq 1.4 \times 10^{20}$ A m^2. This is rather too small (though not ridiculously so). Blackett's idea encounters the same difficulty over the time variations of **B** as did the permanent magnetization theory; it too is incapable of answering Q8. The other planets do not — see Q16 — support (1); neither do stars such as the sun, which reverses its polarity every 11 years. Sensitive experiments performed down deep mines did not support Blackett's hypothesis.

It has been conjectured that the magnetism of some rotating stars is due to charge separation created by centrifugal forces, and the resulting currents about the axis of rotation. Similar explanations have been proffered for the origin of the geomagnetic field. The mechanism fails, however, by many orders of magnitude.

By a process of elimination, it became clear that the geo-

magnetic field **B** is created by electric currents flowing within the earth. It should be recalled here that moving charges inevitably create magnetic field. Displacement currents being negligible, Ampère's law gives

$$\mu_0 \mathbf{J} = \nabla \times \mathbf{B} \tag{2}$$

where **J** is the electric current density. The dominance of the dipole field at the earth's surface suggests a simple answer to Q3: The electric currents creating **B** are remote from that surface, that is, deep within the earth. Recalling that the core, being metallic, conducts electricity far more readily than the (semiconducting) mantle, we may be sure that the electric currents producing **B** flow principally in the core. In what follows we shall treat as an insulator the entire region above $r = b$, the core–mantle boundary (CMB); here $r = |\mathbf{r}|$ where **r** is the radius vector from the geocenter O. (When considering core–mantle coupling, however, the small conductivity of the mantle must be recognized.) Then, by (2)

$$\mathbf{m} = \tfrac{1}{2} \int_{\text{core}} \mathbf{r} \times \mathbf{J} \, d^3x \sim \tfrac{1}{2} b \mathcal{J}(\tfrac{4}{3}\pi b^3) \tag{3}$$

where $\mathcal{J}$ is a typical magnitude of **J** in the core. By (3),

$$\mathcal{J} \sim 3m/2\pi b^4 \tag{4}$$

from which $\mathcal{J} \doteq 3 \times 10^{-4}$ A m^{-2}. We shall see in Section II, B that this is an underestimate.

The core will be modeled by a uniform conductor, with an electrical conductivity σ of 3×10^5 S m^{-1}, about that of iron, of which the core is largely composed. Unless a "battery" exists to maintain the currents, both they and the associated magnetic field will die out through electrical resistance in the *free decay time,*

$$\tau_\sigma = \mu_0 \sigma L^2 \tag{5}$$

where $L = b/\pi$. In making order of magnitude estimates below, we shall adopt L as the typical scale of the geomagnetic field. The entire electromagnetic energy E_m of the earth must be replenished over a time of order $\tau_\sigma \doteq 14{,}500$ yr, where

$$\begin{aligned}
E_m &= \frac{1}{2\mu_0} \int_{\text{all space}} B^2 \, d^3x \\
&= \frac{1}{2} \mu_0 \int_{\text{core}} \int_{\text{core}} \frac{\mathbf{J}(\mathbf{r}) \cdot \mathbf{J}(\mathbf{r}')}{|\mathbf{r} - \mathbf{r}'|} \, d^3x \, d^3x'
\end{aligned} \tag{6}$$

If we use (4) to estimate the right-hand side of (6), we obtain

$$E_m \sim \frac{\mu_0 \mathcal{J}^2}{2b} \left(\frac{4}{3} \pi b^3 \right)^2 \sim \frac{2\mu_0 m^2}{b^3} \tag{7}$$

and this gives $E_m \doteq 10^{20}$ J and an *ohmic heat loss,*

$$Q_\sigma \sim E_m/\tau_\sigma \tag{8}$$

of about 2×10^8 W. These values of E_m and Q_σ will again be seen in Section II,B to be underestimates; the electromagnetic power demand of the geomagnetic field is of order 10^{11}–10^{12} W [See CORE–MANTLE SYSTEM.]

Q2 now raises two further questions:

- Q17. What kind of battery maintains the electric currents?
- Q18. Can that battery supply energy at the necessary rate over geological time?

Several answers have been given to Q17. Taking the word *battery* in its everyday sense, electrochemically produced potential differences over the CMB have been posited, where the iron-rich core meets the silicon-rich mantle. It has also been suggested that significant thermoelectric potential differences are generated on the CMB, due to the differential heating of the CMB by convection currents in the core. Both these ideas suffer from two major drawbacks: First, estimates of the electrochemical and thermoelectric potentials suggest that neither is large enough to produce a field of sufficient strength; second, they do not create the *poloidal* field that is seen at the earth's surface — they generate only *toroidal* field.

We have introduced here an important representation first suggested by H. Lamb in 1881. Since magnetic poles do not exist, we have

$$\nabla \cdot \mathbf{B} = 0 \qquad (9)$$

Lamb showed that *toroidal* and *poloidal* scalars T and P therefore exist such that

$$\begin{aligned}\mathbf{B} &= \nabla \times (T\mathbf{r}) + \nabla \times \nabla \times (P\mathbf{r}) \\ &= \mathbf{B_T} + \mathbf{B_P} \text{ (say)}\end{aligned} \qquad (10)$$

The toroidal field $\mathbf{B_T}$ everywhere lacks a radial component. By (2) and (10),[1]

$$\mu_0 \mathbf{J} = \nabla \times \nabla \times (T\mathbf{r}) + \nabla \times (-\mathbf{r}\,\nabla^2 P) \qquad (11)$$

[1] To derive (11) from (2) and (10) note that

$$\begin{aligned}(\mathbf{B_P}) &= \frac{\partial^2}{\partial x_i\,\partial x_j}(Px_i) - \frac{\partial^2}{\partial x_j\,\partial x_j}(Px_i) \\ &= \frac{\partial}{\partial x_i}\left(x_j\frac{\partial P}{\partial x_j} + 3P\right) - \left(x_i\nabla^2 P + 2\frac{\partial P}{\partial x_i}\right) \\ &= \frac{\partial}{\partial x_i}\left(x_j\frac{\partial P}{\partial x_j} + P\right) - x_i\nabla^2 P\end{aligned}$$

It follows that

$$\mathbf{B_P} = \nabla\left(r\frac{\partial P}{\partial r} + P\right) - \mathbf{r}\,\nabla^2 P \qquad (14)$$

To establish (14)–(16), use (13) and the identity

$$\nabla^2 V = -\nabla^2\left(x_j\frac{\partial}{\partial x_j} + 1\right)P = -\left(x_j\frac{\partial}{\partial x_j} + 3\right)\nabla^2 P$$

Thus, the poloidal field $\mathbf{B_P}$ is created by toroidal currents, and the toroidal field $\mathbf{B_T}$, by poloidal currents.

Any magnetic field can be written as in (10) but, in the current-free exterior of the core, (11) gives

$$T = 0 \qquad \nabla^2 P = 0 \qquad (12, 13)$$

According to (12), the toroidal field is magnetically undetectable at the earth's surface, $r = a$. The field above the core (and in particular at the earth's surface) is $\mathbf{B_P}$, and by (10) and (13), we there have

$$\mathbf{B} = -\nabla V \qquad (14)$$

where V, the *geomagnetic potential*, is given by

$$V = -\frac{\partial(rP)}{\partial r} \qquad (15)$$

and satisfies Laplace's equation:

$$\nabla^2 V = 0 \qquad (16)$$

In the article entitled "Geomagnetism," (14) and (16) led to the basic multipole representation of the main geomagnetic field at the surface of the earth as the sum of a dipole, a quadrupole, an octupole, ... The dipole term $V = \mathrm{O}(r^{-2})$ dominates for large r, and so by (15)

$$P = \mathrm{O}(r^{-2}) \qquad r \to \infty \qquad (17)$$

This condition is very significant. It expresses the absence of any sources of geomagnetic field external to the earth. Any theory that seeks an internal origin for geomagnetism must satisfy (17).

In the absence of inductive effects, potential differences on a spherical CMB generate electrical currents that are purely poloidal, and $\nabla^2 P = 0$ *everywhere* by (16), implying by (11) that P vanishes identically. Electrochemical and thermoelectric potentials generate toroidal magnetic fields only, but (as we shall see) other mechanisms exist that do that effectively. The central difficulty is that of explaining how the earth generates its observed poloidal field, and electrochemical and thermoelectric effects are of no direct help.

Other sources of $\mathbf{B}$ that have been invoked include the Nernst effect and the Hall effect. The former creates an electromotive force (emf) perpendicular to $\mathbf{B}$ and to the gradient $\nabla\Theta$ of the prevailing temperature Θ. Unfortunately the constant of proportionality seems to be far too small to account for the geomagnetic field. This also appears to be true of Hall currents, which are perpendicular to **B** and to the electric field **E**.

C. The Dynamo Hypothesis

Again by a process of elimination, there seems to be only one plausible answer to Q17: The earth's field is maintained by electromagnetic induction — that is, by the emf created when a conductor moves, with velocity **v** (say), across a magnetic field **B**. The emf induced in this way is in the direction **v** $\times$ **B**, perpendicular both to the velocity and the inducing field. The *induced currents* created by this emf generate, according to Ampère's law (2), an *induced magnetic field*. The dynamo theory of geomagnetism, first advanced by Sir Joseph Larmor in 1919, supposes that the inducing and induced fields are one and the same; the field **B** is self-excited. This is not tantamount to hoisting oneself up by one's own boot straps. Although it is true that the electrical resistance of the medium removes magnetic energy (the ohmic heat loss), that energy can be supplied kinetically by the rate of working **v** · (**J** $\times$ **B**) of the mechanism that maintains the velocity **v** against the opposition of the force that acts on any current-carrying conductor, the so-called *Lorentz force,* **J** $\times$ **B**. The dynamo is in fact a machine for generating electrical energy from kinetic energy, in the same way in which electric power is supplied to the home from the kinetic energy of turbines in a commercial power station.

Figure 1a sketches a simple self-excited device, the *homopolar dynamo,* in which a conducting disk D rotates steadily in the magnetic field created by the current I flowing in a surrounding wire loop W. The emf generated in D is the sole source of I, which is drawn off the rim and axle of D by sliding contacts on each. If the angular speed Ω of D is large enough, I and **B** will not diminish, and the dynamo is then self-exciting. The homopolar dynamo is significantly asymmetric; if W is wound in the opposite sense round D, or if the motion of D is reversed, the dynamo will not function. Figure 1b is a more complicated device, T. Rikitake's two-disk dynamo, in which the current created by one disk feeds the wire loop surrounding the other. Its behavior is described later.

For the remainder of this article it will be assumed that the geomagnetic field owes its existence to such a self-excited dynamo operating in the core, though one that lacks the deliberate asymmetry of the dynamos shown in Fig. 1, and which is therefore harder to understand. Initially, we shall suppose that **v** is known; that is, we shall study the *kinematic geodynamo.* Later (in Section III) we shall consider the dynamics that determine **v**.

Although many of Q1–Q18 have not yet been fully answered by the dynamo hypothesis, none seem to be potentially irreconcilable with it. Indeed, replies to some of the questions already seem convincing. For this reason, there is

no other theory that is today in serious contention as an explanation of the earth's magnetism.

II. Kinematic Geodynamo Theory

A. Electromagnetic Induction

Ohm's law for a moving conductor is

$$\mathbf{J} = \sigma(\mathbf{E} + \mathbf{v} \times \mathbf{B}) \tag{18}$$

Substituting this into Faraday's law,

$$\frac{\partial \mathbf{B}}{\partial t} = -\nabla \times \mathbf{E} \tag{19}$$

and using (2), (9), and (18), we obtain the *induction equation:*

$$\frac{\partial \mathbf{B}}{\partial t} = \frac{1}{\mu_0 \sigma} \nabla^2 \mathbf{B} + \nabla \times (\mathbf{v} \times \mathbf{B}) \tag{20}$$

Here and below we measure all variables in a frame co-rotating with the earth.

If **v** = 0 and there are no external sources of **B** [see (12), (13), and (17)], (20) establishes that **B** diffuses away on a time scale which, as the dimensions of the surviving terms in (20) show, is of the order τ_σ defined in (5). If however **v** is large enough, the final term in (20) may overcome this decay. Comparing these two terms we see that, for **B** to survive, it is necessary that the *magnetic Reynolds number,*

$$R = \mu_0 \sigma UL \tag{21}$$

be sufficiently large, where U is a typical speed of core flow.

The angular velocity corresponding to the apparent westward drift of geomagnetic field patterns is about 10^{-11} s^{-1}, corresponding to a velocity of 3×10^{-4} m s^{-1} at the equator of the core. This provides the basis for the usual way of estimating U, which we shall take to be 10^{-4} m s^{-1}. Then, by (21), $R = 40$.

B. Cowling's Theorem, the Omega Effect

Although the necessary condition, $R = O(1)$, for geodynamo action is apparently comfortably satisfied in the core, this condition is not sufficient to ensure that **B** is permanently maintained. It is also necessary that the motions have a sufficiently *low degree of symmetry.* An axisymmetric field, $\overline{\mathbf{B}}$, has a high degree of symmetry and cannot be maintained by dynamo action, according to the celebrated 1933 theorem of T. Cowling. The asymmetric part of the earth's field is essential for the survival of the entire field. In establishing Cow-

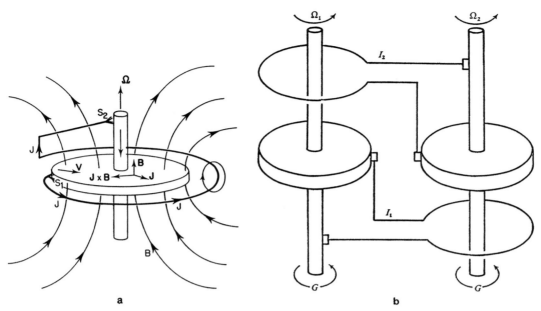

Fig. 1 (a) The homopolar dynamo. The conducting disk turns about its axis in the direction indicated and, by cutting the lines of force of the field **B**, creates an emf that drives a current across the disk and, via the sliding contacts S_1 and S_2, also around the loop **W** in the direction indicated. If the angular velocity of the disk is sufficiently large, the field created by this current may be made identical to the field **B**. The device then acts as a self-excited dynamo. It should be noted that the Lorentz force $J \times B$ acting on the disk is in the opposite direction to the velocity **v** of the disk. The power supplied to the disk by the motor to overcome this force is converted into magnetic energy, which is dissipated ohmically by the resistance of the current path in the dynamo. (b) The Rikitake two-disk dynamo. Two homopolar dynamos are linked together by directing the current generated by one to the encircling wire loop of the other. The disks are driven by the same torque G, but their angular velocities are different, as are the currents they generate.

ling's result below, we shall assume that the $\bar{B}$ is steady, and prove a contradiction. The theorem holds, however, for oscillatory $\bar{B}$ also.

To see why the axisymmetric dynamo fails, we take Oz as the axis of symmetry so that the toroidal field is purely zonal, while the lines of force of the poloidal field lie in meridian planes, as indicated in Figs. 2a and 2b. Since the density at the bottom of the fluid core is only about 10% greater than it is at the CMB, we may assume that, to a good approximation, core fluid is of uniform density ρ, so that

$$\nabla \cdot \mathbf{v} = 0 \tag{22}$$

We may therefore also regard $\bar{\mathbf{v}}$ as the sum of toroidal and poloidal parts, as sketched in Figs. 2c and 2d.

In seeing how toroidal field is created from poloidal field, we are helped by Alfvén's *frozen flux* theorem, which states that if $\sigma = \infty$ [so that the middle term in (20) is absent], the lines of magnetic force move with a conductor, as though

frozen to it. Although R is large in the core, it is not infinite; if it were, the dynamo mechanism would be unnecessary; the field would remain attached to core fluid forever! Alfvén's theorem does not hold exactly; it indicates a tendency, though a strong one. The shearing motion $\bar{v}_T$, shown in Fig. 2c, carries the poloidal lines of force out of the meridional planes and stretches them out along lines of latitude; that is, it creates $\bar{\mathbf{B}}_T$ from $\bar{\mathbf{B}}_P$. This effect is sometimes called the *omega effect* for historical reasons: The velocity shear $\bar{v}_T/s$ has often been denoted by ω. The strength of the ω-effect is measured by an ω-effect magnetic Reynolds number $R_\omega = \mu_0 \sigma U L$, where we now associated U with the strength of the zonal flow $\bar{v}_T$. It is then found that

$$\bar{B}_T \approx R_\omega \bar{B}_P \tag{23}$$

The field created at the poles of the core by the centered dipole m is $\mu_0 m / 2\pi b^3 \doteq 3.7$ gauss. (1 gauss $= 10^{-4}$ T.) In making estimates, we shall assume that $\bar{B}_P = 3.5$ gauss, which

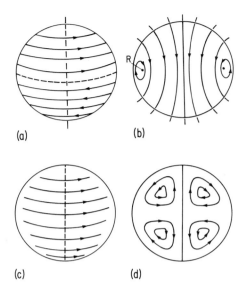

Fig. 2 (a) An axisymmetric toroidal field $\overline{\mathbf{B}}_T$. The lines of force follow latitude circles. Such a field would be generated by a poloidal electric current system $\overline{\mathbf{J}}_P$, that is, one in which the current lines closed in meridian planes. (b) An axisymmetric poloidal field $\overline{\mathbf{B}}_P$. The lines of force lie in meridian planes; R indicates where a ring of O-type neutral points meets the meridian section shown. Such a field would be generated by a toroidal electric current system $\overline{\mathbf{J}}_T$, that is, one in which the current lines followed latitude circles. (c) An axisymmetric toroidal flow $\overline{\mathbf{v}}_T$. The streamlines are latitude circles. (d) An axisymmetric poloidal flow $\overline{\mathbf{v}}_P$. The streamlines close in meridian planes.

is also roughly the magnitude of the field obtained by extrapolating to $r = b$ the **B** observed on $r = a$. Taking $R_\omega = 40$ as typical, (23) shows that $\overline{B}_T = 140$ gauss. The zonal field created from the meridional field by the ω-effect is large. To leading order, both **B** and **v** are toroidal and axisymmetric in the core:

$$\mathbf{B} = \overline{B}_\phi(s, z, t)\hat{\boldsymbol{\phi}} \qquad \mathbf{v} = \overline{v}_\phi(s, z, t)\hat{\boldsymbol{\phi}} \qquad (24)$$

where (s, ϕ, z) are cylindrical coordinates and $\hat{\boldsymbol{\phi}}$ is the unit vector along latitude circles.

Nevertheless, there is no way for the axisymmetric dynamo to make use of the toroidal field to regenerate the poloidal field. Like $\overline{\mathbf{v}}_T \times \overline{\mathbf{B}}_P$ induction just described, the $\overline{\mathbf{v}}_P \times \overline{\mathbf{B}}_T$ emf manufactures only poloidal $\overline{\mathbf{J}}$ and toroidal $\overline{\mathbf{B}}$; since $\overline{\mathbf{v}}_T$ and $\overline{\mathbf{B}}_T$ are parallel they create no emf. The only emf that generates toroidal current is $\overline{\mathbf{v}}_P \times \overline{\mathbf{B}}_P$. But this is zero on the ring R of O-type neutral points (where $\overline{\mathbf{B}}_P = 0$) indicated in meridian projection on Fig. 2(b). At least one such ring must lie in (or on the surface of) the fluid. Thus $\overline{\mathbf{B}}_P$ diffuses away in a time of

order τ_σ, affected in no essential way by the presence of $\overline{\mathbf{v}}$; once $\overline{\mathbf{B}}_P$ has departed, $\overline{\mathbf{B}}_T$ no longer has a source and follows suit. As noted earlier, a difficulty of geomagnetic theory is that of accounting for $\overline{\mathbf{B}}_P$, not that of maintaining $\overline{\mathbf{B}}_T$.

We may now appreciate why our earlier estimates of $\mathscr{J}$, E_m, and Q_σ were too small; (4) is an estimate $\mathscr{J}_T$ of the toroidal current density that generates **m** and $\mathbf{B}_P$, but because of (23) the poloidal current density is much greater. As (2) shows in order of magnitude, $\mathscr{J}_P \sim B_T/\mu_0 L \doteq 10^{-2}$ A m^{-2}, so that $E_m \doteq 6 \times 10^{23}$ J by (7) and $Q_\sigma \doteq 10^{12}$ W.

C. The Alpha Effect

Since an axisymmetric **B** cannot be maintained by dynamo action, it is natural to wonder how the axisymmetric part, $\overline{\mathbf{B}}$, of a general field **B** can persist. Obviously the action of the asymmetric part, $\mathbf{v}' \equiv \mathbf{v} - \overline{\mathbf{v}}$, of **v** must in some way be crucial. It is now generally believed that $\overline{\mathbf{B}}_P$ is created from $\overline{\mathbf{B}}_T$ by an *alpha effect*. Rotation imparts *helicity*, $H = \mathbf{v} \cdot \boldsymbol{\omega}$, to any flow of a rotating stratified fluid, such as that in the core; in other words, it produces a correlation between the flow **v** and its vorticity, $\boldsymbol{\omega} = \nabla \times \mathbf{v}$. If, like the core, the fluid is electrically conducting, this kinetic helicity induces a magnetic helicity, $M = \mathbf{B} \cdot \mathbf{J}$ in the field, that is, a correlation between **B** and $\mathbf{J} = \nabla \times \mathbf{B}/\mu_0$.

The simplest picture of how this correlation arises again makes use of Alfvén's frozen flux theorem. The action of a helical eddy on a field line of $\overline{\mathbf{B}}_T$ may, for simplicity, be thought of in two parts. First (Fig. 3a) the motion $\mathbf{v}'$, in the direction of the arrow and with $\boldsymbol{\omega}'$ somehow removed, creates an Ω-shaped dent in the field line; then the twisting motion of $\boldsymbol{\omega}'$, with now the motion $\mathbf{v}'$ of Fig. 3a magically subtracted, turns the Ω-loop out of the plane of the paper (Fig. 3b). Because of the large field gradients at the base of the Ω, it can ohmically detach from the original field line to produce a closed flux loop (Fig. 3c). By Ampère's law (2), this loop is associated with an emf antiparallel to the field line, and the action of many such loops, as would be generated in a turbulence that possessed a predominance of eddies having the right-handed twist depicted in Fig. 3, is to create an average emf. $\overline{\mathscr{E}} = \overline{\mathbf{v}' \times \mathbf{B}'}$, antiparallel to $\overline{\mathbf{B}}_T$. If, instead of the positive H that we implied by the predominance of right-handed eddies, we had supposed that $H < 0$, we would have obtained a mean emf parallel to $\overline{\mathbf{B}}_T$. The process is called the alpha effect for historical reasons: The constant of proportionality in the simplest example, $\overline{\mathscr{E}} = \alpha\overline{\mathbf{B}}_T$, was chosen, at an influential moment in the development of the subject, to be α. We expect this emf to produce a mean current parallel ($\alpha > 0$) or antiparallel ($\alpha < 0$) to the mean field—that is, to create an M of the same sign as α but of opposite sign to H.

The zonal electric current created by $\overline{\mathscr{E}}$ is, by Ohm's law

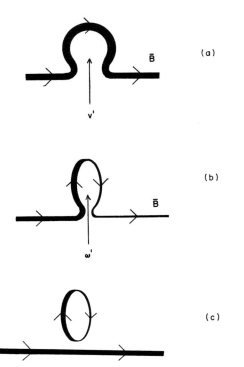

Fig. 3 (a) The motion v' in the direction of the arrow creates an Ω-shaped dent in a field of $\overline{\mathbf{B}}_{\mathrm{T}}$. (b) The twisting motion associated with ω', where ω is the vorticity $\nabla \times \mathbf{v}$, turns the Ω-loop out of the plane of the paper. (c) Because the field gradients are large at the base of the Ω, it can ohmically detach from the original field line to produce a closed flux loop.

(18), of order $\sigma\overline{\mathscr{E}} = \sigma\alpha\overline{\mathbf{B}}_{\mathrm{T}}$. This generates a poloidal field of order

$$\overline{B}_{\mathrm{P}} \approx R_\alpha \overline{B}_{\mathrm{T}} \tag{25}$$

where

$$R_\alpha = \mu_0 \sigma \tilde{\alpha} L \tag{26}$$

is the alpha-effect Reynolds number. [Here $\tilde{\alpha}$ is a characteristic magnitude of α.]

By (23) and (25), the axisymmetric part $\overline{\mathbf{B}}$ of $\mathbf{B}$ can be maintained indefinitely. The α-effect produces $\overline{\mathbf{B}}_{\mathrm{P}}$ from $\overline{\mathbf{B}}_{\mathrm{T}}$, and the ω-effect creates $\overline{\mathbf{B}}_{\mathrm{T}}$ from $\overline{\mathbf{B}}_{\mathrm{P}}$. For this $\alpha\omega$-dynamo to prevent the collapse of $\overline{\mathbf{B}}$, it is necessary only that the *dynamo number*,

$$D = R_\alpha R_\omega \tag{27}$$

be of order unity.[2] Some of the most efficient dynamos (i.e., those for which D is least) are those in which a meridional flow $\overline{\mathbf{v}}_{\mathrm{P}}$ of order $(\mu_0\sigma L)^{-1} = 3 \times 10^{-6}$ m s^{-1} is present to assist the generation of $\mathbf{B}$.

Although it would be naïve to suppose that the flow in a body of fluid as large as the core is completely laminar, there are no indications that it is as highly irregular as the motion in, for example, the solar convection zone. The sun reverses its field every 11 years, implying that τ_σ and therefore σ for the solar convection zone have, through the action of turbulence, been effectively reduced below their molecular values by factors of about 10^5. There is no geomagnetic evidence that the time scales and conductivity of the earth's core are similarly reduced. Unlike the sun, turbulent processes like that depicted in Fig. 3 are probably not the main contributors to α in the core. It is more likely that α is created by large-scale planetary waves, such as those whose existence is suggested by the westward drift; see Section IV. It should also be noticed that the energy emerging per unit area of the solar surface is about 10^{10} times greater than that emerging from the earth's core. The viscous dissipation of energy by the turbulence in the solar convention zone is easily replenished. The energy resources of the earth's core do not, however, permit the luxury of a fully developed turbulence. [*See* CORE.]

The kinematic problem, that of solving the induction equation (20) with external sources of field excluded by conditions (12) and (17), is linear. If the flow $\mathbf{v}$ is not too symmetrical (e.g., if it possesses sufficient helicity) it can, depending on the magnitude of R, regenerate field. Then $\mathbf{B}$ either grows without limit or dies out utterly (with or without oscillation). Between these two extremes lies the knife-edge of criticality ($R = R_c$), for which $\mathbf{B}$ is either steady or oscillates sinusoidally on a time scale of order τ_σ. Many such kinematic dynamo models have by now been constructed. These establish that the dynamo hypothesis has no difficulty answering Q1 and Q2. Kinematic theory is however incapable of answering Q4–Q16. For example, such a linear theory can say nothing about the magnitude of $\mathbf{B}$. Moreover, the period $O(\tau_\sigma)$ of an oscillating kinematic dynamo is far shorter than the observed time scale between reversals of the geomagnetic field, and it cannot explain them. To make progress, it is necessary to understand not only the kinematics governed by (20) but also the dynamics of the fluid motions; and we shall find in Section III,D that axisymmetry again presents difficulties—this time in the force balance.

[2] The phrase "D of order unity" is used, even though D has, for models that have been integrated numerically, invariably exceeded 30. This kind of hazard in making dimensional arguments is well known and has already been met in (5), where *a priori* one might have expected τ_σ to be $\mu_0\sigma b^2$ rather than $\mu_0\sigma b^2/\pi^2$. We shall encounter further examples; see (29).

III. Dynamics of the Geodynamo

A. Magnitude of Forces Acting on the Fluid Core

We may estimate some of the forces (per unit volume) acting on the fluid core as follows:

(a) The *inertial force* is $\rho\, Dv/Dt$, where ρ is core density ($\doteq 10^4$ kg m^{-3}) and D/Dt is the motional derivative. It follows that

$$\rho\, D\mathbf{v}/Dt \approx \rho\mathbf{v}\cdot\nabla\mathbf{v} \approx \rho U^2/b \approx 10^{-10}\ \text{N m}^{-3}$$

where we have taken $U \approx 10^{-4}$ m s^{-1} as before.

(b) The *Coriolis force* is

$$2\rho\mathbf{\Omega}\times\mathbf{v} \approx 2\rho\Omega U \approx 10^{-4}\ \text{N m}^{-3}$$

(c) The *viscous force* is

$$\rho\nu\,\nabla^2\mathbf{v} \approx \rho\nu U/b^2 \approx 10^{-15}\ \text{N m}^{-3}$$

The kinematic viscosity ν of the core is extremely uncertain, and we have here taken $\nu = 10^{-3}$ m^2 s^{-1}, which may be correct within an order of magnitude.

(d) The *magnetic* or *Lorentz force* is

$$\mathbf{J}\times\mathbf{B} \approx \mathbf{J_P}\times\mathbf{B_T} \approx B_T^2/\mu_0 b \approx 10^{-4}\ \text{N m}^{-3}$$

(e) The *Archimedean* or *buoyancy force* is $\rho C\mathbf{g}$, where $\mathbf{g}$ is the acceleration due to gravity; ρC is the density inhomogeneity created by variations in temperature or chemical composition and is hard to estimate. Since $\rho g \approx 10^5$ N m^{-3}, we have

$$\rho C\mathbf{g} \approx 10^{-4}\ \text{N m}^{-3}$$

even if C is as small as 10^{-9}.

A deep question arises here that is related to Q4 and Q5:

- Q19. Why are the Lorentz and Coriolis forces apparently of much the same size in the core, while the inertial and viscous forces are far smaller?

The behavior of a simple example that retains all except the inertial force is illuminating.

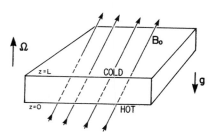

Fig. 4 A horizontal layer of uniform electrically conducting fluid, rotating about the vertical and lying in a magnetic field. Its upper and lower surfaces are maintained at uniform but different temperatures, that of the lower being the greater. If this temperature difference is sufficiently great, the layer will convect.

B. Behavior of a Simple Model of Rotating Magnetoconvection

An infinite horizontal layer of fluid ($0 \leq z \leq L$) rotates about the vertical with angular velocity $\mathbf{\Omega}$ and is at rest in the co-rotating frame; gravity $\mathbf{g}$ is uniform (Fig. 4). The bottom of the layer is maintained at a uniform temperature that exceeds by $\Delta\Theta$ the uniform temperature at which the top surface is held. To simplify study of the key dynamical balances, we shall omit an important electrodynamic condition of the dynamo problem: We shall suppose that the prevailing magnetic field $\mathbf{B_0}$ is uniform—that it is maintained by sources outside the layer.

If $\Delta\Theta$ is slowly increased from zero, a critical stage is reached when $\Delta\Theta = \Delta\Theta_c$ (say), when the state of rest is for the first time linearly unstable to convective overturning, but only on one particular horizontal scale, l_c. The buoyancy force is insufficient to drive convection of smaller l, for its greater vertical shear rate excites a larger viscous opposition; increasingly for large l the buoyancy force is balanced by the pressure gradient, leaving less over to drive the convection. The critical scale l_c represents the best compromise between these two extremes. We shall discuss the dependence of l_c and $\Delta\Theta_c$ on $\mathbf{\Omega}$ and $\mathbf{B_0}$. We introduce dimensionless measures of U, $\mathbf{\Omega}$, $\mathbf{B_0}$, and $\Delta\Theta$, namely the Rossby, Ekman, Chandrasekhar, and Rayleigh numbers:

$$\text{Ro} = \frac{U}{2\Omega L} \qquad E = \frac{\nu}{2\Omega L^2}$$

$$Q = \frac{\sigma B^2 L^2}{\rho\nu} \qquad \text{Ra} = \frac{g\gamma\,\Delta\Theta\,L^3}{\nu\kappa} \tag{28}$$

where γ and κ are the coefficients of volume expansion and the thermal diffusivity of the fluid, respectively. Referring back to forces (a)–(d), we see that Ro, E, and Q, respectively, measure three ratios: inertial force/Coriolis force, viscous force/Coriolis force, and magnetic force/viscous force. Since we are examining only the onset of convection, Ro ≡ 0. We shall, for simplicity, confine attention to convection that is steady in the marginal case Ra = Ra$_c$.

In the absence of rotation and magnetic field, convection starts when

$$l_c = O(L) \qquad \text{Ra}_c = O(1) \qquad (29)$$

that is, when the convection cells are "as broad as they are deep." [As in the case of D—see footnote to (27)—the second of (29) conceals the fact that Ra$_c$ is numerically large, typically about 10^3.]

Suppose now that $\mathbf{B} = 0$ but that the layer is in rapid rotation: $E \ll 1$. In the inertial frame, a rotating layer is packed with vertical vortex lines to which an inviscid fluid would be attached (Kelvin's theorem). In the rotating frame, this gives rise to the Proudman–Taylor theorem: The slow, steady motion of an inviscid, highly rotating fluid is two-dimensional with respect to the rotation axis; that is, $\mathbf{v}$ is independent of z when Ro = E = 0. Since $v_z = 0$ on the two confining surfaces $z = 0$ and L of our layer, two-dimensionality requires that $v_z \equiv 0$. But, if convection is to carry heat from one boundary to the other, v_z is necessarily nonzero between the boundaries!

How can the system "break the constraint" of the Proudman–Taylor theorem? Since we are considering only the onset of convection (Ro = 0), the reply must be, "Through viscosity"; but how can viscosity help when E is so small? The answer is that despite the smallness of E, the condition $E \equiv 0$ for the strict validity of the Proudman–Taylor theorem is violated. If the horizontal scale l is small enough, the effective Ekman number will be increased enough to defeat the Proudman–Taylor theorem and allow convection to occur. Since $l_c \ll L$, the theorem has enforced a strong tendency toward two-dimensionality but has not prevented convection. Small l_c means large viscous dissipation, and therefore a large buoyancy force is required to supply that energy loss—Ra$_c$ must be large. After some analysis, it is found in fact that

$$l_c = O(E^{1/3}L) \qquad \text{Ra}_c = O(E^{-4/3}) \qquad (30)$$

This situation is dramatically changed if now a sufficiently large $\mathbf{B}_0$ is added. It is found that if $Q = O(E^{-1})$, (30) is replaced by

$$l_c = O(L) \qquad \text{Ra}_c = O(E^{-1}) \qquad (31)$$

The reduction in $\Delta\Theta_c$ is astonishing! If $E = 10^{-11}$ (for example) then, according to (30), Ra$_c \doteq 5 \times 10^{14}$ for the non-magnetic layer but, according to (31), Ra$_c \doteq 10^{11}$; that is,

$\Delta\Theta_c$ is reduced by a factor of about 5000 by the magnetic field! As the first of (31) shows, it is no longer necessary for the motions to develop on a small scale to allow viscous forces to defeat the rotational constraint; magnetic forces alone do that. Extremely large diffusive energy losses are therefore avoided, and convection can be initiated with a correspondingly reduced buoyancy force.

With the help of new dimensionless measures, Elsasser and modified Rayleigh numbers,

$$\Lambda = \frac{\sigma B_0^2}{2\Omega\rho_0} \qquad \widetilde{\text{Ra}} = \frac{g\,\gamma\,\Delta\Theta\,L}{2\Omega\kappa} \qquad (32)$$

(31) may be stated in another way:

$$\text{If} \quad \Lambda = O(1), \quad \text{then} \quad l_c = O(L) \quad \text{and} \quad \widetilde{\text{Ra}}_c = O(1) \qquad (33)$$

In fact the smallest value of Ra$_c$, considered as a function of B_0 for given Ω, occurs at an $O(1)$ value of Λ. From the absence of v in (32) and (33), we see that viscosity is completely irrelevant to the convection problem in this range of Λ [See GEOPHYSICAL FLUID MECHANICS.]

C. Relationship of the Simple Model to the Earth's Core

Observing the slow motion of the hour hand of a clock (which in fact turns twice as fast as the earth), one might be forgiven for thinking that the earth rotates slowly. The scale of the core is so enormous, however, that dimensionless measures of rotation such as Ro and E indicate otherwise. Using values given in (a)–(c) above, Ro $\doteq 10^{-6}$ and $E \doteq 10^{-11}$, which again underscores the irrelevance of viscous forces in the core.

According to (b) and (d) above, Λ is the ratio magnetic force/Coriolis force, and is indeed $O(1)$ in the core, in agreement with the first of (33). In the context of a spherical earth, the second of (33) is consistent with the observed planetary scale of the main geomagnetic field. This encourages us to believe that the simple model behaves in some ways like the earth's core, and also that we have uncovered the germ of a geophysically useful idea: By creating a geomagnetic field of the appropriate strength [$\Lambda = O(1)$], the geodynamo permits convection to occur in a core that, in the absence of that field, would be immobilized by the rotational constraint of the Proudman–Taylor theorem. In other words, there is a thermodynamic reason why the geodynamo exists: It helps the earth to cool! This heuristic idea would also explain why the core eschews a perfectly good solution of the magnetohydrodynamic equations, namely the nonmagnetic one, $\mathbf{B} \equiv 0$! A deep answer to Q1 is in the offing.

Several objections may be advanced against this possibly over-ambitious hope. For example,

(i)   By postulating a $\mathbf{B}_0$ maintained by sources outside the layer, we have violated the dynamo condition.
(ii) We have ignored the ohmic requirements of those sources.
(iii) We have studied only the onset of convection, so that α, like Ro, is infinitesimal and cannot assist in field generation.
(iv) The model does not possess the spherical geometry of geophysical reality.

Of these, (iv) does not seem serious. Analyses of models having other magnetic field structures and other (e.g., spherical) geometries invariably lead to qualitatively the same conclusions. In all cases Coriolis forces endow the convective motions $\mathbf{v}'$ with helicity and create an α-effect. Although (iii) is true, we may hope that, since Ro $\ll$ 1 in the core, simple extensions of the analyses will lead readily to the finite α required in the geophysical application. Ideally one would arrange that this α self-consistently maintained the imposed $\mathbf{B}_0$, so creating a convective dynamo. No mathematical model of such a geodynamo has yet been created for the parameter range (33). In response to (i) we would argue that there is no obvious reason why the convection dynamics should depend crucially on the boundary conditions that $\mathbf{B}_0$ happens to satisfy. If this is true, there is no need to worry about (ii); the energy losses of $\mathbf{B}_0$ have already been assessed at the end of Section II,B.

D. Magnetostrophy, the Thermal Wind

We now complicate by returning to the spherical geometry of the core. Our estimates have strongly suggested that we may ignore inertial forces and, except in boundary layers, viscous forces also. Let us therefore explore the possibility of replacing the equation of motion by the *magnetogeostrophic equation*:

$$2\rho\boldsymbol{\Omega} \times \mathbf{v} = -\nabla p + \mathbf{J} \times \mathbf{B} + \rho C\mathbf{g} \qquad (34)$$

As $\mathbf{B}$ evolves according to (20), a state of near magnetostatic equilibrium is, by (34), continually maintained. The pressure gradient must be included in (34) to allow (22) to be satisfied. All forces in (34) are of the same order of magnitude. Only the first has a preferred direction in space, and its presence provides the potential answer to Q6. The directional property of the magnetic compass needle demonstrates that Coriolis forces are important in the core.

The left-hand side of (34) vanishes when we take its scalar product with $\mathbf{v}$; Coriolis forces do no work. The term $-\mathbf{v}\cdot\nabla p[=-\nabla\cdot(p\mathbf{v})$ by (22)] vanishes when integrated over

the core; the pressure gradient does no work *in toto*. The term $\mathbf{v}\cdot(\mathbf{J} \times \mathbf{B}) = -\mathbf{J}\cdot(\mathbf{v} \times \mathbf{B})$ represents the rate at which kinetic energy is converted into magnetic energy through the last term in (18) or (20). It is clearly essential that this lost energy is replenished by the rate of working of the buoyancy force $\rho C\mathbf{g}\cdot\mathbf{v}$. The primary source of C is believed to be not thermal buoyancy but compositional buoyancy—that is, to be caused by the density differences created when light material rises from the inner core boundary, where it has been released during the general cooling of the earth from the heavy ferrous component of the alloy of which the fluid core is composed. Plausibly, this source has been active since the birth of the inner core, and it delivers sufficient power to maintain the geomagnetic field—the answer to Q18 is in the affirmative.

Since (24) holds in a first approximation, (34) gives

$$2\Omega v_* \hat{\mathbf{s}} - \overline{C}_{\mathbf{g}} = \nabla\left(\frac{\overline{p}}{\rho} + \frac{B_\phi^2}{2\mu_0\rho}\right) \qquad (35)$$

where $\hat{\mathbf{s}}$ is the unit vector in the direction of increasing s,

$$v_* = \overline{v}_\phi - v_M \qquad (36)$$

and

$$v_M = \frac{\overline{B_\phi^2}}{2\Omega\mu_0\rho s} \qquad (37)$$

is the *magnetic wind*; $\overline{C}(s, z, t)$ is the axisymmetric part of C. On operating on (35) by $\nabla\times$, we obtain [with the help of $\nabla \times \mathbf{g} = 0$ and $\mathbf{g} \approx -g\hat{\mathbf{r}}$] a more general form of an equation familiar to atmospheric scientists, the *thermal wind equation*:

$$2\Omega \frac{\partial v_*}{\partial z} = \frac{g}{r}\frac{\partial \overline{C}}{\partial \vartheta} \qquad (38)$$

where ϑ is latitude. When integrated, (38) gives

$$\overline{v}_\phi = v_C + v_M + v_G(s, t) \qquad (39)$$

Here v_C is the wind set up by density differences, and v_G is the *geostrophic flow* (which is independent of z and which is, so far in the argument, arbitrary).

It may be shown that, if $v_C = 0$, the $\alpha\omega$-dynamo fails to maintain $\overline{\mathbf{B}}$. Thus v_C is the key contributor to the ω-process described in Section II,B, and our earlier estimate $U = 10^{-4}$ m s^{-1} refers more particularly to v_C than to $v_T(=\overline{v}_\phi)$.

Since v_C is so necessary, some mechanism must be at work to maintain the required pole–equator differences in $\overline{C}$. One way is through the meridional flows $\overline{\mathbf{v}}_P$ such as those sketched in Fig. 2(d). That such flows exist is obvious from the ϕ-component of (34); in the axisymmetric case, neither $-\nabla\bar{p}$ nor $-\rho\bar{C}\mathbf{g}$ contribute, so that

$$\bar{v}_s = \frac{1}{2\Omega\rho}(\overline{\mathbf{J}\times\mathbf{B}})_\phi \qquad (40)$$

Both the axisymmetric field (through $\bar{\mathbf{J}}_P\times\bar{\mathbf{B}}_P$) and the asymmetric field $\mathbf{B}' = \mathbf{B} - \bar{\mathbf{B}}$ (through the ϕ-average of $\mathbf{J}'\times\mathbf{B}'$) contribute to the right-hand side of (40). Together they create $\bar{v}_s$, and then (22) determines the remaining component v_z of the axisymmetric meridional flow $\bar{\mathbf{v}}_P$. This flow slowly stirs the core and creates a ϑ-dependent $\bar{C}$. (Such a meridional flow can also, as we noted earlier, assist the induction process, that is, make the dynamo run more efficiently.)

A central difficulty arises from (40). It may easily be verified that, if an arbitrary $\mathbf{B}$ [together with the $\mathbf{J}$ determined from it by (2)] is substituted into (40) and the indicated average over longitude is taken, the resulting mean flow will, in general, carry fluid across the CMB. In other words, the very necessary condition

$$v_r = 0 \qquad \text{on} \qquad r = b \qquad (41)$$

is not obeyed!

The argument has been advanced that violation of (41) could (and would) happen only if the geostrophic flow, $v_G(s, t)$, which we have so far had no way of determining, were incorrect. It was suggested that a v_G could be found such that, when (20) is solved and the resulting $\mathbf{B}$ is used via (40) and (22) to determine $\bar{\mathbf{v}}_P$, (41) will be obeyed. In principle, this *geostrophic adjustment* is just sufficient to determine v_G. In practice, there is no guarantee that the desired v_G exists; perhaps no adjustment of v_G can bring about (41)! It was further suggested that this might be the case for the geodynamo, and an alternative mechanism for satisfying (41) was proposed, namely core–mantle coupling.

Equation (41) is satisfied if and only if *Taylor's constraint*,

$$\mathcal{T}(s) = 0 \qquad \text{for all} \quad s \qquad (42)$$

is obeyed, where

$$\mathcal{T}(s) = \int_{\mathcal{C}(s)} (\mathbf{J}\times\mathbf{B})_\phi \, dS \qquad (43)$$

and $\mathcal{C}(s)$ is the segment of the cylinder, of radius s with axis Oz, lying within the fluid core; see Fig. 5. This $\mathcal{T}(s)$ is closely related to the electromagnetic torque over $\mathcal{C}(s)$. The other forces in (34) exert no such couple, and failure to satisfy (41) and (42) is tantamount to violating the torque balance on $\mathcal{C}(s)$ that (34) demands. Another way of restoring the balance is through torques acting over the north and south ends of $\mathcal{C}(s)$,

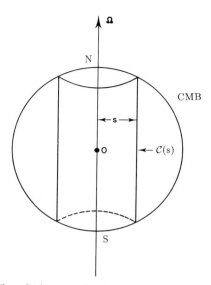

Fig. 5 The cylinder $\mathcal{C}(s)$, of radius s coaxial with the polar axis, which defines the geostrophic torque $\mathcal{T}(s)$ that requires balancing, either by geostrophic adjustment or by a torque created by the mantle on the end of the cylinder.

where $\mathcal{C}(s)$ meets the CMB. There are in reality three such torques, and we have so far ignored them: viscous, electromagnetic, and topographic. By setting $\nu = 0$, we have excluded viscous coupling; by ignoring mantle conduction, we have excluded magnetic coupling; and by supposing that the CMB is axisymmetric, we have excluded topographic coupling.

A study by S. I. Braginsky showed that, by restoring core–mantle coupling, it was possible, both in principle and practice, to bring about the torque balance, to complete the determination of $\mathbf{v}$, and to satisfy all conditions of the problem. The question of whether the Taylor or Braginsky prescription resolves the torque balance takes us right to the contentious forefront of modern geodynamo theory; we shall examine the matter no further here.

IV. Asymmetry and Waves

A. Importance of Asymmetries

Because of the difficulties that they raised, we have so far concentrated mainly on the axisymmetric parts of the geodynamo, that is $\bar{\mathbf{B}}$, $\bar{\mathbf{v}}$, etc. We argued in Section III that the geodynamo is of $\alpha\omega$-type and that therefore $\bar{\mathbf{B}}$ and $\bar{\mathbf{v}}$ are zonal, in a first approximation. The $\bar{\mathbf{B}}$ in (27) is created from $\bar{\mathbf{v}}$ by shearing $\bar{\mathbf{B}}_P$ (the ω-effect); the required $\bar{\mathbf{B}}_P$ relies on the α-effect, and this is provided by the asymmetric part $\mathbf{v}' =$

$\mathbf{v} - \bar{\mathbf{v}}$ of $\mathbf{v}$. It is therefore clear that $\mathbf{v}'$, and the associated $\mathbf{B}'$, play essential roles in the geodynamo mechanism. We may ask:

- Q20. What physical process would create asymmetries in the magnetohydrodynamic state of the core?
- Q21. Can the asymmetries produce the α-effect required to maintain the mean field?

We now attempt what is, in part, an extension of Section III, B to the spherical geometry of the core.

B. Planetary Waves in the Core

In the inertial reference frame, the core rotates almost as a solid body and is therefore threaded by vortex lines parallel to $\mathbf{\Omega}$. If $\mathbf{B}$ (and v) were zero, these lines could transmit *inertial waves* having characteristic frequencies of order Ω. A magnetic field also imparts elasticity to a conducting fluid, through the Faraday–Maxwell tension of the magnetic field lines. If $\mathbf{\Omega}$ (and σ^{-1}) were zero, these lines could transmit *Alfvén waves* having characteristic speeds of V_A, the *Alfvén velocity*,

$$\mathbf{V}_A = \mathbf{B}/(\mu_0 \rho)^{1/2} \qquad (44)$$

These waves resemble those on a plucked string, the tension per unit cross-sectional area of which is the nonhydrostatic Maxwell stress B^2/μ_0, and to which mass ρ per unit area is "frozen." If we take $|\mathbf{B}| \sim B_T \doteq 140$ gauss, we obtain $V_A \doteq 0.12$ m s^{-1}.

When neither $\mathbf{\Omega}$ nor $\mathbf{B}$ are zero, the situation is more complicated. When $V_A \ll \Omega b$, as is the case in the core, two types of waves arise: The *fast* waves are essentially inertial waves, slightly modified by Lorentz forces; they have characteristic periods of a day and are filtered out from the present discussion by omission of the inertial force from (34). The *slow* waves, which concern us, have characteristic frequencies of order

$$\omega_s = \frac{V_A^2}{\Omega L^2} \qquad (45)$$

and periods $\tau_s = 2\pi/\omega_s$ of about 1100 yr. This is encouraging; an answer to Q12 is in prospect.

In the geophysical context, the waves ride on the zonal $\bar{B}_\phi$; see (24). The most unstable disturbances are asymmetric; Q20 is potentially answered. They move longitudinally to the east or west; their frequency is of order (45) but is Doppler shifted by $\bar{v}_\phi$. Two studies have independently suggested that the observed westward drift of the geomagnetic field patterns reflects the motion of these waves; an answer to Q13 is also in prospect.

In the simplest (ideal) case, the slow waves are governed by (34), and by (20) with $\sigma = \infty$. These equations also include the destabilizing effect of the buoyancy force, $\rho C\mathbf{g}$ in (34), which provides an energy source for the magnetic field (see above). In fact, these waves are often named "MAC waves," an acronym that recalls the importance of *Magnetic*, *Archimedean* (buoyancy), and *Coriolis* forces on their dynamics.

The basic concept is simple: The top-heavy density gradient ($\nabla \bar{C}$) in the core is large enough for these waves to exist at a finite amplitude, limited by the rate at which they feed magnetic energy into $\bar{B}_P$ through their helicity and associated α-effect. In this way, they maintain $\bar{B}_P$ and hence, through the ω-effect, they also indirectly maintain the field $\bar{B}_T$ on which they ride. Theoretical implementation of the concept is extremely difficult, but it appears first that the most unstable disturbances are (like the secular variation) of planetary scale. Second, the MAC wave equations distinguish between eastward and westward-moving waves. Because of the Doppler shift in frequency arising from $\bar{v}_\phi$, one cannot go so far as to claim that there is a preference for westward-moving waves, although such indeed is the case for some simple models in which $\bar{v}_\phi = 0$. Unfortunately, when electrical resistance is ignored to simplify the theory, the α-effect either vanishes or is extremely small, so failing to answer Q21. By taking σ to be finite, the model becomes a more complicated version of that studied in Section III,B. The α-effect is restored, but the theoretical obstacles become yet more formidable. It is not surprising that progress has been slow.

C. Asymmetries of Structure

The growing science of mantle tomography has during the past decade assembled abundant evidence that lateral inhomogeneities exist in the lower mantle and D″ layer. Based on their extrapolations of $\mathbf{B}_P$ from the earth's surface to the core surface and on physical arguments, J. Bloxham and D. Gubbins have argued that such inhomogeneities partially lock $\mathbf{B}_P$ at the core surface to the mantle. F. Busse has devised simple theoretical models that illustrate how longitudinal variations in mantle conductivity can create a component of $\mathbf{B}'$ steady in the mantle frame, even though $\mathbf{B}'$ would otherwise have a predominantly westward phase speed. [*See* MANTLE.]

V. Reversals and Other Mysteries, Future Prospects

A. Reversals

MAC wave theory has been held up by technical obstacles; the theory of reversals has, for lack of key physical ideas, never really got off the ground. It is clear only that, if $\mathbf{B}(\mathbf{r}, t)$ obeys the magnetohydrodynamic equations, so does $-\mathbf{B}(\mathbf{r}, t)$;

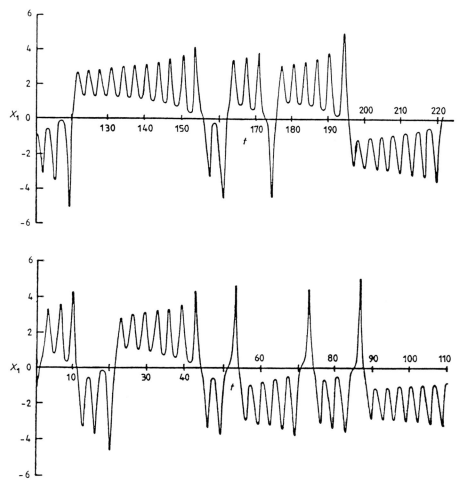

Fig. 6 The time evolution of X_1, the scaled current I_1, carried around one of the wire loops of the Rikitake dynamo; see Fig. 1b. [From Cook and Roberts (1970).]

see (20) and (34). As a corollary, we may conclude, from the lack of marked differences in the paleomagnetic record between the normal and reversed states, that we were right in Section I,B to abandon as insignificant the electrochemical and thermoelectric potentials within the earth, which create such differences.

Perhaps the majority of geomagneticians believe that reversals of the geomagnetic field are autoreversals — that is, the natural behavior, in some parameter ranges, of solutions of the complicated nonlinear magnetohydrodynamic equations (20) and (34). Such autoreversals take place in a number of simple nonlinear magnetomechanical models. For instance, Fig. 6 shows the current as a function of time t in one of the loops of the Rikitake dynamo (Fig. 1b). Although this simple model is remote from geophysical reality, its behavior exhibits some features reminiscent of the geomagnetic field, and these encourage the hope that the geomagnetic dynamo might behave similarly, so answering Q8.

First, while in either polarity state, the record shown in Fig. 6 oscillates with a period of order τ_σ. The earth's field shows similar variations in strength, and there is some evidence that its spectrum contains a *fundamental period* of order τ_σ. Second, reversals occur irregularly in both the model and in the earth; in both cases their statistics are close to Poissonian. Third, like the paleomagnetic record of the earth's field, Fig. 6 exhibits not only long polarity epochs but also short polarity events in which the field explores a new polarity for only one or two cycles. Fourth, though clearly incapable of shedding theoretical light on the structure of the geomagnetic field during a reversal, the Rikitake model apparently like the earth (see Q9) takes a time of order τ_σ to "flip over" from one polarity to the other.

It is harder to understand the long periods that elapse between reversals. Ignoring polarity events, the field maintains one polarity throughout a polarity epoch that is rarely shorter than 2.5×10^5 years. It is hard to identify any time constant in the magnetohydrodynamics of the core that is as long as this. If reversals are auto-oscillations, they may involve processes in the mantle, since only those possess the much longer time scales required. Although there is no concrete information about what these processes might be, it is not totally implausible that the mantle would be involved in the reversal mechanism.

"Theory" becomes even more speculative when one attempts to answer Q10 by explaining why the reversal frequency changes slowly over geologically long periods of time. One should perhaps recall the key role played by the mantle in the dynamics of the core. The mantle must transmit the heat lost by the core in the general cooling of the earth. It therefore acts like a valve. A change in the convective state of the mantle alters the setting of the valve and, because of its greater mobility, the core responds quickly. Perhaps, for some settings of the mantle valve, the core is in a relatively stable magnetohydrodynamic state that is not prone to reversals, whereas for other settings it is less stable and likely to auto-oscillate.

Polarity excursions are often described as attempted, but failed, reversals. It is however still unclear whether they are in any way related to the reversal phenomenon. No answer to Q11 is in prospect.

B. Short-Period Secular Variation

Section IV,B contained, if not concrete answers to Q12 and Q13, at least some plausible suggestions about the origin of the long-period secular variation. Understanding of the short-period secular variation (SPSV) is more rudimentary. For example, no plausible answer to Q15 has yet been proffered; no plausible mechanism has yet been suggested to account for secular change "jerks." [See GEOMAGNETIC JERKS AND TEMPORAL VARIATION.]

Turning next to Q14, perhaps the only component of the SPSV that has been plausibly identified is the so-called *torsional wave*. We saw in Section III,D that the axisymmetric zonal flow $\bar{v}_\phi$ contains a geostrophic part $v_G(s, t)$ that was distinguished by its indifference to the Coriolis force, which could be absorbed into the mean pressure $\bar{p}$. It is however affected by the magnetic field and, if disturbed, it transmits a torsional wave at an Alfvén velocity based on a mean strength of B_s. Such a wave can cross the core in a time of a few decades at most, and the associated changes in **B** therefore form part of the SPSV spectrum. Evidence for the existence of this wave has recently been adduced from an analysis of geomagnetic data and records of variations in the length of

the day. Little is known about other SPSV frequencies, although it has been argued that they are associated with the fluctuations of a thin, stably stratified layer that is believed to exist at the top of the core.

C. The Magnetism of Other Planets

Several authors have attempted to answer Q16 by discovering a general pattern into which the magnetic strength of all planets (and their satellites) fits—a kind of Bode's law of planetary magnetism. So far these attempts amount to little more than dimensional groupings of the parameters describing the dynamo creating region, such as its spatial dimensions b, its density ρ, its electrical conductivity σ, and last but not least, in view of the role of the Coriolis force in creating not only an ω-effect but also the helicity from which the α-effect arises, its angular velocity Ω. For example, J. A. Jacobs has proposed that the field strength B in a planetary core should imply a characteristic Alfvén velocity given, for some function f, by

$$V_A \equiv \frac{B}{\sqrt{\mu_0 \rho}} = \Omega b f(\Omega \tau_\sigma) \tag{46}$$

where τ_σ is the electromagnetic decay time (5). For every planet, even the slowly rotating Venus ($\Omega \approx 3 \times 10^{-7} \text{s}^{-1}$), $V_A/\Omega b$ is very small and $\Omega \tau_\sigma$ is extremely large, suggesting that only the asymptotic form of (46) is relevant. The choice $f(x) \propto x^{-1/2}$ leads to the parameter range discussed in Section III, namely $\Lambda = O(1)$. For Venus, $\Lambda = O(1)$ would predict a core field of order 1 gauss. Although this would be attenuated outside the planet by distance from the core, such a field should certainly have been detected by spacecraft if it existed. The fact that Venus has apparently a very much weaker field points to shortcomings in laws such as (46).

The principal deficiency of (46) is perhaps its lack of dependence on the strength of the mechanism driving the fluid into motion. One would have expected that V_A should involve a second dimensionless grouping such as the Rayleigh number $\widetilde{Ra}$, introduced in (32) for the simple thermal convection problem of Section III,B. Unless Ra achieves an $O(1)$ value, convection and magnetic field generation cannot take place at all. Perhaps the core of Venus is similarly not driven hard enough, so that laws such as (46) become irrelevant. Clearly, all this is highly speculative.

The level of driving also determines the character of a fluid dynamo. For instance, in the strongly driven solar convection zone, Ro = $O(1)$ and the motion and field are highly chaotic. In the weakly driven core of the earth, Ro $\ll$ 1, and the turbulence level is low.

D. What the Future Holds

We have found that every observed facet of the geomagnetic field is potentially explicable by geodynamo theory; none sounded its deathknell. We have seen that some of our 21 questions are satisfactorily answered by the theory, and that there is every prospect of answering the remainder in the course of time. Nevertheless, glaring deficiencies of the theory have been exposed. In some of these, the basic ideas seem to be correct, but their implementation has been hampered by severe theoretical difficulties of a technical nature; in others, the basic mechanisms are unknown or a matter of considerable speculation. It may be confidently expected that, during the last decade of this century, the first type of deficiency will be overcome through the increasing availability of high-speed computers. There is every prospect of exciting developments before the year 2000!

Acknowledgments

I am grateful to S. Braginsky and M. Kono for their advice.

Bibliography

Braginsky, S. I. (1975). Nearly axially symmetric model of the hydromagnetic dynamo of the earth, I. *Geomagn. Aeron.* **15**, 122.

Braginsky, S. I. (1978). Nearly axially symmetric model of the hydromagnetic dynamo of the earth, II. *Geomagn. Aeron.* **15**, 225.

Cook, A. E., and Roberts, P. H. (1970). The Rikitake two-disc dynamo system. *Proc. Cambridge Philos. Soc.* **68**, 547.

Taylor, J. B. (1963). The magneto-hydrodynamics of a rotating fluid and the earth's dynamo problem. *Proc. R. Soc. London, A* **274**, 274.

Glossary

Coriolis force Contribution to the rate of change of momentum of a particle when described in a rotating reference frame.

Dynamo action Process whereby motions within an electrically conducting fluid, interacting with a magnetic field, convert their kinetic energy into magnetic energy, thereby sustaining the magnetic field.

Lorentz force Force experienced by a charge when it moves in a magnetic field.

Magnetohydrodynamics Study of the mutual interaction of the flow of an electrically conducting fluid with the magnetic field through which it moves.

Geomicrobial Transformations

Henry L. Ehrlich

Rensselaer Polytechnic Institute

Ever since they made their first appearance, microbes have played a significant role in shaping the environment at the earth's surface through their physiological activity. This view is supported by study of ancient microfossils. The oldest to date, estimated to have an age of ~3.5 billion years, were found in Western Australia. Because these microfossils resemble modern microbes in shape, inferences about their probable geological activity have been drawn from a comparison with their modern counterparts.

I. Introduction

Geologically significant activity of microbes includes the cycling of organic and inorganic matter, the formation of new minerals through transformation (diagenesis) of preexistent minerals or through *de novo* synthesis from dissolved chemical species (authigenesis), and the breakdown of minerals or rock (weathering). Each of these processes will be considered in more detail in the following sections.

II. Cycling of Organic and Inorganic Matter

Biologically important elements as, for example, carbon, hydrogen, oxygen, nitrogen, phosphorus, sulfur, iron, silicon and others are biologically not available in unlimited quantities in the environment and therefore must be recycled on the death of any organism for the continuance of life. Biological availability of an element does not necessarily correlate with its crustal abundance. Thus, even though in excess of 4.3×10^{16} tons of carbon (inorganic and organic) are present in the earth's crust, less than 0.1% of this is available to life processes. The rest is trapped in carbonates, bitumen, kerogen, petroleum, coal, and natural gas. Some transfer of carbon from the trapped portion to the biologically available portion does take place, but this is nearly balanced by transfer from the biologically available portion to the carbon traps, so that the quantity of trapped carbon remains nearly constant except for the fossil fuel exploited by humans, which constitutes only on the order of 0.1% of the total.

The biological carbon cycle is summarized in Fig. 1. The conversion of inorganic carbon as CO_2 is dependent on *photosynthesis* and in special cases on *chemosynthetic autotrophy*. Photosynthesis is a process in which light is the source of energy that enables the conversion of inorganic to organic carbon with or without the simultaneous release of gaseous oxygen. Chemosynthetic autotrophy, by contrast, is a process dependent on energy derived from the oxidation of reduced inorganic compounds for the conversion of CO_2 to organic carbon from CO_2. It is the only process for biological conversion of inorganic into organic carbon in the dark and is central to the survival of biological communities that exist around hydrothermal vents on the ocean floor. Photosynthetic organisms include purple and green anaerobic bacteria, cyanobacteria, algae, and all plants. Chemosynthetic autotrophs include bacteria that can oxidize substances such as H_2, H_2S, S^0, $S_2O_3^{2-}$, NH_4^+, NO_2^-, Fe^{2+}, and others.

Some of the organic carbon that photosynthetic and chemosynthetic autotrophs release during their lifetime and most of that contained in their remains on their death is returned to the inorganic carbon pool through a biological process called *mineralization*. In this process the organic carbon is

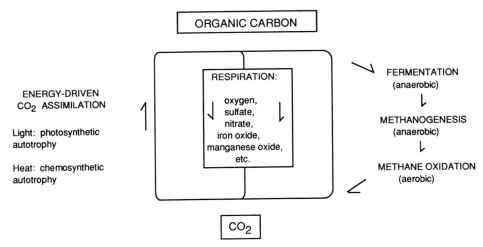

Fig. 1   The biological carbon cycle.

completely oxidized to CO_2 in a series of steps, usually performed by one or more organisms among which microbes, especially bacteria and fungi, play a central role. They represent the *decomposers*. Mineralization occurs aerobically and anaerobically. Under either condition, but especially anaerobically in the presence of external electron acceptors such as sulfate, nitrate, iron oxides, etc., a consortium of different organisms collaborates in stepwise degradation, each member performing a limited series of steps in the degradation. In the absence of sulfate, nitrate, or ferric iron, much of the carbon may be transformed into methane, most of which can be microbiologically converted to CO_2 only on exposure to oxygen in air. The organic carbon that is not mineralized may be assimilated by *heterotrophs* and *mixotrophs*, or it may be conserved in sediment for brief or extended periods. *Predators* obtain their organic carbon by consuming living organisms. The carbon in their excretions and dead remains is subsequently returned to the biologically available carbon pool through mineralization. [*See* PHOTOSYNTHETIC CARBON METABOLISM AND ATMOSPHERIC CO_2.]

As in the case of carbon, biological availability of nitrogen is not unlimited and requires recycling of nitrogen (Fig. 2). It is often the growth-limiting factor in the marine environment. Photosynthetic autotrophs (photosynthetic bacteria, algae, and plants) and chemosynthetic autotrophs (bacteria) require their nitrogen in inorganic form as NO_3^- or NH_4^+. Some bacteria in this group and some heterotrophic bacteria may get their nitrogen from N_2 in the atmosphere when NO_3^- or NH_4^+ are not available. They convert N_2 into ammonia and organic nitrogen. Some of the nitrogen they fix may be excreted by them during their lifetime or the live organisms may be consumed by predators, thereby sharing it with organisms incapable of nitrogen fixation.

The fixed nitrogen that is not shared by excretion or consumed in predation is made available on the death of the nitrogen fixers. Many heterotrophs and some mixotrophs must get their nitrogen from the organic nitrogen pool. For the autotrophs, organic nitrogen must be mineralized (ammonification) to produce NH_4^+, which may then be oxidized by nitrifying bacteria, mostly autotrophs, to NO_3^-. Excess nitrate may be microbially converted to either N_2 or NH_4^+ in respiratory processes known as *denitrification* and *nitrate ammonification*, respectively. While ammonia is formed only anaerobically in this case, N_2 (also NO and N_2O in some instances) may be formed aerobically or anaerobically, but mostly anaerobically. [*See* NITROGEN IN SOIL.]

Although the atmosphere surrounding the earth is a reservoir for gaseous oxygen, which is important for all oxygen-respiring forms of life, the reservoir is not inexhaustible. Indeed, in some environments on the earth's surface into which an excess of biologically oxidizable matter is introduced, oxygen that was initially present may be fully consumed and may be replaced from the atmosphere at too slow a rate to satisfy the demand of the aerobes. Oxygenic photosynthesis is the chief source of oxygen in the atmosphere. The relative constancy of the oxygen concentration in the modern atmosphere is explained by its consumption, mostly by *respiration* of aerobic organisms, at a rate that balances its production by photosynthesis. Aerobic respiration is a process in which oxygen serves as terminal electron acceptor, being reduced to water in the energy-transducing metabolism of microbes and higher organisms having this capacity. In a few instances, oxygen may be directly introduced into an organic molecule by enzymatic action in a process called *oxygenation*. [*See* OXYGEN, BIOGEOCHEMICAL CYCLE.]

The supply of available sulfur, like that of carbon and

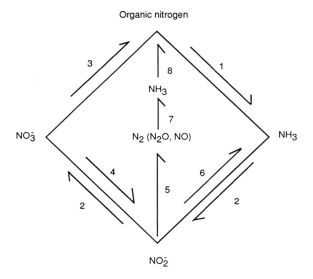

Key:

1, Ammonification
2, Nitrification
3, Nitrate assimilation
4, Nitrate reduction
4+5, Denitrification
4+6, Nitrate ammonification
7, Nitrogen fixation
8, Ammonia assimilation

Fig. 2 The biological nitrogen cycle. The central panel lists possible terminal electron acceptors for the respiratory process.

oxygen, is not inexhaustible in the biosphere. Hence this vital element also needs to be recycled (Fig. 3). Plants generally require their sulfur in the form of sulfate for assimilation. Many microbes also can use sulfate for assimilation. Most animals require their sulfur in the form of organic sulfides (generally sulfur-containing amino acids), however. The sulfur cycle (Fig. 3) ensures that sulfur in the required form is available to all forms of life. In addition to satisfying a nutritional requirement, reduced forms of sulfur serve as a source of energy and reducing power for CO_2 assimilation by chemosynthetic autotrophs and as a source of reducing power for photoautotrophs, which get their energy from sunlight. Fully or partially oxidized forms of sulfur (sulfate, polythionates, thiosulfate, and elemental sulfur) can also serve as terminal electron acceptors in *anaerobic respiration* by some microbes. [See BIOLOGICAL SULFUR, CLOUDS AND CLIMATE.]

Phosphorus is frequently the growth-limiting factor in aquatic environments. It may occur in inorganic or organic form. In general, biological systems can assimilate it only in inorganic form—as orthophosphate. Organic phosphates,

which are usually of biological origin and are mostly but not exclusively phosphate esters, must be hydrolyzed to free the orthophosphate. Controversial claims of microbial phosphate (PO_4^{3-}) reduction to phosphite (PO_3^{3-}), hypophosphite (PO_2^{3-}) and phosphine (PH_3) have been made, but the significance of such reduction, which yields no useful energy, is not understood. On the other hand, bacteria that catalyze the oxidation of hypophosphite and phosphite do occur, and the biochemistry of this oxidation has been studied.

Iron, the fourth most abundant element in the earth's crust, is required by nearly all forms of cellular life. The exception is a small group of lactic acid bacteria. At pH values above 5 under oxidizing conditions, it occurs chiefly in an insoluble, ferric form, mostly iron (III) oxides. Microorga-

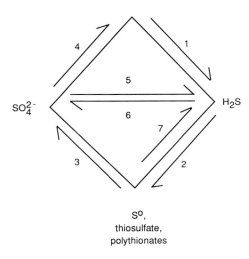

Key:

1, Organic sulfur mineralization
2, H_2S oxidation by aerobic chemoautotrophs and anaerobic photoautotrophs
3, S^0 and partially-reduced-sulfur species oxidation by aerobic chemoautotrophs and anaerobic photoautotrophs
4, Sulfur assimilation by assimilatory sulfate reduction
5, Sulfate reduction, dissimilatory (anaerobic respiration)
6, H_2S oxidation direct to sulfate (aerobic chemoautotrophs and anaerobic photoautotrophs)
7, Sulfur and partially-oxidized-sulfur species reduction (anaerobic respiration)

Fig. 3 The biological sulfur cycle.

nisms that live preferentially at near-neutral pH under oxidizing conditions have consequently had to evolve special iron-scavenging systems consisting of ligands (*siderophores*) that have an extremely high affinity for ferric iron (affinity constants ranging from 10^{30} to 10^{50}) but low affinity for ferrous iron (affinity constant of $\sim 10^8$). These siderophores can dissolve the ferric iron in oxide minerals by forming stable chelates that are then taken up by the iron-requiring microbes, which have specific uptake systems for respective ferrisiderophores. The iron is then reduced by them to ferrous iron and incorporated into the functional groups of iron-based enzymes and electron carriers (*assimilatory iron reduction*).

Some bacteria can use ferric iron, including some oxide minerals, as terminal electron acceptors in respiration (*dissimilatory iron reduction*) by reducing it to ferrous iron. In this process iron from insoluble oxides is solubilized. It remains solubilized as long as conditions continue to be reducing and/or the pH is below 5. Ferrous iron, whether in dissolved form or in mineral form, as in pyrite (FeS_2) for instance, can be oxidized and energy conserved from the oxidation by some aerobic bacteria that live at very acid pH (~ 2.0) and are therefore called *acidophiles*. The product of the oxidation may be dissolved ferric iron or insoluble basic ferric sulfates such as jarosite.

At least one bacterial species, *Gallionella ferruginea*, oxidizes ferrous iron at neutral pH under partially reduced conditions. It accumulates the oxidized iron in its stalk by an as yet unknown mechanism. Under fully oxidizing conditions at neutral pH, ferrous iron autooxidizes too rapidly to be of use as an energy source by *G. ferruginea* or any other potential iron oxidizer. Thus, under certain environmental conditions, microbes play an important role in mobilizing or immobilizing iron.

Silicon, the second most abundant element in the earth's crust, is required for cell protection and structural support by some algae and protozoans, especially diatoms, silicoflagellates, and radiolarians in the sea. It is also taken up by some bacteria and is found in varying amounts in higher forms of plants and animals, but its function in them is unclear. Most silicon is immobilized in silica, silicate, and aluminosilicate minerals. Biologically available orthosilicate amounts to only ~ 2.2 mg kg^{-1} of seawater and ~ 7 mg kg^{-1} of fresh water. To maintain these concentrations, remobilization of the biologically fixed silicon and some weathering of silica, silicate, and aluminosilicate minerals is necessary. Some bacteria and fungi play significant roles in these processes, for instance by forming metabolic products such as acids, bases, and exopolysaccharides, which react with the immobilized silicon and bring it into solution. Only orthosilicate is biologically active. Metasilicate or siloxane species in aqueous systems have to be depolymerized before the silicon in them can be assimilated.

Some bacteria have been shown to participate in such depolymerization.

III. Energy Flow in the Biological Cycling of Matter

The primary driving force of biological cycling of matter in most ecosystems is sunlight, but in some special ones it is geothermal energy. In the former instances, it is the photosynthetic autotrophs that transduce light energy into biologically useful chemical energy by trapping it in organic molecules that enable subsequent consumers to function and to perform their part in the cycling of matter. In the latter instance, it is the chemoautotrophs that transduce chemical energy contained in biologically oxidizable mineral matter that was formed as a result of magmatic activity (examples are volcanic and hydrothermal activity on land, and hydrothermal activity at ocean floor spreading centers). Like photosynthetic autotrophs, the chemosynthetic autotrophs trap chemical energy released in their oxidation of inorganic matter in organic molecules for subsequent utilization by consumers for their function and for the cycling of matter.

In some special environments, photosynthetic autotrophs reside in superficial tissues of certain animals (examples are *Prochloron*, a cyanobacterium-like organism in ascidian worms, a type of primitive chordate; and zoochlorellae, a type of algae, in some marine invertebrates) to share with their hosts for mutual benefit the organic carbon that they produce photosynthetically. In other environments, chemosynthetic autotrophs (such as sulfide- and methane-oxidizers) have become associated with animal hosts, either superficially (as in the gill tissue of molluscs) or internally (as in the body cavity of some vestimentiferan worms), to share the carbon they fix chemosynthetically with their host. In the case of the vestimentiferan worms, this association has become obligatory for the survival of the worms since they lack a gut and cannot survive by utilizing dissolved nutrients in the surrounding water. The chemosynthetic associations are especially notable in communities that exist around submarine, hydrothermal vents from which solutions rich in sulfide or methane are discharged at ocean floor spreading centers and subduction zones; but they occur also in some other marine environments where hydrogen sulfide or methane may be discharged from marine sediment or through submarine vents. They ensure efficient transfer of fixed carbon and energy from the *primary producer* to the *consumer*.

The direction of particular element cycles or parts thereof is determined by environmental conditions (oxidizing or reducing conditions, pH, etc.), which select the organisms and the activity associated with them that are best suited to those conditions. Geothermally driven cycles in which aerobes promote the various reaction steps interact with sunlight-driven cycles for the oxygen needed in their respiration.

IV. Immobilization of Organic and Inorganic Matter in Traps or Sinks

A. Fossil Carbon

Organic matter that is not mineralized by microbial action, although it may be somewhat modified by it, may become buried in sediment and preserved as coal, bitumen, kerogen, or petroleum. Each of these is the product of the action of heat and pressure on corresponding organic residues over time. Although coal is derived mainly from the remains of higher plants, petroleum is derived mainly from the remains of aquatic cyanobacteria and algae, including plankton. Coal, bitumen, and kerogen are generally found at the site of their formation, whereas petroleum after its maturation frequently migrates from the site of its formation and becomes concentrated in a porous rock reservoir. It has been suggested that this migration may be microbially assisted.

B. Mineral Formation

Some bacterial cells may serve passively as nuclei around which specific minerals form. One or more components of a cell envelope may bind specific metal ions that subsequently form metal salts (such as carbonates, phosphates, sulfides) upon which more of the metal salt is deposited, forming a specific mineral crystal. Some other microbes actively form mineral crystals by enzymatic means that become functional and/or structural components in or on their cells. For instance, *magnetotactic bacteria* synthesize magnetite crystals, which are enclosed in special membrane-bound structures called *magnetosomes*. Some calcareous algae such as coccolithophores deposit calcite crystals on their cell surface in special structures called *coccoliths*. Diatoms form cell walls consisting chiefly of silica deposited in species-specific patterns. Radiolarian protozoa form a siliceous support skeleton whose arrangement is species specific.

Microbes may also promote mineral formation external to their cells by nonenzymatic or enzymatic means. In nonenzymatic reactions, they may form metabolic products such as sulfide in sulfate reduction, which can react with heavy metals in solution and precipitate them extracellularly. A good example is the formation of iron pyrite in salt marshes. Some microbes may raise the pH of their environment to a range where sufficient carbonate is formed from respiratory

CO_2 to lead to the extracellular formation of carbonate minerals such as calcite ($CaCO_3$) or rhodochrosite ($MnCO_3$). In enzymatic action, some microbes may transform a dissolved inorganic ion into an insoluble species as, for instance, the oxidation of sulfide to elemental sulfur, the oxidation of ferrous iron to some ferric oxide, or the oxidation of manganous manganese to manganese (III) or (IV) oxide.

Microbes may also form new minerals by promoting *mineral diagenesis*. For instance, in the oxidation of pyrite by acidophilic bacteria like *Thiobacillus ferrooxidans*, the oxidized (ferric) iron may react with the oxidized sulfur (sulfate) to form a basic iron sulfate mineral called jarosite. Similarly, the oxidation of arsenopyrite by *T. ferrooxidans* may lead to the formation of ferric arsenate (scorodite, $FeAsO_4 \cdot 2H_2O$).

V. Remobilization of Mineral Matter in Traps or Sinks

Some bacteria have the ability to solubilize minerals such as oxides or sulfides enzymatically. They may use a mineral as an energy source, as in the case of mineral sulfides, or as a terminal electron acceptor as in the case of oxides. They appear to act on the crystal lattice of the mineral they attack in either instance, but it is still unclear how they do this. In the case of metal sulfide oxidation, some indication exists that the active bacteria attach to specific sites on crystal faces and cause progressive etching around the site of attachment. In the oxidation of metal sulfides, the bacteria involved can be viewed as catalytic conductors that gain biologically useful energy from the transmission of electrons from reduced mineral constituents to oxygen. The oxidized products pass into solution as a result. In the reduction of metal oxides such as MnO_2, the bacteria involved pass electrons actively from a dissolved electron donor to the crystal lattice on the oxide surface, causing the reduction of the Mn(IV) and its resultant solution as Mn(II).

In the case of sulfide mixtures, the oxidizing bacteria may enhance *galvanic interaction* if the sulfide mineral constituents in contact with each other have sufficiently different rest potentials (Fig. 4). A mineral with a low rest potential in such a mixture is the reactive one, whereas a mineral with a high rest potential acts noble by conducting electrons from the reactive mineral to oxygen without permanently changing its oxidation state. Active bacteria such a *T. ferrooxidans* attached to the noble mineral can facilitate the passage of the electrons to oxygen and thereby accelerate the oxidation of the reactive mineral.

Bacteria can also solubilize mineral sulfides or oxides nonenzymatically. In the case of mineral sulfides, *T. ferrooxidans* and other acidophilic iron bacteria can generate acidic ferric

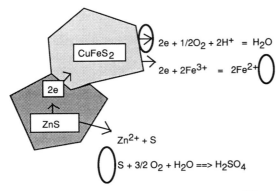

Fig. 4 A model of galvanic interaction between two sulfide minerals, sphalerite (ZnS) and chalcopyrite ($CuFeS_2$), in the presence of *Thiobacillus ferrooxidans*. The sphalerite in this pair is the active mineral and the chalcopyrite the noble mineral. The ovals represent cells of *T. ferrooxidans*. The uppermost cell abstracts electrons directly from the chalcopyrite surface in its respiration. The middle cell reoxidizes ferrous iron to ferric iron. The latter abstracts electrons from the surface of chalcopyrite and is reduced to ferrous iron that is again oxidized by *T. ferrooxidans* to ferric iron. The bottom cell oxidizes sulfur that forms on the surface of sphalerite as a result of oxidation by chalcopyrite. The sulfur passivates the sphalerite if it accumulates on its surface.

sulfate, from pyrite oxidation for instance, which then oxidizes the mineral sulfide chemically, thereby solubilizing it. The ferrous iron in solution from this reaction can be reoxidized enzymatically by the acidophilic iron bacteria for further reaction with the metal sulfide (Fig. 4). Metal oxide minerals can be solubilized by reaction with metabolic products such as formic or oxalic acid that can act as chemical reductants.

Carbonate minerals can be readily solubilized by metabolically produced mineral or organic acids. The former include carbonic, nitric, and sulfuric acids. The latter include formic, acetic, propionic, lactic, succinic, citric, and other acids produced by bacteria and/or fungi.

Al–O bonds in aluminosilicates, unlike Si–O–Si (siloxane) bonds, are susceptible to attack by acid, and thus microbially formed acids can promote their rupture. Siloxane bonds can be broken, especially under acidic conditions, by reaction with exopolysaccharide in bacterial capsules or slime layers such as those formed by *Bacillus mucilaginosus*. Silica (SiO_2), silicates, and aluminosilicates can be solubilized by alkali such as NH_4OH formed by microbes when degrading urea or proteinaceous substances. Finally, silicates and aluminosilicates can be solubilized by extraction of cationic constituents with microbially produced ligands such as 2-ketogluconate and citrate.

VI. Rock Weathering and Pedogenesis

The ability of microbes to generate metabolic products that can attack silicates, aluminosilicates, silica, carbonates, and other minerals contributes significantly in some instances to the weathering of rock. Limestone is especially susceptible to microbial weathering. Indeed, some cyanobacteria and fungi weather limestone by boring beneath its surface to inhabit its interior. *Thiobacillus ferrooxidans* can generate enough sulfuric acid in their oxidation of pyrite to weather the silicate rock in which the pyrite may be embedded in a relatively short time.

The ability of microbes to attack rock minerals also contributes significantly in the development and evolution of mineral soil (*pedogenesis*). In some mature soils in temperate climates, distinct layers (*soil horizons*) can be recognized in a vertical soil profile. Typical layers in spodosols consist, in descending order, of a littler zone or O horizon, a leached layer or A horizon, an enriched layer or B horizon, and a deep layer or C horizon. The activity of microbes in the O and A horizons results in diagenesis of some of the soil minerals in the A horizon. This can lead to the transport of the solubilized mineral constituents to the B horizon by downward percolating soil water. In soils in humid, tropical climates microbial weathering of soil may lead to solubilization of silicate and/or iron from soil minerals and their subsequent reprecipitation on evaporation of soil moisture, causing the cementation of remaining soil particles into nonporous laterite, an infertile soil.

VII. Practical Exploitation of Geomicrobial Transformation

The ability of some microbes to solubilize mineral constituents is being industrially exploited on a worldwide scale in the bioleaching of low-grade copper sulfide ores and to a lesser degree in uranium ores. The ore may be placed in heaps or dumps or treated *in situ*. Naturally associated acidophilic thiobacilli and other iron-oxidizing acidophiles are encouraged to grow by providing environmental conditions (for example, adequate moisture, acid pH, access of air) resulting in the attack of the metal sulfide mineralization as its major energy source. Solubilized copper can be recovered from the solution by treatment with scrap iron (*cementation*), electrolytically and/or by solvent extraction. Solubilized uranium can be recovered on ion exchangers. There is a potential for industrial bioleaching of other mineral sulfides.

Microbial ability to solubilize minerals can also be used in *biobeneficiation* of ores, that is the selective removal of one or more undesirable ore constituents that interfere with recovery of a valuable ore constituent. At the moment, biobeneficiation finds practical application in the processing of pyritic gold ores in which constituents such as pyrite and arsenopyrite interfere in gold recovery by cyanidation. *T. ferrooxidans* can selectively solubilize the interfering pyrite and arsenopyrite, leaving an ore residue from which the gold can then be extracted efficiently by conventional means. A potential industrial application of biobeneficiation is in the desulfurization of coal that is high in pyritic sulfur. Acidophilic, pyrite-oxidizing bacteria have been shown to be able to remove the pyrite effectively from pulverized coal. The microbiological removal of organic sulfur is a problem still in need of an effective process.

The ability of microbes to solubilize minerals on the one hand or to remove solubilized inorganic constituents on the other has potential for *bioremediation* of sites polluted by inorganics. Similarly, the ability of microbes to mineralize a wide range of organic matter has a potential for bioremediation of environments contaminated with organic pollutants. Limitations in the latter case reside in the nature of the organic pollutants, some being easily biodegraded while others are only very slowly degraded or not at all. Some forms of bioremediation have been tried in which microbes with specific metabolic properties have been introduced. A more successful approach often is to stimulate appropriate components of the natural microbial flora by adding nutrients such as phosphate and/or nitrate, whose natural availability is limited, or otherwise creating environmental conditions that favor their activity over that of other microbes.

VIII. Conclusion

Since the beginning of life on Earth, microbes have played a significant role in the evolution of the biosphere portion of Earth's crust. All life has always been and continues to be very dependent on geomicrobial activity. Whether the origin of life and its evolution on the earth was inevitable or a chance happening, given the physical and chemical nature of the planet, is a question to which an answer may be found if and when evidence of life on other planets has been obtained.

Bibliography

Beveridge, T. J., and Doyle, R. J., eds. (1989). "Metal Ions and Bacteria." Wiley, New York.

Ehrlich, H. L. (1990). "Geomicrobiology," 2nd revised and expanded edition. Marcel Dekker, New York.

Ehrlich, H. L., and Brierley, C. L., eds. (1990). "Microbial Mineral Recovery." McGraw-Hill, New York.

Huang, P. M., and Schnitzer, M., eds. (1986). "Interactions of Soil Minerals with Natural Organics and Microbes." SSSA Spec. Publ. No. 17. Soil Science Society of America, Inc., Madison, Wisconsin.

Jannasch, H. W., and Mottl, M. J. (1985). Geomicrobiology of hydrothermal vents. *Science* 229, 717–725.

Glossary

Autotroph Organism that obtains its energy from sunlight or the oxidation of inorganic compounds and its carbon solely from CO_2.

Enzyme Biological catalyst, usually of proteinaceous nature, for specific reactions or groups of reactions.

Exopolysaccharide Polymer of sugars and sometimes sugar acids excreted by some bacteria to form a slime layer or capsule. The polymer can play a role in attachment of bacteria to a surface.

Galvanic interaction Redox reaction between two mineral entities with different rest potential, where the mineral with lower rest potential acts as reductant and the mineral with the higher rest potential acts as oxidant/reductant in conveying electrons to oxygen or some other suitable acceptor.

Heterotroph Microorganism that requires organic matter for its source of energy, for cell synthesis, and for reproduction.

Hydrothermal activity Interaction of an aqueous solution with rock at high temperature and pressure, leading to alteration of the rock and dissolution of some of the rock minerals and formation of others.

Mixotroph Microorganism that obtains its energy and carbon simultaneously from inorganic and organic matter.

Geophysical Fluid Mechanics

G. S. Janowitz
North Carolina State University

Geophysical fluid mechanics is the study of the naturally occurring fluid motions in the atmosphere, oceans, and earth with the objective of understanding the basic processes involved in these motions. This objective is achieved through the use of simplified models and laboratory experiments which incorporate the basic states to be studied. The emphasis on obtaining understanding of the underlying physics in simplified situations distinguishes this field of study from the more applied fields of numerical weather and ocean prediction.

I. Sources of Geophysical Motion

A. Astronomical Gravitational Fields

As the earth moves through space it is acted upon by gravitational forces due to astronomical bodies, principally the sun and the moon. If the earth is viewed as a point mass, its acceleration with respect to an external inertial frame of reference is due to the net astronomical gravitational force. If one now considers a frame of reference of fixed orientation and traveling with the center of the earth, the total acceleration of a parcel of unit mass away from the earth's center is the sum of the acceleration of the parcel relative to the earth-fixed frame of reference and the acceleration of the earth's center. The net astronomical gravitational force on the parcel equals the net force evaluated at the earth's center plus a small correction, the tide-generating force. This tide-generating force, in the absence of other forces, creates the acceleration of the parcel relative to the earth. If we now let

our frame of reference rotate daily about an axis fixed in the earth, the tide-generating force becomes a complex function of time, though a periodic function. Because of the complexities of geometry it becomes nearly impossible to calculate directly motions due to this force, but because of the predictable periodicity of the force, its results, the tides in the ocean, earth, and atmosphere, are relatively easy to observe and predict based on observations. Within the context of geophysical fluid dynamics we can study alterations in a specified tidal motion due to a variety of factors, e.g., variable bottom topography, or stratification. The source of the tidal wave itself need not concern us. [*See* ATMOSPHERIC TIDES; TIDES.]

B. Solar Radiation

The second source of motion in the coupled earth, atmosphere, ocean system is incoming solar radiation. The spatially and seasonally varying incoming solar radiation is the source of nearly all of the nontidal motions in the global system. The coupling of the system and the overall energy balances are discussed below. [*See* SOLAR RADIATION.]

C. Heat Sources in the Earth

Heat released deep within the earth is the third source of geophysical motion and is responsible for large-scale motions in the earth as well as volcanic eruptions and very localized hot spots at the earth's surface.

The overall export of heat from the earth to the atmosphere and ocean is quite small compared to incoming solar radiation and appears to be of limited significance as far as most motions in the atmosphere and oceans are concerned, although in the vicinity of localized hot spots this heating will be of primary importance.

II. Global Energy Balances

A. Atmospheric Balances

Solar radiation is responsible for virtually all nontidal motion. We briefly discuss the processes involved in the coupling of the earth, ocean, and atmosphere. There is an approximate

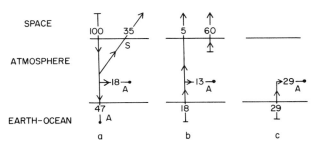

Fig. 1 A sketch of the energy-transferring processes. (a) Short-wave radiation; (b) longwave radiation; (c) mechanical processes. S implies scattering and A implies absorption.

global balance of incoming and outgoing thermal energy in both the atmosphere and the ocean.

If 100 units of shortwave solar radiation were incident at the top of the atmosphere, 18 units would be absorbed by the atmosphere, mostly by ozone and water vapor; 47 units would be transmitted across the lower boundary of the atmosphere to the ocean and earth; and 35 units would be scattered to space, with two-thirds of the scattering due to clouds. The atmosphere would absorb another 13 units of longwave radiation emitted by the earth and ocean, this due to the presence of water vapor and carbon dioxide in the atmosphere. Another 29 units of energy would be mechanically transferred to the atmosphere in the form of sensible and latent heat from the ocean and earth by turbulent eddies in the lower atmosphere. The net gain of 60 units of energy by the atmosphere is lost to space by longwave radiation due to water vapor and carbon dioxide. See Fig. 1 for a sketch of these processes. The numbers given above are only approximate, but a balance must exist because if only 1 unit were retained by the atmosphere it would lead to a 10°C increase in atmospheric temperature in 1 year. [*See* SOLAR–TERRESTRIAL INTERACTION.]

B. Heat Balance in the Ocean

In the earth and ocean the net gain of 47 units is balanced by 18 outgoing units in the form of longwave radiation, 13 of which are absorbed by the atmosphere and the remainder lost to space. The remaining 29 units are transferred to the atmosphere in the form of latent and sensible heat.

Trace elements such as ozone, carbon dioxide, and water vapor play a critical role in both the absorbing and emitting processes, and water vapor condensed in clouds plays a major role in the scattering of incoming solar radiation back to space. Thus changes in the amount of these trace gases can lead to climate change, although because of the many roles played by water vapor quantification of these changes remains difficult. Further, the major source of energy for the

atmosphere is the ocean, through the agency of turbulent transport of latent and sensible heat. This suggests that these two systems are dynamically coupled so that significant changes in one system, say the ocean, will affect the atmosphere, and this in turn will produce further changes in the ocean. As the local heat balances affect the dynamics, these systems are dynamically coupled. [*See* AIR–SEA INTERACTION.]

III. The Equations of Motion

A. Mass Conservation and the Momentum Equation

Latitudinal variations in incoming radiation lead to motions in the atmosphere and oceans which are further complicated by the presence of lateral boundaries of the oceans and a variable bottom topography in both the oceans and the atmosphere. The fluid motions in the atmosphere, oceans, and earth are described by the classical equations of physics applied to a continuum in motion, namely conservation of mass, Newton's second law, and a statement of the first law of thermodynamics which in some forms utilizes the second law to introduce entropy. The fluid motion is viewed from a frame of reference fixed in the earth and rotating with constant angular velocity. Astronomical tide-generating forces are neglected in view of our previous discussion. The equations are first given in vector form. Here $\mathbf{v}$ is the temporally and spatially varying fluid velocity and ρ is the fluid density. The expression for conservation of mass in a continuum, called the continuity equation, is as follows.

$$\frac{1}{\rho}\frac{d\rho}{dt} + \text{div } \mathbf{v} = 0 \tag{1}$$

where $d/dt = \partial/\partial t + (\mathbf{v} \cdot \nabla)$ is the total derivative operator and expresses the time rate of change of a fluid property as experienced by a fluid parcel. A statement of Newton's second law takes the following form in our rotating frame of reference.

$$\rho\frac{d\mathbf{v}}{dt} + \rho 2\mathbf{\Omega} \times \mathbf{v} = \rho\mathbf{g} + \nabla \cdot \boldsymbol{\sigma} \tag{2a}$$

Here the first term on the left-hand side is the particle acceleration relative to the rotating frame of reference, $\mathbf{\Omega}$ is the rotation vector of the earth in radians per second, and the second term is the Coriolis acceleration. The total acceleration of a parcel with respect to an observer in an inertial frame of reference equals the sum of the acceleration relative to the rotating frame, the centripetal acceleration, and the

Coriolis acceleration. The latter two terms arise as follows. A parcel at rest in the rotating frame moves in a circle about the axis of rotation with a rigid-body velocity $\boldsymbol{\Omega} \times \mathbf{r}$ and hence undergoes a centripetal acceleration as seen by the inertial observer. A parcel moving with constant velocity and no acceleration with respect to the rotating frame has a varying direction of motion and changing rigid-body velocity with respect to the inertial observer; these effects together form the Coriolis acceleration. The first term on the right-hand side is the net attraction of the earth's gravitational field less the centripetal acceleration. The vector $\mathbf{g}$ is taken to be of uniform magnitude and to point inward toward the center of the earth. The second term on the right-hand side, the divergence of the stress tensor, is the net force exerted on a parcel by its neighbors. The relation of the stress tensor $\boldsymbol{\sigma}$ to the flow depends on the nature of the particular fluid. This relation is called a constitutive relation. For air, water, and a variety of other common fluids this relation takes the following form in a Cartesian frame of reference, with spatial variables x_1, x_2, and x_3 and velocity components v_1, v_2, and v_3, respectively.

$$\sigma_{ij} = -p\delta_{ij} + \mu \left(\frac{\partial v_i}{\partial x_j} + \frac{\partial v_j}{\partial x_i} \right) \qquad (2b)$$

and j, $i = 1, 2$, or 3. Here p is the mechanical pressure; μ is the absolute viscosity, which we take henceforth as a constant; and δ_{ij} is the Kronecker delta, which is one if the two indices have the same value and is zero otherwise. For geophysical flows with small velocity divergence, the mechanical pressure is identical to the thermodynamic pressure. For the fluid portion of the earth, no constitutive relation is in wide use, and as the viscous force would appear to be a dominant one for internal motions of the earth, we exclude these motions from the remainder of this article. When the expression (2b) is substituted into (2a) we obtain the following momentum equations, with μ a constant.

$$\rho \frac{d\mathbf{v}}{dt} + \rho 2\boldsymbol{\Omega} \times \mathbf{v} = \rho \mathbf{g} - \nabla p + \mu \nabla^2 \mathbf{v} \qquad (2)$$

B. The First and Second Laws of Thermodynamics

An expression of the first law of thermodynamics is required to close the set of equations. For a mass of fluid contained in a box, the change in internal energy between two equilibrium states is equal to the work done on the fluid and the heat added. The internal energy includes the random translational energy of molecules constituting the fluid. For a fluid in motion these molecules also have an ordered translational kinetic energy. We state the first law for a fluid in motion as

follows. The time rate of change of the internal plus kinetic energy of a fluid parcel is equal to the rate at which heat is being added and the rate at which work is done on the parcel. Some of this work is done by gravity forces and some by the fluid surrounding the parcel. A mechanical energy equation can be obtained from Eq. (2) by taking the vector dot product of that equation with the velocity vector. When the mechanical energy equation is subtracted from our statement of the first law, several terms drop out and we obtain the following expression for the first law.

$$\rho \frac{de}{dt} = \frac{p}{\rho} \frac{d\rho}{dt} + 2\mu \sum_{i=1}^{3} \sum_{j=1}^{3} \left[\frac{1}{2} \left(\frac{\partial v_i}{\partial x_j} + \frac{\partial v_j}{\partial x_i} \right) - \frac{\text{div } \mathbf{v}}{3} \delta_{ij} \right]^2$$
$$+ \kappa \nabla^2 T + q_e \qquad (3a)$$

The term on the left-hand side is the time rate of change of the internal energy per unit mass (e). The first term on the right-hand side represents the rate at which pressure forces do work on the parcel; the second term, the dissipation function, is always positive and represents the conversion of mechanical energy into internal energy. The third term represents the addition of heat by molecular conduction of heat, where κ is the thermal conductivity. The final term represents the rate at which heat is added due to the net absorption of radiation and the rate at which latent heat is liberated due to the condensation of water vapor. The latter process, of course, takes place only in the atmosphere when a parcel is saturated and being cooled. Within the context of geophysical fluid dynamics this final term is sometimes specified as a function of space and time. To predict this term explicitly would require that an equation governing the humidity of a parcel, as well as equations governing the emission and absorption of electromagnetic radiation, be specified. As this is well beyond the scope of this article, we will neglect this final term. We can introduce the second law of thermodynamics, which relates changes in the entropy (η) of a system to changes in internal energy and density. Equation (3a) can then be written as follows.

$$\rho T \frac{d\eta}{dt} = 2\mu \sum_{i=1}^{3} \sum_{j=1}^{3} \left[\frac{1}{2} \left(\frac{\partial v_i}{\partial x_j} + \frac{\partial v_j}{\partial x_i} \right) - \frac{\text{div } \mathbf{v}}{3} \delta_{ij} \right]^2$$
$$+ \kappa \nabla^2 T \qquad (3b)$$

If molecular transport of heat and energy dissipation were negligible, then in the absence of already neglected heating due to radiation of condensation this equation would tell us that the entropy of a fluid parcel does not vary as the parcel moves; however, different parcels could have different values for the entropy. Using Eq. (3b) and other thermodynamic relations, the thermodynamic energy relation can be expressed in terms of changes in temperature and pressure or in terms of changes in density and pressure as follows.

$$\frac{d\rho}{dt} = \frac{1}{c^2}\frac{dp}{dt} - \frac{\alpha}{c_p}\left\{2\mu \sum_{i=1}^{3}\sum_{j=1}^{3}\left[\frac{1}{2}\left(\frac{\partial v_i}{\partial x_j} + \frac{\partial v_j}{\partial x_i}\right)\right.\right.$$

$$\left.\left. - \frac{\text{div } \mathbf{v}}{3}\delta_{ij}\right]^2 + \kappa \nabla^2 T\right\} \tag{3c}$$

$$\rho c_p \frac{dT}{dt} - \alpha T\frac{dp}{dt} = 2\mu \sum_{i=1}^{3}\sum_{j=1}^{3}\left[\frac{1}{2}\left(\frac{\partial v_i}{\partial x_j} + \frac{\partial v_j}{\partial x_i}\right)\right.$$

$$\left. - \frac{\text{div } \mathbf{v}}{3}\delta_{ij}\right]^2 + \kappa \nabla^2 T \tag{3d}$$

In these equations c, c_p, and α are the speed of sound, the specific heat at constant pressure, and the coefficient of thermal expansion. The forms of the energy equation are equivalent, but each is useful in a different context. As the thermodynamic equation has introduced a new variable (i.e., temperature), an equation of state is necessary to close the system finally. For the atmosphere, the perfect gas law, $p = \rho RT$, is used, for the ocean the density is frequently taken to increase linearly with increasing pressure and salinity and decrease linearly with increasing temperature. The introduction of salinity in the oceanic case requires a law governing the distribution of salinity, which is as follows.

$$\rho \frac{dS}{dt} = D \nabla^2 pS \tag{3e}$$

C. Boundary Conditions

If the bounding surface of a fluid is given by the equation $F(x, y, z, t) = 0$, under continuum assumptions this surface must always be composed of the same fluid parcels. This poses a constraint on the surface of the fluid, namely the kinematic boundary condition, which can be expressed as $dF/dt = 0$. Across a fluid–fluid interface, an additional condition arises, the dynamic boundary condition, which requires that the pressure be continuous across the interface. The presence of viscous forces also requires that the velocity be continuous across an interface, the "no-slip" condition. This is usually effected in a very thin layer and is dropped as a condition when viscous forces are neglected.

The basic equations governing geophysical motions in the absence of "external" sources of heating have now been covered and the relative importance of the terms in the equations remains to be assessed. Before doing so we must introduce one important new quantity, the vorticity.

D. The Vorticity

When the divergence and curl of a vector field are known and suitable boundary conditions are specified, that vector field can be computed in a straightforward manner. As we shall see in the next section, there are many flows for which the divergence of the velocity field can be taken to vanish. Thus if the curl of the velocity field were known, along with suitable boundary conditions, the velocity field could be calculated. We define a new vector field, the vorticity ω, as follows.

$$\omega = \text{curl } \mathbf{v} \tag{4a}$$

The vorticity can be interpreted as twice the angular velocity, or rate of spin, of a parcel about its own center. Equation (4a) yields the vorticity as seen by an observer in the rotating frame of reference. An observer in an inertial frame would observe the absolute vorticity ω_a, which is given by

$$\omega_a = 2\Omega + \omega \tag{4b}$$

An equation governing the change in the vorticity of a parcel can be derived from Eq. (2) by taking the curl of that equation and utilizing Eq. (1). We obtain the following "vorticity equation."

$$\frac{d}{dt}\left(\frac{\omega_a}{\rho}\right) = \left(\frac{\omega_a}{\rho}\cdot\nabla\right)\mathbf{v} + \frac{1}{\rho^3}\nabla\rho\times\nabla p + \frac{\mu}{\rho^2}\nabla^2\omega \tag{5}$$

The term on the left-hand side can be interpreted as the time rate of change of the absolute vorticity times a particle's volume divided by its mass. The first term on the right-hand side is called the stretching and twisting term and is proportional to the derivative of the velocity in the direction of the absolute velocity. This term is present if the absolute vorticity is not negligible. Variations of the component of velocity in the direction of the absolute vorticity stretch the vortex lines and change the magnitude of the absolute vorticity. Variations of the velocity in the directions perpendicular to that vector change the direction of the absolute vorticity vector. The second term on the right-hand side is called the baroclinic production term. It represents the change in vorticity produced by the torque of the pressure force about the center of mass of the parcel. This term vanishes if surfaces of constant density and pressure coincide. Such flows are called barotropic and come about if the entropy field is uniform at some initial time and heating, which can produce entropy variations, is negligible. The final term is the diffusion of vorticity by molecular viscosity.

Under certain circumstances an additional result can be obtained from the vorticity equation. If for some given flow (1) viscous forces can be neglected; (2) some quantity, say $f(x,$

y, z, t), which is conserved, i.e., $df/dt = 0$, can be found; and (3) either the flow is barotropic or f is a function of p and ρ, then we can obtain Ertel's theorem, which is as follows.

$$\frac{d\pi}{dt} = 0 \qquad \text{where } \pi = \frac{\omega_a \cdot \nabla f}{\rho} \qquad (6)$$

The quantity π is called the potential vorticity. This theorem is of great utility in the study of large-scale flows which satisfy the restrictions stated.

E. Turbulence and the Averaged Equations

The previously stated equations are called the instantaneous equations. In certain regions of the atmosphere and oceans, primarily near horizontal boundaries, rapid variations in flow properties can occur in the direction normal to the boundary. The flow in these regions can be unstable to small disturbances and the result is a fairly random field of vorticity with high-frequency fluctuations. Unstable mass distributions resulting from heating from below or cooling from above can also result in turbulent flows, as can convective overturning due to large-amplitude waves in the fluid. If we performed a large number of experiments holding external conditions fixed—an ensemble of experiments—in turbulent situations each experiment would produce a different result for each flow variable as a function of space and time because of the nonlinear amplification of uncontrollable factors. We can define an average for each variable as a function of space and time over all the experiments in the ensemble, which leads to ensemble-averaged flow properties. If the number of experiments is sufficiently large, the ensemble-averaged properties will be reproducible from one ensemble to the next. The value for any flow variable in any one experiment can be written as the sum of the ensemble average of that variable and the deviation for that particular experiment. The ensemble average of the deviations is, of course, zero. Because the previously stated equations hold for each experiment, each variable can be replaced by the sum of the ensemble average and its deviation in each equation and the resulting equations themselves ensemble averaged. The results are predictive equations for the ensemble-averaged properties. These equations are identical to the previously stated equations with the exception that we have additional terms arising from the nonlinear advective terms. These new terms originate on the left-hand side of the averaged equations but are generally transposed to the right-hand side as they often have a viscous-like effects. For example, in the momentum equation in the x_1 direction the following term appears on the right-hand side.

$$-\bar{\rho} \sum_{j=1}^{3} \frac{\partial}{\partial x_j} \overline{v_i' v_j'}$$

The overbar on the primed quantities indicates an ensemble average of the velocity deviation products. Similar terms appear in the other equations. Relating these nonlinear turbulent terms to mean flow properties constitutes the turbulent closure problem, one of the central problems in fluid mechanics. The equations governing the mean flows require that the flux of momentum and total energy due to the averaged nonlinear terms be conserved across a fluid–fluid interface. A further discussion of turbulent flows is beyond the scope of this article. [See ATMOSPHERE, FLUID DYNAMICS.]

We now consider what information can be derived from our basic equations in the absence of turbulent fluxes and external heat sources.

IV. Simplifications and Scaling Assumptions

A. Common Approximations

Although the dynamics are considerably simplified by remaining outside the limited turbulent regions, they are still quite complex. However, they can be realistically simplified still further by scale analysis. We might imagine a flow field which can be totally described by experimental means so that we know all the flow variables as functions of space and time. We can then compare the relative importance of terms in each equation. We view the flow from a Cartesian coordinate system with an origin on the earth's surface, with the z axis up, the x axis toward the east, and the y axis toward the north. The velocity components in these three directions are w, u, and v, respectively. We can define the scale for each variable as the maximum minus the minimum value of that variable produced by the flow. We can define the length scale for, say, the u variable in the x direction, L_{x_u}, by

$$U/L_{x_u} = \overline{|\partial u/\partial x|}$$

where U is the scale for u and $\overline{|\partial u/\partial x|}$ is an average value for $\partial u/\partial x$. Length scales in the other two directions as well as a time scale can be similarly defined. Generally, the length and time scales for all variables are similar and it is possible that in different regions of the flow different scales may occur, although we exclude this possibility for now. Typically the length and velocity scales in the horizontal, x and y, directions are similar, although the vertical scales may differ from the horizontal scales. Assuming for now that horizontal and vertical scales are similar, we first assess the importance of the frictional terms in the governing equations by using the length and time scales previously defined. If we take the ratio of the nonlinear advective terms to the frictional terms in Eq. (2) we find that

$$\frac{\rho(\mathbf{v} \cdot \nabla)\mathbf{v}}{\mu \, \nabla^2 \mathbf{v}} \sim \frac{\rho UL}{\mu}$$

The dimensionless ratio is called the Reynolds number. If U equals 1 m/s and L equals 1 m, the Reynolds number for flows of air is $O(10^5)$; for the ocean, if U equals 0.1 m/s and L is 1 m, the Reynolds number is about the same. Analysis of advective to diffusive terms in the other equations gives comparable results. We conclude that the effect of molecular transport of momentum and heat is totally unimportant for geophysical flows; these terms are significant in only the smallest of turbulent eddies. Turbulent transport is important in the planetary boundary layers.

We now turn to the continuity equation. From Eq. (3c) we can see that variations in density of a parcel are produced by variations in the pressure of the parcel. Using the momentum equation to obtain an expression for dp/dt and substituting the result into the continuity equation (1), we find that the ratios of the pressure-induced density variations to the scale of the terms in the divergence of the velocity lead to the following three dimensionless numbers:

$$\frac{L}{c\tau}, \quad \frac{U^2}{c^2}, \quad \frac{gL_z}{c^2}$$

The first term is the ratio of the length scale L to the distance a sound wave would travel in time τ. Typically this number is quite small in geophysical problems. The second term is a measure of density variations induced by the conversion of kinetic energy into pressure changes. This term is also quite small. The third term is a measure of density variations induced by pressure variations associated with the change in the vertical location of a parcel. For oceanic flows this term is totally negligible. For atmospheric flows with vertical scale L_z far less than 10 km this term is negligible. Hence, for virtually all oceanic flows and many atmospheric flows the density term in the continuity equation can be neglected. The continuity equation then simplifies to the following.

$$\text{div } \mathbf{v} = \frac{\partial u}{\partial x} + \frac{\partial v}{\partial y} + \frac{\partial w}{\partial z} = 0 \qquad (7)$$

This is the incompressible form of the continuity equation. This simplification is only in this equation, and the energy equation remains as before. Just as the neglect of $\mu \, \nabla^2 \mathbf{v}$ in the momentum equations does not mean that $\mu \, \nabla^2 \mathbf{v} = 0$, so the neglect of dp/dt in the continuity equation does not mean $dp/dt = 0$. The incompressible form for the continuity equa-

tion suggests that the vertical velocity scale is related to the horizontal velocity scale as follows.

$$W/U = L_z/L$$

B. Important Parametric Regimes

We now attempt to understand the balances which might exist for the other equations. Geophysical motions can be influenced by the rotation of the earth through the Coriolis acceleration and by the overall stable increase of the entropy with height, although there are regions in both the atmosphere and oceans which are not stably stratified. The overall stable stratification can be quantified by the Brunt–Väisälä frequency $N(z)$, defined as follows.

$$N^2(z) = \frac{g}{c_p}\frac{d\eta}{dz} = g\left(-\frac{1}{\rho}\frac{d\rho}{dz} - \frac{g}{c^2}\right)$$

The influence of the Coriolis acceleration and the stable stratification depends on the nature of the length, velocity, and time scales which occur in any flow and vary from flow to flow. We first consider the relative importance of the Coriolis acceleration. In the horizontal momentum equations, the ratio of the unsteady acceleration to the Coriolis acceleration and the ratio of the advective acceleration to the Coriolis acceleration are given by the two dimensionless numbers

$$\frac{1}{2\Omega\bar{c}} \quad \text{and} \quad \frac{U}{2\Omega L}$$

If both these numbers are small, the Coriolis acceleration dominates the relative acceleration and nearly balances the horizontal pressure force; such a balance is called geostrophic. For the atmosphere the time scale should exceed 1 day and the length scale 1000 km, and for the ocean the length scale should exceed 10 km for the geostrophic balance to hold. We note that the dimensionless advective ratio is called the Rossby number, which is also the ratio of the vertical component of relative vorticity to the vertical component of Ω. We now discuss the implications of nearly geostrophic balance. Given the large required horizontal length scales, the ratio L_z/L is small compared to one, so the ratio W/U is also small. Further, we can show that if L_z/L is small compared to one the vertical accelerations in the vertical momentum equation are small compared to the pressure gradient terms. Hence, the pressure distribution is hydrostatic in the vertical momentum equation; i.e., the vertical gradient of the pressure associated with the motion balances the disturbance in the density field times the acceleration due to gravity. The ratio of the generation of vorticity in the vertical direction by the baroclinic term to the stretching of planetary vorticity is as follows.

$$\frac{(1/\rho^2 \, \nabla\rho \times \nabla p)_z}{|\omega_a|\partial\omega/\partial z} = \frac{N^2 L_z}{g}$$

This term is small compared to one for both oceanic and atmospheric flows, so the stretching term dominates the vertical component of the baroclinic production term in the vertical component of the vorticity equation. The thermodynamic energy equation was used to estimate the horizontal density gradients and the geostrophic balance was used to estimate the horizontal pressure gradients. One additional term arises on the right-hand side of the vorticity equation from the tilting of the component of the earth's rotation vector parallel to the earth into the vertical direction. This term equals $-2\Omega \cos\theta \, v/\rho R_e$, where R_e is the radius of the earth, and usually is of the same order as the advection of vertical vorticity. This term is usually brought over to the left-hand side of the vorticity equation; it is then called the advection of planetary vorticity. All other terms in the stretching and twisting term are of order L_z/L less than the stretching of planetary vorticity by the vertical velocity. Equating the stretching term to the advection of vorticity term indicates that the vertical velocity scale is a factor of the Rossby number smaller than its initial scaling. Hence rotation in these low Rossby number flows reduces the vertical velocity. An analysis of the horizontal components of the vorticity equation indicates that advection of horizontal vorticity is smaller than the tilting terms by a factor of the Rossby number. The basic balance in the horizontal vorticity equations is thus between the baroclinic production and the tilting terms. These balances constitute the thermal wind equations. We can estimate the importance of the stratification itself on a low Rossby number flow because the appropriate scale for the vertical velocity has been established. From the energy equation in terms of the density we can establish the scale for the density disturbances. With this known, the scale for the density-induced disturbance to the pressure field can be established from the hydrostatic balance. The ratio of the pressure disturbance induced by the stable stratification to that associated with the geostrophic balance is the measure of the relative importance of stratification in flows with specified horizontal and vertical scales. This dimensionless ratio is as follows.

$$N^2 L_z^2/(2\Omega L)^2$$

This dimensionless ratio is known as the Burger number. In many geophysical flows only one of the length scales is specified. The unspecified scale adjusts to make the Burger

number of order 1. Thus the Rossby number, the Burger number, and the ratio L/R_e are the principal determinants of low Rossby number flows. Further analysis of such flows usually proceeds by expanding the scaled variables in a series in the Rossby number. The lowest-order momentum balance is hydrostatic and geostrophic and horizontally nondivergent. The next-order terms in the horizontal momentum equations and continuity equation are manipulated to yield a predictive equation for the lowest-order pressure field.

On the other extreme of geophysical motions, for atmospheric flows with horizontal scales less than 100 km or oceanic flows with horizontal scales less than 1 km or either motion with time scales of less than a few hours, the Coriolis acceleration is of minor importance. However, within the turbulent planetary boundary layer, as the horizontal velocity decreases toward the underlying surface, the Coriolis term does partially balance the turbulent forces. We must reexamine the vorticity equation to establish the appropriate parameters for these flows with shorter length scales. As the earth's rotation is unimportant for such flows, they can be classified as nonrotating. For such flows the vertical length scale can be comparable with the horizontal scale, so that the velocity scales can also be comparable. Such flows can first be classified as either unsteady flows, for which the unsteady terms dominate the advective terms in the total derivative operator, or quasi-steady flows, for which the unsteady terms are of comparable or smaller magnitude than the advective terms. To establish the parameter regime in which the stable stratification is a significant contributor to the vorticity balance, we first estimate the magnitude of the horizontal density gradient from the thermodynamic energy equation and obtain the following result.

$$\nabla_H \rho \sim \frac{N^2 U L_z}{g(1/\tau, \, U/L)_{max}}$$

The baroclinic production of the vertical component of vorticity also requires an estimate for the horizontal pressure gradient. This can be obtained from the horizontal momentum equation. The ratio of baroclinic production to advection of the vertical component of vorticity can be estimated as follows.

$$\frac{[(1/\rho^2)\nabla\rho \times \nabla p]_z}{(d/dt)\omega_z/\rho} \sim \frac{N^2 L_z}{g}$$

This term is small compared to one for both atmospheric and oceanic flows. Changes in the vertical component of relative vorticity are produced only by the stretching and twisting term. The principal portion of the baroclinic production of horizontal vorticity is proportional to the product of the horizontal density gradient with the vertical derivative of the pressure or ρg. The product of the vertical density gradient with the horizontal pressure gradient is of order V^2/gL_z smaller than the principal term and is generally negligible. For quasi-steady flows in which $1/\tau < U/L$ the ratio of baroclinic production to the advection of the horizontal components of vorticity is then

$$N^2 L_z^2/U^2$$

The dimensionless ratio U/NL_z is called the internal Froude number. For unsteady flows in which $1/\tau > U/L$ the ratio of the baroclinic production to the time rate of change of the horizontal components of vorticity is then

$$N^2 \tau^2 L_z^2/L^2$$

Generally the vertical scale is at most a few tenths of the horizontal scale. For time scales less than 10 min for unsteady flows or for quasisteady flows at high Froude number the effects of the stable stratification are unimportant in these turbulent free flows. For these flows, the vorticity of a fluid parcel is changed only by stretching and twisting. For such flows, if no vorticity is present initially, there is no mechanism to produce vorticity, and such flows are called irrotational. For irrotational flows, the velocity vector can be derived as the gradient of a potential function. For divergence-free flows, the potential function satisfies Laplace's equation. The nature of such flows then depends on the boundary conditions applied. The prediction of the nature of surface gravity waves in the ocean represents one of the successful applications of potential theory in fluid mechanics.

Flows for which the Coriolis acceleration is neither dominant nor negligible (i.e., steady flows for which the Rossby number is of order 1) or unsteady flows for which the time scale is of order $1/2\Omega$ are called mesoscale flows. These flows are somewhat more complex to analyze than either quasigeostrophic or nonrotating flows and are usually handled analytically via a small-perturbation approach, an example of which is given in the next section.

V. Wave Motion in Geophysical Fluids

As examples of motions which can occur in geophysical fluids, we consider small-amplitude wave motion in fluids at rest. The velocities that are produced by these waves are considered to be small compared with the phase speed of the waves so that advection of wave-induced perturbation quantities can be neglected in comparison with the unsteady fluctuations. Therefore the dimensionless term $1/2\Omega\tau$ determines the relative importance of rotation.

A. High-Frequency Motions

We first consider an example in nonrotating flows ($1/2\Omega\tau \gg 1$). We consider waves on the surface of the ocean with the origin at the undisturbed surface. We shall also consider time scales small compared with N so that the fluid can be treated as one of uniform density. The depth of the ocean will be taken to be much greater than the horizontal wavelength. As the initial state is without vorticity, the vorticity will remain zero throughout the fluid and the velocity is equal to the gradient of a potential function. For this homogeneous flow the divergence of the velocity is zero so that the potential function satisfies Laplace's equation. We seek a solution of the form

$$\phi(x, y, z, t) = A \cos(kx + ly - \sigma t)e^{\sqrt{k^2+l^2}\,z}$$

This represents a wavelike solution which propagates horizontally. The decay with depth is required by Laplace's equation for the specified horizontal structure of the wave. The kinematic and dynamic boundary conditions can be combined to yield the following condition at the free surface.

$$\frac{\partial^2 \phi}{\partial t^2} + g\frac{\partial \phi}{\partial z} = 0$$

When the assumed form for the solution is substituted into this boundary condition, the following dispersion relation results.

$$\sigma^2 = g\sqrt{k^2 + l^2}$$

This result has been verified experimentally for wavelengths shorter than the fluid depth but longer than a few centimeters. For very short wavelengths the effects of surface tension become important.

B. Intermediate-Frequency Motions

We next consider waves in a fluid of uniform N. The Coriolis effect will play neither a dominant nor a negligible role in the motions considered here. All dependent variables are assumed to vary with space and time as $\cos(kx + ly + mz - \sigma t + \theta)$. When expressions of this form are substituted into the linearized governing equations, the following dispersion relation results.

$$\sigma^2 = \left(\frac{el}{K} + \frac{fm}{K}\right)^2 + N^2 \frac{k^2 + l^2}{K^2}$$

Here $e = 2\Omega \cos \theta$, $f = 2\Omega \sin \theta$, θ is the latitude, and $K =$

$\sqrt{k^2 + l^2 + m^2}$ is the wave number. These waves are called inertial–internal waves.

C. Low-Frequency Motions

As a final example of wave motion in geophysical motions we consider motions for which $1/2\Omega\tau \ll 1$. These flows are quasigeostrophic. We consider motions in which the Burger number is small, bounded above by a free surface, and bounded below by a flat bottom. The assumed motion is then barotropic. The vorticity equation balances the sum of the unsteady change in vertical vorticity and advection of planetary vorticity with the stretching of planetary vorticity associated with the vertical motion of the free surface. The pressure perturbation is assumed to be of the form $\cos(kx + ly - \sigma t)$. With the free surface displacement proportional to the pressure disturbance, all terms in the linearized vorticity equation can be expressed in terms of the pressure. The following dispersion relation results.

$$\sigma = \frac{-2\Omega \cos\theta \, k}{R_e(k^2 + l^2 + f^2/gD)}$$

These waves are called barotropic Rossby waves and appear to play an important role in both oceanic and atmospheric motions.

Bibliography

Batchelor, G. K. (1967). "An Introduction to Fluid Dynamics." Cambridge Univ. Press, London.

Pedlosky, J. (1987). "Geophysical Fluid Dynamics." Springer-Verlag, New York.

Phillips, O. M. (1966). "The Dynamics of the Upper Ocean." Cambridge Univ. Press, London.

Glossary

Baroclinic flows Flows of nonuniform entropy in which the density field produces vorticity.

Barotropic flows Flows in which surfaces of constant density and pressure coincide so that the density field does not produce vorticity.

Geostrophic flows Flows in which the Coriolis acceleration is balanced by the pressure force. Many large-scale flows are in nearly geostrophic balance.

GIS Technology in Ecological Research

Carol A. Johnston
University of Minnesota

A geographic information system (GIS) is a computer-based system for the manipulation and analysis of spatial information in which there is an automated link between the data and their spatial location. A GIS consists of computer hardware and software for entering, storing, transforming, measuring, combining, retrieving, displaying, and performing mathematical operations on digitized thematic data (e.g., soils, vegetation, hydrology) that have been registered to a common spatial coordinate system.

I. Introduction

The spatial distribution of earth resources is of key importance in earth system science. Interactions among air, land, water, and biota most frequently occur when they are juxtaposed in space, so information about their distribution is essential to understanding their interconnectedness. Maps have long been used to represent the location of earth system components, and map overlay techniques can show how they are superimposed in space, but maps are static

and qualitative. Geographic information systems (GISs), computerized systems for manipulating and analyzing mapped data, grew out of the need to derive quantitative information about the distribution of and interactions among earth resources.

A GIS differs from a map in several ways. A map is an analog depiction of the earth's surface, while a GIS records spatially distributed features in digital form. A map simultaneously depicts a variety of landscape features (e.g., topography, vegetation, road networks), while a GIS usually stores each feature as a separate data layer. A map is static and difficult to update, while a GIS data layer can be easily revised. A map is always its own end product, while the end product of a GIS analysis may be a map or data. Although maps can be a form of GIS input or output, a GIS greatly increases the versatility of mapped data because of its wealth of techniques for quantitative analysis.

Although GISs are conventionally used to portray features on the earth's surface, their analysis capabilities can be used with any spatially distributed data. For example, GISs have been used to analyze root systems photographed from behind a vertical glass plate, insect damage to leaves, and underground water movement. GISs can analyze data at a variety of spatial scales, from microscopic to global.

Examples of earth resource questions a GIS can address are:

- How is the distribution of plant community A related to the distribution of plant community B?
- How is the distribution of plant community A related to the distribution of environmental factors, X, Y, and Z?
- How has the distribution of plant community A changed over time?
- How will the distribution of plant community A change if environmental factor X is altered?
- How do ecosystem components X, Y, and Z contribute to material export from watershed Q?
- How will changes in the global distribution of temperature and precipitation affect biomass production?

The development during the 1980s of commercially available GISs operated on personal computers (PCs) or minicomputers led to their increased use for natural resource inventory and planning, where mapping and spatial database management capabilities have been a boon to land use managers. Other reasons for increased GIS use include enhanced analytical capabilities, user-friendliness, and ease of data exchange among different systems. Whatever the reason, GISs are becoming a standard tool in earth systems management and research. This article provides an introduction to GIS technology and its application in ecological research.

II. Basics of GIS

A. GIS Data Structures

There are two basic types of GIS, which differ in the way they store data. Raster-based GISs, also known as grid- or pixel-based systems, portray land surfaces as a matrix of grid cells, each with an individual data value. Vector-based GISs portray land surfaces as points, lines, and polygons. Each of these data structures has advantages and disadvantages, depending on the type of GIS application.

1. Raster Data Structures

Raster-based GISs are in widespread use for spatial analysis because of their computational simplicity and compatibility with remotely sensed data. Each cell in a rasterized database can be assigned only one number, so different geographic attributes must be stored as separate data layers, even when they refer to the same objects. For example, beaver ponds might be described in terms of their age, vegetation, sediment, water depth, or pond identifier. To put all this information into one data layer would require a huge classification system, equal to the product of the possible classes in each category:

50 age classes $\times$ 38 vegetation types $\times$ 10 sediment types $\times$ 40 depth classes $\times$ 837 ponds $\cong$ 636 million classes

Since such a large classification system would be prohibitive, the data for each of the categories are stored in a separate data layer, and the GIS is used to extract and combine information from the layers.

Some raster-based GISs overcome this problem through a relational database, in which a single layer of pixels is used to store integer numbers that refer to items (e.g., points, line segments, polygons) in an attribute table. Each item could have multiple attributes. In the above example, each beaver pond could be classified in an attribute table by age, vegetation, sediment, and depth, reducing the total number of pixel codes needed to 837 (i.e., the number of ponds).

The individual raster is the minimum unit of spatial resolution. Therefore, features that are dimensionless (e.g., points) or smaller than the minimum raster dimension (e.g., wells, streams) are difficult to depict in a raster-based system. At coarse levels of resolution (large rasters), data appear blocky and lines appear stair-stepped. At finer levels of resolution, a raster representation looks more like a map, but data storage requirements increase exponentially.

Finer-resolution raster databases are aesthetically more appealing, but the increased detail they provide may be unnecessary for data analysis. The appropriate raster size should be comparable to the scale at which the ecological process of interest is operating. For example, global circulation models commonly use cells covering degrees of latitude and longitude, subcontinental analyses commonly use square-kilometer cells, and regional analyses commonly use 30 $\times$ 30 m cells (the size of a Landsat Thematic Mapper pixel). Rasters which are too fine can obscure features at coarser scales by providing too much detail. [See ATMOSPHERIC MODELING.]

Quadtrees are a type of raster data storage structure which take into account the inherent resolution of the data. The quadtree is a hierarchical raster data structure based on the successive division of a $2^n \times 2^n$ array into quadrants, in which the data for four quadrants having the same value are aggregated into the next largest block in the hierarchy. As a result, coarse-grained features are stored in large cells and fine-grained features are stored in small cells.

Raster systems are superior to vector systems for depicting continuous data surfaces, such as the reflectance values recorded by a satellite remote sensor. Whereas conventional maps usually group data categorically and spatially (e.g., contour intervals, vegetation patches), raster-based systems can maintain the spatial diversity of the input data. This capability is particularly important for applications involving the analysis of spatial heterogeneity.

2. Vector Data Structures

Vector-based GISs portray land features as points, lines, or polygons. Homogeneous patches are bounded by lines, as they would appear on a conventional map. Because of this similarity, vector-based GISs are preferred for automated cartography applications.

Most commercially available vector-based GISs use an arc–node data structure, in which arcs are the individual line segments defined by each pair of x, y coordinates, and nodes occur at the intersections and ends of lines. Polygons are defined by the arcs which bound them; and topologic relationships (e.g., arc end points, adjacent polygons) are stored in tables with locational data. Attribute data may also be stored with the spatial data, or may be kept in a separate table as a relational database. The tables itemize each point, line, and polygon in the data layer. In GISs with relational databases, all of the items listed are assigned a unique identifier, which is used to link the spatial and attribute information.

Vector-based GISs are superior to raster-based systems for depicting and measuring linear features, such as streams, roads, and boundaries. The use of vector structure is essential in applications where boundary locations must be precise, such as maps which show property ownership or jurisdictional limits.

B. GIS Hardware

GIS hardware consists of the central processing unit (CPU) used to run the software and various peripherals for data input, data storage, visual display, and hardcopy output. Although GISs have been developed for virtually every CPU platform, GISs operated on microcomputers, minicomputers, and workstations have become increasingly popular.

1. Data Entry

Various techniques are used to enter information into a GIS. The most basic means of entering data is manually keying a feature's location and attributes into a computer database. This is relatively simple for point data, such as the location of sampling sites, but is prohibitively slow for detailed maps.

Mapped data are usually entered into a GIS with a digitizing table (Fig. 1). The map is placed on the table and electronically traced using a cursor which converts point locations and line segments into x, y coordinates. Digitizing maps is a tedious, labor-intensive process that can be prone to machine and operator error. It can be very time-consuming to digitize complex databases covering large areas, and GIS users sometimes become so bogged down in data entry that they have little time left for data analysis.

Automated scanning devices provide a more rapid means of map entry (Fig. 2). Scanners can be separated into two types, those that scan lines by following them directly and those that operate in a raster mode, breaking the map image up into tiny squares. Raster scanning devices include flat-bed scanners, drum scanners, scanning cameras, and video digitizers connected to a "framegrabber" that converts the video image into a rasterized digital image.

In addition to their ability to scan maps, video scanners and scanning cameras can create images of objects or aerial photos. When used in conjunction with image analysis pro-grams, they can create a GIS data layer without the necessity of an intermediate map. This approach has been used in innovative applications, such as the particle size analysis of iron ore being dumped from railroad cars.

Although automated scanning has the appeal of reducing the drudgery of manual digitizing, scanning maps require careful checking and editing for erroneous line gaps and overlaps. The output from raster scanners must be processed with a raster-to-vector conversion algorithm before use in a vector-based GIS. Map attributes and labels must be entered manually, so the process is still very labor-intensive. Continued development is needed to advance automated scanning and reduce its cost and labor-intensiveness.

If input data are already in digital form, such as satellite imagery or an existing GIS database, they are usually entered using a floppy disk, tape cassettes, or nine-track tape drive. Although magnetic media have been the standard for data storage and exchange, optical storage devices will undoubtedly become much more important to GIS technology in the future. These devices, known as CD-ROM or WORM (write-once–read-many) devices, have extremely attractive storage capacities: one gigabyte of data can be stored on a single 5.25-inch disk. Additional advantages of CD-ROM include lower reproduction costs and longer stability than magnetic media, consistent with the long-term validity of much geographic data. The drawbacks of this storage medium are that it can be written on only once, and data access is slow in comparison to magnetic disks.

2. Hardcopy Output

In most GIS applications, the desired output is a hardcopy map. As with GIS data structures, both raster and vector plotters exist. The simplest raster plotter is the line printer, and the earliest computer mapping systems used this rudimentary but fast means of hardcopy output. Dot matrix printers, in which tiny pins striking an inked ribbon create gray tones on the paper, represent an improvement over line printers but are still relatively crude. Ink-jet printers can produce hundreds of different colors with varying combinations of cyan, yellow, magenta, and blank inks and are superior for continuous-tone color images (Fig. 3). Laser printers, which are becoming increasingly popular for text printing, can also be used with a GIS for printing page-sized maps (Figs. 4 and 5).

Pen plotters operate in a vector mode, drawing maps as a series of lines on paper held on a flatbed or rotating drum (Fig. 6). Multicolor maps can be drawn using different pen colors, and linear shading patterns can be used to fill in polygons. If the map is to be printed, a scribing tool can be used in place of pens. Electrostatic plotters, in which ink is transferred to electrostatically charged areas of paper, simu-

Fig. 1 Entering a map into a GIS using a digitizing table.

late pen plotters but can produce very high resolution, multi-color maps.

III. Data Sources for Ecological Research Applications of GIS

As with any computer system, a GIS requires suitable input data. Fortunately, a number of U.S. government agencies have developed or are developing digital databases for GIS use (Table I). Federal agencies in the process of digitizing maps include the U.S. Geological Survey (topographic maps, land use maps), the U.S. Soil Conservation Survey (soil surveys), the U.S. Fish and Wildlife Service (wetland inventories), and the U.S. Census Bureau (census data). International organizations, such as the United Nations Environment Programme (UNEP), are developing GIS databases for developing nations and the globe. As GIS use becomes more widespread, these digital databases will become increasingly available.

A base map containing georeferencing information and standard features such as roads and streams is fundamental to most GIS applications. This information can be obtained in vector format from the U.S. Geological Survey's Digital Line Graph (DLG) series). DLG databases are derived from USGS map products and consist of several different thematic layers (Fig. 4). DLG files are available at scales of 1:2,000,000 for state and national applications (Figs. 4 and 5), 1:1,000,000 for state and regional applications, and 1:24,000 for site-specific applications. The 1:2,000,000 and 1:100,000 series are available for the entire United States, but the 1:24,000 series is available for only part of the country.

Another digital product useful in ecological research is the Digital Elevation Model (DEM). Although DEM is used by some as a generic term for any digital representation of the continuous variation of relief over space, it also refers to the Digital Elevation Model databases produced by the U.S. Geological Survey. These are raster representations of elevation, with each pixel assigned an elevation value. USGS 1:1,000,000 DEMs are available for the entire country, and 1:24,000 DEMs are available for portions thereof.

A number of existing tabular databases contain point data

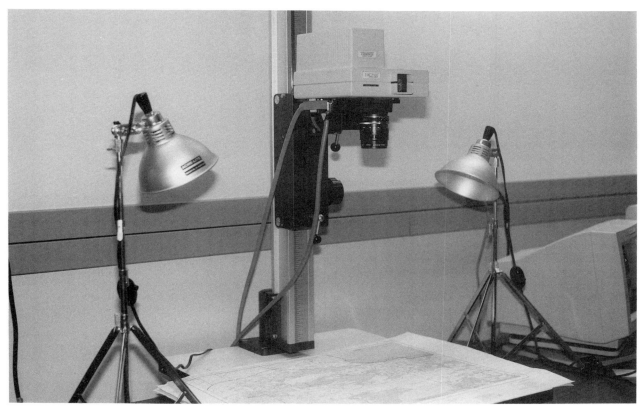

Fig. 2   A scanning camera for automated data entry into a GIS.

which are sufficiently georeferenced for conversion into GIS databases. Examples include precipitation data collected by the National Oceanic and Atmospheric Administration (NOAA) National Climatic Data Center, stream hydrology data in the USGS WATSTORE database, and water quality data in the Environmental Protection Agency's STORET database. The information contained in these databases can be indispensable for regional ecological analyses, but potential users should critically evaluate sample locational coordinates because the quality and/or resolution of locational data may be too poor to warrant conversion. An example is the EPA STORET database, in which much greater attention was paid to the accuracy of the water quality data than to sample location records.

Because ecologists tend to work in sparsely populated areas for which suitable GIS databases are lacking, they must often create their own GIS databases. If suitable maps exist, they can be digitized (see Section II,B,1). Where existing mapped or digital data are lacking, aerial photography can be used to generate the information for many ecological research applications. Air photos are commonly used to map

such features as vegetation, surface water, and land use. The use of air photos in combination with field reconnaissance can provide information about subsurface features such as soils and geology. Historical air photos, taken for most of the country beginning in the late 1930s, can be used to analyze long-term ecological trends. The 50-year period of air photo record has the advantage of covering a time span much longer than most "long-term" field data sets. [*See* PHOTOGRAMMETRY.]

Like aerial photography, remote sensing imagery (Table II) can be analyzed to produce GIS databases of ecological variables, and some commercially available GISs combine both remote sensing and GIS analysis capabilities. There are several advantages to using remote sensing as a source of GIS data. First, remotely sensed images are already in digital form. Second, remote sensing covers extensive areas. Third, remote sensing image analysis can provide quantitative information about ecological properties which cannot easily be derived from aerial photography or field studies. For example, the normalized difference vegetation index (NDVI) which can be calculated from Landsat imagery is an indicator of above-

Fig. 3 Hardcopy output of a satellite image using an ink-jet printer.

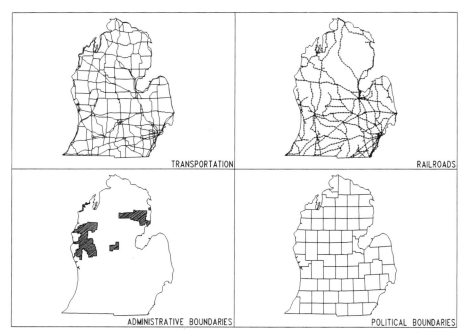

Fig. 4 Selected GIS layers derived from a 1:2,000,000 USGS DLG database. (Courtesy of James Westman, AScl Corporation, U.S. EPA Environmental Research Laboratory, Duluth, Minnesota.)

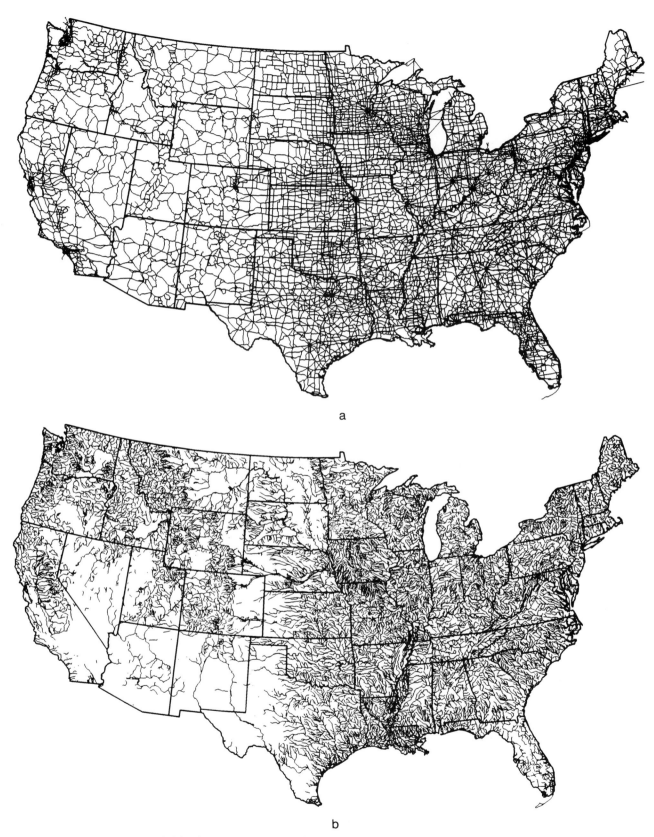

Fig. 5 GIS layers compiled for the conterminous United States from 1:2,000,000 USGS DLG databases. (a) Roads and trails; (b) perennial streams. (Courtesy of James Westman, AScl Corporation, U.S. EPA Environmental Research Laboratory, Duluth, Minnesota.)

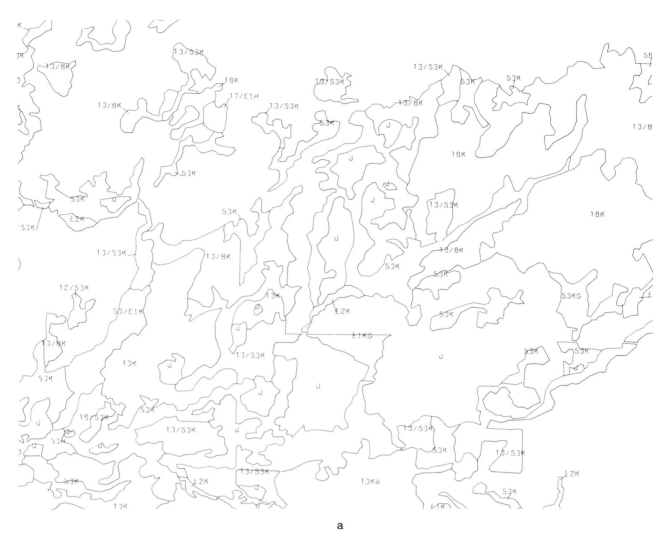

a

Fig. 6 Portion of a computer-plotted Wisconsin Wetlands Inventory overlay. Letter and number codes denote different wetland types. (a) Original format; (b) simplified format developed for regulatory purposes. (*Figure continues.*)

ground green biomass, a measurement which is very difficult and time-consuming in the field. [*See* REMOTE SENSING OF EARTH RESOURCES FROM AIR AND SPACE.]

The major disadvantage of using remote sensing as a source of GIS data is its relatively poor resolution. Although remotely sensed images can provide data about plant communities and environmental conditions, they are unsuitable for studies of individual organisms. Furthermore, the ecological feature of study must be associated with a change in spectral reflectance in order to be detected using a remote sensor, which somewhat limits potential applications of this data source.

Although GIS databases derived from aerial photography or remote sensing are in and of themselves useful, they are combined with field data for many ecological research applications. For example, maps of beaver ponds derived from aerial photos have been combined with field data from aerial counts of beaver colonies to determine how the increase in number of beaver over time affected the landscape. The same maps were combined with field data on nutrient concentrations in different types of beaver ponds to determine how beaver had affected nutrient availability in the landscape.

One obstacle to combining GIS and field-derived data has been the difficulty in obtaining accurate ground locations in

b

Fig. 6   (*Continued*)

IV. GIS for Mapping and Inventory of Earth Resources

GISs are used extensively to map and inventory natural resources, especially by public agencies charged with managing the field, particularly in rural areas with poor ground control. Fortunately, advances such as the use of the global positioning system (GPS) and laser surveying have increased the ease and accuracy of obtaining accurate ground locations. [*See* GLOBAL POSITIONING SYSTEM.]

or protecting them. Just as a warehouse manager needs to know the amount and location of inventory in the warehouse, natural resource scientists need to know the amount and location of sawtimber, ore bodies, wetlands, or other land features before they can effectively manage them. A GIS is a database management system uniquely adapted to this purpose.

The Wisconsin Wetlands Inventory, conducted by the Wisconsin Department of Natural Resources, provides a case study of the benefits of a GIS inventory. The Wisconsin Wetlands Inventory was mandated in 1978 for the purpose of advancing the conservation of wetland resources. Wetlands were located and classified by stereoscopic interpretation of

Table I   Sources of Digital GIS Databases

Database	Source	Scale	Unit area covered
Digital Line Graph (DLG)	USGS	1 : 24,000	7.5′ × 7.5′
Digital Line Graph (DLG)	USGS	1 : 100,000	7.5′ × 7.5′(hydrography)
			30′ × 30′ (other)
Digital Line Graph (DLG)	USGS	1 : 2,000,000	21 regions cover the United States
Digital Elevation Model (DEM)	USGS	1 : 24,000	7.5′ × 7.5′
Digital Elevation Model (DEM)	Defense Mapping Agency	1 : 250,000	1° × 1°
Land Use/Land Cover	USGS	1 : 100,000	30′ × 30′
		1 : 250,000	1° × 2°
Geographic Names Information System (GNIS)	USGS	NA[a]	7.5′ × 7.5′
Soil Surveys (SSURGO)	Soil Conservation Service	1 : 15,840 – 1 : 31,680	Counties
State Soils (STATSGO)	Soil Conservation Service	1 : 250,000	1° × 2°
National Soils (NATSGO)	Soil Conservation Service	1 : 7,500,000	United States
Census Block Group Boundary File	Geographic Data Technology, Inc.	NA	Census urban area
MapBase Digital Street Network (Modified DIME)	ETAK, Inc.	NA	Census urban area

[a] NA, not applicable.

black-and-white infrared aerial photos, delineated on a reproducible 1 : 24,000 photographic base, and digitized using a vector-based system developed in house by the Wisconsin Department of Natural Resources. The software fit each township-centered wetland map to its corresponding 7.5-minute USGS quadrangle by transforming or "rubbersheeting" it using the coordinates of five reference points digitized on both maps. This process helped correct variations in scale on the unrectified photographic bases and provided georeferencing.

After reference information was captured from the USGS quads, wetland boundary lines were digitized and classification codes entered. To process this data into the GIS data layer, polygons were defined by the line segments which circumscribed them. Each wetland classification code was matched with its appropriate polygon, and topological relationships were determined (i.e., the location of a wetland polygon relative to adjacent wetland and upland polygons). Standard computer products include wetland overlay maps (Fig. 6a) and acreage summaries by wetland class for each township in the state. Creation of the wetland data layer for the state's 1719 townships took about 17,190 hours, an average of 10 hours per township.

The computer-generated maps and acreage summaries provided benefits for unanticipated uses. The International Crane Foundation of Baraboo, Wisconsin, used them to estimate the area of suitable crane habitat in two Wisconsin counties by combining their knowledge of crane habitat

Table II   Comparison of Commonly Used Satellite Remote Sensors[a]

	AVHRR	Landsat MSS	Landsat TM	SPOT
Country and agency of origin	U.S., NOAA	U.S., NASA	U.S., NASA	France, SPOT Corp.
First available	1979	1972	1981	1986
Pixel resolution (m)	1100	79	30	20
				(10 in panchromatic)
Overpass frequency (days)	1	18	18	—
Wavelength bands	Red	Green	Blue	Panchromatic
	Reflected IR (two bands)	Red	Green	Green
	Thermal IR	Reflected IR	Red	Red
		Thermal IR	Reflected IR (three bands)	Reflected IR
			Thermal IR	

[a] AVHRR, Advanced Very High Resolution Radiometer; MSS, Multispectral Scanner; TM, Thematic Mapper; SPOT, Système Probatoire d'Observation de la Terre; IR, infrared.

preferences with the wetland vegetation and water classifications on the maps. The Milwaukee Priority Watershed Project used the maps to help develop non–point source pollution abatement plans for watersheds in the Milwaukee River basin. By using wetland overlays and topographic information to identify internally drained areas, they estimated that 10 to 15% of the basin surface area did not contribute non–point source pollution to the river and could be excluded from monitoring, saving the Watershed Project considerable time and money.

Statewide regulatory changes after initiation of the inventory created a major new use for the maps: protective zoning of wetlands. This new application of the Wisconsin Wetland Inventory maps posed a problem that might have been more difficult to overcome without the digital wetland data layer. Standard format wetland maps included wetlands smaller than the minimum size subject to regulation and boundary and classification information unnecessary for zoning. To meet the need for a simplified regulatory version of the wetland maps, computer-plotted overlays were developed to show exterior wetland boundaries, township borders, and only wetlands larger than the minimum size required to be regulated (Fig. 6b). The computerized map base also made it easier to update the regulatory maps as wetlands were rezoned and to monitor resulting changes.

This digital inventory cost more than conventional mapping techniques, but yielded a number of benefits which justified this cost:

- The scale distortion caused by using an unrectified photographic base was minimized.
- Acreage summaries were generated automatically.
- By interfacing different types of data layers, users could address specific management questions which neither data layer could answer alone.
- Map products could be tailored to individual users' needs.
- Maps could easily be updated and used to monitor change.

V. Spatial Analysis Using GIS

The spatial analysis capabilities of GISs distinguish them from computer-aided drafting (CAD) and automated mapping systems. GISs can not only create maps, they can also determine where land features are located relative to the air above (e.g., climate, air quality), the earth below (e.g., soil, bedrock), and other features on the land surface. This capability is essential to our understanding of the interactions among earth resources.

A GIS has numerous analysis capabilities, which can be divided into two general groups: operations performed on individual data layers and operations performed on multiple data layers (Tables III and IV). Despite the differences in data

structure between raster and vector systems, most of these operations can be done on either type of GIS.

A. Operations Performed on Individual Data Layers

1. Point Data

Field methods for studying earth resources often involve the collection of numerical data at individual points, such as water quality monitoring stations and weather stations. If the point data represent linear features (e.g., streams, forest boundaries) or a data surface (e.g., elevation, temperature), a GIS can be used to interpolate information between points, using any of a number of different interpolation algorithms (Table III).

Of the algorithms used for interpolating point data, only kriging takes into account the spatial autocorrelation of the data. It is based on the principle that the observed value of a variable at one location is significantly dependent on values at nearby locations, and it assumes that the spatial variation of any variable can be expressed as the sum of three major components: (1) a structural component, associated with a constant mean value or trend (e.g., a bedrock unit); (2) a random, spatially correlated component; and (3) a random noise or residual error term. Originally developed for ore body estimation, kriging can be applied to any spatially distributed data set. This approach differs from traditional techniques, which interpolate values as a function of the distance between points without accounting for spatial variability.

2. Point or Area Data

Attribute data for points or areas may be used in their original form, or they may be assigned to categorical or ordinal classes using various classification algorithms based on the distribution of data values (Table III). Most image analysis programs and raster-based GISs contain algorithms for assigning data to classes. Vector-based systems generally do not contain these classification algorithms, because the data are usually classified before they are entered.

Recalculation involves the mathematical conversion of data values in a GIS data layer. For example, a data layer of percent soil organic matter can be multiplied by a conversion constant to estimate percent soil carbon. Recalculation is commonly used in modeling applications of GIS. Renumbering also changes data values, but by substitution rather than mathematically. Renumbering is commonly used to combine several classes into a more general class, such as combining granite, basalt, and rhyolite into an igneous bedrock class. It is also used to assign quantitative values to ordinal and categorical data classes, as when codes representing different forest types are substituted with their corresponding pulpwood yields, and to extract specific classes from a data layer

Table III   GIS Operations Performed on Individual Data Layers

Operation	Modifiers	Description
Operations on point data		
Interpolate	B-splines	Interpolate between points to create linear features or a data surface
	Thiessen polygons	
	Trend surface	
	Fourier transform	
	Moving average	
	Kriging	
Nearest neighbor	Distance	Determine the distance to or value of the nearest nonzero point
	Value	
Search	Distance	Select points within a specified distance, range of values, or direction from the target point
	Range	
	Direction	
Operations on point or area data		
Classify	Principal components	Allocate point or cell to a class by analyzing the data with the classification algorithm specified
	Cluster analysis	
	Nearest neighbor	
	Class mean	
	Parallelepiped	
	Maximum likelihood	
	Distribution stretch	
	Level slice	
Recalculate	Add constant	Recalculate values using a weighting constant or mathematical operation
	Subtract constant	
	Multiply by constant	
	Divide by constant	
	Exponentiate	
	Trigonometric function	
	Logarithmic function	
Renumber	—	Reassign new values for old as specified
Buffer (vector systems)	Distance	Identify the area within a given distance of a point, line, or polygon boundary
Spread (raster systems)	Distance	Compute the distance between a given feature and every other cell in the data layer
	Uphill	
	Downhill	
	Resistance	
	Barriers	
Histogram	—	Calculate the frequency distribution (number of points or cells) for each data value
Window	Row, column	Extract a rectangular subset of a file
Clustering operations (raster systems)		
Clump	Distance criterion	Identify contiguous cells with the same value, and assign a unique number to all cells within the cluster (raster data)
Sieve	Number of cells	Delete clusters smaller than a minimum size
Size	—	Classify clumps by area
Report	Lines	Report the length of lines and the perimeter and area of polygons in the data layer
	Polygons	
Border	—	Draw an outline around homogeneous regions of data, classified by location relative to the core region
Edge	Class range	Examine cells adjacent to a homogeneous core region to determine if they meet specified criteria
	Direction	

(continues)

Table III (*Continued*)

Operation	Modifiers	Description
Window operations (raster systems)		
Moving window	Average Minimum Maximum Range Median Mode Majority Minority Class count Deviation Proportion	Summarize values for a square area of cells, and enter the summary value into the central cell
Neighbor analysis	Average Minimum Maximum Range Median Mode Majority Minority Class count Deviation Proportion Location	Summarize values for the eight surrounding cells, and enter the summary value into the central cell (can also be done for polygons in a vector system)
Jumping window	(see moving window)	Aggregate data by entering a summary value into *all* of the cells in a square window
Topographic operations		
Contour	—	Create an isopleth map
Slope	Steepest drop Steepest rise 4-way average 8-way average	Create a gradient map
Aspect	Steepest drop Steepest rise 4-way average 8-way average	Determine slope orientation
Illumination	Sun angle Sun direction	Calculate illumination of an elevation file based on its slope and sun position (produces a shaded relief map)
View	Direction Platform height	Determine which areas can be "seen" from a target point looking over a continuous three-dimensional surface
Routing analyses		
Stream	—	Find shortest route between two points on a continuous surface
Network	Impedance	Find shortest path along network in terms of distance, time, cost

(e.g., extracting lakes from a general land cover map). Although intuitively simple, renumbering is one of the most used operations in GISs.

Buffering operations identify the area within a given distance of a point, line, or polygon and are used to extract a spatial subset from the data layer. For example, a 100-m-wide buffer zone surrounding a beaver pond could be used to determine what vegetation lies within the beaver's foraging radius. Buffering has also been used with streams to investigate interactions between stream water quality and adjacent land uses.

In raster systems, buffering is done by a spreading process which computes the distance between the feature of interest and every other cell within a defined limit, resulting in halos of cells with incrementally larger distances from the central feature. In some raster GIS packages, additional criteria can

Table IV GIS Operations Performed on Two or More Data Layers

Operation	Modifiers	Description
Overlay	Add Subtract (2 layers) Multiply Divide (2 layers) Average Maximum Minimum Median Mode Majority Minority Diversity Deviation	Use the values from cells representing the same location on two or more data layers to compute a summary value, and assign it to the corresponding cell on a new map
Boolean operations	AND OR NOT XOR	Combine two or more data layers using Boolean logic on attributes to yield a third map
Matrix	—	Generate a data matrix comparing two data layers
Cookie cutter	—	Extract a spatial subset of one data layer using a second data layer as a template

be applied to the spread command to control the direction (uphill, downhill), resistance, and barriers to spreading. This allows the spread process to be used in surface routing, such as the overland flow of surface runoff.

Histograms provide the frequency distribution for each data value in the data layer. This may be a count of point occurrences in a vector database, or the number of cells sharing a common class in a raster database. Histograms are the means by which area summaries are generated for a raster database. This capability to reduce the data in a map to a few summary statistics is one of its greatest attractions to ecological researchers, who use the numerical output from GISs to derive mathematical relationships between ecological variables. The data from a histogram report can be exported for use in statistical and modeling programs.

3. Clustering Operations

Because data are stored on a cell-by-cell basis in raster GISs, special operations are needed to aggregate contiguous cells with the same value into individual patches. This is done with a clump procedure, which identifies and numbers cell clusters.

For example, each individual woodlot in a raster data layer could be identified and numbered. In some systems, clumping

is following by a sieving process to eliminate clusters below a minimum size. Once created, the clusters may be classified by size, and reports of area, perimeter, and length of individual features may be generated. The border option in some raster GISs creates a linear boundary around the clusters identified, which is useful in converting the data from raster to vector format.

Although clustering procedures are unnecessary in vector systems because they inherently identify individual polygons, "dissolve and merge" is a similar procedure used in vector systems to combine contiguous polygons which have been reclassified to a common value. The "dissolve" removes the boundaries separating the polygons, and the "merge" combines their contents into a single, larger polygon. This process is used when generalizing information in a data layer, such as combining adjacent corn and soybean fields into a more general "row crops" polygon.

The capability of raster GISs to analyze the cells around the edge of a polygon is of special interest to ecologists studying ecotones, zones of transition between adjacent ecological systems. The edge procedure examines cells adjacent to a homogeneous patch and determines whether they meet specified class and direction criteria. This type of analysis is important for analyzing contiguous habitats in wildlife management applications, and it can be used to determine whether certain plant communities were associated with each other in greater proportion that would be expected from their overall abundance in the landscape as a whole.

4. Windowing Operations

Windowing operations are used in raster GISs to create a new data layer by assigning a value to the central cell of an $n \times n$-dimensional window based on analysis of some or all of the cells in the window. In the moving-window technique, the window is used to scan a raster data layer (Fig. 7). A number of analysis algorithms can be used with a moving window, depending on the desired result (Table III). For example, a majority classification can be used to smooth data, and a class count can be used to determine local diversity. Moving-window analyses of satellite imagery with a range classifier can be used to analyze image texture by measuring the relative degree of difference between the reflectance values of cells in the window. This technique has been used with NDVI, an indicator of net primary productivity derived from satellite image analysis, to detect contrast between areas of high and low productivity. Homogeneous areas appear uniformly dark because of the lack of variation among NDVI values, while areas of contrasting productivity (e.g., aquatic–terrestrial boundaries) appear as bright borders on the scanned NDVI image.

Neighbor analysis is a similar scanning process, but instead of analyzing all cells in the window, it computes a summary value for the eight edge cells in a 3 × 3 cell square and assigns

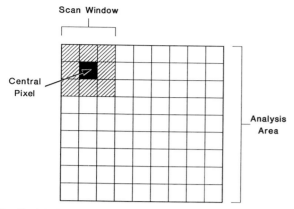

Fig. 7 Moving-window analysis. A summary value derived from the values for each of the nine pixels in the scan window is assigned to the central pixel, and the window is moved until the entire analysis area has been covered.

it to the central cell. In addition to the analysis algorithms used in the moving-window technique, neighbor analysis can be used to determine the location of the eight edge cells relative to the central cell.

The jumping window is a technique for aggregating raster files. All of the values in the original file are assigned the same value and aggregated into a single cell. This technique may be used to reduce file size, or it may be used to detect ecological patterns at a coarser grain than that of the original file. For example, the 30×30 m resolution of Landsat Thematic Mapper (TM) imagery may be suitable for detecting vegetation patterns, while a coarser grain created by aggregating TM cells would be more suitable for analyzing bedrock trends.

5. Topographic and Routing Operations

Topography is related to a number of ecological processes, such as species composition (a function of elevation and aspect), erosion (a function of slope), and insolation (a function of slope and aspect). A number of GISs have topographic operations which can be used with elevation data such as USGS DEM files to generate data layers such as contour intervals, slope, aspect, illumination, and view. These data layers are often used in combination with other GIS data (see Section V,B) to determine how edaphic characteristics control ecological processes.

Whereas most of the operations described above depict static conditions, routing operations can be used to analyze fluxes of materials and organisms across the landscape. Stream analyses can be used with topographic data to determine pathways of air or water flux across the land surface, and network analyses can be used to determine path lengths along defined corridors. For example, network analysis can

be used with a data layer of deer trails to determine the energy cost of traveling between two points along the various possible routes. The trails can be weighted using an impedance factor (e.g., making it more energy expensive to travel through a bog than over land), so that the most efficient solution may be a function of more than just path length.

B. Operations Performed on Multiple Data Layers

1. Overlay

The GIS function most commonly depicted is the overlaying of multiple data layers into a single, comprehensive map. In vector systems, the overlay process merely involves superimposing data layers, so that the linework of two or more layers is displayed simultaneously. Each cell in a raster system can receive only one classification, however, so a combination hierarchy must be specified when an overlay is performed. For example, to superimpose roads on a vegetation layer, the cells classified as roads on the first data layer would have to supersede vegetation cells in corresponding locations on the second data layer. This could be done by assigning roads a higher classification number than any of the vegetation classes and overlaying with a maximum replacement algorithm.

Overlaying can also be used to perform mathematical operations across multiple data layers. For example, the universal soil loss equation developed by the U.S. Department of Agriculture is computed as

$$A = RKLSCP$$

where A is the annual soil loss, R the rainfall erosion index, K the inherent soil erodibility, LS the combination of slope percentage and slope length, C the cover and management factor, and P the conservation practice factor. A number of studies have used GISs to compute soil loss by multiplying data layers representing each of the variables in the equation to create a data layer of annual soil loss (A).

2. Boolean Operations

Boolean operations are used to generate a new data layer from the intersection or union of input data layers. The application of the Boolean operators AND, NOT, OR, and XOR is illustrated in Table V. Boolean operations are typically used to determine where a specific combination of ecological conditions holds. For example, the statement

> If "clay bluff" AND "southwest facing",
> then "erosion = low

could be used with data layers of lakeshore geology and

Table V Truth Table for Boolean Operations[a]

A	B	NOT A	A AND B	A OR B	A XOR B
1	1	0	1	1	0
1	0	0	0	1	1
0	1	1	0	1	1
0	0	1	0	0	0

[a] A value of 1 indicates "true" and a value of 0 indicates "false."

aspect to classify the shoreline erosion hazard of southwest-facing clay bluffs. Boolean operators can also be used in combination:

If "bedrock" OR ("clay bluff" AND "southwest facing"),
then "erosion = low"

Boolean operations may be performed using any number of data layers and conditional statements.

In vector systems with a relational database, Boolean operations can also be performed on a single data layer with multiple attributes. For example, the following statement could be used with a tax parcel map:

If "zone = X" AND "use = commercial" then
"tax = commercial rate for zone X"

Boolean operations are frequently used in ecological research applications of GIS. Boolean analysis can be used with data on home ranges of an individual wildlife species to locate niche overlap, the areas where competition for resources is greatest. Boolean operations can also be used with range distribution maps for multiple plant or animal species to optimize areas for nature preserve siting. Boolean analysis can locate areas having unique combinations of habitat variables required by plants or animals. This technique is commonly used for evaluating site suitability, often in combination with the "recalculate" function for weighting habitat values. When the environmental factors which affect the spatial distribution of biological organisms are unknown or poorly understood, Boolean operations can be used to analyze them by intersecting environmental variables with organism distribution. In this way, a GIS can identify habitat preferences which may be difficult to discern using other methods.

The "matrix" operation is similar to Boolean operations in that it determines spatial coincidence (i.e., the Boolean AND operator), with the difference that it generates a data matrix that is not commutative. For example, the result of the Boolean operation "A AND B" would be the same regardless of whether A is on layer 1 and B is on layer 2, or A is on layer 2 and B is on layer 1. In a matrix analysis, however, the result would be different. This difference is important in temporal

analyses in which data layers must be analyzed sequentially. An example is the transition between meadow and forest: a forest can be converted to meadow over a period of days to months by logging or fire, but a meadow converts to forest over a period of decades to centuries as trees regrow. Because both the mode and rate of conversion differ for these two types of transitions, it is important to keep them distinct.

3. Cookie Cutter

Cookie cutter operations are used to extract a spatial subset of a data layer using a second data layer as a template. For example, the boundaries of a city can be used to extract a spatial subset from a regional land use map. In ecological applications, this capability may be used to extract information about particular ecosystems from a larger GIS or remotely sensed data layer. For example, the vegetation cover in a floodplain area could be extracted by using a map of floodplain boundaries to do a cookie cutter operation on a regional vegetation map.

VI. Temporal Analysis Using GIS

Temporal change is an integral part of community, ecosystem, and landscape functioning: pioneer plant communities are succeeded by secondary plant communities, nutrient availability is affected by litter accumulation, landscape patchiness is altered by disturbance, and ecotone locations are affected by climatic change. GISs provide a powerful tool for studying the magnitude of temporal change and its consequences for ecological systems

Both aerial photography and remote sensing are important data sources for temporal analyses using a GIS. A major advantage of remote sensing is its frequent repeat rate, which is conducive to multitemporal analyses. For example, satellite imagery has been used to analyze the spread of desert land after prolonged drought, the shrinking of the snowpack in mountain ranges, and the loss of tropical rain forest due to land clearing. Although aerial photography has a less frequent repeat rate, it provides a longer historical record for temporal analyses because of its availability since the late 1930s for most of the United States. Historical aerial photography has been used to generate GIS data for temporal analyses such as wetland loss, land use and land cover change, beaver pond establishment, and shoreline erosion.

The GIS operations used for spatial analysis are also applicable in temporal analysis. Overlay functions are used in multitemporal analyses of dynamic ecological variables to compute cumulative totals over a number of observation periods. For example, net annual primary production can be predicted by combining layers of NDVI values from throughout the growing season, and soil moisture surplus or deficit can be predicted by combining layers representing monthly

Table VI Matrix Classification for Two Data Layers[a]

	1	2	3	4
1	1	2	3	4
2	5	6	7	8
3	9	10	11	12

[a] Labels along the side are the class codes for the first data layer; labels across the top are the class codes for the second data layer. Numbers in the matrix are the codes generated by the analysis.

precipitation distribution. These cumulative totals often provide much more meaningful ecological information than any individual data layer.

The matrix operation is essential to temporal analysis of a sequence of events. In this type of analysis, a data matrix is generated in which the number of possible outcomes is equal to the product of the number of classes in the two input data layers. Table VI lists the possible class outcomes for a matrix analysis between data layers containing three and four classes, respectively. The total number of possible class permutations is 12 (3 classes in data layer A $\times$ 4 classes in data layer B). Note that the result of class 1 changing to class 2 (matrix class 2) is different from the result of class 2 changing to class 1 (matrix class 5). This type of analysis is often used to analyze transitions over time, such as vegetation changes caused by disturbance. The information generated by the matrix analysis is useful in itself, or transition probabilities computed from sequential GIS data can be exported to Markov modeling programs designed to predict future changes based on past trends.

VII. Modeling Using GIS

Although the traditional techniques of collecting ecological data at a few points and extrapolating to larger areas have worked well for individual forest stands or abandoned fields, they are less suitable in the growing fields of landscape ecology, biodiversity, and global ecological change. Spatial modeling with a GIS provides this capability.

The simplest GIS models involve the combination of two or more data layers. The layers may be combined mathematically, as illustrated by the universal soil loss equation example (see Section V,B,1) or by using Boolean logic (see Section V,B,2). In either case, the factors which determine the model outcome (i.e., soil loss rates, site suitability) must be known in advance. These relationships are developed subjectively or by empirical testing.

When the empirical relationships needed for model development are not known, a GIS can be used to derive them. GIS-derived data are exported for analysis using multivariate or spatial statistics programs to determine the strength of interaction among the variables. Statistical analysis of GIS data has been used in ecological research to analyze associations among plant communities, between organisms and their environment, between wetlands and downstream water quality, and between coastal characteristics and shoreline erosion, to name a few.

Some ecological phenomena are stochastic and cannot be predicted by environmental variables alone. Examples are fire and climatic events, which occur at unpredictable locations and times. These stochastic influences can be accounted for by incorporating Monte Carlo simulations in GIS models. A GIS can also reduce the spatial uncertainty associated with stochastic events by incorporating disturbance probabilities based on environmental characteristics, since the probability of a stochastic occurrence is not uniform at all locations. For example, some areas are more likely to be affected by fire or lightning strikes than others. A GIS can be used to evaluate landscape characteristics which ameliorate disturbance and to refine stochastic models accordingly. Predictive maps generated by the GIS can be compared with maps of actual occurrences to identify areas of disagreement due to stochastic influences.

One of the most promising areas of GIS application is the marriage of GIS capabilities and simulation modeling. The GIS is used to derive input variables which are exported to the simulation model, and the results of the model runs are returned to the GIS for display. (Fig. 8). This approach has been used with a forest simulation model to predict changes over time in forest composition and biomass as a result of global warming. The model used information about the current distribution of North American forests and predicted climate change to project changes in forest productivity across eastern North America over the next 200 years. The GIS was used to input distributed variables which drive the model and to output model results.

Whether used along or combined with a model, GISs are providing ecologists with the spatial tools needed to study ecological interactions over large areas. Continued development of GIS technology, as well as the integration of ecological science and technology, is essential to solving complex ecological problems at landscape, regional, and global scales.

Acknowledgments

This is contribution 66 of the Center for Water and the Environment and contribution 10 of the Natural Resources GIS Laboratory (NSF DIR-8805437).

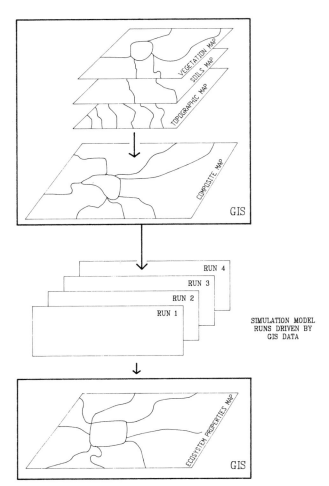

Fig. 8 Integration of a GIS with an ecosystem simulation model.

Bibliography

Burrough, P. A. (1986). "Principles of Geographical Information Systems for Land Resources Assessment." Oxford Univ. Press, Oxford, England.

Johnston, C. A., ed. (1990). "Spatial and Temporal Analysis Using Geographic Information Systems." *Landscape Ecol.* 4(1), special issue.

Mounsey, H., and Tomlinson, R. F. (1988). "Building Databases for Global Science." Taylor & Francis, London.

Peuquet, D. J., and Marble, D. F. (1990). "Geographic Information Systems: An Introductory Reader." Taylor & Francis, London.

Rhind, D., and Mounsey, H. (1990). "Understanding GIS." Taylor & Francis, London.

Ripple, W. J., ed. (1987). "GIS for Resource Management: A Compendium." American Society for Photogrammetry and Remote Sensing, Bethesda, Maryland.

Ripple, W. J., ed. (1989). "Fundamentals of Geographic Information Systems: A Compendium." American Society for Photogrammetry and Remote Sensing, Bethesda, Maryland.

Star, J., and Estes, J. (1990). "Geographic Information Systems." Prentice-Hall, Englewood Cliffs, New Jersey.

Tomlin, C. (1990). "Geographic Information Systems and Cartographic Modeling." Prentice-Hall, Englewood Cliffs, New Jersey.

Glossary

Categorical variables Variables in which the discrete classes are described by name but have no specific order (e.g., plant communities, geologic types, soil series).

Continuous variables Variables in which classes represent a continuous range of quantitative data (e.g., temperature, elevation, population density).

Data layer Digital representation of spatially distributed data having a common theme (e.g., elevation, roads, geology).

Digitize To encode spatially distributed data in digital form.

Ordinal variables Variables in which classes are discrete and have an inherent order (e.g., stream orders, forest size classes).

Overlay Process of combining two or more data layers which have the same spatial extent.

Polygon Multisided, closed figure representing an area.

Raster Data structure which stores information as a regular grid of cells or pixels.

Topology Convectivity or contiguity of geographic features.

Vector Data structure which stores information as points, lines, and polygons.

Glacial Lakes

J. A. Elson
McGill University

A glacial lake is one that receives much water and sediment from glaciers; commonly, part of its margin lakes is glacier ice. Large glacial lakes formed along the margins of retreating Pleistocene ice sheets. Much offshore sediment in these lakes is deposited in "annual" layers (varves). Glacial lakes discharged through very large channels (spillways). Catastrophic flood discharges occurred when retreating ice margins exposed lower outlets, suddenly dropping lake levels. The strandlines of glacial lakes are warped upward toward centers of ice sheets by isostatic rebound of the earth's crust. The distribution of glacial lakes and the evolution of two lake systems are summarized.

I. Introduction

A glacial lake is near or borders a glacier and receives much or most of its water and sediment from the ice. Ice margins commonly impound meltwater against higher land, forming one side of a basin. Pleistocene glacial lakes were as large as 350,000 km². Most were short-lived, lasting tens to hundreds of years, a few more than 5000 years.

Glacial lakes form an ephemeral environment between glaciers and unglaciated land. The thermal inertia of large lakes tempered local climate and prevented the earth beneath from freezing. Glaciolacustrine sediments are mechanically weak and only feebly resist the flow of glaciers compared with the resistance of rock. These conditions contribute to the surging of sectors of the ice sheets into glacial lakes, thus expanding ablation zones and accelerating ice wastage. The surfaces of large lakes are unfrozen most or all of the winter because turbulence generated by wind mixes cold surface water with warmer water below. The open water ameliorates the local climate in winter and may increase snowfall to the leeward. In summer, cold glacial lake water contributes little moisture to the warmer atmosphere. [*See* CLIMATE – ICE INTERACTIONS.]

Glacial outwash normally exported from glacier margins on land by rivers is trapped in glacial lakes; rock flour is deposited as varved clay. Nonglacial lake sediments may be deposited remote from the ice margin in large glacial lakes. Glacial lakes are almost devoid of life near ice margins, but distal parts often contain biota.

II. Morphology and Distribution

A. Shape and Orientation

The distribution of glaciolacustrine sediment depends on the shape of a glacial lake and the length of the glacier margin that it drains. An elongate lake parallel to the ice margin (Fig. 1a) has thick, relatively coarse-grained sediments. In a lake that extends away from the ice margin (Fig. 1b), the sediments grade from coarse and thick near the ice to thin and fine-grained farther away, and nonglacial sediments may occur in the remote parts. Glacial lakes in mountain valleys are deep and usually have thick deposits. Some glaciers and ice sheets eroded deep troughs that initially contained glacial lakes, which became nonglacial when the ice melted. Outlet and valley glaciers extending onto piedmont lowlands erode deep troughs and deposit arcuate moraines that give additional depth.

A few glacial lakes resulted mainly from the residual isostatic depression of the crust after the retreat of an ice sheet. They received much glacial meltwater and sediment, but their basins were not primarily due to damming by glacier ice. Glacial Lake Hitchcock (Fig. 1d) is this type. [*See* GLACIOTECTONIC STRUCTURES AND LANDFORMS.]

Portions of glacial lakes in hilly areas are archipelagos with

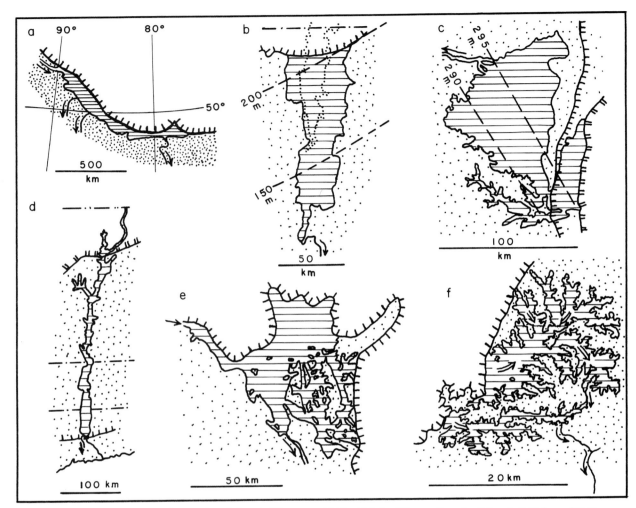

Fig. 1   Maps of type glacial lakes, with scale bars in kilometers. (a) Glacial Lake Ojibway, Ontario. [After A. S. Dyke and V. K. Prest (1987). "Paleogeography of Northern North America, 18,000–5,000 Years Ago." Geological Survey of Canada Map 1703A, with permission of the Geological Survey of Canada.] (b) Glacial Lake Fort Ann, Vermont; dashed lines are contours (isobases) on the warped water plane. [After D. H. Chapman (1937). Late-glacial and postglacial history of the Champlain Valley. *Am. J. Sci.* **34**(8), 98, 114, with permission.] (c) Glacial Lake Wisconsin with isobases of the Elderton Phase. [After L. Clayton and J. W. Attig (1990). "Glacial Lake Wisconsin." Geological Society of America Memoir 173, with permission.] (d) Glacial Lake Hitchcock, New England. [After J. C. Ridge and F. D. Larsen (1990). Re-evaluation of Antev's New England varve chronology and new radiocarbon dates of sediments from Glacial Lake Hitchcock. *Geol. Soc. Am. Bull.* **102**(7) 889, with permission.] (e) Early phase of Glacial Lake Barlow, Ontario. [After J. Veillette (1988). Déglaciation et évolution des lacs proglaciaires Post-Algonquin et Barlow au Temiscamingue. Québec et Ontario. *Géographie Physique et Quaternaire* **42**(1), 7–22, with permission of Les Presses de l'Universite de Montreal.] (f) Glacial Lakes Edmond (north) and Watts (south), Pennsylvania. [After V. C. Shepps *et al.* (1959). "Glacial Geology of Northwestern Pennsylvania." Pennsylvania Geological Survey, 4th series, Bulletin G32, Fig. 11, with permission of the Pennsylvania Geological Survey.]

open basins, as in Fig. 1e. Wave action is mitigated by islands, and strandlines are weakly developed.

During glacial maxima, valleys in unglaciated borderlands were dammed, forming lakes with dendritic outlines (Fig. 1f). These *extraglacial lakes* overflowed through cols into adjacent drainage basins or drained along the ice margin.

B. Distribution of Glacial Lakes

1. Introduction

Extinct glacial lakes are represented by sediments, relict strandlines, and abandoned outlet channels (spillways). The most diagnostic sediments are varved clays, but silt and sand occur near shore and in shallow basins. Deltas were deposited at the mouths of tributary streams, and underflow fans where spillways debouched. Lake sediments (Fig. 2) may not define the extent of a glacial lake at a given time. Many lakes were smaller than the area of sediments attributed to them, as the lakes migrated downslope following the retreating ice margin, abandoning the higher parts of their basins, like Glacial Lake Saskatchewan in Fig. 3. The sediments of Glacial Lake Agassiz occur over 950,000 km², but the area of water never exceeded 350,000 km².

North American glacial lakes usually are studied as complete systems. European research focuses on varved clays, which were used for geochronology for about 60 years before 1950. In Europe glaciolacustrine deposits are extensive only in the Baltic Sea region.

In North America glacial lakes are associated with the Laurentide ice sheet (much of which is shown in Fig. 2), the Cordilleran ice sheet in the western mountains, and several isolated ice caps. Northern Europe and parts of northern Asia were covered by similar ice sheets. The glacial lakes of these ice sheets are discussed below.

2. Glacial Lakes Older Than the Last Glacial Maximum

Varved clays beneath younger glacial sediments are evidence of older glacial lakes; the majority are of mid-Wisconsinan age. Events such as the opening and closing of Hudson Bay and ice-margin oscillations in the corridor between the Cordilleran and Laurentide ice sheets are deduced from them. For example, Glacial Lake Deschaillons in the St. Lawrence valley, southern Quebec, and a predecessor are represented by varved clays separated by nonglacial sediments. The succession shows that the St. Lawrence valley was deglaciated, drained, and then covered by another glacial readvance. Glacial Lake Gayhurst, in southern Quebec on the northwest slope of the Appalachian Mountains, was at least 90 km long and 50 km wide. A high phase (430 m) discharged south across Maine, and a later lower phase (370 m) discharged east into New Brunswick.

3. Zones of Glacial Lakes in Glaciated Regions

To facilitate discussion glacial lakes here are grouped into arbitrary zones according to the relationships of ice margins, drainage divides, and regional slopes (Fig. 2). Zone I is outside the glacial limit and slopes toward the ice. Zone II is within the area glaciated but slopes away from the ice margin. Zone III includes elongate drainage basins parallel or oblique to the ice margin. Zone IV slopes inward toward the ice sheet.

4. Glacial Lakes of the Laurentide Ice Sheets

a. Zone I, Inward Slopes Outside the Area Glaciated

Extraglacial lakes with dendritic outlines were ponded by the Laurentide ice sheet in Montana and North Dakota, where lakes 40 to 80 km wide extended 70 to 160 km up the valleys. They discharged eastward along the ice margin. Repeated glaciations resulted in deepening of the ice-margin channels and diverted parts of the rivers from their northeastward courses southeast to become the Missouri River. Extraglacial lakes also formed in Ohio, Kentucky, and western Pennsylvania (Fig. 1f). Some of these discharged into adjacent valleys through cols at their headwaters. Repeated glaciations deepened the col spillways, and they became part of the Ohio River system.

b. Zone II, Outward Slopes Inside the Glacial Limit

Glacial lakes formed on outward slopes in the following situations: (1) Ice tongues flowed around uplands and dammed valley systems at both ends. Lakes of this origin are uncommon. (2) Deposition of end moraines created lake basins between successive moraine ridges. Lakes of this type occur south of the middle Great Lakes. Lake Douglas in Illinois was 25 km long and 15 km wide. (3) Glacial outwash partly filled trunk valleys, ponding the drainage of tributaries. Most of these *slackwater lakes* were in Indiana and Ohio and had dendritic outlines. Lake Saline, in southern Illinois, was about 70 km across. (4) Preglacial river valleys draining highlands were deepened into troughs by outlet glaciers. Those formed by the Laurentide ice sheet are mostly on the submerged continental shelf off Labrador. (5) Isostatic depression temporarily reversed the regional slope. The slope of the Connecticut Valley in New England was reversed so that the north end was about 260 m lower than the south end, and it was occupied by Glacial Lake Hitchcock (Fig. 1d) for about 4100 years. The lake drained as a result of crustal rebound and the failure of a drift dam across the Connecticut River, which had formed about 20 m of the basin closure.

c. Zone III, Lakes in Drainage Basins Parallel or Oblique to the Ice Margin

Glacial lakes occupied these basins when the downstream parts were obstructed by glacier ice. The Glacial Great Lakes, described in Section VII,B, occupied the upper part of the St. Lawrence River drainage basin.

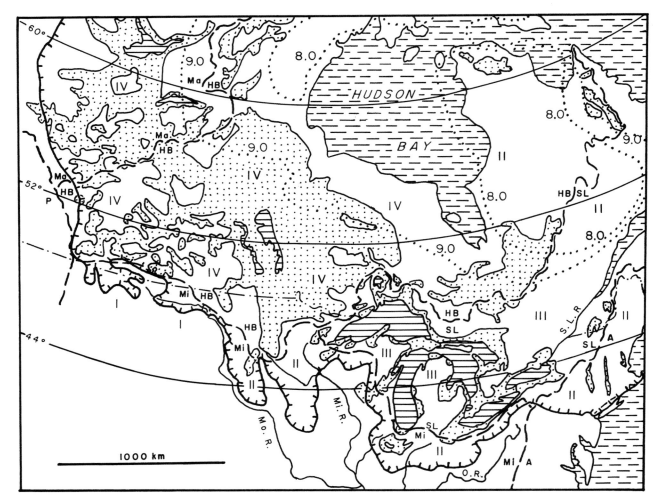

Fig. 2 Map showing glacial lake deposits (stippled) of the southern part of the Laurentide Ice Sheet. Hachured line shows maximum extent of late Wisconsin ice sheet. Dashed lines are drainage divides: Mi, Mississippi; A, Atlantic; SL, Saint Lawrence; HB, Hudson Bay; Ma, Mackenzie; P, Pacific. Rivers: Mo.R., Missouri; Mi.R., Mississippi; O.R., Ohio. Dotted lines are ice margins 9000 and 8000 years ago, marked 9.0 and 8.0, respectively. Roman numerals identify glacial lake zones defined in Section II,B. [After Teller (1987), with permission of the Geological Society of America. Some details from A. S. Dyke and V. K. Prest (1987). "Paleogeography of Northern North America, 18,000–5,000 Years Ago." Geological Survey of Canada Map 1703A, with permission of the Geological Survey of Canada. Glacial lake zones are by the author.]

d. Zone IV, Lakes on Inward-Facing Slopes

Glacial lakes of Zone IV are in the Hudson Bay drainage basin, shown in Fig. 2. More than two-thirds of the area between Hudson Bay and the drainage divide was covered by glacial lakes. Two of these are described here, Lake Saskatchewan in Section II,B,1 and Lake Agassiz in Section VII,A. By 8000 years ago only two residual ice caps remained, one on each side of Hudson Bay. The western ice cap dammed valleys draining toward Hudson Bay, forming glacial lakes to the northwest. The eastern ice cap lay on the drainage divide

in northern Quebec, and the land around it slopes outward (Zone II). Glacial lakes did not form except on inward slopes adjacent to the Torngat Mountains in the east.

5. Glacial Lakes of the Cordilleran Ice Sheet

The terrain at the northern end of the Cordilleran ice sheet drains outward (Zone II). Areas of silt in Alaska, once interpreted as former lakes, now are attributed to other origins. Glacial lakes did occupy the Copper River basin in Alaska and the Old Crow flats of the Porcupine River in northern Yukon

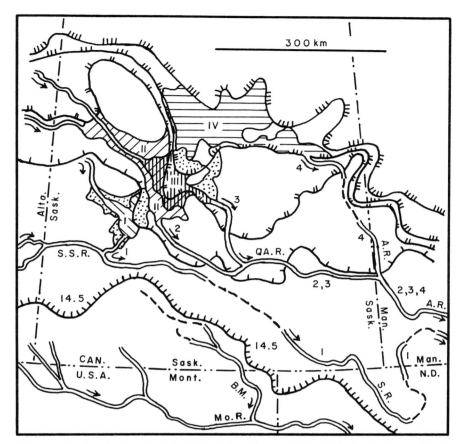

Fig. 3 Map showing deposits (stippled) and four phases of Glacial Lake Saskatchewan (lined, Roman numerals) with the outlet spillways (double lines, numbered in sequence). Hachured lines show ice-sheet margins, the southernmost dated about 14,500 years ago. Valley (spillway) names: Mo.R., Missouri River; B.M., Big Muddy; S.R., Souris River; S.S.R., South Saskatchewan River; QA.R. Qu'Appelle River; A.R., Assiniboine River. Dashed arrows are discharge connections through other glacial lakes, the boundaries of which are omitted for clarity (cf. Fig. 2). [Compiled from E. A. Christiansen (1979). The Wisconsin deglaciation of southern Saskatchewan and adjacent areas. *Can. J. Earth Sci.* **16** (4), 913–938, with permission of the Director, Research Journals National Research Council Canada. Minor detail from Teller (1987) and from A. S. Dyke and V. K. Prest (1987). "Paleogeography of Northern North America, 18,000–5,000 Years Ago." Geological Survey of Canada Map 1703A, with permission of the Geological Survey of Canada.]

Territory at least twice. Small lakes east of the northern Cordillera were dammed by Laurentide ice.

The central part of the Cordillera is a plateau, and narrow glacial lakes existed in the Stikine and Skeena valleys in the west and in the Rocky Mountain trench in the east. In the foothills east of the Rocky Mountains, Glacial Lake Peace extended up the Peace River valley into the eastern Cordillera, dammed by Laurentide ice. In central British Columbia broader glacial lakes occupied basins around Prince George and discharged north into the Peace River. Later, ice recession opened the Fraser valley, and the lakes were drained to

the south. In the southern part of the plateau a reentrant in the retreating margin of the southern Cordilleran ice sheet separated it into highland ice caps on the Coast Range in the west and the Rocky Mountains in the east. In the deglaciated plateau between, the Thompson River valley was occupied by narrow glacial lakes, which discharged southward.

The south end of the Cordilleran Ice sheet was divided into two sections by the Cascade Range. West of the Cascades a piedmont lobe spread into the Puget Sound lowland, damming valleys on the west flank of the Cascades and forming nine small extraglacial lakes. When the ice retreated, larger

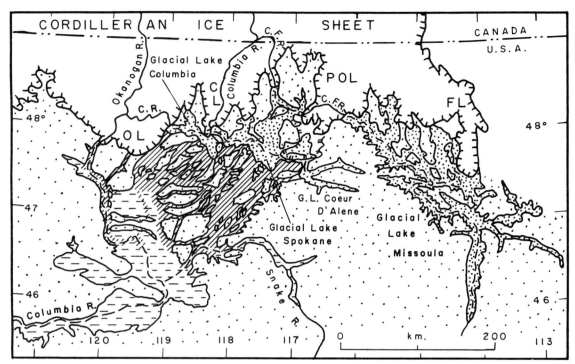

Fig. 4 Sketch map showing glacial lakes at the southern margin of the Cordilleran Ice Sheet during the last glaciation. Glacial lakes stippled; area eroded by catastrophic floods (Scablands) obliquely shaded; area occupied briefly by floodwaters shown by short horizontal dashes; unglaciated and unflooded area dotted. Hachured line shows glacier margin of Pinedale age; line with double hachures north of Glacial Lake Coeur D'Alene is an earlier ice margin. Lobes of the Cordilleran Ice Sheet: OL, Okanagan Lobe; CL, Columbia River Lobe; POL, Pend Orielle Lobe (also called Purcell Trench Lobe); FL, Flathead Lobe. [Sketched from R. B. Waitt, Jr. (1985) Case for periodic, colossal jökulhlaups from Pleistocene Glacial Lake Missoula. *Geol. Soc. Am. Bull.* **96**(10) 1271, Fig. 2. Some details from G. M. Richmond *et al.* (1965). The Cordilleran ice sheet of the northern Rocky Mountains, and related Quaternary history of the Columbia Plateau. *In* "The Quaternary of the United States" (H. E. Wright, Jr. and D. C. Frey, eds.), pp. 232–233. Princeton Univ. Press, Princeton, New Jersey.]

Zone IV lakes were formed in the lowlands. They first drained south to the Chehalis River and then north to the Strait of Juan de Fuca.

East of the Cascade range the Cordilleran ice sheet terminated in several large lobes, occupying major valleys (Fig. 4). From west to east these were the Okanagan, Columbia River, Purcell trench, and the Flathead lobes. All at various times dammed the west-flowing Clark Fork and Columbia rivers and created at least seven deep Zone III glacial lakes. Glacial Lake Missoula in the east was the largest, about 670 m deep. Forty series of varves alternating with coarse flood deposits indicate major fluctuations of its level. It drained repeatedly under the Pend Orielle sublobe that dammed Clark Fork River, each time causing cataclysmic floods that swept southwest across the Columbia Plateau and eroded the Channeled Scablands. Glacial Lake Columbia I, to the west, was ponded in the Columbia Valley behind the Okanagan lobe and was

455 m deep. Its water spilled across the plateau and eroded the upper scabland of Grand Coulee. After several intermediate lakes, Glacial Lake Columbia II formed at a level 215 m below Lake Columbia I and discharged through Grand Coulee. [*See* ICE AGE DYNAMICS.]

The flood discharges (jökulhlaups) of Glacial Lake Missoula affected all the ice lobes and lakes downstream. Estimates of the discharges range from 10.9 to 21.3×10^6 m³/s. There was nearly 2200 km³ of water in Glacial Lake Missoula, and its level fell as much as 500 m in a few days during each outburst.

6. Europe and Asia

a. British Isles

Highlands in the British Isles supported independent ice caps, which merged into piedmont-like glaciers and covered lower areas (Fig. 5). The southern limit was from the Thames

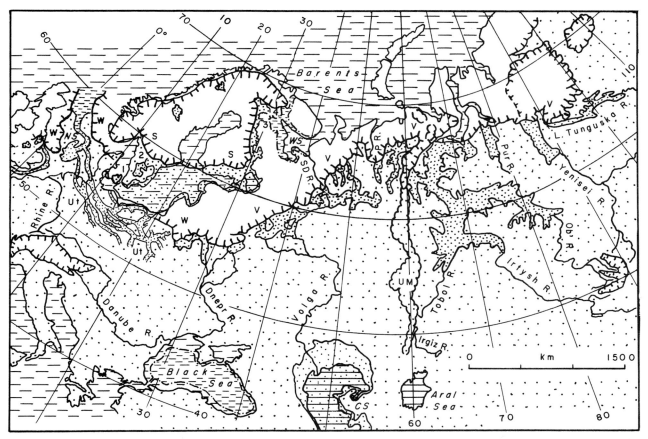

Fig. 5 Sketch map showing major glacial lake basins in northern Europe and northwestern Asia during the maximum extent of the last ice sheets and one later phase of the Scandinavian Ice Sheet. Hachured lines are ice sheet margins: W, Weichsel (in Europe); V, Valdai (in European Russia and Asia); S, Salpausselka (a recessional phase in Scandinavia). The age of the ice boundary between the 'Ob and Yenisei Rivers is uncertain. Glacial lakes and nonglacial lakes of the Caspian Sea and North Sea basins, and Urstromtäler are stippled. Present sea limits shown by long dashes, omitted from the North Sea basin. Quaternary seas in the North Sea, White Sea, and Black Sea basins shown by short close-spaced horizontal dashes. Unglaciated area dotted. NS, North Sea basin; Ut, Urstromtäler; WS, White Sea; SD R., Sev Dvina River; P R., Pechoza River; CS, Caspian Sea. Numbers 1 to 4 around the Baltic Ice Lake indicate successive outlets. The lakes east of the Urals may have been even more extensive than shown when they discharged via the Tobol River to the Aral–Caspian basin. Drainage in the North Sea basin is largely hypothetical. [Compiled and simplified from several sources: U.S.S.R. lakes and ice margins are from A. A. Velichko, ed. (1984). "Late Quaternary Environments of the Soviet Union" (H. E. Wright, Jr. and C. W. Barnowsky, English-language edition eds.). Univ. of Minnesota Press, Minneapolis, Fig. 5-2, by permission of the University of Minnesota; Baltic area after J. J. Donner (1965). The Quaternary of Finland. *In* "The Quaternary" (K. Rankama, ed.), Vol. 1, p. 238. Wiley (Interscience), New York; North sea area adapted from H. Reinhard (1974). Genese des Nordseeraumes im Quartär *Fennia* **129**, 81, Abb.7. Urstromtäler adapted from R. F. Flint (1957). "Glacial and Pleistocene Geology." Wiley, New York, Plate 4.]

Estuary to the Bristol Channel. During deglaciation the ice retreated to the highlands as the snowline rose. A reentrant in the ice margin in central England migrated northward, separating ice of western provenance from eastern ice. Glacial lakes occupied major valleys. Glacial Lake Harrison east of Birmingham was 70 km long by 50 km wide. Near the North Sea coast Glacial Lake Humber, 90 km by 40 km, was south

of York, and to its south and east was the Ouse glacial lake, of similar order. Many small glacial lakes formed as the lowland ice retreated from the dissected slopes of the lower uplands. Outlet glaciers from the highland ice caps disappeared last and left many trough lakes.

Three British glacial lakes are of special interest. The valley of Glen Roy in central Scotland has two strandline terraces ("roads") carved into its sides leading to different cols at the headwaters. In the 1840s glacier dams and shore erosion were agreed to be the only mechanisms that could explain these benches, which required acceptance of the new glacial theory. At the turn of the century P. F. Kendall mapped Glacial Lakes Eskdale and Pickering in the Cleveland Hills in northeast England; he was the first to describe ice marginal drainage and glacial spillways.

b. Northern Europe and Asia

The Scandinavian ice sheet resembled the Laurentide ice sheet, though smaller (Fig. 5). At its maximum extent, meltwater drained into the Black Sea. Glacial lakes formed in the upper Volga valley and in other valleys close to the Baltic–Black Sea divide. Farther east Zone I north-flowing rivers were dammed, and large extra-glacial lakes occupied the valleys of the Sev Dvina and Pechoza rivers. As the ice retreated lakes formed in Estonia, adjacent Russia, and Latvia. Along the southern border of the ice sheet in Poland and Germany, meltwater flowed west through a succession of broad valleys called *urstromtäler*, roughly parallel to the margin. Many were occupied by lakes when they were abandoned for lower channels. They discharged into the Elbe River through glacial lake in the southern North Sea basin, and through the English Channel river to the Atlantic Ocean.

The Baltic Sea basin was occupied by the Baltic Ice Lake, probably the largest glacial lake known. It discharged into the early North Sea at first, but later an outlet to the north on the east side of the ice sheet drained the lake into the White Sea.

As the Scandinavian glacier continued to shrink, in Finland and Sweden many small glacial lakes were impounded by drift dams, especially around the Salpausselka moraine system. Trough lakes were formed in the highlands of Norway and Sweden.

An ice sheet extended across northeastern Europe and Asia as far as Longitude 115° East, obstructing the north flowing drainage (Fig. 5). Extraglacial lakes up to 150 km wide and 1400 to 1800 km long extended up the valleys of the Ob, Pur Yenisei, and Lower Tunguska and discharged southwest into the Aral–Caspian Sea.

7. Highland Ice Caps

Small ice sheets and isolated highland ice caps accumulated on the Alps, southern Andes, Sierra Nevada, Middle and Southern Rocky Mountains, and many mountain systems in central and northeastern Asia. The land around them drains outward (Zone II), and glacial lakes were limited to trough lakes, outlet glacier troughs with end moraines, and valleys dammed by tributary glaciers. Large outlet glaciers of the southern Andes made basins now occupied by lakes typically 20 km long and 5–10 km wide. Four glacial lakes in Tierra del Fuego were about 100 km long and 30 km wide.

III. Outlets of Glacial Lakes

Most glacial lakes discharge through *spillways*. These large channels are troughs 0.5 to 5 km wide and 10 to 150 m deep, now occupied by underfit streams and chains of lakes, and a few by major rivers. The entrance of the spillway usually has a bedrock sill or a resistant lag of boulders and coarse gravel. Spillways do not increase in size downstream like river valleys. Some debouche into other lake basins, as shown in Fig. 3.

Functioning spillways contained water from bank to bank as deep as 12 m during flood flows. Many had equilibrium discharges equal to the water entering the basin from runoff and glacier melt. Others carried only outburst flood discharges and may represent single events.

Estimates of spillway discharges are made from the hydrologic geometry of the channels and the size of sediments in them. A. E. Kehew estimated discharge in the Souris spillway (Fig. 3, no. 1) to have been about 100,000 m³/s, which would have drained its source, Glacial Lake Regina, in five days. The segment of this spillway system crossing from North Dakota into Manitoba carried 250,000 m³/s and could have drained Glacial Lake Souris in nine days. C. L. Matsch estimated Glacial Lake Agassiz discharges through the Minnesota valley as great as 130,000 m³/s. J. T. Teller and L. H. Thorleifson calculated flow through an eastern outlet of Lake Agassiz into Lake Superior to have been 100,000 to 200,000 m³/s during floods, with equilibrium flows of 15,000 to 30,000 m³/s. The discharges of Glacial Lake Missoula and other lakes along the south margin of the Cordilleran ice sheet approached 20×10^6 m³/s (Section II,B,5). [*See* FLOODS.]

IV. Processes in Glacial and Glacier-Fed Lakes

A. Hydrology and Physical Limnology

The nature of the sediments in glacial lakes depends on the temperature and circulation of the water. Lake Malaspina, a glacier-margin lake in Alaska, and glacier-fed lakes in Canada, Norway, and Switzerland have yielded similar information.

Meltwater is produced in seasonal pulses. An initial pulse from melting snow is followed by a general increase to an ice-melt peak in late midsummer; then discharge declines to a trickle in the fall. Smaller peaks due to warm weather or rain are superimposed on the general curve. Lake levels are subdued reflections of the meltwater peaks, with delays depending on the size of the drainage basin and capacity of the lake. Meltwater transports most of the sediment into lakes. The levels of stable glacier-fed lakes fluctuate 1.0 to 2.0 m during a season.

The water of Malaspina Lake is 4–8°C at the surface, declining to 0.5–1.0°C below a depth of about 20 m. Suspended sediment increases from 100–200 mg/liter at the surface to 500–700 mg/liter at 40 m. In the lowest 5 to 20 m of the lake near inlets and outlets the water moves at a velocity of about 10 cm/s (underflows), and near the margins similar velocities occur at intermediate levels (interflows). In other lakes underflow velocities as high as 40 cm/s have been recorded as short pulses. Underflows transport suspended sediment throughout lake basins. Pleistocene glacial lakes were similar to Malaspina Lake, but being much larger they averaged out many of the variations seen in smaller glacial lakes.

B. Flood Discharges (Jökulhlaups)

Flood outbursts (jökulhlaups) are a dynamic interaction of an ice tongue with water almost as deep as the glacier is thick. Flood discharge through an ice dam has been observed at Summit lake in British Columbia. The lake is in a valley dammed by Salmon Glacier, 13 km above its terminus. Summit Lake is 5 km long and 200 m deep at the ice face. In 1961 the lake emptied through a tunnel through the base of the glacier in three days. By 1965 the lake was filled and again emptied suddenly. This is an analog of the flood outbursts of lakes such as Glacial Lake Missoula.

C. Suspended Sediment

The suspended sediment in glacial and glacier-fed lakes settles out as couplets (varves) comprising coarse, light-colored *summer* layers and fine-grained dark *winter* layers. Varves have long been known from Pleistocene sediments and generally have been interpreted as annual layers. Similar couplets occur in existing glacier-fed lakes, and annual deposition has been confirmed by ^{137}Cs dating or by other methods. A coarse layer of sand or silt is deposited during passage of a density flow, which may continue all summer. Clay-size sediment settles from quiet water after the turbulent flow has ended. However, meltwater peaks during the summer may cause a succession of density flows so that as many as five *pseudovarves,* easily misinterpreted as annual, may be deposited. In at least one lake clay from interflows is incorporated into fecal pellets by copepods (small invertebrate animals) and settles out as rapidly as fine sand. Modern researchers are therefore cautious about using varves as annual layers.

V. Sediments of Former Glacial Lakes

A. Introduction

Some englacial debris enters glacial lakes by mud flows or simply falling from ice as it melts. Much more sediment enters from submerged meltwater conduits. These en- and sub-glacial conduits collect debris from drainage areas up to 8 km wide on both sides, and they extend much farther than that into or under the glacier. Conduits are pipe-full channels hydraulically linked to the surface water of the glacier. They flow at high velocities (upper regime) and discharge water and sediment into the lake as a jet. Velocities decrease away from the ice as the jet spreads and coarse material is deposited at the ice front and, with diminishing discharge near the end of the melt season, within the terminal part of the conduit. The lake sediments are finer and thinner with increasing distance from the source. [*See* PALEOLIMNOLOGY; TURBIDITY CURRENTS.]

B. Laminated and Thinly Layered Sediments

The following lake sediments occur in a sequence progressively farther from the ice margin. The same sequence, or part of it, representing increasing distance from the retreating ice margin, may be found from bottom to top at one location.

1. Stony lake clays: Layers 10 to 20 cm thick of muddy silt and sand containing glacially marked pebbles separated by layers of clay about 1 cm thick may occur at the ice margin or the base of a sequence. They are not extensive. They were once thought to be deposited beneath floating ice shelves, but they probably are distal parts of flow tills (Section VC1).
2. Varved sediments, varved clays: Varves of existing lakes are described in Section IV,C. Internal variations in the varves of large glacial lakes tend to fade out with increasing distance from the source. The "clay" in varved clays is clay-size (−4 μm) rock flour; clay minerals occur only where they were derived from older rocks or soils.

Proximal varves are deposited close to the ice margin. They typically are sand layers 10–40 cm thick overlain by clay layers 2–4 cm thick. Farther from the source the two layers are thinner, of nearly equal thickness, and sand is replaced by silt. Beyond this are distal varves 0.2 to 2.0 cm thick in which the clay is thicker than the silt portion. Single thick sandy varves within a sequence of normal varves are *drainage varves*. They represent some change in drainage, introducing a new source of sediment. Changes of sources or lake levels may enable the recognition of distinctive series of varves.

The lowest varves of a series may contain angular clasts ripped up from underlying silt or clay by vigorous underflows. Dropstones from floating glacier ice are common in some varves. Crawling traces (lebenspuren) of invertebrates are found in the silty partings of some varves as close to the ice as 1.0 km.

Varved clay sequences were used for geochronology for the first half of the 1900s, promulgated by G. DeGeer. Varve thickness was plotted against the number of the varve within a sequence, giving graphs similar to tree-ring charts. Several graphs from a basin were averaged to obtain a "normal" curve, which overlapped similar curves from other basins until a complete record since deglaciation was achieved. With the advent of radiocarbon dating in 1950, varve chronology was de-emphasized. DeGeer's chronology for the past 12,000 years recently has been shown to be generally correct. Sophisticated chronological methods now applied to varves include thermoluminescence dating and the measurement of paleomagnetic properties, and interest in them is renewed [See STRATIGRAPHY — DATING.]

C. Other Glaciolacustrine Deposits

1. Subaquatic Flow Tills

Subaquatic flow tills are superimposed layers of silt till 1–10 cm thick with thin lenses or beds of silt or sand in units up to a meter thick. Several units may form masses as thick as 10 m. They are probably deposited by gravity flows from the surface of debris-rich glacier ice standing in a lake. They have been interpreted as tills — and they cause some controversy.

2. Subaquatic Fans and Eskers

Most of the debris carried by continental glaciers is in the lowest few meters of the ice. Most meltwater conduits that transport it to the ice margin debouche at or near the base of the glacier. Sand and silt spread out in fans that grade distally into varved clays. Gravel and sand in the terminal part of the conduit form an esker that grows headward as the ice margin recedes. Gravel and sand may accumulate in a mound at a stable ice margin as well as fill the conduit behind it. The resulting succession of alternating mounds and ridges is a *beaded* esker.

3. De Geer Moraines

De Geer moraines are subparallel, discontinuous, stony sandy till ridges 1–10 m high, 5–40 m wide, spaced 100–300 m apart, in straight or gently lobate patterns convex down-glacier. They occur in glacial lakes and areas of marine submergence. G. DeGeer deduced from associated varved clays that the moraines were deposited annually. Similar ridges in the basin of Generator Lake, Baffin Island form when summer melt raises the lake level, floats the glacier margin, and causes a tension crack at the base of the ice in which till is deposited. Annual formation is likely, but cracks also can form during the runoff peaks within a melt season.

D. Ice-Scour Marks

Curved and straight grooves, many having low lateral ridges, mark the floors of some glacial lakes. Some are many kilometers long, and intersecting patterns are common. Similar patterns occur in areas emerged from the sea. Underwater observations confirm that the marks are made by the keels of drifting icebergs and pressure ridges in sea or lake ice. [See ICEBERGS; SEA ICE.]

E. Deltas and Underflow Fans

Small deltas deposited by nonglacial streams occur on the shores of glacial lakes. They are homopycnal *Gilbert* deltas with distinct topset and foreset beds. Many were incised by the stream when lake level fell, and secondary deltas were deposited at lower levels. These deltas may form quickly, and show former water levels where strandlines are not otherwise apparent.

Where a spillway disgorges into a glacial lake, a large hyperpycnal delta or *underflow fan* is deposited. Some consist mostly of detritus eroded from the spillway. These fans were deposited by sediment-laden underflows. Distal parts are unstructured clayey silt; proximal beds are silt and sand, and an apex of gravel is deposited subaerially. The Assiniboine underflow fan in Glacial Lake Agassiz has an area of about 8000 km^2 and is as thick as 80–90 m. Some underflow fans were deposited by single outburst floods.

F. Fossils

Glacial lakes generally lack life forms near the ice margin. Remote from the ice the water is warmer and contains more nutrients. Fossils of fauna, chiefly mollusks and ostracodes, and flora including peat, wood, and other macro- and micro-fossils, occur in former lagoons behind beaches and in alluviated valleys. A few robust fossils, mostly mollusks and large vertebrate remains, are found in beach deposits. Fossils show that Glacial Lake Agassiz was fully boreal 150–300 km from the ice margin.

VI. Strandlines and Crustal Warping

A. Introduction

In 1884 W. Upham observed that the beaches of Glacial Lake Agassiz rose northward. Surveys of relict strandlines in the Great Lakes and Baltic Sea basins soon followed, and isostatic response of the earth's crust to the growth and disappearance of ice sheets became accepted theory.

B. Strandlines

The development of shore features depends on the fetch and frequency of waves, the slope of the coast, the duration of a particular water level, the resistance of the rocks or soils to erosion, and the climate. Waves are most effective on moderate slopes. In cold climates freeze–thaw weathering aggressively attacks rocks in the wave zone. [*See* EROSION AND WEATHERING.]

Glacial lake strandline forms include

1. Wave-eroded scarps, usually in unconsolidated sediments. Most glacial lakes levels do not last long enough to erode much rock. Scarps up to 10 m high are common in glacial till and areas of silt and clay. Scarps may have a beach at the foot or an eroded berm (terrace) with scattered boulders on it.
2. Beaches. Ridges of sand and gravel are the most abundant form. They are from 1.0 to 8.0 m high, 20 to 100 m wide, and a few hundred meters to hundreds of kilometers long. Small beaches have poorly sorted, angular sand and gravel; but in large ones the sediments are well sorted and the stones usually well rounded, depending on the duration of the water level.
3. Small Gilbert deltas (Section V,E). They are from 10 m to more than 2.0 km across.
4. Terraces. They may be cut in rock by frost action combined with minor wave action. Such benches ("roads" of Glen Roy, Scotland) record water levels even in narrow valleys. Other terraces are formed by waves winnowing away the fines in drift [see (1) above]. Water levels interpreted from strandlines have inherent errors of at least 1.0 to 5.0 m.

C. Crustal Warping

Glacial lake strandlines demonstrate the isostatic uplift of the earth's crust (Figs. 1b and 1c). Studies show postglacial deformation of 1200 m over a distance of 1100 km, southwest of Hudson Bay. The southwest tilt averages 1.1 m/km, and occurred in about 12,000 years.

The oldest, highest water planes that formed when the ice sheet was thickest have slopes increasing from 0.15 m/km distally to 1.1 m/km close to the former ice margin. Younger, lower strandlines formed when the ice sheet was thinner, slope 0.013 to 0.05 m/km; they have straight profiles. The altitudes of points on successive water planes of known ages are used to construct time–altitude curves giving rates of uplift. Uplift rates are 2.5 to 4.0 m/century for up to 4000 years after the oldest strandlines are abandoned. The rates decline to less than 0.3 m/century within several centuries

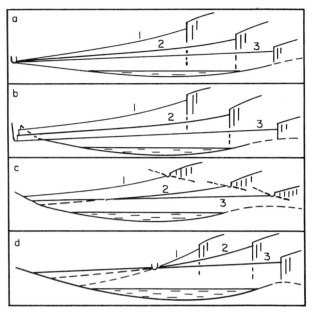

Fig. 6 Schematic diagrams showing the profiles of former glacial lake strandlines in relation to different types of outlets: (a) Distal outlet with a stable sill; (b) distal outlet in erodable materials; (c) proximal outlets; (d) intermediate outlet. Glacier ice is shown by vertical lines. Modern lake shown by short dashes. Numbers show order of strandline formation; dashed strandlines were buried or destroyed by younger lakes.

and then slowly decrease to present values. [*See* CONTINENTAL DEFORMATION.]

D. Lake Outlets and Warped Strandlines

The relationships of relict strandlines are affected by lake outlets, as shown in Fig. 6. In Fig. 6a strandlines hinge upward from the sill of a stable distal outlet, and their ages decrease with lower altitude. Fig. 6b shows an outlet in drift that is eroded until a resistant lag of residual boulders stabilizes the lake level and a strandline forms. An increase in discharge renews erosion and the cycle is repeated. Figure 6c shows strandlines resulting from the opening of proximal outlets. Successive water planes have lower slopes, and a young water level may even submerge the distal parts of older strandlines. Figure 6d shows water planes rotating about an outlet in an intermediate position. Between the outlet and the ice margin the oldest water planes are highest, but beyond the outlet the order is reversed. The latter generally are obliterated by the

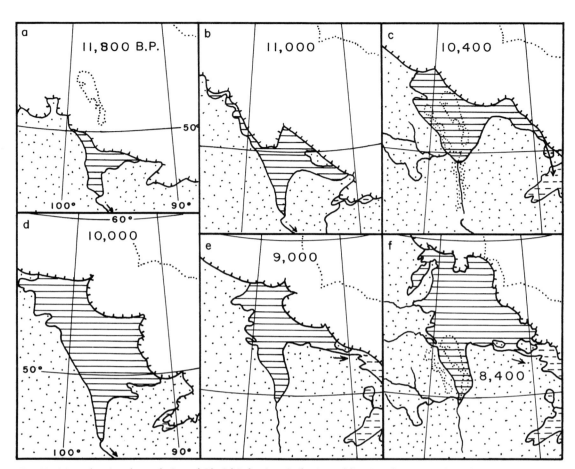

Fig. 7 Maps showing the evolution of Glacial Lake Agassiz (horizontal lines) at the times indicated: 11,800 = 11,800 years before present (= 1950). Ice margins are hachured; other lakes shown by dashed lines; present lakes and the shore of Hudson Bay are dotted where shown. [After A. S. Dyke and V. K. Prest (1987). "Paleogeography of Northern North America 18,000–5,000 Years Ago." Geological Survey of Canada Map 1703A, with permission of the Geological Survey of Canada; and Teller (1987).]

erosion of younger water planes, but sometimes they are preserved under offshore sediments.

VII. Examples of Pleistocene Glacial Lake

A. Glacial Lake Agassiz

Glacial Lake Agassiz was the first glacial lake to be studied as a system, by W. Upham in 1884. The sediments of Lake Agassiz are best known in the Red River valley, one of at least seven sedimentary basins within the lake. There tills deposited during early ice readvances interfinger with the lowest

offshore clays. Subaerial alluvium separates two thick upper clay units. Strandlines are well developed on the western shores of Lake Agassiz and on the east side of the Red River basin.

Lake Agassiz (Fig. 7) existed from 11,800 until about 7500 years ago. The earliest lake, not shown, discharged east along the ice margin into lakes in northern Minnesota and south to the Mississippi River. From 11,800 to 11,000 years ago (Fig. 7a and 7b), the lake expanded northward and discharged south through the Minnesota Valley. About 10,400 years ago outlets north of Lake Superior opened, the water level fell, and the lake decreased in size (Fig. 7c). Alluvium was deposited in the Red River valley in the United States, and shallow-water lake sediments in Canada. Glacier readvance closed the

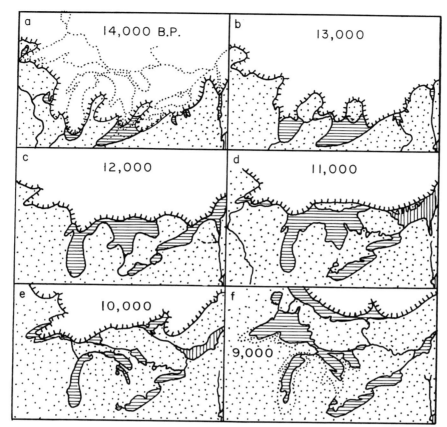

Fig. 8 Maps showing the history of glacial lakes in the Great Lakes basin at 1000-year intervals from 14,000 to 9,000 years before present. Ice margins are hachured; glacial lakes are horizontally lined; marine waters are vertically lined; existing lake outlines are dotted where shown. [After A. D. Dyke and V. K. Prest (1987). "Paleogeography of Northern North America 18,000–5,000 Years Ago." Geological Survey of Canada Map 1703A, with permission of the Geological Survey of Canada.]

eastern outlets about 9900 years ago; the water level rose, and discharge through the Minnesota River valley was reestablished (Fig. 7d). Lake Agassiz was largest, 350,000 km², at this time. Glacier recession reopened eastern outlets (Fig. 7e) through Lake Nipigon into Lake Superior and suddenly lowered the lake level, about 9500 years ago. Continued ice recession in Ontario opened lower outlets into Glacial Lake Ojibway (Fig. 7f). Lake Agassiz drained when Hudson Bay opened about 7500 years ago.

B. Glacial Lakes of the Great Lakes Basins

F. B. Taylor began the study of the Great Lakes before 1895. Before glaciation the Great Lakes basins were valleys of a trellis drainage system on gently dipping Paleozoic rocks.

Repeated glacial erosion of the valley floors into underlying shale formed the present lake basins. The sequence of lakes in the Great Lakes basins is illustrated in Fig. 8. The basins were deglaciated from south to north, and glacial readvances rerouted drainage several times. The Superior, Huron, and Ontario basins had intermediate and proximal outlets resulting in some intersecting standlines.

The early lakes discharged south into the Mississippi system until ice retreat opened outlets east to the Mohawk–Hudson river valley. Because of isostatic depression in the north, further ice recession opened lower outlets from Georgian Bay into Lake Ontario and then east through Lake Nipissing to the Ottawa River. The latter lowered water levels and reduced the lakes in the Michigan and Huron basins to less than half their present sizes. These no longer were true glacial lakes. Uplift of the Nipissing outlet established the present drainage about 4000 years ago.

Bibliography

Dyke, A. S., and Prest, V. K. (1987). Lake Wisconsinan and Holocene history of the Laurentide Ice Sheet. *Geographie Physique et Quaternaire* **41**(2), 237; Geological Survey of Canada Map 1703A.

Jopling, A. V., and McDonald, B. C., eds. (1975). "Glaciofluvial and Glaciolacustrine Sedimentation." Society of Economic Paleontologists and Mineralogists, Special Publication 23, Tulsa, Oklahoma.

Karrow, P. F., and Calkin, P. E., eds. (1985). "Quaternary Evolution of the Great Lakes." Geological Association of Canada, Special Paper 30, St. John's, Newfoundland.

Schluchter, Ch., ed. (1979). "Moraines and Varves. Origin/Genesis/Classification," section titled "Varves and Glaciolacustrine Sedimentation," pp. 279–364. A. A. Balkema, Rotterdam.

Teller, J. T. (1987). Proglacial lakes and the southern margin of the Laurentide Ice Sheet. *In* "North American and Adjacent Oceans during the Last Deglaciation," Vol. K-3 of "The Geology of North America." (W. F. Ruddiman and H. E. Wright, Jr., eds.), p. 39. Geological Society of America, Boulder, Colorado.

Teller, J. T., and Clayton, L., eds. (1983). "Glacial Lake Agassiz." Geological Association of Canada, Special Paper 26, St. John's, Newfoundland.

Glossary

Esker Sand and gravel ridge deposited by a meltwater stream under, within, or on glacier ice close to the margin.

Glaciolacustrine Used to describe sediments deposited in a lake in contact with or receiving most of its water and sediments from a glacier, and also landforms resulting from such deposition.

Homopycnal Describes water entering a lake that has the same density as the lake water.

Hyperpycnal Describes water that is denser than the water in the lake it enters.

Isobase Line of equal altitude on a former water plane.

Jökulhlaup Sudden, often catastrophic, flood discharge resulting from the failure of part of a glacier-ice or glacial-drift barrier that forms part of the basin of a glacial lake.

Varve Layer of sediment in the form of a couplet comprising a summer and a winter layer, representing one year of deposition.

Glaciotectonic Structures and Landforms

J. S. Aber

Emporia State University

Glaciers and ice sheets are parts of the cryosphere, which interacts with other environmental systems at the earth's surface. Glaciotectonism refers to the processes by which glaciers cause deformation of crustal rocks and sediments at scales ranging from microscopic to continental. Various distinctive landforms are related to glaciotectonic structures, which may dominate regions of former glaciation. Repeated episodes of ice sheet glaciation during the Pleistocene (Ice Ages) have created significant, permanent modifications of continental substratum and landscapes in northern Europe and North America.

I. Introduction

Glaciers erode, transport, and deposit rocks and sediments. This knowledge has been the foundation of glacial geology and geomorphology for more than 150 years. The realization that glaciers could also *deform* rocks and sediments came more slowly and is even now not widely appreciated. The study of glaciotectonics has emerged as a significant subdiscipline within glacial geology and geomorphology during the last 40 years. Detailed local studies, regional mapping, and theoretical analyses have been conducted, primarily in North America and Europe.

It is now apparent that glaciotectonic structures and landforms are commonly and widely present in regions of former glaciation. A variety of glacial landforms is now attributed either wholly or partly to glaciotectonic genesis. Hence, ice-pushed structures and landforms must now be included with depositional and erosional features as primary field evidence for past glaciation.

Glaciotectonism may be defined as structural deformation of the earth's crust as a direct result of glacier movement or loading. This definition is fairly simple and not overly restrictive, but it does exclude certain kinds of deformation related to glaciation. Deformation caused by drifting icebergs or by freezing and thawing of dead ice (including permafrost) is not considered to be glaciotectonic. Likewise, deformed structures within glacier ice are excluded. Glaciotectonic structures do include all deformations created in rock or sediment as a consequence of active glacier loading, dragging, or pushing.

II. Glaciotectonic Structures

Glaciotectonic structures range in size from microscopic to continental (Fig. 1). Deformation and disruption of the crust are normally limited to a depth of about 200 m. Isostatic depression and rebound of the crust, as much as 1 km vertically, are also a form of glaciotectonism. Deformed materials range from hard crystalline rocks, to poorly consolidated sedimentary strata, to loose sediments. Deformation may occur in both frozen and thawed materials under either low or high confining pressures. [*See* CONTINENTAL DEFORMATION.]

Ductile and brittle deformations of all kinds are present in glaciotectonic settings. Ductile deformation takes place by plastic or fluid flow of material subjected to glacial stress. Unconsolidated or fine-grained materials, such as clay, silt, shale, or chalk, are most susceptible to ductile deformation. Various kinds of folds, intrusions, diapirs, and contortions result from this style of deformation.

Brittle deformation occurs where rocks or sediments fail by fracturing along discrete planes. Deformation is accomplished by movement or adjustment along fractures with little or no deformation within the rock between fractures. Consolidated or coarse-grained materials, such as sand, gravel, limestone, sandstone, or slate, are most susceptible to brittle deformation. Joints, faults, fissures, brecciation, and other fractured structures result from brittle deformation. Ductile and brittle structures are often intimately associated because

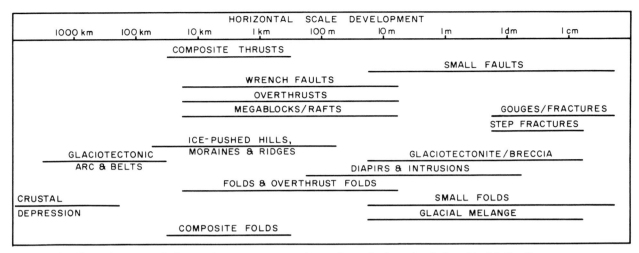

Fig. 1 Chart of common glaciotectonic structures arranged according to horizontal scale (logarithmic). Ductile structures toward bottom; brittle structures toward top. [Reprinted from Aber *et al.* (1989) by permission of Kluwer Academic Publishers.]

of variations in strength within a rock or sediment mass or because of changing conditions of glaciotectonism during deformation.

Glaciotectonic structures may be created in the substratum under thick ice, beneath thin ice, or in the foreland region in front of advancing glaciers. All manner of alpine glaciers, terrestrial ice sheets, and marine-based glaciers are capable of producing glaciotectonic structures during advancing, maximal, or recessional phases of glaciation. In short, glaciotectonic structures may be created wherever glaciers or ice sheets override or push against deformable rocks or sediments.

A striking similarity exists between the structures of ice-shoved hills and those of true mountains. All kinds of structures described from mountains and shields are also present in glaciotectonic settings, although smaller in size. Glaciotectonic structures of ductile style often mimic those seen in igneous or metamorphic rocks. Ice-shoved hills may be regarded as natural scale models of mountains.

III. Glaciotectonic Landforms

A. Categories

Glaciotectonic landforms are the morphologic expressions of subsurface structural deformations brought about by glaciation. Such landforms range from conspicuous ice-shoved hills, to smoothed plains, to anomalous depressions. These landforms may display their original glaciotectonic morphol-

ogy where they are little modified by later events. More commonly, however, subsequent glacial or nonglacial erosion or deposition has altered the initial landform, in some cases obliterating any morphologic expression of the ice-pushed structures. In all cases, some knowledge of subsurface stratigraphy and structure is invaluable for properly interpreting the landforms.

Glaciotectonic landforms may be divided in two general categories on the basis of their morphostructural attributes (Table I):

1. Ice-shoved hills: constructional hills, ridges, or moraines that rise above the general landscape as a result of structural uplift of deformed rock and/or sediment material.

Table I Constructional Glaciotectonic Landforms

Landform	Height (m)	Area (km²)	Morphology
Composite ridges	20–200	1 > 100	Subparallel ridge-and-valley system, arcuate
Hill–hole pair	20–200	1 to > 100	Ridged hill with source basin of similar size
Cupola hill	20–200	1–100	Smoothed dome to drumlin-shaped hill with till cap
Glaciotectonic plain	<20	100 to >1000	Low-relief plain, concealed rafts and diapirs

Includes hill–hole pairs, large and small composite ridges, and cupola hills.

2. Glaciotectonic plains: subdued, streamlined, or smoothed plains. Underlying structures include flat rafts or megablocks, diapirs, and intrusions, as well as buried ice-shoved hills.

These forms represent ideal types within a continuous spectrum of glaciotectonic phenomena. Intermediate, transitional, or mixed landforms exist between these ideal types and are in fact rather common. The materials of which these landforms are constructed may be classified in three groups: (1) pre-Quaternary strata, which are usually consolidated to some degree; (2) preexisting Quaternary strata, both glacial and nonglacial; and (3) penecontemporaneous glacial sediment that was deposited and deformed during the same glaciation. Most glaciotectonic landforms contain all three types of material in varying proportions.

Typical examples of glaciotectonic landforms and structures are described in the following sections using the case-example approach. The examples are selected to illustrate the most common and distinctive types of glaciotectonic phenomena.

B. Hill–Hole Pair

The hill–hole pair is the simplest and most instructive type of ice-shoved landform. It consists of an ice-scooped basin and related hill. Other kinds of ice-shoved hills are variations on this fundamental form. They were often misidentified as kames or bedrock outliers, depending on their internal composition. Hill–hole pairs are now widely recognized in Scandinavia, the north–central United States, western Canada, and elsewhere, both on land and on the continental shelf.

The hill–hole pair consists of a discrete hill of ice-shoved material situated a short distance downglacier from a depression of similar size and shape. The depression is the source of material now in the hill, and ideally the volume of the depression should nearly equal the hill's volume. Depressions are now often the sites of bogs, lakes, or estuaries, and so their apparent sizes are usually reduced by later sedimentation. The ice-shoved hills may also be altered by later erosion or deposition, so an exact volume correspondence does not always exist between the hill and related hole.

The basic morphology of a hill–hole pair is displayed at Wolf Lake, Alberta (Fig. 2). Wolf Lake occupies the source depression from which material was shoved into the large hill immediately to the south. The hill and hole display similar areas and shapes, and their alignment is emphasized by the lineament present on the eastern edge of the hill–hole pair. Characteristic traits of simple hill–hole pairs include:

1. Arcuate or crescentic outline of hill; concave on upglacier side, convex in downglacier direction.
2. Multiple, subparallel, narrow ridges separated by equally narrow valleys following overall arcuate trend.
3. Asymmetric cross profile of hill; higher with steeper slopes on convex (downglacier) side.
4. Topographic depression on concave (upglacier) side; area and shape of hill approximately equal to those of depression.

Where all these morphologic traits are present in a glaciated region with soft bedrock or thick sediment, a hill–hole pair may be identified, even without knowledge of the subsurface.

C. Composite Ridges

The most typical and distinctive glaciotectonic landforms are the ice-shoved ridges found in many glaciated plains. These ridges are composites of upthrust, folded, and contorted sedimentary bedrock that is generally interlayered with and overlain by much glacial sediment. Large composite ridges (>100 m high) usually include a substantial volume of pre-Quaternary bedrock. Small composite ridges (<100 m high) may or may not include deformed bedrock; many, in fact, are composed largely of unconsolidated Quaternary strata.

Composite ridges typically display multiple, narrow ridges separated by steep-sided valleys. The ridges are the upturned ends of thrust blocks or the crests of folds. A close correspondence usually exists between structural features and morphology for composite ridges of the last glaciation. Maximum uplift of thrust blocks is around 200 m.

The large composite ridges of Møns Klint, southeastern Denmark, are undoubtedly the most famous and spectacular of all glaciotectonic sites. Large bodies of deformed chalk along with drift are exposed in a scenic cliff (Fig. 3) and form a rugged landscape inland from the cliff (Fig. 4). The combination of cliff exposures with composite ridge morphology provides a three-dimensional display of Møns Klint's glaciotectonic structure. Chalk bodies have been uplifted as much as 150 m by two phases of glaciotectonic deformation coming from the northeast and south.

The term *push moraine* is commonly and loosely used to refer to ice-shoved ridges. As used here, push moraines are a restricted subset of composite ridges that consist largely or wholly of glaciogenic strata. Brandon Hills are a push moraine, formed between two ice tongues during the final phase of glacier advance, on the Manitoba Escarpment of southwestern Manitoba (Fig. 5). The internal structure consists of tilted blocks of stratified drift that are disrupted by various

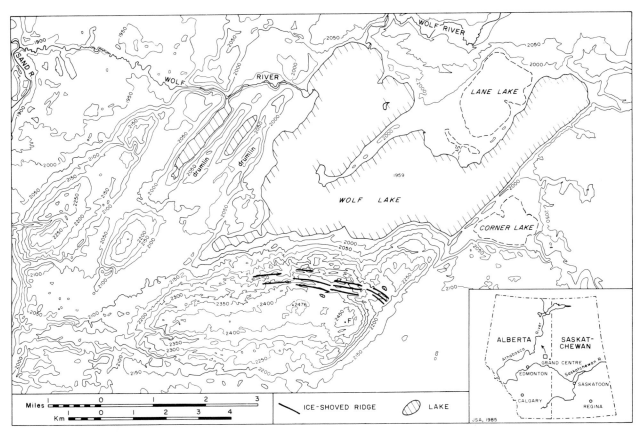

Fig. 2 Topographic map of Wolf Lake vicinity, eastern Alberta. Wolf Lake basin is the source of material now in the hill to the south. Note prominent lineament formed by eastern edges of Wolf Lake and the hill. Ice movement from northeast; elevations in feet; contour interval 50 ft (15 m).

faults, all of which parallel larger blocks and the trends of individual ridges. Composite ridges, covered by a thin veneer of till, resemble a giant fishhook in overall plan. Older stratified drift was deformed, overridden by ice, and covered with discordant till.

D. Cupola Hills

Many glaciated hills have the internal structure of ice-shoved ridges but lack a hill–hole relationship or the typical morphology of composite ridges. Such hills, often displaying a smoothed or rounded morphology, are called *cupola hills*. They have internal structures similar to those of composite ridges, but their external morphologies were substantially modified by overriding ice. Cupola hills have three basic attributes:

1. Interior structure: deformed glacial and nonglacial sediment with or without detached floes of older bedrock.
2. External form: long, even hill slopes with domelike morphology, varying from nearly circular to elongated ovals in plan, 1–15 km long, 20 m to >100 m high.
3. Discordant till: overridden by ice which truncated deformed structures and deposited a basal till cover over hill.

Cupola hills represent the combined effects of ice-shoving and subglacial erosion on the ice-pushed hill. Where subglacial modification is slight, a subdued composite ridge morphology may be preserved. With more modification, a rounded, smoothed cupola hill results. Still greater glacial erosion may create streamlined, drumlin-shaped hills; in fact, many drumlins possess a disrupted core.

The typical characteristics of cupola hills are demonstrated by the Swedish island of Ven in the Øresund strait between

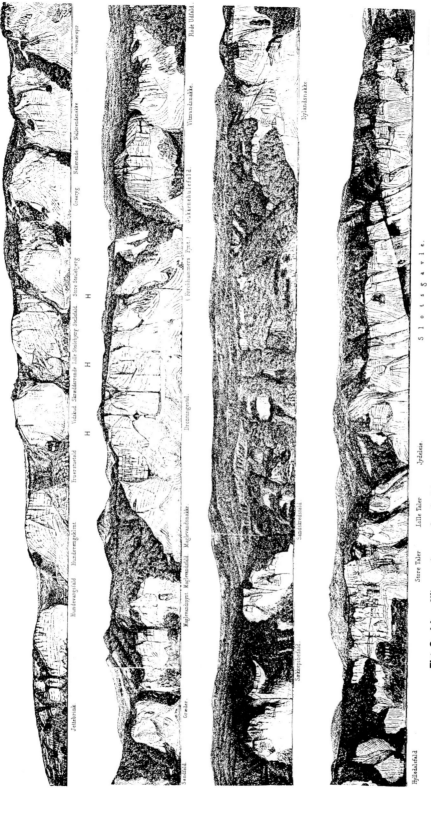

Fig. 3 Møns Klint section, as sketched by Puggaard in 1851, viewed from the east (south at upper left, north at lower right). Chalk masses are detached, folded, and thrust over each other. Black lines within blank chalk masses show deformed flint layers. [Reprinted from Aber *et al.* (1989) by permission of Kluwer Academic Publishers.]

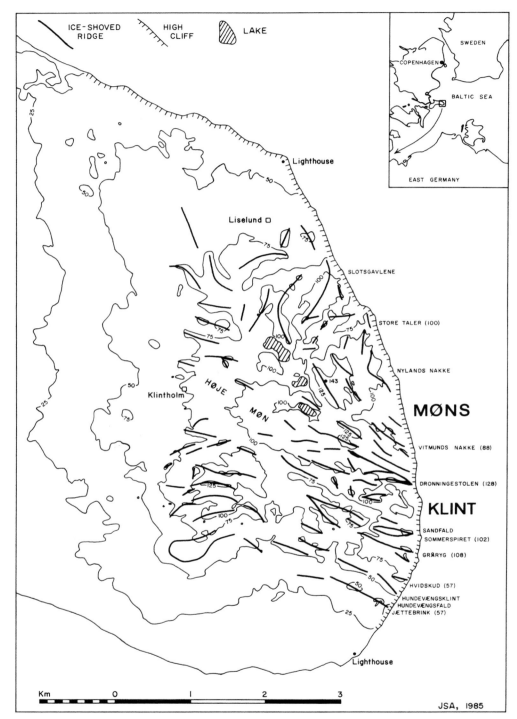

Fig. 4 Topographic map of eastern Møn, Denmark, showing composite ridges. Selected chalk cliffs and interchalk falls are indicated, with heights of chalk cliffs given in parentheses. Contour interval 25 m.

mainland Sweden and Denmark (Fig. 6). The island is an elongated, gently sloping hill bounded by steep sea cliffs. Ven consists entirely of glaciogenic formations that were deposited and deformed by a series of ice advances during the last glaciation.

The major glaciotectonic structures of Ven are gentle synclines, some 400–500 m across and 20–30 m in amplitude, that are disrupted by many lesser faults and folds (Fig. 7). Discordant tills cap the entire island, except where removed by postglacial erosion. The base of structural disturbance seems to coincide with the boundary between glaciogenic strata and underlying fluvial deposits, about 30 m below modern sea level.

E. Glaciotectonic Plains

Glaciated plains underlain by thick drift and having subdued surface morphology cover vast areas in the outer regions of ice sheet glaciation. So-called till plains or drift plains commonly contain subsurface glaciotectonic structures of many types. Ice-pushed structures are locally so abundant, in fact, that such regions could be called *glaciotectonic plains*.

Ice-shoved hills and ice-scooped basins are, in some cases, simply buried beneath younger glacial deposits, as is common in Poland (Fig. 8). Glaciotectonic plains may also contain large intrusive or diapiric structures, as well as large, flat-lying slabs of transported bedrock called rafts or megablocks. These features usually have little or no morphologic expression but may dominate the subsurface structure of large regions.

Megablocks and rafts are flat-lying slabs of transported and deformed bedrock that are usually very thin (<30 m) compared to their areas (often >1 km²). Most known megablocks consist of poorly consolidated sedimentary strata, and most were probably transported only a few kilometers. In some exceptional cases, for example, the Lukow rafts of eastern Poland and the Cooking Lake megablock of central Alberta, transportation of >300 km took place.

The largest known megablock is located near Esterhazy in eastern Saskatchewan, Canada. The megablock covers an area of roughly 1000 km² (Fig. 9). It has no morphologic expression in the drift-mantled plain but is exposed in walls of the Qu'Appelle Valley. Regional bedrock consists of gently dipping, undeformed, sedimentary strata (Fig. 10). The megablock is made up mainly of folded and faulted, siliceous shale. Breccia, slickensides, and mylonite are common microstructures.

The megablock is oval in plan, approximately 38 km long and 30 km wide. Its maximum thickness is about 100 m with an average thickness of around 60 m. From these dimensions, its volume is estimated to be 60 km³. In most places, deformed shale of the megablock rests directly on undisturbed bedrock. Near its western end, however, till is present beneath the megablock. Neither the source nor the direction of movement of the megablock is known. It may not have moved far—perhaps less lateral displacement than its own width—in order to produce the observed structures.

Various kinds of diapirs, intrusions, and wedges are extremely common in glaciogenic sediment. Intrusions include all structures in which one material was injected or squeezed in a mobile state into the body of another material. The intruded material is most usually a clay- or silt-rich sediment, whereas almost any kind of sediment or bedrock may form the host material. The most important question concerns the nature of the stress that led to intrusion; only cases in which active glaciers produced the stress should be considered glaciotectonic.

The Kansas Group stratotype at Atchison, Kansas, displays two large diapirs along with other glaciotectonic structures (Fig. 11). The diapirs were deformed by ice pushing from the northeast, and the intrusive character of the diapirs indicates deformation in a thawed and water-saturated state.

IV. Distribution of Glaciotectonics

A. Continent-Scale Distribution

A general model for continent-scale distribution of glaciotectonic structures and landforms includes three primary landscape zones:

1. Outer zone: all manner of large and small glaciotectonic phenomena are present in thick sediment and soft bedrock on land and on the continental shelf.
2. Intermediate zone: generally thin, patchy drift cover with small, isolated glaciotectonic features found mainly in locally thick drift of last glaciation.
3. Inner zone: widespread, moderately thick drift in which small to medium-sized glaciotectonic features are common.

The overall glacial landscape zonation is well developed both in North America (Fig. 12) and in northern Europe (Fig. 13). The landscape model is, of course, generalized; development of each zone may vary according to local circumstances. Boundaries between zones are sharp in some cases and transitional in others. Each zone may not always be fully developed around the entire area of glaciation. These zones were not created only during the last glaciation, but represent the cumulative results of landscape alteration during multiple episodes of ice sheet glaciation. [*See* ICE AGE DYNAMICS.]

The outer glaciotectonic zone in North America stretches

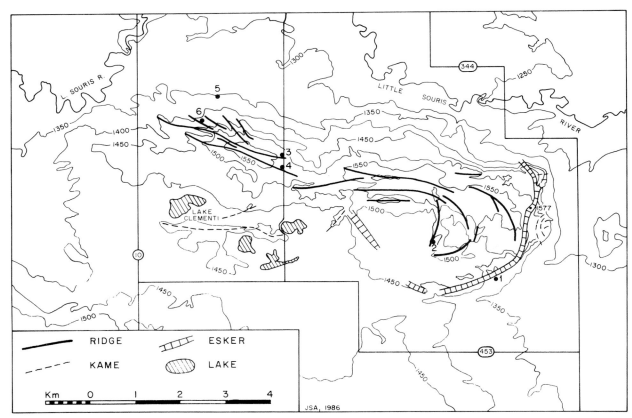

Fig. 5   Topographic map of Brandon Hills vicinity, Manitoba, showing composite ridges and other glacial landforms. Ice movement from north and northeast; elevations in feet; contour interval 50 ft (15 m). [Reprinted from Aber *et al.* (1989) by permission of Kluwer Academic Publishers.]

from the Atlantic Coastal Plain westward into the Great Plains and northward to the Arctic continental shelf. The Hudson Bay Lowlands are also included. In Europe the outer zone extends from Ireland, across the continental shelf and southern Baltic regions, into the Soviet Union. The outer zones are underlain by thick and nearly continuous drift resting on various sedimentary bedrock. Large looped end moraines, drumlin fields, older drift, and multiple till sequences are common. All kinds of large and small glaciotectonic features are abundant throughout the outer zones.

The intermediate zone of glaciation presents a startling contrast to the outer zone. In Europe the intermediate zone includes the Fennoscandian Shield of southern Sweden and Finland and the Caledonian Mountains of Scotland, Norway, and western Sweden. The Canadian Shield and northern Appalachian Mountains make up the intermediate zone in North America. Drift is much thinner and discontinuous. Hard, mostly crystalline bedrock is exposed over vast areas.

Locally thick drift left during the last deglaciation is scattered in end moraines and eskers, but older drift is rarely preserved. Small glaciotectonic features are locally present in the intermediate zone. Small glaciotectonic faults in slate and other consolidated bedrock are widespread.

The inner zone of glaciation in Scandinavia is located in the Caledonian Mountains of south–central Norway and western Sweden and extends across northern Sweden and Finland into the Soviet Union. Drift is again moderately thick and continuous with drumlins and Rogen (ribbed) moraine common. Quaternary deposits predating the last glaciation are commonly preserved beneath till, and multiple till sequences are known in many places. Glaciotectonic structures and small landforms are common in such overridden deposits.

It is clear from the geological context that many of these glaciotectonic features were created in older drift beneath the thick center of the Scandinavian Ice Sheet. Deformation took

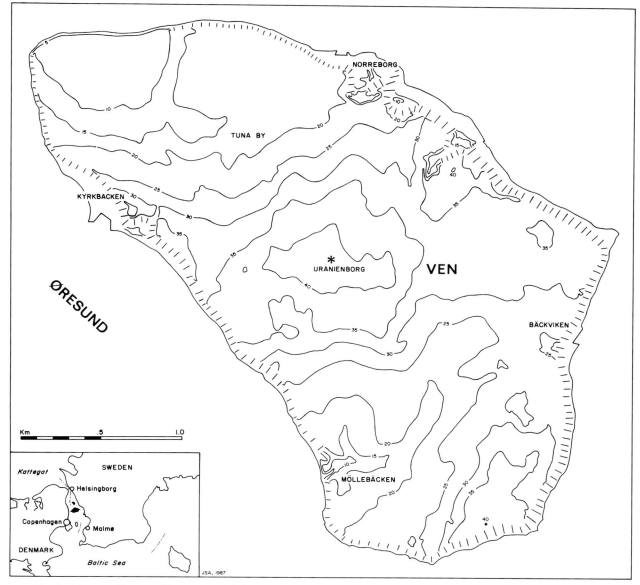

Fig. 6 Topographic map of Ven, a cupola hill in southwestern Sweden. Contour interval 5 m; location of Ven shown on inset map by arrow.

place during all phases of glaciation: advancing, maximal, recessional, and final deglaciation. Two inner zones are present in North America, corresponding to two main sectors of the Laurentide Ice Sheet. These zones are morphologically similar to the inner zone of Scandinavia.

A general impression exists that the inner zones in the regions of ice divides were stagnant areas, in which the ice

sheets had little effect on the landscape, aside from crustal depression and rebound. Crustal depression involved both elastic and plastic deformation in the lithosphere and underlying asthenosphere. As much as 1 km of depression occurred in central Canada and up to 800 m of depression took place in northern Scandinavia. It is now apparent that moderate glacial erosion, deposition, and shallow deformation also

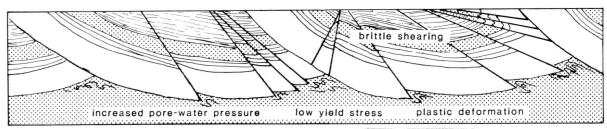

Fig. 7 Schematic model for glaciotectonic structures and conditions of deformation in glaciogenic strata at Möllebäcken, Ven. Large synclines are 300–400 m across and 20–30 m in amplitude. [Reprinted from L. Adrielsson (1984). Weichselian lithostratigraphy and glacial environments in the Ven–Glumsov area, southern Sweden. Doctoral thesis, Lunds Universitet, Sweden.]

took place within the inner zones under dynamic conditions of ice movement.

The distribution of glaciotectonic and other glacial features in North America and Europe follows broadly symmetrical patterns with respect to the Atlantic (Fig. 14). These patterns are related in general to availability of deformable substrata, namely thick sediment or sedimentary bedrock. The three-zone model thus reflects long-term modification of

the continental substratum during repeated ice sheet glaciations.

B. Regional Distribution

The regional distribution of glaciotectonic features may be evaluated for the outer landscape zone, in which such features are abundant and their pattern of distribution may be

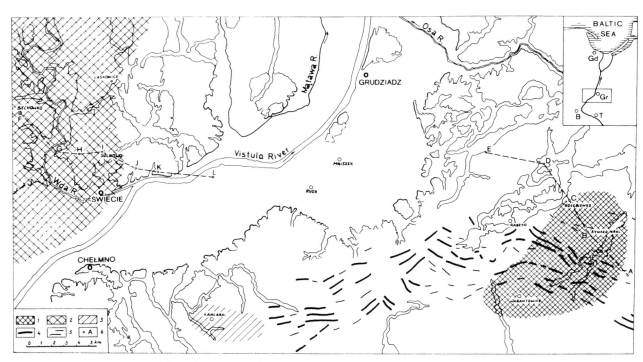

Fig. 8 Buried ice-shoved hills in the lower Vistula region, Poland. 1, 2, and 3 are glaciotectonic disturbances involving uplifted sedimentary bedrock; 4 and 5 are major and minor end moraines; 6 shows borehole sites. [Reprinted from Aber *et al.* (1989) by permission of Kluwer Academic Publishers.]

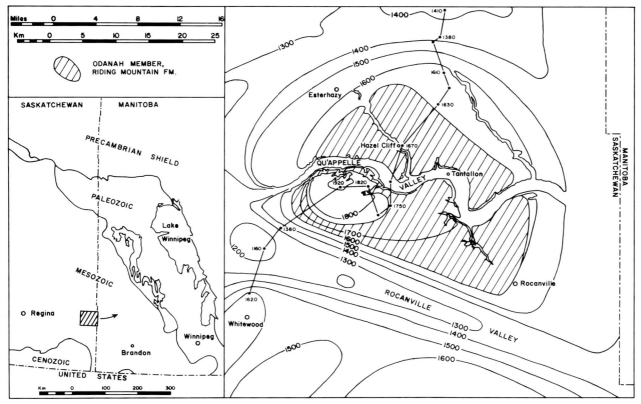

Fig. 9 Bedrock contour map of Esterhazy vicinity, Saskatchewan. The 1600-ft contour between Esterhazy and Rocanville defines the position of the megablock. Elevations in feet; contour interval 100 ft (30 m). [Reprinted from Aber *et al.* (1989) by permission of Kluwer Academic Publishers.]

compared with other glacial phenomena. Two regional distribution patterns are recognized for glaciotectonic features:

1. Random, sporadic distribution of megablocks, rafts, diapirs, and other features that have little or no morphologic expression, along with small cupola hills. These features were presumably created in subglacial settings, and their locations are primarily related to local conditions of substratum materials.
2. Ice-marginal distribution of prominent ice-shoved hills. These features were created at or near active ice margins, and their locations are closely related to development of ice lobes or tongues.

The basic geometry of ice lobes was recognized more than a century ago and is controlled primarily by preglacial, bedrock topography. In central North Dakota, ice-shoved hills are arrayed in belts from 2–3 km to 3–5 km wide immediately inside ice margin positions. These belts define the outlines of larger ice lobes (Fig. 15). Individual hills within a belt often display their own looped shapes reflecting development of local ice tongues. Behind these belts, the glaciotectonic hills become progressively smaller and smoothed, giving way upice to streamlined terrain of low relief.

This pattern could be interpreted as the result of simultaneous creation of streamlined and ice thrust features beneath and behind a stationary ice margin. Thrusting presumably took place in a narrow frozen-bed zone, whereas streamlining occurred under thawed-bed conditions. Permafrost was probably involved in many situations, but thrusting of unfrozen material could also take place, particularly above confined aquifers. It is equally plausible that ice shoving took place during advance of ice lobes that subsequently overrode the ice-pushed hills.

The location of ice lobes, and thus ice-shoved hills, was generally controlled by bedrock topography. However, this was not the case in the central Netherlands, where the Saalian glaciation advanced over a flat alluvial plain. Nonetheless,

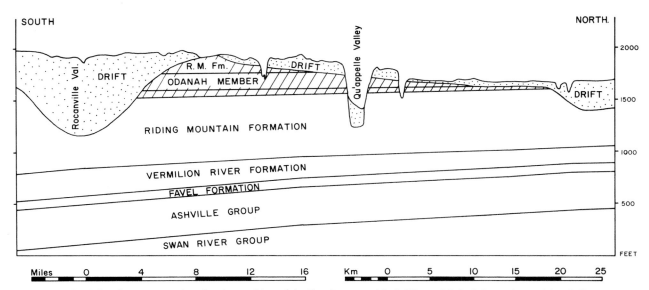

Fig. 10 Subsurface cross section showing position of the Esterhazy megablock (diagonal lining). It consists mainly of siliceous shale of the Odanah Member of the Riding Mountain Formation. See Fig. 9 for location; vertical exaggeration 42×. [Reprinted from Aber *et al.* (1989) by permission of Kluwer Academic Publishers.]

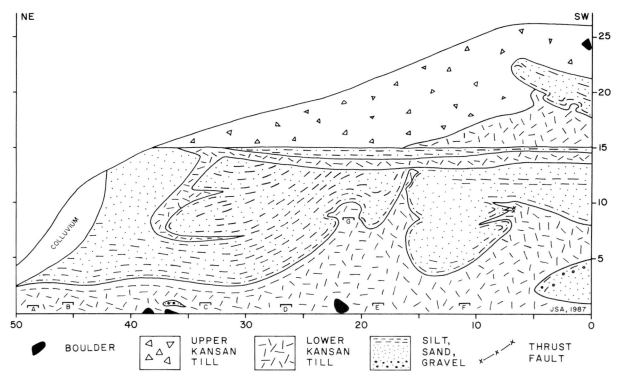

Fig. 11 Kansas Group stratotype. Lower Kansas till intrudes up into Atchison Formation sand, which is capped by upper Kansas till.

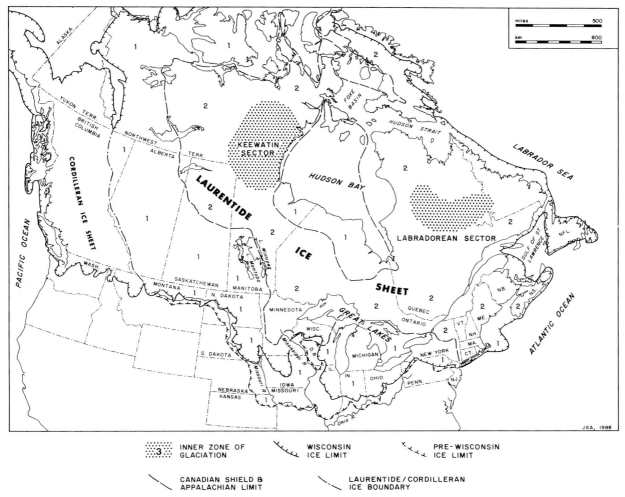

Fig. 12 Three major glacial landscape zones in North America: 1, outer zone of thick sediment and conspicuous ice push deformation; 2, intermediate zone with thin glacial sediment and widely exposed basement rock; 3, inner zone of moderate glacial sediment.

composite ridges display conspicuous lobate distributions around deep glacial basins that are now partly filled with younger sediment. The basins were apparently created by a combination of subglacial meltwater erosion and glaciotectonic thrusting.

A model for glaciotectonism associated with ice lobes consists of two stages: (1) proglacial thrusting of ice-shoved hills, followed by (2) subglacial modification of overridden hills (Fig. 16). Initial proglacial thrusting takes place along a décollement that may be controlled by any of several features: lower boundary of permafrost, lithologic or stratigraphic boundary, position of confined aquifer, etc. High

groundwater pressure is presumably developed along the décollement.

Subglacial meltwater may either erode tunnel valleys or deposit eskers, and proglacial meltwater may cut spillways across the ice-pushed ridge and deposit outwash sediment on the distal side of the hill. Small, temporary lakes may form, and superglacial sediment moving down the ice front may become incorporated in the ice-shoved hill. Such features would ideally outline the frontal and lateral margins of the ice lobe at the time of thrusting and may be preserved where ice did not later overrun the hill.

With continued glacier advance, the hill may be overrid-

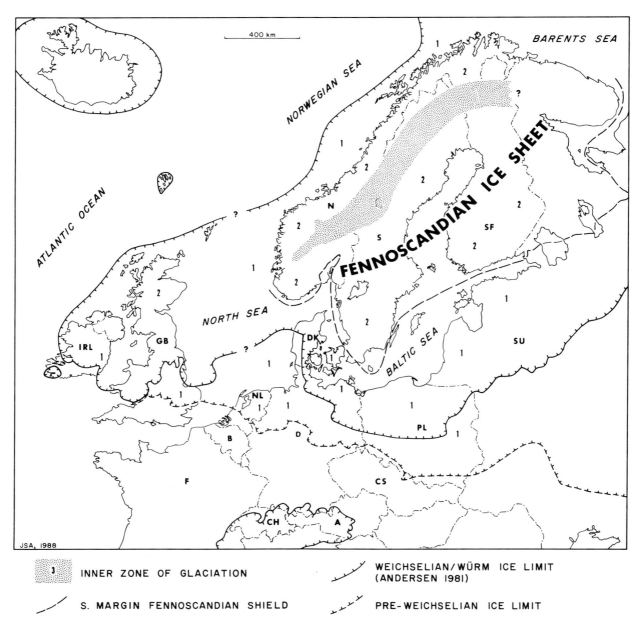

Fig. 13 Map of northern Europe showing major glacial landscape zones. Moderately thick and continuous drift containing many glaciotectonic features is present in the inner zone. Numbered zones same as in Fig. 12.

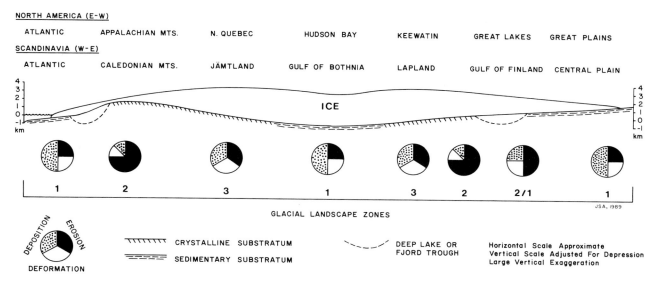

Fig. 14 Glacial landscape symmetry of North America (east–west) and Europe (west–east) with schematic ice sheet profile and substratum geology. Pie diagrams represent the local, relative contributions of glacial deposition, erosion, and deformation in modifying the three landscape zones. Numbered zones same as in Figs. 12 and 13.

den, at which point a penetrative subglacial style of deformation may occur. Erosion of the hill provides reworked sediment for a discordant till cover, and gradually a cupola hill morphology begins to develop. Further subglacial molding may create a streamlined, drumlin-shaped hill, and ultimately the ice-shoved hill may be greatly reduced in size.

Under ideal circumstances, a series of ice-pushed hills may be created during ice lobe advance and manage to survive throughout the glaciation. This results in a *glaciotectonic landscape*, in which wide, low basins alternate with narrow belts of ice-shoved hills. Such is the case in the Suwalki Lakeland of northeastern Poland, where a festoon pattern of composite ridges was created during progressive ice lobe advances (Fig. 17).

C. Applied Glaciotectonics

Glaciotectonic terrain is typically anomalous in its texture, lithology, geochemistry, and engineering properties compared with the undeformed surroundings. The main effects of glaciotectonism are (1) disruption of normal stratigraphy, (2) reduction of rock and/or sediment strength, and (3) increased rock and/or sediment variability. Such anomalies are potentially important for many human activities that involve excavating in, building on, or other modification of the land.

Planning and operation of mines; prospecting and mineral exploration; construction of roads, bridges, and buildings; and soil mapping are all applied aspects of glaciotectonics.

Formations that are normally buried may be brought to the surface by ice thrusting. Ice scooping may, conversely, remove the normal formations from an area. Failures in mine walls and highway excavations have taken place in disturbed overburden. Knowledge of the distribution and types of glaciotectonic features is necessary for proper human development in such situations.

V. Mechanisms of Glaciotectonism

A. Fundamental Cause of Glaciotectonism

Glaciotectonism is the result of interaction between the cryosphere, lithosphere, and hydrosphere. Glaciotectonic deformation may occur in a wide variety of settings: in front of glaciers, beneath ice margins, or under thick centers of ice sheets. Deformation may take place during all stages of glaciation and affect all manner of well-consolidated rocks and loose sediments. Topographic situations vary from mountains to continental shelves. In short, glaciotectonic features may be expected wherever sedimentary strata were overridden or pushed by glaciers.

Many factors have been cited as or presumed to be important or necessary conditions for glaciotectonism: elevated groundwater pressure, permafrost, topographic obstacles, substratum lithology, compressive or surging ice flow, subglacial meltwater or proglacial lakes, etc. Most of these factors are related to local conditions that often vary considera-

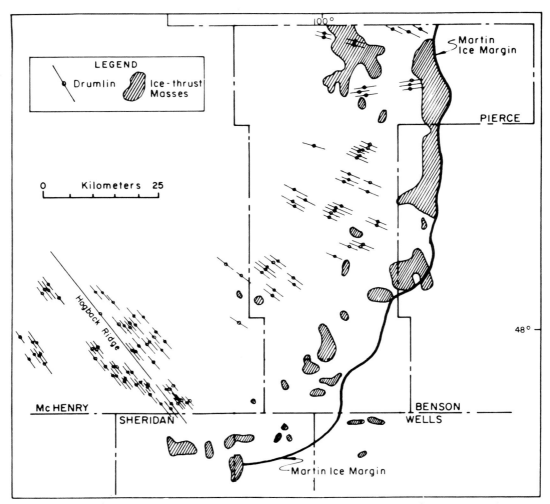

Fig. 15 Glacial features associated with the Martin ice margin in central North Dakota. Note zonation of ice-pushed and streamlined features in relation to ice lobe. [Reprinted from Aber *et al.* (1989) by permission of Kluwer Academic Publishers.]

bly over short distances or times. Only one factor—*lateral pressure gradient*—operates everywhere beneath a glacier, regardless of local conditions.

The lateral pressure gradient is simply a result of unequal loading of the substratum beneath a glacier, in which thickness decreases toward the ice margin. This is the fundamental cause of glaciotectonism. A glacier imposes two kinds of stress on its bed: vertical stress due to static weight of ice column and drag or shear stress due to ice movement. The vertical stress is independent of ice movement. Wherever ice thickness changes over a distance, the inequality of loading generates a lateral pressure gradient regardless of the velocity or direction of ice flow.

The lateral stress created by unequal ice loading is typically several times greater than shear stress due to glacier drag. Lateral stress is greatest near the ice margin, where ice thickness changes are most rapid, but also exists to a lesser degree beneath the centers of thick ice sheets. Proglacial thrusting of ice-shoved hills is essentially a mechanism of squeezing out material from beneath the ice margin.

Much glaciotectonism should be viewed as the result of hydraulic pressure rather than mechanical deformation. Substantial uplift of thrust blocks forming large ice-shoved hills can occur only near the ice margin, where the ice load is light. Beneath thick ice, subhorizontal thrusting of rafts and diapirism are likely, but not major structural uplift.

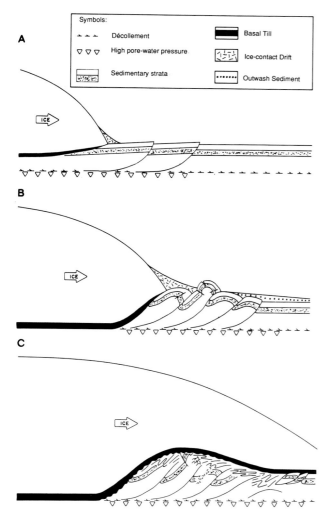

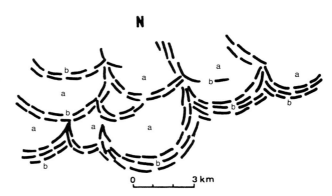

Fig. 17 Sketch map showing festoon pattern of ice-shoved ridges (b) and intervening basins (a) for Suwalki Lakeland, northeastern Poland. [Reprinted from Aber *et al.* (1989) by permission of Kluwer Academic Publishers.]

Fig. 16 Schematic models of proglacial thrusting and subglacial modifications of an ice-shoved hill during advance of an ice lobe. [Reprinted from Aber *et al.* (1989) by permission of Kluwer Academic Publishers.]

B. Analogous Nonglacial Deformation

Many structures within the earth's crust are produced in a manner analogous to glaciotectonism. Wherever an advancing mass imposes an increasing load on weak substratum material, the conditions for deformation may develop beneath or in front of the mass. Examples range from soft-sediment deformations associated with deltas to thrusting of mountain ranges at plate boundaries. An essential ingredient in all cases is displacement above a décollement in which high hydrostatic pressure is developed. Glaciotectonic deformation is not unique from this point of view; it is similar to many

other crustal disturbances, the only significant differences being temporal and spatial scales.

Convergent plate boundaries illustrate certain similarities with ice-shoved hills. Two tectonic arcs are commonly developed in connection with subduction zones (Fig. 18). The outer arc, located near the trench, is characterized by high-pressure, low-temperature deformation and metamorphism. Sedimentary volcanoes fed by cool diapirs of mud, mélange, or salt are typical of the deformed accretionary wedge. These deformations are created by lateral (hydraulic) pressure generated within and below the leading edge of the overriding plate. This hydraulic arc is analogous to deformation during glacier advance and overriding of the substratum. [*See* LITHOSPHERIC PLATES; PLATE TECTONICS, CRUSTAL DEFORMATION AND EARTHQUAKES.]

Continental collisions produce mountains in which huge thrust sheets may be pushed hundreds of kilometers horizontally. Such thrust sheets contain much fluid; some of the fluid remains within the mountain structures, some escapes in springs, and some is expelled into permeable sedimentary strata of the foreland beyond the mountains. The action of the thrust sheet is comparable to that of a giant squeegee or roller that drives fluid ahead as it advances.

This situation is analogous to the outer zone of glaciation, where advancing ice thrusts poorly consolidated and saturated sedimentary strata. Fluid is driven out of compacted sediments, as well as derived from subglacial melting and infiltration from the ice surface. Where trapped in confined aquifers or beneath permafrost or stagnant ice, groundwater migrates into the proglacial environment before escaping at springs. Thrusting of ice-shoved hills is highly dependent on such high-pressure fluid within décollements, just as thrusting of mountains requires fluid support. From a mechanical point of view, creation of major thrust mountains along convergent

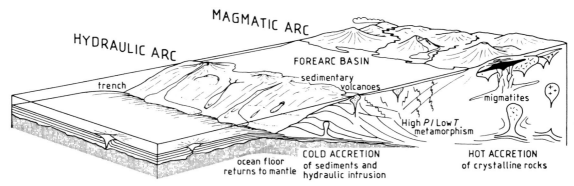

Fig. 18 Hydraulic and magmatic arcs associated with subduction zones. Cold deformation in the hydraulic arc is analogous to glaciotectonism. [Reprinted from Talbot and Brunn (1989).]

plate boundaries is really no different from thrusting of ice-shoved hills.

Bibliography

Aber, J. S., Croot, D. G., and Fenton, M. M. (1989). "Glaciotectonic Landforms and Structures," Glaciology and Quaternary Geology Series. Kluwer Academic, Dordrecht, The Netherlands.

Bluemle, J. P., and Clayton, L. (1984). Large-scale glacial thrusting and related processes in North Dakota. *Boreas* **13**, 279–299.

Croot, D. G., ed. (1989). "Glaciotectonics: Forms and Processes." Balkema, Rotterdam.

Meer, J. J. M. van der, ed. (1988). "Tills and Glaciotectonics." Balkema, Rotterdam.

Moran, S. R., Clayton, L., Hooke, R. LeB., Fenton, M. M., and Andriashek, L. D. (1980). Glacier-bed landforms of the Prairie region of North America. *J. Glaciol.* **25**, 457–476.

Ruszczyńska-Szenajch, H. (1985). The origin of glacial rafts: detachment, transport, deposition. *Boreas* **16**, 101–112.

Talbot, C. J., and Brunn, V. (1989). Melanges, intrusive and extrusive sediments, and hydraulic arcs. *Geology* **17**, 446–448.

Wateren, D. F. M. van der (1985). A model of glacial tectonics, applied to the ice-pushed ridges in the central Netherlands. *Geol. Soc. Denmark Bull.* **34**, 55–74.

Glossary

Cryosphere The frozen portion of the earth's surface, including glaciers, ice sheets, sea or lake ice, perennially frozen ground, and snow.

Glacier Mass of ice formed by the accumulation and recrystallization of snow into solid ice (density > 0.8 g/cm³) that moves due to internal plastic creep, basal sliding, or deformation of the substratum.

Glaciotectonism Structural deformation of the earth's crust as a direct result of glacier movement or loading.

Global Cyclostratigraphy

M. A. Perlmutter and M. D. Matthews

Texaco Exploration and Production Technology Department

Global cyclostratigraphy is a process-response model that predicts stratigraphic patterns by integrating climatic and tectonic history on global, regional, and basinal scales. Long-term and short-term global climate and climatic change, topography, bathymetry, uplift rates, subsidence rates, provenance and sea or lake level are all integrated into the model. The generalized stratigraphy of a basin can be forecast by determining the effects of these factors on sediment flux and accommodation space over time.

I. Conceptual Framework

Stratigraphic cycles are predicted by integrating those local processes and conditions that control sediment flux to a basin and those that control the distribution, deposition, and preservation of sediment within a basin (Fig. 1). Basin-related processes and conditions are constrained by geologic and climatic systems acting on global and regional scales over varying time frames. These broader-scope systems are categorized as either long-term (effects are observable for a million years or more) or short-term (effects are observable for thousands to hundreds of thousands of years). See Fig. 2.

Long-term systems that affect depositional processes include (a) the overall global climate for a geologic interval; (b) the tectonic conditions that control continental drift, sea-floor spreading, topography, bathymetry, the existence of faults, uplift rates, subsidence rates, and some eustatic changes; and (c) the petrology of the provenance areas. Short-term systems that affect sedimentation include those climatic changes caused by oscillations of the earth's orbit and axis of rotation (Milankovitch cycles). [*See* Cyclic Sedimentation, Climatic and Orbital Insolation Changes in the Cretaceous.]

By grouping the controls on deposition in this manner, long-term systems provide a relatively stable framework for analyzing the effects of short-term changes on deposition. The long-term climatic framework is interpreted or taken directly from published information. The effects of Milankovitch cycles are then superimposed. These cycles, which occur at periods of about 400,000 yr and 100,000 yr (orbital eccentricity), 40,000 yr (obliquity), and 20,000 yr (longitude of the perihelion), constructively and destructively interfere, forcing the long-term base climate to vary between predictable end-member conditions. Milankovitch cycles are modeled using a set of cosine functions.

Climate has a direct effect on potential sediment production and transport. Changes in sediment texture, mineralogy, and yield to a basin can be estimated by superimposing climatic variations on the topographic framework and provenance of a particular region. Occurrence and distribution of depositional environments and their preservation potential within the basin then become a function of sediment flux (controlled by short-term climate) and accommodation space (a function of long-term tectonic processes). Thus, the stratigraphic record can be forecast by integrating the tectonic and climatic histories of a basin.

II. Global Climatic Patterns

To understand how regional climate can change during a Milankovitch cycle, the present global climatic pattern must be understood in relationship to atmospheric dynamics. This is done by evaluating zonal and azonal characteristics of atmospheric circulation and seasonal changes in these properties. Climatic change is then modeled by estimating the responses of the circulation pattern to changes in insolation. [*See* Climate.]

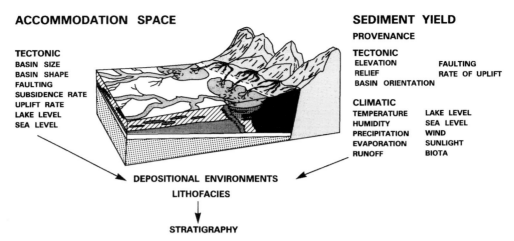

Fig. 1 Controls on basin fill. Factors are separated into those that affect sediment weathering and transport and those that affect accommodation space. [After Perlmutter and Matthews (1990).]

A. Zonal Circulation Patterns

Hadley circulation is a simple representation of the zonal pattern of atmospheric dynamics. Circulation in each hemisphere is divided into three cells: Hadley (tropical), Ferrel (temperate), and Polar (polar). See Fig. 3. Temperature and the positions of these cells control the zonal distribution of humidity, runoff, and evaporation. Humid (high runoff) zones are located proximal to atmospheric upwelling, and dry to arid conditions occur near downwelling. Cell position and size change seasonally as a function of the latitude of the thermal equator. [*See* ATMOSPHERIC CIRCULATION SYSTEMS.]

B. Azonal Circulation Patterns

Perfectly zonal, atmospheric climatic patterns are disrupted by azonal processes related to the distribution of land, sea,

and topography. Azonal effects include the seasonal migration of the Intertropical Convergence Zone (ITCZ), circulation around mid-latitude highs, proximity to oceans, and orographic effects (Fig. 4).

A latitudinal shift of the ITCZ away from the equator is generated by a combination of the tilt of the earth's axis, continental size and distribution, and the differences in heat capacity of land and sea. This shift tends to be small over oceans and large over large continents (Figs. 4 and 5). The actual climatic effect of this shift in any particular area depends on the width of the envelope described by the annual migration of the ITCZ, position within the envelope, the seasonal pressure variation between continent and ocean, wind fetch over land and water, and topography. Equatorial regions affected by narrow envelopes tend to undergo small annual climatic changes. Regions affected by wide envelopes tend to experience large, annual climatic swings known as monsoons. These areas undergo seasonal changes in tempera-

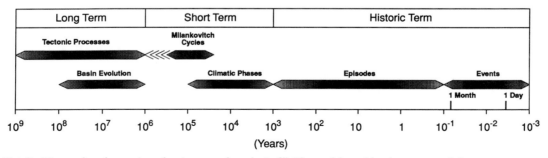

Fig. 2 Time scales of operation of major controls on basin fill. The model considers long-term and short-term processes and conditions. Historic-term episodes and events are below the level of resolution of the model.

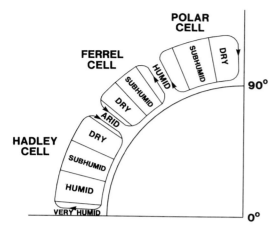

Fig. 3 Idealized latitudinal distribution of Hadley circulation and associated zones of humidity. [After Perlmutter and Matthews (1990).]

ture, rainfall, and prevailing wind direction. Near the equator, large shifts of the ITCZ can limit the occurrence of tropical rain forests by causing extended dry periods. Away from the equator, effects can produce an increase or decrease in expected precipitation. Climatic zones outside the ITCZ envelope also appear to shift or "buckle" poleward.

The mid-latitude high associated with the junction of the Hadley and Ferrell cells is not truly zonal. Instead it consists of a series of disconnected high-pressure regions resulting

from the differential heating of continents and oceans. Circulation around these mid-latitude high-pressure areas transfers heat and moisture onshore or offshore, depending on the latitude and the coast of the landmass. The amount of climatic modification is a function of wind fetch, orientation of the coastline, and ocean circulation. Wind is generally onshore in equatorward areas of mid-latitude eastern coasts (east coast effect), increasing precipitation in a zone subject to desertification when unmodified (Figs. 4 and 5). Winds are generally onshore in poleward areas of mid-latitude western coasts (west coast effect), moderating temperatures and increasing precipitation. [See MID-LATITUDE CONVECTIVE SYSTEMS.]

Ocean currents tend to transport warm equatorial water poleward and cold polar water equatorward. In combination with prevailing winds, western (ocean) boundary currents like the Gulf Stream provide heat and moisture to eastern and western coasts of continents as indicated above. Eastern (ocean) boundary currents can cool and increase relative humidity on western coastlines. Upwelling produced by wind direction and coastline orientation may also raise relative humidity on mid-latitude western shorelines by cooling air and producing fog. [See OCEAN CIRCULATION.]

Orography can also modify regional and local climates. Orographic effects are a function of initial temperature and humidity of an air mass, change in elevation, lapse rate, and wind direction. Rising air adiabatically cools increasing relative humidity. Descending air warms adiabatically, decreasing relative humidity. As a result, windward sides of mountain

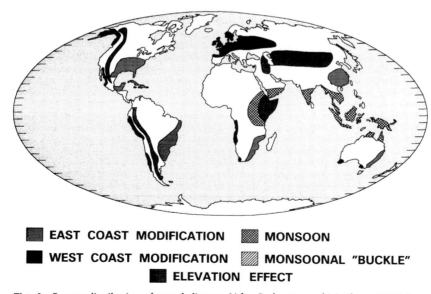

EAST COAST MODIFICATION   MONSOON

WEST COAST MODIFICATION   MONSOONAL "BUCKLE"

ELEVATION EFFECT

Fig. 4 Present distribution of azonal climates. [After Perlmutter and Matthews (1990).]

Fig. 5 Atmospheric circulation map for July, present day. The position of the ITCZ is indicated by the broken line. Large arrows indicate surface winds. Small arrows indicate surface ocean currents. [After Perlmutter and Matthews (1990).]

ranges tend to have more runoff and leeward sides less run-off. [*See* MOUNTAINS, EFFECT ON AIRFLOW AND PRECIPITATION.]

III. Milankovitch Cycles and Climatic Succession

A. Zonal Climatic Succession during a Milankovitch Cycle

The position and size of Hadley circulation cells respond to changes in the amount and seasonal distribution of insolation caused by the interaction of eccentricity, tilt, and precession. Therefore, the mean annual and seasonal positions and sizes of the six Hadley circulation cells vary over the course of a Milankovitch cycle (Figs. 5, 6, and 7). Compared to the climatic minimum, the boundaries of the cells at the climatic maximum are displaced poleward. The Hadley cell is larger and the Polar cell smaller. As these cells migrate over a Milankovitch cycle, zones of humidity migrate with them, defining the climatic extremes or end-members a region is likely to experience.

B. Zonal Shift of Hadley Circulation during a Milankovitch Cycle

There is an excellent published database of the distribution of paleoclimate indicators that allows the pattern of global cli-

matic change to be mapped from the end of the Pleistocene through the present day. These data, which include geographic positions over time of high, moderate, and low stand lakes, peats, dune fields, paleowind vectors, and glaciers, clearly indicate the environmental migration caused by shifts in atmospheric circulation that occurred during the last climate cycle.

The latitudinal variation of temperature and humidity has been idealized for changes in Hadley circulation from the last climatic minimum (phase C), about 18,000 yr B.P., to the last climatic maximum (phase A), about 8,000 yr B.P. (Fig. 8). End-member climates and their latitudinal boundaries are indicated on the right-hand side of this figure. The positions of these end-member conditions delineate 13 zones or cyclostratigraphic belts that, to a first approximation, undergo similar climatic successions during a Milankovitch cycle. The position and succession of each of these belts can be applied to the entire Pleistocene. [*See* CLIMATIC CHANGE: RELATIONSHIP BETWEEN TEMPERATURE AND PRECIPITATION.]

Of particular interest are mid-latitude Belts 4B, 5, and 6. Interplay of the upwelling and downwelling arms of the Ferrel cell cause humidity (runoff) to be highest at or near the climatic minimum. This condition is opposite that of other belts. Belt 3, for example, is tropical and humid at the climatic maximum and temperate and arid at the minimum, while Belt 5 is temperate and arid at the maximum and temperate and humid at the minimum.

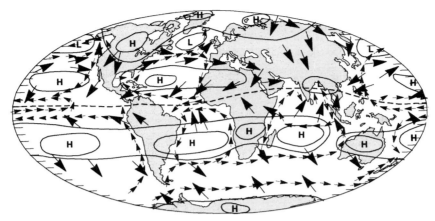

Fig. 6 Atmospheric circulation map estimated for July, last climatic minimum. The position of the ITCZ is indicated by the dashed line. Symbols are the same as for Fig. 5. [After Perlmutter and Matthews (1990).]

C. Nonzonal Climatic Modifications to Cyclostratigraphic Belts

To go from a global-scale approximation to a regional-scale application, the idealized zonal climatic succession indicated by a particular cyclostratigraphic belt must be modified by the factors that cause climates to be azonal: the seasonal shift of the ITCZ, mid-latitude high pressure systems, ocean currents, and the lapse rate. Effects of these factors can migrate latitudinally, intensify, or diminish during a Milankovitch cycle (refer to Figs. 4, 5, and 6).

At the climatic maximum, seasonal migration of the ITCZ is enhanced because increased insolation increases pressure contrasts between land and sea. Deeper continental low pressure combined with higher sea-surface temperatures can cause an increase in precipitation over land. At the climatic minimum, migration of the ITCZ is less as a result of the reduced temperature and pressure differential between land

and sea. Precipitation over land may also be less because of the lower pressure contrast and lower sea-surface temperatures.

Anticyclonic circulation around mid-latitude high-pressure systems moves poleward as temperatures increase between the climatic minimum and the maximum. This causes the positions of the east and west coast effects to move poleward as well. Summertime coastal effects at the climatic maximum may be enhanced as a result of greater pressure differences between sea (high pressure) and land (low pressure). Expansion of high-pressure systems at the maximum may cause desertification to increase on the equatorward portions of mid-latitude western coastlines where offshore or alongshore winds occur. Anticyclonic circulation migrates equatorward from the maximum to the minimum. Coastal effects also migrate equatorward and may be diminished. High-pressure systems shrink in size.

The pattern of changes in atmospheric circulation over the span of a Milankovitch cycle can also produce regular variations in the positions of mid-latitude oceanic gyres. Boundary currents warm and shift poleward at the climatic maximum, combining with onshore winds to further increase the landward transport of heat and moisture. Upwelling zones may also shift poleward. Boundary currents cool and move equatorward as the climate moves toward minimum conditions, reducing landward transfer of heat and moisture. Upwelling zones may also shift equatorward at the climatic minimum.

The increase in lapse rate between the climatic maximum and minimum can cause significant environmental changes in elevated areas. Low lapse rate at the maximum permits warmer climates to occur at greater altitudes. Higher lapse rate during the minimum forces environments that once could exist at high altitudes to shift toward sea level.

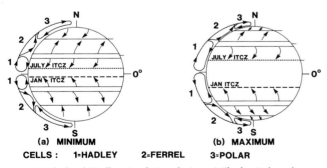

Fig. 7 Idealized Hadley circulation during a Milankovitch cycle. Mean annual position and size of each cell is indicated for (a) the climatic minimum and (b) the climatic maximum. [After Glennie (1984).]

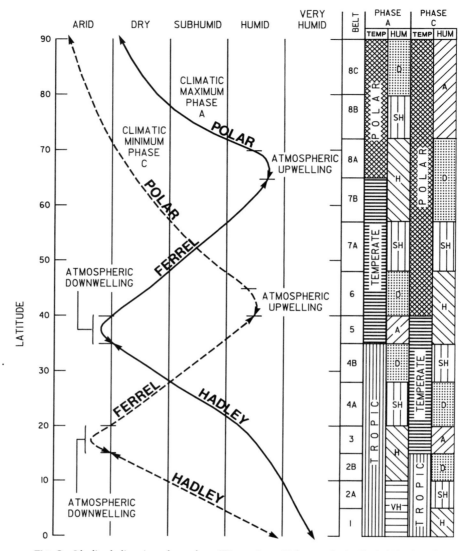

Fig. 8   Idealized climatic end-members, Wisconsin to Holocene. Latitudinal shift of Hadley circulation, temperature, and humidity between maximum and minimum conditions is mapped. Humidity and latitude are plotted on the x and y axes, respectively. Climatic maximum, phase A, is indicated by the solid line, and climatic minimum, phase C, by the dashed line. The range of the Hadley, Ferrel, and Polar cells is indicated along these lines. By convention, the area of downwelling between the Ferrel and Hadley cells is included in the Ferrel cell, and the area of upwelling between the Ferrel and Polar cells is included in the Polar cell. Columns A and C record the ranges of temperature and humidity for phases A and C, respectively. Temperature has been related to cell for this time period. Thirteen cyclostratigraphic belts (1–8C) indicate regions affected by similar climatic extremes. Key abbreviations: VH is very humid, H is humid, SH is subhumid, D is dry, and A is arid. As an example, note that Belt 3 shifts from tropical/humid (phase A) to temperate/arid (phase C) while Belt 5 shifts from temperate/arid (phase A) to temperate/humid (phase C). [After Perlmutter and Matthews (1990).]

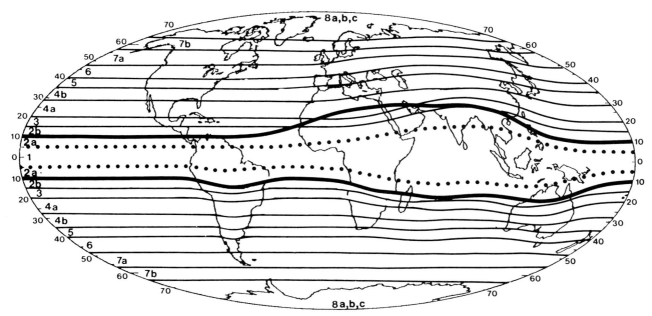

Fig. 9 Cyclostratigraphic belt map for 0–3 mybp. This map represents a zonal first approximation of the distribution of climatic end-members. Refer to Fig. 8 for climatic range for each belt. The January/July positions of the ITCZ at the climatic maximum are indicated by the solid lines and its positions at the climatic minimum are indicated by the dotted lines. Note that within the envelope described by a wide shift of the ITCZ (where it is not parallel to latitude) climate may be monsoonal. As a result, climatic conditions forecast for Belt 1 will not be strictly tropical/humid or very humid. The specific climate is a function of the factors described in the text. [After Perlmutter and Matthews (1990).]

D. Paleoclimatic Maps

Paleoclimatic maps vary with land/sea distribution, regional topography, global sea level, and ocean circulation. Therefore, maps must be derived for each new set of global and regional conditions. A rigorous treatment of climatic change requires the generation of seasonal atmospheric circulation maps at the different climatic phases of a Milankovitch cycle. For example, Figs. 6 and 7 are maps of atmospheric circulation representing two phases of the most recent climate cycle, the climatic minimum (phase C) and the present (phase B2).

If construction of seasonal maps for all phases is not practical, the idealized global distribution of cyclostratigraphic belts can be interpreted from seasonal maps of the climatic maximum and minimum (see Fig. 9). Climatic transitions between these end-members must then be interpolated and azonal modifications made to each phase. Climatic indicators should be used to constrain paleoclimate maps and the climatic succession for a region. These indicators include the depositional sequence, lake levels, evaporites, red beds, dune fields, coals, mineral assemblages, sediment grain characteristics, paleowind directions, phosphorites, and floral and faunal assemblages.

IV. The Effects of Climate and Elevation on Sediment Yield

The dynamics of sediment yield to a basin is a complex, nonlinear interaction of processes that include climate, slope, elevation, vegetation, and rock type. Traditionally, these processes and their responses are simplified, linearized, and treated as steady state. For example, although short-term climate is known to change rapidly, climate is usually considered to be stable and the influence of climate on short-term changes in yield is commonly ignored. The Global Cyclostratigraphic model, however, treats the dynamics of sediment yield in a conceptual nonlinear, time-variant framework, thereby permitting yield to be estimated as a function of climate. This allows a more realistic evaluation of stratigraphy that includes the effect of short-term climatic change.

A. Controls on Sediment Yield

The rate of sediment production from bedrock is controlled by the rate of weathering and the rate of exposure of bedrock to weathering. Weathering rate (and the characteristics of the produced sediment) is primarily determined by the interaction of climate with bedrock. This interaction also controls vegetation, which accelerates and modifies the weathering process. The rate of erosion from the provenance area is a function of the energy of the transporting medium (air, water, or ice) and is determined by the interaction of climate with slope. Vegetation affects erosion through root binding and modification of raindrop impact, infiltration rate, and runoff. The rate of erosion determines the rate of exposure. The effect of elevation on yield occurs through its positive correlation with slope and its relationship with climate through lapse rate (adiabatic heating and cooling), which can enhance precipitation or evaporation. [*See* EROSION AND WEATHERING.]

Most current paradigms of erosion and transport are derived from studies of recent rates of denudation, transport, and deposition. For the short duration of these studies, the existing climate is time invariant. Although weather may change, the climate is steady state. Under this condition, the rates of sediment production and transport (removal) from the provenance area should be in equilibrium. However, short-term climatic changes can cause an imbalance between production and transport rates, causing a sediment surplus or deficit in the provenance area. In surplus conditions, the transport system cannot remove all the sediment being created. In deficit conditions, the transport system has the capacity to remove more than what is being produced. As a result, sediment yield may be higher or lower than the equilibrium value for a specific climate, depending on the climatic succession. [*See* SEDIMENT TRANSPORT PROCESSES.]

B. Sediment Flux: The Time Rate of Change of Sediment Yield

As indicated in the conceptual framework, Milankovitch cycles are modeled using cosine functions. As a result, the rate of climatic change of end-members conditions (A and C) is slower than transition climates (B and D). In terms of sediment yield, this may permit production and erosion in phases A and C to approach an equilibrium, while forcing production and erosion in phases B and D to be out of balance.

For example, as conditions evolve from a more humid to a more arid climate, the rate of sediment removal in the provenance area will decrease more rapidly than the rate of production. This change is also accompanied by a decrease in both the amount of fine-grained material being produced and the amount of mineral alteration because of the reduced role of chemical weathering. The net effect is "storage" (production surplus) of relatively unaltered, coarse-grained sediment in the provenance region or on basin margins. This storage will continue through the arid end-member until a point in the subsequent climatic shift (from arid back to humid) when the

rate of erosion increases faster than sediment production. At this stage of change, "stored" sediment may be "flushed" into the depositional basin. The increase in moisture also accelerates chemical weathering, increasing both the fine-grained component of the sediment and the alteration of mineralogy. The concurrent increase in vegetation prolongs the process. As the flush proceeds toward the wet end-member and the available stored sediment is exhausted, the rate of production and removal may return to a dynamic equilibrium. [*See* CHEMICAL WEATHERING CONTROLS ON SAND COMPOSITION.]

The phenomenon of sediment storage and flush is best developed in regions whose climates shift from an arid phase to a dry or subhumid phase, and in regions that experience glaciation and deglaciation. Glacial flushes are not restricted to polar latitudes but may occur at any latitude where there is altitude sufficient to cause alpine glaciers, such as the equatorial Andes. In addition, river systems can bypass meltwaters out of polar latitudes. For example, the Mississippi River drainage area flooded the Gulf of Mexico with meltwater and sediment during the last deglacial event.

Repetitive episodes of storage and flush cause sediment flux to be cyclic. The general stratigraphy of an area can then be modeled by evaluating flux in context with accommodation space. In lacustrine basins, maximum yield occurs as lake-level rises. Keep in mind, however, that lake-level rise and maximum yield can occur during any climatic phase, depending on the geographic position of the basin. For example, in Belt 3 lakes rise and flux is highest in phase D1 and D2; in Belt 4A lakes rise and flux is highest in phases D2 and A; in Belt 4B lakes rise and flux is highest in phases B2 and C; and in Belt 5 lakes rise and flux is highest in phases B1 and B2. [*See* SEDIMENTARY GEOFLUXES, GLOBAL.]

The association of flux and climatic phase is critical in the evaluation of fluvial systems and in the timing of flux to continental margins relative to sea level. Although maximum sediment flux can occur at any climate phase, short-term sea-level low stands always occur at or near the climatic minimum (phase C). Thus, timing of flux to marine basins can occur at any stage of sea level and is, therefore, a function of paleogeography, as well as the base level associated with sea-level changes. This may have an effect on the interpretation of the record of long-term sea level because changes in timing of sediment delivery to continental margins can be caused by continental drift through different climatic belts. [*See* SEA LEVEL FLUCTUATIONS.]

V. Climatic Succession, Sediment Flux, and Stratigraphy: An Example

As indicated above, sediment flux is estimated by superimposing Milankovitch-induced climatic changes on the topography produced by the long-term tectonic conditions. The distribution of likely depositional environments, and therefore the stratigraphy, can be modeled over the history of a basin by balancing sediment flux with accommodation space. Space available for deposition is estimated from topography, bathymetry, and subsidence rates. As an example of the potential effects of climatic succession, depositional processes and stratigraphic history have been modeled for a hypothetical extensional basin filling with sediments in Belt 3. Climatic end-members in this belt range from tropical/humid in phase A to temperate/arid in phase C (Fig. 8). Transitional climates in phases B1 and B2 are tropical/subhumid and tropical/dry, respectively. Phases D1 and D2 show the same transitional climates in reverse order. The depositional environments and stratigraphy should exhibit changes that represent these climatic extremes and the transitional climates between them.

A. Tectonic Framework

The tectonic framework of this hypothetical basin has been described in the form of the end-member conditions that tend to occur at different stages of rift evolution: (1) high rift shoulders with accommodation space greater than sediment influx and (2) low rift shoulders with accommodation space less than sediment influx. End-member (1) represents conditions that tend to occur during the middle stages of rifting, and end-member (2) represents very early and later stages of rift evolution.

Rift architecture is treated as a series of linked half-grabens. Rift shoulder elevation is assumed similar on either side of the basin. This eliminates differences in the weathering climate caused by lapse rate, permitting textural differences between fault and hinge margins to be thought of in terms of transport distance and slope. The provenance area on the hinge margin is farther from the depocenter and has a lower average slope than the provenance area on the fault margin, which tends to be steep and local. Therefore, sediments transported to the basin from the hinge side are finer grained despite producing the same range of grain sizes as the fault side. Hinge margins have greater sediment yield because drainage areas are larger. [See CONTINENTAL RIFTING.]

B. Cyclostratigraphy of a Rift

End-member tectonic condition (1)—high margins, subsidence rate greater than sedimentation rate—produces excess accommodation space. This is expressed as a surface depression that will fill with water forming a lake, if precipitation exceeds evaporation. End-member tectonic condition (2)—low margins, subsidence rate less than sedimentation rate—more than fills the available accommodation space. Runoff forms an axial river system, resulting in the bypass of sedi-

ment. In Belt 3, the climatic succession forces runoff to vary from high to low, causing lake levels, river discharge, and sediment flux to vary significantly during a Milankovitch cycle.

At the climatic maximum (phase A) (Figs. 10a and 11a), the climate is tropical and humid. Rainfall and runoff are high, resulting in large, deep lakes or meandering axial rivers. There is significant solution of bedrock. Assuming a granitic provenance, sediment is predominantly produced by biochemical weathering, with grain sizes mostly in the range of silt and clay. Clastic material sand-sized or larger is estimated at 15% of the volume of weathered sediment. In tectonic condition (1), efficient sediment transport occurs in upland drainage areas. In either tectonic condition, lowlands are highly vegetated, trapping coarse material and slowing transport.

In the first stages of cooling, phase B1, the climate becomes tropical and subhumid (Figs. 10b and 11b). Rainfall and runoff are less than at the climatic maximum. Lakes and axial rivers shrink. Biochemical solution of bedrock decreases. Physical weathering becomes more important, resulting in the formation of more coarse clastic sediment, estimated at up to 30% of the volume of weathered sediment. Even though runoff decreases, coarse clastic sediment influx to the basin increases because production is higher. This causes alluvial fans to expand. Lower lake level reduces base level, causing downcutting and locally increasing sediment yield from drainage basins along the fault, adding to the expansion of the alluvial fans. Downcutting within the basin itself serves only to rearrange sediments already there and does not increase sediment influx to the basin as a whole. Basinward redistribution of sediment already in the basin occurs as a result of erosion at the apex of alluvial fans, in river channels, and along lake margins.

In the later stages of cooling, phase B2, the climate becomes tropical and dry (Figs. 10b and 11b). Rainfall is low, with runoff much less than in the previous phase. Overall, evaporation exceeds precipitation, forcing lakes to continue to contract and rivers to shrink and become underfit. The lower volume of water in the hydrologic cycle and the restricted level of plant growth severely reduce solution. As a result, physical processes dominate weathering. The reduced availability of water, however, now decreases the rate of sediment production but increases the relative amount of coarse material, estimated to comprise about 40% of the total. Sediment yield to the basin is less than in phase B1 because of lower rates of weathering and runoff. Basinward transfer of sediment as a result of falling lake level will also occur at a reduced rate. As the cycle approaches the climatic minimum (phase C), runoff decreases further and becomes flashy. Transport becomes even less efficient and sediment begins to be stored in the drainage basin and in alluvial fans along basin margins.

At the climatic minimum, phase C, the environment be-

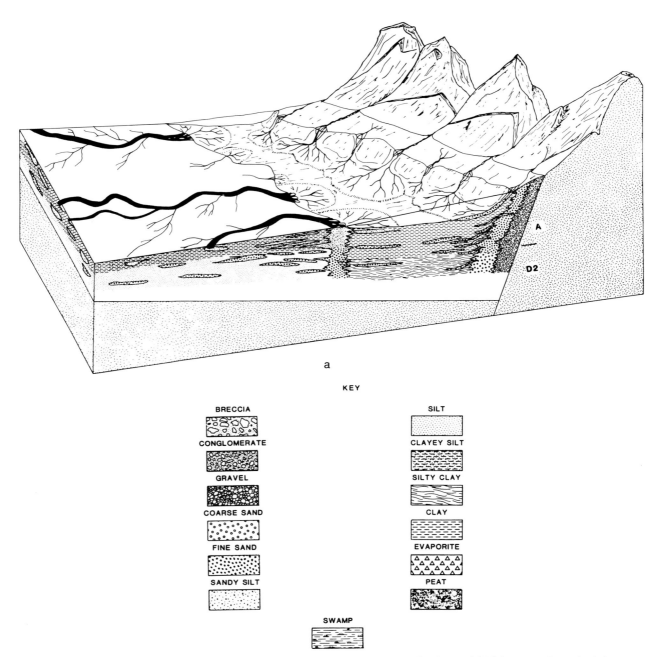

a

KEY

BRECCIA

CONGLOMERATE

GRAVEL

COARSE SAND

FINE SAND

SANDY SILT

SILT

CLAYEY SILT

SILTY CLAY

CLAY

EVAPORITE

PEAT

SWAMP

Fig. 10 Depositional/stratigraphic scenario for Belt 3. Tectonic framework: half-graben with high basin margins and subsidence rate greater than sedimentation rate. (a) Climate: phase A, climatic maximum, tropical/humid conditions. (b) Climate: phases B1 and B2, cooling transition, tropical/subhumid conditions in B1 and tropical/dry conditions in B2. (c) Climate: phase C, climatic minimum, tropical to temperate/arid conditions. (d) Climate: phases D1 and D2, warming transition, tropical/dry conditions in D1 and tropical/subhumid conditions in D2. [After Perlmutter and Matthews (1990).] (*Figure continues.*)

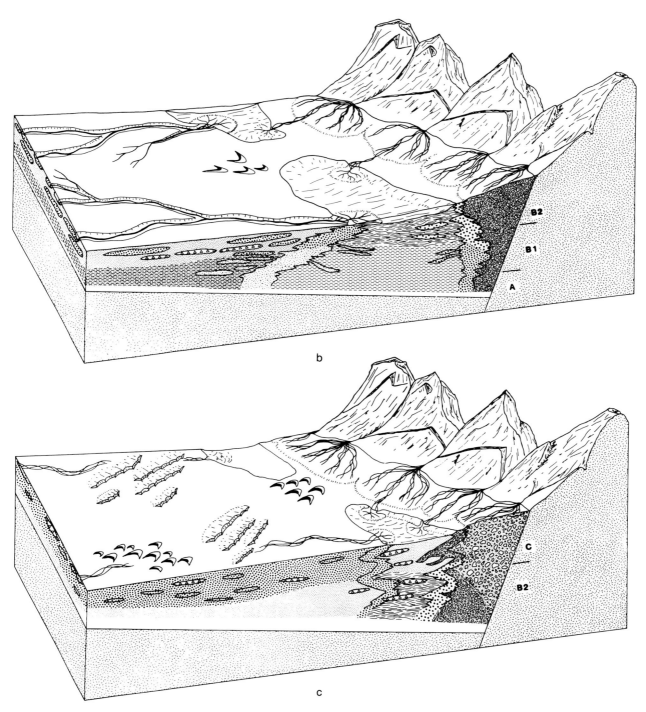

b

c

Fig. 10 (*Continued*)

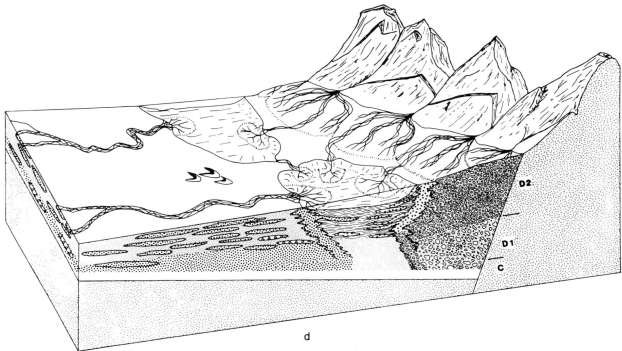

d

Fig. 10   (*Continued*)

comes arid (Figs. 10c and 11c). The area defined by Belt 3 is now under the influence of the atmospheric downwelling that occurs near the boundary of the Hadley and Ferrel cells. Rainfall is very low and sporadic. Runoff occurs as flash floods. Locally sourced lakes become hypersaline, playas, salt pans, or dry up completely. Rivers are ephemeral. Biochemical weathering is minimal. Physical processes govern weathering, and the rate of weathering is low. However, up to 70% of all the sediment produced is estimated to be coarse. In general, sediment transport to the basin is inefficient, and most sediment is left in the upland provenance areas or deposited in alluvial fans. Aeolian processes rework sediment deposited on basin margins toward the center of the basin.

In the early stages of warming, phase D1, the climate becomes tropical and dry (Figs. 10d and 11d). Rainfall and runoff increase. Lakes and river systems begin to expand and freshen. Physical weathering is still the dominant sediment-producing process, although the proportion of coarse material is reduced to 40%. This is caused by the increase in biochemical solution that accompanies the increase in moisture in this phase. Sediment yield to the basin increases as a result of higher runoff. In addition to the sediment weathered during this phase, sediment left in the provenance area or deposited in a metastable position along basin margins in the

preceding arid phase can now be transported. This produces the pulse of deposition termed a flush, which persists until this excess sediment is completely transported. The additional bed load being transported at this time will cause river systems that previously meandered to become braided. A flush may extend through a dry phase into the subsequent subhumid phase.

In the later stages of warming, phase D2, the climate becomes tropical and subhumid (Figs. 10d and 11d). Rainfall and runoff continue to increase and lakes and rivers continue to expand. Weathered sediments are finer grained (30% coarse) due to the increase in biochemical processes caused by the increase in moisture. Yield to the basin increases due to higher rates of sediment production and runoff and the cached sediments from phase C not flushed in phase D1. As the flush wanes, transported sediment becomes balanced with or limited by the material actually being produced. Rivers begin to meander again at this time.

VI. Summary

Global cyclostratigraphy provides a consistent, global framework that allows the prediction of the general stratigraphic

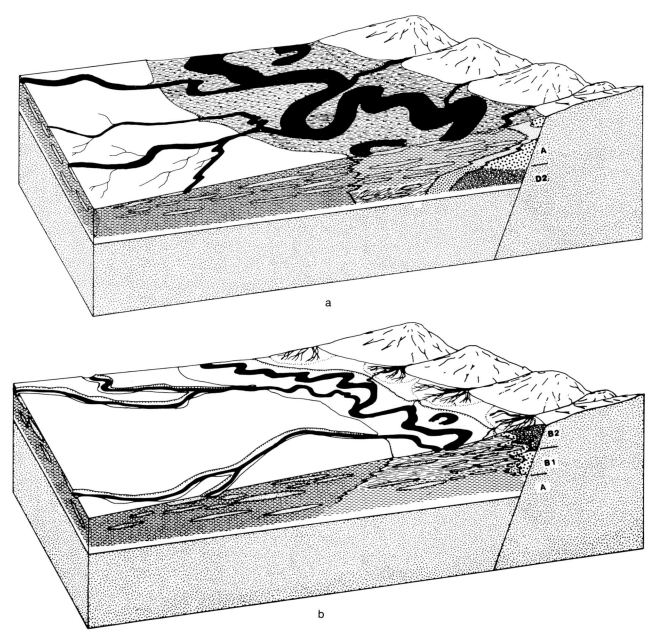

a

b

Fig. 11 Depositional/stratigraphic scenario for Belt 3. Symbols are the same as for Fig. 10. Tectonic framework: half-graben with low basin margins and subsidence rate less than sedimentation rate. (a) Climate: phase A, climatic maximum, tropical/humid conditions. (b) Climate: phases B1 and B2, cooling transition, tropical/subhumid conditions in B1 and tropical/dry conditions in B2. (c) Climate: phase C, climatic minimum, tropical to temperate/arid conditions. (d) Climate: phases D1 and D2, warming transition, tropical/dry conditions in D1 and tropical/subhumid conditions in D2. [After Perlmutter and Matthews (1990).] (*Figure continues.*)

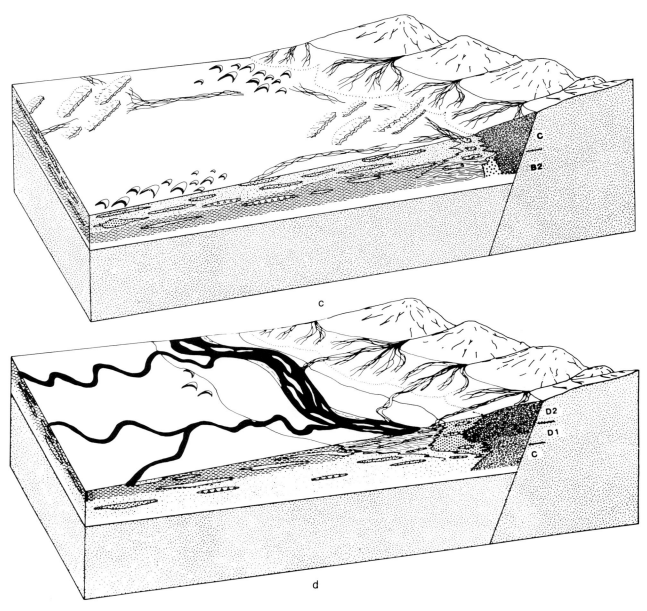

Fig. 11 (*Continued*)

patterns as a basin evolves. Stratal architecture is constrained by integrating short-term Milankovitch-driven variation in sediment yield and type with the long-term tectonic history of a basin. Thus, the tectonic evolution of a basin and its adjacent highlands acts as a slowly changing framework on which the shorter-term sediment yield cycles are superimposed.

Milankovitch oscillations cause variations in the amount and distribution of insolation. The earth's atmospheric circulation pattern shifts in response to this change in heating. Zones of atmospheric upwelling (humid areas) and downwelling (arid areas) migrate and change as a result. Mapping the migration in the position of these cells permits the climatic end-members and succession to be estimated for an area. A zonal first approximation of climatic change suggests 13 separate successions of cyclostratigraphic belts.

The direct relationship of climate with sediment yield causes yield to vary over these same short-term time scales. Climatic cycles, combined with the complex, nonlinear forces that control yield, cause yield to vary between conditions of dynamic equilibria and disequilibria. Sediment yield may vary by only a small amount to as much as an order of magnitude over the span of a cycle, depending on the belt and the interaction of the various Milankovitch frequencies.

The same cyclic climatic processes also affect weathered grain properties and the depositional systems within a basin. Thus, after defining the general history of subsidence and uplift, we can relate the quantity and type of sediment received by a basin and the distribution of depositional environments within the basin to its paleolatitudinal position.

Bibliography

Berger, A., Imbrie, J., Hays, J., Kukla, G., and Saltzman, B., eds. (1964). "Milankovitch and Climate," Parts 1 and 2. NATO ASI Series. D. Reidel, Boston.

Glennie, K. W. (1984). Early Permian-Rotliegend. *In* "Introduction to the Petroleum Geology of the North Sea" (K. W. Glennie, ed.), pp. 41–60. Blackwell Scientific, Boston.

Perlmutter, M. A., and Matthews, M. D. (1989). "Global Cyclostratigraphy: Effects on the Timing of Sediment Delivery to Continental Margins Relative to Sea Level." *Proceedings of the AGU Chapman Conference on Long Term Sea Level Changes. Snow Bird, Utah, April 17–20, 1989.*

Perlmutter, M. A., and Matthews, M. D. (1990). Global cyclostratigraphy — A model. *In* "Quantitative Dynamic Stratigraphy" (T. A. Cross, ed.), pp. 233–260. Prentice-Hall, Englewood Cliffs, New Jersey.

Street, F. A., and Harrison, S. P. (1985). Lake levels and climate reconstruction. *In* "Paleoclimate Analysis and Modeling" (A. D. Hecht, ed.), pp. 291–340. Wiley (Interscience), New York.

Glossary

Climate cycle Climatic succession associated with Milankovitch cycles.

Climatic maximum Warmest phase of a climate cycle.

Climatic minimum Coolest phase of a climate cycle.

Cyclostratigraphic belts Regions of the earth's surface undergoing similar climatic cycles.

Milankovitch cycles Variations in the orbital eccentricity, axial obliquity, and longitude of the perihelion that cause differences in the seasonal and latitudinal distribution of insolation.

Global Positioning System

Robert C. Dixon
Omnipoint Data Co. Inc.

The Global Positioning System (GPS) is a military system designed to allow completely passive (no transmission required on the part of a user) position location, using only signals from a set of earth-orbiting satellites. (See Fig. 1.) Provision is made for civilian use of the system, however, since GPS is intended to supplant many other position-location systems currently in use. Position-location resolution is intended to be much better for the military users than for civil users. Nevertheless, it is expected that civil uses for GPS for surveying, vehicle location, maritime navigation, and other applications will be widespread by the mid 1990s. The system is planned to have full operational capability by 1992, and user receivers of all kinds have been and are currently in development. This includes receivers that are large enough to require a twin-engine aircraft for transportation as well as those that are small enough to be carried in a pocket. Current receivers can resolve differential position to within millimeters and absolute position to within a few meters.

I. Global Positioning System

The Global Positioning System, or GPS, is a general-use position location and timing system consisting of three subsystems, or segments.

A. Space Segment

The space segment is made up of the constellation of 24 satellites (which is not yet complete but is expected to be operational in late 1991). These satellites are identical to one another, and will be in six orbital planes, each inclined at 55° relative to the earth's equator, as illustrated in Fig. 2. Originally, the space sector was intended to include 28 satellites, but this number was reduced to 21 because of funding limitations and then restored to 24 in 1989. In all cases, a portion of the satellites are considered to be spares. That is, the system is designed to operate with a fraction of the number that are actually in orbit. However, the spares are "hot" spares, that is, they are continually in use, along with "operational" satellites.

The space segment is designed to provide a probability of better than 99% that, for a 10-year period, four or more satellites with "good" geometry will be in view of users over most of the earth's surface.

"Good" geometry is defined to mean that the geometric dilution of precision (called GDOP) is less than an arbitrarily chosen amount (six). Geometric dilution of precision amounts to that effect that occurs when triangulation is attempted using a short baseline, as illustrated in Fig. 3. Ideally, the GPS user will have three or more satellites that are well above the horizon and yet far enough from one another that they provide baselines that are adequate for accurate position measurement. Figure 4 shows the predicted coverage for 21 satellites, with "outages" defined as those times in which GDOP is greater than six. Note that for these areas, which are comparatively small, outages are less than one hour per day, and that even then the outages are split into two parts of one-half hour maximum duration each.

Satellite orbits will have a duration (or period) of approximately 12 hours at 10,900 nautical miles (22,200 km), which causes every satellite to overfly every position on the earth once in every 24 hours. Statistically, one can expect to have 12 satellites, on the average, in view, no matter what one's location on the earth. The trick is to choose those satellites that provide the best signal as well as satisfactory geometry.

Characteristics of the satellites are

- Weight   1721 pounds
- Power supply   700 W at end of life
- Frequency standards   Two cesium, two rubidium

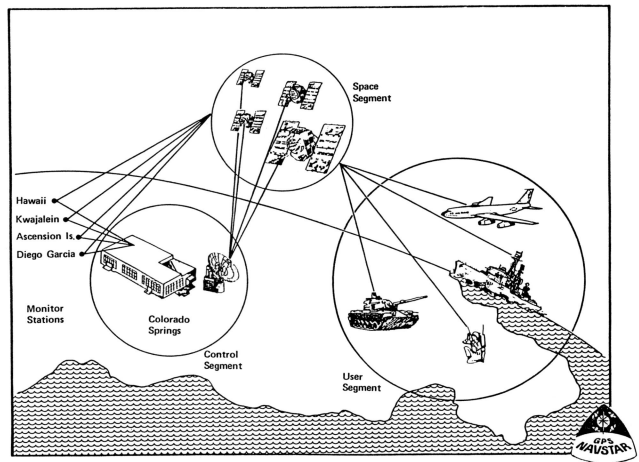

Fig. 1 GPS overall system.

- Transmitted signals   Ranging signals, almanac, handover information, spacecraft housekeeping and health information, satellite position and time, and data
- Received signals   Almanac, data, correction factors, and satellite ephemerides

The GPS system is intended for use by military and civilian users and is already being used by both, even though the space segment is not yet complete. An advantage of the system is that only the number of hours of coverage per day decreases when the number of satellites is reduced. The accuracy of the system is the same, as long as four or more satellites, with adequate geometry, are in view.

B. Control Segment

The control segment consists of the control facility, originally at Vandenberg Air Force Base in California but since 1988 in

Colorado Springs, Colorado, and four monitoring stations. These monitoring stations are located at surveyed sites in Diego Garcia, Hawaii, Kwajalein, and Ascension islands.

The control facility monitors all satellites' positions, their timing and health, and uses this information, together with information garnered from the monitoring stations and other services, such as the National Weather Service, to generate correction factors. These correction factors are then transmitted to the satellites to correct their ephimerides as well as the other information they transmit to users for correction of the positions they calculate.

It should be noted that the overall GPS system is designed to continue to operate without supervision from the control segment for a period of up to 18 days while providing the stated accuracy. This is the reason for the multiple atomic standards on the satellites, and it is also the reason that an almanac of satellite position is maintained on every satellite, under the supervision of the control facility.

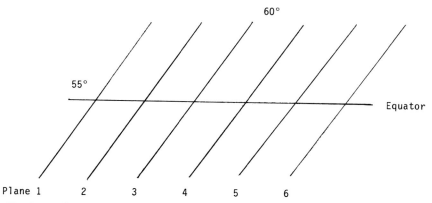

Fig. 2 Satellite constellation geometry: 24 satellites altogether, in six planes, four per plane. Orbital period 12 hours.

The monitoring stations, as well as the control facility, being at accurately surveyed sites, do not need to receive signals from multiple satellites to determine satellite ranging or timing accuracy. They derive the satellite's accuracy by simply comparing their known position to the range measured with the position that the satellite reports, to measure any satellite error. Similarly, the monitoring station can measure the satellite's frequency and timing offset, if any.

C. User Segment

The GPS user segment is the most varied of all. The system has already been put to many uses that the designers never intended, and many more will be conceived in the future. The

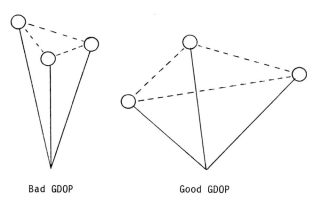

Bad GDOP   Good GDOP

Fig. 3 Illustration of satellite configurations producing good and bad GDOP (geometric dilution of precision). When distance from satellite to satellite is small, angular resolution is poor. Best GDOP is produced when volume of tetrahedron is greatest. Definition is

$$\text{GDOP} = (\sigma_x^2 + \sigma_y^2 + \sigma_z^2)^{1/2}$$

user segment includes a full range of configurations designed to serve myriad uses, from handheld through those that cannot be carried in any vehicle smaller than a medium-sized twin-engined aircraft.

Anyone or anything that employs the satellite downlink signal or signals can be considered part of the user segment. In the future, almost everyone will fit in this category.

II. GPS Signal Structure (Satellite to Ground)

GPS signals are of a class called "spread spectrum" signals that employ bandwidths much greater than necessary to send the information conveyed (see Section VI). Very little information, considering the bandwidth used, is included in the satellite downlink signals, other than that necessary to the task of position location. This information is transmitted at a low 50 bits per second, and little room is left for any other information to be transmitted. The low information rate is vital to the system, however, to enable the user to take advantage of the system's spread spectrum processing gain and jamming margin. For example, at any time a given user can expect to see signals from an average of 12 satellites. These signals are all simultaneously incident on the receiver's antenna. The only method that the receiver has available to effectively separate these signals, winnowing just the one wanted from all the others, is to employ the spread-spectrum code from the desired satellite. That is, each satellite will transmit a code that is different from all the other satellites, and the user must recognize the desired signal by looking for that code and rejecting all other codes.

GPS downlink signals (for positioning) are transmitted at two frequencies, designated L1 and L2, that are derived from a single source and are therefore coherent with one another

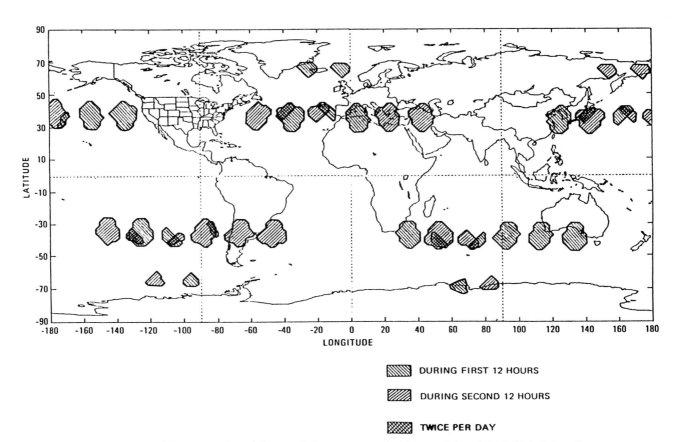

▨	DURING FIRST 12 HOURS
▨	DURING SECOND 12 HOURS
▨	TWICE PER DAY

Fig. 4 GPS coverage, full operational capability (unaided user set operation). Zones of degraded (PDOP > 6) 4-satellite coverage for 21-satellite constellation using 5° mask angle. Maximum duration is one-half hour.

on transmission. The accuracy of the frequency source is of the order of parts in 10^{12} to 10^{13}, as each satellite contains four atomic standards—two rubidium and two cesium. These four standards are included to insure reliability and to combine the long-term stability of cesium with the short-term stability of rubidium. These standards operate at a frequency of 5.115 MHz and are combined to produce a single 5.115 MHz reference signal. The L1 and L2 carrier frequencies generated from the reference frequency are 1575.42 MHz for L1 and 1227.6 MHz for L2. (There is a third, L3, carrier frequency generated from the reference signal, but it is not intended for GPS users.) The L1 frequency is 308 times the 5.115 MHz reference, and the L2 frequency is 240 times the reference.

All other frequencies used for GPS signal generation are also derived from the 5.115 MHz reference signal. These are listed in Table I.

The spread spectrum approach used by GPS is direct sequence modulation, in which the carriers are directly modulated, using biphase phase-shift keying (the same as is used in

many data transmission systems, except that here a code instead of data is used to modulate the carrier). Two codes are employed, one of which is called the C/A code, which stands for coarse/acquisition or clear/acquisition. This code is used by military users for initial synchronization and subsequent transfer to another code, called the P code, using information carried as data within the C/A code signal. Civilian or "unauthorized" users employ only the C/A code, which cannot provide the same accuracy in position measurements as the P code.

The C/A code has a length of 1023 chips (one clock interval of the code) and a chip rate of 1.023 megachips per second. This code therefore repeats 1000 times per second with a period of one millisecond. The C/A code is a Gold code, generated by a dual 10-stage linear maximal shift register sequence-generator. The number of codes available from Gold code generators of this length, without changing the base pair of codes, is 1025. Only 37 of the codes available are designated for use at any given time, however. The GPS program office has stated that the C/A codes may be changed at any time; therefore, those who intend to employ the downlink signals must design their receivers so that their codes may also be changed readily.

Each satellite transmits a different C/A code from the same set of Gold codes. That is, they are one of the same set of 1025 codes.

A second code, or P code (which stands for either precision or protected code) also modulates a carrier at the L1 frequency. The rate of the P code is exactly 10 times the C/A code rate, or 10.23 megachips per second. The P code is much more complex than the C/A code, however, even though it is generated in a similar manner. That is, both the C/A code and the P code are generated by modulo-two addition of linear maximal shift register-generated sequences. The C/A code uses only two 10-stage registers, whereas the P code is generated by four 12-stage registers. These registers are also truncated so that they do not output their normal code length, which is $2^{12} - 1$, and the truncation lengths used are also not the same. Therefore, when these registers and their codes are modulo-two added together, they generate a code whose length is 2.3546×10^{14} chips and whose period (at 10.23 Mchip per second) is 266.4 days.

The P code, however, is not allowed to complete its cycle. Rather, it is offset by one week in each satellite and started again weekly at the same point. That is, all satellites transmit the same P code, but operate in a different section of the code that is offset from all other satellites by at least one week. Also, it is expected that the P code will be encrypted for use by authorized users only. The P code, because of its 10-times-higher chip rate, can provide 10 times better range resolution than the C/A code; this is one reason why the military GPS users can expect better position accuracy than the civilian or nonauthorized users. (Authorization in this context means simply that the user has access to the P code key; it has no reference to a users permission to use the GPS downlink. Permission is not required, although in the future a fee may be charged for access to knowledge of the correct connections for either the P code or the C/A code.)

Two L1 carriers are used, but both are at the same frequency; 1575.42 MHz. After generation, by multiplying the reference signal (5.115 MHz) by a factor of 308, the carrier signal at 1575.42 MHz is divided in two, and one of the two is phase shifted by 90° with respect to the other (see Figure 5). Then these two carriers are individually biphase-modulated by the P and C/A codes and added back together, with the P-code-modulated carrier suppressed by 3 dB. The overall effect of this process is to produce a dual-carrier, doubly modulated signal at the L1 frequency. This also gives the C/A-code-modulated signal a small (3 dB) advantage to make up for the fact that not only is its code shorter, but its processing gain is 10 times less than the P-code-modulated carrier.

The L2 carrier is modulated by only one code, which can be either the P code or the C/A code. (It is not anticipated that the C/A code would ever be used, since the primary reason for having the second carrier is to allow calculation of a compensation factor for ray-bending due to the signal passing through the ionosphere.) As this correction factor is needed only by authorized users to produce the higher accuracy provided to them, it is unlikely that the C/A code would be transmitted on the L2 carrier. Biphase modulation is employed on the L2 carrier, as on the two L1 carriers.

Biphase, or BPSK modulation, is a simple modulation format in which the carrier is shifted by either plus 90° or minus 90° from its unmodulated condition, depending on whether the data, or code, is a ONE or a ZERO. When four phases are used (implying that two data bits or two code chips are present) the signal is called "quadriphase" or four-phase PSK (QPSK). One can consider the L1 signal to be an unbalanced quadriphase signal, since the P-code-modulated carrier is at one-half the power level of the C/A-code-modulated carrier. Alternatively, it can be considered to be two independent carriers that just happen to be at the same frequency, phase shifted by 90°, and with one carrier at half the power level of the other. (This is, in fact, the way GPS receivers usually operate. They independently despread and demodulate the P-coded and C/A-coded signals, and a C/A-code-only receiver must ignore the fact that the P-code-modulated signal is present.)

A simple BPSK signal can be expressed as

$$f(t) = A \cos \omega_c t \pm 90°$$

where $+90°$ represents a ONE and $-90°$ represents a ZERO. A QPSK signal is represented as

$$f(t) = A \cos \omega_c t \pm 90° \pm 45°$$

where $+90°$ represents a ONE and $-90°$ represents a ZERO of one data bit or code chip, and $+45°$ represents a ONE and $-45°$ represents a ZERO of another data bit or code chip.

In the GPS system, the actual data is embedded in the two codes, by modulo-two addition of both codes with the data. The modulo-two addition process is one in which $1 = AB' + A'B$ ("+" here is the logical OR). That is, when two inputs are the same, the output is a ZERO, and when they are different, the output is a ONE. Now, remembering that the code rates in GPS are at much higher rates than the data (data is at 50 bits per second, while the P and C/A code rates are 10.23 Mchip per second and 1.023 Mchip per second, respectively), if we add the code and data together modulo-two we see that the data causes the code to be inverted when the data is a ONE and noninverted when the data is a ZERO. Also, in the case of GPS, the data rate is an exact submultiple of both code rates, so there is always an exact number of code chips per data bit. For the C/A code, in fact, there are exactly 20

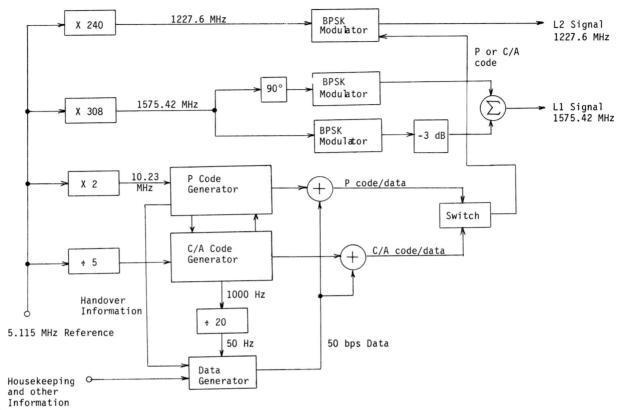

Fig. 5   Block diagram showing carrier generation and modulation structure.

repetitions of the C/A code per data bit, for a total of $20 \times 1023 = 20{,}460$ chips per bit of information. The P code, on the other hand, is much longer in period than the data bit period; so we can only say that there are 204,600 chips per data bit.

It is important to realize that exactly the same data is added to both the C/A code and the P code. That is, even though it is expected that the P code may be encrypted, the data transmitted will not be protected by that encryption process, as it will also be transmitted on the C/A code that is not encrypted.

Data content in the GPS system consists of a handover word, used to allow authorized users to synchronize more quickly to the P code after acquiring C/A code synchronization; an almanac, which tells a user where to expect the satellites to be for the next 18 days; and correction factors for tropospheric anomalies, satellite problems, or other factors known to the master station that can be transmitted on the satellite uplink and relayed to users on the satellite downlink. See Table I.

III. Satellite Carrier Generator and Modulator Architecture

The GPS satellite architecture was shown in Fig. 5. This satellite is interesting, and somewhat unusual in that all GPS-related signals are generated from the same 5.115 MHz reference frequency, which makes all of the downlink (except for the command link) signals coherent at the satellite. For this reason, it has been necessary to degrade the C/A-code-modulated channel to prevent it from being used to provide position accuracy similar to that provided by the P-code-modulated channel. The signals are generated as follows:

- P code clock   5.115 MHz multiplied by two (10.23 MHz)
- C/A code clock   5.115 MHz divided by five (1.023 MHz)
- Data clock   5.115 MHz divided by 1,023,000 (50 Hz)

Table I Summary of Satellite Downlink Signal Characteristics

Frequency of operation	L1 1575.42 MHz (308 × 5.115 MHz)
	L2 1227.6 MHz (240 × 5.115 MHz)
	L3 1381.05 MHz (270 × 5.115 MHz)
	Sband 2227.5 MHz (command downlink)
Signal strength at user (0 dB gain antenna)	L1 −160 dBW (C/A code)
	−163 dBW (P code)
	L2 −166 dBW (C/A or P code)
	L3 −166 dBW (one of C/A or P code)
	−169 dBW (each of two signals)
P code	10.23 × 10⁶ chip per second
	266.4-day period, truncated to one week
	All satellites transmit different delay of same code.
C/A code	1.023 × 10⁶ chip per second
	1.0-millisecond period Gold code
	All satellites transmit different code.
	Code length 1023 chips
Modulation	L1 Unbalanced QPSK, using both P code and C/A code. P and C/A code carriers shifted 90°. P code carrier reduced 3 dB compared to C/A code carrier.
	L2 BPSK, using either P code *or* C/A code.
	L3 BPSK, using either P code *or* C/A code, *or* balanced QPSK using both codes.
Data	Rate 50 bits per second, modulo-two added to both P code and C/A code.
	Data frame (or page) = 1500 bits/30 seconds
	300 bits per subframe/6 seconds
	25 pages total data/12.5 minutes
	Handover word transmitted every 6.0 seconds
Transmitted RF power (E.R.P.)	L1 +26.8 dBW (C/A) +23.8 dBW (P)
	L2 +19.7 dBW (C/A *or* P code)
	L3 +20.1 dBW (one code)
	+17.1 dBW (each of two codes)
Nominal path loss	L1 186.8 dB
	L2 185.7 dB
	L3 186.1 dB

- L1 carrier   5.115 MHz multiplied by 308 (1575.42 MHz)
- L2 carrier   5.115 MHz multiplied by 240 (1227.6 MHz)
- L3 carrier   5.115 MHz multiplied by 270 (1381.05 MHz)

Vector diagrams showing the P code and C/A code modulated BPSK signals interpreted as two independent signals and as a single unbalanced QPSK signal are shown in Fig. 6. Spectra of the L1 and L2 signals are given in Fig. 7.

IV. Position and Time Measurement Using GPS

A GPS user can derive both position and time information, as well as velocity, from the satellite downlink signals. This is accomplished through a method known as multilateration, which is the generic term for any process in which position is determined by measurement and processing of ranging measurements. GPS receivers do not require directional antennas for positioning, although some GPS receivers do employ them for other purposes.

Three satellite signals are required by a receiver to derive position, and a fourth satellite signal is required to derive time. Of course, the quality of these signals is important, but a receiver is likely to have 12 signals to choose from at any given time. Therefore, it would choose those with the best signal-to-noise ratio and the best geometry.

On receipt of three satisfactory signals and range measurement from the satellites to the receiver,[1] the receiver employs a set of three simultaneous equations in three unknowns from which it calculates its position. Assume the receiver is at position $x_{unknown}$ (X_u), $Y_{unknown}$ (Y_u), and $Z_{unknown}$ (Z_u) with satellites at $X_1Y_1Z_1$, $X_2Y_2Z_2$, and $X_3Y_3Z_3$ (which are precisely known, because the satellite tells the user's receiver where it is, and the control facility tells the satellite where the satellite is). The range from a satellite to the user receiver can then be expressed as (using the three-dimensional form of the Pythagorean theorem):

$$R^2 = (X + X_u)^2 + (Y + Y_u)^2 + (Z + Z_u)^2$$

where R is measured pseudo-range; X, Y, and Z are the satellite's known coordinates; and X_u, Y_u, and Z_u are the receiver's unknown coordinates.

The position of the receiver with respect to three satellites whose positions are known allows a set of three equations in three unknowns to be generated. The receiver can solve this set of equations to determine its position. These equations are

$$R_1^2 = (X_1 + X_u)^2 + (Y_1 + Y_u)^2 + (Z_1 + Z_u)^2$$
$$R_2^2 = (X_2 + X_u)^2 + (Y_2 + Y_u)^2 + (Z_2 + Z_u)^2$$
$$R_3^2 = (X_3 + X_u)^2 + (Y_3 + Y_u)^2 + (Z_3 + Z_u)^2$$

[1] In the GPS system, the user is completely passive and does not need to transmit for any reason to determine position. Range measurements are one way measurements and are therefore called "pseudo-range" measurements.

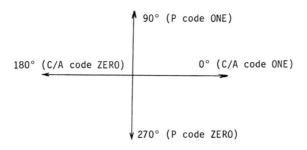

(a) L1 signal structure as two independent BPSK signals in phase quadrature.

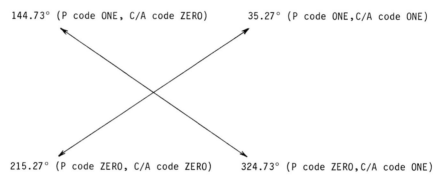

(b) L1 signal structure as an unbalanced quadriphase signal.

Fig. 6 Illustration of the L1 signal as independent BPSK and combined as unbalanced QPSK signals.

where R_n is the pseudo-range measurement on satellite N; X_n, Y_n, and Z_n are the coordinates of satellite N; and X_u, Y_u, and Z_u are the coordinates of the GPS receiver. These equations can be solved for the three unknowns, usually in three to five iterations, by a modern microcomputer.

Thus far, the assumption has been that the user receiver has a good, accurate assessment of satellite time (which is maintained by the control facility). If so, the receiver can make a range measurement by setting its code generator to coincide with system time and searching until it synchronizes with the downlink code from the satellite it chooses. By counting the number of code chips it must search, it derives an accurate estimate of the range to the satellite. If time is not accurately known, however, the receiver's measurement of range will be in error by an amount $C\Delta t$, where C is the code rate in chips per second and Δt is the time error.

This error can be resolved by measuring the range to a fourth satellite and forming a fourth equation, which gives a set of four simultaneous equations in four unknowns — which can be solved as before. These are

$$R_1^2 = (X_1 + X_u)^2 + (Y_1 + Y_u)^2 + (Z_1 + Z_u)^2 + (C\Delta t)^2$$
$$R_2^2 = (X_2 + X_u)^2 + (Y_2 + Y_u)^2 + (Z_2 + Z_u)^2 + (C\Delta t)^2$$
$$R_3^2 = (X_3 + X_u)^2 + (Y_3 + Y_u)^2 + (Z_3 + Z_u)^2 + (C\Delta t)^2$$
$$R_4^2 = (X_4 + X_u)^2 + (Y_4 + Y_u)^2 + (Z_4 + Z_u)^2 + (C\Delta t)^2$$

The fourth equation allows a solution for the time offset Δt, which allows the user receiver to be set accurately to the system time used by the satellites and maintained by the control facility. On subsequent position measurements, assuming the receiver has a stable clock, only three satellites need to be in view with good geometry.

No attempt will be made here to demonstrate a solution to either set of simultaneous equations, as this task is beyond the skill and patience of most humans. We see, therefore, that the GPS system is an ideal application of microcomputers,

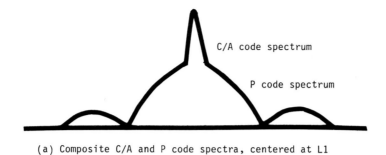

(a) Composite C/A and P code spectra, centered at L1

(b) P code spectrum, centered at L2

Fig. 7 Spectra typical of GPS transmitters at L1 and L2. Envelope of (b) has $(\sin x/x)^2$ power distribution. Envelope of (a) is composite of two $(\sin x/x)^2$ spectra, where C/A-code-modulated carrier has twice the power of the P-code-modulated carrier, and its code is at one-tenth the rate.

which make practical use of small, portable receivers for position location by multilateration, where they were never practical before.

As GPS system time can be derived from the system itself, by measuring range to four satellites whose position is known and whose time is maintained by the control facility, a user can turn on the receiver from a cold start, having no idea of present position or the correct system time, and still find both within a reasonable amount of time. The time specified for military systems to accomplish this task from a cold start is 7.0 to 10.5 minutes. Nonmilitary receivers vary greatly in their capabilities. There are more than 100 GPS receivers already developed or in the process of being developed, so no attempt will be made here to define their abilities. The time required to obtain a position measurement from a cold start consists of

1. Time for the receiver's clock to warm up.
2. Time to search through the satellite downlink C/A codes, acquiring synchronization, and measuring range on each satellite in view.
3. Time to synchronize the P code, using the handover word (as applicable).

4. Time to perform the solution of the set of simultaneous equations for time and position.

Of course, the more information the receiver has that reduces the degree of uncertainty, the quicker these tasks can be performed. For example, if the receiver knows (through the almanac) what satellites should be in view, it is not necessary to search through the C/A codes of those that are not available. In fact, if the receiver knows which satellites will provide the best geometry for a measurement *a priori*, it can go directly to the codes transmitted by those satellites and reduce measurement time even more. Some systems can, in fact, synchronize directly to the P code, given the proper timing information through outside aiding, without using the time-consuming process of C/A code synchronization and then transferring to the P code.

The code search process is one of slipping the receiver's code in time, with respect to the downlink code, and counting the number of chips slipped. The technique of slipping the receiver's code is a simple one, accomplished by delaying the receiver's code clock (but not its timing reference) some fraction of a chip at a time and counting the number of delay increments. A typical GPS receiver might delay its code clock by a tenth of a chip time for each increment, which effectively increases its range resolution by a factor of 10. Search rate in a receiver is limited by the receiver's bandwidth, which is in turn tailored to the data rate for which the receiver is designed. As GPS receivers are designed for a data rate of 50 bits per second, a typical GPS receiver can be expected to search for code synch at a similar rate. Therefore, one can expect a simple, unaided GPS receiver to search through a single C/A code in approximately $1023 \div 50 = 20.5$ seconds.

Searching blindly through the P code, with a mere one-second time uncertainty, would require some $10.23 \times 10^6 \div 50 = 204,600$ seconds, or almost 57 hours. This is the primary reason for the existence of the C/A code. Since it is there, however, it is made available for use by those who are not authorized to use the P code.

V. Resolution of the Ionospheric Propagation Problem

A significant problem in any system that must transmit through the atmosphere, and measure distance from transmitter to receiver, is the fact that the path length of the signal is not always a straight line. The signal's path length is a function of the angle at which the signal enters the ionosphere and the density of the electrons in the volume through which it passes. Therefore, a satellite directly overhead may have less error in the range measurement made using its signal than a satellite near the horizon. This problem is mitigated for GPS purposes by measuring range from a satellite using both the L1 and L2 frequencies and generating a correction factor that allows the true range to be determined. [See IONOSPHERE.]

Let us assume that the signal coming through the ionosphere is delayed by an amount K/f^2 in code chips, where K is a constant which is a function of the electron density and the angle of the signal path, and f is the signal frequency. (Remember that signal frequency is known to parts in 10^{12} or better.) A range measurement could then be expressed as

$$R = R_{\text{true}} + K/f^2$$

and if range measurements are made at both L1 and L2, we have

$$R_1 = R_{\text{true}} + K/f_1^2$$
$$R_2 = R_{\text{true}} + K/f_2^2$$

and

$$\text{(a) } R_1 = R_{\text{true}} + K/f_1^2$$
$$\text{(b) } R_2 = R_{\text{true}} + K/f_2^2$$

Table II GPS Error Sources

Error contributor	Type of error	Cause
Pseudo-range noise	Random	Poor signal-to-noise ratio; bandwidth to clock tracking loop
Pseudo-range quantization	Random	Random error in range count
Code clock drift	Bias	Code clock-rate error
Computation	Random	Algorithms; computer limitations
Ephemeris uncertainty	Bias	Inability to measure and/ or predict exact position
Satellite clock delay	Bias	Variation in clock correction
Group delay	Bias	Variation in transmitter and/or receiver RF delay
Ionospheric delay	Bias	Inability to predict exact correction factor
Tropospheric delay	Bias	Inability to predict exact correction factor
Multipath delay	Random	Signal arriving from more than one direction
Doppler shift of carrier and code clock	Bias	Relative velocity of satellite and user; uncertainty in position

Dividing (a) by (b), we get

$$\frac{R_1 - R_{\text{true}}}{R_2 - R_{\text{true}}} = \frac{K/f_1^2}{K/f_2^2} = \frac{f_2^2}{f_1^2}$$

and solving for R_{true}

$$R_{\text{true}} = \frac{(f_2^2/f_1^2)(R_2 - R_1)}{(f_2^2/f_1^2) - 1}$$

which results in the true range, in terms of the two range measurements made (at the two frequencies) and the frequencies themselves. Since we know f_1 and f_2 accurately, this reduces to

$$R_{\text{true}} = \frac{0.60718\,(R_2 + R_1)}{0.39281}$$

and we see that the ionospheric delay can be estimated by measuring the range at the L1 and L2 frequencies, allowing the true range to be determined.

Table II examines GPS error sources, and Fig. 8 presents graphic analyses of navigation performance results for comparison of position errors.

Table III is a survey of GPS measurement capability.

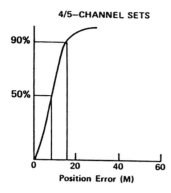

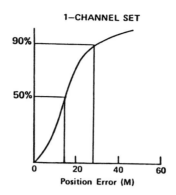

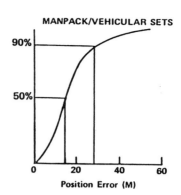

6 Host Vehicles
76 Missions
3 and 4 Satellites

Fig. 8 Phase 1 navigation performance results (3-D position error).

VI. Spread Spectrum Signals

Spread spectrum signals have special properties that are vital to making the GPS system viable. These are

- *Selective addressing* This property allows a GPS receiver to synchronize to a single satellite's signal while ignoring all of the others.
- *Code division multiple access* This property allows multiple GPS signals to be sent at the same time, with minimum interference and negligible crosstalk between them.
- *Good range resolution* Any spread spectrum system can measure range to another spread spectrum system with a resolution that is never worse than a distance equal to that traveled at the speed of light during one code-chip time. As GPS has two code-chip rates (1.023 Mchip/s and 10.23 Mchip/s), it can resolve range with a resolution never worse than approximately 1000 feet (C/A code) or 100 feet (P code). In practice, spread spectrum systems routinely measure range to one-tenth their chip time, and sometimes to one-thousandth their chip time. (A chip time is the reciprocal of the code rate.)
- *Inherent privacy/security* The code used in a spread spectrum system determines its security level. In GPS, the two codes are linear (C/A) and encrypted (P).
- *Multipath resistance* Spread spectrum systems treat any signal that arrives with a delay greater than one chip time behind the signal to which the receiver is synchronized

as an unwanted interfering signal, and rejects it. This is a very valuable property for a ranging system.
- *Interference rejection* GPS makes use of this property in two ways: First, it employs the interference rejection to operate in the presence of both friendly and unfriendly interference. Second, GPS makes use of this property to allow a receiver to operate when there are typically 12 or more GPS signals coming into the receiver at the same time.
- *Low power density* This characteristic of a spread spectrum signal (that it has low average power density, i.e., its transmitted power is spread over a wide bandwidth) helps to reduce interference to other non-GPS systems.

Two main types of spread spectrum systems are in general use. These are "frequency hopping" and "direct sequence." Of the two, the code rates used in direct-sequence systems are much more amenable to high resolution ranging than those used in frequency hopping. (Frequency-hopping systems typically use codes whose chip rate is no higher than a few thousand chips per second; therefore, their range resolution is usually poor.)

GPS is a direct-sequence system that, like many other spread spectrum systems, uses BPSK modulation. This is only one form of modulation used by such systems; the others are QPSK and MSK. (Note that these are the same types of modulation commonly used in data transmission systems.) BPSK is the most common, however.

An advantage offered by spread spectrum systems is that they offer a processing gain that is equal to the ratio of the RF

Table III GPS Measurement Capability[a]

Specification	Capability

Position location accuracy	Using P code
	  16-meter SEP[b], 50%
	  25-meter SEP, 90%
	  Unstated, 10%
	Using C/A code
	  100-meter CEP[c], 50%
	  (Sometimes stated as 40-meter CEP; this is adjustable and has been stated at various times as 40 to 200 meters)

Velocity measurement accuracy	Using P code
	  0.1 meter per second
	Using C/A code
	  Not stated

Satellite position accuracy	Better than 10 meters; limited by wobble due to gravitational anomalies and prediction capability
Satellite time accuracy	Better than 10 nanoseconds
Time to first fix, cold start	7.0 to 10.5 minutes (military systems)
Time measurement accuracy	100 nanoseconds; independent measurements
Differential time measurement accuracy	25 nanoseconds; two receivers simultaneously using the same four satellites

[a] This is a general listing; some designs may be better.
[b] SEP, spherical error probability. The probability that a position measurement will be within a sphere with the stated radius (three-dimensional).
[c] CEP, circular error probability. The probability that a position measurement will be within a circle with the stated radius (two-dimensional).

bandwidth used divided by the data rate. This processing gain produces an improvement in signal-to-noise ratio similar to that produced by the deviation ratio improvement in FM systems. The processing gain provided by a GPS receiver is approximately 42.5 dB for the C/A code, and 52.5 dB for the P code. Because of the very low signal strength of the GPS signal, as seen at the receiver (about 0.03 μV) the processing gain is needed to bring the signal above the thermal noise. In the GPS (P code) signal bandwidth, the thermal noise is near -105 dBm. The signal, on the other hand, is at -135 dBm. Therefore, the signal-to-noise ratio at the receiver input is -30 dB. The 52.5-dB processing gain brings the signal-to-noise ratio after spread spectrum processing by the receiver to $+22.5$ dB, which provides an excellent bit error rate to the 50 bit per second data channel.

Bibliography

Dixon, R. C. (1984). "Spread Spectrum Systems." Wiley, New York.

Navstar GPS Joint Program Office (1988). "User's Overview." El Segundo, California.
Spilker, J. J., Jr. (1978). GPS signal structure and performance characteristics. *J. Inst. Navigation* 25(2), 29.
VanDierendonck, A. J., Russell, S. S., Kopitske, E. M., and Birnbaum, M. (1978). The GPS navigation message. *J. Inst. Navigation* 25(2), 29.

Glossary

Almanac Downlink information that defines the position of the GPS satellites.

Biphase modulation Process of causing a carrier to assume one of only two phases, depending on the status of a binary modulating signal (example: zero degrees for a One and 180 degrees for a Zero).

C/A code Coarse/acquisition code. Sometimes referred to as a *clear/acquisition code*. Used by military users for initial synchronization and subsequent transfer to P code.

Chip rate Reciprocal of a code chip time, or the code clock rate.

Code chip One clock interval (or bit) of a spectrum-spreading code.

Code period Time required for a code to repeat.

Constellation Overall configuration of all GPS satellites and their relative position with respect to one another.

Control segment That portion of the GPS system consisting of the ground monitoring, satellite command and control, and satellite downlink correction facilities.

Downlink Any signal transmitted from a satellite to the earth, in the GPS context.

GPS coverage Percentage of time that a particular location has four or more satellites in view with good angular resolution, or geometric dilution of precision (GDOP).

Handover word Information transmitted on the GPS downlink to aid an authorized GPS user in receiving the P code modulated signals.

Housekeeping Information transmitted as part of a downlink to describe the satellite's status.

L1 One of a pair of frequencies at which the GPS downlink signal is transmitted. This frequency is 1575.42 MHz.

L2 One of a pair of frequencies at which the GPS downlink signal is transmitted. This frequency is 1227.6 MHz.

Multilateration Act of determining position through the use of multiple range measurements.

P code Precision code. Sometimes referred to as a *protected code*.

Pseudo-range One-way range measurement (as opposed to a measurement made over a round-trip signal path).

Range Distance from a satellite (in the GPS case) to a receiver.

Space segment Satellite portion of the GPS system.

Spread spectrum Type of modulation format that employs bandwidths much wider than the information being conveyed.

Triangulation Act of determining position through the use of multiple angle measurements.

Truncation In the GPS context, this is the act of shortening the spectrum-spreading code to a length less than its natural length.

Uplink Any signal transmitted from the earth to a satellite, in the GPS context.

User segment Overall conglomerate user portion of the GPS system consisting of all downlink receivers, both military and nonmilitary.

Global Runoff

B. L. Finlayson and T. A. McMahon
University of Melbourne

Runoff is a component of the water balance and is that part of precipitation not returned to the atmosphere by evaporation or transpiration, together known as evapotranspiration. On the continents, runoff is returned to the sea predominantly as flow in rivers and to a lesser extent as flow in groundwater aquifers. Over the oceans, there is a term in the water balance which is conceptually equivalent to runoff on the continents. Runoff is the component of the water balance which is readily available for use by the human population; it supports terrestrial aquatic biota and transports natural weathering and erosion products, as well as the waste and pollutants of civilization, to the sea. The moisture status of an area can be described by the runoff ratio, the proportion of precipitation which becomes runoff. It is not possible to estimate global runoff directly from measurements of flow in rivers because the network of gauging stations is inadequate. Mean values are estimated using models of the water balance based on measured rainfall and measured or estimated evapotranspiration and usually the models are calibrated with available streamflow data. These estimates of runoff are used to describe its spatial distribution. Measurements of runoff in rivers provide information about the way runoff varies through time and its statistical characteristics such as variability and serial correlation. Measurements of runoff in streams also provide information about the magnitude and frequency of floods. Water not evaporated is stored on the continents in river channels, lakes, soil, and groundwater aquifers and as ice and snow, and is gradually released to become runoff. These various delays in storage, superimposed on the seasonal distribution of precipitation and evaporation, combine to produce seasonal runoff regimes.

I. Introduction

The continental water balance may be stated simply as

$$\text{Precipitation} = \text{evapotranspiration} + \text{runoff} + \text{change in storage}$$

Provided the calculation of this balance is carried out over a sufficiently long period of time, it can be assumed that there is no change in storage and that term can therefore be ignored. Precipitation and evaporation are observed or estimated at single points in space, and reasonable extrapolations can be made between the data points to provide global coverage. Runoff can be modeled as the difference between precipitation and evaporation, as is done in published global water balance studies, and estimates of runoff from continental-scale regions are long-term averages calculated from this water balance model. Runoff calculated in this way is referred to as climatic runoff. [*See* EVAPORATION AND EVAPOTRANSPIRATION; WATER CYCLE, GLOBAL.]

On the continents the amount of water in the water balance is limited to the volume of precipitation, and the proportion which becomes runoff (the runoff ratio) decreases as the climate becomes more arid, as shown in Fig. 1. The relationship between precipitation and runoff is nonlinear. Figure 1 also illustrates the nature of the relationship between precipitation and runoff over a range of values of the radiation balance of the land surface. The water balance can also be calculated over the oceans, but here water is always available for evaporation, so "negative runoff" is possible. This has important implications for the characteristics of the ocean surface in different parts of the globe, particularly surface salinity. Of all the world's oceans, only the Arctic and Pacific oceans have a positive water balance, and this excess is transported by currents to make up the deficits in the other oceans.

By contrast, any measurement of runoff refers to an area, the drainage basin above the point of measurement, and many of the properties of runoff are area dependent. Mea-

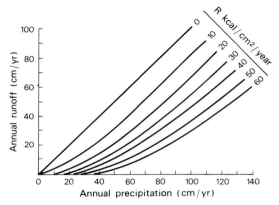

Fig. 1 Annual runoff versus annual precipitation as a function of the radiation balance of the land surface (R). [From Nemec (1983).]

sured runoff does not describe conditions at the point of measurement, as do precipitation and evaporation, but represents the integration of many factors over the drainage basin area. Measured runoff is not necessarily the "natural" flow from the drainage basin, but is affected by artificial storages in the catchment, interbasin transfers of water, and the modification of natural runoff by land use practices. There are now probably very few drainage basins of moderate or large size which could be strictly described as natural. It is difficult to obtain adequate lengths of runoff data from representative areas of the globe, and this, more than anything else, has inhibited comparative regional studies of runoff at the global scale. Direct measurements of streamflow are necessary to provide information on the seasonal regime of runoff, runoff variability, flood behavior, and relationships between runoff and other environmental variables.

Mean annual runoff, as derived from the water balance model, provides only part of the information required about runoff for water resources development. Areas with similar mean annual runoffs will have very dissimilar water resources potential if the variability of flow is different. The mean and variability themselves are also not sufficient to describe water resources potential. For various reasons, such as the carryover of soil moisture from one year to the next, storage delays in the groundwater, water storage in natural lakes, or the influence of anomalies in the global circulation, runs of years of above and below-average runoff, longer than would be expected to occur randomly, are sometimes experienced, and these are described by the serial correlation of the annual flows. The global average storage delay time of runoff is about 2 months, though clearly this is related to catchment size such that the larger the catchment area, the longer the storage delay time of runoff.

Runoff is normally measured in water years, not calendar years, where a water year begins at the end of the period of lowest flow in the annual runoff regime. Although there are no firm rules regarding terminology and units of measurement, *runoff* is normally used to refer to depth per unit time such as millimeters per year or volume per unit time per unit area such as cubic meters per second per square kilometer; *discharge* is used to refer to rates of flow in, for example, cubic meters per second; and *volume* is expressed per unit time as in millions of cubic meters per year or megaliters per year.

II. Depth and Distribution of Global Runoff

Mean annual global runoff (in volume units) from the continents and islands is in the order of 39,700 km³. Of this total, only some 2200 km³ occurs as groundwater discharge and the rest, 37,500 km³, is runoff in rivers. The continental surfaces may be divided into areas of external, or exoreic, drainage where runoff is into the oceans and internal, or endoreic, drainage to inland basins. Some areas of the globe have no appreciable surface runoff (areic) and therefore lack well-developed river networks. These areic areas are usually the product of extreme aridity, though in some cases, such as the Nullarbor Plain in Australia, the effects of aridity are exacerbated by underlying permeable rocks so that virtually all of the runoff is as groundwater. These areas are occasionally crossed by large rivers carrying runoff from moister climatic zones, termed exotic rivers. Examples include the Nile in Egypt, the Euphrates in Syria and Iraq, and the Murray in western New South Wales and South Australia.

The water balance for the land areas of the globe by individual continents is shown in Table I. On the basis of depth of runoff, Australia is clearly the driest continent, drier even than Antarctica, though only because evaporation in Antarctica approaches zero. When the water balance terms are considered as volumes (Table I), the size of the continents also becomes an important factor. In these terms, Africa and Europe, which have such different runoff depths, are similar. Runoff as a percentage of precipitation (the runoff ratio) is also shown in Table I for each continent. As the climate becomes drier, the runoff ratio decreases because delays in storage provide opportunity for the stored water to be evaporated before it becomes runoff. Most of the continents have runoff ratios in the range 30–40%. The runoff ratios of Australia and Africa are considerably lower, and that of Antarctica is unusually high.

If runoff is taken to be a measure of the available water resource, it is interesting to consider its per capita availability,

Table I   Precipitation (*P*), Evaporation (*E*), and Runoff (*D*) as Depth and Volume and the Runoff Ratio (*D/P*) for the Continents

Continent	Area (10⁶ km²)	Volume (10³ km³)			Depth (mm)			Runoff ratio (%)	Runoff per capita (megaliters/yr)
		P	*E*	*D*	*P*	*E*	*D*		
Asia	44.1	30.7	18.5	12.2	696	420	276	40	4.6
North America	24.1	15.6	9.7	5.9	645	403	242	38	15.8
South America	17.9	28.0	16.9	11.1	1564	946	618	40	46.3
Europe	10.0	6.6	3.8	2.8	657	375	282	43	4.1
Africa	29.8	20.7	17.3	3.4	696	582	114	16	7.2
Australia	7.6	3.4	3.2	0.2	447	420	27	6	132.5
South Pacific islands	1.3	3.7	1.5	2.2	2846	1153	1692	59	343.8
Antarctica	14.1	2.4	0.4	2.0	169	28	141	83	—
World	148.9	111.1	71.4	39.7	746	480	266	36	8.9

Source: Runoff data from Baumgartner and Reichel (1975); population from United Nations, *World Population Statistics,* 1985.

listed in Table I. The distribution of available water does not approximate the population distribution and the discrepancies shown in Table I at the continental scale are much greater at the regional and local scales; at these scales, the continental values can be very misleading.

The major controls on the overall pattern of runoff are climate and topography, and, of course, these two factors are not independent. The concentration of high levels of runoff along the major mountain chains is a notable feature of the distribution shown in Fig. 2. This concentration is in response to the combination of orographically induced precipitation and the reduction in evapotranspiration with increasing elevation, a function of lower temperatures and increased cloudiness. There is clearly a very strong association between runoff and rainfall, and the description that follows highlights the association between runoff and mechanisms in the atmosphere for inducing uplift and therefore precipitation.

The equatorial belt, lying between 20°N and 20°S latitude, experiences generally high levels of runoff despite the high temperatures there. This is the area where rainfall is generally produced by uplift along the intertropical convergence zone. Poleward of this zone, to about 35° latitude, large areas of very low runoff, the tropical and subtropical deserts, are associated with the subtropical high-pressure cells which form under the descending limb of the Hadley circulation. These features of the global circulation dominate the climates of Australia and nonequatorial Africa. Through a combination of the effects of latitudinal range and major mountain chains on the other continents, these subtropical high-pressure cells have a less dominant effect on their overall climatic patterns. [*See* ATMOSPHERIC CIRCULATION SYSTEMS; MOUNTAINS, EFFECT ON AIRFLOW AND PRECIPITATION.]

Between 35°N and 35°S, zones of relatively high runoff are found on the eastern margins of the continents, while low-

runoff desert conditions extend to the coast on the western margins. This is a response to the easterly air flows associated with the Hadley circulation (the trade winds and the equatorial easterlies). Uplift is mainly orographic or convective. Poleward of these latitudes, in the Ferrel cell, the circulation is from the west and higher-runoff areas are generally found on the western margins of the continents. In the Ferrel cells, in addition to the orographic effect, uplift occurs along fronts which form where warm tropical airmasses and cold polar airmasses meet. [*See* ATMOSPHERIC FRONTS; MID-LATITUDE CONVECTIVE SYSTEMS.]

Figure 2 shows surface salinity for the oceans. Where the water balance is positive, salinity is relatively low, and high salinity is found in areas of negative water balance. These high-salinity areas are the oceanic equivalent of the tropical and subtropical deserts on the continents and are important net source areas of water vapor to the atmosphere. Significant areas of low salinity are also found near the mouths of major river systems such as the Amazon and the Mississippi.

In addition to the natural storage of runoff on the continents in lakes, rivers, and groundwater, the construction of dams has also led to large-scale storage. It has been estimated that about 15% of global annual river discharge is artificially regulated and this proportion is increasing. Scientists concerned with global sea level rise estimate that this artificial storage on the continents has been sufficient to suppress the rise by about 0.75 mm/yr over the past 30 years. Although the storage of increased amounts of runoff on the continents in reservoirs and in inland basins seems to offer a potential means of averting at least part of the global sea level rise predicted to accompany atmospheric warming caused by the accumulation of "greenhouse gases," such storage may have unwanted environmental consequences. The use of large amounts of runoff for irrigation on the continents has led to

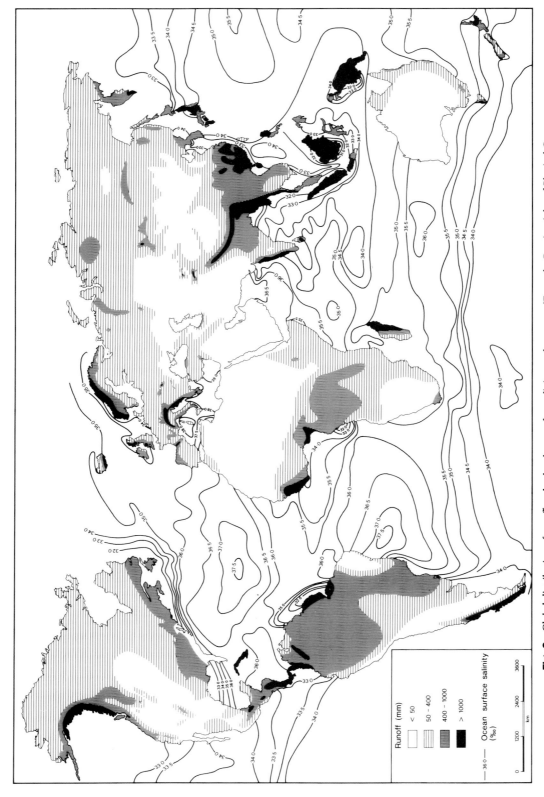

Fig. 2 Global distribution of runoff on land and sea surface salinity on the oceans. [From the Russian Atlas of Physical Geography (1964).]

Runoff (mm)

  < 50

  50 – 400

  400 – 1000

  > 1000

Ocean surface salinity (‰)

  —36.0—

0   1200   2400   3600

km

rising water tables and waterlogging and salinization of soils in the irrigation areas; the construction of dams leads to permanent flooding of large areas of valuable agricultural land; and the reduction of flow in major river systems is detrimental to in-stream biota and to the productivity of estuarine and deltaic ecosystems.

III. Characteristics of the Annual Runoff of the Continents

This section and those which follow are based on runoff records from nearly 1000 rivers worldwide. These data cannot be used for the water balance calculations discussed above, but they enable the time-dependent aspects of runoff to be investigated and in this way are complementary to the water balance approach. [*See* RIVERS.]

Some key hydrologic parameters (based on annual flows) calculated from the streamflow data are presented in Table II. The values of mean annual runoff given in this table are averages of stations in the data set and do not represent the actual values for the continental areas for which they are given. They should not be confused with the continental values of mean annual runoff shown in Fig. 2 or Table I, which are taken from water balance studies. The world trend is for mean annual runoff to decrease as basin size increases. This trend is apparent in all continental regions except northern Africa, where there is a slight increase in mean annual runoff as basin area increases. This is probably an artifact of the small amount of data available for that region. Note, however, the very low mean annual runoff values for large catchments in Australia and, to a lesser extent, southern Africa.

Three other parameters are included in Table II, namely the coefficient of variation of annual flows, C_{vr} (defined as the standard deviation divided by the mean), the maximum observed annual flow volume divided by mean annual flow, $V_{max}/\overline{V}$, and the storage capacity (required to meet a reservoir yield equal to 80% of the mean annual runoff at 95% reliability) divided by mean annual flow, τ_{80}. A striking feature of these parameters is that Australian and southern African streams exhibit variabilities about double those for the rest of the world and extremes are larger, as are storage needs.

A. Volume

Figure 3 shows the relationship for each continental region between annual flow volume and catchment area. In all cases, the correlations are significant at the 1% level and catchment area explains a high proportion of the variance, the lowest being Australia at 60%. Most of the relationships cluster around the world one, with minor exceptions for South

Pacific islands, Australia, and southern Africa. Lower flow volumes for Australia and southern Africa, especially for the larger catchments, are as would be expected for continents in their latitudinal positions. North Africa lies close to the world line, even though it is a predominantly arid area, because the data for North Africa come mainly from the humid zone bordering the Gulf of Guinea and the Nile system. No runoff data are available for the arid parts of North Africa.

B. Variability

The coefficient of variation of annual flow volumes, C_{vr}, is used here as the most appropriate measure of variability because it enables direct comparisons to be made between streams with very different mean annual flows. In Fig. 4 the relationship between C_{vr} and mean annual runoff is shown for each continent. Mean annual runoff is a climatic indicator representing the aridity or humidity of the climate. Australia and southern Africa, while following the world trend of decreasing variability with increasing humidity of the climate, are notable in having higher C_{vr} values than the other continents over the whole range of mean annual runoff. [*See* CLIMATE.]

Values of variability stratified by continental region and catchment area are summarized in Table II. The world mean is 0.43 and the values vary over a wide range. Combining all data by regions, Australia and southern Africa have C_{vr} approximately twice those for other regions. On statistical grounds it would be expected that variability should decrease as catchment area increases. This is the case for Asia, North Africa, and the Pacific islands. The other continents, with the exception of Australia, show no strong trend. Australia is anomalous in having a marked increase in variability with increase in catchment area, which can be explained in terms of the distribution of climates on the Australian continent. Humid climatic zones parallel the coast in a relatively thin strip around the northern, eastern, southeastern, and southwestern coasts. Any large catchment in Australia must extend into parts of the drier interior, and this contributes to the increase in C_{vr} as catchment area increases.

C. Distribution Types

Prediction in hydrology depends to a large extent on knowing the underlying statistical frequency distribution of the data. Of the world streams, 60% have annual flows which are not significantly different from Normal at the 5% level of significance. However, only 31% of Australian and southern African streams are Normally distributed, compared with 71% for the other continents. This observation is important; in the main, for streams in Australia and southern Africa analyses involving statistical techniques are more difficult than else-

Table II Some Hydrologic Parameters Based on Annual Streamflow Volumes[a]

Continent	Area ranges (km²)	Number of streams	Average record length (yr)	Average catchment area (km²)	Average mean annual runoff (mm)	Average C_{vr}	Average $\dfrac{V_{max}}{\overline{V}}$[b]	Average τ_{80}[c]
World	$0-10^3$	434	28	322	820	0.45	2.2	0.79
	10^3-10^4	273	35	3,376	540	0.48	2.5	0.99
	10^4-10^5	180	36	37,675	410	0.37	2.0	0.54
	$>10^5$	87	47	527,312	230	0.33	2.0	0.46
	All	974	33	55,153	610	0.43	2.2	0.77
Northern	$0-10^3$	0						
Africa	10^3-10^4	3	11	3,836	170	0.54	2.2	0.86
	10^4-10^5	5	34	48,168	190	0.37	2.0	0.39
	$>10^5$	15	30	301,913	210	0.25	1.6	0.23
	All	23	29	207,872	200	0.31	1.8	0.35
Southern	$0-10^3$	55	26	298	280	0.81	3.6	2.38
Africa	10^3-10^4	30	31	3,315	100	0.78	3.5	1.82
	10^4-10^5	11	30	31,235	180	0.70	3.2	1.61
	$>10^5$	4	38	299,254	70	0.54	2.5	0.98
	All	100	28	16,565	210	0.78	3.5	2.07
Asia	$0-10^3$	42	16	242	900	0.47	2.1	0.81
	10^3-10^4	31	21	4,637	790	0.45	2.4	1.06
	10^4-10^5	45	14	32,163	400	0.30	1.6	0.31
	$>10^5$	25	36	734,462	310	0.28	1.7	0.29
	All	143	20	139,600	620	0.38	2.0	0.62
North	$0-10^3$	83	23	347	1,690	0.31	1.7	0.30
America	10^3-10^4	54	47	3,142	720	0.39	2.2	0.54
	10^4-10^5	33	45	36,491	510	0.38	2.2	0.63
	$>10^5$	19	61	403,496	150	0.35	2.1	0.42
	All	189	37	47,984	1,050	0.35	2.0	0.44
South	$0-10^3$	11	34	516	640	0.39	2.2	0.47
America	10^3-10^4	17	34	3,538	730	0.33	1.9	0.34
	10^4-10^5	21	38	39,614	670	0.34	2.0	0.42
	$>10^5$	4	47	1,983,646	450	0.41	3.7	0.96
	All	53	36	166,647	670	0.35	2.1	0.44
Europe	$0-10^3$	103	34	344	520	0.30	1.7	0.28
	10^3-10^4	79	33	3,234	510	0.27	1.6	0.24
	10^4-10^5	61	47	42,500	350	0.31	1.8	0.35
	$>10^5$	17	65	342,758	250	0.25	1.8	0.19
	All	260	39	33,497	460	0.29	1.7	0.28
South	$0-10^3$	40	20	299	1,170	0.26	1.5	0.24
Pacific	10^3-10^4	10	18	2,433	1,750	0.22	1.4	0.15
	10^4-10^5	0						
	$>10^5$	0						
	All	50	20	725	1,290	0.25	1.5	0.22
Australia	$0-10^3$	100	35	323	540	0.59	2.6	1.10
	10^3-10^4	49	42	3,213	210	0.88	4.0	2.53
	10^4-10^5	4	46	30,275	81	0.98	4.3	3.09
	$>10^5$	3	30	120,333	37	1.12	4.8	3.50
	All	156	37	4,307	420	0.70	3.1	1.65

[a] The "All" category in this table does not give a measure of the mean value for the continent, only of the data used to compile this table.

[b] Maximum observed annual flow divided by mean annual flow.

[c] Storage capacity required to meet 80% of draft at 95% reliability divided by mean annual flow.

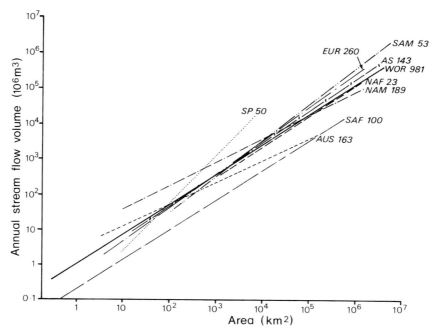

Fig. 3   Relationship between annual discharge volume and catchment area for each continental region.

D. Persistence

where because non-Gaussian methods must be employed. [*See* HYDROLOGIC FORECASTING.]

D. Persistence

Persistence (dependence, nonrandomness) is the nonrandom characteristic of a hydrologic time series. For example, if sufficient soil moisture in a catchment is carried over from one year to the following year such that the flows in the second year are greater than they otherwise would be, or vice versa, these effects are regarded as persistence. Persistence may also arise if there is a serial correlation in the rainfall which gives rise to the runoff.

The most common measure of short memory effects is lag one autocorrelation, where annual flows are correlated with the flow from the previous year. There appear to be no significant differences between continents or climatic zones in terms of short memory effects. The most important control on the lag one autocorrelation coefficient is catchment size. Large catchments, which therefore have longer storage delays, have the strongest short memory effects.

A British engineer, H. E. Hurst, demonstrated that many long time series of natural phenomena exhibit long-term memory. This has become known as the Hurst effect. Most streamflow records show this effect, but with few exceptions, one of which is the Nile, records are too short to be reliable indicators of long-term memory. In many water resource system analyses, the Hurst effect is of little consequence. However, for streams with high variability or in storage–yield analysis of regulated flows, with either high draft or high reliability, the Hurst effect has a significant impact on the results. For example, for a large reservoir (with capacity three times mean annual flow) and an 80% draft rate, the system time reliability is 89% for flows following a Markovian time series, compared to 81% for flows which exhibit the Hurst effect.

E. Reservoir Storage and Yield

The required storage for a constant draft of 80% and a reliability of 95%, τ_{80}, is calculated from the coefficient of variation of annual flow by

$$\tau_{80} = 2.76 C_{vr}^2$$

Storage sizes estimated by this method do not include provision for net evaporation losses or seasonal effects in demand or inflow. From this equation, it can be seen that storage is proportional to the square of C_{vr} and hence, as the mean storage values in Table II show, there are some important regional disparities in this regard. In particular, Australia and

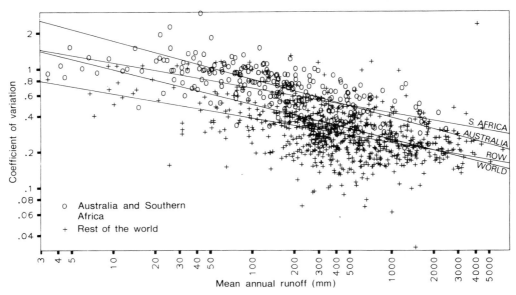

Fig. 4 Relationship between coefficient of variation of annual flows and mean annual runoff. Stations in Australia and southern Africa and the rest of the world are separately identified.

southern Africa require considerably larger reservoirs than the rest of the world for the same degree of river regulation.

Interregional differences observed in C_{vr}, as between Australia and southern Africa and the rest of the world, are considerably magnified by τ_{80} as shown in Table II. This has important economic consequences because storages in regions such as Australia and southern Africa must be more than four times larger than those on other continents to provide the same degree of regulation on the basis of flow variability alone. High evaporation rates in these two continents compound this problem. In these areas of high irrigation need, the cost of providing water storage is therefore relatively high.

IV. Characteristics of the Annual Flood Series

The annual flood series consists of the instantaneous maximum discharge of the largest flood in each water year. Analysis of floods occupies a central place in all studies of rivers: for the design of structures in engineering hydrology, for channel formation and sediment transport in fluvial geomorphology, and as an important component of ecosystem dynamics in river ecology. Key parameters of the annual flood series calculated from the authors' global data set are listed in Table III by continental regions. As is the case with annual runoff, the annual floods of Australia and southern

Africa are more variable than those of other continents. [*See* FLOODS.]

A. Annual Peak Discharge

A fundamental control on the size of the mean annual flood ($\bar{q}$, in cubic meters per second) is catchment area, and the relationship between $\bar{q}$ and catchment area is remarkably similar for catchments in all continents (Fig. 5). If the mean annual flood is expressed in per unit area terms ($\bar{q}_s$, m/s/km²), it decreases as catchment area increases. It has already been shown that Australia and southern Africa are significantly different from the other continental regions with regard to certain aspects of annual flow. These differences can also be found in the peak discharge data. Figure 6 shows the relationships between $\bar{q}_s$ and mean annual runoff. Over the range 10–1000 mm of mean annual runoff, Australia and southern Africa have specific mean annual floods considerably greater than those in the rest of the world.

B. Variability

Two measures of the variability of the annual flood series are shown in Table III. They are the index of variability (I_v) and the ratio of the 100-year return period flood to the mean annual flood ($q_{100}/\bar{q}$). The index of variability is the standard deviation of the logarithms of the annual flood series and has been used as an index of the flash flooding potential of

Table III Some Hydrologic Parameters Based on Annual Peak Discharges[a]

Continent	Area ranges (km²)	Number of streams	Average record length (yr)	Average catchment area (km²)	Average $\bar{q}_s$ (m³/s km²)	Average I_v	Average $\dfrac{q_{100}}{\bar{q}}$
World	$0–10^3$	423	27	302	0.81	0.32	4.5
	$10^3–10^4$	244	32	3,535	0.23	0.29	5.1
	$10^4–10^5$	176	31	36,271	0.08	0.25	3.9
	$>10^5$	88	42	548,994	0.03	0.17	2.9
	All	931	30	59,812	0.44	0.28	4.4
Northern Africa	$0–10^3$	0					
	$10^3–10^4$	7	15	3,721	0.12	0.34	16.8
	$10^4–10^5$	10	30	34,065	0.04	0.23	2.8
	$>10^5$	15	35	566,690	0.02	0.08	1.4
	All	32	29	277,095	0.05	0.18	5.2
Southern Africa	$0–10^3$	72	26	210	0.47	0.49	5.6
	$10^3–10^4$	25	31	3,482	0.15	0.43	13.6
	$10^4–10^5$	10	29	28,250	0.08	0.39	12.4
	$>10^5$	4	27	152,020	0.01	0.40	5.7
	All	111	27	8,944	0.34	0.46	8.1
Asia	$0–10^3$	40	16	262	0.58	0.32	7.8
	$10^3–10^4$	31	19	4,111	0.40	0.23	3.2
	$10^4–10^5$	43	14	31,832	0.11	0.22	3.3
	$>10^5$	24	42	691,434	0.04	0.16	3.4
	All	138	21	131,168	0.30	0.24	4.6
North America	$0–10^3$	111	24	266	1.36	0.26	4.0
	$10^3–10^4$	36	34	4,182	0.29	0.26	4.6
	$10^4–10^5$	29	32	32,833	0.10	0.26	3.9
	$>10^5$	19	39	482,158	0.02	0.16	2.2
	All	195	28	52,592	0.85	0.25	3.9
South America	$0–10^3$	2	17	808	0.53	0.18	5.7
	$10^3–10^4$	6	25	3,772	0.23	0.16	2.1
	$10^4–10^5$	13	29	46,768	0.11	0.14	2.0
	$>10^5$	4	48	1,557,193	0.04	0.10	1.6
	All	25	30	274,440	0.16	0.14	2.3
Europe	$0–10^3$	61	36	438	0.21	0.16	2.2
	$10^3–10^4$	76	36	3,302	0.12	0.17	2.4
	$10^4–10^5$	56	41	42,857	0.06	0.18	2.8
	$>10^5$	16	57	363,555	0.03	0.14	2.1
	All	209	39	40,644	0.12	0.17	2.4
South Pacific	$0–10^3$	40	23	296	1.32	0.24	3.0
	$10^3–10^4$	12	22	2,284	0.86	0.16	2.7
	$10^4–10^5$	0					
	$>10^5$	0					
	All	52	23	755	1.21	0.22	2.9
Australia	$0–10^3$	97	32	333	0.68	0.39	4.9
	$10^3–10^4$	50	40	3,323	0.16	0.51	5.8
	$10^4–10^5$	16	42	29,597	0.05	0.56	6.5
	$>10^5$	6	41	233,668	0.01	0.50	7.7
	All	169	36	12,272	0.45	0.45	5.4

[a] The "All" category in this table does not give a measure of the mean value for the continent, only of the data used to compile this table.

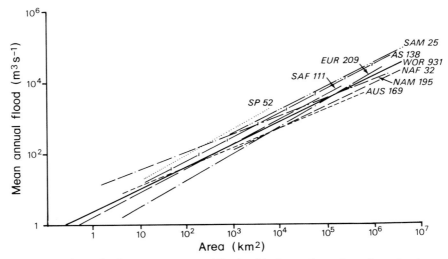

Fig. 5   Relationship between mean annual flood and basin area for each continental region.

drainage basins. In Fig. 7 the indices of variability of annual peak discharges for the world data set are stratified by Köppen climatic zones and for each climate the world distribution is included for ease of comparison. Climate types B and C differ significantly from the world pattern, with proportionally fewer stations with low I_v, and climate type B has many more stations with high I_v than is the case elsewhere. It should be noted that these two climate types include the

tropical and subtropical deserts and their subhumid margins. Characteristics of the major climate types in the Köppen system are given in Fig. 7.

Average values of I_v for the continental areas are included in Table III. Clearly, Australia and southern Africa exhibit values of flood variability considerably greater than the rest of the world. South American streams are the least variable with an average I_v of only 0.14. This repeats the

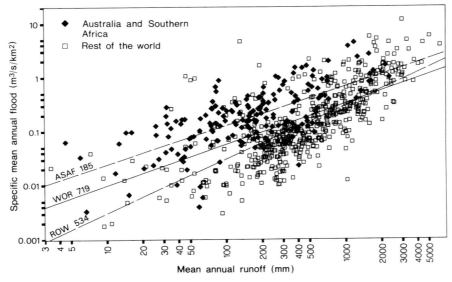

Fig. 6   Relationship between specific mean annual flood and basin area. Stations in Australia and southern Africa are identified separately from those in the rest of the world.

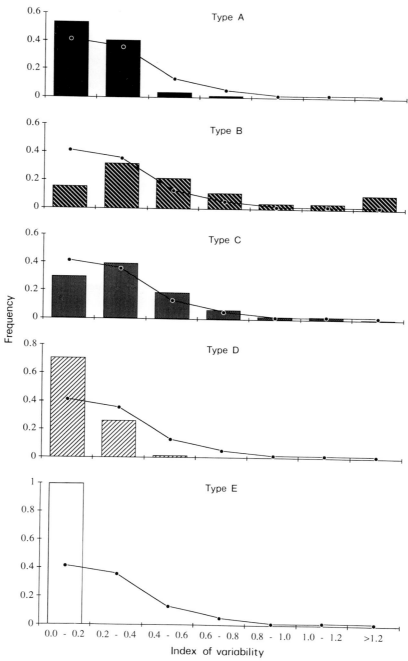

Fig. 7 Frequency distributions of the standard deviation of the logarithms of annual peak discharges for major climatic zones. Type A, equatorial humid climates; type B, semiarid and arid climates; type C, humid temperate climates; type D, snow climates; type E, polar climates.

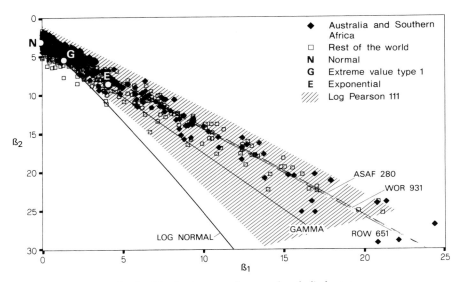

Fig. 8 Distribution types for annual peak discharges.

C. Distribution Types

The selection of an appropriate frequency distribution to fit to an annual flood series is an important aspect of practical hydrology because of the need to predict the size of floods with a given probability of occurrence as the basis for design. Figure 8 is a moment ratio diagram for annual peak flows, where β_1 is the square of the coefficient of skewness and β_2 the kurtosis. On this graph, the Normal distribution which is represented by a point does not fit the data but the log-Pearson type III distribution, which plots as an area, fits these data best.

V. Seasonal River Regimes

The pattern of variation of discharge through the year constitutes the seasonal regime of a river. It represents the outcome of the interaction of the seasonal distribution of precipitation and evaporation with storage delay times, which are affected by catchment size, the presence of groundwater aquifers, and the presence of winter storage as snow. The seasonality of flow is an important consideration in water resource planning. Natural flow regimes have been operating long enough for aquatic ecosystems to have become adapted to them. It is becoming increasingly evident that where the natural flow regime is modified by river regulation, there are detrimental effects on river ecology. [*See* RIVER ECOSYSTEMS.]

Classifications of river regimes have usually been based on assumed associations with climatic zones, though this ignores the influences of the catchment through such characteristics as size, topography, and geology. The streamflow database discussed above has been used to generate a classification of global river regimes based only on flow data. The classification was generated using cluster analysis and consists of 15 groups. The average flow regime pattern for each group is shown in Fig. 9 together with a band 1 standard deviation wide. Figure 10 is a world map of these regime groups, though its reliability is limited by poor data coverage in many areas. Regime types are clearly related to climatic zones, but there is not a perfect correlation because of the importance of catchment characteristics as controls on runoff generation.

Postscript

These controls are still poorly understood and the mechanisms operating are not sufficiently well known to allow us to explain the differences observed between the continents, especially as regards variability. The science of hydrology has largely been developed in the northern hemisphere continents of North America and Europe, which have streamflow characteristics quite different from those of the southern hemisphere continents of Australia and southern Africa. The concepts developed in North America and Europe form the basis of understanding of the subject, and textbooks and hand-

pattern found for the variability of annual streamflow volumes (Table II).

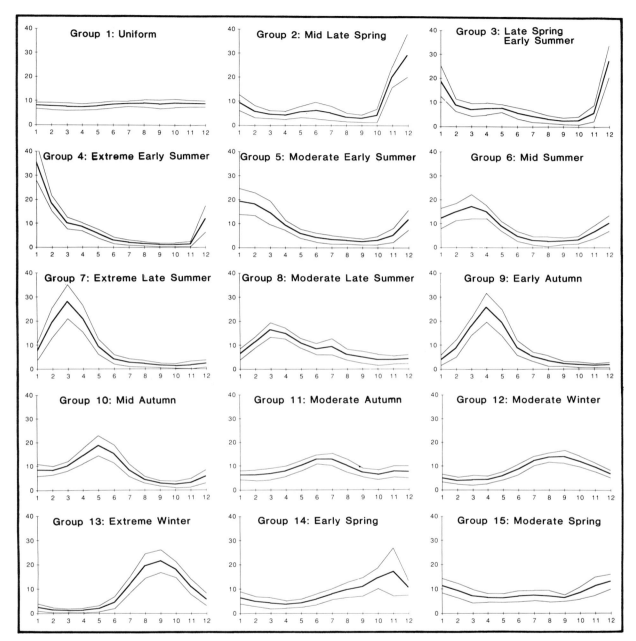

Fig. 9 Seasonal regime types. The band shown is plus or minus 1 standard deviation. [From Haines *et al.* (1988).]

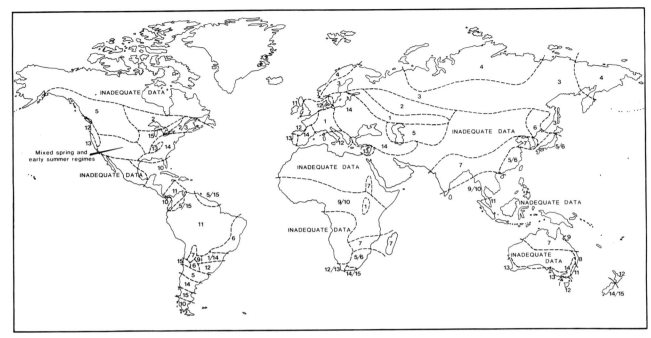

Fig. 10 World distribution of the seasonal regime types shown in Fig. 9. [From Haines *et al.* (1988).]

books written there are used in hydrological education and practice throughout the world. A better understanding of differences between regions at the continental scale could lead to the regionalization of the globe for the purposes of model transferability.

Bibliography

Baumgartner, A., and Reichel, E. (1975). "The World Water Balance." Elsevier Scientific Publishing, Amsterdam.

Haines, A., Finlayson, B. L., and McMahon, T. A. (1988). A global classification of river regimes. *Appl. Geogr.* **8**(4), 255–272.

Nemec, J. (1983). The concept of runoff in the global water budget. *In* "Variations in the Global Water Budget," (A. Street-Perrot, M. Beran, and R. Ratcliffe, eds.), Reidel, Dordrecht, Netherlands.

Glossary

Annual flood series Maximum instantaneous discharge in each year of record.

Coefficient of variation Standard deviation of the annual flows divided by the mean annual flow.

Index of variability Standard deviation of the annual flood series in the log domain.

Runoff ratio Proportion of precipitation which becomes runoff, usually calculated on an annual basis and expressed as a percentage.

Seasonal runoff regime Distribution of flow in a stream throughout the year.

Water year Twelve-month period beginning at the end of the annual low-flow period.

Global Temperatures, Satellite Measures

Roy W. Spencer

NASA Marshall Space Flight Center

Since 1979, satellite-borne radiometers have been capable of monitoring global deep-layer mean atmospheric temperatures. This provides a valuable supplement to ground-based thermometer and weather-balloon measurements, especially in remote regions of the earth. The satellite measurements allow the monitoring and investigation of climate change events, their causes, and thus their role in, and response to, the changing Earth.

I. Introduction

Satellites are beginning to be utilized for monitoring global and regional atmospheric temperature variations associated with climate change events. Although providing only an indirect measure of temperature through the spaceborne measurement of thermally emitted radiation by the atmosphere, some satellite instruments have revealed useful long-term stability and thus sensitivity to small temperature changes. Combined with the global coverage that satellites can provide, these advantages make satellite monitoring an important supplement to earth-based thermometer measurements for not only monitoring changes in the earth system, but quantitatively understanding them. Global atmospheric temperature changes during the decade 1979–1988 are presented and discussed.

II. The Role of Atmospheric Temperature in Global Change

The temperature of the atmosphere is intricately linked to a wide variety of processes and circumstances. In a globally averaged sense, it is the result of the balance between (1) the amount of solar energy received by the surface and atmosphere and (2) how much of that energy is lost to space by infrared radiational cooling. Because the surface of the earth curves away from the sun toward the poles, more energy is gained than is lost in the tropics, whereas more is lost than gained in the polar regions. The excess thermal energy in the tropics is transported poleward within atmospheric and oceanic circulation systems that form in response to the energy imbalance. At the same time, cooler polar air is transported to the tropics to complete a continuous cycle of energy transport. The characteristics of the atmospheric circulation systems performing this transport depend on the geographic distribution of ocean, deserts, vegetation, snow cover, and other features. But also, some of these surface features are themselves a result of the circulation systems and their temperature and moisture characteristics. Therefore, rather than being randomly superimposed upon each other, each often reflects characteristics of the other. Any change in the earth system (e.g., rain forest depletion, radiation balance changes from greenhouse gas production) can result in temperature changes that in turn can impact the earth system, even in a region different from where the original surface change took place. [*See* ATMOSPHERIC CIRCULATION SYSTEMS; CLIMATE; OCEAN CIRCULATION.]

III. Global Monitoring from Space

A. Satellite Radiometry of the Earth

Since the 1960s, various satellites have carried earth-viewing radiometers for the retrieval of information regarding various atmospheric and oceanic characteristics. Some missions have been proof of concept, and others have provided daily operational information, such as is needed for weather forecasting and monitoring purposes. Most of these satellites have been in near-polar, sun-synchronous orbits that provide global coverage at the same local sun time each day. Due to increasing awareness of global environmental problems, combined with the global monitoring satellites can provide, these data have only recently been examined for their utility in understanding the roles of the oceans, land, and atmosphere in the changing earth system.

B. Radiometric Temperature Measurements

Everything emits very low levels of radiation at an intensity dependent on its temperature and its ability to emit thermal radiation (emissivity). Measurements of thermally emitted radiation by infrared or microwave radiometers on spacecraft are usually termed brightness temperatures (T_b). Measurements of the thermal radiation emitted by oxygen (using microwave radiometers) and of that emitted by carbon dioxide (using infrared radiometers) have been used since 1979 to routinely monitor atmospheric temperature conditions to improve weather forecasting in regions where there is little ground-based temperature data. Now these data are being examined more critically to determine whether they can be used in the demanding role of an earth system monitoring tool. Because of the long wavelength of radiation (approaching 1 cm) compared to cloud particle sizes (typically 0.001 to 0.01 cm), the microwave measurements are only weakly affected by clouds. This allows temperature measurements to be made through many clouds, especially cirrus (ice) clouds. Also, since the amount of oxygen in the atmosphere is very large (20.95% by volume) it is essentially unaffected by climate changes (such as the seasonal cycle in northern hemisphere vegetation cover), and so provides a stable temperature tracer for long-term monitoring. However, unlike ground-based thermometer or weather-balloon measurements, which are made at a specific selected level in the atmosphere, the satellite measurement can only be made for fairly deep layers of the atmosphere. The intensity of the received radiation varies within a layer according to a "weighting function" which is a characteristic of the frequency chosen for the radiometer to measure. Different frequencies have different weighting function positions in atmospheric altitude, allowing different layers' temperatures to be monitored.

C. Results from the TIROS-N Microwave Sounding Units

The Microwave Sounding Units (MSUs) have been flown since 1979 on the TIROS-N series of National Oceanic and Atmospheric Administration (NOAA) satellites for daily weather monitoring and forecasting purposes. The MSUs have four channels, each receiving a different frequency of microwave radiation in the 50 to 60 GHz oxygen absorption band, to allow temperature measurements of four different, but overlapping, layers in the atmosphere. Here we will examine results from MSU channel 2, which at 53.74 GHz measures the temperature of a layer of air extending throughout the troposphere (Fig. 1). To the extent that the weighting function intersects the surface of the earth, variations in surface emissivity can slightly contribute to the T_b measurements so that they do not correspond exactly to air tempera-

ture measurements. Some clouds, precipitation, and atmospheric water vapor variations also contribute to this contamination. Fortunately, comparisons to weather balloon measurements have revealed that these contaminations, at least over the United States where the comparisons were performed, are very small compared to the air temperature signature. Therefore, in the following discussion we will consider temperature and MSU channel 2 T_b to be largely interchangeable, but the reader is reminded that they are not the same.

A total of seven of the MSUs were flown through 1988, and their measurements have been combined to examine the 10-year variability of atmospheric temperatures over the globe as well as for specific regions. Individual MSUs have an instrumental accuracy no better than about 1°C, which would appear to not be sufficient to match the usual accuracy of 0.1°C for climate monitoring demands. However, it has been determined from intercomparisons between MSUs operating at the same time on different spacecraft that the differences between them remain constant to a remarkable level of precision (0.01°C). Therefore, since climate monitoring of temperature puts a much greater emphasis on the variability of temperature rather than its absolute magnitude, the MSUs can provide useful climate information. This is also strong evidence for instrument stability, which is necessary for monitoring of (potentially small) global changes over time from space.

1. The Annual Cycle in Temperature

The average annual variability of the global and hemispheric MSU channel 2 temperatures is shown in Fig. 2, as well as the annual variability as a function of Earth latitude band. Note that the globally averaged temperatures are somewhat higher in the northern hemisphere summer than in the southern hemisphere summer. This is due to the greater amount of land surface in the northern hemisphere, which warms more readily from solar heating than does the ocean-dominated southern hemisphere. From the latitude distribution of the annual cycle we see that there is only a very small cycle near the equator, since this region is in perpetual summer. The polar regions, however, experience large seasonal temperature fluctuations due to the cycle between day-long darkness in the winter to day-long sunlight in the summer. Humans have adapted their agriculture, commerce, clothing, housing, and so on to the average environmental conditions associated with these average temperature conditions.

2. Temperature Anomalies during 1979–1988

When the environmental conditions to which we have become accustomed deviate too much from what is normal for a particular time of year, disruption of normal activities can result. The deviations from the average seasonal temperature cycle are usually termed "anomalies." Although they are sometimes seemingly non-temperature-related (e.g., drought

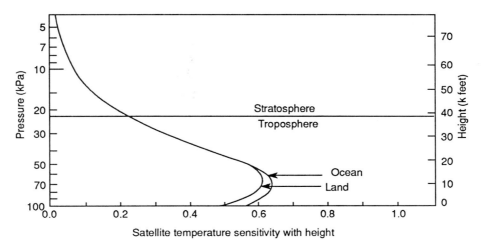

Fig. 1 Atmospheric weighting function of MSU channel 2 as a function of height (thousands of feet) and pressure (kilopascals). The profiles are somewhat different for ocean versus land because the ocean, due to an emissivity of 0.5, reflects some of the downwelling atmospheric radiation back upward.

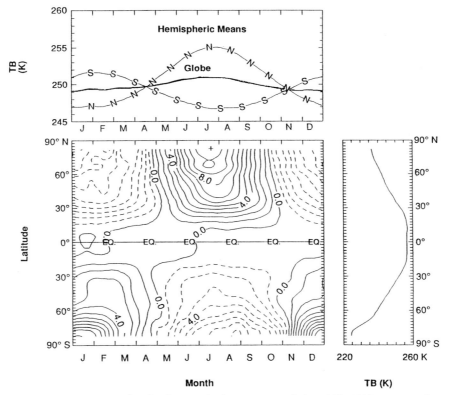

Fig. 2 Average annual cycle of tropospheric temperature during 1979–1988 as measured by MSU channel 2 for the hemispheric and global averages (top), and zonal (within latitude band) averages (bottom). The zonal variations (bottom left) are relative to the average zonal brightness temperature (bottom right).

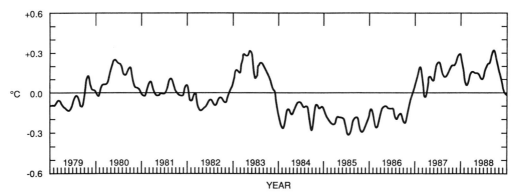

Fig. 3 Globally averaged temperature anomalies during 1979–1988 as measured by MSU channel 2.

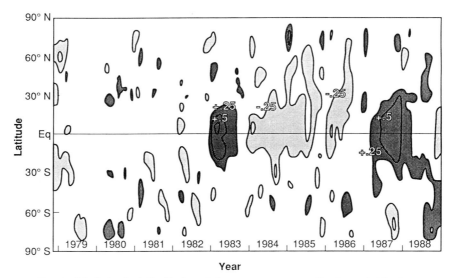

Fig. 4 The latitudinal distribution of the temperature anomalies shown in Fig. 3.

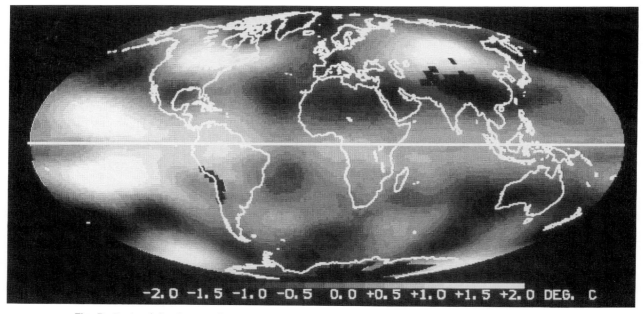

Fig. 5 Regional distribution of temperature anomalies during early February 1983 measured by MSU channel 2.

or flood), in almost all cases they are the result of atmospheric circulation changes that have associated air temperature anomalies. (As discussed earlier, the atmospheric circulation changes can themselves be the result of some other anomalous change in surface conditions.)

The globally averaged temperature anomalies measured by the MSUs during 1979–1988, the result of subtracting the average annual cycle from the original measurements, are illustrated in Fig. 3. Note that they are quite small, with magnitudes always less than 1°C. However, the global consequences of these changes can be dramatic. For instance, the warm anomalies of 1982–1983 and 1987–1988 were associated with particularly strong El Niño/Southern Oscillation (ENSO) events. During these events unusually warm water builds up in the eastern Pacific during a weakening of easterly trade winds. On a time scale of months, oceanic thermal energy is transferred to the atmosphere, which can disrupt weather conditions worldwide. During this 1982–1983 event, Australia and southern Africa experienced extreme drought, western South America had devastating floods, and the Pacific coast of the United States was repeatedly battered by strong cyclones. [See EL NIÑO AND LA NIÑA; SEA SURFACE TEMPERATURE SIGNALS FROM SATELLITES.]

Some of the regional character of these events is seen in Fig. 4, where the temperature anomalies are averaged in Earth latitude bands. We see that the 1983 ENSO warm event was concentrated in the tropics. If we further investigate which part of the tropics is responsible for this warming, we find that the tropical east Pacific (Fig. 5) experienced most of the warming, a direct response to the ocean heating by the sun. Also note, however, that this heating resulted in a band of tropical warmth circling the earth. Thus, we see some of the global consequences of the regional event.

IV. The Future

Global satellite surveillance of atmospheric temperature will continue throughout the 1990s with improved sensors. The results and examples presented here represent only the beginnings of the relatively new field of global temperature monitoring with satellites.

Bibliography

Khalsa, S. J. S., and Steiner, E. J. (1988). A TOVS dataset for study of the tropical atmosphere. *J. Appl. Meteorol.* **27**, 851–862.

Spencer, R. W., and Christy, J. R. (1990). Precise monitoring of global temperature trends from satellites. *Science* **247**, 1558–1562.

Spencer, R. W., Christy, J. R., and Grody, N. C. (1990). Global atmospheric temperature monitoring with satellite microwave measurements: Method and results, 1979–1985. *J. Clim.* **3**, 1111–1128.

Glossary

Brightness temperature Apparent temperature of an object based on its radiative brightness. It equals the thermometric temperature when the object emissivity equals 1.

Emissivity Ability of an object to emit thermal radiation, ranging from 0 for a nonemitter (perfect reflector) to unity for a perfect emitter (nonreflector).

Radiometer Instrument that measures visible, infrared, or microwave radiation at one or more frequencies.

Troposphere Lowest major layer of the earth's atmosphere, extending from the surface to an average altitude of 12 km, through which the temperature generally decreases with height.

Weighting function Vertical distribution of the intensity of radiation received by a spaceborne radiometer, that is, what levels in the atmosphere the radiometer is sensitive to.

Groundwater

William Back
U.S. Geological Survey

The discipline of groundwater has both scientific and engineering aspects. Groundwater is a resource whose occurrence, movement, and chemical character must be understood in order to exploit and manage the resource through proper techniques of well construction and well-field operation. Groundwater can also present technological problems to be solved during construction of engineering projects. A third aspect of the discipline of groundwater is the study of its influence on geologic processes.

I. Introduction

Hydrology is the science of all phases of water (gas, liquid, and solid) of the earth — its occurrence, distribution, circulation, chemical and physical properties, and how it responds to and affects the environment. "Hydrology" includes the past history of water on the earth, its present occurrence, and its future changes.

Groundwater is a fundamental resource of the earth. As far as we know, the existence of free water as liquid, solid, and vapor differentiates the earth from all other planets; it is the primary consideration and basis of all human activity. Billions of dollars are expended annually throughout the world to develop groundwater as a resource, to protect aquifers from contamination, or to remediate those that have inadvertently become contaminated by some otherwise beneficial activity. In addition, impressively large sums are expended annually on engineering projects to ameliorate the deleterious effects of ground water, such as failure of dams, inefficiency of reservoirs, seepage of water into tunnels and mines, seepage of water during highway construction, and loss of stability of railroad and highway cuts. However, despite the extreme economic and societal importance of these aspects of groundwater, this chapter emphasizes another aspect — its interrelation with and influence on geologic processes.

II. Hydrologic Cycle

The hydrologic cycle is a continuum of processes in which water evaporates from oceans, lakes, and rivers; sublimates from snow and ice; and transpires from plants into the atmosphere; condenses to precipitate as snow, sleet, and rain; and ultimately returns to oceans as streamflow or discharge of groundwater from aquifers. The cycle is driven by solar and planetary energy, wherein the sun evaporates the water; the earth's rotation contributes to ocean currents and movement of water vapor; and the earth's gravity controls water runoff and infiltration. Just as water transforms from phase to phase, so too does energy. When solar energy causes evaporation, that energy is conserved and released again when the vapor condenses. This latent heat of vaporization helps control weather and climate. The simplicity of this concept can be deceptive, however, because only a small amount, about 1%, of the earth's total water supply moves within the hydrologic cycle each year. Most of the world's water supply, more than 97%, is salt water of the oceans. Most of the fresh-water supply is in ice caps and glaciers. [*See* WATER CYCLE, GLOBAL.]

The hydrologic cycle is an extremely important concept. We rarely see the complete idealized cycle, because mountains and oceans do not exist everywhere, deserts experience infrequent rainfall, and groundwater discharge to oceans is largely unobservable. The hydrologic cycle provides the scientific paradigm of hydrology, and scientists trained in different disciplines work on different segments of the cycle. Separation of the hydrologic cycle into various disciplines can be done on the basis of differences in the time scale for various processes involved in the cycle. Rainfall and overland flow can occur in hours, effects on streamflow can be measured in days, infiltration to the water table is measured in days and weeks, and the movement of groundwater from areas of recharge to areas of discharge can range from years to thousands of years. Within each of these segments of the cycle, dominant physical processes are different enough to warrant differentiation of the discipline of hydrology into subdisciplines.

In practice, hydrologists limit themselves to that part of the hydrologic cycle that occurs on the land. Studies of the oceans are the science of oceanography and studies of water vapor are the science of meteorology. The land part of hydrology can be subdivided into disciplines that include glaciology (the study of glaciers and ice); limnology (the study of freshwater and saline lakes); soil physics and chemistry (study of soil moisture); surface-water hydrology (the study of streams, lakes and estuaries); and groundwater (which includes the study of subsurface water).

The sources of groundwater can be meteoric, marine, metamorphic, and magmatic water, and water of diagenetic origin. Groundwater of meteoric origin is water that has "recently" been a part of the atmospheric part of the hydrologic cycle. Groundwater of marine origin is sea or ocean water that has intruded into coastal aquifers. Water of diagenetic origin is water that has been in the geologic formations that comprise aquifers for a significant period of time, perhaps millions of years. This definition thus includes "connate" water, which hydrogeologists do not recognize as existing in the sense of its original definition. "Connate water" was meant to specify ocean water that filled pore spaces between sediment particles at the time of their deposition. However, rock–water reactions during this residence time change the chemical composition of groundwater, and remove the original water, so that truly connate water does not exist. Waters of metamorphic and magmatic origin are released from rocks that are undergoing metamorphism or crystallization from a magmatic melt. [*See* CRUSTAL FLUIDS.]

Within the concept of the hydrologic cycle are some minor flow processes that are unimportant in terms of water resources but of great geologic importance. These include thermally induced free water convection near mid-ocean ridges and continental areas of high heat-flow, and expulsion of pore fluids during sediment compaction and diagenesis. Also, plutonic rocks, shales, and evaporites once considered to be essentially impermeable are now known to have permeabilities that are significant over geologic time. The importance of groundwater in the geologic cycle (Fig. 1) was emphasized by B. Hitchon in 1976, who stressed the ubiquity of water in processes within the "secondary environment" of weathering, erosion, diagenesis, and metamorphism. Groundwater also is of vital important in the "primary environment." For example, convecting groundwater is often the major factor in the cooling of magmatic bodies. [*See* GROUNDWATER FLOW AND GEOTEMPERATURE PATTERN.]

In traditional concepts of hydrogeology, a groundwater

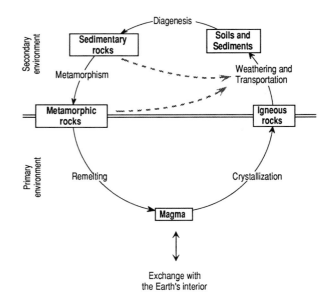

Fig. 1 Rock cycle. Groundwater processes are ubiquitous in the secondary environment. [After Hitchon (1976) and Sharp and Kyle (1988).]

basin commonly has shallow, intermediate, and deep regional flow systems that comprise zones of recharge, lateral flow, and discharge. When considering the role of groundwater in geologic processes, it is desirable to broaden the definition of recharge to include all sources of water for the particular strata of interest (Table I)—that is, dehydration and compaction of clays and fluid escaping from crystallizing magmas can be considered "recharge" from internal fluid sources. Thus, "recharge" includes the source of the subsurface water that forms mineral deposits, and "discharge" includes the manner by which that water leaves the deep subsurface. The zone of "lateral flow" can be considered to include the processes that circulate water from the recharge to the discharge zone, or the processes that recirculate the water.

III. Historical Developments of Hydrogeologic Concepts

Hydrology is a relatively young science that rests on a foundation of physical and chemical principles and processes identified during the past 100 years. Valid hydrogeologic concepts that originated in the latter part of the nineteenth century and the early decades of this century remain accepted and are actively applied today. Modern contributions have added significant advances, but the basic foundation of these early fundamental concepts, principles, and processes still determine the nature of the science and its applications.

Table I   Zones of Groundwater Flow[a]

Sources of water
   Rainfall and snow
   Marine
   Diagenetic
   Compaction
   Mineral dehydration
   Metamorphic
   Magmatic
Mechanisms of flow
   Gravity drive
   Overpressures
   Free convection
   Thermally driven
   Concentration driven
   Osmotic mechanisms
Losses of water
   Evapotranspiration
   Mineral hydration
   Discharge to surface water
   Discharge to oceans
   Burial in sediments
   Subduction

[a]Modified from Sharp and Kyle (1988).

Hydrogeology is an interdisciplinary science dependent on many branches of the physical, chemical, and biological sciences. It is a merging of geology and hydrology and could develop only after establishment of these two basic sciences. Several geologic phenomena were first identified and understood as a consequence of groundwater investigations. A number of distinguished early North American geologists worked on topics of hydrogeology to identify the fundamental controls exercised by lithology and structure on the continuity of aquifers and the flow of groundwater. In 1885, T. C. Chamberlain outlined the geologic and hydrologic conditions necessary for distant recharge and artesian flow of groundwater. His clear explanation of regional occurrence of groundwater and the hydrology of artesian systems can be considered the beginning of formal hydrogeology in North America. Chamberlain described water-bearing and confining units and their control on groundwater, and made the important observation that totally impermeable confining units do not exist. In 1897, N. H. Darton provided additional fundamental concepts during his classical field studies of the Dakota sandstone, which is an extensive artesian aquifer system of the north-central plains region of the United States. Darton identified the recharge area of the Dakota as being the topographically high Black Hills of southwestern South Dakota, and the discharge areas as being 300 to 500 km to the east.

Field studies similar to those early ones have continued to the present and provide the basic understanding of geologic controls on the occurrence, movement, and chemical behav-

ior of groundwater. Scientific and economic interests in artesian systems have continued throughout the twentieth century. The Dakota sandstone served as a model for study and debate long after publication of Darton's early explanation of artesian flow, which rested on the relative differences between the slope of the land surface and the slope of the potentiometric surface. For example, in 1928, O. E. Meinzer based his explanation of artesian pressure on the concept of compressibility and elasticity of artesian aquifers, and developed the fundamental concept of storativity during a study of the Dakota artesian system. This aquifer system continues to serve as an experimental field site to determine hydrologic properties of shales and confining beds with low permeability.

Based on these early concepts was the realization that much of the water pumped from an artesian aquifer is derived from compaction of the lithologic framework. Recognition of such compaction and compressibility provided the explanation for subsidence of the land surface that occurs in many places where oil, gas, and water have been pumped. At places in the San Joaquin Valley, California, the land surface has subsided more than 10 m. Subsidence has caused numerous environmental and engineering problems in many parts of the world, such as in Houston, Texas; Mexico City; Venice, Italy; Shanghai, China; and areas of Japan.

The chemical aspects of hydrogeology developed somewhat later after the basic geologic and hydrodynamic components were understood. By the 1950s, many of the chemical principles and processes of fundamental importance to the understanding of groundwater chemistry had been identified. These include the primary chemical reactions and their controls on the chemical character of groundwater, the concept of mineral equilibrium in natural water environments, oxidation–reduction reactions pertaining to groundwater environments, geochemistry of regional flow systems, shale–membrane phenomena, and heat flow. The marked advances in the principles of groundwater geochemistry that occurred in the 1960s and later were attributable predominantly to the growing appreciation of the fundamental role of chemical thermodynamics. Principles of thermodynamics have provided the basis for calculations of mass balance, mass transfer, mineral equilibrium, and kinetics of reactions in groundwater systems. Combining isotope geochemistry with hydrogeology has provided an additional dimension to understanding geochemistry of groundwater. [*See* WATER GEOCHEMISTRY.]

IV. Hydrogeologic Development of Aquifers

Because groundwater is present in all lithologies, essentially every known geologic process can contribute to formation of aquifers and other fluid-containing strata. All rocks in the crust above the zone of rock flowage contain water, albeit some rocks contain far smaller quantities than others. Lithology is a primary control on the mode of occurrence of groundwater. For example, in unconsolidated sand, water is present in interstitial pore space among the grains; in limestone and dolomite, it is present primarily in dissolution channels; in consolidated sedimentary rocks, it is present along fractures, joints, and bedding planes; in igneous and metamorphic rocks, it is present in weathered zones and along fractures.

Deducing the geologic evolution of aquifers in a region is comparable to deducing the evolution of geologic formations in that region, except that there is a greater emphasis on processes that generate pores, voids, and other openings within the rock, rather than on the fabric, texture, and mineral genesis of the rock matrix itself. Throughout the world, geologic processes that generated aquifers are not equally significant at every place or throughout geologic time. However, the science of geology has now progressed to the point where the fundamental processes are well understood, and it is rather straightforward to identify those processes that have either controlled or influenced development of the characteristics of a particular aquifer.

For example, in the West Indies, several processes have controlled the formation of the aquifers. As the Caribbean Plate moved from the Pacific Ocean, gabbro and low-density basalt were extruded onto the Plate, giving it a thick, buoyant nature. Because of differences in density between the light crust of the Caribbean plate and the surrounding oceanic crust, subduction of the denser material occurred; this gave rise to the Antillean arc structure that moved northeastward from its original position near present-day Central America. Major tectonic and magmatic events culminated with the collision of the northward-moving greater Antillean arc with the Florida–Bahama platform. The present position of the Caribbean islands results from east–west strike–slip movements along faults parallel to the arc. The cessation of volcanism and movement permitted accumulation of great thicknesses of calcareous mud and sediments that have been lithified through various processes of diagenesis and comprise major carbonate aquifers. The tectonism generated fractures and faults in the igneous rocks that permitted infiltration of water. Water is obtained from the relatively thick zones of saprolite that developed through weathering of the igneous rocks. Still other aquifers formed from the deposition of alluvium, either along stream channels or as alluvinal fans. [See FAULTS; LITHOSPHERIC PLATES.]

In other parts of the world, such as the Dinaric Mountains along the Adriatic Sea, major aquifers formed by dramatic tectonic uplift of limestone, producing extensive fractures that permitted extensive dissolution. The solution-enlarged fractures serve as conduits and reservoirs in the otherwise impermeable limestone. In Hawaii and parts of the northwestern United States, volcanism produced basalt flows that are interlayered with stream channel deposits. These sediments, combined with permeable weathered zones at tops and bottoms of basalt flows, make the entire volcanic sequence a permeable, productive aquifer system. The largest springs in the world issue from volcanic rocks and from limestone. Throughout much of north-central North America and northern Europe, coarse-grained material deposited by meltwater from retreating glaciers forms productive aquifers.

V. Groundwater in Geologic Processes

The role of groundwater in performing geologic work has long been a topic of interest in the earth sciences. The spectrum of diverse phenomena, ranging from processes deep within the crust to those that occur at or near the land surface, demonstrates the wide range of groundwater influence and control. Common to all these processes is the understanding that subsurface fluids do not exist in passively porous rock; instead, there exists a complex coupling between the rock, the moving water, heat, and constituents dissolved in the water.

This coupling, commonly referred to as water–rock interaction, can be categorized into three broad groups. First, there is the coupling between stress, strain, and pore fluids, most frequently in elevated-temperature environments. This coupling is of importance for understanding development of abnormal pressures in active depositional basins, earthquake generation, dissipation of frictional heat, rock strength and faulting, and pore-pressure development in response to molten intrusions. The second and third major types of water–rock interactions can be termed mass and energy transport and transfers occurring within saturated porous rocks.

"Transport" is defined as the processes that move mass or energy from one point to another in a porous medium. The movement of mass (water and solutes) is referred to as diffusion, dispersion, or advection of a chemical substance, whereas the energy (heat) transport is referred to as conduction, thermodispersion, or advection. "Transfer" means the manner in which, and commonly the rate at which, mass or energy contained in one phase (liquid, gas, or solid) is transferred to another phase. This indicates another type of coupling between transport and transfer, in that the physical phenomena (transport) often intrudes upon the chemical or thermal dynamic phenomena (transfer). Under these categories are included the temperature-dependent, liquid-forming, phase transformations, such as conversion of montmorillonite to illite, or kerogen to hydrocarbon; the generation, movement, and entrapment of hydrocarbons; the mobilization and migration of ore-forming fluids and their ultimate precipitation to form ore bodies; and a host of competing diagenetic reactions and transporting environments in clastic and non-clastic rocks where groundwater is the primary fluid.

Many of the important water–rock interactions involved in geologic processes can be viewed as problems in mass and energy transport and transfers in deformable (either physical or chemical) bodies, where the deformation is contemporaneous with deposition; occurs before, during, or after uplift; or does not occur. If momentum is considered to be the exertion of a force on a body, it too can be regarded as a transport process, suggesting that the role of groundwater in geologic processes can be viewed from the perspective of momentum, mass, and energy transport and transfers within geologic environments. In most cases, these processes occur simultaneously and sometimes interfere with each other. Although the individual phenomena are generally separated to facilitate study, some geologic processes are the consequences of the coupling of these many processes and reactions.

A. Geomorphology

Groundwater affects near-surface processes and landforms in a wide variety of ways. Its role in weathering, soil development, slope failure, and development of karst topography has long been acknowledged; however, its importance in other aspects of geomorphology has begun to be recognized only in the last few decades.

Karst landscapes are the result of dissolution of limestone and dolomite by runoff, infiltration, or deep-circulating groundwater. The processes of karst formation are such that the transport of mass in solution is of greater importance than transport by other processes. Karst terrains evolve through time, and the overall rate of the processes producing them depends on chemical and hydrodynamic processes. The dissolution of carbonate rocks entails equilibrium controls and kinetic controls. Thermodynamic factors determine the maximum amount of dissolved carbonate that can be carried in a given volume of water, and this, combined with the flow rate, determines the maximum rate at which karst landscapes will develop. The evolution of karst landscapes can be considered to result from a set of coupled processes, some of which depend on mass-transport rates and others that depend on dissolution kinetic rates.

Karst processes are influenced by the strong relation between climate and chemistry of groundwater. The influence of climate on the chemistry of groundwater results primarily from variations in temperature, precipitation, and concentration of CO_2, all of which enter into equilibrium controls and kinetic controls. Dissolution of carbonate rocks by groundwater forms an underground system of conduits and smaller openings that are fundamental to karst development; these conduits are comparable to stream channels in the fluvial system. Such features as sinkholes, poljes, and deep karren cannot develop until cave systems have been established

beneath them. These cave systems integrate the drainage from many sources and concentrate the discharge at large karsitic springs. Landforms at spring points include erosional features such as gorges, created by headward sapping of a cave roof or a cliff foot, and constructional features, such as travertine terraces, cascades, and dams.

B. Diagenesis of Carbonate and Clastic Rocks

Two of the major controls on the occurrence and movement of groundwater are the distribution of permeability and porosity. Although diagenesis is a major control on these properties, other geologic factors influence the distribution of permeability. For example, the porosity and permeability of carbonate rocks can be affected by more than 60 different processes and controls (Table II).

The term "diagenesis" refers to changes that occur in sediments following deposition. These changes are the outcome of diverse and commonly complex physical, chemical, and biological processes. This section emphasizes the role of groundwater in the diagenesis of clastic and carbonate rocks through the mechanism of mass transport. Processes that mechanically deform clay minerals, chemically alter aquifer framework grains, and form authigenic minerals can produce numerous diagenetic features. Diagenetic processes are related not only to chemical phenomena but also to hydrodynamic phenomena. Considerable quantities of mass must be transported in a flow system to develop the diagenetic features. Models of groundwater flow in regional geologic basins and in smaller flow systems are an integral part of many diagenetic studies.

The processes of advection and diffusion are responsible for the physical transport of mass, and chemical and biological reactions are responsible for diagenetic changes. These reactions and processes partition the mass among the water, mineral, gas, and organic phases, and are controlled by thermodynamic principles. The direction and rate of advective transport are in the direction and at the rate of groundwater flow, and flow lines trace the pathways for mass movement.

Patterns of flow are determined by gradients in hydraulic potential, distributions of permeability and, to a lesser extent, by fluid density and viscosity. Potential gradients that provide the driving forces for flow are generated mainly by gravity, compaction, or dilation of a body of sediments, free convection, and physical and chemical processes. These last processes include (1) phase changes such as gypsum to anhydrite, or smectite to illite; (2) the solution or precipitation of minerals; (3) fluid expansion caused by temperature changes; (4) osmotic or membrane phenomena; and (5) metamorphic reactions (Fig. 2).

Groundwater moves in a gravitational field because of

Table II Processes and Controls That Affect Porosity and Permeability of Carbonate Rocks

Geologic factor	Processes	Controls	General influence
Diagenetic	Compaction Cementation Pressure solution Solution[a] (includes recrystallization, inversion, micritization)	Original porosity and permeability Original mineralogy Grain size/surface area Proximity to sea level (uplift or burial) Volume and rate of water movement Fluid chemistry: pH, pCO_2, salts in solution; temperature, pressure	Influences initial distribution of porosity and permeability of indurated rock mass; many of these are geochemical in nature; they occur very early in the history of the rock
Geochemical	Solution[a] (dissolution) Dolomitization Dedolomitization Precipitation Sulfate reduction Redox	Groundwater flux Original porosity and permeability Mineralogy Fluid chemistry: pH, pCO_2, salts in solution, temperature, pressure, mineral-water saturation	Influences later development of porosity and permeability; influences water chemistry
Lithologic-stratigraphic		Layer thickness Sequence thickness Variability in texture (vertical) Variability in permeability (vertical) Original porosity and permeability inherited from diagenesis Bulk chemical purity Grain size	Influences anisotropy of rock mass, thereby resulting in zones potentially more permeable if other geologic factors are favorable
Structural-tectonic	Uplift Tilting Folding Jointing Faulting Metamorphism	Fracture density Openness of fractures Layer (permeability) orientation	Influences orientation of permeability zones Influences integrity of confining layers In extreme instances (metamorphism), influences existence of permeability zones
Hydrologic	Dynamic groundwater flow[a]	Climatic–temperature Climatic–precipitation Depth of circulation Location of boundaries Existence of complete flow systems Flux Initial anisotropy-vertical variation Springs Surface-water/groundwater relation Recharge Hydraulic gradient Size of groundwater basin	Influences existence of flow systems Influences rate of flow system evolution
Weathering-geomorphic	Infilling (fluvial and glacial) Unloading	Topography Relief Soil development Cap rock Degree of karstification Base level Surface slope	Influences development of flow systems Influences destruction of permeability by sedimentation Influences shallow porosity-permeability development
Historical geologic–chronologic		Sequence of events Duration of events	Influences stage of development of specific permeability zones

Source: Brahana *et al.* (1988).
[a]Most important.

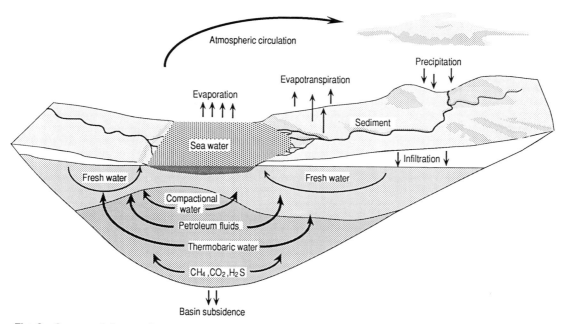

Fig. 2   Conceptual diagram illustrating various hydrologic regimes in a large, active sedimentary basin. [After Galloway and Hobday (1984) and Schwartz and Longstaff (1988).]

potential gradients created when water recharges a basin in topographically high areas. These gradients result in the flow of groundwater from regions of high potential to regions of lower potential, shown as fresh-water flow (Fig. 2) and gravity flow (Fig. 3). The actual direction of groundwater flow is determined not only by the potential gradient but also by the distribution of permeability.

Groundwater flow also occurs in response to the compaction of saturated clay-rich sediments. The theory of consolidation explains how the total stress or load of a given volume of sediment is borne by both the solid matrix and the water. Incremental loading caused by continued sedimentation directly increases water pressure in the pores. This increase in pore pressure provides the potential for water flow, shown as "compactional water" in Fig. 2. Sediments consolidate when drainage occurs and pore pressures dissipate. The rate of drainage compared to the rate of sedimentation determines whether the water pressure in sediments will be normal or overpressured (fluid pressures in excess of hydrostatic pressure) at a given depth. Thermobaric water (Fig. 2) is water released by dehydration reactions of clays and other minerals; it exists in the deepest part of the basin where temperatures and pressure are high.

Water flow by convection is another mechanism for advective mass transfer in geologic basins (Fig. 3). Convection cells can develop in thick sandstones or even in fractured shale units under normal geothermal gradients. Free convection tends to occur more frequently with increase in depth and in

formation thickness. Certain mineralogical phase changes transfer water from the mineral to the liquid phase. Potential for flow is generated because the addition of even small quantities of water to a pore results in an increase in pressure. Solution or precipitation of minerals and water expansion caused by heating, as shown in Fig. 3, have much the same effect — both alter the volume relation between water and pores.

Osmotic or membrane effects contribute to flow in a different way. Highly compacted sediments, rich in clay, can act as a semi-permeable membrane. When waters of different salinity are on opposite sides of the membrane, the difference in chemical potential causes a flow of water through the membrane from the fresh-water side, where the thermodynamic activity of water is comparatively great, to the saline-water side. Groundwater flow systems exist at a hierarchy of scales, ranging from deep basinal systems, which are the site for burial diagenesis, to localized systems such as an island or a sabkha setting, which are commonly the sites for early diagenesis of carbonate rocks.

C. Formation of Ore Deposits

The role of groundwater in the mobilization, transport, and deposition of ore deposits has long been recognized, but the lack of understanding of geochemical controls and the absence of viable transport models precluded application of groundwater principles that is now possible.

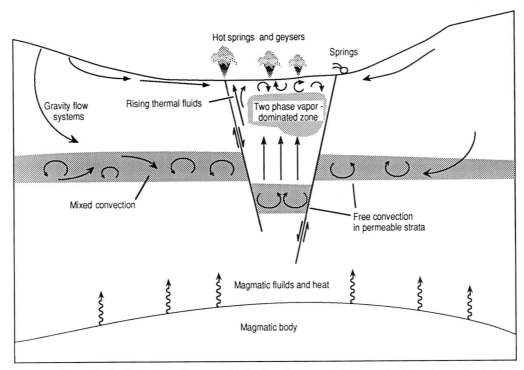

Fig. 3 Flow paths for hydrothermal systems. Predominantly meteoric fluids can be driven by gravity, by free convection, or by a combination of the two. Flow systems can be both deep and shallow. Vapor-dominated and hot-water-dominated systems can coexist. [After Sharp and Kyle (1988).]

Groundwater is of paramount importance in a variety of ore-forming processes, but the hydrodynamics of these processes have only recently been the subject of quantitative investigations. Research topics of common interest to economic geologists and hydrogeologists include fluid flow on both local and regional scales, the mixing of groundwater, and water–rock interactions. Hydrogeologic models capable of estimating timing and rates of fluid flow, when coupled with transport and geochemical modeling, can lend insight into the genesis of mineral deposits (Fig. 3). Furthermore, the analysis of these fossil-hydrogeologic systems will ultimately increase understanding of the subsurface part of the hydrologic cycle by providing paleo-hydrologic data.

Four necessary aspects for the generation of ore deposits by ground water are (1) a source for the mineral constituents, (2) dissolution of the minerals in water, (3) migration of the ore fluids, and (4) precipitation of the minerals in response to physical and chemical changes in the fluid or rocks. The relation of hydrogeology to economic geology has been demonstrated for many types of ore bodies. These include carbonate-hosted lead–zinc deposits; porphry copper deposits; epithermal deposits associated with shallow hydrothermal systems; stratiform sulfide deposits; uranium roll-front depos-

its; weathering-zone deposits such as lateritic ores or iron, nickel, aluminum, and gold; supergene sulfide deposits; and metamorphic vein deposits. [See ORE FORMATION.]

The specific requirements for mineral deposition in a hydrogeologic system are (1) a source of ore-forming constituents for transport in solution or as colloids; (2) water for transport; (3) existence of a potential field for water flow; the most probable driving mechanisms are hydraulic forces, (gravity and compaction) but thermal, chemical, osmotic, or electroosmotic fields also exist; (4) sufficient permeability, either primary or secondary, to permit water flow; (5) proper thermodynamic conditions to precipitate the minerals; (6) adequate formation dispersivity to permit differentiation and distribution of minerals; and (7) sufficient permeability contrast to focus the mineralizing fluid in sites of ore formation.

D. Oil and Gas Accumulation

The role of groundwater in migration and entrapment of oil and gas has been accepted since the classic work of M. King Hubbert in the 1940s and 1950s. The chief forms in which hydrocarbons migrate are (1) oil droplets and gas bubbles, (2) a continuous oil–gas phase, (3) colloidal or micellar solu-

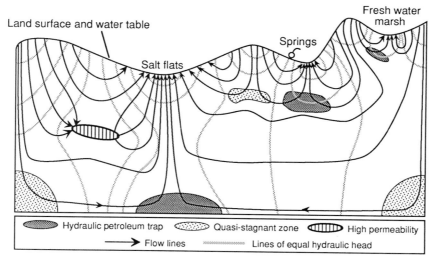

Fig. 4 Regionally unconfined gravity flow of formation water and related hydrogeologic phenomena. [Modified from Toth (1988).]

tions, (4) true solutions, (5) petroleum in a framework of organic matter, and (6) soluble organic matter. Various energy sources give rise to forces that cause migration, and they are not mutually exclusive. These forces can act singly or in combination, and their relative importance can change with place and time in an evolving basin. The sources of energy include (1) compaction and elastic rebound of the rocks, (2) buoyancy, (3) gravitation in either confined carrier beds or aquifers or in a regionally unconfined rock framework, (4) gas expansion, (5) thermal expansion of liquids, (6) molecular diffusion of hydrocarbons, and (7) osmosis.

The basic dynamic requirement for entrapment is the existence of a minimum fluid potential at the site for hydrocarbon accumulation (Fig. 4). Minimum fluid potential can develop in the following three essentially different situations: (1) with no water flow, (2) in the direction of water flow, and (3) in the direction opposite to water flow. In the absence of water flow, the force that transports hydrocarbons is buoyancy, and, at the upper boundary of a coarse-grained bed, a minimum potential occurs with respect to the buoyant petroleum particles. At that location, additional energy would be required to force the hydrocarbons into the overlying fine-grained rock against the resistance of micropore filtration and a capillary barrier; therefore, the oil accumulates. Various versions of dynamic entrapment occur by certain lithological and structural features that act as traps for oil that moved in the direction of water flow. In quasi-stagnant regions between flow systems, minimum fluid potentials can develop without particular lithological or structural requirements and from "hydraulic traps" (Fig. 4).

In groundwater-discharge areas, flow converges, and the potentiometric contours form closed zones of head; minimal lateral escape of hydrocarbons is not possible (Fig. 4). Upward flow of water is possible if the rocks are saturated with water despite reductions in pore size across boundaries of different strata. Micropore filtration and capillarity, however, will present an energy barrier for hydrocarbons at these boundaries, thus completing a three-dimensional sink, or trap, for petroleum. In regions where descending flows diverge, such as beneath recharge areas, fluid potentials decrease radially outward. However, the lateral gradients are so slight, if they exist at all, in the center of these mounds that descending hydrocarbons may be unable to overcome the resistance of facies changes and other barriers along the bedding and can accumulate between the diverging flow systems. These accumulations are unstable and relatively sensitive to shifts in flow fields; they can thus be expected to be less common in nature than accumulations associated with converging flows. The modes, forces, and patterns of hydrocarbon migration and their temporal and spatial variations depend directly on the evolutionary history of a sedimentary basin. Therefore, if migration is to be understood, it must be analyzed in the context of basin evolution. Numerical models have been successful in depicting and explaining migration of petroleum to provide rational working hypotheses that could aid exploration. [See BASIN ANALYSIS METHODS.]

VI. Effects of Global Climatic Change

The occurrence, movement, and chemical character of groundwater can be significantly affected by changes in climate. These changes include such aspects as redistribution of rainfall patterns, which would (1) lead to relocation of recharge and discharge areas; (2) affect the base flow and sedimentation rates of streams; and (3) affect the type and distribution of vegetation, which can be a major control on the infiltration capacity of an area.

Perhaps less frequently considered is the effect of sea-level fluctuation on the groundwater regime. The fresh-water flow system in many coastal aquifers terminates at a chemical rather than a physical boundary. This chemical boundary results from the encroachment of salt water from the ocean into the aquifers. Sea-level elevation is a major control on both the hydraulic gradient and the nature and position of the down-gradient boundary of fresh-water flow. The position of the fresh-water–seawater interface and the thickness of the fresh-water lens in an aquifer are controlled by the relation of fresh-water head to sea level. The magnitude and distribution of this fresh-water head in an undeveloped aquifer is controlled by the size and elevation of the recharge area, transmissivity of the aquifer, permeability of overlying material, and amount of precipitation. Pumping large volumes of water from the aquifer can have local or regional effects on head. These factors, in turn, control the recharge rate, altitude of the potentiometric surface, and degree of confinement—all of which determine the amount and location of groundwater discharge, hydraulic gradient, and physical characteristics of the groundwater–salt-water mixing zone.

As groundwater flows toward discharge areas along a coast, it mixes with seawater that has encroached into the aquifer. This mixing generates brackish water, which is less dense and, therefore, more buoyant than seawater; this brackish water rises and discharges near the shore. Numerous submarine and near-shore brackish-water springs represent discharge from this mixing zone. This discharge requires a replacement of seawater, thereby establishing a cyclic flow system in which water moves from the sea, through the coastal part of aquifers, upward along the interface, through the mixing zone, and back to the sea.

The mixing zone associated with this interface is a geochemically reactive area that has been a major influence on the development and distribution of porosity and permeability, particularly in carbonate rocks. In addition, in many coastal carbonate areas, carbonate dissolution that occurs in the mixing zone has a pronounced effect on formation of geomorphic features, such as coastal caves, karren karst, bays, lagoons, and crescent-shaped beaches.

Sea-level fluctuations have also been a contributing factor in the formation of many alluvial-filled valleys in coastal areas. At times of sea-level decline, rivers incised the valleys to the base level of the sea. After a subsequent rise of sea level, these valleys were drowned, and the rivers progressively filled the valley with alluvial and estuarine deposits of clay, sand, and gravel. Emergence of these deposits permits them to be filled with fresh water, and they can become significant

aquifers. A further future rise in sea level will flood these valleys and cause the fresh-water aquifers in the alluvium to be intruded by the encroaching sea water.

Acknowledgments

I wish to express my appreciation to the many authors (1988) of *Hydrogeology*, published by the Geological Society of America as Volume 0-2 in their series, *Decade of North American Geology*, from which much of the material presented here was excerpted. I also wish to thank James A. Miller, Charles A. Appel, and David A. Aronson of the U.S. Geological Survey for their review and comments on the manuscript.

Bibliography

Back, W., and Freeze, R. A., eds. (1983). Chemical hydrogeology. *Benchmark Papers in Geology* **73**, Hutchinson and Ross, Stroudsburg, Pennsylvania.

Back, W., Rosenshein, J. S., and Seaber, P. R., eds. (1988). "Hydrogeology: The Geology of North America" Vol. 0-2 in the series "Decade of North American Geology." Geological Society of America, Boulder, Colorado.

Brahana, J. V., Thrailkill, J., Freeman, T., and Ward, W. C. (1988). Carbonate rocks. *In* "Hydrogeology: The Geology of North America" (W. Back, J. S. Rosenshein, and P. R. Seaber, eds.), Vol. 0-2 in the series "Decade of North American Geology," pp. 333–352. Geological Society of America, Boulder, Colorado.

Domenico, P. A., and Schwartz, F. W. (1990). "Physical and Chemical Hydrogeology." Wiley, New York.

Driscoll, F. G. (1986). "Groundwater and Wells." Johnson Division, St. Paul, Minnesota.

Fetter, C. W. (1988). "Applied Hydrogeology." Merrill Publishing Co., Columbus, Ohio.

Freeze, R. A., and Back, W., eds. (1983). Physical hydrogeology. *Benchmark Papers in Geology* **72**, Hutchinson and Ross, Stroudsburg, Pennsylvania.

Freeze, R. A., and Cherry, J. A. (1979). "Groundwater." Prentice-Hall, Englewood Cliffs, New Jersey.

Galloway, W. E., and Hobday, D. K. (1983). "Terrigenous Clastic Depositional Systems: Applications to Petroleum, Coal, and Uranium Exploration." Springer-Verlag, New York.

Hitchon, B. (1976). Hydrogeochemical aspects of mineral deposits in sedimentary rocks. *In* "Handboook of Stratabound and Stratiform Ore Deposits," Vol. 2. (K. H. Wolf, ed.), pp. 53–66 Elsevier, Amsterdam.

La Fleur, R. G., ed. (1984). "Groundwater as a Geomorphic Agent." Allen & Unwin, Inc., Winchester, Massachusetts.

Narasimhan, T. N., ed. (1982). "Recent Trends in Hydrogeology." Special Paper 189, Geological Society of America, Boulder, Colorado.

Schwartz, F. W., and Longstaff, F. J. (1988). Ground water and clastic diagenesis. *In* "Hydrogeology: The Geology of North America," (W. Back, J. S. Rosenshein, and P. R. Seaber, eds.), Vol. 0-2 in the series "Decade of North American Geology," pp. 413–434. Geological Society of America, Boulder, Colorado.

Sharp, J. M., Jr., and Kyle, J. R. (1988). The role of groundwater processes in the formation of ore deposits. *In* "Hydrogeology: The Geology of North America," (W. Back, J. S. Rosenshein, and P. R. Seaber, eds.), Vol. 0-2 in the series "Decade of North American Geology," pp. 461–483. Geological Society of America, Boulder, Colorado.

Toth, J. (1988). Groundwater and hydrocarbon migration. *In* "Hydrogeology: The Geology of North America" (W. Back, J. S. Rosenshein, and P. R. Seaber, eds.), Vol. 0-2 in the series "Decade of North American Geology," pp. 485–502. Geological Society of America, Boulder, Colorado.

White, W. B. (1988). "Geomorphology and Hydrology of Karst Terrains." Oxford Univ. Press, New York.

Glossary

Advection Process by which dissolved constituents (solutes) are transported by bulk motion of flowing groundwater.

Diffusion Mixing of a solute in response to molecular movement.

Dispersion Resulting distribution of concentrations of solutes by advection and diffusion.

Hydraulic conductivity Ability of rocks to transmit water, taking into account the viscosity and density of the water.

Hydraulic potential Commonly referred to as fluid potential, which is related to head, the summation of forms of energy (elevation, pressure, and velocity) that cause water to move.

Permeability Measure of the ability of soil or rocks to transmit fluids or gases.

Porosity Ratio of the volume of void space to the total volume of rock or soil.

Transport model Mathematical equation that accounts for the sources, movement, and fate of dissolved constituents.

Groundwater Flow and Geotemperature Pattern

Kelin Wang

Pacific Geoscience Centre, Sidney, B.C., Canada

The interdisciplinary study of groundwater flow and geotemperature patterns focuses on the interactions between the hydrological and thermal regimes of the earth's lithosphere, especially the upper crust. These interactions are subsurface energy transport and redistribution processes controlled by the thermal and hydraulic properties of the subsurface formations. Hydrodynamic concentration of thermal energy in limited volumes of the shallow earth is of considerable economic and scientific value to human society.

I. Introduction

The earth is unique in the solar system for its characteristic and precious hydrosphere. The highly mobile bodies of water form a physical cycle that involves migration, evaporation, condensation, and precipitation, and a chemical cycle that involves hydration and dehydration of rock minerals facilitated by the recycling of crustal materials into the deeper interior. Both cycles contribute substantially to energy transport processes in the outer earth. Although groundwater does not make as direct a contribution as surface water to the balance of the atmospheric energy budget, it plays many important roles in reshaping the geothermal regime in the earth's lithosphere, especially the upper crust. [*See* CRUSTAL FLUIDS.]

Groundwater systems are maintained by infiltrating meteoric water from ground surface and to a minor extent water released from rock minerals in the earth. The majority of the groundwater systems considered in this article are those mainly of meteoric origin. In tectonic zones, such as plate boundaries, water supply by compaction of sediments and dehydration of metamorphosed rocks is of tremendous importance. Release of water by cooling magmas intruded into the earth's crust is observed in hydrothermal systems. Other sources of groundwater are of very limited quantity and have negligible thermal consequences. [*See* GROUNDWATER; WATER CYCLE, GLOBAL.]

The principal source of the earth's internal heat are the decay of long-lived radioisotopes, such as ^{40}K, ^{238}U, and ^{232}Th, and probably the dissipation of the earth's primordial heat, namely the heat of accretion and differentiation. Exothermic chemical and biochemical reactions and mineral phase changes, tectonic movements, and tidal friction, etc., also generate heat but are of negligible significance on a global scale. The earth's internal heat is transferred to the ground surface at a global rate of about 4.2×10^{13} W mainly by means of conduction, but wherever there is groundwater flow, the process is augmented or hindered by convection. Manifestations of geothermal energy such as hot springs and hydrothermal fields are the extreme cases of such convective heat transfer.

The economic and scientific value of the thermal effects of groundwater flow have long been appreciated. Use of thermal water for medical and recreational purposes can be found in historical records of any culture. The relation of ore deposits to thermal waters was observed and recorded early in the sixteenth century. Serious consideration of the influence of groundwater flow accompanied the first group of systematic measurements of terrestrial heat flow in the early 1930s. In the early 1980s, extensive saline water with strong circulation was observed at a depth as great as 5 to 9 km in the world's deepest well drilled in Kola Peninsula, U.S.S.R. The thermal effects of groundwater on large-scale geological and tectonic processes in the earth's crust is attracting further attention. [*See* HEAT FLOW THROUGH THE EARTH.]

For ease of elucidation, it is necessary to distinguish between local scale and regional scale groundwater flow systems, although there is a complete spectrum of problems of different scales in between. In this article, the term local scale flow system implies not only a small distance scale (a few kilometers) and depth scale (a few hundred meters), but also a relatively simple hydrogeologic setting, such as a single aquifer. In local scale systems, the velocity of water flow along major paths can often be considered as nearly uniform. The term regional scale flow system is used for a distance scale of tens of kilometers or more, and a depth scale of a few kilometers or more, such as in a sedimentary basin.

II. Conductive and Convective Heat Transfer in Porous Media: Basic Concepts

A. Heat Conduction in the Presence of Groundwater

The heat flow density (HFD, W m^{-2}) by conduction $\mathbf{q}$ is determined by the temperature gradient Γ (K m^{-1}) and the thermal conductivity λ^r (W m^{-1} K^{-1}) of the porous medium and is given by Fourier's law

$$q_i = -\lambda_{ij}^r \Gamma_j = -\lambda_{ij}^r \frac{\partial T}{\partial x_j} \qquad (1)$$

where T (K) is temperature, x_j (m) is the jth coordinate of the Cartesian coordinate system, with $x_3 = z$ being elevation (if upward) or depth (if downward). Subscripts denote components of vectors (represented by boldface characters) or tensors in the Cartesian system; repeated subscripts imply summation. The thermal conductivity is generally a tensor, but reduces to a scalar for isotropic media. If the temperature difference ΔT and the thermal conductivity λ^r over a depth interval Δz are measured, the vertical conductive HFD can be determined as

$$q = -\lambda^r \frac{\Delta T}{\Delta z} \qquad (2)$$

The heat conduction equation derived from (1) is

$$\frac{\partial}{\partial x_i} \lambda_{ij}^r \frac{\partial T}{\partial x_j} = \rho^r c^r \frac{\partial T}{\partial t} - Q^h \qquad (3)$$

where t (s) is time, ρ^r (kg m^{-3}) and c^r (J K^{-1}) are the density and specific heat of the porous medium, respectively, and Q^h (W m^{-3}) is the heat source/sink in the medium. Since the variations of ρ^r and c^r with temperature tend to cancel each other's effects, the quantity $\rho^r c^r$ (J m^{-3} K^{-1}) is usually consid-

ered as one parameter with its temperature dependence neglected, and is termed the specific thermal capacity. The thermal conductivity divided by the specific thermal capacity is called the thermal diffusivity, $\alpha^r = \lambda^r/(\rho^r c^r)$ (m^2 s^{-1}). It is the thermal diffusivity α^r that determines the transient behavior of heat conduction in rocks. The geothermal field can often be considered to be at steady state; processes with many short-period variations are often approximated as steady state over a long time in the average sense. In such cases, it is the thermal conductivity λ^r that matters.

The bulk thermal conductivity of a fluid–solid complex such as a water-saturated porous rock depends on both the fluid and solid components. One of the expressions for such a bulk conductivity (λ^r) is the geometric mean

$$\lambda^r = (\lambda^s)^{1-\phi} \lambda^\phi, \qquad (4)$$

where ϕ (dimensionless) is porosity, the ratio of the total volume of pore spaces to that of the whole rock. The superscript s denotes solid matrix. No superscript is used for the physical properties of water, but a superscript f can be added to represent fluid in general. Since water has a thermal conductivity (0.6 W m^{-1} K^{-1}) lower than that of most rock minerals, a saturated porous rock with a higher porosity is a poorer conductor of heat.

The bulk specific thermal capacity of a water-saturated porous medium is given by

$$\rho^r c^r = \rho c \phi + \rho^s c^s (1 - \phi) \qquad (5)$$

B. Convective Heat Transfer in Porous Media

For nonisothermal subsurface flows, a reference hydraulic head h is defined as

$$h = \frac{P_f}{\rho_0 g} + z \qquad (6)$$

where P_f is fluid pressure, ρ_0 is water density at a reference temperature T_0, g is gravity, usually taken as a constant, and $z = x_3$ is elevation (not depth). The magnitude and direction of water flow in a porous medium is represented by darcy velocity or specific discharge $\mathbf{v}$ (m s^{-1}) which is given by Darcy's law,

$$v_i = -\frac{\kappa_{ij}}{\mu} \rho_0 g \left(\frac{\partial h}{\partial x_j} + \frac{\rho - \rho_0}{\rho_0} \delta_{3j} \right) \qquad (7)$$

where κ (m^2) is the permeability of the porous medium, μ (kg s^{-1} m^{-1}) is the dynamic viscosity of water, and ρ is the water

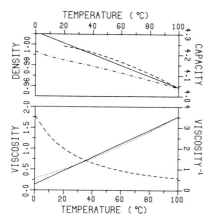

Fig. 1 The specific thermal capacity ρc (J cm^{-1} K^{-1}), density ρ (g cm^{-3}), and dynamic viscosity μ (kg s^{-1} m^{-1}) of water as functions of temperature. The actual ρ and μ are represented by dashed lines, and ρc by dot-dashed line; dotted line is μ^{-1}. Straight solid lines are possible linear approximations for ρ and μ^{-1}.

density at temperature T. κ is generally a tensor, but reduces to a scalar for isotropic media. The values of ρ and μ vary with temperature and slightly with pressure; in most practical cases, only the temperature dependence needs to be considered (Fig. 1). For moderate temperature ranges (~ 100 K), such temperature dependence is often approximated with linear functions (Fig. 1). In cases where the temperature dependence of ρ can be neglected (e.g., in local flow systems) the buoyancy term involving ρ vanishes and Eq. (7) becomes analogous to Fourier's law of heat conduction. The product $(\kappa_{ij}/\mu)\,\rho_0 g$ is termed the hydraulic conductivity, usually denoted by K (m s^{-1}); K depends on the properties of both the porous medium and the flowing fluid, while the permeability κ is a characteristic of the porous medium only.

The partial differential equation for conservation of momentum derived from (7) is

$$\frac{\partial}{\partial x_i}\left[\frac{\kappa_{ij}}{\mu}\,\rho_0 g\left(\frac{\partial h}{\partial x_j}+\rho_r\delta_{3j}\right)\right]=s\frac{\partial h}{\partial t}-Q^{\mathrm{f}} \qquad (8)$$

where s (m^3 m^{-1}) is the specific storage of the fluid–solid complex, depending on the porosity and the compressibilities of the solid matrix and water; Q^{f} (m^3 s^{-1}) is a term representing water source or sink; s and Q^{f} are analogous to ρc and Q^{h} in the heat conduction equation (3).

The amount of thermal energy in a unit mass of water at temperature T is $\rho c(T-T_0)$, where the thermal energy at the reference temperature T_0 is arbitrarily taken to be zero. Similar to those of rocks, the specific thermal capacity of water ρc remains nearly a constant with changing temperature (Fig. 1). Although water is a relatively poor heat conductor, it carries

and exchanges heat with the surrounding rocks when it flows. Due to its large heat capacity and high mobility, water can transfer heat very efficiently. With a darcy velocity **v**, the HFD caused by water flow is $\rho c\mathbf{v}(T-T_0)$. The total HFD by conduction and convection is therefore

$$q_i^{\mathrm{total}}=-\lambda_{ij}^{\mathrm{r}}\frac{\partial T}{\partial x_j}+\rho c v_i\,(T-T_0) \qquad (9)$$

Here conduction takes place in both the solid matrix and the water-filled pore spaces. By assuming that thermal equilibrium between the pore water and the solid matrix is achieved instantaneously, the heat conduction–convection equation is given as

$$\frac{\partial}{\partial x_i}\lambda_{ij}^{\mathrm{r}}\frac{\partial T}{\partial x_j}-\rho c v_i\frac{\partial T}{\partial x_i}=\rho^{\mathrm{r}}c^{\mathrm{r}}\frac{\partial T}{\partial t}-Q^{\mathrm{h}} \qquad (10)$$

Equations (8) and (10) can be solved analytically in a few simple cases; whenever available, we prefer them for their handiness and lucidity. Occasionally, analog models are employed to simulate the problem in laboratory using other physical variables such as electrostatic fields. Numerical modeling methods have been increasingly used to solve these equations with more complicated and realistic parameter structures and boundary/initial conditions.

C. Free Convection

We are dealing with the process of forced convection or advection, that is, the fluid flow caused by gravitational or compactional forces. When the geothermal gradient is high enough and the geometry of the hydrogeological system is favorable, the buoyancy force may render the water system unstable and result in free convection. Ocean water can circulate through the permeable basaltic ocean floor by means of free convection and affect the observed oceanic HFD pattern. On the continent, however, natural convection is likely to be important only in the hydrothermal areas where terrestrial heat flow is abnormally high.

III. Thermal Effects of Local Scale Flow Systems

A. Vertical Flow in a Permeable Formation

When groundwater flows vertically in a permeable rock formation, the temperature and HFD perturbations can be studied using a simple, steady-state, one-dimensional model. Consider water flowing at constant velocity v through such a rock

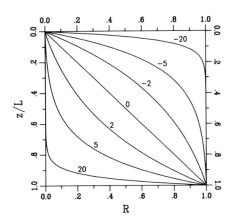

Fig. 2 Relative temperature R in a permeable formation in the presence of vertical groundwater flow, as defined in Eq. (11). The numbers on the curves are Peclet numbers; negative values indicate upward flow.

formation with thickness L and a uniform bulk thermal conductivity λ, with the upper boundary $z = 0$ and lower boundary $z = L$ kept at constant temperatures T_0 and T_1, respectively. The HFD in the absence of water flow, $q_1^* = \lambda(T_1 - T_0)/L$, is defined as a characteristic conductive HFD, and $q_2^* = \rho c v(T_1 - T_0)$ as a characteristic advective HFD.

By solving the appropriate form of Eq. (10), the relative temperature in the medium is

$$R(z) = \frac{T(z) - T_0}{T_1 - T_0} = \frac{\exp(Pz/L) - 1}{\exp(P) - 1} \qquad (11)$$

where P is the Peclet number of the system, defined as

$$P = q_2^*/q_1^* = \rho c v L/\lambda \qquad (12)$$

P tells us the relative importance of convective heat transfer; in a given formation, it depends mainly on the flow velocity v. R, as a function of z/L for different Peclet numbers, is plotted in Fig. 2. As z changes from 0 to L, T varies from T_0 to T_1, and R from 0 to 1. Without a hydraulic disturbance ($P = 0$), the temperature–depth (T–z) profile is linear. When water flows downward ($P > 0$), it cools the upper portion of the system, reducing the geothermal gradient, causing a lower conductive HFD; but the gradient in the lower portion is enhanced, and hence yields a higher conductive HFD.

Equation (11) is often used to interpret the curvatures observed in borehole T–z profiles, and even to estimate the darcy velocity v. However, Eq. (11) can be applied to borehole data only when the local flow system is not significantly disturbed by the existence of the borehole at the time of measurement. Water also flows between the borehole and the formation, unless the borehole section through the for-

mation is cased. More important, many of the borehole T–z data interpreted using Eq. (11) might be subject to the effects of water flow along the hole itself, which will be discussed in a subsequent section.

B. Flow in a Thin Aquifer

1. A Horizontal Aquifer

Consider a cross section of a horizontal aquifer with thickness H at a depth D from the upper surface which is kept at a constant temperature, as illustrated in Fig. 3a. The thermal conductivity of the isotropic water–rock complex of the aquifer is λ^a, and that of the overlying and underlying rocks is λ^b. Water at temperature T_0 flows with velocity v from $x = 0$ into the section. Assume that water is well mixed in the thin aquifer and hence the vertical temperature variation within the aquifer is negligible; the relative temperature of the water as a function of horizontal distance x is then

$$R(x) = \frac{T(x) - T_n}{T_0 - T_n} = \exp\left(-\frac{1}{\beta}\frac{x}{D}\right) \qquad (13)$$

in which

$$\frac{1}{\beta} = \frac{\rho c v D}{2\lambda^a}\left[\sqrt{1 + \frac{4\lambda^a\lambda^b}{(\rho c v)^2 HD}} - 1\right] \qquad (14)$$

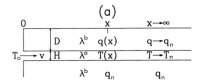

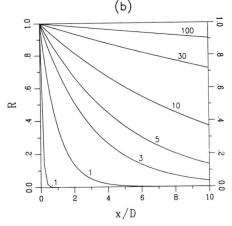

Fig. 3 Thermal effects of water flow in a horizontal thin aquifer. (a) The schematic model. (b) The relative temperature R of the aquifer as defined in Eq. (13). The numbers on the curves are values of the parameter β.

where ρc is the specific thermal capacity of water, and T_n is the undisturbed temperature of the aquifer (without water flow or as x tends to infinity). R, as a function of x/D, is plotted in Fig. 3b for various β values.

The magnitude of the temperature disturbance depends on the temperature of the recharging water. If $T_0 = T_n$, the water flow does not affect the geothermal regime. If $T_0 > T_n$ or $T_0 < T_n$, depending on where the recharging water originates, the thermal regime near the recharge area is disturbed, with the disturbance decaying exponentially with distance. The rate of the decay depends on the dimensionless number β, which is determined by many parameters of the flow system as seen in Eq. (14). In the case of pure conduction $(v = 0)$, β depends only on the thermal and geometrical parameters, $\beta = \lambda^a H/\lambda^b D$. In the case of fast water flow, $v > 10^{-8}$ m s^{-1} for an HD value of about 1000 m^2, β reduces to the Peclet number for this system $P = \rho c v H/\lambda^b$. The larger P is, the slower the decay of the disturbance with distance. $R(x)$ also represents the relative vertical HFD in the medium above the aquifer

$$R(x) = \frac{q(x) - q_n}{q_0 - q_n} \tag{15}$$

where q_0 and $q(x)$ are the vertical HFDs in the medium above the aquifer at distances 0 and x, respectively, and q_n is the undisturbed HFD, or the HFD below the aquifer. With this expression, the parameter β or P can be estimated, and v inferred, from HFD data.

2. A Dipping Aquifer

In a pervious formation or a fracture zone with a gentle dip angle, water flowing along the dip direction affects the local geothermal regime in a fashion similar to that of a horizontal aquifer discussed in the previous section. Larger water velocity and aquifer dip angle will give rise to more prominent thermal disturbances. Figure 4 shows two numerically simulated cross sections of temperature fields in the presence of a thin aquifer or fracture zone. In both cases, the thermal conductivity of the whole medium is homogeneous and the HFD from depths in the crust is a constant 60 mW m^{-2}, so that without water flow the isotherms would be equally spaced horizontal lines. In one case, an aquifer is dipping at an angle of 30°, with water inflow from the shallower end at a darcy velocity of 10^{-7} m s^{-1}. In the other case, groundwater flows upwards along a vertical path at the same velocity. With a much larger water velocity, the latter would resemble the local thermal regime near a fault-controlled hot spring.

When a dipping aquifer is very thin, the water velocity and the aquifer thickness can be lumped into a volumetric flow rate f(m^3 s^{-1} m^{-1}), that is, the volume of water flowing across a stripe of unit width in the dip direction per unit time. If the aquifer has a small dip angle and the section under

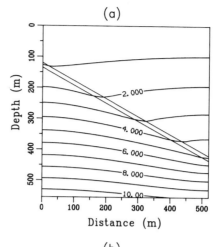

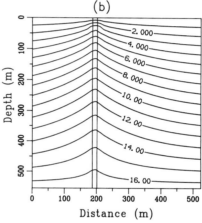

Fig. 4 Geotemperatures in the presence (a) of downward groundwater flow along an aquifer dipping at 30° and (b) of upward flow along a vertical fracture zone. In both cases the background HFD is 60 mW m^{-2}, the thermal conductivity of the whole medium is 2 W m^{-1} K^{-1}, and the darcy velocity of water flow is 10^{-7} m s^{-1}.

consideration is away from any thermal or hydrological boundaries, the difference in the HFD values below and above the aquifer can be approximated by

$$\Delta q = q_{\text{below}} - q_{\text{above}} = \rho c f \Gamma \sin \theta \tag{16}$$

where Γ is the normal geothermal gradient assumed to exist below the aquifer and θ is the dip angle. This formula is often used to estimate the volumetric flow rate f or mass flow rate ρf as Δq can be measured in the field. Since the conditions for this expression to hold are impossible to satisfy in the field, the flow rates thus obtained should be regarded as order-of-magnitude estimates.

C. Water Flow near an Active Fault

The frictional motion of a fault generates heat. The accumulation and dissipation of heat in and near the fault zone are complicated processes, and depend on a number of parameters including the thermal and mechanical properties of the rocks, the geometry of the fault, the local stress field, and the rate of fault movement. One of the most important influencing factors is the water initially contained in the rocks as pore fluid. As the fault moves, frictional heating will cause a temperature rise in the fault zone. The thermal expansion of the water in the rocks produces a higher pore fluid pressure. The increased pore fluid pressure has an effect of reducing the effective resistive stress of the fault plane, acting like a "lubricant." The amount of heat generated through friction is thus limited. However, if the water under higher pressure within and near the fault zone can flow away so that the pore pressure is unable to build up, the frictional heating may raise the temperature high enough to cause local melting of the rocks. Although the advective heat transfer by flowing water also has a cooling effect, the reduction of pore fluid pressure may play more significant roles in the whole process. Here the controlling parameter is the permeability of the rocks. A higher permeability allows water to drain faster and therefore reduces the pore fluid pressure more effectively. On a larger scale, the inability of an active fault to limit resistive stress because of water drainage may be partly responsible for some earthquakes.

D. Drilling-Induced Flow

1. Flow along a Borehole

When measuring temperatures in a borehole, heat flow scientists and petroleum workers often encounter fast water flow within the borehole, with an order of magnitude of centimeters per second. This type of flow occurs when the hole provides a pathway between two aquifers, one of which has a higher hydraulic head than the other, or when the hole opens an exit for a confined aquifer. Although such flow has a very limited range of influence on the geothermal regime, it hinders the correct determination of actual rock temperatures.

If water at temperature T_0 continuously enters a borehole of radius R at a depth $z = 0$, and flows along the hole at a constant velocity v, the temperature of the flowing water as a function of depth z and time t can be calculated using the equation

$$T = T_0 + \Gamma z - \Gamma R F(t) \exp\left\{1 - \left[-\frac{1}{F(t)}\frac{z}{R}\right]\right\} \quad (17)$$

in which $F(t)$ (dimensionless) is a measure of the rate of heat

transfer; for a time range of a few days to several years after the onset of the drilling-induced flow, it is given by

$$F(t) = \frac{P}{2}\left[\ln\left(\frac{2\sqrt{\alpha^r t}}{R^2}\right) - 0.29\right] \quad (18)$$

where Γ is the undisturbed geothermal gradient; P is the Peclet number of the flow system, defined as $P = \rho c v R / \lambda^r$; λ^r and α^r are the thermal conductivity and thermal diffusivity of the surrounding rock formation, respectively; and ρc is the specific thermal capacity of water. The sense of the curvatures in the borehole T–z profile in response to flow direction is the same as that caused by vertical water flow in a rock formation (Fig. 2); but the shape is quite different, since the temperature restores rather abruptly to that of the undisturbed rocks as water leaves the hole either at another aquifer or at the ground surface.

2. Intrusion of Drilling Fluid into a Permeable Formation

Fluids (drilling mud) circulate in a borehole at the time of drilling to cool the grinding bit and carry away the cuttings. Some of the drilling fluids may intrude into a permeable rock formation and disturb the local temperature field. The effect of such disturbance on the borehole T–z profile would be a conspicuous curvature in the section of intrusion. The sense of the curvature depends on the temperature of the intruding drilling fluid relative to the rock formation temperature. Such a local temperature anomaly dies away rather quickly if there is water flow along the formation, but it may last for a number of years if the water is stagnant in the formation so that the dissipation of the temperature anomaly is only by heat conduction.

IV. Thermal Effects of Regional Scale Flow Systems

A. Flow Patterns of Groundwater and Heat in Sedimentary Basins

1. Qualitative Interpretation of Field Data

A sedimentary basin is formed as a result of continued crustal subsidence accompanied by the deposition of sedimentary sequences. For an understanding of the tectonic processes that cause the crustal subsidence, it is important to know the crustal HFD pattern beneath the basin. The geothermics and hydrogeology of sedimentary basins are of particular economic interest to the petroleum industry.

The basement of a sedimentary basin usually consists of low permeability crystalline igneous or metamorphic rocks. The much more permeable sedimentary rocks in the basin such as sandstones and limestones form a hydraulically continuous environment. In such an environment, the topographic relief of the water table, which indicates the lateral variation of hydraulic head, is commonly a subdued replica of the land surface. Topography affects local atmospheric circulation, causing more precipitation in upland areas, and the groundwater system is recharged in these areas by infiltrating meteoric waters. In lowland areas, deficient precipitation is compensated by discharge of groundwater as well as convergence of surface waters. In a topographically asymmetric basin, a large-scale gravity-driven groundwater flow system may develop, on which is superimposed many smaller-scale localized flow systems.

The thermal effects of a basin-scale flow system can be qualitatively understood by regarding it as consisting of many interrelated local flow systems. In recharge and discharge areas, where the vertical component of water flow is prominent, the geotemperature pattern can be greatly distorted. According to Fig. 2, the downward flow in the recharge region will cause an apparent low HFD near the ground surface; the upward flow in a topographic depression will cause an apparent high HFD near the surface. At large distances from the discharge and recharge areas, the conductive geothermal regime may not be perturbed. The case of water flow along a single horizontal aquifer, Fig. 3, can be used as an illustration. The relative importance of convective heat transfer in each portion of the regional flow system is represented by a Peclet number that can be defined for the local flow system of that portion.

These ideas prove to be very useful in the interpretation of field hydrological and thermal data. An example is the study of the Western Canada sedimentary basin, also called the Prairies basin, which is a vast foreland basin situated between the Canadian Shield and the Cordillera. The Precambrian basement is mostly the western extension of the Canadian Shield. The sedimentary sequence comprises Paleozoic rocks unconformably overlain by Mesozoic rocks, with a maximum total thickness exceeding 6 kilometers in the western part of the basin near the foothills of the Rocky Mountains.

A large number of bottom-hole-temperature (BHT) data were obtained from thousands of petroleum wells drilled in the basin. BHTs are measured a few times in a well during drilling, with appropriate corrections made for the disturbances to the local geotemperature pattern caused by the drilling activities. The inaccuracy of individual measurements of this kind and the paucity of data from each well prohibits detailed analysis of HFD at any single location, but the multitude of these data covering a large area provides reliable information on the variation of the HFD pattern in the statistical sense. Based on these data, it is found that the conductive HFD in the sedimentary rocks of the Prairies basin varies both laterally and vertically. The lateral variations of the near-surface HFD values, being lower in the southwest

upland areas and higher in the northeast, are generally negatively correlated with the elevation of the topographic surface. Such an apparent (surface) HFD pattern could be due either to a general trend of increase in the basal HFD toward northeast, or to the disturbances caused by regional-scale groundwater movements. However, there is little evidence, from the measurements of basal rock heat productivities and seismic surveys, for the former.

If the apparent HFD variation is indeed caused by basal HFD variations, the HFD should be constant in a column of sedimentary rocks at one location. However, vertical HFD variations also appear to exist in the basin. It is interesting to compare the HFD values in the Mesozoic and Paleozoic sedimentary sequences. The difference between the HFD in the Mesozoic formations q_m and in the Paleozoic formations q_p is denoted by $\Delta q = q_m - q_p$. The average Δq over a small area can be obtained. The contours of these local Δq values are shown in Fig. 5, which can be regarded as an enhanced expression of the lateral variations of the apparent HFD since the Paleozoic HFD has a slight trend of decrease toward north. The Δq values are generally negative in the southwest, increasing northeastward. This pattern is best explained by regional-scale groundwater movements. The Rocky Mountains to the west of the basin provide ample meteoric water to recharge the groundwater system, and the downward flow of the cold water causes a lower apparent HFD at the land surface and a negative Δq. As water discharges in the lower elevation areas in the northeast, a positive Δq is observed.

Similar topography-correlated heat flow patterns have been observed in sedimentary basins in different parts of the world, such as the Rhine graben in Germany, the Bohemian Cretaceous basin in Czecholovakia, the Liaohe basin in North China, the Great Artesian basin in Australia, the Uinta basin of the Western United States, the Western North Sea basin and the Great Plains in the United States.

2. Identification of Regional-Scale Permeability and Basal HFD

Although in many cases a hydraulically modified geotemperature pattern in a basin appears evident, it is difficult to conclude the extent of the thermal effects of water flow. The principal reasons age (1) direct permeability measurements often do not suggest flow velocities large enough to cause substantial thermal disturbance in a basin, and (2) the real lateral distribution of basal HFD is not known. To solve these remaining problems, further theoretical studies are needed.

Rock permeabilities are very heterogeneous and may vary by orders of magnitude over a short distance. Values determined in laboratories on rock samples are seldom representative of the formation permeabilities. Values determined in the field through well tests include the averaged effects of the varying permeability values and of small fissures and fractures

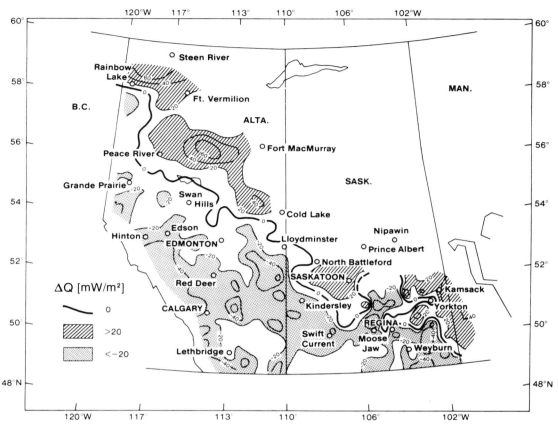

Fig. 5 Contour map of the HFD difference between the Mesozoic and Paleozoic sedimentary sequences, $\Delta q = q_m - q_p$, in the Western Canada sedimentary basin. [From Majorowicz *et al.* (1986).]

in a local area, and are preferred in the study of local flow systems. On the regional scale, however, effects of fractures on the orders of hundreds of meters make the effective permeabilities even larger. Such regional-scale values are impossible to measure experimentally—they have to be inferred from related geological and geophysical information. The basal HFD values can mainly be obtained through deep borehole measurements, but boreholes drilled into the basement rocks are not often available. Other parameters, such as the thermal conductivities of the rock formations, are not well known and therefore often need to be inferred as well.

When the parameter values are known and we solve Eq. (8) and (10) for temperature and hydraulic head, we have a forward problem. The numerical modeling techniques have been used to solve such forward problems. When we use available data to infer some poorly known parameters such as the permeabilities and the basal HFD, we have an inverse problem. For the purpose of obtaining unique and stable solutions with error estimates for the inverse problems, geo-

physical inverse theories can be applied. Inverse numerical modeling refers to methods that incorporate the geophysical inverse theories and the forward numerical modeling techniques. Inverse numerical modeling techniques are widely used in groundwater hydrology for the identification of aquifer parameters such as hydraulic conductivity and specific storage, using hydrological data. Methods that invert simultaneously the thermal and hydrological data to resolve regional-scale permeability and basal HFD have recently been developed.

B. Flow in Fractured Rocks

In the study of the thermal effects of groundwater flow in fractured areas, two cases have to be differentiated. In one case, often in basaltic terrains, fractures are densely and evenly distributed over a large volume of rock, and the lengths of the fractures are small compared to the scale of the flow system under consideration. Thus, the fractured rocks can be approximated as porous media. Darcy's law, Eq. (7), and the relevant conclusions and methods in the previous section can be applied in such a situation, provided the flow velocity is small. Therefore, this case is not considered here. In the other case, as for most fractured igneous and metamorphic bedrock terrains, fractures with much larger separations divide the low-permeability rock formations into large blocks, and the flow system is largely controlled by a relatively small number of fracture zones. Very often, these fracture zones are so permeable that water flows along them at a speed several orders of magnitude faster than what we most frequently encounter in porous rocks. Obviously, Darcy's law would be an invalid approximation for such flow systems. Heat transfer in the rock blocks is nearly entirely by conduction, but in the fracture zones it is principally by convection.

It is difficult to define the flow pattern in such areas, due to the scarcity of hydrological and thermal data, and also due to the complex configuration of flow paths. Flow paths depend on the separation between, the aperture of, and the hydraulic properties of the filling materials in the fractures. Usually, the best we can do is to examine the disturbance to the T–z profiles caused by local flow systems of individual fractures using available borehole data, and to estimate the HFD values at depths and the volumetric or mass rate and the influence depth of groundwater flow in fractures. The thermal effects of dipping aquifers discussed in Section III,B apply to this situation. Drilling-induced flows, especially those along the boreholes, are common phenomena and also deserve special attention.

An example is the severely hydraulically perturbed terrestrial heat-flow data in Finland, which geologically is in the central Baltic Shield, with a typical average HFD value of 35 mW m^{-2}. The locations of 17 boreholes selected for their relatively large depth are indicated in the geological map of the country in Fig. 6a. The measured T–z data from each hole are plotted in Fig. 6b; they were reduced by subtracting a graphically determined average geothermal gradient of the hole mainly for the purpose of saving plotting space. Some of the curvatures on these profiles are due to other effects such as the paleoclimatic changes at the ground surface and the spatial variations of thermal conductivities of the rock formations. On careful corrections for these effects, the disturbances caused by water flows were isolated, with the flow types identified on the T–z profiles. Disturbance caused by an along-fracture type flow, typically characterized by a gradient change (also see Section III,B), is indicated by an asterisk; disturbance caused by an along-borehole type flow, appearing as a temperature offset on the profile (also see Section III,D), is shown by vertically connected arrows. In most cases, water flows occur between 0 to 400 meters depth. However, in hole Kolari R197, water was observed to flow continuously out of the collar of the hole, driven by a higher pressure

at depths; and in the Lavia hole, water flows down the hole from a depth of 650 to 950 meters. The apparent mean HFD value of a single borehole ranges from 18 to 73 mW m^{-2}, which are believed to deviate from the value at depths by several tens of percent. The volumetric rates of water flow along fractures were estimated to be of the order of 10^{-8} to 10^{-6} m^3 s^{-1} m^{-1}, using the approximate method of Eq. (16) with the dip angles of fractures obtained from geologic logs or seismic studies. In unfilled fractures, this corresponds to a flow velocity of 10^{-6} to 10^{-4} m s^{-1} for an aperture of 1 cm. In fractures filled with sediments, the effective area that water flows across is inversely proportional to the porosity of the filling material, so that lower porosity implies higher flow velocity for a given flow rate. Significant along-fracture flows were observed in both igneous and metamorphic rock types. In the granitic and other felsic and intermediate intrusive rocks, which are commonly regarded as hydraulically tight, such flows also exist. Plutonic rocks are generally more "permeable" than metamorphic rocks due to their larger fracture lengths and apertures.

The chemical contents and temperature of groundwater are good indicators of the rates and depths of circulation in fractured areas, and usually contain information on the origin and path of flows. In the Precambrian metamorphic terrains in the northeastern part of Sao Paulo, Brazil, a deep (several kilometers) flow system is identified mainly through the spatial distribution and chemical characteristics of thermo-mineral springs. At shallow depths, fresh water often implies faster water exchange between the groundwater system and the atmosphere; more saline water is normally more stagnant.

C. Tectonically Induced Flows

1. Compaction Driven Flow in Orogenic Zones

When porous rocks undergo deformation caused by compressional forces, the fluids trapped in the pore spaces are forced to flow by a pressure gradient accompanied by a reduction of rock porosity. Such compressional forces can be the consequences of continuous sedimentation or tectonic movements. The former does not cause perceptible perturbation to the geothermal regime according to results of numerical modeling with typical sedimentation rates. Therefore only the latter is considered.

At continental collision zones, such as the Alpine–Himalaya, and Appalachian orogenic belts, continental margins loaded with oceanic sediments are buried underneath thrust sheets (Fig. 7). As pore waters, referred to as tectonic brines in Fig. 7, trapped in the sediments of the underplate are expelled by the heavy load of the thrust sheets, they will affect the thermal regime of the thrust belt and the foreland basin. Compaction-driven water flow of this type in orogenic zones has recently been modeled using numerical techniques.

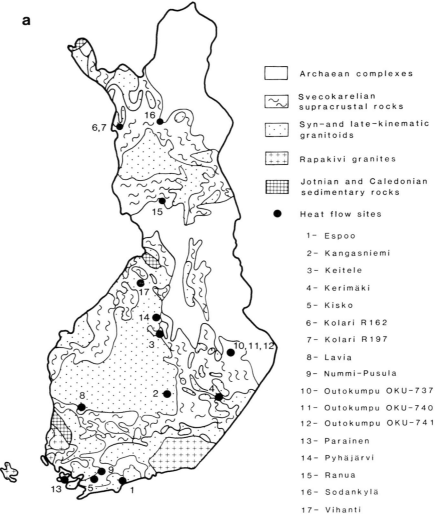

a

Archaean complexes

Svecokarelian supracrustal rocks

Syn- and late-kinematic granitoids

Rapakivi granites

Jotnian and Caledonian sedimentary rocks

● Heat flow sites

1- Espoo

2- Kangasniemi

3- Keitele

4- Kerimäki

5- Kisko

6- Kolari R162

7- Kolari R197

8- Lavia

9- Nummi-Pusula

10- Outokumpu OKU-737

11- Outokumpu OKU-740

12- Outokumpu OKU-741

13- Parainen

14- Pyhäjärvi

15- Ranua

16- Sodankylä

17- Vihanti

Fig. 6 Thermal effects of groundwater flow in fractured areas. (a) A simplified geological map of Finland, central Baltic Shield, and the locations of 17 boreholes. (b) T-z profiles from these boreholes reduced by subtracting a graphically determined gradient shown below each profile (mK m^{-1}); disturbance caused by along-fracture water flow is indicated by an asterisk, and that caused by along-borehole flow by vertically connected arrows. [From Kukkonen (1988).] (*Figure continues.*)

In addition to temperature and water flow as described by Eq. (8) and (10), other factors such as porosity variation in space and time and the speed of thrust have to be considered, with appropriate simplifying assumptions invoked. For a homogeneous sediment model with horizontal and vertical permeability values of 10^{-15} and 10^{-16} m^2, respectively, 1 million years of continuous thrusting at a typical rate of 5 cm/yr results in a 20% vertical compression of the 5 km thick sediments under the thrust sheets, but the thermal disturbance in the foreland basin caused by water expulsion from

the compressed sediments is found to be insignificant. However, in the presence of higher permeability flow paths such as aquifers, the thermal effects of the compaction-driven flow will be enhanced. In an extreme but quite realistic case, in which a highly permeable vertical fault zone 600 m wide exists in the foreland basin as a conduit between a deeper aquifer and the land surface, temperatures near the fault zone may rise as high as 40°C above normal. Since shear failure of rocks frequently occurs during the convergence processes at destructive plate boundaries, such localized thermal anoma-

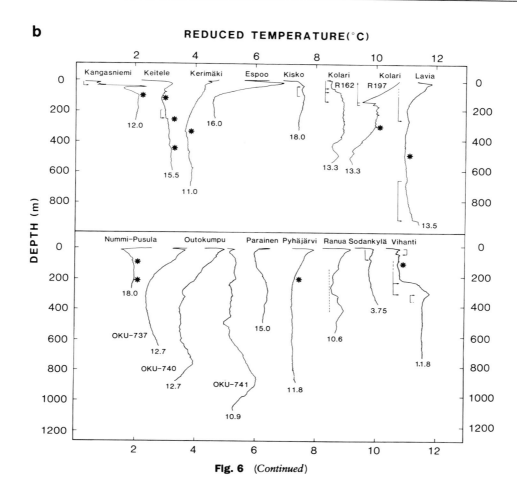

Fig. 6 (*Continued*)

lies may well account for some of the observed ore deposits, hydrocarbon deposits, hot springs, and remagnetization of ancient rocks in such areas. Fluids characteristic of Paleozoic origins observed in tiny inclusions of precipitated minerals in the North American Appalachian basin appear to be, at least partially, direct supporting evidence for such processes. [*See* PLATE TECTONICS, CRUSTAL DEFORMATION AND EARTHQUAKES.]

2. Metamorphic Water

Of paramount significance from a geochemical and geomechanical point of view but currently less investigated in the studies of terrestrial heat flow and groundwater flow is fluid flow from greater depths that is related to regional metamorphism in tectonically active regions. As subduction and collision take place, dewatering of the underplate is not only caused by compactional forces expelling pore waters, as discussed in the previous section, but also by the changes in rock minerals at more extreme temperature–pressure (T–P) conditions. In subduction zones, prograde metamorphism takes

place as the underplate is heated and pressurized. Such prograde metamorphism is characterized by loss of large amounts of volatiles, mostly water, initially bonded in the structure of the rock minerals. On the regional scale, massive fluid flow along appropriate pathways must occur and have a profound influence on the lithospheric thermal regime.

For example, in a recent study of the heat flow data in relation to the subduction of the Juan de Fuca plate west of Vancouver Island, Canada, it was proposed that, as a result of the metamorphic dehydration of the subducting plate, water might be transporting heat up from lower crustal depths to within 10 km of the surface, producing an apparent shallow heat source near the volcanic axis. Impressive thermal consequences of regional metamorphism related water flow in the past are often recorded in the geology of such regions. In the big Maria Mountain metamorphic terrain, southeastern California, for example, the high temperature of the metamorphism that took place in Late Cretaceous cannot be explained by mechanisms such as abnormal radioisotopic heat generation or magma intrusion into the crust, but only by the

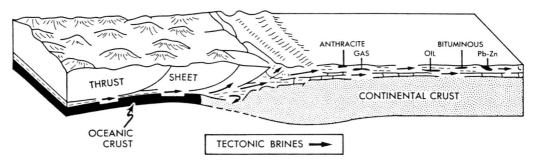

Fig. 7 Schematic model of thrust zone, showing the possibility of compaction-driven groundwater flow. [From Deming *et al.* (1990).]

presence of hot fluid produced in the devolatizing metamorphic process. Enormous water flux at the time of metamorphism is evidenced by the formation of massive wollastonite in that area.

V. Practical Aspects

A. Groundwater as Thermal Energy Resources

Terrestrial heat as an energy resource is of little economic value unless it is concentrated in a limited volume of the earth by some natural agents. Three conditions must exist to form a geothermal energy reservoir: (1) an abnormally high local HFD value, usually related to magma intrusion into the crust; (2) a permeable zone that contains hot fluid such as water; and (3) a much less permeable covering formation to prevent rapid loss of thermal fluid.

Thermal water or steam at temperatures higher than 150°C may be useful for electricity generation; these are often referred to as high-enthalpy fluids. Those below 150°C are low-enthalpy fluids and may be used for domestic and industrial heating. Thermal springs at a temperature comfortable to human bodies are often tourist attractions. In tectonically active areas, such as Italy, New Zealand, Japan, and western North and South America, hydrothermal systems are important energy resources. The long duration of supply, relatively low cost, and low pollution potential make geothermal energy a feasible partial solution to the energy shortage problem that is gradually but increasingly faced by human society.

B. Groundwater as Waste Thermal Energy Storage

Water is often used as an economical coolant fluid in industry. As the heated water infiltrates the ground, groundwater becomes a part of the cooling system. Such a dumping

of heat may be potentially hazardous to the local ecology and causes a problem known as thermal pollution. Heat transport processes in the groundwater system are therefore sometimes numerically modeled in advance as a part of the policy-making procedure to predict the potential effects to the environment. Another way of making groundwater a waste thermal energy storage is the burying of industrial nuclear wastes, the radioactive leftovers of nuclear reactions. Although nuclear disposal sites are chosen in impermeable crystalline rocks, fractures in these rocks may still form a groundwater flow system. In such cases, and in most other cases of industrial waste disposal, the transport of harmful solutes dissolved in the groundwater is often a much more serious problem than thermal pollution.

C. Thermal Water and Mineral Resources

The chemical contents of thermal waters, especially those directly related to magmatic activities, are of great economic value. Thermal waters may carry various minerals dissolved in them in complex ionic forms and precipitate these materials in certain places to form ore deposits. Local thermal anomalies caused by groundwater flow may create a suitable environment for the formation of certain ore bodies. For example, the formation of the Mississippi Valley-type lead–zinc deposits are believed to be associated with brines at a temperature of 100 to 150°C. Numerical modeling of the thermal and hydrological history of anomalous thermal fields will help in the exploration of these mineral resources.

D. Thermal Water and Hydrocarbon Deposits

Petroleum, that is, oil and natural gas, are hydrocarbon deposits generated from dissolved organic matter buried in sedimentary basins. Similar to ore deposition processes, groundwater contributes to the formation of hydrocarbon deposits in two ways — transporting the dissolved materials and creating an appropriate temperature environment. Cross-formational groundwater flow in sedimentary basins is the principal mechanism of migration of hydrocarbons, and petroleum deposits are frequently found in locations where water flow is forced to slow down or change course. Temperature condition is essential to petroleum generation, and the range 70–160°C is usually believed to be optimal for the hydrocarbon maturation process. Numerical modeling of the thermal and hydrological history of sedimentary basins is widely performed in petroleum exploration. To constrain the subsurface thermal conditions in the geological history, various paleotemperature indicators, such as vitrinite reflectance, $^{40}K/^{39}K$, fission scar tracks in apatite, pollen translucency, etc., are often used.

Bibliography

Beck, A. E., Stegena, L., and Garven, G., eds. (1989). "Hydrogeological Regimes and Their Subsurface Thermal Effects," AGU Monograph 47. American Geophysical Union, Washington, D.C.

Deming, D., Nunn, J. A., and Evans D. G. (1990). Thermal effects of compaction driven groundwater flow from overthrust belts. *J. Geophys. Res.* **95**(B5), 6669.

Haenal, R., Rybach, L., and Stegena, L., eds. (1988). "Handbook of Terrestrial Heat Flow Density Determination." Kluwer Academic Publishers, Dordrecht, The Netherlands.

Kukkonen, I. T. (1988). Terrestrial heat flow and groundwater circulation in the bedrock in the central Baltic Shield. *Tectonophysics* **156,** 59.

Majorowicz, J. A., Jones, F. W., and Jessop, A. M. (1986). Geothermics of the Williston Basin in Canada in relation to hydrodynamics and hydrocarbon occurrences. *Geophysics* **51,** 767.

Mase, C. W., and Smith, L. (1987). Effects of frictional heating on the thermal, hydrologic, and mechanical response of a fault. *J. Geophys. Res.* **92**(B7), 6249.

Smith, L., and Chapman, D. (1983). On the thermal effects of groundwater flow: 1. Regional scale systems. *J. Geophys. Res.* **88**(B1), 593.

Wang, K., and Beck, A. E. (1989). An inverse approach to heat flow study in hydrologically active areas. *Geophys. J. Int.* **95,** 69.

Glossary

Conductive heat transfer Flow of heat in a medium due purely to a temperature difference.

Convective heat transfer Heat transfer by flow of fluids, also called advective heat transfer in the case of forced convection.

Geothermal gradient Rate of temperature increase with depth in the earth, or the vertical component of the geotemperature gradient.

Heat flow density (HFD) Flow of heat across a unit area per unit time, regardless of the means of heat transfer.

Groundwater Transport of Contaminants

Jean M. Bahr
University of Wisconsin

Industrialized nations throughout the world are increasing their efforts to identify and control sources of groundwater contamination. With these efforts have come new research initiatives to gain a better understanding of the processes that control groundwater flow and contaminant transport in the subsurface. Results of this research are being used to develop models to predict the effects of contaminant releases to groundwater and to design effective measures for aquifer restoration. This article will review the sources and mechanisms of groundwater contamination, the physical and chemical processes that control contaminant migration and transformation, and some of the measures currently in use or in development to mitigate problems of groundwater contamination.

I. Sources and Mechanisms of Groundwater Contamination

A. Background

Groundwater contamination resulting from human activities is not a new phenomenon. Links between contamination of wells and outbreaks of diseases such as cholera were recognized in the 1800s. However, the variety and volumes of contaminants introduced to groundwater have increased dramatically with the growth of the synthetic chemical industry since the 1940s. Many of the same fossil fuel products and chlorinated organic compounds that have been implicated in global climatic change have also been detected in groundwater. In the United States, public awareness of the problems

of groundwater contamination was catalyzed by the discovery in the mid-1970s of an abandoned disposal site at Love Canal in Niagara Falls, New York. Since that discovery, thousands of such sites have been identified in North America, many of them with more extensive groundwater contamination than that found at Love Canal.

Groundwater is an important source of water for human and livestock consumption, irrigation, and industrial processes throughout the world. Contamination of groundwater can directly limit any of these water uses. At the same time, it can also limit use of surface water supplies that interact with the groundwater system. Base flow in streams and rivers is supplied by discharge from aquifers. Groundwater also discharges to many lakes and wetlands, where evaporation and transpiration lead to concentration and accumulation of dissolved species. Thus, contaminants transported to the surface by groundwater can pose a significant threat to fragile ecosystems. In extreme cases of evaporation and volatilization of contaminants at the water table, groundwater contaminants may also become a source for atmospheric pollutants such as nitrous oxide. [*See* GROUNDWATER.]

B. Classification of Contaminants

In the broadest sense, any dissolved or immiscible substance in the groundwater system that is found in concentrations exceeding those in the natural system can be considered a contaminant. Identifying the presence of contaminants and quantifying the extent of contamination, therefore, requires careful characterization of the "natural" background water quality. Such characterization may not be easy in areas that have a long history of contaminant releases to groundwater. In a more practical sense, the terms contamination or pollution may be applied to situations where concentrations reach levels that are objectionable from the standpoint of human health or other criteria. Using this definition, not all releases by human activities are necessarily sources of contamination. On the other hand, there are many groundwater systems, such as deep sedimentary basins or coastal aquifers, which contain saline water that is unsuitable for domestic consumption. The problems associated with this "natural" pollution can be exacerbated by upward or inland movement of the salt water in response to nearby groundwater development.

Groundwater contaminants can be classified in a number

Table I Classes of Groundwater Contaminants

Class	Examples
Inorganics	
Metals and cations	Lead, chromium, sodium
Nonmetals and anions	Nitrate, sulfate, chloride
Radionuclides	Tritium, radium, radon
Microorganisms	Coliform bacteria, viruses
Organics	
Petroleum hydrocarbons	Gasoline, diesel, jet fuels
Halogenated solvents	Trichloroethylene, chloroform
Wood preservatives	Creosote, pentachlorophenol
Pesticide compounds	Aldicarb, dibromochloropropane
Other industrial organics	PCBs, chlorobenzenes

of ways including their chemical nature, their sources, and the mechanisms by which they enter and migrate in the groundwater system. In terms of chemical characteristics, the principle contaminants of concern can be divided into the classes listed in Table I. In general, inorganic contaminants enter the groundwater system dissolved in a water-based, or aqueous, phase. Although many organic compounds are introduced to groundwater as aqueous-phase solutes, many of these compounds can also enter the subsurface and persist as a separate nonaqueous-phase liquid (NAPL). In the case of petroleum hydrocarbons, the NAPL is lighter than water. Chlorinated organics, either alone or in mixtures, form NAPLs that are denser than groundwater.

Migration paths of NAPL contaminants within the groundwater system can differ significantly from those of dissolved contaminants. Dissolution of organic compounds where water and NAPL are in contact can provide a continuous source for additional dissolved groundwater contaminants. The subsurface NAPL sources may be located at considerable distance from the site of a surface spill or leak, which can complicate the process of identifying the original source of contamination. Because drinking water standards for many NAPL compounds are in the range of parts per billion, very small volumes of NAPL can ultimately contaminate very large volumes of groundwater. Microorganisms and some larger organic molecules are too large to truly dissolve in groundwater but will remain in groundwater as suspended colloids. Transport of these colloids in aquifers may be inhibited by filtration in materials with small pores or accelerated by preferential flow through larger pores.

C. Contaminant Sources

A variety of sources for groundwater contamination are illustrated in Fig. 1. Many cases of groundwater contamination can be attributed to planned releases of wastes through septic

tanks, injection wells, and land application of waste water or sewage sludge. Certain other planned activities, such as irrigation, pesticide application, animal feeding operations, and application of de-icing salts, also necessarily result in releases to groundwater. Other important sources are those designed to store or transport wastes and hazardous materials but which sometimes result in leaks or spills. These include landfills, surface impoundments, fuel pipelines, and underground storage tanks. Finally, contaminants in the atmosphere or in surface water may enter the groundwater system through the natural processes of groundwater recharge by precipitation and infiltration.

Mechanisms for transport from these sources to groundwater include infiltration through the unsaturated zone to the water table, as is the case for landfill and septic tank leachate, and direct introduction via injection wells or leaks in tanks or pipelines located below the water table. Poorly constructed or abandoned wells can provide conduits for rapid exchange between aquifers at different depths, leading to contamination in zones that would normally be isolated from the source of contamination. Changes in directions or rates of groundwater flow caused by pumping can in some cases induce or accelerate infiltration from surface or subsurface sources.

II. Processes Controlling Groundwater Contaminant Transport

A. Physical Processes

1. Transport of Dissolved Contaminants

Observations of plumes of dissolved contaminants emanating from continuous or finite sources indicate that contaminants migrate in the general direction of groundwater flow while at the same time they tend to spread to occupy a larger volume of the aquifer. These two features of contaminant transport are often attributed to two separate physical processes: advection, or transport with the average groundwater flow; and dispersion, or spreading of contaminants along and transverse to the principal directions of groundwater flow. In fact, both advection and dispersion can be considered as macroscopic manifestations of complex transport paths and nonuniform velocity at smaller scales that are due to heterogeneity of the geologic media through which groundwater is flowing. Recognizing that advection and dispersion are intimately related, it is still useful to discuss them separately along with a brief description of methods that are used to quantify the relevant parameters. [*See* GROUNDWATER FLOW AND GEOTEMPERATURE PATTERN.]

a. Advection

Conservative solutes are those that do not react with the aquifer solids or degrade or decay during transport. These

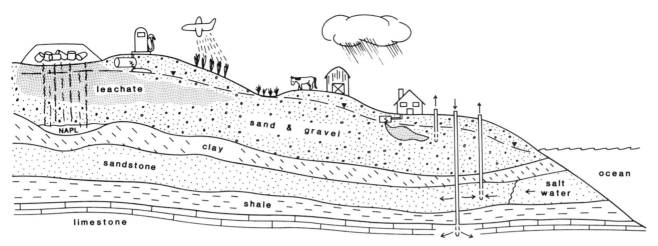

Fig. 1 Sources of groundwater contamination. The dashed line labeled with a solid triangle represents the water table.

solutes move in the direction of groundwater flow at an average velocity corresponding to the average linear, or advective, velocity of the groundwater. Groundwater flow is governed by Darcy's law, which can be written as

$$q = -K \, dh/dl \qquad (1)$$

where q is the volumetric flow rate per unit cross-sectional area of aquifer perpendicular to the direction of flow (units of length/time); dh/dl (dimensionless) is the hydraulic gradient that can be measured as the change in head (water level elevation) h between two points separated by distance l; and K is the hydraulic conductivity (units of length/time), a measure of the geologic medium's capacity to transmit water. The flow rate q is referred to by a variety of terms including the specific discharge, the Darcy flux, and the Darcy velocity.

It is important to recognize that the specific discharge does not correspond to the average velocity expected for contaminants moving with the water because the portion of the cross-sectional area of a geologic medium that is filled with solids is not available to groundwater flow. A better estimate of the average groundwater velocity can be obtained by dividing the specific discharge by the porosity n, the volumetric fraction of the medium that is occupied by pores. The advective, or average linear, velocity is therefore defined as

$$v = \frac{q}{n} = \frac{-K}{n} \, dh/dl \qquad (2)$$

For porous media that do not contain large numbers of unconnected and deadend pores, the porosity provides a good approximation of the percentage of open area perpendicular to flow. In fact, the porosity used in Eq. (2) should actually be the "effective" porosity, which can be considerably smaller than the total porosity for very fine-grained materials. Estimating the effective porosity can be particularly difficult for media in which fractures make up the dominant pathways for fluid flow.

Knowledge of the advective velocity in an aquifer affected by contamination allows prediction of average travel times for conservative contaminants to move from a source to critical discharge points such as domestic wells, streams, lakes, or wetlands. A number of techniques have been developed to measure or estimate the advective velocity in groundwater systems. The simplest and most commonly employed method involves computing the advective velocity by Eq. (2), using estimates of the hydraulic gradient, hydraulic conductivity, and effective porosity. The magnitude and direction of the hydraulic gradient is determined from measurement of water levels in wells, using these measurements to construct maps of the water table and/or the potentiometric surface at various depths. Hydraulic conductivity can be measured by field pumping tests, by single-well response tests known as slug tests, or by laboratory permeameter measurements. For certain types of unconsolidated sediments, reasonable correlations have been obtained between hydraulic conductivity and grain size distributions, allowing indirect estimation of hydraulic conductivity for these materials. Porosity can be measured in the laboratory by a variety of techniques, assuming that it is possible to obtain an undisturbed sample of the soil or rock.

Direct methods of determining advective velocity have also been used at a number of sites. These methods make use of naturally occurring or artificially introduced "tracer" solutes. Movement of these solutes with the groundwater is monitored over some distance or time and the observed travel times and distances are used to compute velocities. A technique called borehole dilution employs only a single well. An easily detected solute is introduced at a known concentration. As water flows through the well it dilutes the tracer, the rate of dilution depending on the rate of groundwater flow. Concentrations in the well water are determined continuously or periodically, preferably using a measuring device that can remain submerged in the borehole. Results of a borehole dilution test can be used to determine the magnitude of the advective velocity, but they do not provide information on the direction of flow.

Tracer tests involving multiple sampling points can be used to obtain more detailed information on both direction and velocity of groundwater flow. Tests using introduced conservative tracers such as chloride and conducted under natural groundwater flow conditions generally require a very large number of sampling points and months or years of monitoring. Environmental isotopes of oxygen, hydrogen, carbon, and other elements, for which concentrations have varied over time due to climatic conditions or radioactive decay, can be used as natural tracers to provide information on groundwater age and, hence, travel time from the recharge area. As an example, groundwater sampling for tritium, which was introduced to the atmosphere in high concentrations by atmospheric nuclear weapons tests during the 1950s and 1960s, has been used successfully in a number of settings to estimate ages of groundwater that entered the subsurface after 1953. Use of environmental tracers requires detailed knowledge of the input history of the tracer as well as location of recharge areas and general directions, both vertical and horizontal, of groundwater flow. Contaminants themselves can also serve as tracers to determine the advective velocity, provided the location and timing of their introduction to the system are known and the contaminants can be considered conservative solutes.

b. Dispersion

The advective velocity, which describes the average transport behavior of a conservative contaminant, provides sufficient information for predicting mean arrival times of contaminants. However, if more accurate information on expected concentrations or distributions of contaminants within an aquifer is required, it is also necessary to consider the deviations from average velocity that give rise to spreading, or dispersion, of a plume. Figure 2 illustrates processes at the microscopic scale that are responsible for dispersion. The first of these processes (Fig. 2a) is molecular diffusion, in which contaminants move from areas of high concentration to lower concentrations following Fick's second law

$$\frac{\partial c}{\partial t} = D^* \frac{\partial^2 c}{\partial x^2} \tag{3}$$

where c is the solute concentration, D^* is the coefficient of molecular diffusion in the porous medium, t is time, and x is

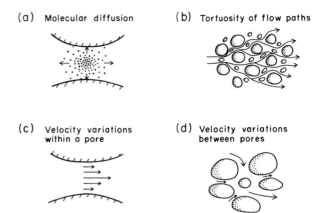

Fig. 2 Microscopic processes causing dispersion during contaminant transport.

distance. D^* will be less than the diffusion coefficient in free water because diffusion of solutes in a porous medium will be inhibited by the presence of the solids and must, therefore, take a tortuous path through the pore space. Molecular diffusion will cause dispersion of solute in all directions, thus spreading a plume both longitudinally (along the direction of groundwater flow) and transverse to the flow direction. Spreading due to molecular diffusion will take place even in the absence of groundwater flow.

The tortuous geometry of flow paths in a porous medium give rise to additional spreading of contaminants, as illustrated in Fig. 2b. Like molecular diffusion, tortuosity causes both longitudinal and transverse dispersion. However, in contrast to molecular diffusion, which depends only on concentration gradients, the rate of dispersion resulting from flow path tortuosity increases with the advective velocity of groundwater. The variability in velocity within pores (Fig. 2c) and between pores of different sizes (Fig. 2d) are additional causes of spreading that depend on the advective velocity. Effects of these velocity variations will be manifested primarily in longitudinal dispersion.

The effects of tortuosity and variations in microscopic velocity are usually described together as "mechanical" dispersion. Mechanical dispersion has been approximated by a model which follows the form of Fick's law. For a two-dimensional problem in which flow is in the x direction this model can be written as

$$\frac{\partial c}{\partial t} = \frac{\partial}{\partial x} D_{\mathrm{L}} \frac{\partial c}{\partial x} + \frac{\partial}{\partial y} D_{\mathrm{T}} \frac{\partial c}{\partial y} \tag{4}$$

where D_{L} and D_{T} are, respectively, the coefficients of mechanical dispersion in the longitudinal and transverse directions x and y. The coefficients of mechanical dispersion are further assumed to depend on the advective velocity as

$$D_L = \alpha_L v \qquad (5)$$

and

$$D_T = \alpha_T v \qquad (6)$$

where the α's are called the longitudinal and transverse dispersivities, assumed to be properties of the porous medium.

Models combining advective transport, molecular diffusion according to Eq. (3), and mechanical dispersion following the relationships expressed in Eq. (4) through (6) have been found to be in good agreement with results of laboratory experiments using columns on the order of tens of centimeters to a meter in length. However, rates of plume spreading observed in the field for travel distances of tens to hundreds of meters are much greater than rates that would be predicted using values of dispersivity determined from laboratory experiments. A number of recent experiments indicate that this greater dispersion at the field scale is the result of spatial variability of hydraulic conductivity due to aquifer heterogeneity. Such variability of hydraulic conductivity has been observed even in relatively homogeneous media such as glacial outwash sands.

A variety of approaches have been suggested to improve models of solute transport and dispersion at the field scale. The deterministic approach, which has been successfully applied to aquifers with well-defined stratification, involves explicitly incorporating known variations in hydraulic conductivity (and consequently variations in advective velocity) into the transport model. The drawback of this approach is that it requires detailed knowledge of the spatial distribution of aquifer heterogeneity.

An alternative approach involves incorporating statistical descriptions of the hydraulic conductivity distribution into "stochastic" transport models to predict average displacement and spreading of a plume. These stochastic models require less specific information on aquifer heterogeneity than deterministic models, but they require that the statistical distribution of hydraulic conductivity does not change significantly in the direction of groundwater flow—in other words, that there are no gradual or abrupt facies changes. Stochastic theories of dispersion also suggest after some, possibly very long, travel distance, the dispersion process can again be approximated by a Fick's law model. These models of dispersion at long distances employ a "macrodispersivity" that is considerably larger than the dispersivity used to model mechanical dispersion resulting from pore-scale velocity variations.

The concepts of fractal geometry are providing a new tool to better understand dispersion at the field scale, particularly

in systems that cannot be considered "uniformly heterogeneous" as required by the stochastic models. Models of transport in porous media assuming a self-similar or fractal geometry have yielded results that appear to approximate observed plume spreading. This approach may also lend itself to modeling transport in fractured rocks, where flow paths and mixing processes are even more complex than in sediments or rocks with random, interconnected pores.

2. Transport of Nonaqueous-Phase Liquids

Migration of organic phase contaminants in the subsurface is also governed by Darcy's law, which for these fluids can be written as

$$q_0 = -K_0 \, dh_0/dl \qquad (7)$$

In Eq. (7) q_0 is the volumetric flow rate of the organic fluid, K_0 is the specific fluid conductivity, and dh_0/dl is the gradient in organic fluid head.

Two main fluid properties, density and viscosity, affect fluid conductivity and fluid head, causing these to differ from the corresponding groundwater values. The fluid conductivity K_0 depends on these properties according to the relationship

$$K_0 = \frac{k \rho_0 g}{\mu_0} \qquad (8)$$

where k is the intrinsic permeability of the aquifer, ρ_0 the fluid density, g the gravitational constant, and μ_0 the fluid viscosity. Fluid head is defined as

$$h_0 = z + \frac{p}{\rho_0 g} \qquad (9)$$

where z is the elevation above a datum and p the fluid pressure. The resulting patterns of NAPL migration can vary significantly from those of dissolved contaminants. NAPLs with densities less than that of water will have smaller gradients in the vertical direction, resulting in accumulation and spreading at the water table following infiltration through the unsaturated zone (see Fig. 3). Dense NAPLs, in contrast, will have greater vertical gradients and will accumulate along basal surfaces where low-permeability materials underlie the aquifer. Fluid viscosity affects fluid conductivity, with less viscous fluids having higher K_0 and thus greater mobility under a given gradient.

When both groundwater and NAPLs coexist in the aquifer pores, the conductivity to each fluid is reduced by the presence of the other. The fluid conductivities in Eqs. (1) and (7) must then be modified using a coefficient of relative permeability. A number of recent experimental and theoretical

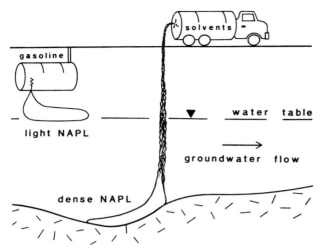

Fig. 3 Subsurface distribution of nonaqueous-phase liquid (NAPL) contaminants.

studies have been directed toward improving estimates of relative permeabilities for various combinations of fluids and aquifer materials. These studies have shown that at some level of fluid saturation the relative permeability drops to zero and the fluid becomes trapped at a "residual saturation." Although these immobile NAPLs will not migrate as a separate phase, they can dissolve in groundwater as it flows by, thus serving as a continuing source for dissolved contaminants.

3. Transport of Colloids

Particles with diameters of 10 μm or less can remain suspended in water even at the low velocities common in groundwater systems. Some of these colloids, such as bacteria, viruses, and some organic macromolecules, may be considered contaminants by themselves. Others, such as clays, mineral precipitates, and natural humic compounds, may facilitate contaminant transport by scavenging heavy metals or organics that might otherwise be retarded by interactions with aquifer solids.

Colloids can move by advection as long as the aquifer medium contains pores with diameters sufficiently larger than those of the colloids. Because colloids are likely to move preferentially through the larger pores, the average velocity of mobile colloids may be significantly greater than the average velocity of a dissolved contaminant in the same aquifer. In the presence of smaller pores, some colloids will become trapped by the process of straining filtration. Filtration is enhanced by electrostatic interactions or other processes that lead to coagulation of the particles. Sorption onto aquifer solids will also serve to immobilize or retard groundwater colloids.

Quantitative prediction of colloid mobilization and filtration requires improved understanding of the rates and mechanisms of physiocochemical interactions among colloids and between colloids and aquifer solids. Laboratory and field studies of these interactions are just beginning. Particularly challenging to researchers is the problem of identifying and predicting preferential flow paths resulting from fractures and the arrangement and distribution of large pores.

B. Chemical Processes

Chemical reactions involving groundwater contaminants can occur entirely within the liquid phase (homogeneous reactions) or can result in transfer of mass between groundwater and the aquifer solids (heterogeneous reactions). A further useful distinction is between reactions that proceed in one direction only, referred to as irreversible reactions, and reversible reactions, which reach an equilibrium limit when rates of the forward and reverse steps are balanced. Homogeneous reactions will affect the form of a contaminant, but by themselves have no effect on the net mobility of the contaminant or its transformation products. In contrast, heterogeneous reactions can cause retardation—that is, a decrease in the rate of contaminant migration—if they cause contaminants to be temporarily held in the immobile solid phase. The extent of this retardation depends in part on whether the rates of reversible reactions are sufficiently fast, relative to rates of groundwater flow, to maintain local equilibrium in the groundwater system. Irreversible homogenous and heterogeneous reactions ultimately result in a decrease in the contaminant mass in groundwater, an effect referred to as attenuation. Figure 4 provides a comparison of concentration profiles and breakthrough curves for conservative, retarded, and attenuated contaminants.

The following sections briefly describe the important classes of chemical processes and their effects on contaminant migration and distribution in groundwater. It should be recognized that groundwater contaminants are affected by multiple reactions that lead to simultaneous retardation and attenuation. Although recent modeling advances have permitted researchers to explore some systems of coupled reactions, prediction of the potentially synergistic effects of large numbers of simultaneous reactions remains a theoretical and computational challenge.

1. Decay and Degradation

One of the simplest reactions to incorporate into models of groundwater contaminant transport is radioactive decay, which follows the rate law

$$\frac{dc}{dt} = -\lambda c \qquad (10)$$

where λ is a first-order decay constant. Products of radioac-

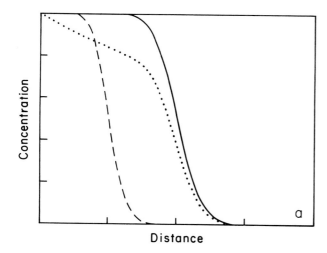

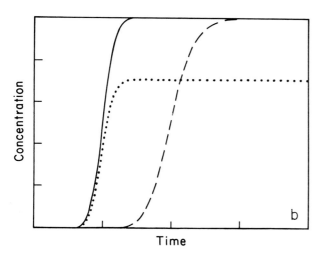

Fig. 4 Concentration profiles and breakthrough curves for conservative (——), retarded (– – –), and attenuated (. . . .) contaminants. (a) Concentration profiles represent contaminant distributions measured at a single time, with a continuous source located to left. (b) Breakthrough curves illustrate changes in concentration with time a fixed distance from the continuous source.

tive decay may themselves be radioactive, such as in the case of radium decay to radon. Thus, while decay causes attenuation of the original contaminant, it may generate a new contaminant that can continue to migrate in the groundwater system. This potential problem of merely substituting one contaminant for another is a common feature of many decay and degradation reactions.

Integration of Eq. (10) yields an expression for the half-life of a decaying contaminant — that is, the time required for its mass in the system to decrease to one-half is initial value. The importance of accounting for this type of decay depends on the length of the half-life relative to the transport time of interest. Thus, ^{14}C with a half-life of 5730 years could be considered conservative for residence times in groundwater on the order of several hundred years. Tritium (^3H), with a half-life of 12.3 years, could be treated as a conservative solute in laboratory column experiments where travel times are on the order of a few days. However, simulations of field-scale tritium transport for periods of several years or more would overestimate concentrations if they did not explicitly account for radioactive decay.

A number of other irreversible reactions are commonly modeled using a first-order decay law. These include nitrification of ammonia and hydrolysis of a variety of organic contaminants. Strictly speaking, the rate laws for these transformations are not first-order; but in many cases a "pseudo-first-order" approximation works well under conditions of constant pH, temperature, and other chemical characteristics of the solution. Simulation of these reactions during groundwater transport requires knowledge of the appropriate rate constants for ambient groundwater conditions.

Microorganisms in the groundwater system also play an important role in attenuation of contaminants. In some cases, the microbes break down contaminants in order to use them as energy sources. In other cases, changes in groundwater chemistry due to microbial activity may enhance rates of abiotic transformations. Microbial degradation has also been successfully modeled as a pseudo-first-order reaction. In situations where contaminant concentrations are high and microbial degradation is limited by the availability of an electron acceptor such as oxygen, or where the microbial population changes over time, more complex models are required. In some cases, there may be a threshold concentration of contaminants or other nutrients below which microbial populations cannot be maintained. In such cases, biodegradation will cease and, in contrast to the case of radioactive decay or hydrolysis, attenuation of contaminants will reach a finite limit.

2. Sorption

Sorption reactions involve transfer of contaminants between groundwater and the surface of aquifer solids. If a sorption reaction is reversible and sufficiently fast to assume local equilibrium in a flowing groundwater system, and if the sorbing solute does not compete with other solutes for surface sites, concentrations in the solid phase $\bar{c}$ and in groundwater c can be related by an expression known as a sorption isotherm, having the general form

$$\bar{c} = f(c) \qquad (11)$$

Equilibrium-controlled sorption delays contaminant transport because mass is continually being transferred between the mobile fluid phase and an immobile solid phase. The magnitude of this delay can be quantified by the "retardation factor" R_f, which equals the ratio of the average linear velocity of a conservative solute to the average linear velocity of a sorbing solute. R_f depends on the sorption isotherm through the relation

$$R_f = 1 + \frac{\rho}{n}\frac{d(\bar{c})}{dc} = 1 + \frac{\rho}{n}f'(c) \qquad (12)$$

where ρ is the aquifer bulk density and n the porosity. It can be seen from Eq. (12) that, depending on the form of the isotherm. R_f may be a function rather than a constant. Thus, the velocity of sorbing solute can vary as a function of the solution phase concentration. This implies that retardation of a given contaminant can change during transport as a result of dispersion or attenuation processes that cause changes in concentration.

Sorption reactions referred to as exchange reactions involve competition between solutes for surface sites. In these cases, the equilibrium concentrations of dissolved and sorbed species depend not only on the contaminant of interest but also on concentrations of any competing species, either dissolved or in the solid phase. Retardation caused by competitive sorption then varies with the concentrations of all exchanging species. Sorption or exchange reactions that are not sufficiently fast to approach local equilibrium during transport can also cause contaminant retardation. Nonequilibrium retardation will vary with both time (or travel distance) and concentration, increasing to the equilibrium limit at some, possibly very long, distance from the contaminant source.

Many sorption or exchange reactions involve charged or polar solutes interacting with surfaces of opposite charge. Most organic contaminants are neither charged nor polar and thus would not be expected to sorb in this manner. Nevertheless, partitioning of many of these nonpolar, or "hydrophobic," contaminants between groundwater and aquifer solids has been successfully modeled using an isotherm of the form of Eq. (11) with the function $f(c)$ equal to the product of the solute concentration c and a constant partition coefficient K_p. This leads to a constant retardation factor defined by

$$R_f = 1 + \frac{\rho}{n}K_p \qquad (13)$$

Attempts to relate K_p to the hydrophobicity of the contaminant and the organic carbon content of the aquifer have met with considerable success, suggesting that solid organic carbon is the primary aquifer sorbent for dissolved organics. However, significant retardation of organic contaminants has also been observed in aquifers with very low concentrations

of organic carbon, indicating that partitioning and retardation of organic contaminants proceeds by other mechanisms as well.

3. Complexation

Interactions between species in solution can generate combined species known as ion-pairs, solution phase complexes or chelates. Although these homogeneous complexation reactions do not affect the rate at which the dissolved contaminants migrate, they may have important consequences for heterogeneous or degradation reactions involving the constituent species. For example, if a charged solute such as a heavy metal or radionuclide forms a strong complex with a solute of opposite charge, the resulting charge neutralization will make the radionuclide less likely to sorb onto a charged solid surface in the aquifer. The net mobility of the metal is therefore increased because retardation is reduced. A number of documented cases of enhanced mobility or remobilization of radioactive contaminants have been attributed to the formation of complexes with natural and anthropogenic organic solutes. In other cases, the complex may actually be more likely to sorb than the uncomplexed contaminant, if the complexing agent has a strong affinity for the aquifer solids. In such a case, formation of the complex could lead to greater retardation.

Another potential consequence of homogeneous complexation is reduced attenuation by degradation. This would occur if organic contaminants held in complexes are less readily utilized as energy sources for microorganisms or are less subject to abiotic reactions such as hydrolysis. Accounting for possible homogeneous reactions is important to the development of accurate predictive models of contaminant transport in groundwater.

4. Precipitation–Dissolution

Precipitation–dissolution reactions constitute a fourth class of reactions that can affect the transport of inorganic contaminants. Like sorption reactions, precipitation and dissolution involve reversible mass transfer between the groundwater and aquifer solids. In contrast to sorption reactions, for which the maximum concentration in solution is unlimited given sufficient contaminant input, precipitation–dissolution reactions provide constraints on the maximum concentrations of dissolved species through the equilibrium "solubility product" relation

$$K_{sp} = \prod_i c_i^{\nu_i} \qquad (14)$$

In Eq. (14) K_{sp} is the equilibrium constant for a

precipitation–dissolution reaction, the c_i's are concentrations of all dissolved species that are produced upon dissolution, and the v_i's are the stoichiometric coefficients of the species in the reaction. It should be noted that many precipitation–dissolution reactions occur relatively slowly in groundwater systems and, therefore, may not reach the equilibrium limit defined by Eq. (14).

Given continued input of a precipitating contaminant, this type of reaction can lead to attenuation of the plume. Attenuation will reach a finite limit, however, when contaminant concentrations in solution reach their equilibrium values. Contaminants held in the solid phase following precipitation can be remobilized if the input of contaminants ceases or if changes in groundwater chemistry such as an increase or decrease in pH alter the equilibrium groundwater concentration. Because of this possibility for remobilization, attenuation is not permanent, as in the case of decay or degradation. In this sense, the effects of precipitation–dissolution reactions on transport can also be viewed as a form of retardation. It is important to recognize, however, that this retardation will differ significantly from retardation resulting from sorption reactions and that a retardation factor model of the form of Eq. (12) does not apply.

III. Control of Groundwater Contamination

A. Physical Containment

When a localized source of groundwater contamination such as a buried tank or soil contaminated by a spill can be identified, the first step in a clean-up program should be removal of the source. If the source is not readily identifiable or localized, or if a groundwater plume already extends from the source, the next step is often an attempt to contain the plume and limit further migration. Construction of a subsurface vertical barrier to surround the plume or to divert groundwater flow away from the contaminated area provides one means of physical containment of shallow contamination. Barrier walls constructed by excavating a trench and backfilling with a mixture of soil and bentonite clay have been installed to depths of over 30 m. Other types of subsurface barriers include driven sheet piles and injected grout curtains. To prevent underflow of contaminated groundwater, barrier walls should extend to a low-permeability zone below the plume. In the absence of a low permeability zone, shallow pumping within the walled area is necessary to limit migration of the plume beneath the wall. Surface sealing to reduce rainfall infiltration will also improve the effectiveness of physical containment structures.

B. Hydraulic Control

In cases where conditions preclude the construction of a physical barrier, such as in deeper aquifers or areas of shallow bedrock that cannot be readily excavated, containment is best achieved by hydraulic controls. One or more wells can be used to modify hydraulic gradients and limit plume migration. These wells can be used simply to contain the plume or they can serve simultaneously as extraction wells to remove contaminated groundwater for surface treatment. Combinations of injection and extraction wells can be used in some cases to flush clean water through a contaminated zone in order to accelerate clean-up.

The volumes of water involved in most pump-and-treat systems can be very large, particularly when the contaminant plume has migrated and dispersed to fill large portions of an aquifer. In addition, removal of contaminated water is usually accompanied by removal of some water from uncontaminated zones, further increasing the treatment volumes. When contaminants are initially held in the solid phase by sorption or precipitation reactions, it may be necessary to remove a volume of water that is many times the initial volume of the plume. Thus, the same mechanisms that serve to limit migration rates and contaminant concentrations during groundwater transport can also inhibit rapid and efficient removal from the subsurface. Treatment and disposal of extracted groundwater presents additional environmental problems. The trade-offs between mitigation of groundwater contamination and generation of new sources of atmospheric or surface water pollutants have not always been clearly recognized.

C. *In-Situ* Biorestoration

Perhaps the most promising approach to aquifer restoration is enhancement of natural degradation processes by injection or infiltration of water containing nutrients and other substances that can stimulate growth of subsurface microbial populations. The feasibility of such biostimulation has been demonstrated at a number of sites affected by spills of petroleum products. Field experiments using several chlorinated solvents have also yielded encouraging results. Questions remain concerning the nature of the biodegradation products, which may themselves be contaminants. Better methods of estimating the optimal mix of nutrients and more efficient injection techniques are also needed. A major limitation of biorestoration, however, is that it cannot remove NAPL contaminants because it relies on an electron transfer process that must take place in the water phase. Injection of steam and/or surfactants may aid in the removal of NAPL contaminants. Continued research and experimentation should, in the near future, lead to improved design of treatment systems.

Bibliography

Bear, J., and Verruijt, A. (1987). "Modeling Groundwater Flow and Pollution." Reidel, Dordrecht, The Netherlands.

Dagan, G. (1989). "Flow and Transport in Porous Formations." Springer-Verlag, Berlin.

Lee, M. D., Thomas, J. M., Borden, R. C., Bedient, P. B., Wilson, J. T., and Ward, C. H. (1988). Biorestoration of aquifers contaminated with organic compounds. *CRC Crit. Rev. Environ. Control* **18**(1), 29–89.

Mackay, D. M., and Cherry, J. A. (1989). Groundwater contamination: Pump-and-treat remediation. *Environ. Sci. Technol.* **23**(6), 630–636.

McCarthy, J. F., and Zachara, J. M. (1989). Subsurface transport of contaminants. *Environ. Sci. Technol.* **23**(5), 496–502.

National Research Council, Water Science and Technology Board (1990). "Ground Water Models, Scientific and Regulatory Applications." National Academy Press, Washington, D.C.

Naymik, T. G. (1987). Mathematical modeling of solute transport in the subsurface. *CRC Crit. Rev. Environ. Control* **17**(3), 229–251.

Ronen, D., Magaritz, M., and Almon, E. (1988). Contaminated aquifers are a forgotten component of the global N$_2$O budget. *Nature (London)* **335**, 57–59.

Schwille, F. (1988). "Dense Chlorinated Solvents in Porous and Fractured Media" (translated by J. F. Pankow). Lewis Publishers, Chelsea, Michigan

Wheatcraft, S. W., and Tyler, S. W. (1988). An explanation of scale-dependent dispersivity in heterogeneous aquifers using concepts of fractal geometry. *Water Resour. Res.* **24**(4), 566–578.

Glossary

Aquifer Saturated, permeable geologic unit that can store and transmit significant quantities of groundwater.

Discharge area Region in which there is a net loss of water from the groundwater system (discharge) to surface water bodies or by evapotranspiration.

Plume Volume of contaminated groundwater that occupies a continuous region of an aquifer and emanates from a single source.

Potentiometric surface Imaginary surface defined by the levels to which water will rise in wells that are open at the same elevation. The slope of the potentiometric surface determines the horizontal direction of groundwater flow.

Recharge area Region in which there is a net addition of water to the groundwater system (recharge) as a result of infiltration from surface water bodies or an excess of precipitation over evapotranspiration and runoff.

Solute Inorganic or organic species that is dissolved in groundwater.

Unsaturated zone Zone between the land surface and the water table in which fluid pressures are less than atmospheric.

Water table Surface along which fluid pressure is equal to atmospheric pressure.

Gulf Stream

P. Cornillon
University of Rhode Island

This article explores the details of the Gulf Stream System and the relationship of this current system to basinwide, hemispherical, and global processes. Of particular interest is the coupling between the atmosphere and the oceans. The prevailing winds over the North Atlantic give rise to an asymmetric current system in the ocean, with intense poleward currents on the western edge of the basin (the Gulf Stream System) and a broad weak equatorward flow in the eastern half of the basin. The intense current on the western boundary dissipates the bulk of the kinetic energy put into the system by the wind and results in a large northward transport of thermal energy to approximately 35°N, and, downstream of Cape Hatteras (the point at which the Gulf Stream leaves the continental margin), a large transfer of heat from the ocean to the atmosphere. Coupled with the subtropical gyre is a subpolar gyre in which water moves farther north, loses more thermal energy, increases in density, sinks, and flows to the south as North Atlantic Deep Water. This water type crosses the equator and the South Atlantic, eventually being caught up in the Atlantic Circumpolar Current, by which it is transported to the rest of the major ocean basins.

The modulating effect of the ocean on the spatial and temporal variability of the surface temperature resulting from the processes outlined above in turn plays a critical role in atmospheric circulation. The complexity of this coupling — the ocean to the atmosphere and vice versa — in conjunction with the difficulty in measuring oceanic circulation over the appropriate spatial and temporal scales has resulted in relatively slow progress in understanding the system as a whole. Satellites, long-term moorings, cross-basin sections [undertaken as part of the World Ocean Circulation Experiment (WOCE)], and deployment of "smart," disposable instruments, combined with faster computers and increased personnel are expected to contribute dramatically over the next decade toward a better understanding of the ocean and the coupled land–atmosphere–ocean system as a whole.

I. Overview of the Gulf Stream System and Connecting Currents

The major ocean basins, with the exception of the South Pacific, show a marked asymmetry in their currents, with a strong narrow poleward flow on the western side of the basin and a broad weak equatorward flow on the eastern side. The Gulf Stream System comprises the currents on the western side of the North Atlantic (Fig. 1). The Gulf Stream's counterpart in the North Pacific is the Kuroshio and Kuroshio Extension, in the South Atlantic is the Brazil Current, and in the Indian Ocean is the Agulhas.

Strong poleward currents in the western North Atlantic were first described by Ponce de León in 1513 and have received a great deal of attention over the past four and three-quarters centuries. Although the large thermal gradient marking the shoreward side of the Gulf Stream, one of its major features, was described as early as 1612 by M. Lescarbot, Benjamin Franklin and, independently Charles Blagden, in the late 1700s, were the first to describe the use of the stream's strong thermal signal as a navigational aid. Franklin was also the first to print a chart of the stream (1769) and to describe attempts to measure its vertical thermal structure (1785). Over the next century, it became apparent that the Gulf Stream was not the simple steady current envisioned but rather one that varied in position as a function of time (1814–1825) and appeared to enclose, within its warm boundaries, masses of cold water (1832, most likely cold core rings) and cold veins (1845, possibly entrainment features or shingles along the cold edge of the stream).

By the early 1900s knowledge of the Gulf Stream had grown to the point at which it became clear that the "current" under study could logically be divided into several regions (currents) of different properties and that a consistent nomenclature would contribute to communication associated with this current system. To this end, C. O'D. Iselin proposed the following use of terms in 1936 (see Fig. 1 for a geographic description of the currents): the *Gulf Stream System* refers to the strong western boundary currents in the

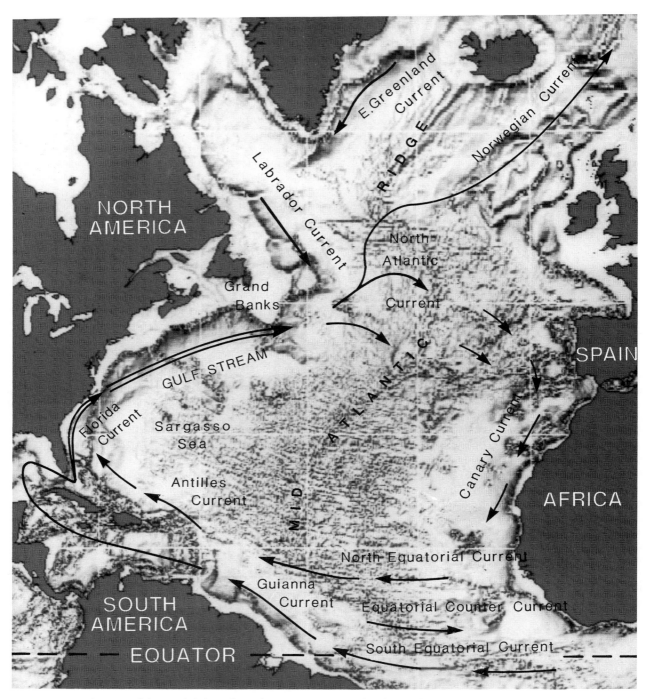

Fig. 1   Bathymetry of the North Atlantic basin shown in perspective. The white grid is in 15° increments in both latitude and longitude beginning at the equator (0°N) and at the prime meridian (0°W).

North Atlantic extending from the Florida Straits past the tail of the Grand Banks of Newfoundland, the Grand Banks being a shallow area (less than 200 m) located approximately between 53° and 49°W and 48° and 43°N, with the southernmost extent at 50°W, 43°N. The *Florida Current* is the portion of the current from the Florida Straits to immediately south of Cape Hatteras, North Carolina (75°W, 35°N). The *Gulf Stream* is that part of the Gulf Stream System extending from Cape Hatteras to the tail of the Grand Banks, where it is joined by the Labrador Current from the north. In the literature, there is still significant variability in the use of these terms. The Florida Current is, on occasion, defined as the current from the Florida Straits to the point at which the current joins with the Antilles Current. Downstream of this point, the current is called the Gulf Stream. The term Gulf Stream is also used in place of the Gulf Stream System; hence it overlaps with the Florida Current. In this article, Iselin's definitions are used.

A. Currents Feeding the Florida Current

The Florida Current is fed by the Loop Current, the Antilles Current, and inflow from both the shelf side and the Sargasso Sea (the water in the center of the North Atlantic subtropical gyre) side along most of its length. The Loop Current is the current that flows from the Yucatan Straits through the Gulf of Mexico, joining the Florida Current at its origin immediately to the south of Key West, Florida. The transport of the Loop Current at this point is on the order of 30 Sv (1 Sv = 10^6 m^3 s^{-1}). The Antilles Current is less well defined, although recent work suggests a northwestward-flowing current between approximately 300 and 1000 m from about 22°N to the Florida Current at about 28°N. The transport of this current is on the order of 4 Sv. Of some significance in terms of water properties is that the Antilles Current appears to be an extension of the Gulf Stream recirculation, and not a branch of the North Equatorial Current (NEC) as originally thought; the contribution from the NEC to the Gulf Stream System comes via the Loop Current.

B. Currents Emanating from the Gulf Stream

Downstream of the Grand Banks the precise nature of the flow, which is referred to as the North Atlantic Current or the North Atlantic Drift, is poorly understood both observationally and theoretically. Early theories (1930–1970) suggested two quite different scenarios. In one scenario, based on hydrographic sections, the Gulf Stream splits into two main branches immediately downstream of the Grand Banks. The southern branch contributes to the interior portion of the transport of the subtropical gyre. The northern branch, the North Atlantic Current, flows to the northeast past

Newfoundland, at which point it turns to the east crossing the Mid Atlantic Ridge at about 30°W, 50°N. It splits several more times before crossing the Mid-Atlantic Ridge, with some of its transport contributing to the interior of the subtropical gyre, some to the eastern boundary current (the Canary Current), and some to the subpolar gyre.

In the other scenario, based on the distribution of dissolved oxygen in the water, no part of the Gulf Stream rounds the Grand Banks, and the Southeast Newfoundland Rise acts to separate the flow into two anticyclonic gyres. In this model, the North Atlantic Current is actually part of an entirely separate "Northern Gyre." Although the water property distribution was satisfied by this model, the proposed flow lines differ dramatically from dynamic height contours.

More recent work, based on satellite-tracked buoys and hydrographic sections, supports neither of these scenarios, i.e., neither a permanent branching nor confinement of the subpolar gyre to the east of Newfoundland. These data suggest that the North Atlantic Current spreads over the entire North Atlantic as a rather broad easterly drift with a maximum at the Subarctic Front. Furthermore, this model does not point to the currents feeding the Canary Basin and the Portugal Current, and, more important, it suggests that the North Atlantic Current contributes little to the circulation of the southern gyre, a much larger percentage of water being transported to the north than previously thought.

The foregoing discussion covers, in a cursory fashion, the various currents connecting to the currents of the Gulf Stream System, both at its origin in the Florida Straits and at its termination downstream of the Grand Banks. In the following, the focus will be on the Gulf Stream System, i.e., the Gulf Stream and the Florida Current. Furthermore, the major emphasis will be on the currents' physical properties, only passing reference being made to their biological properties. The remainder of this article is divided into two major parts: one in which specific properties of the currents are detailed, and the other in which the larger-scale (oceanographic and atmospheric) implications of the current system are explored.

II. The Gulf Stream Close Up — the Details

A. Bathymetry of the Gulf Stream System

Figure 2 provides a three-dimensional perspective of the bathymetry as it is related to the Gulf Stream System. The mean path of the stream from the Florida Straits to the Grand Banks, as determined from 5 years of satellite data, is also indicated. Topographically, the path of currents in the Gulf Stream System may be divided into three regions: the Florida Straits to Cape Hatteras, Cape Hatteras to the New England seamounts, and the New England seamounts to the Grand

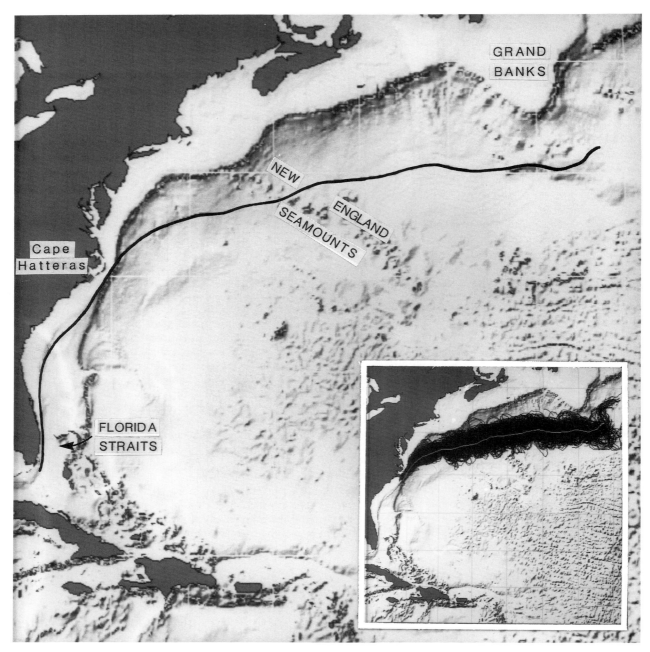

Fig. 2 Detail of the western North Atlantic bathymetry shown in perspective with the mean path of the Gulf Stream, derived from satellite data, shown as the solid black line. The inset shows all satellite-derived paths drawn on the same bathymetric perspectives.

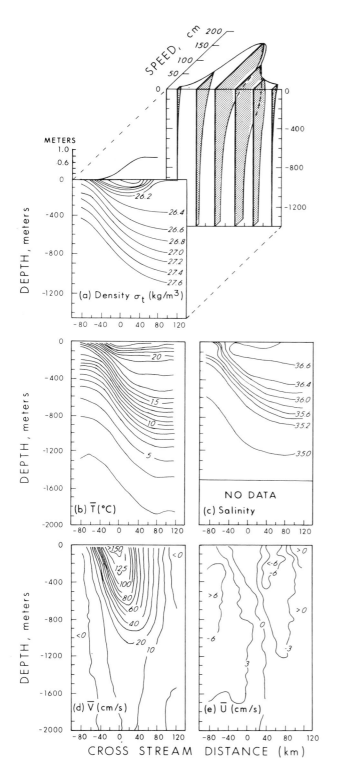

Fig. 3   Cross-stream properties of the Gulf Stream at 73°W.
(a) Density contoured versus depth and cross-stream distance,
change in surface elevation relative to an arbitrary level surface, and
perspective showing the downstream current. (b) Temperature. (c)
Salinity. (d) Downstream component of the current. (e) Cross-
stream component of the current. The cross-stream distance is
measured from the location of the maximum surface velocity, the
positive direction being offshore. [Panels (b), (d), and (e) redrawn
from Fig. 10 of Halkin and Rossby (1985).]

Banks. As will become apparent in the following, these regions appear to represent dynamically different regimes.

B. Cross-Stream Variability of Physical Oceanographic Properties

An extensive study of the Gulf Stream approximately 150 km
downstream of Cape Hatteras (≈ 73°W) by H. T. Rossby in
the early 1980s yielded the first detailed (and the most complete to date) view of the cross-stream (normal to the mean
path of the current) current structure as a function of depth
in this region. We will therefore begin with a discussion of the
cross-stream properties of the Gulf Stream at this location.
The properties investigated are sea surface elevation, temperature, salinity, density, downstream (or alongstream—parallel
to the mean path of the current) current, and cross-stream
current, the last five as a function of depth as well as of
cross-stream position.

Although not measured until recently (by satellite altimeter), one of the more telling characteristics of the Gulf Stream
is the approximately 1-m increase in surface elevation observed when moving from its shoreward to its seaward side.
The change in elevation, plotted in Fig. 3a, gives rise to a
cross-stream pressure gradient on level surfaces beneath the
surface. (A level surface or geopotential surface is one everywhere normal to the local gravity vector. Were this surface to
be solid, an object on it would feel no horizontal force due to
gravity. The cross-stream elevation of Fig. 3a is plotted relative to such a surface.) This pressure gradient exists because
the weight of the overlying water increases from left to right
(in Fig. 3a) on such surfaces, and it results in a shoreward-
directed force on fluid parcels. Counterbalancing this is an
apparent force, called the Coriolis force, which is proportional to the horizontal velocity and results from the fact that
the ocean, as we view it, is not fixed in an inertial coordinate
system but rather in one that is rotating. Basically, we live on
a rotating sphere. The Coriolis force points to the right
(seaward in this plot) of fluid parcel velocities in the northern
hemisphere. From the surface elevation plot of Fig. 3a, it is
evident that the cross-stream pressure gradient near the sur-

face varies from zero on the shoreward side of the stream to a maximum at 0 km and then back to zero on the seaward side. Coincident with this, the downstream current must increase from zero to a maximum and then return to zero if the Coriolis force is to balance the pressure gradient. This is shown schematically in the rear panel of Fig. 3a. Also shown in Fig. 3a is the cross-stream density structure, which is obtained from temperature and salinity data similar to those contoured in Fig. 3b and c. Because the mean density of the water above any given level surface tends to decrease seaward (with the exception of a thin layer near the surface), the cross-stream pressure gradient also tends to decrease with depth. Again, keeping in mind the balance between the Coriolis force and the horizontal pressure gradient, the decreasing pressure gradient must result in decreasing velocity with depth. This is also shown schematically in the rear panel of Fig. 3a. Figure 3d shows contours of the downstream velocity in which the basic features outlined above are clearly evident.

The foregoing discussion focuses on the downstream velocity and its relation to the surface elevation and density fields. Less well understood are the mechanisms forcing the cross-stream velocity structure shown in Fig. 3e. Of particular interest is the fact that these data show that at 73°W the Gulf Stream, at least over the depth range indicated, is a region of convergence; i.e., there is a net influx of fluid from the Sargasso Sea and from the Slope Water regions. As a result of this inflow, the transport of the stream increases with downstream distance in this region.

Rossby's observations focused on the structure of the current in the upper 2000 m of the water column, leaving unanswered the question of the depth to which the influence of the Gulf Stream might be observed. The first unambiguous evidence that the current downstream of Cape Hatteras extends to the seafloor (> 4000 m) was provided by a single long-term (approximately 1 year) mooring at 68°W, 37°37′N, the historical mean position of the stream at that longitude. Cross-stream profiles of the downstream current obtained from this mooring at a number of depths as it meandered (this term will be defined later) past the mooring are shown in Fig. 4. The downstream component of the current, although quite weak, is still evident at the bottommost current meter (688 m above the bottom) on the mooring.

C. Along-Stream Variability of Physical Oceanographic Properties

A number of cross-stream sections of the Florida Current have been made over the past 25 years. Contours of the downstream isotachs on these sections show the evolution of the structure of the current as it flows through the Florida Straits, where it is completely constrained bathymetrically,

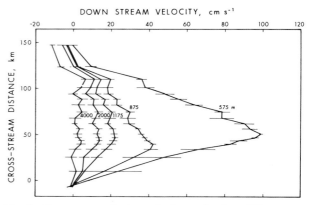

Fig. 4 Downstream velocity at depths ranging from 575 to 4000 m (688 m above the seafloor) as a function of cross-stream distance. [Redrawn from Fig. 3 of M. M. Hall and H. L. Bryden (1985). *Geophys. Res. Lett.* **12**, 203–206.]

out onto the ocean shelf. Figure 5 is a plot of a number of such sections. As the current passes through the Florida Straits it encounters an increasingly smaller cross-sectional area, which decreases from 60 km² immediately south of Key West to 36 km² at the northern extreme of the Little Bahama Bank. As a result, the current increases in speed (maximum surface velocity from 140 to 180 cm/s) while at the same time becoming more and more asymmetric. After leaving the Florida Straits, the current decelerates at the surface, broadens, and deepens, while the transport increases. Despite the bathymetric constraint of the shelf at this point, the cross-stream isotach contours (Fig. 5, section VII) look quite similar to those in deep water downstream of Cape Hatteras (Fig. 3d): the maximum in the current shifts seaward with depth; the width of the current is on the order of 100 km; the maximum surface current is on the order of 150 cm/s; and the maximum current falls to approximately 20% of its value at about 900 m. Other measurements of the cross-stream structure of the downstream velocity farther downstream, e.g., at 68°W (Fig. 4), suggest similar characteristics. It therefore appears that there is little change in the baroclinic structure of the current of the Gulf Stream System from the point at which it leaves the Florida Straits to as far downstream as reliable cross-stream measurements exist.

The relatively small change in the baroclinic structure of the stream can also be seen in the downstream evolution of the cross-stream thermal structure. Although it is the density that is important in calculations related to the dynamics of the current, the density is to a large extent determined by the temperature in the Gulf Stream region, the effect of changes in salinity being relatively small compared to those in temperature. Figure 6 shows two cross-stream thermal sections, one at 53°30′W and the second at 73°W. It is quite apparent

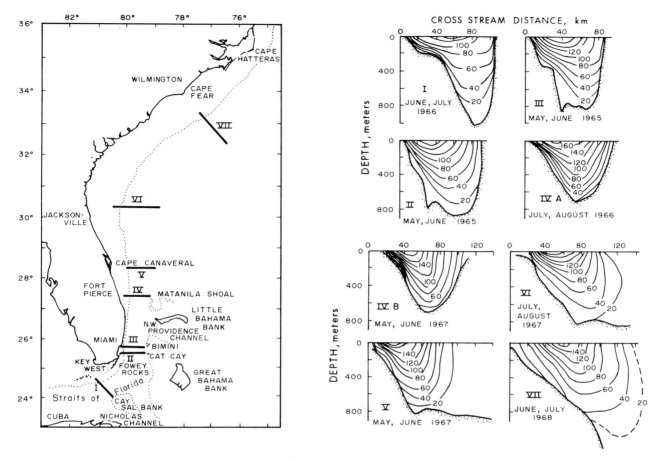

Fig. 5 Contours of the downstream velocity of the Florida Current on the sections shown in the accompanying map of the region. [Redrawn from Figs. 1 and 3 of W. S. Richardson, W. J. Schmitz, Jr., and P. P. Niiler (1969). *Deep-Sea Res.* **16**(Suppl.), 225–231.]

from these data that the cross-stream thermal gradient, and by inference the cross-stream density structure, remains largely unchanged with downstream distance; only the distance over which these gradients persist increases slightly.

Given that the baroclinic structure of the Gulf Stream System changes little from 33°N to the Grand Banks, one might expect the same to be true of the transport. This is, in fact, not the case; the transport increases dramatically from 30 Sv in the Florida Straits to ≈ 160 Sv at 65°W (Fig. 7, a compilation of available observations of Gulf Stream transport). The transport of the current may be viewed as consisting of two components: a baroclinic component, the part of the current that goes to zero near the bottom, and a barotropic component, the part of the current that is independent of depth. The bulk of the change in transport of the Gulf Stream results from a change in the barotropic component. Because this component of the velocity is integrated over the entire water column, small changes in it give rise to large

changes in the transport. From the Florida Straits to the New England seamounts the Gulf Stream System is a region of convergence (Fig. 3e), hence the increase in transport. Downstream of the seamounts, the current is thought to be divergent with a resulting decrease in transport. There is a suggestion of this in the few measurements available (Fig. 7), but the convoluted nature of the stream's path makes precise measurements in this region difficult, hence the large variability in these data.

D. Instabilities Associated with the Gulf Stream System

The discussion of the relationship between the current and the cross-stream pressure gradient suggests that the Gulf Stream might be viewed as a boundary or even as a barrier between warmer, lighter water to the south and east and

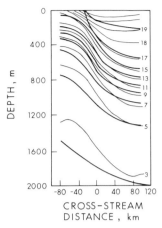

Fig. 6 Contours of temperature as a function of depth and cross-stream distance at two different locations. Thick lines correspond to a section at 58°30′W and thin lines to a section at 74°W, repeated from Fig. 3b. For clarity, only odd temperature contours have been retained.

colder, heavier water to the north and west. (Recent work using heat, oxygen, and potential vorticity as tracers suggests that a view of the Gulf Stream as a barrier is quite appropriate even in the region in which the stream is divergent.) It is the Coriolis force resulting from the current that maintains the barrier separating the two regions. The notion of the stream as a barrier is quite real in the sense that warmer water on the Sargasso side is actually higher in a geophysical sense. The warmer water may therefore be viewed as applying pressure on the barrier. If for some reason the barrier is weakened (a reduction in the strength of the current), this pressure would result in the barrier being distorted. Instabilities of this type are called baroclinic instabilities and arise because the ocean is stratified with a significant vertical shear in the horizontal currents and because the ocean is on a rotating sphere. Instabilities of the current can arise equally well from the fact that a strong current in a fluid at rest gives rise to large horizontal velocity shears that are unstable. These instabilities are referred to as barotropic instabilities. Both types of instabilities are present in the currents associated with the Gulf Stream System, giving rise to large temporal fluctuations in the stream's path referred to as meanders. A number of meanders are evident in the sea surface temperature (SST) field (Fig. 8) derived from a satellite-borne thermal infrared sensor called the Advanced Very High Resolution Radiometer (AVHRR). In this image of the Gulf Stream System from the Florida Straits to the Grand Banks, the Florida Current and Gulf Stream appear as the lighter-colored water. [*See* SEA SURFACE TEMPERATURE SIGNALS FROM SATELLITES.]

Infrared images obtained from satellite-borne sensors between 1975 and 1989, as well as *in situ* instrumentation and multiple ship surveys, have been used to characterize meanders. The most rapidly growing meanders are approximately 400 km in length, propagate downstream at approximately 8 km/day, and double in amplitude at first every 50 km and farther downstream every 400 km. As the amplitude of a meander becomes comparable to its wavelength, it becomes increasingly convoluted, often either detaching completely or shedding large volumes of water.

Meanders that pinch off are called Gulf Stream rings and are the largest instabilities resulting in cross-stream displacement of fluid, a momentary puncturing of the barrier characterized above. Gulf Stream rings arise when a meander folds back on itself and then detaches; hence, their vertical structure is very nearly axisymmetric. The geometry of the process and their dynamics require that rings breaking off to the south of the stream spin cyclonically (counterclockwise when viewed from above) and those breaking off to the north spin anticyclonically. Because meanders breaking off to the south enclose cold slope water at their center, they are called cold core rings. Similarly, those breaking off to the north enclose warm Sargasso Sea water and are called warm core rings. Both

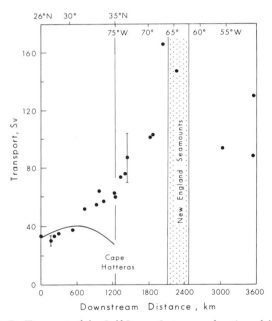

Fig. 7 Transport of the Gulf Stream System as a function of distance from Miami, Florida. The solid line is the transport calculated with climatological winds assuming Sverdrup dynamics in the ocean interior. For the two locations at which multiple measurements were made, the standard deviation of these measurements is indicated.

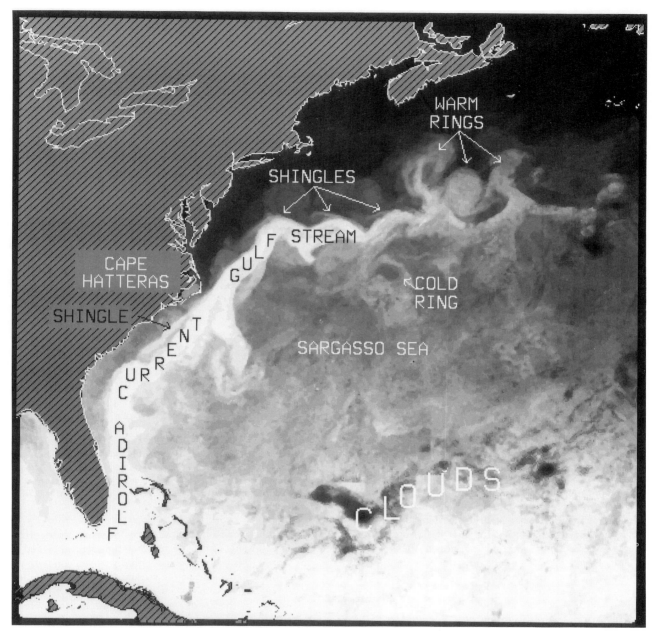

Fig. 8 Sea surface temperature of the western North Atlantic for mid-April 1985 derived from the infrared channels of the satellite-borne AVHRR.

types of rings propagate at approximately 6 km/day (≈ 6.9 cm/s) to the west. Warm core rings are constrained in their movement by the continental shelf (see Fig. 1) to the north and the Gulf Stream to the south. This constraint often results in their being reabsorbed into the Gulf Stream or in their properties being dramatically modified during their life through interactions with Gulf Stream meanders. Both warm and cold core rings have been studied extensively by satellites and *in situ* instrumentation and none have been observed to decay freely; all have been reabsorbed by the Gulf Stream, warm core rings either through an interaction with a meander downstream of Cape Hatteras or in the apex formed by the stream and the continental shelf immediately downstream of Cape Hatteras, and cold core rings reabsorbed by meanders downstream of Cape Hatteras or in the Florida Current between the Grand Bahama Banks and Cape Hatteras. Warm core rings form preferentially downstream of the point at which the stream crosses the New England seamounts. Figure 9 shows the location between 75°W (Cape Hatteras) and 45°W at which all observed warm core rings were formed from, or reabsorbed by, the Gulf Stream for the period 1980–1985. In a separate study dealing extensively with warm core ring statistics for the period 1975–1986, the mean annual rate of formation was found to be 8.7 per year; the distribution of ring lifetimes was distinctly bimodal with 44% surviving more than 20 weeks, 32.7 weeks on the average, and the remainder surviving 7.7 weeks on average. Short-lived rings were those reabsorbed as a result of a meander–ring interaction downstream of 70°W. The rings that survived westward of 70°W generally survived to the stream–shelf apex. Not as much is known about the life history of cold core rings because they are more difficult to track in satellite-derived infrared imagery once they have moved away from the stream. The formation rate and speed of westward propagation are thought to be quite similar to those of warm core rings, but their life span is thought to be much longer, some surviving as long as 2 years, and the distribution of life spans is not thought to be bimodal in that there is no constriction in their path as there is for warm core rings at 70°W.

Gulf Stream rings are quite important in the transport and dynamics of the stream and the cross-stream transport of heat, salinity, and other properties. Ejection of a ring is one way in which very large instabilities are removed from the stream's path, and reabsorption of a ring appears to be a source of instability, as is an encounter between a meander and a ring that does not result in reabsorption. Based on the formation of approximately 18 rings (warm and cold) per year and a mean radius of 75 km, rings recycle approximately 25% by volume of Gulf Stream water; i.e., averaged over a year, 25% of the time Gulf Stream water is removed from the stream by rings downstream of the seamounts and reintroduced into the stream upstream of the seamounts, the bulk of this occurring at or upstream of Cape Hatteras. In that Gulf

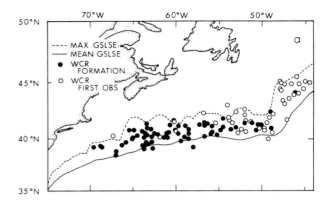

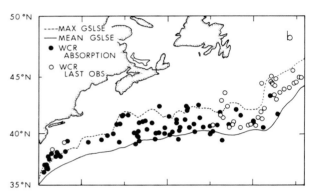

Fig. 9 (a) Location of observed formation (●) or first sighting (○) of 115 warm core rings (WCRs) for the period May 29, 1980 through May 16, 1985. The solid line is the 5-year mean of the Gulf Stream landward surface edge (GSLSE). The dashed line shows the northward extreme of this edge. (b) Locations of the absorption (●) or last sighting (○) of the 115 rings shown in (a). Solid and dashed lines as in (a). [Redrawn from Figs. 12 and 13 of Auer (1987).]

Stream rings generally enclose a core of water coming from the opposite side of the stream, when they are reabsorbed by the stream, this water is mixed with the stream and with the water on the ring side of the stream. For example, a warm core ring carries Sargasso Sea water in its core. When it is reabsorbed by the stream, say at Cape Hatteras, this water is mixed into the stream and into the Slope Water.

As suggested earlier, there are instabilities of the Gulf Stream which result in the ejection of stream water into the surrounding fluid without the detachment of a meander. These are smaller in magnitude than Gulf Stream rings, do not appear to carry fluid from one side of the stream to the other, and appear to be confined to loss of fluid above the main thermocline. To the north, two classes of such losses have been characterized — *shingles* and *overwashes*. To the south, all such losses have, for the present, been included in one group called *warm outbreaks* (of the Gulf Stream into the Sargasso Sea), although at least two different mechanisms seem to be responsible for the ejection of fluid to the south. Shingles, identified in Fig. 8, occur at the crest of meanders and are long and narrow and approximately parallel to the stream. Fluid within the shingle appears to have a velocity toward the west, i.e., in a direction opposite to that of the stream. The western end of the shingle either has a cyclonic swirl or extends anticyclonically to the fringes of a warm core ring (downstream of Cape Hatteras), should one be present. Expendable bathythermograph (XBT) and conductivity–temperature–depth (CTD) sections through shingles show that they are relatively shallow, extending only to approximately 50 m, and quite narrow, on the order of 1–10 km across. Work with isopycnal floats deployed in the main thermocline at the approximate center of the Gulf Stream off Cape Hatteras shows that the floats when moving into the crest of a meander will "slide" up isopycnal surfaces toward the shoreward or cyclonic side of the stream. When the isopycnal surface on which the floats are located reaches approximately 50 m, the floats are often ejected from the stream, moving in a path similar to that traced out by shingles. Although shingles appear to be relatively small in volume, they are ubiquitous, generally evident at a number of locations between the Florida Straits and the New England seamounts in any single satellite image (Fig. 8). Of particular significance is the fact that shingles are responsible for the transport of Gulf Stream water into the Slope Water in a region (Cape Hatteras to the New England seamounts) in which the Gulf Stream is generally convergent, as indicated by the cross-stream isotachs shown in Fig. 3e. Also of significance is that upstream of Cape Hatteras, where the transport of the current is significantly less than that downstream of this point (Fig. 7) and where warm core rings do not play a role in cross-current fluid exchange, shingles are the dominant mechanism for moving water from the current to adjacent water on the shoreward side of the current.

The biological importance of the loss of Gulf Stream water to the surrounding fluid has been highlighted by a recent event immediately upstream of Cape Hatteras. Introduction of *Ptychodiscus brevis* (a red tide organism) resulted in significant mortality of the local shellfish community, numerous cases of human illness, and the closing of the shellfish grounds (with an estimated loss of $25 million to the fishery). *Ptychodiscus brevis* is abundant between September and February in the Gulf of Mexico, hence presumably in the Gulf Stream. Prior to this event, however, there had been no reported outbreaks of red tide caused by *P. brevis* on the shelf in the vicinity of Cape Hatteras, although three such events have been documented along the east coast of Florida, each following a similar outbreak on the west coast of Florida. For these, a link from the western coast to the eastern coast via the Loop Current and the Florida Current was

established. The event observed on the shelf in the vicinity of Cape Hatteras in November 1987 followed an outbreak on the east coast of Florida and appears to have resulted from anomalous winds that occurred at the same time that a shingle was present. The shingle mixed Gulf Stream water with the adjacent shelf water and the wind moved this water onshore carrying the microorganism to the shellfish areas.

Overwash water is the name recently given to relatively large amounts of Gulf Stream surface water that simply appears to have been left behind by a meander. There is no well-defined structure to these events. As in the case of shingles, they are quite shallow and, thus far, their existence has been documented only immediately downstream of Cape Hatteras. One peculiarity of this water is that it has a density very close to that of the adjacent Shelf or Slope Waters and floods onto the shelf with a fair degree of regularity. For an 11-year period surveyed for such events, there was an indication of overwash water moving onto the shelf between 3 and 9% of the time. In that the discharged Gulf Stream water is relatively rich in nutrients and that this is one of the few mechanisms identified for moving offshelf water onto the shelf, the impact of such events on the biology of the shelf may be quite significant.

To the south of the Gulf Stream large patches of warm water are seen in satellite imagery leaving the stream at the trough of meanders. These are called warm outbreaks and appear to be driven by two mechanisms. In one case they are drawn off by the interaction of a cold core ring with a meander trough, much as some shingles are to the north. In other cases, they are part of an anticyclonic cell upstream of the trough of the meander, possibly the analog of shingles (seen to the north) that are not associated with a warm core ring. The geometry of warm outbreaks is, however, fundamentally different from that of shingles. Outbreaks have a small aspect ratio, rarely appear to be organized, more like overwash water discussed above, and are very large in comparison to shingles — with dimensions on the order of 100 km in both directions. They are therefore closer in horizontal extent to cold core rings, but when any organized motion is evident, it appears to result from anticyclonic circulation as opposed to the cyclonic circulation of cold rings. There is also no indication, at the surface, of Slope Water enclosed in these features. The only vertical information available for warm outbreaks indicates that they extend into the main thermocline, i.e., to depths of at least 700 m; hence they involve loss of a large volume of fluid from the stream. Finally, they appear (again at the surface) to be extremely unstable, disintegrating in approximately 10 days. This means that the fluid which they contain is rapidly mixed into the adjacent Sargasso Sea.

The last mixing mechanism that we discuss is the very small scale mixing associated with the large thermal and salinity gradients observed on the shoreward side of the stream, again downstream of Cape Hatteras. The cross-stream gradients are at a maximum when very cold fresh Shelf Water is entrained along the edge of the stream. The Shelf Water filaments are remarkable in that they have been observed to extend more than 500 km downstream of their first point of contact, with much of the filament being less than 10 km (order 1 km) across. XBT and CTD sections through such entrainment features show that, beneath the surface, they may be entrained at several points into the Gulf Stream, resulting in an interleaving of Shelf Water and Gulf Stream Water. These conditions result in double diffusive mixing (mixing resulting from the fact that the rate for molecular diffusion of thermal energy differs substantially from that for salt), and vertical thermal profiles in these regions indicate the characteristic staircase structure associated with such mixing.

III. The Gulf Stream from Afar — the Big Picture

In this section we focus on the role that the Gulf Stream System plays in the circulation of the North Atlantic. The general circulation of the North Atlantic (Fig. 1) is divided into two large gyres: the subtropical gyre with the North Equatorial Current on its southern extreme, the Guiana, Antilles, and Florida currents on its western edge, the Gulf Stream and the North Atlantic Drift on its northern edge, and the Canary Current closing the circulation on its eastern edge; and a northern gyre, the subpolar gyre, with the North Atlantic Current to its south, the Norwegian Current to its east, and the Greenland and Labrador currents to its north and west. The subpolar gyre is important when considering the circulation of the subtropical gyre because the North Atlantic Current feeding the northern gyre represents one of the mechanisms for transferring relatively warmer (lighter) fluid from the upper layer (the water above the main thermocline) of the subtropical gyre to the lower layer of this gyre. Basically, the fluid leaving the gyre by this path is cooled in the Norwegian and Labrador seas (which results in an increase in its density), sinks, and then moves to the south in the Deep Western Boundary Current (DWBC, also referred to in the literature as the Western Boundary Undercurrent), crossing the path of the Gulf Stream with a transport of about 27 Sv at a depth of approximately 4000 m in the vicinity of Cape Hatteras. This water continues on to the south, eventually crossing both the North and South Atlantic oceans. To replenish the water lost to the subpolar gyre, some of the water in the lower layer, i.e., beneath the main thermocline, is carried up through the thermocline in the tropics and fed back into the subtropical gyre. There are of course other paths that upper layer fluid may take to reach the lower layer. These, although not in this context, were discussed in detail

in the previous section on mixing. In addition to this mixing, cross-thermocline mixing takes place over the remainder of the gyre, but, because of the relatively lower levels of kinetic energy available for such mixing and the relatively lower thermal and salinity gradients required for double diffusive mixing, the contribution from the remainder of the gyre is thought to be small. [See OCEAN CIRCULATION.]

The Gulf Stream System can then be viewed as a set of currents that (1) closes the circulation of the subtropical gyre, (2) provides several mechanisms for moving upper layer fluid into the lower layer, and (3) moves thermal energy from the tropics to the polar region.

A. Closing the Gyre — Mass Transport

The forcing for the horizontal circulation of the subtropical gyre is dominated by the prevailing winds over the North Atlantic basin. The atmospheric circulation over the North Atlantic is roughly symmetric and centered close to the middle of the basin. The bulk of the forcing is from the westerlies to the north, centered at about 38°N, and the easterlies to the south, centered at about 20°N. As indicated in the introduction, the oceanographic circulation resulting from these winds is markedly asymmetric with a broad diffuse current on the eastern half of the basin and a very narrow strong current on the western edge, this despite the fact that the atmospheric circulation forcing the gyre is roughly symmetric. The asymmetry in the currents is a consequence of the variation of the Coriolis parameter with latitude, which in turn results from the fact that we are dealing with currents on a rotating sphere. This is dealt with in detail in the oceanographic literature and will not be addressed here. [See AIR–SEA INTERACTION; ATMOSPHERIC CIRCULATION SYSTEMS.]

Because the North Atlantic is closed to the north, the net transport across any east–west section crossing the entire Atlantic is, for all practical purposes, equal to zero. This is simply a result of the conservation of mass taken together with the fact that very little water is stored for long periods of time on land or in the atmosphere and the net north–south transport of water in the atmosphere is relatively small, as is the net transport through the Bering Straits. The fact that the net north–south transport for the entire water column is zero does not mean that the same is true for any given layer, e.g., the upper 1000 m of the ocean. But if we consider the ocean to consist of two layers, water above the main thermocline and water below it, the transport to the north of water in and above the main thermocline by the western boundary current at the latitude of separation, 35°N, is very nearly equal to the transport of water in and above the main thermocline to the south integrated across the remainder of the basin. There is not exact equality because some fluid is lost to the lower layer either vertically or laterally, as described in the previous section. The transport to the south across the remainder of the ocean, which results from the wind, can be estimated from the Sverdrup relation (derived from the equations of motion in which pressure, Coriolis force, and wind stress are retained). From the Florida Straits to approximately 30°N the measured transport of the Florida Current is very nearly equal to that estimated from the known winds (Fig. 7), both increasing nearly linearly. To the north of 30°N, the wind-induced southward transport in the ocean interior decreases, while the northward transport in the Florida Current continues to increase. This increase continues downstream of Cape Hatteras to approximately the New England seamounts, after which the transport appears to decrease, although measurements in this region are scarce. Between Cape Hatteras and the Grand Banks, the Gulf Stream forms the northern edge of a tight *recirculation gyre*, sometimes referred to as the *Worthington gyre*. This tight recirculation gyre differs from the subtropical gyre in that water leaves the stream downstream of the seamounts and reenters the stream upstream of the seamounts *without* penetrating the interior of the gyre, i.e., water that is not part of the wind-driven Sverdrup circulation. The mechanisms driving this gyre (and a weaker one thought to exist to the north of the stream between the seamounts and the Grand Banks) are poorly understood at present. Warm and cold core rings clearly contribute to the mass of water removed from the current downstream of the seamounts and returned upstream, their movement being determined in part by their internal dynamics. But even assuming nine of each (warm and cold rings) per year with radii of 75 km, this accounts for only one-quarter or less of the transport. The other mixing mechanisms discussed above contribute significantly less than rings to the recirculation. At least one-half of the increase in transport must come from another source. Rectification of the intense eddy field described in the following section with the attendant mass flux has been suggested as a possible contributor. Measurements required to test this hypothesis are, however, extremely difficult to make in that they deal with second-order terms in the equation of motion. Suffice it to say that, at present, the physics associated with the increase in transport between 30°N and the New England seamounts and the subsequent decrease downstream of that point is poorly understood.

B. Turbulence and the Gulf Stream System

As previously discussed, satellite-derived imagery of the Gulf Stream System, especially downstream of the separation point near Cape Hatteras, shows a significant exchange of Gulf Stream water with the surrounding fluid. Associated with this exchange of mass is a lateral diffusion of kinetic energy. This region is the primary sink for the energy introduced into the

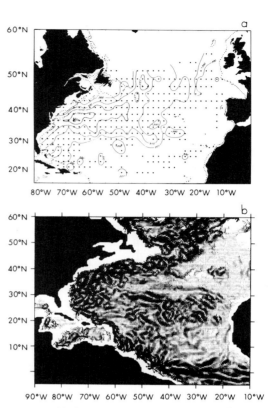

Fig. 10 (a) Eddy kinetic energy (centimeters squared per second squared) in the western North Atlantic. [Redrawn from Fig. 10 of Richardson (1983).] (b) Instantaneous horizontal velocity field at 1375 m from the WOCE community model. [From Fig. 2 of P. Müller and G. Holloway (1989). *EOS* **70**, 818–830.]

gyre by wind and heating, the quantities forcing the circulation of the gyre. The importance of the Gulf Stream System in this regard is indicated in Fig. 10. Figure 10a shows contours of eddy kinetic energy (EKE, the kinetic energy remaining in the fluid after removal of the mean flow) determined from float data. Figure 10b is one realization of the instantaneous horizontal velocity field determined by the WOCE community general circulation model. There are, not surprisingly, differences in these fields, resulting from the different points of view which they represent. The similarity in the Gulf Stream region, of a well-defined maximum following the mean path of the stream with a moderately rapid decrease to either side, is quite striking, however. The maximum, centered on the stream's mean path, results primarily from the meandering of the stream; a given location in the envelope of stream paths (inset in Fig. 2) appears to move in and out of the stream as meanders pass, resulting in dramatic fluctuations of the current. The dynamic variability outside the envelope of stream paths is, however, still substantially larger than it is in the center of the gyre. The *e*-folding distance for the dynamic variability on the seaward side of the stream is approximately 300 km; i.e., the EKE has fallen to $1/e$ of its maximum value approximately 300 km to the south of the stream's mean path. The fact that the energy level is still relatively large, say 200 km from the mean path, is due in part to the passage of rings, which contribute to the variability much as meanders in the Gulf Stream do. Studies of float trajectories in which estimates of the EKE were made, excluding floats in cold core rings, still show energy levels substantially above those in the center of the gyre. Given that the EKE level decreases away from the Gulf Stream, it appears that the stream itself is the source of these elevated levels. This is thought to be caused either directly by energy radiated from the stream, indirectly by local instabilities of the recirculation gyre (which exists only because of the stream), or by rings (ejected from the stream as described above) transferring and radiating energy as they migrate to the west. Evidence exists for all three of these mechanisms, but the relative contribution of each has not yet been clearly established.

The large EKE values associated with the stream are important for two reasons: (1) eddy mixing is roughly proportional to eddy kinetic energy, hence large values of eddy kinetic energy translate to large values of mixing, of importance physically, biologically and chemically; and (2) eddies are thought to contribute to the general circulation of the gyre both by rectification, through which they enhance the mean flow, and by turbulent dissipation, through which they remove energy input to the gyre by wind and heat. [See OCEAN EDDIES AND GLOBAL CLIMATE.]

On the shoreward side of the stream, the falloff in EKE is much more rapid than it is on the seaward side because of the lateral constraint of the continental shelf. The more rapid falloff in EKE, a measure of friction between the current and the surrounding fluid, may be viewed as replenishing the potential vorticity removed from the system by the negative wind stress curl forcing the gyre.

C. Heat Transport

To a very close approximation, the earth is in thermodynamic equilibrium, absorbing solar radiation primarily in the visible portion of the electromagnetic spectrum; converting the absorbed energy into thermal energy in the atmosphere, land, cryosphere, and ocean; and emitting radiation, primarily in the thermal infrared portion of the spectrum. (It is the emitted thermal infrared radiation as measured by satellite-borne sensors that is used to obtain the SST fields discussed earlier). Absorption of solar radiation in the equatorial region is substantially larger (per unit area) than in the polar regions. Were the absorbed energy not redistributed horizontally,

local thermal balances (absorption and reradiation) would dominate. This would give rise to colder temperatures in the polar regions than we observe, warmer temperatures in the equatorial regions, and temperatures significantly more variable in time at midlatitudes. It is the redistribution of thermal energy in both space and time that tends to modulate the extremes that would otherwise be seen. The ocean plays a critical role in both the spatial and temporal modulation of the thermal energy. With regard to the storage and release of thermal energy on seasonal scales *(temporal modulation)*, the ocean dominates the other components (atmosphere, land, and cryosphere). Because the total heat content of the atmosphere is roughly equal to that of the upper 3 m of the ocean, the rate of heat storage of the atmosphere is severely constrained, being approximately an order of magnitude less than that of the oceans (in the northern hemisphere). There is very little variability in the temperature of the cryosphere, hence little variability in the stored thermal energy. Finally, the ocean covers three times the area covered by land and the apparent heat capacity of the ocean resulting from wind-induced vertical mixing at the surface is substantially larger than that of land, hence land also contributes relatively little to the storage of thermal energy. [See SOLAR RADIATION.]

Coupled with the relatively greater overall heat capacity of the ocean is the large redistribution of mass by currents, carrying with it the thermal energy associated with this mass *(spatial modulation)*. The atmosphere also contributes significantly to the spatial redistribution of thermal energy, the relative fraction being a strong function of latitude: at low latitudes the heat transport is dominated by the oceans, while at high latitudes it is dominated by the atmosphere. This large latitudinal variability in the fraction of the net poleward heat flux resulting from the oceanic contribution is a consequence of the importance of western boundary currents in the overall equation. As a result of the large vertical gradients both in the horizontal velocity and in the temperature of western boundary currents, they carry a larger volume of thermal energy to the north than is returned to the south by deep currents, the eastern boundary current, and the ocean interior combined.

In the northern hemisphere, heat transport in the North Atlantic is somewhat more important than that in the North Pacific despite the fact that the Atlantic is smaller. This is a result of the fact that the northern North Atlantic extends farther to the north, with a substantially larger area of water north of 50°N. As discussed earlier, it is in the Labrador and Norwegian seas that surface water cools, sinks, and flows to the south, crossing the entire North and South Atlantic. (There is no comparable formation of deep water in the North Pacific.) This southerly flow gives rise to a northward flow of equal volume and, more important from the heat transport perspective, of relatively warmer water. This has three important consequences: (1) warmer water is moved farther to the north by the North Atlantic Current, giving rise

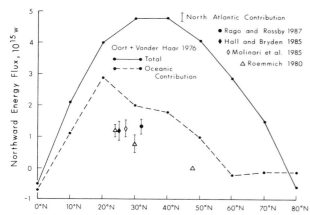

Fig. 11 Poleward heat transport in the northern hemisphere. [Data from A. H. Oort and T. H. Vonder Haar (1976). *J. Phys. Oceanogr.* **6**, 781–800; T. A. Rago and H. T. Rossby (1987). *J. Phys. Oceanogr.* **17**, 854–871; M. M. Hall and H. L. Bryden (1985). *Geophys. Res. Lett.* **12**, 203–206; R. L. Molinari, W. D. Wilson, and K. Leaman (1985). *Science* **227**, 295–297; D. Roemmich (1980). *J. Phys. Oceanogr.* **10**, 1972–1983.]

to a larger heat flux from the ocean to the atmosphere at high latitudes than in the North Pacific; (2) the poleward heat transport in the Atlantic is somewhat larger at midlatitudes than in the Pacific (e.g., estimates, using sections at approximately 24°N, indicate a poleward flux in the Atlantic of $1.22 \pm 0.32 \times 10^{15}$ W compared with $0.85 \pm 0.25 \times 10^{15}$ W in the Pacific); and (3) there is a net northward flux of heat across the equator in the Atlantic that is not thought to exist in the Pacific.

Figure 11 shows estimates of the total northward heat transport in the northern hemisphere, the oceanic contribution to this total, and the contribution of the Atlantic Ocean. The solid and dashed curves for the total and oceanic contributions, respectively, were determined by A. H. Oort and T. H. Vonder Haar using satellite observations of infrared and reflected solar radiation, radiosonde data, and mechanical bathythermograph (MBT) and XBT data. Although the errors associated with these curves, especially the oceanic fraction, are large, the general conclusions of their work, i.e., the latitudinal partition of heat transport, are thought to be correct. The heat transport estimates of the North Atlantic were obtained from a number of studies. It is clear from Fig. 11 that the North Atlantic plays a major role in the redistribution of thermal energy in the northern hemisphere, especially at low to midlatitudes.

Between 25° and 35°N, the Florida Current transports thermal energy to the north. The DWBC, the Canary Current, and the ocean interior transport thermal energy to the south. The difference, the net poleward transport of thermal energy,

increases only slightly between the Florida Straits and Cape Hatteras from $1.22 \pm 0.32 \times 10^{15}$ W at 24.5°N to $1.34 \pm 0.19 \times 10^{15}$ W at 32°N (Fig. 11), this despite a doubling in the volume transport of the Florida Current in the same distance. This is because (1) there is a concomitant increase in the southward volume transport of the remainder of the ocean (recall that the net volume flux across a line of constant latitude is nearly zero because the North Atlantic is, for all practical purposes, closed to the north) and (2) the increase in transport is dominated by the barotropic component.

As indicated previously, at Cape Hatteras the northward-flowing Florida Current turns into the east-northeastward-flowing Gulf Stream. Between Cape Hatteras and the New England seamounts, the stream's volume transport increases, although there is a loss of warm surface water to both sides of the stream via warm outbreaks, shingles, and overwashes. Downstream of the seamounts, the Gulf Stream loses mass primarily through the detachment of Gulf Stream rings. All of these processes tend to remove warm surface water from the stream. Some of this water is reintroduced into the stream upstream of the seamounts, but not after a significant loss of thermal energy to the atmosphere. Furthermore, the water in the stream itself spends a significant fraction of time at a nearly constant latitude, exposed on the average to dry cool air blowing off the North American continent. Because of the combination of these factors, the region downstream of Cape Hatteras is one of the oceanic regions with the greatest flux of thermal energy into the atmosphere. This loss of thermal energy to the atmosphere continues well beyond the Grand Banks as the North Atlantic Current carries the still relatively warm surface water farther north and east. This loss of thermal energy to the atmosphere, primarily through evaporation of the surface waters, coupled with the substantial decrease in northward-flowing water in general, gives rise to the precipitous drop in the northward transport of heat by the ocean and the relative increase in the fraction carried by the atmosphere.

The preceding discussion has emphasized the importance of oceanic heat transport in the global climatic balance and, in particular, the role played by the Gulf Stream System. It would be inappropriate to close this section without briefly mentioning the importance of this heat transport to the circulation of the North Atlantic itself, hence ultimately to the Gulf Stream. Admittedly, the dominant forcing of the subtropical gyre in the North Atlantic is wind induced. The deep water flow, formation of deep water properties, maintenance of the main thermocline, and net northward transport of upper ocean waters are, however, all directly affected by the surface heat flux. These properties are in turn related to the upper ocean circulation. For example, it has been argued that the separation of the Gulf Stream from the continental shelf in the vicinity of Cape Hatteras is controlled in part by the DWBC. Whether or not this is true, there is little doubt that the thermohaline (driven by thermal and salinity differences rather than by the wind) circulation driven by the equator–polar imbalance in surface irradiance is coupled with the upper ocean circulation. It is equally evident that the atmospheric circulation is coupled with the oceanic circulation, both through the wind-induced stress on the sea surface and through the air–sea heat flux.

Bibliography

Auer, S. J. (1987). A 5 year climatological survey of the Gulf Stream System and its associated rings. *J. Geophys. Res.* **92**(C11), 11,709–11,726.

Bower, A. S., Rossby, H. T., and Lillibridge, J. L. (1985). The Gulf Stream—barrier or blender? *J. Phys. Oceanogr.* **15**, 24–32.

Halkin, D., and Rossby, T. (1985). The structure and transport of the Gulf Stream at 73°W. *J. Phys. Oceanogr.* **15**, 1439–1452.

Richardson, P. L. (1983). Eddy kinetic energy in the North Atlantic from surface drifters. *J. Geophys. Res.* **88**(C7), 4355–4367.

Tracey, K. L., and Watts, D. R. (1986). On Gulf Stream meander characteristics near Cape Hatteras. *J. Geophys. Res.* **91**(C6), 7587–7602.

Glossary

Baroclinic   Function of depth; generally refers to the velocity field.

Barotropic   Independent of depth; generally refers to the velocity field.

Continental shelf   Region from the coast to the abrupt change in seafloor depth, which generally occurs at a water depth of about 100 m.

Continental slope   Region over which the seafloor depth changes rapidly from approximately 100 to more than 4000 m.

Coriolis force   Fictitious force resulting from the fact that we view the ocean from a rotating coordinate system.

Cyclonic   Spinning in the direction in which the earth rotates when viewed from the pole—counterclockwise in the northern hemisphere.

Geostrophic relation   Equation expressing the balance between the Coriolis force and the horizontal pressure gradient.

Isopycnal   Line or surface on which density is constant.

Isotach   Line or surface on which water velocities are constant.

Isotherm   Line or surface on which temperature is constant.

Section   Number of stations made along a line.

Station   Set of vertical measurements of oceanographic properties, such as temperature or salinity, made at one location, generally by ship.

Subpolar gyre   Circulation system poleward of the Subtropical Gyre.

Subtropical gyre   Wind-driven basinwide circulation system occurring between approximately 20° and 40° latitude.

Thermocline   Part of the water column in which the temperature changes rapidly in the vertical; can be viewed as dividing the ocean into a two-layer system.

Hail Climatology

Wayne M. Wendland

Illinois State Water Survey and University of Illinois – Urbana

Hail consists of soft or hard roundish balls of ice generated within convective clouds. The stones consist of concentric spherical shells of ice formed about a nucleus. Generally, diameters are on the order of a centimeter or less but can reach 10 cm or more. Hail severely damages crops and some structures, particularly if blown by strong, gusty winds.

I. Introduction

Hailstones are concentric spheres of ice frozen about some nucleus, formed within a cumulonimbus cloud. The stones tend to be unevenly round but may be noticeably oblong with pronounced ice knobs. Hailstones are composed of many roughly concentric spheres of ice which increase the size of the stone by growth first on a nucleus, then on previously formed ice. Young, building cumulus and cumulonimbus develop upward due to atmospheric instability (i.e., when the ambient vertical decrease in density is greater than that which results if a parcel of air is lifted and exposed to reduced pressure). Instability is enhanced when a lower layer of warm, moist air is overlain by cool, drier air.

The primary causes of severe thunderstorms and hail are one or more of the following: (1) strong, fast-moving fronts, (2) intense surface heating, and (3) orographic lifting, all of which encourage overturning. Within the United States, the first two tend to be the primary causes in the upper Midwest, whereas the third dominates along the eastern slopes of the Rockies.

Hail events, i.e., hail from one spatially and temporally continuous storm, tend to be short-lived (a few tens of minutes) and usually affect only a few hundred square kilometers. Forecasting that hail will occur at a given location several hours in the future is presently not possible. Short-term forecasts can be made if hail is already occurring and if a likely trajectory is known. Hail forecasts several hours in advance of the event (termed "watches" in the United States) are now limited to statements that areas of hail are likely within an area on the order of several hundred thousand square kilometers within a 6-h window.

In the United States, at present, the National Weather Service (NWS) issues severe weather "watches" when hail is expected within the next few hours. "Warnings" are issued only after the phenomenon has already occurred and a likely trajectory is known. They are usually valid for the next hour or so for an area of 2000 to 4000 km². Currently, watches precede hail events by several hours. Presumably, the next generation of Doppler radars (NEXRAD), to be installed by the mid-1990s, will provide somewhat longer lead times and more accurate spatial forecasts. [*See* DOPPLER RADAR.]

II. Means of Sensing Hail

NWS uses no unique instrument to record the occurrence or size of hail. Observers at NWS first-order and cooperative stations visually observe and record the times of hail occurrence and generally include comments on size. First-order sites observe 24 h each day; cooperative stations take an observation only once each 24 h, although they generally record extreme events which occur at times other than the standard time of observation, if noted.

Defining areas of hailfall from NWS observations alone presents a crude record at best because (1) the circa 40-km spacing between observing sites is larger than the typical size of an area of hailfall and (2) hail occurrence cannot be assumed to be included in all cooperative observations because the observer is required to only record temperature and precipitation at one time each day. Hailfall at other times, particularly at night, may well escape detection.

Special hail networks have been established from time to time with special instrumentation for specific, usually short-term studies. *Hailpads* have been so used with rather favorable results. The pad consists of a square slab of Styrofoam about 30 cm on a side and 3 to 5 cm thick, covered with aluminum foil. Indentations record the size distribution and concentration of the hailstones during the time interval of exposure. Energy expended by the stones can easily be calculated.

Cameras have been used to function automatically when hailfall is sensed. Stones which fall between the lens and a black background are recorded, to be counted and measured after film development.

In addition, cadres of volunteer observers have been trained and used for special studies. The observers not only record size distribution and relative concentration (percentage of ground cover, depth) but also often gather and preserve specimens for later analysis.

III. Quality of Hail Observations and Analyses

Because the average area of a hailstorm is usually only a few tens of square kilometers and the average distance between most NWS cooperative weather stations is on the order of 40 km, many hail events are never recorded by the National Weather Service network. Certainly, the areal extent, trajectory, and hence damage estimates of these storms cannot be determined from standard weather observations alone.

Detailed meso–network scale characteristics of hail events have been studied since the 1950s, primarily in Alberta, Colorado, and Illinois, by incorporating (1) routine weather service observations, (2) special dense hailpad networks (spacing of 2 km or so), (3) networks of volunteer observers (similar spacing), and (4) crop losses determined from hail insurance data. These have done much to increase knowledge of temporal and spatial characteristic of hail events but are too few and discontinuous in space and time to add any information to our understanding of hail climatology.

IV. Morphology of Hail Events

Hail is generated within cumulonimbus clouds (convective cells) averaging 1 or 2 km in diameter and moving at forward speeds of 35–60 km/h. A hail event from such a thunderstorm typically has a tenure measured in minutes. In addition, the hail does not necessarily emanate from the entire cloud base, but falls from only portions of the base and may start and stop over the lifetime of the cumulonimbus cloud. [*See* CLOUD DYNAMICS.]

The resulting pattern of hail on the surface is therefore fragmented on both temporal and spatial scales. The smallest area of a continuous event is a hailstreak (the hailfall area affected by one cumulonimbus), several of which become hailswaths, relatively concentrated areas of hailfall with some spatial breaks. Median hailstreak areas in Illinois and North Dakota are about 20 km² (range 2.5–2040 km²), and mean speed is about 30 km/h. Average duration of a hail event is about 10 min. Eighty percent of all hailstreaks occupy areas of less than 25 km². The streaks tend to be separated from each other by about 25 km. The longest hailstreak measured was about 80 km in length and 15–30 km in width.

At any point on the surface within a hailstreak, the average duration is 3–6 min, with some 43% of all hail observations occurring between 1400 and 1800 CST. On average, each square meter within a hailstreak receives about 240 stones during an event.

The effects of a particularly devastating hailstorm of June 13, 1984 near Denver (golfball- to softball-sized hail) was monitored by a mesonetwork of surface observing sites, the NWS cooperative stations in the area, dual Doppler radars, wind profilers, radiosondes, lightning detectors, and photography. The duration of the storm exceeded 1½ h with a general accumulation in depth of about 0.25 m and drifts to 1 m. This was the most costly and destructive hailstorm to date, with an estimated cost of $350 million (1984$).

Extreme hailstones of 290 g were reported near Cedoux, Saskatchewan on August 27, 1973, and another of 249 g near Wetaskiwin, Alberta on July 6, 1975.

The average energy imparted to the surface being hit by hail varies greatly over small and large scales. Even on the average, however, studies in Illinois and Colorado claim mean impact energy is about 0.22 J per 1000 cm².

V. Causes of and Development of Hail

A. Synoptic Situations Conducive to Hail Formation

For thunderstorms and hail to occur, the instability of the atmosphere must first exceed some threshold. Instability is enhanced when dry and potentially colder air overlies relatively warm, humid air. If this vertical profile is disturbed by

(1) the passage of a cold front, (2) surface convergence, or (3) strong surface heating, the resulting overturning promotes vertical growth of cumulonimbus and associated (strong) vertical currents. Since the diameter of a single cumulonimbus cloud is on the order of only a few kilometers and the cloud exists for about an hour, the area affected by hail from one cloud is necessarily limited to the same scales. However, if thunderstorms group along a squall line and new growing cumulonimbus clouds replace those spent, a hail event can affect substantially larger areas and continue (with hail-free areas and varying intensities) for hours. [*See* ATMOSPHERIC FRONTS; METEOROLOGY, SYNOPTIC.]

Sixty-four percent of hail events in the upper Midwest from 1900 to 1959 were associated with cold fronts, 18% with stationary fronts, 8% were air mass storms, 6% were associated with low pressure centers, and 4% were associated with warm fronts. Those along the Front Range in Colorado are primarily orographically induced. [*See* MOUNTAINS, EFFECT ON AIRFLOW AND PRECIPITATION.]

B. Development of Cumulonimbus and Hail within Cumulonimbus

Cumulonimbus clouds typically exhibit lifetimes of an hour or so and often reach altitudes of 10 km (occasionally 20 km) during that time interval. In mid latitudes in summer, the portion of the cloud beneath about 5 km (portion of the atmosphere where temperature lapses to 0°C, i.e., the freezing level) is virtually all liquid, whereas the part above the freezing level is composed of ice crystals and supercooled water (i.e., liquid water, but with a temperature below 0°C). Vertical motions within the cloud propel particles (natural or human-made minerals) upward, where they serve as a nucleus about which supercooled water freezes into an embryo of one of two types: graupel or a frozen drop. Over Colorado, 80–85% of all hail embryos tend to be primarily graupel, suggesting that the ice (Bergeron) process is the primary mechanism for precipitation formation in that area. However, the ratio of the two types varies greatly from case to case and location to location.

As the embryo moves up and down within the cloud (direction may shift several times during the lifetime of a hailstone) it may encounter more supercooled water, the latter freezes to the former (due to the higher vapor pressure of the supercooled water), and the size of the stone grows. Each additional freezing tends to coat the stone with a new shell of ice, much like the morphology of an onion.

Hailstones move up within the cloud until the vertical currents decrease, or the stone is blown out the top of the cloud, or the stone becomes sufficiently large and falls to earth by gravity. Stone diameters are most commonly less than about 2 cm, but stones of several times that have been observed.

VI. Spatial and Temporal Distributions of Hail

A cursory inspection of climatological literature yielded only one generalized map purporting to show the worldwide mean distribution of hail days (Fig. 1). Areas with the greatest frequency of hail days are generally inland, mid-latitude sites, including:

- United States (from the Rockies to the Appalachians, and the Great Basin)
- Central Mexico
- Alberta
- Mid-Central America
- Windward side of the Andes and a few isolated locations in Brazil and Argentina
- Eastern Europe
- Southwestern Soviet Union
- Coastal Soviet Union from 40° to 50°N
- Mongolia
- China
- India
- Burma
- Central and eastern equatorial Africa
- South Africa
- Malagasy

Oceanic frequencies are doubtless under-represented due to incomplete observations.

A. Distribution within the United States

The spatial diversity of hail day frequency for the central United States is shown in Fig. 2. This map was prepared from observations made at 1285 NWS first-order and cooperative stations from 1910 to 1960. Frequencies and gradients are greatest in mountains (the latter from 60 to 180 days per 20 years over a distance of about 80 km) and least in the southeasternmost seven states of the study area (data density was limited to only three sites per state in these latter states). Monthly hail frequency studies for the 17 states have been prepared. From November through February the pattern is relatively smooth, with frequencies generally of 2 to 5 hail days per 20 years. The area of maximum frequency is found in Oklahoma and Missouri. Beginning in March, the frequencies throughout the 17-state region increase dramatically and the pattern becomes much more discontinuous, with the scale size of anomalies being on the order of only about 650 km². From March through May the core of high-frequency hail days moves northward to a poorly defined area, including Oklahoma, Missouri, Texas, Kansas, Nebraska, Iowa, and Minnesota. In June the area with maximum frequency shifts to the High Plains and lee Rockies, including Montana,

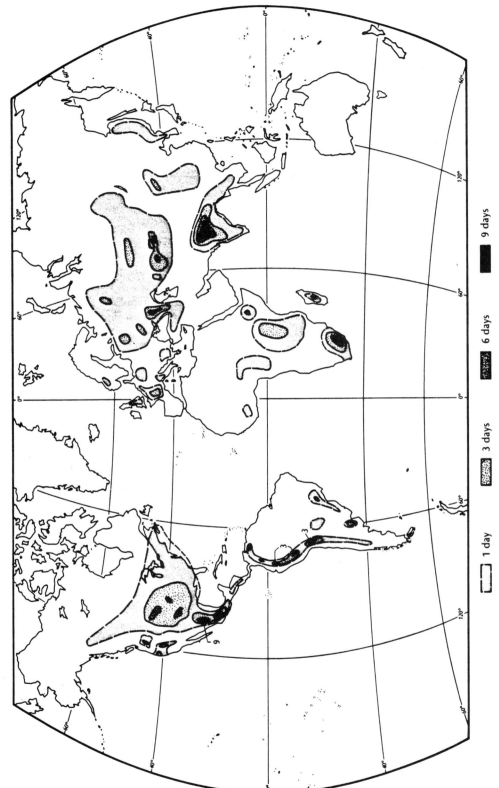

Fig. 1 Worldwide hail areas presented as mean annual number of days with hail. [From Griffiths and Driscoll (1982). Reproduced with permission of Chas. E. Merrill Publ. Co.]

☐ 1 day   ▦ 3 days   ▨ 6 days   ■ 9 days

Fig. 2 Hail day frequency in the United States presented in number of hail days in 20 years. [From Stout and Changnon (1968). Reproduced with permission of the Illinois State Water Survey.]

Wyoming, North Dakota, South Dakota, Nebraska, Kansas, and Colorado, and the maximum core frequency is about 25 hail days per 20 years. From July through October, the maximum band becomes spatially discontinuous and collapses to the south, with average frequencies being 3 to 4 days per 20 years but with a strong positive anomaly of 12 days in New Mexico.

The temporal patterns were not found to be constant in these studies. From 1901 to 1960, the southernmost states tended to exhibit maximum frequency in the first two or last two decades, whereas the more northern states exhibited their monthly maxima in the 1920s and 1930s.

Pentad frequencies of hail from 1901 through 1970 have been examined from Montana, Texas, Iowa, Minnesota, Illinois, Georgia, South Carolina, and North Carolina. Of these, only those from Montana, Texas, Iowa, and Kansas indicated a trend, with maximum hail activity in the first or second half of the 1950s.

Because hail events are strongly correlated with the position of cold fronts and because of the northward movement of the mean frontal position in spring, hail events, too, exhibit a northward movement in spring, resulting in a maximum frequency in southern Illinois in May and moving to northern Illinois in midsummer.

Since thunderstorms are much more frequent during the warmer half-year and temperatures over the United States (and the northern hemisphere and the world) generally increased from the late 1800s to about 1940 or 1950, one may wonder whether the northward shift in maximum frequency and the increase in frequency to about 1950 were the result of the general warming.

On a smaller scale, there is evidence that large industrialized urban areas provide a temperature, atmospheric stability, and particulate environment which enhances hail frequency over and downstream of major cities for several tens of kilometers. Increased hail frequencies were noted in and particularly downstream of Kansas City, Omaha, Chicago, St. Louis, Cleveland, Houston, and Washington, D.C.

In the late 1950s, the Crop–Hail Insurance Actuarial Association (CHIAA) of Chicago sponsored research at the Illinois State Water Survey to better determine the climatology of hail in the United States. This work continued for over 20 years, resulting in a compendium of hail data. These data provide the basis for virtually all spatial and temporal conclusions in this article.

Hail is most frequently observed on the High Plains east of the Rockies and within the upper Midwest. Hail events of the High Plains are typically 5–15 times more intense than hail events of the Midwest. Such intense hailstorms tend to increase in frequency from spring to September. Damage to corn, however, is greatest in July because of the vulnerability of the plant at that time of growth.

B. Distribution in Areas outside the United States

Quantitative hail climatologies of good quality from countries other than the United States are few and certainly not areally complete. Only a few fragmented statements are permitted. Hail has been observed in northern Zinjiang, China to be larger than that in the south. Mean annual frequency is 21 days, primarily from March through December, with 80% falling between May and August. The hail "year" in southern China continues from February through May. Mideastern and far northern China then becomes the major frequency area from April through July. The northern Yellow River basin observes its maximum frequency from May to September, whereas far northeastern China and northwestern Sichaun Province receive their maximum frequency in May and June.

Maximum hail frequency is from January through June in India, with about 80% of all hail days occurring during that interval.

VII. Crop Damage from Hail

Hail damage is not precisely documented, again because of the relatively small area of effect and because of the relatively less dense network of weather observing sites. What is known has been gleaned from the broad-scale storms in the United States that were observed by the National Weather Service network or from intense, meso-scale field studies carried out over relatively short periods of time. Such small-scale studies have been completed in Alberta, Colorado, and Illinois. The number of extremely damaging hailstorm days is directly proportional to the total number of summer hailstorm days, a relationship which permits extrapolation of relative damage into areas with frequency data only.

Figure 3 shows the potential for damage from hail in the United States based on hail size and frequency and the susceptibility of crops. The greatest potential exists in Colorado, Kansas, Nebraska, Oklahoma, and Texas.

Within the United States the Great Plains suffers the greatest hail damage, primarily focused on the corn belt. North Dakota experiences the greatest crop damage of any state. In 1972 the average annual crop loss from hail was $284 million, about 1% of national crop production.

VIII. Suppression of Hail by Active Weather Modification Techniques

Neither structures, possessions, nor commercial crops can be *protected* from hail damage. About 2% of the U.S. crop

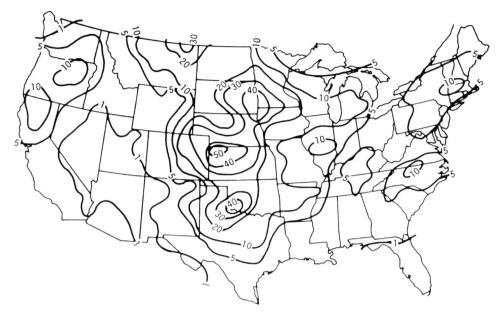

Fig. 3 Potential for hail damage to property in the United States. Units are one-thousandth of 1% of residential property value per year. [From Changnon (1977b). Copyright © 1977 American Meteorological Society.]

production is lost to hail damage annually. Even if hail forecasts were vastly improved (i.e., including better timing, with appropriate lead time and intensity of the event), little improvement in protection would be noted. Currently, one can only *mitigate* against losses with hail insurance, distributing local losses over a greater number of policies from a larger area. It is estimated that about 15% of U.S. crops are so insured. Since hail damage can substantially decrease a farmer's income for the entire year, there is considerable use of crop insurance. However, interest in the potential for hail suppression continues, waxing and waning with recent experience.

Attempts at hail suppression thus far have focused on modifying the temperature or other physical characteristic of the cloud of interest by the introduction of silver iodide or, to a lesser degree, dry ice. These agents apparently act as nuclei about which hail may form, enhancing the formation process. The effectiveness of hail suppression is equivocal, in part depending on the modification method, experimental design, field data collected, and interpretation of those data. Research has found that hail suppression is perceived as being more effective by men than by women, and by more highly educated persons. [*See* WEATHER AND CLIMATE MODIFICATION.]

Even disregarding these perception factors, the effectivity of hail suppression is still equivocal. Many seeding studies have been completed with differing results. Some atmospheric scientists strongly tout the capability of seeding to either enhance precipitation or suppress hail, but a general conclusion, representing the consensus of atmospheric scientists, is still equivocal. In spite of the inconclusive nature of statements on hail suppression within the scientific community, operational programs continue.

Whereas precipitation enhancement seeding programs continue in several western states of the United States (funded by state agencies and local governments), few hail suppression programs are ongoing. Rather, they are short-term projects, focused on areas and temporal windows exhibiting a high potential for damaging hail. A hail suppression program continues in west Kansas, near the Colorado border, sponsored by 16 Kansas counties. More than a decade of seeding experience in that area suggests a 30–35% reduction in the occurrence of hail, relative to two control areas. A decrease in hail would seem to be supported by the 11% reduction in insurance premiums in the seeded area over the last 15 years.

Hail suppression apparently enjoys a more favorable status in some other countries. Some 5 million hectares of farmland were reportedly protected by means of hailstorm seeding in the USSR, although there were no quantitative data to support the claim. Another 350,000 ha were seeded by cannon, and 730,000 ha in Moldavia by rocket.

Table I   Reductions in Hail Losses to Crops from Several Studies

Site	Length of study (yr)	Reduction (%)
Argentina[a]	3	34
Texas[b]	4	48
North Dakota[b]	15	31
North Dakota[c]	4	60
South Africa[d]	3	40
South Dakota[e]	4	18–40
Colorado (NHRE)[f]	3	30
Gansue Province, China[g]	?	50
Gansue Province, China[h]	9	+29 (increase)
Greece[i]	5	43–93
Bulgaria[j]	?	36–56
Mancheng, China[k]	?	71
Yugoslavia[l]	20	?
Greece[l]	3	?

[a] Not statistically significant.
[b] Based on insurance premium changes.
[c] Based on crop hail loss reductions.
[d] Ratio of area of loss to value of loss.
[e] Rainfall increased 4–12%.
[f] Reduction of hail mass (rainfall increased 25%).
[g] Reduction of hail damage.
[h] Based on hailpad surveillance.
[i] Preliminary results; no more specifics given.
[j] Based on crop loss changes.
[k] Reduction in area of hail damage.
[l] Experiments conducted; no results given.

Table I presents results of hail suppression projects from several studies and areas. Each of these was an experiment, with both seeding and control areas. Seeding was done only on days when sufficient clouds were present in the operational area and when the atmosphere displayed certain characteristics, e.g., instability and humidity. In some cases clouds were double-blind seeded; i.e., the controller in the seeding aircraft introduced silver iodide in some clouds and a placebo in others, the identity of each agent being unknown to the controller. Reductions are not to be compared or combined per se because the individual results were determined by many different methods (see Table I footnotes).

In part because of the uncertainty of seeding efficacy, weather modification is generally regulated at the state level in the United States, the federal government requiring only that the activity be reported to the National Oceanic and Atmospheric Administration.

IX. Summary and Conclusions

Known characteristics of hail and the formation of hail have been gleaned from (1) vertical atmospheric soundings of temperature, humidity, and pressure in and outside hail areas; (2) aircraft observations within and about hail areas; (3) radar data; (4) data from vertical profilers in areas with hailfall; (5) studies of synoptic situations which produced hail; (6) examination of scores of hailpad data; and (7) hail observations by the National Weather Service, state agencies, and specially trained observers. Although the data are many, much is yet to be learned.

Hail is formed within cumulonimbus (thunderstorm) clouds. Such clouds primarily form in warmer environments, where warm, humid air near the earth's surface is overlain by cooler, drier air. Cumulonimbus clouds are triggered by fast-moving, strong cold fronts, often along squall lines, and by intense surface heating within airmasses.

In most, but not all instances, high probability of hail within an area of a few hundred thousand square kilometers can currently be forecast several hours before the expected event. The U.S. National Weather Service issues a severe weather *watch* in those instances. A *warning* follows for a much smaller area and shorter temporal window after hail has been observed and an expected trajectory can be made.

Hail suppression experiments in which nucleating agents were added to cumulonimbus clouds have been undertaken in countless instances in numerous countries. The model of cloud seeding to stimulate weather modification is generally accepted, though certainly not by all. Experimental results suggest a hail decrease of perhaps some 15–30% within the seeded area. However, since experiments are carried out in the free atmosphere, a nagging question persists as to whether the atmosphere was "ready to begin hail of its own accord." Indeed, there is less agreement about the degree to which the seeding modified the "natural" hailfall.

In spite of uncertainties, several routine hail suppression programs have continued for more than 10 years, funded by individuals and state agencies, obviously in the belief that the cost is well spent. At present, the most general, active method for the mitigation of hail loss remains hail insurance.

Bibliography

Changnon, S. A., Jr. (1960). "Relations in Illinois between Annual Hail Loss Cost Insurance Data and Climatological Hail Data." Crop–Hail Insurance Actuarial Association, Chicago.

Changnon, S. A., Jr. (1960). "25 Most Severe Summer Hailstorms in Illinois." CHIAA Research Report No. 4. Illinois State Water Survey, Champaign.

Changnon, S. A., Jr. (1969). Insurance-Related Hail Research in Illinois during 1968. CHIAA Research Report No. 40. Illinois State Water Survey, Champaign.

Changnon, S. A., Jr. (1977a). On the status of hail suppression. *Bull. Am. Meteorol. Soc.* 58, 20–28.

Changnon, S. A., Jr. (1977b). The scales of hail. *J. Appl. Meteorol.* **16,** 626–648.

Changnon, S. A., Jr. (1978). "The Climatology of Hail in North America," pp. 107–128. Monograph No. 38. American Meteorological Society, Boston.

Changnon, S. A., Jr., and Stout, G. (1967). Crop-hail intensity in central and northwestern United States. *J. Appl. Meteorol.* **6,** 542–548.

Changnon, S. A., Jr., Davis, R. J., Farhar, B. C., Haas, J. E., Ivens, J. L., Jones, M. V., Klein, D. A., Mann, D., Morgan, G. M., Jr., Sonka, S. T., Swanson, E. R., Taylor, C. R., and Van Blokland, J. (1977). "Hail Suppression, Impacts and Issues." Final Report to National Science Foundation, ERP75-09980. Illinois State Water Survey, Champaign.

Farhar, B. C., Changnon, S. A., Jr., Swanson, E. R., Davis, R. J., and Haas, J. E. (1977). "Hail Suppression and Society." Summary of Technology Assessment of Hail Suppression. Illinois State Water Survey, Champaign.

Griffiths, J. F., and Driscoll, D. M. (1982). "Survey of Climatology." Chas. E. Merrill, Columbus, Ohio.

Huff, F. A., and Changnon, S. A., Jr. (1959). "Hail Climatology of Illinois." Report of Investigation No. 38. Illinois State Water Survey, Champaign.

Stout, G. E., and Changnon, S. A., Jr. (1968). "Climatography of Hail in the Central United States," CHIAA Research Report No. 38. Illinois State Water Survey, Champaign.

Glossary

Airmass Large mass of air (thousands of kilometers in diameter and a few kilometers thick) with similar temperature and humidity characteristics; forms over a large surface with common temperature, albedo, heat content, etc.

Bergeron process Process by which ice crystals grow at the expense of supercooled water droplets, coexisting in a cloud with temperatures below 0°C. Because vapor pressure over water is greater than over ice at a common temperature, if supercooled water is sufficiently available, ice grows at the expense of water.

Convergence Movement of air on a plane toward a common point or common line (resulting in rising currents); the result of friction unbalancing the forces of geostrophic flow.

Cumulonimbus Cloud generally 10 or more kilometers tall, consisting of ice and water, appearing like towering, billowing cotton, generally with a white top (anvil) extending downstream from the cloud. Commonly called thunderstorm or thunderhead. Cloud turret can be white, gray, or very dark. Contains rain, supercooled water, graupel, ice crystals, and hail and is turbulent.

Graupel Snow pellets, i.e., white, opaque, approximately round ice particles having a snowlike structure and 2–5 mm in diameter.

Hail Roundish ice mass consisting of concentric layers of ice formed on a mineral nucleus; hard, may be lumpy; diameters of 5 cm or more, though most of the hail measured at the earth's surface is less than 0.5 cm in diameter.

Instability Property of a system such that certain disturbances or perturbations introduced into the steady state will increase in magnitude; tendency of an atmosphere to overturn, enhanced by some disturbance or perturbation; specifically when a parcel of air, being lifted, encounters ambient air cooler than itself and thereafter rises of its own accord.

Heat Flow through the Earth

Kevin P. Furlong
Pennsylvania State University

The solid earth is a heat engine that drives plate tectonics, produces mantle convection, and controls the rates and locations of surface deformation. Heat is transferred through the earth by several mechanisms—conduction, convection (advection), and radiation. Near the surface conduction dominates as the heat transport mechanism, while within the mantle convection and radiation play important roles in heat transfer. By studying the thermal regime in the solid earth we can determine the energy budget of the earth, the rates at which crust and lithosphere are produced and destroyed, the factors controlling mantle convection, and the interaction between the dynamic processes of the earth's interior and processes at the earth's surface.

I. Overview

A. Global Heat Budget

The integrated heat budget of the solid earth (surface heat flux derived from within the earth) is approximately 42 TW (terawatts) (42×10^{12} W). Although this flux is substantially less than the solar-driven heat flux from the surface through the atmosphere (1.9×10^5 TW), it is the heat from the earth's interior that drives mantle convection, plate tectonics, and crustal formation processes. Of the total energy budget of 42 TW, the largest fraction is transferred into the oceans. The oceanic lithosphere (the quasi-rigid outer shell of the earth incorporating the crust and uppermost 50–100 km of

the mantle) represents approximately 60% of the earth's surface but accounts for 72% of the heat flux. The largest fraction of the oceanic heat budget is associated with the formation of new crust and lithosphere at mid-ocean ridges. The nature of heat flow in the oceans is linked to the age of the oceanic lithosphere, with a substantial decay in the heat flow during the first 30 Ma (million years) after formation at the mid-ocean ridge. [*See* OCEANIC CRUST.]

The heat flow within the continental lithosphere is less closely related to the age of the lithosphere but rather is controlled by the nature of tectonic activity in the region and the amount of radiogenic material (primarily within the continental crust), which produce heat. The mean heat flow in the continental lithosphere (including the flooded continental shelves) is approximately 57 mW/m², substantially less than the mean of approximately 99 mW/m² for the oceanic lithosphere. The complex, heterogeneous structure of the continental crust leads to substantial variability in heat flow in the continents, with differences of tens of mW/m² over distances of 10–100 km. Additionally, because much of the surface heat flow in continental regions is produced by the decay of radiogenic material within the crust, the magnitude of heat flow decreases with depth in the continental crust. It is estimated that as much as 60% of the observed surface heat flow results from heat produced within the crust, with only 40% transferred from the earth's mantle. [*See* CONTINENTAL CRUST.]

B. Internal versus External Heat Flux

As mentioned above, the internal heat flux of the earth pales in comparison with the daily flux of heat into the earth system from the sun. The solar flux however is essentially matched by the heat radiation in space, and thus the solar heating (ignoring the role of the atmosphere in trapping heat) would do little more than maintain the earth's surface at a mean annual temperature of approximately 13°C. The solar flux does play a dominant role in climate, atmospheric temperatures, and the oceanic thermal regime; but deeper than a few meters to tens of meters in the solid earth, the internal heat flux dominates.

The internal heat flux is derived from several sources: radiogenic heat production, secular cooling of the earth, crystallization processes within the core, and tidal energy dissipation. At present, the radiogenic component dominates. Within the continental crust, a substantial fraction of the heat flux is locally derived from the decay of heat-producing elements (notably K, Th, and U), but even though the concentration of these primary heat-producing elements is substantially reduced in the mantle and core of the earth, they still dominate in the heat budget.

Another important attribute of the internal heat flux is the various modes of heat transfer that occur. Within the lithosphere heat transfer is primarily solid-state conductive — that is, through processes acting at the crystal lattice level. Within the main portion of the earth's mantle and within the outer core, heat transfer is primarily advective — heat transfer with mass circulation. This process is an extremely efficient mechanism for heat transfer resulting in substantial heat flow with very small temperature differences.

C. Heat Sources and Sinks

The primary heat sources are the decaying radiogenic elements within the earth. Potassium (K), thorium (Th), and uranium (U) provide the bulk of the heat within the earth. For rocks typical of the continental crust, the volumetric heat production is in the range of $0.5 - 3.0 \ \mu W/m^3$. The oceanic crust (and perhaps the deeper levels of the continental crust) is more mafic in composition with volumetric heat production typically in the range of $0.1 - 0.7 \ \mu W/m^3$. Mantle compositions are substantially reduced in these heat-producing elements, with values in the range of $0.01 - 0.05 \mu W/m^3$.

The interior of the earth is constantly undergoing change. Mineralogical variations with temperature and pressure within the mobile mantle involve reactions that may be endothermic or exothermic. Although these reactions and their thermodynamic character are important to our understanding of the convective processes within the mantle, their role as important heat sources or sinks is minor. Another potential heat source/sink pair is the melting and subsequent crystallization of magmas. Producing a magma is a substantial heat sink, as approximately 3.3×10^5 J/kg (8.25×10^8 J/m^3) is required to melt typical earth materials. This corresponds to 10 million years of the radiogenic heat output of a typical crustal rock of comparable volume. However, the process of melting is not a net heat sink because the same quantity of heat is released when the magma crystallizes. Thus absorption of heat into the melting process and its subsequent release on crystallization should be seen not as a heat sink but rather as a process by which heat may be efficiently carried by a volume of rock. That is, the transport of magma within the earth is a very efficient heat transfer mechanism, since the material transported carries this large latent-heat component.

II. Global Heat Flux

A. Mode of Heat Transfer in the Earth's Interior

Beneath the lithosphere, heat conduction no longer serves as the primary mechanism of heat transfer in the mantle. Rather, as a result of convection within the mantle, heat is efficiently transferred by advection or as a result of the mass transfer that occurs during convective flow. As a result, the temperature gradient within the convecting mantle is substantially different from that in the conduction-dominated lithosphere. In particular, since heat is very efficiently transferred by advection the temperature gradient is significantly less than in the lithosphere. Typical temperature gradients in the convecting mantle are approximately $0.5°C/km$ as compared to gradients of $5-20°C/km$ typical for the lithosphere. These low thermal gradients allow temperatures in the interior of the mantle to remain reasonably low, keeping mantle rocks at temperatures below melting conditions for most of the mantle. The convection that occurs in the mantle involves the flow of solid, viscous material. [See MANTLE.]

B. Heat Sources in the Earth's Interior

The actual distribution of heat sources within the mantle and core of the earth is not completely understood. Based on samples of mantle material brought to the earth's surface by diatremes or in associated with some volcanic activity, upper mantle rocks are believed to have radiogenic heat production at least one tenth of lower crustal rocks, at values of approximately $0.01-0.05 \ \mu W/m^3$. Although this value is relatively small, the large volume of the mantle, if at this level of heat production, would produce a substantial amount of heat — greater than that thought to be generated in the mantle. Thus the average radiogenic heat production in the mantle may be less than that observed in upper mantle rocks that have been exposed on the surface.

Heat is also being generated in the earth's core, both from radioactive decay and heat released during slow crystallization of the liquid outer core. The heat generated in the core allows the heat flow at the core–mantle boundary (CMB) to be slightly less than the heat flow at the earth's surface. However, since heat flow is defined by the heat transfer per unit surface area, and the surface area of the core is only about 30% of the area of the earth's surface, the heat flux across CMB represents approximately 20% of the surface flux. Since heat flux across the CMB is of the same order of magnitude as the surface flux, temperature gradients across the thermal boundary layer at the CMB are similar to temperature gradients within the lithosphere. [See CORE; CORE–MANTLE SYSTEM.]

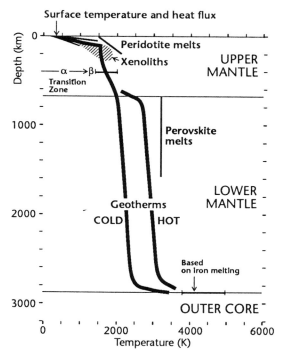

Fig. 1 Estimates of average temperature structure (geotherm) for the crust and mantle of the earth. These estimates are constrained by melting relations for mantle and core compositions, consequences of mantle convection, and surface heat flux. [Modified from Jeanloz and Morris (1986).]

C. Global Geotherm

Taking into account the various modes of heat transport and levels of heat production within the earth's interior, an estimate can be made of the temperature structure, or the geotherm, for the earth. Such a geotherm for the mantle and outer core is shown in Fig. 1, constrained by a variety of data. In addition to estimates of heat production in the interior, seismological data can be used to constrain the locations within the earth where partial melting may occur. This information can be combined with experimental and theoretical estimates of the melting behavior of earth materials as further constraints on the geotherm.

In the upper 100–200 km the temperature gradient is quite high (averaging 10–15°C km), while through the bulk of the mantle and the outer core the gradient is only approximately 0.5°C/km. This reflects the consequences of conductive versus advective heat transfer. In the conductive thermal boundary layer, which is the lithosphere, the high gradients are necessary to accommodate the magnitude of the heat

flux. Within the bulk of the mantle and the outer core, the vigor of the convecting regime can vary to accommodate the heat flux, and thus the temperature gradients are similar to adiabatic gradients. A second boundary layer exists at the CMB. This compositional and thermal boundary layer separates the convecting outer core from the convecting mantle. The depth extent of this thermochemical boundary layer and the temperature jump across it are not completely constrained, although the best evidence links it to the D″ layer of the lower mantle, implying a thickness of approximately 200 km. The temperature jump across the boundary, initially thought to be relatively small, may in fact be comparable in magnitude to the lithospheric thermal boundary layer with a temperature change of approximately 1000°C across the 200 km D″ boundary.

III. Thermal Regime of the Lithosphere

A. Thermal Evolution of the Oceanic Lithosphere

Often seen as a major success of the plate tectonic model, the evolution of the oceanic lithosphere can also be seen as a major example of the interaction of thermal processes with tectonic processes. The plate tectonic paradigm of new oceanic crust and lithosphere forming at mid-ocean ridges and subsiding as it moves away from the ridge crests has been directly linked to the thermal structure and evolution of the oceanic lithosphere. In addition, not only is the thermal regime of the lithosphere correctly predicted by the thermal models applied to the oceanic lithosphere, but also the bathymetry, gravity field, and lithospheric structure are well described. Thermal processes in the oceanic lithosphere can be placed into two complementary categories. First is the formation and thermal evolution of the lithosphere as a consequence of conductive processes. Second is the modifications of the thermal regime by the advection of heat associated with fluids circulating within the young oceanic crust. [See LITHOSPHERIC PLATES.]

Near the mid-ocean ridges, observed heat flow is extremely variable, but it can be quite high (>300 mW/m²). Away from the ridge crests the heat flow generally decreases as a function of the age of the lithosphere, reaching values of approximately 65 mW/m² for 50 Ma oceanic lithosphere and <50 mW/m² for 100 Ma lithosphere. This systematic decrease of heat flow with age is well described by the conductive cooling of the lithosphere after formation at the ridge, while the large variability near the ridge crests is linked to processes of fluid circulation within the young oceanic crust.

1. Processes at Mid-Ocean Ridge Crests

Mid-ocean ridges form the boundary between diverging oceanic plates. At these locations the passive upwelling of mantle material results in the enhanced partial melting of the asthenospheric mantle. The segregation and accumulation of this melt forms the 5–10 km thick oceanic crust bounded above by ocean water and below by the beginnings of the oceanic lithosphere. As the lithosphere moves away from the ridge crest it cools, additional lithospheric mantle is added to its base by cooling, the lithosphere subsides, and the heat flux through its surface decreases. These processes can be quantified by a simple yet very physical description of the cooling process.

The thermal evolution of the oceanic lithosphere is well described by the processes of a cooling half-space. The thermal regime at zero age at the ridge crest can be thought of as a constant (asthenospheric) temperature from the ocean bottom into the asthenospheric mantle. This constant half-space temperature structure will cool though its top (base of the ocean), providing large heat flux near the ridge crest that decays over time (distance from the ridge). The process is mathematically described by the error function [erf(x)] which, for the case of the cooling oceanic lithosphere, is given by

$$T(z, t) = T_0 \, \mathrm{erf}[z/(4\kappa t)^{1/2}]$$

where κ describes the thermal diffusivity of the lithosphere, z is depth below the crust/water interface, and t represents time after formation at the ridge crest. Thermal diffusivity is a measure of the rate at which the system can lose heat, in units of length squared per unit time. T_0 is the initial temperature of the upwelling mantle (temperature of the asthenosphere) and is normally assumed to be in the range of 1200–1350°C.

An example of the predicted thermal structure for various ages of lithosphere is shown in Fig. 2. The lithosphere is the region where there is a significant thermal gradient, and temperatures are less than the solidus (initial melting) temperature for mantle peridotites. This temperature condition for the base of the lithosphere is approximately 90% of the asthenospheric temperatures near the ridge crest, and thus the increase in thickness of the lithosphere with age (distance from the ridge) is given approximately by

lithosphere thickness (km) $= 13 \times [\text{Age (Ma)}]^{1/2}$

This systematic cooling and thickening of the oceanic lithosphere also causes a systematic (isostatic) subsidence of the oceanic lithosphere. As a result ocean water depth increases also with the square root of the age, away from the ridge crests, which have an average water depth of approxi-

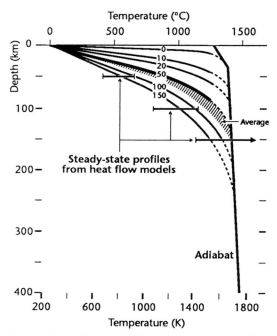

Fig. 2 Geotherms for oceanic lithosphere, parameterized by age of oceanic lithosphere (time since formation at mid-ocean ridge). [Modified from Jeanloz and Morris (1987).]

mately 2.7 km. The average age-dependent water depth in the ocean is given by

water depth (km) $= 2.7 + (1/3) \times [\text{Age (Ma)}]^{1/2}$

Such a systematic variation in water depth with age provides a tool for evaluating the history of oceanic conditions through a technique known as Backtracking. Using predictions of water depth at prior times for a particular marine deposit or fossil, estimates of paleoenvironment, ocean chemistry, or sea-level variations can be made. In addition areas of the oceanic lithosphere that deviate from the systematic age–depth relation can be identified as regions affected by processes beyond the simple cooling that normal oceanic lithosphere undergoes. These anomalous regions have been related to hotspot heating and active dynamic processes within the mantle (e.g., mantle convection).

2. Role of Fluid Circulation

Initial measurements of the heat-flow field in young oceanic lithosphere showed extreme variability, with many sites indicating very low conductive heat flow. However, in these same regions of low heat flow, the subsidence history and lithospheric thickness were consistent with the simple cooling model. In addition, the deviation in observed (conductive) heat flow from that predicted by the cooling model was negative and approached the predicted values as the lithosphere aged and the sediment thickness increased. When heat flow measurements were made along transects with short distances between measurement sites, this problem was resolved. The deviation from predicted heat flow is a consequence of fluids circulating in the crust and sedimentary layer of young oceanic lithosphere. Thus heat transfer in these regions has a large component of advection. The variation of observed heat flow with position indicated that the circulation cells have a dimension of approximately 5–10 km, consistent with a characteristic 5–10 km length for the roughness of the young oceanic crust. In regions of the young oceanic lithosphere blanketed by sediments, a shorter-wavelength thermal signature associated with convection within the sedimentary layer is also seen. Much of the heat transferred by advection is released into the ocean by the submarine equivalent of hot springs and the spectacular black smokers, with their unique biota.

The magnitude of this advective component in the oceans is quite large. Detailed heat flow measurements indicate that approximately 3–10 TW of heat is transferred by this process. This represents between 10% and 25% of the total global heat budget and as much as 1/3 of the oceanic component of the heat flux. It is important to recognize that this heat advection by water circulating in the oceanic crust and sediments is a near-surface phenomenon. The circulation typically extends to depths of a few kilometers or less and ceases when the fractures within the crust are sealed by hydrothermal deposits.

Fluid circulation plays an important part in the thermal structure and evolution of the sedimentary layer and the upper oceanic crust, producing large lateral variations in temperature and heat flux. At depths of 2–3 km within the crust, temperatures may vary by as much as 200°C over distances of <5 km. The thermal output of a single black smoker can be quite large, on the order of 10 MW. One potential outcome of this large fluid circulation process is the development of hydrothermal ore deposits. This fluid circulation does not play a substantial role in the overall cooling history of the lithosphere, however. Since the water circulates only to depths of a few kilometers, the process that controls the heat loss from the lithosphere is still conduction. The circulation of fluids and the associated heat advection serves to efficiently redistribute the heat and produce localized zones of large heat flux. [See HYDROTHERMAL SYSTEMS.]

B. Thermal State of the Continental Lithosphere

The variation of heat flow with lithospheric age in the oceans is well resolved. Whether similar behavior occurs in the

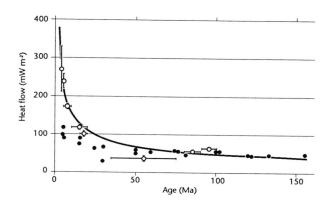

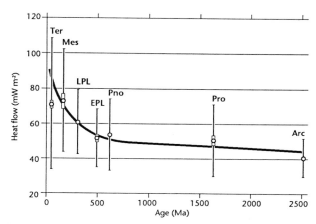

Fig. 3   Relationship between surface heat flow and lithospheric age for oceanic (upper) and continental (lower) regions. In continental regions, age is determined as time of last major tectono-thermal event.

continents is less clear. In Fig. 3 the relationship between heat flow and age are shown for both oceanic and continental regimes. In both cases there is an apparent (nonlinear) decrease of heat flow with age. In the case of the continents however this decrease occurs over a period of time approaching 500 Ma as compared to the characteristic time of 50 Ma for the oceans.

The thermal regime in the continents is more complex than the oceanic lithosphere. This is a result of the combination of large variability in crustal structure and lithology in the continents and the diverse tectonic histories experienced by the continental lithosphere. In contrast with the oceanic thermal regime, where the temperature structure changes with time, many areas of the continental crust are in a thermal steady state. In addition, the rock types typically constituting the continental crust contain enough radiogenic elements to provide a significant fraction of the heat flux. Also, the complex structure of the continental crust leads to large spatial variability in both temperature and heat flow.

The basic equation describing the steady-state thermal structure of the continental lithosphere is

$$\nabla^2 T = -A/k$$

where A represents the volumetric heat production and k is thermal conductivity. This equation is integrated (twice) under appropriate boundary conditions to evaluate the temperature structure as a function of position. Although normally this is done with respect to only one dimension (depth), there are times when it is appropriate to evaluate the thermal structure in two or three dimensions. To solve the above equation, the distribution of heat production and thermal conductivity must be known.

1. Heat Production in the Continental Crust

The details of the distribution of heat-producing elements in the crust are not known. We can be sure that heat production of near surface rocks cannot be representative of the entire crust and that there is substantial variability in heat production with rock type. Typical near-surface continental rocks have heat-production values of approximately $1.5-2.5$ μW/m³. If such values continued throughout the entire continental crust (typically ~ 40 km thick), heat production within the crust would account for $60-100$ mW/m² of surface heat flow—more than is normally observed!

During the 1960s several discoveries helped to place constraints on the possible distribution of heat production in the continental crust. In studies where both detailed heat-flow measurements and measurements of the heat production of the crustal rocks at the heat-flow sites were made, a correlation was observed between the value of surface heat flow and heat production (Fig. 4). This apparent linear relationship can be interpreted in several ways, but one interpretation that helps to constrain the distribution of heat production in the crust is that the intercept value of heat flow at zero heat production (reduced heat flow) corresponds to a background heat flux from the deeper crust or uppermost mantle. The linear increase of heat flow as heat production increases can then be used to infer a depth scale for the region of enhanced heat production in the crust.

The slope of the $q-A$ (heat flow–heat production) relation is in units of length. The simplest interpretation is that the heat production is constant from the surface to the depth given by the slope of the $q-A$ relation (typically ~ 10 km), with very low values of heat production below that depth. It can be shown that any heat-production distribution that, when integrated over depth, gives the equivalent integral to

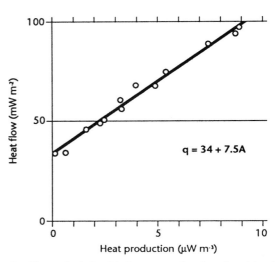

Fig. 4 Observed relationship between surface heat flow (q) and near-surface heat production (A) in the northeastern United States. [Modified from R. F. Roy, D. D. Blackwell, and F. Birth (1968). Heat generation of plutonic rocks and continental heat flow provinces. *Earth Planet. Sci. Lett.* **5**, 1–12.]

this constant depth function is also acceptable. In addition to the constant depth function, an exponential decrease in heat production with depth is also often used in constructing crustal geotherms. This exponential function has the added property of being unaffected by erosional variations. Using this model of an exponential decrease of heat production with depth, the resulting geotherm for the continental crust is given by

$$T(z) = T_0 + (q_0 z)/k - (A_0 b^2/k)(z/b + e^{-z/b} - 1)$$

where T_0 is surface temperature, q_0 is surface heat flow, A_0 is surface heat production, and b is the length scale determined by the $q-A$ relation. This equation describes the (steady-state) temperature as a function of depth (z). The temperature gradient as a function of depth is given by the first derivative of the above equation, which is

$$dT/dz = q_0/k + (A_0 b/k)(e^{-z/b} - 1)$$

and heat flow as a function of depth is given by

$$q(z) = q_0 + A_0 b(e^{-z/b} - 1)$$

The concentration of heat-producing elements in the upper crust serves to reduce the variation in heat flow in the lithosphere. Below the radiogenic-enriched zone, heat flow varies only slightly with depth at values $40-60\%$ of surface values. Although variations in conductivity also play a role, temperature gradients in the lower crust and upper mantle are also only 40–60% of near-surface temperature gradients. As an example of this, under typical crustal conditions of surface temperature (T_0) of 10°C, a surface heat flow (q_0) of 75 mW/m², near-surface heat production (A_0) of 2 μW/m³, a q–A length scale (b) of 10 km, and thermal conductivity (k) of 2.5 W/m-K, surface temperature gradients are approximately 25°C/km. At a depth of 20 km (two depth-scale lengths), the temperature is approximately 250°C, the temperature gradient is approximately 8°C/km, and the heat flow at that depth is approximately 20 mW/m². If the effects of heat production had not been included, the temperature estimate would be 500°C and heat flow and gradient would be the same as surface values. Although these calculations assume a simple structure of the crust, they point up the importance of including the effects of heat production in evaluating the thermal regime of the continental crust. Estimates of temperatures within the crust that ignore these effects are normally substantial overestimates.

2. Lithologic Variability in the Continents

Most continental crust has experienced several episodes of tectonism since it initially was formed. This tectonism has the effect of producing a heterogeneous crust, where lithologic variations may occur with length scales of 5–50 km. Associated with the lithologic variation is variation in thermal properties (conductivity, heat production), which will affect the steady-state thermal regime.

The thermal effects of lithologic variability are easily seen in the vicinity of salt domes and salt diapers. In such regions the salt body has substantially higher thermal conductivity than the surrounding rocks and thus conducts heat more efficiently than the surroundings. Heat is thus channeled into the salt, and the heat flow above the salt is enhanced. Since the heat energy must be conserved in the region, this enhancement of heat flow in the salt is at the expense of the surrounding rocks, resulting in depressed heat flow outside the salt diaper. Although the conductivity contrasts are usually somewhat smaller in typical continental crust, these effects can still be important.

When the crustal heterogeneity involves both thermal conductivity and heat production, the consequences can have great significance for the thermal structure of the crust. For many of the igneous rocks in the continental crust there is a positive correlation between thermal conductivity and heat production. As a result the heat-channeling effects of thermal-conductivity contrasts are enhanced by increases in heat production in the high thermal-conductivity units. If this heterogeneity is exposed at the surface, there will be a positive correlation between observed surface heat flow and near-surface heat production. Much of the observed q–A linear relationship may be simply a consequence of this

crustal heterogeneity rather than providing information on the depth distribution of heat production in the continental crust. At present we are unable to resolve this ambiguity, but care must be used in the interpretation of heat-production distributions in continental crust. This uncertainty does not alter the conclusion that heat production is generally enhanced in the upper crust and depleted in the lower continental crust.

3. Transient versus Steady-State Thermal Regimes

Much of the continental crust is near thermal equilibrium and thus is well described by stead-state models of the temperature structure. In association with crustal deformation and large-scale plate tectonic activity, some regions of the continental lithosphere undergo transient thermal events that must be considered in evaluating the temperature structure. In contrast to the thermal evolution of the oceanic lithosphere, a variety of transient thermal regimes must be considered for continental crust. Many of these primarily affect the near-surface thermal regime (described in a later section), but large-scale tectonic events thermally perturb much of the crust and lithosphere. Large-scale extensional events and thrusting can dramatically alter the temperature structure throughout the crust and affect the metamorphic history, mechanical properties, and occurrence of partial melting and volcanism.

Extension in the crust and lithosphere can occur by several mechanisms, for example pure shear (stretching) or simple shear (displacement along discrete shear/fault zones). In either case there is a transfer of material both vertically and horizontally. The thermal regime is most affected by the vertical component of mass transfer, resulting in the advection of heat. Since heat advection is significantly more efficient than heat conduction, the temperature structure that develops during the extensional event is not in equilibrium with the conductive thermal regime appropriate for the region. As a result there will be a temporal change in the temperature field during and after the extension.

There is an apparent enigma in the thermal behavior during extension of the crust and lithosphere. During extension there is an increase in the heat flow through the surface. This is a result of heat advection increasing the thermal gradient, enhancing the conductive heat flow. This effect can be quite pronounced with heat flow increasingly by a factor of two or more during a large extensional event. Extension is also a cooling event for the crust. At first this appears contradictory to the increase in heat flow during extension, but it can be easily understood by considering what happens to a particular volume of rock during extension. As the rock is brought closer to the surface it cools, and it is the heat it releases on cooling that helps to increase the heat flow. Thus one must not confuse the response of the heat flow with the temperature behavior of a particular volume of rock. The underlying behavior is that the crust involved in extension is cooled by the event. [*See* CONTINENTAL RIFTING.]

Extension as a cooling process creates another paradoxical situation—volcanism is often associated with large-scale lithospheric extension. This volcanism is derived by two processes associated with the extensional tectonics. First, the deeper levels of the crust and the lithospheric mantle initially cool very little during the extension, since they are far removed from the earth's surface where the heat is released. As a result, if there is a large amount of extension, there will be a significant decrease in the pressure conditions for these deep rocks as they are brought closer to the surface. The conditions under which crustal and mantle rocks partially melt, the solidus, depends on both the temperature and the pressure. If there is sufficient decrease in pressure during extension, with only minor temperature decrease, then partial melting may be initiated. This process is most effective near the base of the lithosphere where the mantle rocks are very close to their melting conditions prior to extension. The second process that produces some of the volcanism during extension is melting of the lower crust as a result of heat advected by the transport of melt. As long as the solidus temperature for the lower or middle crust is significantly lower than that for the melt derived from the mantle, some melting of the crust may occur. The volume of crustal melt produced by this process is normally limited to be slightly less than the introduced mantle melt. The specific volume will depend on ambient pressure and temperature regimes in the lower crust and the fluid or volatile content of the crustal rocks. [*See* CRUSTAL FLUIDS.]

With the exception of the mid-ocean ridge environment, geologic settings where melt is generated are non-steady-state. During these melting events the temperature regime may reach a point where there is little or no change over a period of time. This dynamic steady-state regime, where the temperatures do not change as long as the process is actively proceeding, plays an important role in other environments as well. During large-scale crustal extension a point may be reached where the near-surface heat flow does not change with time. During this period the temperature at a specific depth may also be constant. However the temperature of a particular volume of rock is not constant but rather will be decreasing.

C. Families of Geotherms

The thermal regime in the lithosphere can be highly variable. Under assumptions of the heat transport processes and the composition, reasonable estimates of the temperature versus depth function, the geotherm, can be made. These geotherms are parameterized differently for oceanic and continental regions. In oceanic terrains the geotherm is parameterized by

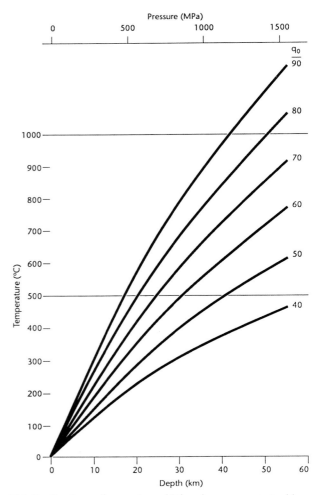

Fig. 5 Geotherms for continental lithosphere, parameterized by surface heat flow (q_0).

the age of the lithosphere, while for continental regions the normal parameterization uses the surface heat flow. Families of these geotherms can be constructed that provide an initial estimate of the temperature regime within the lithosphere. An example of such families of geotherms is shown in Fig. 5. In this case assumptions are made about the relationship between the total amount of crustal heat production and heat flow, thermal conductivity, and the behavior of the geotherm as the solidus temperature is approached.

IV. Crustal and Near-Surface Thermal Regimes

The temperature structure of the interior of the earth is dominated by processes only slightly related to the evolution of the surface of the earth. Although the temperature gradients within the upper crust are controlled by the heat flow from the interior and the thermal conductivity of the crustal rocks, tectonic and climatic processes can significantly alter the near-surface thermal regime. In addition, the deviations of the upper-crustal thermal structure from simple conductive geotherms can be used to place limits on climate variability, erosion rates, upper-crustal fluid flow, and glaciation.

A. Effects of Periodic Variations in Surface Temperature

Changes in surface temperature approximate simple periodic functions on several time scales. Over the period of one day temperatures vary approximately as a sine wave, as does the temperature variation over a complete year. Temperatures may change in such a systematic way on longer time scales, but that is not confirmed.

In cases where the heat flow is constant and the surface temperature changes approximately as a sine wave, the transient effect of the surface variation on the steady-state geotherm is given by

$$T(s, t) = A_0 e^{-kz} \sin(\lambda t - kz - \epsilon)$$

where A_0 is one-half of the amplitude of the temperature variation, λ is the angular frequency of the temperature variation (e.g., $2\pi/\text{day}$), ϵ is a lag term used to align the sine wave with the time frame used (e.g., days of year), and k is the wave number defined as

$$k = \sqrt{(\lambda/2\kappa)}$$

where κ is the thermal diffusivity. This function can be applied over a variety of time scales. Since this is the variation from the normal geotherm, the effects are simply added to the appropriate value of temperature at a given depth. This function decreases rapidly with depth, and the temperature effect is approximately $1/500$ of the surface effect at a depth of one thermal wavelength. The thermal wavelength (γ) is equal to $(2\pi/k)$. For typical values of thermal diffusivity, the thermal wavelength for a temperature variation of one day is approximately one meter!

1. Daily Variation

The daily surface-temperature variation affects subsurface temperatures only in the upper 1 m. In fact it is only in the upper few tens of centimeters that any perceptible effects are seen. Additionally, other processes such as water soaking the soil and evaporation play a much more important part in the transient temperature regime in the upper few meters on time scales of a few days.

2. Seasonal (Yearly) Variation

In contrast to the small importance of the daily variation, the yearly (seasonal) variation plays an important part in the thermal regime of the near surface. The appropriate thermal wavelength for the yearly variation is approximately 20 m. In Fig. 6 an example of the temperature perturbation with depth for the yearly cycle is shown. These results are shown as a temperature change from the average temperature at each depth. Clearly seen in this figure is the combined effects of a decrease in amplitude with depth and the progressive time lag in maximum or minimum temperatures with depth. What this means is that the coldest time of the year at the surface does not correspond to the coldest time at all depths, but rather the effect is akin to a thermal wave moving downward with a velocity given by λ/k. For the yearly variation this velocity is approximately one meter/20 days. The coldest time of the year at 3 m depth is approximately two months after that at the surface. It is for this reason that buried pipes tend to freeze during the February–March period rather than in early January.

B. Effects of Secular Variations in Surface Temperature

Longer term variations in surface temperature perturb the geotherm to depths substantially greater than the daily or seasonal variation. The processes at work are similar to those producing the thermal evolution of oceanic lithosphere. In general, climatic changes and emplacement or removal of ice sheets acts as a step change in the surface temperature. The effects of this step change propagate downward, perturbing the background geotherm.

Although the effects of climatic variations on the thermal regime near the earth's surface have been understood for many years, it is only recently that attempts have been made to utilize the temperature structure as a means for placing constraints on past climate. Temperatures measured within a few hundred meters of the earth surface may provide a means to place constraints on climate variations going a few hundred years into the past. In the past, deviations in the thermal regime in the upper 100 m of a borehole were considered the result of climate variability or fluid circulation, but further interpretation was not attempted.

During the past few years, deviations from the appropriate temperature–depth profile for a site with well-constrained heat flow and thermal conductivity structure have been utilized in an attempt to unravel past climate. One of the first of these studies was conducted in the permafrost of northern Alaska, where the temperature structure of the permanently frozen crust would be unaffected by water movement. Attempts are now being made to extend this approach to heat-flow measurement sites in general.

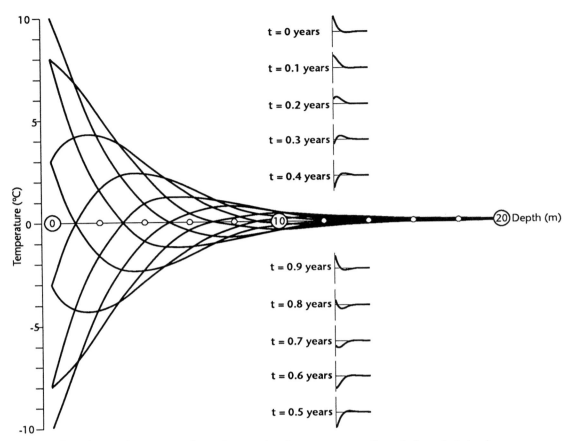

Fig. 6   Effect of seasonal temperature fluctuation on subsurface temperatures. Curves indicate deviation from average subsurface temperatures during 10 intervals throughout the year. Amplitudes of the curves are based on assumed ±10°C variation at surface.

The possibility of using borehole temperatures to study climate variability over the past few centuries is exciting for several reasons. First, the study site can be essentially anywhere on the continents. The earth has been recording the effects of surface temperature changes continuously. It is now simply a matter of deciphering that record. Second, the borehole temperature history is most sensitive to the past one to two hundred years. Information on global climate during this period is critical for the proper evaluation of the more recent trends in global temperatures. Third, with this approach the actual temperature history is being measured. Most other tools require the use of a surrogate (e.g., isotopes) to record the climate history.

This approach, to utilize borehole temperature regimens for climate studies, is still in its infancy. Certainly the assumption of a step change or sine wave variation in surface temperatures is overly simplified. With higher resolution temperature logging and detailed thermal-conductivity logs, more sophisticated temperature histories can be determined.

Another promising locale for these studies is within the ice sheets of Greenland and Antarctica. Within the ice sheets no water is moving, and the thermal properties of the ice are relatively constant below the upper few meters. The situation is complicated by the burial of the ice by new surface accumulations, but that effect can be monitored by dated horizons within the ice sheet. The polar regions are potentially most sensitive to changes in global climate, and borehole temperature measurements may provide a direct means to monitor climate variability in those regions during the past few centuries. [*See* POLAR REGIONS, INFLUENCE ON CLIMATE VARIABILITY AND CHANGE.]

C. Effects of Fluid Flow, Sedimentary Basins

Water moving through the earth's crust can efficiently transport significant quantities of heat. The general effect of this advection of heat is to depress the thermal regime in the recharge area of the fluid flow regime while enhancing temperatures and heat flow in the discharge area. The role of fluids in perturbing the temperature structure in basins is most pronounced in basins adjacent to highlands or mountains. Such foreland basins, which form in response to the loading effects of the mass of the mountains, are prime sites for thermal perturbations as a result of the topographic relief associated with the mountains. These highlands serve as the hydraulic potential, or head, to drive the flow regime. The result is a depression of isotherms (lower geothermal gradient) where the fluids are moving dominantly downward (the recharge area) and an elevation of geotherms where the fluids move upward, typically near the edge of the basin. The effects of this fluid flow can alter temperatures by 10–20°C, and amount that can have dramatic effects on the rates of geochemical reactions within the basin.

Fluid flow may also play a role in areas where relatively wet, porous sediments are being compacted. In such cases the fluids are being driven out of the collapsing pore spaces, and they normally would flow in a lateral and upward direction. As a result the primary effect of this regime is to raise isotherms and enhance heat flow. The degree to which this mechanism operates is unclear; but in regions where it has occurred, it may play a role in the development of sediment-hosted hydrothermal ore deposits.

D. Uplift and Erosion, Subsidence and Burial

Unroofing a rock unit via erosion or burial by sedimentation have pronounced effects on the near-surface thermal regime. In addition the change in the position of the rock strata relative to the earth's surface produces a variation in the temperature history during and after the period of erosion or burial. It is important to separate the processes of uplift and subsidence from the processes of erosion and burial, as the thermal effects may differ. Uplift implies changing the rock strata position relative to the center of the earth (or some other suitable datum such as sea level), and erosion or unroofing means bringing the strata closer to the earth's surface. Similarly subsidence refers only to the position of the strata relative to the datum, and burial implies moving the strata relative to the surface. Uplift and subsidence may be caused by a variety of thermal processes, but they do not necessarily cause any changes in the thermal regime of the earth or the thermal history of the rock units.

In a fashion to similar to extension, which also tends to bring strata closer to the surface, erosion has the effect of enhancing near-surface heat flow while causing the cooling of a specific rock unit. Erosion rates can vary greatly, and it is normally considered that a significant increase of heat flow will be observed when erosion exceeds approximately

1 mm/yr. Although the change in surface heat flow is negligible for erosion rates less than this figure, the temperature–depth path followed by a rock unit may deviate from the background geotherm even for quite low erosion rates. This effect can be exploited in studies of the tectonic evolution of the crust, as much of this thermal history may be recorded in isotopic, mineralogic, or organic systems within the strata.

Burial by sedimentation is effectively the inverse of erosion. The near-surface heat flow is depressed and the temperatures of the buried rock units increase. Burial rates for which observable surface heat-flow effects are generated are similar to erosion rates. As for erosion, even when burial rates do not cause a significant heat-flow anomaly, the temperature history may substantially differ from that for burial along the background geotherm.

Burial by sedimentation most typically occurs within developing sedimentary basins, which are also prime targets for petroleum exploration. The specific burial and erosion history experienced by a potential hydrocarbon source can have dramatic effects on the timing and rate of hydrocarbon maturation and production of economic oil and gas deposits. [See BASIN ANALYSIS METHODS.]

Another geologic locale for dramatic thermal effects from burial and erosion is in association with thrust belts. These are typically associated with regions of plate convergence, with dramatic examples from the European Alps. During thrusting, sections of the crust 1–10 km thick may be moved over similar strata. This overthrusting occurs quite rapidly with these thrust sheets moving at rates approaching a few centimeters per year. This causes extremely rapid burial of the strata below the thrust sheet, and the ensuing mountains are subjected to rapid erosion. In this way the thermal history of the buried strata will be the juxtaposition of burial followed by erosion, and the thrust sheet will experience erosion. The thermal history for the thrust sheet is not one of simply cooling however, as the thrust sheet will initially cool rapidly as a result of being placed on top of the cooler near-surface rocks. If there were no subsequent erosion, there would be gradual warming of these strata back to the normal geothermal regime. However, normally there will be substantial erosion, so the resulting thermal history will record the competing effects of thermal rebound after thrusting and cooling associated with erosion. The effects of this tectonism on the thermal regime can often be determined by studying the temperature record contained in fluid inclusions and organic markers within the strata both overthrust and buried. [See OROGENIC BELTS.]

E. Magmatic Activity

The emplacement of magmas on or near the earth's surface has a dramatic effect on the thermal regime near the intrusion or lava flow. These effects are in addition to any larger scale increase in heat flow (and crustal temperatures) associated with the underlying cause of the magmatic activity. These events would normally take one of four forms: plutons, sills, dikes, or flows. Flows are emplaced on the surface and are normally no more than a few meters in thickness. In a few cases lava flows may approach 100 m in thickness. Plutons are intrusions, most typically spherical or teardrop shape, and may have radii in the 3–5 km range. Sills are horizontal, relatively flat intrusions one to a few hundred meters in thickness, and dikes are sheetlike near-vertical intrusions of similar size.

In most cases the thermal evolution of these magmatic events is dominated by thermal conduction. Even in cases where the intrusive itself is convecting, the rate at which heat can be removed from the boundary of the intrusion is thermal conduction in the surrounding crust. This effect is sometimes modified by fluid circulation in the surrounding crust, but in most cases this does not appear to be the case. In a system dominated by conduction, the boundary between the intrusion and the surrounding crust does not exceed the mean temperature of the intrusion and surrounding crust prior to the emplacement of the magmatic body. This holds generally true for sills, dikes, and plutons. Local areas may deviate from this temperature constraint as a consequence of complex intrusion geometry or complicated crustal lithologies. The major thermal perturbation occurs over a distance of approximately two pluton radii, or equivalently a distance of the dike/sill thickness. The cooling history of plutons and dikes/sills are shown in Fig. 7. [*See* MAGMATIC PROCESSES IN SILLS.]

For lava flows emplaced on the surface, similar conditions apply, although they are somewhat modified by being at the earth's surface. First, a lava flow can lose heat much more easily through its surface than its base since the atmosphere efficiently removes the heat by advection. Second, in the near surface, it is much more likely that fluid flow will modify the cooling history, particularly after the flow has cooled substantially. The history of a cooling lava flow is shown in Fig. 8. Because of the rapid cooling of surface flows, they are an ideal medium to record the earth's magnetic field and its variations. However, in the few cases where lava flows exceed 5–10 m in thickness, care must be exercised in using them for this purpose, as the time period over which the flows cool may allow the recording of a complex variation in magnetic field.

V. Tectonism and Thermal Processes

Tectonic activity such as the formation of mountain ranges, movements along faults, and the eruptions of volcanoes can have a dramatic effect on the thermal regime of the crust. This is a two-way street, as the principal energy source for such tectonism is heat from the earth's interior. But here we

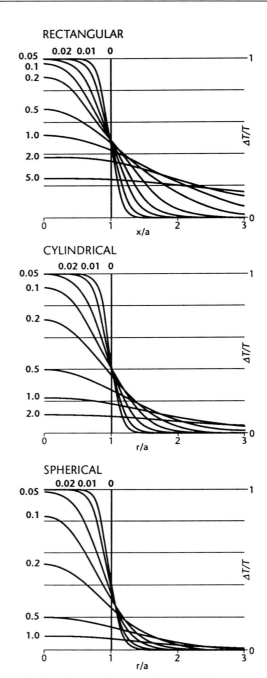

Fig. 7 Cooling history of intrusives for assumed geometries of a rectangular pluton (sills/dikes), cylindrical pluton, and spherical pluton. Temperatures are relative to temperature of the intrusive. Distance axis is scaled relative to the half-width or radius of the intrusion. Curves are labeled by nondimensional time after intrusion.

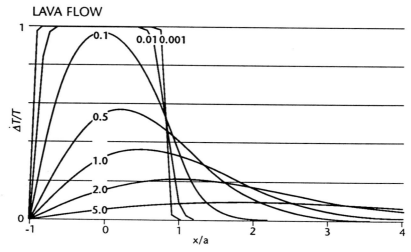

LAVA FLOW

Fig. 8 Cooling history of a surface lava flow. Distances are scaled by the half-width of the flow; temperature is scaled by the temperature of the flow.

will be concerned primarily with the consequences of tectonic activity on the thermal regimen and the methods by which those effects can be observed and exploited to understand the nature of the tectonic activity.

A. Heating along Faults

Movement along faults, whether abruptly during earthquakes or slowly and continuously as a result of fault creep, produces heat. This heat is produced as a result of the energy expanded to overcome the resistance (friction) to fault movement. This heat is similar to the heat produced by the brakes on an automobile or rubbing your hands together. The mechanisms are similar, and as such the physics are reasonably well understood. If the heat produced along a fault can be determined, the physical conditions of that fault motion can be evaluated. For example the heat produced will be directly proportional to the shear stress necessary to overcome the friction along the fault. Thus if one can measure the resulting fault-generated heat flow, estimates on the frictional strength of the fault zone can be determined. As a simple rule of thumb, a shear stress of 100 MPa (1 kbar) will generate a heat flow from the fault surface of approximately 30 mW/m². [*See* FAULTS.]

For the San Andreas fault in central California, the lack of a heat-flow anomaly local to the fault zone indicates that the fault moves during earthquakes or by creep with levels of shear stress less than 20–40 MPa (200–400 bars). Whether such low levels of shear stress are also characteristic of other fault zones is not known, although investigations of some exposed thrust faults have found some cases of local melting

on the fault surface indicative of significant shear heat generation during movement on those faults.

B. Rheology and Temperature Structure

The manner by which rocks within the earth deform, or the rheology of the rocks, is a function of composition, temperature, and the stress regime acting on the rocks. We normally assume that rocks will deform by whatever mechanism can occur at the lowest stress level. Temperature plays a very important part in controlling the mode and rate of deformation. Within the crust, this temperature effect appears to play an important part in controlling the depth of earthquakes. [*See* ROCK RHEOLOGY AND MASS TRANSPORT IN THE SOLID EARTH.]

Earthquakes occur in the cooler upper parts of the crust, where rocks permanently deform most easily by frictional sliding or brittle failure. As the temperatures in the crust increase, a point is reached where the deformation is most easily accomplished by ductile processes. At this depth, earthquakes are not generated. Determining this depth of transition from brittle faulting to ductile creep requires some knowledge of lithology and, very important, the geotherm or temperature structure of the crust. For the oceanic lithosphere in which the crust is relatively thin (~5–10 km), earthquakes occur in both the crust and uppermost mantle. For most regions of the continental crust with typical thicknesses of 25–40 km, the brittle–ductile transition occurs within the crust, and the lower crust appears to deform by ductile creep mechanisms. [*See* PLATE TECTONICS, CRUSTAL DEFORMATION AND EARTHQUAKES.]

The temperature-dependent rheologic behavior of the lithosphere also determines how the lithosphere deforms when a load is placed on it. This load may be a volcanic mountain or sediments deposited in a basin. In all cases the way in which the lithosphere flexes is controlled by the temperature structure. Hotter lithosphere deforms more easily, flexing into a narrower but deeper basin than colder lithosphere with the same load.

C. Metamorphism

The temperature regime in the crust will determine what the equilibrium mineralogic assemblages are for a particular rock type. Changes in this mineralogy, or metamorphism, will occur if the temperature conditions change. Although the metamorphic reactions are complex and will occur at different rates depending on other conditions such as pressure, fluid content, and stress, the occurrence of specific metamorphic rock types can be used to evaluate the temperature history of a suite of rocks. Using detailed determinations of the metamorphic conditions, the Pressure–Temperature–time (or $P-T-t$) history of the rocks can be determined. Such $P-T-t$ evaluations are providing insight into the processes that have built mountain ranges such as the Himalaya, and the way in which mountains such as the Appalachians are eroded.

Metamorphism also occurs when magmas are intruded into the crust. This contact metamorphism forms as a rind surrounding the intrusive; and its size, shape, and mineralogy can provide information on the temperature history of not only the metamorphic rocks but also the intrusive itself. [See METAMORPHIC PETROLOGY.]

VI. Environmental Issues

The temperature regime in the earth plays an important role in how the crust deforms, its strength, and how fluids move through the crust. If we change the thermal regime in the crust by burying heat-producing radioactive waste we may alter other processes of importance. The potential of changes in the temperature regime at the earth's surface may also propagate into the interior. Additionally, the large flux of heat from the interior of the earth, if harnessed, could provide a source of energy to supplement our dwindling fossil fuel resources.

A. Radioactive Waste Disposal

Radioactive waste disposal is clearly a process that alters the thermal regime. Depending on the volume and heat-producing strength of the buried waste, temperatures may rise to the point of concern. The heating effect has several primary consequences. First, the rise in temperature may cause the surrounding material to deform more easily. This is especially

a potential problem if the waste is buried in rock types that plastically deform at temperatures less than $500°C$. Second, the rise in temperature of the surrounding rocks may generate thermal stress sufficient to cause failure or faulting in the rocks. Third, the temperature gradients near the waste may drive fluid convection, which may disperse waste beyond the burial region. Finally, the increase in temperatures will cause most chemical reactions involving crustal fluids to proceed at a faster rate. [See NUCLEAR WASTE DISPOSAL.]

B. Effects of Global Warming

The principal consequence of global temperature changes on the thermal regime within the earth will be to change the upper boundary condition. As described earlier these effects will be recorded in the temperature regime within the upper crust. The effects of change of only a few degrees in surface temperature will have little effect on most of the crustal thermal processes. Global warming or other climate modification may affect the thermal regime in sedimentary basins in two small but potentially important ways. First, the change of a few degrees in temperature can significantly change the rates of hydrocarbon maturation reactions. Second, and potentially most important, since the temperature regime in many basins is substantially affected by fluid flow within the basin, the availability of water in the recharge area becomes important. If global warming or other climate change reduces the water availability, it could result in a significant reduction in the flow of water in the basin and the associated advective heat transfer. Increases in water flux could have the opposite effect.

C. Geothermal Energy

Tapping the heat flux from the interior is not simple. Hot springs efficiently bring heat to the surface, but rarely are hot springs located where they are needed. Geothermal heat can be used in a variety of forms—electric power generation, space heating, or farming. Using geothermal sources for the generation of electricity is unfortunately at present of limited use. Most geothermal systems are not at sufficiently high temperatures for generation. Space heating and farming uses are potentially more useful. Geothermal space heating of buildings can be accomplished earlier by direct use of the heat or using the geothermal source as a heat reservoir tapped via a heat pump. The usefulness of geothermal energy in farming can be in the space heating of greenhouses and in the extension of the growing season by the use of warm waters in irrigation.

The effective exploitation of geothermal energy is not trivial. Many technological problems remain in economically using the energy. In addition the problem of geothermal sources not necessarily coinciding with the potential energy users will continue to be a problem.

VII. Summary

The thermal regime in the interior of the earth is produced primarily by heat generated with the decay of radiogenic elements within the crust and mantle of the earth. The heat is transferred by two principal mechanisms — conduction and advection. Thermal conduction is the dominant heat transfer mechanism within the lithosphere and perhaps within the D″ region of the core–mantle boundary. Within the bulk of the earth's mantle and the outer core, convection dominates and, hence, advection is the dominant mode of heat transfer.

The overall thermal structure of the earth is modified by a variety of processes that cause both transient and steady-state variations to the background thermal processes. Compositional heterogeneity can produce regions of elevated or depressed heat flow as well as deviations from the normal vertical orientation of heat flow. Tectonism such as thrusting, erosion, burial, and volcanism produces substantial transient effects in the lithospheric thermal structure. The large-scale plate-tectonic processes of sea-floor spreading is a mechanism for the bulk of global (surface) heat loss.

The detailed thermal structure in the upper crust is affected by variations in near-surface temperatures, and thus it records temperature variations from climatologic and tectonic processes. Detailed analyses of near-surface temperature regimes may be a powerful tool to investigate the record of global change.

Bibliography

Furlong, K. P., and Chapman, D. S. (1987). Thermal state of the lithosphere. *Rev. Geophys.* **25**(6) 1255.
Furlong, K. P., and Edman, J. D. (1989). Hydrocarbon maturation in thrust belts: thermal considerations. *In* "Origin and Evolution of Sedimentary Basins and Their Energy and Mineral Resources." *Geophys. Monogr., Am. Geophys. Union* **48**, 137.
Jeanloz, R., and Morris, S. (1986). Temperature distribution in the crust and mantle. *Annu. Rev. Earth Planet. Sci.* **14**, 377.

Glossary

Geotherm Function describing temperature in the earth as depth increases.

Heat advection Transfer of heat content of a material with mass transfer.

Heat flow Rate at which heat energy crosses a specified unit area, given in units of mW/m^2.

Lithosphere Strong outer layer of the earth composed of the crust and uppermost mantle, normally approximately 100 km thick.

Radiogenic heat Heat energy produced by the decay of radioactive materials in the earth. The principal heat-producing elements in the earth are uranium (U), thorium (Th), and potassium (K).

Thermal conductivity Rock property that describes the efficiency of heat transfer.

Heavy Metal Pollution of the Atmosphere

E. A. Livett
Manchester University

Atmospheric fluxes of heavy metals have been enhanced by human activity to a considerable degree. Urban and industrial activities generate heavy metals in both particulate and gaseous phases. These metals are dispersed efficiently on a local, regional, and global scale; a significant part of the atmospheric heavy metal load around the world can now be attributed to pollution. Historical trends suggest that, following a continual increase in heavy metal pollution for at least the last 3000 years, levels are beginning to decrease as awareness grows. Environmental levels of heavy metals in the future depend not only on the controlled, efficient use of fossil fuel and mineral reserves but also on the residual burdens of metals in the global reservoirs.

I. Introduction

From earliest times, humanity has modified the global cycles of the elements. Indeed, air pollution is inextricably linked with human origins, as the use of fire distinguishes *Homo sapiens* from their nearest anthropoid kin. Nevertheless in some respects the term *pollution* is misleading. The natural cycles of the elements involve all phases of the earth—atmosphere, geosphere, hydrosphere, and biosphere—and as human beings are part of the biosphere, it could be argued, that they are also a part of the natural cycles. All attempts to define pollution thus involve ambiguity; especially as we can never be sure of the composition of the pre-human environment. Definitions, however artificial, are nevertheless convenient; and for the purposes of this article pollution is defined as the man-made, or anthropogenic, redistribution of substances.

The metals we consider—antimony (Sb), arsenic (As), cadmium (Cd), chromium (Cr), cobalt (Co), copper (Cu), lead (Pb), manganese (Mn), mercury (Hg), nickel (Ni), selenium (Se), tin (Sn), vanadium (V), and zinc (Zn)—have diverse chemical and geochemical properties. Lead and tin are Group IVB metals; antimony is in Group VB; arsenic and selenium are semi-metals of Groups VB and VIB, respectively; and the remainder are Transition Group metals. Nevertheless, one important shared attribute is the binding preferences of their ions for sulfur groups, rather than for oxygen-donor atoms in ligand formation in organic systems; it is this attribute that underlies the environmental importance of these metals. Although many of the metals (excepting As, Cd, Hg, and Pb) are essential plant and animal micronutrients at the trace concentrations at which they are found in an *unpolluted* environment ($\sim 1.0 - 1000\ \mu g\ g^{-1}$), all have distinct toxic properties at higher, pollutant, concentrations, prompting widespread concern as their global abundance has increased as a result of human activities.

II. Fluxes and Cycles

The atmosphere is one compartment of the global system in which chemical elements cycle continuously, passing through reservoirs (air, water, and organisms) and accumulating in or being depleted from sinks (soils and sediments). The principal atmospheric pathways are shown in Fig. 1.

The immediate source of a greater part of the atmospheric trace-metal burden is the geosphere, whence the metals are released by the processes of fragmentation, volatilization, or combustion. All emissions, whether natural or anthropogenic, result from one of these processes. Over the range of 14 metals considered here, atmospheric cycles are of two basic types: for the so-called *lithophile* elements (Co, Cr, Mn, Ni, V), release to the atmosphere via rock and soil erosion constitutes a significant fraction (between 40% and 80%) of natural emissions; whereas for the more volatile *chalcophile* elements (As, Cd, Cu, Hg, Pb, Sb, Se, Sn, Zn) continental crustal inputs to the atmosphere are relatively less important, and vulcanism can make a significant contribution to the natural atmospheric metal load ($> 60\%$ in the case of Cd). It is for the second group of metals that human activities have made the most significant perturbation of the natural cycles,

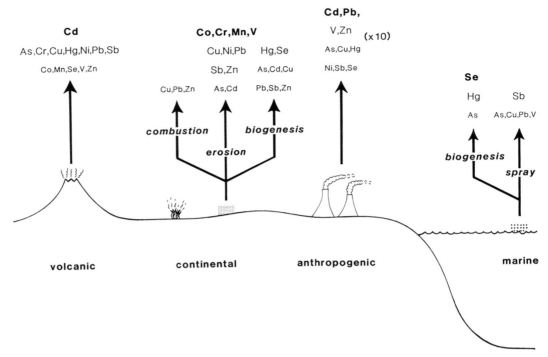

Fig. 1 Principal sources of heavy metal emissions to the global atmosphere. Contributions are shown as a percentage of total natural emissions for each metal. Bold type, > 50%; light type, > 25%; and small type, > 10%. For anthropogenic emissions, the quantities are > 500%, > 250%, and > 100%, respectively.

enhancing emissions (and deposition) by up to 30-fold. [*See* ATMOSPHERIC EFFECTS OF VOLCANIC ERUPTIONS.]

A. Emissions

1. Sources and Quantities

An inventory of the principal sources of anthropogenic emissions is given in Table I. Combustion processes (columns 1–5), which essentially separate the relatively volatile trace-metal constituents from the more refractory matrix constituents of mineral ores, fossil fuels, and secondary materials, account for at least 80% of the emissions of most metals. The estimates of the amounts of metal released by these processes and the forms in which they are released, however, are imperfectly known and involve such factors as the initial concentration of the metal in the parent material, its relative volatility, the temperature and efficiency of combustion, the presence and type of control devices used, and the total amount of material processed. All such inventories, especially for diffuse processes and for a broad geographical scale, inevitably incorporate a wide margin of error and so should be interpreted with reservation.

Coal combustion generates the widest spectrum of heavy metal emissions. The average trace metal composition of coal ranges from 0.5 $\mu g\,g^-$ to 50 $\mu g\,g^{-1}$ in the order: Cd < Hg < Sn < Se < As,Co < Sb < Cu,Ni < Cr < Pb,V < Mn,Zn; but as much as 30–40% (Zn, Ni, Pb, and Sb) or as little as 5–10% (Mn and As) of the original content is released by combustion. Coal combustion is the major source of Co emissions on a worldwide scale.

Heavy metal emissions associated with the combustion of heavy oils for heat and electricity generation are dominated by vanadium, zinc, and nickel. The average heavy metal contents of oils are 50 $\mu g\ g^{-1}$, 30 $\mu g\ g^{-1}$, and 10 $\mu g\ g^{-1}$, respectively, although values as high as 1400 $\mu g\ g^{-1}$ (V) and 345 $\mu g\ g^{-1}$ (Ni) have been reported for particular deposits. Concentrations of the other trace metals are all below 1 $\mu g\ g^{-1}$. Fuel oils, in contrast, are purified, and their content and emission rates of trace metals would be negligible were it not for the presence of additives. Most notably, the addition of up to 0.27 g 1^{-1} Pb to gasoline makes automotive emissions the most important single contributor (> 60%) to the anthropogenic atmospheric lead burden. Other metals (Cd, Cr, Hg, Ni, Sb, Se, Sn, and Zn) are emitted in trace amounts in car exhausts, by virtue of their inclusion in lubricants and as a result of engine and tire wear.

Nonferrous metal smelting is a localized but nevertheless

Table I   Global Atmospheric Emissions Inventory for Heavy Metals[a]

Metal	Source[b]								Total
	1	2	3	4	5	6	7	8	
				(10^9 g yr^{-1})					
As	1.98	0.06	—	12.32	1.42	0.53	0.49	2.02	18.82
Cd	0.53	0.14	—	5.43	0.16	0.44	0.87	—	7.57
Cr	11.27	1.41	—	—	15.63	1.33	0.84	—	30.48
Cu	5.18	1.96	—	23.86	1.49	0.41	2.47	—	35.37
Hg	2.09	—	—	0.13	—	—	1.34	—	3.56
Mn	10.72	1.39	—	3.17	14.73	—	8.26	—	38.27
Ni	13.77	27.07	—	8.78	3.57	0.90	1.56	—	55.65
Pb	8.16	2.42	248.03	49.85	7.63	7.29	4.47	4.50	332.35
Sb	1.30	—	—	1.53	<0.01	—	0.67	—	3.51
Se	3.06	0.80	—	2.36	<0.01	<0.01	0.1	—	6.32
Sn	1.00	3.27	—	1.06	—	—	0.81	—	6.14
V	7.92	76.12	—	0.06	0.75	—	1.15	—	86.00
Zn	11.1	2.16	—	72.43	19.54	13.90	9.5	3.25	131.88

[a] No reliable data for cobalt.

[b] (1) Coal combustion; (2) oil combustion; (3) gasoline consumption; (4) nonferrous metal smelting; (5) ferrous metal smelting; (6) other industry; (7) other agricultural/urban activities; (8) miscellaneous.

major source of As, Cd, Cu, Se, and Zn emissions, and an important source of Pb and Ni emissions on a global scale. The high emission factors of trace metals from the smelting process (e.g., 1 g kg^{-1} metal produced) reflect the often high purity of the parent ore (e.g., lead-bearing minerals contain up to 80% Pb). The emissions spectrum from smelters also depends on the geochemical associations in the ore-bearing minerals; galena (PbS) and sphalerite (ZnS) frequently occur together, as do chalcopyrite ($CuFeS_2$) and pentlandite (($FeNi)_9S_8$), while the rarer elements occur principally as substitutes for the more abundant ones (e.g., Cd for Zn and Co for Cu). The semi-metals As and Se occur mainly as metal arsenates and selenides in sulfide deposits — indeed lead ore is the major commercial source of arsenic.

Ferrous metal production, like nonferrous smelting, is a locally important source of atmospheric emissions of certain metals and accounts for the greater part (60–80%) of chromium and manganese emissions on a worldwide scale. Zinc, lead, and manganese, in that order, are the major pollutants from iron production; emissions factors are highest when the parent material is siderite or magnetite, but these factors are still three orders of magnitude lower than for nonferrous metal smelting. Steel and alloy manufacture generates a wider emissions spectrum, comprising Cr, Ni, Cu, Cd, and V in addition to the major pollutants — Pb and Zn.

The contributions of cement and fertilizer manufacture (column 6), low-temperature combustion processes (e.g., waste incineration and wood combustion — column 7) and noncombustion processes (column 8) to global emissions are particularly difficult to estimate as these activities are widespread globally. Also, the distinction between some human activities and natural phenomena is often blurred (e.g., erosion of contaminated soil versus erosion of natural ore deposits; the agricultural practice of slash-and-burn versus forest wildfires).

2. Formation and Physicochemical Characteristics

The morphology, size, and chemical speciation of heavy metals released to the atmosphere are of prime importance in determining the subsequent fate and impact of these pollutants on global ecosystems. Figure 2 illustrates the size range of some particles generated by anthropogenic and natural processes. Anthropogenic emissions are characteristically dominated by small (<1 μm) particles, and the high-temperature processes (fossil fuel combustion, ferrous and nonferrous smelting), which account for the major fraction of anthropogenic emissions, give rise to particles of characteristic morphology and chemistry. In all these processes, a complete separation of the elemental trace metals from the parent material by volatilization is achieved by high temperatures. Formation of metal-rich particles subsequently occurs by a variety of mechanisms in the cooling flue gas stream.

In coal combustion, a significant fraction of the vaporized metal condenses onto the surface of spherules (MMED 1–100 μm) that are formed from a melt of the more refractory matrix elements in the coal, resulting in a strongly inverse proportional relationship between heavy metal concentration and particle size. In car exhaust fumes, lead and other metals become tightly bound to carbon and hydrocarbon material in

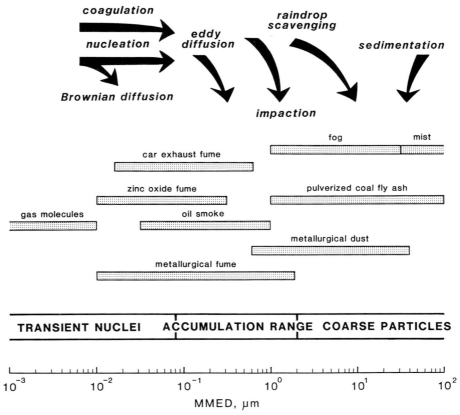

Fig. 2 Sizes of some natural and pollutant particles and the processes that effect their deposition.

tiny ($<$0.15 μm), electron-dense spheres, which subsequently coalesce to form irregular chain-like clusters. The size of these aggregates ranges from 1 to 5 μm (exceptionally to 20 μm).

In ferrous and nonferrous metal production a number of separate processes are involved, and these are reflected in the particulate size range. The lowest temperature stage—roasting and sintering—typically generates particles of MMED $<$1–40 μm, while the hotter reduction and purifying processes generate a fume comprising particles of MMED $<$1 μm.

The chemical speciation of anthropogenic emissions and the solubility of the compounds depends on the pollutants present, the solubility products of the individual chemical species, the surface area-to-volume ratio of the particles, the composition of the rainwater, and the presence of protective coatings. Thus, sub-micron particles and highly irregular aggregates (e.g., auto-exhaust particles) are relatively soluble, as are the nitrates, sulfates, and ammonium compounds frequently present in combustion-related emissions. Generally,

the solubility of anthropogenic heavy metal particles ranges from 80–50%, in the order V $>$ Cr $>$ Mn $>$ As $>$ Co $>$ Ni $>$ Cu $>$ Zn$=$Cd $>$ Pb. Mercury is unique in that a large proportion ($>$90%) of the total emissions from both combustion-related and cold processes is in a stable, elemental, gaseous form (Hg^0) and as such is insoluble. No more than 10% of the lead in vehicle exhaust is in a gaseous form (Pb(CH$_3$)$_4$); Ni(CO)$_4$ is also emitted but decomposes rapidly to the carbonate on exposure to water vapor.

B. Dispersal and Deposition

A mixed population of pollutant particles (see Fig. 2) is continually being injected into the atmosphere from sources scattered nonuniformly over the earth's surface. These pollutants are subjected to mixing and sorting, solution and aggregation, and dispersal and deposition, so that local, regional and global aerosols are formed. Meteorological factors are the primary agents of pollutant dissemination, but ultimately it is the characteristics of the individual particles that determine their fate.

1. Meteorological Factors and Pollutant Dispersal

Most combustion-related processes generate plumes (localized, concentrated streams of particles), whether issuing from industrial stacks, car exhaust pipes, or cigarettes. The first stage of dispersal involves the complete mixing of the plume material with the local air mass by turbulent mixing and molecular diffusion. Vertical transport of the plume often occurs initially and is most effective when the emission is buoyant and has an appreciable efflux velocity and conditions are calm with a positive temperature lapse rate. As the plume rises, frictional eddies around its edge achieve some mixing (resulting in entrainment, or enlargement of the plume). Little or no plume rise will occur, however, when the lapse rate is zero or negative, or when down-drafts around neighboring high buildings exceed the upward velocity of the plume (hence the rationale for building tall stacks). In these situations, the emissions may be trapped at ground level.

When the buoyancy and efflux velocity of the plume are equalled by the air pressure, the predominant motion becomes horizontal, and further mixing with the local air mass is achieved principally by molecular diffusion and eddy diffusion (turbulent mixing). These processes are mathematically analogous, but the coefficient of eddy diffusion is several orders of magnitude greater than the coefficient of molecular diffusion. Essentially, the Gaussian, or normal, distribution of pollutant concentrations across the plume is preserved while it is progressively diluted, the rate of dilution being proportional to the wind speed. Eventually, the pollutant concentrations within the plume approximate those of the surrounding air mass.

Mixing of pollutants on a regional and global scale is achieved both by the mean global air circulation systems and by turbulence (deviations in the mean wind speed and direction) within these systems. [See ATMOSPHERIC CIRCULATION SYSTEMS.]

The prevailing direction of global air circulation is west–east and is most strongly expressed in the upper troposphere around latitudes of 30°N and 30°S (the subtropical jet stream). These movements originate directly from the earth's movements and can transport a particle or a *parcel* of air that has been carried to this altitude around the globe in two to three weeks. Circulation in the lower atmosphere is more complex: At mid-latitudes in both hemispheres fluctuating westerly winds encircle the globe, while at low latitudes northeasterly (in the northern hemisphere) or southeasterly (in the southern hemisphere) winds (the Trade Winds) prevail. Superimposed on these latitudinal wind systems and their

associated air masses is a series of north–south (meridional) circulating currents, with velocities considerably less than those of the latitudinal systems. These meridional circulating systems, or cells, are driven by differences in the heat energy budgets between poles and equator, and cause the north–south transport of pollutants. [See ZONAL AND MERIDIONAL WINDS IN THE EARTH'S ATMOSPHERE.]

The so-called Hadley cells in the north and south tropics dominate the system and are separated at or near the equator by the interhemispheric tropical convergence zone (ITCZ), which extends in altitude to the tropopause. It is in this zone, mainly during times of maximum deviation from its mean position, that pollutant exchange between hemispheres, and from troposphere to stratosphere, can occur. The descending limb of these cells in each hemisphere is associated with rain.

Two more analogous, but weaker, cells occur in each hemisphere, in the polar regions and at the mid-latitudes. Here, they are associated with the frontal systems developed between adjoining air masses that are important in the local-to-regional transport and rainout of pollutants (see below). A low-pressure center (cyclone) can trap pollutants over a source region for up to several weeks, whereas the high pressures occurring with anticyclones can cause their rapid dispersal. The southerly encroachment of the polar air mass over northwestern Europe during the winter is the means by which pollutants are transported to the arctic region.

Particle residence times in the atmosphere vary widely, from around a week at low altitudes to a month or more in the path of the strong westerlies at high latitudes, and are longest—one to two years—in the stratosphere.

2. Size and Deposition of Particles

The large-scale transport of pollutants discussed above concerns air masses, but the size of individual particles in these aerosols is equally important in determining how far these pollutants are transported and how quickly and effectively they are removed from the atmosphere by dry and wet deposition.

For particles of MMED > 5 μm the force of gravity eventually overcomes the buoyancy of hot effluent gases, and the particles will tend to sediment out of stack emissions and car exhaust fumes relatively near to the source. The distance traveled prior to deposition is directly proportional to the wind speed and inversely proportional to the falling velocity (which obeys Stokes' law). Therefore, the distribution of pollutants deposited around a point source reflects the prevailing wind and is graduated according to particle size. The remaining fraction of particles that remains airborne is a true aerosol. For this fraction, dry deposition is much slower and is achieved either by surface impaction and eddy diffusion (particles of MMED 1–5 μm) or by molecular diffusion (particles of MMED < 1 μm, which includes gases); surface characteristics and microtopography thus assume importance for the deposition of aerosols. The efficiency of dry deposition is empirically defined as deposition velocity V_g.

$$V_g = \frac{\text{flux of particles to the surface}}{\text{concentration of particles in air}}$$

V_g does not relate to a specific mechanism but to the net deposition process, and usually stipulates the substrate and altitude. Maximum values of V_g (e.g., 8–20 cm s^{-1} for 5-μm zinc smelter emissions) have been recorded for large particles deposited by gravity, decreasing to values of, for example, 0.1–3.0 cm s^{-1} for smaller particles. V_g is lowest for particles of MMED 0.1–1 μm, for which the processes of eddy diffusion and molecular diffusion are least effective due to high aerodynamic resistance. For particles of <1 μm, V_g increases again to ~0.3 cm s^{-1}, as molecular diffusion becomes the effective deposition mechanism. The further the particle is transported from its source, the more likelihood there is of atmospheric reactions, such as aggregation, taking place. The aggregation of primary motor exhaust particles of 0.15 μm was described above; similar processes occur widely as an aerosol ages, resulting in an increase in frequency of particles in the 0.1–1.0 μm size-class (the *accumulation range*), which decreases the net efficiency of deposition of the aerosol.

The association of pollutant particles with water occurs when the aerosol is transported to rain-forming altitudes of 1–3 km. Two distinct processes are involved: in-cloud scavenging and below-cloud scavenging. In the first, mist droplets form around hygroscopic or soluble pollutant particles of MMED <0.1 μm (nucleation), and grow to ~1–60 μm. In upland regions this heavy-metal-enriched mist can be transferred directly to vegetation by impaction *(occult deposition)*. Below-cloud scavenging involves the capture of heavy metal particles at lower altitudes by falling raindrops (MMED <1000 μm); its effectiveness varies with particle size, with minimum rates occurring for particles of 0.1–1.0 μm, and with maximum rates for particles of 4–10 μm. The resulting water droplets (e.g., fog, >1.0 μm; rain ~1000 μm) are deposited by impaction or sedimentation. The concentrations of pollutants in rain are inversely proportional to raindrop size, and the rate of wet deposition decreases logarithmically with the duration of a rain event.

The overall efficiency of wet deposition, or washout factor W, is defined most simply as

$$W = \frac{\text{concentration of element in precipitation}}{\text{concentration of element in air}}$$

Washout efficiency depends on many variables, including chemical speciation and particle size of emissions as well as

the likelihood of particular pollutants to be carried to rain-forming altitudes. Generally, the lowest values of W (<500) are recorded for the crustal elements Al, Fe, and Sc and the heavy metals Cr, Pb, Sb, and Se; intermediate values (500–1000) are recorded for V, Mn, As, and Co; and maximum values (1000–3000) for Ni, Cu and Zn and 1000–10000 for Hg. The overall contribution of wet and occult deposition to total deposition ranges from 20 to 80%.

The size distribution of an anthropogenic aerosol is thus increasingly modified as it is transported away from its source. The broadest size spectrum occurs in urban/industrial regions, while the predominance of the 0.1–1.0-μm size range is characteristic of remote regions.

III. Distribution and Status

It is known that earth-to-air fluxes of many heavy metals resulting from human activities now equal or exceed the natural fluxes, and that over 90% of these emissions originate in the northern hemisphere. Following emission, however, the pollutants are diluted by the global air mass as a result of the complex processes of mixing and transport, and eventually they may be deposited at great distance from their source. We shall examine the distribution and pollutant status of heavy metals in the regional and remote air masses and the relative importance of the sources of the pollutants in these contexts.

A. The Status of Pollutants

During the last few decades, the establishment of national and international air-sampling networks has provided, for the first time, records of trace metal concentrations in air and in dry and wet deposition covering all the geographical regions of the world. The range of air concentrations of 14 metals presented in Fig. 3 illustrates the tripartite grouping of the sites into *polluted* or *urban* (near to industrial or urban areas); *rural* (at least 20 km from major pollution sources, where the atmospheric composition typifies the region or country); and *remote* (high-altitude or high-latitude sites where the air composition is representative of the continental, hemispheric, or even global aerosol). Superimposed on this geographical gradient is the approximately 10^3-fold difference in average air concentrations between the most and the least abundant metals. The pollutant status of the metals on a global scale can be assessed by ranking the metals according to average air concentration and according to anthropogenic emissions (in parentheses), as follows: Pb 1(1), Zn 2(2), Mn 3(5), Cu 4(6), Ni 5(4), V 6(3), As 7(8), Cr 8(7), Hg 9(11), Se 10(10), Cd 11(9), Sb 12(12). The ranks generally show good agreement; that is, air concentration reflects the degree of anthropogenic enhancement. A higher rank with respect to concentration than to

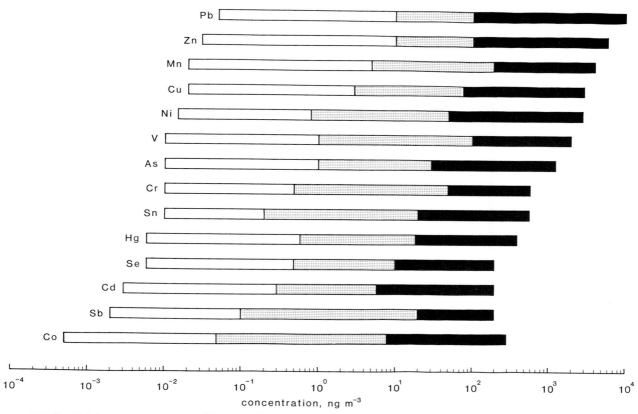

Fig. 3 Global concentration ranges of heavy metals in air at remote (unshaded), rural (stippled), and polluted (shaded) localities.

emission rate implies that the metal is poorly dispersed globally, while a lower concentration rank implies that natural sources rival pollutant sources in their global importance (e.g., Mn).

Another way of assessing the impact of human activities on the atmospheric cycles of the metals is to assume that the main source of the natural metal flux is crustal—derived from weathered rock and soil. The enhancement of total heavy metal concentration or deposition above the estimated crustal input—the crustal enrichment factor, EF_{crust}—is thus

$$EF_{crust} = \frac{[X/Al]_{air}}{[X/Al]_{crust}}$$

where X is the metal and Al (aluminum) is the crustal reference element (scandium or silicon can also be used). The concentration of crustal material in air varies considerably, for example from <1.0 ng Al m^{-3} in the snow-covered Antarctic to >1000 ng Al m^{-3} in arid desert regions. Values of EF_{crust} vary correspondingly, depending on the geochemical affinities of the element as well as its source strength. For

example, values for urban regions range from 3800 for Pb, 940 for Cd, 300 for Zn, and 149 for Cu (the chalcophile elements) down to 3.2 for Mn and 2.2 for Fe (the lithophile elements). Values of EF_{crust} greater than 10 have been detected in the Antarctic aerosol for the metals As, Cd, Cu, Hg, Pb, Sb, Se, and Zn (the *anomalous enrichments*), engendering much debate, as there is no recent increase in heavy metal deposition, signifying pollution, in this region. Other sources postulated for these enrichments are vulcanicity, forest fires, and low-temperature volatilization.

B. Source Apportionment

Many techniques have been developed to assess the relative contributions of different anthropogenic activities to the atmospheric metal load in rural and remote regions.

1. Elemental Signatures

The emission spectra of pollutant sources comprise characteristic ratios of heavy metals; these ratios tend to survive unchanged through the processes of transport, mixing, and scavenging to produce an identifiable spectrum in heavy metal deposition in rural or remote regions. Specific signatures include $V:Se = 1500-42,000$ for oil-fired power plants; $V:As = 10$ for coal fired plants; and $Br:Pb = 0.32$ for motor exhaust. In North America the regional signature $Mn:V = 2-3$ is characteristic of the midwestern aerosol (dominated by oil-burning sources), and $Mn:V = 0.2$ identifies the northeastern air mass. Elemental signatures are most effective for short-time resolution sampling (when meteorological factors are constant).

2. Receptor Models

These methods seek to mathematically resolve the observed variation in heavy metal concentration at a site into its separate components. In the chemical element balance method, the emission factors of all the likely pollution sources in the region are used to break down the total particulate concentration C of element i in a sample into its constituents, as follows:

$$C_i = \sum_j m_j X_{ij}$$

where m_j is the fraction of the particulate material originating from an individual source j and X_{ij} is the concentration of element i in material from source j. The concentrations C_i and X_{ij} are obtained by analysis, and the unknown m_j is derived by a least-squares fit method using a subset of selected marker elements such as Pb (auto-exhaust), V (heating-oil combustion), or As (coal combustion). Multisource elements (e.g., Cd, Cu, Zn) forming the residual subset serve to test the success of the analysis: A poor fit for these elements means that some important source(s) may have been omitted from the equation. An example of source apportionment in the rural northeastern United States is soil dust (53%), coal burning (25%), motor exhaust (12%), sea salt (5%), refuse incineration (3%), and oil burning (2%).

Factor analysis is another receptor model, but one that does not require prior knowledge about numbers or emissions characteristics of sources and does not provide quantitative apportionments. Large data sets are analyzed to determine the minimum number of factors, which accounts for the greater part of the observed variation. Three sources were thus revealed for the South Pole aerosol: marine (Na, Mg); crustal (Al, Ca, Fe, K, Mn); and a third, independent source (Ag, Cd, Cu, Pb, and Zn).

3. Stable Lead Isotope Analysis

This approach is based on the occurrence in rocks and minerals of four stable isotopes: ^{206}Pb, ^{207}Pb, and ^{208}Pb, which are all radiogenic, and ^{204}Pb, which is nonradiogenic. The respective ratios of these four isotopes in lead-bearing minerals are specific to the age and type of rock: The proportion of radiogenic lead and the concentration of ^{206}Pb are higher the younger the rock. This enables lead from coal combustion emissions (of Carboniferous age) to be distinguished from anti-knock lead, which is manufactured from lead deposits dating back to the Middle Precambrian and the Ordovician Periods. Isotopic analyses of atmospheric lead in remote continental areas of North America reveal that automobile emissions are making a major contribution ($>50\%$) to the atmospheric lead burden in these regions at the present day.

IV Past, Present, and Future

Human insight into its impact on the global environment and atmosphere has lagged far behind mastery of the tools of despoliation. Atmospheric pollution has been recognized as a problem since at least the times of the Ancient Romans, but it is only in the last few decades that the development of sophisticated sampling devices (e.g., Cascade Impactors) and sensitive analytical techniques [e.g., instrumental neutron activation analysis (INAA) and proton induced X-ray emission (PIXE)] has enabled the whole spectrum of trace metals to be determined in wet and dry deposition and in individual particles. Our knowledge of the historical development of pollution prior to the 1960s is fragmentary; but more recent trends have been well-documented, and we are now equipped to assess the efficacy of pollution-control devices and the results of changes in energy requirements and tentatively to predict trends and formulate strategies for the future.

A. The Historical Perspective

Since it is only in recent decades that direct pollution monitoring has been possible, we must use indirect methods to assess the impact of human activities on past environments. These methods include geochemical monitoring (the analysis of separate dated layers of stratified deposits such as ice and snow, peat and aquatic sediments), and biological monitoring (the analysis of growth-incremental layers of long-lived organisms such as corals and trees; or whole preserved specimens that lived at different times in the past such as herbarium specimens of mosses and lichens). The information derived from these materials is intrinsically variable, being subject to the accumulation characteristics of the medium and the possible subsequent mobility of the metals. Furthermore, the records are retrospective and time-averaged, and are highly dependent on the occurrence or availability of the materials. At best, therefore, these methods furnish a skeleton of records for disjunct episodes in time and space. [See HISTORICAL GEOLOGY METHODS.]

There are no pollution records for the early millennia of industrial history. The first records of charcoal in Australia date to ~100,000 yr B.P., the discovery and use of native

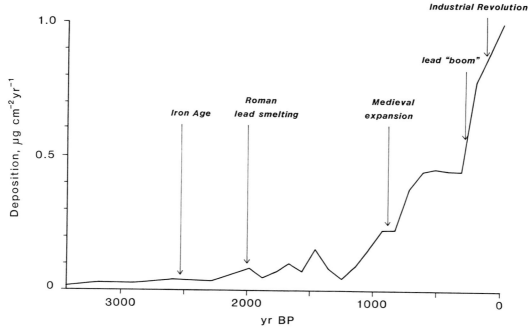

Fig. 4 Retrospective record of atmospheric lead deposition in rural Britain from 3500 B.P. to the present day.

copper in the Near East to before 5000 yr B.P., and the smelting of copper in Egypt and of lead in Palestine to the third and second millennia B.C., respectively. The longest quantitative sequences of heavy metal deposition are from Northwest European peat cores. The generalized pollution history depicted in these records can be summarized by the profile of lead deposition, as a wide-spectrum pollutant, shown in Fig. 4. The earliest enhancements of heavy metal (lead, zinc, and copper) deposition in these cores date back to the late Bronze Age/Early Iron Age at ~2500–3000 B.P. and are attributable to increased soil mobilization due to tillage, charcoal and wood burning, and iron smelting. Later, during the Roman occupation of Northern Europe, lead mines were worked at many sites in Britain, and a two- to fivefold increase in deposition rates of lead (and possibly also of zinc, cadmium, and copper) associated with these workings would have been widespread on a local scale. Lead deposition rates calculated from the concentrations in the peat nearby were ~0.25 μg cm^{-2} yr^{-1}.

Following the departure of the Romans the population stagnated, and agriculture and industry declined; there was a corresponding decrease in the atmospheric deposition of metals. The early Medieval Period, however, saw renewed growth and a revival of agriculture and industry; this period was associated in many regions of Britain with increased deposition of metals associated with diverse activities. Hence-

forth, atmospheric lead pollution was probably a regional rather than a local phenomenon, affecting remote areas for the first time.

In Britain the modern industrial era was ushered in by the first large-scale exploitation of coal and the invention of the steam engine and the coke furnace. Coal consumption (which rose 10-fold during the first half of the nineteenth century) and metal smelting predominated as pollution sources. Heavy metal (Pb, Zn, Cd, Cu, Ni, Hg, and V) deposition at semi-rural sites in Britain soared to 10 or 20 times previous levels. Following the wave of industrialization that spread to continental Europe in the nineteenth century, heavy metal deposition in remote areas rose two- or threefold. The rate of lead deposition at semi-urban sites in Britain in the 1850s was ~6 μg cm^{-2} yr^{-1}.

There is good geochemical documentation of the growth in atmospheric heavy-metal pollution in North America, too, over the 150–200 years following the industrialization of the subcontinent in the nineteenth century. Increases in deposition on a local scale closely followed the westward and northward migration of the European settlers. Typically, lead, zinc, copper, cadmium, nickel, and mercury in atmospheric deposition increased two- or threefold over presettlement levels (although there is no assessment of existing pollution levels attributable to the indigenous Indian populations). By as early as A.D. 1850 anthropogenic lead was detectable in deposition records from remote, high-altitude sites, showing that the subcontinental air mass was already polluted.

The first detectable increase in lead deposition in ice deposits from Greenland is dated to A.D. 1750, implying that pollution from industrial Northwest Europe, particularly Britain, had been efficiently dispersed northward in the polar air mass. There is no conclusive proof, however, that Antarctic snows are yet enriched with lead or other metals—and there is still ongoing debate as to whether there has been significant interhemispheric exchange of pollutants.

B. Recent Trends

The present century has seen, initially, continued industrial growth and fuel consumption and, concomitant with this, increases in total suspended particulates and heavy metal deposition in most of the developed countries. Leaded gasoline was introduced in 1923, and consumption has increased threefold since 1950. These developments have been mitigated, however, by the imposition of emissions controls following the implementation of the Clean Air Acts in the United Kingdom and the United States in 1956 and 1963, the replacement of coal by oil as the major world energy source between 1950 and 1970, and the enforced reduction in the lead content of gasoline since 1970 in the United States and since 1972 in Western Europe. The detection of environmental responses in heavy metal deposition to these changes is achieved both by long-term programs of real-time air monitoring and by analyses of time-averaged deposition—for example in the seasonal snowpack.

Controls on industrial emissions have certainly resulted in an amelioration of air quality in the immediate vicinity of industry. Reductions of up to 95% in total suspended particulates and in concentrations of associated chalcophile metals (As, Cd, Sb, Co, and Zn) have been reported in North America and Europe following the installation of electrostatic precipitators, the construction of taller stacks, and the conversion of many power plants from coal to oil burning. Flue gas "scrubbers" are reported to collect Ni and V as well as SO_2 emissions. In many urban areas of the United States and Western Europe air concentrations of heavy metals, particularly lead, were rising by ~5–10% per year until 1970, but have recently declined by up to 60% as the enforced reduction of lead in gasoline has taken effect. Improved incineration practice has also brought about a decline of >50% in air concentrations of Cd.

Regional and continental trends are less clear-cut, as there are fewer records than for polluted areas. In the United Kingdom there has been a significant decline in air concentrations of only Co, Cr, Mn, Pb, Sb, and V at rural sites from 1972 to 1981, which has been attributed to the decline in industry following the economic recession, as well as to the efficacy of emissions controls, and to a lesser extent to the

recent reductions of Pb in gasoline. Concentrations of Cu and Zn have remained stable. In North America, while some semi-rural districts have experienced an amelioration in air quality, there has been a net increase in heavy metal deposition to some of the more remote regions. This has been attributed to the more efficient dispersal of industrial emissions to these distant regions by taller stacks (e.g., the 381-m-high stack at Sudbury, Ontario), which has offset any beneficial results of emissions controls.

It has not proved possible to detect recent trends in anthropogenic heavy metals in the global air mass, as short-term and medium-term variations in deposition in the remote polar regions are primarily a function of variation in the transport of pollutants from the mid-latitudes, brought about by fluctuating meteorological conditions.

C. Future Prospects

In all scenarios for atmospheric pollution in the future, energy and transportation, the major contributors to global atmospheric emissions of heavy metals and gaseous pollutants, are the two main protagonists.

1. Energy

Traditionally, fossil fuels have supplied the world's energy needs, from the development of coal-based technology in the eighteenth and nineteenth centuries to the rise to preeminence of oil in the mid-twentieth century. In 1970s projections of energy supply and demand by the year 2000, nuclear fission was to have supplied 40% of the market, with fossil fuels supplying 30% and renewable and miscellaneous sources making the balance. However, development and innovation in the nuclear industry has been hampered by socioeconomic drawbacks; and oil supplies are precarious, and reserves are dwindling rapidly. With new coal deposits continually being discovered and world reserves standing at approximately 7725×10^9 metric tons (230 years' supply at the current rate of consumption), expanded coal utilization in the twenty-first century provides a more realistic solution to the energy gap.

Optimistically, further improvements in particulate and gaseous emissions controls in the conventional coking and electricity-generating industries during the next few decades will offset the environmental impact of the increased consumption of coal. Eventually, however, these traditional processes will be superseded by new technologies, whereby coal will be converted efficiently into liquid and gaseous fuels. Most important, operating temperatures in these processes will be much lower, resulting in substantially reduced vaporization of the chalcophile elements and hence a reduction in emissions via the flue stream.

Although the chemical conversion process is cleaner, the numerous stages involved in liquefaction and gasification, as well as in the consumption of their products, will diversify the pollution risks. These include the disposal of spent catalysts (containing Co, Cr, and V) and heavy-metal-rich residues from the liquefaction and gasification stages. It may prove difficult to control emissions of the volatile elements As, Hg, and Se from the liquefaction process and from the combustion of the gas product.

2. Transportation

The introduction of tetraethyl lead in gasoline in 1923 to prevent preignition, or "knocking," resulted in an immediate and widespread response of atmospheric lead levels; but three decades elapsed before public awareness was roused. Since the 1970s, the lead content of gasoline was progressively reduced in the United States and later in Western Europe, and U.S. legislation now dictates that new cars be adapted to utilize lead-free fuel. But is this feasible and environmentally desirable?

Several options exist if lead is to be excluded forthwith. First, the engine can be redesigned to enable it to consume low-octane fuel without "knocking," but this will result in higher fuel consumption. Second, gasoline can be further refined to increase its octane rating, but this will increase its carcinogen content. Third, an alternative anti-knock agent can be substituted for lead, but this is likely to introduce a different pollution risk (e.g., emissions of Mn_3O_4 from the agent methylcyclopentadienyl tricarbonyl—MMT). The last option is the use of an entirely different fuel, but this still remains a distant prospect.

The simplest short-term, though not complete, solution to the problem is to filter the lead from the exhaust fumes. Lead traps have a good retention rate (98%) for particle sizes $>9\ \mu m$ but only 50% efficiency for the modal, $1\ \mu m$, size fraction of the emissions. An environmental advantage is that the lead residue can be recycled, thus conserving a nonrenewable resource.

3. Conclusion

The provision of power, whether for heating, transport or countless other activities is at the crux of the pollution problem at the present day. No future energy technology can be "pollution-free," relying on expendable materials for fuels, energy conversion, storage, or conduction. If emissions were to be substantially reduced in the future, there would be a rapid decline in atmospheric flux rates initially, due to the quick turnover time (<1 month) of most heavy metals in the atmosphere. However, enhanced heavy metal deposition over hundreds of years in the northern hemisphere, and particularly of lead during this century, has loaded the global ocean, sediment, and soil reservoirs with heavy metals (e.g., 10 μg cm^{-2} Pb), accumulated preferentially in the surface organic layers. Residence times in these reservoirs are of the order of 10^3 to 10^5 years, and there is thus considerable potential for recycling (see Fig. 1). Mercury, notably, and possibly also As, Pb, and Se, undergo biological methylation, whereby they can be returned to the atmosphere from soils and sediments. Eolian remobilization of organic and crustal materials is also a significant part of the atmospheric cycle of some metals, as is enrichment of oceanic aerosols by salt spray.

In effect, therefore, human activities have raised the heavy metal baseline for the whole global system, and it is likely that enhanced atmospheric fluxes will continue indefinitely.

Bibliography

Chadwick, M. J., and Lindman, N., eds. (1982). "Environmental Implications of Expanded Coal Utilization." Pergamon Press, Oxford.

Coughtrey, P. J., Martin, M. H., and Unsworth, M. H., eds. (1987). "Pollutant Transport and Fate in Ecosystems." Blackwell, Oxford.

Galloway, J. N., Thornton, J. D., Norton, S. A., Volchok, H. L., and McLean, R. A. N. (1982). Trace metals in atmospheric deposition: a review and assessment. *Atmos. Environ.* **16,** 427.

Harrison, R. M., and Laxen, D. P. H. (1981). "Lead Pollution, Causes and Control." Chapman and Hall, London.

Henderson-Sellers, B. (1984). "Pollution of Our Atmosphere." Adam Hilger, Bristol, England.

Legge, A. H., and Krupa, S. V., eds. (1986). "Air Pollutants and their Effects on the Terrestrial Ecosystem: Current Status and Future Needs of Research," *Adv. Environ. Sci. Technol.* **18.** Wiley (Interscience), New York.

Livett, E. A. (1988). Geochemical monitoring of atmospheric heavy metal pollution: theory and applications. *Adv. Ecol. Res.* **18,** 65.

Lodge, J. P., Jr. (1989). "Methods of Air Sampling and Analysis." Lewis Publishers, Chelsea, Michigan.

Nriagu, J. O., and Davidson, C. I., eds. (1986). "Toxic Metals in the Atmosphere," *Adv. Environ. Sci. Technol.* **17.** Wiley (Interscience), New York.

Warneck, P. (1988). "Chemistry of the Natural Atmosphere." Academic Press, San Diego.

Glossary

Aerosol Two-phase system comprising a gas with solid or liquid particles in suspension.

Anthropogenic Correctly, "giving rise to humanity"; but more commonly used to signify "human-induced."

Chalcophile Associated with the sulfide phase in minerals.

Ligand Chemical species bonded to the central atom in a complex ion.

Lithophile Associated with the silicate phase in minerals.

MMED Mass median equivalent aerodynamic diameter; where the equivalent aerodynamic diameter of a particle is the diameter of a sphere of density 1 g cm^{-3} that has the same falling velocity.

Stratosphere Atmospheric layer extending from the tropopause to an altitude of ~ 50 km.

Troposphere Atmospheric layer nearest to the earth's surface and showing a linear temperature lapse to the tropopause at 10–15 km altitude.

Historical Geology Methods

A. J. Boucot
Oregon State University

Historical geology is essentially an early twentieth century term. Textbooks and courses entitled Historical Geology, as contrasted with Physical Geology, first began to proliferate after World War I. However, contrary to what the layman might think, historical geology does not in the strict sense deal with the earth's history. The bulk of the earth, the mantle and core regions situated well below the ever-so-thin crust, are and have been essentially unavailable from a historical viewpoint. Precious little material even of upper mantle origin is recognized in the earth's crust. Such exceptional materials include some of the xenoliths present in diamond pipes containing mineral grains that have been dated by radiometric means. Historical geology, then is actually the history provided for us from the crustal rocks preserved at the earth's surface. A valuable complement of crustal rocks is also obtained from boreholes made largely in the search for oil and gas, but these too penetrate crustal rather than mantle rocks. JOIDES boreholes from the ocean basins provide us with a picture of ocean basin crustal history, supplemented by dredge hauls and geophysical data. Still, when all is said and done the bulk of the information from which the history of the earth's crust is reconstructed is obtained from surface exposures, including manmade excavations of one kind or another (quarries, road cuts, railroad cuts and the like).

Implicit in all historical geology is the fact that the further back one goes in time the less reliable are the conclusions. Information about increasingly older rocks becomes more and more sketchy, and the conclusions based on same increasingly more diagrammatic. This is true for both the physical and biological aspects of historical geology. The reasons for the progressively poorer record back in time are twofold: (1) proportionally greater areas of younger rocks are exposed at the earth's surface, with the Quaternary being best exposed; (2) the older time units have had progressively greater opportunities for having had a larger and larger part of their original extent destroyed by erosion and also to have been seriously altered by metamorphism that commonly destroys most fossil remains as well.

I. Purpose

As pointed out above, historical geology deals chiefly, almost entirely, with the earth's crust. Information about the initial formation of the earth is provided almost entirely by astronomers, geophysicists, and geochemists dealing largely with physical concepts rather than with information obtained directly from the rocks. We have no rocks recognized from the initial stages of the earth within the 4.5 b.y. range and very few for the first two billion years or so, in accord with the comments in the previous paragraph.

A. Classical Approach

The classical approach to historical geology, embodied in text after text and course after course bearing the label, has been to provide a brief account of thoughts concerning the earth's formation, something about absolute age determination and data relevant to the age of the universe and the solar system including the earth, a chapter on Precambrian history, and then the remaining 90% or so devoted to the final 15% or less of earth history, the interval during which well-skeletonized fossils are available as contrasted with the paucity of organic remains in the Precambrian. This approach made some sense to the geologist, particularly in the first half of this century, owing to the difficulty and expense of obtaining reliably dated (absolute ages) rocks from which a history could be constructed for the largely unfossiliferous Precambrian rocks as contrasted with the Cambrian and younger beds with their relative abundance of well-skeletonized fossils. Also involved here is the fact that the bulk of the geological information accumulated for the land areas involves the younger, post-Precambrian rocks. Since World War II the Precambrian rocks have been subjected to far more critical scrutiny and study, but there is still a very disproportionate amount of attention devoted to the younger beds. Additionally, on land the built-up regions adjacent to most of the world's universities and research centers are not situated close to regions of Precambrian rocks. The bulk of the exposed Precambrian rocks occur in such relatively remote regions as the Canadian Shield, large parts of central Africa, remote parts of the Guayanas and Brazil, and so forth. Also involved is the fact that most practicing geologists are not working with Precambrian rocks. In summary, the lack of concern shown for Precambrian history in most historical geology texts and courses is an indication of several factors. Exceptional in these regards is P. Cloud's magisterial "Oasis in Space," which turns the ordinary textbook emphasis upside down by devoting a far more appropriate number of pages to Precambrian history.

The classical approach has emphasized the historical geology of a particular region, normally that of the writer's backyard. North American books commonly deal almost entirely with North American geological history, with little attention given to the rest of the world; European volumes are Europe oriented; and so forth. This type of provincialism was justified formerly by the assumption that the geological history of one part of the world was largely similar to that of the remainder. We now realize that this assumption is badly in error. For example, the geological histories of Arabia and eastern North America have remarkably little in common.

The classical approach has also emphasized a format in which individual units are devoted to the Cambrian and younger periods, particularly to their paleogeography (of course, only for the region being considered, such as North America), to the common fossils occurring in the beds of each period, and to evidences for such things as local mountain building for the region being considered, as well as evidences through time of local volcanism and climate. The classical approach has been largely descriptive, with little attention given to problems raised by the data.

B. Modern Approach

The modern approach is more problem oriented. It is not enough merely to bring late Ordovician, Pennsylvanian–Lower Permian, southern hemisphere Oligocene to present, and northern hemisphere Quaternary continental glaciation to the student's attention. One must, additionally, consider what factor(s) might be involved in causing continental glaciation. The modern approach recognizes that a global account is necessary because not all parts of the earth share a common history. In fact, a case can be made that the geological history of every region is in large part unique.

The purpose, then, of modern historical geology is to go beyond the mere recording of facts about the earth's crust through time and space to a consideration of causality. Because of our relatively low level of understanding this modern approach requires that many, commonly conflicting, conclusions be considered.

II. Physical Objectives

The physical objectives of historical geology are many. The following summary briefly considers a few of the more outstanding possibilities.

A. Paleogeography through Time, Including Positions of the Continents

Since C. Schuchert's day an attempt to at least reconstruct the ever-changing positions of the major continental shorelines has been a goal of historical geology. Since Schuchert more attention has been devoted to trying to reconstruct the position of ancient topography, largely from the evidence of preserved montane roots, since the actual mountains themselves are rapidly worn down.

The phase following the heading refers to the paleogeographical revolution in the geologist's thinking beginning in the mid-1960s. Prior to that time most geologists thought of the positions of the present continents as having been relatively fixed since the beginnings of the rock record, with Alfred Wegner's thoughts concerning continental drift given little serious regard by most. Following the popularization of the plate tectonic *cum* continental drift concept beginning in the mid-1960s there was a total change in geological thought on this subject. It is fair to say that today most geologists view some type of major change in position of the continents through times as a reality, although there is little overall agreement on the exact configuration of that ever-changing paleogeography until one reaches the Cenozoic, with its almost virtual identity to modern geography in many regards. This is not the place to review the varied reasons behind the different measures of paleogeographical disagreement, but the nongeologist should be warned that they exist. Considering the disagreements that now exist it is unlikely that any high level of paleogeographical agreement will be reached in this century.

1. Terrestrial

In the terrestrial environment an attempt is now made, particularly for the younger strata where we have a better preserved sedimentary record, to work out the location of lakes and rivers, as well as of topographic features such as mountains. Varied evidence is employed in this effort, largely taken from the purview of the sedimentologist.

2. Marine

In the marine environment an effort is made to locate the shoreline region and then the offshore slope, and the width of the continental shelf itself is estimated. Absolute depths cannot be obtained directly from either physical or biological data, but reliable approximations can be made.

B. Mountain Building through Time

The geologist is well equipped to consider the occurrences of mountain building in time and space. The record of the sedimentary rocks makes it clear whenever there is a gap in sedimentation in one region, adjacent to which pre-gap rocks in another region have suffered tilting, folding, faulting, and metamorphism and intrusive activity, following which sedimentation is again recorded, that mountain building may have been involved. As stated earlier, only the roots of earlier-occurring mountains are preserved until one reaches the latter part of the Cenozoic. Still, by plotting the geographic positions and geologic ages of these mountain roots one can obtain a fair view of the changing intensity and location of mountain building though time. This activity has always been an integral part of historical geology.

C. Volcanism through Time

Volcanism through time is partly connected with mountain building through time since many montane regions are areas of active volcanism both during their period of formation and well before their formation. In addition to the volcanism associated with mountain building there is the volcanism associated with continental rifting—of the African Rift Valley sort. This type of rift volcanism can be tracked far back in geological time. Interestingly enough there has not yet been a serious effort at carefully compiling the total volcanic record —the nature of the volcanics—as well as the preserved volumes and estimated original volumes on a paleogeographical base, except at the most elementary textbook level. [See CONTINENTAL RIFTING.]

D. Climate through Time

Climate through time has been a primary concern of historical geology from the beginning. Geologists early on recognized that the overall global climatic gradient, as well as the climate present at any one locale, changed markedly back and forth through geological time. The term *Ice Ages*, plural, has been in use for many years. Likewise, the amazing fact that palm trees were formerly present in west-central Greenland, crocodiles in northern Ellesmereland, nummulitic limestone beneath the Hatton–Rockall Plateau, and evidence favoring the presence of mangrove swamps in the London and Paris regions reminds us that the presently high climatic gradient is not a permanent feature.

The geologist uses physical and biological data for determining the distribution of past climates, with proper concern being taken as well for the changing, different positions of the continents. Marine salt beds and calcrete-type soils, for example, are taken as good evidence for relatively arid regions, normally present at lower latitudes, whereas tropical-type coals and bauxites are indicative of low-latitude tropical to subtropical humid regions (with arid belts possible to either side), whereas tillites, striated pavements, roche moutonées, and similar phenomena are indicative of former continental glaciation. [See BAUXITE AND LATERITE SOIL ORES; CLIMATE; PALEOCLIMATIC EFFECTS OF CONTINENTAL COLLISION AND GLACIATION.]

E. Atmosphere and Ocean Composition through Time

Samples of ancient ocean water and air have not been preserved in the geological record (recent discussions of "air" in ancient amber notwithstanding). There are no reliable geochemical techniques for determining the salinity of the water formerly associated with sedimentary rocks. Recourse, therefore, must be made to indirect evidences of ancient seawater and atmospheric composition. The prime evidence is had from the physiologies of modern organisms and their evident environmental conservatism back in time. For example, modern cephalopods are a stenohaline group; that is, they are unable to tolerate seawater compositions deviating very much up or down from the average 35 parts/thousand dissolved salts present in the modern seas. Fossil cephalopods, back into the Upper Cambrian when they first appeared in skeletonized condition, all occur in marine beds that the geologist would conclude represent "normal" marine conditions—none in hyper- or hyposaline beds as indicated by the presence of evaporites or of fresh-water proximity. The oxygen requirements of modern animals and plants are such that any major deviations from the present tenor of that gas seem unlikely. For example, any major reduction in the tenor of oxygen in the atmosphere since the advent of mammals in the later Triassic would certainly have totally eliminated that group; this is also the case for many other groups of organisms on land and sea. [See EVAPORITES.]

III. Biological Objectives

The biological objectives of historical geology are varied; they depend in largest part on the evidence provided by fossils, the remains of once living organisms, for their solution. It needs to be kept in mind that the fossil record is not an unbiased sample of past life. Quite the contrary. Most fossils consist of skeletonized remains only. Soft tissues and soft-bodied organisms are very rarely preserved. We are undoubtedly in total ignorance of many major soft-bodied groups or of lightly skeletonized groups. For example, in the modern seas the copepods are the major element in the zooplankton, forming the bulk of the animal biomass for higher trophic level creatures to subsist on. Yet there is not a single fossil normal marine copepod, not even one in the Pleistocene, owing to the poorly mineralized nature of their exoskeletons, which must be recycled by scavengers and other processes with amazing rapidity. The same is true for the euphausids, another group of most abundant zooplankton (the krill favored by whalebone whales), particularly in the polar seas. On land we find that herbaceous plants are almost never fossilized, in contrast to woody plants. On land the fossil record of both plants and animals consists almost entirely of low-elevation organisms; our knowledge of pre-Pleistocene montane biotas is minor at best. Obligate cave dwellers have very little chance of fossilization. Most of our fossil birds come from the shorebird and marine category; fossil parrots and finches are few. In general, the marine fossil record is far more likely to preserve its organisms than is the case with the far more oxidizing and erosive terrestrial environment (freshwaters are intermediate in this regard between the marine and the terrestrial). For example, the desert environment is very hard; we have not even a single fossil cactus.

A. Origin(s) of Life, Geological Evidences

In the strict sense the fossil record tells us nothing about the actual origin(s) of life. The rocks provide some circumstantial evidence bearing on the early environments present on land and sea but no direct evidence about actual origins. Our first evidence of life consists of Archaeozoic procaryotes, stromatolitic structures and cell membranes suggestive of procaryotic cyanobacterial and bacterial life capable of photosynthesis. Primitive as these organisms are, they are still far distant from the actual origin(s) of life. The origin of life appears to be a problem for the biochemist and molecular biologist rather than for the geologist–paleontologist at this time.

B. History of Life: Deductions from the Fossil Record, Deductions from the Modern Biota

In contrast to the origins of life the history of life is an area where the fossil record provides a wealth of information, once the biases of that record are taken into account. There is a surprisingly useful, reliable record for the times of appearance and disappearance of the more abundant, well-skeletonized organisms, although the record for the increasingly less common well-skeletonized organisms is correspondingly less reliable. We can even deduce a number of biotic interrelations from this well-skeletonized record. For example, an ancient lake yielding a fish with sharp, pointed teeth clearly points to the existence of animal prey of one sort or another, even in the absence of the prey organism(s), as well as of some type of plant or fungal food for that prey. This is what might be termed ecological thinking. The absence of normal marine copepods from the fossil record does not deprive us completely of knowledge concerning their possible first appearance. Their first appearance was very likely at least by the earlier Jurassic, at which time we find crab carapaces bearing the swollen deformities caused today by certain parasitic isopods, which themselves today use a normal marine copepod today as the intermediate host. This is ecological–evolutionary thinking. We lack much information about the marine plankton of the Cambrian; but one assumes based on that of the modern seas that such a plankton, even if unknown, probably existed.

1. Significance of Biostratigraphic Record

The fossil record does not consist of a homogeneous mass of skeletal debris derived from once living organisms. Rather, it consists of a definite, time-sequence set of biotic units. Each of these time-unit biotas consists of a set of biotas, each one evolved to exist in a particular environment. The biostratigrapher, the stratigraphic paleontologist, works at developing a highly refined set of globally and regionally applicable biotic units. Most of these units persist for many fives and even tens of millions of years before being abruptly replaced geologically by the next unit. Earlier in the last century many paleontologists viewed the nature of the fossil record as consistent with intermittent catastrophes that totally eliminated the biota, following which life was generated *de novo*. Today we are in possession of far more information, which enables us to adopt a neocatastrophist viewpoint; we are aware that following each of these major and minor catastrophes there was a varying percentage of surviving organisms that gave rise through adaptive radiation to the contents of the next unit. [*See* STRATIGRAPHY—DATING.]

a. Extinctions, Crises in the History of Life

It has been evident since the first half of the last century that major (and minor) eliminations of significant percentages of the existing biota, on land and sea, have taken place from time to time. The cause or causes of these crises are poorly understood. There is now good evidence that the earth was impacted by an extraterrestrial body at the Cretaceous–Tertiary boundary. Whether or not this impact event had a significant, minor, or no effect on the biota is still a subject of contention, as well as whether or not similar impacts occurred contemporaneously with other extinction events. There does not yet seem to be good evidence for such impacts at any extinction event horizon other than the Cretaceous–Tertiary. Our uncertainty concerning the cause(s) of the extinctions is understandable when considered against the fact that informed specialists are unable to agree about whether or not the relatively recent extinctions affecting many of the large mammals of the New World and Australia were largely caused by human activity, by climatic change, or by a combination of both. In other words, if we are unable to satisfactorily solve this relatively modern puzzle involving animals whose close relatives are still extant, what chance do we have for understanding the demise tens and hundreds of millions of years ago of organisms about whose habits we have a very poor understanding?

b. Adaptive Radiations

Because the history of life, from the point of its first recognition in the earlier Precambrian, is not a smooth-flowing record in which one species gradually evolves into another, but is rather a sequence in which lengthy intervals of biotic fixity are interspersed with brief intervals of revolutionary change (adaptive radiations, commonly but not invariably following after extinctions), it is necessary to consider their history. Because the best-known part of the fossil record, the bulk of the fossil record, has to do with the marine benthic environment, it is not surprising that most of the well-documented radiations are described from that environment. They range from major ones—such as the appearance of soft-bodied eumetazoans in the Ediacaran, skeletonized metazoans at the beginning of the Cambrian, the change from Paleozoic to Mesozoic marine organisms at the Permian–Triassic boundary—to far lower-level, third- and fourth-order adaptive radiations of concern mostly to specialists. Similar, although largely diachronous, adaptive radiations are documented within the nonmarine environment. For example, the major Phanerozoic subdivisions, based on marine invertebrates, are Paleozoic, Mesozoic, and Cenozoic, with boundaries at the Permian–Triassic and Cretaceous–Cenozoic boundaries, whereas the major Phanerozoic subdivisions based on higher land plants are Paleophytic, Mesophytic, and Cenophytic, with boundaries situated within the Late Permian and the Lower Cretaceous—times when nothing equally revolutionary happened within the marine benthic world.

C. Evidence for the Theory of Organic Evolution

One of the crowning achievements of modern science is the concept of organic evolution. This concept is based in no small part, both in Darwin's day and our own, on evidence from the geologic record. The history of life is an integral part of historical geology. The components making up the history of life need to be considered.

1. Taxonomic *cum* Morphologic

From Darwin's day the time sequence of fossil organisms has been an integral part of the concept of organic evolution. In group after group, at one taxonomic level or another, the fossil record provides evidence of earlier groups of organisms occurring prior to later ones. For example, procaryotic autotrophs occur alone in the earlier Precambrian—to which are added calcareous, eucaryotic algae in the Cambrian, lower embryophytes in the Middle Ordovician, vascular plants in the later Silurian, flowering plants in the earlier Cretaceous—whereas with vertebrates, we find jawless fishes alone in the Upper Cambrian—to which are added jawed vertebrates in the earlier Silurian, amphibians in the Late Devonian, reptiles in the Mississippian, mammals in the Upper Triassic, and placentals in the earlier Cretaceous—and so on for group after group of organisms. All of this is based on a consideration of morphology-taxonomy through geologic time. For the well-sampled groups the time sequence is reliable; but for the more poorly sampled groups (including the soft-bodied organisms and such lightly skeletonized groups as insects, mites, and spiders, among others) the sequence may be temporally unreliable.

2. Behavioral and Coevolutionary

Paralleling the evolutionary sequences built up from the evidence of morphology and taxonomy are the conclusions about behavioral evolution and coevolution. Most of the information about behavior is derived from a consideration of functional morphology; in other words, when one finds a vertebrate jaw containing sharp, spiky teeth it takes little imagination to conclude that its owner was a carnivore. The evidence of functional morphology, arrayed in a time-sequenced evolutionary manner, is most reliable when applied to taxa that belong to extant groups. We can provide reliable conclusions about the behavior of modern crabs and lobsters that can be directly applied to fossil crabs and lobsters, whereas conclusions about the behavior of an extinct subclass such as the trilobites are far less reliable, although their position as arthropods permits us to make some generalizations that are undoubtedly reliable. Additional to the evidence of functional morphology is that derived from what I term *frozen behavior*. For example, when one finds a pair of ants *in copulo* it is easy to compare their posture with that of modern ants belonging to the same group in order to see whether behavioral changes have occurred (in general, behavior at the family and lower levels tends to have many conservative aspects).

Coevolution may involve such things as the relation between parasite and host. The fossil record does provide evidence about certain kinds of coevolution. This evidence also occurs in a time-sequenced manner. The evidence we possess suggests that coevolved relations are very conservative once they appear. Mother Nature does not permit an arms race between disease organisms and their hosts nor between predators and their prey (the popular nineteenth century concept of an arms race between fossil cats and horses has turned out to be a gross misinterpretation of the record).

3. Community Ecologic

An integral part of the fossil record concerns itself not just with when varied taxa first and last appear (adaptive radiations and extinctions) but also with their abundances (both the fixed and changing aspects) through time *within* communities. One may define a community as a regularly recurring association of organisms within which each organism has a characteristic abundance—some very common, others less abundant, some rare. For example, the famous rhipidistian crossopterygian *Latimeria*, first found in the 1930s in very small numbers off the Seychelles, is presently so rare and geographically restricted as to lack a fossil record following the Cretaceous, whereas prior to the Cenozoic the rhipidistians were fairly abundant in the fossil record back as far as the Devonian. Brackish-water oysters, those that commonly occur in oyster banks in estuaries, first appear in the Cretaceous in great abundance, together with some co-occurring invertebrates common to the reef-oyster community, and persist in that way to the present.

Community history, as might be expected, is historically arrayed in a manner tightly parallel with the extinctions and adaptive radiations of the taxa. In other words, an extinction tends to largely eliminate or at least largely restructure previously existing community types. An adaptive radiation, on the other hand, tends to generate new community types or at least to greatly restructure previously existing community types. All of these community changes are grist for the historical mill.

4. Biogeographic

The physical geologist is concerned with the ever-changing aspect of the earth's geography. The paleontologist is concerned with the ever-changing biogeography. The descriptive aspect of biogeography deals with the distribution of organisms, pure and simple. Historical biogeography deals with the distribution of fossil organisms in time and space.

5. Biostratigraphic

Biostratigraphic evidence supporting the concept of organic evolution is provided by means of a time sequence of organisms, present for many groups, indicating a change from one level of organization to another level. For example, with photosynthetic organisms we first find Archaeozoic procaryotes, cyanobacteria, persisting for several billions of years, following which the Cambrian sees the first appearance of eucaryotic calcareous algae in the sea. On land we find evidence of cyanobacteria back into the Precambrian, with hepatics, well-developed embryophytes, first appearing in the Middle Ordovician, followed in the Upper Silurian by vascular plants. Among the vascular plants there is a regular progression from more primitive to more advanced, culminating in the earlier Cretaceous with the appearance of the flowering plants. Turning to the vertebrates one finds jawless fish present from the Upper Cambrian, jawed fish from the Lower Silurian, amphibians from the Upper Devonian, reptiles from the Mississippian (a very recent discovery pushed their lower range back from the Pennsylvanian), followed finally in the Upper Triassic by the mammals, with placental mammals not appearing until the earlier Cretaceous. Similar sequences are available for many other of the well-skeletonized animal groups.

D. Historical Biogeography

Historical biogeography deals with the distribution of organisms in time and space, both on land, fresh-water and terrestrial, and sea. Biogeographic units may be defined as areas, aquatic or terrestrial, in which there is potentially free reproductive communication between all of the organisms. Within each biogeographic unit there is, of course, the problem that each organism occurs within a specific community or communities. This is merely an enunciation of the obvious that each organism within any biogeographic unit can exist only within a specific local environment suited to its own requirements.

The hierarchy of biogeographic units, highest to lowest, is Realm, Region, Province, Subprovince. Synthesis of the distribution of fossils from the Cambrian to the present indicates that biogeographic units have been ever changing. This is not surprising when one considers that the basic paleogeography has always been in a state of flux, that global climatic gradients have always been increasing and decreasing in a sporadic manner, and that major extinctions and adaptive radiations have been taking place through time as well.

Concern for the causal in historical geology now makes it obvious that the factor(s) involved in causing major biogeographic changes, such as that from high cosmopolitanism to high provincialism, correlate poorly in time with those responsible for the major extinctions and adaptive radiations. Such a conclusion is based on such facts as the presence in the mid-Upper Devonian of a truly major extinction within the marine benthic environment during a time of globally high cosmopolitanism; whereas the major extinction at the end of the Permian was preceded by high provincialism but followed by high cosmopolitanism; and that at the end of the Ordovician was both preceded and followed by moderately high provincialism.

E. Community History

Community history deals with the first appearance, last appearance, and change in communities of organisms through time. Community history is concerned with whether or not the contents of a community type show evidence for change. Study of their history indicates that there is an inverse relation during the existence of a community type for rapid phyletic evolution on the part of the rare genera but conversely slower evolution for the more abundant forms. This is an area where a great deal remains to be learned. We still, for example, are unclear about whether immediately prior to a major extinction horizon it is the less abundant, more stenotopic organisms that drop out first, or whether during an adaptive radiation it is these same organism types that appear last.

IV. Conclusion

There are many additional possibilities for inquiry within the historical geology sector. The items summarized here are chiefly the traditional items. For example, we have yet to seriously consider the evidence provided by the fossil record pertaining to the history of disease or the rate of changes involved. Overall, the geological record is our only reliable source for assessing rate phenomena of all types, physical and biological. A variety of geochemical, including isotopic, items of a basically historical nature are currently being considered, although conclusions based on the currently available data are still in the contentious stage for the most part. Historical geology is potentially useful to planners because it provides one means of assessing the long-term effects of major environmental change. For example, would a sea-level canal through Lake Nicaragua materially affect the near-shore marine biotas of the adjacent eastern Pacific and Gulf of Mexico? Should we conduct the experiment before ascertaining the result? Examination of the fossil record—that is, what happens when a major biogeographic barrier in the shallow marine environment is removed—has been done. The results indicate that taxa from either side of the former barrier become extinct in what appears to be an almost random manner rather than having the biota of one or the other side totally decimated. Likewise, we are aware from an examination of the geological record that when a cold-climate biogeographic unit is removed, owing to global lowering of the climatic gradient, the former cool-climate biota is almost totally eliminated. In modern terms, goodbye polar bears, snowshoe rabbits, and Arctic foxes and their ecosystem. We have a lot to learn.

Bibliography

Cloud, P. (1988). "Oasis in Space." W. W. Norton, New York.

Gray, J., and Boucot, A. J. (1979). "Historical Biogeography, Plate Tectonics, & the Changing Environment." Oregon State Univ. Press, Corvallis, Oregon.

Hallam, A. (1973). "Atlas of Palaeobiogeography." Elsevier, Amsterdam.

Holland, H. D. (1984). "The Chemical Evolution of the atmosphere and Oceans." Princeton Univ. Press, Princeton, New Jersey.

Kauffman, E. G., and Walliser, O. H. (1990). "Extinction Events in Earth History: Lecture Notes in Earth Sciences, No. 30." Springer-Verlag, Berlin and New York.

Larwood, G. P. (1988). "Extinction and Survival in the Fossil Record." Systematics Assoc. Special Vol. No. 34, Oxford Science Publ., Oxford, England

Schopf, J. W. (1983). "Earth's Earliest Biosphere: It's Origin and Evolution." Princeton Univ. Press, Princeton, New Jersey.

Glossary

Adaptive radiation Evolutionary diversification of a taxon's morphology, ecology, physiology, behavior, and other characteristics over a geologically short time interval, leading to the *appearance* of a number of new taxa.

Biostratigraphy Relative (not absolute, i.e., number of years) age dating and time correlation of stratified rocks (mostly sedimentary plus some volcanics) by means of fossils contained in those rocks, based on our knowledge of their time sequence locally, regionally, and globally— all of this ultimately being a function of evolutionary change.

Historical biogeography Discipline that deals with the ever-changing distribution of organisms, marine and nonmarine, over the earth's surface through geologic time; i.e., the historical counterpart of present-day biogeography, which discusses the modern distribution of all living organisms.

Paleogeography Working out by means of geologic criteria (chiefly climatically sensitive sedimentary rocks and environmentally sensitive rocks and structures, such as nonmarine, shoreline, or deep-water types) and structural geologic evidence for the displacement of rock bodies that originally were continuous, geophysical criteria (chiefly the evidence of remanent magnetism, which provides data regarding the original latitudinal position of rock bodies, although not whether that latitude is north or south), and paleontologic criteria (chiefly the evidence of historical biogeography that helps to establish which rock bodies contain fossils, suggesting original reproductive communication between same) of the ever-changing geographies of the past.

Hydrogen Peroxide Dynamics in Marine and Fresh Water Systems

William Cooper
Florida International University

David Lean
Environment Canada

Knowledge of the mechanisms of formation and decay of hydrogen peroxide (H_2O_2) in fresh and marine waters is a prerequisite to insight on its distribution in the water column. With this information it will then be possible to assess how H_2O_2 may shape ecosystem biogeochemistry. As a strong oxidizing agent, H_2O_2 may alter many chemical and biological processes. The state of our current understanding will be reviewed and the possible impact of H_2O_2 on the hydrosphere will be discussed.

Throughout this review, examples are provided from experiments recently conducted at the research station of the National Water Research Institute of Environment Canada. Here, on the edge of the Canadian Shield, lakes have a wide variety of chemical features. For consistency, most of the experiments discussed below were conducted on Sharpes Bay, Jacks Lake.

I. Formation of H_2O_2 in Surface Waters

Hydrogen peroxide has been found in marine, estuarine, and fresh water ecosystems but until recently no overall predictive relationships have emerged. Typical concentrations of H_2O_2 (Table I) are in the 80 to 400 nM range; in general, values for marine system are lower than for lakes and rivers. Diel patterns of H_2O_2 have been obtained. Concentrations generally increase during the daytime and decline at night, but until recently patterns have appeared complex and at times inconsistent. The formation of H_2O_2 could potentially result from either photochemical or biological processes as well as chemical oxidation–reduction reactions.

A. Photochemical Formation

Any process that involves an electron transfer in oxygenated waters can potentially lead to H_2O_2 formation, but the major source occurs when sunlight strikes dissolved organic compounds. All natural waters contain yellow organic compounds originally called "Gelbstoffe" or humic substances, which absorb sunlight energy in the ultraviolet (UV) region. This results in the formation of excited-state (highly energetic) molecules of humic substances (HS*) that transfer their energy to oxygen, resulting in the formation of H_2O_2.

Sunlight that penetrates the atmosphere and reaches the water has a minimum wavelength of approximately 300 nm. The lower portion of the solar spectrum has been conveniently divided into different regions. The range from 280 to 320 nm is referred to as UV-B, and UV-A is from 320 to 400 nm. The region from 400 to 750 nm is the visible range and corresponds to that referred to as photosynthetically active radiation (PAR). In natural waters, light in the PAR region is absorbed by microscopic plants called algae; it is converted to biomolecules, which are used as energy not only for the plants but also to support the bacteria and animal community. Infrared light exists above 750 nm, but we cannot see it, nor does it play a significant role in surface-water photochemical reactions. [*See* SOLAR RADIATION.]

Light in the lower wavelength region (UV-B) is damaging to all bacteria, plants, and animals. Life on earth as we know it today is protected by absorption of UV-B light by ozone (O_3) in the stratosphere. Therefore, if stratospheric ozone is depleted, an increase in light energy available for photochemical reactions will occur partly as a consequence of (a) the increase in the intensity of the UV-B wavelengths and (b) an increased proportion of lower wavelengths with higher energy output

Table I   Reported H_2O_2 Surface Concentrations in Natural Waters

Water source	H_2O_2 Natural levels (nM)
Fresh water	
Volga River area, Russia	1300–3200
Reservoir, Russia	700–1300
Southeastern United States	90–320
Jacks Lake, Ontario, Canada	10–800
Lake Erie	50–200
Lake Ontario	100
Sewage	
Russia	1200–8500
Seawater	
Texas coastal waters	14–170
Biscayne Bay, Florida coast	80–210
Gulf of Mexico	120–140
Coastal	100–240
Offshore	90–140
Bahama Bank	50–190
Peru coastal and offshore	8–70
Estuarine	
Chesapeake Bay	3–1700

per unit quanta. At 300 nm, 90 kcal per mole quanta (einstein) can break most carbon bonds if absorbed. Increased incidence of skin cancer is only one of the consequences. The quantitative consequences of this increase in available energy are not known, but there is little doubt that the photochemical processes resulting in the formation of H_2O_2 will increase. [*See* OZONE, ABSORPTION AND EMISSION OF RADIATION.]

For a photochemical reaction to occur, light must be absorbed by a molecule. The light absorbance characteristics of several lake waters is illustrated on Fig. 1. The highest curve was obtained from an acid bog, which contains higher levels of humic substances. The lower curves were for water samples obtained from Brookes Bay (Jacks Lake), Anstruther Lake (a soft-water lake), Sharpes Bay (Jacks Lake), and Lake Erie, respectively. The predominant absorbing materials in natural waters are operationally defined as humic substances (HS), which represent a significant part of the so-called dissolved organic matter (DOM) fraction and are also referred to as dissolved organic carbon (DOC). This substance has defied complete identification by the most modern chemical techniques. It consists of a complex mixture of small (molecular weights up to 500) and large (molecular weights up to and in excess of 10,000) molecules with both aliphatic and aromatic character. A generalized mechanism for the photochemical formation of H_2O_2 from light absorption by humic substances (HS) is as follows. The ground state (not excited) of the humic substances (HS) is the most common form

found in natural waters; but when it absorbs energy from sunlight, it is excited to the singlet state ($^1HS^*$). Through energy transfer, intersystem crossing, it goes to the excited triplet state ($^3HS^*$). In both cases the singlet and triplet states have more energy than the ground state, and both are reactive. One major difference between the $^1HS^*$ and the $^3HS^*$ is that the lifetime of the $^3HS^*$ is much longer than the lifetime of the $^1HS^*$; therefore, it is much more likely to react with another molecule. In other words, there is a longer time for a collision to occur. This results in an energy transfer producing what is called a caged complex ($[HS^+\cdot + e^-]$) that splits apart to give an aqueous electron (e^-_{aq}). These reactions are illustrated here.

$$HS + light \rightarrow {}^1HS^* \text{ or } {}^3HS^* \tag{1}$$

$$^1HS^* \text{ or } {}^3HS^* \rightarrow [HS^+\cdot + e^-] \tag{2}$$

$$[HS^+\cdot + e^-] \rightarrow HS^+\cdot + e^-_{aq} \tag{3}$$

Most molecules are found in their ground state. The oxygen molecule is an exception in that the ground state is the triplet state ($^3O_2$). Thus, ground state O_2 can accept energy either from an aqueous electron or the $^3HS^*$. Either reaction results in the formation of superoxide ion ($O_2^-\cdot$), which is a weak acid with a pKa of 4.8. In other words, an equilibrium exists in aqueous solution that depends on the pH or hydrogen ion concentration of the system. Disproportionation or dismutation (reaction with itself) of the $O_2^-\cdot$ occurs, resulting in the formation of H_2O_2. These reactions and other potential side reactions are illustrated here.

$$O_2 + e^-_{aq} \rightarrow O_2^-\cdot \tag{4}$$

$$^3HS^* + {}^3O_2 \rightarrow HS + {}^1O_2 \tag{5}$$

$$^3HS^* + {}^3O_2 \rightarrow HS^+\cdot + O_2^-\cdot \tag{6}$$

$$HS + {}^1O_2 \rightarrow HS^+\cdot + O_2^-\cdot \tag{7}$$

$$^3HS^* + RNH_2 \rightarrow HS^-\cdot + RNH_2^+ \tag{8}$$

$$HS^-\cdot + O_2 \rightarrow HS + O_2^-\cdot \tag{9}$$

$$2O_2^-\cdot + 2H^+ \rightarrow H_2O_2 + O_2 \tag{10}$$

$HS^+\cdot$ and $HS^-\cdot$ are the cation and anion radicals of humic substances. RNH_2 represents nitrogen-containing compounds. [*See* ORGANIC MATTER DECOMPOSITION IN THE MARINE ENVIRONMENT; OXYGEN, BIOGEOCHEMICAL CYCLE.]

Although details of the above mechanisms have not been confirmed, several pathways have been studied. First, experi-

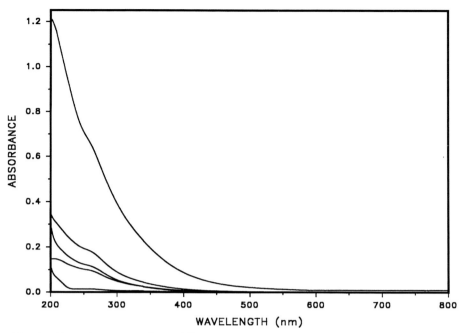

Fig. 1 Absorption spectra of several filtered (0.2 μm) waters at a variety of DOC concentrations. Distilled water does not show any significant absorption.

ments have been conducted that provide evidence that the intermediate $O_2^-\cdot$ does play an important role in the overall reaction. The enzyme called superoxide dismutase (SOD) was added to samples of seawater and lake water before exposing the sample to sunlight. The production of H_2O_2 was compared to a sample of the same water with no SOD. SOD excludes other side reactions and quantitatively converts $O_2^-\cdot$ to H_2O_2. Relatively few natural waters have been studied using this technique, and until additional experiments are conducted on a variety of natural waters no generalizations are possible. Our preliminary observations suggest that much of the $O_2^-\cdot$ formed is converted to H_2O_2 leaving little to react with other substances.

Second, studies to determine the relative importance of the reduction of O_2 by an aqueous electron (e_{aq}^-) or direct energy transfer of the triplet excited state $^3H^*$ to ground state O_2 have been reported. However, no conclusive results have been obtained, and both reactions are shown in the mechanism.

The third area, the effect of concentration of HS in Eq. (3) on the formation of H_2O_2 has received more intensive investigations than the other two. The efficiency of light in initiating a chemical reaction is expressed as a quantum yield. This is a dimensionless number which, if multiplied by 100, gives the efficiency (in percent) for the overall process. To describe a sunlight-induced photochemical reaction, the quantum

yield must be determined at several wavelengths throughout the solar spectrum. The resulting curve is an action spectra. Quantum yield is the ratio of chemical energy produced per unit of sunlight energy (1 einstein is equal to 1 mole of quanta). In the case of photochemical reactions involving humic substances where the exact structure, molecular weight, and variability from one location to another are unknowns, an exact number is impossible. However, apparent quantum yields for several natural waters containing humic substances have been reported. Examples of spring water, groundwater, and lake water are shown in Fig. 2. The important feature of these apparent quantum yields is that the wavelengths responsible for most of the H_2O_2 formation are between 300 and 400 nm. These UV-A wavelengths are typically filtered out within the first meter, but in low humic waters they may penetrate up to 10 m.

From the action spectra shown in Fig. 2, it is also apparent that light at wavelengths lower than 300 nm are more efficient in producing H_2O_2. From this observation it is possible to speculate that if stratospheric ozone is depleted, increased formation of H_2O_2 would be expected in natural waters.

Humic substances represent a quantity of carbon from 1 to 20 times more abundant than the living carbon component in lakes and oceans; they can exist at concentrations from < 1 to > 30 mg/liter. As such, the measure of DOC or DOM is generally assumed to be related to the quantity of humic

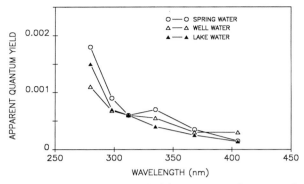

Fig. 2 Apparent quantum yield of four water samples measured in the range from 280 to 405 nm.

substances present rather than such compounds as simple sugars or amino acids. Organic compounds that are easily degraded through microbial activity represent an insignificant fraction of the DOC pool, and only refractory organic compounds that are resistant to microbial breakdown persist. Studies have been reported that show a positive correlation of H_2O_2 formation rate with increasing DOC. In other studies, the formation rate of H_2O_2 correlates well with fluorescence (excitation at 365 nm and emission at 490 nm) of the waters. It is believed that the humic substances present in the waters are responsible for this fluorescence. Thus, the rate of formation measured per unit volume is related to the concentration of dissolved organic carbon or humic substances but, as will be discussed below, on an areal basis the formation lakes in most natural waters is similar. Only the depth distribution of the formation changes with the concentration of humic substances.

Humic substances are not the only photochemically active compounds that may lead to the formation of H_2O_2 in natural waters. The amino acids tryptophan and methionine, when exposed to light, have been shown to produce H_2O_2 in oxygenated waters. It is thought that this results from the reduction of O_2 by e_{aq}^-. Dissolved flavins (e.g., riboflavin) have also been shown to produce H_2O_2 in oxygenated waters exposed to sunlight. In general, simple molecules such as those above are easily metabolized and broken down by aquatic bacteria and are rarely found at high concentrations in natural waters. Therefore, the relative importance of this formation pathway has not been established.

The formation of H_2O_2 can be determined by incubating water samples in quartz glass tubes suspended in the water column at fixed depths. An example of one such experiment is shown on Fig. 3. To exclude any influence of microorganisms, samples were filtered through 0.2 μm Nuclepore filters and incubated *in situ* from 1200 to 1400 h on 12 and 13 September 1990. Additional experiments on the changes

in water column [H_2O_2] were also conducted on these days and will be discussed in Section II,C. Formation rates declined exponentially with depth but even at 1 m, there was still measurable production even though little UV-A light penetration extends to this depth. Formation by the longer wavelengths is the likely explanation even though the quantum yield is low. This illustrates the need for reliable spectral quality estimates of light in the deeper waters.

B. Biological Formation

It is well known that H_2O_2 is associated with metabolism in aerobic microorganisms and algae. Therefore, it is possible that biological processes may result in the formation of H_2O_2 in natural waters. Extracellular enzyme systems, such as L-amino acid oxidases, have been identified in phytoplankton that are used for the oxidation of amino acids to form ammonia (the ammonia is subsequently utilized by the organism). Another reaction by-product is H_2O_2. It is very possible that this process may contribute significant quantities of H_2O_2 to natural waters below the depths at which photochemical processes are important. However, in the near surface, the photochemical processes are quantitatively more important than the biological processes that lead to the formation of H_2O_2.

C. Chemical Formation

The last potential pathway for H_2O_2 formation in natural waters is as a consequence of chemical reactions. One example of this for copper redox cycling in marine waters:

$$Cu^+ + O_2 \rightarrow Cu^{2+} + O_2^- \cdot \qquad (11)$$

This general process will be discussed in Section III.

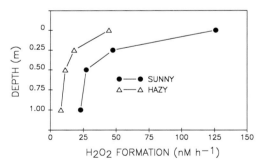

Fig. 3 Formation rate of H_2O_2 in filtered Sharpes Bay water contained in quartz tubes held at several depths in the water column from 1200 to 1400 on 12 and 13 September 1990.

II. Decay or Loss of H_2O_2 in Natural Waters

The loss or decay of H_2O_2 in natural waters can result from either biological and/or chemical decomposition.

A. Biological Decay

Although as an oxidizing agent hydrogen peroxide may potentially react with inorganic compounds such as reduced species of the oxides of sulfur and nitrogen, it has been established that the reaction of H_2O_2 with catalase and/or peroxidase enzyme systems associated with heterotrophic bacteria is principally responsible for the losses of H_2O_2 observed in natural waters. These organisms are responsible for breaking down organic materials; they exist at concentrations from 500,000 to 8,000,000 organisms per milliliter.

To determine a decay rate, samples of unfiltered lake water are incubated in the dark and subsamples removed at intervals for up to 24 h. The decline follows first-order kinetics, so the slope of the natural logarithm of the concentration as a function of time provides the rate constant. It represents the fraction of the H_2O_2 lost per unit time. Concentrations of added H_2O_2 give similar decay-rate constants in the range from 50 to 500 nM. This is important to the development of simple models developed from extrapolation of laboratory-derived rate constants.

The addition of microbial inhibitors, sterilization, filtration through various pore size filters, and the development of kinetic expressions for the decomposition of H_2O_2 using pure strains of bacteria provide evidence as to the importance of biological rather than chemical decay of H_2O_2. In one experiment, H_2O_2 was added to the samples of reconstituted soil suspensions and natural waters. Typical decay rates resulted in half-lives of H_2O_2 of from 1 to 5 h. Insignificant H_2O_2 losses were observed over 7 h, when the same samples were treated with 0.025 M formaldehyde or 0.50 mM Hg(II). Samples that were autoclaved or boiled for 10 min showed no loss of H_2O_2 over 24 h.

Another experimental approach, used to establish that bacteria are important in the loss of H_2O_2 in natural waters, involves determining the decay rates in waters filtered through various pore sizes. A 64-μm Nitex screen removed major zooplankton, the 12-μm Nuclepore filter removed large algae, the 1-μm filter removed almost all algae and about 10% of the heterotrophic bacteria, and the 0.2-μm filter removed almost all the heterotrophic bacteria. For example, the decay rate expressed as half-life for H_2O_2 in a sample from Sharpes Bay obtained on 11 September 1990 (Table II) demonstrates that organisms < 1 μm but > 0.2 μm are responsible for much of the decay. This fraction contains ca. 90% of the bacteria. When all organisms and other particulate mate-

Table II The Loss of H_2O_2, as Expressed by Half-Life, in Lake Water and Filtrates of Lake Water through Various Pore-Size Filters

Sample	Organisms	Half-life, $t_{1/2}$ (h)[a]
Lake water (unfiltered)	Natural assemblage	4.4
64 μm	Zooplankton removed	4.7
12 μm	Large algae removed	6.4
1.0 μm	Small algae removed	19.1
0.2 μm	Bacteria removed	58.7

[a] The half-life was determined as it would be for a first-order kinetic process (i.e., the ln 2 divided by the decay rate constant).

rial is removed, H_2O_2 is quite stable in the water sample. Without decay, concentrations of H_2O_2 would increase. During the night or dark phase when little or no H_2O_2 is formed, the concentration of H_2O_2 is lowered by the decay processes. With a half-life of 60 h, a 12-h dark period would result in the loss of only approximately 20% of the H_2O_2.

The third approach that has been used to provide evidence that the decay processes are microbially mediated were experiments with pure bacteria cultures. *Vibrio alginolyticus* was grown in pure culture and the kinetics of the loss of H_2O_2 determined. The loss of H_2O_2 was shown to be second order overall, first order in both H_2O_2 concentration and bacteria numbers. The equation is

$$-d[H_2O_2]/dt = k_2[H_2O_2][\text{cell number}] \qquad (12)$$

where $k_2 = 1.6 \times 10^{-9}$ ml cell^{-1} min^{-1}. A similar result was obtained for the fresh water bacterium *Enterobacter cloaceae* where $k_2 = 1.5 \times 10^{-9}$ ml cell^{-1} min^{-1}.

Having established that microbial processes are important in the decay of H_2O_2 in natural waters, then several variables would affect the rate of decomposition. These variables would include not only bacterial numbers, but also the kind of bacteria and their physiological state. Other variables that may be important are temperature and H_2O_2 concentration. Other than the data given above, no data have been reported relating bacterial numbers to decay rates, at least at H_2O_2 concentrations normally found in natural environments. While the decay rate is a function of concentration, the decay rate constant is independent of concentration throughout the range of 50–500 nM. We have found that decay rates in lakes are related to total bacterial numbers (Fig. 4). To determine bacteria numbers, a sample is treated using a DNA-specific fluorescent dye, then counted using a high-power microscope illuminated using light of a wavelength that causes the dye to be visible. In the example, the decay rate constant is related to total bacteria numbers for Sharpes Bay and for the Eastern, Central, and Western basins of Lake Erie. The decay rate constant can be considered to represent the fraction of the

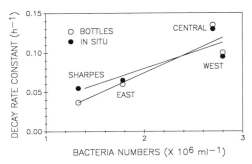

Fig. 4 Hydrogen peroxide decay-rate constant in water samples from Sharpes Bay, Jacks Lake, Ontario, Canada and from the East, Central, and West basins of Lake Erie plotted as a function of bacteria numbers.

H_2O_2 lost per hour. Times 100, the range is from 3 to 14% per hour. The results were obtained in two ways. H_2O_2 concentrations measured in water samples were incubated in bottles kept in the laboratory in the dark and compared with H_2O_2 concentrations monitored in the lake water column from 9 p.m. to 5 a.m. the next morning. The two methods gave similar values. Nevertheless, the relationship was not perfect and we suspect that some of the discrepancy occurred because of errors in bacteria counts and the presence of a class of bacteria capable of photosynthesis also called blue-green "algae." Clearly, more work is required, but the results are encouraging.

Decay rate constants were also determined for a water sample incubated at several temperatures (Fig. 5). For every 10°C increase in temperature the rate constant increased by about 0.05 h^{-1}. Freshly collected lake water was stored for a 12-h acclimation period at the temperatures shown. H_2O_2 was added and the decay rate determined by measuring the change in concentration with time. These results show that temperature directly affects the decay rate of lake waters unless some physiological adaptation occurs after longer acclimation. Changes in bacteria abundance was not measured. Nevertheless, it would seem that the cooler the water, the slower the decomposition of H_2O_2.

B. Chemical Decay

As mentioned above, H_2O_2 is a strong oxidant. Thermodynamically, it is one of the most highly oxidizing chemicals, but kinetically it is often limited. That is, the rate at which it oxidizes chemicals is often slow. A generalized reaction that would lead to the loss of H_2O_2 would be

$$H_2O_2 + M^{n+} \rightarrow M^{(n+1)+} + OH \cdot + OH^- \qquad (13)$$

where M^{n+} is the reduced form of a metal and $M^{(n+1)+}$ is the

more oxidized form; $OH \cdot$ is the hydroxyl radical (the strongest oxidizing substance in aqueous solution); and OH^- is a hydroxyl ion. The hydroxyl radical is simply the hydroxyl ion minus one electron in the outer ring. In marine environments, studies of Cu and Fe have shown that these reactions do occur. However, because of the low concentrations of the metals in the ocean, these reactions would be unlikely to occur in the upper water column where H_2O_2 is present. In fresh water the reactions of H_2O_2 and metals has not been studied, but the formation of Fe(II) from Fe(III) has been observed in samples exposed to sunlight.

C. Distribution of H_2O_2 in Natural Waters

The distribution of H_2O_2 results from the balance between formation, decay, and the physical process of mixing. We have concluded above that the formation of H_2O_2 is due to abiotic photochemical processes, but decay results from microbial degradation. To understand the factors that influence the distribution of H_2O_2 through the upper mixed layer, wind- and temperature-driven mixing processes must be considered.

In general, few depth profiles of H_2O_2 exist and even fewer diel (day–night) measurements have been made. However, there are some general trends. The wide variation in H_2O_2 concentration in surface waters is remarkable (Table I). The upper water column is at a constant temperature and has been referred to as the mixed layer. In lakes during the summer, the mixed layer or epilimnion is usually less than 7 m but may extend to 20 m in large lakes like Superior. In oceans, it is not uncommon to have the mixed layer as deep as 100 m. Since sunlight is absorbed on an areal basis, complete depth H_2O_2 concentration profiles are necessary so that the total quantity can be quantified.

To illustrate the combined influence of wind and sunlight,

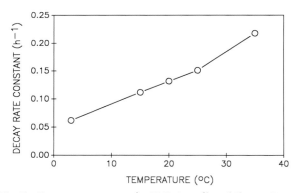

Fig. 5 Decay-rate constants for H_2O_2 in unfiltered Sharpes Bay lake water contained in glass bottles in the dark at temperatures from 3 to 35°C. Initial concentrations were about 270 nM.

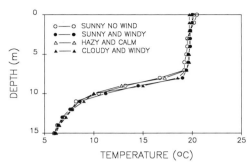

Fig. 6 Temperature profiles in Sharpes Bay, Jacks Lake, Ontario, Canada measured on 11 September 1990 when conditions were sunny with no wind and on subsequent days that were sunny and windy, hazy and fairly calm, and cloudy and windy.

Table III Areal Daily Formation of H_2O_2

Location	Date	mg H_2O_2 m^{-2} day^{-1}
Gulf of Mexico		
Near-shore	May 82	21.4
Coastal	August 82	20.4
Cariaco trench	March 86	21.6
Jacks Lake		
Sharpes Bay	June 88	54.0–61.9
Lake Erie		
East basin	July 88	27.1–48.2
Central basin	July 88	25.2–28.8
West basin	July 88	23.0–39.1

we conducted experiments on 11–14 September 1990 that illustrate features common to all systems. Temperature (Fig. 6) and H_2O_2 concentration profiles (Fig. 7) were measured each afternoon at 1600 h. Early morning H_2O_2 concentrations were near detection limits (10 nM). The temperature profiles were similar on all four days except that surface temperatures were slightly higher and deep epilimnetic water slightly cooler on calm days. Furthermore, the water was warmer in the metalimnion during windy days. In summary, the temperature profiles provided little information on subtle differences in water column mixing.

Recalling that most of the H_2O_2 formation occurred in the top meter (Fig. 3), the *in situ* H_2O_2 concentration profiles (Fig. 7) illustrate that H_2O_2 is an excellent tracer for vertical mixing. The first day was sunny with no wind. Although the temperature indicated a fairly well mixed epilimnion to 7 m (Fig. 6), the H_2O_2 concentration reflected rapid formation at the surface with little mixing. The next day was windy and sunny and the H_2O_2 was mixed down to 4 m with elevated concentrations even to 7 m. Nevertheless, the areal H_2O_2 formation rate on these two days was almost identical. The

next day was hazy (diffuse cloud cover) with no wind; lower concentrations of H_2O_2 were observed but concentrations were higher near the surface. The fourth day, under cloudy and windy conditions, the little H_2O_2 that was formed was more evenly distributed.

The major difference between these H_2O_2 concentration profiles and those observed in oceanic environments is that H_2O_2 is found down to greater depths and in regions like the Sargasso Sea, as deep as 100 m. Since the formation occurs only in surface waters, the deep mixing zone results in concentrations generally 100 nM H_2O_2 or less.

The half-lifes for H_2O_2 decay for the coastal region are 24–30 h and exceed 100 h in the oligotrophic ocean compared to the 4.4 h in Sharpes Bay (Table II). This gives the impression that both production and losses are slower in the ocean. However, expressed on an areal basis total production corrected for decay and the total 24-h decay are in balance and are similar to the range observed in lakes (Table III).

Although the total amount of H_2O_2 formed, corrected for decay, is the same as that measured in the quartz flasks, the H_2O_2 is distributed throughout the water column, revealing the recent pattern of vertical mixing. In marine environments such as the Sargasso Sea, this mixing may result in H_2O_2 being observed to greater than 100 m, whereas in fresh water the distribution is limited to the 2 to 20 m depth of the epilimnion.

A potentially significant source of H_2O_2 in surface waters is rain. A summary of the H_2O_2 concentration in rain is provided in Table IV. The concentration of H_2O_2 is given in μM whereas for surface water, all concentrations are given in nM. In many cases, the H_2O_2 concentration in rain is 100 to 1000 times that found in surface water. Therefore, a 1-cm rainfall could result in a significant increase in the H_2O_2 in the upper water column.

In oceanic environments, the upper mixed layer is much deeper than in lakes. However, the decay rates are such that a rain event may increase the H_2O_2 to 10 m and remain at elevated levels for several days. In lakes, a rain event would

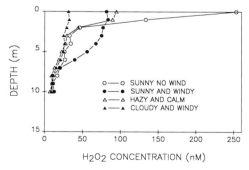

Fig. 7 Hydrogen peroxide profiles as a function of depth.

Table IV Concentration of H_2O_2 in Rain from Marine Environments and in Ontario, Canada[a]

Location	H_2O_2 (μM)
Gulf of Mexico	11.4–82.0
Western Atlantic	8.4–20.6
Florida Keys	24.3–31.9
Jacks Lake, Canada	1.3–34.0

[a] Units of concentration are μM and are 1000 times nM concentrations on Table I.

cause a much greater increase in the upper water column $[H_2O_2]$; however, it would not persist as long because of the higher rate of decay.

III. Impact of H_2O_2 on Hydrosphere

Many details of the cycling of H_2O_2 in natural waters are not known and almost no studies have been conducted that examine the question of its impact on the ecosystem. Hydrogen peroxide plays an important role in oxidation–reduction (redox) processes in natural waters. Another potentially important reaction pathway related to H_2O_2 is the peroxidase enzyme systems. These reactions may be important in both inorganic and organic reactions where oxidized species or free radicals are formed and react to form other products. We can speculate that it may restrict the distribution of organisms not able to protect themselves from such exposures.

Redox metal cycling in natural waters is of interest for several reasons.

1. Metal toxicity to organisms is often related to the oxidation state.
2. Metal bioavailability is often related to the oxidation state.
3. Geochemical cycling of the metals involves redox chemistry.

The redox chemistry of metals in seawater is very complicated, and a simplified description will be presented here. The focus is how H_2O_2 affects this chemistry. It should be pointed out that there are few similarities between fresh and marine systems regarding metal speciation. The principle reason for this is the presence of salts in the marine waters, which form ionic and nonionic (ion pairing) complexes with different oxidation states of the metals. These bound metals are stable

on time scales that are comparable to formation and decay of H_2O_2. A generalized mechanism has been proposed.

$$M^{n+} + H_2O_2 \rightarrow M^{(n+1)+} + OH\cdot + OH^- \quad (14)$$

$$M^{n+} + OH\cdot \rightarrow M^{(n+1)+} + OH^- \quad (15)$$

$$H_2O_2 \leftrightarrow H^+ + HO_2^- \quad (16)$$

$$M^{(n+1)+} + HO_2^- \rightarrow M^{n+} + HO_2 \quad (17)$$

$$HO_2 \leftrightarrow H^+ + O_2^-\cdot \quad pK_a = 4.8 \quad (18)$$

$$M^{(n+1)+} + O_2^-\cdot \leftrightarrow M^{n+} + O_2 \quad (19)$$

where HO_2^- is the peroxide base and HO_2 is the conjugate acid of superoxide.

Mechanisms such as the one proposed above are helpful in defining the overall chemistry of chemicals involved. However, without a detailed understanding of the individual reaction rate constants, the relative importance of these reactions cannot be evaluated. For example the oxidation of Fe(II) by H_2O_2 has been shown to be a first order with respect to $[OH^-]$, while no pH dependence was observed for the oxidation of Cu(I) in the range of 6.2 to 8.3. The Cu(I) oxidation H_2O_2 was shown to be highly dependent on $[Cl^-]$, with the rate increasing as the $[Cl^-]$ decreased. The details of these reactions are beyond the scope of this article. [See WATER GEOCHEMISTRY.]

As mentioned above, although H_2O_2 is thermodynamically a strong oxidant, it is often limited kinetically in reactions with organic compounds. As shown in the above mechanisms, the presence of metals results in the formation of more highly reactive species (superoxide and hydroxyl radicals). No detailed studies have been reported that assess the influence of these reactions in the fate of organic compounds (e.g., pollutants); however, it appears that these processes may serve to degrade organic compounds in natural waters.

In summary, the redox cycling of both Cu and Fe may be affected by the presence of H_2O_2. It has also been shown that the presence of these two metals in natural waters may affect the formation and decay of H_2O_2. These observations cannot be generalized because little research has been conducted in the area. One example is Cr where H_2O_2 appears to affect the redox chemistry in marine systems. Quantitatively, biological decay appears to be more significant in the overall cycling of H_2O_2 in the upper water column when compared to losses due to metal cycling.

It has been shown that biological decay is important in accounting for the loss of H_2O_2 in aquatic systems. Two enzymes are responsible for the removal of H_2O_2: catalases and peroxidases. Catalases decompose H_2O_2 with evolution of oxygen and water. Peroxidases decompose H_2O_2 to water with no evolution of oxygen; however, it does require a chemical in its reduced state that acts as an electron or hydrogen donor. The reduced chemical is converted to its oxidized state. The oxidized chemical in some cases may be a highly reactive free radical, which can then initiate a series of reactions that may lead to the formation of different chemicals and/or the loss of the original molecule.

One implication of the peroxidase-mediated decomposition of H_2O_2 is the fate of organic pollutants. These reactions may occur in the dark or in the light and require the presence of H_2O_2 and extracellular peroxidases. More specifically, *p*-anisidine was stoichiometrically removed when added to a natural water merely by adding H_2O_2. In control solutions without peroxidase, H_2O_2 did not react with *p*-anisidine.

The importance of this mechanism in aquatic systems cannot be assessed at this time but it is known that many compounds (e.g., phenols) serve as H donors in peroxidase-mediated reactions. This area of research has been largely overlooked, but it is possible to speculate that these reactions may be involved in the *in situ* formation of complex molecules that resemble humic substances and possibly the formation of surface microlayers.

The influence of H_2O_2 on the organisms that live in lakes and oceans has rarely been studied even though concentrations in the water exceed intracellular concentrations by about four orders of magnitude. Bacteria growth rates are inhibited at 500 nM, but the effect on algae, zooplankton, and fish is not known. Perhaps, the vertical migration of zooplankton may be related to presence of H_2O_2 and other oxygen radicals.

Bibliography

Cooper, W. J., Zika, R. G., Petasne, R. G., and Fischer, A. M. (1989). Sunlight-induced photochemistry of humic substances in natural waters: major reactive species. *In* "Aquatic Humic Substances: Influence on Fate and Treatment of Pollutants." Advances in Chemistry Series No.

219, pp. 333–362. American Chemical Society, Washington, D.C.

Cooper, W. J., Lean, D. R. S., and Carey, J. (1989). Spatial and temporal patterns of hydrogen peroxide in lake waters. *Can. J. Fish. Aquat. Sci.* **46**, 1227–1231.

Moffett, J. W., and Zika, R. G. (1987). Reaction kinetics of hydrogen peroxide with copper and iron in seawater. *Environ. Sci. Technol.* **21**, 804–810.

Palenik, B., Zafiriou, O. C., and Morel, F. M. M. (1987). Hydrogen peroxide production by a marine phytoplankter. *Limnol. Oceanogr.* **32**, 1365–1369.

Zika, R. G., Saltzman, E. S., and Cooper, W. J. (1985). Hydrogen peroxide concentrations in the Peru upwelling area. *Mar. Chem.* **17**, 265–275.

Glossary

Dissolved organic carbon (DOC) Concentration of dissolved organic compounds, expressed as carbon, in natural waters. DOC is operationally defined by the equipment used for analysis.

Fluorescence Emission of a certain wavelength light, resulting from and occurring only during the absorption of light of another wavelength.

Humic substances (Gelbstoffe) Ill-defined organic compounds that give waters their brown/yellow color. It is a complex mixture of negatively charged polymers, with a molecular weight distribution that is more or less continuous from less than 500 to greater than 20,000. The molecules have both aliphatic and aromatic characteristics.

Photosynthetically active radiation Portion of the solar spectrum from 400 to 750 nm.

Quantum yield Efficiency of conversion of light energy (sunlight energy in the case of natural waters) to chemical energy.

Superoxide ion ($O_2^-\cdot$) Reduced form of dioxygen that is an anion free radical.

UV-A Portion of the solar spectrum from 320 to 400 nm.

UV-B Portion of the solar spectrum from 280 to 320 nm.

Hydrologic Forecasting

Soroosh Sorooshian
The University of Arizona

The survival of life on this planet depends not only on the presence of adequate amounts of water on the land, in the oceans, and in the oceans, and in the atmosphere but also on the continuous exchange of water between these three phases. This redistribution of water occurs through several processes, the major ones of interest to human activity being precipitation, evaporation, and surface and subsurface runoff. Increases in the demand for water due to population growth and industrialization have resulted in a need to properly manage the quantity and quality of the available water. The development of effective management policies requires us to estimate how the various states (e.g., surface water storage volumes, groundwater levels) and fluxes (i.e., precipitation rates, snowmelt rates, streamflow, reservoir evaporation rates) of the hydrologic cycle will vary with time. The process of obtaining these estimates is known as hydrologic forecasting. Today, such forecasts are typically generated with the help of computer-based models.

I. Introduction

Most of the techniques for hydrologic forecasting are designed to aid in the management of river basins. A comprehensive management plan includes components such as con-sideration of flood protection, energy generation, water supply, irrigation, navigation, fisheries, recreation, and preservation of the aquatic system (Fig. 1). Central to all of these is the ability to forecast how the quantity and quality of water in rivers and lakes, groundwater reservoirs, and glaciers and snowpacks will evolve over time. The nature of the problems associated with forecasting depends on the temporal and spatial scales over which the forecasts are required. Hydrologic forecasts typically fit into one of the following three temporal classifications: (1) short term (1 hour to 2 days), medium term (2–10 days), and long term (more than 10 days). [*See* WATER CYCLE, GLOBAL.]

For example, for local authorities to implement floodplain evacuation measures during flash floods, short-term forecasts on the order of 15 min to several hours (depending on the size of the upstream watershed) may be required. On the other hand, the operation of major reservoirs for flood control and power generation requires medium-term streamflow forecasts. Finally, irrigation and domestic water supply allocations require long-term forecasts of surface and subsurface water storage volumes.

Hydrologic forecasts are issued by various local, state, and federal agencies. The type of forecast most frequently issued by these agencies is the short-term streamflow forecast. These are of particular importance when the danger of flooding is high. Flash floods, which kill more people nationwide than any other natural disaster, rip through river channels during heavy rains, surge over the channel banks, and sweep everything before them. Medium-term forecasts are often associated with seasonal floods, caused by the spring rains and snowmelt, that fill narrow tributaries with too much water too quickly. [*See* FLOODS.]

In the United States, the agency responsible for generating most of the flood and river flow forecasts is the National Weather Service. The NWS operates 13 regional River Forecast Centers (RFCs) around the United States. These centers prepare river and flood forecasts and warnings for approximately 3000 communities. Other federal agencies involved in hydrologic forecasting include the U.S. Army Corps of Engineers, the U.S. Forest Service's Soil Conservation Service, the U.S. Bureau of Reclamation, and the U.S. Geological Survey. In addition, various local agencies at the city, county, and state levels often generate their own local area forecasts. Such forecasts are typically generated with the help of computer-

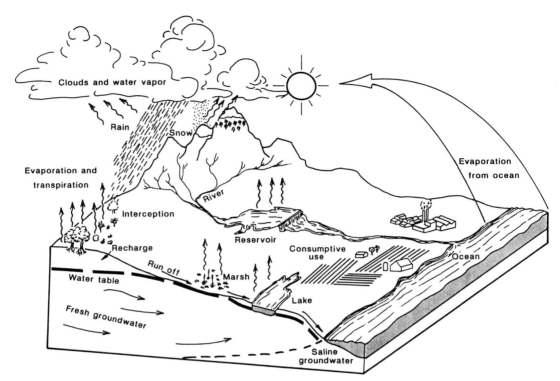

Fig. 1 The hydrologic cycle.

based mathematical models. This discussion, therefore, focuses on the usage of models for hydrologic forecasting.

II. Present State of the Art

Until the early part of this century, hydrology remained essentially an empirical science. The 1930s and 1940s brought with them the introduction of a more theoretical, quantitative approach through the work of people such as L. K. Sherman and R. E. Horton. The rapidly expanding fields of mathematical systems theory, computer science and technology, and automated remote data acquisition systems (e.g., radio, radar, satellites) have given rise to more sophisticated forecasting methods. Computerized techniques are now indispensable because of the large quantities of data which must be processed in a timely manner. The coming decade will see further developments in data acquisition and handling procedures with the implementation of advanced radar and satellite monitoring systems.

Hydrologic forecasting technologies in use today are based largely on the systems approach to decision making. The systems approach provides a logical way by which to describe any physical system so that a mathematical model of that system can be developed and programmed into a computer.

Such a computer model consists of three major components (Fig. 2). The main component is a simplified computer representation, $F(I(t), S(t), \Theta)$, of the physics of the watershed. It describes the mechanisms by which precipitation and other inputs $I(t)$ are transformed into surface and groundwater flows and storages $S(t)$ and into streamflow, evaporation losses, and other outputs from the watershed $O(t)$. The specific physical characteristics of each watershed (slope, soil type, vegetation density, etc.) are described by model parameters Θ.

A great number of different hydrologic forecast models have been developed using the systems approach. These models use different simplification strategies and were developed for different purposes (Table I). The following sections describe in more detail those forecast purposes which directly address the major hydrologic fluxes. The exception is evapotranspiration, which has not, to date, received much attention as a forecast issue.

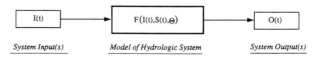

Fig. 2 System representation of a hydrologic model.

Table I Modeling Requirements for Various Hydrologic Purposes

Forecast purpose	System input(s)	Hydrologic systems	System output(s)
Flood forecasting	Rainfall rate	Watershed	Flood hydrograph (peak flood stage and time to peak)
Streamflow routing	Upstream inflows	River reach	Downstream hydrograph
Reservoir operation	Reservoir inflows	Reservoir	Reservoir level and discharge
Snowmelt forecasting	Snowpack extent and depth, solar radiation, rainfall	Snowpacks and glaciers	Snowmelt runoff hydrograph
Low-flow forecasting	Subsurface recharge, upstream reservoir releases	River and groundwater aquifer	Streamflow hydrograph
Groundwater recharge	Infiltration from lakes and rivers, leakage from other aquifers, infiltration from irrigation, pumping stresses, etc.	Aquifer	Groundwater levels, groundwater capture
Streamflow quality	Streamflow, point and non–point source contaminant concentrations	River reach ecosystem	Dissolved oxygen, temperature, and concentration
Groundwater contamination	Spatial and temporal distribution of contaminant sources	Aquifer	Location and evolution of contaminant plumes

III. Short-Term Streamflow and Flood Forecasting

The transformation of a tranquil river basin into a raging flood zone area can result in serious loss of life and property. Each year in the United States, approximately 75,000 people are driven from their homes, about 200 people lose their lives, and over $2 billion in damages are sustained. Accurate short-

term streamflow forecasts are an important aid in mitigating the damage that can be caused by flooding.

Streamflow forecasts are based on measurements and estimates of the amount of rain which has fallen in the upstream watershed during the past few days. Often, estimates of soil moisture values and evaporation rates are also required in order to make a forecast. Rainfall which is not absorbed by soil and plants will take a certain amount of time to travel through and over the soil and down stream channels until it

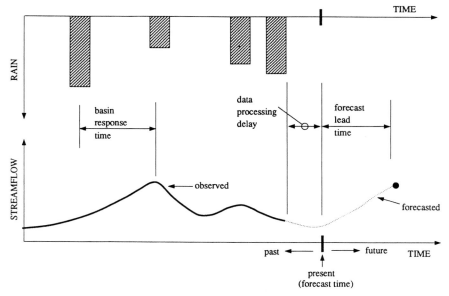

Fig. 3 Flood forecast scenario.

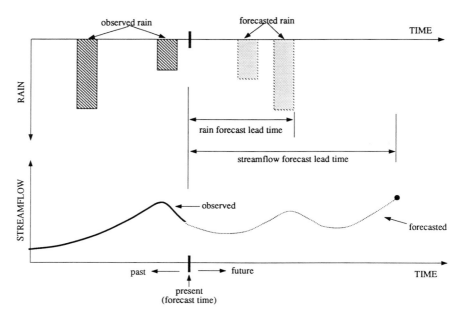

Fig. 4 Flood forecast scenario with precipitation forecast.

reaches the river location where the forecast is desired. That amount of time is called the watershed response time or *basin lag*. In most cases, the forecast lead time (length of time between the present and the forecast event) can be no longer than the watershed response time (Fig. 3). In fact, it is usually shorter than the watershed response time because a certain amount of time is required for transmission and processing of the rainfall data and for computation and dissemination of the forecast. The forecast lead time can be extended beyond the watershed response time if precipitation forecasts are available (Fig. 4). [*See* GLOBAL RUNOFF; RIVERS.]

Two approaches are commonly used to develop short-term streamflow and flood forecasting models: the conceptual rainfall runoff (CRR) model approach and the time series (TS) model approach. The CRR approach models the internal physical processes of a watershed through mathematical descriptions of the storage reservoirs and other nonlinear processes by which water is routed from the inputs to outputs through the storages. In contrast, the TS approach models the causal linkages between inputs and outputs without using detailed representations of the intervening physical processes. Each approach has advantages and disadvantages.

An example of an operational CRR model is the Soil Moisture Accounting (SMA) Model component (Fig. 5) of the U.S. National Weather Service River Forecast System (NWSRFS). The National Weather Service is responsible for short-term streamflow and flood forecasts throughout the United States. The SMA model is capable of generating accurate forecasts for medium- to large-scale watersheds (typically larger than 100 square miles) in various climatic and hydrologic regimes. In order to do so, two requirements must be satisfied: (1) The model parameters must be properly calibrated to the watershed, and (2) accurate measurements of precipitation over the watershed must be obtained

IV. Precipitation Forecasting

A. Precipitation Measurements

Short-term hydrologic forecast models depend on recent observations of hydrometeorological variables such as precipitation and streamflow. Precipitation varies quite significantly in time and space, making it difficult to obtain estimates of adequate quality. Data collection networks must, therefore, be designed very carefully. The most widely used procedure for measuring precipitation is to place rain gauges at several scattered locations. Because rain gauges measure precipitation only at a point, interpolation and averaging methods must be used to compute estimates of precipitation over the surrounding areas. More recently, radar-based precipitation estimation procedures have come into use. Although radar measurements have the advantage of providing spatial averages, the current state of the art is unable to provide precipitation estimates of good accuracy.

Hydrometeorological data can be collected either manually or automatically. The traditional manual approach can be very time consuming, which reduces forecast lead time. Mod-

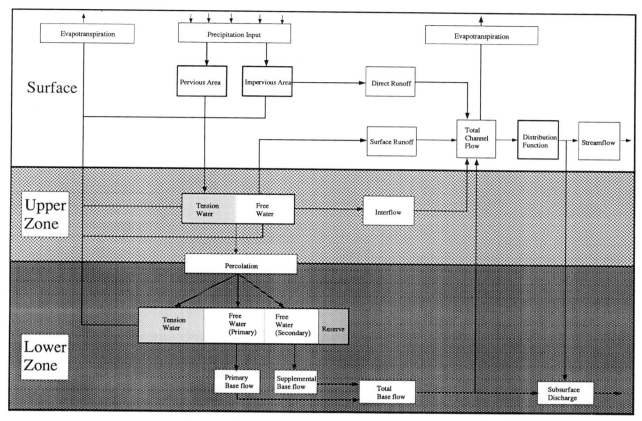

Fig. 5 Schematic representation of SMA-NWSRFS model.

ern data collection systems are based on automatic measuring instruments, radio transmission of data, and collection and processing by computers. By reducing the time required to transmit and process data to a few seconds, automated data collection systems have made flash flood forecasting (short basin response times) feasible.

B. Precipitation Forecasting

As mentioned earlier, precipitation forecasts can extend the forecast lead time beyond the watershed response time. Because current weather prediction capabilities are limited to 3–4 days, precipitation forecasts can, at most, add a few hours to 1–2 days to the forecast lead time. A precipitation forecast must be quantitative and based on a spatial and temporal time scale compatible with the streamflow forecast model. Such a forecast is known as a quantitative precipitation forecast (QPF).

Quantitative precipitation forecasts are routinely issued by the U. S. National Weather Service. These forecasts are made using numerical weather prediction (NWP) models which are

based on dynamic and thermodynamic equations describing the atmosphere. Conventional NWP models predict only the larger-scale features of the weather and are best suited for providing QPFs for about 12–36 h. Mesoscale NWP models can provide more detailed, shorter-term forecasts ranging from 1 to 18 h. "Nowcasting," which is based on linear extrapolations of the current weather, can provide forecasts for up to 2 h ahead. [*See* WEATHER PREDICTION, NUMERICAL.]

Another approach to providing QPFs that is currently under development and testing is the station precipitation model (SPM). The SPM approach is based on surface observations of pressure, temperature, dew point temperature, and precipitation (for forecast updating). The technique appears promising, but its implementation for operational forecasting requires more research.

At present, precipitation forecasts are not widely used in hydrological forecasting. One major problem is that QPFs are typically not available at the space and time scales and locations needed for hydrologic forecasting. A second is that precipitation forecast procedures can be expensive. However, important research is taking place, and we can expect greater usage of precipitation forecasting in the future.

V. Snowmelt Runoff Modeling and Forecasting

A. Meltwater (from Snowpacks and Glaciers) Forecasting

Snowpacks and glaciers are a significant form of water storage in some areas. Meltwater runoff from these storages provides a major portion of streamflow in such areas and thus plays an important role in their hydrology. Reliable predictions of the rates at which meltwater will be released from the snowpack are required for the optimal design and operation of water resource systems and for the issuance of reliable river forecasts and flood warnings. For example, 75% of the streamflow in the western United States originates from snowmelt. In contrast to areas where the predominant source of streamflow is rainstorms, snowmelt-fed and streams exhibit a relatively even distribution of flow from season to season or from year to year.

Meltwater runoff forecasts depend on the purpose for which they are required and on the physical attributes of the watershed. One or several of the following quantities may be required:

- Accumulation rate (percentage of snow in the precipitation)
- Areal extent of snow-covered areas and glaciers; water equivalent of snow cover
- Heat flux to the snow and ice surfaces
- Climatic parameters and thermal–physical properties

B. Snow Cover Extent Measurements

1. Ground Measurements

Reliable real-time operational forecasts of snowmelt runoff require, in most cases, direct observations of the snow cover. To date, manual point measurements have been the most reliable method of obtaining information on snow depth, water equivalent, snow density, and temperature. In many situations (particularly in hard-to-reach mountainous regions), practicality, safety, and economic considerations dictate the installation of automated recording instruments. In the United States, an automated data collection and telemetry network system called SNOWTEL has been installed to cover 11 western states.

2. Aircraft Surveys

Terrestrial and aircraft surveys are also used to collect information for meltwater runoff forecasting. Aircraft surveys provide information on the area extent of the snow cover (photographic methods) and estimates of the area distribution of its depth (stereophotogrammetry methods). [See PHOTOGRAMMETRY.]

A method of estimating snow water equivalent involves the use of natural gamma radiation. In this technique, the terrestrial gamma radiation is measured over a flight line from a low-flying aircraft. Flight lines are calibrated by measuring the terrestrial radiation without snow cover. The water mass in the winter snow cover attenuates the terrestrial radiation signal. The NWS Airborne Gamma Radiation Snow Survey program measures the snow water content at over 1000 flight lines in 16 states and five Canadian provinces. The radiation data collected on snow-covered calibrated flight lines give some measure of the degree to which the snow water equivalent has attenuated the radiation signal. Consequently, it is possible to infer the amount of snow water equivalent along the flight line with a high degree of accuracy.

3. Satellite Surveys

In remote and large areas where good ground information is unavailable, satellite observations are probably the only source of regularly available information of the extent of snow cover. In the United States, NOAA has a mapping program covering 271 drainage basins in the West and Midwest using Geostationary Operational Environment Satellite (GOES) imagery and digital interpretation techniques to produce basin snow cover area estimates. For regions with frequent cloud cover, new sensors and methods (e.g., microwave systems) are being developed which promise to provide information on the extent and water equivalent of snow cover under any weather conditions. These techniques are, however, still experimental and under development.

C. Snowmelt Runoff Modeling and Forecasting

After estimates of the extent of snow cover and its water equivalent have been obtained, melt rate forecasts can be produced. Because direct measurements of melt rate are not possible, the forecasting problem is somewhat different from that of rainfall runoff forecasting, and theoretical physics-based methods have been the main focus of research and development. Various approaches have been suggested for snowmelt forecasting.

The procedure most widely used for flood warning and reservoir operations is the degree-day method. A degree-day is defined as a departure of 1° in mean daily temperature above 32°F (0°C). The degree-day factor is the depth in inches (millimeters) of water melted per degree-day and is determined by dividing the volume of streamflow produced by melting snow within a given time period by the total degree-day for the period. The degree-day method performs well for areas where the basin is covered with a fairly uniform depth of snow.

The energy balance (or energy budget) method is the most theoretically sound method for determining snowmelt rates. Recently developed models can simulate temperature, albedo, density, water equivalent, and other characteristics during both accumulation and ablation. These models work on the principle of the heat transfer within the snowpack and the changes in energy storage as required in the energy balance equation. Although the energy budget approach is recommended for snowmelt forecasting purposes whenever possible, numerous complications may prevent its use. Among the more obvious complications are the difficult-to-model influences of variations in elevation, slope, aspect, and forest canopy cover on radiation exchange, temperature, and wind. The development of more sensitive remote sensing sensors and geographic information system (GIS) technologies has stimulated new research that is expected to result in operationally practical energy balance models in the near future. [See GIS TECHNOLOGY IN ECOLOGICAL RESEARCH.]

Several snowmelt runoff models are currently in operational use, ranging from simple regression-type degree-day relationships to the more complex energy budget models. The Soil Conservation Service (SCS) uses a multiple-regression procedure for generating seasonal snowmelt forecasts. The NWSRFS model of the National Weather Service uses a relatively complex snow accumulation and ablation model. The Corps of Engineers' Streamflow Synthesis and Reservoir Regulation (SSARR) model provides two options, a temperature index method and an energy budget method, for snowmelt estimation.

VI. Low-Flow Forecasting

The forecasting of "low flows" is of special operational importance in many river basins. Sustained periods of low flow can have serious economic impacts with potentially damaging ecological effects. The type of information that must be forecast (river level, temperature, dissolved oxygen concentration, etc.) during low-flow conditions varies, depending on the river basin and the range of purposes for which the river is used. The forecast lead times range from medium- to long-term. Medium-term predictions of low streamflow conditions are of particular interest for river navigation, power plants using cooling water and discharging thermal water, industries discharging waste water, management of multipurpose reservoirs, and various ecological considerations such as fish population maintenance (oxygen content, etc.). In addition, long-term forecasts of low streamflow

conditions are critical to urban and agricultural water supply industries.

Two kinds of low-flow conditions concern hydrologic forecasters. In temperate regions, a several-week dry spell may lower the groundwater table sufficiently to result in low flow levels in the rivers. In arid and semiarid regions, however, dry spells may persist for several years, resulting in the depletion of mountain snowpacks and glaciers, which are a primary source of river flow.

Each of these climatic conditions will require a different approach to low-flow forecasting. In the case of rivers fed primarily by subsurface sources, the hydrologic forecasting approach will rely heavily on the use of numerical groundwater models to simulate various aquifer recharge and depletion scenarios. These scenarios will be based on long-term meteorological forecasts generated by the NWS and other agencies.

In arid and semiarid regions, such as the Colorado River basin, long-term forecasts of low-flow conditions are strongly tied to extended meteorological forecasts of rain and snow. Under drought conditions (several years of well below average precipitation), it is not uncommon for the snowpack to be so severely depleted that there is no measurable flow. Present hydrologic methods for the study of droughts are based on statistical analyses searching for trends in historical records. Recent advances in climate modeling, however, hold the potential for more reliable predictions of drought conditions based on representations of global atmospheric circulation patterns and meteorological conditions. [See ATMOSPHERIC MODELING; DROUGHT.]

The problem of hydrologic forecasting under extreme conditions (such as droughts) illustrates the need for a greater understanding of the processes coupling the surface phase of the hydrologic cycle to the atmosphere and the groundwater storages. Research in this area is expected to receive much attention in the coming decades, especially as significant amounts of high-quality data become available through sources such as the NEXRAD weather radar modernization program of the NWS, the Earth Observing System (EOS) satellite to be placed in orbit by NASA, and the weather satellite program of NOAA.

VII. Surface Water Quality Modeling

Rivers, lakes, and estuarine regions have historically been the most common locations for human settlements and have, therefore, long been receptacles for various forms of waste discharges. In the early stages of such settlements, the characteristics of such discharges did not exceed the assimilative neutralizing capacity of the local ecosystem, and the quality of water in the rivers, lakes, and estuaries remained high.

However, with the industrial revolution and the tendency for human settlements to become larger and larger, the quality of surface waters began to deteriorate rapidly. In the last three decades, a great deal of emphasis has been placed on the development of strategies for remediating surface water quality problems and their severe ecological consequences.

Various models have been developed for this purpose. These models typically include the following features:

1. They accept as inputs various discharge constituents as functions of time at points of entry to the system.
2. They simulate the mixing reaction kinetics of the system at different locations and different times (e.g., bacterial biodegradation, chemical hydrolysis, physical sedimentation.
3. They synthesize a space- and/or time-distributed output at the system outlet (e.g., dissolved oxygen or nutrients, temperature, turbidity, dissolved solids).

Water quality constituents input to the system may include thermal, biological, organic, inorganic, and radiological wastes. Wastes are typically classified as conservative or non-conservative. Pollutants which do not decay significantly with time or disappear from the stream or lake system by settling, adsorption, or other means are classified as conservative wastes. Examples of such pollutants include total dissolved solids and pollutants such as pesticides and herbicides which decay only slowly with time. Unstable pollutants which have a time-dependent decay, such as biochemical oxygen demand (BOD), heat, and radioactive wastes with short half-lives, are classified as nonconservative wastes.

Several nonconservative-waste models have been proposed for lakes and rivers and are in operational use throughout the world. The classical dissolved oxygen depletion model developed by H. W. Streeter and E. B. Phelps (and its many later versions) is possibly the most widely used river water quality model in the world. Many comprehensive simulation models have also evolved over the years. Best known among them are QUAL-I, DOSAG-I, and the Environment Protection Agency (EPA) Storm Water Management Model (SWMM) designed to simulate the quantity and quality of storm water flow from urbanized areas.

Various lake water quality models are available. The simpler of these models are developed by assuming that the body of water is completely mixed. This assumption is applicable to lakes in which the extent of turbulence due to wind, tides, or other forces is great enough that, for all practical purposes, the concentration of the pollutant in the system can be considered to be completely homogeneous. The more complex lake water quality models assume that the lake has several vertically stratified layers, with each layer assumed to be completely mixed.

VIII. Forecast Problems in Groundwater Systems

Groundwater is an important source of water throughout the world, and its use in irrigation, industries, municipalities, and rural homes continues to increase. In the United States, 39% of the total water used comes from groundwater sources. Shortages of groundwater due to excessive withdrawals emphasize the need for accurate forecasts of the availability of groundwater and the importance of proper planning to ensure continued availability of water supplies. Furthermore, in the last few decades, groundwater contamination has emerged as the major threat to future groundwater supplies. It is estimated that more than 5 million contamination problems need to be addressed in the United States. [See GROUNDWATER.]

Groundwater systems, unlike surface water systems, are completely concealed from view; consequently, conceptual, physical, or mathematical models are the only way to achieve an understanding of their potential yields and responses to natural or human-related stresses.

Forecast problems in groundwater hydrology are essentially long-term in nature. Individual precipitation events rarely have a significant impact on groundwater levels; the major factors affecting the distribution and movement of groundwater are pumping (output) and recharge (input). In order to forecast future spatial and temporal variations in aquifer conditions, sophisticated numerical modeling procedures must be employed.

Numerous groundwater flow and contaminant transport computer algorithms have been developed based on finite-element and finite-difference approximation procedures (e.g., USGS MODFLOW and USGS MOC).

Growing public concerns regarding the fate of environmentally toxic contaminants in groundwater systems have resulted in increased emphasis on scientific research in this area. At present, however, few flow and transport problems can be modeled with a great deal of confidence. The most satisfactory results to date have come from models involving only the flow of water or the transport of a single nonreactive contaminant in a saturated porous medium. More sophisticated models accounting for partial saturation, the presence of several mobile fluids, fracturing, and the presence of reacting contaminants are necessary in order to predict the behavior of many existing contaminated aquifer systems. [See GROUNDWATER TRANSPORT OF CONTAMINANTS.]

In the United States, two agencies, the Nuclear Regulatory Commission (NRC) and the Environmental Protection

Agency (EPA), are particularly concerned with the predictive capabilities of groundwater modeling in support of their many regulatory activities. These two agencies deal with contaminant transport and recognize the need to evaluate present conditions and predict potential migrations.

The EPA is concerned primarily with predictive modeling in connection with the Superfund, hazardous waste management, and underground injection programs. For these purposes, long-term forecast capabilities to estimate plume extents and rates of movements over several months to years are critical.

The NRC, on the other hand, is concerned primarily with regulating the disposal of low- and high-level radioactive wastes at suitable sites. Because of the extremely long half-life of these contaminants (hundreds to thousands of years), disposal sites must be selected and designed to minimize the risks associated with leakage into the environment. This requires that potential pathways by which radionuclides may migrate into and through groundwater systems be accurately identified and evaluated. Because such movement is expected to be very slow, transport models with the capability of providing forecasts with lead times on the orders of hundreds to thousands of years are required. [*See* NUCLEAR WASTE DISPOSAL.]

Because of the high risks associated with radioactive leakage, it is not surprising that existing models are being scrutinized extremely carefully and have been subject to severe criticisms. This, in turn, is motivating very rapid research and development in the field of contaminant transport modeling.

IX. The Future of Hydrologic Forecasting

The science of hydrologic forecasting is still relatively young and is evolving rapidly. Along with progress in the understanding of fundamental principles governing hydrologic processes, advances in implementation of new systems-theoretic modeling approaches are resulting in improved forecast capabilities. Hydrologic forecasts must improve in two ways: forecast lead times must be extended, and forecast confidence must be improved. All of this requires improvements in methods for collection, transmission, processing, and storage of large quantities of data representing the high degree of spatial and temporal variability of hydrologic processes. Future advances in the performance of on-line monitoring schemes will be achieved partly through the exploitation of on-site microprocessor technology and partly through exploitation of remote sensing technologies.

A major initiative toward providing improved surface water forecast services on a national basis is NOAA's proposed Water Resources Forecasting Services (WARFS) program. The beginning infrastructure for WARFS is the current National Weather Service's River Forecast System, relying strongly on the NEXRAD weather radar modernization program, the Automated Surface Observing System (ASOS) program, and the Advanced Weather Interactive Processing System (AWIPS).

The NEXRAD and ASOS programs will provide much of the necessary technology for observing precipitation amounts on a nationwide basis at the temporal and spatial resolutions required. The data and computer systems provided by the NWS modernization programs, along with the technology of advanced hydrologic and climate forecast models, will be used to (1) support forecast service requirements of government and quasigovernment water managers; (2) provide water resources forecasts to private sector intermediaries, who will serve specific industries; (3) satisfy needs for forecast services at near-, mid-, and long-term time scales for a wide variety of water use situations nationwide; (4) provide critical information on forecast reliability; and (5) incorporate improved weather and climate forecast information into hydrologic models.

This last area is of critical importance. One of the earliest and most serious impacts of climatic changes will be on water resources. The operational focus of the WARFS initiative will provide information essential to assessing actual or potential climate change impacts on our nation's rivers and streams. Similarly, information from WARFS may be used to identify the hydrologic component in large-scale atmospheric models and to evaluate the validity of climate analyses and long-term experience and technology needed to enable the United States to manage the consequences of global climate change.

In the groundwater area, the future is expected to bring significant improvements in the technology for three-dimensional computer modeling of multiple contaminant transport in multiphase, multimedia, variable-saturation aquifers. The effectiveness of such models is contingent on our ability to validate them. In recognition of this, several intensively monitored experimental sites are being developed and supported by various government agencies throughout the country.

Bibliography

Anderson, M. G., and Burt, T. P. (1985). "Hydrological Forecasting." Wiley, Chichester, England.

Bras, R. L. (1990). "Hydrology, an Introduction to Hydrologic Science." Addison-Wesley, Reading, Massachusetts.

Burges, S., ed. (1986). "Trends and Directions in Hydrology," special issue of *Water Resour. Res.* **22**(9).

Collier, C. G. (1989). "Application of Weather Radar Systems, a Guide to Uses of Radar Data in Meteorology and Hydrology." Wiley, New York.

Hudlow, M. D. (1988). Technological developments in real-time operational hydrologic forecasting in the United States. *J. Hydrol.* **109**, 69–92.

Kraijenhoff, D. A., and Moll, J. R., eds. (1986). "River Flow Modeling and Forecasting." Reidel, Dordrecht, The Netherlands.

National Research Council. (1990). "Ground-Water Models, Scientific and Regulatory Applications." National Academy Press, Washington, D.C.

Singh, V. P. (1988). "Hydrologic Systems," Vols. I and II. Prentice-Hall, Englewood Cliffs, New Jersey.

Sorooshian, S. (1987). The impact of catchment modeling on hydrologic reliability. *In* "Engineering Reliability and Risk in Water Resources" (L. Duckstein and E. Platé, eds.), pp. 365–390. NATO ASI Series E, Applied Sciences No. 124. Martinus Nijhoff, Dordrecht, The Netherlands.

Thomann, R. V., and Mueller, J. A. (1987). "Principles of Surface Water Quality Modeling and Control." Harper & Row, New York.

Glossary

Forecast lead time Length of time between the moment at which a forecast becomes available and the time at which the event is forecast to take place.

Hydrologic forecasting Estimation of future states of hydrologic phenomena.

Watershed response time Time required for rainfall which is not absorbed by soil and plants or evaporated to travel over and through the land surface to the streamflow location at which the forecast is desired.

Hydrothermal Systems

Robert P. Lowell
Georgia Institute of Technology

Hydrothermal systems encompass the interconnected processes of heat, chemical, and aqueous fluid transport in the earth's crust. These systems redistribute heat energy in the crust and give rise to zones of concentrated heat energy storage and transfer and, often, mineral deposition. Hydrothermal systems occur in both terrestrial and submarine settings. Hydrothermal systems are often categorized as high-temperature or low-temperature systems; low-temperature systems typically have subsurface temperatures of less than 150°C. Low-temperature continental systems have little subsurface boiling. Continental two-phase systems are categorized as hot-water or vapor-dominated systems. As a result of high pressures at the seafloor, submarine systems do not boil until the temperature is well above 350°C. High-temperature submarine systems have discharge temperatures in excess of 150°C. This article is concerned with the earth science aspects of hydrothermal systems; treatments of the engineering and utilization aspects of the systems can be found elsewhere.

I. Introduction

Hydrothermal systems are one component of a complex, global convective heat engine that drives the dynamic planet Earth. Heat is transported from the interior by vigorous convection in the mantle to the base of the lithosphere. The lithosphere is broken into a number of plates. The relative motions of these plates, volcanic and tectonic activity, and the release of stored elastic energy as earthquakes are manifestations of the underlying convection system. [*See* LITHOSPHERIC PLATES; PLATE TECTONICS, CRUSTAL DEFORMATION AND EARTHQUAKES.]

Heat is transported through the lithosphere mainly by thermal conduction. Approximately 40% of the thermal flux through the continental lithosphere can be attributed to heat produced in crustal rocks as a result of the decay of long-lived radioactive isotopes of uranium, thorium, and potassium; the remaining 60% is conducted from below. Direct transfer of heat through the continental lithosphere by volcanic and hydrothermal activity accounts for less than 1% of the total heat flux. [*See* CONTINENTAL CRUST; HEAT FLOW THROUGH THE EARTH.]

The thermal regime of the oceanic lithosphere is dominated by the process of lithospheric creation along the ocean ridge system (Fig. 1). Lithospheric plates move apart along the ridge axis. The rate of movement is different along different segments of ridge axis but is nearly constant in time along any particular segment. As the plates separate, hot upwardly convecting mantle material fills the void at the trailing edges. Decompression of the ascending mantle results in partial melting; the melt segregates from the residual mantle and ascends buoyantly toward the seafloor. Some of the magma erupts onto the seafloor, but most crystallizes at depth of one to several kilometers. The new hot lithosphere that is created at the ridge axis cools as it moves laterally away. [*See* OCEANIC CRUST.]

Figure 2 shows a theoretical curve of conductive heat flux as a function of the age of the lithospheric plate, together with the observed mean conductive heat flow for a number of ocean ridges. The observed values are substantially lower than the theoretical curve to a considerable age. The discrepancy between the theoretical and observed data is due to heat transfer by hydrothermal activity. This component of heat transfer is difficult to measure directly.

Table I depicts the components of heat loss through the earth's surface. A striking feature of the table is that nearly 25% of the global heat loss is due to hydrothermal systems in the oceanic crust. Approximately 80% of the hydrothermal heat loss is due to low-temperature off-axis circulation; the remainder is due to high-temperature discharge in the neo-volcanic zone within a few kilometers of the ridge axis.

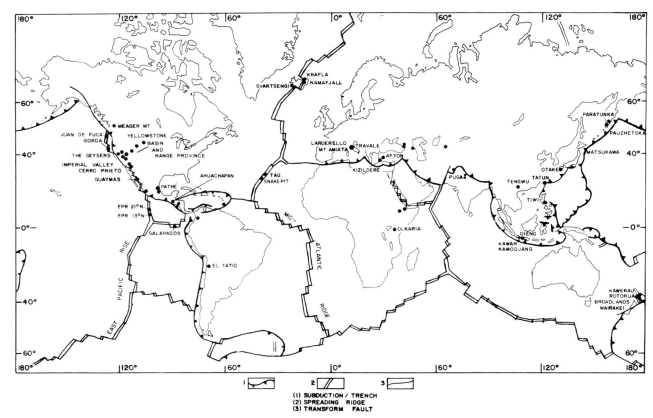

Fig. 1 World map showing major plates, with ridges, transform faults, subduction zones, and plate boundaries. Several major hydrothermal systems are also noted by dots. [Adapted from Rybach (1981).]

II. Ingredients

Hydrothermal systems consist of a heat source and an aqueous fluid circulation system. Chemical reactions occur between the fluid and rock, and heat is transferred to the surface by the circulation. The temporal variability of the systems is a fundamental aspect of their behavior. Each of these ingredients is described below. Figures 3a and 3b, respectively, show schematic models of a single-pass magma–hydrothermal system and a low-temperature warm-spring system. Although these figures specifically depict terrestrial systems, the same main ingredients are present in submarine systems.

A. Heat

1. Geothermal Gradient

Conductive heat flux, Q, is defined by $Q = \lambda \, dT/dz$, where λ is the thermal conductivity of the rock and dT/dz is the rate at which the temperature T increases with depth in the earth, that is, the geothermal gradient. The thermal conductivity of crustal rocks ranges from approximately 1.8 to 5.0 W/m °C, with most igneous and metamorphic rocks in the neighborhood of 2.5 W/m °C. The geothermal gradient ranges from approximately 10°C/km in tectonically stable continental regions to more than 100°C/km in tectonically active areas and in young oceanic crust. Fluids circulating to depth in the crust can be heated by the geothermal gradient and flow out at the earth's surface as warm or hot springs.

2. Magmatic Heat

Whereas many low-temperature hydrothermal systems appear to be related to heating of circulating water by the geothermal gradient, most high-temperature hydrothermal activity is located in regions of recent volcanism. The intrusion of magma to shallow levels of the crust provides additional heat to be tapped by the hydrothermal system. Magmatic heat consists of the latent heat of crystallization as well as sensible heat. Heat loss from crystallizing and cooling

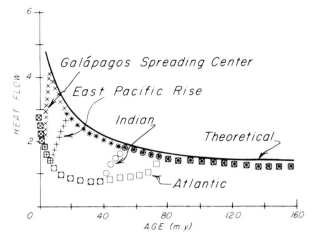

Fig. 2 Observed mean heat flow for oceanic spreading centers compared with the theoretical curve. [From Anderson *et al.* (1977).]

magma raises the temperature of the adjacent and overlying rock and increases the geothermal gradient. Regions of high geothermal gradient are often associated with recent volcanism; therefore the heat source for hydrothermal systems operating in regions of high geothermal gradient may be ultimately of magmatic origin. Ocean ridge crests dominate the earth's volcanic output, so it is understandable that high-temperature hydrothermal activity is concentrated there. Figure 1 shows several areas of known high-temperature hydrothermal activity. The continental systems are all associated with areas of recent volcanism.

B. Fluid

Hydrothermal systems are characterized by the transport of heat in the crust by an aqueous fluid. In continental

Table I Rate of Heat Loss from the Earth

Component	Heat loss ($\times 10^{13}$ W)
Continents	
Radiogenic heat production in crust	$\simeq 0.46$
Conductive heat flux from below crust	$\simeq 0.68$
Extrusion of lavas	$\simeq 0.003$
Hydrothermal losses	$\simeq 0.01$
Total	$\simeq 1.15$
Oceans	
Conduction	$\simeq 2.03$
Extrusion of lavas	$\simeq 0.01$
Hydrothermal losses	$\simeq 1.01$
Total	$\simeq 3.04$
Global heat loss	$\simeq 4.2$

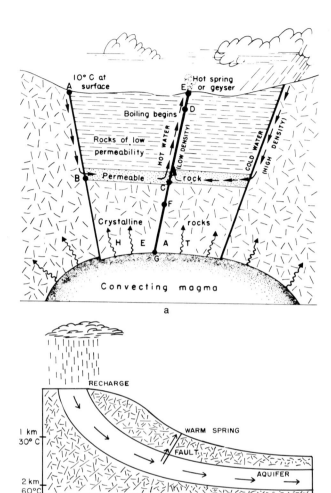

Fig. 3 Schematics of (a) a single-pass terrestrial magma–hydrothermal system and (b) a terrestrial low-temperature system. [Part (a) from White (1973).]

systems, the fluid may be of meteoric, connate or metamorphic, or magmatic origin, or a combination of the three types. In principle, the origin of the fluid can be distinguished on the basis of its hydrogen and oxygen stable isotope geochemistry. In submarine systems, the fluid is seawater; although some fluid inclusion studies suggest the presence of magmatic fluids at depth, there is no conclusive data to suggest a magmatic component in active submarine systems. [*See* CRUSTAL FLUIDS.]

1. Meteoric Waters

Meteoric waters are surface waters such as rivers and lakes, ocean water, water contained in ice caps, and ground waters that have recently been in contact with the atmosphere. These waters have a characteristic relationship between their oxygen and hydrogen stable isotope ratios. Waters from hydrothermal systems with subsurface temperature exceeding approximately 200°C typically show increased oxygen isotopic ratios relative to local meteoric waters. The enrichment can be accounted for by isotopic exchange between the local meteoric water and the rocks through which it passes. Acidic thermal waters show an enrichment in the hydrogen isotopic ratio as well as in the oxygen isotopic ratio.

2. Connate and Metamorphic Waters

Connate waters are waters trapped in the interstices of sedimentary rock at the time of its formation. Metamorphic waters are those released during dehydration reactions in metamorphic rocks at depth. Both of these waters may have been of meteoric origin, but long-time burial and reaction with their host minerals may have changed their chemical signature. Low-temperature hydrothermal systems in sedimentary basins and geopressured geothermal systems may contain connate waters. It may be difficult in practice to distinguish between connate, metamorphic, and meteoric source waters, but this is not generally a critical issue.

3. Magmatic Waters

Magmas generally contain dissolved volatiles such as H_2O and CO_2. As the magma crystallizes, water may exsolve as a distinct vapor phase. Such a vapor phase in isotopic equilibrium with the magma would have a distinct relatively narrow range of oxygen isotopic ratios. Observations of the isotopic variations of a number of different hydrothermal systems lie along nearly parallel lines intersecting the meteoric trend. There is no suggestion that these waters contain a magmatic component. Because of the limitations on the data, the presence of a small amount of magmatic fluid, at most a few percent, cannot be ruled out. Nearly all the fluid in active high-temperature continental hydrothermal systems is of meteoric origin.

C. Permeability

For a hydrothermal fluid to transfer heat, it must circulate through the rock. The rate of fluid flow through permeable rock is generally described by an empirical relationship called Darcy's law:

$$q_i = K_{ij}(\partial H/\partial x_j) \tag{1}$$

where q_i is the volume of fluid crossing a unit area per unit time, $\partial H/\partial x_j$ is the gradient of the hydraulic head, and K_{ij} is the hydraulic conductivity tensor. The hydraulic conductivity is often written in terms of a permeability tensor k_{ij}, where

$$K_{ij} = gk_{ij}/\mu$$

where g is the acceleration due to gravity and μ is the dynamic viscosity. For most applications the permeability is treated as a scalar k. The permeability depends only on the properties of the rock and has units of (length)^2. These units are appropriate because permeability is found to be related to the square of the average grain size as shown in Eq. (2). Permeability values are often given in millidarcies (md); 1 md $\simeq 10^{-15}$ m^2. The permeability of rocks is due to interconnected interstitial, or pore, spaces in the rocks or to fractures. The permeability due to interconnected pore spaces is termed the primary permeability; that due to fractures is termed secondary permeability. The permeability of a rock in the natural setting is a function of several parameters, including some that are related to hydrothermal flow itself. Therefore the permeability may be a complex function of time and space in a given hydrothermal system. The permeability is the most important physical property governing the behavior of a hydrothermal sytem and its temporal evolution. It is also the most difficult to measure *in situ* on a system scale. It is often inferred from limited data and from modeling calculations.

1. Porous Media

The ratio of the volume of interstitial space to the total volume of a rock is the porosity of the rock. Porosity is often expressed as a percent. Igneous and metamorphic rocks typically have low porosity; a few percent or less is common. Sedimentary rocks typically have much higher porosity; 20–30% is not unusual.

Effective or interconnected porosity is related to permeability. The relationship is not a straightforward one because the fluid takes an irregular, or tortuous, path through the solid. The permeability is a function of the size and number of fluid pathways as well as their tortuosity. Many attempts have been made to derive mathematical relationships between porosity and permeability for idealized arrangements of pore structures. The results suggest a relationship of the form

$$k = Cb^2\phi^n \tag{2}$$

where k is the permeability, b an average grain size of the medium, ϕ the porosity, and C a numerical constant. The power n is 2 or 3 in most formulations. Equation (2) is often found not to hold well in practice. It is used because it is generally easier to obtain estimates of porosity than of permeability for a hydrothermal sytem, and this is the best relationship available. The failure of the relationship is probably due to fracture control of the permeability in natural systems.

2. Fractures and Faults

In virtually all hydrothermal systems the permeability is controlled by discrete fractures, fracture networks, and faults zones rather than by interconnected porosity. This is easily understood for hydrothermal systems in igneous and meta-

Table II Fracture Permeability [from Eq. (3)]

b (m)	d (m)	ϕ (%)	k (m²)
0.01	0.00001	0.1	8.3×10^{-15}
	0.0001	1.0	8.3×10^{-12}
0.1	0.0001	0.1	8.3×10^{-13}
	0.001	1.0	8.3×10^{-10}
1.0	0.001	0.1	8.3×10^{-11}
	0.01	1.0	8.3×10^{-8}

morphic rocks because of their low porosity and low primary permeability. That fractures and fracture networks greatly enhance the permeability can be seen by considering a simple model consisting of an arrangement of parallel planar fractures of uniform width d separated by a distance b is an impervious rock matrix. The permeability for this arrangement is

$$k = d^3/12b \qquad (3)$$

Table II shows the permeability derived from Eq. (3), as well as the fracture porosity $\phi = d/b$, for a number of fracture widths and spacings. The table shows that fractured rocks with low porosity can be highly permeable. The fracture permeability is typically several orders of magnitude greater than the permeability due to interconnected pore spaces. For example, the permeability of unfractured granite determined from laboratory measurements is typically $10^{-18} - 10^{-20}$ m². *In situ* measurements of permeability in igneous and metamorphic rocks often indicate the existence of zones of permeability several orders of magnitude greater than the low values found in laboratory measurements on intact samples. This is a further indication of the importance of fractures in controlling *in situ* permeability in igneous and metamorphic rocks. It is interesting to note that estimates of bulk permeability for both terrestrial systems in igneous rocks and submarine hydrothermal systems often lie in the neighborhood of $10^{-12} - 10^{-15}$ m². Such an estimate points to the dominance of fracture permeability in igneous rocks.

Fractures are not uniformly distributed throughout the rock because stresses are inhomogeneous. Both thermal and tectonic stresses may operate to concentrate fractures. In Icelandic hydrothermal systems, zones of high permeability are associated with magmatic dikes which tend to intrude into regions of tensional stress. Hydrothermal discharge in both continental and submarine hydrothermal systems is focused by tectonic faults.

Although it is apparent that faults and fractures dominate the permeable pathways for hydrothermal fluids, it is often not necessary to consider the flow in individual fractures. Mathematical models of flow in fractured rock often treat the fracture permeability as an equivalent porous medium. [*See* FAULTS.]

3. Formation and Evolution of Permeability

Rock permeability varies not only spatially but also temporally as a result of inhomogeneous stress distribution. Tectonic activity and thermal contraction tend to create permeability, whereas thermal expansion, ductile deformation, and chemical deposition tend to seal permeability.

The solubility of many minerals in hydrothermal fluids is strongly temperature dependent. For example, the solubility of silica attains a maximum between 350 and 400°C. If the fluid is heated above 400°C or cooled below 350°C, silica will precipitate from the solution and clog the fractures and pore spaces. Amorphous silica and quartz are common vein minerals in hydrothermal systems.

When hydrothermal fluid moves at depth through rock that is hotter than the fluid, the fluid extracts heat from the rock. As the rock cools, it contracts and new fractures may be created. As the heated fluid rises toward the surface, it moves through rock that is cooler. As the fluid transfers heat to the rock, the rock expands thermally and permeable paths may close.

The deformation of rock under stress is a function of temperature. At temperatures less than $\simeq 350-450$°C brittle failure under stress is the rule. At higher temperatures, up to $\simeq 750$°C, rocks may fracture under stress, but quasiplastic flow may seal the fractures. At a temperature greater than $\simeq 750$°C, rock deformation is largely ductile. Fluid inclusion data from some deep crustal oceanic rocks suggest the existence of trapped hydrothermal fluid at temperatures as high as 600°C. The temperature dependence of rock deformation suggests that hydrothermal fluid does not have direct access to a magma chamber. The sealing of fractures by quasiplastic flow at temperatures between 350 and 450°C may explain why hydrothermal circulation may be generally limited to that temperature range.

The relative importance of the various thermal and chemical effects in creating and sealing permeability is not well known. It is generally thought that chemical precipitation is the primary cause of sealing of permeability in the shallow parts of hydrothermal systems. Thermal expansion may play a minor role, but the effect of thermal expansion on temporal variability in hydrothermal systems is largely unexamined. Thermal contraction and ductile deformation may play dominant roles in the temporal variability in the deep parts of hydrothermal systems, but the effects have not been quantified in detail.

D. Heat Transfer Mechanisms

The fundamental mechanism of heat transfer between rock and the hydrothermal fluid is conduction. If the permeability can be described in terms of a porous-medium model, the fluid is in intimate contact with the rock mass. In this case, thermal equilibrium is assumed; the rock and fluid at any point are at the same temperature. If, however, the permeability is controlled by discrete, widely separated fractures, the fluid in a fracture may be at a different temperature than the rock at some distance from the fracture wall. Heat transfer between the impermeable rock and the fluid in the fracture is by conduction. A zone of conductive heat transfer also exists between the base of hydrothermal circulation and the underlying magma chamber.

The hydrothermal fluid transfers heat by convective circulation. In low-temperature systems, the circulation may be largely forced by topography. In terrestrial systems, ground waters from regions of high topography descend, are heated by the geothermal gradient, and discharge as warm springs at lower elevations. The topographic head drives the circulation. In oceanic systems, ridge crest topography gives rise to lateral temperature gradients within the crust that drive circulation. As a result of ridge crest topography, fluid tends to ascend beneath topographic highs and descend at topographic lows.

In many systems, the convective movements are largely a result of buoyancy differences between the colder descending fluid and warmer ascending fluid. The vigor of buoyancy-driven convective motions is related to a dimensionless parameter called the Rayleigh number and defined as $Ra = g\alpha \Delta Tkh/va$, where α is the coefficient of thermal expansion, g the acceleration due to gravity, k the permeability, ΔT the temperature difference, h the thickness of the layer across which the temperature difference is applied, a the effective thermal diffusivity of the water-saturated medium, and v the kinematic viscosity of the fluid. In vigorously convecting hydrothermal systems, the heat flux at the surface may be one to two orders of magnitude greater than in the absence of convection. [*See* GEOPHYSICAL FLUID MECHANICS.]

E. Water–Rock Reaction

Hydrothermal fluids react chemically with the rocks through which they pass. As a result, the chemical and isotopic compositions of both the water and the rock are changed. Rock type, fluid composition, permeability, duration of activity, pressure, and temperature all affect the formation of hydrothermal minerals, although the original rock type is much less significant at temperatures above 280°C. [*See* WATER GEOCHEMISTRY.]

1. Geochemical Thermometers

Chemical reactions, the partitioning of various isotopes between mineral and fluid phases, and the temperature dependence of the concentration of certain constituents in hydrothermal fluids have led to the development of a number of geochemical thermometers that can be used to estimate subsurface temperatures from surface samples. For example, the solubility of silica in hydrothermal waters is strongly tempera-

ture dependent, but the rate of precipitation decreases dramatically as the temperature decreases. Consequently, if a fluid saturated with silica at high temperature is brought to the surface, the concentration of dissolved silica will often provide a good measure of the temperature at depth. Other common geothermometers include (1) Na/K, which makes use of the temperature dependence of partitioning of these elements between aluminosilicate rocks and solution; (2) Na–K–Ca, which includes the role of calcium, and (3) stable isotope ratios, such as $^{13}C/^{12}C$, $^{18}O/^{16}O$, D/H, and $^{34}S/^{32}S$, which make use of the temperature-dependent partitioning of these isotopes between rock and fluid.

Various factors affect the performance of each of these geothermometers, so often many are used. Careful comparison of the results obtained with several geothermometers can provide useful information on subsurface conditions.

2. Ore Deposits

A consequence of water–rock reactions is the incorporation of ore metals that are present in the rock as trace constituents (such as copper, zinc, lead, tin, gold, and silver) into the hydrothermal fluid as dissolved constituents. Most common ore minerals are sulfides that are extremely insoluble in water; solubility is achieved by the formation of bisulfide or chloride ion complexes. If mechanisms are present to cause local precipitation of the metals in appreciable concentrations, an ore deposit is formed. Local precipitation can result from a rapid drop in temperature—for example, because of mixing with cooler near-surface waters—and from boiling. Many types of ore deposits that have been exploited are attributed to hydrothermal systems. The study of active hydrothermal systems for the purpose of understanding the formation of ore deposits is of great importance. [*See* MARINE MINERAL RESOURCES; ORE FORMATION.]

Some active high-temperature submarine hydrothermal systems are presently forming massive sulfide ore deposits on the seafloor. This type of deposit is exposed on land, incorporated in rocks of ancient as well as relatively young geologic age. The deposits are particularly well exposed in ophiolite complexes in Cyprus and Oman. The study of sulfide deposits in ophiolites, together with their feeder and reaction zones, provides a window into processes actively occurring at depth in submarine hydrothermal systems. [*See* MAGMATIC SULFIDE DEPOSITS.]

F. Time Dependence

Hydrothermal systems are fundamentally transient phenomena. The temporal variability may have a range of time scales. Heat input may vary as magma chambers solidify and cool, or are periodically replenished. Thermal conduction from hot rock decays as time $t^{-1/2}$. Permeability may change as a result of tectonic, thermal, or chemical effects. Fluid recharge pat-

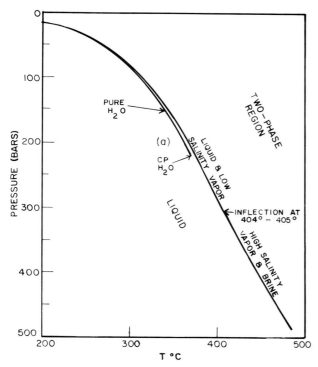

Fig. 4 Boiling point curve for pure water and for seawater. [From Bischoff and Rosenbauer (1984).]

terns may change because of climatic changes (precipitation patterns, glacial episodes, etc.).

Large continental hydrothermal areas such as Yellowstone Park may last for several tens of thousands of years, although individual subsystems may have a shorter lifetime. High-temperature hydrothermal activity at the trans-Atlantic geotraverse (TAG) system on the Mid-Atlantic Ridge at 26°N appears to have undergone several episodes over the past 40,000 years, with individual episodes lasting several hundred years. Individual black smoker vents at ocean ridge crests may last for only a few tens of years.

III. Boiling and Phase Separation

As heated fluids rise toward the surface in continental hydrothermal systems, they may undergo boiling. Figure 4 shows the boiling point curve for pure water and for seawater. Pure water has a critical point (P_c, T_c) of 220 bars, 374°C; at temperature and pressure conditions above the critical point only a single fluid phase exists. Suppose fluid is heated at some depth to a temperature $T < T_c$ (e.g., point a in Fig. 4). If during its ascent its temperature and pressure place

it on the two-phase boundary, the fluid will begin to boil. As the fluid continues to rise its temperature will be constrained to lie on the two-phase boundary; when it reaches the surface the fluid will be a water–stream mixture. [See WATER, PHYSICAL PROPERTIES AND STRUCTURE.]

A. Saline Waters

Meteoric hydrothermal fluids usually have dilute concentrations (i.e., <1%) of dissolved salts. Seawater contains approximately 3% dissolved salts by weight. Magmatic fluids may contain an order of magnitude more salt than does seawater. The presence of dissolved salts in hydrothermal fluid changes the critical pressure and temperature and the composition of the phases in thermodynamic equilibrium along the two-phase boundary. In contrast to the case for pure water, the phase boundary in saline fluid systems extends beyond the critical point (Fig. 4 shows seawater as an example). For saline systems, therefore, one considers subcritical and supercritical boiling separately. At $P < P_c$ a saline fluid may undergo subcritical boiling. In this case, the salt remains in the liquid phase; the vapor phase is almost pure water. At $P > P_c$, a saline fluid may undergo what is termed supercritical boiling. In this case, boiling produces a dense brine and a vapor phase of reduced salinity.

For seawater, P_c and T_c are approximately 300 bars and 400°C, respectively (Fig. 4). Seafloor pressures at ocean ridge crests are typically 200–250 bars, and the pressure at the top of a magma chamber, approximately 2 km below the seafloor, is approximately 200 bars higher. Both supercritical boiling and subcritical boiling are possible in seafloor hydrothermal systems. The data for some active seafloor systems show salinity variations that are indicative of phase separation, but many active high-temperature vents are in the neighborhood of 360°C and appear to be single-phase systems. At a seafloor pressure of 250 bars, a fluid would have to be ≃ 390°C to boil.

B. Brines

Studies of fluid inclusions from fossil continental hydrothermal systems and from oceanic crustal rocks show the occurrence of highly saline brines, and although some of these fluid inclusions may contain fluid of magmatic origin, many studies suggest that the brines may have evolved from relatively dilute waters. The evolution of brines from dilute hydrothermal waters and the implications of brine formation for the physics and longevity of hydrothermal systems are of considerable interest.

Boiling of saline waters causes the formation of a dense brine that will tend to sink to the bottom of the system. Double-diffusive convection, in which a convecting layer of cooler, dilute hydrothermal fluid overlies a convecting layer of hot, saline brine, may occur. The brine layer may also act as a thermally conducting barrier between the hydrothermal system and a magma chamber.

IV. Types

A. Continental Systems

1. Warm Water

Warm-water systems are those in which subsurface boiling does not occur. Temperatures are usually less than 100°C. Warm-water systems include geopressured systems. Many low-temperature hydrothermal systems appear to derive their heat from the regional geothermal gradient. Examples include many isolated warm springs in the eastern United States and elsewhere, some systems in the Basin and Range Province of the western United States, and low-temperature systems in Iceland.

2. Liquid Dominated

In liquid-dominated two-phase systems a layer of two-phase liquid overlies a single-phase fluid. The phases are intermingled in the two-phase region, and the pressure distribution through the system is nearly hydrostatic. The surface discharge of liquid-dominated systems is characterized by boiling hot springs and geysers; the fluids are generally neutral to alkaline in pH and chloride rich. The vast majority of high-temperature continental hydrothermal systems are liquid dominated. Well-known examples include the geyser basins of Yellowstone National Park, United States; Wairakei and Broadlands, New Zealand; and Ahuachuapan, Mexico (Fig. 1). Temperatures of these systems are typically in the range 200–300°C.

3. Vapor Dominated

Although most hydrothermal systems are liquid dominated, a few well-known and exploited systems are characterized by extensive regions in which vapor is the dominant phase (e.g., Geysers, United States; Lardarello, Italy; Kamodjang, Indonesia; and Matsukawa, Japan). Vapor-dominated zones indicate the occurrence of phase separation upon boiling. Systems with extensive vapor-dominated regions act as a heat pipe; heat is carried upward by the vapor phase to the base of a relatively impermeable caprock, where small amounts of liquid water condense and flow downward. Within vapor-dominated zones, the pressure distribution is usually much less than hydrostatic and is often essentially uniform over the thickness of the vapor-dominated zone. Vapor-dominated zones may also occur "parasitically" in liquid-dominated systems. Figure 5 shows some conceptual models of hydrothermal systems that have vapor-dominated zones. Systems with extensive vapor zones appear to be similar to model I.

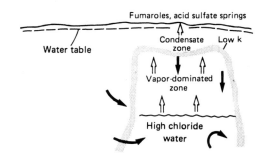

A. MODEL I

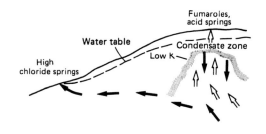

B. MODEL II

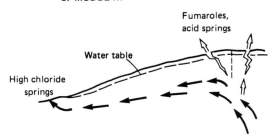

C. MODEL III

Fig. 5 Some conceptual models of vapor-dominated zones. Solid arrows are for liquid, open arrows for vapor. (A) Representation of model I; the vapor-dominated zone is large and there is limited throughflow. (B and C) Models II and III, respectively. Both models II and III involve lateral flow that links acid–sulfate features at higher elevations with high-chloride springs at lower elevations. [From Ingebritsen and Sorey (1988).]

Such systems have little through flow, and the vapor-rich zone is sealed from the surrounding liquid-dominated regions by a low-permeability barrier. The surface discharge from these systems tends to be acidic, sulfate rich, and chloride poor compared to the discharge from liquid-dominated systems. Models II and III represent examples of vapor-dominated zones that may occur parasitically within liquid-dominated systems. Models II and III suggest that the formation of vapor-dominated zones occurs as a result of significant topographic relief, where the vapor zone occurs at high elevations and more typical chloride-rich hot spring discharge occurs at lower elevations. Both models II and III involve lateral flow that connects the acid–sulfate surface features that typify vapor-dominated conditions at high elevations with high-chloride springs at lower elevations. Lassen Volcano, California and the Mirror Plateau and Mud Volcano regions of Yellowstone Park may be examples of model II.

B. Submarine Systems

1. Low Temperature

As evidenced by Fig. 2, hydrothermal activity persists in oceanic crust of considerable age. Most of this activity occurs at low temperature. Limited data on *in situ* permeability recently obtained from boreholes indicate that the upper few hundred meters of the crust may have a permeability in the neighborhood of $10^{-13} - 10^{-14}$ m² but that at greater depths the permeability drops to $10^{-17} - 10^{-18}$ m². Ocean sediments are generally at least one order of magnitude less permeable than the upper crust. If the limited data on crustal permeability are indicative of the general permeability condition in the oceanic crust outside the axial zone, low-temperature hydrothermal circulation in the oceanic crust is likely to be confined to a thin upper crustal layer. The Rayleigh number would probably be subcritical and the circulation patterns controlled by topography of the crustal rocks beneath the sediment cover. Fluid discharge in these systems would be concentrated at topographic highs and recharge at topographic lows. Local outcrops of the permeable layer through the sediment cover may also control sites of fluid recharge and discharge.

2. High Temperature

Active high-temperature systems characterized by the venting of black smoker plumes near ocean ridge axes are among the most spectacular hydrothermal phenomena (Fig. 6). Sulfide-laden black smoker fluids typically have a temperature in the neighborhood of 350°C; precipitates from black smoker discharge form sulfide chimneys and larger sulfide deposits on the seafloor. The discharge from several individual smoker vents forms plumes in the oceanic water column that may affect oceanic circulation patterns. Geochemical and tempera-

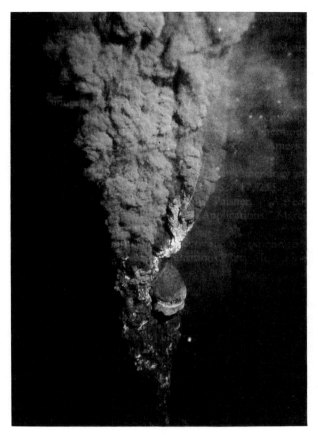

Fig. 6 Black smoker vent at the East Pacific Rise 21°N hydrothermal field. (Courtesy of Woods Hole Oceanographic Institution, Woods Hole, Massachusetts.)

ture signals of hydrothermal plumes have been detected thousands of kilometers from the ridge axis. The heat source of these systems is probably a shallow subaxial magma chamber. Locations of several known active high-temperature submarine systems are shown in Fig. 1.

a. Mass and Geochemical Flux

The total hydrothermal mass flux through the oceanic crust is estimated to be $10^{17} - 10^{18}$ g/yr. At this rate the entire mass of the oceans would be recycled through the oceanic crust in a few million years. Chemical reactions between hydrothermal seawater and oceanic crustal rocks effectively remove Mg and SO_4 from the circulating fluid and enrich it in Ca, K, Si, Fe, Mn, Ba, Li, and Rb. Variable depletions and enrichments of Na, Cl, and Sr are seen in different vents. Enrichments in Li and Rb exceed those of river input by factors of 5 to 10. Enrichments in Ba, K and Si are one-third to two-thirds of the river input values. Enrichment in Ca and depletions in Mg and SO_4 are approximately equal to those of river input. Submarine hydrothermal systems thus have a significant impact on the global geochemical budget of the oceans.

Exchange of magnesium for calcium is an important factor in the global CO_2 cycle. The additional calcium is balanced by the precipitation of $CaCO_3$, which in turn produces CO_2. It is thought that 14–22% of atmospheric CO_2 is due to seafloor hydrothermal activity. Tectonic plate reorganization approximately 40–50 million years ago may have led to increased hydrothermal activity, which in turn may be responsible for an increase in CO_2 and atmospheric temperatures during that time.

b. Biological Activity

The discovery of diverse biological communities at ridge crest vents has revolutionized scientific thinking about biological processes in the deep ocean and has led to reevaluation of the understanding of biological structures and processes from the molecular to the ecosystem level. Among the distinctive features of vent communities are the presence of (1) chemosynthetic bacteria at the base of the food chain that derive their energy from hydrogen sulfide and other compounds in the hydrothermal vent fluid (as opposed to all previously known ecosystems, which used solar energy), (2) novel symbiotic relationships; and (3) great temporal and spatial variability of vent environments compared to environments elsewhere in the deep ocean. Vent animals are largely different from the surrounding deep-sea animals, suggesting extensive evolutionary adaptations to the unique vent environment.

The great spatial and temporal variability of vent environments on a very restricted spatial scale, combined with the close connection of vent biological communities with the geochemistry of vent fluids, allows vents to serve as excellent natural laboratories for a variety of biological studies. The recognition that black smoker vents probably have lifetimes of the order of a few decades opens interesting questions regarding the life histories of vent organisms, how they colonize, how genetic differentiation and speciation occur, and how organisms disperse and find new vents when old ones die. Studies of vent biology may even shed light on the origin of life on earth.

V. Methods of Study and Future Developments

The study of hydrothermal systems requires an integrated approach that incorporates field, laboratory, and theoretical investigations of both active and fossil systems. Both geophysical and geochemical techniques are needed. Fossil systems provide critical information about the end products of hydrothermal activity, such as rock alteration, zones of mineralization, and ore deposition, and their spatial relationships. It is more difficult to sort out aspects of the temporal evolution from the study of fossil systems, and dissection and

overprinting of the system by subsequent geologic processes lead to complexities. The study of active systems provides a snapshot view of the system behavior. Long-term monitoring can elucidate some aspects of temporal behavior, but not those that operate on geologic time scales. Moreover, the study of active systems is hampered by an inadequate knowledge of the subsurface, particularly of the deep parts of the system.

A. Continental Systems

Although continental hydrothermal activity has been recognized and utilized since antiquity, commercial exploitation and detailed scientific investigation are quite recent. Today, geothermal resources are used in a number of countries throughout the world, for the generation of electricity and space heating as well as for a variety of industrial, agricultural, and balneological uses. They are still much underutilized resources, however, Only in the past 40 years have intensive field and laboratory investigations been carried out and theoretical models of continental hydrothermal systems developed. Studies of active hydrothermal systems in connection with the formation of ore deposits are still rather primitive.

The association of continental hydrothermal activity with volcanism also suggests the importance of hydrothermal activity in connection with the volcanic environment. The hydrothermal regime of volcanoes bears on volcanic hazards related to landslides and edifice instability as well as on processes related to eruptions themselves. Studies of volcano–hydrothermal systems are just beginning to receive serious attention. [*See* VOLCANOES.]

High-temperature systems respond to changes in heat input due to magma emplacement and subsurface temperature, and circulation patterns can be affected by climatic changes such as glacial ice cover. To the extent that these local conditions reflect changes of a global nature, the system may record global change on time scales of thousands to tens of thousands of years. Warm-spring systems in which the fluid throughput is rapid—i.e., the residence time of the fluid in the subsurface is less than 100 years—may provide insight into recent changes. Relatively little work has been done on relating temporal changes in continental hydrothermal systems to global change.

B. Submarine Systems

While continental hydrothermal systems have little more than a local impact on the earth's energy and geochemical budgets, albeit an important impact from a human perspective, submarine systems have a global impact. Ridge crest hydrothermal systems are a central component of a large dynamic system that transfers a significant fraction of the earth's internal energy from the interior to the surface along the ridge axis. The enormous mass flux of seawater through the system exerts a control on certain geochemical mass balances that greatly influences the composition of the global ocean and the CO_2 cycle in the atmosphere on geologic time scales. The geochemistry of the exiting solutions controls the vent biological ecosystems. High-temperature plumes may affect the circulation of the oceans at middepths. The effects of recently discovered megaplumes on ocean circulation are still unknown but megaplumes have been hypothesized to be related to El Niño events. [See EL NIÑO AND LA NIÑA.]

Studies of seafloor hydrothermal systems require an integrated approach. Hydrothermal systems cannot be studied without an understanding of the spatially and temporally dependent processes by which magma is transported to the ridge axis and forms oceanic crust; nor can these systems be studied without considering the dynamical relationships between biological processes and the geochemical, geological, and physical processes involved in hydrothermal venting and the transport of hydrothermal solutions into the oceans.

The study of active submarine hydrothermal systems at ocean ridge crests is scarcely more than a decade old and hence is still in its infancy. It is hampered by almost complete ignorance of conditions at depth. There are no boreholes in active ridge crest systems and hence no subsurface samples of rock or fluid. Long-term monitoring of seafloor hydrothermal systems is needed; only a few vent systems have been sampled more than once over the past decade, and few attempts have been made to examine short-term temporal variability.

Active high-temperature systems have been found in a variety of tectonic settings at both sedimented and unsedimented ridge crests and over the full range of spreading rates. The distribution of hydrothermal systems and ore deposits along ridge crests is unknown. Less than 1% of the axis has been explored systematically.

The study of active systems must be complemented by the study of fossil systems as exposed in ophiolites. The study of fossil seafloor systems suffers from the same problems as the study of fossil terrestrial systems, but three-dimensional information is critical to the eventual unraveling of the complex interrelated physical, geological, geochemical, and biological processes.

Bibliography

Anderson, D. N., and Lund, J. W. (1987). Geothermal resources. *In* "Encyclopedia of Physical Science and Technology" (R. A. Meyers, ed.), Vol. 6, p. 184. Academic Press, Orlando, Florida.

Anderson, R. N., Langseth, M. G., and Sclater, J. G. (1977). The mechanism of heat transfer through the floor of the Indian Ocean. *J. Geophys. Res.* **82**, 1828.

Berger, B. R., and Bethke, P. M., eds. (1985). "Geology and Geochemistry of Epithermal Systems." Volume 2 in the series "Reviews in Economic Geology." Society of Economic Geologists, El Paso, Texas.

Bischoff, J. L., and Rosenbauer, R. J. (1984). The critical point and two-phase boundary of seawater, 200–500°C. *Earth Planet. Sci. Lett.* **68**, 172.

Elder, J. (1981). "Geothermal Systems." Academic Press, London.

Fournier, R. P. (1989). Geochemistry and dynamics of the Yellowstone National Park hydrothermal system. *Ann. Rev. Earth Planet. Sci.* **17**, 13.

Grassle, J. F. (1986). The ecology of deep-sea hydrothermal vent communities. *Adv. Mar. Biol.* **23**, 302.

Ingebritsen, S. E., and Sorey, M. L. (1988). Vapor-dominated zones within hydrothermal systems: evolution and natural state. *J. Geophys. Res.* **93**, 13,635.

Ocean Studies Board. (1988). "The Mid-Oceanic Ridge—a Dynamic Global System." National Academy of Sciences, Washington, D.C.

Rona. P. A. (1988). Hydrothermal mineralization at oceanic ridges. *Can. Mineral.* **26**, 431.

Rybach, L. (1981). Geothermal systems, conductive heat flow, geothermal anomalies. *In* "Geothermal Systems: Principles and Case Histories" (L. Rybach and L. J. P. Muffler, eds.), pp. 3–35. Wiley, New York.

White, D. E. (1973). Characteristics of geothermal resources. *In* "Geothermal Energy" (P. Kruger and C. Otte, eds.), p. 69. Stanford Univ. Press, Stanford, California.

Glossary

Black smokers Seafloor hydrothermal fluids jetting from chimneys typically about 10 cm in radius. Flow rates are $\simeq 1$–5 m/s; temperature is 350–400°C. Black coloration is due to sulfide mineral precipitates.

Fluid Inclusions Tiny cavities, 1.0–100.0 μm in diameter, containing liquid and/or gas, formed by the entrapment of fluid in crystal irregularities. Fluid inclusions provide information on the temperature, pressure, and chemical composition of the fluids and/or gases trapped at the time of mineral formation.

Geopressured systems Hydrothermal systems contained in sedimentary rock in regions of normal geothermal gradient. They represent a special case in which the pore fluids are under pressure exceeding the hydrostatic pressure.

Lithosphere Semirigid upper platelike layer of the earth. Its thickness is near zero at oceanic ridge axes, but it increases with age to about 100 km. Oceanic lithosphere consists of a crustal rock layer $\simeq$ 5 km thick of basaltic composition underlain by mantle rocks. Continental lithosphere consists of a layer of crustal rock 30–60 km thick of low density and variable composition underlain by mantle rocks.

Magma   Mobile molten or partially molten rock generated within the earth.

Magma chamber   Reservoir of magma within the upper crust from which volcanic materials can be extruded.

Massive sulfide ore deposit   Occurrence of a concentrated mass of sulfide minerals such as pyrite, sphalerite, or chalcopyrite in one place, as opposed to their being disseminated or occurring in veins.

Ophiolites   Layered suite of basic and ultrabasic silicate rocks consisting of pillow basalts, basaltic dikes, gabbros, and peridotites that are thought to represent segments of oceanic lithosphere that have been thrust onto land at convergent plate margins.

Ice Age Dynamics

Kirk A. Maasch
Yale University

Ice ages, or glacial epochs, are periods of time during which the earth experiences notably colder climatic conditions on a global scale. Such times are characterized by the growth of large continental glaciers into mid-latitude regions, covering a much larger total area than the present-day ice sheets. Ice ages have occurred at widely spaced intervals of geologic time (approximately 200 Myr, 1 Myr is 1 million years) since the Precambrian, lasting for around 1–10 Myr. During the latest glacial epoch, commonly referred to as the Pleistocene ice age, there have been major global oscillations between glacial and interglacial climatic states on a time scale of 10–100 kyr (1 kyr = 1 thousand years). Although the exact causes for ice ages, and the glacial cycles within an ice age, have not been proven, they are most likely the result of a complicated dynamical interaction among many components of the earth system.

I. Introduction

The existence of ice ages on earth has been pondered for well over a century now and still has no complete explanation.

Since the early nineteenth century, when it was first recognized that vast glaciers had once covered much of Europe and North America, countless theories have been proposed with continuing controversy regarding external versus internal causes. During the past decade external causes due to variations in the earth's orbital geometry, as proposed by M. Milankovitch, has become widely accepted as an explanation for the Pleistocene ice ages. This "Milankovitch revival" is largely due to the availability of continuous paleoclimate records, made possible by advances in deep-sea drilling technology, which appear to contain periodicities near those of orbital variations. In particular, these continuous records contain a dominant periodicity in the late Pleistocene of roughly 100 kyr, near which rather weak forcing due to eccentricity variations occurs. Additional variations are observed in these records at periods of approximately 41 and 22 kyr, corresponding to change in obliquity and precession of the equinoxes, respectively. However, the mere presence of variability at these periods is not necessarily evidence of a fundamental cause. In essence we do not really know that there would not still have been ice ages on earth even in the complete absence of any external earth orbital forcing. [See PALEOCLIMATIC EFFECTS OF CONTINENTAL COLLISION AND GLACIATION.]

In fact, analysis of oxygen isotope records from deep-sea sediment cores (a proxy for global ice mass) spanning the entire Pleistocene reveals that 100-kyr periodicity is virtually absent before 900 ka (ka≡thousands of years before present). Consequently, a theory based solely on external earth-orbital causes, which has never been able to explain satisfactorily the presence of a dominant 100-kyr cycle, is also faced with explaining its sudden origin in the mid-Pleistocene. To further complicate the issue, there is evidence that rapid transitions toward cooler mean climatic conditions occurred at roughly 900 ka and 2.4 Ma (Ma≡millions of years before present). Since the character of the earth's orbital variations has not changed significantly during the past several million years, a purely external astronomical cause for these changes seems unlikely.

The idea that atmospheric carbon dioxide may play an important role in glacial–interglacial cycles during the Pleistocene has also been revived. At the end of the last century it had been hypothesized that lowering the amount of CO_2 in the atmosphere could cause an ice age, although at the time there was no evidence for such a decrease. It is now possible to measure the composition of tiny air bubbles trapped in polar ice, thus providing a paleo-CO_2 record. These measurements indicate that CO_2 levels in the atmosphere were approximately 90 ppm (parts per million) lower during a glacial time than during an interglacial. The CO_2 record from Vostok, Antarctica, which spans the last 160 kyr, shows the characteristic 100-kyr periodicity also seen in the proxy record for global ice mass. Exactly why such variations in atmospheric CO_2 occur and what (if any) the connection is with growth and decay of large continental ice sheets is not precisely known. Because of uncertainties in the time sales for both marine and ice cores, the phase relationship between atmospheric CO_2 and continental ice mass as inferred from the marine oxygen isotope records remains questionable, although some inferences have been made using a marine proxy for CO_2 based on carbon isotopes.

It seems likely that a combination of internal feedback mechanisms involving the cryosphere, atmosphere, and ocean, along with external forcing due to orbital variations, are responsible for the glacial cycles that dominated the Pleistocene. Thus, it should be possible to describe climatic variation with a dynamical system containing internal feedbacks (capable of unstable oscillatory as well as stable steady-state behavior) and externally forced by earth-orbital variations. A constant challenge to the development of a complete theory for climatic variations is provided by continual accumulation of paleoclimatic evidence. Further motivation, in addition to the geological aspects of the problem, is a concern over the impact on climate of burning fossil fuels. Recent speculation on possible climatic effects due to anthropogenic modification of atmospheric composition (mainly greenhouse gases like CO_2) point toward a need to understand the effects of natural variations to help assess effects of anthropogenic forcing of the climate with CO_2. [*See* ATMOSPHERIC TRACE GASES, ANTHROPOGENIC INFLUENCES AND GLOBAL CHANGE.]

The importance of an interplay between the collection and interpretation of data and the development of a theory capable of explaining all of the observations cannot be overemphasized at this point. The remainder of this article is broken into two parts: (1) a summary of the geologic evidence for global climatic variation over the Pleistocene; and (2) a discussion of progress toward the development of a complete theory of the ice ages, conforming as closely as possible to all aspects of the evidence.

II. Global Climatic Variation during the Pleistocene

During the last few decades there has been rapid expansion of our knowledge on paleoclimates, primarily due to advancements in technology. Deep-sea sediment cores that are on the order of a few hundred meters in length, the source from which a large fraction of this knowledge is derived, may now be obtained with little or no loss of sediment. This means that nearly continuous records spanning several million years, and having resolution of $\sim$1000 years, are becoming available.

The oxygen and carbon isotopic composition of foraminifera shells, relative abundances of temperature-sensitive microfossils, the amount of calcium carbonate, and detritus of continental origin (carried by either ice or wind) that make up the sediment are among numerous quantities measured in these cores. After obtaining the cores and information that they contain, we are left with climatological interpretation as the central problem. To address this problem it is necessary to examine each of many possible influences on the measured quantities and evaluate their significance in the context of the deep-sea sediment cores. In this way it is possible to narrow down the cause of fluctuations in these quantities to only a few. For instance, the oxygen isotope record may be used as a proxy for the evolution of global ice mass through the Pleistocene, and using the carbon isotopes it is possible to trace the paths of deep water and perhaps deduce long-term variations in the levels of atmospheric CO_2.

In addition to these marine records, the terrestrial record also provides important information about the magnitude of ice mass and its areal extent, along with numerous other climatically sensitive variables. Although most of the terrestrial evidence is fragmented and often difficult to correlate with the deep-sea records, some continuous indicators of global climatic variation have been obtained.

In the remainder of this section we will expand on the paleoclimatic evidence for changes in global ice mass and surface temperature, both of which represent measures of the climatic state used to define an ice age. In addition, evidence for variations in atmospheric CO_2 and deep-ocean temperature and circulation will also be described because of the potential connection between these and the time evolution of both temperature and ice on a global scale. Since data from a core is measured as a function of depth, another important question that needs to be addressed is how to put it on a time scale. The progress toward establishing a global climatic history for the Pleistocene relies largely on the correlation of isotopic, paleomagnetic, and biostratigraphic events in deep-sea sediment cores. Relatively few of these events have been reliably dated, and hence caution must be used when constructing a time scale.

A. Global Ice Volume

Terrestrial evidence in the form of moraines, striations, erratics, and other glacial features of the landscape can be used to reconstruct the limits of ice at the time of the last glacial maximum. As early as the late 1700s it was hypothesized that there had once been significantly more ice on the earth than at the present time. Although much of the evidence for glaciations prior to the last one have been largely erased, enough has survived for the geologists of a century ago to realize that the northern hemisphere had undergone not one, but multiple glaciations during the Pleistocene. The incompleteness of the continental record led to the notion that there were four major glaciations during the Pleistocene, an idea that prevailed up to the late 1950s. With the availability of more continuous records, mainly from the deep sea, we now recognize that there were many more glacial–interglacial cycles than was originally thought (at least 10 during the last one million years). What has come to be the primary evidence for these glacial cycles during the Pleistocene is the marine oxygen isotope record derived from planktonic foraminifera.

1. Marine $\delta^{18}O$ Record

When water evaporates from the sea surface, it is isotopically lighter than the water left behind. As precipitation that falls in the form of snow at high latitudes gets locked up in the glaciers during an ice age, the oceans become progressively heavier in oxygen isotopic composition. These variations in the $\delta^{18}O$ (delta notation is defined in the glossary) of ocean water in turn affect the $\delta^{18}O$ content of $CaCO_3$ in foraminiferal shells. Unfortunately, other factors also affect the $\delta^{18}O$ in these shells, including the temperature of the ocean, change in evaporation minus precipitation as measured by salinity, vital effects of individual species of foraminifera, and differential dissolution. However, although any or all of these effects could be important, if cores and the foraminifera to be isotopically analyzed are chosen carefully, the dominant effects are the oceanic composition (global ice volume) and temperature. Heavy (light) values of $\delta^{18}O$ correspond to cold (warm) temperature and/or high (low) volume of continental ice.

The first systematic measurements of $\delta^{18}O$ in deep-sea cores were made by C. Emiliani in the mid-1950s using planktonic foraminifera. At the time he believed temperature to be the dominant control on fluctuations in $\delta^{18}O$ records that he obtained from Caribbean cores; he concluded that equatorial sea-surface temperatures were as much as 5–6°C cooler during glacial times than during an interglacial. Slightly more than a decade later, N. J. Shackleton argued that the ice volume effect, not temperature, was the main cause for fluctuations in the $\delta^{18}O$ record. Based on a comparison of planktonic and benthic $\delta^{18}O$ curves, measured in the same core,

Shackleton and N. D. Opdyke concluded that the planktonic record had a slight temperature overprint, while the benthic record was more representative of the ice volume alone. This argument was based on the assumption that the temperature of bottom water, which today is near freezing, could not have been much colder during a glacial period. As detailed benthic $\delta^{18}O$ records became available, the idea that they were the best proxy for continental ice volume became the commonly accepted view.

Some complications arise, however, when estimates of the magnitude of sea-level change based on the interpretation of marine $\delta^{18}O$ as an ice record are compared with other independent evidence for sea-level change. Late Pleistocene sea-level changes have been reconstructed using several methods of analyzing coastal terraces and emergent shoreline deposits. In particular, the coral reef terraces in tropical regions of tectonic uplift (e.g., Barbados, Haiti, and New Guinea) have been studied in considerable detail. The estimated amplitude of sea-level change due to ice volume changes, inferred from the coral terraces, conflicts with estimates based on benthic $\delta^{18}O$. It has been hypothesized that this conflict may partially reflect some need to understand the role of floating ice masses that do not affect sea level. More than likely, though, it mainly points back to our lack of knowledge concerning exactly what portion of the variance in the $\delta^{18}O$ record is actually due to changes in oceanic temperature versus isotopic composition. [See SEA LEVEL FLUCTUATIONS.]

One way in which this discrepancy can be accounted for is by a reconsideration of the earlier assumption that deep-water temperature remained constant through a glacial cycle, thus invalidating the idea that the benthic $\delta^{18}O$ is the best available proxy for global ice volume. Ten years ago, R. K. Matthews and R. Z. Poore had suggested that the low-latitude planktonic $\delta^{18}O$ record is dominated by changes in global ice volume. This was based on work by R. E. Newell, who proposed that tropical sea-surface temperature probably remains almost constant, the result of a balance between radiation and evaporative latent heat flux. Thus, to a first-order approximation the planktonic $\delta^{18}O$ record measured in low-latitude deep-sea sediment cores might represent the best available proxy for fluctuations in the global ice volume during the Pleistocene.

2. $\delta^{18}O$ Stratigraphy

In his early work on deep-sea cores, Emiliani referred to the apparent succession of warm and cold periods as isotopic stages designated by numbers increasing with depth. The warm, interglacial stages (isotopically light) were assigned successive odd numbers alternating with successive even numbers representing the cold, glacial stages (isotopically heavy). Stages within which more detail existed were further broken into lettered substages. Emiliani recognized 17 stages in many cores from both the equatorial Atlantic Ocean and the Carib-

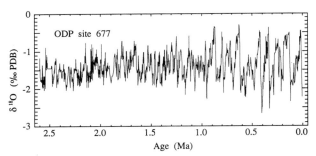

Fig. 1 A 2.6 Myr planktonic $\delta^{18}O$ record from ODP site 677 in the eastern equatorial Pacific (Data courtesy of N. J. Shackleton.)

bean Sea. W. S. Broecker and J. van Donk referred to the past six major deglaciations as "terminations." Termination I, II, III, IV, V, and VI correspond to the 1/2, 5/6, 7/8, 9/10, 11/12, and 13/14 stage boundaries, respectively. With the analysis of the first Pacific cores in the early 1970s it became readily apparent that these isotopic stages really were globally recognizable events. Shackleton and Opdyke defined 22 stages in western equatorial Pacific core V28-238 that correlated well with previously studied cores from the Caribbean Sea and Atlantic Ocean. As more cores were collected and analyzed for $\delta^{18}O$, and as the resolution of these records improved, a more detailed stratigraphy has developed.

The numbering scheme originally proposed by Emiliani has subsequently been modified to include a better description of substage structure within the last 750 kyr (stages 1–20). Instead of using a stage number followed by a letter to describe these substage events, a tenths digit was added—even for positive substage excursions and odd for negative ones. This system affords considerable flexibility in that if more detailed structure is ever recognized, another decimal place could be added to describe it. Subsequent extensions to the isotope stratigraphy back into the late Pliocene have been proposed with numbered stages reaching more than 100.

Figure 1 shows the equatorial Pacific planktonic $\delta^{18}O$ record from ODP site 677. This high-resolution record is currently the most detailed first-order estimate of variations of global ice volume over the past 2.6 Myr. What is readily apparent in Fig. 1 is that the last 1 Myr is dominated by 10 glacial–interglacial cycles (around 100-kyr period) upon which are superimposed higher-frequency fluctuations having periods near 40 and 20 kyr. Prior to this time there is no such dominant 100-kyr cycle, although a dominant near-40-kyr periodicity seems to remain.

B. Surface Temperature

Paleotemperature fluctuations have been estimated by numerous methods at widely distributed locations over the surface of the earth. Many of these estimates, particularly on land,

only extend back into (sometimes through) the last glacial cycle. Sea-surface temperature (SST) estimates derived from the marine sedimentary record, on the other hand, can often be extended back over 10^5–10^6 years. Although paleotemperature records are often dominated by local or regional temperature fluctuations, there is an underlying cold–warm cycle common to many of these that reflects the global signal of glacial–interglacial cycles clearly recorded in the marine $\delta^{18}O$ record.

1. Marine SST Record

It has long been recognized that planktonic organisms could be used to reconstruct the temperature history of the surface oceans. Around the turn of the century, J. Murray cataloged warm- and cold-water species of planktonic foraminifera; although he lacked access to deep-sea cores, he knew that they would contain information on temperature fluctuations of the geologic past. In the 1930s, W. Schott demonstrated that indeed foraminifera could be used to distinguish glacial from post-glacial conditions. F. B. Phleger found several warm–cold cycles in cores obtained by the Swedish Deep Sea Expedition.

Subsequently, several methods were developed by which quantitative estimates of surface ocean temperatures can be made using foraminiferal assemblages. The presence/absence of warm/cold species or species groups is the simplest of these. Another, the method of total faunal analysis, sums up the percentages of cold- and warm-water species, which are then used to estimate temperature based on empirically determined relationships between present-day observed temperatures and faunal assemblages. In the early 1970s, J. Imbrie and N. G. Kipp developed a more sophisticated method using multivariate factor analysis to write transfer functions to estimate paleotemperatures. At high latitudes, where planktonic foraminifera cannot be used, transfer functions based instead on either radiolaria or diatoms have been developed. Based on such quantitative methods, SST estimates have been made down many cores throughout the world oceans. Not surprisingly, during glacial times it is colder than during interglacials.

2. Terrestrial Record

At high latitudes, estimates of the surface air temperature can be made based on the oxygen and hydrogen isotopic composition of ice cores drilled in the polar ice caps. Such cores recovered in Greenland and Antarctica provide a record of climatic variation over the last glacial cycle. Oxygen and hydrogen isotopic ratios ($\delta^{18}O$ and δD) can be measured in the ice downcore with resolution as high as 100 years. Both $\delta^{18}O$ and δD are strongly a function of the temperature at which the precipitation condensed (colder temperatures correspond to lower delta values). Although the isotopic composition of the ice may also depend on other effects, including changes in ice thickness with time, changes in ice-flow pattern, changes in the source area of precipitation, and changes in the seasonal distribution of precipitation, it appears that the $\delta^{18}O$ and δD of polar snow does correlate with the air temperature at the site of deposition. In East Antarctica the isotope–temperature relationship has an observed slope of 6‰/°C for δD and 0.75‰/°C for $\delta^{18}O$. The temperature at Vostok during glacial times (estimated using δD) is approximately 12°C colder than present-day values. In Greenland the observed isotope–temperature slope is slightly lower; 6.62‰/°C for $\delta^{18}O$. At Dye 3, Greenland, this implies that the glacial–interglacial temperature contrast is about 13°C.

In nonglaciated mid-latitude areas it is possible to estimate surface temperature several ways. Loess, which is a wind-blown silt, is indicative of a cold, dry (glacial) climate. Pleistocene loess deposits cover large areas of the mid-latitude northern hemisphere. During the intervening warm, more humid interglacial times, soils can form in these areas. The alternating layers of loess and soils can be used to deduce both temperature and humidity variations back in time. In central Europe, loess deposits indicate that there were eight glacials and eight interglacials during the latter half of the Pleistocene (the last 800 kyr)—the same number of cycles recognized in the deep-sea $\delta^{18}O$ record. Loess deposits in China extend back even farther, providing a detailed chronology of climate change through the entire Pleistocene. The Chinese loess records the inception of northern hemisphere glaciation at around 2.4 Ma as well as the increased amplitude of the glacial–interglacial cycle in the mid-Pleistocene. Glacial times correlated with oxygen isotope stages 2, 6, 12, and 16 are strong in the Chinese loess record, as they are in the marine $\delta^{18}O$ record. While it is not quite clear why, variations in the magnetic susceptibility of loess from China closely resemble the fluctuations observed in the marine $\delta^{18}O$ record, leading to a tentative conclusion that it is in some way indirectly related to global ice mass.

Pollen records recovered from bogs or lake beds reflect both local and regional climatic variation, as well as global change. Climatic variations inferred from pollen data from Europe, Asia, and South America (spanning 100 kyr to 1 Myr) are similar to those derived from the marine $\delta^{18}O$ record. For example, a late Pleistocene record of climatic variation in Greece reconstructed from pollen reflects the glacial–interglacial cycles seen in both the deep-sea and loess records. Similarly, pollen-derived paleoclimate histories from Japan and Taiwan also exhibit the same kind of glacial–interglacial pattern over the late Pleistocene. It has also been suggested on the basis of a longer (3 Myr) pollen record from South America that there is a connection between the long-range altitudinal shifts of the forest line in Eastern Cordillera, Colombia and fluctuations of global temperature; 48 isotopic events identified in the marine $\delta^{18}O$ record can be visually

correlated with temperature changes inferred from this record.

Lake sediments and sometimes calcite deposits on land can be isotopically analyzed to obtain continuous records reflecting paleotemperature variations. One such calcite vein, deposited by groundwater moving through the Devils Hole fault zone at Ash Meadows, Nevada has been analyzed for $\delta^{18}O$ providing a continuous record spanning from around 50 ka to beyond 500 ka. The $\delta^{18}O$ values here are presumed to represent fluctuations of mean surface temperature in the recharge area of the Ash Meadow basin. A comparison of the Devils Hole $\delta^{18}O$ record with $\delta^{18}O$ measured at Vostok, Antarctica and in the deep-sea sediments reveals a marked similarity suggesting that this record does indeed reflect changes in global paleoclimatic conditions. Calcite deposits in caves have long been used as a monitor of past climates; however, because they cease to grow during glacial times these records are discontinuous—they only record climate change during interglacials.

C. Atmospheric CO_2

It is now widely recognized that the amount of trace constituents in the earth's atmosphere have indeed changed with time. Most noted are long-term variations in the so-called "greenhouse" gases, particularly CO_2. Correlation between these changes in CO_2 and global ice mass point to a possible CO_2–climate connection. Observed variations in other constituents of the atmosphere, such as the greenhouse gas methane, as well as dust and other particulates, may also be related to climate change.

1. Ice Core CO_2 Record

The polar ice sheets in Greenland and Antarctica provide an opportunity to obtain direct measurements of the composition of ancient air. During the formation of glacial ice, tiny air bubbles become trapped as the snow compacts. This air is younger than the surrounding ice, and the exact age difference depends on the temperature as well as on the accumulation rate. Refined techniques for extraction and analysis of these bubbles provide direct measurements of the CO_2 concentrations in this air. It then becomes possible to reconstruct a time history of atmospheric CO_2. Potential sources for error in the CO_2 record include melting and refreezing, which will increase the amount of CO_2 in ice, and interactions of CO_2 with impurities in the ice, which can lower the amount of CO_2. Additionally, as with the deep-sea sediment cores, estimating age as a function of depth presents a major problem. Dating the ice cores can be done several ways, including counting annual layers, identifying marker horizons (i.e., ash layers), using radioactive dating, correlation with deep-sea records, and using ice flow models.

Analysis of air bubbles in six different polar ice cores, two

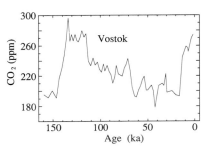

Fig. 2 The Vostok ice core estimate of atmospheric CO_2 variation.

from Greenland and four from Antarctica, indicate that atmospheric CO_2 was 30% lower during the last glacial than during the present interglacial. The cores from Camp Century and Dye 3 in Greenland, and Byrd Station and Dome C in Antarctica provide a record spanning approximately the last 30–50 kyr. More recently, a core from Vostok, Antarctica has made it possible to extend this record back to 160 ka (Fig. 2). The Vostok CO_2 record shows a high correlation with $\delta^{18}O$ and δD measured in the ice, including the dominant period of oscillation near 100 kyr. High levels of CO_2, near 280 ppm, correspond with interglacials whereas an average level of around 190 ppm is observed during glacial times. High peaks in CO_2 concentration occur close to the time of rapid deglaciation. However, because of uncertainties in the time scale for the CO_2 record from the ice cores, and the fact that the enclosure process acts to smooth the record, reliable measurements of periodicities or phase relationships between global-ice mass and atmospheric CO_2 are difficult to make with any certainty. Recent measurements of $\delta^{18}O$ made on the O_2 of air bubbles in the Vostok core, which have tentatively been linked with global ice volume, may provide a means by which this phase can be more accurately determined. Preliminary results show the CO_2 maximum preceding the ice minimum by several thousand years.

2. Marine $\delta^{13}C$ Record

The time evolution of atmospheric CO_2 has also been estimated indirectly by using $\Delta\delta^{13}C$ (the difference between planktonic and benthic $\delta^{13}C$ measured in foraminifera from deep-sea cores) as a proxy. The use of carbon isotopes in this manner stems from the proposal made by Broecker that the observed changes in atmospheric CO_2 must be caused by alterations in the chemistry of seawater. A comparison of the $\Delta\delta^{13}C$ record from eastern equatorial Pacific core V19-30 with the Vostok CO_2 record shows similarity in the general trend over the past 160 kyr, including the rapid increase in CO_2 associated with the last two deglaciations. There are, however, some differences between the two records as well, suggesting that effects other than CO_2 changes are also re-

flected in the $\Delta\delta^{13}C$ record. Since $\delta^{18}O$ and $\Delta\delta^{13}C$ were both measured in the same core, they can be placed on a common time scale and the phase between them can be estimated. In doing this it appears that the minimum in $\Delta\delta^{13}C$ (high-atmospheric CO_2) leads the minimum in $\delta^{18}O$ (low ice volume).

D. Deep Ocean Properties

Changes in the circulation and chemistry of the oceans contribute significantly to global climatic variability. Deep ocean circulation patterns and residence times of the deep waters play an important role in redistributing heat as well as controlling atmospheric CO_2. While it is possible to estimate sea surface conditions using faunal abundances of planktonic fossils, the ability to infer deep-water properties in a similar manner has proven to be difficult. Investigations of geochemical properties of benthic species, on the other hand, have been more successful in revealing past variations in the behavior of the deep oceans. Among the geochemical quantities measured in foraminifera that provide information useful in reconstructing paleoceanographic conditions are $\delta^{18}O$, $\delta^{13}C$, and ^{14}C, and the cadmium (Cd) content of $CaCO_3$. [*See* CARBON DIOXIDE TRANSPORT IN OCEANS; OCEAN CIRCULATION.]

1. Temperature

Since the $\delta^{18}O$ of benthic foraminifera is a function of both oceanic composition and temperature, an estimate of global ice volume (and its isotopic composition) can lead to an estimate of deep-ocean temperature. Such an estimate for global ice volume can be made from sea-level changes deduced from coral terraces (as discussed above). If the amount of $\delta^{18}O$ attributed to changes in oceanic composition is then subtracted from a benthic $\delta^{18}O$ record, the residual $\delta^{18}O$ represents an approximate deep-water temperature. Making these assumptions, it appears that for the past 150 kyr the mean deep-ocean temperatures were similar to those of the modern ocean only during peak interglacial times, and during glacial times the deep sea was about 2°C cooler. The main uncertainty in reconstructing deep-water paleotemperature in this manner is the possibility that regional isotopic variability may not always be negligible.

2. Circulation

The $\delta^{13}C$ and the Cd/Ca ratio measured in foraminiferal shells can be used to reconstruct deep-ocean circulation patterns. In the modern ocean, the distribution of $^{13}C/^{12}C$ ratios are related to biological activity. Living organisms preferentially take up ^{12}C, leaving the ocean surface enriched in ^{13}C (high $\delta^{13}C$). At the same time these organisms also take up nutrients (phosphorus and nitrogen) and, although it is not understood why, Cd. As biological debris sinks to the bottom, most of its decays, leading to deep water lower in both

δ^{13}C and O_2, but enriched in nutrients and Cd, relative to the surface waters.

Individual deep-water masses have a characteristic geochemical signature when they form. Subsequently, these water masses will then accumulate different amounts of decaying debris along their flow path as they age. Knowing the distribution of δ^{13}C and Cd, the path of a water mass can then be traced and the strength of mixing between different source components can be estimated.

North Atlantic Deep Water (NADW) plays an important role driving the thermohaline circulation of the world oceans and, hence, in the ventilation of the deep waters. NADW is enriched in δ^{13}C and depleted in nutrients. As NADW ages, flowing through the South Atlantic and Indian oceans, and ending up in the deep Pacific, it gradually becomes depleted in δ^{13}C and enriched in nutrients. The δ^{13}C gradient between the deep Atlantic Ocean and the deep Pacific Ocean is in some sense a measure of the relative strength of NADW production.

The δ^{13}C content of some species of benthic foraminifera appear to reflect the δ^{13}C of the bottom water. Also, Cd in solid solution in the $CaCO_3$ shells of benthic foraminifera is representative of the Cd content of the bottom water. Thus, by measuring these quantities downcore at several locations, it becomes possible to reconstruct the distribution of these geochemical tracers in previous times. In doing so, the indications are that that NADW production was significantly reduced during glacial times relative to the present. [See DE-GLACIATION, IMPACT ON OCEAN CIRCULATION.]

Additional information on the residence time of deep waters can be obtained by measuring ^{14}C in the shells of foraminifera. Comparison of the ^{14}C age of benthic foraminifera with those of planktonic foraminifera provides a measure of the ventilation rate of the bottom waters. From the few available estimates, it appears that the ^{14}C age of NADW decreased from 650 years at the last glacial maximum to 350 years today. The age of Pacific bottom waters also decreased by several hundred years to a present-day value of 1600 years.

E. Pleistocene Time Scales

Although the precise timing of glacial events throughout the Pleistocene is difficult to determine, the development of an accurate time scale is necessary to quantitatively estimate periodicity and phase relationships between climatic variables—information crucial to understanding why the earth experiences ice ages. With particular concern for the marine δ^{18}O record, a wide variety of ideas have been employed in the struggle to assign age to depth in deep-sea sediment cores. Among the many tools that have been instrumental in the development of the Pleistocene time scale are radiometric dating, correlation based on isotope stratigraphy, magnetic stratigraphy and biostratigraphy, and

spectral methods for tuning deep-sea records to variations in the earth's orbital parameters.

1. Interpolated Time Scales

The first time scales developed for the late Pleistocene δ^{18}O records were based on the assumption of a linear sedimentation rate between the recognizable stratigraphic events for which age estimates were available. Dates on material directly in the sediments (namely, ^{14}C) are restricted to the most recent part of the record. However, older age estimates not made directly on the actual core sediments can come from radiometrically dating an event, which can then be correlated to some depth in the core. [See RADIOCARBON DATING.]

The raised coral terraces used to estimate sea level, which have been U-Th dated at 80, 105, 125 ka, are correlated with the last interglacial (oxygen isotope stage 5). Magnetic reversals, which have been K-Ar dated in basalt flows all over the world and shown to be globally synchronous events, can be readily detected in deep-sea sediments. During the last three million years there were two long periods (epochs) of normal polarity and one reversed. The transition from the Gauss normal to the Matuyama reversed (M/G) occurred at $\sim$2.43 Ma and from Matuyama reversed to Brunhes normal (B/M) at $\sim$0.73 Ma. Within the Matuyama there are two shorter normal vents, the Olduvai (1.88–1.66 Ma) and the Jaramillo (0.98–0.91 Ma). [See MAGNETIC STRATIGRAPHY.]

Biostratigraphic zonations based on the first and last appearances of marine microorganisms can also be used to correlate deep-sea cores as well as to aid in the construction of a time scale. These datum levels are based on calcareous nannoplankton, diatoms, planktonic foraminifera, and radiolarians. The most important of these are the ones that have been shown to be globally synchronous. [See STRATIGRAPHY —DATING.]

Other attempts have been made to construct interpolated time scales based on the assumption that a certain part of the total sediment accumulation has remained constant, namely that the influx of terriginous material has not varied significantly with time. In the early 1950s G. Arrhenius suggested that during the Pleistocene accumulation rates of terriginous material were constant in the open Pacific Ocean. Thus, he attributed fluctuations in TiO_2 concentrations in the sediments to variable accumulation rates (particularly of $CaCO_3$) in the deep sea. Subsequently this idea has been used with the assumption that other terriginous components of the sediment (aluminum, clay, and fine silt) accumulate at a constant rate.

2. Tuned Time Scales

Spectral analysis of paleoclimate time series has become a routine practice during the last 20 years. In addition to the spectra of individual records, estimates of cross spectra can be used to determine coherency and phase relationships between variables such as $\delta^{18}O$, CO_2, and SST. Obtaining this information is an important step toward understanding possible dynamic interactions among components of the climate system.

The first attempts at spectral analysis of deep-sea paleoclimate records were made to test the Milankovitch hypothesis of ice ages. These early efforts, in the late 1960s, were hampered by large uncertainties in the time scale. By the early 1970s longer records, which extended back through the B/M magnetic transition, became available and the initial interpolated time scale improved enough to allow detection of peaks in the spectrum near 100, 40, and 20 kyr. Although the interpretation of these spectral estimates as evidence for cause remain a subject of debate, the one thing clear is that spectra of the Pleistocene paleoclimate record contain significant peaks near these periodicities (sometimes along with additional peaks that are also significant).

Spectral estimates were then used to attempt to refine the interpolated time scales by assuming a relationship between the $\delta^{18}O$ record and variations in the geometry of the earth's orbit. The first of these was the TUNE-UP time scale of J. Hays, Imbrie, and Shackleton. For the $\delta^{18}O$ record from RC11-120/E49-18 they estimated that between 70 and 300 ka the component of $\delta^{18}O$ in a band centered on 41 kyr lags obliquity by 9 ± 3 kyr, and for the portion of the same record spanning 0–150 ka the component in a band centered on 23 kyr lags precession by ~3 kyr (although when looking at the record from 0–300 ka these are approximately in phase). Using these results, they went on to assume that in fact these phase relationships would be constant over the entire record if their time scale was correct. They then proceeded to tune the time scale by adjusting the age of the 11/12 stage boundary from 440 to 425 ka and the 7/8 boundary from 251 to 247 ka. With these changes the phase relationships then appeared to remain constant over the last 425 kyr in the 41 kyr (obliquity) band and the last 340 kyr in the 23 kyr (precessional) band.

Subsequently, several other tuned time scales have been developed, mainly variations on the same basic theme. The one most widely used is a stacked oxygen isotope chronology developed by Imbrie and others (commonly referred to as the SPECMAP record). The approach taken to build this record may be summarized as follows: First, five $\delta^{18}O$ records were stratigraphically analyzed and placed on an initial interpolated time scale. Then they were orbitally tuned by an iterative procedure aimed at achieving an optimum phase lock between the data and two target curves — the variations in obliquity and precessional index. The phase lag between obliquity and the 41 kyr component of the data was assumed to be 8 kyr, and between the precessional index and the 22 kyr component of the record, 5 kyr. The data were filtered with bandpass filters centered on 41 and 22 kyr, designed to pass >50% of the variance at periods from 35–50 kyr and

17–27 kyr, respectively. The phase of the filtered records in each band was then compared with the phase of the target curves, and adjustments were made to the age estimates of one or more control points before the filtering process was repeated. The SPECMAP time scale is the result of approximately 120 such iterations. Finally, the records were normalized to zero mean and unit variance and averaged together to produce the stacked SPECMAP $\delta^{18}O$ curve (which has come to be the most widely quoted time scale in the literature). In fact, in many cases it has become almost blindly accepted as the "most accurate" and simply applied to many diverse forms of paleoclimate data, which in turn are subjected to spectral analysis. This, quite naturally, acts to accentuate any Milankovitch periodicities that may be in a record by implicitly making the assumptions that were made in constructing the SPECMAP time scale.

III. Ice Age Theories and Climate Models

To build a complete theory for the climatic variations during the Pleistocene (or any other time), all of the available observational evidence must be used to provide the constraints that guide the formulation of such a theory. This article does not provide a detailed historical account of the development of every theory for ice ages; however, three relevant ideas that when combined, can provide a credible account for the Pleistocene ice ages are outlined here.

A. ''Greenhouse'' Forcing of Climate Change

The idea of "greenhouse" warming in connection with paleoclimatic variability goes back at least to the middle of the nineteenth century. In the early 1860s J. Tyndall measured absorption of infrared radiation by CO_2 and H_2O vapor, concluding that their presence results in a warming of the atmosphere. Tyndall was probably the first to suggest that ice ages may have been caused by decreased atmospheric CO_2. In 1896 S. Arrhenius calculated the radiative effects of doubling atmospheric CO_2, concluding that the average temperature would rise by around 5°C. He postulated that this warming would be more pronounced due to a positive feedback involving an increase in the amount of water vapor (the dominant greenhouse gas) caused by the increased temperatures, and furthermore, higher latitudes would warm by an amount greater than low latitudes. He also did a half-CO_2 calculation and arrived at a 4°C decrease in temperature. Arrhenius suggested that variable volcanic input of CO_2 followed by re-equilibration with weathering processes could cause variation of atmospheric CO_2, which in turn could cause the ice ages. Around the turn of the century, T. C. Chamberlin constructed a geological scenario for CO_2 as a cause of the ice ages which at that time probably made this theory the most widely accepted explanation for climate change.

Since any actual evidence for such changes in CO_2 remained unavailable, the theories invoking CO_2 as a possible cause of the ice ages, although not completely forgotten, did for some time take a back seat to more popular ideas such as the earth orbital theories, among others. A brief revival of the belief that lowering atmospheric CO_2 could cause ice ages came about in the 1950s. G. N. Plass recalculated the effect that a doubling and halving of atmospheric CO_2 would have on surface temperature using a computer to solve the radiative transfer equations. His results showed an increase in the earth's average surface temperature of 3.6°C for a CO_2 doubling and a decrease of 3.8°C if CO_2 were halved. Plass went on to postulate that if the amount of CO_2 in the atmosphere–ocean system could be drawn down to a critically low value, atmospheric CO_2 along with large continental ice sheets could oscillate at the periodicity observed in the climate records.

Once again, however, due to lack of any evidence for lower CO_2 and with all of the data made available from deep-sea cores in the 1960s and 1970s, the Milankovitch theory once again became the center of attention. It was not until the late 1970s that the potential importance of CO_2 was once again recognized. At this time B. Saltzman was including feedbacks involving CO_2 as a source of instability of the climate system in his effort to develop a more complete theory to explain ice ages. Also at this time, the first measurements from polar ice cores were made, which showed that indeed the concentration of atmospheric CO_2 was lower during glacial times than it is at the present.

B. Orbital Forcing of Climate Change

In the late eighteenth century J. L. Lagrange and P. S. Laplace developed the general theory for secular variations of the earth's orbital geometry. These so-called secular changes, which are due to the gravitational interaction between the earth and other planets, include variations in eccentricity, obliquity, and the precession of the equinoxes. The eccentricity (e), which measures how eliptical the earth's orbit is, ranges between approximately 0.00 and 0.06. Several periodicities have been identified (dominated by periods near 100 and 413 kyr). Variations in eccentricity change the total amount of solar radiation received in a year at the top of the atmosphere (more radiation when e is high). The obliquity (ϵ), which is the angle between the earth's equatorial plane and the plane of the orbit, varies roughly between 22° and 24.5° with a period of around 41 kyr. Changes in ϵ latitudinally redistribute incoming solar radiation—high values of ϵ are associated with larger seasonal extremes of radiation at the

poles. Changes in the precessional index ($P \equiv e \sin \omega$) result from precession of the equinoxes, measured by the longitude of perihelion (ω) modulated by eccentricity. An eliptical orbit means that the earth–sun distance of each season changes as a function of ω and e. Minimum values of P are associated with maximum values of summertime radiation in the northern hemisphere (the southern hemisphere is 180° out of phase).

By the middle of the nineteenth century several theories for the cause of ice ages based on these changes in the geometry of the earth's orbit had been proposed (for example, by J. A. Adhemar in 1842 and J. Croll in 1864). It was originally believed by Adhemar that the precession of the equinoxes would result in ice ages alternating between the northern and southern hemispheres every 11 kyr. Later, recognizing that the eccentricity would modulate the precessional effect, Croll hypothesized that during times when the orbit was nearly circular there would be no ice ages.

The actual solar radiation received at the top of the atmosphere is a function of latitude, time, eccentricity, obliquity, longitude of perihelion, and the solar constant. In the 1920s Milankovitch calculated the incoming solar radiation from 25° to 75° north and south (at every 10° of latitude) for the last 600 kyr (values which have since been refined and extended). Milankovitch proposed that these changes in radiation were enough to cause ice ages, and that the major factor in determining when an ice age would occur was the summer temperature at high northern latitudes, which controls the melting of winter snow accumulation. [See SOLAR RADIATION.]

C. Internal Instability

It is possible that even with constant external forcing, the internal dynamical interactions among relevant variables within the climate system can give rise to oscillatory behavior driven by instabilities in the system. This in fact is a plausible way to account for the as yet unexplained dominance of 100-kyr periodicity during the late Pleistocene. In addition, the phase relationships between climate variables (such as ice, CO_2, and SST) as well as the emergence of dominant 100-kyr periodicity can be accounted for in this way.

The observations of rapid transitions and strong variability at frequencies where there is only weak forcing imply that the internal climate system contains instability and nonlinearity. The system does not appear to be in quasi-static equilibrium with respect to external forcing due to orbital variations. This is evidenced by the fact that at times during the past when values of eccentricity, obliquity, and precessional index were nearly identical the amount of global ice mass could range from full glacial to interglacial conditions. In addition, based on a more detailed picture of the events during the last 18 kyr (constructed using both marine and continental evidence) it appears that the deglaciation began well ahead of the orbitally induced insolation maximum and was not unidirectional, as are changes in isolation, but instead was characterized by several shorter-period oscillations. Also, as noted, the dominant periodic response in the late Pleistocene is near 100 kyr in spite of relatively small external forcing at this period; and other climatic variability exists at periodicities far removed from any known forcing (although many of these have been severely supressed by using tuned time scales). Furthermore, even though the character of variations in the geometry of the earth's orbit have remained virtually unchanged over the last several million years, there was the rapid transition in the mean and variance (including the introduction of the 100-kyr cycle) of global ice mass at roughly 900 ka. In other words, the response of the climatic system to external forcing changed during the Pleistocene, implying that internal dynamics are important.

D. Modeling the Ice Ages

It must be recognized that modeling is not simply curve fitting, but instead represents an effort to discover the form by which the principles of conservation of mass, momentum, and energy can be expressed to account for the behavior of the complex climatic system and perhaps predict the consequences of changed boundary and initial conditions. When confronted with the difficult problem of actually formulating a statement of the governing laws, an inductive (rather than deductive) approach to the problem is necessary in our case, mainly because the magnitude of the net mass and energy fluxes involved in the Pleistocene ice-mass changes are far too small to calculate or measure. Vertical flux parameterizations (e.g., of sensible heat and latent heat) are accurate to no more than ~ 1 W/m², whereas fluxes required to account for observed long-term changes in ice volume (or sea level) are only $\sim 10^{-1}$ W/m² or less (accumulation minus ablation of Pleistocene ice sheets equals 1 cm/yr in net global evaporation). Other slowly changing quantities pertinent to climate change present a similar problem. Changes in atmospheric CO_2 occurred naturally during the Pleistocene at a rate of less than 10^{-1} ppm/yr (anthropogenic increase is on the order of 1 ppm/y). To change the deep-ocean temperature by 2°C over 10 kyr only requires the order of 10^{-1} W/m². The fact that these fluxes are so small and can be accomplished by so many potentially relevant feedbacks effectively renders impossible a deduction of climate variability beginning with so-called first principles.

Put in the most simple terms, ice can accumulate in high northern latitudes if the snow that falls in the winter can survive through the following summer. This means that all of the slowly varying factors that can lead to slow changes in surface temperature and precipitation must be considered in the development of an ice age model. With the use of equilib-

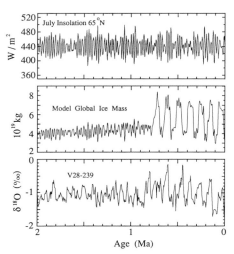

Fig. 3 Time evolution for the past 2 Myr of July insolation at 65°N taken as the orbital forcing (top), the model ice mass response (middle), and the ice mass variation inferred from the $\delta^{18}O$ record obtained from western equatorial Pacific core V28-239 (bottom). [From Saltzman and Maasch (1990).]

rium climate models (such as atmospheric general circulation models) it is possible to estimate the relative importance of these factors. It turns out that the surface temperature at high latitudes is at least as sensitive to varying CO_2 (within the range observed in the ice cores) as it is to radiative forcing due to earth-orbital variations. Hence, if earth-orbital variations are going to be considered relevant to the cause of ice ages, then CO_2 cannot be neglected. [*See* ATMOSPHERIC MODELING.]

As an example, Fig. 3 shows the results of a model that embodies each of the three ideas expanded above. This model is based on the hypothesis that free and forced variations in the concentration of atmospheric carbon dioxide, coupled with changes in the deep-ocean state and global ice mass, under the additional influence of earth-orbital forcing, are primary determinants of the climatic state. The variations of incoming July solar radiation at 65°N due to orbital variations, which is assumed to be the main external periodic forcing acting on the system over the late Pleistocene, are shown in the top panel of Fig. 3. Note that the general character of this forcing does not change over the last 2 Myr, while the model solution for global ice mass (shown in the middle panel of Fig. 3) includes the transition to the near-100-kyr major ice age oscillations of the late Pleistocene observed in the paleoclimate record. For comparison, the $\delta^{18}O$ measured in western equatorial Pacific core V28-239 (a proxy for ice mass) is shown in the bottom panel of Fig. 3. In obtaining this solution it was assumed that variations in tectonic forcing lead to a reduction of the equilibrium CO_2

concentration (perhaps due to increased weathering of rapidly uplifted mountain ranges over this period). As a consequence of this CO_2 reduction the model dynamical system can bifurcate to a free oscillatory ice-age regime that is under the "pacemaker" influence of earth-orbital (Milankovitch) forcing.

An important point to understand is that, given the lack of knowledge about the actual set of laws that govern long-term climate variability, models must remain flexible enough to accommodate new observational information as it becomes available. Model evolution is an iterative procedure making use of both improved observational data and the increased theoretical knowledge acquired from previous efforts. While fitting the available observations is necessary, model solutions also include information that we are currently unable to directly compare with observation. These predictions ultimately provide the most important test of a model and the hypothesis on which it rests; model predictions will either be verified, in which case we can continue to consider it a valid explanation, or they will prove to be incorrect, prompting modification and/or reinterpretation of the proposed variables and the effects of their feedbacks upon one another.

IV. Summary

A vast amount of evidence has been collected that indicates dramatic oscillations in climatic conditions of global proportion throughout the Pleistocene. It has been well documented that global ice mass has undergone huge fluctuations on the order of 10^{19} kg (equivalent to more than 100 meters of sea level) during the last million years. In addition, it has been demonstrated that the concentrations of atmospheric CO_2 vary by at least 30% through a glacial cycle, with glacial periods characterized by low pCO_2 (~ 190 ppm) and interglacials by high pCO_2 (~ 280 ppm). Ocean temperature and circulation have also undergone glacial–interglacial changes, with colder conditions generally prevailing in glacial times. These temperature fluctuations are more pronounced in the surface waters at high latitudes, where decreases on the order of 5–$10°C$ are believed to have occurred. Ocean deep water may have been as much as $2°C$ cooler during glacial times. North Atlantic Deep Water production seems to have been considerably reduced during glacial times relative to the present. It seems almost inevitable that the cryosphere, atmosphere, and ocean must be dynamically linked in ways that must be understood before we can hope to know why the

earth has ice ages. External earth orbital forcing alone is not sufficient to account for the observations, although it has certainly influenced the climate. To develop a complete theory for the ice ages it is necessary to account for all of the observed climatic variations (not just ice or temperature); and furthermore, in principle the theory should take into consideration the full record, not only the late Pleistocene. That is, the transition into an ice age from a warmer climatic state must also be accounted for by any complete theory. Although models have been constructed that can account plausibly for many of the main variations, a universally acceptable theory for the ice ages remains an ongoing challenge.

Bibliography

Bradley, R. S. (1985). "Quaternary Paleoclimatology." Allen & Unwin, Boston.

Hecht, A. D., ed. "Paleoclimate Analysis and Modeling." Wiley, New York.

Saltzman, B., and Maasch, K. A. (1988). Carbon cycle instability as a cause of the late Pleistocene ice age oscillations: Modeling the asymmetric response. *Global Biogeochem. Cycles* **2**, 177–185.

Saltzman, B., and Maasch, K. A. (1990). A first-order global model of late Cenozoic climatic change. *Trans. Proc. Soc. Edinburgh: Earth Sci.* **81**, 315–325.

Wright, H. E. (1989). The Quaternary. *In* "The Geology of North America — An Overview," The Geology of North America, Vol. A (A. W. Bally and A. R. Palmer, eds.), pp. 513–536. Geological Society of America, Boulder, Colorado.

Glossary

Benthic Forms of marine life that live on the bottom of the ocean.

Delta (δ) notation Isotopic ratio in a sample compared to the same ratio in a standard defined as

$$\delta = \frac{R_{sample} - R_{standard}}{R_{standard}} \times 1000$$

where R is the ratio of the heavy to the light isotope and δ is in parts per thousand (‰).

Planktonic Floating organisms that live in the surface waters of the ocean.

Icebergs

Stephen F. Ackley

Cold Regions Research and Engineering Laboratory

Icebergs are the floating remnants of ice produced on the continents in glaciers, ice sheets, and ice shelves that originate as snow compressed into ice. They differ from sea ice, which also floats in the ocean in cold regions, in that sea ice is produced by the direct freezing of seawater. Icebergs are a unique part of the earth's hydrologic cycle. The cycle is initiated as oceanic evaporated moisture, which is condensed, frozen, and deposited as snow precipitation on the continents and compressed into ice by a 100 m or so of overburden snow. The ice flows to the ocean in ice streams or glaciers that are driven by gravity and that densify and deform as the ice migrates (they behave as slowly moving rivers of ice). The cycle is completed by the ice breaking off into the sea as icebergs, which then melt slowly in the ocean. For the dominantly ice portion of the hydrologic cycle the time scales are typically from decades to thousands of years, compared with the annual or slightly longer time scales more typical of the water-based hydrologic cycle in lower latitudes.

I. Origin of Icebergs

There are three types of iceberg-producing formations: glaciers, ice sheets, and ice shelves. Glaciers that are grounded at their termini (not afloat) calve very small icebergs whose longest dimension is generally less than the ice thickness (Fig. 1). Most of the glacier fronts in the northern hemisphere are grounded and produce icebergs of this type. Ice streams of glaciers that have a floating length of at least several times their width calve icebergs that have a width on the order of the ice thickness and, consequently, a length that may be

several times the thickness. Ice shelves (Fig. 2), very large masses of floating ice primarily in Antarctica, are fed by both glaciers and direct precipitation and produce icebergs that have a long dimension from hundreds of meters to more than 100 km. Icebergs are produced by fracturing of the ice mass by one or a combination of forces: movement of the glacier (essentially the propagation of stress by gravity acting on the material); flexure induced by tides, waves, storm surges, and other rapid or periodic changes in the underlying water level; undermining of the glacier by runoff producing a shear failure near the face of the glacier; and unbalanced hydrostatic forces on the front face of the ice sheet (since it sticks out of the water). Particularly large icebergs may be produced by collision of an iceberg with an existing ice shelf or ice tongue, causing a long deep fracture that can generate a new iceberg. For example, the Trolltunga iceberg in Antarctica, which measured 50 by 100 km was apparently produced by collision of an iceberg from the Amery Ice Shelf against the Trolltunga (a floating ice tongue or shelf extending out from the Antarctic continent at longitude 1°W). The Trolltunga iceberg, after spending several years grounded on a shoal, later collided with the Larsen Ice Shelf, producing an additional berg of 90 by 35 km. [See WEST ANTARCTIC ICE SHEET.]

A. Antarctica

Estimates of the annual production of icebergs are related to the unknown mass balance quantity for the Antarctic ice sheet and vary by almost a factor of 10. The middle portion of this range for Antarctica, $0.5-1.2 \times 10^{15}$ kg/yr (about 1000 km^3), is from two to nearly six times the estimate of the Greenland iceberg mass per year. This is commensurate with the size and volume of the ice sheet, which is equivalent to 21×10^6 km^3 of water, or almost 10 times that of Greenland. Ice reaches the sea along approximately 20,000 km of coastline, of which about 11,000 km consists of ice shelves. These ice shelves account for the major differences in size and character between the Antarctic and Greenland icebergs. The total number of icebergs per year is estimated to be about 100,000 for Antarctica. [See ANTARCTICA.]

The major ice shelves—the Ross, Filchner-Ronne, Larsen, and Amery—as illustrated earlier, can produce enormous icebergs and can account for a tremendous share of the iceberg production in Antarctica. Perhaps 60–80% of the

ONE OF THE GLACIERS WE CROSSED

Photograph by F. H

Fig. 1 A glacier terminus on South Georgia Island. The glacier is probably grounded underwater at the terminus. Small icebergs (foreground) fracture off as the glacier pushes over the grounded area. [Photograph from F. Hurley in E. Shackleton (1920). "South, the story of Shackleton's last expedition, 1914–1917." Macmillan, New York.]

Fig. 2 Edge of one of the Antarctic ice shelves in the Atlantic sector of the Antarctic. The ice shelf is floating but attached to the continental ice sheet.

Fig. 3 Tabular iceberg in Antarctic waters. These flat-topped icebergs typically break off from ice shelves (Fig. 2).

Fig. 4 Tabular iceberg (long dimension about 1 km) floating in the Weddell Sea. Melting pack ice floes are seen in front and back of the iceberg. The dark spot near the center of the iceberg marks the place where a flock of seabirds landed. Birds typically use the icebergs as floating islands in these waters.

icebergs by volume are generated from the ice shelves, and therefore tabular icebergs (Fig. 3) are quite common. One reported iceberg was 75×110 km in area and contained $1000–2000$ km³ of ice, i.e., a year or more of estimated ice sheet average calving loss in one iceberg. The Ross and Filchner-Ronne ice shelves are roughly 0.5×10^6 and 0.4×10^6 km² in area (roughly the size of France or Texas). Icebergs measuring many tens of kilometers are generated from these ice shelves at intervals of $10–100$ years, and kilometer-size icebergs calve off annually or more frequently. The wide variation in iceberg production can be illustrated by the behavior of the Amery Ice Shelf. It is estimated that the average annual output is about $2–5\%$ of the iceberg production of the continent. In 1962, however, one-fifth of the ice shelf broke off. This one calving incident produced about 8.2×10^{17} g of ice (800 km³ or $20–50$ times its annual average), or again about 1 year's production for the entire continent in one breakoff from an ice shelf.

B. Greenland

The largest producer of icebergs in the northern hemisphere is the Greenland Ice Sheet. It generally calves about 240 km³ of ice into the neighboring seas. The Jacobshavn Glacier in west Greenland—probably the fastest-moving ice stream in the world (up to 8 km/yr)—accounts for about 7% of the total iceberg production in west Greenland. The average iceberg calved from Greenland has a mass of 5×10^9 kg (0.005 km³). According to one estimate, about 20,000 icebergs are produced annually from west Greenland. However, these estimates are rough, and another value cited is $10,000–34,000$ annually from all of Greenland. North of Jacobshavn Glacier, Rinks Glacier produces icebergs over 600 m thick. Icebergs here and at Jacobshavn are affected by being in deep long fjords with sills at their seaward end that may inhibit the movement of the icebergs. Farther north, a group of glaciers discharges into Melville Bay. These glaciers (Streenstrup, Dietrichson, Nansen, Kong Oscar, and Gade) produce most of the icebergs that eventually reach the open sea. In Kane Basin, Humboldt Glacier, 100 km across its face, appears to produce many large icebergs. These are grounded in front of it for many years and may be freed up by unusually high tides. The east coast of Greenland does not produce as many icebergs as the west coast. The glaciers there discharge into long fjords that are sometimes blocked by small islands. A further impediment on the east coast is the sea ice cover, which can block the seaward movement of the icebergs.

C. Alaska and Other Northern Hemisphere Source Regions

Alaskan glaciers are generally grounded and icebergs are rarely encountered in areas very far from their source. They account for a small percentage of the total icebergs in the northern hemisphere. They are mainly of interest because of their proximity to shipping lanes; for example, the Columbia Glacier is near Valdez at the end of the Alaska pipeline. The Columbia Glacier undergoes highly varying rates of iceberg discharge as it is a tidewater glacier that is periodically pinned and unpinned on shoals near its terminus. All other source regions, such as the Canadian Arctic islands, Svalbard, Novaya Zemlya, and Alaska, together contain only about 13% of the quantity of ice held in the Greenland Ice Sheet and have roughly the same percentage of discharge (20 km³). The only sources of sizable Arctic icebergs (called ice islands in the Arctic Basin) are the five ice shelves of northern Ellesmere Island, the largest of which is the Ward Hunt Ice Shelf. These have periodically calved (Ward Hunt in 1962) and have produced ice islands that are sometimes used for manned bases (Fletcher's Ice Island, T-3). The ice islands were identified as a potential collision hazard to offshore drilling structures in the region off Alaska and Canada when oil exploration efforts were taking place in the early 1980s. [See ARCTIC.]

II. Characteristics of Icebergs Relative to Origin

Icebergs are classified as follows:

Tabular: length-to-height ratio $> 5 : 1$ (Fig. 4)
Blocky: length-to-height ratio about $2.5 : 1$ (Fig. 5)
Drydock: eroded icebergs with twin columns or pinnacles (Fig. 6).
Dome: smooth rounded icebergs with low sides (Figs. 7a and 7b)
Pinnacle: icebergs with a central spire or pyramid with one or more spires dominating the shape (Fig. 8)

Tabular icebergs are the largest, and the other shapes are often the result of tabular iceberg deterioration. As they diminish in size, tabular, blocky, and pinnacle icebergs become dome icebergs if they are unstable and roll or drydock icebergs if they remain upright. A summary of production, median and maximum dimensions, and other characteristics of Antarctic and Arctic icebergs as compiled by J. R. Keys from several sources is shown in Table I. Tabular icebergs are formed from ice shelves, and the dominance of ice shelf coastline in Antarctica accounts for their common sighting there. Relatively little ice shelf exists in the northern hemisphere, so tabular icebergs are rarely sighted in northern waters. Size distributions of icebergs have been reported for Antarctica based on satellite imagery. A modal length within sea ice was $0.65–0.75$ km. For a sample at the edge of the

Fig. 5 Blocky iceberg that is slightly tilted because of imbalance induced by calving events. The iceberg prow has raised out of the water, creating a sloping beach that penguins and seals are using as a resting place between feeding forays into the surrounding waters.

Fig. 6 Drydock iceberg floating with pack ice floes. The twin spires with a submerged region of the iceberg between them are characteristic of these icebergs.

Fig. 7 (a) Dome iceberg consisting of a clear ice lump with bands of white bubbly ice. This iceberg has overturned and appears green because of refraction of light through the clear portions from the surrounding waters. (b) Dome iceberg of white bubbly ice. The vertical lines on the iceberg are ice strata from the annual accumulations of snow, which were originally laid down horizontally. The iceberg has turned about 90° from its original orientation because of imbalance induced by calving events.

Fig. 8 Pinnacle iceberg with a pyramidal spire. The orientation of this iceberg lies between the slight tilt of a blocky iceberg (Fig. 5) and the overturning of a dome iceberg (Fig. 7b).

pack it was 0.35–0.45 km. Thickness of the ice shelves at their seaward ends, where most icebergs are generated, is typically about 200 m. Because most icebergs are generated from the ice shelves, which have an upper layer composed of precipitation, the mean density is estimated to be 720–880 kg/m³. Where the equilibrium line of the glacier is well above sea level, as it is on some outlet glaciers in Antarctica, the density rises to 900 kg/m³. Bubbles constitute around 3–7% by volume of the ice in icebergs. [*See* Sea Ice.]

For northern hemisphere icebergs, bubbles similarly constitute 3–8% of their volume. Bubbles are either round or tubular, and the tubular bubbles (length 4 mm compared to a diameter of 0.02–0.18 mm) are thought to result from flow stress. Some small tabular icebergs are generated from Greenland but have width-to-thickness ratios close to 1 : 1 (blocky icebergs). These are thought to have low stability and are probably likely to roll, which also contributes to the rare sitings of tabular bergs in northern waters. In nearly all cases, northern hemisphere icebergs have a density of up to 900 kg/m³ because the equilibrium line of the source glacier is well above sea level, in contrast to most areas of Antarctica. This means the calving front of the glacier is in the ablation zone of summer melt, which removes the low-density mate-

rial from the upper surface of the glacier before calving and leaves only ice that has been increased in density at the deeper portions of the glacier to form the icebergs.

Dirt or ash sometimes appears in icebergs (Fig. 9). The dirt arises from contact with the basal moraine of the glacier, and through melt–freeze processes bands of dirt are interleaved with clear ice in the bottom of the ice. These bands are seen when erosional processes in the ocean cause the icebergs derived from these glaciers to tilt or roll over. The source regions for these icebergs in Antarctica are the valley glaciers of the Antarctic Peninsula or the island arcs, such as the South Orkneys and South Sandwich Islands. Volcanic ash is sometimes seen as well and results from deposition onto the glacier surfaces from volcanoes such as that on Deception Island.

III. Distribution of Icebergs and Fate in the Ocean

A. Southern Hemisphere Drift Patterns

In the Antarctic the drift near the coast or source region of the icebergs is generally from east to west (East Wind Drift). It

Table I   Annual Production, Median and Maximum Dimensions, and Other Characteristics of Antarctic and Arctic Icebergs

	Antarctic	Arctic
Mean annual iceberg production		
Total mass (tonnes)	$10-16 \times 10^{11}$	2.2×10^{11}
Total volume (km³)	1,200–2,000	240
Total numbers (rough estimate)	6,000–70,000	10,000–50,000
Median width (m)		
All icebergs	Less than 100	85[a]
	Less than 200	130[b]
	300	(length of bergs longer than 50 m)
Tabular bergs only	690 (mean of 24)	
	600	
Median visible heights (m)		
All icebergs	30–50	23[a]
Main range for tabular bergs in Ross Sea	12–36	8 (median mean height)[b]
Mean maximum draft (m)	210 (mean of 18)	95[a]
Mean density (kg/m³)	840	910
Estimated median mass (tonnes)	10^5-10^7	$\begin{cases} 1-5 \times 10^6 \text{ (mean)}^b \\ 4 \times 10^5 \text{ (median)}^{a,b} \end{cases}$
Maximum recorded length (km)	185	50
	170	
Maximum recorded thickness (m)	390	600?
Maximum recorded height (m)	140	103
Maximum recorded draft (m)	330	450
Estimated maximum mass (tonnes)	10^{12}	5×10^{10}
Maximum concentration	~ 90	230–910[d]
bergs per 1000 km², i.e., within 17.8 km of observation point	1500[c]	
in Ross Sea or Canadian drilling areas or prospects at least 20 km	20 at least	40[b], 5[a]
offshore		20–60[b]
Maximum recorded daily drift (km)	46	81

Source: From Keys (1984) as summarized from various references.
[a] Labrador Sea.
[b] Eastern Lancaster Sound–northern Baffin Bay.
[c] Bay of Sails, western McMurdo Sound, Ross Sea.
[d] Disco Bay, West Greenland.

has been estimated that icebergs remain in this 100-km coastal zone for about 13 years. However, that may be a high estimate and may include time spent aground. The average lifetime is about 6 years for drifting bergs. Large tabular bergs have been tracked in the East Wind Drift for 1–5 years. Consistent with this lifetime estimate, the half-life of icebergs south of the Antarctic Convergence is estimated to be 2–4 years. Strong northward drift is noted in the western Weddell Sea in the cyclonic Weddell Gyre, along the coast of the Antarctic Peninsula. This is the strongest region of offshore drift, but other regions of northward iceberg drift have been identified, especially north of the Ross Sea and off East Antarctica (80–100°E). Once icebergs cross about 65°S latitude, they fall into the influence of the westerlies and begin to travel toward the east (opposite to the coastal flow). They also continue to drift northward at a decreasing rate until their movement becomes purely east–west at 45°S. The effective northern limit of the drift is the Polar Front, where the surface water temperature changes from freezing values south of the front to 6–14°C north of it.

Icebergs drift in an erratic fashion because of current eddies, wind, and waves. Large icebergs may take several hours to approach an equilibrium drift velocity after an abrupt change in currents and winds, but a small berg can take less than an hour. Icebergs move at rates between 60 and 75% of the current speed. Measured average speeds vary from 4 to 13 km/day (5–15 cm/s), and the maximum recorded daily drift is 46 km.

Fig. 9 Dirt or (probably) volcanic ash band in an iceberg in the Weddell Sea.

Fig. 10 Slightly tilted iceberg with seals on the sloping prow at the left-hand side. The smoothed lower portions near the water line are due to melting by waves; the rounded lumps seen higher up are probably accumulations of spray ice from breaking waves. The hanging block in the right foreground was probably created by a calving event induced by wave undercutting at the iceberg water line.

Fig. 11   Small tabular iceberg with a V notch in the left center. Snow has partially filled and bridged the upper portions of a crevasse; the dark place at the base of the V shows the unfilled portion of the crevasse. Several of these dark spots are seen on the sides of the iceberg, indicating that it is riddled with crevasses and is likely to break up further along these lines of weakness.

B. Northern Hemisphere Drift Patterns

The major area of iceberg drift is Baffin Bay and the Labrador Current between Greenland and eastern Canada. Icebergs produced on the west coast of Greenland enter the northward-moving West Greenland Current, which has branches that move westward and join the southward-moving Baffin Island Current. These icebergs then continue southward, joining the Labrador Current, which has a southward extension to 40°N, the latitude of New York and other population centers on the East Coast of North America. These icebergs, in both number and southern extension, present a hazard because they cross the major shipping lanes of the North Atlantic. Other areas of iceberg drift are less noticeable, in terms of both proximity to population centers and the total number (and mass) of icebergs involved. Small numbers of icebergs are generated off glaciers in Franz Josef Land and Severnaya Zemlya and drift westward with the ice and water, eventually entering the East Greenland Current, where they are swept southward to their eventual demise in the frontal zones between the East Greenland Current and northward-moving warm waters in the Greenland Sea and the North Atlantic. In the polar basin, there are few occurrences of icebergs, and these are mainly generated from the Arctic Archipelago ice shelves such as the Ward Hunt Ice Shelf. These bergs are carried into the clockwise circulation of the Beaufort Gyre, where they may circulate for several years, as T-3 did. Some of these icebergs (ice islands in the polar basin) may ground on the shelves off Alaska and Canada, or they may end up entering the Transpolar Drift Stream, pass through Fram Strait and into the East Greenland Current, and eventually reach the warm, north-flowing Atlantic waters, where they are destroyed.

C. Destruction of Icebergs in the Ocean

If icebergs melted as an entirety, it would take an extremely long time for them to melt because their surface-to-volume ratio is quite small. Breakup effects accelerate their deterioration by initially breaking large bergs into small ones, which are then rapidly destroyed by the following mechanisms.

Chief among effects leading to breakup is the wave field in

the open ocean. Wave flexure and fatigue cause icebergs to break up into pieces that are more or less uniform in size. Once the iceberg is smaller than about 1000 m, other deterioration mechanisms begin to dominate. The most severe deterioration mechanism, once they enter waters even slightly above freezing, is wave action, which enhances melting (rather than flexure). Waves melt away ice in a band about the water line (Fig. 10), eventually causing calving by a shear failure of the ice above, which is then subjected to rapid deterioration by being almost completely immersed in the warm water. The remaining berg may become unstable and roll, resulting in further melting and a new wave erosion potential at the water line. Continued calving, rolling, and melting quickly reduce the size of the iceberg.

Icebergs usually disappear quickly when they encounter warm waters—for example, off Newfoundland, when they hit the warm (12°C) North Atlantic Current. It is possible for icebergs to be transported across the warm currents in cold-core eddies, so occasionally they survive as far east as the Azores and as far south as Bermuda.

The transfer of heat from the water to the iceberg occurs in a boundary layer that is generally turbulent and is made more complex by the phase change from water to ice and by the dissolved salts that are in the water. Although a short laminar region can exist in the oceanic environment, turbulence dominates over the vertical faces of an iceberg. The melt rate is about 30% higher when turbulence is present, as it usually is in the open ocean. The rate of melt is also dependent on the vertical length of the wall. For example, the melt rate of a 1-m face is three times as great as that of a 100-m face. Iceberg deterioration therefore accelerates as the berg calves into smaller pieces.

IV. Interaction of Icebergs with Nature and Humans

A. Contributions to Ice Sheet Mass Balance

Early estimates place the net accumulation or input to the Antarctic Ice Sheet at about 2×10^{15} kg/yr. The mass loss is mainly due to basal melting of the ice shelves (by ocean water) and iceberg calving. The basal melting contribution is estimated to be about 0.2×10^{15} kg/yr, with large uncertainties. Calving loss estimates have been as low as 0.6×10^{15} kg/yr, higher at 1×10^{15} kg/yr, and as high as $2-3 \times 10^{15}$ kg/yr. The uncertainty in the calving loss is probably considerably higher than all other uncertainties in the mass balance combined. The combined observational data set (ship sitings internationally) indicates that the rate of iceberg calving from Antarctica has been underestimated. Newer estimates, based on a lifetime figure of 2–4 years and the observations of

icebergs, place the calving loss at 2.3×10^{15} kg/yr. Although these estimates are difficult to make, the more recent complete observations suggest that the rate of calving is three to four times larger than the smaller of the earlier estimates. Because of the mechanics of iceberg calving in Antarctica, as indicated earlier, these values have a wide range of annual or even decadal uncertainty, since two of the larger events within 10 years can produce a wide anomaly in the effect of iceberg calving on the mass balance. That this can happen was illustrated within a 3-year period in the 1980s on the Filchner–Ronne and Ross ice shelves. One calving event on each ice shelf generated a few icebergs equivalent to the annual estimate of iceberg calving loss. Many tens of thousands of additional icebergs were generated around the 20,000-km coast of Antarctica at the same time, confounding a well-behaved annual average calving loss estimate.

B. Iceberg Effects on Human-Made Structures

The most highly publicized interaction of iceberg and humans was the sinking of the *Titanic* in 1912 after it collided with an iceberg. This sinking led to the establishment of the International Ice Patrol in 1914 and its assignment to the U.S. Coast Guard. Ship sinkings and collisions with icebergs diminished rapidly with the tracking and reporting of these hazards, and also with the advent of radar detection of icebergs in the modern era. However, collisions with icebergs are still a hazard and have caused ship damage as recently as 1990 in the North Atlantic. Positive use of icebergs was made in the last century by whalers in Baffin Bay, who often tied up in the lee of an iceberg and used it as a floating breakwater. The iceberg protected the ship from both waves and storm-driven pack ice.

Icebergs also present a threat to undersea pipelines and storage facilities being developed on the margins of Baffin Bay and the Labrador Sea. On these coasts, towboats have been employed to alter slightly the trajectory of icebergs that present a danger to drill ships, offshore platforms, and the like.

C. Towing Icebergs as Freshwater Sources in Arid Regions

The idea of using icebergs as a source of fresh water for arid regions has recently been considered. The problem was divided into (1) locating a suitable supply of icebergs, (2) calculating the power required to transport the iceberg to a location where fresh water is needed, (3) calculating the amount melted in transit, and (4) estimating the economic feasibility of the venture.

It was concluded that a suitable supply of tabular icebergs

existed in Antarctica and that, if the icebergs could be towed with only melting losses, the other parts of the problem were also viable. The feasibility of this usage, however, was limited to southern hemisphere arid areas such as the deserts of Chile, South Africa, and western Australia. However, difficult engineering problems were posed by the integrity of icebergs including the presence of cracks and crevasses (Fig. 11), the ability to tie up or on to them, and the combination of breakup and other forces. In 1977 the First International Conference on Iceberg Utilization was held in Ames, Iowa to review the state of knowledge on icebergs and propose techniques for towing them to arid regions. Several authors addressed some of the more engineering-related aspects of iceberg towing. These considerations include the initial size and shape (availability of suitable bergs), roll stability, wave flexure and mechanical breakup (a serious problem in the open ocean portion of the tow), tensile towing stresses where the iceberg is attached to the tow vessel, tug power, and, finally, handling and processing at the final destination.

Primarily because of the technical difficulties revealed, this effort has not progressed beyond the conceptual ideas presented at this conference and a subsequent one held in Cambridge, England in 1980. An interesting footnote on the economic considerations came later. In the initial stages, the economic questions had centered on trade-offs between the costs of delivery and the value of the water delivered for irrigation. Subsequent calculations showed that the iceberg, if used as the cold side of a conventional power plant cooling cycle, would essentially pay for its transportation costs in increased efficiency of the power plant. The fresh water, after the iceberg melted and warmed, would be available for irrigation and municipal and industrial usage essentially free of charge.

Bibliography

Hult, J. L., and Ostrander, N. C. (1973). "Applicability of ERTS for Surveying Antarctic Iceberg Resources." R-1354-NASA-NSF. RAND Corporation, Santa Monica, California.

International Glaciological Society (1980). "Proceedings of the Conference on the Use of Icebergs." *Ann. Glaciol.* **1.** International Glaciological Society, Cambridge, England.

Jacobs, S. S., MacAyeal, D. R., and Ardai, J. L., Jr. (1986). The recent advance of the Ross Ice Shelf, Antarctica. *J. Glaciol.* **32**(112), 464–474.

Keys, J. R. (1984). "Antarctic Marine Environments and Offshore Oil." Commission for the Environment, Wellington, New Zealand.

Robe, R. Q. (1980). Iceberg drift and deterioration. *In* "Dynamics of Snow and Ice Masses" (S. C. Colbeck, ed.), pp. 211–259. Academic Press, New York.

Schwerdtfeger, P. (1979). On icebergs and their uses. *Cold Reg. Sci. Technol.* **1**(1), 59–79.

Swithinbank, C., McClain, P., and Little, P. (1978). Drift tracks of Antarctic icebergs. *Polar Rec.* **18**(116), 495–501.

Wadhams, P., and Josberger, E., eds. (1983–1987). "Iceberg Research," Nos. 1–13. Scott Polar Research Institute, Cambridge, England.

Walton, D. W. H., ed. (1987). "Antarctic Science." Cambridge Univ. Press, London.

Weeks, W. F., and Campbell, W. J. (1973). Icebergs as a fresh water source: An appraisal. *J. Glaciol.* **12**(65), 207–233.

Weeks, W. F., and Mellor, M. (1978). Some elements of iceberg technology. *In* "Iceberg Utilization" (A. A. Husseiny, ed.), pp. 45–98. Pergamon, New York.

Glossary

Ablation zone Region of an ice body where there is net annual loss of ice by melt, evaporation, or iceberg calving.

Calving Breaking off of an iceberg from a glacier, ice sheet, ice shelf, or a larger iceberg by fracturing of the snow and ice allowing the iceberg to separate from the parent ice body.

Equilibrium line Division between accumulation and ablation zones on a glacier, ice sheet, or ice shelf. The annual net mass balance is zero at the equilibrium line.

Glaciers Flowing bodies of ice that are driven by gravity acting on a snow and ice mass. Glaciers are divided into an accumulation zone, where, on average, snow and ice mass is increasing, and an ablation zone, where snow and ice mass is decreasing.

Ice sheets Largest ice masses, usually associated with the snow and ice covering Antarctica and Greenland.

Ice shelves Floating masses of ice that are fed by glaciers or ice streams that flow from the large ice sheets and by direct precipitation of snow onto the surface.

Igneous Petrogenesis

University of British Columbia

Igneous rocks are those that have solidified from the liquid state. Igneous petrogenesis is the study of the origin of such rocks by characterization of the source regions, conditions of melting, and the processes that acted during solidification. Recognition of individual petrogenetic processes is based on textural features and chemical variations preserved in the rocks. Volcanic rocks (extrusives) solidified at or near the surface, while plutonic rocks (intrusives) solidified at depth below the surface. During the process of petrogenesis, an original system which may be solid or liquid, divides into two or more parts of which at least one is mainly liquid, and has a differential chemical composition than the original starting material. The starting material and the separation mechanism by which the system splits into parts distinguish the various theories of petrogenesis. An important problem of petrogenesis is distinguishing primary magmas from differentiated or evolved magmas.

I. Introduction

A. Magmatism and the Earth

Igneous rocks comprise an important part of the crust and upper mantle of the earth. Over the past 4 billion years igneous activity has transported significant amounts of energy and material (including water and carbon dioxide) from the mantle to the crust, thereby adding essential material to the continental crust, hydrosphere, and atmosphere. As a result of this interaction the continents, for example, have the average composition of andesite, an intermediate volcanic rock. Today, Mid-Ocean Ridge Basalt (MORB) is the most voluminous rock in the crust of the earth. It underlies the ocean floor and is generated by volcanic activity at the ocean ridges (e.g., the Mid-Atlantic Ridge). Oceanic islands (Hawaii, Iceland) owe their existence exclusively to igneous volcanic activity. Igneous activity is prominent along island arcs (Japan, Lesser Antilles) and active continental margins (Chile, Cascades), areas well known for agriculture, mining of volcanogenic ore deposits, and volcanic hazards. [*See* CONTINENTAL AND ISLAND ARCS; OCEANIC ISLANDS, ATOLLS, AND SEAMOUNTS; VOLCANOES.]

The energy source for most igneous activity is the inner heat of the earth (geothermal energy), but some rare igneous bodies have been generated by meteorite impact. The Sudbury *lopolith* with its associated nickle ores is a prime example, formed by an impact 1.7 billion years ago. [*See* IMPACT CRATER GEOLOGY.]

The tools used by igneous petrologists to study petrogenesis are varied. Examination of field and tectonic relations and tectonic association is the initial step, which may be followed by hand specimen and microscopic petrography, major and trace element chemistry (including rare earth elements), stable and radioactive isotopes (such as Nd, Sm, Sr, K, Pb, U), and consideration of the results of melting and other laboratory experiments on natural and/or synthetic materials. [*See* ANALYTICAL PETROLOGY.]

The remainder of this article discusses the various mechanisms and materials involved in magma generation and differentiation within the earth's crust and mantle. Major problems are still to be solved in igneous petrogenesis. For discussions of energy interchange and volatile and condensed material transport between igneous and other earth systems, consult other articles in this encyclopedia.

B. Igneous Material

From a chemical standpoint, almost every major type of natural material may form a melt and then solidify as an igneous rock: silicates (by far the most common, e.g., basalts), carbonates (Na-rich and Ca-rich carbonatites), or oxides (El Laco magnetite flow, Peru). Sulfides are also known as discrete liquid droplets in some igneous intrusives but are not known to form large discrete igneous bodies. If we consider only terrestrial examples, temperatures of extrusion at the surface range from less than 200°C for recent African carbonatite lavas (Na_2CO_3-rich) to more than 1300°C for Archean Mg-rich flows known as komatiites. [*See* MAGMATIC SULFIDE DEPOSITS.]

II. Partial Melting

A. Introduction

Melting is the primary petrogenetic process ultimately involved in the origin of all igneous rocks. Various schemes of melting and interaction of melt with crystals have been developed to explain different features of igneous rocks. The local source region being melted may be *closed* and not exchange material with its surroundings, or it may be *open* to such exchange. In equilibrium batch melting (a closed process) the melt remains in contact with the solid phases, thus maintaining equilibrium between solid and liquid within the batch. In Rayleigh or perfect fractional melting (an open process) the melt is continuously removed from the solid as it is produced, so that it never has an opportunity to react with the solid residue. Natural processes probably lie between these two extremes.

B. Closed System Melting

The upper mantle consists of peridotite, a rock comprising mainly the Fe- and Mg-bearing minerals: olivine and lesser amounts of enstatite orthopyroxene and diopside clinopyroxene. Peridotite also contains lesser amounts of Al-bearing minerals, spinel, plagioclase, and garnet. Plagioclase (a low-pressure phase) and garnet (a high-pressure phase) never occurs together, but spinel may accompany either. A typical spinel peridotite (lherzolite) might contain by volume: olivine 80%, enstatite 10%, diopside 8%, and spinel 2%. Observed specimens of upper-mantle material vary greatly in the proportions of minerals present. The mantle is, therefore, heterogeneous on a hand specimen scale (centimeters). Other evidence from isotope studies suggest it is heterogeneous on a scale of kilometers and varies locally on a global scale, while still remaining a variety of peridotite. For example, each volcano in Hawaii has a slightly different isotopic signature that is distinct from that of Mid-Ocean Ridge Basalt (MORB). Mantle heterogeneity may be due to previous partial melting episodes, which may be of great antiquity and represent both refractory residuum and/or melt that failed to escape to the surface as volcanic material. This heterogeneity may be layered or veined and presents great difficulties to petrogenetic studies. [*See* MANTLE.]

The normal state of the peridotite of the earth's mantle at the present time (and probably for at least the past billion years) is solid. However, experiments show that addition of as little as 0.4% H_2O may cause melting of about 1% in fertile mantle (not having produced melt previously) for depths of 100 km to 250 km (probably corresponding to the seismic low-velocity zone). Any increase in temperature above the normal temperature at depth (the geotherm temperature) will cause more melt to form within the low-velocity zone.

If a large fraction of the mantle melts (say 30 to 50%), the resulting magma must be extremely MgO-rich (> 30% MgO by weight) and give rise to olivine-rich eruptives such as komatiites. Such melts were common in the Archean about 2.4 billion years ago but not known from recent terranes. There is no current consensus among petrologists as to the viability of such large degrees of mantle melting in recent time.

Partial melting (leaving a solid refractory residuum) in the mantle at shallow depths as close to the surface as 50 km (under Hawaii, for example) is claimed by some petrologists to produce tholeiitic basaltic magmas, the most common volcanic product in Hawaii. However, other researchers consider it probable that the primary melt, which is stable with mantle mineralogy, is picritic (olivine-rich basalt) and must, therefore, differentiate to tholeiite by crystal fractionation. Crystallization and accumulation of olivine form olivine-rich cumulate rocks near the base of the oceanic crust (so-called underplating), and the resulting residual magma erupts as tholeiite lava flows at the surface. Similar processes are envisioned for continental tholeiites such as the Columbia River basalts, although some involvement with continental material appears necessary to explain their chemistry. [*See* CONTINENTAL FLOOD BASALTS.]

At ocean ridges, MORB is erupted. If we assume a picritic parental magma in equilibrium with mantle material due to partial melting, olivine and plagioclase fractionation appears to be necessary to explain the chemistry of the lavas. An additional complication is magma mixing (see Section III,J). Most MORB does not, therefore, represent primary magma.

Not all primary magmas are derived by partial melting of mantle peridotite. In the earth's continental crust, primary melts form from the metamorphosed equivalents of shales and clastic rocks (i.e., sandstones) in high-grade metamorphic terranes (upper amphibolite and granulite grade). These melts are usually granitic to dioritic in composition, their solid equivalents comprising mainly plagioclase, alkali feldspar, quartz, and mafic minerals (biotite, hornblende). Rocks of granitic composition begin to melt at about 650°C at depths of 10 to 20 km (3 to 7 kbar pressure). In contrast, rhyolite (equivalent in composition to granite) is erupted at the surface at about 950°C to 1000°C. Hydrous materials melt at lower temperatures than anhydrous material and are probably responsible for the onset of melting. However, water-saturated melts cannot move upward to lower pressures without crystallizing. Only effectively water-free magmas can reach the surface (in a thermodynamically stable state) and erupt as effusive flows. Any magma containing water erupting quietly at the surface does so under disequilibrium conditions. It is likely that most water-charged magmas have a tendency to erupt explosively (as, for example, the dacite magma at Mt. St. Helens in 1980). The depth of intrusion and/or style of eruption, therefore, have petrogenetic significance. [*See* CONTINENTAL CRUST.]

It is currently thought that, within the continental crust, intrusion of basaltic or picritic magma derived from the mantle could provide a heat source to elevate temperatures to about 1000°C, necessary for a "dry" (i.e., water-free) rhyolite or dacite magma to erupt at the surface.

Many volcanoes (especially andesitic) are associated with volcanic arcs overlying zones of earthquake epicenters (Benioff zones) at destructive plate margins. It is certainly possible for a subducting slab of oceanic crust to undergo melting. It is currently thought, however, that the slab is not hot enough to melt; instead, water is released (by dehydration of amphibolite). This released water is mobile and moves upward, acting as a flux and thus facilitating melting in the mantle above the slab (see open sytem petrogenesis, below). [*See* CRUSTAL FLUIDS; OCEANIC CRUST.]

C. Open System Partial Melting

When partial melting takes place, the initial melt is characterized by high concentrations of incompatible elements. These elements are less easily accommodated in the solid phases; thus, on melting they partition preferentially into the liquid phase. In an open system, with progressive loss of melt, partial melting may yield changing primary melts. We may view such partial melting as occurring in stages; with each melting episode the melts are extracted from the local system and thus are not present to interact with the residuum. Examples might be (with increasing temperature): in the crust, granite, granodiorite, diorite; in the mantle, nephelinite, alkali basalt (basanite), olivine basalt, picrite. The isotopic and trace-element signature of such progressive melting of the same original starting material should be distinctive and help to identify the process.

Zone melting is a process observed in engineering practice as a mechanism for increasing the concentration of impurities in part of an experimental charge. It involves movement of a molten zone through a solid by combined melting and crystallization. As a naturally occurring process it is problematical.

D. Open System with Metasomatism

Recent work on isotope geochemistry suggests that more than one source is necessary to explain the chemistry of ocean island basalts and possibly MORB. A somewhat speculative mechanism used to explain this multisource feature is metasomatism of the mantle immediately prior to the melting vents by a hydrous, incompatible-element-rich, fluid. Veining observed in some mantle nodules may be taken as indirect evidence of this process. [*See* MANTLE METASOMATISM.]

The effect of metasomatism on continental crustal melts has yet to be explored systematically. Crustal rocks usually contain enough water to act as a flux (bearing in mind that completely water-saturated melts cannot move to zones of lower pressure). The progressive de-watering of sedimentary sequences during burial releases water to higher levels in the crust and thus does not provide water to the lower part of the crust, where temperatures are high enough for partial melting.

III. Magma Differentiation Processes

A. Introduction

After melting of the source region and collection of the melt phase, the magma may modify its composition through one or more physical and chemical processes called differentiation. Recognition of the individual processes responsible for differentiation is based on textural and chemical features preserved in the rocks as well as the geological and tectonic features of the rock occurrence.

Crystal fractionation is essential to most petrogenetic theories. As with melting, two extreme cases may be identified: perfect fractional or Rayleigh crystallization (in which crystals never react with the melt, usually due to physical removal from the system) and equilibrium crystallization (in which the crystals remain in contact with the liquid and continuously react with it). The natural cases probably lie between these extremes. Liquid fractionation curves describe the changes in liquid chemistry during crystallization (the liquid line of descent). This important theoretical concept was developed by N. L. Bowen and co-workers in the 1940s and explicitly derived for olivine much later (1978).

Olivine, $(Fe,Mg)_2SiO_4$, a solid solution of fayalite (Fa) the Fe-bearing component and forsterite (Fo) the Mg-bearing component, is an essential mineral in discussions of mantle melting. In perfect fractional crystallization, the relation between melt composition (say amounts of FeO and MgO) and the changing composition of the fractionating solid (say olivine) is explicit.

$$\left(\frac{\Delta FeO}{\Delta MgO}\right)_{solid} = K_d \left(\frac{FeO}{Mg}\right)_{liquid} \qquad (1)$$

where K_d is the distribution coefficient. P. L. Roeder and R. F. Emslie showed that for olivine K_d is nearly constant ($K_d = 0.3$) over a wide range of conditions. Assuming K_d is constant, an explicit equation exists for the liquid line of descent, thus:

$$FeO = A \cdot MgO^{K_d} \qquad (2)$$

where the FeO and MgO of the liquid are measured in

extensive amounts (i.e., grams) rather than weight percentage, and *A* is a constant that is a function of the parental composition. This expression can be solved for the FeO–MgO liquid line of descent for any primitive magma if the value of K_d and parental composition are known. These relationships are extremely useful for evaluating the primary nature of magmas.

B. Crystal Fractionation Separation Mechanisms

Gravitational settling/flotation as a petrogenetic mechanism is currently central to modern petrogenetic theory and dates back at least as far as Charles Darwin, in the 1840s. It is based on the observation that there is generally a difference in density between crystals of a given mineral species and the liquid from which they crystallized. Usually the crystals are denser than the liquid and will tend to sink. Stokes's law allows an estimate of the sinking (or floating) velocity if the densities, size of crystal, and liquid viscosity are known. For example, a 1-mm diameter crystal of olivine would settle at about 250 m/yr in a basaltic liquid. Augite would sink at a slightly slower rate for a given size of crystal.

It should be noted that convective velocities of liquid may be orders of magnitude faster than crystal settling velocities, thus transporting crystals around a magma chamber under circumstances where a simple calculation of settling velocities might suggest it is unlikely. Crystals need only settle out of the convection current (a much smaller distance compared with the size of the magma chamber) in order to produce the chemical effects of crystal settling. Strictly speaking, this is crystal settling aided by convection currents (themselves also due to density differences). [*See* MAGMA CHAMBER DYNAMICS.]

Minerals such as sodalite, leucite, and analcite have such a low density (<2.5 g/cm^3) that they will commonly have a tendency to float. Plagioclase is capable of either sinking or floating depending on the density of the liquid and the composition of the feldspar. The iron content of the liquid is important in determining if plagioclase will sink or float.

There is excellent evidence from field descriptions of picritic and komatiitic flows that olivine concentrations at the bottom of the flows resulted from crystal settling (flowage differentiation would produce accumulation at the top as well as the bottom). Crystal settling has also been observed in experimental charges, and evidence from alkaline sills suggests that sodalite and leucite have accumulated by flotation. Notwithstanding the field evidence, there is a difficulty with universally invoking crystal fractionation. If natural silicate liquids behaved like Newtonian liquids—that is, if they had no yield strength—then even the tiniest crystal would sink if it were denser than the surrounding liquid. However, liquids

of the *Bingham* type possess a yield strength and in an unstressed state behave similarly to solids; they do not begin to flow like liquid until a certain minimum force has been applied. It has been suggested that natural silicate liquids behave like Bingham liquids; consequently, large crystals settle more slowly than might be expected. Furthermore, small crystals below a certain minimum size will not settle at all! Some natural compositions such as granite would yield liquids with such high calculated viscosities that unrealistic times may be required for crystals to sink a significant amount. It is clear, therefore, that crystal fractionation by settling, while important, must be examined in every case and cannot be routinely accepted as the major cause of fractionation.

In the above cases, the liquid is viewed as stationary and the crystals are considered to move through it. In Sections III,C and III,D the crystals are considered stationary and the liquid flows past them, thus allowing more complex combinations of liquids and crystals.

C. Flow Differentiation

Flow differentiation of a magma containing suspended crystals is essentially a hydrodynamic process resulting from the velocity of flowing magma near a wall or chamber floor. In a vertical magmatic body, for example, crystals may be either entrained in rapidly flowing magma in the center or trapped in the stationary boundary layer at the sides, depending on velocity, type of flow (turbulent or laminar), crystal size, and relative density contrast of crystal and liquid. Note that in this process, if the flow is vertical, then crystal size is more important than density contrast.

D. Flow Crystallization

Flow crystallization results when crystals are fixed to the wall of a magma chamber and the liquid flows past, continually presenting the crystal with new material for growth. Flow may be in any direction from the top or bottom of the magma chamber.

E. Filter Pressing

This process requires more than 50% crystallization of a parent liquid in a quiet environment. The crystals form an interlocking mesh through which the coexisting liquid is squeezed upward (it is usually buoyant because it is less dense than the crystal network). Liquids derived by filter pressing are believed to form the alkaline or granophyric streaks and pods in some large basaltic sills such as the Palisade sill in New Jersey. Segregation vesicles in which late-stage liquid oozes into gas-filled vesicles in a lava flow are a microscopic variant of this process. Observations of this type are important, for they show the ability of crystal fractionation to produce evolved or differentiated liquids from basalts magma purely by crystallization. Extraction of the melt from the crystal mesh is the most difficult part of this mechanism, and it is not known if it occurs on a large scale. [*See* MAGMATIC PROCESSES IN SILLS.]

F. Other Crystal Fractionation Processes

Layered intrusions are large bodies of igneous rock which, unlike volcanoes, preserve a record of crystallization within the earth's crust. They are characterized by rhythmic layering whose origin is still problematical. Observations from such layered bodies also indicate that some perplexing forms of crystallization occur which, among other features, produce monomineralic layers whose melting points are too high for the layers to have been completely liquid.

G. Liquid Immiscibility

Liquid immiscibility involves the separation of a parental homogeneous liquid into two liquids of contrasting chemical composition (oil and water, for example, are two immiscible liquids). Naturally occurring liquids are known to exhibit such immiscibility. Silicate–oxide immiscibility may occur in iron-rich silicate liquids and is in rare instances responsible for some iron-ore deposits (Kiruna district, Sweden) and an unusual magnetite–apatite-rich oxide lava flow in Peru. Silicate–sulfide immiscibility is also known from the Kilauea lava lake and possibly some magmatic Ni-bearing sulfide deposits. Silicate–carbonate immiscibility has been suggested as a petrogenetic process in the origin of carbonatites, some of which contain important ore deposits.

The most common (even if largely unrecognized) form of immiscibility involves silicate–silicate immiscibility, which was first documented from lunar material sampled by the Apollo astronauts. Subsequently this phenomenon was described in terrestrial basalts and andesites. It commonly results in an iron-rich liquid coexisting with a silica-rich liquid. These two liquid phases are thermodynamically stable but have radically different chemical and physical properties (e.g., density). The two liquids typically occur in a globular texture (resembling oil in water) between plagioclase laths in rocks representing a high degree of crystallization. Since the silica-rich liquid is similar in composition to rhyodacite (a silica-rich lava), it is possible that this process may be more common and might be a major petrogenetic process.

H. Liquid Fractionation

The term liquid fractionation involves all mechanisms in which a parent homogeneous liquid separates into liquids of different composition without the direct influence of crystal-

lization or liquid immiscibility. The Soret effect, in which a chemical gradient is imposed on a liquid by a thermal gradient, is well known as a laboratory demonstration and is capable of fractionating isotopes as well as chemical species. Laboratory thermal gradients are typically several orders of magnitude greater than natural gradients, and so it is not clear that this process is important in nature.

The gravitational gradient within a vertical magma column is theoretically capable of producing a chemical gradient, but the effect is small and requires a large vertical extent. Water is the one major constituent that might conceivably be concentrated in the upper part of a magma chamber by gravitational effects.

Since liquids of different composition generally have different densities, a buoyant effect will accompany any compositional gradient, no matter how it is produced, and denser compositions (typically more iron-rich) will tend to sink while less dense liquids (more silica-rich) will tend to float. This effect tends to produce vertical stratification in magma chambers with more silica-rich liquids at the top. It is also known that convective velocities are likely to be orders of magnitude greater than settling velocities of crystals. It is possible, therefore, that most of the mass movement in bodies of magma is due to movement of liquid rather than crystals.

I. Gaseous Transfer Processes

Various processes involving a separate gaseous phase (possibly supercritical) have been proposed. The gas bubbles may attach themselves to crystals, thus making them buoyant (a process familiar to amateur wine makers), or may directly transport material in solution. The efficacy of these mechanisms is not known.

J. Open System Differentiation; Magma Mixing

Magma mixing involves mixing of two different parental magmas to yield a hybrid daughter magma of intermediate composition. It is currently recognized as an important mechanism of fractionation in the genesis of MORB and andesite. Much of the complexity of MORB chemistry can be explained if MORB magma chambers are periodically injected with pulses of primitive basalt, which mixes with fractionated basalt in the chamber. Studies of layered intrusives suggest that periodic injection of basalt is not limited to ocean ridges. This type of mixing may be called *back mixing*, as evolved liquids are mixed back with their more primitive precursor.

The above mixing involves similar magmas that may even be cogenetic, that is, from the same liquid line of descent. Another type of mixing involves unrelated, chemically distinct magmas such as basalt and rhyolite. These magmas may have temperatures as much as 200°C apart (basalt 1200°C and rhyolite 1000°C), and the viscosity difference may exceed four orders of magnitude (with basalt the more fluid). While mixing of equal amounts of these magmas could produce a magma of intermediate composition such as andesite, the vast differences in temperature and viscosity preclude simple mixing. Mixing of the fluid, high-temperature basalt with the viscous, low-temperature rhyolite will chill the basalt, heat the rhyolite, and produce complex turbulent mixing. The exact mechanism by which these two magmas mix to form a third is an ongoing problem of petrogenetic theory.

K. Open System Differentiation, Assimilation

Assimilation (sometimes called contamination) involves the complete or partial digestion of solid material by a liquid magma. The solid may be the *wall rock* containing the magma or material picked up by the magma on its way toward the surface. Since magmas are always in contact with wall rock (with which they are usually unstable), assimilation should be ubiquitous.

Assimilation may involve diffusion, melting, chemical reaction, and simple mechanical disaggregation. Field evidence of the efficacy of this mechanism comes from partially assimilated material in which the physical and chemical processes were arrested in varying degrees of completion. In any given case the exact mechanism is quite complex and probably involves all of these processes. Some of the best evidence for assimilation comes from partially digested xenoliths (foreign rocks) in granitic bodies. In spite of its low temperature (650–1000°C), granite is capable of assimilating basaltic rocks, and it is thought that the water content of the granite assists in the process.

Except in obvious cases, it is not clear to what extent magmas are affected by assimilation. It is considered doubtful if this process accounts for the production of large quantities of evolved melts in the crust; however, it may be an important process for modifying trace elements and isotopes of primitive magmas, thus giving the illusion of a complex history.

IV. Primary Magmas

One of the main reasons for studying igneous rocks is to learn about planetary interiors. Since igneous rocks are the products of fusion, their compositions must carry information about the source material from which the melt was extracted. The earth's lower crust and mantle are currently inaccessible, and igneous rocks are one of the few samples of the earth's interior. Not all igneous rocks offer the same insights into the earth's interior. Evolved magmas, since they are modified from precursor magmas, have their primary parental characters obscured by differentiation processes. Primary magmas,

in contrast, are not differentiated; therefore, they can provide the most information about the earth's interior.

Defining the theoretical concept of a primary magma is fairly simple, but determining if a given rock solidified from a primary magma is a difficult problem. Positively identified primary magmas have proven remarkably elusive. It is not even generally accepted by all petrologists that primary magmas can erupt at the surface without becoming fractionated.

How many different parental magmas are there? In past years it was commonly thought that the igneous rocks in a given geographical area had but one parental magma. All other magmatic rocks were considered to have been derived from this primary magma by the process of differentiation. It is now recognized that this concept, although valid locally, is somewhat oversimplified. The quest for primary magmas has occupied petrologists for a generation and is an ongoing problem.

Although no single characteristic is diagnostic of primary status, there are several current criteria for recognizing primary melts from the mantle. In addition to a high melting point (> 1200°C), the melt should be stable with mantle olivine (Fo89 $+$), have a high magnesium number (> 70, or 100 Mg/(Fe $+$ Mg) in molecular units), and have relatively high contents of nickle (> 250 ppm) and chromium (> 500 ppm). Some possible candidates for primary liquids stable with upper-mantle peridotite are shown in Table I. These

Table I Hypothetical Primary Liquids[a]

	1	2	3	4	5	6
SiO_2	44.0	56.8	48.0	49.0	44.1	32.27
TiO_2	0.3	0.3	1.3	0.8	3.7	4.59
Al_2O_3	7.2	11.9	11.0	15.7	12.1	2.44
Fe_2O_3	0.1	1.7	1.0	0.6	0.3	3.00
FeO	13.1	7.1	11.0	7.5	10.9	10.99
MnO	0.2	0.1	0.1	0.1	0.1	0.19
MgO	27.1	12.6	15.0	11.1	13.1	30.33
CaO	6.2	7.9	9.0	13.0	10.1	7.26
Na_2O	0.7	1.0	2.0	1.9	3.5	0.28
K_2O	0.6	0.4	0.4	0.17	1.3	1.58
P_2O_5	0.03	0.03	0.2	0.08	0.8	3.52
H_2O	0.1	0.01	0.1			0.32
CO_2	0.1	0.01	0.1			2.94
Ni	1500	300	400	250	300	1030
Cr	320	475	1000	950	540	1900
Mg#	79	74	71	71	68	83
Fo%	92	91	89	89	88	94

[a] Major oxides in weight %; trace elements in parts per million; Mg# in molecular units (see text), Fo in mole %. Magma types from the following location: (1) komatiite (MunroTwp., Ontario); (2) boninite (Pacific island arc); (3) olivine tholeiite (hypothetical parent, Hawaii); (4) primitive MORB; (5) basanite (Antarctica); (6) kimberlite (West Greenland).

hypothetical magmas range from komatiite, the Precambrian rock type, to kimberlite, a rare rock possibly associated with carbonatite and sometimes containing diamonds thought to originate at depths as great as 200 km. These magmas are representative of various areas and have had their iron oxide ratios adjusted to indicate stability with mantle olivine. Bearing in mind that the quest for primary liquids is an ongoing research problem, the compositions in Table I should be regarded as model primary liquids in keeping with the philosophy of this article.

Bibliography

Basaltic Volcanism Study Project (1981). "Basaltic Volcanism on the Terrestrial Planets." Pergamon Press, New York.

Hess, P. C. (1989), "Origins of Igneous Rocks." Harvard Univ. Press, Cambridge, Massachusetts.

McBirney, A. R. (1984). "Igneous Petrology." Freeman Cooper & Co., San Francisco.

Pearce, T. H. (1978). Olivine fractionation equations for basaltic and ultrabasic liquids. *Nature (London)* **276**, 771–774.

Philpotts, A. R. (1990). "Principles of Igneous and Metamorphic Petrology." Prentice-Hall, Englewood Cliffs, New Jersey.

Wilson, M. (1988). "Igneous Petrogenesis." Unwin Hyman, Boston.

Glossary

Cogenetic (comagmatic) Igneous rocks related by differentiation to a common parental source.

Differentiation (fractionation) Process by which one magma gives rise to another of different composition. Differentiated or fractionated magmas are said to be evolved from a more primitive parent.

Incompatible element Element that partitions strongly into the liquid when a solid melts.

Liquid line of descent In a graph of chemical composition, a line connecting magmas related by differentiation to a common parent. The chemical expression of magmatic evolution.

Magma Naturally occurring liquid rock often containing volatiles such as water or carbon dioxide. Igneous rocks form by solidification of magma, usually by freezing.

Magma chamber Holding chamber for magma within the earth.

Primary magma Magma produced by melting a solid precursor. Magma only slightly differentiated in composition from primary magma is called primitive.

Impact Crater Geology

H. J. Melosh

University of Arizona

The surfaces of most solid planets and moons in the solar system are scarred with circular craters produced by the impacts of smaller objects. More than 100 impact craters ranging in size from tens of meters to 140 km have also been recognized on Earth. Fresh impact craters are roughly circular rimmed depressions surrounded by hummocky blankets of debris. They form when an extraterrestrial body strikes Earth or another planet at a velocity exceeding a few kilometers per second. Crater formation is an orderly, although rapid, process that begins when the projectile first strikes the planet's surface and ends after the debris around and within the crater comes to rest. The crater is excavated by strong shock waves created as the projectile plunges into the surface. These shock waves cause diagnostic high-pressure mineralogical changes in the rocks surrounding the crater. The size of the final crater is a function of the speed and mass of the projectile that created it, as well as such other factors as the angle of impact and acceleration due to gravity. A freshly formed crater on Earth, with its characteristic morphology, is eventually altered and degraded by erosion or tectonic processes. Impact cratering played a major role in the formation and early Archean history of Earth and the moon and probably caused at least one major biological extinction at the end of the Cretaceous era.

I. History of Investigation

Craters were discovered in 1610 when Galileo pointed his first crude telescope at the moon. Galileo recognized their raised rims and central peaks but described them only as circular "spots" on the moon. Although Galileo himself did not record an opinion on how they formed, astronomers argued about their origin for the next three centuries. The word "crater" was first used in a nongenetic sense by the astronomer J. H. Schröter in 1791. Until the 1930s most astronomers believed the moon's craters were giant extinct volcanoes; the impact hypothesis, proposed sporadically over the centuries, did not gain a foothold until improving knowledge of impact physics showed that even a moderately oblique high-speed impact produces a circular crater, consistent with the observed circularity of nearly all of the moon's craters. Even so, many astronomers clung to the volcanic theory until the high-resolution imagery and direct investigation of the Apollo program in the early 1970s firmly settled the issue in favor of an impact origin for nearly all lunar craters. In the current era spacecraft have initiated the remote study of impact craters on other planets, beginning with Mariner IV's unexpected discovery of craters on Mars on July 15, 1965. Since then, craters have been found on nearly every other solid body in the solar system.

The first terrestrial structure shown unambiguously to be created by a large impact was Meteor Crater, Arizona. This 1-km-diameter crater and its associated meteoritic iron were investigated in detail by D. M. Barringer from 1906 until his death in 1929. Since Barringer's work a large number of small impact structures resembling Meteor Crater have been found. Impact structures larger than about 5 km in diameter were first described as "cryptovolcanic" because they showed signs of violent upheaval but were not associated with the eruption of volcanic materials. J. D. Boon and C. C. Albritton in 1937 proposed that these structures were really caused by impacts, although final proof had to wait until the 1960s, when the shock-metamorphic minerals coesite and stishovite proved that the Ries Kessel in Germany and subsequently many other cryptovolcanic structures are the result of large meteor impacts.

Finally, theoretical and experimental work on the mechanics of cratering began during World War II and was extensively developed in later years. This work was spurred partly by the need to understand the craters produced by nuclear weapons and partly by the fear that the "meteoroid hazard" to space vehicles would be a major barrier to space exploration. Computer studies of impact craters were begun in the early 1960s. A vigorous and highly successful experimental program to study the physics of impact was initiated by D. E. Gault at NASA's Ames facility in 1965.

These three traditional areas of astronomical crater studies, geological investigation of terrestrial craters, and the physics of cratering have blended together in the post-Apollo era. Traditional boundaries have become blurred as extraterrestrial craters are subjected to direct geologic investigation, the earth's surface is scanned for craters using satellite images, and increasingly powerful computers are used to simulate the formation of both terrestrial and planetary craters on all size scales. The recent proposals that the moon was created by the impact of a Mars-size protoplanet with the proto-Earth 4.5 Gyr ago and that the Cretaceous era was ended by the impact of a 10-km-diameter asteroid or comet indicate that the study of impact craters is far from exhausted and that new results may be expected in the future.

II. Crater Morphology

Fresh impact craters can be grossly characterized as *circular rimmed depressions*. Although this description can be applied to all craters, independent of size, the detailed form of craters varies with size, substrate material, planet, and age. Craters have been observed over a range of sizes varying from 0.1 μm (microcraters first observed on lunar rocks brought back by the Apollo astronauts) to the more than 2000-km-diameter Hellas basin on Mars. Within this range a common progression of morphologic features with increasing size has been established, although exceptions and special cases are not uncommon.

A. Simple Craters

The classical type of crater is the elegant bowl-shaped form known as a *simple crater* (Fig. 1a). This type of crater is common at sizes less than about 15 km in diameter on the moon and 3 to about 6 km on Earth, depending on the substrate rock type. The interior of the crater has a smoothly sloping parabolic profile and its rim-to-floor depth is about one-fifth of its rim-to-rim diameter. The sharp-crested rim stands about 4% of the crater diameter above the surrounding plain, which is blanketed with a mixture of ejecta and debris scoured from the preexisting surface for a distance of about one crater diameter from the rim. The thickness of the ejecta falls off as roughly the inverse cube of distance from the rim. The surface of the ejecta blanket is characteristically hummocky, with mounds and hollows alternating in no discernible pattern. Particularly fresh simple craters may be surrounded by fields of small secondary craters and bright rays of highly pulverized ejecta that extend many crater diameters away from the primary. Meteor Crater, Arizona, is

a slightly eroded representative of this class of relatively small craters. The floor of simple craters is underlain by a lens of broken rock, *breccia,* which slid down the inner walls of the crater shortly following excavation. This breccia typically includes representatives from all the formations intersected by the crater and may contain horizons of melted or highly shocked rock. The thickness of this breccia lens is typically one-half to one-third of the rim-to-floor depth.

B. Complex Craters

Lunar craters larger than about 20 km in diameter and terrestrial craters larger than about 3 km have terraced walls, central peaks, and at larger sizes may have flat interior floors or internal rings instead of central peaks. These craters are believed to have formed by collapse of an initially bowl-shaped *transient crater,* and because of this more complicated structure they are known as *complex craters* (Fig. 1b). The transition between simple and complex craters has now been observed on the moon, Mars, Mercury, Venus, and Earth, as well as on some of the icy satellites in the outer solar system. In general, the transition diameter scales as g^{-1}, where g is the acceleration due to gravity at the planet's surface, although the constant in the scaling rule is not the same for icy and rocky bodies. This is consistent with the idea that complex craters form by collapse, with icy bodies having only about one-third the strength of rocky ones. The floors of complex craters are covered by melted and highly shocked debris, and melt pools are sometimes seen in depressions in the surrounding ejecta blanket. The surfaces of the terrace blocks tilt outward into the crater walls, and melt pools are also common in the depressions thus formed. The most notable structural feature of complex craters is the uplift beneath their centers. The central peaks contain material that is pushed upward from the deepest levels excavated by the crater. Study of terrestrial craters has shown that the amount of structural uplift h_{su} is related to the final crater diameter D by

$$h_{su} = 0.06 D^{1.1}$$

where all distances are in kilometers. The diameter of the central peak complex is roughly 22% of the final rim-to-rim crater diameter in craters on all the terrestrial planets.

Complex craters are generally shallower than simple craters of equal size, and their depth increases slowly with increasing crater diameter. On the moon, the depth of complex craters increases from about 3 km to only 6 km while crater diameter ranges from 20 to 400 km. Rim height also increases rather slowly with increasing diameter because much of the original rim slides into the crater bowl as the wall collapses. Complex craters are thus considerably larger than the transient crater from which they form: estimates suggest that the crater diameter may increase as much as 60% during collapse.

As crater size increases, the central peaks characteristic of smaller complex craters give way to a ring of mountains (Fig. 1c). This transition takes place at about 140-km diameter on the moon and about 20-km diameter on Earth, again following a g^{-1} rule. The central ring is generally about 0.5 of the rim-to-rim diameter of the crater on all the terrestrial planets.

The ejecta blankets of complex craters are generally similar to those of simple craters, although the hummocky texture characteristic of simple craters is replaced by more radial troughs and ridges as size increases. Fresh complex craters also have well-developed fields of secondary craters, including frequent clusters and "herringbone" chains of closely associated, irregular secondary craters. Very fresh craters, such as Copernicus and Tycho on the moon, have far-flung bright ray systems.

C. Multiring Basins

The very largest impact structures are characterized by multiple concentric circular scarps and are hence known as *multiring basins*. The most famous such structure is the 930-km-diameter Orientale basin on the moon (Fig. 1d), which has at least four nearly complete rings of inward-facing scarps. Although opinion on the origin of the rings still varies, most investigators feel that the scarps represent circular faults that slipped shortly after the crater was excavated. There is little doubt that multiring basins are caused by impacts: most of them have recognizable ejecta blankets characterized by a radial ridge-and-trough pattern. The ring diameter ratios are often tantalizingly close to multiples of $\sqrt{2}$, although no one has yet suggested a convincing reason for this relationship.

Unlike the simple–complex and central peak–internal ring transitions discussed above, the transition from complex craters to multiring basins is not a simple function of g^{-1}. Although multiring basins are common on the moon, where the smallest has a diameter of 410 km, none at all have been recognized on Mercury, with its two times larger gravity, even though the largest crater, Caloris basin, is 1300 km in diameter. The situation on Mars has been confused by erosion, but it is difficult to make a case that even the 1200-km-diameter Argyre basin is a multiring structure. A very different type of multiring basin is found on Jupiter's satellite Callisto, where the 4000-km-diameter Valhalla basin has dozens of closely spaced rings that appear to face outward from the basin center. Another satellite of Jupiter, Ganymede, has both Valhalla-type and Orientale-type multiring structures. Since gravity evidently does not play a simple role in the complex crater–multiring basin transition, some other factor, such as the internal structure of the planet, may have to be invoked to

explain the occurrence of multiring basins. The formation of multiring basins is a topic of active research.

D. Aberrant Crater Types

On any planetary surface a few craters can always be found that do not fit the simple size–morphology relation described above. These are generally believed to be the result of unusual conditions of formation in either the impacting body or the planet struck. Circular craters with symmetric ejecta blankets or elliptical craters with "butterfly-wing" ejecta patterns are the result of very low impact angles. Although moderately oblique impacts yield circular craters, at impact angles less than about 6° from the horizontal the final crater becomes elongated in the direction of flight. Small, apparently concentric, craters or craters with central dimples or mounds on their floors are the result of impact into a weak layer underlain by a stronger one. The ejecta blankets of some Martian craters show petal-like flow lobes that are believed to indicate the presence of liquid water in the excavated material. Craters on Ganymede and Callisto develop central pits at a diameter where internal rings would be expected on other bodies. The explanation for these pits is still unknown. In spite of these complications, however, the simple size–morphology relation provides a simple organizing principle into which most impact craters can be grouped.

III. Cratering Mechanics

The impact of an object moving at many kilometers per second on the surface of a planet initiates an orderly sequence of events that eventually produces an impact crater. Although this is really a continuous process, it is convenient to break it up into distinct stages that are dominated by different physical processes. This division clarifies the description of the overall cratering process, but it should not be forgotten that the different stages really grade into one another and that a perfectly clean separation is not possible. The most commonly used division of the impact cratering process is into contact and compression, excavation, and modification.

A. Contact and Compression

Contact and compression is the briefest of the three stages, lasting only a few times longer than the time required for the impacting object (referred to hereafter as the "projectile") to traverse its own diameter, $t_{cc} \approx L/v_i$, where t_{cc} is the duration of contact and compression, L the projectile diameter, and v_i the impact velocity. During this stage the projectile first contacts the planet's surface (hereafter, "target") and transfers its energy and momentum to the underlying rocks. The specific kinetic energy (energy per unit mass, $\frac{1}{2}v_i^2$) pos-

Fig. 1 Impact crater morphology as a function of increasing size on the moon. (a) Simple crater: 2.5-km-diameter crater Linné in western Mare Serenitatis (Apollo 15 Panametric Photo strip 9353). (b) Complex crater with central peak: 102-km-diameter crater Theophilus (Apollo 16 Hasselblad photo 0692). (c) Complex crater with internal ring: 320-km-diameter crater Schrödinger (Lunar Orbiter IV medium resolution frame 094). (d) Multiring basin: 620-km-diameter (of most prominent ring) Orientale basin (Lunar Orbiter IV medium resolution frame 194). *(Figure continues.)*

Fig. 1   *(Continued)*

Table I   Linear Shock–Particle Velocity Equation of State Parameters

Material	ρ_0 (kg/m³)	c (km/s)	S
Aluminum	2750	5.30	1.37
Basalt	2860	2.6	1.62
Calcite (carbonate)	2670	3.80	1.42
Coconino sandstone	2000	1.5	1.43
Diabase	3000	4.48	1.19
Dry sand	1600	1.7	1.31
Granite	2630	3.68	1.24
Iron	7680	3.80	1.58
Permafrost (water saturated)	1960	2.51	1.29
Serpentinite	2800	2.73	1.76
Water (25°C)	998	2.393	1.333
Water ice (−15°C)	915	1.317	1.526

sessed by a projectile traveling at even a few kilometers per second is surprisingly large. A. C. Gifford, in 1924, first realized that the energy per unit mass of a body traveling at 3 km/s is comparable to that of TNT. Gifford proposed the *impact–explosion analogy,* which draws a close parallel between a high-speed impact and an explosion. During contact and compression the projectile plunges into the target, generating strong shock waves as the material of both objects is compressed. The strength of these shock waves can be computed from the Hugoniot equations, first derived by P. H. Hugoniot in his 1887 thesis, relating quantities in front of the shock (subscript 0) to quantities behind the shock (no subscripts):

$$\rho(U - u_p) = \rho_0 U$$
$$P - P_0 = \rho_0 u_p U$$
$$E - E_0 = \tfrac{1}{2}(P + P_0)(1/\rho_0 - 1/\rho)$$

In these equations P is pressure, ρ density, u_p particle velocity behind the shock (the unshocked material is assumed to be at rest), U the shock velocity, and E energy per unit mass. These three equations are equivalent to the conservation of mass, momentum, and energy, respectively, across the shock front. They hold for all materials but do not provide enough information to specify the outcome of an impact by themselves. The Hugoniot equations must be supplemented by a fourth equation, the equation of state, that relates the pressure to the density and internal energy in each material, $P = P(\rho, E)$. Alternatively, a relation between shock velocity and particle velocity may be specified, $U = U(u_p)$. Since this relation is frequently linear, it often provides the most convenient equation of state in impact processes. Thus,

$$U = c + S u_p$$

where c and S are empirical constants. Table I lists the measured values of c and S for a variety of materials. These equations can be used to compute the maximum pressure, particle velocity, shock velocity, etc. in an impact. A rough estimate of these quantities is obtained from the planar impact approximation (sometimes called the impedance-matching solution), which is valid as long as the lateral dimensions of the projectile are small compared with the distance the shock has propagated. This approximation is thus valid through most of the contact and compression stage. Unfortunately, there is no simple formula for this approximation. The simplest expression is for the particle velocity in the target, u_t (the particle velocity in the projectile is $v_i - u_t$):

$$u_t = \frac{-B + \sqrt{B^2 - 4AC}}{2A}$$

where

$$A = \rho_{0t} S_t - \rho_{0p} S_p$$
$$B = \rho_{0t} c_t + \rho_{0p} c_p + 2\rho_{0p} S_p v_i$$
$$C = -\rho_{0p} v_i (c_p + S_p v_i)$$

and the subscripts p and t refer to the projectile and target, respectively. This equation can be used in conjunction with the Hugoniot equations and equations of state to obtain any other quantities of interest. Thus, the pressure behind the shock is given by $P = \rho_{0t} u_t (c_t + S_t u_t)$. The pressures in both the target and projectile are the same by construction of the solution.

As the projectile plunges into the target, shock waves propagate both into the projectile, compressing and slowing it, and into the target, compressing and accelerating it downward and outward. At the interface between target and projectile the material of each moves at the same velocity. This equals one-half the impact velocity if they are composed of the same materials (note that in the above equation $A = 0$ in this case, but the numerator also vanishes and the right-hand side of the equation approaches $-C/B$, which equals $v_i/2$). The shock wave in the projectile eventually reaches its back (or top) surface. At this time the pressure is released as the surface of the compressed projectile expands upward, and a wave of pressure relief propagates back downward toward the projectile–target interface. The contact and compression stage is considered to end when this relief wave reaches the projectile–target interface. At this time the projectile has been compressed to high pressure, often reaching hundreds of gigapascals, and upon decompression it may be in the liquid or gaseous state due to heat deposited in it during the irreversible compression process. The projectile generally carries off 50% or less of the total initial energy if the density and compressibility of the projectile and target material do not differ too much. The projectile–target interface at the end of contact and compression is generally less than a projectile diameter *L* below the original surface.

Contact and compression is accompanied by the formation of very high velocity *jets* of highly shocked material. These jets form where strongly compressed material is close to a free surface, for example, near the circle where a spherical projectile makes contact with a planar target. The jet velocity depends on the angle between the converging surface of the projectile and the target, but may exceed the impact velocity by factors as great as 5. Jetting was initially regarded as a spectacular but not quantitatively important phenomenon in early impact experiments, where the incandescent streaks of jetted material amounted to only about 10% of the projectile's mass in vertical impacts. However, recent work on oblique impacts indicates that in this case jetting is much more important and that the entire projectile may participate in a downrange stream of debris that carries much of the original energy and momentum. Oblique impacts are still not well understood and more work needs to be done to clarify the role of jetting early in this process.

B. Excavation

During the excavation stage the shock wave created during contact and compression expands and eventually weakens into an elastic wave while the crater itself is opened by the much slower *excavation flow*. The duration of this stage is roughly given by the period of a gravity wave with wavelength equal to the crater diameter *D*, $(D/g)^{1/2}$, for craters whose excavation is dominated by gravity *g* (this includes craters larger than a few kilometers in diameter, even when excavated in hard rock). Thus, Meteor Crater was excavated in about 10 s, whereas the 1000-km-diameter Imbrium Basin on the moon took about 13 min to open. Shock wave expansion and crater excavation, although intimately linked, occur at rather different rates and may usefully be considered separately.

The high pressures attained during contact and compression are almost uniform over a volume roughly comparable to the initial dimensions of the projectile. However, as the shock wave expands away from the impact site the shock pressure declines as the initial impact energy spreads over an increasingly large volume of rock. The pressure in the shock *P* as a function of distance *r* from the impact site is given roughly by

$$P(r) = P_0(a/r)^n$$

where a $(=L/2)$ is the radius of the projectile, P_0 is the pressure established during contact and compression, and the power *n* is between 2 and 4, depending on the strength of the shock wave (*n* is larger at higher pressures—a value $n = 3$ is a good general average).

The shock wave, with a release wave immediately following, quickly attains the shape of a hemisphere expanding through the target rocks. The high shock pressures are confined to the surface of the hemisphere: the interior has already decompressed. The shock wave moves very quickly, as fast as or faster than the speed of sound, between about 6 and 10 km/s in most rocks. As rocks in the target are overrun by the shock waves, then released to low pressures, mineralogical changes take place in the component minerals. At the highest pressures the rocks may melt or even vaporize upon release. As the shock wave weakens, high-pressure minerals such as coesite or stishovite arise from quartz in the target rocks, diamonds may be produced from graphite, or maskelynite from plagioclase. Somewhat lower pressures cause pervasive fracturing and *planar elements* in individual crystals. Still lower pressures create characteristic cone-in-cone fractures called *shatter cones* (Fig. 2) that are readily recognized in the vicinity of impact structures. Indeed, many terrestrial impact structures were first recognized from the occurrence of shatter cones. Table II lists a number of well-established shock metamorphic changes and the pressures at which they occur.

The expanding shock wave encounters a special condition near the free surface. The pressure at the surface must be zero at all times. Nevertheless, a short distance below the surface the pressure is essentially equal to *P*, defined above. This situation results in a thin layer of surface rocks being thrown upward at very high velocity (the theoretical maximum velocity approaches the impact speed v_i). Since the surface rocks are not compressed to high pressure, this results in the ejection of a small quantity of unshocked or lightly shocked rocks at speeds that may exceed the target planet's escape velocity. Although the total quantity of material ejected by this *spall* mechanism is probably only 1–3% of the total mass excavated from the crater, it is particularly important scientifically as this is probably the origin of the recently discovered meteorites from the moon and of the SNC (Shergottite, Nakhlite, and Chassignite) meteorites which are widely believed to have been ejected from Mars.

The weakening shock wave eventually degrades into elastic waves. These elastic waves are similar in many respects to the seismic waves produced by an earthquake, although impact-generated waves contain less destructive shear wave energy than earthquake waves. The seismic waves produced by a large impact may have significant effects on the target planet, creating jumbled terrains at the antipode of the impact site. This effect has been observed opposite Caloris basin on Mercury and opposite Imbrium and Orientale on the moon. The equivalent Richter magnitude *M* caused by an impact of energy E $(=\frac{1}{2}m_p v_i^2)$ is given approximately by $M = 0.67 \log_{10} E - 5.87$.

Target material engulfed by the shock wave is released a short time later. Upon release, the material has a velocity that

Fig. 2 Shatter cones are characteristic features in rocks shocked to pressures from 2 to 6 GPa. Shown here are shatter cones in the Mid-Proterozoic Pentecost orthoquartzite from the Spider impact structure, Western Australia. (Photo courtesy of George Williams.)

is only about one-fifth of the particle velocity in the shock wave. This *residual velocity* is due to thermodynamic irreversibility in the shock compression. It is this velocity field that eventually excavates the crater. The excavation velocity field has a characteristic downward, outward, then upward pattern that moves target material out of the crater, ejecting it at angles close to 45° at the rim. The streamlines of this flow cut across the contours of maximum shock pressure, so that material ejected at any time may have a wide range of shock levels. Nevertheless, the early, fast ejecta generally contain a higher proportion of highly shocked material than the later, slower ejecta. Throughout its growth, the crater is lined with highly shocked, often melted, target material.

Inside the growing crater, vaporized projectile and target may expand rapidly out of the crater, forming a vapor plume that, if massive enough, may blow aside any surrounding atmosphere and accelerate to high speed. In the impacts of sufficiently large and fast projectiles some of this vapor plume material may even reach escape velocity and leave the planet, incidentally also removing some of the planet's atmosphere. Such *impact erosion* may have played a role in the early history of the Martian atmosphere. Even in smaller impacts the vapor plume may temporarily blow aside the atmosphere, opening the way for widespread ballistic dispersal of melt droplets (tektites) above the atmosphere and perhaps permitting the formation of lunarlike ejection blankets even on planets with dense atmospheres, as has been observed on the Soviet Venera 15/16 and U.S. Magellan images of Venus.

Table II Peterographic Shock Indicators

Material	Indicator	Pressure (GPa)
Tonalite (igneous rock)	Shatter cones	2–6
Quartz	Planar elements and fractures	5–35
	Stishovite	15–40
	Coesite	30–50
	Melting	50–65(?)
Plagioclase	Planar elements	13–30
	Maskelynite	30–45
	Melting	45–65(?)
Olivine	Planar elements and fractures	5–45
	Ringwoodite	45
	Recrystallization	45(?)–65(?)
	Melting	>70
Clinopyroxene	Mechanical twinning	5–40(?)
	Majorite	13.5
	Planar elements	30(?)–45
	Melting	45(?)–65(?)
Graphite	Cubic diamond	13
	Hexagonal diamond	70–140

The growing crater is at first hemispherical in shape. Its depth $H(t)$ and diameter $D(t)$ both grow approximately as $t^{0.4}$, where t is time after the impact. Hemispherical growth ceases after a time of about $(2H_t/g)^{1/2}$, where H_t is the final depth of the transient crater. At this time the crater depth stops increasing (it may even begin to decrease as collapse begins), but the crater diameter continues to increase. The crater shape thus becomes a shallow bowl, finally attaining a diameter roughly three to four times its depth. At this stage, before collapse modifies it, the crater is known as a *transient* crater. Even simple craters experience some collapse (which produces the breccia lens), so that the transient crater is always a brief intermediate stage in geological crater formation. However, because most laboratory craters are "frozen" transient craters, much of our knowledge about crater dimensions refers to the transient stage only and must be modified for application to geological craters.

Laboratory, field, and computer studies of impact craters have all confirmed that only material lying above about one-third of the transient crater depth (or about one-tenth of the diameter) is thrown out of the crater. Material deeper than this is simply pushed downward into the target, where its volume is accommodated by deformation of the surrounding rocks. Thus, in sharp contrast to ejecta from volcanic craters, material in the ejecta blankets of impact craters does not sample the full depth of rock intersected by the crater, a surprising fact that has led many geologists astray in their estimation of the nature of the ejected debris.

The form of the transient crater produced during the excavation stage may be affected by such factors as obliquity of the impact (although the impact angle must be less than about 6° for a noticeably elliptical crater to from at impact velocities in excess of about 4 km/s), the presence of a water table or layers of different strength, rock structure, joints, or initial topography in the target. Each of these factors produces its own characteristic changes in the simple bowl-shaped transient crater form.

C. Modification

Shortly after the excavation flow has opened the transient crater and the ejecta has been launched onto ballistic trajectories, a major change takes place in the motion of debris within and beneath the crater. Instead of flowing upward and away from the crater center, the debris comes to a momentary halt, then begins to move downward and back toward the center whence it came. This collapse is generally attributed to gravity, although elastic rebound of the underlying, compressed rock layers may also play a role. The effects of collapse range from mere debris sliding and drainback in small craters to wholesale alteration of the form of larger craters in which the floors rise, central peaks appear, and the rims sink down into wide zones of stepped terraces. Great mountain rings or wide central pits may appear in still larger craters.

These different forms of crater collapse begin almost immediately after formation of the transient crater. The time scale of collapse is similar to that of excavation, occupying an interval of a few times $(D/g)^{1/2}$. Crater collapse and modification thus take place on time scales very much shorter than most geologic processes. The crater resulting from this collapse is then subject to the normal geologic processes of gradation, isostatic adjustment, infilling by lavas, etc. on geologic time scales. Such processes may eventually result in obscuration or even total obliteration of the crater.

The effects of collapse depend on the size of the crater. For transient craters smaller than about 15 km in diameter on the moon, or about 3 km on Earth, modification entails only collapse of the relatively steep rim of the crater onto its floor. The resulting simple crater (see Fig. 1a) is a shallow bowl-shaped depression with a rim-to-rim diameter D about five times its depth below the rim H. In fresh craters the inner rim stands near the angle of repose, about 30°. Drilling in terrestrial craters (Fig. 3) shows that the crater floor is underlain by a lens of broken rock (mixed breccia) derived from all of the rock units intersected by the crater. The thickness of this breccia lens is typically one-half the depth of the crater H. Volume conservation suggests that this collapse increases the original diameter of the crater by about 15%. The breccia lens often includes layers and lenses of highly shocked material mixed with much less shocked country rock. A small volume of shocked or melted rock is often found at the bottom of the breccia lens.

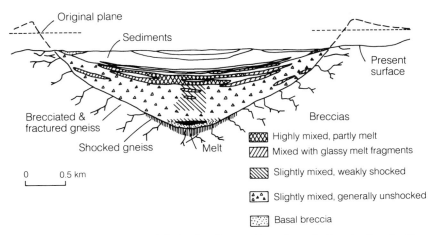

Fig. 3 Geologic cross section of the 3.4-km-diameter Brent Crater in Ontario, Canada. Although the rim has been eroded away, Brent is a typical simple crater that formed in crystalline rocks. A small melt pool occurs at the bottom of the breccia lens and more highly shocked rocks occur near its top.

Complex craters (Figs. 1b and 1c) collapse more spectacularly. Walls slump, the floor is stratigraphically uplifted, central peaks or peak rings rise in the center, and the floor is overlain by a thick layer of highly shocked impact melt (Fig. 4). The detailed mechanism of collapse is still not fully understood because straightforward use of standard rock mechanics models does not predict the type of collapse observed. The current best description of complex crater collapse utilizes a phenomenological strength model in which the material around the crater is approximated as a Bingham fluid, a material which responds elastically up to differential stresses of about 3 MPa, independent of overburden pressure, and then flows as a viscous fluid with viscosity on the order of 1 GPa-s at larger stresses. In a large collapsing crater the walls slump along discrete faults, forming terraces whose widths are controlled by the Bingham strength, and the floor rises, controlled by the viscosity, until the differential stresses fall below the 3-MPa strength limit. A central peak may rise, then collapse again in large craters, forming the observed internal ring (or rings). Figure 5 illustrates this process schematically. The rock in the vicinity of a large impact may display such an unusual flow law because of the locally strong shaking driven by the large amount of seismic energy deposited by the impact.

The mechanics of the collapse that produces multiring basins (Fig. 1d) is even less well understood. Figure 6 illustrates the structure of the Orientale basin on the moon with a highly vertically exaggerated cross section derived from both

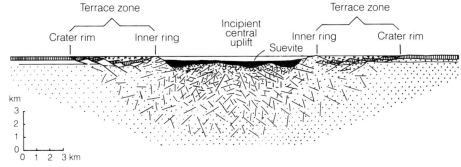

Fig. 4 Geologic cross section of the 22-km-diameter Ries Crater in Germany. Drilling and geophysical data suggest that this is a peak ring crater. Its central basin is filled with suevite, a mixture of highly shocked and melted rocks and cold clasts.

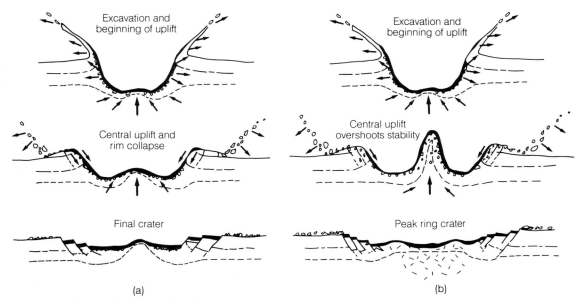

Excavation and
beginning of uplift

Central uplift and
rim collapse

Final crater

(a)

Excavation and
beginning of uplift

Central uplift
overshoots stability

Peak ring crater

(b)

Fig. 5 Schematic illustration of the formation of (a) central peak and (b) peak ring craters. Uplift of the crater floor begins even before the rim is completely formed. As the floor rises, rim collapse creates a wreath of terraces surrounding the crater. In smaller craters the central peak "freezes." In larger craters the central peak collapses and creates a peak ring before motion ceases.

geological and geophysical data. Note that the ring scarps are interpreted as inward-dipping faults above a pronounced mantle uplift beneath the basin's center. One idea that is currently gaining ground is that the ring scarps are normal faults that develop as the crust surrounding a large crater is pulled inward by the flow of underlying viscous mantle material toward the crater cavity (Fig. 7). An important aspect of this flow is that it must be confined in a low-viscosity channel by more viscous material below; otherwise the flow simply uplifts the crater floor and radial faults, not ring scarps, are the result. Special structural conditions are thus needed in the planet for multiring basins to form on its surface, so that a g^{-1} dependence for the transition from complex craters to multiring basins is not expected (or observed). This theory is capable of explaining both the lunar-type and Valhalla-type multiring basins as expressions of different lithosphere thicknesses.

IV. Scaling of Crater Dimensions

One of the most frequently asked questions about an impact crater is "how big was the meteorite that made the crater?" Like many simple questions, this has no simple answer. It should be obvious that the crater size depends on the meteorite's speed, size, and angle of entry. It also depends on such factors as the meteorite's composition, the material and composition of the target, surface gravity, and presence or absence of an atmosphere. The inverse question, of how large a crater will be produced by a meteorite of a given size with known speed and incidence angle, is in principle much simpler to answer. However, even this prediction is uncertain because there are no observational or experimental data on the formative conditions of impact craters larger than a few tens of meters in diameter, and the impact structures of geologic interest range up to 1000 km in diameter. The traditional escape from this difficulty is to extrapolate beyond experimental knowledge by means of scaling laws.

The first scaling laws were introduced in 1950 by C. W. Lampson, who studied the craters produced by TNT explosions of different sizes. Lampson found that the craters were similar to one another if all dimensions (depth, diameter, depth of charge placement) were divided by the cube root of the explosive energy W. Thus, if the diameter D of a crater produced by an explosive energy W is wanted, it can be computed from the diameter D_0 of a crater produced by a (generally smaller) explosive energy W_0 using the proportion

$$D/D_0 = (W/W_0)^{1/3}$$

An exactly similar proportion may be written for the crater depth H. This means that the ratio of depth to diameter,

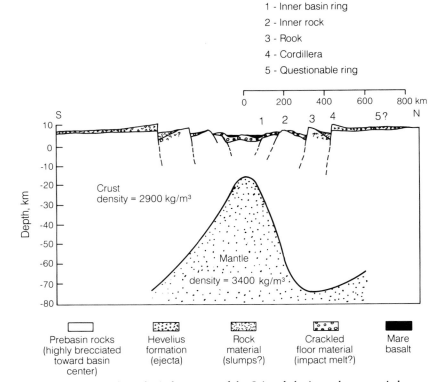

1 - Inner basin ring
2 - Inner rock
3 - Rook
4 - Cordillera
5 - Questionable ring

Fig. 6 Geologic and geophysical structure of the Orientale basin on the moon. A dense mantle plug underlies the center of the basin. The crustal thinning above the plug is due to about 40 km of crustal material ejected from the crater that formed the basin. The great ring scarps formed during collapse of the crater. Note the 10× vertical exaggeration.

H/D, is independent of yield, a prediction that agrees reasonably well with observation. In more recent work on large explosions the exponent 1/3 in this equation has been modified to 1/(3.4) to account for the effects of gravity on crater formation.

Although impacts and explosions have many similarities, a number of factors make them difficult to compare in detail. Thus, explosion craters are very sensitive to the charge's depth of burial. Although this quantity is well defined for explosions, there is no simple analog for impact craters. Similarly, the angle of impact has no analog for explosions. Nevertheless, energy-based scaling laws were very popular in the older impact literature, perhaps partly because nothing better existed, and many empirical schemes were devised to adapt the well-established explosion scaling laws to impacts. [See Natural Detonations.]

This situation has changed rapidly in the last decade, however, thanks to more impact cratering experiments specifically designed to test scaling laws. It has been shown that the great expansion of the crater during excavation tends to decouple the parameters describing the final crater from the parameters describing the projectile. If these sets of parameters are related by a single, dimensional coupling parameter (as seems to be the case), it can be shown that crater parameters and projectile parameters are related by power-law scaling expressions with constant coefficients and exponents. Although this is a somewhat complex and rapidly changing subject, the best current scaling relation for impact craters forming in competent rock (low porosity) targets whose growth is limited by gravity rather than target strength (i.e., all craters larger than a few kilometers in diameter) is given by

$$D_{at} = 1.8\rho_p^{0.11}\,\rho_t^{-1/3}\,g^{-0.22}\,L^{0.13}\,W^{0.22}\,(\cos\theta)^{1/3}$$

where D_{at} is the diameter of the transient crater at the level of the original ground surface, ρ_p and ρ_t are densities of the projectile and target, respectively, g is surface gravity, L is projectile diameter, W is impact energy $(=\pi/12\,L^3\rho_p v_i^2)$, and θ is the angle of impact from the vertical. All quantities are in SI units.

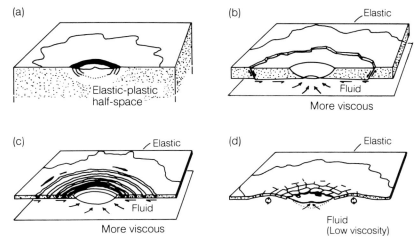

Fig. 7 Ring tectonic theory of multiring basin formation. Drawings (a)–(c) illustrate the effect of decreasing lithosphere thickness, from a very thick lithosphere (a) that prevents rings from forming, to a moderately thick lithosphere (b) that produces one or a few lunar-type rings, to a thin lithosphere (c) in which Valhalla-type systems develop. Drawing (d) illustrates the hypothetical case of a very fluid mantle underlying a thin lithosphere in which a radial-plus-concentric fracture pattern develops. Patterns of this kind have been observed around large explosions on floating ice.

Crater depth H appears to be a constant times the diameter D_{at}. Although a few investigations have reported a weak velocity dependence for this ratio, the experimental situation is not yet clear.

V. Impact Cratering and Planetary Evolution

Over the last few decades it has become increasingly clear that impact cratering has played a major role in the formation and subsequent history of the planets and their satellites. Aside from their scientific interest, impact craters have also achieved a modest economic importance as it has become recognized that the fabulously rich Sudbury nickel deposit in Ontario, Canada, is a tectonically distorted 140-km-diameter impact crater. Similarly, the Vredefort structure in South Africa is probably a large old impact crater. Oil production has been achieved from a number of buried impact craters, such as the 10-km-diameter Red Wing Creek crater in the Williston Basin and the 3.2-km-diameter Newporte structure in North Dakota. On a more homely level, the 60-km-diameter ring-shaped depression in the Manicouagan crater in eastern Quebec is currently used as a reservoir supplying New York City.

Modern theories of planetary origin suggest that the planets and the sun formed simultaneously 4.6×10^9 years ago

from a dusty, hydrogen-rich nebula. Nebular condensation and hydrodynamic interactions were probably capable of producing only planetesimals about 10 km in diameter that accreted into planetary-scale objects by means of collisions. The time scale for accretion of the inner planets by mutual collisions is currently believed to be between a few tens and 100 million years. Initially rather gentle, these collisions became more violent as the random velocities of the smaller planetesimals were increased during close approaches to the larger bodies. The mean random velocity of a swarm of planetesimals is comparable to the escape velocity of the largest object, so as the growing planetary embryos reached lunar size, collisions began to occur at several kilometers per second. At such speeds impacts among the smaller objects were disruptive, whereas the larger objects had sufficient gravitational binding energy to accrete most of the material which struck them. Infalling planetesimals bring not only mass but also heat to the growing planets. In the past it was believed that the temperature inside a growing planet increased in a regular way from near zero at the center to large values at the outside, reflecting the increase in collision velocity as the planet became more massive. However, it now seems probable that the size distribution of the planetesimal population was more evenly graded between large and small objects and that each growing planetary embryo was subjected to many collisions with bodies comparable in size to the embryo itself. Such catastrophic collisions would deposit heat deep within the core of the body, wiping out any regular law for temperature increase with increasing radius and making the thermal evolution of growing plants rather stochastic.

The origin of the moon is now attributed to a collision between the proto-Earth and a Mars-size protoplanet near the end of accretion 4.5×10^9 years ago. This theory has recently supplanted the three classical theories of lunar origin (capture, fission, and coaccretion) because only the giant impact theory provides a simple explanation for the moon's chemistry, as revealed in the lunar rocks returned by Apollo. One view of this process is that a grazing collision vaporized (by jetting) a large quantity of the proto-Earth's mantle along with a comparable quantity of the projectile. Most of the mass of the projectile merged with Earth (incidentally strongly heating Earth; if Earth was not molten before this impact it almost certainly was afterward), but one or two lunar masses of vapor condensed into dust in in stable Keplerian orbits about Earth and then later accumulated together to form the moon.

Sometime after the moon formed and before about 3.8×10^9 years ago the inner planets and their satellites were subjected to the *late heavy bombardment,* an era during which the cratering flux was orders of magnitude larger than at present. The crater scars of this period are preserved in the lunar highlands and the most ancient terrains of Mars and Mercury. A fit to the lunar crater densities using age data from Apollo samples gives a cumulative crater density through geologic time of

$$N_{cum}(D > 4 \text{ km}) = 2.68 \times 10^{-5} [T + 4.57 \times 10^{-7} (e^{\lambda T} - 1)]$$

where $N_{cum}(D > 4 \text{ km})$ is the cumulative crater density (craters per square kilometer) of craters larger than 4 km in diameter, T is the age of the surface in gigayears ($T = 0$ is the present), and $\lambda = 4.53 \text{ Gyr}^{-1}$. The current cratering rate on the moon is about 2.7×10^{-14} craters with $D > 4 \text{ km/km}^2/$ yr. On the earth the cratering rate has been estimated to be about 1.8×10^{-15} craters with $D > 22.6 \text{ km/km}^2/\text{yr}$, which is comparable to the lunar flux taking into account the different minimum sizes, since the cumulative number of craters $N_{cum}(D) \sim D^{-1.8}$. There is currently much debate about these cratering rates, which might be uncertain by as much as a factor of 2.

These high cratering rates in the past indicate that the ancient Earth should have been heavily scarred by large impacts. Based on the lunar record, it is estimated that more than 100 impact craters with diameters greater than 1000 km should have formed on Earth. Although little evidence of these early craters has yet been found, it is gratifying to note the recent discovery of thick impact ejecta deposits in 3.2–3.5 Gyr Archean greenstone belts in both South Africa and Western Australia. Since rocks have recently been found dating back to 4.2 Gyr, well into the era of heavy bombardment, it is to be hoped that more evidence for early large craters will eventually be discovered. Heavy bombardment also seems to have overlapped the origin of life on Earth. It is possible that impacts had an influence on the origin of life, although whether they suppressed it by creating global climatic catastrophes (up to evaporation of part or all of the seas by large impacts) or facilitated it by bringing in needed organic precursor molecules is unclear at present. The relation between impacts and the origin of life is an area of vigorous speculation.

The idea that large impacts can induce major volcanic eruptions is one of the recurring themes in the older geologic literature. This idea probably derives from the observation that all of the large impact basins on the moon's nearside are flooded with basalt. However, radiometric dates for Apollo samples made it clear that the lava infillings of the lunar basins are nearly 1 Gyr younger than the basins themselves. Furthermore, the farside lunar basins generally lack any lava infilling at all. The nearside basins are apparently flooded merely because they were topographic lows in a region of thin crust at the time that mare basalts were produced in the moon's upper mantle. Simple estimates of the pressure release caused by stratigraphic uplift beneath large impact craters make it clear that pressure release melting cannot be important in impacts unless the underlying mantle is near the melting point before the impact. Thus, it is probably safe to say that, to date, there is no firm evidence that impacts can induce volcanic activity. Impact craters may create fractures along which preexisting magma may escape but themselves are probably not capable of producing much melt. Nevertheless, both the Sudbury and Vredefort structures had associated massive igneous intrusions whose genesis is sometimes attributed to the impact, although in this case they may have been triggered by the uplift of hot lower crust. Further study of these issues is needed.

The most recent major impact event on Earth seems to have been a collision between Earth and a 10-km-diameter comet or asteroid 65 Myr ago that ended the Cretaceous era and caused the most massive biological extinction in recent geologic history. Evidence for this impact has been gathering from many sites over the last decade and now seems nearly incontrovertible. First detected as an enrichment of the siderophile element iridium in the ~3-mm-thick K/T (Cretaceous/tertiary) boundary layer in Gubbio, Italy, the iridium signature has now been found at more than 100 locations worldwide, in both marine and terrestrial deposits. Accompanying this iridium are other siderophiles in chondritic ratios, shocked quartz grains, coesite, stishovite, and small (100–500 μm) spherules resembling microtektites. All these point to the occurrence of a major impact at the K/T boundary. In addition, soot and charcoal have been found at a number of widely separated sites in abundances which suggest that the entire world's standing biomass burned within a short time of the impact. An impact of this magnitude should have produced a crater between 200 and 300 km in diameter. At the time of this writing, a prime candidate is a 180-km-diameter structure buried 1 km beneath the north shore of the Yucatan Peninsula. Southward thickening of a terrestrial ejecta deposit that seems to underlie the iridium-rich layer in North America suggests an impact site somewhere in the Caribbean, as does a 1-m-thick layer of tektite-like ejecta discovered on Haiti. Although many details of the impact, and especially of the extinction mechanism, still have to be worked out, the evidence for a great impact at this time is becoming overwhelming, although a few geologists still adhere to some kind of volcanically induced extinction crisis. Future work should be able to resolve the mysteries surrounding this striking demonstration of the importance of impact craters in geology.

Bibliography

Carr, M. H., Saunders, R. S., Strom, R. G., and Wilhelms, D. E. (1984). "The Geology of the Terrestrial Planets," NASA SP-469. U.S. Government Printing Office, Washington, D.C.

Grieve, R. A. F. (1987). Terrestrial impact structures. *Ann. Rev. Earth Planet. Sci.* **15**, 245.

Mark, K. (1987). "Meteorite Craters." Univ. of Arizona Press, Tucson.

Melosh, H. J. (1989). "Impact Cratering: A Geologic Process." Oxford Univ. Press, New York.

Taylor, S. R. (1982). "Planetary Science: A Lunar Perspective." Lunar and Planetary Institute, Houston.

Glossary

Breccia Broken, angular rock debris.

Ejecta blanket Circular apron of debris surrounding an impact crater.

Impact crater Circular, rimmed depression produced by the impact of a solid body traveling at more than a few kilometers per second.

Shock metamorphism Characteristic changes in rock mineralogy caused by the compression of a passing shock wave.

Shock wave Moving wave of strong compression that usually travels faster than sound.

Indian Ocean, Evolution in the Hotspot Reference Frame

Robert A. Duncan

Oregon State University

The Indian Ocean basin was formed by the breakup and disintegration of the eastern portion of the supercontinent of Gondwana, beginning about 160 million years ago and continuing today. Africa, Antarctica, India, and Australia were once joined in a landmass that covered much of the southern hemisphere. In a series of rifting episodes, first Africa, then India, then Australia separated from Antarctica and drifted northward to equatorial latitudes, while new ocean floor filled in the opening spaces. The islands of Madagascar, the Seychelles, and Sri Lanka are continental fragments left behind in the spreading process. The history of the development of this region can be deciphered through the bathymetric and magnetic record of sea-floor spreading and linear volcanic trails related to hotspot activity.

I. Introduction

The major topographic features of the floor of the Indian Ocean are three active spreading ridge systems (Fig. 1) and a widely distributed collection of islands, plateaus, and linear volcanic chains (Fig. 8). The spreading ridges are the sites of lithospheric plate separation and progressive addition of new ocean floor through upwelling and melting of the upper mantle. The Southwest Indian ridge separates the African plate from the Antarctic plate; the Southeast Indian ridge divides the Australian plate from the Antarctic plate; the Central Indian ridge marks the boundary between the African

and Indian plates. As in other ocean basins the spreading-ridge morphology correlates with the rate of plate separation: The unusually slow-spreading Southwest Indian ridge (1.0 cm/yr opening) is extremely rugged, narrow, and heavily dissected by closely spaced transform faults, which offset short spreading-ridge segments. In contrast, the moderately fast-spreading Southeast Indian ridge (7.5 cm/yr opening) is smoother and broader, with longer ridge segments. [*See* LITHOSPHERIC PLATES.]

With the exception of the islands, plateaus and linear chains, the ocean floor deepens away from the spreading ridges in proportion to the square root of its age. This simple relationship is seen worldwide and results from conductive cooling and contraction of the oceanic lithosphere. The ridge crests, where new crust is forming today, lie between 2500 and 3000 m water depth, while 100 million year old ocean floor is about 5800 m deep. Thus, water depth is a rough but useful guide to the age distribution of ocean floor in the Indian Ocean. In some regions, such as the Somali Basin (between East Africa and Madagascar) and the Indus and Bengal abyssal fans (west and east of peninsular India), the weight of enormous accumulations of biogenic and terrigenous sediments has depressed the lithosphere in excess of the normal thermal subsidence.

A more precise means of determining the age of the ocean floor beneath various parts of the Indian Ocean has been through the mapping and identification of sea-floor magnetic anomalies. The basaltic rocks that constitute the oceanic crust have formed from magmas that erupted at the spreading ridges. As the basalts crystallize and cool they acquire a weak magnetization parallel to the earth's magnetic field. From time to time the direction of the earth's magnetic field reverses polarity, so that ocean crust formed during a reversed interval has a magnetization opposite to crust formed during a normal interval like the present. Thus, a stripe of sea floor of uniform magnetization is formed all along the spreading ridge system during a single magnetic polarity interval. The width of the stripe is the product of the duration of the magnetic polarity interval and the spreading rate. A pattern of normally and reversely magnetized stripes parallel to the spreading ridges can then be mapped from magnetometer

Fig. 1   Bathymetric features and plate boundaries of the Indian Ocean basin. Arrows indicate current directions of relative plate motion.

measurement of the sea-floor magnetic anomalies produced by these rocks of alternating magnetic polarity. Finally, the ages of the magnetic reversals over the last 160 million years have been determined from radiometric dating. Hence, this *magnetostratigraphic time scale* for the history of magnetic reversals can be used to determine the age of the ocean floor for which sequences of magnetic anomalies have been identified. [*See* DEPOSITIONAL REMANENT MAGNETIZATION; OCEANIC CRUST; PALEOMAGNETISM.]

The opening of the Indian Ocean can thus be deciphered through a series of spreading events recorded in the bathymetry and magnetic anomaly patterns on the seafloor. This history of lateral motion of separate pieces of the lithosphere is termed *relative plate motions*. Other systems for determining past plate positions are the (1) paleomagnetic and (2) hotspot reference frames. Because the inclination of the magnetic direction recorded in rocks uniquely determines only the latitude at which the magnetization was acquired, any north–south motion of a plate can be measured. East–west motion, however, cannot be resolved this way. Some linear volcanic chains are related to stationary thermal anomalies in the upper mantle, called *hotspots*. These are regions of unusu-

ally high melt production, maintained over long periods (up to 100 million years or more) by convective upwelling of *plumes* of hot material from the deep mantle. According to this model the direction of plate motion over the hotspots is simply the orientation of the lineaments, and the velocity is found from the rate of migration of volcanism along the chains. Thus, plates can be "backtracked" along these lineaments to the sites of current hotspot activity for ages of interest. Both these schemes provide *absolute plate motions*, in the sense that motion is measured in a reference frame that is fixed; that is, the magnetic pole (the spin axis) and the mantle, respectively.

We can follow the main features of Indian Ocean evolution in each of these systems after we review the current plate geometry and motions.

II. Current Plate Boundaries

The Indian Ocean is now bounded by four major plates: the African, Indian, Australian, and Antarctic, each of which is

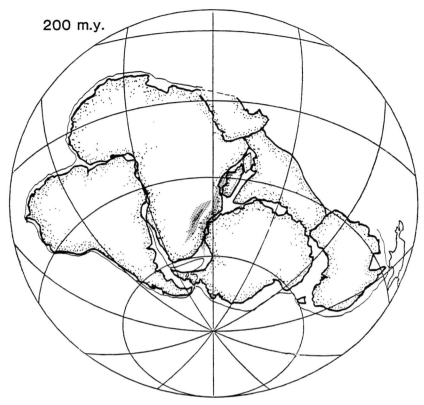

Fig. 2 Paleogeographic reconstruction of the Indian Ocean region for 200 million years ago, in the hotspot reference frame. The Karoo flood basalts erupted about this time in southern Africa. Continental rifting began between Africa, Madagascar, and Antarctica at about 160 million years ago.

part oceanic and part continental (Fig. 1). The Antarctic plate is separated from the African and Australian plates by the Southwest and Southeast Indian spreading ridges. The African–Indian plate boundary is the Central Indian spreading ridge. African–Arabian separation also occurs along the Red Sea and Gulf of Aden spreading systems, and the East African Rift may develop into a new ocean basin that further dismembers the African plate in the future. The Indian plate moves northeast relative to the Arabia along the Owen Fracture Zone (transform fault), which connects with the Himalayan zone of collision between the Indian plate and Asia. A long transform fault to the east of India allows northward motion of the Indian plate relative to Southeast Asia.

Differential subduction along the northern boundary of the Indian–Australian plate pair is now causing compression in the oceanic lithosphere between the two plates along an east–west zone lying roughly between latitude 10 and 20°S. Evidence for this new boundary is in warping of the Indian plate, thrust faulting, and seismicity. The Australian plate is

bounded to the north by the Sumatra–Java–Sunda subduction zone and, further east, a complex collisional boundary of subduction zones and transform faults. In the southeasternmost corner of the Indian Ocean, south of Australia, is the Macquarie Ridge, a transform-thrust fault boundary between the Australian and Pacific plates.

Measurements of stress in the crust can be made in boreholes on land and at sea, and the orientation of faults and dikes can reveal stress conditions in the plates. These data are used to interpret the direction and magnitude of forces acting on the plates in the Indian Ocean region. Generally, compressional stresses are found perpendicular to the spreading ridges and indicate a push from the elevated ridge crests. Tensional stresses are seen near the subduction boundary south and east of Java–Sumatra, indicating a pull from the descending oceanic slab beneath the Indonesian archipelago. Compressional stresses are common throughout peninsula India and south of Sri Lanka, aligned in the direction of Indian–Asian plate collision.

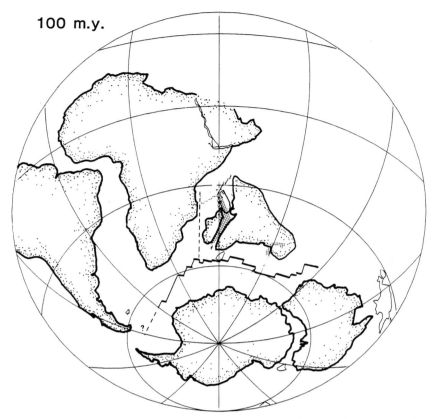

Fig. 3   Paleogeographic reconstruction of the Indian Ocean for 100 million years ago, in the hotspot reference frame. Sea-floor spreading shifted to separation between India and Antarctica at about 130 million years ago.

III. Relative Plate Motions

A. 160 – 130 Million Years

Some rifting between Africa and Antarctica may have begun as early as 200 million years ago, concurrent with enormous eruptions of the Karoo flood basalts in South Africa (Figs. 2 and 8). However, this did not develop into a successful spreading event. By mid-Jurassic time (about 160 million years) true oceanic spreading was established along an axis that now defines the east coast of Africa. In this episode Madagascar, which has nestled against the coast of Africa east of Kenya, moved southward with eastern Gondwana (Antarctica, India, and Australia) away from Africa. This opened up the Somali and Mozambique Basins between 160 and 130 million years ago. Within this same period new ocean floor was being formed in the Argo Basin north of Australia, by spreading in the Tethys Ocean. (This ocean lay to the north of India and has been largely consumed by collision of

Australia, India, and Africa with the Asian plate, and only small fragments such as the Argo Basin remain.) [*See* CONTINENTAL FLOOD BASALTS; CONTINENTAL RIFTING.]

B. 130 – 100 Million Years

Slow spreading between the India – Madagascar plate and the Australia – Antarctica plate began at about 130 million years and connected to the west to the new spreading ridge separating Africa from South America (Fig. 3). Thus the Indian and South Atlantic Oceans have grown within the same period. Plate separation in the east was initially northwest – southeast, forming the Wharton, the Bengal, and African – Antarctic basins. At about 117 million years much of the Kerguelen Plateau, the Naturaliste Plateau and the Rajmahal Traps (Fig. 8) were formed on the new ocean floor by rapid eruption of enormous quantities of flood basalts signaling the birth of the Kerguelen hotspot. Subsequent reorganization of the spreading geometry rifted and dismembered this large

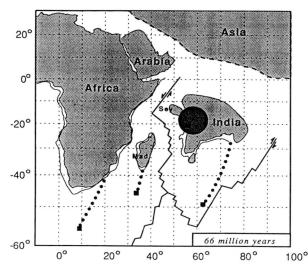

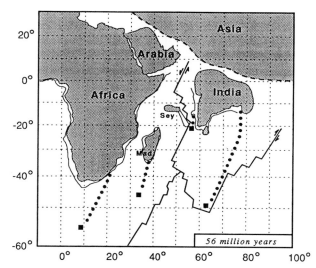

Fig. 4 Paleogeographic reconstruction of the Indian Ocean for 66 million years ago, in the hotspot reference frame. Dotted path describes the earlier path of the Kerguelen and other hotspots. The Reunion hotspot burst into life at this time, covering much of western India in several kilometers of flood basalts.

Fig. 5 Paleogeographic reconstruction of the Indian Ocean for 56 million years ago, in the hotspot reference frame. Heavy dotted paths describe the tracks of hotspots.

oceanic plateau. To the west separation of Africa from South America occurred along several long transform faults that transported the Falkland plateau to the west and left open-deep-water circulation between Africa and Antarctica.

C. 100 – 40 Million years

Beginning at about 100 million years the Indian plate started a spectacular drive to the north, which has culminated in its collision with Asia. A new spreading geometry developed, with a western triple junction separating Madagascar from India from Antarctica, and an eastern triple junction separating India from Australia from Antarctica. Fast spreading along the Indian – Antarctic and Indian – Australian ridge sections allowed rapid northward motion (up to 18 cm/yr) of India during this period, opening the Madagascar Basin and the sea floor south of Sri Lanka. At 66 million years the Reunion hotspot burst into activity with rapid eruption of the Deccan flood basalts, covering much of western India (Fig. 4). Probably in concert with this event the spreading ridge separating India from Madagascar jumped to the north and severed the Seychelles bank from a position adjoining Pakistan. This new spreading axis then began to open the Arabian Sea (Fig. 5) and continues today as the Carlsberg ridge portion of the Central Indian ridge. To the east, plate separation between Australia and Antarctica began to accelerate at about 55 million years.

D. 40 Million Years to Present

From the initial breakup of Gondwana, Antarctica has moved very little over the mantle. This means that all spreading ridges separating it from neighboring plates have had to move slowly to the north as new ocean floor has been added. At about 40 million years the Central Indian and the Southeast

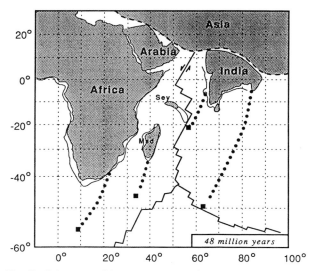

Fig. 6 Paleogeographic reconstruction of the Indian Ocean for 48 million years ago, in the hotspot reference frame. Heavy dotted paths describe the tracks of hotspots.

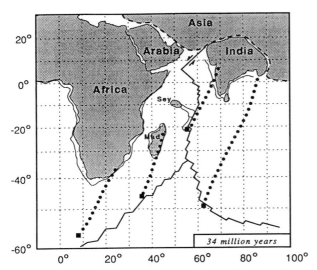

Fig. 7 Paleogeographic reconstruction of the Indian Ocean for 34 million years ago, in the hotspot reference frame. Migration of the spreading ridge system over the hotspots severed the tracks on the Indian plate at this time. Heavy dotted paths describe the tracks of hotspots.

Indian ridges migrated northward over the Reunion and Kerguelen hotspots. This interrupted the previously continuous trail of volcanism left by these two thermal anomalies on the Indian plate. Subsequent tracks have marked the African and Antarctic plates, respectively.

North of India and Australia the Tethys Ocean closed along a series of subduction plate boundaries. Elevated fragments (e.g., plateaus, island arcs) of the Indian plate began to collide with Asia at least as early as 55 million years (Fig. 5). This is termed a *soft* collision because the Indian plate was not slowed significantly until about 45 million years, when the *hard* collision began as continental India arrived against central Asia (Fig. 6). Rather than subduct into the mantle like oceanic lithosphere, the continental lithosphere thickened by shortening, accordion-style, through thrust faulting. The Tibetan Plateau owes its elevation to crustal thickening, which accommodated about 500 km of crustal shortening. This time also coincided with a global reorganization in plate motions (for example, the bend in the Hawaiian-Emperor hotspot track) and subduction directions. [*See* PLATE TECTONICS, CRUSTAL DEFORMATION AND EARTHQUAKES.]

At about 34 million years the tracks of the hotspots on the Indian plate were cut off by migration of the spreading ridge over the hotspots (Fig. 7). Initial uplift of the Himalayan Mountains occurred in early Miocene time, or about 23 million years. The large abyssal fans of the Indus and Ganges rivers have grown since this time. Subduction of old sea floor formed in the Tethys Ocean continues northwest of Austra-

lia, along the Java Trench. Continental rocks of the Australian plate are resisting subduction from Timor eastward and *microplates* are developing in this complex plate boundary. At about 10 million years ago an east–west compressional boundary between the Indian and Australian plates (Fig. 1) started to develop, probably as a result of increasing resistance to plate convergence in the Himalayan region. Differential motion between Australia and India is also occurring along a transform fault bounding the Ninetyeast Ridge. [*See* MICROPLATES.]

The northward separation of Australia from Antarctica during this period had a profound effect on global circulation, climate, and biotic evolution. By about 25 million years deep water could move freely past the South Tasman Rise, the Drake Passage (between South America and Antarctica) opened, and the Circum-Antarctic Current was established. This created the thermal isolation of Antarctica by dividing the warm subtropical circulation from the cold subpolar circulation. The decrease in trans-latitudinal heat transport led to the development of Antarctic ice sheets in the middle Miocene (about 10 million years) and growth of sea ice. The Red Sea started opening at about 20 million years, at the same time as spreading in the Gulf of Aden. Continental rifting in eastern Africa began at about 30 million years and is developing only slowly, with probably 100 km extension and continental plate thinning. This may ultimately connect southward with the Southwest Indian ridge to become a more active spreading boundary in the future, or it may fail to go much farther. [*See* PALEOCLIMATIC EFFECTS OF CONTINENTAL COLLISION AND GLACIATION.]

IV. The Hotspot Reference Frame

The global constellation of persistent, long-lived, stationary sites of excessive mantle melting, called hotspots, forms a convenient frame of reference for plate motions, through the orientations and age distributions of volcanic trails left by these melting anomalies. Hotspots appear to mark the tops of narrow plumes of upwardly convecting, warm mantle material. It is likely that these mantle plumes originate deep in the mantle (perhaps 3000 km down, at the core–mantle boundary) and are not much deflected by horizontal flow of the upper mantle. [*See* CORE–MANTLE SYSTEM; MANTLE.]

In the Indian Ocean, two major hotspots have been active during much of the opening of this basin. The island of Reunion is the present manifestation of the hotspot that earlier produced the island of Mauritius, the volcanic ridge underlying much of the Mascarene Plateau, the Chagos Bank, the Maldive and Laccadive Islands, and the massive accumulation of the Deccan flood basalts, western India. This volcanic trail is parallel with the Ninetyeast Ridge, a remarkably linear, submarine volcanic chain linked to hotspot activity

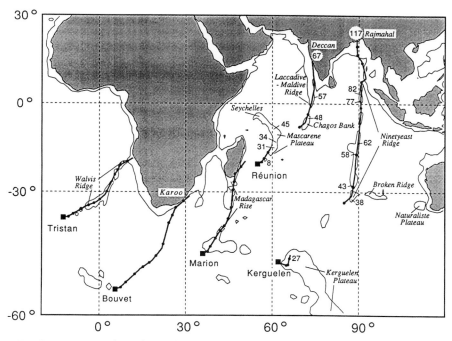

Fig. 8 Hotspot tracks in the south Atlantic and Indian Ocean basins. Solid squares show current locations of hotspots and heavy lines are computer-generated tracks that should mark the volcanic trails left on passing plates if these hotspots have been stationary. (Dots are 10 million year increments in the predicted age of volcanic activity.) Locations of elevated features related to hotspots are indicated and the age of volcanic activity at sampled sites is shown (numbers are millions of years). The Karoo, Deccan, and Rajmahal are flood basalt provinces that mark the catastrophic beginnings of hotspot activity.

now centered near the Kerguelen Islands, on the Antarctic plate (Fig. 8).

Through a combination of surface sampling and deep sea drilling these volcanic lineaments have been studied in detail and provide an unusually complete record of the activity of mantle plumes, from their initiation in catastrophic flood basalts events, through maturity in continuous volcanic ridges, into decline as discrete islands marking intermittent volcanic activity. The Reunion hotspot burst to life with eruption of as much as 1.5 million cubic kilometers of basaltic lava flows throughout a broad region of the Indian subcontinent at 66 million years. Radiometric dating and magnetic stratigraphy have established that the eruptions occurred in less than one million years. Such a phenomenal eruption rate is an order of magnitude greater than today's most vigorous volcanic systems (e.g., Hawaii). Following this flood basalt initiation event the eruption rate fell by a factor of one hundred.

As India drifted northward over the hotspot between 66 and 40 million years, a north-to-south volcanic trace was generated that is now the Laccadive–Maldive–Chagos ridge.

At about 36 million years the Central Indian spreading ridge crossed over the hotspot, and subsequent hotspot activity was recorded only on the African plate. Thus, between 36 million years and the present, northeastward motion of Africa over the hotspot has formed the Mascarene Plateau and the islands of Mauritius (starting 7 million years ago) and Renion (starting 2 million years ago). A large seamount to the west of Reunion may be the future site of the next island in this chain. The future site of this hotspot activity should be west to southwest of Reunion Island. [*See* OCEANIC ISLANDS, ATOLLS, AND SEAMOUNTS.]

In the eastern Indian Ocean the Kerguelen hotspot has left a similarly complete record of the direction and rate of plate motions. This hotspot also began with a flood basalt event, at about 117 million years, which formed a huge oceanic plateau between India, Australia, and Antarctica. This plateau was subsequently broken up by later spreading ridges, into the Kerguelen Plateau, Naturaliste Plateau, and the Rajmahal Traps, eastern India. The products of hotspot activity between 117 and 100 million years are not well identified, but possibly include portions of the Ninetyeast Ridge buried under the Bengal fan sediments. The Broken Ridge was built at 88 million years and the Ninetyeast Ridge marks the trail of the Indian plate over the Kerguelen hotspot from about 100 to 40 million years. The southern end of this 500-km-long volcanic ridge was terminated by the Southeast Indian spreading ridge migrating across the hotspot. From 40 million years until the present the hotspot has built up the Kerguelen Islands on the slow-moving Antarctic plate.

A. Stationary Melting Anomalies

The idea that hotspots are stationary volcanic sources over periods as long as 100 million years is supported by a simple graphical test. The African plate is first prescribed to follow hotspot tracks in the South Atlantic basin (e.g., Walvis Ridge, Madagascar Rise). The motions of the Indian, Australian, and Antarctic plates relative to Africa are then added from the known spreading history preserved on the Indian Ocean sea floor. The resulting calculated motions for the plates surrounding the Indian Ocean should match the orientations and age distribution of sites along the Reunion and Kerguelen hotspot tracks—if there has been no significant interhotspot drift. Figure 8 shows such calculated hotspot tracks plotted over the elevated ridges and plateaus of the Indian Ocean. These compare well with the actual lineaments and sample site ages and demonstrate that Atlantic and Indian Ocean hotspots have remained fixed with respect to one another for this period. Hence backtracking plates along hotspot tracks is a simple yet reliable method for reconstructing plate positions with respect to the mantle reference frame. Figures 2–7 show a series of reconstructions of the Indian Ocean in the hotspot reference frame.

B. True Polar Wander

Finally, plate motions can also be described relative to the earth's spin axis, in the paleomagnetic reference frame. Differential motion between the mantle and the spin axis can then, in principle, be determined by comparing plate motions recorded by the hotspot and paleomagnetic reference frames. Such motion has been termed true polar wander (distinguished from the apparent wander of the magnetic axis inferred from time sequences of paleomagnetic directions recorded in rocks from a given plate). True polar wander could occur through mass redistribution (e.g., mantle convection and plate motions) sufficient to shift the entire figure of the earth relative to its spin axis.

In the absence of true polar wander, mantle plumes would not move with respect to the spin axis. Hence, every volcano generated along a given hotspot track would record the same magnetic inclination—that of the site of hotspot activity. If such paleolatitudes are not constant, it follows that the man-

tle reference frame moves with respect to the spin axis, which is true polar wander. Measurements from the two Indian Ocean hotspot tracks show that the hotspots moved about 5°, or 500 km, to the north at about 40 million years. This coincides with a southerly motion of Pacific hotspots (i.e., Hawaii) of similar magnitude. While this is a small amount, the indications are consistent with a rotation of the entire earth at about the same time as the hard collision of India with Asia. [See POLAR MOTION AND EARTH ROTATION.]

Bibliography

Duncan, R. A. (1981). Hotspots in the southern oceans—an absolute frame of reference for motion of the Gondwana continents. *Tectonophysics* **74,** 29–42.

Morgan, W. J. (1981). Hotspot tracks and the opening of the Atlantic and Indian Oceans. *In* "The Sea," Vol. 7 (C. Emiliani, ed.), pp. 443–475. Wiley (Interscience), New York.

Norton, I. O., and Sclater, J. G. (1979). A model for the evolution of the Indian Ocean and the breakup of Gondwanaland. *J. Geophys. Res.* **84,** 6803–6830.

Richards, M. A., Duncan, R. A., and Courtillot, V. E. (1989). Flood basalts and hotspot tracks: Plume heads and tails. *Science* **246,** 103–107.

Glossary

Absolute plate motions Direction and velocity of plate movements inferred from volcanic lineaments related to hotspot activity or paleomagnetic measurements.

Hotspots Stationary, long-lived thermal anomalies in the upper mantle that form volcanic trails on passing plates.

Magnetostratigraphic time scale Ages have been assigned by radiometric dating to each of the reversals of the earth's magnetic field, in sequence, for the last 160 million years.

Mantle plumes Upwardly convecting, narrow columns of the mantle that supply hotspots with warmer, deep-mantle material.

Microplates Small pieces of lithosphere that have approximately equal horizontal and vertical dimensions.

Plates Large, rigid, pieces of the earth's outermost shell that are in horizontal motion and separate, collide, and slide against one another.

Relative plate motions Direction and speed of horizontal displacements of plates, inferred from sea-floor magnetic anomalies and transform fault orientations.

True polar wander Rotation of the whole earth with respect to its spin axis, due to a change in the principal moments of inertia.

Infrared and Raman Spectroscopy in the Geosciences

P. F. McMillan

Arizona State University

Infrared and Raman spectroscopies are the two major techniques used to study the vibrational properties of materials. Vibrational spectroscopy is applied in a variety of ways in the earth sciences; as a versatile analytical tool, in detailed studies of structure and bonding of minerals and other earth materials, and as a technique for remote sensing of atmospheres and planetary surfaces.

I. Introduction

A. Vibrations of Molecules and Solids

The atoms in a molecule or condensed phase are held together by chemical bonds, which determine the equilibrium structure. If the atoms are displaced from their equilibrium positions, the chemical forces tend to restore the structure to its original state. At any temperature, the atoms are in constant motion within the molecule, undergoing slight excursions from their equilibrium state. The interatomic force field determines that these excursions take the form of small periodic displacements about the equilibrium position of each atom, known as the normal modes of vibration of the molecule or material. Because each atom can move in three independent directions in space, there are $3N - 6$ normal modes of vibration for a system with N atoms. The remaining 6 degrees of freedom correspond to translation (3) and rotation (3) of the system as a whole. Each normal mode is

associated with a characteristic vibrational frequency, determined by the strength of the interatomic forces, the masses of the atoms involved, and their relative positions in the molecule. In a crystalline solid, the atoms are arranged in a regular periodic array, and the vibrational modes take the form of vibrational waves traveling through the crystal. These are characterized not only by the oscillation frequency of the atoms about their equilibrium positions but also by the wavelength of the lattice vibrational wave. For both molecules and crystals, the number and type of vibrational modes observed in infrared or Raman spectroscopy are determined by the symmetry of the molecule or crystal.

B. Infrared Spectroscopy

Vibrational frequencies lie in the range $10^{12}–10^{14}$ s^{-1}, which corresponds to light in the infrared region of the electromagnetic spectrum (vibrational frequencies are often expressed in wave number units, with 1 $cm^{-1} \equiv 2.998 \times 10^{10}$ s^{-1}). In infrared spectroscopy, a beam of infrared light is passed through the sample, and the absorption spectrum gives the vibrational frequencies of the material. For solid or liquid samples, the infrared beam can be reflected from the surface to give an infrared reflectance spectrum. Materials at a given temperature also emit infrared radiation, with an emission spectrum which reflects the thermally excited vibrations of the material.

C. Raman Spectroscopy

This technique is named after Sir C. V. Raman, one of its codiscoverers (C. V. Raman and K. S. Krishan in Calcutta and G. Landsberg and L. Mandelstam in Moscow, both in 1928). An intense beam of light in the visible region of the spectrum is passed through the material, and small frequency shifts due to interaction with vibrational modes of the sample are observed. In earlier work, sunlight or light from a mercury lamp was used as the source of the visible radiation; now lasers are used to provide the incident beam. Raman scattering is complementary to infrared spectroscopy for studying molecular vibrations. In general, not all vibrational modes of a given sample are active in the infrared or the Raman spectrum (this is determined by the molecular symmetry), and obtaining both gives a more complete record of the vibrational spectrum.

II. Applications to Mineralogy

A. Phase Identification and Analysis

Because the vibrational spectrum of a substance is determined by the details of its structure and bonding, each material has a characteristic vibrational spectrum, which can be used as a "fingerprint" for identification of that phase in an unknown sample. This type of qualitative analysis is complementary to other laboratory techniques in common use, especially x-ray diffraction, but vibrational spectroscopy can have certain advantages over these. For example, some minerals, such as the Al_2SiO_5 polymorphs andalusite and sillimanite, or the chain (α) and ring (β) polymorphs of $CaSiO_3$ wollastonite, have almost identical x-ray patterns but very different infrared and Raman spectra. In addition, the vibrational spectra are often extremely sensitive to changes in structural order, even on a distance scale of several tens of angstrom units (10^{-10} m), and can be used to distinguish minerals in slightly different disordered states. Because vibrational spectroscopy does not depend on the existence of long-range structural order, it is also extremely useful for studying highly disordered systems such as glasses, liquids, and dense fluids. Finally, x-ray diffraction is relatively insensitive to light elements such as hydrogen, lithium, or boron, which are important in the earth sciences. Vibrations involving these atoms can usually be easily observed in infrared or Raman spectra. [See POWDER DIFFRACTION TECHNIQUES.]

With suitable calibration, both infrared and Raman spectroscopic analyses can be made highly quantitative. In general, the absolute intensity of a particular infrared- or Raman-active vibrational mode cannot be calculated from first principles with any degree of certainty, so empirical correlations based on series of standards containing known concentrations of the species or phase of interest must be established. Quantitative infrared and Raman studies which have had a major impact in the earth sciences include the determination of total water content and H_2O/OH speciation in silicate glasses via infrared absorption spectroscopy and the *in situ* nondestructive analysis of fluid inclusions by Raman spectroscopy.

B. Structural Studies

Most of the vibrational spectroscopic studies to date in mineralogy have been carried out to investigate some structural feature of the mineral in question. This generally requires an interpretation of the vibrational spectra observed in terms of a structural model, hence a correlation between the dynamic (vibrational) structure and the static structure of the mineral concerned. This is often obtained by carrying out a vibrational calculation, constrained by the experimental spectra, which gives the atomic displacements corresponding to each vibrational mode. These calculations can be extremely useful, but the results must be treated with some care because they can be highly model dependent. Major progress is being made in determining vibrational properties from first principles via solid-state *ab initio* calculations.

Useful structural information can be gained from partial interpretation of the vibrational spectra. For example, some types of vibration fall in characteristic frequency ranges, such as O—H stretching (normally 3000–3700 cm^{-1}) or Si—O stretching (800–1200 cm^{-1}) vibrations. The Si—O stretching frequency depends on the polymerization state of the tetrahedral SiO_4 groups present (i.e., the number of SiOSi linkages per tetrahedron), or the coordination state (tetrahedral SiO_4 or octahedral SiO_6) of the silicate unit, so infrared and Raman spectroscopy can be a useful complement to other methods of structure determination. The O—H stretching frequency is highly dependent on the immediate OH environment and the presence or absence of hydrogen bonding. Vibrational studies can be invaluable in helping locate the hydrogen atoms in OH-containing minerals. Both infrared and Raman spectroscopy can be carried out with plane-polarized light, and the resulting polarized spectra as a function of crystal orientation give valuable information on the symmetry of vibrational modes, hence the symmetry of the structural groups involved (for example, distortion of CO_3^{2-} groups in carbonate minerals from trigonal planar symmetry), and their orientation (for example, the orientation of H_2O molecules within channels in natural cordierites).

C. Thermodynamic Studies

Most rock-forming minerals have vibrational frequencies in the range 50–1500 cm^{-1}, which corresponds to vibrational energies of approximately 0.5–15 kJ/mol or to thermodynamic temperatures up to 2000 K (1727°C). This means that over the temperature range of the earth's crust and much of the mantle, changes in internal energy of minerals with temperature are mainly due to excitation of vibrational modes. It follows that the heat capacity of minerals over the temperature range of the crust and mantle can be calculated from a knowledge of their vibrational spectra. Methods for carrying out these calculations have been developed and applied to a wide range of rock-forming minerals. Despite some ambiguities in the heat capacity calculations, due mainly to the lack of complete vibrational data, they form a useful complement to experimental determinations of mineral heat capacities. This is especially true for phases stable only at high pressure such as $MgSiO_3$ garnet, ilmenite, and perovskite, which are probably major constituents of the earth's mantle and for which direct determinations of high-temperature heat capacities are difficult to carry out in the laboratory. Heat capacity calculations of this type have been useful in constraining high pressure–high temperature phase diagrams for minerals in the system (Mg,Fe)O-SiO$_2$, resulting in a better understanding of the likely mineralogy of the earth's mantle. [*See* MANTLE.]

D. Phase Transitions

A major fraction of vibrational spectroscopic studies in mineralogy has been devoted to *in situ* studies of temperature- and pressure-induced phase transitions. One of the classical studies was carried out by Raman and T. M. K. Nedungadi in 1940, when they followed the Raman spectrum of quartz as a function of temperature through the transition between the α and β forms at 573°C. They observed that one vibrational mode, at 207 cm^{-1} at room temperature, decreased markedly in frequency with increasing temperature and disappeared at the α–β phase transition temperature. This type of behavior is known as *soft mode behavior*. A "soft" vibrational mode is one whose atomic displacement pattern mimics the structural changes occurring during the phase transition, so that exciting the vibrational mode "drives" the phase change. Many such thermal phase transitions driven by soft modes have now been studied by infrared and Raman spectroscopy, including a number relevant to mineralogy, such as the α–β quartz transition and phase transitions in minerals with the rutile and perovskite structures.

With the advent of the diamond anvil cell coupled with infrared and Raman spectroscopic techniques, a growing number of pressure-induced soft mode phase transitions have been studied. In the diamond cell, the sample is compressed between two opposed diamond planar faces. Because of the small surface area of the diamond faces, enormous pressures (in the kilobar to megabar range) can be generated by application of even modest forces to the back plates of the diamonds. Diamond is an ideal material for this type of cell because of its great strength, enough to withstand the large strains generated in the anvil. In addition, it is transparent to much of the visible and infrared spectrum, permitting a wide range of spectroscopic studies to be carried out *in situ* at high pressure. Some studies are already being carried out under combined high pressure and high temperature conditions within the diamond anvil cell, and soon it will be possible to investigate a wide range of phase transitions under the pressure–temperature conditions of the deep earth. Such studies will be important for understanding both the detailed mineralogy of the mantle and core and the thermal budget of the earth, since displacive phase transitions at high pressure and high temperature in minerals such as (Mg,Fe)SiO$_3$ perovskite and SiO$_2$ stishovite could be important in transporting energy within the earth.

There is currently also much interest in the structures and vibrational properties of molecular gases such as hydrogen, nitrogen, and methane solidified in the diamond cell. These gases form the major constituents of the giant outer planets,

and this work is particularly important for gaining an understanding of the structure of the planetary interiors. Intense interest has focused on the possibility of metallization in nitrogen, oxygen, and even hydrogen at high pressures, associated with a transition from the molecular solid to an atomic solid state. In this case the deep interiors of the Jovian planets could be composed of metallic condensed gases, giving rise to a planetary magnetic field. [*See* CORE–MANTLE SYSTEM.]

III. Applications to Geochemistry

A. Structure of Silicate Glasses and Melts

Vibrational spectroscopy has been used extensively to study the structures of silicate glasses and melts in an effort to understand the nature and properties of natural magmas. In most of these studies, systematic changes in the spectra with composition (or some other variable) are noted, and the glass spectra are interpreted by comparison with the spectra of corresponding crystals or by carrying out vibrational calculations. For simple alkali and alkaline earth silicate glass series, the variations of the spectra with composition and temperature have given valuable insights into the number and distribution of different silicate polymer species, which have been correlated with variations in the melt and glass viscosity, density, and immiscibility behavior. The spectra of aluminosilicate glasses have been much more complex to interpret, but a better understanding of the structural role of aluminum in aluminosilicate glasses and melts is beginning to emerge. These studies have been particularly powerful when coupled with other structural techniques, such as nuclear magnetic resonance spectroscopy, Mössbauer spectroscopy, or x-ray absorption spectroscopy and diffraction. Of considerable interest are recent studies of melts *in situ* at high temperatures (which give insight into not only structural properties but also dynamic transport and relaxation processes) and at high pressures in the diamond anvil cell. It is already possible to carry out vibrational and other spectroscopic studies on silicate melts under the pressure and temperature conditions of magma generation in the earth's crust and mantle. These studies are certain to lead to a better understanding of melt processes and crystallization phenomena within the earth and terrestrial planets in terms of the physics and structural chemistry of the melt phase. [*See* MAGMA CHAMBER DYNAMICS; NMR SPECTROSCOPY IN THE EARTH SCIENCES; SILICATES.]

B. Dissolution of Water and Carbon Dioxide in Molten Silicates

Water and carbon dioxide are the two most important volatile species which dissolve in natural aluminosilicate melts, in terms of both their abundance and their effect on melt behavior. Both infrared and Raman spectroscopies have been used to study the dissolution mechanisms of these volatile species. Quantitative infrared absorption studies have shown that water dissolves in silicate glasses and melts as both molecular H_2O and bound hydroxyl (OH) species, and the relative proportions of these species vary with temperature, pressure, and total water content. The OH/H_2O speciation reaction can be correlated with the effect of dissolved water on melt properties such as viscosity and electrical conductivity and aids in understanding the relationship between hydrated aluminosilicate melts and aqueous fluids containing dissolved aluminate and silicate species. Infrared and Raman studies have shown that carbon dioxide also dissolves in silicate melts both as molecular CO_2 and via reaction with the silicate species to form carbonate (CO_3^{2-}) groups. These investigations have led to an understanding at the molecular level of how dissolution of carbon dioxide should increase the viscosity of silicate melts at depth, and they are beginning to yield insights into the mechanisms for separation of carbonate from silicate magmas at high pressure within the earth.

C. Speciation in Aqueous Fluids

Water is the most abundant volatile species in the earth system, and an area of intense interest in geochemistry concerns the structure of aqueous fluids and liquids and the speciation of metals and gases dissolved in them. Because the infrared absorption of water itself is so strong, it is difficult to carry out infrared spectroscopic studies on species dissolved in water, and Raman spectroscopy has been the technique of choice in most studies of this type. The structure of water and the interpretation of its vibrational spectrum remain topics of current debate. Some features of the spectra can be understood in terms of the gas-phase molecule H_2O, but others cannot, and the roles of hydrogen bonding and rapid proton exchange are subjects of continuing interest and discussion. The results of Raman spectroscopic experiments with increasing temperature have been interpreted as indicating a reduction in the number of hydrogen bonds. At low temperatures, application of pressure appears to have little effect on either the Raman spectrum or the structure, but at temperatures in the supercritical regime increasing pressure further decreases the importance of hydrogen-bonded structures and water begins to behave as a normal polar fluid. There have been several studies of the effects of dissolved gases and electrolytes on the Raman spectrum of water, and Raman spectroscopy has been used to investigate the structures and equilibria of species dissolved in water as a function of pressure and temperature, but much work remains before a systematic pattern of the behavior of geochemically important species can begin to emerge. [See CRUSTAL FLUIDS; WATER, PHYSICAL PROPERTIES AND STRUCTURE.]

D. Studies of Fluid Inclusions

Fluid inclusions are microscopic bubbles of gas or liquid trapped during formation of a mineral or glassy material, and

they provide a record of the fluid phases with which the mineral was in contact within the earth. Many techniques have been developed for qualitative and quantitative analysis of the fluid and any dissolved species, including infrared spectroscopy of bulk samples, but one of the most successful has been micro-Raman spectroscopy. Because the incident light beam for Raman spectroscopy is in the visible region of the spectrum, it can easily be focused to a small spot by an ordinary microscope. In the Raman microprobe, developed independently by P. Dhamelincourt in France and G. Rosasco in the United States, the laser light is focused to a small spot $(1-5 \mu m)$ by a microscope objective and the Raman scattered light is also collected by the same objective before being sent into the spectrometer. The result is that a Raman spectrum can be obtained from a region as small as $1 \mu m$ in diameter at the focus of the incident and scattered beams. If the sample containing the inclusions is transparent to visible radiation, it is easy to focus the beam inside micrometer-sized inclusions and obtain the characteristic spectrum of any molecular species inside. This technique provides a rapid, *in situ*, nondestructive means of obtaining qualitative and quantitative analyses of the major species in natural fluid inclusions, and it has had a major influence on our understanding of the nature of fluids in the earth's crust and mantle. In addition, micro-Raman spectroscopy combined with thermodynamic observations of fluid inclusions in minerals from ore deposits has led to a better understanding of the role of fluids in deposition and transport processes in ore-forming bodies. Micro-infrared instruments are now also being developed, using reflection (Cassegrain) objectives to focus the infrared radiation, and are beginning to provide information complementary to results of micro-Raman studies on fluid inclusions. [See ORE FORMATION.]

IV. Other Applications in Earth and Planetary Science

A. Infrared Emission Spectroscopy: Remote Sensing of Planetary Surfaces

In infrared absorption or reflection spectroscopy, the infrared beam incident on the sample is generated by a source and then passed through a spectrometer or interferometer to the detector. In infrared emission spectroscopy, the sample itself acts as the source, and the thermal emission spectrum in the infrared region is characteristic of the vibrational spectrum and temperature of the material. This technique is useful in the laboratory for studying samples which are opaque to infrared radiation and hence not amenable to absorption studies, such as materials adsorbed on metals and samples with rough, irregular surfaces with low reflectivity. This is the case of soils, and infrared emission studies in the laboratory have been useful in determining the mineral constituents and adsorbate species in natural soil samples. In addition, aircraft-mounted infrared emission spectrometers have been used both to measure thermal profiles of terrains and to identify the water content and mineralogy of soils from their characteristic vibrational signatures. Along with other teledetection techniques, airborne infrared emission spectroscopy is a valuable tool in terrestrial prospecting and resource evaluation. In an exciting extension of this technique, infrared emission spectrometers mounted on spacecraft in the near future will provide valuable new information about the surface mineralogy of the moon, Mars, and perhaps other terrestrial planets. [See REMOTE SENSING OF EARTH RESOURCES FROM AIR AND SPACE.]

B. Raman Lidar: Remote Sensing of Atmospheric Gases

Lidar stands for light detection and ranging and refers to a family of laser-based remote sensing techniques which involve sending a powerful laser beam into the atmosphere and collecting and analyzing the backscattered signal. In Raman lidar techniques, the characteristic vibrational or rotational Raman lines of gas-phase molecules in the atmosphere are monitored to provide a three-dimensional map of their concentration profiles. In addition, the relative intensities of the rotational Raman lines of O_2 and N_2 molecules give a useful measure of the temperature profile in the atmosphere. Vibrational Raman lidar is particularly useful for monitoring SO_2, CO_2 and water vapor and has been used to study atmospheric pollution by power plants and aircraft and automobile emissions. The technique has begun to be used to measure the concentrations and spatial distribution of gases in volcanic gas plumes. Two of the major advantages of Raman lidar are that it can be carried out remotely throughout all phases of an eruption, eliminating the need to fly aircraft through the plume, and that several gas species can be monitored simultaneously, which is not possible with other remote sensing methods in current use. [See REMOTE SENSING OF THE UPPER ATMOSPHERE, INSTRUMENTATION AND OBSERVATIONAL TECHNIQUES.]

Bibliography

Calas, G., ed. (1986). "Méthodes Spectroscopiques Appliquées aux Minéraux." Société française de Minéralogie et Cristallographie, Paris.

Farmer, V. C., ed. (1974). "The Infrared Spectra of Minerals." Mineralogical Society, London.

Hawthorne, F., ed. (1988). "Spectroscopic Methods in Mineralogy and Geology." Mineralogical Society of America, Washington, D.C.

Hemley, R. J., Bell, P. M., and Mao, H. K. (1987). Laser techniques in high-pressure geophysics. *Science* **237**, 605.

Kieffer, S. W., and Navrotsky, A., eds. (1985). "Microscopic to Macroscopic: Atomic Environments to Thermodynamic Properties." Mineralogical Society of America, Washington, D.C.

McMillan, P. (1988). Raman spectroscopy in mineralogy and geochemistry. *Ann. Rev. Earth Planet. Sci.* **17**, 255.

Radziemski, L. J., Solarz, R. W., and Paisner, J. A., eds. (1987). "Laser Spectroscopy and Its Applications." Marcel Dekker, New York.

Scott, J. F. (1974). Soft-mode spectroscopy: experimental studies of structural phase transitions. *Rev. Mod. Phys.* **46**, 83.

Glossary

Fluid inclusion Bubble of liquid or gas trapped inside a solid mineral-phase, usually formed during crystallization of the mineral in the presence of a fluid phase.

Infrared emission Light in the infrared region of the spectrum emitted by a heated sample.

Infrared radiation Light with wavelength between approximately 0.001 and 1 μm, in the infrared region of the spectrum.

Infrared spectroscopy Interaction of infrared radiation with a sample, usually to excite vibrational modes of the material.

Normal modes of variation Natural resonances of the coupled periodic displacements of atoms in a material about their equilibrium positions.

Raman lidar *Light* detection *and* ranging technique, based on the Raman effect.

Raman scattering Scattering of visible light accompanied by a shift in wavelength due to interaction with molecular vibrations.

Raman spectroscopy Analysis of Raman scattered light to give information on vibrational modes of the sample.

Soft mode Vibrational mode for which the atomic displacements track the structural changes followed during a displacive phase transition. The frequency of the soft mode goes to zero at the transition temperature or pressure.

Ionosphere

Kenneth Davies

*National Oceanic and Atmospheric Administration**

The ionosphere is of interest both as a medium that affects radio waves and as a plasma that absorbs solar ionizing radiations. The ionized component of the upper atmosphere, between heights of approximately 50 and 2000 km, is closely coupled to the neutral upper atmosphere and to the magnetosphere. The ionosphere is affected by a variety of natural and anthropogenic disturbances at or near the ground.

The ionization varies by orders of magnitude depending on geographic and/or geomagnetic location, time of day, season, and solar activity (e.g., flares, sunspot cycle). It is this variability that makes the ionosphere interesting as a plasma laboratory and gives rise to problems for users of ionospheric radio systems.

I. Ionospheric Regions

A. Production and Loss of Ionization

The upper atmosphere is divided into the D, E, and F regions, above which lies the protonosphere or plasmasphere. The letters E and F were used by Sir Edward Appleton to denote the electric fields of the waves reflected from the lower layer and upper layer, respectively (Fig. 1). It happens that the various regions can be distinguished by different ion compositions and the ion productions in each region result from different solar wavelengths.

1. Production of Ions

The production of ionization by a broad spectrum of solar radiation absorbed in a multiconstituent atmosphere is rather

* Retired.

complicated. Qualitatively we see that, at the top of the atmosphere, the gas is so tenuous that little energy is absorbed. Low in the atmosphere, where the density is high, most of the energy has already been absorbed at higher levels. Hence, there is a maximum of energy absorption (and hence a maximum of ion production) at some intermediate height. [*See* SOLAR RADIATION.]

The ion production rate at a height h in a single-constituent, isothermal atmosphere for solar zenith angle χ is given by

$$q(z, \chi) = q(0) \exp[1 - z - \sec \chi \exp(-z)] \qquad (1)$$

where $q(0)$ is the maximum rate of ion production when the sun is overhead ($\chi = 0$) and the reduced altitude z is measured from the height (h_0) at which the optical depth is unity, i.e.,

$$z = (h - h_0)/H \qquad (2)$$

where H is the pressure scale height. The total production rate is a summation over the different wavelength bands and is, therefore, best calculated by a computer. The function $q(z, \chi)$ is often called the Chapman function.

2. Loss of Ionization

The electron density N_e is a balance between the three processes of production (q), loss (L), and movement (M). Hence, the equation of continuity is

$$\frac{dN_e}{dt} = q - L - M \qquad (3)$$

The loss is primarily by dissociative recombinations of a molecular ion (XY^+):

$$XY^+ + e \rightarrow X + Y \qquad (4)$$

with a recombination coefficient α so that $L = \alpha N_e^2$. Under the assumption of quasiequilibrium, $dN_e/dt \approx 0$, and no movement, the electron density profile is

$$N_e(h, \chi) = N(0) \exp \tfrac{1}{2}[1 - z - \sec \chi \exp(-z)] \qquad (5)$$

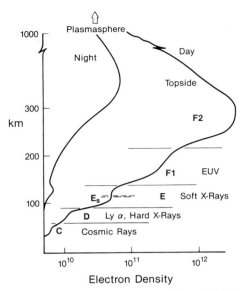

Fig. 1 Ionosphere structures on a summer day and night in middle latitudes and the main bands of solar and cosmic ionizing radiations.

The maximum electron density, N_{max}, depends on χ as follows:

$$N_{max} = N_{max}(0) \cos^{1/2} \chi \tag{6}$$

and the height (h_m) of maximum electron density depends on χ as

$$h_m = h_0 + H \ln \sec \chi \tag{7}$$

where h_0 is the height of maximum density when the sun is overhead. Although the foregoing equations do not hold in detail for the real ionosphere, they have provided an invaluable guide to the organization and representation of data for the D, E, and F layers. The maximum electron (el.) density is often represented by the maximum plasma (or critical) frequency f_c (in megahertz), where

$$N_{max} = 1.24 \times 10^{10} f_c^2 \text{ (el. m}^{-3}) \tag{8}$$

B. Region Formation

1. The D Region

In the D region ($\approx 50-90$ km) ionization is largely due to x rays with wavelengths between about 0.1 and 1.0 nm, cosmic rays, and Lyman α radiation (121.6 nm). The Lyman α penetrates to the D region because the overlying atmosphere is transparent at this wavelength. X rays ionize O_2 and N_2, and Lyman α ionizes the trace constituent NO and other gases with low ionization potentials. During solar disturbances x rays are major sources of ionization of O_2 and N_2. Some ionization is produced by solar ultraviolet ionization of excited molecular O_2 ($^1\Delta_g$). Below about 70 km the C layer is produced by galactic cosmic rays. Since these charged particles are affected by the earth's magnetic field, there is a geomagnetic control of the D region.

The loss processes in the ionosphere, and especially in the D region, can proceed via complicated chains of ion-atom interchange reactions. Thus positive ions are lost either by dissociative recombination with electrons ($\alpha_D \approx 10^{-12}$ m^3 s^{-1}) or by neutralization with negative ions ($\alpha_i \approx 10^{-12}$ m^3 s^{-1}) formed by electron attachment to a neutral molecule (e.g., O_2^-). Negative ions can also be destroyed by photodetachment by visible light. The ratio (λ) of negative ion density to positive ion density (N^-/N^+) is normally less than 1 below about 75 km during the day. At night this ratio increases because photodetachment is not effective. In the presence of negative ions the rate of increase of electron density is

$$\frac{dN_e}{dt} = \frac{q}{1+\lambda} - (\alpha_D + \lambda\alpha_i)N_e^2 \tag{9}$$

Positive and negative ions also play important roles in the D region. Below about 82 km the dominant positive ions are water clusters of the type $H^+(H_2O)_n$. These ions produce a large increase in the effective electron loss rates, and this leads to a "ledge" near 82 km at which the electron density increases rapidly with altitude.

2. The E Region

The E region ($\approx 90-140$ km) is formed by radiations in the range 10–3.1 nm and radiations with $\lambda > 80$ nm. The Lyman β radiation (102.5 nm) and C(III) (97.7 nm) are of special importance because they both ionize O_2, while the Lyman continuum is responsible for some ionization of atomic oxygen. A source of ionization at night is the night sky radiations of Lyman α, Lyman β, and helium II (30.4 nm).

In the E region there are few, if any, negative ions and the dominant electron loss process is dissociative recombination with O_2^+, e.g.,

$$O_2^+ + e \rightarrow O + O \tag{10}$$

Thus the peak of the E region follows approximately Eqs. (6) and (7). In the E region $\alpha_D \approx 10^{-13}$ m^3 s^{-1} and $N_e \approx 5 \times 10^{11}$ m^{-3}, so that the time constant τ [$=(2\alpha_D N_e)^{-1}$] ≈ 10 s and the electron density follows closely the changes in production rate.

3. The F Region

The F region is ionized by radiations in the wavelength range 10–80 nm, which ionize O, O_2, and N_2.

In the F_1 region (≈ 140–220 km) atomic oxygen is important. Direct radiative recombination between an electron and an atomic ion is very slow. Hence, electrons are lost via a two-stage process in which the positive oxygen ion is lost by ion–molecule interchange followed by dissociative recombination, for example,

$$O^+ + N_2 \rightarrow NO^+ + N \tag{11a}$$

$$NO^+ + e \rightarrow N^* + O^* \tag{11b}$$

in which the N and O atoms are left in excited states, denoted by asterisks. Because the molecular density decreases with height, the effective loss rate will decrease with height.

There are no peaks in absorption of solar radiations in the F_2 region. The F_2 region (> 220 km) peak is caused by plasma transport rather than production and loss. Transport includes, for example, diffusion, winds in the neutral atmosphere, and electromagnetic drift.

The vertical diffusion speed ω is given by

$$\omega = -D(h) \sin^2 I \left(\frac{1}{N_e} \frac{dN_e}{dh} + \frac{Mg}{2kT} \right) \tag{12}$$

where N is electron density, h height, M the ion mass, g the acceleration due to gravity, k Boltzmann's constant, T temperature, I magnetic dip, and $D(h)$ the ambipolar diffusion coefficient (because the electrons and ions are electrically coupled). Plasma diffusion is maximal in the direction of the geomagnetic field, which leads to field-aligned diffusion and, hence, field-aligned irregularities. Because the diffusion coefficient increases inversely as the neutral density, diffusion increases rapidly with height. Even including diffusion, the F_2 region profile is approximately a Chapman layer. Diffusive separation in a plasma differs from that in a neutral gas because, as the light electrons try to separate from the heavy positive ions, an electric field (E) is established so that the electrons and ions move together (ambipolar diffusion). For an electron–proton gas in the F_2 region $E \approx 0.05$ μV/m, and for an electron–oxygen ion gas $E \approx 0.8$ μV/m. Because of this coupling, the mean (electron–ion) mass is half the ion (gas) mass and the plasma scale height is twice that of the neutral gas.

Neutral winds result from heating of the neutral air by solar radiation and/or energy input into the auroral zones by electron bombardment and Joule power dissipation. Normally, the air is hottest at the subsolar point, and the resulting pressure gradients drive horizontal winds from the dayside to the nightside. In the D region collisions are sufficiently numer-

ous that the plasma motion is determined by the neutral motions. In the E region the ions tend to move with the neutrals but the electrons are tied to the magnetic field so that a polarization electric field is established that can affect ion motions in the F_2 region. In the F_2 region, however, the collision frequency is relatively low and the horizontal neutral wind cannot move the ions across the earth's magnetic field. The ions adhere to the magnetic field like rings on a bar; they can slide along the field easily but cannot move across it. At a middle-latitude location, an equatorward wind lifts a plasma, whereas a poleward wind lowers the plasma. Because the loss rate decreases with height, the equatorward wind increases the electron densities, whereas the poleward wind decreases the electron densities. In low magnetic latitudes, transequatorial winds transport plasma from one hemisphere to another, producing hemispherical asymmetry in the F_2 region.

Electromagnetic ($\mathbf{E} \times \mathbf{B}$) drifts are important in the F_2 region because they are the main mechanism for moving plasma across magnetic field lines. Electric fields are particularly important near the dip equator, where vertical motions are important, and in the winter polar cap, where horizontal movement is important. In the northern hemisphere a west-to-east electric field will increase the height of the layer and, hence, the peak electron density, and the converse holds for an east-to-west electric field.

C. Plasma Physics

The ionosphere is a weakly ionized plasma whose properties can depart appreciably from those of the neutral gas in which it is embedded. For example, the free electrons can absorb, refract, and scatter radio waves, and in the presence of the earth's magnetic field the ionosphere behaves as a birefringent medium, splitting incident radio waves into ordinary and extraordinary waves with opposite senses of rotation of their electric and magnetic fields. The radio refractive index is anisotropic, so the direction of energy flow can differ from the direction of wave (front) propagation. The movement of plasma is governed by electric and magnetic fields and (through collisions) by the motions of the neutral air. Further, the energy distribution can be nonthermal owing to the presence of energetic electrons produced by photoemission. Neutral winds and electric fields produce horizontal and vertical motions of plasma that affect the global structure of the ionosphere and ensure the maintenance of the nighttime ionosphere, especially during the long polar night.

II. Spatial and Temporal Structures

A. Structure of the D Region

Electron density profiles in the day and night D region (50–90 km) are highly variable from one day to the next. On an average day electron concentration follows roughly a $\cos^n \chi$ law, where χ is the solar zenith angle and n is about unity.

The electron collision frequency in the D region is related to atmospheric gas pressure p (in newtons per square meter) by

$$\nu_m = 8 \times 10^5 p \text{ s}^{-1} \tag{13}$$

Here ν_m is the collision frequency of electrons with thermal energy kT, where $k = 1.38. \times 10^{-23}$ J/K is Boltzmann's constant and T is temperature in kelvins. The neutral composition in the D region is essentially the same as that at the ground. However, minor constituents such as NO, O_3, and O_2 $(^1\Delta_g)$ play important roles.

Within the D region are two layers, the D layer and the C layer (see Fig. 1). The former is produced by solar ultraviolet and x rays and the latter by galactic rays. In the D region between 50 and 85 km, the neutral gas temperature decreases from about 260 K to about 180 K. In the thermosphere, above 85 km, the temperature increases steadily. The neutral scale height varies between about 8 km (at 50 km) and about 6 km (near 90 km). Of course, individual daily values vary considerably about the average.

The D region has been studied by measuring the absorption of radio waves passing through it. The absorption depends on the integral of the product of the electron density N_e and the electron collision frequency ν ($\int N_e \nu \, dh$) throughout the D region. The global distribution of absorption in Fig. 2 shows, for example, that in middle latitudes the absorption is high in winter and low in summer. This is called the winter anomaly. Near the dip equator the absorption is maximal around the equinoxes. The winter anomaly shows up as groups of days of enhanced D region electron densities in the winter months. It is due to (1) storm aftereffects, whose magnitude is increased in winter, and (2) days on which there are anomalous enhancements (e.g., radio absorption) associated with meteorological phenomena, such as stratospheric warmings. The latter increases are associated with increases in nitric oxide and O_2 $(^1\Delta_g)$ levels and decreases in water cluster ions resulting from increases in temperature.

Ion composition in the D region is rather complicated. Both positive and negative ions are present, and the latter are particularly important at night. The two primary positive ions are O_2^+ and NO^+. Via a number of chain reactions these produce such ions as O_4^+, H_3O^+, and $H_3O^+(H_2O)_2$. Negative ion profiles show that above 80 km O_2^- is the dominant ion and below 80 km NO_3^- is dominant with smaller contributions ($\approx 20\%$) from O_3^-, NO_2^-, CO_3^-, and CO_4^-.

B. Structure of the E Region

1. The Normal E Region

In the E region (90–140 km) the temperature increases rapidly with height, and we pass from the region in which the gases are essentially mixed (the homosphere) to a region in which the constituents are separated by diffusion (the heterosphere). For modeling purposes the lower boundary of the heterosphere is set at 120 km. Above about 80 km there is appreciable dissociation of O_2 molecules into atomic oxygen. The level of maximum O concentration lies between 85 and 100 km, and the peak density can be of the order of 10^{19} m^{-3}.

The E region is important for geomagnetism because the currents that give rise to some geomagnetic variations — for example, the equatorial electrojet and the auroral electrojet — flow at heights between 100 and 150 km.

The ionization in the E region can be broadly divided into a regular E layer and sporadic E; ledges near the level of maximum electron density give rise to a subdivision of regular E into E_1 and E_2. The regular E layer tends to follow ion production by solar radiation, whereas sporadic E may be present at night or at any time of day. One type, middle-latitude sporadic E, is thought to be produced by the action of strong neutral wind shears on long-lived metallic ions (e.g., Na^+, Mg^+, Ca^+).

The dominant positive ions in the E region are O_2^+ and NO^+. A peak in O_2^+ density around 110–115 km is produced by the absorption of C(III) and Lyman β radiation. Nitric oxide is thought to be an important molecule in the E region because of its rapid charge interchange with O_2^+ to give NO^+.

To a first approximation, the daytime critical frequencies (in megahertz) of the E region are given by

$$f_o E = 0.9[(180 + 1.44R_{12})\cos \chi]^{0.25} \tag{14}$$

where R_{12} is the 12-month smoothed sunspot number.

During the night $f_o E$ falls to a minimum given by

$$f_o E_{min} = 0.36(1 + 0.005R_{12}) \tag{15}$$

More detailed models for the nighttime $f_o E$ have been derived by M. L. Muggleton. Global maps of $f_o E$ based on observations, for equinox, summer, and winter, for sunspot maximum (1958) have been given by M. Leftin.

2. Sporadic E

Whereas the critical frequency of the regular E layer varies little from day to day (with a standard deviation < 5%), the maximum frequency of sporadic E ($f_o E_s$) is highly variable in both space and time. The critical frequency can vary from less than 2 MHz to over 30 MHz at a given location. In middle latitudes, E_s is largely a summer day phenomenon and auroral sporadic E is essentially a nighttime phenomenon. Some types of E_s, such as that encountered near the magnetic equator, are relatively well behaved. On some occasions, E_s is opaque and blankets the overlying layers. On other occasions, the upper layers can be seen through the E_s, which indicates either that the reflection is partial or that E_s is cloudy and the

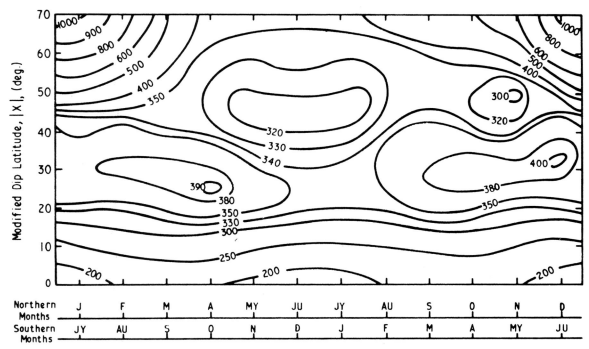

Fig. 2 Absorption in decibels of radio waves of frequency 1 MHz for an overhead sun and a smoothed $x = \tan^{-1}(I/\sqrt{\cos \phi})$ where I is the magnetic dip in radians and ϕ is geographic latitude. [From P. A. Bradley (1976). A new computer-based method of HF sky-wave signal prediction using vertical-incidence ionosonde measurements. *In* "Radio Systems and the Ionosphere," AGARD Conference Proceeding No. 173 (W. T. Blackband, ed.), p. 11. Advisory Group for Aerospace Research and Development, North Atlantic Treaty Organization.]

radio waves penetrate through gaps. Sporadic E usually occurs over the height range 90–120 km. One characteristic of sporadic E, its partial transparency, gives rise to a "blanketing" frequency $f_b E_s$ below which it is opaque and above which it is transparent.

C. Structure of the F_1 Region

In the F_1 region (140–220 km), the most important ionizable constituent is molecular nitrogen, which is ionized by solar extreme ultraviolet radiation. The neutral temperature in the F_1 region at solar maximum increases rapidly with height from around 500 K at 140 km to over 1000 K at 220 km. For ionospheric physics the F_1 region is important because the altitude at which the ion production is a maximum for an overhead sun [see Eq. (1)] is in the F_1 region, whereas the height of maximum electron density usually occurs in the F_2 layer.

To a first approximation, the critical frequency $f_o F_1$ of the F_1 region is given by

$$f_o F_1 = (4.3 + 0.01 R_{12}) \cos^{0.2} \chi \qquad (16)$$

A more comprehensive expression has also been derived by

E. D. Ducharme and co-workers. At night the F_1 region merges with the F_2 region to form a single F region.

D. Structure of the F_2 Region

The largest electron densities in the ionosphere usually occur near the peak of the F_2 region. The F_2 region lies between about 220 and 2000 km. The electron density is maximal there because the molecular density and therefore the electron loss rate decreases rapidly with height.

The dominant (positive) ion in the F_2 region is O^+, which gives way to H^+ above a transition height. The density of O^+ reaches a peak around 300 km and has values ranging from about 10^{10} m^{-3} at night in high latitudes near sunspot minimum to about 5×10^{12} m^{-3} near the equator at sunspot maximum. The dominant neutral gas is O, but N_2 and O_2 play important roles in the loss of plasma. The height of the peak and the maximum density are highly variable, depending on the neutral wind and electromagnetic drift. In the absence of diffusion and other effects, the ion density would increase steadily, owing to the rapid decay of the loss rate, until a height was reached at which there were insufficient neutrals to ionize. However, O^+ ions produced in the topside diffuse downward through the peak of the F_2 region with typical

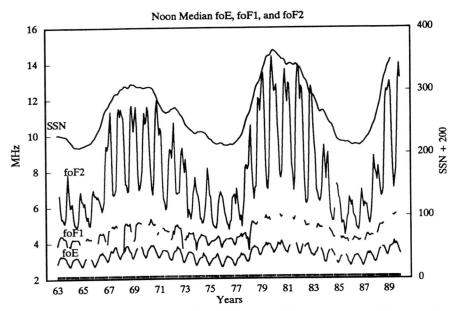

Fig. 3 Long-term variations of F_oE, f_oF_1, and f_oF_2 at Boulder, Colorado and of the 12-month smoothed sunspot numbers (SSNs). (Courtesy of R. Conkright.)

downward velocities near the F_2 peak of 10–20 m s^{-1}. Unlike f_oE and f_oF_1, f_oF_2 does not follow a cos χ law either diurnally or seasonally. This is illustrated in Fig. 3, which brings out the following features:

1. The seasonal variations of f_oE and f_oF_1 are in phase with the solar zenith angle, whereas f_oF_2 is in antiphase. The latter has been called the F_2 winter anomaly.
2. The F_1 layer disappears in some winters.
3. There is a marked increase in the critical frequencies in sympathy with the 11-year sunspot cycle.

The winter anomaly occurs in the daytime only, and it is due to a large summer electron loss rate caused by an increase in the molecular-to-atomic composition of the neutral atmosphere; thus it is more appropriately described as a summer anomaly.

In middle latitudes in winter the diurnal maximum of f_oF_2 occurs near midday, but during the summer it occurs in the late afternoon. In low latitudes the diurnal maximum can occur in the evening.

The 12-month smoothed F_2 noon maximum (or critical) frequency f_oF_2 is approximately linearly dependent on the 12-month smoothed sunspot number R_{12}. The intercepts and slopes depend on geographic location. On average, f_oF_2 saturates for sunspot numbers greater than about 150.

The worldwide distribution of f_oF_2 at a given universal time (UT) is illustrated in Fig. 4, which reveals the marked geomagnetic control of the F_2 region. The most distinctive features on the map are the two regions of high f_oF_2 lying around

±20° dip latitude. This feature is called the equatorial (or Appleton) anomaly. The "highs" lag behind the subpolar point but move around with the sun. Because of the geomagnetic control, the critical frequencies of the F_2 layer exhibit a marked longitudinal effect. The highs move from east to west and also vary in geographic latitude, tending to remain at a constant magnetic dip. In middle to high magnetic latitudes, the night F_2 layer is characterized by roughs of low f_oF_2 lying between about 50° and 70° of geomagnetic latitude. The electron density inside a rough is drastically reduced, e.g., by as much as a factor of 2 at 1000 km altitude and as much as an order of magnitude at the F_2 peak.

Spread F is an indication of the occurrence of relatively small-scale irregularities in the F_2 region. The phenomenon is of global dimensions and is particularly severe near the magnetic equator and in the auroral zones. In middle and high latitudes, these irregularities appear to occur in patches with horizontal extents up to several hundred kilometers and drift speeds of about 100 m s^{-1}. Spread F can change rapidly, e.g., within a minute. Equatorial spread F occurs after sunset and near the equinoxes. This type of spread F tends to disappear in large ionospheric storms. The most likely time of occurrence is between 2100 and 0100 local time (LT) and is earlier at sunspot maximum.

E. The Topside

The region above the height of maximum electron density (approximately 300–2000 km) is called the topside and is

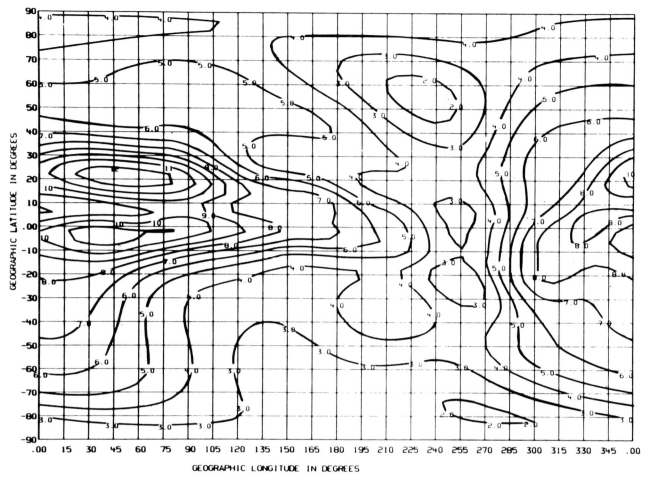

Fig. 4 Contour map (in megahertz) of the global distribution of the monthly median value of f_oF_2 for April 1976 at 1200 UT. (Courtesy of C. M. Rush.)

inaccessible to high-frequency radio soundings from the ground. Knowledge of the topside has come from several techniques, including topside (iono-) sounders, incoherent scatter radars, and *in situ* satellite detectors. The dominant positive ion in the topside is O^+ and, at heights greater than about 100 km above the peak, the electron density decays exponentially with height with a scale height appropriate to an O^+–electron plasma. Above a transition height h_T, H^+ becomes the dominant ion and the scale height increases markedly. At about 1000 km the electron density is of the order of 10^{10}–10^{11} el. m^{-3}.

F. The Protonosphere

The protonosphere is the part of the outer ionosphere that is bounded on the lower side by the transition height, at which the concentrations of O^+ and H^+ are equal, and on the upper side by the plasmapause, at which the plasma density drops by one or two orders of magnitude. The protonosphere may be thought of as the part of the (inner) magnetosphere that corotates with the earth. Within the protonosphere the neutral densities are small and the plasma profiles are determined by transport. Protons are produced at the base of the protonosphere by charge exchange between ionospheric O^+ ions and hydrogen atoms:

$$O^+ + H \rightleftharpoons H^+ + O \tag{17}$$

The position of the transition height varies from around 800 km on a winter night at low sunspot numbers to several thousand kilometers (e.g., 3000 km) during a summer day at high sunspot numbers. Under quiet conditions the plasmapause is located on an L shell of approximately 6. During a large disturbance it can contract to an L shell of 2 or 3. Inside the plasmapause there is thermal plasma at temperatures of 1000–3000 K. Beyond the plasmapause the ions have superthermal energies ($\approx$ 1000 eV).

There is essentially no plasma production in the protonosphere. By day plasma diffuses up from the F_2 layer, where the relative densities are high. At night, when the plasma densities in the F_2 layer fall, plasma diffuses downward. The protonosphere acts like a reservoir; it takes plasma from the ionosphere by day, stores in it a loss-free environment, and returns it to the ionosphere at night, thus helping to maintain the nighttime F region.

G. Winds and Drifts

1. Winds in the D and E Regions

Several mechanisms give rise to the horizontal motion of plasma in the ionosphere—for example, neutral winds, waves in the neutral atmosphere, and electromagnetic drift. Neutral winds result from unequal heating of the atmosphere. Gravity waves are disturbances in the atmosphere with periods longer than the buoyancy, or Brunt–Väisälä, period, which varies from about 5 min at 100 km to about 15 min at 300 km. In the lower ionosphere the electron–neutral collision frequency is sufficiently large to ensure that the electrons move with the ambient wind. In the height range 90–100 km, wind speeds are of the order of 10 m s^{-1}.

2. Winds and Drifts in the F Region

Meridional winds are important in the F region. Neutral winds have been calculated, taking into account the restraining effect of ion drag. In this model, winds flow from the subsolar region over the high-latitude regions to the dark hemisphere. Near 1400 LT at a middle-latitude station the winds flow poleward, and near 0200 LT they flow equator-. ward. The net effect of this wind system on the ionospheric plasma is to lower the height of maximum density by day and raise it at night. Thus by day f_0F_2 is lower than it would be in the absence of wind, whereas at night f_0F_2 is higher than it would otherwise be because the plasma is lifted to heights at which the loss rate is reduced. During ionospheric storms there is strong heating of the auroral neutral gas by deposition of energetic electrons and by Joule (electrical) heating. This heating is most pronounced on the nightside, so that at night in middle latitudes the wind due to auroral heating reinforces the diurnal wind, whereas on the dayside they are in opposition. Zonal winds are effective in modifying f_0F_2. Winds have a diurnal variation, ranging from around 50 m s^{-1} during the day, when ion drag is large, to over 300 m s^{-1} at night, when ion drag is small. [*See* ZONAL AND MERIDIONAL WINDS IN THE EARTH'S ATMOSPHERE.]

In high latitudes and when strong electric fields are present, large horizontal drifts of plasma have been observed ($\approx$ 1 km s^{-1}). Plasma is convected over the polar cap from day to night, thus maintaining the F_2 ionization during the long winter night. The motion of the ions through the neutral gas sets up a wind in the neutral gas which modifies that produced by the sources enumerated above.

In low latitudes, vertical ($E \times B$) drifts play a significant role in establishing the distribution of ion density in the F_2 region. The height of maximum electron density, h_mF_2, reaches a very high level by about 1900 LT and then falls so that at midnight it is about 100 km lower than at noon. This evening uplift of the equatorial F_2 layer results from upward electromagnetic ($E \times B$) drift caused by a west-to-east electric field generated in the E region. This is called the fountain effect. This height variation is especially noticeable at Huancayo, Peru, where at 1900–2000 LT the F layer peak can be above 500 km. The vertical speed at Huancayo has a diurnal variation with a maximum of about 30 m s^{-1}, which occurs around 2000 LT.

3. Traveling Ionospheric Disturbances

The movement of ionospheric irregularities can be produced (1) by the advection of irregular plasma structures by winds in the neutral atmosphere and/or electromagnetic drifts and (2) by the passage of acoustic gravity waves in the neutral atmosphere. In the lower ionosphere the neutral gas motion is imparted to the plasma, but in the F_2 region only the component parallel to the geomagnetic field is imparted to the plasma. Thus, as the wave passes over a given location, the height of a surface of fixed ion density oscillates, and this can be seen as a traveling ionospheric disturbance (TID). TIDs are relatively large-scale wavelike structures (100–1000-km horizontal scales) which travel with speeds between about 50 and 1000 m s^{-1} and with wave periods from about 5 min to over an hour. TIDs can be classified as very-large-scale waves and medium-scale waves. [*See* ATMOSPHERIC GRAVITY WAVES AND TRAVELING IONOSPHERIC DISTURBANCE.]

Observations at several radio frequencies show that these disturbances consist of long fronts which are tilted forward so that the disturbances appear first at high heights and move downward. Very-large-scale disturbances, which originate in the auroral zone and travel to great distances, are associated with magnetic storms. They have wavelengths of the order of 1000 km and periods of 1 h or more. Medium-scale disturbances, with wavelengths of the order of 100–200 km and periods ranging from 5 to 45 min or so, do not appear to travel great distances and may originate in local tropospheric weather.

H. The Auroral Zones

The auroral zones are defined as the regions of the earth where visible aurora occurs overhead. The north and south auroral zones are relatively narrow rings situated between magnetic latitudes of about 64° and about 70°. Aurora is the result of excitation of atmospheric atoms and molecules by energetic ions and electrons released from the overlying magnetosphere. With sufficiently energetic electrons (e.g., 10 keV) intense ionization of the neutral air can occur at heights from around 100 km to over 150 km. This ionization can have important consequences for radio surveillance, such as the presence of spurious echoes in over-the-horizon radars. Auroral structures can range from small-scale, sharply defined, narrow columns, which are aligned along magnetic field lines, to large-scale structures oriented in the magnetic east–west direction. The aurora is characterized by substorms in which the auroral displays undergo large-scale expansions and contractions and the auroral zones move equatorward. During substorms the energy of the precipitating electrons can be sufficient to penetrate to the D region and cause high "auroral" absorption of radio signals. [*See* POLAR AURORA; SUBSTORMS.]

III. Ionospheric Disturbances

A. Types of Disturbances

The term *ionospheric disturbance* is used to cover a wide variety of ionospheric conditions that depart from the usual or quiet state. However, it must be recognized that even in the quiet ionosphere things are not static. For example, sporadic E may appear, fluctuate wildly, and disappear; small ripples are almost always present; and the F_2 region is marked by fluctuations in critical frequency (f_oF_2), height of maximum density (h_mF_2), and total electron content (TEC) from day to day and within a day. Because the causes of these quiet-time variations cannot always be identified, they are called "geophysical noise."

There is a set of disturbances that are associated, either directly or indirectly, with events on the sun. These include the following:

1. Sudden ionospheric disturbances
2. Ionospheric storms
3. Polar cap absorption events
4. Traveling ionospheric disturbances (large scale)
5. Associated disturbances in the geomagnetic field (solar flare effect) and auroral and magnetospheric substorms.

These effects are shown in Fig. 5.

B. Solar Disturbances

Ionospheric disturbances are often associated with various solar features, such as solar flares, disappearing filaments, and coronal mass ejections, which are related to enhancements in ionizing radiations and charged particles. The D region is affected by x rays in the range 0.1–1.0 nm. The E region is affected by soft x rays (1–10 nm), and the F region is affected by extreme ultraviolet. A flare can last from a minute to several hours, with an average duration of about 30 min.

Flares are assigned an "importance" based on area at maximum brightness (0 to 4) and brightness (faint, F; normal, N; and bright, B). An optical flare of importance greater than 1 N is accompanied by x ray and extreme ultraviolet emissions and radio bursts. X ray flares are rated C, M, and X according to the fluxes (E) in the 0.1–0.8-nm range as measured by instruments on satellites in earth orbits, where for C, $10^{-6} < E < 10^{-5}$ W m^{-2}; for M, $10^{-5} \leq E \leq 10^{-4}$ W m^{-2}; and for X, $E > 10^{-4}$ W m^{-2}. An additional digit is added to the C, M, and X to indicate the coefficient of the flux; e.g., C8 = 8 × 10^{-6} and M3 = 3 × 10^{-5}. Most class X and some class M flares produce sudden ionospheric disturbances that affect radio communications (solar radio bursts), and some produce solar protons.

Coronal holes are large, magnetically unipolar solar regions that do not emit x rays but do emit low-energy charged particles into interplanetary space; these particles form the solar wind plasma, which has an average speed of about 400 km s^{-1}. During some disturbances this speed can exceed 1000 km s^{-1}. Such great speed often produces shocks that impinge on the earth's magnetosphere and under certain conditions may produce geomagnetic and ionospheric storms.

During some large flares, protons with energies in the range 1–400 MeV are emitted, enter the earth's polar regions, and produce intense D region absorption of radio waves called polar cap absorption. During quiet magnetic conditions there is a lower cutoff latitude for these protons of about 65° (for 10-MeV particles).

C. Sudden Ionospheric Disturbances

These are ionospheric effects that start essentially at the onset of the visible flare and are caused by ionization due to x rays and extreme ultraviolet rays. In some sudden ionospheric disturbances (SIDs), the electron densities in the D region can increase by more than a factor of 10. These enhancements lead to several radio propagation disturbances, including (1) shortwave fadeout due to enhanced absorption of radio signals, (2) sudden phase anomaly due to the lowering of the reflection height of very-low-frequency (3–30 kHz) waves, (3) sudden cosmic noise absorption of galactic radio noise on radio frequencies near 30 MHz, and (4) sudden enhancements of atmospheric noise or sudden decreases of atmospheric noise on radio frequencies in the range 1–75 kHz. The 12-month averaged count of the occurrence of sudden ionospheric disturbances in the D region varies with the sunspot cycle as seen in Fig. 6. Sudden ionospheric distur-

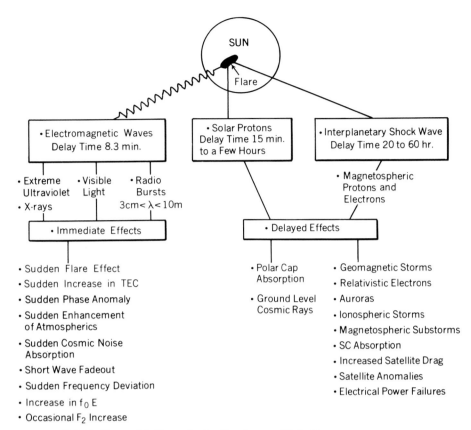

Fig. 5 Terrestrial effects of solar disturbances. SC, Sudden commencement.

bances caused by ionization enhancements in the E and F regions include (5) sudden frequency deviation of radio waves in the frequency range 3–30 MHz caused by ionization increases in the upper E region and (6) sudden increases in total electron content caused by enhancements in the F region.

D. Polar Cap Events

When a solar proton enters the atmosphere, it produces a pair of ions for every 35 eV of energy. A complicated ion chemistry is associated with this ionization. The intense ionization, created in the height range $\approx 50–90$ km, produces polar cap absorptions (PCAs) at high radio frequencies and phase advances at very low radio frequencies that last from 1 h to more than 60 h. The solar proton fluxes extend over the polar cap including the auroral zone, and they are not closely correlated with magnetic disturbance except during the later stages.

The occurrence of PCAs tend to follow the sunspot cycle.

The history of a PCA is roughly as follows. The absorption appears first at high latitudes and expands equatorward. It is high during daylight but decreases at night because electrons are removed by attachment to form negative ions. To summarize:

1. PCAs are almost always preceded by a major solar flare. The time delay may be as short as $\frac{1}{2}$ h and the average delay is 3–4 h.
2. The duration varies from less than 1 day to about 10 days.
3. Magnetic activity is usually absent during the first day or so of the event, after which a sudden commencement magnetic storm may occur. Most, but not all, PCAs have their maxima before the start of the magnetic storm.
4. Nearly every PCA is preceded by a broadband centimeter solar radio burst.

E. Relativistic Electron Precipitation

Relativistic electron precipitation (REPs) are essentially auroral zone events and result in increases of D region ionization.

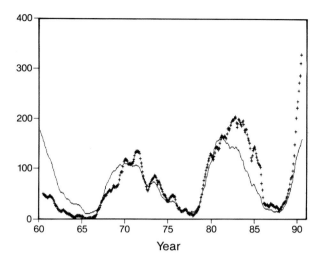

Fig. 6 Long-term variation of the occurrence of D region sudden ionospheric disturbances and smoothed sunspot numbers. The occurrence is low prior to about 1965 because the numbers of observers were low.

They have a large latitudinal extent (≈ 2000 km) and can last about 1 h to more than 6 h. The seasonal variation of occurrence has peaks at the equinoxes. REPs are associated with enhanced geomagnetic disturbance and are essentially daytime phenomena, with maximum occurrence before noon. Electron energies are on the order of 500 keV, and the stoppage of these electrons by atmospheric molecules gives rise to x ray bursts called bremsstrahlung.

F. Ionospheric Storms

From the point of view of society, these are by far the most important ionospheric disturbances because of their relatively long duration (several days) and their adverse effects on radio systems in the populated middle latitudes. Ionospheric storms are prime examples of the interction of the ionosphere with the magnetosphere and the neutral atmosphere. These storms are associated with auroral and geomagnetic storms that have various effects on human activities, including breakdown of electric power systems, spacecraft charging, reduction of satellite lifetimes, and interference with geophysical exploration.

A typical magnetic storm lasts from a few hours to a few days and begins with a sudden increase in the earth's horizontal magnetic field (or sudden commencement, SC). This SC is followed by a decrease below normal, followed by a recovery (2–3-day) phase. Magnetic storms are most intense near the auroral zones, which move equatorward in severe storms. Some storms have gradual commencements.

Magnetic storms occur when solar wind with a sufficiently large southward magnetic field reaches the magnetosphere. The magnetic field of the solar wind "reconnects" with the geomagnetic field, and the convection of these magnetic field lines over the polar caps constitutes a gigantic dynamo that generates dawn-to-dust voltages across the magnetosphere of up to 10^5 V and electric currents of 10^7 A which generate 10^{12} W of power. It is thought that during quiet periods ions from the solar wind, and especially from the ionosphere, fill the magnetosphere with plasma. Magnetospheric substorms accelerate ions and electrons on the nightside from the plasma sheet; some follow magnetic field lines into the auroral zones, causing visible auroras, auroral ionization, radio absorption, neutral atmosphere heating, and so forth. Thus ionospheric ions and electric fields affect the magnetosphere, which, in turn, modifies the ionosphere and atmosphere; this process is called atmospheric–magnetospheric coupling. [*See* MAGNETOSPHERIC CURRENTS.]

Auroral storm ionization and heating are caused mostly by the precipitation of electrons with energies in the range from about 1 keV to about 20–30 keV. Auroral heating of the neutral gas can change completely the global pattern of winds in the thermosphere. Large equatorward winds are established on the nightside and large-amplitude gravity waves are generated that help to transport energy to lower latitudes; both of these modify the ionosphere on a worldwide scale.

In the middle latitudes, the disturbed D region is characterized by a storm "aftereffect" or "poststorm" effect in which the D region disturbance persists for some days after the magnetic storm is over. This is thought to be caused by electron precipitation from the magnetosphere. At times of some sudden commencements, there are brief absorption enhancements with durations of the order of 5–10 min. These are essentially auroral zone phenomena and appear on both the dayside and the nightside.

The most conspicuous feature of the disturbed E region is the appearance of auroral or storm E_s, particularly during nighttime. Auroral E_s is produced by energetic electron bombardment.

The most prominent feature of the F region storm is the variation of the electron density and total electron content. The effects of a storm on N_mF_2 at a given location depend in a complicated way on the local time, season, latitude, and storm time, which is the time measured from the onset of the sudden commencement. In general, there is an initial positive stage in which N_mF_2 is enhanced, followed by a negative stage in which N_mF_2 is depressed. Individual storms can depart markedly from this statistical pattern. The positive stage is most pronounced on the afternoon of day 1, on which the geomagnetic storm started, and the negative stage lasts for some 3 days during which conditions return slowly to normal.

The positive phase of a storm is the result of increased

equatorward winds produced by heating of the auroral zone by particle precipitation and Joule heating by auroral electric currents. The wind lifts the plasma up the geomagnetic field, and the lower electron loss rates lead to increased electron densities. The negative phase results from changes in atmospheric composition that increase the loss rate. There are two aspects of the movement of storms: (1) the equatorward motion of the auroral zone and (2) the movement of the disturbance away from the auroral zone. It appears that a disturbance moves from about 70° geomagnetic latitude to about 50° geomagnetic latitude in about 10 h. To mitigate the effects of storms on various human systems, forecasting and warning centers are in operation in several countries.

IV. Ionospheric Modifications

Besides the solar control of the ionosphere, there are numerous natural and anthropogenic (transient) effects that modify the ionosphere.

A. Atmospheric Sources

Ionospheric traveling disturbances have been detected which originated from the following sources at or near the earth's surface: thunderstorms, weather fronts, hurricanes, and solar eclipses.

Although electrical coupling exists between the troposphere and the conducting ionosphere, the more noticeable effects are those transmitted mechanically in the form of acoustic–gravity waves. Thunderstorms produce acoustic waves with periods of 3–5 min by mechanical oscillation of the cloud. There is evidence for waves resulting from the cooling of the troposphere and ozonosphere during solar eclipses.

B. Terrogenic Effects

Terrogenic effects are those in which the ionosphere is influenced by the electrical conductivity structure of the underlying surface of the earth. Electric currents are induced in the earth which react on the ionosphere, and these are particularly important in auroral latitudes, where there are large primary ionospheric currents. There is coupling of thunderstorm electric fields and currents from the troposphere to the ionosphere. The thunderstorm generator keeps the electrical potential of the ionosphere at about 2×10^5 V; this potential can maintain an electric field of about 200 V m^{-1} near the earth's surface, and this global coupling is affected by orographic features. Traveling ionospheric disturbances are generated by earthquakes and volcanoes. Earthquakes produce both short-period (≈ 1 min) and long-period (≈ 1 h) waves. The short-period waves caused by earthquakes come from

propagation of Rayleigh waves near the earth's surface followed by coupling into the atmosphere. The long-period (earthquake) waves travel from the source in the atmosphere. Disturbances from the eruption of Mt. St. Helens on May 18, 1980 produced global atmospheric and ionospheric effects. [See ATMOSPHERIC EFFECTS OF VOLCANIC ERUPTIONS.]

C. Anthropogenic Sources

1. Explosions

A number of ionospheric effects of chemical and nuclear explosions have been detected. At long distances, the more pronounced effects are the short-period (≈ 1 min) and long-period (≈ 1 h) waves originating in nuclear explosions.

Near a nuclear fireball, marked changes in the ion densities are caused by x rays and radioactive emissions. The ionized debris moves along the geomagnetic field to the conjugate region, where the electron density is enhanced. The trapped electrons bounce between conjugate ionospheres and drift eastward, encircling the earth within a few hours. The ionization of the D region produced by the precipitation of energetic electrons results in high absorption of radio signals. At higher altitudes, increased ionization forms field-aligned irregularities resulting from the high mobility of plasma parallel to the magnetic field.

2. Radio Heating

High-power radio waves are used to modify the ionosphere. There are two types of radio heating: (1) nonresonant or resistive heating, in which the frequency of the radio wave is far different from the plasma frequency, and (2) resonant heating, in which the wave frequency is close to the plasma frequency.

Nonresonant heating results from the absorption of radio energy in the D region, which raises the electron temperature and hence the collision frequency, thus increasing the absorption coefficient of the D region. Examples of this type of heating include wave interaction or cross modulation, gyrointeraction, and self-distortion.

In the E region, powerful radio signals modulate the conductivity so that the ionosphere behaves as a very large antenna in which currents oscillate at the modulation frequency (e.g., ≈ 10 kHz) and radiate radio waves on this frequency.

Resonant heating of the F region produces a variety of effects. The immediate effects are plasma resonance excitations followed (in ≈ 5 s) by field-aligned irregularities and anomalous absorption. The energy extracted from the heating (or pump) wave raises the electron temperature and increases the volume of the heated region. Also observed during heating are airglow at 630 nm, the plasma line, spread F, and ion–acoustic waves.

3. Chemical Modification

Chemical seeding may be deliberate or inadvertent. Deliberate modification provides tracers (e.g., barium releases) for the study of ionospheric parameters; induces instabilities in plasma irregularities in order to simulate natural disturbances; and creates "holes" in the ionosphere, leading to "field-aligned" ducts for possible radio communications. Inadvertent modification can be produced by exhaust gases from rockets and chemicals (e.g., chlorofluorocarbons) from the ground.

Bibliography

Baker, D. N. (1986). Statistical analyses in the study of solar wind–magnetosphere coupling. *In* "Solar Wind–Magnetosphere Coupling" (Y. Kamide and J. A. Slavin, eds.), Vol. 17. Terra Scientific Publishing, Tokyo.

Banks, P. M., and Kockarts, G. (1973). "Aeronomy," Parts A and B. Academic Press, New York.

Barr, R., Stubbe, P., Reitveld, M. T., and Kopka, H. (1986). ELF and VLF signals radiated by the "Polar Electroject Antenna": experimental results. *J. Geophys. Res.* **91**, 4451.

Birkmeyer, W., and Hagfors, T. (1986). Observational technique and parameter estimation in plasma line spectrum observations of the ionosphere by chirped incoherent scatter radar. *J. Atmos. Terr. Phys.* **48**, 1009.

Brasseur, G., and Solomon, S. (1986). "Aeronomy of the Middle Atmosphere," 2nd ed. Reidel, Dordrecht, The Netherlands.

Davies, K. (1990). "Ionospheric Radio." Peter Peregrinus, London.

Kato, S. (1980). "Dynamics of the Upper Atmosphere." Reidel, Dordrecht, The Netherlands.

Kelley, M. C. (1989). "The Earth's Ionosphere." Academic Press, San Diego.

Migulin, V. V., and Gurevich, A. V. (1985). Investigation in the USSR of nonlinear phenomena in the ionosphere. *J. Atmos. Terr. Phys.* **47**, 1181.

Muldrew, D. B. (1983). Alouette-ISIS radio wave studies of the cleft, the auroral zone, and the main trough and of their associated irregularities, *Radio Sci.* **18**, 1140.

Popov, L. N., Krakovetzkiy, Yu. K., Gokhberg, M. B., and Pilipenko, V. A. (1989). Terrogenic effects in the ionosphere: a review. *Phys. Earth Planet. Inter.* **57**, 115.

Reid, G. C. (1986). Solar energetic particles and their effects on the terrestrial environment. *In* "Physics of the Sun" (P. A. Sturrock, T. E. Holzer, D. M. Mihalas, and R. K. Ulrich, eds.), Chapter 22. Reidel, Dordrecht, The Netherlands.

Rush, C. M., PoKempner, M., Anderson, D. N., Perry, J., Stewart, F. G., and Reasoner, R. (1984). Maps of f_oF_2 derived from observations and theoretical data. *Radio Sci.* **19**, 1083.

Stubbe, P., Kopka, H., Rietveld, M. T., Frey, A., Høeg, P., Kohl, H., Nielsen, E., Rose, G., LaHoz, C., Barr, R., Derblom, H., Hedberg, A., Thide, B., Jones, T. B., Robinson, T., Brekke, A., Hansen, T., and Holt, O. (1985). Ionospheric modification experiments with the Tromsø heating facility. *J. Atmos. Terr. Phys.* **47**, 1151.

Whitehead, J. D. (1989). Recent work on mid-latitude and equatorial sporadic E. *J. Atmos. Terr. Phys.* **51**, 401.

Glossary

Ambipolar diffusion Diffusion of electrons and positive ions which are coupled by a polarization electric field.

Bremsstrahlung X rays produced by the stoppage of energetic electrons in the earth's upper atmosphere.

Dissociative recombination Neutralization of a positive molecular ion by recombination with an electron to produce two neutral particles (atoms). These atoms may be in excited states.

Equatorial anomaly Structure of the low-latitude ionosphere in which there are two electron density maxima, at about $\pm 15°$ magnetic dip latitude, rather than one maximum at the subsolar point.

Geophysical noise Fluctuations in a geophysical parameter that are of a quasirandom nature and cannot be attributed to fluctuations in known causative phenomena.

Incoherent scatter radar Sensitive radar operating on a frequency well above the critical frequency that detects radio signals scattered from refractive index irregularities in the ionosphere and/or neutral atmosphere.

L shell Surface of the geomagnetic field given by the longitudinal drift of an L line along which trapped charged particles bounce between northern and southern hemisphere. The L value is a length, in units of earth radius, which reduces to the equatorial radius of a field line in the case of a dipole field.

Stratospheric warming Disturbance of the winter polar, middle atmosphere lasting for several days and characterized by a warming of the stratosphere (13–50 km altitude) by some tens of degrees.

Terrogenic effects Structures in the ionosphere attributable to the electrical (conductivity) structure of the ground.

Land–Atmosphere Interactions

Roni Avissar

Rutgers University

Land–atmosphere interactions are the physical, chemical, and biological processes and feedbacks that occur, at different spatial and temporal scales, between the land and the atmosphere. They involve various mechanisms such as the exchange of radiation, heat, gases, and particles, at the earth's surface. The hydrologic cycle and the carbon dioxide cycle are but two typical examples of such interactions.

I. Introduction

Land–atmosphere interactions attract interest from most earth science disciplines. This is because they involve a large number of physical, chemical, and biological processes that take place in the biosphere at different spatial and temporal scales. Of particular importance are the exchanges of energy, mass, and momentum between the surface and the atmosphere. These processes have a strong impact on the atmospheric planetary boundary layer (PBL) and on the surface heat fluxes. Therefore, land–atmosphere interactions play a

major role in studies and applications related to climate, hydrology, ecology, agriculture, and the like.

Solar radiation is the only significant source of energy for the land–atmosphere system. As shown in Fig. 1, which represents an annual average global energy balance for the atmosphere, much of the radiation retained by this system is actually absorbed at the earth's surface. About one-third of the total incoming solar radiation at the top of the atmosphere is reflected to space. From the remaining amount, about one-third is absorbed by the different atmospheric components and the other two-thirds are absorbed by the earth's surface. There, it is redistributed into infrared radiation (about one-third) and sensible and latent heat flux. It is therefore not possible to explain (let alone to predict) the long-term behavior of land–atmosphere interactions without a proper understanding of the mechanisms that convert the solar radiation absorbed at the surface into local heat storage, emission of infrared radiation, and latent and sensible heat releases. [*See* SOLAR RADIATION.]

II. Land–Surface Energy Balance

Conservation of energy at the earth's surface implies that radiative (Q_R), convective (Q_C), and conductive (Q_G) contributions of energy to the surface are balanced. Thus, the earth's surface energy balance can be expressed with the following equation:

$$Q_R + Q_C + Q_G = 0 \qquad (1)$$

In the soil, in general $|Q_G| \gg |Q_C|$ and $|Q_G| \gg |Q_R|$, so that only ground conduction is retained as the principal contribution to the heat balance (when rain falls or snow melts and percolates into the soil, substantial heat can be transferred so that these inequalities may not be satisfied). In the atmosphere these inequalities are reversed (i.e., $|Q_C| \gg |Q_G|$ and

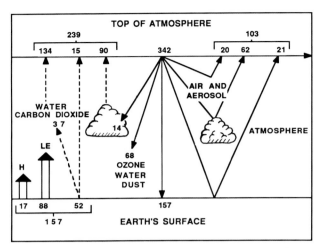

Fig. 1   Annual average global energy fluxes into and out of the atmosphere, in W m⁻². (H is sensible heat flux, LE is latent heat flux.)

$|Q_R| \gg |Q_G|$) since molecular transfers of heat in turbulent air are ineffective compared with radiation and convection. [*See* SOIL KINETICS.]

For a bare soil surface, Eq. (1) can be explicitly written as

$$\underbrace{\overbrace{-\rho c_p \overline{w'\theta'}}^{H_G} \overbrace{-\rho L_h \overline{w'q'}}^{LE_G}}_{Q_C}$$
$$+ \underbrace{(1 - \alpha_G)R_{sw}^{\downarrow} + \varepsilon_G R_{lw}^{\downarrow} - \varepsilon_G \sigma T_G^4}_{Q_R}$$
$$\underbrace{- \lambda \frac{\partial T_S}{\partial z}}_{Q_G} = 0, \qquad (2)$$

where H_G and LE_G are sensible and latent heat flux, respectively; ρ is air density; c_p is air specific heat at constant pressure; L_h is latent heat of evaporation; α_G and ϵ_G are soil surface albedo and emissivity, respectively; σ is the Stefan–Boltzmann constant; $R_{sw}^{\downarrow}$ and $R_{lw}^{\downarrow}$ are solar (shortwave) and infrared atmospheric (longwave) radiation, respectively; λ is soil thermal conductivity; T_S is soil temperature; z is depth in the soil; and T_G is soil surface temperature (i.e., T_S at $z = 0$).

A. Convective Heat Fluxes

The terms w', θ', and q' are perturbations (or turbulent parts) from the mean vertical component of the wind, from the potential temperature, and from the specific humidity, respectively. The covariances $\overline{w'\theta'}$ and $\overline{w'q'}$ (overbar indicates averaging) are obtained from the conservation of heat and mois-

ture equations, respectively, by expanding the dependent variables into mean and turbulent parts. An averaging of the different terms over time and space is then performed, assuming that the average of the turbulent parts is zero (i.e., a Reynolds averaging). The mathematical procedure involved in the representation of these conservation principles will not be described here; it is provided in most introductory books in atmospheric science and fluid dynamics.

By analogy with the molecular flux of heat, the covariance $\overline{w'\theta'}$ can be represented by the product of a turbulent-transfer coefficient and the gradient of potential temperature above the ground surface or, alternatively, the product of a drag coefficient and the difference of potential temperature above the surface. Similarly, for the covariance $\overline{w'q'}$, the transfer (or drag) coefficient is multiplied by the gradient (or the difference) of humidity. These representations are often called *first-order closure* representations because the mean correlations are specified as functions of one or more of the averaged dependent variables.

When explicit equations are developed for the fluxes (which include triple correlation terms involving variables that must be represented in terms of the covariances or the averaged dependent variables or both) the representation is referred to as *second-order closure*. Although theoretically more satisfying, this more expensive approach with its greater degrees of freedom has not improved simulations of the evolution of the resolvable dependent variables in the PBL over those obtained using the best first-order representations.

Turbulence in the PBL is mostly generated by forcings from the ground. For example, solar heating of the ground during sunny days causes thermals of warmer air to rise. Frictional drag on the air flowing over the ground causes wind shears to develop, which frequently become turbulent. Obstacles like trees and buildings deflect the flow, causing turbulent wakes adjacent to, and downwind of the obstacles.

Turbulence is several orders of magnitude more effective at transporting heat (or other quantities) than is molecular diffusivity. As a result, typically about 10 to 20% of the net radiation (Q_R) at the soil surface is conducted into the soil. The remaining part is transferred into the atmosphere by turbulent heat fluxes.

B. Radiative Heat Fluxes

The solar radiation that reaches the ground is influenced by the slope and the orientation of the ground surface with respect to the sun. This can be described with the following formula:

$$R_{sw}^{\downarrow} = R_{sw}^{\perp}[\cos \alpha \cos Z + \sin \alpha \sin Z \cos(\beta - \gamma)] \qquad (3)$$

where $R_{sw}^{\perp}$ is direct solar radiation at the ground on a unit cross-sectional area perpendicular to sun's rays; Z is the

Table I Albedos of Shortwave Radiation and Emissivities of Longwave Radiation for Representative Types of Land Surfaces

Land surface	Albedo[a]	Emissivity[a]
Fresh snow	0.70–0.95	0.99
Old snow	0.45–0.55	0.82
Ice	0.20–0.40	0.98
Soils	0.05–0.45	0.90–0.98
Asphalt	0.05–0.20	0.95
Concrete	0.10–0.35	0.70–0.97
Tar and gravel	0.08–0.18	0.92
Grass	0.16–0.26	0.90–0.97
Agricultural crops	0.18–0.30	0.90–0.95
Deciduous forests	0.15–0.20	0.95–0.98
Coniferous forests	0.10–0.15	0.97–0.98
Urban areas	0.10–0.27	0.85–0.95
Water	0.03–1.00	0.99

[a] Ranges indicate minimum and maximum values found in the published literature.

zenith angle, which depends on day of the year, latitude, and hour; α is the slope of the terrain; and $\beta - \gamma$ is orientation of the sun's azimuth β with respect to the azimuth of the terrain slope γ.

The term $\epsilon_G \sigma T_G^4$ is derived from the Stefan–Boltzmann law. It represents the infrared radiation emitted by the soil surface. As indicated, this radiation is proportional to the fourth power of the surface temperature.

Surface albedo and emissivity have an important impact on the amount of radiative energy received and emitted by the surface. As indicated in Table I, which provides a few typical values of these photometric properties, a relatively broad range of values is found at the surface of the earth. It is interesting to note the particularly high albedo of snow and ice and the relatively low albedo of forest. Thus, in addition to the limited amount of radiation received at high latitudes (i.e., near the poles) due to the large angle between the sun's rays and horizontal surfaces [see Eq. (3)], ice and snow, which mostly cover these locations, reflect a large part of this energy. On the contrary, the low albedo of forest usually found at low latitudes (i.e., near the equator) indicates a large absorption of energy at these locations.

C. Conductive Heat Flux

On bare, dry land there is no evaporation (i.e., no latent heat flux) and the net input of radiative energy received at the soil surface during daytime hours raises its temperature. The stronger the solar radiation, the higher the soil surface temperature. The sun's zenith angle is minimum at noon and is 90° at sunrise and sunset. Thus, under cloud-free atmospheric conditions, the solar radiation and the soil surface tempera-

ture are maximum at noon [see Eq. (3)]. Typically, in such condition, the soil surface temperature is higher than the air temperature above it and also higher than the subsurface soil temperature, at least until mid-afternoon. As a result, heat propagates into the atmosphere and the ground from high temperature to low temperature. [See HEAT FLOW THROUGH THE EARTH.]

The partition of heat between the ground (i.e., soil heat flux) and the atmosphere (i.e., sensible heat flux) depends, on the one hand, on the gradient of temperature between the soil subsurface and the soil surface and between the atmosphere and the soil surface. On the other hand, it depends on the ability of the soil and the atmosphere to transfer heat from the soil surface. This ability is expressed by the soil thermal conductivity and, as discussed above, by the atmospheric turbulent-transfer (or drag) coefficient(assuming, for the sake of the present discussion, a first-order closure approximation), respectively.

A typical soil consists of minerals (e.g., quartz, aluminosilicates), organic matter, water, and air. The solid phase and the porosity (i.e., the relative pore volume) of the soil are usually assumed to be constant, although mechanical pressure and chemical reactions within the soil might modify the porosity. But the liquid and gaseous phases, which together constitute the porosity, vary continuously, depending on precipitation, evapotranspiration, and water percolation at the bottom of the soil layer.

The thermal conductivity of each soil constituent differs very markedly, as is shown in Table II. Hence, the space-average (macroscopic) thermal conductivity of a soil depends on its mineral composition and organic matter content, as well as on the volume fractions of water and air. One complicating factor is that thermal conductivity is sensitive not merely to the volume composition of a soil but also to the sizes, shapes, and spatial arrangements of the soil particles (soil particle size can vary widely from on the order of less than a micrometer for clay to a few millimeters for sand). Since the thermal conductivity of air is very much smaller than that of water or solid matter, a high air content (or, correspondingly, a low

Table II Thermal Conductivities of Soil Constituents (at 10°C) and of Ice (at 0°C)

Constituent	Thermal conductivity (W m^{-1} K^{-1})
Quartz	8.8
Other minerals (average)	2.9
Organic matter	0.25
Water (liquid)	0.57
Ice	2.2
Air	0.025

Table III Average Thermal Conductivities of Soils

Soil type	Porosity	Soil water content	Thermal conductivity ($W\ m^{-1}\ K^{-1}$)
Sand	0.4	0.0	0.29
	0.4	0.2	1.75
	0.4	0.4	2.17
Clay	0.4	0.0	0.25
	0.4	0.2	1.17
	0.4	0.4	1.58
Peat	0.8	0.0	0.06
	0.8	0.4	0.29
	0.8	0.8	0.50

water content) corresponds to a low thermal conductivity. Moreover, since the proportions of water and air vary continuously, the soil thermal conductivity is time variable.

The thermal conductivity of quartz, which is the major component of sand, is higher than that of other minerals. Moreover, sandy soils are typically associated to a relatively low porosity. Therefore, the more sand is present in a soil, the better its conductivity.

Table III provides the average thermal conductivity of different soils at various soil water content. Clearly, both water/air content and soil type (characterized by a particle size distribution) strongly affect this conductivity.

Heat conducted within the soil is stored there, according to the soil heat capacity. This other soil thermal property is also dependent on the soil constituents; it is usually calculated by addition of the heat capacities of the different soil components, weighted according to their volume fractions. As indicated in Table IV, heat capacity of air is about three orders of magnitude smaller than that of the other soil components and, therefore, it is usually omitted in this calculation. The more water in the soil, the larger its heat capacity and, as a result, the more heat is stored in the soil.

During nighttime, more heat is lost at the soil surface (by emission of infrared radiation) than is received from the atmosphere. As a result, the soil surface is usually colder than the subsurface soil layer and the air above it. This generates a heat flux from within the soil (where it has been stored during the day) to the surface. The larger the soil heat conductivity, the better the conduction of heat to the cooling surface.

On wet land, the net input of radiative energy received at the soil surface is used mainly for evaporation. As a result, both the sensible heat flux to the atmosphere and the soil heat flux are usually very small. When the ground is covered by a dense vegetation, water is extracted mostly from the plant root zone by transpiration. In that case, latent heat flux is dominant even if the soil surface is dry, so long as there is enough water available in the plant root zone. Because large quantities of energy are required in the change of phase from

liquid to vapor ($\approx 2.45\ MJ\ kg^{-1}$ at 20°C), evapotranspiration provides an efficient mechanism for the dissipation of heat from the ground surface. As a result, completely different land–atmosphere interactions are obtained with dry soils or with wet soils.

D. Soil Water Balance

Thus, the water held in the soil affects not only the thermal properties of the soil but also (and mainly) the energy redistribution at the soil surface. To evaluate correctly how much water is available in the soil for evapotranspiration, a water balance can be estimated from the following equation:

$$\underbrace{\Delta S}_{\text{storage}} = \underbrace{(P + I + U)}_{\text{gains}} - \underbrace{(R + D + E + T)}_{\text{losses}}, \qquad (4)$$

wherein ΔS is change in soil moisture storage, P is precipitation, I is irrigation, U is upward flow into the soil layer from underground water, R is runoff, D is downward drainage out of the soil (percolation), E is direct evaporation from the soil surface, and T is plant transpiration. [See SOIL WATER: LIQUID, VAPOR, AND ICE.]

The largest composite term in the "losses" is generally the evapotranspiration ($E + T$). Thus, obviously, the water balance is intimately connected with the energy balance and both must be considered simultaneously to evaluate the surface heat fluxes.

Flow and storage of water in the soil depend on the hydraulic properties of the soil, which are related, among other parameters, to the particle size distribution of the soil. In general, the smaller the soil particles, the stronger the water is held in the soil. That means that water will flow and percolate more easily in a sandy soil than in a clayey soil. Correspondingly, water storage will be more effective in a clayey soil. Yet it is important to emphasize that this is only a very schematical description of the mechanism. As a matter of fact, complex physical, chemical, and even biological processes are involved in the soil–plant–water relationships.

Table IV Volumetric Heat Capacities of Soil Constituents (at 10°C) and of Ice (at 0°C)

Constituent	Heat capacity ($W\ m^{-3}\ K^{-1}$)
Quartz	2.0×10^6
Other minerals (average)	2.0×10^6
Organic matter	2.5×10^6
Water (liquid)	4.2×10^6
Ice	1.9×10^6
Air	1.25×10^3

Their detailed analysis is, however, beyond the scope of this article and can be found in most introductory books in soil physics.

In vegetated surface, different energy balance equations must be written for the vegetation layer and the soil surface. Equation (2) must be modified to account for the exchange of radiation between the vegetation and the soil surface. The turbulent heat fluxes between the soil surface and the atmosphere are affected by the presence of the plant canopy and, therefore, require a specific consideration. Usually, for the vegetation layer, it is assumed that the storage of heat by the plants is negligible. Thus, Eq. (1) can be simplified to

$$Q_R + Q_C = 0 \qquad (5)$$

which indicates that the net radiative energy absorbed by the plants (i.e., the radiation received from the sun, the atmosphere, and the soil surface less the radiation emitted by the plants) is redistributed into convective fluxes only.

The treatment of these terms is more complicated than for a bare land surface. This is due to the complex interactions between the soil, the canopy, and the atmosphere. From a physical point of view, however, the same conservation principles are assumed.

The earth's surface is rather rough, especially over continents (orography), and provides the major sink of momentum for the atmosphere. That is, in the absence of any other mechanism, wind speed would decrease due to friction at the surface. The maintenance of an active circulation, therefore, requires a continuous supply of kinetic energy to the atmosphere. On the global scale, this is realized by the large difference of solar radiation absorbed at the equator and at the poles, which results in a persistent temperature gradient between these locations. Similarly, regional atmospheric circulations will be determined by the juxtaposition of different landscapes characterized by a different ability to release sensible and latent heat into the atmosphere. In fact, a more detailed discussion of the energetics of the atmosphere would show that the space and time distribution of the temperature field largely controls the dynamics of the atmosphere both at global and at regional scales.

III. Methods of Investigation

There are three fundamental methods for studying land–atmosphere interactions — *observations, physical models,* and *mathematical models.*

A. Observations

Our fundamental understanding of the processes involved in the interactions between the land surface and the atmosphere comes from simultaneous measurements of atmospheric and surface characteristics. Over the past years, quality and accuracy improvements of sensors, data acquisition systems, and measuring methods has led to remarkable progress in our understanding of the basic mechanisms of these interactions.

There are two categories of sensors that can be used to measure the different air and ground variables relevant to land–atmosphere interactions — *direct sensors* and *remote sensors*. Direct sensors are ones that are placed in the ground or on some instrument platform (e.g., tower, aircraft) to make *in-situ* measurements at the location of the sensor. Remote sensors measure emissions at various wavelengths that are generated (or modified) by the land or the atmosphere at locations distant from the sensor. Active remote sensors produce their own emissions (sound, light, microwave), and passive remote sensors measure waves produced by the earth (infrared, microwave) or the sun (visible).

Disadvantages of direct sensors include modification of local environmental conditions by the sensor or its platform and the fact that it gives a potentially unrepresentative point value. Disadvantages of remote sensors include the inability to measure certain atmospheric characteristics; their small signal-to-noise ratio, which necessitates averaging over relatively large volumes (i.e., point values are not reliable); and the fact that the emissions may be modified in unknown ways as they propagate to the sensor. At present, satellite-based remote sensors do not have adequate vertical resolution to provide much data in the atmospheric layer close to the earth's surface. However, they provide detailed information about the earth's surface characteristics and their use in conjunction with other methods offers an attractive solution to study land–atmosphere interactions.

Although field measurements represent "truth" by definition, it is a truth that is composed of the superposition of many simultaneous effects and processes. For a few situations, we can attempt to isolate or focus on one specific process by the careful selection of a field site or weather pattern. However, we can never fully isolate any one process, and the weather is rarely reproducible. In addition, it is difficult to do sensitivity studies by systematically altering certain physical parameters and then measuring the effect. Thus, observation methods by themselves occupy a limited place in studies of land–atmosphere interactions, especially when considered from a global change perspective. However, it is important to emphasize that their contribution to validate (even partly) the involved processes is unrivalled.

B. Physical Models

With physical models, scale-model replicas of observed ground surface characteristics (e.g., topographic relief, trees) are constructed and inserted into a controlled-flow laboratory chamber, such as a wind tunnel. The fluid flow through the chamber is adjusted so as to best represent the larger-scale atmospheric flow. However, complete equivalence of the laboratory model and atmospheric prototype flow fields requires geometric, kinematic, dynamic, and thermal similarity. In addition, boundary conditions upstream, downstream, at the lower surface, and near the top of the physical model must be similar to those at the corresponding boundaries of the modeled atmospheric domain. Thus, flows in scaled physical models can at best be only partially similar to their atmospheric prototype and cannot at present include all processes present in the atmosphere such as Coriolis acceleration, exchange of energy by radiation, moist processes, and atmospheric circulations. [*See* ATMOSPHERIC MODELING.]

The multiple similarity requirements, as well as the characteristics of the wind tunnels and their instrumentation, limit their application to the study of processes whose horizontal scales range between 2 m and, at most, 50 km, and temporal scales of up to a few hours. As a result, their use to simulate land–atmosphere interactions in the context of global change is also limited. It is worth noting, however, that their contribution to our understanding of land–atmosphere interactions in complex terrain (e.g., emission of pollutants in densely built-up areas) is appreciable.

C. Mathematical Models

Mathematical models can incorporate our best understanding of the physical and chemical processes believed to be important. Atmospheric scientists have developed a rather large set of such models, from simple zero-dimensional to complex three-dimensional models, so that different issues may be investigated at the appropriate scale and with the relevant amount of detail. These models are usually composed of one or more differential equations relating variables and parameters. Diagnostic models are made of one or more time-independent equations, while prognostic models contain at least one time-dependent equation, whose purpose is to predict the value of a variable some time in the future, on the basis of current and past values of this and other variables.

Dynamical models of the atmosphere are based on a set of conservation principles (usually conservation of motion, mass, heat, water, and other gaseous and aerosol materials) and, therefore, include prognostic equations for the three components of the wind field, for the temperature field, and possibly for the moisture field, as well as diagnostic relations such as the equation of state for ideal gases. To simulate appropriately land–atmosphere interactions, prognostic equations for the soil temperature and for the soil moisture are sometimes also included. Since such a set of nonlinear equations cannot be integrated in time analytically, these models are converted into a different but related set of equations that require only a finite number of operations for their solution: For example, a differential equation might be replaced by a finite difference equation. This process requires that continuous variables (such as the temperature or pressure fields) be reduced to a finite set of values at given locations (usually a square or a rectangular grid on the horizontal plane, with multiple levels in the vertical) and times.

A typical mesoscale (i.e., regional) atmospheric model will cover an area of 10^4 to 10^6 km^2, with a horizontal resolution (i.e., the real domain represented by one grid element in the numerical model) of 1 to 20 km. In the vertical direction, such models use 5 to 20 layers to describe the structure of the first 3 to 15 km of the atmosphere. General circulation models (GCMs) used in global climate studies typically represent the entire earth's surface with a horizontal resolution of 3° to 8° of latitude and longitude, and the vertical structure of the atmosphere with 2 to 11 layers. Those used for short-term numerical weather prediction have a somewhat higher resolution, for example 1° horizontally and 18 layers in the vertical.

These new equations are then further converted into computer programs for processing. Clearly, there is a trade-off between the amount of detail that is desirable to address the issues at hand (and, in the case of prognostic models, the length of time over which the prediction is to occur), and the resources required to carry out the computation.

Three types of limiting factors are inherent in atmospheric numerical models. One type arises from the approximations and assumptions made in choosing the form of the mathematical system of equations to be used to describe atmospheric behavior. The second type results from the properties of the numerical scheme used to solve the governing set of simultaneous partial differential equations. The third type depends on the choice of model grid resolution and model domain size. Despite these limits, numerical modeling offers the most flexible method to study land–atmosphere interactions and is the only available method to evaluate the magnitude of these interactions at the global scale.

IV. Land and Atmosphere Parametrization in Numerical Models

A. Land Parametrization

One of the main difficulties in representing the surface processes in atmospheric numerical models is related to the number and complexity of the nonlinear processes involved. For instance, the albedo of a soil depends on the soil surface moisture content, which itself varies as a function of atmospheric input (precipitation) as well as soil processes (infiltration, runoff) and the rate of evaporation. The latter depends in turn on the availability of water at the surface, the gradient of humidity between the surface and the atmosphere, and the wind profile. When vegetation is present, the system is even more complex. This is because the transpiration of plants, which usually constitutes the most significant part of the total evapotranspiration, is biophysically controlled by valve-like structures on the leaf surface called stomata. These stomata interact strongly and rapidly with different components of their environment (e.g., solar radiation, carbon dioxide concentration, air humidity and temperature, soil moisture in the root zone), and their mechanism is apparently specific to each plant type. Also, the plant canopy modifies the roughness of the surface and, therefore, influences the pattern of wind and the turbulent fluxes of heat and moisture above and within the canopy.

In addition to this fundamental problem, the spatial resolution of atmospheric models (typically from 10 km² in mesoscale models and up to 5×10^5 km² in GCMs) is also generally much larger than the characteristic spatial scale of the relevant surface processes. Even on the smallest scales, large heterogeneities of soil type and soil wetness, of vegetation type and vegetation density, of topography, of urbanization, etc., are frequently observed. Since the fluxes of mass, momentum, and energy between the land surface and the atmosphere depend heavily on the nature, structure, composition and evolution of this surface, one would ideally like to be able to represent explicitly these complex processes at the spatial and temporal scales at which they occur. This cannot be done, however, because of the computational cost that would be involved and the tremendous amount of data that would need to be processed for each simulation. On the other hand, it may not be necessary to include every detail of the surface either, to predict the general behavior of the atmosphere. For practical reasons, it is therefore desirable to approximate the highly variable contributions to the surface fluxes in terms of the larger-scale processes that can be described explicitly in the model: The small-scale processes are said to be *parametrized* in terms of variables describing the large-scale processes. This difficult but necessary step in modeling this system is not specific to surface processes; similar procedures must be applied to represent cloud and precipitation processes in the atmosphere, for example.

The current state-of-the-art parametrizations of land surface in atmospheric models are based on the "big leaf—big stoma" concept. With this concept, the vegetation is represented as a single leaf with negligible heat capacity. Its density is characterized by a ground cover parameter, σ_f, which is the fractional area over which the foliage prevents solar radiation from reaching the ground. Thus, σ_f varies between 0 (bare ground) and 1 (full canopy cover). The surface energy fluxes (sensible and latent) are calculated from two energy budget equations, one for the vegetation layer and one for the soil surface (which is partly bare and partly covered by vegetation). No storage of heat is allowed in the canopy layer and, therefore, the net radiative energy absorbed by the plant canopy is released back into the atmosphere as sensible and latent heat. The contribution of the vegetation and the bare soil to the global energy fluxes from each grid element of the model are linearly extrapolated using σ_f. The leaf has one single stoma that, in the most sophisticated parametrization, reacts to the various components of the plant environment known to have an impact on its mechanism.

In fact, however, the microenvironment of each leaf in the canopy is somewhat different from that of the other leaves in the same canopy, and it varies in time. This is due to the particular inclination, orientation, and shading of the different leaves which, as a result, absorb different amounts of solar radiation and, therefore, have different temperature, humidity, and heat exchange (sensible and latent) with the ambient air. In addition, the wind constantly modifies these conditions. Also, plants change the orientation of their leaves in response to nyctinastic (sleep movements), seismonastic (movements in response to shaking), and heliotropic (movements in response to changes in illumination) stimuli. Obviously, because of the relatively fast reaction of the stomata to the microenvironment, a large variability of stomatal conductances and microenvironmental conditions is expected even in otherwise homogeneous canopies, generating gradients of temperature and humidity inside the canopy. Such variability and gradients are indeed observed in field experiments.

Thus, because the surface fluxes are strongly nonlinearly related to the gradients of temperature and humidity, the "big leaf—big stoma" concept is not appropriate to represent the landscape heterogeneity of the real world. Moreover, a large number of settable empirical constants (sometimes as many as 50) must be provided to implement this type of parametrization. Because these constants might be different for different growing periods and for all the different plants found in the domain represented by a single grid element of the atmospheric models, a gigantic amount of data would need to be collected, archived, and compiled to provide the "representative" constants for the "big leaf—big stoma" of each grid element of the model. An extremely sophisticated micrometeorological field experiment allowing complete control of the plant environment would be required to provide a reasonable value for these constants. Therefore, there is no realistic way to provide such observations and it is not yet clear how such constants should be combined to describe the real world.

In addition, with this parametrization, the land is assumed to consist only of bare soil and vegetated surface. Except for the crude representation of the stomatal mechanism, none of the biological or chemical interactions between the land and the atmosphere is parametrized. Thus, built-up areas that require a particular treatment to correctly represent the land roughness, the release of heat fluxes, and the emission of pollutants, are not considered.

Therefore, the current state-of-the-art land-surface parametrizations for atmospheric models are poorly equipped to simulate land–atmosphere interactions. Consequently, the best the atmospheric models can do is a sensitivity analysis of the various components that affect the ideal "big leaf—big stoma." It must be conceded, though, that such analysis is a useful first step in that it helps to bracket the relative importance of some of the parameters involved in the system.

B. Atmosphere Parametrization

The parametrization of numerical models used to simulate land–atmosphere interactions includes not only the land aspects discussed above but, obviously, includes also atmospheric aspects. For instance, the hydrologic cycle cannot be simulated accurately without an appropriate parametrization of atmospheric turbulence (which is a key process for a correct evaluation of the surface heat fluxes) and of cloud physics. The problems involved in the parametrization of these complex processes require an extensive discussion that is beyond the scope of this article. These parametrizations, however, can be studied in detail from the many textbooks that have been published on this subject.

V. Numerical Modeling of Land–Atmosphere Interactions

Various numerical experiments have been realized with microscale, mesoscale, and global-scale atmospheric models to better understand the extent of land–atmosphere interactions at the local, regional, and global scales. Obviously, because of the limit of current parametrizations, these experiments have concentrated mostly on the impact of dry and moist (vegetated) land on the hydrologic cycle, probably the most important cycle in the climate system. A few typical cases are discussed here.

A. Local Interactions

The Colorado State University mesoscale numerical model is used for the simulations presented here. Starting from a given initial state, the conservation equations for mass, momentum, moisture, and energy, expressed in finite-difference are used in the model to calculate the evolution of the fields of wind, temperature, and water vapor. Because the domain simulated with this model generally does not exceed a few hundred kilometers, the curvature of the globe is neglected and a horizontal grid is adopted. In the version of the model used for this study, the atmosphere is represented by 19 levels between the surface and an elevation of 10 km, with a horizontal resolution of 3 km. The prescribed land conditions are topography, surface roughness, and soil and vegetation type.

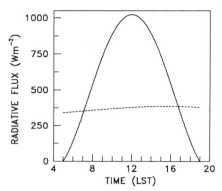

Fig. 2 Daytime variation of solar radiation (——) and infrared atmospheric radiation (– – –), simulated for a cloudless mid-summer day (August 15) at latitude 37° N.

Figure 2 illustrates the daytime variation of the radiative fluxes (solar and atmospheric) simulated for a cloudless mid-summer day at latitude 37° N. The maximum solar radiation for this time of the year and this latitude is slightly over 1000 W m^{-2}, and sunrise and sunset occur at 5:24 a.m. and 6:36 p.m. local standard time (LST), respectively. The average infrared atmospheric radiation is about 375 W m^{-2}. This energy input on a dry, bare soil surface and on a densely vegetated, moist surface results in the surface temperature and the sensible and latent heat fluxes presented in Fig. 3. Figure 4 shows the profiles of potential temperature and specific humidity at 2 p.m., obtained above these surfaces up to a height of 3000 m in the atmosphere.

Clearly, the temperature as well as the heat fluxes are extremely sensitive to the ability of the earth's surface to evaporate and transpire water. At noon, for instance, the surface temperature, sensible heat flux, and latent heat flux differences between the two simulated cases are as large as 26 K, 425 W m^{-2}, and 600 W m^{-2}, respectively. The faster heating rate of the dry land surface generates a vigorous turbulent mixing and an unstably stratified PBL, as can be seen from the profile of potential temperature in Fig. 4a (the height of the PBL is given by the atmospheric level at which the potential temperature starts to increase with altitude—at about 2000 m in this case). On the contrary, the slower heating rate of the vegetated surface limits the development of the PBL to a height of about 750 m, but the plant transpiration provides a supply of moisture that significantly increases the amount of water in the shallow PBL, as evident from the profile of specific humidity in Fig. 4b.

For these simulations a "clay loam" soil type (as classified by the United States Department of Agriculture) was considered. This soil is characterized by relatively low hydraulic and thermal conductivity. Thus, water does not percolate easily from the plant root zone and, therefore, remains available for

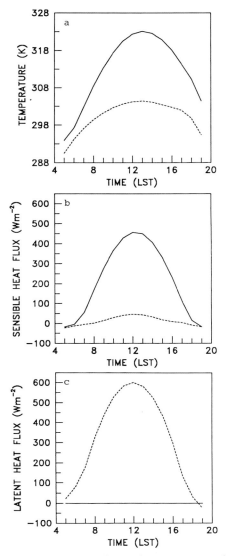

Fig. 3 Daytime variation of (a) surface temperature; (b) sensible heat flux; and (c) latent heat flux, simulated for a cloudless mid-summer day (August 15) at latitude 37°N for a bare, dry land (——) and for a densely vegetated, moist land (---).

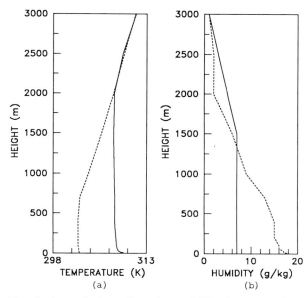

Fig. 4 Atmospheric profile, at 2 p.m. (LST), of (a) potential temperature; and (b) specific humidity, simulated for a cloudless mid-summer day (August 15) at latitude 37°N over a bare, dry land (——) and over a densely vegetated, moist land (---).

It is interesting to note that the solar and atmospheric radiation absorbed at the earth's surface has a considerable impact on the atmosphere up to a height of about 2000 m (at 2 p.m.). However, it affects at most only the upper 30 cm of the soil layer (when transpiring plants cover the ground, soil temperatures are influenced hardly at all).

A sensitivity analysis of the different parameters involved

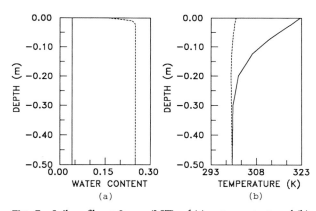

Fig. 5 Soil profile, at 2 p.m. (LST), of (a) water content; and (b) temperature, simulated for a cloudless mid-summer day (August 15) at latitude 37°N for a bare, dry land (——) and for a moist soil covered by a dense plant canopy (---).

transpiration during a relatively long time period. This is illustrated by the moisture profile of the wet soil presented in Fig. 5a, which indicates that only the upper soil level has lost water by the combined effect of evaporation and percolation at this time of the day (2 p.m.). The relatively low thermal conductivity of a dry soil results in a strong temperature gradient of approximately 1.5 K cm⁻¹ in the upper level of the bare, dry soil at the same time (Fig. 5b).

in the land – atmosphere interactions consists of increasing and decreasing substantially the value of one of the parameters of the system (e.g., solar radiation, wind speed, surface albedo, surface roughness) and noting how these changes affect the diurnal variation of the energy fluxes and the temperature and humidity profiles in the atmosphere and the soil. Such an analysis shows that the major factors affecting land – atmosphere interactions are the input of energy (radiative and kinetic) to the system and the water availability for evapotranspiration. The former is mainly controlled by solar and atmospheric radiation, surface albedo, wind speed, and surface roughness. The latter, of course, depends on the soil water content, but is also strongly affected by the soil hydraulic properties that control the water movement to the soil surface, where it is evaporated. In addition, when vegetation is present, the transpiration of plants is controlled by the plant stomatal mechanism. Therefore, the credibility of a numerical model used to simulate land – atmosphere interactions depends on its ability to represent appropriately these processes.

B. Regional Interactions

Continental surfaces are very heterogeneous. This can be readily appreciated by looking at maps of soil, vegetation, topography, or land use patterns, for instance. This heterogeneity results in variations in water availability for evapotranspiration and, consequently, in differential heating for different surfaces subjected to the same solar radiation input.

A direct consequence of this differential heating is that the air immediately above these surfaces also heats up in proportion to the temperature of the surface (Fig. 4a). As the air above dry land becomes warmer and less dense than the air above moist land, a pressure gradient is created at some distance above the ground between these different areas. Pressure gradients produced in this way generate a movement of air from the relatively cool (high pressure) to the relatively warm (low pressure) area. This wind enhances the surface heat fluxes and results in a stronger heating of the air above dry land and in a larger supply of water in the atmosphere above moist and vegetated surfaces. Thus, this process has a positive feedback on the intensity of the wind and on the surface heat fluxes.

The full set of physical processes that control and affect the land – atmosphere interactions at the regional scale can be studied in great detail with mesoscale atmospheric models, once they are properly set up to represent appropriate surface conditions. As a demonstration, two numerical experiments (two-dimensional) were carried out with the Colorado State University mesoscale numerical model. In the first experiment, the development of an irrigated agricultural area in an arid region that consisted essentially of bare, dry land was

simulated. This type of human activity has been observed at various geographical locations (e.g., the Sacramento Valley and the area adjacent to the Salton Sea in California; the San Luis Valley in Colorado; along the Nile River in Egypt and the Sudan; near Lake Chad in Africa; the northern Negev Desert in Israel). In the second experiment, the process of deforestation of a densely vegetated tropical area was simulated. This other type of human activity occurs at an alarming pace in tropical regions such as the Amazon basin in Brazil and is also frequent in mid-latitudes.

The simulated regions were 180 km wide, consisting of three adjacent areas (bare or vegetated) each 60 km wide. For the arid land simulation, the agricultural area was located in the middle of the domain and was flanked by two dry-land areas. Correspondingly, for the tropical numerical experiment, the deforested area was in the middle of the domain and was flanked by two densely vegetated areas. These horizontal domains are schematically represented at the bottom of each graph in Fig. 6 and 7, where vegetated areas are represented by a dark underbar.

The numerical integration for the simulations was started at 6 a.m. (LST), which corresponds to the time when the sensible heat flux becomes effective in the development of the convective PBL on sunny summer days (see Fig. 3b). The horizontal distribution of surface heat fluxes, surface temperature, and air temperature (at a height of 5 m above the surface) obtained after 8 hours of simulation (i.e., at 2 p.m.) are presented in Fig. 6a for the arid case and in Fig. 6b for the tropical case. Figure 7 presents the meteorological fields (u is the west – east horizontal component of the wind parallel to the domain; w is the vertical component of the wind; θ is the potential temperature; and q is the specific humidity) obtained at the same time for both the arid (Fig. 7a) and the tropical (Fig. 7b) cases. Only the lower 3 km of the atmosphere (i.e., roughly the PBL) are displayed in this figure.

Even though the model was initialized with spatially homogeneous atmospheric and land conditions (in each of the land – surface subdomains separately), after 8 hours of simulation all the meteorological fields as well as the surface fluxes and the temperatures depict a considerable spatial variability. Of course, the major variability of surface heat fluxes and temperatures is due to the transition from dry land to vegetated areas. As expected, dry land is characterized by high sensible-heat fluxes, high temperature, and low latent-heat fluxes. On the contrary, vegetated land is characterized by relatively low sensible-heat fluxes, low temperature, and very high latent-heat fluxes. But superimposed on the variability due to landscape heterogeneity, there is another variability caused by the feedback of the wind, the humidity, and the temperature of the lower atmosphere. In particular, the impact of the wind intensity on the sensible heat flux and the surface temperature is clearly noticeable in the dry land: At

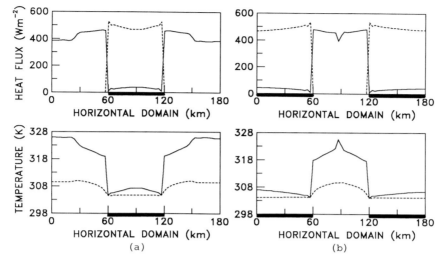

Fig. 6 Spatial distribution of sensible heat flux (——), latent heat flux (---), surface temperature (——), and air temperature at a height of 5 m above the surface (---), simulated at 2 p.m. (LST) for (a) a 60-km wide agricultural area developed in an arid region (latitude 37°N); and (b) a 60-km wide deforested area in a densely vegetated tropical region (latitude 0°). Vegetated areas are indicated by dark underbars.

these locations where relatively strong winds have developed, the sensible heat flux is about 20% higher than that obtained in the calm areas of the dry land. Correspondingly, the surface temperature is about 8 K cooler in the dry land exposed to high winds.

The regional circulation (depicted by the horizontal and vertical components of the wind in Fig. 7) for both simulations is quite evident. As one can see, two wind cells have developed in opposite directions in the two considered cases. The intensity of the wind is mainly governed by the contrast of sensible heat flux between the dry, bare land and the vegetated areas. Thus, the horizontal wind intensity of each wind cell in both simulations is approximately similar. However, the vertical component of the wind is about four times stronger in the deforestation case than it is in the arid one. This is due to the convergence of the flow above the deforested area as compared to the divergence of the flow from the agricultural region to the surrounding arid areas. These circulations (often referred to as "vegetation breezes") are comparable in intensity and extent to sea and lake breezes, which are generated by the contrast of sensible heat flux between dry land and bodies of water.

It is interesting to note that in the deforestation simulation, the relatively strong horizontal winds converging toward the center of the deforested area (which is also the center of the simulated domain) create, at their meeting point, a small zone where the horizontal component of the opposing winds is

virtually eliminated. As a result, the sensible heat flux drops significantly and the surface temperature increases substantially in this small zone. This is illustrated by the peak values in the curves of sensible heat flux and temperature in Fig. 6b.

As schematically illustrated in Fig. 8, these circulations have a considerable impact on cloud formation and the hydrologic cycle. For instance, in the deforestation simulation, the forest evapotranspiration provides a supply of moisture that significantly increases the amount of water in the shallow PBL that has developed above it. This PBL is significantly forced by the relatively large aerodynamic roughness of the forest, which produces an important mechanical mixing. The moisture is advected by the thermally induced flow toward the deforested area, where it is well mixed within the relatively deep convective boundary layer. Humid air masses raised to the higher and cooler atmospheric levels by the relatively strong vertical velocity may condense and generate clouds under appropriate atmospheric conditions.

A few observational studies have been carried out during recent years to try to validate the regional land–atmosphere interactions described here. For instance, measurements of temperature, humidity, and wind taken from a small airplane flying at different altitudes above the boundary between dry, arid areas and intensively irrigated areas have shown, at least qualitatively, the existence of such processes. Satellite infrared temperature images have clearly shown high temperature contrasts between such dry and irrigated areas.

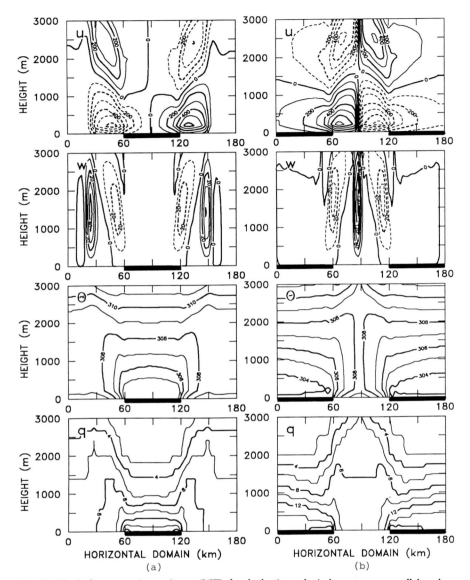

Fig. 7 Vertical cross section, at 2 p.m. (LST), for the horizontal wind component parallel to the domain (u) in cm s^{-1}, positive from left to right; the vertical wind component (w) in cm s^{-1}, positive upward; the potential temperature (θ) in K; and the specific humidity (q) in g kg^{-1}. Comparative data are shown for (a) a 60-km wide agricultural area developed in an arid region (latitude 37°N) and (b) a 60-km wide deforested area in a densely vegetated tropical region (latitude 0°). Vegetated areas are indicated by dark underbars. Solid contours indicate positive values; dashed contours indicate negative values.

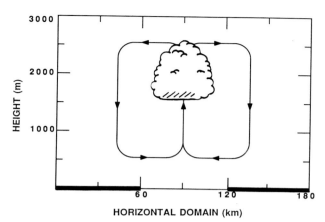

Fig. 8 Schematic representation of cloud formation resulting from a regional circulation generated by deforestation in a tropical region. Arrows indicate wind direction and forested areas are indicated by dark underbars.

Another type of regional circulation generated by an extended contrast of sensible heat flux is the so-called "urban circulation." These circulations, which usually converge toward the "urban heat island," have an important impact on the advection, diffusion, and mixing of various pollutants emitted within the cities. These pollutants modify the chemical composition of the atmosphere and may affect considerably the radiative energy input to the land surface. They may also react with different chemical and biological systems. Depending on the intensity of the contrast of sensible heat flux obtained between the urban and their neighboring rural areas, circulations even stronger than vegetation breezes can develop.

During daytime hours, the highest air temperature is usually obtained near the earth's surface, and it decreases with height in the atmosphere. As a result, topographical features might create horizontal gradients of temperature, as schematically illustrated in Fig. 9. The air temperature in A (at a distance ΔZ_1 from the earth's surface) is colder than the air in B, which is closer to the surface ($\Delta Z_2 \ll \Delta Z_1$). This type of gradient of temperature might also generate circulations known as "thermally induced upslope circulations."

These circulations, which are also important aspects of land–atmosphere interactions, have been studied extensively with mesoscale atmospheric models (in the limit of the models' parametrization). Because of the essential similarity of their mechanism to that of a vegetation breeze, no further example is provided here.

C. Global Interactions

Atmospheric GCMs have been used occasionally to study the impact of land surface interactions on the weather and cli-

mate at the global scale. Perhaps the most representative of such numerical experiments was performed by Shukla and Mintz in 1982, using the Goddard Laboratory for Atmospheric Sciences' GCM.

Starting from a given initial state, the conservation equations for mass, momentum, moisture, and energy, expressed in finite-difference form for a spherical grid, are used in the model to calculate the evolution of the pressure field at the earth's surface and of the fields of wind, temperature, and water vapor at nine levels between the surface and an elevation of 20 km. The fields of the convective clouds and precipitation, and the large-scale upglide clouds and precipitation, are also calculated, with a horizontal resolution of 4° of latitude and 5° of longitude over the globe and a nearly continuous variation in time. The prescribed surface boundary conditions are the ocean surface temperature, the large-scale topography and surface roughness, and the surface albedo. The detailed structure and properties of this model and its ability to simulate the broad characteristics of the observed climate of the earth are described in a number of publications.

In Shukla and Mintz's numerical experiment, two 60-day integrations starting from identical conditions on 15 June, with two different constraints placed on the land–surface evaporation, were performed. In one case, referred to as the "wet-soil case," the actual rate of evaporation was always set equal to its potential (maximum) value under the existing climatological conditions. In the other case, referred to as the "dry-soil case," the actual rate of evaporation was permanently set to zero. Such a study, sometimes called the "global irrigation experiment," is of course a very crude and unrealistic experiment; but it represents a useful first step in that it brackets the behavior of the model between two extreme cases.

Figures 10, 11, and 12 show the precipitation, surface temperature, and surface pressure reduced to sea level, respec-

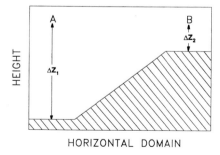

Fig. 9 Schematic representation of the horizontal gradient of temperature created by the topographical feature represented at the bottom of the graph. The points A and B are at the same altitude above sea level, but point B is closer than point A to the warm surface.

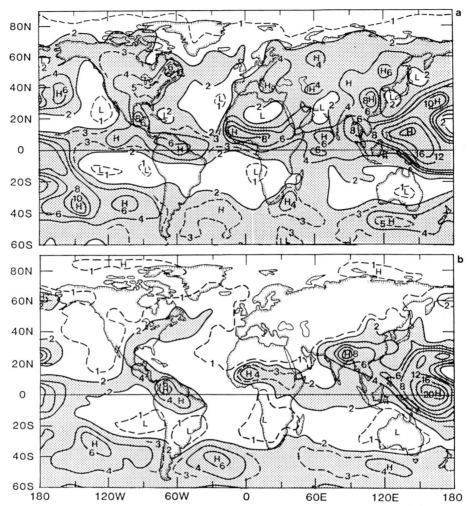

Fig. 10 Average precipitation, in mm/day, for the month of July, as simulated with the Goddard Laboratory for Atmospheric Sciences' general circulation model. Evapotranspiration from the continental surfaces is always set equal to (a) potential evapotranspiration and (b) zero. Shaded areas receive more than 2 mm/day. [Reproduced from Shukla and Mintz (1982); copyright 1982 by the American Association for the Advancement of Science.]

tively, obtained from these two numerical experiments. These results are time-averaged for July, the month when the northern hemisphere's extratropical region exhibits the maximum potential evapotranspiration. As one can notice, considerable differences in the patterns of precipitation, surface temperature, and surface pressure were observed in these simulations.

A detailed analysis of the results shows that over most of North America, for example, precipitation is between 3 and 6 mm/day in the wet-soil case and less than about 1 mm/day in the dry-soil case. The ground surface temperature is 15 to

25°C warmer in the dry-soil case as compared to the wet-soil case. This is due to the combined effect of evaporative cooling and increased cloudiness (which reduce the amount of solar radiation absorbed by the land surface) on the wet continents. Similarly, surface pressure is about 5 to 15 mbar lower over most of the land areas in the dry-soil case, suggesting enhanced cyclonic circulations over the continents.

Similarly to the atmospheric circulations generated by the contrast of dry and moist (vegetated) areas, the contrast between the ocean (which maintains an approximately constant surface temperature on a daily time scale) and dry land

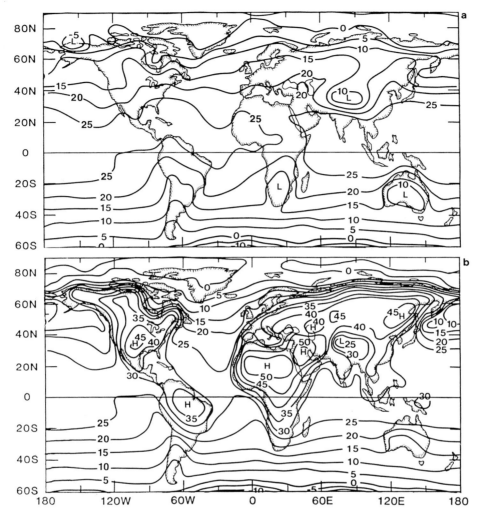

Fig. 11 Same as Fig. 10 but for surface temperature, in °C. [Reproduced from Shukla and Mintz (1982); copyright 1982 by the American Association for the Advancement of Science.]

produces strong circulations on the regional scale (i.e., sea breezes) and global scale. At the regional scale, such circulations advect moist air from the ocean over the dry land, where it is expected to increase the convective precipitation activity. Because of the poor resolution of the GCM, however, this type of regional precipitation, which would wet the land and modify the surface evaporation and the hydrological cycle, is not simulated. A better grid resolution (i.e., a smaller domain represented by one grid element in the numerical model) would probably have provided different results for the dry-soil case.

But the sensible heat flux obtained over the ocean is quite similar to that obtained over wet land. Thus, the contrast between ocean and wet land cannot produce the convective activity obtained from the contrast of ocean and dry land. Therefore, considering only this particular aspect, a better grid resolution would not modify significantly the results for the wet-soil case. This emphasizes the problem of resolution and/or parametrization of the mathematical models.

VI. Summary and Conclusions

Land–atmosphere interactions involve many physical, chemical, and biological processes that occur, at various spatial and

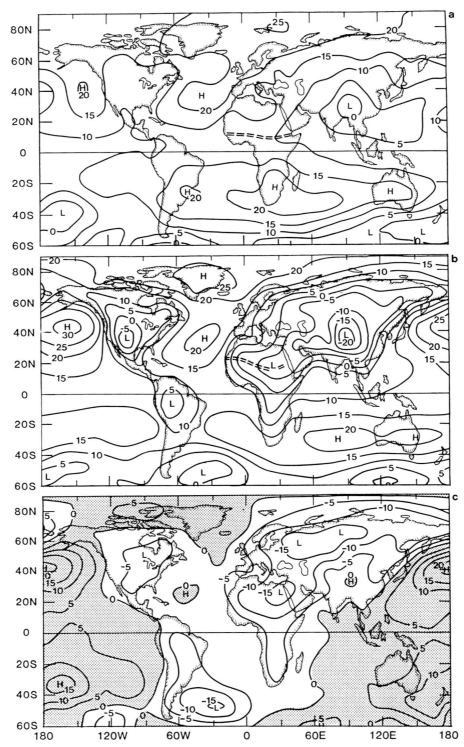

Fig. 12 Same as Fig. 10 but for surface pressure reduced to sea level (mbar minus 1000). In addition, (c) shows the difference of surface pressure between case (a) and case (b). Shaded areas indicate negative differences between the two cases. [Reproduced from Shukla and Mintz (1982); copyright 1982 by the American Association for the Advancement of Science.]

temporal scales, between the land and the atmosphere. Only limited information on the extent and magnitude of these interactions can be obtained from physical models and/or directly from observational methods. Mathematical models, if correctly parametrized and initialized, offer potentially a more appropriate alternative. These models, in conjunction with observational methods used to validate processes and provide adequate initialization data, have the potential ability to provide a reliable estimation of these interactions. In addition, these models can be used to simulate different scenarios to forecast the expected impact of human activities on the landscape and the atmosphere and, therefore, help to establish policies.

Water availability for evapotranspiration at the earth's surface has a considerable impact on land temperature, on the atmospheric planetary boundary layer (PBL) development, and on atmospheric temperature and humidity. High temperature, low humidity, and a well-developed PBL are obtained over dry land surfaces. On the contrary, low temperature, high humidity, and a shallow PBL are obtained over moist (vegetated) land surfaces. The contrast between dry and vegetated domains, at the regional scale, may generate atmospheric circulations that have a considerable impact on the local hydrologic cycle. These regional effects superimposed on global processes generated by land–ocean–atmosphere interactions probably have a significant impact on the global hydrologic cycle, one of the most important cycles affecting global climate. Unfortunately, due to the current limit of available computer resources, global climate simulations cannot be run with the grid resolution required to represent these regional effects. A reasonable parametrization of such effects is not yet available and, therefore, needs to be developed.

Obviously, the parametrization of land surface processes in atmospheric models plays a major role on the simulation of land–atmosphere interactions. During the past years, a few research groups have provided important contributions to this difficult task, but state-of-the-art parametrizations currently in use are still far removed from representing the landscape heterogeneity found in the real world. Therefore, current models can be used, at best, to provide sensitivity analysis of some aspects of the hydrologic cycle and cannot be expected to provide accurate climatic forecast.

Clearly, an important research objective would be to improve these parametrizations to account for landscape heterogeneities and to include other processes believed to have an important impact on land–atmosphere interactions and the climate. For instance, an appropriate representation of the biophysical processes involved in the exchange of carbon dioxide between the land and the atmosphere is greatly desirable. This element, which is released at an alarming pace in the atmosphere as a result of human activity, is suspected of having a significant impact on the radiative cooling of the earth (a "greenhouse effect") and, therefore, on the global climate.

Most recent studies indicate a tendency to drop the deterministic approach adopted so far to represent the land in atmospheric models for a statistical-dynamical approach. With this new approach, probability density functions are used for the various parameters of the soil–plant–atmosphere system rather than theoretical representative values. This apparently more promising parametrization is, however, still in its infancy and a strong research effort is still required to produce a reliable parametrization.

Finally, the present article has presented a discussion of the basic mechanisms involved in land–atmosphere interactions, as well as a brief description of the tools available to study these interactions. This is, however, not more than the "tip of the iceberg." Many important numerical and observational studies that have been carried out on these interactions and their impact on our day-to-day lives have not been discussed here. The reader interested in a detailed treatment of the subject is referred to more technical reviews (e.g., Avissar and Verstraete, 1990).

Bibliography

Avissar, R., and Verstraete, M. M. (1990). The representation of continental surface processes in atmospheric models. *Rev. Geophys.* **28,** 35–52.

Avissar, R., Moran, M. D., Wu, G., Meroney, R. N., and Pielke, R. A. (1990). Operating ranges of mesoscale numerical models and meteorological wind tunnels for the simulation of sea and land breezes. *Boundary-Layer Meteorol.* **50,** 227–275.

Dickinson, R. E., ed. (1987). "The Geophysiology of Amazonia: Vegetation and Climate Interactions." Wiley, New York.

Hansen, J. E., and Takahashi, T., eds. (1984). "Climate Processes and Climate Sensitivity," M. Ewing Series 5. American Geophysical Union, Washington, D.C.

Hillel, D. (1982). "Introduction to Soil Physics." Academic Press, London.

Houghton, J. T., ed. (1984). "The Global Climate." Cambridge Univ. Press, Cambridge, England.

Pielke, R. A. (1984). "Mesoscale Meteorological Modeling." Academic Press, New York.

Shukla, J., and Mintz, Y. (1982). Influence of land–surface evapotranspiration on the Earth's climate. *Science* **215,** 1498–1501.

Stull, R. B. (1988). "An Introduction to Boundary Layer Meteorology." Kluwer Academic Publishers, Dordrecht, The Netherlands.

Wood, E. F., ed. (1990). "Land Surface-Atmospheric Interactions for Climate Modeling: Observations, Models and

Analysis." Kluwer Publishing Co., Dordrecht, The Netherlands.

Glossary

Advection Primarily used to describe predominantly horizontal motion of properties (e.g., energy and mass) in the atmosphere.

Albedo Ratio of the amount of solar radiation reflected by a body to the amount incident upon it.

Convection Mass motions within a fluid resulting in transport and mixing of properties (e.g., energy and mass). Usually restricted to predominantly vertical motion in the atmosphere.

Emissivity Ratio of the total radiant energy emitted per unit time per unit area of a surface at a specified wavelength and temperature to that of a black body (i.e., a hypothetical body that absorbs all of the radiation striking it) under the same conditions.

Evapotranspiration Combined transfer of water from the earth's surface to the atmosphere by evaporation (i.e., the process by which water is transformed into water vapor) and transpiration (i.e., the evaporation from plants, which is controlled by a plant biophysical mechanism).

Infrared atmospheric radiation Energy received from the atmosphere by radiation in the range 3.0–100.0 μm of the electromagnetic spectrum.

Latent heat Heat released or absorbed per unit mass by a system in changing phase.

Planetary boundary layer That part of the atmosphere that is directly influenced by the presence of the earth's surface and that responds to surface forcings with a time scale of about an hour or less.

Potential temperature Temperature a parcel of dry air would have if brought adiabatically (i.e., without gaining or losing energy) from its present position to a standard pressure of 100 kPa.

Sensible heat That heat energy able to be sensed (e.g., with a thermometer).

Solar radiation Energy received from the sun by radiation in the range 0.15–3.0 μm of the electromagnetic spectrum.

Large Marine Ecosystems

Kenneth Sherman

National Oceanic and Atmospheric Administration
National Marine Fisheries Service

Large marine ecosystems (LMEs) are extensive regions of the world ocean, generally on the order of $\geq$ 200,000 km^2, characterized by distinct bathymetry, hydrography, productivity, and trophically related populations. Global change, as a result of ozone depletion and greenhouse warming, may become a source of stress on the biomass production of LMEs. The dramatic fluctuations in marine biomass yields over the past two decades, when considered in light of growing concerns over coastal pollution and global change, are serving to accelerate the movement toward the adoption of LMEs as regional units for management of living marine resources. Coastal nations can now apply an ecosystem approach to the management of large marine ecosystems, based on customary international law.

I. The Large Marine Ecosystem Concept

Large marine ecosystems (LMEs) are regional units of ocean space that are the subject of research and management aimed at the maintenance and sustained development of marine resources at risk from overexploitation, pollution, and environmental change. Studies of LMEs and their management are based on the relationships among organisms, their environment, and the ecological and socioeconomic pressures controlling growth, survival, mortality, and recruitment of the multispecies assemblages supporting biomass yields of LMEs.

The concept of regional marine ecosystem research and management has been growing slowly within the international community. This trend began with the deliberations of the International Council for the Exploration of the Sea (ICES). The first attempts by ICES to deal with the research and management of large marine ecosystems took place in Kristiana, Norway, in 1901, where representatives from Denmark, Germany, Norway, Russia, and the United Kingdom initiated the planning of joint biological and hydrographical studies of the North Sea and the Baltic Sea. The meeting was prompted by the realization that the capacity of the oceans to produce commercially desirable fish species was finite and that "overfishing" could result in serious depletion of the stocks. During the intervening 90 years, ICES has provided a fertile ground for the development of joint international studies of marine ecosystems. In the process, generations of scientists participating in the work of ICES helped to focus attention on the advantages of coordinated multidisciplinary studies of the Baltic Sea, North Sea, Norwegian Sea, Barents Sea, Icelandic Shelf, Iberian coastal waters, and more recently the Arctic Ocean. Much of the new knowledge of ecosystem structure and function is included in the series of publications produced by ICES. Among the most informative are symposium volumes dealing with the principal forces driving variability in biomass yields of marine ecosystems, including those focused on (1) natural environmental perturbations; (2) pollution; and (3) a combination of overexploitation, pollution, and environmental perturbations.

A. New Era of Ocean Management: Law of the Sea

International law is supportive of an ecosystem-based approach to ocean resource management as codified in the United Nations Convention of 1982 on the Law of the Sea (UNCLOS). The UNCLOS designated the areas of coastal nations extending up to 200 miles from the baselines of territorial seas as exclusive economic zones (EEZs) wherein coastal states could exercise sovereign rights to explore, exploit, conserve, and manage the natural resources of the zones.

Nearly 95% of the annual global biomass yields of living marine resources (e.g., fishes, crustaceans, mollusks, algae) is produced within the boundaries of the EEZs. Under the terms of UNCLOS, coastal nations assumed responsibilities for

1. Ensuring through proper conservation and management measures that the maintenance of the living resources in the EEZ is not endangered by overexploitation.
2. Promoting the optimal utilization of the living marine resources in the EEZ without prejudice to the need for conservation of those resources.
3. Protecting and preserving the marine environment and taking all measures necessary to ensure that activities under their jurisdiction or control are so conducted as not to cause damage by pollution to other states and their environment from any source.

B. The Evolving LME Approach to Ocean Research and Management

It is within the EEZs that initial focus on the importance of large marine ecosystems as entities for the management and conservation of living marine resources was directed. Although the designation of LMEs is an evolving scientific and geopolitical process, sufficient progress has been made to allow for useful comparisons to be made of the different processes influencing large-scale changes in the biomass yields of living marine resources in LMEs. To facilitate the comparisons, a series of symposia have been convened at the annual meetings of the American Association for the Advancement of Science (AAAS). Reports presented to the AAAS symposia over the past several years (1984, 1987, 1988, 1990) argued that LMEs are tractable global units for the conservation and management of living marine resources and demonstrate the utility of the comparative ecosystem approach for determining the principal sources of variability in biomass yields.

By matching sampling effort to the time and space scales of the processes that are of most direct influence to growth and survival of living marine resource populations, forecasts of biomass yield trends among the species can be improved for LMEs. Studies of changes in abundance and population renewals of resource species in general and fish stocks in particular on a large marine ecosystem scale are in agreement with the proposition by Robert Ricklefs of the University of Pennsylvania that ecologists should begin to address critical community processes on a regional basis. Ricklefs argues that it is with studies on the regional scale of events that ecologists should be challenged to broaden the concepts of regulation of local community structure and incorporate historic, systematic, and biogeographic information into the principles of

community ecology, in an effort to help unite local and regional perspectives.

C. Research Strategies for the Future

As the trend for the management of living marine resources moves from single species to multispecies assemblages, it becomes increasingly important to encompass entire ecosystems as management units. This approach will ensure that management measures designed to optimize the natural productivity of target species will also include consideration for related competitor/predator populations and their environments.

Greater emphasis has been focused over the past decade within the National Marine Fisheries Service of NOAA on approaching fisheries research from a regional ecosystem perspective in seven LMEs within the exclusive economic zone of the United States—the Northeast Continental Shelf, the Southeast Continental Shelf, the Gulf of Mexico, the California Current, the Gulf of Alaska, the East Bering Sea, and the Insular Pacific including the Hawaiian Islands. These ecosystems, in 1989, yielded 9.0 billion pounds of fisheries biomass valued at approximately $17 billion to the economy of the United States.

A description of the sampling programs providing the biomass assessments within the U.S. exclusive economic zone has been published in Folio Map 7 produced by the Office of Oceanography and Marine Assessment of NOAA's National Ocean Service. The map depicts the six ecosystems under investigation (Fig. 1). Sampling programs supporting biomass estimates in LMEs within the EEZ of the United States are designed to (1) provide detailed statistical analyses of fish and invertebrate populations constituting the principal yield species of biomass, (2) estimate future trends in biomass yields, and (3) monitor changes in the principal populations. The information obtained by these programs provides managers with a more complete understanding of the dynamics of marine ecosystems and how these dynamics affect harvestable stocks. Additionally, by tracking components of the ecosystems, these programs can detect changes, natural or human-induced, and warn of events with possible economic repercussions. Although sampling schemes and efforts vary among programs (depending on habitats, species present, and specific regional concern), they generally involve systematic collection and analysis of catch-statistics; the use of NOAA vessels for fisheries-independent bottom and midwater trawl surveys for adults and juveniles; ichthyoplankton surveys for larvae and eggs; measurements of zooplankton standing stock, primary productivity, nutrient concentrations, and important physical parameters (e.g., water temperature, salinity, density, current velocity and direction, air temperature, cloud

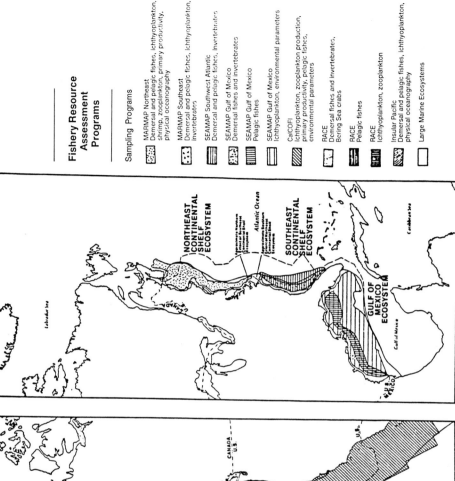

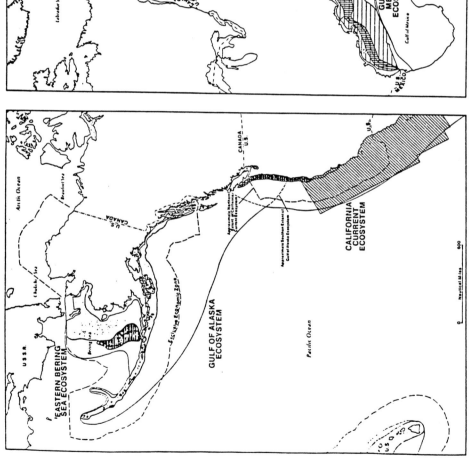

Fig. 1 Large marine ecosystems of the United States. [This figure is a modified version of Folio Map No. 7, "A National Atlas: Health and Use of Coastal Waters, United States of America." U.S. Department of Commerce, National Oceanic and Atmospheric Administration, National Ocean Service, Office of Oceanography and Marine Assessment, Washington, D.C. (1988). Modified and reproduced with permission. Figure drawn by Lyn Indalecio.]

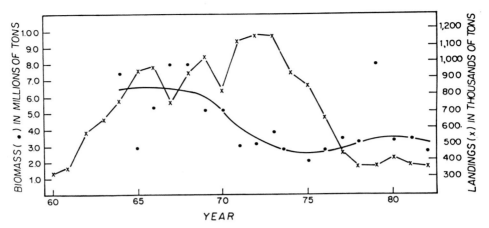

Fig. 2 Annual catch trends, excluding menhaden and large pelagic species such as large sharks and tuna, and estimated biomass of "exploitable" fish and squid of the Northeastern Continental Shelf ecosystem, 1960 to 1982. [From M. P. Sissenwine (1986). Perturbation of a predator-controlled continental shelf ecosystem. *In* "Variability and Management of Large Marine Ecosystems" (K. Sherman and L. M. Alexander, eds.), pp. 55–85. AAAS Selected Symposium 99, Westview Press, Boulder, Colorado.]

cover, light conditions); and, in some habitats, measurements of contaminants and their effects.

A critical feature of each program is the development of a consistent long-term database for understanding interannual changes and multiyear trends in biomass yields for each of the LMEs. For example, during the late 1960s and early 1970s, when there was intense foreign fishing within the Northeast Continental Shelf ecosystem, marked alterations in fish abundances were recorded. Significant shifts among species abundances were observed. The finfish biomass of important species (e.g., cod, haddock, flounders, herring, and mackerel) declined by approximately 50% (Fig. 2). This was followed by increases in the biomass of sand lance (Fig. 3) and elasmobranchs (dogfish and skates) and led to the conclusion that the overall carrying capacity of the ecosystem for finfish did not change. However, excessive fishing effort on highly valued species allowed for low economic valued species to increase in abundance. Analysis of catch-per-unit-effort and fishery-independent bottom trawling survey data were critical sources of information used to implicate overfishing as the cause of the shifts in relative abundance among the species of the fish community within the shelf ecosystem.

The NMFS–NOAA regional programs concerned with forecasting trends in biomass yields are being conducted in relation to the dynamics of large marine ecosystems. The studies are now yielding information on large-scale, between-year changes in ecosystem productivity and fish abundances. They will continue to provide an improved information base for understanding and managing the nation's living marine resources. Recent programs, including GLOBEC, initiated by

the National Science Foundation will expand research on problems of recruitment and provide critical knowledge of the principal factors controlling productivity within marine ecosystems. There is also a need to understand better the relationships among the boundary areas of LMEs, particularly the wetlands and estuaries, many of which provide critical habitat for the early life stages of fishes and invertebrates found in the LMEs as adults. Research and monitoring efforts that include the wetlands and estuaries at the landward margin of LMEs will provide important information on the cause and effect of perturbation in marine biomass yields and greatly improve the management of marine resources in LMEs. In an effort to monitor long-term global change, NOAA, along with the National Science Foundation and National Aeronautics and Space Administration, agreed in 1986 to develop and implement a program aimed at studying the globe as an ecosystem. The effort, to be conducted as part of the International Geosphere–Biosphere Program (IGBP), will encourage the use of a permanent network of satellites and buoys, new instruments for measuring changes on land and in the sea, acoustic systems, electronic pattern-recognition systems, and other technical advances that will allow for more systematic measurements to monitor biological and physical properties of LMEs.

II. A Global Perspective of LMEs

The 1990s represent a period of political and social transition that is likely to increase pressures for ocean development. As

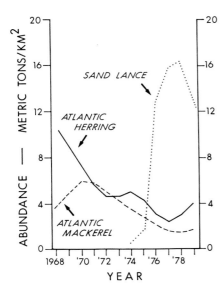

Fig. 3 Decline of Atlantic herring and Atlantic mackerel and apparent replacement by the small, fast-growing sand eel in the Northeast Continental Shelf ecosystem (measured in metric tons per square kilometer, 1968–1979).

global economies and populations expand (e.g., Western Europe, developing coastal countries), so do the potentials for resource depletions and environmental stress on regional seas. Coincident to the economic expansion is a greater awareness of global environmental issues (warming, sea-level rise, ozone shield depletion) and the need for improvements in the management of the oceans and their resources. Recent reports describe the alteration of non-steady-state marine ecosystems by natural and anthropogenic driving forces resulting in negative economic impacts.

The fish component of LMEs have adapted reproductive, growth, and feeding strategies to the unique environmental conditions within the LME. Changes to the fish components of these systems can trigger a cascade effect involving higher trophic levels (e.g., birds, marine mammals), lower trophic levels (zooplankton, phytoplankton), and the economies dependent on the resources of the LME.

A. Atlantic Rim LMEs

Increasing attention has been focused over the past few years on synthesizing available biological and environmental information influencing the natural productivity of the fishery biomass within LMEs in an effort to identify the principal, secondary, and, where important, the tertiary driving forces causing major shifts in the species composition of biomass yields.

Changes in the ocean climate of the northern North Atlan-

tic during the late 1960s and early 1970s have been considered by some marine scientists as the dominant cause of change in the food chain structure and biomass yields of at least three northern North Atlantic LMEs. Large-scale declines in the population levels of important fish stocks (e.g., capelin, cod) within the Norwegian Sea ecosystem, Barents, Sea ecosystem, and West Greenland Sea ecosystem have been observed.

In the West Greenland Sea ecosystem, cod stocks were displaced southward since 1980, attended by a decrease in the average size and abundance of cod. Biomass yields declined from about 300,000 metric tons (mt) per year in the mid-1960s to less than 15,000 mt in 1985. Both changes appear to have been the result of short-term cooling negatively affecting the growth and survival of early developmental stages of cod, causing a reduction in recruitment by influencing larval physiology, stability of the water masses, and the dynamics of the plankton community. Since the 1920s, the annual biomass yield of cod has been related to changes in temperature with catches increasing during warm periods and declining during cool periods. The effects of fishing mortality on the decline of the cod are of secondary importance, following the major influence of climatic conditions over the North Atlantic.

To the east, changes in the temperature structure of the Norwegian Sea ecosystem appear to be the major driving force controlling the recruitment of the important cod stocks of the system. Strong and medium sizes of cod biomass are related to warm years. The conditions for growth and survival of early developmental stages of cod are enhanced during warmer years when the larval cod are maintained for longer periods within coastal nursery grounds. There, the cod's most important prey organism, the copepod, *Calanus finmarchicus*, swarms in very high densities under conditions of well-defined thermocline structure and consequently provide optimal feeding conditions on the abundant phytoplankton production.

The changes in biomass yields of the Barents Sea ecosystem have been attributed primarily to changes in hydrographic conditions and secondarily to excessive fishing mortality. The average annual biomass yield of the ecosystem in the 1970s was about two million metric tons (e.g., fish, crustaceans, mollusks, algae). However, by the 1980s annual yields declined to approximately 350,000 mt (Table I). The decline of inflow of warm Atlantic water into the Barents Sea ecosystem, coupled with excessive levels of fishing effort led to (1) the collapse of the major fisheries of the region (cod, capelin, haddock, herring, redfish, shrimp); (2) subsequent disruption in the structure of the food chain; and (3) an increase in the abundance levels of the shrimp-like euphausiids representing a significant amount of biomass that is underutilized in relation to the potential sustained yield of this ecosystem body. Given the depressed state of the fish stocks, any restoration

Table I Nominal Catches (Tons) of Fish, Crustaceans, and Mollusks in the Barents Sea

Years	Bulgaria	U.K.	Denmark	Norway	West Germany	U.S.S.R.	Others	Total
1975	3,288	107,341	18,951	1,040,164	23,780	748,711	84,076	2,026,311
1976	—	74,886	5,071	1,569,403	35,440	885,456	25,971	2,596,227
1977	3,207	58,013	3,957	1,630,222	6,851	1,089,664	13,983	2,805,897
1978	—	26,108	13,542	1,044,837	—	859,256	2,835	1,946,578
1979	6,956	15,988	3,074	569,511	—	617,087	1,939	1,214,561
1980	1,539	7,728	9,618	445,242	—	584,448	3,139	1,051,714
1981	779	3,446	17,683	557,633	406	851,629	2,834	1,434,410
1982	385	3,703	11,488	548,478	136	592,427	615	1,157,232
1983	265	4,251	924	347,822	—	400,934	8,320	762,516
1984	—	1,413	5,103	295,741	21	360,716	—	662,994
1985	27	724	—	349,698	440	473,023	—	823,912
1986	—	1,730	370	211,521	220	220,781	—	434,622
1987	—	1,014	—	157,573	—	197,675	—	356,262

Source: V. M. Borisov (1990). The state of the main commercial species of fish in the changeable Barents Sea ecosystem. Paper presented to AAAS sponsored symposium on Food Chains, Yields, Models, and Management of Large Marine Ecosystems, convened by K. Sherman and B. D. Gold, 16 February 1990, New Orleans, Louisiana.

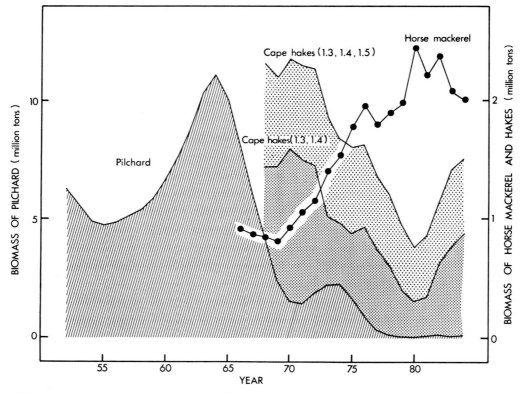

Fig. 4 Estimates of the biomass of pilchard *(Sardinops ocellatus)*, Cape horse mackerel *(Trachurus capensis)*, and Cape hakes *(Merluccius capensis* and *M. paradoxus)* of the Benguela Current ecosystem showing expansion of the horse-mackerel resource following collapse of the pilchard resource and the opposite trends in biomasses of horse mackerel and Cape hakes during the 1950s through the 1980s. [From R. J. M. Crawford, L. V. Shannon, and P. A. Shelton (1989). Characteristics and management of the Benguela as a large marine ecosystem. *In* "Biomass Yields and Geography of Large Marine Ecosystems" (K. Sherman and L. M. Alexander, eds.), pp. 169–219. AAAS Selected Symposium 111, Westview Press, Boulder, Colorado.]

management would need to consider significant reduction in fishing effort by fisheries of Norway and the USSR, the coastal nations that share the resources of the Barents Sea ecosystem.

In the North Sea ecosystem, important species have "flipped" from a position of dominance to a subordinate position within the ecosystem. A biomass flip occurs when the population of a dominant species rapidly drops to a very low level and is replaced by another species. The finfish stocks of the North Sea ecosystem have been subjected to intensive fishing mortality. The biomass flip in the North Sea occurred over the decade of the 1960s. The pelagic herring and mackerel yields flipped downward from 5 million metric tons (mmt) to 1.7 mmt, whereas small fast-growing and commercially less desirable sand lance, Norway pout, and sprat increased by 1.5 mmt along with an approximate 36% increase in gadoid yields. The causes for the biomass flips are not very well understood. Several arguments have been put forward that correlate the "flip" with changing oceanographic conditions as the principal driving force. Other explanations support overexploitation as the major source of the flip. However, none of the arguments can be considered more than speculative at this time, pending the rigorous analysis of more recent information.

Further to the south, the Iberian Shelf ecosystem has recently been examined in relation to variability of biomass yield. The alternation in abundance levels of horse mackerel and sardine within the Iberian Coastal ecosystem is attributed to changes in natural environmental perturbation of its thermal structure rather than to any density-dependent interaction among the two species. Similarly, in the Benguela Current ecosystem of the southwest coast of Africa, the long-term fluctuations in the abundance levels of pilchard, horse mackerel, and hakes are attributed to changes in the oceanographic regime (Fig. 4). The Benguela LME is bounded at both the equatorward and polarward extremities by warm water within the ecosystem. Cold nutrient-rich water is upwelled at moderate intensity in the central section and more intensely in the northern and southern areas. Environmental conditions favor either the epipelagic or the demersal species, never both simultaneously, and have been the principal driving force for large-scale shifts in abundance among the fish species. The effects of the fisheries on changes in species abundance are secondary. Changes in abundance of pilchard stocks have led to detectable effects in the abundance level of dependent predator species, particularly marine bird populations.

B. Pacific Rim LMEs

The greatest increases in biomass yields in the Pacific have been reported at the area of confluence between the Oyashio Current ecosystem and Kuroshio Current ecosystem off

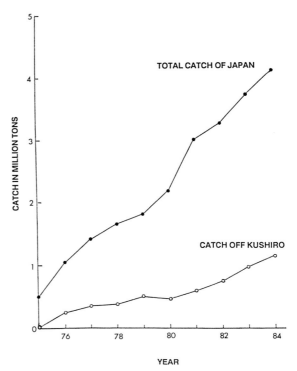

Fig. 5 Catch increases in Japanese sardine during the decade from 1975 to 1984 from the area of confluence between the Oyashio Current ecosystem and the Kuroshio Current ecosystem off the coast of Japan. [From T. Minoda (1989). Oceanographic and biomass changes in the Oyashio Current ecosystem. *In* "Biomass Yields and Geography of Large Marine Ecosystems" (K. Sherman and L. M. Alexander, eds.), pp. 67–93. AAAS Selected Symposium 111, Westview Press, Boulder, Colorado.]

Japan and in the Humboldt Current ecosystem off Chile. In the Oyashio and Kuroshio Current ecosystems the yield of Japanese sardines increased from less than 500,000 mt in 1975 to just over 4 mmt in 1984 (Fig. 5). The changes in yield of the Chilean sardine in the Humboldt Current ecosystem also increased from about 500,000 mt in 1974 to 4.3 mmt in 1986. The increased yields have been attributed to density-independent processes involving an increase in lower food-chain productivity, made possible by coastward shifts in the boundary areas of the Oyashio and Kuroshio systems, and water mass shifts in the Humboldt Current ecosystem. The effects of fishing on the sardines in both areas was of secondary importance to the enhanced productivity of the phytoplankton and zooplankton components of the ecosystems, providing an improved environment for growth and recruitment. Studies are underway to determine the extent of the teleconnection between the Pacific-wide El Niño events of the past decade and both (1) the multimillion metric-ton

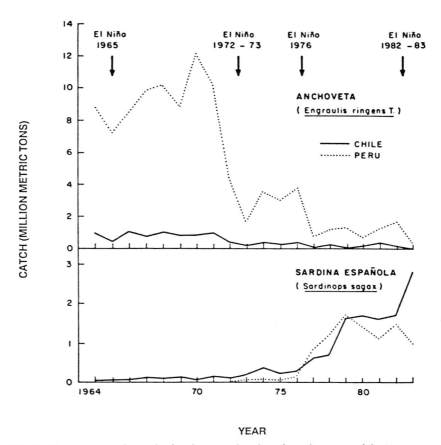

Fig. 6 Fluctuations in the catch of anchovies and sardines from the waters of the Humboldt Current ecosystem off the coasts of Chile and Peru. [From J. R. Canon (1986). Variabilidad ambiental en relacion con la pesqueria neritica pelagica de la zona Norte de Chile. *In* "La Pesca en Chile" (P. Arana, ed.), pp. 195–205. Escuela de Ciencias del Mar, Facultad de Recursos Naturales, Universidad Catolica de Valparasio.]

increases in yields of sardines occurring nearly simultaneously in the northern and southern hemispheres and (2) the dramatic decline in the biomass yields of anchovy in the northern areas of the Humboldt Current ecosystem in the early 1970s from about 12 mmt in 1970 to less than 2 mmt by 1976 (Fig. 6). [*See* EL NIÑO AND LA NIÑA.]

Although less dramatic, the long-term shift in abundance levels between sardines and anchovies within the California Current ecosystem (Fig. 7) is considered the result primarily of natural environmental change and secondarily from intensive fishing, rather than from any density-dependent competition between the two species.

Changes in biomass yields of two other Pacific rim LMEs have been the result of overexploitation. The introduction of highly efficient modern trawlers to the Gulf of Thailand ecosystem, in an effort to increase fishing efficiencies, led to

excessive fish mortality and a marked reduction in annual yields of biomass of fish for human consumption between 1977 and 1982 (Fig. 8).

Intensive fishery effort resulted in the depletion of the demersal fish stocks and dramatic reductions in the biomass yields of the Yellow Sea ecosystem. Between 1958 and 1968 fisheries' yields declined from 180,000 mt to less than 10,000 mt. The fishery then shifted to pelagic stocks reaching a level of 200,000 mt in 1972, followed by a reduction to less than 20,000 mt in 1981. The fisheries of the Yellow Sea in 1982 were shifted principally to anchovy and sardine with a total annual yield of all species 40% lower than the 1958 level. The demersal fishery remained in a depleted state (Fig. 9). In an effort to maximize economic yield, high-value species are being considered for introduction into the coastal waters of the Yellow Sea to enhance the demersal fisheries. Initial

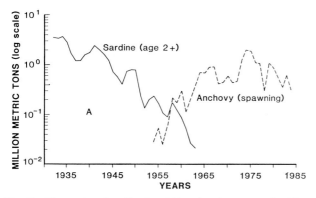

Fig. 7 Time series of sardine (age 2+) and anchovy spawning biomass (log scale) of the California Current ecosystem. "A" denotes approximate anchovy spawning biomass in 1940–1941. [From A. D. MacCall (1986). Changes in the biomass of the California Current ecosystem. *In* "Variability and Management of Large Marine Ecosystems" (K. Sherman and L. M. Alexander, eds.), pp. 33–54. AAAS Selected Symposium 99, Westview Press, Boulder, Colorado.]

attempts to introduce juvenile prawns for grow-out have been successful. Management strategies to enhance the recovery of depleted demersal stocks are contingent on reaching agreements among four countries involved in the fisheries of the Yellow Sea ecosystem: China, North Korea, South Korea, and Japan.

The importance of a natural predator driving an ecosystem is evidenced in the studies Australian scientists conducted on the large-scale changes in the community structure of the Great Barrier Reef ecosystem that extends over 230,000 km² of the Queensland continental shelf. The predation of the crown-of-thorns starfish in the 1960s and 1970s resulted in a shift in the biomass of corals, community structure of the benthos, and a decoupling of energy transfer to several fish stocks in the system.

To the north and west of Australia lies the relatively pristine Banda Sea ecosystem, where no large-scale fisheries are presently being conducted. The ecosystem is under the influence of monsoon-induced seasonal periods of large-scale upwelling and downwelling. Biological feedback to these environmental signals is reflected in the changes in phytoplankton, mesozooplankton, micronekton, and fish. During upwelling events, productivity of the ecosystem is enhanced by a factor of 2 to 3. The biomass of pelagic fish resources is also higher during the upwelling period. The fish biomass of the ecosystem is estimated at between 600,000 mt to 900,000 mt in the peak upwelling season (August), and 150,000 mt to 250,000 mt in the downwelling period (February). The estimated sustained annual biomass yield of the ecosystem is approximately 30,000 mt of pelagic fish.

C. Indian Ocean LMEs

Information on the biomass yields of the LMEs of the Indian Ocean has been largely limited to the reports of the Food and Agricultural Organization (FAO) of the United Nations.

The developing maritime states around the rim of the Indian Ocean have recently become active in the application of ecosystem principles to management of the living resources within their EEZs. The total biomass yield of the rim countries amounted to approximately 3.5 mmt in 1986, or 4.3% of the global total. Development of the LME approach to research and management is under consideration by the Indian government for the living marine resources in the Bay of Bengal ecosystem and West Indian Shelf ecosystem. The total annual yield reported by India from the two ecosystems is approximately 1.6 mmt or 48.2% of the total yield reported for the Indian Ocean. Other maritime nations reporting substantial annual biomass yields are Bangladesh with 6.7% of the total Indian Ocean yield and Burma with 15.5% of the total (Table II). Both Bangladesh and Burma participate in the fisheries of the Bay of Bengal ecosystem.

The next tier of states utilizing significant amounts of fisheries biomass from the Bay of Bengal ecosystem are Pakistan with 9.6% of the total biomass yield, Sri Lanka with 4.4%, Iran reporting catches for the Indian Ocean amounting to 3.4% of the total, and Oman with 3.3%. The United Nations Environmental Program and the FAO Fisheries Department are involved in joint studies with the maritime nations around the Indian Ocean rim fostering research and management programs aimed at implementing a balanced strategy for ensuring sustained yields of the living marine resources within the regional LMEs.

D. Polar LMEs

Common to most polar LMEs, according to a recent review by Gotthilf Hempel, are seasonal or permanent ice covers, year-round low temperatures, and intense seasonality in irradiance. The Arctic Ocean is characterized by a meridional current system and the Antarctic by latitudinal circular currents. Large gyres are features of both the Arctic Sea LME and the Antarctic ecosystem. The Arctic LMEs have broad shelf areas, are well stratified, and are characterized by seasonal depletions of nutrients. In contrast, the Antarctic ecosystem is for the most part a well-mixed, deep-water upwelling system. The Arctic LMEs support large populations of pelagic fish. The fish fauna of the Antarctic are limited to a few species in relatively low abundance. The shrimp-like euphausiids are the keystone species to the Antarctic food chain, whereas they are not as numerous in the Arctic LMEs. [See ANTARCTICA; ARCTIC.]

The polar ecosystems support large populations of birds and mammals. Major living marine resources of the Arctic

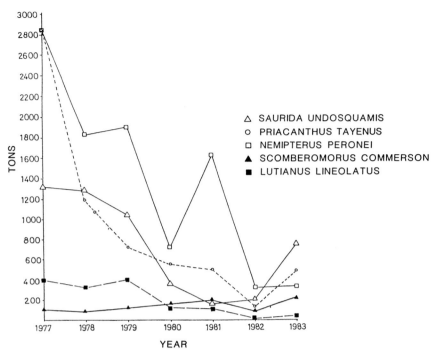

Fig. 8 Decline in the total catch of carnivorous feeding species of fish from the Gulf of Thailand ecosystem. [From T. Piyakarnchana (1989). Yield dynamics as an index of biomass shifts in the Gulf of Thailand ecosystems. *In* "Biomass Yields and Geography of Large Marine Ecosystems" (K. Sherman and L. M. Alexander, eds.), pp. 95–142. AAAS Selected Symposium 111, Westview Press, Boulder, Colorado.]

LMEs are the fishes inhabiting the shallow areas of the Bering Sea ecosystem and the Barents Sea ecosystem.

The finfish of the Antarctic ecosystem have been overexploited and are in a depleted state. The principal biomass yield is the euphausiid, *Euphausia superba*, known as "krill." The annual standing stock of krill is estimated at several hundred million metric tons. Present levels of catches are about 500,000 mt annually.

In the Antarctic, the Weddell Sea ecosystem has been studied rather intensely by scientists of the Alfred Wegener Institute for Polar Research, Federal Republic of Germany. The abundant fish species is a relatively small pelagic species, *Pleurogramma antarcticum*. In the Weddell Sea ecosystem, it is the principal prey of marine birds and mammals, rather than krill. The permanent ice cover of the Weddell Sea precludes any potential for significant fishing or whaling activities.

In contrast, in the Arctic subpolar area of the Bering Sea the fishery resources are among the most productive in the world. The sustained biomass yield reached 2.5 mmt in the early 1970s and declined to about 1.7 mmt in 1984. The estimated standing stock of biomass was 15 mmt in the early

1980s and is on a downward trend, declining to about 13 mmt in 1984, reflecting the impacts of intensive fishing effort. The principal species caught by the fishery is the Alaskan pollock. From an ecosystem perspective, important research questions presently being addressed concern pollock stock structure, the role of cannibalism on recruitment, the dependence of marine birds and mammals on pollock as a principal prey species, and the determination of total allowable catch levels that will ensure sustained yields of the resource.

III. LMEs and Biomass Yields

On a global scale, the experts have been off the mark in earlier estimates of annual sustained global yield of usable biomass. Projections given in *Global 2000 Report* of 1980, prepared by the U.S. Council on Environmental Quality and the Department of State, were reported as expected to rise little, if at all, by the year 2000 from the 60 mmt reached in the 1970s. In contrast, estimates given by J. Wise in the rebuttal volume by J. Simon and H. Kahn, entitled *The Resourceful Earth*, argue

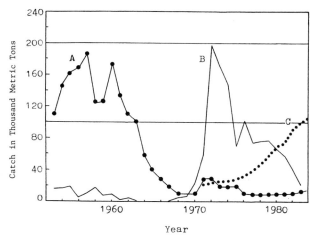

Fig. 9 Annual catch of dominant species: (A) small yellow croaker and hairtail; (B) Pacific herring and Japanese mackerel; (C) *Setipinna taty,* anchovy and scaled sardine of the Yellow Sea ecosystem, 1953 through 1984. [From Q. Tang (1989). Changes in the biomass of the Yellow Sea ecosystem. *In* "Biomass Yields and Geography of Large Marine Ecosystems" (K. Sherman and L. M. Alexander, eds.), pp. 7–35. AAAS Selected Symposium 111, Westview Press, Boulder, Colorado.]

for an annual sustained yield level of 100 to 120 mmt by the year 2000. The trend has continued upward since the 1970s; the 1986 level of usable marine biomass yields on a global basis was approximately 80 mmt. The lack of clear definition of actual and or potential global yield is not unexpected, given the rather limited efforts previously directed to the improvement of the information base on living marine resources on a global basis.

A. Global Change and LMEs

Although the annual global yield of living marine resources has been increasing since the 1960s, most of the gain has been from herring and herring-like fishes whose populations are responsive to changes in non-steady-state oceanographic conditions exemplified by El Niño type events. Demersal fish stocks are heavily exploited. Against this background of increasing stress on the genetic diversity and adaptability of resource stocks is a growing concern for the potential effects of global climate change on ocean productivity. The nature of contemporary biosphere feedback to climate change is an area of science that is only now being addressed. The possible doubling of atmospheric levels of carbon dioxide, methane, chlorofluorocarbons, and nitrous oxide will likely result in global temperature increases over the next 100 years. Climatologists report the decade of the 1980s to be the warmest on

record. Climate models suggest that different physical equilibrium states between the ocean and atmosphere may rise as the climate changes. The changes will likely be accompanied by human-induced effects, including additions of nutrients from agricultural practices and runoff and depletion of high-altitude ozone levels. Considerable uncertainty surrounds the effects of human-induced changes on the global ecosystems. The processes linking changes in greenhouse gas concentrations to climate change are complex. Although the average global temperature will increase, the regional rates of temperature change are not well understood. Major uncertainties remain regarding the effects of climate change on ocean productivity, biomass yields, and the related socioeconomic consequences. Programs are now being developed that will focus on LMEs as global regions, where responses of ecosystem productivity in relation to global change and other natural and anthropogenic sources of change can be assessed, monitored, and predicted. [*See* Atmospheric Trace Gases, Anthropogenic Influences and Global Change; Biodiversity and Human Impacts; Climate.]

The predictions of physical change, based on models linking global ocean circulation to temperature, precipitation, sea level, and greenhouse gas increases, estimate a mean temperature rise of 1.8°C by the year 2020; by 2070, the estimate of increase is 3.5°C for temperature and 7% for precipitation. Sea level rise by 2070 is estimated to be about 45 cm.

Given the expected magnitude of change, it is likely that the structure and productivity of LMEs will be significantly altered. A summary of the principal driving forces controlling biomass yields in LMEs from around the globe is depicted in Fig. 10.

B. Predation Driven LMEs

There are well documented examples of excessive fishing mortality and natural predation as the principal sources of decline in the biomass of important resource species within LMEs. The biomass of commercially important fish stocks of the Northeast Continental Shelf ecosystem of the United States declined by approximately 50% between 1968 and 1975. It was replaced by species with low economic value. The principal cause of the loss of biomass was excessive fishing mortality on juvenile and adult stocks of fish. Recent estimates of the impact of natural mortality on the fish component within one of the Northeast Shelf subsystems — Georges Bank — were determined. Large fish preying on smaller-sized fish accounted for consuming approximately 75% of the annual fish production; 10% of the fish production was utilized as biomass yield for food and industrial purposes; 10% was consumed by marine mammals; and 5% by marine birds.

In the Western Pacific, the 40% decline in the demersal

Table II Indian Ocean Biomass Yields, Fish, Crustaceans, Mollusks—Nominal Catch by Country, 1987

Marine fishing areas	Amount of biomass yield in thousands of metric tons	Percent of Indian Ocean catch	Percent total
Western Indian Ocean Rim			
Tanzania	47,775	1.37	
Somalia	17,000	0.49	
Kenya	6,875	0.2	
Mauritus	17,952	0.5	
Reunion	1,521	0.04	
	91,123		2.6
Northwestern Indian Ocean Rim			
Saudi Arabia	45,500	1.30	
Yemen Arab Republic	22,254	0.64	
People's Democratic Republic of Yemen	48,492	1.39	
Oman	115,011	3.3	
Iran	120,000	3.4	
Pakistan	336,129	9.6	
	687,386		19.6
West Indian Continental Shelf Ecosystem and Bay of Bengal Ecosystem			
India	1,681,485	48.2	
Sri Lanka	153,537	4.4	
Bangladesh	232,858	6.7	
Burma	540,873	15.5	
Australia	100,000	2.9	
	2,708,753		77.7
Indian Ocean total	3,487,262.0		100.0
World total	80,501,200.0		4.3

Source: United Nations Food and Agricultural Organization Annual Fishery Statistics Yearbook, 1989.

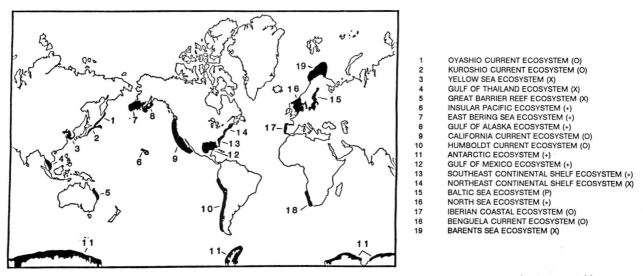

1	OYASHIO CURRENT ECOSYSTEM (O)
2	KUROSHIO CURRENT ECOSYSTEM (O)
3	YELLOW SEA ECOSYSTEM (X)
4	GULF OF THAILAND ECOSYSTEM (X)
5	GREAT BARRIER REEF ECOSYSTEM (X)
6	INSULAR PACIFIC ECOSYSTEM (+)
7	EAST BERING SEA ECOSYSTEM (+)
8	GULF OF ALASKA ECOSYSTEM (+)
9	CALIFORNIA CURRENT ECOSYSTEM (O)
10	HUMBOLDT CURRENT ECOSYSTEM (O)
11	ANTARCTIC ECOSYSTEM (+)
12	GULF OF MEXICO ECOSYSTEM (+)
13	SOUTHEAST CONTINENTAL SHELF ECOSYSTEM (+)
14	NORTHEAST CONTINENTAL SHELF ECOSYSTEM (X)
15	BALTIC SEA ECOSYSTEM (P)
16	NORTH SEA ECOSYSTEM (+)
17	IBERIAN COASTAL ECOSYSTEM (O)
18	BENGUELA CURRENT ECOSYSTEM (O)
19	BARENTS SEA ECOSYSTEM (X)

Fig. 10 Predominant variables influencing changes in fish species biomass in large marine ecosystems. Predominant variable—predation (X); environment (O); pollution (P); inconclusive information (+).

finfish yields of the Yellow Sea ecosystem is attributed to overfishing, considered as fishing predation. Within the Yellow Sea ecosystem, Chinese scientists are experimenting with the "grow-out" of fleshy prawn. They have experienced modest success in harvesting these introduced prawns in an ecosystem where the natural predator field has been reduced through excessive fishing "predation."

The case study of the Gulf of Thailand is yet another example of excessive fishing effort (e.g., human predation) from the introduction of high-technology fishing methods, reducing the biomass of bottom fish to a fraction of the sustained abundance level.

C. Environmentally Driven LMEs

Some large-scale changes in abundance levels of marine resources result from changes in ocean productivity driven principally by environmental changes. Examples of the multimillion metric-ton response in fish production of LMEs to favorable environmental conditions include the boundary region between the Kuroshio and Oyashio Current ecosystems off Japan and the Humboldt Current ecosystem off Chile.

The impact on global fishery yields from responses of these LMEs to ocean physics and productivity is significant. In 1974, the total annual yields of world fisheries was approximately 60 mmt; by 1986 global biomass yields increased to 80.9 mmt. The additional biomass resulted from increases in yields (landings) of sardines and anchovies. The yields of these herring-like fishes increased from an annual level of 10 mmt in 1974 to 23.9 mmt in 1986. Of this increase, Japanese sardines from the Oyashio and Kuroshio Current ecosystems increased from 400,000 mt to 5.2 mmt during this period and the yield of Chilean sardines from the Humboldt Current ecosystem increased from 500,000 mt in 1974 to 4.3 mmt in 1986. The increases have been attributed to density-independent processes involving an increase in phytoplankton and zooplankton food-chain productivity made possible by coastward shifts in the mixing area of the Oyashio and Kuroshio Current ecosystems off Japan and water mass shifts in the Humboldt Current ecosystem off Chile.

Changes within the California Current LME, causing the declines in abundance levels of Pacific sardine and subsequent increase in anchovy biomass, are considered the result of natural environmental change rather than any density-dependent competition between the two species.

In the Atlantic the LME shift in abundance between horse mackerel and sardines, within the Iberian Coastal ecosystem, is attributed to changes in natural environmental perturbations rather than to any density-dependent interaction between the two species. Similarly, in the Benguela Current ecosystem, the long-term fluctuations in the abundance of horse mackerel and pilchard are attributed to environmental changes.

D. Pollution Driven LMEs

Changes in the productivity of the Baltic Sea ecosystem, one of the LMEs for which considerable effort has been expended in the synthesis of long-term biological, physical, and chemical databases, have been attributed to the effects of coastal pollution. Within the Baltic Sea ecosystem, according to the comprehensive study of Gunnar Kullenberg, anoxic conditions are occurring with increasing frequency due, in part, to an increased input of organic material from land-based sources resulting in higher levels of primary production and a change in ecosystem balance. The nutrient levels have increased over the past several decades, and a considerable input of various substances has been reported, including organochlorines. A change from oligotrophic to eutrophic conditions has occurred in the coastal zone and in parts of the offshore waters of the Baltic. There are indications that oligotrophic fish are being replaced by eutrophic species, including bream and roach. The increased primary production has led to an increasing sedimentation changing hard bottom to soft bottom, contributing to a decline and disappearance of some fauna and a change in benthic community structure.

The eutrophication events may have been beneficial to some species. The catch of Baltic cod, herring, and sprat contributed to a doubling of the biomass yield of these species during the past two decades to an annual yield level of approximately 900,000 mt. However, in recent years the cod stocks have declined in abundance and the incidence of fish disease has also increased. Although it is not possible to generalize, at least in some cases the diseases are related to industrial impacts and the reduction in cod may be attributed to the low oxygen levels of bottom waters on the spawning ground. A rigorous program to monitor key ecosystem components of the Baltic is underway by several international scientific organizations in an effort to reduce coastal pollution and increase the sustained biomass yield of the ecosystem.

The Adriatic Sea ecosystem is under stress from increasing pollution. This LME is the focus of a number of important studies by the International Council for the Scientific Exploration of the Mediterranean Sea and the United Nations Regional Seas projects, conducted by the U.N. Environmental Program. The results of the studies should provide a sound basis for remedial actions to reduce pollution levels within the ecosystem.

The offshore waters of other LMEs examined have shown little evidence of serious pollution. However, the coastal margins represent threatened environments. The ecosystems and resources of the coastal zone are under stress from expanding population pressures in both developed and underdeveloped coastal nations. Development encroaching on the coastal zone includes urban, industrial, commercial, and agricultural activities that exert pressures on delicate ecosystem habitats, including wetlands, coral reefs, and sea grass beds. Pollution from shipping, oil spills, dumping, and offshore mining are in need of tighter controls to limit contaminant inputs. Excessive inputs of nutrients into coastal waters have been implicated as the source of the growing frequency and extent of plankton blooms and associated episodes of biotoxin damage to mollusks, other living marine resources, and to human health. Other serious pollutants include waste floatables and plastics that entangle marine organisms and litter shallow water areas and beaches. [*See* COASTAL POLLUTION; MARINE POLLUTION.]

IV. Management of LMEs

The international legal framework for oceans management is evolving toward an ecosystem approach to the management of LMEs. If the evolution is to be realized, it will be necessary for coastal nations to adopt an ecosystem approach within the boundaries of their exclusive economic zones. Customary international law doctrines call for comprehensive and interrelated treatment of the resources of the oceans and adjacent coasts. Government officials can use these doctrines and incorporate total ecosystem management in the application of laws related to oceans and coastal management. Furthermore, they can apply these doctrines in negotiations with foreign countries involving bilateral and multilateral fisheries agreements. Arrangements for the protection and sustained use of all marine resources can be premised on total ecosystem management.

According to Martin Belsky, Dean of the Albany Law School, legal scholars have long argued that the oceans are a "commons" and that a comprehensive ecosystem approach to oceans management is both essential and the evolving rule. They point to the 1982 United Nations Convention on the Law of the Sea (UNCLOS) as strong evidence that the evolution has now been completed. The UNCLOS indicates that the ecosystem approach is now binding international law. By definition, a treaty is binding on those states that have agreed to it (e.g., the doctrine of *pacta sunt servanda*). Therefore, for those nations that have ratified UNCLOS, a comprehensive approach is mandated.

A. The Antarctic Ecosystem Model

The Convention for the Conservation of Antarctic Marine Living Resources (CCAMLR) is an international agreement that supports an ecosystem approach to the conservation and management of living resources found in ocean areas surrounding Antarctica. The convention mandates a management regime committed to applying measures to ensure that harvesting of Antarctic species, such as finfish and krill, is conducted in a manner that considers ecological relationships among dependent and related species. The implementation of CCAMLR is carried out against a background of enlightened international activities in Antarctica, that in recent decades have been concerned with scientific research and cooperation, demilitarization, denuclearization, resource utilization, and environmental protection. The parties to the convention have conducted their activities under the system of legal, political, and scientific relationships established by the Antarctic Treaty of 1959.

The CCAMLR was negotiated from 1977 to 1980, entering into force in 1982. The CCAMLR Convention Area includes the marine area south of the Antarctic Convergence, the boundary between 48 and 60 degrees South separating the cold Antarctic waters and the warmer subantarctic waters. South of this boundary is defined as the Antarctic marine ecosystem. The convention applies to "the populations of finfish, mollusks, crustaceans, and all other species of living organisms, including birds, found south of the Antarctic Convergence."

Members of the Commission for the Conservation of Antarctic Marine Living Resources are Argentina, Australia, Belgium, Brazil, Britain, Chile, the European Community, Germany, France, India, Japan, New Zealand, Norway, Poland, South Africa, South Korea, the Soviet Union, Spain, and the United States.

The conservation approach adopted by the Commission seeks to

1. Prevent any harvested population from falling below the level that ensures the greatest net annual increment in yield.
2. Maintain the ecological relationships between harvested, dependent, and related populations of Antarctic living marine resources.
3. Restore depleted populations.
4. Prevent or minimize the risk of changes in the marine ecosystem that are not potentially reversible over two or three decades.

The Commission has produced conservation measures for depleted stocks of finfish. Among the management measures adopted are limits on catch-levels, mesh-regulation measures for aiding the recovery of fish stocks, and regulations prohibiting all directed fisheries for the bottom-living species of Antarctic cod, *Notothenia rossii*, in the waters of South Georgia, the South Orkneys, and the Antarctic Peninsula.

B. The Northern California Current Ecosystem Model

A multidisciplinary team of scientists and policy experts have prepared a plan for managing the Northern California

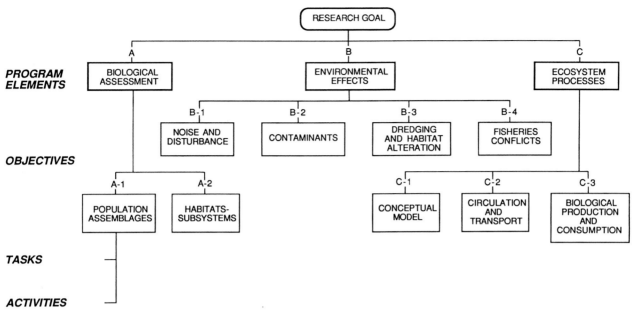

Fig. 11 Principal components of environmental risk. Program elements of this plan represent categories of information needed to evaluate each of the components at risk to the Northern California Current ecosystem. [From D. L. Bottom, K. K. Jones, J. D. Rodgers, and R. F. Brown (1989). "Management of Living Marine Resources. A Research Plan for the Washington and Oregon Continental Margin." Oregon Dep. Fish. Wildl., Publ. No. NCRI-T-89-004, Newport, Oregon.]

Current as a large marine ecosystem. The objective of the plan is to minimize the risks of human disturbance that could lead to large-scale and irreversible changes to the ecosystem.

According to D. L. Bottom of the Oregon Department of Fish and Wildlife, management activities within the northern California Current region presently are scattered among a large number of state and federal agencies whose jurisdictions divide the ecosystem along geographic and resource boundaries. Different agencies are responsible for managing marine birds, marine mammals, sport and commercial fisheries, pollutant discharges, dredging and disposal, and oil, gas, and mineral development. Preparation of a resource management plan for the Northern California Current ecosystem represents an important step toward integrated management. Ultimately, a regional structure or process will be necessary to coordinate management activities in Oregon with those in northern California and Washington (Fig. 11).

Regional management is necessary to direct local resource uses in a manner that will minimize the risks of single or cumulative effects on an entire ecosystem. Species composition and oceanographic conditions indicate that the northern California Current region (Cape Mendocino, California, to Vancouver Island, British Columbia) represents an ecological unit that is appropriate for regional planning and management.

Research activities in the Pacific Northwest are also segregated among many agencies and institutions. An integrated program of research is needed to assure that the sum of individual environmental studies yields the understanding that is required to manage the entire ecosystem. The Oregon Plan addresses several scales of information to support ecosystem management in the northern California Current region. Areawide surveys are necessary to understand large-scale variability and to direct local development activities in a manner that will minimize risks to the ecosystem. Studies of single and local environmental effects are needed to develop lease stipulations, make siting decisions, and monitor performance. Considering the high cost of oceanographic studies, wherever possible, individual research activities should also consider the broader regional goals that are a mutual concern of many research agencies and institutions. The plan will be used to encourage cooperation among the many research interests in the northern California Current ecosystem.

C. The Northeast U.S. Shelf Model

For three-and-a-half centuries the Northeast Shelf ecosystem has supported large, important common-resource fisheries. Utilization of the biomass yields from the fisheries extends from the export trade in salted cod of the early colonial period to the intensive exploration of total finfish biomass in the late 1960s and early 1970s. Heavy fish mortality imposed on the resources by European factory fleets precipitated the passage of legislation by the United States in 1976, creating a Fishery Management Zone (FMZ) from the territorial seas of the United States seaward 200 miles. In 1983, the FMZ was declared by Executive Order of the President as part of the U.S. Exclusive Economic Zone (EEZ).

The LME encompasses 260,000 km² extending from the Gulf of Maine in the north to Cape Hatteras in the south. The shelf ecosystem is among the most productive in the world. The annual biomass yields (e.g., crustaceans, mollusks, fish, algae) contribute one billion dollars annually to the economy of the coastal states from Maine to North Carolina.

Uses of the Northeast Shelf ecosystem as a source of petrogenic hydrocarbons and as a repository for wastes has heightened concerns for the "health" of the ecosystem and its capacity for sustainable production of usable biomass. Added to the concern is the loss of wetlands, aerosol fallout, and runoff of nitrogenous particulates contributing to coastal eutrophication.

The offshore waters of the ecosystem do not show any adverse impacts of pollution. Impacts have been limited to periodic shellfish closures in small embayments to protect human health from pathogens; mortalities of benthic mollusks and crustaceans from anoxic events associated with unusual phytoplankton blooms; secondary effects of algal blooms causing mortalities of salmonids reared in large coastal holding nets; disease outbreaks among species being cultured for market (e.g., mussels, oysters); periodic mortalities of cetaceans and pinnipeds on coastal beaches, apparently caused by the ingestion of prey feeding on biotoxic dinoflagellates; and the periodic incidence of biotoxin-bearing dinoflagellates causing shellfishing closures as a protection to human health. Measured against increasing pollution-induced losses of marine resources, it is clear that the major impacts on living resources of the shelf ecosystem are the result of excessive fishing mortality. The structure of the fish community has been changed significantly by overfishing over the past two decades. The gadoids (cod, haddock, hake, pollock) have been significantly overfished and are in a depleted state. They have been replaced in part by a growing mackerel and herring biomass undergoing recovery from the overfishing of the early 1970s and a large increase in small, low-valued elasmobranchs (spiny dogfish; skates)—the latter component having increased within the Northeast Shelf ecosystem on Georges Bank from about 24% of the fish biomass in 1963 to 74% in 1986.

Management measures adopted by two regional management councils have been ineffective in enhancing the recovery of the highly valued gadoids. The major difficulty in achieving improved management results is the lack of selectiv-

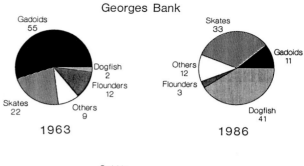

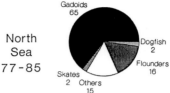

Fig. 12 Species shifts and abundance of small elasmobranchs (dogfish and skates) on Georges Bank within the Northeast Continental Shelf ecosystem of the United States compared with the North Sea ecosystem. [From K. Sherman, E. B. Cohen, and R. W. Langton (1991). "The Northeast Continental Shelf: An Ecosystem at Risk." Proceedings of the Conference: Gulf of Maine, Sustaining our Common Heritage, Convened by The Gulf of Maine Governors and Premiers (in press).]

ity of trawling and the large number of vessels engaged in the fisheries. Consideration is being given to the adoption of adaptive management strategies aimed at reducing the losses of long-term biomass yield of highly valued species through a controlled manipulation of predatory stocks (e.g., small elasmobranchs) when it can be quantitatively demonstrated that this is the best decision among an array of plausible options.

Comparative ecosystem studies in relation to the development of management strategies can be instructive. Spiny dogfish and skates are important predators of young gadoids, including cod and haddock. Their increase in abundance adds to the predation stress on gadoids on Georges Bank and in other areas of the Northeast Shelf ecosystem. The present fishery for small elasmobranchs in the Northeast Shelf ecosystem is rather limited. However, in the North Sea ecosystem where dogfish and skates are regularly fished, they represent only 4% of the fish biomass. Therefore, their predation impact on gadoids in the North Sea is diminished and the percentage composition of gadoids to other fish components of the ecosystem is significantly higher (Fig. 12). Considerations in any management protocol for the Northeast Shelf ecosystem will need to account for not only the multispecies fish interrelationships, but also the impacts of pollution on the nursery grounds of fish stocks and the interactions between fish and protected species, including pinnipeds, ceta-

Table III   Key Spatial and Temporal Scales and Principal Elements of a Systems Approach to
the Research and Management of Large Marine Ecosystems

Spatial – Temporal Scales

Spatial	*Temporal*	*Unit*
Global (world ocean)	Millennia-decadal	Pelagic biogeographic
Regional (exclusive economic zones)	Decadal-seasonal	Large marine ecosystems
Local	Seasonal-daily	Subsystems

Research elements
  Spawning strategies
  Feeding strategies
  Productivity, trophodynamics
  Stock fluctuations/recruitment/mortality
  Natural variability (hydrography, currents, water masses, weather)
  Human perturbations (fishing, waste disposal, petrogenic hydrocarbon impacts, aerosol contaminants, eutrophication effects)
Management elements: options and advice — international, national, local
  Bioenvironmental and socioeconomic models
  Management to optimize fisheries yields
Feedback Loop
  Evaluation of ecosystem status
  Evaluation of fisheries status
  Evaluation of management practices

ceans, and sea turtles. No significant changes in the lower end of the food chain (e.g., phytoplankton, zooplankton) have been detected over the past 30 years, suggesting that the major driving force of the biomass yields of the Northeast Shelf ecosystem is the human predator, thereby providing maximum options for the management of biomass yield from an ecosystem perspective.

D. The Global Model

Although the designation of LMEs as global management units is an evolving scientific and socioeconomic process, sufficient progress has been made to allow for useful comparison of the processes influencing changes in biomass yields in different LMEs. Effective management strategies will be contingent on the identification of the major driving forces causing large-scale changes in biomass yields. Management of species responding to strong environmental signals will be enhanced by improving the understanding of the physical factors forcing biological changes. Whereas in other LMES, where the prime driving force is predation, options can be explored for implementing adaptive management strategies. Remediation actions are required to ensure that the "pollution" of the coastal zone of LMEs is reduced and does not become a principal driving force in any LME.

There is a growing awareness of the utility of the LME approach to resource management among marine scientists,

geographers, economists, government representatives, and lawyers. Concerns remain regarding the socioeconomic and political difficulties in management across national boundaries, as is the case of the Sea of Japan ecosystem, for example, or the North Sea ecosystem, or the 38 nations sharing the resources of the Caribbean Sea ecosystem. For at least one LME, the Antarctic, a management regime has evolved based on an ecosystem perspective in the adoption and implementation of the Convention for the Conservation of Antarctic Marine Living Resources. Efforts are underway to implement ecosystem management within the LMEs of the United States' exclusive economic zone, and at least one consortium of states (Oregon, Washington, California) is considering the management of an LME designated as the Northern California Current ecosystem. The Northeast Coast ecosystem of Australia is presently under an ecosystem-oriented management regime.

A hierarchical approach to the management of LMEs is depicted in Table III. The hierarchy allows for the LMEs to serve as the link between local events (e.g., fishing, pollution, storms) occurring on the daily-to-seasonal temporal scale and their effects on living marine resources and the more ubiquitous global effects of climate changes on the multidecadal time scale. The regional, temporal focus of season to decade is consistent with the evolved spawning and feeding migrations of the fishes—the keystone species of most large marine ecosystems. These migrations are seasonal and occur

Table IV   Selected Hypotheses Concerning Variability in Biomass Yields of Large Marine Ecosystems

Ecosystem	Predominant variables	Hypothesis
Oyashio Current Kuroshio Current California Current Humboldt Current Benguela Current Iberian Coastal	Density-independent natural environmental perturbations	*Clupeoid population increases:* Predominant variables influencing changes in biomass of clupeoids are major increases in water-column productivity resulting from shifts in the direction and flow velocities of the currents and changes in upwelling within the ecosystem.
Yellow Sea U. S. Northeast Continental Shelf Gulf of Thailand	Density-dependent predation	*Declines in fish stocks:* Precipitous decline in biomass of fish stocks is the result of excessive fishing mortality, reducing the probability of reproductive success. Losses in biomass are attributed to excesses of human predation expressed as overfishing.
Great Barrier Reef	Density-dependent predation	*Change in ecosystem structure:* The extreme predation pressure of crown-of-thorns starfish has disrupted normal food chain linkage between benthic primary production and the fish component of the reef ecosystem.
East Greenland Sea Barents Sea Norwegian Sea	Density-independent natural environmental perturbations	*Shifts in abundance of fish stock biomass:* Major shifts in the levels of fish stock biomass within the ecosystems are attributed to large-scale environmental changes in water movements and temperature structure.
Baltic Sea	Density-independent pollution	*Changes in ecosystem productivity levels:* The apparent increases in productivity levels are attributed to the effects of nitrate enrichment resulting from elevated levels of agricultural containment inputs from the bordering land masses.
Antarctic Marine	Density-dependent perturbations	*Status of krill stocks:* Annual natural production cycle of krill is in balance with food requirements of dependent predator populations. Surplus production is available to support economically significant yields, but sustainable level of fishing effort is unknown.
	Density-independent natural environmental perturbations	*Shifts in abundance in krill biomass:* Major shifts in abundance levels of krill biomass within the ecosystem are attributed to large-scale changes in water movements and productivity.

over hundreds to thousands of kilometers within the unique physical and biological characteristics of the regional LME to which they have adapted. As the fisheries represent most of the usable biomass yield of the LMEs and fish populations consist of several age classes, it follows that measures of variability in growth, recruitment, and mortality should be conducted over multiyear time scales. This is necessary to interpret the effects of environmental, biological, and fishing effects on changing abundance levels of the year class to the populations of the species constituting the fish community, their predators and prey, and physical environment.

Consideration of the naturally occurring environmental events and the human-induced perturbations affecting demography of the populations within the ecosystem is necessary. Based on a firm, scientific understanding of the principal causes of variability in abundance and with due consideration to socioeconomic needs, management options can be considered for implementation from an ecosystems perspective. The

Table V List of 30 Large Marine Ecosystems and Subsystems for Which Syntheses Relating to
Principal, Secondary, or Tertiary Driving Forces Controlling Variability in Biomass Yields Have
Been Completed for Inclusion in LME Volumes Through February 1991

Large marine ecosystem	Volume number[a]	Authors
U.S. Northeast Continental Shelf	1	M. Sissenwine
	4	P. Falkowski
U.S. Southeast Continental Shelf	4	J. Yoder
Gulf of Mexico	2	W. Richards and M. McGowan
	4	B. Brown *et al.*
California Current	1	A. MacCall
	4	M. Mullin
	5	D. Bottom
Eastern Bering Shelf	1	L. Incze and J. Schumacher
West Greenland Shelf	3	H. Hovgård and E. Buch
Norwegian Sea	3	B. Ellertsen *et al.*
Barents Sea	2	H. Skjoldal and F. Rey
	4	V. Borisov
North Sea	1	N. Daan
Baltic Sea	1,5	G. Kullenberg
Iberian Coastal	2	T. Wyatt and G. Perez-Gandaras
Mediterranean-Adriatic Sea	5	G. Bombace
Canary Current	5	C. Bas
Gulf of Guinea	5	D. Binet and E. Marchal
Benguela Current	2	R. Crawford *et al.*
Patagonian Shelf	5	A. Bakun
Caribbean Sea	3	W. Richards and J. Bohnsack
South China Sea-Gulf of Thailand	2	T. Piyakarnchana
Yellow Sea	2	Q. Tang
Sea of Okhotsk	5	V. Kusnetsov *et al.*
Humboldt Current	5	J. Alheit and P. Bernal
Indonesia Seas-Banda Sea	3	J. Zijlstra and M. Baars
Bay of Bengal	5	S. Dwivedi
Antarctic Marine	1,5	R. Scully *et al.*
Weddell Sea	3	G. Hempel
Kuroshio Current	2	M. Terazaki
Oyashio Current	2	T. Minoda
Great Barrier Reef	2	R. Bradbury and C. Mundy
	5	G. Kelleher
Gulf of California	5	L. Mee
South China Sea	5	D. Pauly and V. Christensen

[a] Vol. 1, "Variability and Management of Large Marine Ecosystems," edited by K. Sherman and L. M.
Alexander, AAAS Selected Symposium 99, Westview Press, Boulder, Colorado (1986). Vol. 2, "Biomass Yields
and Geography of Large Marine Ecosystems," edited by K. Sherman and L. M. Alexander, AAAS Selected
Symposium 111, Westview Press, Boulder, Colorado (1989). Vol. 3, "Large Marine Ecosystems: Patterns, Processes, and Yields," edited by K. Sherman, L. M. Alexander, and B. D. Gold. AAAS Symposium, AAAS Publishers, Washington, D.C. (1990). Vol. 4, "Food Chains, Yields, Models, and Management of Large Marine
Ecosystems," Edited by K. Sherman, L. M. Alexander, and B. D. Gold, AAAS Symposium, Westview Press,
Boulder, Colorado (in press). Vol. 5, "Stress, Mitigation, and Sustainability of Large Marine Ecosystems,"
edited by K. Sherman, L. M. Alexander, and B. D. Gold, AAAS Publishers, Washington, D.C. (in press).

final element in the hierarchy is the feedback loop that allows for evaluation of the effects of management actions at the fisheries level (single species, multispecies) and the ecosystem level, with regard to the concept of resource maintenance and sustained yield. It will be necessary to conduct supportive research on the processes controlling sustained productivity of LMEs. Within several of the LMEs, important hypotheses concerned with the growing impacts of pollution, overexploitation, and environmental changes on sustained biomass yields are under investigation (Table IV). By comparing the results of research among the different systems, it should be possible to accelerate an understanding of how the systems respond and recover from stress. The comparisons among LMEs should allow for narrowing the context of unresolved problems and capitalizing on research efforts underway in the different ecosystems. A list of reports describing the effects of biological and physical perturbations on the fisheries biomass yields of 30 large marine ecosystems is given in Table V.

Global change in the form of ozone depletion, warming, and the greenhouse effect may become a source of stress on the biomass production of the oceans. The rather dramatic decadal fluctuations in marine biomass yields, when considered in light of the growing concerns over global change, should serve to accelerate the movement toward adoption of LMEs as regional units for the conservation and management of living marine resources under existing maritime law.

Bibliography

Abelson, P. H. (1986). The International Geosphere–Biosphere Program, *Science* **234** (4777), 657.

Alexander, L. M., Allen, S., and Hanson, L. C. (1989). "New Developments in Marine Science and Technology: Economic, Legal and Political Aspects of Change." 22 L. Sea Inst. Proc., Univ. Hawaii, Honolulu.

Beddington, J. R., May, R. M., Clark, C. W., Holt, S. J., and Laws, R. M. (1979). Management of multispecies fisheries. *Science* **204** (4403), 267–277.

Belsky, M. H. (1989). The ecosystem model mandate for a comprehensive United States ocean policy and law of the sea. *San Diego Law Review* **26**(3), 417–495.

Burger, J., ed. (1988). "Seabirds & Other Marine Vertebrates: Competition, Predation and Other Interactions." Columbia Univ. Press, New York.

Gulland, J. A., ed. (1988). "Fish Population Dynamics, The Implications for Management," 2nd ed. Wiley, Chichester.

International Council for the Exploration of the Sea (ICES) (1978). "North Sea Fish Stocks—Recent Changes and Their Causes" (G. Hempel, ed.). *Rapp. P.V. Reun., Cons. Int. Explor. Mer* **172**.

ICES (1978). "Marine Ecosystems and Fisheries Oceanogra-

phy" (T. R. Parsons, B.-O. Jansson, A. R. Longhurst, and G. Saetersdal, eds.). *Rapp. P.V. Reun., Cons. Int. Explor. Mer* **173**.

ICES (1984). "The Biological Productivity of North Atlantic Shelf Areas" (J. J. Zijlstra, ed.) *Rapp. P.V. Reun., Cons. Int. Explor. Mer* **183**.

ICES (1986). "Contaminant Fluxes through the Coastal Zone" (G. Kullenberg, ed.). *Rapp. P.V. Reun., Cons. Int. Explor. Mer* **186**.

ICES (1989). "Oceanography and Biology of Arctic Seas" (G. Hempel, V. Alexander, G. Dieckmann, J. Jakobsson, J. Meincke, and M. Spindler, eds.). *Rapp. P.V. Reun., Cons. Int. Explor. Mer* **188**.

MacCall, A. D. (1986). Rethinking research for fishery and ecosystem management. *In* "Rethinking Fisheries Management" (J. G. Sutinen and L. C. Hanson, eds.), pp. 173–193, Proceedings from the Tenth Annual Conference Held June 1–4, 1986. Center for Ocean Management Studies, The University of Rhode Island, Kingston.

May, R. M., ed. (1984). "Exploitation of Marine Communities." Springer-Verlag, Berlin.

Morgan, J. R. (1988). Large marine ecosystems, an emerging concept of regional management. *Environment* **29** (10), 4–9 and 29–34.

Postma, H., and Zijlstra, J. J., eds. (1988). "Ecosystems of the World 27: Continental Shelves." Elsevier Science Publishers B.V., Amsterdam.

Ricklefs, R. E. (1987). Community diversity: relative roles of local and regional processes. *Science* **235**, 167–171.

Rothschild, B. J., ed. (1988). "Toward a Theory on Biological-Physical Interactions in the Ocean." Kluwer Academic Publishers, Dordrecht, The Netherlands.

Sainsbury, K. J. (1988). The ecological basis of multispecies fisheries, and management of a demersal fishery in tropical Australia. *In* "Fish Population Dynamics, The Implications for Management," 2nd ed. (J. A. Gulland, ed.), pp. 349–382. Wiley, Chichester.

Scientific Committee for the Conservation of the Antarctic Marine Living Resources (SC-CAMLR-VII) (1988). "Report of the Seventh Meeting of the Scientific Committee, Hobart, Australia, 24–31 October 1988."

Sherman, K., and Alexander, L. M., eds. (1986). "Variability and Management of Large Marine Ecosystems." AAAS Selected Symposium 99, Westview Press, Boulder, Colorado.

Sherman, K., and Alexander, L. M., eds. (1989). "Biomass Yields and Geography of Large Marine Ecosystems." AAAS Selected Symposium 111, Westview Press, Boulder, Colorado.

Simon, J. L., and Kahn, H., eds. (1984). "The Resourceful Earth." Basil Blackwell, Inc., New York.

Sutinen, J. G., and Hanson, L. C., eds. (1986). "Rethinking Fisheries Management." Proceedings from the Tenth Annual Conference Held June 1–4, 1986. Center for Ocean Management Studies, The University of Rhode Island, Kingston.

U.S. Council on Environmental Quality and the Department of State (Gerald O. Barney, director). (1980). "The Global 2000 Report to the President: Entering the Twenty-First Century," Vols. I–III, U.S. Government Printing Office, Washington, D.C.

Valiela, I. (1984). "Marine Ecological Processes." Springer Advanced Texts in Life Sciences (D. E. Reichle, series ed.). Springer-Verlag, New York.

Wise, J. P. (1986). Fisheries statistics—boring stuff until you need them for monitoring. *In* "Oceans '86—Conference Proceedings" Vol. 3—Monitoring Strategies Symposium, pp. 904–907. Marine Technology Society, CH2363-0/86/0000-0904, IEEE, Washington, D.C.

Glossary

El Niño Occasional extreme warming of the tropical Pacific Ocean.

Elasmobranchs Taxonomic category of cartilaginous fishes including small sharks and skates.

Euphausiids Small, shrimp-like crustaceans, important as prey to fishes and whales, now the focus of an expanding fishery in Antarctic waters by Japan and the Soviet Union, resulting in new products for human consumption.

Eutrophic Waters with high nutrient levels and high productivity.

Fish recruitment Process of population renewals in the ocean characterized as the life stage when fish first become the targets of a fishery.

Gadoid Taxonomic category of bony fishes that are cod-like, including cod, haddock, hake, and pollock.

GLOBEC Global Ocean Ecosystem dynamics program sponsored by the U.S. National Science Foundation to support studies of the factors controlling the productivity levels of selected ecosystems around the globe.

Keystone species Species within an ecosystem that are critically important as prey supporting large numbers of other species, as for example krill in the Antarctic that are the principal food source of whales, seals, and penguins.

Oligotrophic Waters with low nutrient levels and low productivity.

Teleconnection Very large-scale interaction (e.g., ocean-wide) between the atmosphere and ocean having an impact on regional conditions.

Lithospheric Plates

M. K. McNutt
Massachusetts Institute of Technology

The surface of the earth is covered by a mosaic of rigid, drifting lithospheric plates that are thin (typically less than a few hundred kilometers) compared to their lateral dimensions (typically a few thousand kilometers). The relative motion of these plates, at rates of tens of millimeters per year, leads to earthquakes, volcanism, rifting, and mountain building along plate boundaries, but relatively little deformation within plate interiors. It is principally through the creation of lithosphere at midocean ridges and its reassimilation into the mantle at convergence zones that elements and energy are exchanged between the earth's mantle and its surficial crust, oceans, and atmosphere.

I. Origin of the Concept of a Lithosphere

The term *lithosphere* was originally coined by Joseph Barrell in 1914 to refer to the outermost, mechanically strong layer of the earth that is capable of supporting large surface loads, such as volcanoes. Its roots come from the Greek *lithos* and *sphere*, which combine to denote the earth's rocky shell. The concept took on a much increased importance in the late 1960s when the theories of continental drift and sea-floor spreading were combined under the paradigm of plate tectonics, which explains geologic activity on the earth's surface as the consequence of the interactions along the boundaries of a finite number of lithospheric plates that drift laterally relative to one another. Whereas the earth's crust, mantle, and core can be distinguished by fundamental changes in rock chemistry and mineralogy, the lithosphere differs from the underlying asthenosphere principally in its deformation properties. Because the viscosity of the earth's mantle depends strongly on temperature, the lithosphere in the upper ~100 km of the mantle is highly viscous and resistant to deformation compared to the softer, lower-viscosity asthenosphere. Lithosphere is usually classified as either oceanic or continental lithosphere, depending on whether its surficial layer consists of thin (~6 km), dense (~2800 kg/m^3) oceanic crust, or thick (~40 km), more buoyant (~2700 kg/m^3) continental crust.

II. The Lithosphere as a Drifting Plate

A. Plates and Plate Boundaries

According to the theory of plate tectonics, the earth's surface is currently composed of approximately seven major plates and six minor ones (Fig. 1). A plate will typically include both continental and oceanic types of lithosphere, and the geometry of lithospheric plates is constantly evolving as new plate boundaries are created and old ones cease to be active in response to changing stresses driving plate motion. Boundaries between any two plates consist of three types: (1) constructive boundaries, where the two plates diverge; (2) destructive boundaries, where plates converge; and (3) conservative boundaries, where two plates slide past each other with no component of convergence or divergence (Fig. 2).

Boundaries of the first type most commonly occur in the ocean basins along mid-ocean ridges, which are globe-encircling chains of volcanoes. As the two plates drift apart, volcanism along the mid-ocean ridge accretes new lithospheric material (basalts and peridotites) onto the trailing edges of the plates. On the continents, divergent plate boundaries are manifest in the form of continental rifts which, if they remain active, eventually lead to the formation of a mid-ocean ridge and ocean basin between two continental fragments. Characteristics of constructive plate boundaries include tholeiitic to alkalic volcanism, earthquakes with focal mechanisms indicating normal faulting, and high topography caused by hot, buoyant asthenosphere filling in the void left in the wake of the separating plates. A classic example of a

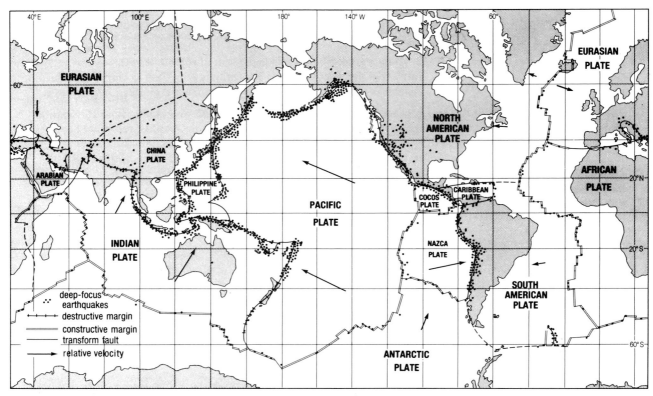

Fig. 1   The world pattern of plates, ocean ridges, trenches, and transform faults in relation to earthquakes epicenters indicated by dots. Tentative positions of plate margins are given by dashed lines. The length and direction of arrows indicate the relative velocities of the plates averaged over the past few Ma, using the African fixed frame of reference. The arrow in the key corresponds to a relative velocity of 50 mm/yr. [From Open University Course Team (1989).]

constructive plate boundary is the Mid-Atlantic Ridge. Rifting of a supercontinent approximately 200 million years ago led to the formation of the Atlantic Ocean as the Americas drifted away from Eurasia and Africa. [*See* CONTINENTAL RIFTING; OCEANIC CRUST; SUPERCONTINENTS.]

Averaged over the globe, lithosphere is consumed along the destructive, or convergent, plate boundaries at the same rate that it is created along the constructive plate boundaries so as to keep a constant earth radius. Convergent plate boundaries take the form of subduction zones if the lithosphere being consumed is oceanic. In this case, the consumed plate sinks down into the earth's mantle beneath the overriding plate, which may consist of continental or oceanic lithosphere, to form narrow arcuate deeps, called oceanic trenches, in the sea floor. This process can lead to thrust faulting at shallow depth and other earthquakes as deep as 700 km beneath the earth's surface as stress is relieved within the brittle core of the downgoing plate. Some components of the subducting plate, including some marine sediments, are melted off at depths of approximately 125 km. The rising melt interacts

with the overlying asthenosphere to form arcs of basaltic and andesitic volcanoes on the overriding plate that often erupt in violent, gas-rich explosions. Much of the Pacific basin is bounded by subduction zones, such as the Marianas trench on the west where the sea floor reaches the greatest depth anywhere on the earth, in excess of 11 km below sea level. [*See* CONTINENTAL AND ISLAND ARCS; VOLCANOES.]

Although the volcanic arcs above some subduction zones lie directly on continental crust, others erupt on the edge of a small ocean basin between the volcanic arc and a continent. These marginal or back-arc basins form by either presently active or previously active sea-floor spreading, which at first seems incongruous with the compressional nature of plate convergence zones. One hypothesis that can explain the extensional nature of back-arc basins supposes that in these locations the motion of the plates is such that the locus of the trench is migrating seaward relative to the overlying continent as the dense subducting slab founders into the asthenosphere. The back-arc basin then forms in the gap that would otherwise exist between the continent and the subducting oceanic

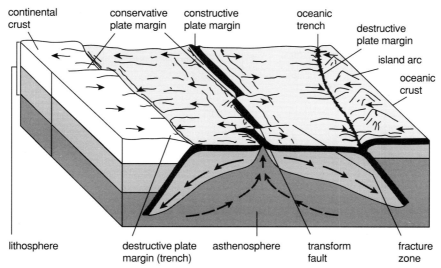

Fig. 2   Vertically exaggerated diagram showing the basic concepts of plate tectonics. Plates of rigid lithosphere (which include oceanic or continental crust and uppermost mantle) overlie a layer of relatively low strength, the asthenosphere. Mantle material rises at constructive plate margins (ocean ridges), and plate material descends into the mantle at destructive plate margins (trenches). At conservative plate margins (transforms), plates merely slide past each other. [From Open University Course Team (1989).]

lithosphere. The western Pacific contains numerous examples of presently and previously active back-arc basins, such as the Fiji and Marianas basins.

Even though convergence zones probably involve subduction of oceanic lithosphere initially, continental lithosphere eventually encounters the trench axis for plates containing both continental and oceanic types of lithosphere. The thick, buoyant continental crust renders such lithosphere more difficult to subduct than the oceanic type, and significant density reductions may extend into the subcontinental upper mantle as well. The response of the plates to continuing convergence is to deform along their margins into folded mountain belts, often sweeping up any intervening island arcs and pieces of back-arc basins into a suture zone between the colliding continents. The Alpine–Himalayan belt, extending more than a quarter of the circumference of the earth, is a spectacular example of a fold belt resulting from the continental collision following the subduction of Tethys, an earlier ocean basin that lay along the southern margin of Europe and Asia. [*See* OROGENIC BELTS.]

Debate still continues in the earth science community concerning what fraction of the continental crust and its underlying mantle lithosphere can be recycled into the mantle. The fact that the average age of the continental crust is about 2 billion years, whereas the greatest age of oceanic crust is only 200 million years (with the exception of some volumetrically minor, more ancient ophiolite outcrops), precludes the possibility that continental crust participates in the subduction recycling to the same extent as does oceanic crust. However, continent-derived sediments deposited in the ocean basins evidently are subducted, because their distinct geochemical fingerprints can be detected in the andesitic volcanics derived from slab melts.

Along conservative, or transcurrent, boundaries, lithosphere is neither created nor consumed. Therefore, volcanism is a very minor factor along such plate margins, but shallow earthquakes are frequent as the plates slip past each other along the earth's surface. In the ocean basins, transcurrent boundaries take the form of transform faults that offset discrete spreading segments (Fig. 2). Although these faults cease to accommodate differential lateral plate motion beyond the ridge-transfusion intersections, their morphologically prominent off-ridge extensions, called fracture zones, bound lithosphere of different ages. Transcurrent boundaries are also found on continental lithosphere, notable examples being the San Andreas fault in California and the Anatolian fault in northern Turkey and Iran.

B. Plate Motions

The motions of the lithospheric plates on the earth's surface are described in several different ways. Relative motion models describe only the direction and rate of motion of one plate with respect to another. Absolute motion models at-

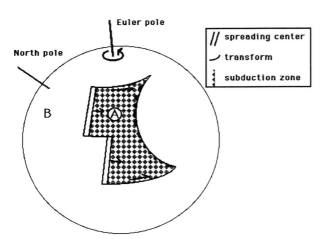

Fig. 3 Illustration showing how the relative motion of plate A with respect to plate B can be described by a rigid-body rotation about a Euler pole. The spreading ridges lie along lines of longitude, and the transforms correspond to line of latitude (small circles) for the Euler pole.

tempt to find some frame of reference external to the plates themselves from which to measure the velocities of the plates. In both cases, the motions of the plates are most easily expressed in terms of Euler poles and either angles or angular rates of rotation about those poles (Fig. 3). Euler poles can be classified as either instantaneous poles of rotation, in which case each pole is associated with an instantaneous rotation rate that describes the velocity of a particular plate with respect to another plate or the external reference frame at a particular instant in time, or as finite rotation poles, in which case the associated rotation angle describes the total rotation necessary to bring the plate from its initial position at time t_1 to its final position t_2 with respect to another plate or the external reference frame. In the latter case, the calculated rotation may not correspond to the path actually taken by the plate between times t_1 and t_2 because the location of the instantaneous Euler pole may have varied over that interval of time.

It has only been extremely recently that it has been possible to directly measure the geologically instantaneous plate motions using precise geodetic positioning. Fixed benchmarks placed on opposite sides of a plate boundary are periodically resurveyed to determine the relative motion of the two plates. To resolve plate motions with only years to decades of survey data, positions must be accurate to the centimeter level. Some of the most promising techniques for attaining this accuracy involve the latest in space-age technology, such as very long baseline interferometry (VLBI), which utilizes the phase lag between ratio signals from distant quasars received at two different positions on the earth's surface to precisely measure changes in their separation.

The instantaneous plate motions determined geodetically are in rather good agreement with those derived from using geological markers to measure plate motion over the past 1 to 5 million years. These geologically determined Euler poles and angles are technically finite rotation parameters, but they can be converted to instantaneous rotation poles and angular rates on the assumption that motion has remained constant during the past few million years. Directions of plate motion are most easily derived from the strike of the transform faults, since plate velocity must be locally tangent to their trend. The orientation of ridge crests is usually normal to the direction of plate motion, although well-documented cases of oblique spreading make this method less reliable. Once the azimuth of plate motion is determined, the rate is readily calculated from the amount of offset of dated markers (e.g., stream gravels, volcanic ash flows, etc.) across transform faults.

In the ocean basins, magnetic lineations in the sea floor (Fig. 4) have provided almost all of the information on rates of relative plate motion. These magnetic lineations are caused by periodic reversals of the earth's magnetic field. Magma upwelling at the mid-ocean ridge acquires a direction of remanent magnetization in the direction of the earth's field at the time of volcanism. This signal is frozen into the oceanic crust as its spreads away from the ridge, forming a magnetic stripe in the sea floor of either normal or reversed polarity and with a width dependent on the rate of plate divergence and the duration of the given polarity interval. Because durations of different polarity intervals over the past few tens of millions of years have been well calibrated by radiometric dating of volcanic lavas, the age of oceanic lithosphere and rates of plate divergence can be easily determined by comparing the magnetic signal observed by towing a magnetometer above the sea floor with the known magnetic reversal chronology.

It is more difficult to determine the azimuth and rate of relative motion between two plates sharing only a subduction zone boundary (e.g., the Philippine Plate with any of its neighbors). In such settings, the directions of first motions for earthquakes along plate boundaries are often the only available indicator of direction of motion, and can be used to get a rough estimate of rate based on earthquake frequency and size.

A number of reference frames have been used to convert relative plate motions between two plates to absolute plate motions between each plate and some external reference. The following are the most commonly used:

1. The *hotspot* reference frame assumes that sites of active midplate volcanism, such as the youngest volcanoes in the Hawaiian, Society, and Cape Verde island chains, represent the surface expression of mantle plumes anchored in the deep mantle. By measuring the motion of

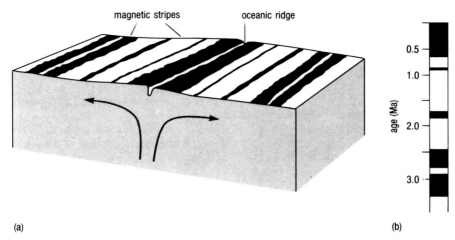

magnetic stripes   oceanic ridge

age (Ma)

0.5
1.0
2.0
3.0

(a)   (b)

Fig. 4 (a) Asthenospheric material rises under an oceanic ridge to produce new lithosphere and "freezes in" the polarity of the earth's magnetic field at the time of volcanism. As the plate drifts away from the ridge, a symmetric pattern of magnetic stripes is produced. (b) A simplified time scale for polarity of the earth's magnetic field over the past few million years, showing alterations of black (normal) and white (reversed) polarity. [From Open University Course Team (1989).]

the plates past these plumes fixed to the mantle frame, one obtains their velocity with respect to the deep mantle.

2. The *no-net-torque* reference frame requires that the net effect of all plate rotations is to produce no uncanceled torques on the lithosphere or underlying mantle. Given any set of plate boundaries and relative velocities, one can always find a unique reference frame such that the total torque exerted by all of the plates on the layer below is zero. Conceptually, because this frame of reference moves with the mean velocity of the deep mantle, it should lead to absolute plate velocities in agreement with those derived from the hotspot frame of reference.

3. The *Africa fixed* reference frame assumes that the African Plate does not drift. The locus of hotspot volcanism on the African Plate over the past few tens of millions of years is consistent with the hypothesis that Africa has not drifted much with respect to the hotspot frame of reference.

4. The *geomagnetic* reference frame refers plate motions with respect to the earth's geomagnetic field, on the assumption that it is an axial dipole field when averaged over several tens of thousands of years. Magnetic minerals in molten volcanic rocks align in the direction of the local strike and dip of the earth's magnetic field, and lock in that signal as they cool. Even after the plate containing those rocks has drifted to a new location, extremely sensitive laboratory instruments can recover the orientation of the paleomagnetic field of volcanic rock samples and thus determine the location of the plate with respect to the geomagnetic pole at the time of volcanism. The dis-

advantage of this reference frame is that it is insensitive to the purely longitudinal component of plate motion, along which the strike and dip of the magnetic field vector for an axial dipole is constant; but it is extremely useful for determining the motions of continents prior to the Cretaceous.

Both the hotspot and no-net-torque reference frames are of little use in the Jurassic because we do not know the plate boundaries and most of the hotspot tracks have been subducted. [*See* OCEANIC ISLANDS, ATOLLS, AND SEAMOUNTS.]

Despite these different conventions for referencing absolute plate motions, they all produce approximately the same absolute velocity field, especially for rapidly moving plates for which the uncertainties are smaller. The general features of such solutions are that mostly oceanic plates that are attached to subducting slabs (e.g., the Pacific Plate) are drifting very rapidly, while largely continental plates drift more slowly (Fig. 1). Attempts to determine whether this relationship between plate velocity and the distribution of continents and oceans has also held in the past are plagued by large uncertainties, but they suggest that this correlation may not apply in the Cretaceous (144–66 Ma) when the geometry of plate boundaries was rather different from what it is today.

C. Forces Driving Plate Motions

For many years earth scientists have sought clues as to what forces drive the motion of the plates. External factors have been considered, such as torques from the sun and moon, but in general most researchers favor internal sources. The relationship between the lithosphere and the underlying asthenosphere has been particularly problematic. Does convection in the asthenosphere, powered by internal energy sources such as radioactive heat generation and chemical differentiation, drive the motion of the overlying plates, or is motion of the plates largely decoupled from mantle flow by a very low-viscosity cushion beneath the plates? Most earth scientists favor the latter model because the current plate geometry does not allow simple cells of mantle upwelling beneath ridge crests and downwelling at subduction zones. For example, for a plate entirely rimmed by spreading centers, such as the Antarctic Plate, if upwardly diverging mantle flows drive spreading on the surrounding mid-ocean ridges, a problem of mass balance results under the center of the plate where there is no subduction zone. In addition, the absolute velocities of plate motions today are not correlated, either positively or negatively, with plate areas. Therefore, it seems unlikely that flow in the asthenosphere either drives or greatly resists plate motion.

A number of other forces within the lithosphere itself have been proposed as possible drivers for plate motion. A likely factor is the excess mass of subducting slabs. Because they are cooler than the surrounding asthenosphere, they tend to pull the overlying attached plate into the mantle by their own negative buoyancy. This mechanism has been proposed as the explanation for the rapid velocity of the Pacific Plate toward the subduction zones in the northwestern Pacific where extremely old, cold lithosphere is currently being subducted. Another possible contributor to plate motions is termed *ridge push*. Because hot asthenosphere upwells into the void left by diverging plates at mid-ocean ridges, the ridges are hotter and stand higher than the surrounding sea floor. Therefore, a force exists that tends to drive the lithosphere "downhill" from the point of elevated potential energy. Seismologists have examined the focal mechanisms of relatively rare midplate earthquakes for evidence of a state of stress in oceanic lithosphere consistent with the ridge push hypothesis. Although a number of other factors influence the stress state of the lithosphere, such as thermal stress from plate cooling, it does seem possible that the stress state of plate interiors is consistent with the predictions of ridge push. However, this factor is probably secondary to slab pull in determining the motion of the plates.

III. Thickness of the Oceanic Lithosphere

The previous discussion has focused on the lithosphere as a thin shell moving across the earth's surface. Observations of plate motions alone do not provide information as to the depth to which rocks near the earth's surface participate in the drift pattern observed geologically and geodetically.

Other geophysical observations do bear on the thickness of the plates, particularly for oceanic lithosphere that is relatively easy to interpret compared to continental lithosphere on account of its younger average age, more uniform crustal structure, and simpler geologic history.

A. The Thermal Lithosphere

One extremely common criterion for distinguishing the lithosphere from the asthenosphere is to define the lithosphere as including that section of the earth's outermost layer where heat transfer is governed by conduction rather than convection. It then becomes possible to calculate the apparent thickness of the thermal plate by comparing observations, such as heat flow through the surface of the lithospheric plate and its subsidence as it contracts and cools, with the predictions from mathematical models of thin plates conductively cooling with time. The asthenosphere is assumed to be well mixed by convection such that it can be modeled as isothermal, whereas the overlying lithosphere is characterized by a steep thermal gradient. [*See* HEAT FLOW THROUGH THE EARTH.]

Two different models have been proposed for the base of the thermal lithosphere. For the half-space model, it is assumed that conductive cooling can penetrate to arbitrarily great depth in the mantle given sufficient cooling time. The base of the lithosphere is defined as the depth to an isotherm T_m (~1350°C), which is some fraction of the melting temperature of the mantle. According to the predictions of this model, the plate is continually cooling and thickening with time since formation at the mid-ocean ridge. An alternative to the half-space model is the plate model, which assumes that there is a maximum thickness L for the lithosphere below which conductive cooling does not penetrate. For time that is short compared to $L^2/(\pi^2\kappa)$, in which κ is the thermal diffusivity of the lithosphere, both half-space and plate models predict identical thermal structures for the lithosphere. At greater time, however, the plate model gradually approaches an equilibrium conductive gradient at depths less than L such that the thickness of the lithosphere never exceeds L.

Observations of subsidence of the oceanic lithosphere have been interpreted as favoring the plate model because at lithospheric ages exceeding 70 Ma, the depth of the seafloor fails to increase and the heat flow fails to decrease as fast as the rate predicted for half-space cooling. The data are more consistent with conductive cooling of a plate only 125 km thick. Although this value has been adopted as the average thickness of oceanic lithosphere, there remain some unresolved questions concerning the applicability of the cooling plate model. The first problem is that the observations of depth and heat flow on old oceanic lithosphere that have been used to calibrate the model may be corrupted by the thermal effects of midplate (hotspot) volcanism. For example, the old Pacific lithosphere just east of the western Pacific trenches passed over the volcanic centers of French Polynesia at 80 to 100 Ma. It is thought that these hotspots may have reheated the lithosphere, leading to the departures from the half-space cooling model that are observed today. A second problem is that for the plate model to be physically valid, a heat source must be present to prevent the lithosphere from thickening beyond 125 km. Small-scale convection, shear heating, and radioactive heat generation at the base of the lithosphere have all been suggested as possible heat sources, but only the first is still considered to be a viable mechanism.

B. The Seismic Lithosphere

The base of the lithosphere does not correspond to a major seismic discontinuity, such as the Moho or the core–mantle boundary; but because the velocity and attenuation of seismic waves are sensitive to temperature and melt content, the lithosphere can be detected with seismic data. In this context, the lithosphere can be defined as a high-velocity "lid" lying above a low-velocity asthenosphere with high attenuation for shear waves. Based on analysis of surface waves, the depth to the low-velocity zone beneath the old ocean basins lies at approximately 100 km, although there remains some debate as to whether anisotropy in the propagation of surface waves might cause an apparent low-velocity zone. The observed increase in phase velocities of surface waves propagating through the lid with increasing age of the plate is consistent with the decrease in temperature predicted by the thermal plate model.

The seismic lithosphere defined above is distinct from the *seismogenic lithosphere*, which is the layer that can generate earthquakes. The seismogenic lithosphere represents only the uppermost part of the thermal plate where rocks are cool enough (less than 600–800°C) to rupture by brittle failure, rather than ductile flow. Because the depth to the brittle/ ductile transition depends strongly on temperature as well as pressure, the maximum depth of earthquake hypocenters can be used to estimate the geothermal gradient in the lithosphere. [See SEISMOLOGY, THEORETICAL.]

C. The Elastic Lithosphere

The elastic lithosphere is that region of the lithosphere that is capable of elastically supporting stresses in excess of 100 MPa caused by plate loading over time scales exceeding 1 million years. In response to the applied loads, such as volcanoes, mountain belts, or other mass anomalies applied above, within, or below the plate, or by lateral thrusts applied to its end, the lithosphere flexes in the manner of an elastic plate. The wavelength of the flexure is indicative of the effective elastic thickness of the plate—a stiff plate produces a long-wavelength moat and arch around the applied load. Based on

observations in the ocean basins of the wavelength of the moats and arches surrounding seamount loads and the distance to the outer rise seaward of trenches where slabs are bent down into the mantle, the effective elastic thickness increases from 5 km for young lithosphere near the mid-ocean ridges to 40 km for lithosphere greater than 80 Ma. The thickness of the elastic plate as a function of age coincides with the depth to the 450–600°C isotherm in the cooling plate model. This result is consistent with laboratory experiments on rock deformation that predict an exponential dependence of viscosity on temperature. Thus the 450–600°C isotherm marks a sharp transition between a largely elastic layer in the upper lithosphere underlain by a more fluid zone where large stresses are rapidly relaxed by ductile creep.

The thickness of the plate that supports an applied load decreases from a value more typical of the thickness of the thermal lithosphere to the thickness of the elastic plate during the first 10^4–10^5 years after loading as high stresses relax in the lower lithosphere. For this reason, the elastic plate thickness for very short-term loads, such as Pleistocene ice sheets, is much larger than the value appropriate for longer term loads, such as volcanoes. Although the region of the oceanic upper mantle between the base of the elastic plate and 125 km is not sufficiently strong to sustain stresses exceeding 100 MPa over long time scales, it is apparently viscous enough to transfer heat by conduction rather than by convection and to drift as a plate.

D. The Chemical Lithosphere

In addition to representing a mechanical and thermal boundary layer, the lithosphere is also a chemical boundary layer in that its uppermost layer consists of the earth's crust, which has differentiated from the mantle. There is less consensus as to whether the lithospheric upper mantle is also chemically distinct from the asthenosphere. Analyses of oceanic xenoliths point to an upper mantle depleted in the constituents of basalts. If the entire 6 km of oceanic basaltic crust were extracted from the underlying upper mantle, the thickness of the depleted zone would be approximately 40 km. Because the extraction of basalt leaves the source rock depleted in aluminum, iron, and other elements that crystallize in the extremely dense garnet phase, this depleted layer would be more buoyant than undepleted lithosphere. It is thought that the buoyancy from a depleted upper mantle might be an important factor in explaining the high elevation of oceanic plateaus, which have crustal thickness, and presumably depleted layers, 2 to 3 times the values for normal oceanic lithosphere. There is little direct evidence that this depleted mantle layer is a pervasive feature of typical oceanic upper mantle; but if it is, it could have important geodynamic implications.

IV. Similarities and Differences between Continental and Oceanic Lithosphere

We know that continental lithosphere is not mechanically equivalent to oceanic lithosphere because continental nuclei are more than 3 Ga old, whereas the ocean floor does not exceed 200 Ma in age. One of the major questions in earth science is whether this difference in the plate-tectonic recycling history of continental versus oceanic lithosphere can be explained solely by the difference in composition and thickness of continental crust, or whether it requires fundamental differences in the thermal and chemical structure of the subcrustal upper mantle. At present, evidence favors the hypothesis that beneath the old continental cratons, at least, the lithospheric plate is at least a factor of two thicker than the 125-km value determined for oceanic lithosphere.

A. Thickness of the Thermal Lithosphere

The thickness of the oceanic thermal lithosphere has been calibrated using the observed subsidence and heat flow as a function of lithospheric age. It has been more difficult to estimate the thickness of the thermal lithosphere beneath the continents for many reasons. First of all, because oceanic crust is uniform in thickness, except near plateaus and seamounts, the depth of the sea floor largely reflects the cooling and contraction of the lithosphere. On the continents, elevation variations are dominated by changes in the composition and thickness of the crust that are largely uncorrelated with lithospheric age, except for the Precambrian cratons and platforms where the sedimentary record indicates that they have been approximately at sea level for the past 2500 Ma. Either the thermal thickness of continental lithosphere must be sufficiently small (~125 km) such that the platforms had already reached thermal equilibrium by that time, or some other process must offset the subsidence from thermal contraction and cooling of a thicker plate.

A second problem concerns the contribution of radioactive heat generation to the thermal evolution of the lithosphere. Because oceanic crust is thin and mafic in composition, the contribution of radioactive heating to heat flow is minor. Continental crust, on the other hand, has concentrated the long-lived radionuclides. Surface heat flow is highly correlated with the radiogenic productivity of surface rocks, but lack of information on the composition of the continental lower crust leads to large uncertainty in the component of heat flux from the cooling mantle lithosphere. To the extent that this signal can be extracted, it appears that the subcontinental mantle lithosphere requires several hundred million years to reach thermal equilibrium, compared to less than 100 million years for that beneath the ocean basins; and thus

continental lithospheric thickness could be 250 km or greater.

B. Seismic Velocity Structure

Seismic data has provided some of the most provocative evidence that lithospheric roots beneath old continental cores extend below 125 km. The technique of seismic tomography has resolved lateral variations in seismic velocity in the upper mantle with respect to a radially stratified velocity model. The fact that high-velocity structures beneath the continents extend to depths of 250 km or greater has been used to argue that continental lithosphere must be at least that thick, otherwise plate motion would offset these structures laterally. Additional evidence on the nature of these deep continental roots has come from the analysis of the difference in travel times between a direct ScS wave and the first multiple, ScS_2, that has an extra two-way traverse through the mantle and an extra surface reflection beneath either an old ocean basin or an old continental craton. If the mantle lithosphere beneath old continental cratons were thermally and chemically identical to that beneath old ocean basins, the travel-time difference $ScS_2 - ScS$ should be the same regardless of whether the surface bounce point for ScS_2 is beneath the continent or the ocean, after correcting for the differences in crustal structure. In fact, the difference is several seconds smaller for the continents, indicating that the continental lithosphere has much higher velocity. The magnitude of the difference may be too large to be explained by thermal processes alone, even if one assumes a thicker thermal plate beneath the continents.

C. Chemistry of the Mantle Lithosphere

One intriguing explanation for the $ScS_2 - ScS$ travel time residuals is that the lithosphere beneath the cratons is chemically different from that beneath old ocean basins, in addition to being thermally thicker. Analysis of mantle xenoliths from the deep continental lithosphere point to a depletion in Fe relative to (Fe + Mg), perhaps caused by the extraction of large quantities of basaltic melt in the process of forming the thick continental crust. This depletion has the effect of increasing seismic velocities, in agreement with the $ScS_2 - ScS$ travel-time residuals; but their magnitude may require that the depleted layer beneath the continental cratons extend to depths of 400 km. Although considerable debate continues as to whether continental lithospheric roots extend to such great depth, the added advantage of this hypothesis is that it leads to a density reduction via garnet depletion that can offset the density increase from thermal cooling, thus stabilizing the cratons near sea level. Furthermore, geodynamic arguments suggest that lithospheric thickness greater than

125 km will not be convectively stable unless the density increase from lithospheric cooling is offset by such a density reduction. [*See* Mantle.]

D. Rheological Behavior

The elastic plate thickness for old continental lithosphere provides particularly persuasive support for differences in the lithospheric thermal structure between continents and oceans. Whereas elastic plate thicknesses for oceanic plates do not exceed 40 km, values in excess of 100 km have been determined using several different techniques to analyze gravity data over surface and buried loads on the continents. The largest values are found for continental cratons such as the Canadian and Indian shields. Lower elastic plate thicknesses, more typical of the oceanic values, are found for the younger continental margins. The differences in maximum values for elastic plate thickness cannot be attributed to the rheological behavior of continental crust. Laboratory experiments on feldspar, plagioclase, and quartz, typical constituents of the continental crust, indicate that the upper 30 to 40 km of continental lithosphere should be extremely weak compared to the upper oceanic lithosphere. Therefore, the subcontinental lithosphere must be much stronger than the suboceanic lithosphere, either by virtue of its temperature or its chemistry. If the plate is more than 250 km thick, elastic plate thicknesses of 100 km and greater could be achieved for lithosphere older than 600 Ma through deepening of the elastic/ductile boundary as marked by the depth to the 600°C isotherm. In addition, if the lower continental lithosphere is depleted in volatile elements, perhaps again in the process of extracting melts to form continental crust, a hotter, and therefore deeper isotherm (~800°C) would mark the elastic/ductile transition because of the higher energy needed to activate ductile flow in "dry" rocks.

E. How Differences in Lithospheric Structure Have Affected the Geologic Evolution of Continents versus Ocean Basins

One obvious difference between continental and oceanic lithosphere is the degree to which the rigid plate hypothesis of plate tectonics applies to the two different kinds of lithosphere. Deformation in the ocean basins is confined to narrow plate boundaries, with relatively undeformed plate interiors. Plate boundaries in the continents are often extremely diffuse and involve processes virtually unknown in the oceans, such as splaying of transform faults, doubling of the crustal thickness in convergence zones via thrust faulting, lateral extrusion of thickened crust along strike-slip faults,

and delamination of the upper crust from the rest of the lithosphere. [*See* Continental Deformation.]

The difference in behavior of the two types of lithosphere in plate boundary zones can be explained by the fact that the continental crust has a zone of weakness at precisely the depth of maximum strength for oceanic lithosphere. Figure 5 shows examples of strength envelopes proposed for oceanic and continental lithosphere. Their shapes are derived from insight gained from laboratory experiments on rocks with the appropriate composition for either continental or oceanic lithosphere. The envelopes indicate the maximum stress that can be sustained by the lithosphere before it fails either by brittle sliding along faults in the upper lithosphere or by ductile creep in the lower lithosphere. Compared to the oceanic case, the continental yield envelope is more complicated in structure, with an additional ductile zone predicted to lie in the relatively weak lower continental crust if the geotherm is steep enough to activate ductile flow at that depth for sialic rocks. Failure in this zone is thought to be responsible for the internal deformation of continental plates.

The base of the yield envelopes in Fig. 5 approximately coincides with the elastic thickness of the lithosphere, although for highly bent plates, failure at its top and bottom can lead to an effective elastic thickness that underestimates the true depth to the base of the yield envelope. The fact that the mantle lithosphere beneath the continents is so much stronger may account for the stability of continental platforms and shields, particularly if the local geotherm is too cold to allow the development of a ductile zone in the lower continental crust.

The fact that most continental lithosphere remains at the surface of the earth for billions of years despite continuing plate subduction is not easily explained by just the great strength of old continental plates. Their buoyancy must prevent them from recycling back into the mantle. Most of the buoyancy can be attributed to the thick, light continental crust. Additional buoyancy may come from a depleted subcrustal lithosphere, although because this is only well developed beneath the old continental cratons, it cannot account for long-term preservation of the younger continental margins, which are also, on average, much older than oceanic lithosphere.

V. Importance of the Lithosphere

The importance of the lithosphere can hardly be overemphasized from either a scientific or a societal perspective. From a purely scientific standpoint, the lithosphere preserves the only record of the earth's geologic, biologic, and climatic

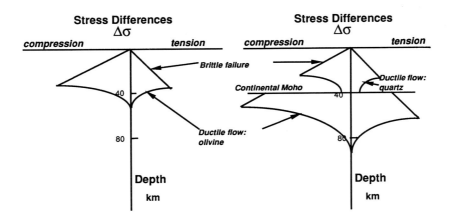

Fig. 5 Examples of strength envelopes for oceanic and continental lithosphere. In both cases, strength is limited in the upper lithosphere by brittle failure as stresses exceeding the limits of the envelope are relieved by sliding along preexisting faults. At greater depth, stresses are relieved by ductile flow. The base of the yield envelope is twice as deep for continental lithosphere as compared with old oceanic lithosphere because the continental thermal plate is assumed to be twice as thick. The continental yield envelope has an additional region of low strength just above the Moho caused by ductile failure of quartz-rich rocks in the lower crust. The horizontal stress axes are poorly calibrated, but maximum stress levels within the envelopes may reach 1000 MPa.

history. The magmatic and metamorphic processes occurring along plate boundaries are responsible for transporting energy from the earth's interior to the surface and cycling elements into and out of the mantle. From a societal standpoint, the lithosphere is the dynamic layer responsible for both natural hazards and resources, as well as the surface upon which we live.

A. Geologic Hazards along Plate Boundaries

Certainly one of the spectacular successes of plate tectonics has been its ability to explain the global distribution of earthquakes, and in fact a map of seismicity is almost equivalent to a map of plate boundaries. Earthquakes occur along plate boundaries because friction resists the ability of the plates to move relative to one another. The nature of this friction is highly variable, depending on rock type, the depth to the brittle/ductile transition, the heterogeneity or roughness of the fault surface, and the mode of failure. For example, rocks typically fail more easily under extension than compression. Mid-ocean ridge earthquakes thus tend to be shallow (<10 km) and small to moderate in magnitude. They

rarely occur near continents, so pose little seismic risk. Earthquakes along transform boundaries are deeper (<20 km) and of moderate to large magnitude. When they occur in populated areas, such as the 1906 San Francisco earthquake, they can be very destructive. The largest and the deepest (<700 km) earthquakes are generated in convergence zones. Shallow subduction zone earthquakes, such as the 1964 Alaskan earthquake, can be particularly damaging, both by virtue of the ground shaking they produce and the tsunamis they can generate. [*See* PLATE TECTONICS, CRUSTAL DEFORMATION AND EARTHQUAKES; TSUNAMIS.]

Attempts to mitigate earthquake hazards involve developing better methods of forecasting when events are likely to occur and designing structures to withstand them. While the state of the art of our understanding of nucleation of earthquakes does not yet permit short-term forecasting, knowledge of plate motion rates and the size and frequency of historic earthquakes have enabled long-term forecasting (decades to centuries) or seismic probabilities along the more active plate boundaries. Buildings and other structures in earthquake-prone areas are now designed to "give" in such a way that they dampen out the ground shaking, without either collapsing in catastrophic failure or amplifying the natural frequency of the strong ground motion. Better design of buildings deserves most of the credit for the low fatality rate in the 1989 Loma Prieta earthquake (~40 deaths), compared to the large death toll of the 1989 Armenian earthquake (~25,000 deaths) of similar magnitude. [*See* EARTHQUAKE PREDICTION TECHNIQUES.]

Volcanic eruptions and landslides triggered by both earthquakes and volcanoes also pose substantial risk along plate boundaries. Although volcanic eruptions are easier to predict than earthquakes because they are commonly preceded in the days-to-hours time frame by small-magnitude earthquake tremors marking the movement of magma beneath the volcano, volcanic eruptions can be violently explosive and deadly (e.g., the 1980 Mount St. Helens eruption). [*See* VOLCANIC TREMOR.]

B. Resources Stored in the Lithosphere

Traditionally, most earth science activities have been directed toward the discovery, assessment, and exploitation of nonrenewable (over human time scales) resources via the oil and mining industries. Processes occurring within lithospheric plates and along their boundaries, such as magmatism, metamorphism, and sediment burial are responsible for creating and concentrating most of earth's resources, such as coal, oil, gas, metals, and gems. As earth scientists have better understood the nature of plate boundary interactions today and the geometry of plate boundaries in the geologic past, it has become easier to focus exploration activities by predicting where new resources wait to be discovered and, just as important, where they are extremely unlikely to occur. [*See* BASIN ANALYSIS METHODS; ORE FORMATION.]

The long-term stability of the continental lithosphere has also allowed the accumulation and storage of nonrenewable resources over several billion years. Industrial society is exhausting many of these resources, especially fossil fuels, over the period of only a few centuries. Clearly we must become more efficient at finding new deposits and accelerate development of alternative energy sources, including geothermal energy.

C. Lithospheric Plates and Climate

The motion of the lithospheric plates is constantly changing the size, shape, and distribution of continents and ocean basins. This in turn controls the pattern of ocean and atmospheric circulation and swings in sea level, both of which have been important factors in changing the global environment by modifying the zonal belts of temperature and precipitation.

A record of changing ocean circulation patterns and climate is preserved in the skeletal remains of marine organisms buried in sea-floor sediments. The distribution and type of ancient warm-water and cold-water species mark the onset of the Gulf Stream pattern of circulation, as the width of the Atlantic Ocean increased by spreading on the Mid-Atlantic Ridge, and the initiation of the Antarctic Circumpolar Current with the opening of the Drake Passage between South America and Antarctica. The uplift of the Himalaya Mountains and the Tibetan Plateau within the past 40 million years as India collided with Asia has greatly modified the climate of central Asia, and the weather systems are now dominated by the monsoons. The record demonstrates that local changes in atmospheric or oceanic circulation patterns in one area can lead to major environmental changes worldwide. Thus studies of the global response to plate tectonic–induced changes in climate provide critical constraints on predicting the ultimate impact on the earth's climate of more immediate changes to the environment caused by human activities. [*See* PALEOCLIMATIC EFFECTS OF CONTINENTAL COLLISION AND GLACIATION.]

In the past few hundred thousand years, changes in sea level have been caused by changes in volume of continental ice sheets. In the past few hundred million years, however, even larger sea-level fluctuations have been the direct result of movement of the lithospheric plates. Shallow-water marine sediments deposited on continental margins and platforms at 510 Ma and again at 90 Ma indicate sea-level rises of 400 to 600 m relative to where the sea stands today. Plate motions affect sea level by changing the total volume of the ocean basins. Because the depth of the sea floor increases with increasing age of the lithospheric plate, periods of accelerated spreading at mid-ocean ridges lead to a reduction in the average age, and therefore the total volume, of the ocean basins. The rifting of continents has a similar effect. Extended continental lithosphere takes up more area on the surface of the earth compared to its unextended state, and it therefore displaces more seawater volume. Conversely, periods of either enhanced subduction of young sea floor or continental collision can increase the volume of the ocean basins, leading to a fall in sea level. Although tectonically controlled sea-level changes occur very slowly compared to those from glaciation, they do impose a long-term trend on the higher-frequency fluctuations. [*See* SEA LEVEL FLUCTUATIONS.]

D. Transfer of Mass and Energy between the Atmosphere, Hydrosphere, and Mantle

The lithospheric plates are an integral part of a global cycle in which chemical elements and energy are exchanged with the underlying mantle and the overlying oceans and atmosphere (Fig. 6). Volcanic activity along plate boundaries is responsi-

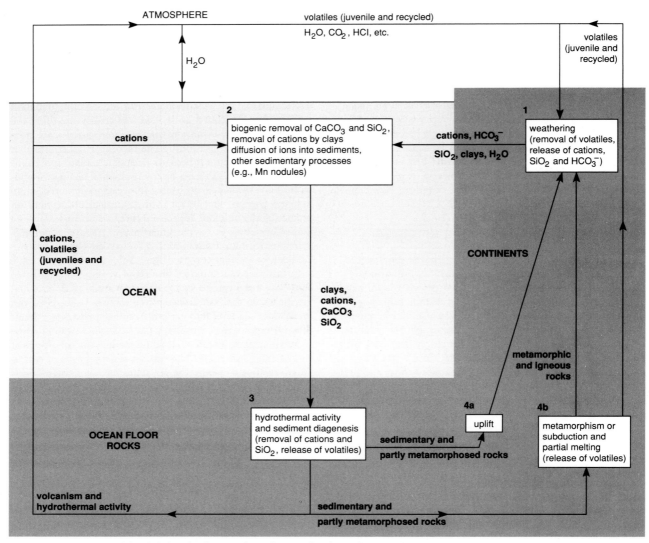

Fig. 6 Details of the global cycle, showing the pathways of exchange between the lithosphere, oceans, and atmosphere. [From Open University Course Team (1989).]

ble for outgassing components such as water, methane, CO_2, HCl, and SO_2, etc., from the mantle into the oceans and atmosphere. Weathering of crustal rocks on the continents and sea floor releases cations (Ca^{2+}, Na^+, K^+, and Mg^{2+}) into rivers and oceans. These dissolved constituents can later be redeposited in oceanic sediments via biogenic precipitation of calcium carbonate and silica or sedimentation of clays removed from the seawater. The elements are finally recycled back into the mantle in inclusions or chemical bonds with sediments and crustal rocks transported down subduction zones. Both gravity and seismic data suggest that even conti-

nental crust and lithosphere can be subducted to some depth in the mantle, and thus might provide an important mechanism for renewing heterogeneity in isotopic systems and minerals in the convectively mixed earth's mantle. Overall, the system appears to be in long-term steady state such that rates of input and removal of dissolved constituents in the oceans and atmosphere are balanced.

Although the rate at which the sea floor is recycled through the system is slow (100 Ma), in only 10 Ma the entire volume of the ocean water can be cycled through the hydrothermal systems operating along the mid-ocean ridges where hot volcanic rock directly interacts with seawater. The time scales involved with an individual volcanic system are even shorter and do impact the earth on human time scales. For example, temperatures in some mid-ocean ridge hydrothermal systems exceed 300°C, which means that any magnesium dissolved in seawater reacting in the system can be completely removed and deposited in hydrothermal veins in about a week. The subaerial eruption of a major volcano can occur in a matter of minutes and can release gases and particulate matter into the atmosphere which over weeks to years profoundly affect incoming solar radiation and therefore climate. [*See* Hy-DROTHERMAL SYSTEMS.]

Bibliography

Cox, A., and Hart R. B. (1986). Plate Tectonics: How It Works." Blackwell Scientific Publications, Palo Alto, California.

DeMets, C., Gordon, R. G., Argus, D. F., and Stein, S. (1990). Current plate motions. *Geophys. J. Int.* **101**, 425–478.

Lerner-Lam, A. L., and Jordan, T. H. (1987). How thick are the continents? *J. Geophys. Res.* **92**, 14007–14026.

Open University Course Team (1989). "The Ocean Basins: Their Structure and Evolution." Pergamon Press, Oxford.

Glossary

Asthenosphere More mobile region of the mantle immediately beneath the higher viscosity lithosphere that does not drift with the plates.

Euler pole Pole passing through the center of the earth about which a plate can rotate as a rigid body from any initial to any final position on the earth's surface.

Ophiolite Assemblage of rock units including pillow basalts, diabase sheeted dikes, gabbros, and serpentinized peridotites lying unconformably on continental lithosphere, which likely represent sections of oceanic lithosphere accreted to the continent during plate collision.

ScS Seismic phase consisting of a shear wave traveling down through the mantle that is reflected at the core–mantle boundary and returns to the earth's surface as a shear wave. ScS_n denotes a multiple ScS wave that is further reflected $n-1$ times at the earth's free surface.

Suture zone Lineation in continental lithosphere marking the location of an earlier plate collision that led to the amalgamation of formerly separate continental masses.

Xenolith Rock sample with mineralogy foreign to the igneous host in which it occurs. Thought to be a fragment of wall rock from volcanic conduits in the deep lithosphere rapidly transported to the surface during eruptions.